Systems Engineering for the Digital Age

Systems Engineering for the Digital Age

Practitioner Perspectives

First Edition

Edited by
Dinesh Verma

Published by John Wiley & Sons, Inc., Hoboken, New Jersey.
Published simultaneously in Canada.

For general information on our other products and services or for technical support, please contact our Customer Care Department within the United States at (800) 762-2974, outside the United States at (317) 572-3993 or fax (317) 572-4002.

Wiley also publishes its books in a variety of electronic formats. Some content that appears in print may not be available in electronic formats. For more information about Wiley products, visit our web site at www.wiley.com.

Library of Congress Cataloging-in-Publication Data:
Names: Verma, Dinesh, 1964- editor.
Title: Systems engineering for the digital age: practitioner perspectives / edited by Dinesh Verma.
Description: First edition. | Hoboken, New Jersey: Wiley, [2024] | Includes index.
Identifiers: LCCN 2023035119 (print) | LCCN 2023035120 (ebook) | ISBN 9781394203284 (cloth) | ISBN 9781394203291 (adobe pdf) | ISBN 9781394203307 (epub)
Subjects: LCSH: Systems engineering.
Classification: LCC TA168 .S8854 2024 (print) | LCC TA168 (ebook) | DDC 620.001/171–dc23/eng/20230808
LC record available at https://lccn.loc.gov/2023035119
LC ebook record available at https://lccn.loc.gov/2023035120

Cover Design: Wiley
Cover Image: © Dinesh Verma, Executive Director, on behalf of Systems Engineering Research Center, Issarawat Tattong/Getty Images

Set in 9.5/12.5pt STIXTwoText by Straive, Pondicherry, India

SKY10055988_092223

Contents

List of Contributors

Adam M. Ross
Massachusetts Institute of Technology
Cambridge, MA, USA

Alejandro Salado
University of Arizona
Tucson, AZ, USA

Andres Sousa-Poza
Engineering Management/System Engineering
Old Dominion University
Norfolk, VA, USA

Art Pyster
George Mason University
Fairfax, VA, USA

Azad M. Madni
Astronautics, Aerospace, and Mechanical Engineering
University of Southern California
Los Angeles, CA 90089, USA

Barry M. Horowitz
University of Virginia
Charlottesville, VA, USA

Benjamin Kruse
Stevens Institute of Technology
School of Systems and Enterprises
Hoboken, NJ, USA

Brian Chell
Stevens Institute of Technology
School of Systems and Enterprises
Hoboken, NJ, USA

Brian Duffy
University of Southern California
Information Sciences Institute
Marina del Rey, CA, USA

Brian Sauser
Department of Logistics and Operations Management
University of North Texas
Denton, TX, USA

Cesare Guariniello
School of Aeronautics and Astronautics
Purdue University
West Lafayette, IN, USA

Cihan Dagli
Missouri University of Science and Technology
Engineering Management and Systems Engineering Department
Rolla, MI, USA

Cody Fleming
Iowa State University
Mechanical Engineering
Ames, IA, USA

Craig Charlton
University of Southern California
Information Sciences Institute
Marina del Rey, CA, USA

Dalia Bekdache
School of Aeronautics and Astronautics
Purdue University
West Lafayette, IN, USA

Daniel Browne
Georgia Tech Research Institute
Systems Engineering Research Division
Smyrna, GA, USA

Dan DeLaurenits
Purdue University
Aeronautics & Astronautics
West Lafayette, IN, USA

Daniel Dunbar
Stevens Institute of Technology
School of Systems and Enterprises
Hoboken, NJ, USA

David Fullmer
Georgia Tech Research Institute
Systems Engineering Research Division
Smyrna, GA, USA

Dinesh Verma
Stevens Institute of Technology
School of Systems and Enterprises
Hoboken, NJ, USA

Donna H. Rhodes
Massachusetts Institute of Technology
Cambridge, MA, USA

Eileen Van Aken
Virginia Tech
Blacksburg, VA, USA

Ellins Thomas
Booz Allen Hamilton
El Segundo, CA, USA

Frank Patterson
Georgia Tech Research Institute
Systems Engineering Research Division
Smyrna, GA, USA

Gary Witus
College of Engineering
Wayne State University
Detroit, MI, USA

Gregg Vesonder
Stevens Institute of Technology
Hoboken, NJ, USA

Hector Saunders
US Space Force Space Systems
Command (SSC)
El Segundo, CA, USA

James D. Moreland
Virginia Tech University
Grado Department of Industrial and Systems
Blacksburg, VA, USA

Jiang Li
Electrical Computer Engineering
Old Dominion University
Norfolk, VA, USA

Joel S. Patton
Virginia Tech Applied Research Corporation
Arlington, VA, USA

John Dzielski
Stevens Institute of Technology
School of Systems and Enterprises
Hoboken, NJ, USA

John M. Colombi
Department of Systems Engineering and
Management
Air Force Institute of Technology
OH, USA

Jon Wade
University of California
San Diego, CA, USA

Joseph Bradley
Leading Change LLC
Melbourne, FL, USA

Judith Dahmann
The MITRE Corporation
MITRE Labs
McLean, VA, USA

Kaitlin Henderson
Virginia Tech
Blacksburg, VA, USA

Karen Marais
School of Aeronautics and Astronautics
Purdue University
West Lafayette, IN, USA

Kristin Giammarco
Department of Systems Engineering
Naval Postgraduate School
Monterey, CA, USA

Lirim Ashiku
Missouri University of Science and Technology
Engineering Management and Systems Engineering Department
Rolla, MI, USA

Lu Xiao
School of Systems and Enterprises
Stevens Institute of Technology
Hoboken, NJ, USA

Mark R. Blackburn
Stevens Institute of Technology
School of Systems and Enterprises
Hoboken, NJ, USA

Megan M. Clifford
Stevens Institute of Technology
School for Systems and Enterprises
Hoboken, NJ, USA

Michael Orosz
University of Southern California
Information Sciences Institute
Marina del Rey, CA, USA

Michael Sievers
Systems Architecting and Engineering
University of Southern California
Los Angeles, CA 90089, USA

Michael Shih
Booz Allen Hamilton
El Segundo, CA, USA

Mustafa Canan
Information Sciences Department
Naval Postgraduate School
Monterey, CA, USA

Navindran Davendralingam
Research Science, Amazon Inc.
Seattle, WA, USA

Nicole Hutchison
Stevens Institute of Technology
Hoboken, NJ, USA

Paul T. Grogan
Stevens Institute of Technology
School of Systems and Enterprises
Hoboken, NJ, USA

Peter A. Beling
Virginia Polytechnic Institute and State University
Hume Center for National Security and Technology
Blacksburg, VA, USA

Payuna Uday
Stevens Institute of Technology
Systems Engineering Research Center
Hoboken, NJ, USA

Richard Threlkeld
Missouri University of Science and Technology
Engineering Management and Systems Engineering Department
Rolla, MI, USA

Rob Cloutier
University of South Alabama
Mobile, AL, USA

Roshanak R. Nilchiani
Stevens Institute of Technology
School of Systems and Enterprises
Hoboken, NJ, USA

Roger Jones
Stevens Institute of Technology
School of Systems and Enterprises
Hoboken, NJ, USA

Samuel Kovacic
Engineering Management/System Engineering
Old Dominion University
Norfolk, VA, USA

Santiago Balestrini-Robinson
Georgia Tech Research Institute
Systems Engineering Research Division
Smyrna, GA, USA

Shawn Dullen
Stevens Institute of Technology
School of Systems and Enterprises
Hoboken, NJ, USA

Thomas McDermott
Stevens Institute of Technology
School for Systems and Enterprises
Hoboken, NJ, USA

Tim Sherburne
Virginia Polytechnic Institute and State University
Hume Center for National Security and Technology
Blacksburg, VA, USA

Timothy D. West
Stevens Institute of Technology
School of Systems and Enterprises
Hoboken, NJ, USA

Tom Hagedorn
Stevens Institute of Technology
School of Systems and Enterprises
Hoboken, NJ, USA

Tom McDermott
Stevens Institute of Technology
School for Systems and Enterprises
Hoboken, NJ, USA

Waterloo Tsutsui
School of Aeronautics and Astronautics
Purdue University
West Lafayette, IN, USA

Valerie B. Sitterle
Electronic Systems Laboratory
Georgia Tech Research Institute
NW Atlanta, GA, USA

William B. Rouse
McCourt School of Public Policy
Georgetown University
Washington, DC, USA

William Shepherd
Stevens Institute of Technology
Hoboken, NJ, USA

Ye Yang
Amazon.com Inc.
New York, NY, USA

Preface

The Systems Engineering Research Center (SERC) was established in the Fall of 2008 as a government-designated University Affiliated Research Center (UARC). The SERC has produced 15 years of research, focused on an updated systems engineering toolkit (methods, tools, and practices) for the complex cyber-physical systems of today and tomorrow. The principal sponsor of the SERC is the Office of the Under-Secretary of Defense for Research and Engineering in the U.S. Department of Defense. A unique governance element of the SERC is its establishment as a national network of preeminent universities in the United States focused on systems research.

The impact of the SERC on the state of systems engineering practice and education has been profound. Examples include the Body of Knowledge for Systems Engineering (SEBoK) – the most popular systems engineering website in the world with over 50,000 unique visitors every month; *GRCSE* – the first graduate reference curriculum for systems engineering; the *Mission-Aware Security* methodology for assessing system architectures for cyber-resilience; and a diversity of research projects to support the transition of *digital engineering* from strategy into practice including advanced tradespace analysis methods, modeling methods for digital project workflows, digital engineering measurement methods, and digital engineering competencies. Therein lies the focus of this book – translating some of the mature research inspired by the SERC into a compendium of chapters, organized into topical clusters, for the benefit of practicing engineers in industry and government.

This book features a total of 41 chapters, organized into eight topical clusters. Each of these chapter clusters was organized and edited by the following cluster leaders. Every one of the cluster leaders has been a principal investigator on multiple research projects within the SERC:

1	Transforming Engineering Through Digital and Model-Based Methods	Mark Blackburn (Stevens Institute of Technology)
2	Executing Digital Engineering	Jon Wade (University of California – San Diego)
3	Tradespace Analysis in a Digital Engineering Ecosystem – Context and Implications	Val Sitterle (Georgia Tech Research Institute)
4	Evaluating and Improving System Risk	Nicole Hutchison (Stevens Institute of Technology)
5	Model-Based Design of Safety, Security, and Resilience Systems	Tom McDermott (Stevens Institute of Technology)
6	Analytic Methods for Design and Analysis of Missions and Systems-of-Systems	Dan DeLaurentis (Purdue University)
7	Applying Systems Engineering to Enterprise Systems and Portfolio Management	Dan DeLaurentis (Purdue University)
8	Systems Education and Competencies in the Age of Digital Engineering, Convergence, and AI	Nicole Hutchison (Stevens Institute of Technology)

The chapter cluster leaders engaged with over 60 authors and coauthors to bring this initiative to completion - and all of us were managed by Dr. Payuna Uday, a chapter coauthor and the book's overall project manager. But for her patience and grace, combined with firm deadlines, this project may have never completed. I cannot thank her enough.

Given the context of this book, I would be remiss if I did not acknowledge my colleagues who have served in the role of SERC's Chief Technology Officers (Jon Wade and Tom McDermott); and Chief Scientists (Barry Boehm and Dan DeLaurentis) over the course of the last 15 years. As many of you know, we lost Professor Barry Boehm in 2022. Each of these four leaders have nurtured the networked model and collaborative ethos of the SERC. Their collective efforts have nucleated a national network of stellar research faculty with a collective interest in systems research.

Finally, this book has been written with the practitioner community of engineers and systems engineers in mind. As such the chapters have a pragmatic and utilitarian orientation. We have kept each chapter succinct and balanced in terms of depth and breadth of the materials included. I hope you find it useful. To support more in-depth reading and interest, this book has an accompanying website with additional materials organized in a manner consistent with the organization of this book – www.digitalse.org).

Acknowledgment

This book is inspired by research supported by the U.S. Department of Defense through the Systems Engineering Research Center (SERC) under Contracts HQ003419D0003, HQ003413D0004, and W15QKN18D0040. SERC is a federally funded University Affiliated Research Center managed by Stevens Institute of Technology. Any opinions, findings, and conclusions or recommendations expressed in this material are those of the authors and do not necessarily reflect the views of the U.S. Department of Defense.

Acronyms

Acronym	Meaning
6DOF	Six degrees of freedom
A&S	Acquisition and sustainment
AADL	Architecture analysis and design language
AARP	American Association of Retired Persons
AAS	Advanced automation system
ABM	Agent-based models
ACAT	Acquisition category
ADP	Approximate dynamic programming
AFD	Assessment flow diagram
AFRL	US Air Force Research Lab
AFSIM	Advanced framework for simulation integration and modeling
AGE-MOEA	Adaptive evolutionary algorithm based on non-Euclidean geometry for many objective optimization
AHP	Analytical hierarchy process
AI	Artificial intelligence
AI4SE	Artificial intelligence methods and tools within systems engineering
AIA	Aerospace Industries Association
AIAA	American Institute of Aeronautics and Astronautics
AIRC	Acquisition Innovation Research Center
ALT	Acquisition, logistics, and technology
AM	Additive manufacturing
ANSYS	Analysis system
API	Application programming interface
APPEL	Academy of Program/Project and Engineering Leadership
APS	Active protection system
AR	Augmented reality
ARDEC	Armament Research Development and Engineering Center
ART	Agile release train
ASA	Assistant Secretary of the Army
ASEE	American Society for Engineering Education
ASME	American Society of Mechanical Engineers
ASOT	Better accessibility of information
AST	Authoritative source of truth

ASuW	Anti-surface warfare
ASW	Anti-submarine warfare
AT	Assistive technologies
AT&L	Acquisition Technology and Logistics
ATR	Automatic target recognition
AugI	Augmented intelligence
AV	Architectural view
AWB	Analytic workbench
AZ	Azimuth
B.Eng. or BE	Bachelor of Engineering
BS	Bachelor of Science
B2B	Business-to-business
BA	Barabási-Albert
BATNA	Best alternative to a negotiated agreement
BFO	Basic formal ontology
BPMN	Business process modeling notation
BWA	Biological warfare agent
CA	Competitive advantage
CAD	Computer-aided design
CAPEC	Common attack pattern enumeration and classification
CAS	Complex adaptive systems
CASA	Center for Adaptive Systems Applications
CASE	Complex adaptive situations environment
CASoS	Complex adaptive system of systems
CATIA	Computer-aided three-dimensional application/analysis
CBA	Cost–benefit analysis
CBM	Condition-based maintenance
CBN	Causal Bayesian Network
CBT&E	Capability-based test and evaluation
CCO	Common core ontologies
CD	Continuous delivery
CDD	Capability development document
CDR	Critical design review
CDRLs	Contract data requirements lists
CEP	Circular error probability
CESUN	Council of Engineering Systems Universities
Cf	Consequence of failure
CFD	Computational fluid dynamics
CFM	Cryogenic fluid management
CI	Configuration items
CI	Continuous integration
CID	Continuous iterative development
CM	Configuration management
CNC	Computer numerical control
COCOTS	Constructive cost model for commercial-off-the-shelf
COMPASS	Comprehensive modelling for advanced system of systems

CONOPs	Concept of operations
CorBoK	Core body of knowledge
COSYSMO	Constructive systems engineering cost model
COTS	Commercial-off-the-shelf
CPS	Cyber physical systems
CPT	Classical probability theory
CPU	Central processing unit
CR	Change request
CRACK	Collaborative representative authorized committed knowledgeable
CRWS	Cyber resilient weapon system
CSAR	Combat search and rescue
CSP	Constraint satisfaction problem
CSRM	Cyber security requirements methodology
CSV	Comma separated variables
CTD	Contribution to design
CTEs	Critical technology elements
CTIs	Critical technology integrations
CTO	Chief technology officer
CV	Capability viewpoint
CVaR	Conditional value at risk
CVE	Common vulnerability exposures
CWE	Common weakness enumeration
CYBOK	Cyber body of knowledge
D.Eng.	Doctor of Engineering
DALs	Design assurance levels
DANSE	Designing for adaptability and evolution in system of systems engineering
DARPA	Defense Advanced Research Projects Agency
DAS	Defense acquisition system
DASD	Deputy Assistant Secretary of Defense
DAU	Defense Acquisition University
DbC	Design by contract
DCTO(MC)	Deputy CTO for mission capabilities
DE	Digital engineering
DECF	Digital Engineering Competency Framework
DEE	Digital engineering environment
DEFII	Digital engineering framework for integration and interoperability
DESEP	DE systems engineering plan
DESM	Digital engineering success measure
DESOW	Digital engineering statement of work
DETAF	Digital engineering tradespace analysis framework
DEVCOM	Combat capabilities development command
DHS	Department of Homeland Security
DIDs	Data item descriptions
DIKW	Data information knowledge wisdom
DIM	Disruption impact matrix
DLA	Defense Logistics Agency

DoD	Department of Defense
DoDAF	Department of Defense Architecture Framework
DoDI	Department of Defense Instruction
DOE	Design of experiments
DOE	Department of Energy
DoS	Department of State
DPG	Defense planning guidance
DSF	Decision support framework
DSM	Design structural matrix
DT	Digital thread
DT	Digital transformation
DT	Digital twin
DTA	Digital twin aggregate
DTI	Digital twin instances
EA	Evolutionary algorithm
EC	European Commission
ECO	"easy, common, or old"
EFFBD	Enhanced functional flow block diagram
EFV	Expeditionary fighting vehicle
EOI	Expressions of interest
EPIRB	Emergency position indicating radio beacon
ER	Erdõs-Rényi
ESA	European Space Agency
ESRD	End-stage renal disease
ETA	Event tree analysis
EVM	Earned value management
F2T2EA	Find fix track target engage assess
FAA	Federal Aviation Administration
FAM	Fuzzy associative memory
FAR	Federal Acquisition Regulation
FBI	Federal Bureau of Investigation
FC	Flexible contract
FDA	Food and Drug Administration
FDM	Fused deposition modeling
FEA	Finite element analysis
FILA	Flexible and Intelligent Learning Architectures
FILA-SOS	Flexible and intelligent learning architectures for system of systems
FIS	Fuzzy inference system
FLIR	Forward looking infrared radar
FLS	Fuzzy logic system
FMEA	Failure modes and effect analysis
FMECA	Failure mode effects and critically analysis
FOREST	Framework for operational resilience in engineering and system test
FOSS	Free and open-source software
FRC	Flexible resilience contract

FTA	Fault tree analysis
GA	Genetic algorithm
GAO	General Accounting Office
GDS	Graph data structure
GE	Graph energy
GM	General motors
GOGO	Government owned government operated
GOTS	Government-off-the-shelf
GPA	Grade point average
GPSR	Global positioning system receiver
GPUs	Graphic processing units
GQM	Goal question metric
GR	Gyro rate
GRA	Government reference architecture
GRCSE	Graduate reference curriculum for systems engineering
GT	GPS time
GUI	Graphic user interface
HAI	Human artificial intelligence
HIPPA	Heath Insurance Portability and Accountability Act
HITL	Hardware in the loop
HMI	Human machine interface
HMM	Hidden Markov model
HMO	Huckel molecular orbital
HMS	Human–machine system
HMT	Human–machine teaming
HoQ	House of quality
HPC	High-performance computing
HW	Hardware
I&T	Integration and testing
IAPRs	Integrated acquisition portfolio reviews
IBD	Internal block diagram
ICD	Initial capabilities document
IDEF5	Integrated definition for ontology description capture method
IDIQ	Indefinite delivery indefinite quantity
IEEE	Institute of Electrical and Electronics Engineers
IFDs	Informational flow diagrams
IISE	Institute of Industrial and Systems Engineers
IMC	Interactive Markov chain
IME	Integrated modeling environment
IMU	Inertial measurement unit
INCOSE	International Council on Systems Engineering
IOC	Initial operational capability
IoIF	Interoperability and integration framework
IoT	Internet of things
IP	Intellectual property

IR	Infrared
IRBs	Institutional review boards
IRIs	Internationalized resource identifiers
IRL	Integration readiness level
ISEDM	Integrated systems engineering decision management
ISO	International Standards Organization
ISR	Intelligence surveillance and reconnaissance
ISS	International Space Station
IT	Information technology
JCIDS	Joint capabilities integration and development system
JDMS	Joint data management system
JFF	Joint fact finding
JIT	Just in time
JPEO-CBRND	Joint Program Executive Office for Chemical, Biological, Radiological, and Nuclear Defense
JPL	Jet Propulsion Lab
JSON	JavaScript object notation
JST	Japan Science and Technology Agency
JUONs	Joint urgent operational needs
KDPI	Kidney donor profile index
KPAs	Key performance attributes
KPIs	Ker performance indices
KPPs	Key performance parameters
KSA	Knowledge, skills, abilities
KSAB	Knowledge, skills, abilities and behaviors
KSABC	Knowledge, skills, abilities, behaviors, cognition
KSAs	Key system attributes
L&R	Launch and recovery
LCS	Littoral combat ships
LE	Learning enabled
LE	GPS location
LGE	Laplacian graph energy
LNC	Laplacian natural connectivity
LogIT	Logistics information technology
LP	Linear programming
LTL	Linear temporal logic
M&S	Modeling and simulation
M.Eng. or ME	Master of Engineering
MS	Master of Science
M2M	Machine to machine
MA	Mission aware
MADM	Multiple attribute decision making
MAE	Mission assured engineering
MAPS	Modular active protection systems
MATLAB	Matrix Laboratory

MAUT	Multi-attribute utility theory
MAVE	Modeling analytical and visualization environment
MAVT	Multi-attribute value theory
MBA	Model-based assurance
MBSE	Model-based systems engineering
MBSEVE	Model-based systems engineering in virtual environments
MBTD	Mission-based test design
MCDM	Multiple-criteria decision-making
MCE	Model-centric engineering
MCM	Mine counter measures
MD	Mahalanobis distance
MDAO	Multidisciplinary design analysis and optimization
MDAP	Major Defense Acquisition Program
MDK	Model development kit
MDO	Multidisciplinary design optimization
MDP	Markov decision problem
MDS	Multidimensional scaling
ME	Mission engineering
MECF	Mission engineering competency framework
MF	Membership functions
MFD	Modular function deployment
MHACO	Multi-objective hypervolume based ant colony optimizer
MIM	Mission integration management
MISD	Model interface specification diagram
MISDP	Mixed integer semi-definite programming
MIT	Massachusetts Institute of Technology
ML	Machine learning
MMP	Minimal marketable products
MMS	Model management system
MOB	Main operating base
MOD	UK Ministry of Defense
MOEA/D	Multi-objective evolutionary algorithms based on decomposition
MOEAs	Multi-objective evolutionary algorithms
MoEs	Measures of effectiveness
MOLP	Multi-objective linear programming
MOMINLP	Multi-objective mixed integer nonlinear programming
MOO	Multi-objective optimization
MOOM	Multi-objective optimization methods
MOPs	Measures of performance
MOSA	Modular open systems approach
MP	Monterey Phoenix
MPL	Mars polar lander
MPT	Methods, processes, and tools
MPVIP	Monterey Phoenix Virtual Internship Program
MRAP	Mine resistant ambush protected

MRDB	Market research database
MS	Multi-stakeholder
MS	Motor speed
MSA	Material solution analysis
MSFC	Marshall Space Flight Center
MSL	Mission system library
MSTSE	Multi-stakeholder tradespace exploration
MUSTDO	Multi-stakeholder dynamic optimization
MVP	Minimum viable products
NAS	National airspace system
NASA	National Aeronautics and Space Administration
NAVAIR	Naval air systems command
NAVSEA	Naval sea systems command
NAVSEM	NAVAIR systems engineering method
NBC	Nuclear, biological, and chemical
NBI	Normal boundary intersection
NC	Natural connectivity
NCO	Non-combatant operations
NDI	Non-development items
NDIA	National Defense Industrial Association
NDS	National Defense Strategy
NFRs	Non-functional requirements
NIST	National Institute of Standards and Technology
NLGE	Normalized Laplacian graph energy
NLNC	Normalized Laplacian natural connectivity
NLP	Natural language processing
NLS	Natural language system
NOV	Net option value
NPS	Naval postgraduate school
NPV	Net present value
NRP	Naval research program
NSA	National Security Agency
NSGA	Non-dominated sorting genetic algorithms
NSS	National Security Strategy
NSWC-CR	Naval Surface Warfare Center-Crane, Division
NTEs	Non-critical technology elements
NTIs	Non-critical technology integrations
NUD	"new, unique, and difficult"
NWS	Naval warfare scenario
ODASD(SE)	Office of the Deputy Assistant Secretary of Defense for Systems Engineering
OEM	Original equipment manufacturer
OIR	Optimality influence range
OI	Orbital inclination
OODA	Observes orients decides and acts
OOSE	Object-oriented systems engineering
OPCAT	Object-process CASE tool

OpenMBEE	Open model based engineering environment
OPH	Organ procurement hospitals
OPM	Object process methodology
OPOs	Organ procurement organizations
OR	Operations research
ORS	Operationally responsive space
OSD	Office of the Secretary of Defense
OSD ME	Office of the Secretary of Defense Mission Engineering
OSMA	Orbital semi-major axis
OTA	Other transaction authority
OTS	Off-the-shelf
OUSD	Office of the Under Secretary of Defense
OUSD(R&E)	Office of the Undersecretary of Defense for Research and Engineering
OV	Operational viewpoint
OWASP	Open Web Application Security Project
OWL	Web ontology language
PBD	Point-based design
PCR	Platform configuration registers
PDF	Portable document format
PDF	Probability distribution function
PDR	Preliminary design review
PEO	Program Executive Office®
PERT	Project/Performance/Program Evaluation and Review Technique
Pf	Probability of failure
PFL	Pareto front learning
PGM	Probabilistic graph model
PhD	Doctor of Philosophy
PI	Pragmatic idealism
PIF	US Army Prototype Integration Facility
PLM	Product lifecycle management
PM	Program/Project management
PMBOK	Project management body of knowledge
PMI	Project Management Institute
PNG	Portable network graphics
POM	Program objective memorandum
POMDP	Partially observable Markov decision process
POs	Project owners
POSG	Partially observable stochastic games
PPBE	Planning, programming, budgeting, and execution
PRA	Probabilistic risk assessment
PSM	Practical software measurement
PSSM	Practical software and systems measurement
PuCC	Pugh controlled convergence
QAs	Quality attributes
QFD	Quality function deployment
R&D	Research and development

RA	Refactor adjustment
RAM	Reliability, availability, and maintainability
RAST	Recovery assist, secure and traverse system
RCM	Reliability-centered maintenance
RCO	Air Force Rapid Capabilities Office
RCV	Remote controlled vehicles
RDF	Resource description framework
RDP	Reality, domain, perspective
RDT&E	Research development test and evaluation
REST	Representational state transfer
RF	Rework fraction
RFC	Request for change
RFI	Request for information
RFP	Request for proposal
RMP	Risk management plan
ROI	Return on investment
RPO	Robust portfolio optimization
RT	Research task
RV	Rebuild value
S&T	Science and technology
SA	System agent
SA	Situation awareness
SA	System architecture
SAM	Surface to air missiles
SAMs	Stakeholder analysis model
SAR	Synthetic aperture radar
SAT Solvers	Boolean satisfiability solvers
SAW	Simple additive weighting
SBCE	Set-based concurrent engineering
SBD	Set-based design
SBT	Set-based thinking
SD	System dynamics
SDCI	System disruption conditional importance
SDDA	Systems developmental dependency analysis
SDI	System disruption importance
SE	System engineering
SEBok	Systems engineering body of knowledge
SEI	Software Engineering Institute
SEP	Systems engineering plan
SERC	Systems Engineering Research Center
SET	Systems engineering transformation
SETR	Systems engineering technical review
SF	Strategic fit
SHACL	Shapes constraint language
SIMs	System importance measures
SMAD	Space mission analysis and design

SMA	SoS Manager Agent
SMART	Specific measurable achievable relevant time-bound
SME	Subject-matter experts
SMs	ScrumMaster
SMS-EMOA	S-metric selection evolutionary multi-objective optimization algorithm
SNRs	Signal-to-noise ratios
SOA	Service-oriented architecture
SOCOM	Special operations command
SODA	Systems operational dependency analysis
SOF	Special operations forces
SOGA	Self-organizing genetic algorithm
SORDAC	SOCOM Research and Development Acquisition Center
SoS	System of systems
SoSE	System of systems engineering
SoSWG	System of Systems Working Group
SOW	Statement of work
SPARQL	Simple protocol and RDF (resource description framework) query language
SPAWAR	Space warfare
SQL	Structured query language
SQOTA	System qualities ontology tradespace and affordability
SQs	System qualities
SRA	System readiness assessment
SRL	System readiness level
SSM	Six step model
STAMP	Systems-theoretic accident model and process
STEM	Science technology engineering math
STL	Stereolithography
STPA	Systems theoretic process analysis
STPA-Sec	Systems theoretic process analysis for security
StSTL	Stochastic signal temporal logic
SUV	Suburban vehicle
SUW	Surface warfare
SV	Space vehicle
SV	Strategic value
SV	Systems viewpoint
SvcV	Services viewpoint
SW	Software
SWaP	Size weight and power
SWCM	Software configuration management
SWEBOK	Software engineering body of knowledge
SWFTS	Submarine warfare federated tactical systems
SWRL	Semantic web rule language
SWT	Semantic web technologies
SysML	System modeling language
TA	Threat agent
TARDEC	Tank and Automotive Research Development and Engineering Center

TAT-C	Trade-space analysis tool for designing constellations
TBD	To be determined
TBI	Traumatic brain injury
TCs	Transplant centers
TD	Technical debt
TDD	Test driven design
TDM	Technical debt management
TE	Test and evaluation
TF-IDK	Term frequency inverse document frequency
TIMs	Technical interchange meetings
TM	Traditional manufacturing
TMRR	Technology maturation risk reduction
TOGAF	The open group architecture framework
TOPSIS	The technique for ordered preference by similarity to an ideal solution
TOTP	Time-based one-time password
TPM	Trusted platform module
TRA	Technology readiness assessment
TREEs	Testable resilience efficacy elements
TRL	Technology readiness level
TRRB	Technical readiness review boards
TSE	Tradespace exploration
TTPs	Tactics, techniques, and procedure
TWC	Teamwork cloud
TxFx	Transfer functions
UARC	University Affiliated Research Center
UAS	Unmanned aircraft system
UAV	Unmanned aerial vehicle
UCARS	UAV common automatic recovery system
UML	Unified modeling language
UNOS	United network for organ sharing
UPSTAGE	Universal platform for simulating tasks and actors with graphs and events
UQ	Uncertainty quantification
USAF	United States Air Force
USD	US Dollars
USSF	US Space Force
UV LIDAR	Ultraviolet light detection and ranging
UWV	Underwater vehicles
V&V	Verification and validation
VR	Virtual reality
VV&A	Verification validation and accreditation
WBS	Work breakdown structure
XML	Extensible markup language

About the Companion Website

This book is accompanied by a companion website:

www.wiley.com/go/verma/systemsengineering

Part I

Transforming Engineering Through Digital and Model-Based Methods

Mark Blackburn

Chapter 1

Fundamentals of Digital Engineering

Mark R. Blackburn and Timothy D. West

Stevens Institute of Technology, School of Systems and Enterprises, Hoboken, NJ, USA

Background

Between 2013 and 2021, the Systems Engineering Research Center (SERC) partnered with the Naval Air Systems Command (NAVAIR) to explore the viability of transforming systems engineering through holistic model-centric system engineering. The vision driving this exploration was to leverage mission and system-based analyses and engineering to reduce the development and integration cycle time by at least 25% for large-scale air vehicle systems (Blackburn et al. 2021b). NAVAIR had a rigorous Systems Engineering Technical Review (SETR) process that incorporated over 6000 analysis and design checks. SETRs utilized a time-consuming manual, paper-based approach that made it hard to conduct asynchronous reviews as the design of different aspects of the system was maturing at different rates. In addition, the increasing complexity of missions and systems often yielded anomalous or contradictory requirements that prolonged system delivery by months. After one such situation, which was identified by the contractor, a NAVAIR lead engineer requested that the team model subsystem interactions that had resulted in the contradictory requirements. They discovered the underlying contradiction in the model and updated the requirements. This effort took approximately seven months. Similar cases shared with the research team became the motivation to develop a model-based approach that would transform systems engineering through digital engineering.

The eight-year SERC research task for NAVAIR consisted of three major phases. During Phase One, from 2013 to 2015, the research team worked with the sponsor to perform a global scan of holistic and advanced approaches to Model Centric Engineering (MCE)/DE, with over 30 organizations providing their best examples and demonstrations. We met at engineering and manufacturing locations of industry (e.g., large DoD contractors and automakers) and government (e.g., NASA/JPL, Sandia National Labs); we usually spent a day seeing demonstrations and learning about the most advanced technologies being used by these organizations. Phase One culminated in late in 2015, when NAVAIR leadership decided to move quickly to keep pace with other organizations that have adopted DE. NAVAIR leadership desired to transform, not simply evolve, in order to perform effective oversight of prime contractors that are using modern modeling methods for mission, system, and discipline-specific engineering (Blackburn et al. 2015).

Systems Engineering for the Digital Age: Practitioner Perspectives, First Edition. Edited by Dinesh Verma.

Companion website: www.wiley.com/go/verma/systemsengineering

Phase Two began in early 2016, when the NAVAIR sponsor introduced the Systems Engineering Transformation (SET) Framework concept discussed in the New Approach Section. There was also a change in leadership who decided to accelerate SET (Blackburn et al. 2017). NAVAIR had used systems engineering rigor for many years and applied that rigor to systematic planning that develops six (6) functional areas, including SERC Research as shown in Figure 1.1. The SERC Research eventually contributed to many other facets of the SET plan, which are shown with round-edge boxes with dashed lines, which many examples captured as models for a number of Surrogate Experiment use cases.

Surrogate Pilot Experiment Called Skyzer

The Surrogate Pilot experiments were conducted by a combination of SERC researchers, NAVAIR engineers, contractors, and a surrogate contractor over three phases starting in late 2017. Phase one (1) developed a Search and Rescue mission use case called Skyzer that culminated at the end of 2018 with the development of mission and systems models that linked to the request for proposal (RFP) response model from the surrogate contractor. The focus was on characterizing how to run a model-based acquisition for a program that is based on the SET Framework concept. The results produced an implementation that demonstrated how to fully link models in an authoritative source of truth (AST); the initial thinking was to establish a single source of truth, but as Skyzer developed into a distributed set of linked models, AST was deemed to be more descriptive of the linked Skyzer models. The team demonstrated the art-of-the-possible by doing "everything" in models, simulating a new operational paradigm between government and industry in a collaborative AST using the NASA JPL's Open Model Based Engineering Environment (OpenMBEE) combined with No Magic SysML model authoring tools and a model repository called Teamwork Cloud. The surrogate contractor's RFP response refined the mission and system models, creating subsystem models with detailed design and analysis information using multiple physics-based and discipline-specific models. We also introduced Digital Signoffs for the Source Selection technical evaluation, completed directly in the surrogate contractor's RFP response model using the View Editor and DocGen, which are components of OpenMBEE. We leveraged the unclassified surrogate pilot results and models to facilitate workforce development training that is being used by Defense Acquisition University to update its DE curriculum.

Based on the success of Phase One of the Skyzer surrogate pilot, NAVAIR commissioned a Phase Two (2) effort starting in 2019, but with new objectives from evolving SET priorities (Blackburn et al. 2021b). The most important priority was to align the Skyzer mission and system models with the new and evolving NAVAIR Systems Engineering Method (NAVSEM). During this phase, we demonstrated how Digital Signoffs in AST can transform Contract Data Requirements Lists (CDRLs) and Data Item Descriptions (DIDs). We created a Capability Based Test & Evaluation (CBT&E) model based on a Model-Based Testing Engineering process that linked to the mission and system models. Additionally, we demonstrated the use of models for capturing Statement of Work (SOW) language, which furthered the workforce development and training effort. We also joined our sponsors in presenting the models, methods, digital signoffs, and DE environments results to several contractors to get feedback on the DE approach that the sponsor would use on future programs.

Surrogate Pilot Phase Three (3) began in 2020 and ended in August of 2021 (Blackburn et al. 2021b) and focused on presenting results and providing demonstrations to encourage program deployment. We continued aligning mission, system, and surrogate contractor RFP models with the evolving NAVSEM modeling method. We also created Stakeholder Analysis Model (SAMs) for Airworthiness (criteria to get flight clearance) derived from MIL-STD-516C (2014) that aligns

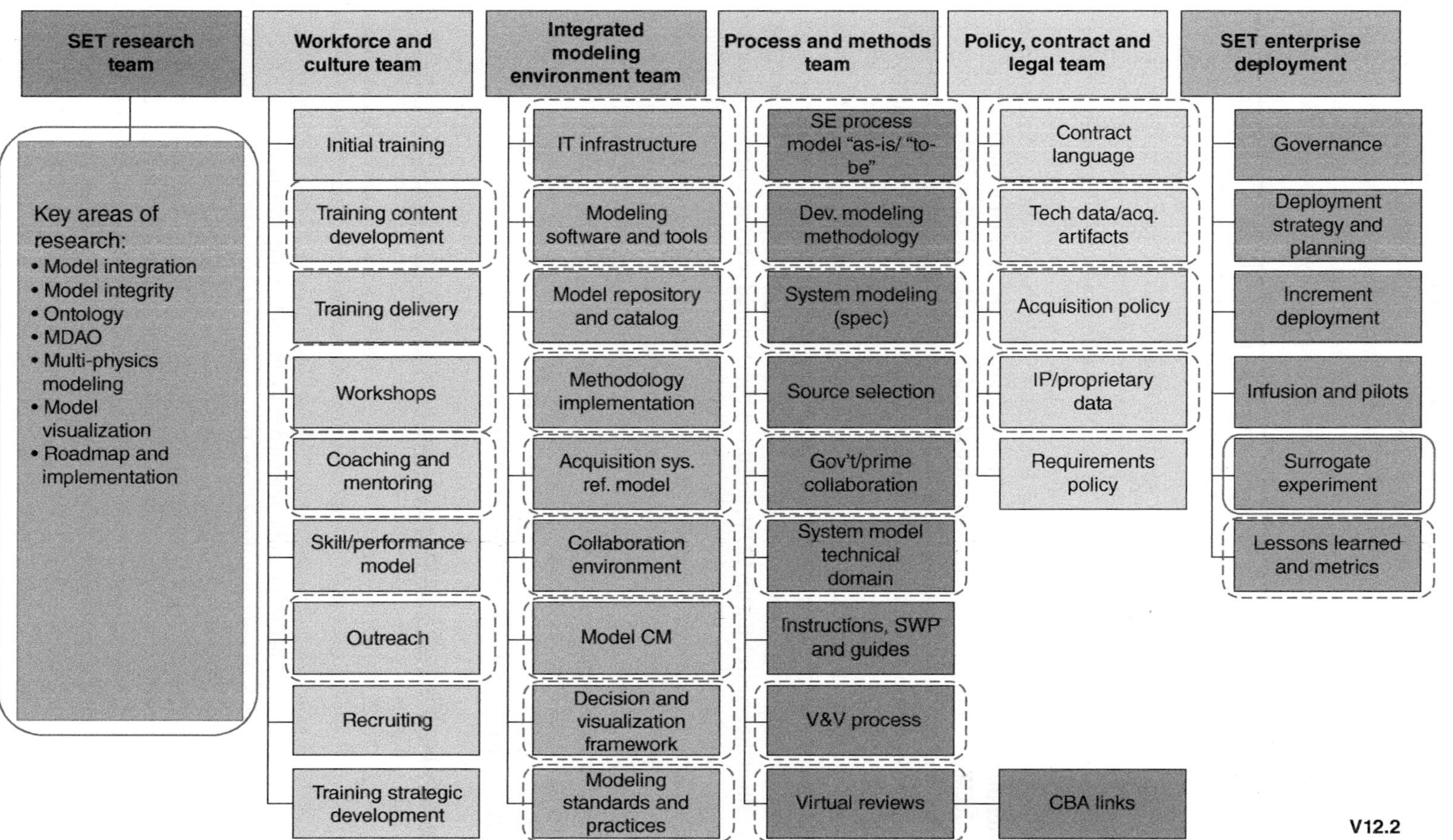

Figure 1.1 SET functional areas with impacts on SET research and surrogate pilot.

and links with Skyzer technical data in mission and systems models. We created a SAM for a Cost Model derived from MIL-STD-881E that aligns and links with the Skyzer technical data in mission and systems models, as well as link to the CBT&E and Airworthiness models to factor in cost from those efforts. We created an example using Model Curation criteria applied to Skyzer models. We used the SAM models to demonstrate how to derive mission and system-specific model information from reference models.

During Phase Three, the SERC researchers also supported additional efforts, such as (1) performing a Cyber Ontology Pilot using ontologies that have been transformed into training material for US Army DEVCOM sponsor, (2) investigating how Digital Signoffs (Blackburn and Kruse 2023) can contribute to a new form of technical Baselines Progress Measures to transform away from monolithic reviews, and (3) correlating Digital Engineering Success Measure (DESM) categories with lessons learned benefits observed during the NAVAIR Surrogate Pilot to further explain the benefits demonstrated by Skyzer (McDermott et al. 2020).

In 2016, we started a similar SERC research task for the US Army, focused on DE approaches that could better inform science and technology decision making, which often requires the fabrication of pre-production prototypes. Our research included modeling and simulation of discipline-specific, multi-physics analyses to help them understand system and mission requirements and tradeoffs. We developed the interoperability and integration framework (IoIF), a computationally based framework that uses ontologies (a.k.a., information models) and semantic web technology to link different model data at different levels of abstraction with multi-physics and discipline-specific model data, facilitating greater cross-domain model integration for analysis and design. We also developed two IoIF training course based on the Cyber Ontology use case, and a Catapult use case to help transition the research to DEVCOM, and this is being shared with the US Space Force (USSF).

In 2022, we started another SERC research task with the USSF. They have similar challenges like NAVAIR; we have developed DE training material from the combined efforts of NAVAIR and DEVCOM. We are using the modeling pattern developed for Skyzer to create another surrogate set of models aligned with the domain of USSF called Spacer, with fire-monitoring satellites that communicate with ground systems.

Problem

There are many possible ways to describe the challenges faced by the research sponsors. We characterized a simplified perspective on "traditional mission and systems engineering" process phases (Blackburn et al. 2018a, 2018b), as shown in Figure 1.2. We created a few surrogate use cases (i.e., fictious, but hypothetically relevant to the sponsor mission and systems). The information includes various types of linked modes and associated model artifacts that represent information for these different processes. Model artifact can be mapped into an underlying linked information model that enables a decision framework based on the Integrated Systems Engineering Decision Management (ISEDM) process for comparing different mission, system, and subsystem tradeoffs (Cilli 2015), which map back to the Concept of Operations (CONOPs). The fundamental problem faced by most organization was that the vast amount of information needed to make decisions was not integrated across the domains (e.g., mechanical, electrical, software, etc.) nor at different levels of abstraction (e.g., missions, systems, subsystems). Traditionally data and information are captured in disparate forms and documents. The relationships between these artifacts are often informal and must be understood, created, updated, and managed manually. The research sponsors, as well as industry contractors, understood that significant advances in computationally based

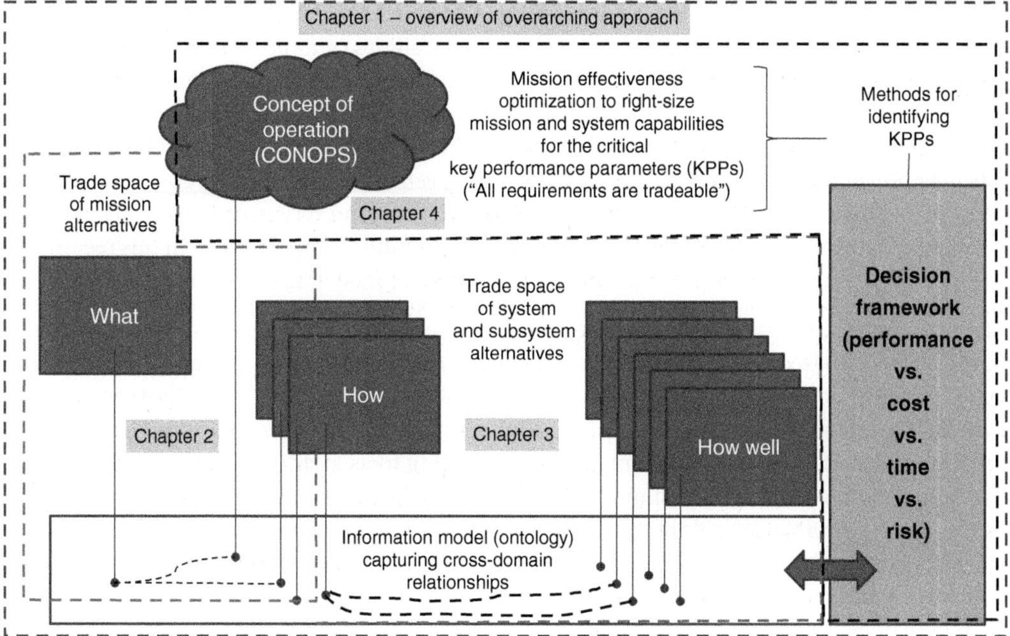

Figure 1.2 Context of systems engineering of challenge areas.

methods, models, and tools had occurred, but they had been unable to transform these advances into a comprehensive digital engineering ecosystem. Therefore, we use the term digital engineering in the broadest sense to factor in all facets of analysis, design, and configuration management using models and simulations appropriate for the purpose and need. In addition, if any information is changed, it should trigger the re-analysis of other related information to understand the potential impacts of the changes.

We notionally defined the relationships between information captured in a traditional "Vee" approach as it relates to information presented in five book chapters for DE.

Chapter 1: **Fundamentals of Digital Engineering** provides an overview of the overarching approach for formalizing the processes shown in Figure 1.2 using different use case examples, which are discussed in more detail in the other four book chapters.

Chapter 2: **Mission and Systems Engineering Methods** discusses "What" we want – requirements and constraints, as well as supporting structural and behavioral analyses, which refine one level of requirements (e.g., mission) down through lower levels of requirement (e.g., systems).

Chapter 3: **Transforming Systems Engineering Through Integrating Modeling and Simulation and Digital Threads** discusses more about "How" (1 or more) – designs to achieve the "What," as well as "How well" (usually many) to assess the "How" using analysis, testing, reviews, and assessing how the design satisfies the requirements, given the constraints to achieve the mission concept and mission measures. This is done using the formalization of the underlying "Information Model" that links the data or metadata from many different domains at different levels of abstraction using IoIF.

Chapter 4: **Digital Engineering Visualization Technologies and Techniques** discusses how the formalization of different models discussed in Chapters 1–3 fit into a continuous loop using enabling technologies and IoIF. We transformed from a static, document-based CONOPs into a dynamic graphical CONOPs enabled by gaming technologies to assist stakeholders and developers reach a shared mental model of the required capabilities of a future system. The Decision Framework enabled by IoIF demonstrates how data from the information model data can be used to populate the Decision Framework to determine the Key Performance Parameters (KPPs) and other mission measures of the various stakeholders based on various types of analyses and help optimize capabilities for mission effectiveness given time, cost, and risk considerations.

Chapter 5: **Interactive Model-Centric Systems Engineering** discusses how humans use or should use models to support decision making and communication of results. In this regard, the information in the chapter therefore applies to all types of modeling shown in Figure 5.2.

Our research team formalized this conceptional characterization of mission, system, subsystem, multiple physics-based, and discipline-specific engineering models and linked interoperable models to demonstrate the art-of-the-possible. The formalized Information Model concept is done using IoIF combined with domain and application ontologies; ontologies can be thought about as a type of schema for linking data from different modeling tools at different levels of abstraction. This formalization of models and simulations allowed us to demonstrate how to do integration and testing on the proverbial left side of the "Vee." Historical evidence indicates that system integration and testing often uncover many defects (Chilenski and Ward 2014) that can be traced back to contradictory requirements, as discussed above.

The remainder of this chapter provides an overview of the NAVAIR and US Army DEVCOM research. The subsequent chapters provide additional details for each of the major contributions to Digital Engineering.

New Approach

Skyzer Pilot Context

In 2017, the NAVAIR leadership decided to conduct surrogate pilot experiments using an evolving set of use cases to simulate the execution of the new SET Framework concept, shown in Figure 1.3[1] as part of the SET Enterprise Deployment. The notional concept was to move away

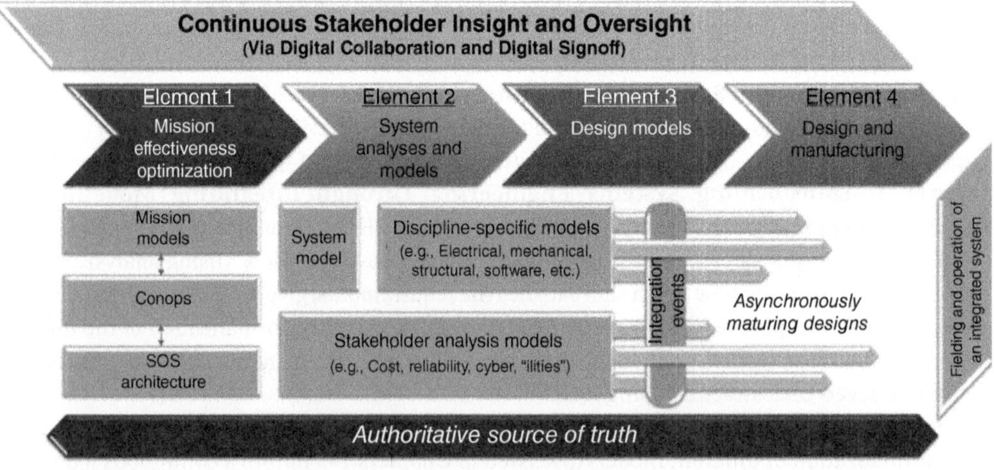

Figure 1.3 NAVAIR systems engineering transformation framework.

from the traditional "Vee" process approach into a more iterative approach. This notional framework represents a new operational paradigm between government and industry, but can apply equally well with contractors and subcontractors. The research provided analyses into the needed enterprise capabilities and built on efforts for cross-domain model integration, model integrity, ontologies, semantic web technologies, multi-physics modeling, and model visualization addressing evolving needs and priorities of SET (SERC RT-157/170/195/1008/1036) (Blackburn et al. 2021b). The research continuously demonstrated the benefits of modeling methods enabled by computational automation, and a Digital Engineering Environment that enabled collaboration in an AST.

The SET Framework looked for an iterative approach for considering the mission needs and measure for tradespace analyses using Multidisciplinary Design Analysis and Optimization (MDAO) at mission (Element 1), system (Element 2), and subsystem levels as extended by the contractor (Element 3), with continuous asynchronous reviews in a digital collaborative environment and interactions, such as digital signoffs within the AST (Elements 3 and 4). The concept of a collaborative AST should enable the acquisition organization's subject matter experts (SMEs) to have insight and oversight by seeing the model information, as well as the modeling methods used to produce analysis results to support for making decisions (Kruse et al. 2020). The results of the decisions could be captured as digital signoffs directly in the models by replacing traditional SETRs. Such digital signoffs should be done continuously and asynchronously as the design matures.

Our SERC team created both linked models and established an implementation of a digital engineering environment (DEE) that simulated the execution of the SET Framework concept up to and including some of Element 3. The results demonstrate improved **consistency, traceability, collaboration, and information sharing**, and align with the five goals of the DoD Digital Engineering Strategy (Deputy Assistant Secretary of Defense (Systems Engineering) (2018)):

1) Formalize the development, integration, and use of models to inform enterprise and program decision-making.
2) Provide an enduring, authoritative source of truth.
3) Continuously incorporate technological innovation to improve the engineering practice.
4) Establish a supporting infrastructure and environments to perform activities, collaborate, and communicate across stakeholders.
5) Transform the culture and workforce to adopt and support digital engineering across the lifecycle.

One of the best early decisions in the surrogate pilot experiments was to "model everything," not because one would normally do that, but to demonstrate the art-of-the-possible. This approach made everything accessible in the context of descriptive models using the System Modeling Language (SysML). These descriptive models formalize information about the system structure, behaviors, interfaces, and requirements and demonstrated how such an approach can completely replace documents. We used NASA/JPL developed OpenMBEE, which provided collaborative access to the simulated government team members as well as the industry surrogate contractor. OpenMBEE also provided the DocGen capabilities (Kruse and Blackburn 2019), which permitted all stakeholders access to the model information using a web-based representation of the model. DocGen creates stakeholder-relevant views extracted directly from the modeled information, so SMEs can visualize modeled information in the OpenMBEE View Editor and perform digital signoffs, even if they lack SysML training or tools. The View Editor is a component of OpenMBEE that allows users to edit or comment on information generated from the model using a web browser, where edits in the View Editor can be synchronized back into the model if desired.

We developed a modeling modularization method enabled by Magic Draw's Project Usage model import capability, demonstrating the concept of and AST (Goal #2). We modeled everything demonstrating how one can transform many traditional document-centric activities, for example, conducting certain types of SETRs in the DEE and eliminating some types of SETR checks that are subsumed into the modeling process. More importantly all models were linked together in the AST to promote **collaboration/info sharing**, **information access**, **reduce defects**, improve **consistency**, **improve traceability**, and eliminate some types of work for greater **efficiencies**.

We used descriptive modeling methods to extend beyond processes guidelines (the "what") and developed the analysis and design artifacts (the "how") that should be modeled to have sufficient and relevant information to make decisions (Goal #1). Descriptive modeling languages should include structure (data properties, parts), behavior, interfaces, and requirements. Descriptive **modeling methods** are needed for different abstraction levels such as mission, system, contractor refinement of the system model, subsystem, and discipline-specific models. A method also defines the types of relationships between the artifacts, which often provides information about cross-domain relationships and dependencies (Goal #1). Technology features that complement methods are the use of View and Viewpoints, which are inputs to OpenMBEE DocGen. A View and Viewpoint can be used to define the needed model artifacts that are associated with the desired modeling method. Methods define the required types of artifacts, which again leads to **consistency**, better **understanding of the system architecture**, **standardization**, as well as to more effectively assess **completeness** of the generated "specification."

The new approach also employed digital signoffs in the AST to provide an example for how to transform from document deliverables to support asynchronous reviews enabled by **collaborative information sharing** in the AST. Digital signoffs link criteria traditionally required in a document at various program review points to link with model artifacts as evidence. We determined an approach to use OpenMBEE View and Viewpoints as means for placing a digital signoff directly with model information that provided the needed evidence. Digital signoffs can be updated in the OpenMBEE View Editor, with the signoff information (e.g., approval, risk, approver, comments) added that gets synchronized back into the model. We also established a basis for automating digital signoff metrics. If a piece of information associated with a digital signoff is changed, we demonstrated how the signoff can be automatically reverted to a value such as "to be determined" (TBD) for triggering the need for another review using the View Editor. This should **reduce cost** by transforming/eliminating documents that take on a new form in the model providing greater **efficiency**, **consistency**, **automation**, and **standardization**.

These examples represent the most important lessons learned through Skyzer such as Consistency, Collaboration/Info Sharing, Automation, Information Access, Improve Traceability, and Collaboration Environment in an AST, which are part of the DE technologies. We do know that there might be some perception that modeling takes longer, but we also know that the **increased rigor reduces defects**, especially cross-domain, or level-to-level (mission to system), because all the models are linked together (i.e., **traceability**) using enabling technologies such as model imports. We are also able to render and edits these models to **collaborate** and **share information** directly in a "cloud-like" way. The models **increased rigor** using formal **standardized** languages enabled more **automation** to increase **productivity** and provide greater **efficiencies**; these should result in **reduced time**.

Figure 1.4 shows the DE for Systems Engineering Roadmap[2] that depicts the DE Strategic Goals and some enabling technologies that have been demonstrated by surrogate models for our research sponsors, where the semantics underlying the DE information and technologies will

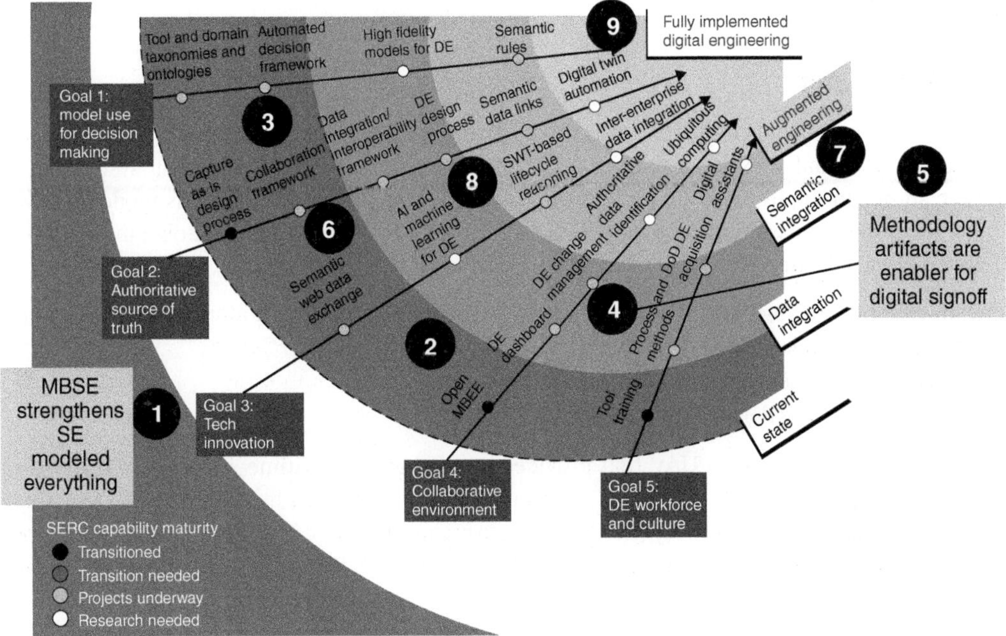

Figure 1.4 Digital engineering for systems engineering roadmap.

evolve and align leading to advancements in DEE. The numbered items characterize one scenario on the roadmap:

1) We refer to Model Based Systems Engineering (MBSE), which is supported by descriptive modeling tools such as Magic Draw as a Technical Innovation (Goal 3) that has been strengthening traditional System Engineering as it has been evolving over the past 15 years.
2) OpenMBEE played an important role in the surrogate pilot experiments to support a Collaborative Environment (Goal 4), where many of these same capabilities are part of the No Magic (now 3D Catia) suite of tools.
3) OpenMBEE also provided support for an AST (Goal 2), but there are several methodological rules that must be followed to support the AST as a Collaboration Framework. (Several enabling technologies are part of No Magic toolset.)
4) NAVAIR established NAVSEM modeling method that characterized modeling artifacts (how) for the different process steps (what).
5) Those modeling method artifacts provided the basis for enabling digital signoffs, because a signoff is associated with a digital artifact.
6) Semantic data exchange is another technical innovation (Goal 3) that provides a basis for greater tool interoperability needed for collaboration that has been demonstrated with the DEVCOM use cases using IoIF.
7) We envision that improving semantics integration will lead to artificial intelligence (AI) as part of domain-based knowledge representation.
8) The use of artificial intelligence (AI) and machine learning (ML) with the advancements of digital assistants in the future can promote more automation with modeling method compliance supporting augmented intelligence for engineering.
9) All these lead to a more fully integrated digital engineering environment supporting richer decision-making across the system domains under various mission scenarios.

Results

NAVAIR and Skyzer

The Skyzer experiments conducted over three (3) surrogate pilot phases evolved and elaborated mission, system and subsystem analyses, and requirements using a hypothetical system called Skyzer, shown in Figure 1.5. In addition, three (3) different stakeholder analysis models (SAMs) link directly to the mission, system, and RFP response (contractor) models as shown in Figure 1.6; those SAM models include:

1) Capability-Based Test & Evaluation (CBT&E), based on NAVAIR modeling approach for Mission-Based Test Design (MBTD)
2) Airworthiness Model derived from MIL-STD 516C
3) MBSE Cost Models derived from MIL-STD 881E

Skyzer's CONOPs is for an UAV that provides humanitarian maritime support for search and rescue use cases as reflected in Figure 1.5. This use case was extended in Phase Two (2) to include a ship-based Launch and Recovery (L&R) system to create another capability to research methods for CBT&E, based on a NAVAIR modeling approach for MBTD. Phase Three (3) developed a deep dive for the landing gear to examine scenarios for modeling information related to Airworthiness (the process used to get a flight clearance). In addition, Phase Three developed an MBSE Cost Model, which used the Skyzer Airframe characteristics for the cost modeling basis; the Airframe characteristics are derived from the Phase One contractor model. The scope of the UAV design included multi-physics design considerations that are based on computational fluid dynamics (CFD) to analyze the airflow over the surface of the vehicle, structural integrity, weight, and vehicle packaging. The surrogate pilot team simulated a request for information (RFI) before officially releasing the RFP in the form of models concluding the Phase One effort. Performance requirements such as a cruise speed of 170 knots forced the design of Skyzer to be something other than

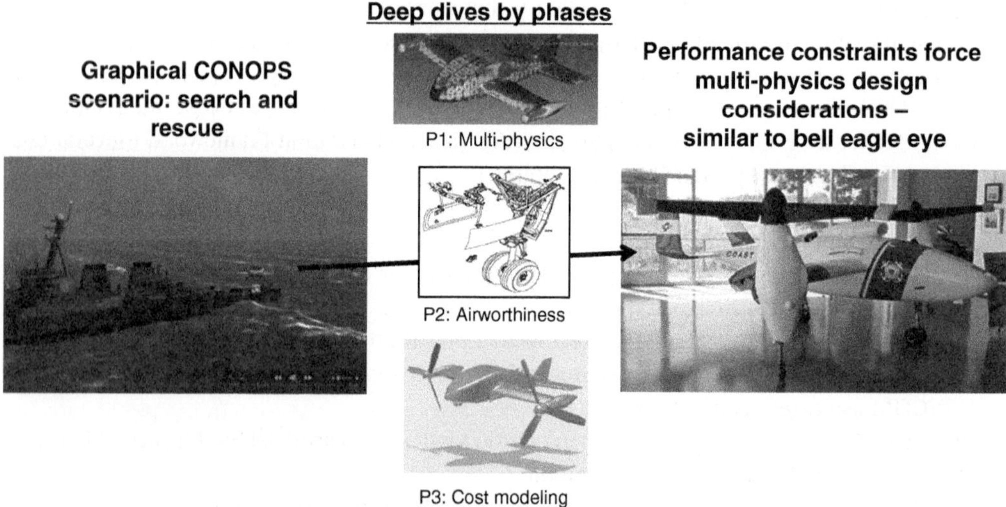

Figure 1.5 Graphical CONOPS for Skyzer UAV. *Source:* Lestocq/Wikimedia Commons/CC BY-SA 4.0.

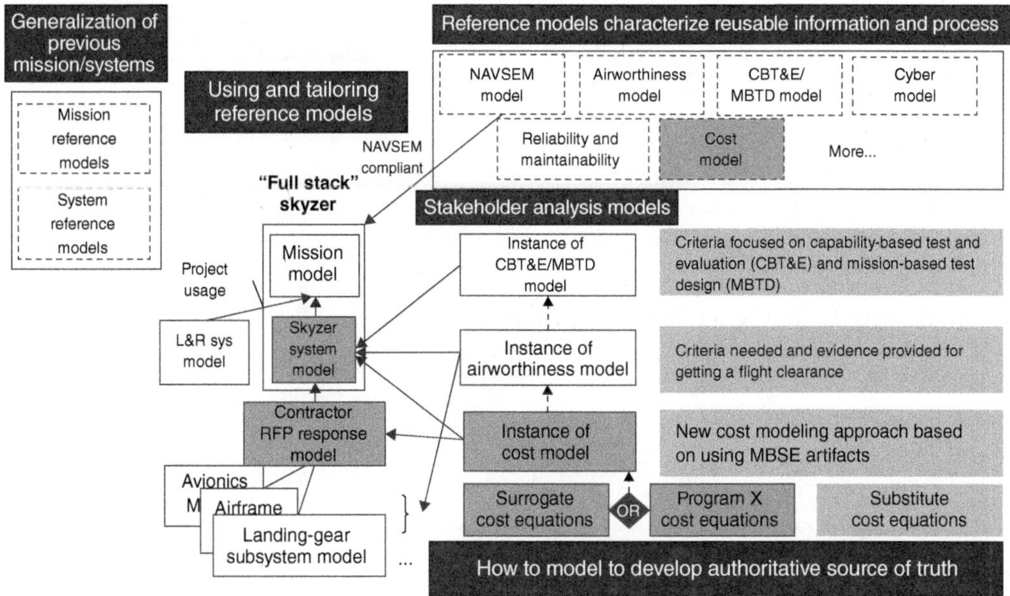

Figure 1.6 "Full Stack" of modularized models.

a traditional helicopter and ultimately a design similar to the Bell Eagle Eye ("Bell Eagle Eye" 2022), which was proposed in the surrogate contractor RFP response models. This proposed design was evaluated in a surrogate source selection that was embedded as part of the contractor RFP response by the sponsor team (a retired NAVAIR SME who had no SysML training) using digital signoffs in the OpenMBEE View Editor. The efforts during Phase Three, in addition to the Airworthiness and Cost Model, included updates to digital signoffs, models to align with the evolution of NAVSEM, and recommendations for adding guidance to NAVSEM based on needs identified in the evolution of the modeling for Skyzer.

The team discovered several important contributions as Skyzer was used to explain and demonstrate what it means to have an AST. These include how modeling methods characterize the needed modeling artifacts that enable the digital signoff concept, methods for model modularization to ensure separation of concerns, classification, acquisition, and how technical models at the mission and system model link to SAMs for Test and Evaluation, Airworthiness and Cost models. These SAM models link back to the "Full Stack" of technical models as reflected in Figure 1.6. Additionally, the SAM Cost Model links to the "Full Stack," but can also link directly to the Airworthiness and CBT&E models. The surrogate contractor model was linked to the government models of the AST. The NAVAIR sponsors believed that these types of model examples provide demonstrations to justify program adoption as well as support workforce development.

Three types of reference models shown in Figure 1.6. The first, shown in the upper left corner are reference models that generalize prior designs. These "reference architectures" can be used as a starting point (starter models) for a new design and may contribute to 60% or more of the design to accelerate the analysis and design process. Reference architectures often capture best practices and optimal designs from prior tradeoff analyses and actual deployments; such reference architectures may also bring verification support such as testing, which can also be used to reduce the time for developing a new system. Notice that reference architectures can also be used to represent prior missions such as Search and Rescue.

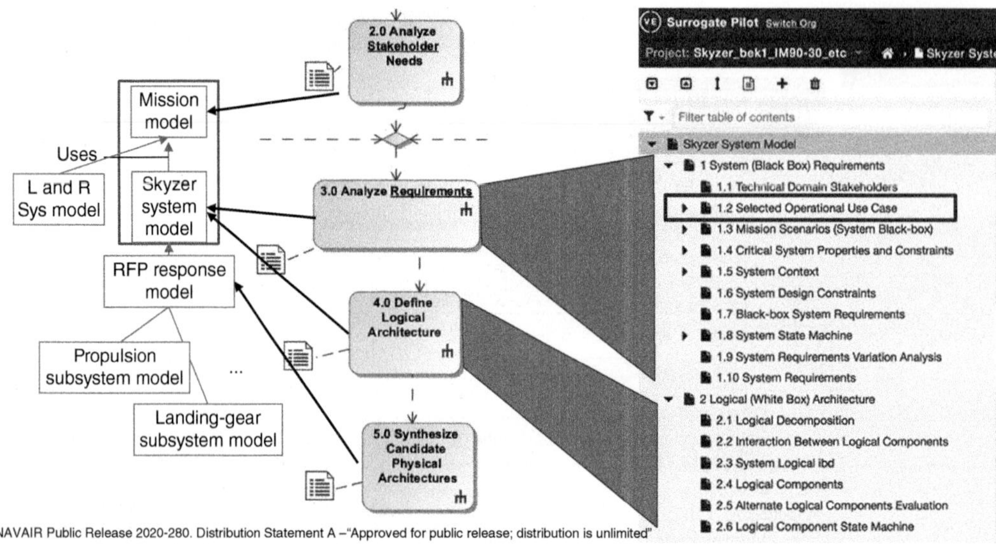

Figure 1.7 Update View and Viewpoint for Skyzer system model.

The second type of reference model is a process model. NAVSEM is a reference model that defines the process guidelines (what) and artifacts (how) that are captured in the Skyzer "Full Stack" of technical models. The NAVSEM method is formalized primarily using a SysML activity diagram, as shown in Figure 1.7. NAVSEM also describes the artifacts associated with each process such as the Selected Operational Use Case, which is represented using a SysML use case diagram. Skyzer also defined digital signoffs for each type of artifact.

The third type of reference model defines the superset of all criteria that may be used for any modeling effort. For example, the Airworthiness reference model includes all the criteria from MIL-STD 516C (2014) along with best practice guidance from years of use by SMEs. An "Instance of Airworthiness" model for Skyzer as shown in Figure 1.6 uses a subset of the criteria that relates to the actual technical data of Skyzer's "Full Stack." For example, the Airworthiness reference model includes criteria for helicopters and air vehicles with a pilot, which would not apply to Skyzer, because it is an unmanned tilt-rotor aircraft. The Cost Model is a reference model based on MIL-STD 881E but also incorporates other methodological guidance captured by NAVAIR SMEs over years of doing cost analysis on system programs. The Cost Model links to the "Full Stack" of technical models but also links to the Airworthiness and CBT&E models to factor in costs associated with getting a flight clearance, which could involve flight testing. The Cost Model was developed with generic cost equations (surrogates), and based on the selection of a contractor, NAVAIR has captured cost equations for the different contractors based on prior programs. This allows NAVAIR to incorporate vendor-specific cost equations based on historical cost data into their source selection analyses. Once the program starts, costs such as Total Ownership Cost can be continuously tracked during project execution as the design matures. In turn, this information can be used to refine the historical cost data for each contractor.

In addition to the models shown in Figure 1.6, the research team developed other models, including (1) a Surrogate Project/Planning Model that characterizes the objectives for the surrogate pilot and research, (2) the Systems Engineering Technical and Management Plan model, a type of DE Systems Engineering Plan, (3) a Surrogate Acquisition Model for Skyzer to support Source Selection Evaluation and Estimation, and (4) View and Viewpoints for DocGen and other

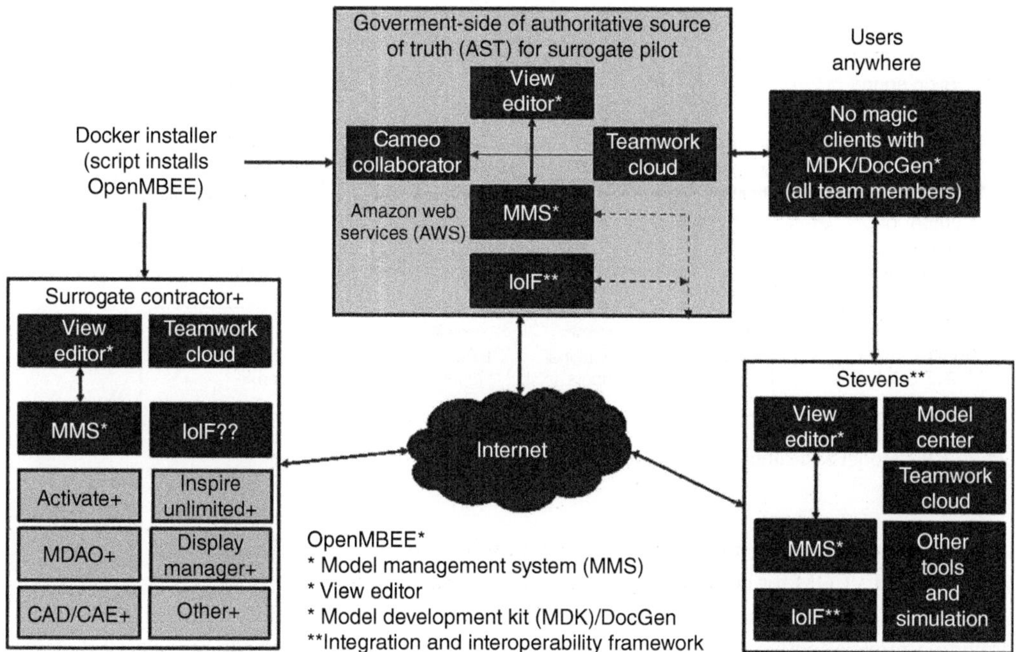

Figure 1.8 Digital Engineering Environment (DEE) elements of authoritative source of truth.

model Libraries that are used in conjunction with DocGen to generate the specifications from the models based on stakeholder views. This research produced over 60 products that include models, presentation, reports, videos, and links to the surrogate pilot autogenerated models.

Figure 1.6 represents the data that were formalized as models using the DEE shown in Figure 1.8.[3] The DEE capabilities support modeling for the Skyzer AST, model management system (MMS) (like configuration management), DocGen, and collaboration through web-based browser to view the information generated from the model, as well as discipline-specific and multi-physics model and simulations. This shows that the models for the AST shown in Figure 1.6 are produced by a distributed set of tools in various environments; this perspective best illustrates why the AST is not a single source of truth. These and other details are covered in Chapter 2.

Interoperability and Integration Framework (IoIF)

Our research with US Army DEVCOM started in 2016 (Blackburn et al. 2017, 2018b) and revealed several key challenges involving the use of many types of modeling and simulation tools. In general, the tools were not integrated. Subject matter experts (SMEs) would often receive document-based requirements that would be used to create some type of analysis models (e.g., aerodynamics, stress, thermal) and run various simulations to produce results that were captured in a report that is given to another SME (e.g., geometry), who would then repeat the process in their area of expertise. The typical result was a series of stove-piped efforts that lacked the model integration needed for true model-based design and development. DEVCOM needed a way to better integrate models for simulations across the various domains at different levels of abstraction to support an iterative process for tradespace analysis and decision-making. Figure 1.9 below, excerpted from a SERC

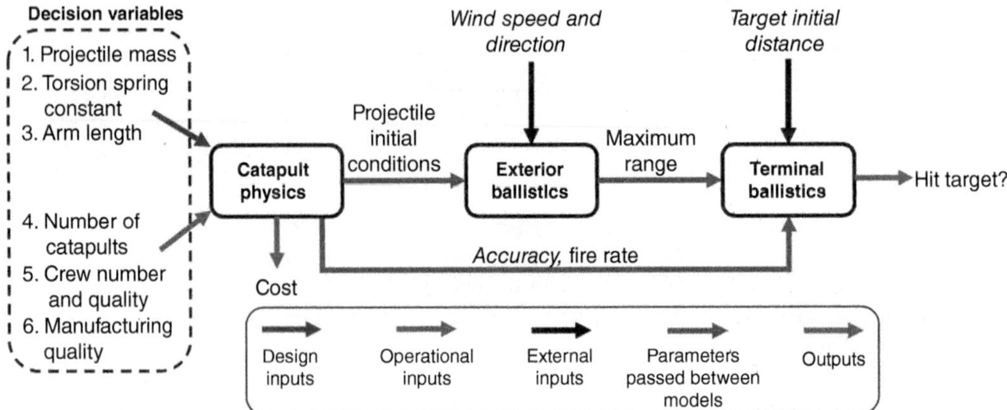

Figure 1.9 Model and simulation interactions for mission level optimization.

researcher's dissertation on Mission Level Optimization (Chell 2021), offers an approach to addressing DEVCOM for better model integration across disciplines.

Our early research demonstrated how ontologies and semantic web technologies (SWT) could provide a more effective way to "integrate" models using an interoperable repository that was formalized using ontologies (a type of information model) using IoIF. IoIF leverages SWT, including using an ontology to support a Decision Framework as reflected in Blackburn et al. (2018a, 2018b). IoIF is a framework that has evolved since 2017. The early results and demonstrations made IoIF and DEVCOM relevant domain ontologies the focus of the subsequent three-year research task conducted under ART-002 (Blackburn et al. 2022a) and an additional two-year research task conducted under ART-022. Specifically, we investigated various scenarios focused on five major IoIF demonstrations. Related efforts investigated the development of graphical CONOPs and DEVCOM-relevant mission, system, subsystem and analysis models linked to co-simulations workflows coordinated by the IoIF. IoIF uses a central interoperable repository to input and output various types of model information from different modeling tools and transforms or maps ontology aligned data from various types of models relevant for analysis use cases.

In April 2018, the three Navy system commands (NAVAIR, NAVSEA and SPAWAR) initiated a plan to build mission and system-relevant interoperable ontologies (Blackburn et al. 2021b). The initial effort focused on using an ontology architecture to scope the identified need, enforce interoperability, creating common terminology across domains, and serve as an enabler for DE. A Cyber Ontology pilot was conducted, where our SERC team supported the effort to develop and demonstrate how to use a SysML model of a computer architecture that aligns with a Cyber Vulnerability ontology using IoIF. IoIF provided the computational means for semantic reasoning, which associated potential vulnerabilities with elements of a computer network model and synchronized the information back in SysML, linking the vulnerability with the elements of the system model architecture. The Cyber Ontology was public domain and was used to develop an IoIF training course for our Army sponsor that is now being shared with NAVAIR, US Air Force and US Space Force.

We also conducted research on using AI/ML Design Patterns for Digital Twins and Model-Centric Engineering (Blackburn et al. 2021a) using SWT, leveraging the resulting synergies derived from that research such as IoIF. We contributed our Skyzer and DEVCOM results to Digital

Engineering Competency Frameworks (WRT-1006) (Pepe et al. 2020), because the Surrogate Pilot efforts illustrated new types of DE Competencies (e.g., Digital Signoffs). We correlated analysis on DESM categories with lessons learned from the Skyzer Surrogate Pilot under WRT-1001 (McDermott et al. 2020). The Skyzer models are also being used for workforce development in the new Digital Acquisition University DE Curriculum upgrade. The team also developed Cyber Ontology Case Study Training Course, which was delivered virtually to DEVCOM attendees to support the transition of the research results to DEVCOM.

The objective of this course was to educate attendees on the concept of a "full stack" of models that are "integrated" using the IoIF by extending the cyber ontology use case. The course guides attendees in how IoIF uses ontologies and SWT to map domain information to different types of models. Course exercises involve extending a computer network model developed in SysML to learn how to add stereotypes that tag objects in a SysML model that are mapped to a cyber ontology. The course has exercises for attendees to develop tool proxies using the IoIF service for exchanging data between IoIF with different modeling tools (e.g., MATLAB) and dashboards. The class exercises were designed to extend a SysML computer network model and ontology with a new type of component seeded with a vulnerability. The exercises help attendees learn how to use a triplestore data repository as reflected in Figure 1.10.

The research team developed four use cases to investigate methods and technologies for modeling the "full stack" for an AST. Figure 1.11 depicts the "full stack" of SysML models developed for DEVCOM's munitions-centric mission and systems. This image shows the various elements that contribute to an "integrated" set of models and simulations. Starting at the top, we developed evolving ontologies that align with the DEVCOM munitions domain, as represented in Reference Architectures. We defined the mission model that characterized mission objectives and measures in the SysML descriptive modeling language. We defined a Graphical CONOPs, which was used to elicit requirements. The CONOPs was developed in the Unity gaming engine, which allowed us to validate model requirements through simulated operations. The system model and analysis models were also defined in SysML. The analysis model characterized the metadata returned from the various discipline-specific co-simulations using Computational Fluid Dynamics (CFD),

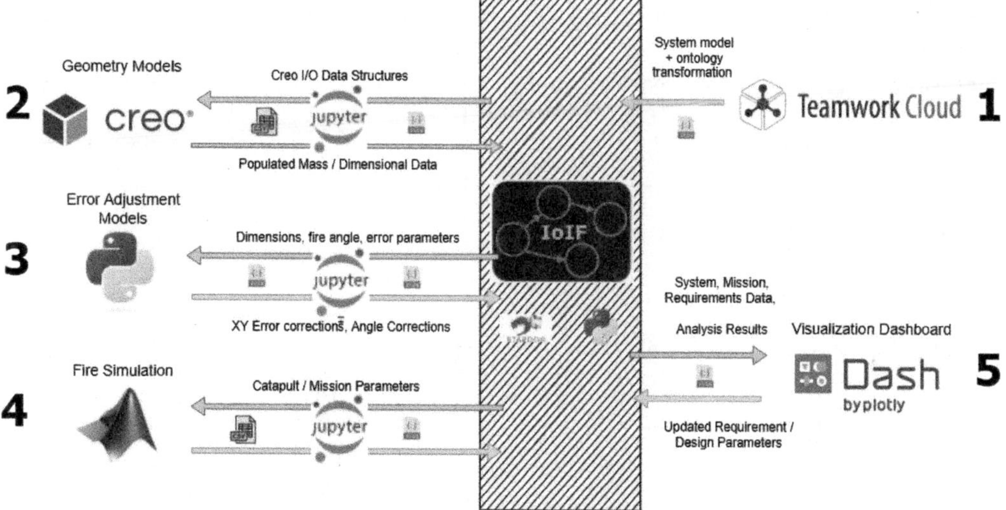

Figure 1.10 IoIF cyber ontology-based use case used in course and exercise.

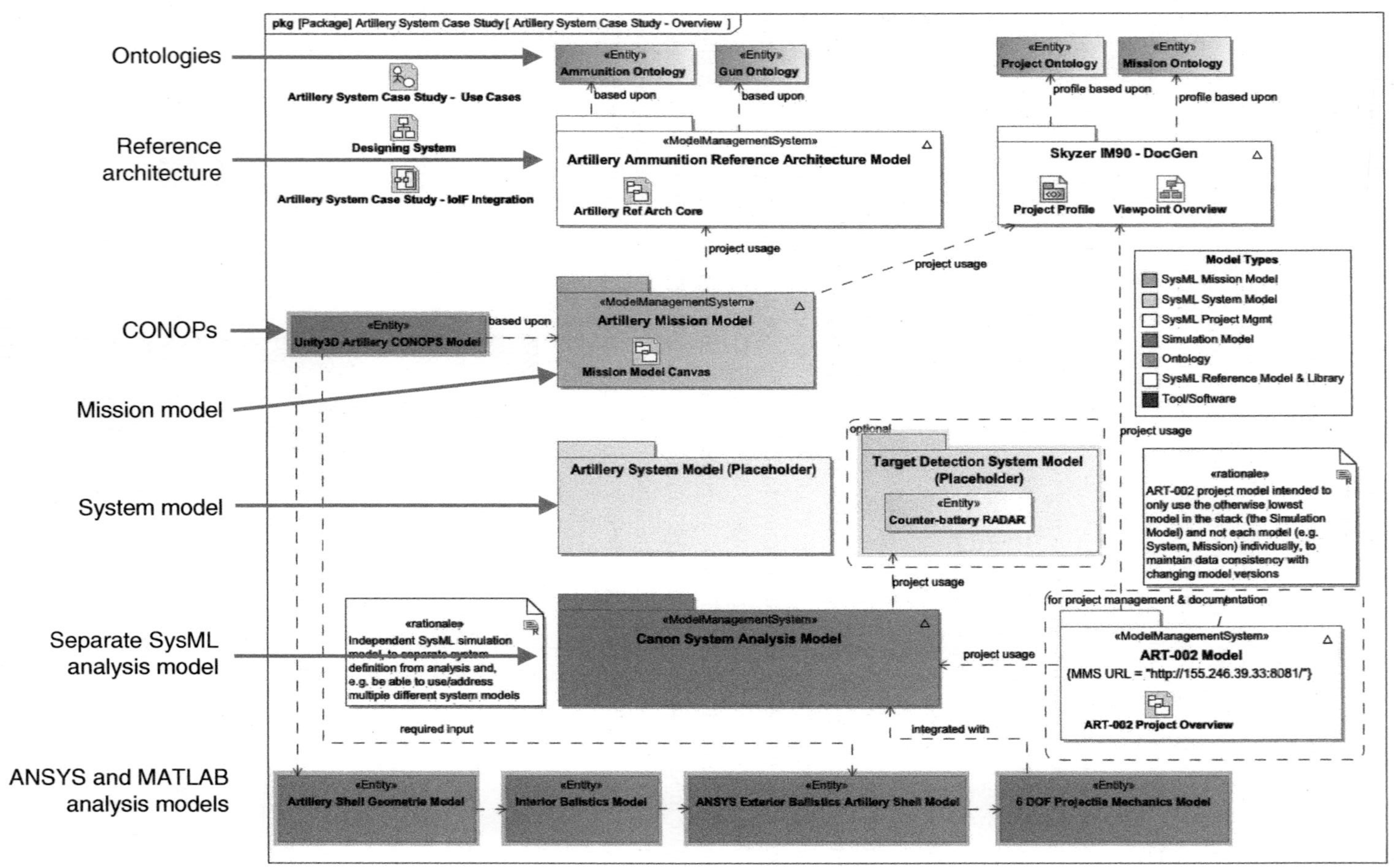

Figure 1.11 "Full Stack" of models for project research including designing system tools.

Finite Element Analysis (FEA), Six Degree of Freedom (6DOF) analyses, thermal, and geometry analyses; this was all coordinated through IoIF. The team also used the IoIF to develop several dashboards that could visualize various design tradeoffs. We also developed an IoIF Service capability to eliminate file-based data exchanges between tools. Instead, IoIF uses a REpresentational State Transfer (REST)-type interface for distributed data exchanges between IoIF and the modeling tools for a workflow. The various tools, including IoIF, are shown in green around the outside of the figure. Subsequent chapters provide additional details on the capabilities and limitations of these tools.

Related Chapters

Chapter 2: Mission and Systems Engineering Methods

This chapter provides information on how the NAVSEM modeling methods was used to develop models for the Skyzer surrogate pilot. In addition, it discusses how model modularization should be done using enabling technologies within some descriptive modeling tools to link models at different levels of abstraction to establish a collaborative authoritative source of truth (AST). This chapter explains the use and approach for linking Stakeholder Analysis Models that are instantiated from reference models such as the Cost model, which is used as a case study in Chapter 2. The chapter also discusses the tools used to establish the AST and explains how modeling methods enable Digital Signoffs using OpenMBEE, DocGen, MMS, View Editor, and View and Viewpoints. The use of DocGen and Digital Signoffs demonstrated how to transform the way that we can do CDRLs and can be performed continuously as different parts of the models mature at different rates. This chapter also provides some guidelines on performing model management, which is similar to configuration management, but subtly different using models. Both sponsored research tasks developed a Digital SE Management and Technical Plan, which is a type of DE Systems Engineering Plan (DESEP) (Kruse et al. 2020). Finally, this chapter discusses how Digital Signoff can support the Source Selection process using an example from Skyzer.

Chapter 3: Transforming Systems Engineering Through Integrating M&S and Digital Thread

This chapter builds on Chapter 2 and goes into additional detail about how descriptive models typically defined in SysML, can be linked to discipline-specific and multi-physics models that often have their own simulations that is relevant to a particular need (e.g., Computational Fluid Dynamic to understand airflow of a surface, aerodynamics, stress, thermal, etc.). This chapter discusses a novel approach for formalizing how to characterize Digital Threads that link information from high-level mission objectives down through parameters in the mission, system and discipline-specific models using the concept of an Assessment Flow Diagram (AFD) (Cilli 2015).

This chapter discusses challenges of tool-to-tool integration that was overcome using domain/application interoperable ontologies that are aligned and configured with the computational framework called the IoIF. This chapter discusses new techniques to enhance design and analysis tasks that incorporate data integration and interoperability across various engineering tool suites spanning multiple domains at different abstraction levels. SWT offers data integration and interoperability benefits as well as other opportunities to enhance knowledge represented by disparate models. This chapter discusses a methodology for incorporating SWT into engineering design and analysis tasks. The methodology includes a new interface approach that provides a tool-agnostic model representation enabled by SWT that exposes data stored for use by external users through standards-based interfaces. Use of the methodology results in a tool-agnostic authoritative source of information spanning the entire project, system, mission, enterprise domain, and using SWT to link descriptive models to discipline-specific and multi-physics models. This chapter also explains the benefits associated with using ontologies and SWT for engineering.

Chapter 4: Digital Engineering Visualization Technologies and Techniques

This chapter discusses how enabling technologies of gaming engines transformed the approach for developing Concept of Operations (CONOPs). There were several different use cases developed through two research tasks RT-168 and ART-002 that showed how to computationally enable Mission-level concerns. Graphical CONOPs have been used to help elicit requirement to support model validation (getting the "right" mission and system objectives/requirements) using visualization, and have been used as way to provide confidence of the simulations by comparing results from 1D (i.e., one-dimensional) simulation with 2D simulation (in the graphical CONOPs). This independence provides a way to verify that both the 1D and 2D are accurate, when they both converge to the same results. This further enables a way to run the gaming engine simulation through 1D inputs to simulate 1000s of trades vs. 10s when run manually. This chapter shows DE visualization technologies that derive data from the IoIF such as Decision Framework Dashboard and Digital Thread analysis.

Chapter 5: Interactive Model-Centric Systems Engineering

This chapter discusses how humans work or should work with models. Models are increasingly used for decision making in systems engineering, yet while human users are essential to a model's success, research into human-model interaction has been lacking. Models represent an abstraction of reality and can come in a variety of forms and formats. Humans use models to augment their ability to make sense of the world and anticipate future outcomes. The idea that "humans use" models, highlights human interaction as a necessary factor in model-centric engineering.

Future Work

The emphasis of future efforts should include support for transitioning the research to programs, such as the IoIF training course and the DAU DE curriculum using publicly available examples such as Skyzer to support workforce development. We did develop training material that required hands on exercises using models, code, tools and other technologies where the sponsors actually conducted demonstrations to present to their senior leadership. However, we need more support and workforce development for both the technical as well as the acquisition professional that support program solicitation, developing and responding to the DE Request for Proposals (RFP), DE Systems Engineering Plan and DE Statement of Work. We investigated the concept of Baseline Progress Measure to continuously assess and measure maturity of analysis and design in the context of models. Digital signoffs are an enabling technology that provide measures and metrics to support the concept of Baseline Progress Measures.

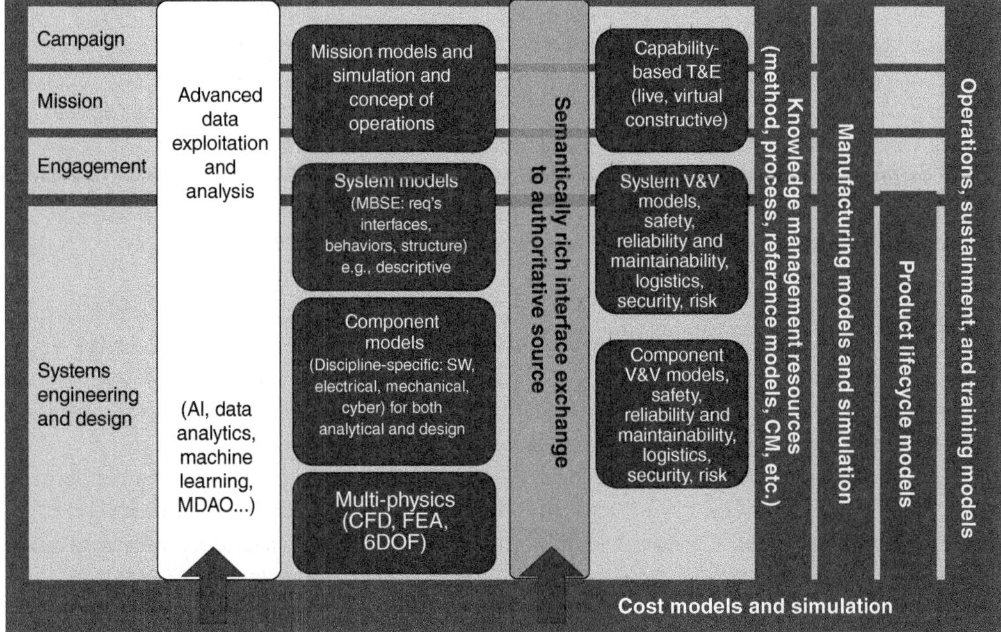

Figure 1.12 Digital Engineering Environment reference model.

Finally, we provide a reference model abstraction of the different models that are associated with DE enabling technologies as shown in Figure 1.12. Most of the NAVAIR and DEVCOM case studies discussed in the following chapters focus on the early part of the lifecycle, primarily up to about contract award. There is a need to address the later stages of the lifecycle that are notionally reflected on the right side of this figure.

Notes

1 NAVAIR Public Release 2017–370. Distribution Statement A – "Approved for public release; distribution is unlimited."
2 https://sercuarc.org/wp-content/uploads/2020/01/ROADMAPS_2.3.pdf
3 NAVAIR Public Release 2019–443. Distribution Statement A – "Approved for public release; distribution is unlimited."

References

Bell Eagle Eye (2022). In Wikipedia. https://en.wikipedia.org/w/index.php?title=Bell_Eagle_Eye&oldid=1111743865 (accessed 22 September 2022).

Blackburn, M.R. and Kruse, B. (2023). Conducting design reviews in a digital engineering environment. *INCOSE Insight* 25 (4): 42–46.

Blackburn, M. R., Bone, M., and Witus, G. (2015). Transforming System Engineering Through Model-Centric Engineering. *Systems Engineering Research Center. Technical Report SERC-2015-TR-109.*

Blackburn, M. R., Blake, R., Bone, M., et al. (2017). Transforming Systems Engineering Through Model-Centric Engineering. *Systems Engineering Research Center. Technical Report SERC-2017-TR-101.*

Blackburn, M. R., Bone, M., Dzielski, J., et al. (2018a). Transforming Systems Engineering Through Model-Centric Engineering. *Systems Engineering Research Center. Technical Report SERC-2018-TR-103.*

Blackburn, M. R., Verma, D., Giffin, R., et al. (2018b). Transforming Systems Engineering Through Model-Centric Engineering. *Systems Engineering Research Center. Final Technical Report SERC-2017-TR-110.*

Blackburn, M. R., Austin, M., and Coelho, M. (2021a). Using AI/ML Design Patterns for Digital Twins and Model-Centric Engineering. *Systems Engineering Research Center. Final Technical Report SERC-2021-TR-007.*

Blackburn, M. R., Dzielski, J., Peak, R., et al. (2021b). Transforming Systems Engineering Through Model-Centric Engineering. *Final Technical Report SERC-2021-TR-012.*

Blackburn, M. R., Grosse, M. I., Gabbard, J., et al. (2022a). Transforming Systems Engineering Through Model-Based Systems Engineering. *Final Technical Report. Systems Engineering Research Center.*

Blackburn, M.R., McDermott, T., Kruse, B. et al. (2022b). Digital engineering measures correlated to digital engineering lessons learned from systems engineering transformation pilot. *Insight* 25 (1): 61–64.

Chell, B. (2021). *Multidisciplinary System and Mission Design Optimization*. Stevens Institute of Technology.

Chilenski, J. and Ward, D. (2014). 2014 Modeling and Simulation (M&S) Subcommittee Report.

Cilli, M. (2015). Seeking improved defense product development success rates through innovations to trade-off analysis methods. Doctoral thesis. Stevens Institute of Technology.

Deputy Assistant Secretary of Defense (Systems Engineering) (2018). *Digital Engineering Strategy*. US Department of Defense. https://ac.cto.mil/wp-content/uploads/2019/06/2018-Digital-Engineering-Strategy_Approved_PrintVersion.pdf.

Kruse, B. and Blackburn, M. R. (2019). Collaborating with OpenMBEE as an authoritative source of truth environment. *17th Annual Conference on Systems Engineering Research (CSER)* (3–4 April 2019). Washington, DC: National Press Club.

Kruse, B., Hagedorn, T., Bone, M., and Blackburn, M. R. (2020). Collaborative management of research projects in SysML. *18th Annual Conference on Systems Engineering Research (CSER)* (8–10 October 2020).

McDermott, T., Van Aken, E., Hutchison, N., et al. (2020). Digital Engineering Metrics. *Systems Engineering Research Center. Technical Report SERC-2020-SR-002.*

MIL-HDBK-516C (2014). *Department of Defense Handbook. Airworthiness Certification Criteria.* U.S. Department of Defense. http://everyspec.com/MIL-HDBK/MIL-HDBK-0500-0599/MIL-HDBK-516C_52120/

Pepe, K., Hutchison, N., Blackburn, M. et al. (2020). Preparing the acquisition workforce: a digital engineering competency framework. *INCOSE International Symposium* 30: 920–934. https://doi.org/10.1002/j.2334-5837.2020.00763.x.

Biographical Sketches

Mark R. Blackburn, PhD is a Senior Research Scientist with Stevens Institute of Technology since 2011 and principal at KnowledgeBytes. Dr. Blackburn has been the Principal Investigator (PI) on 17 System Engineering Research Center (SERC) research tasks for both US Navy NAVAIR,

US Army DEVCOM and US Space Force on Digital Engineering Transformation Research Tasks. He has also been PI on a FAA NextGen and National Institute of Standards and Technology projects and has received research funding from the National Science Foundation. He develops and teaches a course on Systems Engineering for Cyber Physical Systems. He is a member of the SERC Research Council, OpenMBEE Leadership Team and INCOSE Pattern Working Group focused on the Semantic Technologies for Systems Engineering initiative. Prior to joining Stevens, Dr. Blackburn worked in industry for more than 25 years. Dr. Blackburn holds a PhD from George Mason University, a MS in Mathematics (emphasis in CS) from Florida Atlantic University, and a BS in Mathematics (CS option) from Arizona State University.

Timothy D. West currently serves as the Deputy Director of the U.S. Air Force Research Laboratory's Operations Directorate. In this capacity, he provides leadership and oversight of the Laboratory's air and space flight test program, demonstrating combat readiness of the military's latest science and technology development programs. Prior to his current assignment, Mr. West served for nearly 30 years as a military officer in the U.S. Air Force, where he held command positions in science and technology, test and evaluation, and program management. He is a graduate of the US Air Force Test Pilot School, the Air Command and Staff College, and the Air War College. Mr. West holds MS degrees in Aerospace and Industrial Engineering from the University of Tennessee Space Institute, and a BS in Mechanical Engineering from the University of Kentucky. Mr. West is also a Systems Engineering PhD Candidate at Stevens Institute of Technology.

Chapter 2

Mission and Systems Engineering Methods

Benjamin Kruse, Brian Chell, Timothy D. West, and Mark R. Blackburn

Stevens Institute of Technology, School of Systems and Enterprises, Hoboken, NJ, USA

Problem

As noted in Chapter 1, this initial research began in 2013 because the sponsor (NAVAIR) was concerned about the cycle time required to perform their mature Systems Engineering Technical Review (SETR) process for analyzing a mission, developing system requirements, performing system analysis and design, and conducting system testing in order to produce an integrated air vehicle. Their traditional "Vee" process relied on milestone-centric document review as the primary mechanism to perform insight and oversight, which drove longer development cycles and stop-work situations, which contributed to cost and schedule growth. They needed a faster, cheaper approach that was conducive to continuous design review and real-time decision making, one that leveraged a comprehensive Digital Engineering Ecosystem to not only perform these tasks but also to capture the rationale behind them. They also wanted to leverage technology to identify contradictory requirements earlier in the analysis and design phases when the design was still fluid, rather than during system integration and testing of a near-final design, where design changes are significantly more expensive. NAVAIR also wanted 25% reduction in development time from that of the traditional large-scale systems, and they believed a more holistic Model-Based Systems Engineering (MBSE) approach leveraging Mission-based Analysis and Engineering could achieve this reduction.

Background

NAVAIR chartered the SERC to perform an evaluation of emerging system design through computationally-enabled models. NAVAIR also tasked the SERC to begin collecting and structuring evidence to assess the technical feasibility of moving to a "complete" model-driven lifecycle. The initial emphasis in 2013 was on the "technical feasibility" of such a Vision. At NAVAIR's direction, nontechnical hurdles (e.g., organizational adoption, training, usability, etc.) were ignored during early research. NAVAIR recognized the growing popularity of MBSE across government, industry, and academia. NAVAIR also understood that MBSE's multilevel, multi-domain

Systems Engineering for the Digital Age: Practitioner Perspectives, First Edition. Edited by Dinesh Verma.

Companion website: www.wiley.com/go/verma/systemsengineering

model, analysis, and simulation capabilities were steadily increasing to take advantage of high-performance computing, and that emerging tool environments could now support various degrees of fidelity that connect to different and complementary views of the system under analysis and design.

At NAVAIR's behest, we conducted a structured global scan (Blackburn et al. 2014) of the most advanced and holistic approaches to performing model-centric engineering (MCE), which is now more generally referred to as digital engineering (DE). The global scan revealed that MCE/DE was in use, and adoption was accelerating. MCE[1] can be characterized as an overarching digital engineering approach that integrates different model types with simulations, surrogates, systems, and components at different levels of abstraction and fidelity across disciplines throughout the lifecycle. Industry is trending toward more integration of computational capabilities, models, software, hardware, platforms, and humans-in-the-loop. The integrated perspectives provide cross-domain views for rapid system-level analyses allowing engineers from various disciplines using dynamic models and surrogates to support continuous and often virtual verification and validation for trade-space decisions in the face of changing mission needs.

As part of the global scan, we performed site visits to various organizations. Figure 2.1 correlates some of the organizations visited with the MCE topics discussed during the global scan, including the lifecycle perspectives, shown in Figure 2.2, which are relevant to many cross-cutting topics (Blackburn et al. 2015). We subsequently conducted additional research on most of these topics, as described in subsequent chapters of this book.

In 2015, NAVAIR proposed a new framework for the systems engineering transformation (SET), depicted in Figure 1.3 of Chapter 1, which would make the mission, system, and subsystem processes more iterative and agile. We assessed and refined the evolving framework in order to support a new operational paradigm to mission engineering, analysis, and acquisition, which would be jointly led by the government and research team with a collaborative, industry-led design effort. As Figure 1.1 of Chapter 1 depicts, we binned research roadmap into six functional lanes to better align with NAVAIR's priorities in the accelerated SET and to address some of the non-technical hurdles listed above.

The research, which continued through 2021, yielded a modeling framework that leveraged high-performance computing (HPC) to enable the AST and the integration of multi-domain and multi-physics models, and to provide a method for model integrity. Completing the modeling and infrastructure for the DE environment was a critical step to enabling an AST, as was selecting a modeling method. Despite the availability of thousands of software tools,[2] no single tool or family of tools offered a truly comprehensive and integrated solution to DE. Every organization we interviewed had to architect and engineer their own DE environment to meet their unique needs. Most organization used a federated network of commercial tools, often developing the integrating fabric between the different tools, models, simulations, and data. Some organizations have encoded historical knowledge in reference models, model patterns to embed methodological guidance to support continuous orchestration of analysis through new modeling metrics, and automated workflows.

The scope of the research looked at the cross-cutting relationships associated with the research needs, as shown in Figure 2.3, including the evolving OpenMBEE as the experimental integrated modeling environment. The use of OpenMBEE proved to be very valuable, as discussed in the next section. Five use cases were employed in this research.

UC00: Conduct automated model reasoning: This use case employed ontologies and semantic web technologies to perform reasoning about completeness and consistency across cross-domain models to achieve model integration through interoperability. As discussed in more

	Instances where discussed (not exhaustive)												Characteristics						From kickoff briefing					
Discussion topics (not exhaustive)	NASA/JPL	A	B	C	Altair	GE	Sandia	DARPA META (VB)	DARPA META (BAE)	Model center	Automotive	CREATE	Performance	Integrity	Affordability	Risk	Methodology	Single source of tech truth	Prioritization and Trade-off analysis	Concept engineering	Architecture and design analysis	Design and test reuse and synthesis	Active system characterization	Human-system integration
Modeling CONOPS	x															x	x	x	x	x			x	
Modeling patterns	x								x					x		x	x	x		x	x	x		
Multi-physics modeling and simulation		x	x	x	x			x	x		x	x	x	x					x	x	x	x	x	
Multi-discpline/domain analysis and optimization	x	x	x	x	x	x	x	x	x	x			x	x	x	x		x	x		x	x	x	
Mission-to-system-level simulation integration	x	x	x													x		x	x	x	x	x	x	x
Affordability analysis			x				x						x	x	x	x			x		x	x	x	
Quantification of margins			x				x						x	x	x	x	x		x		x	x	x	
Requirement generation (from models)	x		x					x									x	x	x		x	x		
Tool agnostic digital representation	x	x			x				x								x	x	x		x	x	x	x
Model measures (thru formal checks)	x		x			x		x	x							x	x	x			x	x		
Modeling and sim for Manufacturability			x			x		x					x	x	x	x	x	x	x	x	x	x	x	
Process automation (workflows)	x				x				x	x							x	x			x			
Iterative/agile use of MCE	x	x	x							x							x				x	x		
High performance computing	x	x	x		x		x	x			x	x	x	x	x				x	x	x		x	
Platform-based and surrogates	x	x	x								x									x	x	x	x	
3D environments and visualization	x	x	x	x	x	x	x	x			x		x	x					x	x		x	x	x
Immersive environments		x	x								x									x			x	x
Domain-specific modeling languages	x	x	x	x	x	x	x	x	x		x		x	x			x			x	x	x		
Set-based design		x				x							x	x	x	x			x	x	x			
Model validation/qualification/trust							x					x		x		x	x		x		x	x		
Modeling environment and infrastructure	x	x	x	x	x	x	x	x	x	x	x	x	x	x	x	x	x	x	x	x	x	x	x	x

Figure 2.1 Traceability and scope of data collection of MCE relevant topics.

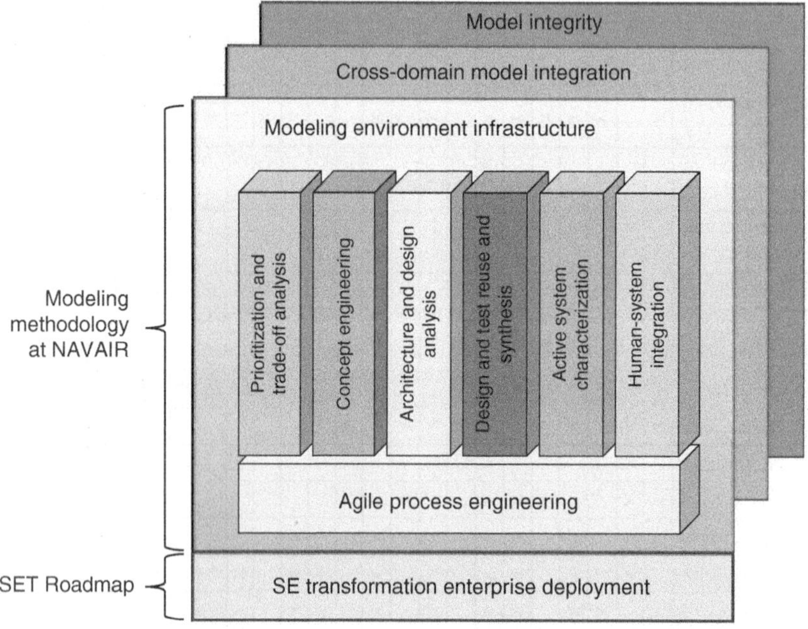

Figure 2.2 SE transformation research areas (SERC).

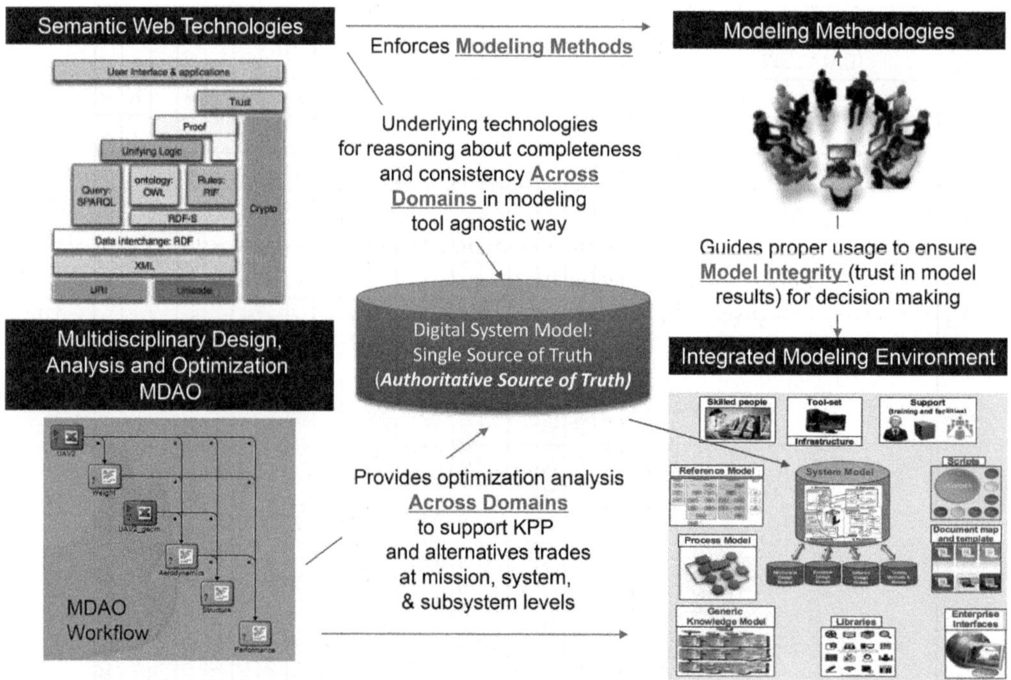

Figure 2.3 Cross-cutting relationships of research needs.

detail in Chapter 4, this use case demonstrated an authoritative source of truth (AST) using tool-agnostic approaches to methodology enforcement and conformance that also support model integrity. Information about the systems and their various components were converted into explicitly defined, machine-readable subject/predicate/object statements that complied with Resource Description Framework (RDF) and Web Ontology Language (OWL) taxonomies. Tools such as SPARQL Protocol and RDF Query Language (SPARQL) were then employed to enrich the provided assertions about the system by extracting and graphing logical implications that could be derived from the explicit statements.

UC01: Research multidisciplinary design, analysis, and optimization (MDAO): This use case employed MDAO at the mission, system, and subsystem levels to enable continual assessment of design tradeoffs, as described in Chapter 4. This use case also investigated methods for tracing system capabilities through the performance of cross-domain trade space analyses using MDAO techniques. MDAO is an approach for calculating optimal designs and understanding design trade-offs in an environment that simultaneously connects many different models, evaluates them in a consistent and efficient manner, and optimizes for one or many objectives. For example, when designing a vehicle, there is typically a trade-off between maximizing performance and maximizing efficiency. Calculating either of these objectives requires multiple disciplinary models (geometry, weight, aerodynamics, propulsion). MDAO prescribes ways to integrate these models and explore the necessary trade-offs among the objectives to make a design decision. While the theoretical foundations of MDAO are well established, several barriers to practical implementation exist. Chief among these is the lack of model integration, which prevents designers of one subsystem from easily assessing how changing a design variable affects the results of other subsystems' models or simulations – the stovepipes or siloes which many systems engineering efforts strive to overcome.

UC02: Formalize an authoritative source of truth (AST) using an integrated modeling environment (IME): As the title implies, this use case employed an IME to develop an AST for use by NAVAIR and its industry partners. This not only is essential to provide a unified and ideally standardized access to the consistent and correct, i.e., true, system, and mission data for, e.g., MDAO or semantic reasoning, but also impacts all related workflows, which has implications on both technologies and workforce development. Especially in respect to allowing industry contractor access, methods for model modularization are also crucial to ensure separation of concerns, classification, and acquisition.

UC03: Develop and manage models using an IME: The methodology for all these technologies in the context of the IME workflows, such as methods for the creation of system models, including interrelated subsystems and system of systems, related mission models, e.g., based on a digital mission model canvas, and methods for MDAO modeling (discussed in more detail in Chapter 4). Additionally, in order to handle such various models, there are methods for model management needed as well as for model modularization, to support constraints needed for developing an AST with many different users and roles, thereby relating to many other use cases. This handling of models also includes required methods for representing and organizing reference models, model libraries, process models, and discipline-specific models. Having interconnected models in an AST also enables methods for tracing change impacts and developing and tracing capability measures to KPPs; more details are discussed in Chapter 4.

UC04: Integrate physics-based modeling with SysML models: This use-case-enriched basic SysML models with the underlying physics-based models. It included interconnecting MATLAB models and other physics-based models with Cameo-based SysML models, as well as MDAO-derived physics models, to assess model integrity and design risks and uncertainties, as discussed in Chapter 3.

New Approach

Skyzer Surrogate Pilot Experiment

As noted in Chapter 1, Skyzer is a notional aircraft acquisition program consisting of a tilt-rotor remotely controlled unmanned air vehicle (UAV) to conduct ship-based search and rescue for the Navy. The fictitious program was crafted by NAVAIR to evaluate and demonstrate the SET Framework (Figure 1.3 in Chapter 1), without the security constraints that accompany an actual weapon system acquisition program. Skyzer became the centerpiece of several Surrogate Pilot experiments, conducted by a combination of SERC researchers, NAVAIR engineers, contractors, and a surrogate contractor over three phases starting in late 2017 (Blackburn et al. 2017). Figure 1.5 in Chapter 1 depicts the Concept of Operations (CONOPs) for Skyzer's humanitarian maritime search and rescue mission.

Phase one (1) culminated at the end of 2018 with the development of mission and systems models that linked to the Request for Proposal (RFP) Response model from the Surrogate Contractor. The focus was on characterizing the process for running a model-based acquisition based on the SET Framework concept. We successfully developed fully linked models in an AST. The Surrogate Pilot team officially released the RFP in the form of models, concluding the Phase One effort.

The surrogate contractor's RFP response refined the mission and system, creating subsystem models with detailed design and analysis information using multi-physics and discipline-specific models. Performance requirements such as a cruise speed of 170 knots forced the design to be something other than a traditional helicopter and ultimately a design like the Bell Eagle Eye ("Bell Eagle Eye" 2023) was proposed in the surrogate contractor RFP response models. This proposed design was evaluated in a surrogate source selection that was embedded as part of the contractor RFP response by the sponsor team (a retired NAVAIR SME who had no SysML training) using introduced digital signoffs in the OpenMBEE (2023) View Editor, which are discussed below.

The use case was extended in Phase Two (2) to include a ship-based Launch and Recovery (L&R) system in order to create another capability, where we can research methods for Capability Based Test & Evaluation (CBT&E), based on a NAVAIR modeling approach for Mission-Based Test Design (MBTD). Phase Three (3) developed a deep dive related to the landing gear in order to examine some scenarios for modeling information related to Airworthiness (the process used to get a flight clearance). In addition, Phase Three developed an MBSE Cost Model, which used the Skyzer Airframe characteristics for the cost modeling basis; with the Airframe characteristics being derived from the Phase One contractor model. The efforts during Phase Three, in addition to the Airworthiness and Cost Model, included updates to Digital Signoffs, updates to the models to align with the evolution of NAVSEM, and recommendations for adding guidance to NAVSEM based on needs identified in the evolution of the modeling for Skyzer.

OpenMBEE, DocGen, and View Editor

NASA/JPL developed OpenMBEE was used as part of the IME in order to provide collaborative access to the models for both the government team members and the industry surrogate contractor (Hutchison et al. 2020). OpenMBEE has three main components: the Model Development Kit plugin (MDK) (Kruse and Blackburn 2019), the Model Management System (MMS), and the View Editor. To serve as an AST, MMS stores the model data in an open and accessible way while providing versioning, workflow management, and access control to enable multi-tool integration across disciplines. For instance, it stores all model elements of all incorporated SysML models, including their histories and branches.

The MDK is a plugin for the NoMagic modeling tool that synchronizes between MMS and the SysML models. It also includes DocGen, which is a language for model-based document creation based on the view and viewpoint paradigm. DocGen allows a model-based document creation following ISO-42010 (ISO/IEC/IEEE 2011), where views are defined as representations of a system from the perspective of a viewpoint. The viewpoints hereby provide the information needed to build a view out of thc available data to address stakeholder concerns. DocGen permits all stakeholders access to stakeholder-specific model information using a web-based representation of the model using the View Editor. An example of View and Viewpoint hierarchy is provided in the Digital Signoff subsection below.

The View Editor offers live access to web-based model data in the form of DocGen views for agile virtual reviews and real-time collaboration to surpass static documents. It enables a shift from document-centric to model-based approaches. As a component of OpenMBEE, the View Editor allows users to edit or comment on information generated from the model in a web browser, with most edits in the View Editor being synchronized back into the model. Skyzer SMEs used the View Editor to successfully visualize modeled information in the web browser and perform digital signoffs (as further explained below), without any SysML training or tools.

The surrogate pilot effort demonstrated the value of having a set of generic viewpoints defined in a viewpoint library, as done for the RFP (Hutchison et al. 2020). With standardized viewpoints, only a few modelers need to know how to create them. Additionally, standardized viewpoints allow the modeler to pre-plan view hierarchies to define the document structure and the type of required model content prior to the modeling, thereby guiding the modeling and design process by specifying what information is needed. This can be enhanced by including warning messages in the viewpoints and by checking the validity of model inputs. We employed a template view hierarchy in the NAVSEM process detailed below, being ever mindful of model-based document creation best practices, such as limiting the size of diagrams and using their documentation and employing specific modeling considerations to enable a successful processing of model elements through the viewpoints.

NAVSEM

Chapter 1 provides a general overview of how the Skyzer models align with the NAVSEM method. NAVSEM defines both the process (what) as well as the method artifact (how), which are important to the digital signoff process. NAVSEM is formally described in a SysML as shown in Figure 2.4. There are 11 main process steps represented as an SysML activity diagram, and each of the process steps are further decomposed as shown in Figure 2.5. The approach fosters compliance with NAVSEM. The outline of the DocGen-generated "document" is consistent with the NAVSEM process steps. If information is missing from an element within the outline, then this alerts reviewers that the model information is missing and thus not yet fully compliant with NAVSEM.

Model Management and Modularization

As part of the surrogate pilot experiment to "model everything" to demonstrate the art-of-the-possible, we created multiple distributed models that were linked together in an AST to enable information sharing and access, better collaboration, improved consistency, enhanced traceability, reduced defects, and greater efficiencies through the elimination of certain types of work. These models captured both system information and linked project-relevant data, to be used by DocGen and the View Editor.

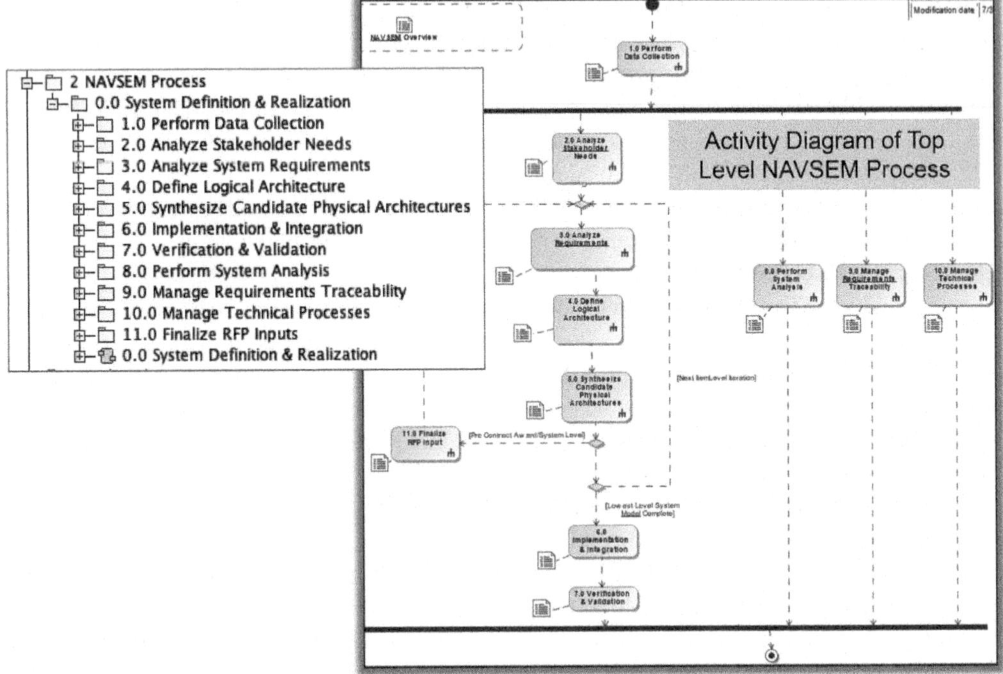

Figure 2.4 NAVSEM high-level process.

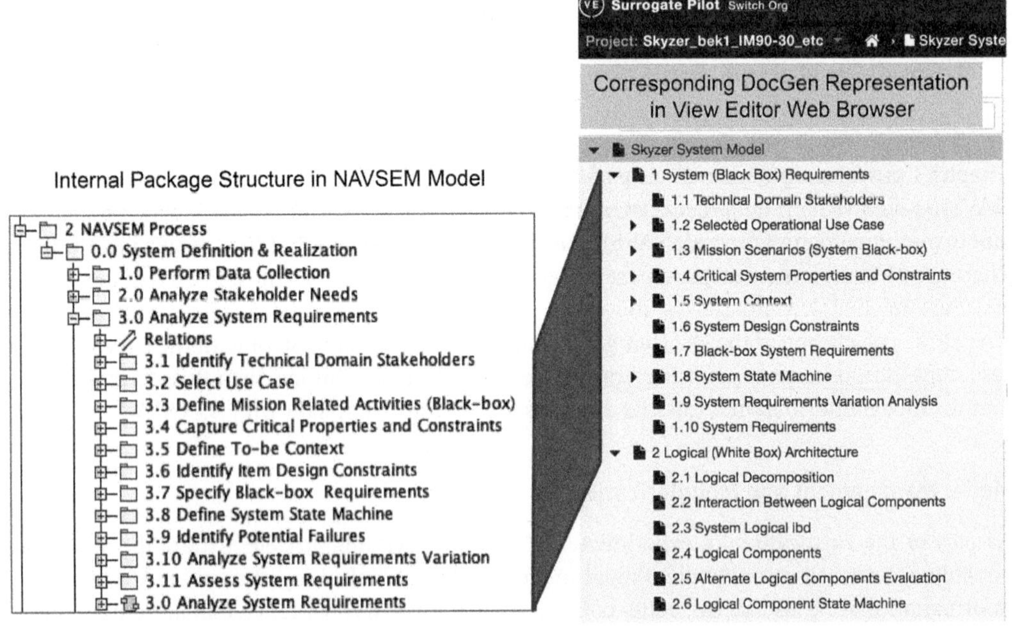

Figure 2.5 Internal package structure maps to DocGen-generated structure.

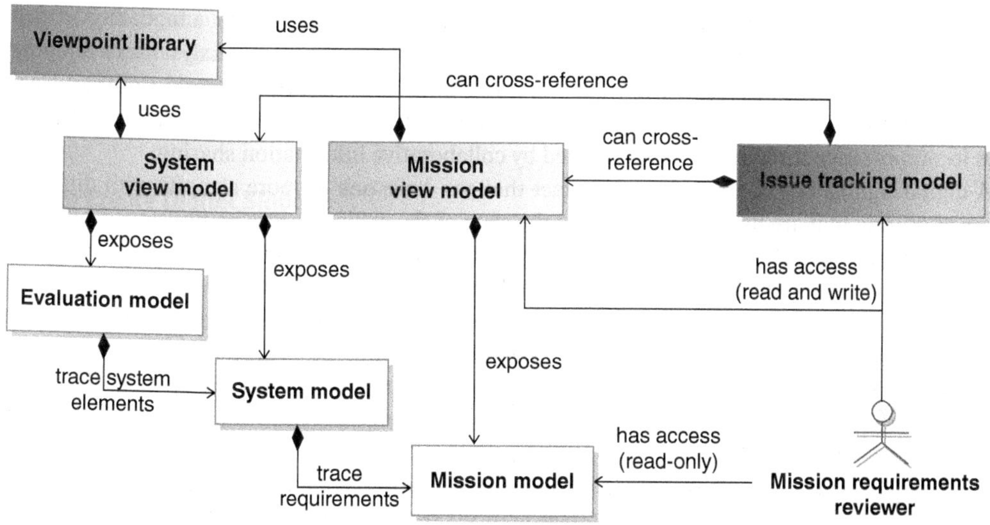

Figure 2.6 Five usage and user permission examples for mission requirements review.

To effectively design and manage these distributed models, we split up the various models into manageable models and employed a modeling modularization method enabled by MagicDraw's project usage capability. Project usage provided a means for accessing shared elements of the used project from within another model, e.g., to allow traceability links from system elements back to specific mission requirements. This approach facilitated model library reuse, where the library is separated from the rest and protected from accidental changes, yet still available for use. Project usage also allowed us to separate mission, system, subsystem, analysis models or even SysML models that only include a view hierarchy to allow the resulting document to be editable in the web browser, while keeping the exposed model content read-only. The project usage mechanism also enabled refined user-access management, e.g., specific or partially restricted access.

Figure 2.6 depicts an example structure of distributed and modularized models, where project usage relations are displayed as compositions (Kruse and Blackburn 2019). The white elements on the bottom represent connected domain models that allow traceability, e.g., of requirements from the mission model or system elements from the system model. In addition, these models are used by the predefined view hierarchies in the view models described above. The viewpoints come from the Viewpoint Library in the top-left corner. The model for Issue Tracking on the right uses the two view models. This structure, together with specific access and editing rights, allows the reviewer to edit the Issue Tracking model and the Mission View Model. A reviewer can, for example, enter comments into the document that expose the mission requirements, or directly add comments as new issues in the Issue Tracking Model, while not being able to edit the actual requirements in the Mission Model.

Digital Signoff

During Phase Three of the surrogate pilot, we investigated the concept of digital signoffs embedded within the models (Blackburn and Kruse 2023). A digital signoff is a means to capture an evaluating intent (e.g., an approval or rejection), in a dependable, and legally binding way that does not require paper or electronic documents (e.g., PDFs), but is instead part of the model

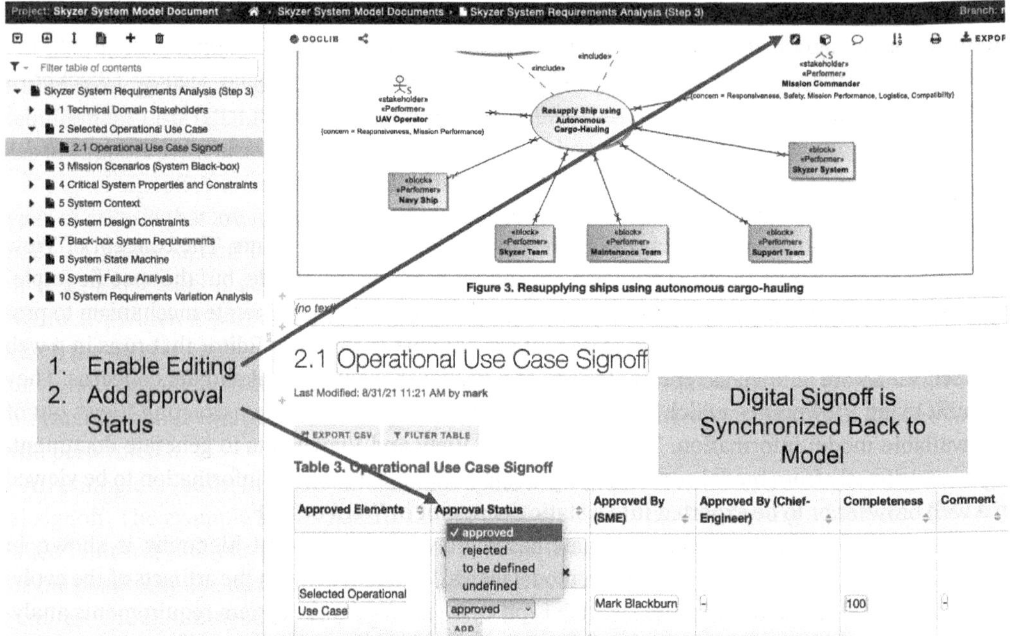

Figure 2.9 View Editor showing digital signoff for operational use case for system.

Currently, we develop one or more View and Viewpoint hierarchies as shown in Figure 2.8. Each view can hereby be understood as a section of the live document that is automatically generated from the exposed model element(s) based on the DocGen instructions of the viewpoint for which it conforms. For instance, for signoffs we place a view (e.g., Operational Use Case Signoff) link to the model artifact to be signed off (e.g., Selected Operational Use Case) using an ***expose*** relation and have the View ***conform*** to a fitting Viewpoint (e.g., Dual Signoff Table) in order to create a table containing the signoff(s), which in this case requires two persons to sign off as shown in Figure 2.9, from an image of the View Editor. Authorized users can ***Enable Edits*** and select the ***Approval Status*** and provide the name of the approver.

This type of technology enables and formalizes decision-making, and directly associates each decision with specific model information related to that decision. Like DocuSign, the model and digital signoff facilitates a digital document for a contract. The digital signoff can (and should) be located in one, and only one place to constitute decisions in an AST. Requesting specific model elements through digital signoffs that are part of standardized view hierarchies can also enforce the use of specific modeling methods such as the NAVSEM. This reduces time, because the signoff can be performed as soon as the associated artifacts are "ready for review." The digital signoff is a DE construct, which means its state can be changed (computationally) if anything associated with the signed off artifacts (models) is changed. This can eliminate work and mistakes that occur when using documents, because it is often difficult to trace such relationships within one or more documents. This leads also to digital signoff metrics, as shown in Figure 2.10 that can be automatically generated to guide management and assessment of risk. These measures and metrics are automatically calculated and can again be viewed from a web browser.

By having the View and Viewpoint hierarchies in separate SysML projects as their exposed model content, it is possible to assign editing rights for only the views with the signoffs and not the

Date	Number of Sign Offs	Number of High Risk Sign Offs	Ratio of Approved Sign Offs	Ratio of Rejected Sign Offs	Average Risk
2023.04.17 13.44	5	1	0	0	55
2023.04.29 16.54	5	1	0.6	0.2	40

Figure 2.10 Digital signoff metric example.

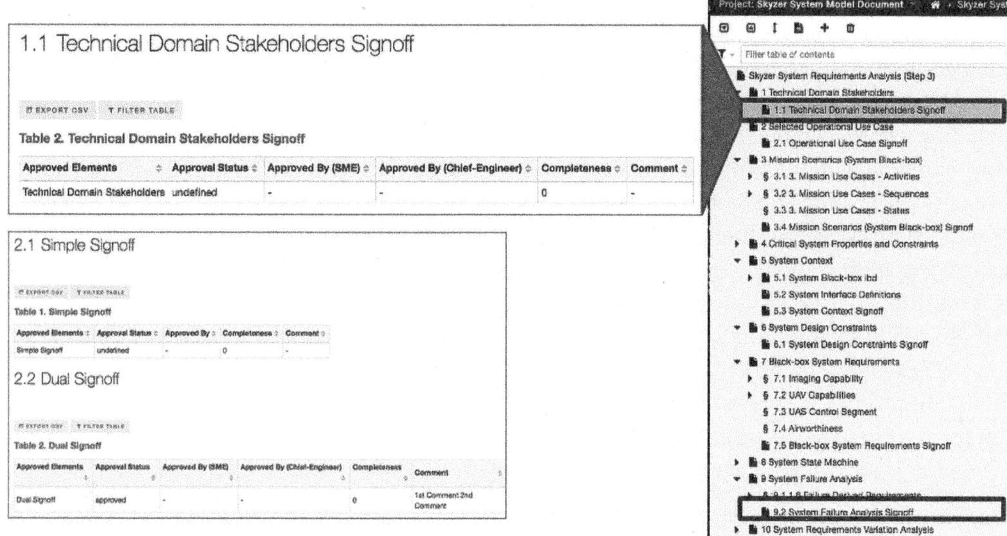

Figure 2.11 Template based examples as seen via View Editor in the web browser.

model content or vice versa. This can for instance prevent modelers from signing off their own work or alternatively reviewers from accidentally changing the underlying mission or system model. In addition to the examples in Figures 2.7 and 2.9, there are three examples shown in Figure 2.11, which reflects on how different digital signoff templates can be used depending on the needs of the particular signoff (e.g., single person vs. dual person signoff).

Multidisciplinary Design Analysis and Optimization (MDAO)

As part of our MDAO-related research task objectives, we investigated methods for assessing the impacts of design changes on system capabilities, supporting conceptual design trade-off analyses while running complex, multi-physics models efficiently. We performed design optimization on models ranging from the mission level all the way down to subsystem levels. Specific MDAO examples included:

- Generic multidisciplinary models of NAVAIR-relevant system examples, including analyses of the geometry, structure, aerodynamics, propulsion, stress, thermal, and performance capabilities, to be used as an example case
- MDAO architectures such as multidisciplinary feasible and interdisciplinary feasible to compare simulation results when searching for optimized solutions (Chell 2021)
- Using systems representations (e.g., SysML, Domain Specific Models) to map inputs (parameters and variables) and outputs (objectives, constraints, intermediate parameters) among the individual models

- Conducting trade studies on the UAS design using established approaches and tools for MDAO, exploring different approaches, tools, and visualization techniques to display information and uncertainty most effectively for decision-makers
- Exploring ways that previous trade study results on detail-phase product design can be useful toward new conceptual design of products with varying mission capability requirements
- Using the surrogate pilot to understand the barriers to implementing this type of MDAO, culturally and practically/theoretically
- Exploring the ways that MDAO and MBSE tools can work together

Optimizing complex models is a difficult problem, as solutions generally improve with the number of function evaluations within the optimization loop, and these models can have long run times. We investigated methods for enhancing MDAO workflow efficiency on a fixed-wing model including using a combination of MDAO architectures and multi-fidelity MDAO. MDAO architectures combine a multidisciplinary model with a solution algorithm, and multi-fidelity MDAO leverages the fast run times of lower-fidelity models with the accuracy of higher-fidelity models.

Another thrust within the MDAO research was to apply MDAO to mission-level optimization using a graphical CONOPs. In this case, surrogate models were applied to speed up the run times of a multi-physics quadcopter search mission model. This scenario contained high uncertainty, and the resulting search times could vary by several orders of magnitude. To reduce this uncertainty, a design of experiments (DOEs) approach was used to identify the inputs for a Monte Carlo simulation. Then, the outputs from those simulations were used to create surrogate models. Optimizing these surrogates found a more optimal design than any of the failure-free designs within the DOE. More information on graphical CONOPs is presented in Chapter 4.

Collaborative Management of Research Projects in SysML

We created a model for the System Engineering Technical and Management Plan (Kruse et al. 2020). We used a similar approach for the research use cases and pilot case study to demonstrate how the management of systems engineering can be done in SysML models within OpenMBEE. Figure 2.12 provides an overview of the packages and model elements contained in this model. We created these elements and provided continuous updates from its DocGen view hierarchy to manage the project through supported web-based collaboration, model-based report generation, and enabled semantic reasoning.

We have an underlying ontology that provides for semantic reasoning, which is seen as a key enabler and accomplished using a SysML profile that is aligned to an underlying project ontology. This results in not only using the advantages of a DE engineering environment for managing the project, but also demonstrates the benefit of semantic enabled reasoning that is a focus of research of the research project. As such it is not only used for the project model, but also throughout most of the various domain models, e.g., shown in Figure 2.6. Excerpts of the ontology as well as the profile for the project model are shown in Figure 2.13b (Kruse et al. 2020). The figure shows the ontology terms "Agent," as the bearer of a "Role of Responsibility," which gets prescribed by an "Assignment" that is to accomplish further things. On the SysML profile side there is the "perform" dependency with its tagged value, called "role of responsibility." This relation is used below to specify that an "Agent" called Researcher performing the role of responsibility of the task lead for the "Assignment" Research Task 1. This example shows that there is not a one-to-one mapping between the terms of the project ontology and the matching project profile. For example, the term "Role of Responsibility" gets realized in form of a subsidiary property and the relations "bearer of" and "prescribes" are only realized implicitly through the "perform" dependency.

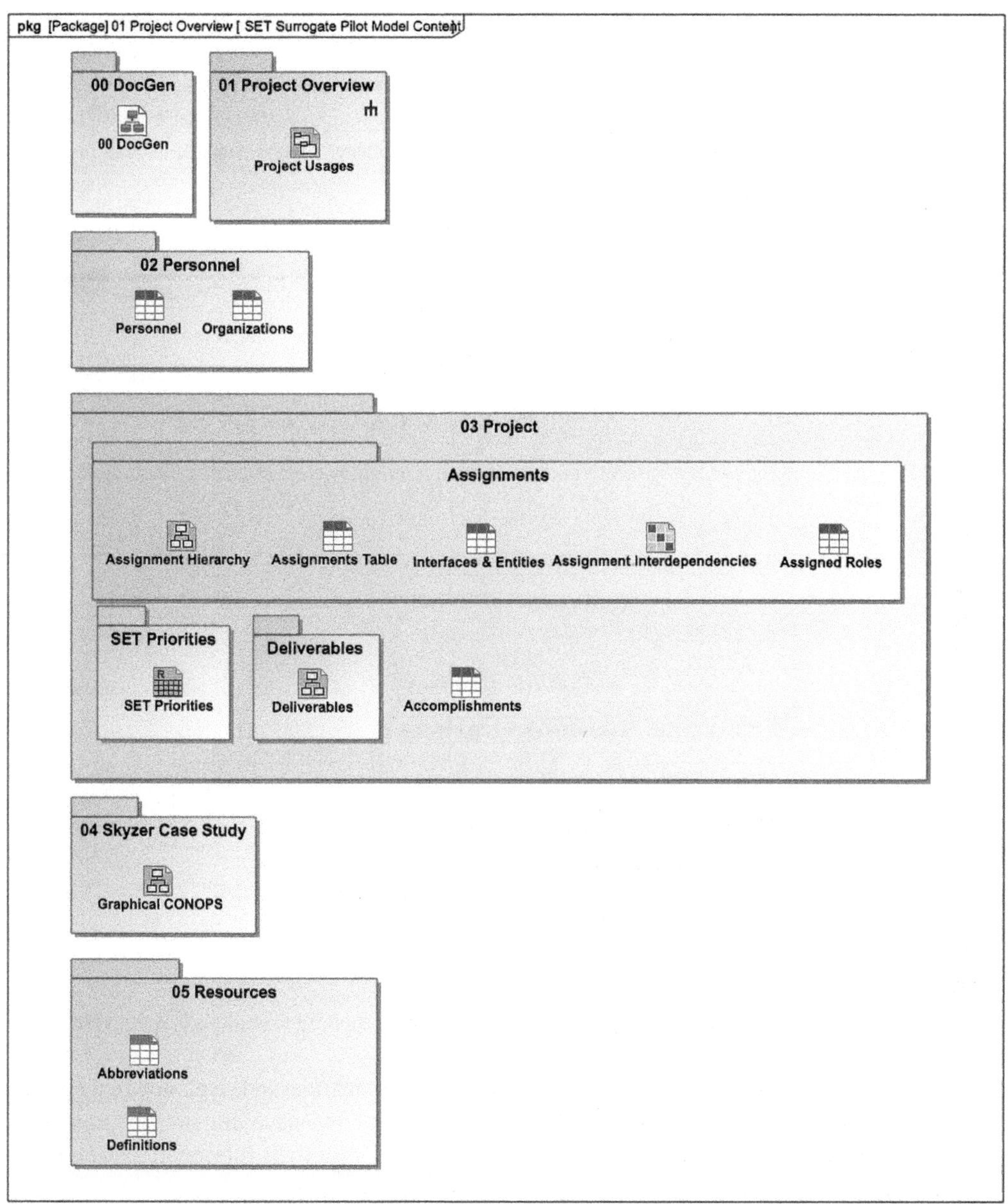

Figure 2.12 Model packages and elements for the system engineering technical and management plan model.

The content of the project model in SysML includes a hierarchy of assignment elements. Each assignment has a property for its status and can use its documentation for a textual description that also becomes part of the auto-generated documents as project reports. Linked to the assignments are researchers and other stakeholders as "agents" that perform certain roles of responsibility, as shown in Figure 2.13b. The interrelations between the different assignments as well as their required inputs and outputs, i.e., the deliverables, are modeled using internal block diagrams, as shown Figure 2.14. This also shows an assignment to align and refactor the "Skyzer Mission Model" and "Skyzer System Model" according to the "NAVSEM Starter" process model, while investigating

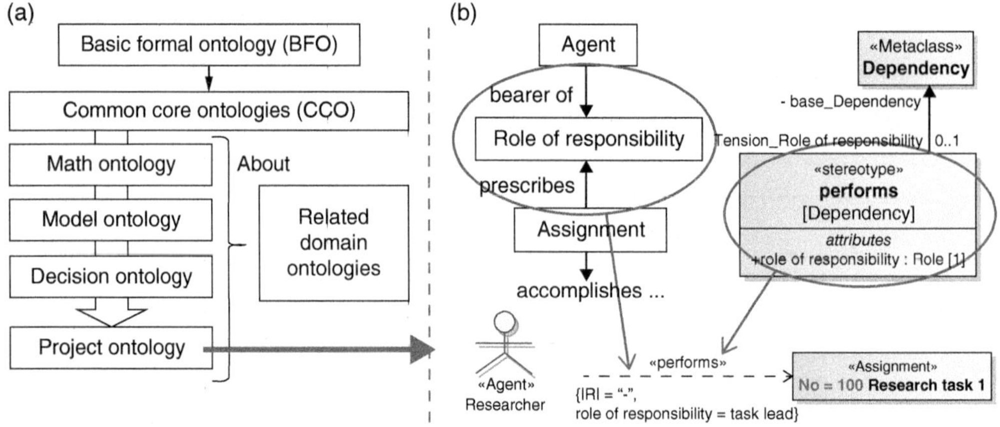

Figure 2.13 Project ontology ecosystem with excerpt of project ontology elements corresponding to SysML.

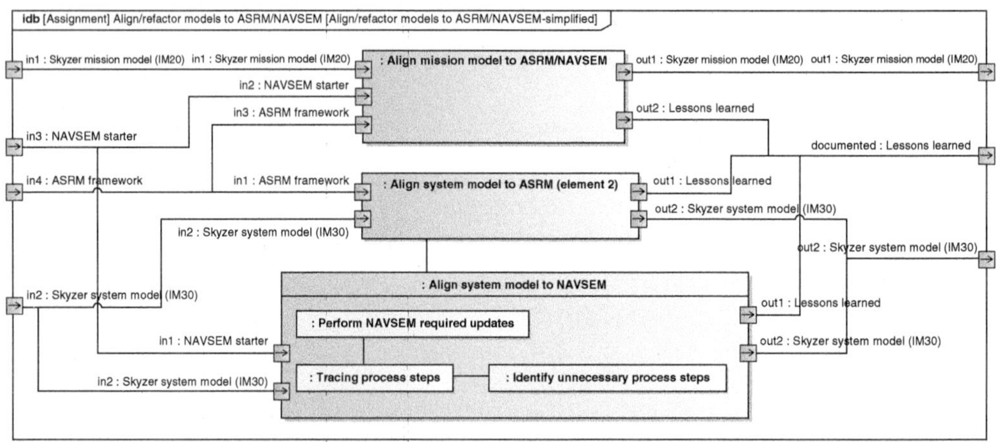

Figure 2.14 Simplified internal block diagram of assignments with their interrelations and deliverables.

their use and applicability as well as documenting any lessons learned, e.g., about identified unnecessary process steps, as shown as a subsidiary assignment. By specifying the assignments as specialized class elements in the profile, they can be modeled with their interrelations and deliverables, in contrast to, e.g., extended requirements or activity elements.

The accomplishments of the project are modeled as shown on top of Figure 2.15 using a stereotyped dependency with comment, date, and status properties. The dependency relates the accomplished entity with the achieving assignment. Having a project usage relation in the modeling tool gives direct access to all other used SysML models of the project. This allows to directly refer to the used models, their content or their documents when capturing accomplishments. Examples are given on the bottom of Figure 2.15 with an excerpt from the View Editor showing an accomplished addition to the mission model in the form of an added diagram for the ongoing alignment to NAVSEM/ASRM and the completed change of the mission model document's view hierarchy. The representation in the View Editor allows researchers to edit the date, status, comment, as well as the names of the accomplished entity and the assignment in the table in a web browser, without a SysML modeling tool. It is also possible to adapt generic placeholder elements into new

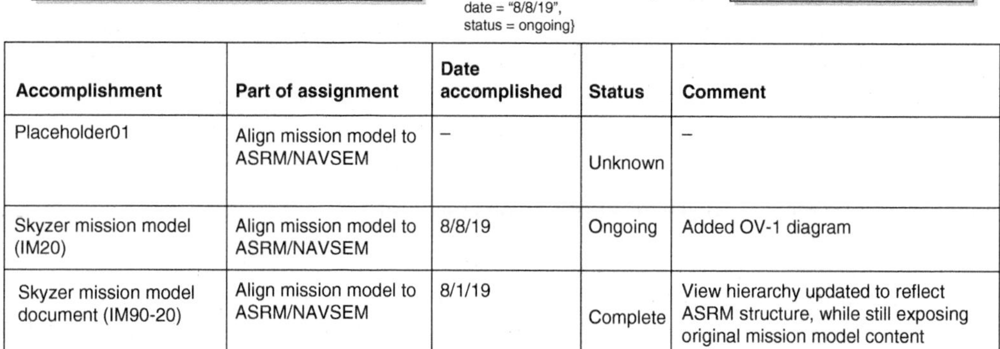

Accomplishment	Part of assignment	Date accomplished	Status	Comment
Placeholder01	Align mission model to ASRM/NAVSEM	–	Unknown	–
Skyzer mission model (IM20)	Align mission model to ASRM/NAVSEM	8/8/19	Ongoing	Added OV-1 diagram
Skyzer mission model document (IM90-20)	Align mission model to ASRM/NAVSEM	8/1/19	Complete	View hierarchy updated to reflect ASRM structure, while still exposing original mission model content

Figure 2.15 Example accomplishment in SysML (top) and derived View Editor table (bottom).

accomplishments. Similar placeholder elements also exist for assignment elements in the project backlog. Yet, to properly integrate the renamed placeholder elements, additional work in the SysML modeling tool is required.

Finally, the project model contains additional resources about the project goals, a brief overview over the investigated Skyzer case study, and a glossary with a list of used acronyms. The model data yields several metrics that are calculated and exposed within the generated documents. For example, for the number and status of accomplishments. This demonstrates that a System Engineering Technical and Management Plan can be modeled as a descriptive model and be potentially included as part of the AST.

Ontologies

In 2016, we started a similar SERC research task for the US Army, focused on science and technology (S&T), which often requires the fabrication of pre-production prototypes. Therefore, our research included modeling and simulation of discipline-specific, multi-physics analyses that help them understand system and mission requirements and trade-offs. We developed the Armaments Interoperability and Integration Framework (IoIF), a computationally based framework that uses ontologies (a.k.a. information models) and semantic web technology to link different model data at different levels of abstraction with multi-physics and discipline-specific model data, facilitating greater cross-domain model integration for analysis and design. An example of the resulting ontology ecosystem under the Basic Formal Ontology (BFO) can be seen in Figure 2.13a. We also developed two IoIF training courses based on a Cyber Ontology and Catapult use cases to help transition the research to the Army sponsor. See Chapters 3 and 4 for more details on these use cases.

Results

The Skyzer experiments conducted over three (3) surrogate pilot phases evolved and elaborated mission, system and subsystem analyses and requirements using a hypothetical system called Skyzer and demonstrated how to systematically apply different types of modeling methods at the mission, system, and subsystem levels. These methods aligned with the NAVSEM process model, but also created artifacts to be used in DocGen and to created Digital Signoffs as a way to transform away from using document-based signoff, by embedded the signoff with the decision evidence directly in the model. Methods define the required types of artifacts, which again leads to **consistency**, better **understanding of the system architecture**, **standardization**, as well as to more effectively assess **completeness** of the generated "specification." The methods also included guidance for modularity that linked all models to support an AST as shown in Figure 1.6 in Chapter 1.

Application of DE Metrics to Skyzer Pilot Lessons Learned

We performed an analysis to correlate DE benefit categories with lessons learned benefits observed during the Surrogate Pilot that applied DE methods and tools using an AST by creating models for everything to demonstrate the art-of-the-possible (McDermott et al. 2020). The analysis correlated rating from 17 lesson learned (Blackburn et al. 2021) categories to 22 DE benefit areas grouped into four metrics. The metrics categories include:

- Measure people *adoption* and enterprise process adoption (**adoption**)
- Analyze breadth of *usability* and issues with usability (**user experience**)
- Measure *productivity* indicators (**velocity/agility**)
- Generate *new value* to the enterprise (**quality and knowledge transfer**)

The correlated analysis uses a rating system to correlate the strength of each key lessons learned benefit against the benefit categories. We used the lessons learned in this analysis, because they directly rely on DE practices, methods, models, and tools that should enable efficiencies, and contribute to productivity. The DE approach integrated methods and tools with enabling technologies: Collaborative DE Environment (DEE) supporting an AST not just for the Government but also for the contractor. It also required the use of DEE technology features (e.g., Project Usage [model imports], DocGen, View Editor, Digital Signoffs) and methods to accomplish those lessons learned. The efforts demonstrated a means for a new operational paradigm to work directly and continuously in a collaborative DEE to transform, for example, how Contract Data Requirement List (CDRLs) can be subsumed into the modeling process using Digital Signoff directly in the model that is accessed through a collaborative DEE.

Figure 2.16 shows correlated analysis of 22 DE Success Measure Categories with top 6 of 17 lessons learned benefits observed during NAVAIR Surrogate Pilot that applied DE methods and tools using an AST that modeled everything to demonstrate the art-of-the-possible (McDermott et al. 2020).

We used a scoring/weight of: blank (0), three (3), five (5), and nine (9), where 9 has a strong relationship from underlying aspects of the lesson learned/benefits to the benefits categories. We create a total weighting across the benefits categories (row 2 has score for each measure) and similarly for each lesson learned (final column computes score for each lesson learned by row). The highest-ranking DE/MBSE benefit areas across the lessons learned are summarized below. The numbers in the parentheses reflect the rankings from Figure 2.16.

- [Knowledge Transfer] Better Communication/Info Sharing
- [Quality] Increased Traceability
- [Velocity/Agility] Improved Consistency
- [Knowledge Transfer] Better Accessibility of Information
- [User Experience] Higher Level of Support for Automation
- [Adoption] Quality and maturity of DE/MBSE Tools

This analysis is attempting to relate the lessons learned from the surrogate pilot to the DE metrics categories	Quality						Velocity/agility							User experience					Knowledge			Other						Total
	Reduce errors/defects (16)	Improve traceability (61)	Improve system quality (21)	Reduce risk (22)	Increased rigor (0)	Reduce cost (33)	Consistency (44)	Reuse (37)	Efficiency (13)	Improve Standardization	Collaboration/info sharing (68)	Integration/V&V (11)	Reduce time (24)	Automation (0)	Reduce SE task burden (0)	Manage complexity (48)	Productivity (14)	System understanding (24)	Information access (27)	Knowledge capture/sharing (13)	Architecture/sys understanding (23)	Planning	Priorities	Methods	Collaboration Env and AST	Workforce development		
Total	58	108	87	80	95	62	117	77	91	99	111	51	60	111	59	71	91	76	101	90	79	84	62	93	116	77		
Establish infrastructures for IME tools and AST as early as possible	3	9	9	5	5	3	9	5	5	5	9	5	5	9	5	5	9	5	9	3	9	5	5	3	9			153
Technically feasible to develop everything as a model	5	9	9	5	9	3	9	5	5	5	9	3	5	9	5	5	5	9	9	5	9	3	3	5	9	3		160
Use digital signoffs as a means for evolving from CDRLs	5	9	3	5	5	9	9	5	9	9	5	5	5	9	5	3	5	3	5	5	3	5	5	5	5	5		146
Establish and align modeling with methods & guidelines	5	9	9	5	5	5	9	5	5	9	5	5	5	5	5	3	5	5	5	5	9	5	3	9	5	9		154
Surrogate pilot demonstrated a new operational paradigm for collaborations in AST	3	9	5	5	9	3	9	3	5	5	9		3	9	3	5	5	9	9	9	5	3	3	6	6	9		149
Technology enables collaborative capabilities in MCE	3	9	5	5	5	3	9	5	9	9	9	3	3	9	5	5	5	5	9	5	5	1	1	9	9	5		150

Figure 2.16 DE digital engineering measures correlated to DE lessons learned (Top 6).

These align quite closely with the highest ranked metrics categories in the literature review and survey (McDermott et al. 2020). As this analysis was developed independently of the literature review and survey results, it provides additional validation of the rankings listed in Figure 2.16. Of note in this example, which is more advanced than a number of other DoD acquisition pilots, is the focus on automation. Reducing workload via automation is a key aspect of User Experience in the DE/MBSE implementation.

Primary lessons learned are:

- It is technically feasible to develop everything as a model
- Must establish and align modeling with methods and guidelines
- Establish infrastructures for IME tools and AST as early as possible
- Technology enables collaborative capabilities in model centric engineering

Both DE/MBSE are tightly coupled to quality of systems engineering methods, processes and workforce capabilities. However, the digital transformation of SE in much more tightly coupled with technology. The quality and maturity of the **DE/MBSE tools**, particularly integration of the Collaboration Environment and the AST is critical.

Future Work

Model-based systems engineering (MBSE) has been around for at least 20 years and the process, methods and tools have matured. Many organizations have been actively using MBSE on programs for at least 10 years and some 20 years. There has been system engineering transformations for many organizations as well as the Government who acquires large scale systems of system. The acquisition process is moving away from documents and instead using tools and technologies using integrated models. Even the late adopters are aware of the benefits and while there is still a need for broader work force development for technical people performing MBSE using tools and language such as SysML, there is a new need for broader understanding by Subject Matter Experts (SME) that now must provide information to Stakeholder Analysis Models (e.g., Cost, Reliability, Risk, etc.) that link with the technical models. It is well understood that the models should be integrated to establish an authoritative source of true, which also leverages rigorous model/configuration management. SMEs must be able to find, read, analyze and update information within the AST. However, we should not expect them to use the MBSE authoring tools. There are other approaches for using Tool Proxies, such as DocGen that generate views of model information, where SMEs can access, analyze and update information that can be synchronized back into the model using appropriate model/configuration management practices.

There is also a need to develop other types of program-related information such as a Digital Engineering Statement of Work (DESOW) that specifies how the work should be created in the form of models, such as Skyzer, which is an unclassified type of reference model that can be used to illustrate method-compliance modeling. This needs to consider how the models should be evolved and maintained as programs go beyond manufacturing and production, to operations and sustainment.

Notes

1 DASD has increased the emphasis on using the term Digital Engineering. A draft definition provided by the Defense Acquisition University (DAU) for DE is: **An integrated digital approach that uses authoritative sources of systems' data and models as a continuum across**

disciplines to support lifecycle activities from concept through disposal. This definition is similar to working definition used throughout our prior research task RT-48/118/141 for Model Centric Engineering (MCE).

2 Certain commercial software products are identified in this material. These products were used only for demonstration purposes. This use does not imply approval or endorsement by Stevens, SERC, or NAVAIR, nor does it imply these products are necessarily the best available for the purpose. Other product names, company names, images, or names of platforms referenced herein may be trademarks or registered trademarks of their respective companies, and they are used for identification purposes only.

3 NAVAIR Public Release 2019-443. Distribution Statement A – "Approved for public release; distribution is unlimited."

References

Bell Eagle Eye (2023). https://en.wikipedia.org/wiki/Bell_Eagle_Eye (accessed 4 June 2023).

Blackburn, M. and Kruse, B. (2023). Conducting design reviews in a digital engineering environment. *Insight* 25 (4): 42–46.

Blackburn, M., Cloutier, R., Hole, E., and Witus, G. (2014). Transforming System Engineering Through Model Based System Engineering. *Systems Engineering Research Center, Final Technical Report, Research Task 48.*

Blackburn, M., Bone, M., and Witus, G. (2015). Transforming System Engineering Through Model-Centric Engineering. *System Engineering Research Center, Research Task 141. Technical Report SERC-2015-TR-109.*

Blackburn, M., Blake, R., Bone, M. et al. (2017). Transforming Systems Engineering Through Model-Centric Engineering Research Task 157. *SERC-2017-TR-101.*

Blackburn, M., Kruse, B., Stock, W., and Ballard, M. (2020). Digital engineering modeling methods for digital signoffs. *NDIA Systems and Mission Engineering Conference. 2020 Virtual Systems and Mission Engineering Conference* (10 November–13 December 2020).

Blackburn, M., Dzielski, J., Peak, R. et al. (2021). Transforming Systems Engineering Through Model-Centric Engineering. *Final Technical Report SERC-2021-TR-012, WRT-1036 (NAVAIR).*

Chell, B. (2021). Multidisciplinary system and mission design optimization. Dissertation. Stevens Institute of Technology.

Delp, C., Lam, D., Fosse, E., and Lee, C. (2013). Model based document and report generation for systems engineering. In: *IEEE Aerospace Conference* (02–09 March 2013). Big Sky, USA. IEEE.

Hutchison, N., Pepe, K., Blackburn, M. et al. (2020). Developing the Digital Engineering Competency Framework (DECF). *Technical Report SERC-2020-TR-010.*

ISO/IEC/IEEE (2011). *Systems and Software Engineering – Architecture Description.* ISO/IEC/IEEE 42010:2011(E) (Revision of ISO/IEC 42010:2007 and IEEE Std 1471-2000), pp. 1–46. https://doi.org/10.1109/IEEESTD.2011.6129467.

Kruse, B. and Blackburn, M. (2019). Collaborating with OpenMBEE as an authoritative source of truth environment. *Procedia Computer Science* 153 (C): 277–284.

Kruse, B., Hagedorn, T., Bone, M., and Blackburn, M. (2020). Collaborative management of research projects in SysML. *18th Annual Conference on Systems Engineering Research (CSER)* (8–10 October 2020).

McDermott, T., Van Aken, E., Hutchison, N. et al. (2020). Digital Engineering Metrics. *Technical Report SERC-2020-SR-002.*

OpenMBEE: Open Model-Based Engineering Environment (2023). www.openmbee.org (accessed 4 June 2023).

Biographical Sketches

Benjamin Kruse acquired his diploma in mechanical engineering in the field of aerospace at the Technical University Munich in Munich (2012, Germany) where he also started his doctorate before finishing his Dr.sc. at ETH Zurich in Zurich (2017, Switzerland). There he investigated formal and reuse-based support for multi-disciplinary concept design using the model-based systems engineering language SysML. Afterward, he became a Research assistant professor at Stevens Institute of Technology in Hoboken, NJ (2017, USA), working on the model-based systems engineering side of the systems engineering transformation through model-centric engineering as part of implementing the Department of Defense's digital engineering strategy. This work was continued as an ISE Post-Doctoral Associate for the Virginia Polytechnic Institute and State University in Blacksburg, VA (2021, USA) before switching to the e:fs TechHub GmbH in Gaimersheim (2022, Germany) in order to work as a developer for model-based systems engineering in the Volkswagen Group.

Brian Chell, PhD is a postdoctoral research associate with the Systems Engineering Research Center at Stevens Institute of Technology. His research interests are in distributed space systems and applying multidisciplinary design optimization techniques to mission-level analysis. Brian received a PhD in systems engineering and an ME in space systems engineering from Stevens, and a BS in aerospace engineering sciences from the University of Colorado Boulder.

Timothy D. West currently serves as the Deputy Director of the U.S. Air Force Research Laboratory's Operations Directorate. In this capacity, he provides leadership and oversight of the Laboratory's air and space flight test program, demonstrating combat readiness of the military's latest science and technology development programs. Prior to his current assignment, Mr. West served for nearly 30 years as a military officer in the U.S. Air Force, where he held command positions in science and technology, test and evaluation, and program management. He is a graduate of the US Air Force Test Pilot School, the Air Command and Staff College, and the Air War College. Mr. West holds MS degrees in aerospace and industrial engineering from the University of Tennessee Space Institute, and a BS in mechanical engineering from the University of Kentucky. Mr. West is also a Systems Engineering PhD Candidate at Stevens Institute of Technology.

Mark R. Blackburn, PhD is a senior research scientist with Stevens Institute of Technology since 2011 and principal at KnowledgeBytes. Dr. Blackburn has been the principal investigator (PI) on 17 System Engineering Research Center (SERC) research tasks for US Navy NAVAIR, US Army DEVCOM, and US Space Force on Digital Engineering Transformation Research Tasks. He has also been PI on a FAA NextGen and National Institute of Standards and Technology projects and has received research funding from the National Science Foundation. He develops and teaches a course on Systems Engineering for Cyber Physical Systems. He is a member of the SERC Research Council, OpenMBEE Leadership Team and INCOSE Pattern Working Group focused on the Semantic Technologies for Systems Engineering initiative. Prior to joining Stevens, Dr. Blackburn worked in industry for more than 25 years. Dr. Blackburn holds a PhD from George Mason University, a MS in mathematics (emphasis in CS) from Florida Atlantic University, and a BS in mathematics (CS option) from Arizona State University.

Chapter 3

Transforming Systems Engineering Through Integrating Modeling and Simulation and the Digital Thread

Daniel Dunbar, Tom Hagedorn, Timothy D. West, Brian Chell, John Dzielski, and Mark R. Blackburn

Stevens Institute of Technology, School of Systems and Enterprises, Hoboken, NJ, USA

Introduction

While discipline-specific modeling and simulation have enabled immense gains in efficiency and computation, broader design and analysis tasks that must reach across disciplines and individual models to a broader system context often rely on manual transfer of data or direct tool-to-tool integration, which is brittle in nature (Bone et al. 2019). The chapter discusses digital engineering (DE) in the broadest way, and the accompanying authoritative source of truth (AST) seek to enable better collaboration between all types of models not only for individual system design and analysis tasks but across the entire life cycle of the system, including mission, system, and subsystem levels of abstraction. This chapter details efforts using graph data structures and ontology aligned data as an AST to enable cross model collaboration in a way that addresses current challenges in the process while building a foundation for future advances.

The structure of the chapter is as follows: the Background and Motivating Use Case gives some background on modeling, simulation, and the concept of the Digital Thread and introduces a catapult use case that is used throughout the chapter. The Integration Methodology section discusses a high-level framework enabled by ontology-aligned graph data structures in a DE context, and a python-based implementation of the framework is discussed in relation to the motivating use case. The Discussion section looks at the framework's application in light of the needs DE presents, and the Conclusion provides a summary as well as avenues for future expansion.

Background and Motivating Use Case

Modeling and Simulation

Modeling and simulation in engineering is so ubiquitous that the existence of tools enabling sophisticated analysis to be performed with a mouse click may be taken for granted. The use of the singular verb in the previous sentence was deliberate, intended to imply that modeling and simulation capabilities are often so tightly coupled in modern tools that the boundary is indistinguishable to the user, and that both may be done nearly simultaneously. An Internet search will return many different definitions for a model and for a simulation, and these definitions often reflect some

Systems Engineering for the Digital Age: Practitioner Perspectives, First Edition. Edited by Dinesh Verma.

Companion website: www.wiley.com/go/verma/systemsengineering

aspect of the individuals or organizations formulating the definition. In the context of DE, we need to recognize that all the definitions that might be conceived may fit some specific use case at some time. For this reason, it is worth taking a moment to consider what these two terms could mean, how they are different, and how they are related. These ideas will be important when we discuss the concept of the "full-stack" of models that includes mission, system, physics-based and discipline-specific models, and a new construct called the assessment flow diagram (AFD). The full-stack places the system of interest within the world of possibilities of its use, and at the same time may consider aspects of the system at the resolution of individual components and parts. The computationally formalized AFD (Cilli 2015) models the interconnection of simulation capabilities associated with one or more analysis or design processes.

A model is a representation of something constructed for the purpose of informing about some aspect or aspects of a thing, mission, or system. Models, like the systems they are intended to describe, must have well defined boundaries so that it is clear what is and is not part of the system being modeled. Models may be physical representations, but often they are abstractions of some characteristics of a system or object. In the most general sense, models simply enumerate the features and characteristics of a system or thing. This is the nature of a reference model whose purpose is to describe the class of elements it represents. Reference models can be generalized to represent more specific types of elements in a class, or they can be extended to include information about how the enumerated features or characteristics of the system are related. These enumerable features may be divided into those that can be selected or specified individually, and those that can only be determined after some subset of the individual elements have been defined. Models may also contain information about how these features are related.

A better way to motivate what constitutes a model is to consider some examples. The examples help to establish a broad view of what might be considered a model. We use a catapult example, derived from publicly available information as a "system of interest" throughout this chapter, and start by considering models of the features and characteristics of a catapult. A reference model for a catapult might have placeholders for its purpose in general terms of the type of target it is intended to attack, the kind of projectile, where it is to be used, the crew that operates it, and other characteristics. This model may be generalized into different and distinct models based on the energy storage mechanism and type of projectile. At this level of resolution, it is possible to consider the physical characteristics of the catapult and how it performs. These types of models can be characterized using the System Modeling Language (SysML) and its semantics. SysML is considered a "descriptive" modeling language because its purpose is to provide a capability to describe aspects of a system from many different viewpoints. While it can be argued that SysML can be extended to describe anything, at some point it becomes necessary to use modeling tools specifically developed to represent and characterize something so that it can be fabricated. Examples of engineering discipline or domain-specific tools are parametric solid modeling for mechanical engineering and circuit layout and design for electrical engineering.

The catapult system used throughout this chapter is represented by a SysML block definition diagram shown in Figure 3.1. The catapult is part of a larger Artillery System and Catapult Mission and operates in a mission environment with defined targets. It has its own physical architecture defined in Figure 3.2 and represented as a project usage (i.e., a type of model import) in the Mission SysML model. Multiple levels of abstraction in the Mission and System architecture contain information and parameters that may be used in relevant modeling and simulation activities.

A simulation is a realization of a model that can be used to compute outputs or dependent variables based on a set of inputs or independent variables. In the context here, the model can be thought of as a description of a thing and the rules that characterize it. The simulation exercises

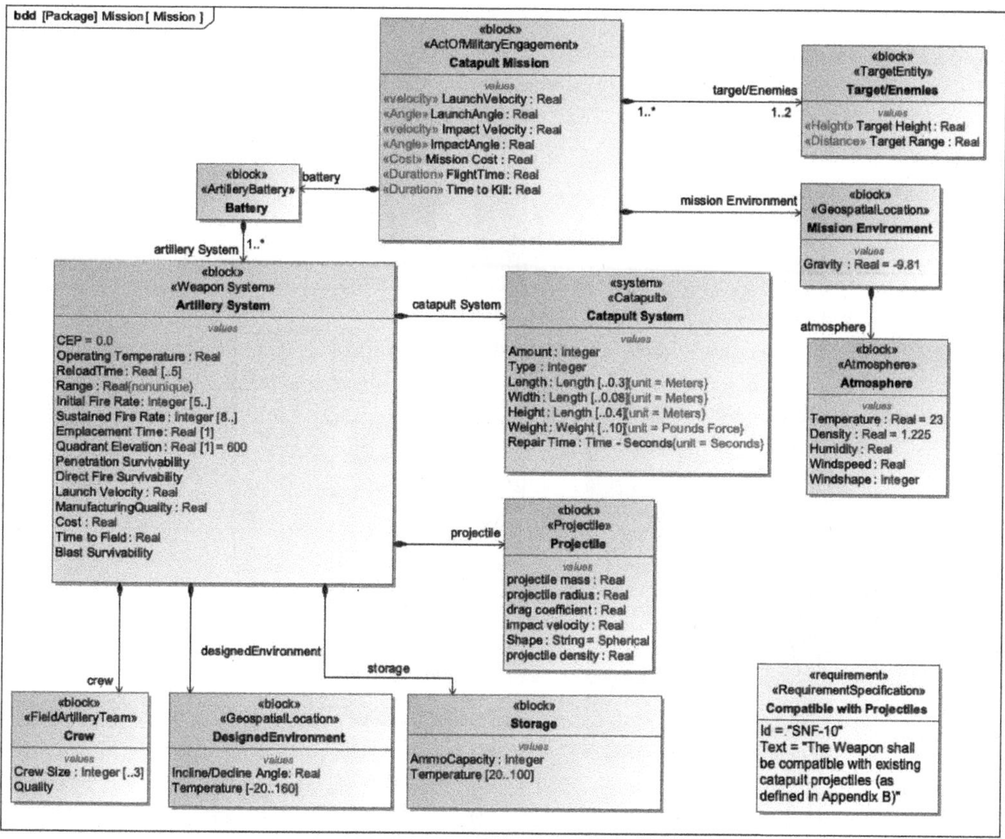

Figure 3.1 Mission level block definition diagram.

the rules that complete the description of the system. In this sense, a rendering of a part by a solid modeling tool is a simulation. The purpose of the rendering is for the user to examine the part and perhaps see if it fits properly with renderings of other parts. Given a set of mathematical formulas, spreadsheets can be used to implement simulations. In our catapult example, we use a fire simulation to simulate the use of the catapult in the field.

Digital Thread

The digital thread (DT) is a common term within engineering that indicates a movement from individual models to a series of models working together in an integrated computing environment working toward a common goal within the engineered system's life cycle. Where two or more models contribute parameters, design constraints, etc. to a larger design or analysis task, the implementation of a DT creates an explicit connection that enables more automated and semi-automated performance of complex analyses while reducing errors caused by manual connection of parameters across models. DT representations are discussed in Chapter 4.

Whereas domain-specific modeling and simulation is fairly limited in its scope through the limits of the software tool implementing the model simulation or the limited set of tool-to-tool integrations developed and maintained to connect one model and tool to another, the use of a centralized AST gives greater freedom to implement a DT that connects multiple tools and

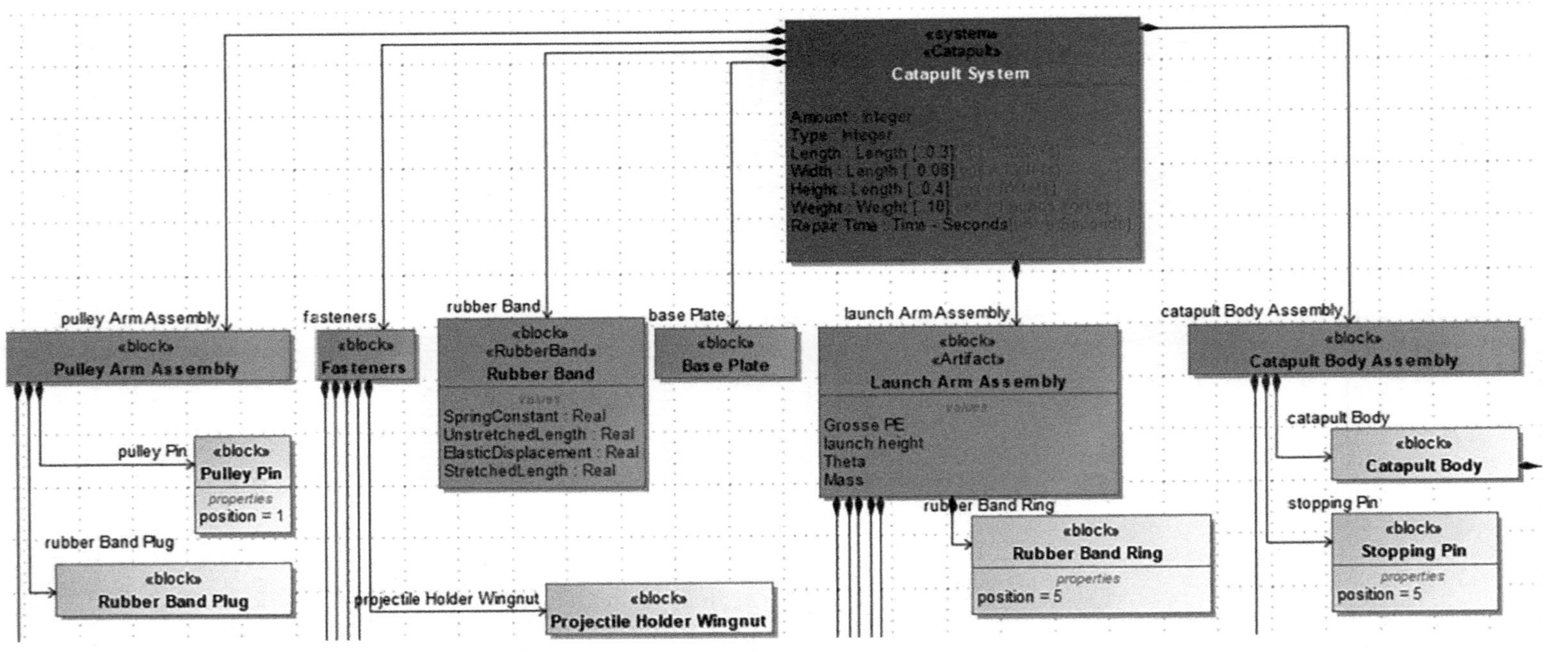

Figure 3.2 Partial catapult system physical architecture – imported as project usage.

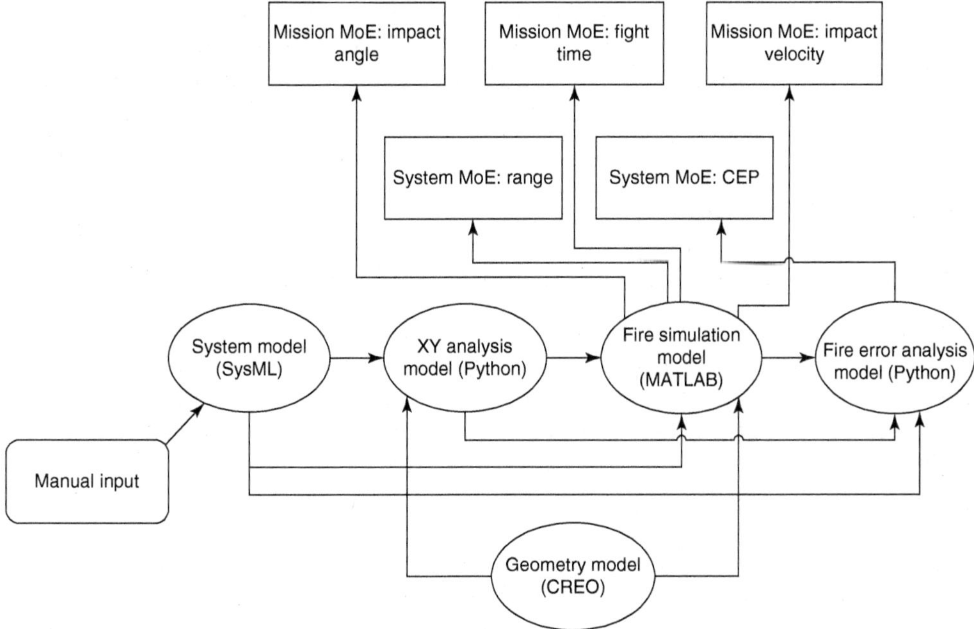

Figure 3.3 Abstract digital thread showing interconnections between models and MoEs.

viewpoints of a system during design or analysis. This AST acts as a common source of information that all models within a system under design or a system of analysis can pull from to ensure consistent data usage.

An abstraction of the catapult example (Figure 3.3) shows five models needed to be integrated in a DT to inform mission and system objectives. The System Model is parametrically informed by the Geometry Model and manual input. It in turn informs the other three models – the XY Analysis Model, the Fire Simulation Model, and the Fire Error Analysis Model. The XY Analysis Model informs the Fire Simulation Model and the Fire Error Analysis Model. The Fire Simulation Model informs the Fire Error Analysis Model. Two system level Measures of Effectiveness (MoEs) are defined – Circular Error Probability (CEP) and Range. The Fire Simulation Model produces Range, and the Fire Error Analysis Model produces CEP. All three mission-level MoEs defined are provided by the Fire Simulation Model.

As can be seen in Figure 3.3, the individual models used in the analysis are highly interconnected. A higher-level analysis that seeks MoEs at the system and mission level requires the use of all of the models. Traditionally, the transfer of parameter information from model to model would be done either manually by designers and analysts or through direct tool-to-tool integration. The solution here enables a standards-based approach to configuring interfaces that allows for automated and semi-automated computer transfer of information without the burden of individual tool-to-tool interfaces being built and maintained.

In this chapter, multiple modeling and simulation tools are used, primarily to show the flexibility of the methodology. It is understood that the number of modeling and simulation tools available and used by the broader community is larger than one chapter can address. Therefore, the chapter focuses on methodology and standards-based integration protocols and data formats, such as the REpresentational State Transfer (REST), Application Programming Interface (API), and JavaScript Object Notation (JSON). This chapter serves as a foundation for building an ecosystem that fits many different design and domain contexts by describing a methodology for representing knowledge captured by models and simulations in a tool and domain agnostic graph data structure with the use of standards-based techniques for interfacing with external tools.

Integration Methodology

Digital Engineering Framework for Integration and Interoperability

The Digital Engineering Framework for Integration and Interoperability (DEFII) provides a conceptual foundation for integrating ontologies and other semantic technologies into a DE context (Dunbar et al. 2023). It lays out an approach for transforming data from tool-specific sources into a tool-agnostic graph representation that can be accessed and analyzed in multiple ways to yield unique insight into the system under design and enable DT applications.

Ontology Development

The foundation of the DEFII framework relies on ontology-aligned data (Figure 3.4). Ontologies have been demonstrated to be useful potential candidates for representing ASTs in a DE context. They have their root in philosophy but have seen evolution in computer science and the natural sciences as a way of capturing knowledge in a manner that is both human and computer readable (Arp et al. 2015). Modern ontologies can be written in languages like the Web Ontology Language (OWL)[1] that enforce formal mathematical logic on the ontologies and enable automated reasoning to infer new information embedded in the logic of the explicitly asserted data. They are part of a suite of technologies referred to as Semantic Web Technology (SWT). This suite of technologies

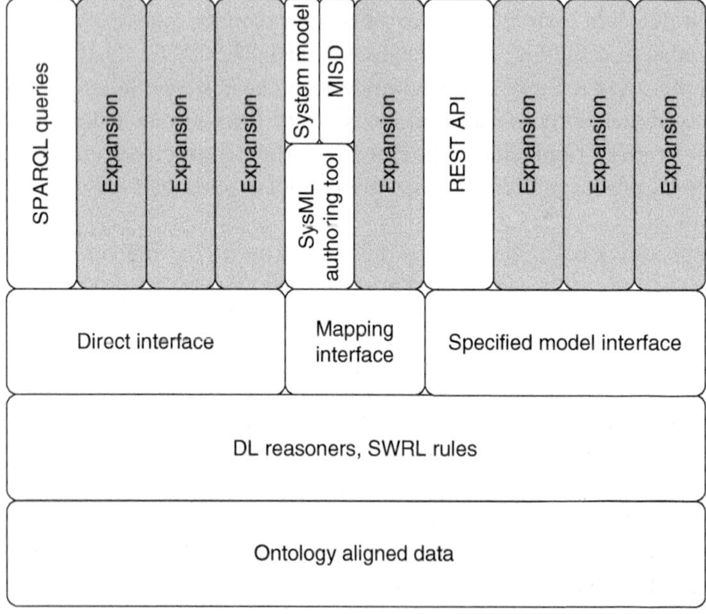

Figure 3.4 Digital engineering framework for integration and interoperability (DEFII). *Source:* Adapted from Dunbar et al. (2023).

includes things like Description Logic reasoners, rule languages, query languages, and constraint languages that all operate on linked data captured as a graph to align with ontologies.

Model integration is specifically aided by the use of ontologies because they enforce a common, controlled vocabulary. The same word can have very different meanings in different domains and tool sets. For example, a "layer" in a diagram or design application may refer to a collection of drawing components that is grouped together and can be made visible or invisible relative to other components in the drawing. However, a "layer" in the Additive Manufacturing domain may refer to an individual layer of filament in the 3D printing process. Thus, the same word used in two different ways introduces confusion when attempting to perform model integration. Ontologies make explicit what terms have what definition in the model integration context. This enforcement of common terminology enables effective semantic integration.

Realistically, the ontologies needed for most model integration are relatively small and could be captured in an Excel spreadsheet, or even in an old-school steno notebook. However, we believe such an approach is short-sighted, as it would create a new set of disciplinary "stovepipes" that would likely result in model integration issues downstream. Additionally, the logic, richness, and depth of a well-designed ontology often gets lost in these rudimentary approaches. Instead, we recommend the use of tools such as Stanford University's Protégé,[2] an open-source graphical user interface (GUI) for building ontologies (Noy et al. 2003). To better facilitate cross-compatibility between related application ontologies, we also recommend aligning ontological terms with a domain-neutral, top-level ontology, such as the Basic Formal Ontology (BFO) developed by Barry Smith and Pierre Grennon (Ruttenberg 2020), and ideally under a mid-level "bridging" ontology that offers common semantic pathways to connect the domain-neutral ontology with the domain/application-specific ontology being developed. One such "bridging" construct that is gaining popularity is the Common Core Ontologies (CCO), a family of open-source ontologies that "represent and integrate taxonomies of generic classes and relations across all domains of interest" (CUBRC, Inc. 2020). This approach not only enhances cross-domain compatibility, but it also optimizes ontology extensibility, shareability, and usability – critical attributes for "future proofing" any ontology development. For additional insight on this subject, see Arp et al. (2015), which offers 24 best practices for ontology development.

Data aligned to ontologies form a special kind of Graph Data Structure (GDS). GDS, in general, include a series of nodes connected by vertices. This data structure has been shown to be a useful representation for performing system design analysis (Cotter et al. 2022; Medvedev et al. 2021) and appropriate for use as an AST (Mordecai et al. 2021). Thus, ontology aligned data represented in a GDS enables the use of both reasoning based on the formal representation of the ontology and the graph-based algorithms and research being performed on GDS more broadly. Thus, the advantages are threefold:

1) **Enforces a common vocabulary:** the use of ontologies enforces a common vocabulary, at least within the AST. This common vocabulary ensures that all tools and users interacting with the data connected to the AST are clear about what data they are accessing.
2) **Enables semantic reasoning and querying:** the formal nature of ontologies allows for semantic expansion of the underlying data. This utilizes relationships and axioms stored in the ontologies to infer additional facts about the data. In addition, querying using standard languages like SPARQL, a query language similar to SQL but for triplestore repositories instead of relational databases, enables precise extraction of data according to patterns.
3) **Access to graph-based algorithms and analysis:** mathematical graph theory has enabled the development of useful algorithms to analyze a graph data set. Analyses such as shortest path first can give unique insights to a system under design and are enabled by the graph data structure underlying the ontology-aligned data.

Reasoning and Rules

The second layer of the DEFII framework is automated reasoning. As mentioned above, the use of ontologies creates opportunities for automated expansion of the underlying data by inferring additional facts about the system based on relationships and other axioms built into the ontologies themselves. In addition, use of rule languages like the Semantic Web Rule Language (SWRL) allows for more automated expansion of the data based on context-specific rules. Both DL reasoning and SWRL rule application can happen in an automated fashion without any action by users of the framework. During the setup process of the triplestore repository used to house the ontology-aligned data, reasoning profiles can be configured to automatically reason on data within the repository. Thus, this layer is below any interface layer to represent its automated nature.

Interfaces

The DEFII framework allow for the advantages of graph data structures, and more specifically ontology-aligned graph data structures, without the need for all users of the framework to understand the details of ontologies or the broader SWT stack. It does this by abstracting away many of the details involved in the SWT stack and exposing data through controlled interfaces. There are three notional interfaces as part of the DEFII Framework to address this concern:

1) **Direct interface:** the Direct Interface focuses on the SWT stack. While things like ontologies and DL reasoners are part of the foundation of the framework, the Direct Interface allows directed usage of other SWT tools. For example, SPARQL is a query language designed to query graph data structures and can be used to explore ontology aligned data.
2) **Mapping interface:** the Mapping Interface is a tool or model-dependent interface that can translate a tool or model data external to the framework into terms aligned with its underlying ontologies. This process ***begins*** with the tool or model in question and moves toward ontology-aligned data.
3) **Specified model interface:** the Specified Model Interface is a tool-agnostic interface that exposes ontology aligned data to users and toolsets outside of the ontological context. This interface ***begins*** with the ontology-aligned data and moves outward.

The Specified Model Interface must be configured to package models of interest together in a manner consistent with designers/analysts' needs. This is accomplished through the use of specifications within the system model. The use of the Model Interface Specification Diagram (MISD) (Figure 3.5) enables parameters found within a system model to be packaged together into a single interface.

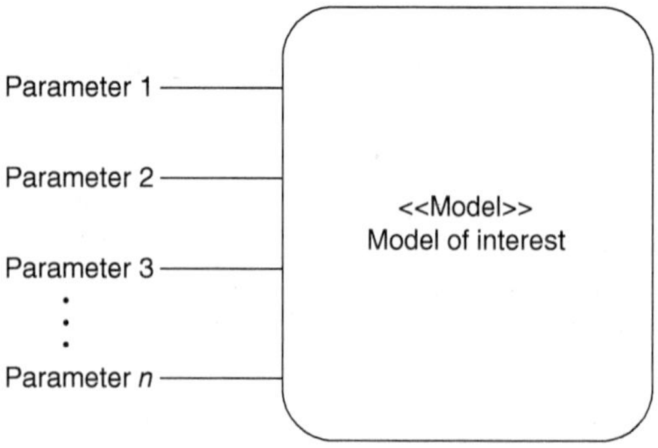

Figure 3.5 Abstract model interface specification diagram (MISD).

After a model has been specified in an MISD, the model can then be exposed through the Specified Model Interface. The notional interface can be implemented in many ways, but the use of a REST API allows for a standards-based interface to the packaged model. REST APIs are becoming ubiquitous in software development and interfacing and use the JSON format to transfer data between tools. For example, see Figure 3.6 where a simply physics model has packaged three different elements into a single model (MISD 1).

Many software tools offer access to REST APIs using http services, so connecting model information to various software tools is typically straightforward. In the cases where a software package does not provide http services, simple middleware can be used to convert a packaged model from a JSON structure outputted by a REST API into a format that is readable by the tool in question, such as XML or comma separated variables (CSV).


```
"MISD 1":{
    "Mass": "27 kg",
    "Gravity": "9.81 m/s/s",
    "Velocity": "40 m/s"
},
```


Figure 3.6 Sample JSON representation of an MISD.

An additional advantage to packaging the models in a standard format like JSON is that this allows the models to be used by multiple tools for different purposes. For example, the fire simulation analysis model in the catapult example is run by MATLAB. A web application dashboard could also access the same information via the same REST API endpoint to provide visualizations on what MATLAB is using to perform its analysis. The specified model is tool-agnostic – it is simply a collection of parameters grouped together in a way that is deemed useful to relevant stakeholders.

Assessment Flow Diagram

An AFD groups individual MISDs into a larger analysis context, where multiple models share parameters or influence each other. This concept is based on computationally formalizing Cilli's work on assessment flows (Cilli et al. 2015) and enables higher level analyses. Whereas an MISD packages particular parameters together in a way that makes them easily accessible to external tools, an AFD shows how the models themselves interact.

Consider the catapult use case and the abstract DT shown in Figure 3.3. The series of model interact with each other. This can affect things like sequencing and may also lead to necessary trade off analyses as a common parameter affects models in different ways. The inclusion of all models into a common AFD allows explicit, formal representation of the relationships between models to be considered during analysis. The use of a diagram also enables the communication of those connections in a human readable format. Figure 3.7 shows three MISDs aggregated to form a higher-level AFD, and there is an implicit flow between the model's inputs (bottom blocks of MISD) and outputs (top blocks of MISD) to reach the high-level system or mission objectives (Dunbar et al. 2023).

Interoperability and Integration Framework

The DEFII framework gives a conceptual method for implementing an AST in a functionally useful way in a Digital Engineering context. However, it must be instantiated in software. The Armaments Interoperability and Integration Framework (IoIF) (Bone et al. 2018) is a python-based implementation of the framework. Using this implementation in the Catapult use case produces an instantiation of the framework with implementation that is detailed in Figure 3.8.

This implementation can use a number of different triplestores to store ontology aligned data. Ontotext's GraphDB[3] is shown above. GraphDB provides functionality to perform SPARQL queries on the data directly using its graphical user interface (GUI). Dassault Systemes' SysML authoring suite, including Catia Teamwork Cloud[4] is used to represent the system model. MATLAB[5] and Creo[6] are shown using middleware to connect the tools to the Specified Model Interface, and MATLAB and a Web Dashboard are shown with direct access to the REST API. Python libraries have been built to enable the Mapping Interface and the Specified Model Interface. In addition to the System Model using the Mapping Interface, four different analysis models are represented.

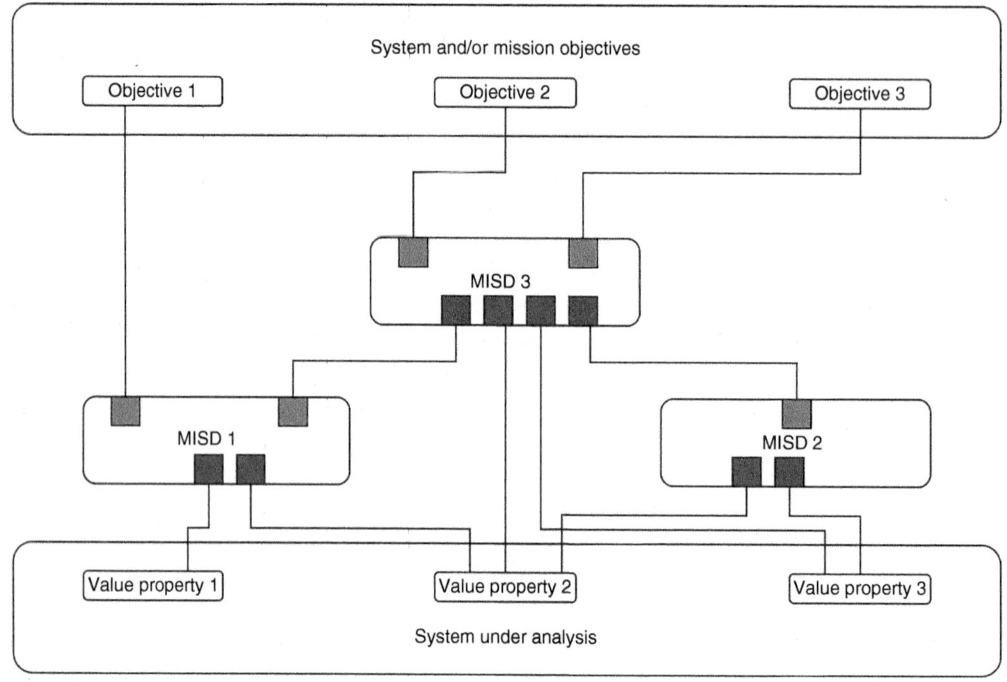

Figure 3.7 Abstract assessment flow diagram connecting system parameters, modeling, simulation, control, and visualization Dunbar et al. (2023).

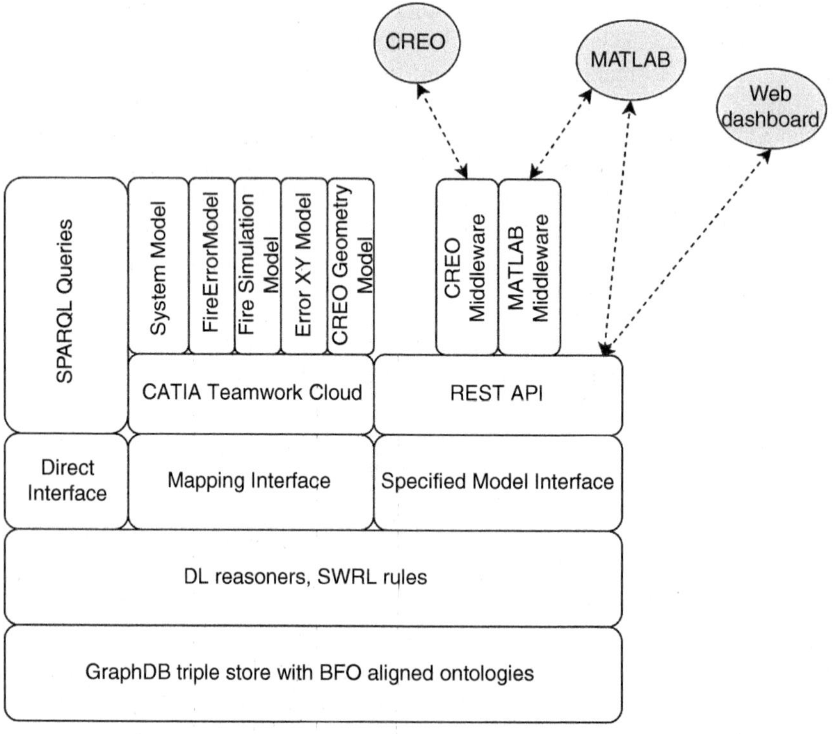

Figure 3.8 Instantiated framework of catapult use case – IoIF pythonic implementation.

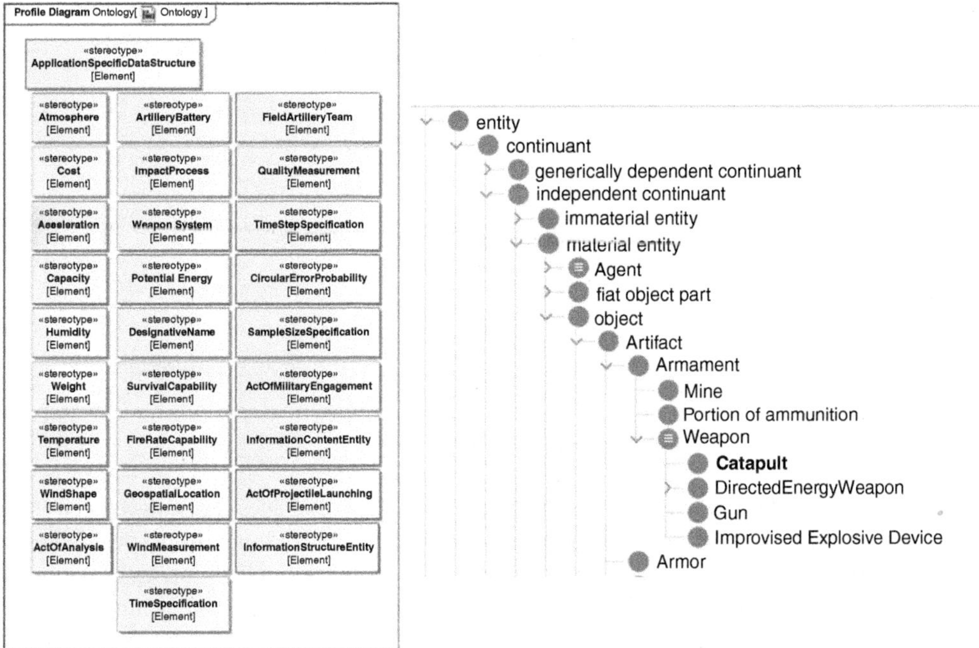

Figure 3.9 Ontological elements used in SysML profile and protégé editing software.

Mapping is performed on the SysML model and uses SysML stereotypes to tag the system model with ontologically relevant terms. The model can be read programmatically, and components with stereotype tags that correspond to classes present in the ontologies can be matched to create an ontology aligned representation of the system model. A profile can be created in SysML to capture all the custom, ontology linked stereotypes included in a model for tagging (Figure 3.9 left). This Profile corresponds to ontologies represented the OWL ontology language.

Mapping is then implemented in IoIF by extracting model elements that have been tagged with custom SysML stereotypes corresponding to classes in the ontologies being used for a specific project. Figure 3.10 from Dunbar et al. (2023) shows pseudocode that explains the mapping process from an RDF representation of the SysML system model (created by accessing the TWC REST API) to an ontology-aligned AST.

Ontologies are, by definition, extensible. This means that they can be expanded in the future. Ontologies only represent data explicitly stated and offer no insight using reasoning based on omission. This is not to be confused with reasoning based on mathematical logic. For example, a reasoner could properly infer from an ontology on family relations that a mother's son is her father's grandson, even if this relationship is not explicitly stated based on axioms defining the relationship between the terms mother, son, and father, and grandson. However, a reasoner could not infer from an ontology that a particular frog is NOT green simply because the ontology does not label the frog with the color green. This concept is connected to the idea of ontological realism and the use of the Open World Assumption, a philosophical assumption that has implications on engineering analysis beyond the scope of this chapter. However, incomplete ontologies are still usable and in most cases evolve over time.

Partial implementations of ontologies are acceptable. Ontologies are based on a never-ending expansion of the body of knowledge and thus are never fully "done." They are always capable of

```
"Step 1":    "Execute SPARQL Query for stereotyped
elements;  store results for class and name",
"Step 2":    "Loop through the results evaluating for
'name' variable existence in defined ontologies",
"Step 3":    "Where 'name' exists as a class in the
ontology, perform the following three tasks",
"Step 3a":   "Create a triple in form ([class]_entity,
rdf:type, [name])",
"Step 3b":   "Create a tiple in the form ([class]_spec,
rdf:type: DirectiveInformationContentEntity)",
"Step 3c":   "Create a tiple in the form ([subject from
Step 3b], Comm:prescribes, [subject from Step 3a])",
"Step 4":    "Put newly created triples into the graph
database",
"Step 5":    "Add a link between relevant elements in
mapped repository and original data repository"
```

Figure 3.10 Mapping pseudocode. *Source:* Adapted from Dunbar et al. (2023).

being expanded in the future. In practical terms, this means that ontologies can be developed as far as they are useful without being overly concerned with capturing everything. Ontologies should be developed in a principled, consistent manner to promote interoperability (for more information on ontology development, see Arp et al. (2015)), but they need not be exhaustive beyond the usefulness of the project. If more detail is needed in the future, the ontologies can be extended without fear that the extensions will break the current functionality. An example of this is seen above. In Figure 3.1, all blocks are supplied with custom stereotypes that indicate an ontological class associated with the blocks. However, Figure 3.2 shows mixed results – the Catapult itself and multiple components within the Launch Arm Assembly are stereotyped, but many components in the diagram are not. The particular analysis being performed does not need representation of these components and subsystems, so the ontologies were not developed. This does not prevent successful execution of the Digital Thread as presented, nor it does prevent expansion of the ontological understanding of the catapult system in the future should further analyses need it. It is not an "all or nothing" approach.

After a model has been mapped using the SysML stereotypes as tags to connect components to ontology classes, analysis can be performed on the graph representation. The two methods relevant to the DT shown above are the MISD and the AFD. Figure 3.11 shows a list of MISDs (tagged with the <<model>> stereotype) that are to be used in the System of Analysis. Note the Catapult Analysis block (tagged with the <<ActOfAnalysis>> stereotype) includes both the MISDs and the Catapult Mission block.

An MISD also includes a parametric portion that shows how the model connects to the system under analysis. Figure 3.12 shows the constraint block associated with the ErrorXY MISD in a parametric diagram. The bottom ports indicate inputs to the model, and the top port indicate outputs. These can be connected to other components represented in the system model.

When multiple MISDs are aggregated, an AFD can be formed. This displays not only multiple models of interest but how those models interact with each other. Figure 3.13 shows the full AFD for the Catapult Analysis. The system model parameters (at the bottom) are informed by the geometry model (on the right) and they inform additional models above them. These models are interrelated and produce outputs that correspond to System and Mission level objectives. The ports on the constraint blocks indicate direction of the links and show which ports act as inputs and which act as outputs. This formal representation may look like Figure 3.3 with more color and detail, but it brings much more value by being machine readable. IoIF can interpret the AFD to discover

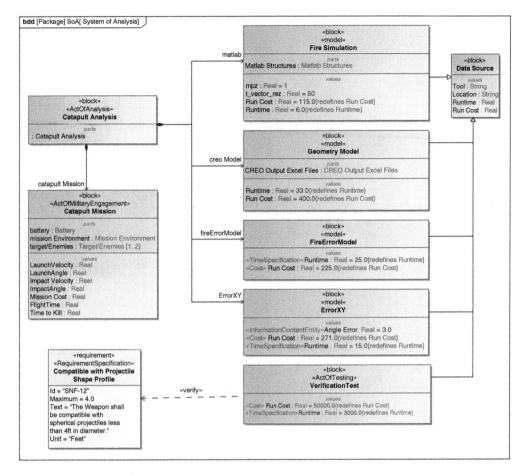

Figure 3.11 Assessment flow diagram of analysis elements.

connections between models and impact analysis of changed parameters. Thus, the diagram is a way of configuring the way the AST works and how it packages information together for discipline-specific tools to interact with the data and provide meaningful progress on design and analysis goals. Note – the AFD in Figure 3.13 is condensed to capture information in a format that is readable on letter paper. An AFD need not be so condensed in a computer model and expansion may make the diagram easier to read and understand depending on the context.

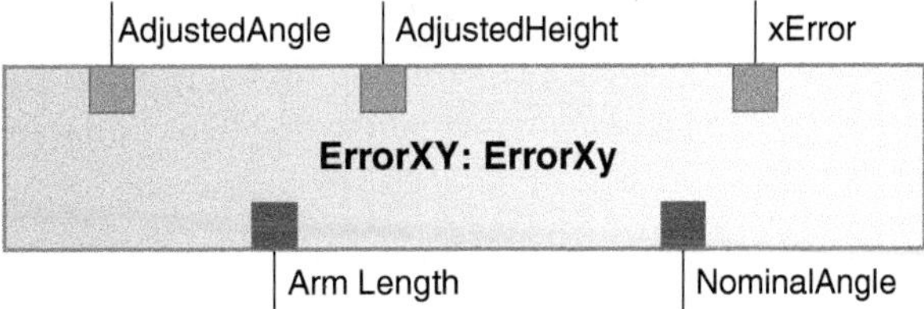

Figure 3.12 Error model – MISD.

The AFD in Figure 3.13 by itself would create four top level model segments in the JSON document for each instance represented in the SysML model: Geometry, ErrorXY, Fire Simulation, and FireErrorModel. A partial JSON output can be seen in Figure 3.14.

Additional structure within a model may be needed and can be represented as an Internal Block Diagram in SysML and transformed to the REST API JSON structure. For example, the

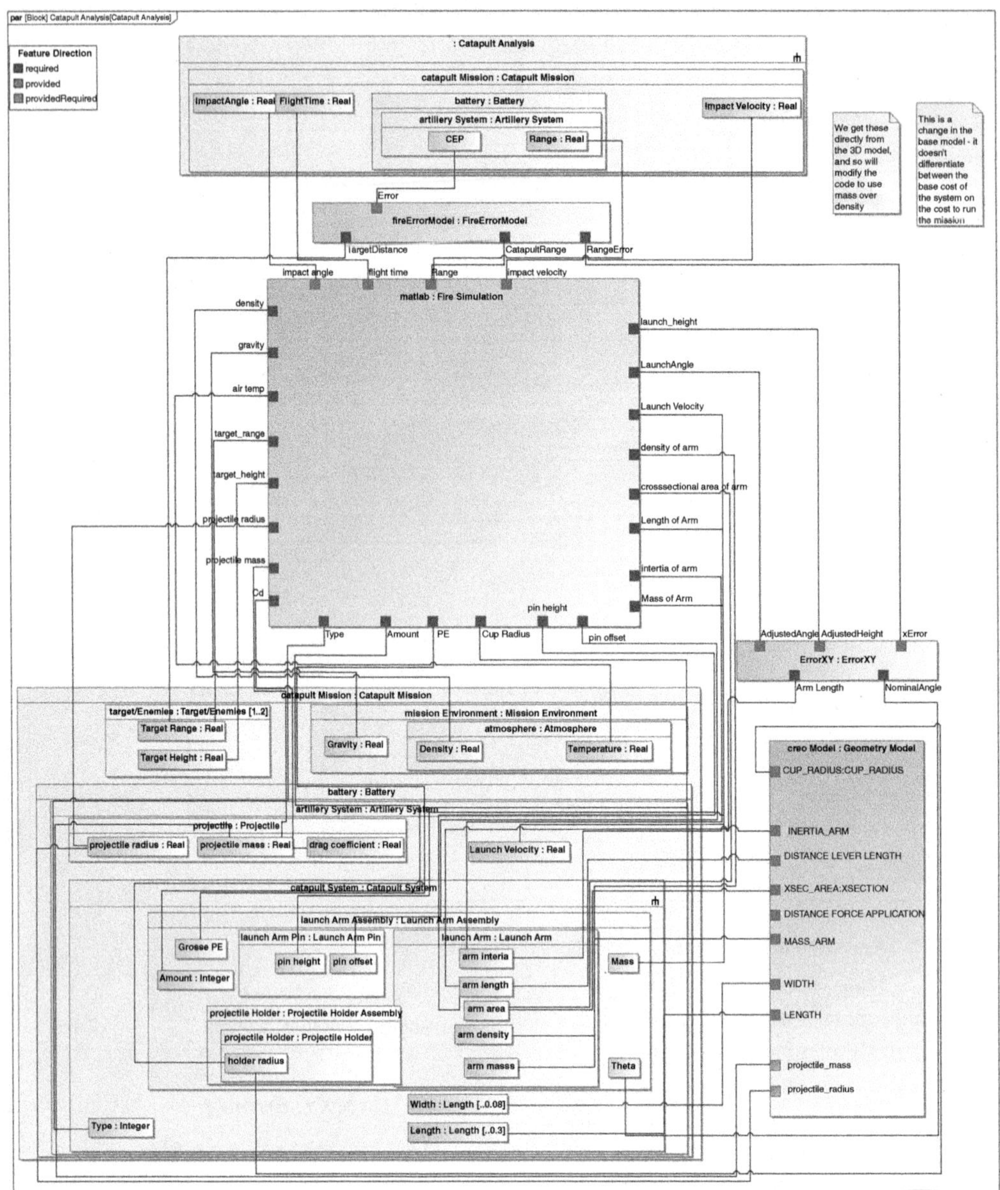

Figure 3.13 Catapult assessment flow diagram.

implementation of the DT in the Catapult use case uses MATLAB to run a Fire Simulation model (Figure 3.15). This model requires data to be parsed into multiple Microsoft Excel files. The use of middleware to transform JSON from the REST API to spreadsheets could take care of parsing the data into the proper spreadsheets. However, pulling the structure into the SysML model allows the JSON structure itself to guide

```
"Geometry Model": {
  "XSEC_AREA:XSECTION": 12,
  "MASS_ARM": 5,
  "INERTIA_ARM": 5,
  "Run Cost": 400,
  "DISTANCE LEVER LENGTH": 2,
  "Band Elastic Displacement": 0,
```

Figure 3.14 Partial AFD JSON output.

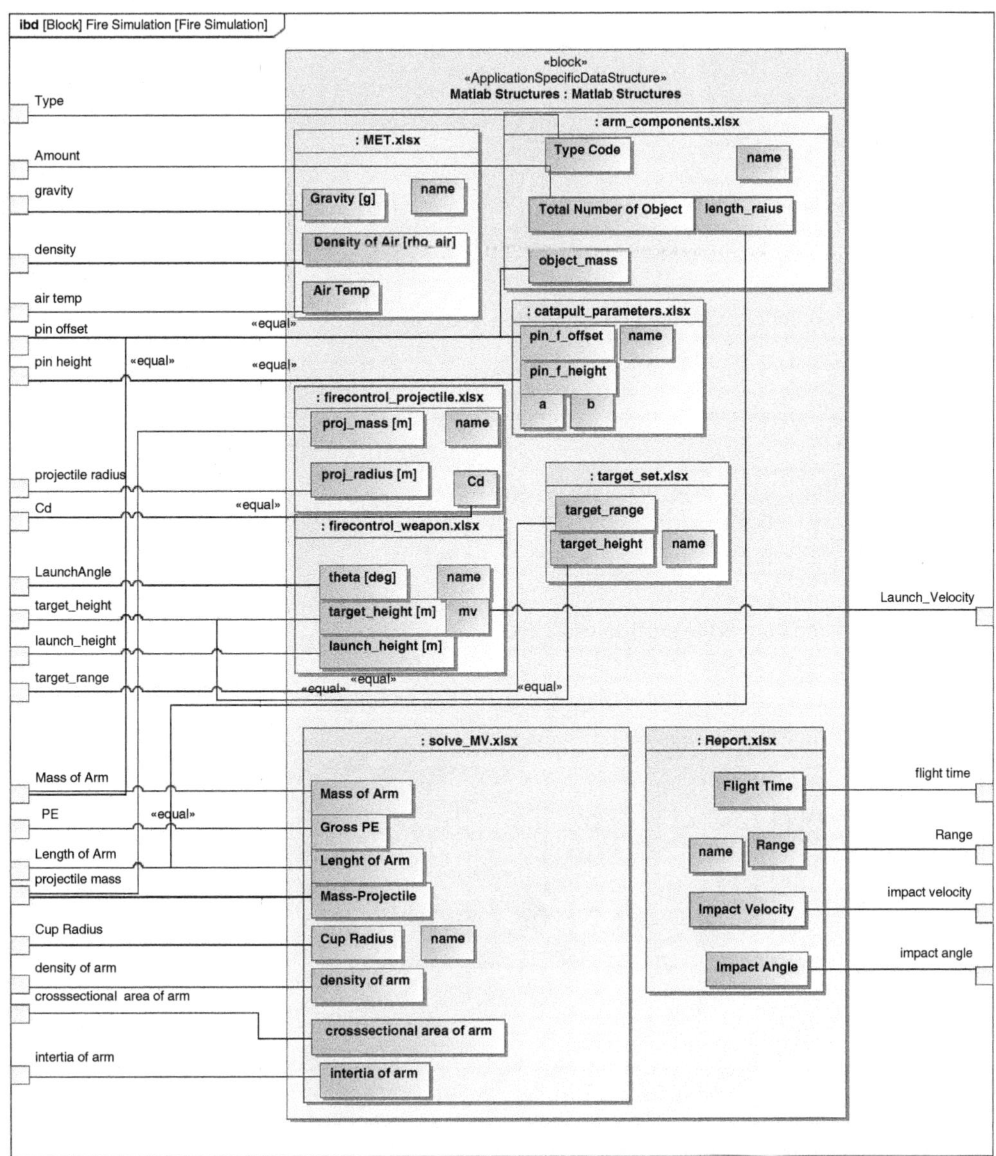

Figure 3.15 Internal block diagram for fire simulation model.

the parsing. This would allow other tools accessing the same REST API endpoint to see the groupings of parameters associated with the Fire Simulation model. An example of the JSON output is in Figure 3.16, and an example of the resulting excel file is in Figure 3.17.

Digital Thread Enacted

With the AST established and the various MISDs aggregated into a broader AFD, the DT is enabled (Figure 3.18). This DT relies on the triplestore to exchange parameter values, which ensures that all models are accessing the same information. Sequencing is shown and can be automated.

```
"firecontrol_projectile.xlsx": {
  "id" : "http://www.kpdm_dev.edu/kpdm_mapped.owl#ef388258-4535-4f6c-96a3-15db42cd4d7a_entity",
  "projectile radius": 0.7882,
  "proj_radius [m]": 0.7882,
  "name": "firecontrol_projectile.xlsx",
  "Cd": 0.5,
  "projectile mass": 0.541504,
  "proj_mass [m]": 0.541504
},
"firecontrol_weapon.xlsx": {
  "id": "http://www.kpdm_dev.edu/kpdm_mapped.owl#e5d9d2b6-9272-487e-82ba-378c5f90b6dd_entity",
  "target_height": 7,
  "target_height[m]": 7,
```


Figure 3.16 JSON output of segmented fire simulation model using IBD.

	A	B	C
1	proj_radius [m]	proj_mass [m]	Cd
2	0.02	0.0027	0.5
3	0.02	0.0027	0.5

Figure 3.17 Fire control projectile excel spreadsheet.

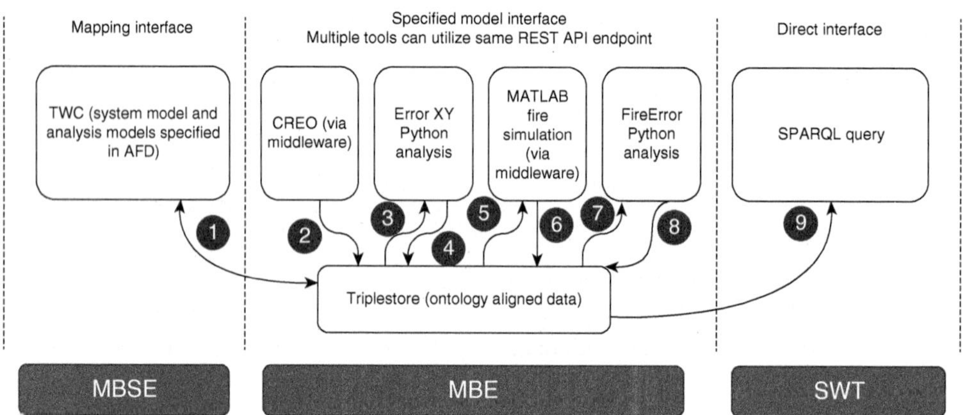

Figure 3.18 Digital thread associated with interfaces and disciplines.

1) The system model and associated analysis models (all in SysML) are imported to the Triplestore (functioning as the AST) via the mapping interface.
2) CREO model results are written to parameters associated with the catapult system via a standard PUT call on the REST API. A custom middleware is used to convert the CSV report generated by CREO to the JSON structure setup by the Specified Model Interface.
3) Parameters associated with the Error XY model are pulled from the AST via a GET request on the REST API. These are used to run a Python analysis script to produce a range of height adjustments on the catapult.
4) The results of the Error XY analysis are written back to the AST using a PUT call on the REST API.

5) Parameters associated with the MATLAB Fire Simulation model are pulled from the AST via a GET request on the REST API. A custom middleware is used to convert the JSON results of the GET request into a series of Microsoft Excel spreadsheets MATLAB expects to perform the simulation.
6) The results of the Fire Simulation are written back to the AST. A custom middleware is used to convert the Excel report generated by MATLAB to the JSON structure setup by the Specified Model Interface.
7) Parameters associated with the Fire Error model are pulled from the AST via a GET request on the REST API. These are used to run a Python analysis script to produce an error value.
8) The results of the Fire Error analysis are written back to the AST using a PUT call on the REST API.
9) A SPARQL query can be performed directly on the AST to list all the instances of the Catapult analysis currently in the AST. This is done via the Direct Interface. This is not needed for the specific DT discussed in this chapter, but it is an important capability for analysts to have and is displayed here to show how it could be used.

Discussion

Use of the DEFII framework addresses the DE need to establish an AST that can be used to implement DTs. It does so by representing system data in an ontology-aligned graph data structure. This provides a few unique advantages as an approach that may not be readily apparent.

The DEFII framework and the accompanying IoIF python-based implementation enable implementation of a DT in a way that allows users unfamiliar with graph data structures (GDS), ontologies, or coding a way to configure their portion of the DT for use in a broader analysis context. The ability to configure the AST via diagrams in SysML allows system modelers to specify the models of interest. Discipline-specific engineers can utilize the packaged models through a standard-based access point like a REST API endpoint while continuing to do their primary design and analysis tasks in tools familiar to them.

Because the AST is a graph data structure, implementing a DT as described in this chapter is only a portion of the capabilities and analyses that can be performed on the underlying system data. Graph-based algorithms can provide additional insight into the structure of the system under design, DL reasoners can provide greater semantic expansion of the underlying data, and additional SWT tools can be deployed to provide value to designers and analysts. Thus, the DEFII framework addresses the known DT problems of today while setting up an ecosystem for more advanced applications in the future.

Conclusion

Future Expansion

While the current implementation of the DEFII framework utilizes ontologies to tag models and develop the graph-based representation of the system under design or analysis, many advantages that come with the formal semantics of an ontology remain underutilized. The five use cases demonstrated to date do not take much advantage of formal reasoning, but it is an objective for future demonstrations and projects. Formal reasoning allows for the expansion of data based on axioms and relationships defined in ontologies, and this expansion could enable additional applications. Hennig et al. use the formal nature of ontologies to reason on engineering design to automatically determine certain aspects of the system under design such as connection of engineering disciplines to appropriate subsystems and the identification of Critical Item Lists (Hennig et al. 2016). These types of reasoning tasks could be incorporated into the various implementations of the DEFII framework.

Additionally, use of ontology-aligned data allows for further utilization of the SWT suite that has been built up over time to serve other industries and applications. Dunbar et al. (2022) looks at the use of the Shapes Constraint Language (SHACL) as a mechanism for verifying certain aspects of a system as defined by the ontology-aligned data, such as whether certain parameters are present in the system under analysis or whether fields correspond to a defined set of regular expressions. This research sets the stage for more complex verification tasks such as checking whether a system of analysis is well formed in construction and in use.

Final Thoughts

Implementation of a Digital Thread (DT) in an automated or semi-automated fashion is crucial to the future of systems engineering, because any change to any of the interrelated models, as reflected in the AFD, will likely change the DT. The multidisciplinary nature of complex projects means that shared parameters in a system across multiple engineering disciplines are a reality and a complicating factor in performing design and analysis tasks efficiently and consistently. Increasing project complexity often brings on the need for multidisciplinary modeling and simulation. The inputs to some models are the outputs to other models and this chapter discusses how to formalize these connections and build tools and tool proxies that can traverse the larger system of analysis using graph-based interoperability as a means to achieve cross-domain model integration. The DEFII framework gives a theoretical foundation to a graph-based approach of the DT, and IoIF is an implementation of this framework, which has been used for several use cases. This discussion should to provide enough detail to readers and organizations to implement their own versions of the DEFII framework while general enough to expand to fit their own use cases.

Notes

1 https://www.w3.org/OWL/
2 https://protege.stanford.edu
3 https://graphdb.ontotext.com
4 https://www.3ds.com/products-services/catia/products/no-magic/teamwork-cloud/
5 https://www.mathworks.com/products/matlab.html
6 https://www.ptc.com/en/products/creo

References

Arp, R., Smith, B., and Spear, A.D. (2015). *Building Ontologies with Bsic Formal Ontology*. MIT Press.

Bone, M., Blackburn, M., Kruse, B. et al. (2018). Toward an interoperability and integration framework to enable digital thread. *System* 6 (4): 46. https://doi.org/10.3390/systems6040046.

Bone, M.A., Blackburn, M.R., Rhodes, D.H. et al. (2019). Transforming systems engineering through digital engineering. *The Journal of Defense Modeling and Simulation: Applications, Methodology, Technology* 16 (4): 339–355. https://doi.org/10.1177/1548512917751873.

Cilli, M. V. (2015). Improving defense acquisition outcomes using an integrated systems engineering decision management (ISEDM) approach [PhD Thesis]. In ProQuest Dissertations and Theses. Doctoral dissertation. Stevens Institute of Technology. http://ezproxy.stevens.edu/login?url=https://www.proquest.com/dissertations-theses/improving-defense-acquisition-outcomes-using/docview/1776469856/se-2?accountid=14052.

Cilli, M., Parnell, G.S., Cloutier, R., and Zigh, T. (2015). A systems engineering perspective on the revised defense acquisition system. *Systems Engineering* 18 (6): 584–603. https://doi.org/10.1002/sys.21329.

Cotter, M., Hadjimichael, M., Markina-Khusid, A., and York, B. (2022). Automated detection of architecture patterns in MBSE models. In: *Recent Trends and Advances in Model Based Systems Engineering* (ed. A.M. Madni, B. Boehm, D. Erwin, et al.), 81–90. Springer International Publishing https://doi.org/10.1007/978-3-030-82083-1_8.

CUBRC, Inc. (2020). *An Overview of the Common Core Ontologies.* Buffalo, NY: CUBRC. https://github.com/CommonCoreOntology/CommonCoreOntologies/blob/master/documentation/An%20Overview%20of%20the%20Common%20Core%20Ontologies%201.3.docx.

Dunbar, D., Hagedorn, T., Blackburn, M., and Verma, D. (2022). Use of semantic web technologies to enable system level verification in multi-disciplinary models. In: *Advances in Transdisciplinary Engineering* (ed. B.R. Moser, P. Koomsap, and J. Stjepandić). IOS Press https://doi.org/10.3233/ATDE220632.

Dunbar, D., Blackburn, M., Hagedorn, T., and Verma, D. (2023). Graph representation of system of analysis in determining well-formed construction. *2023 Conference on Systems Engineering Research* (16–17 March 2023). Hoboken, NJ: Stevens Institute of Technology.

Dunbar, D., Hagedorn, T., Blackburn, M. et al. (2023). Driving digital engineering integration and interoperability through semantic integration of models with ontologies. *Systems Engineering* 21662. https://doi.org/10.1002/sys.21662.

Hennig, C., Viehl, A., Kämpgen, B., and Eisenmann, H. (2016). Ontology-based design of space systems. In: *The Semantic Web – ISWC 2016*, vol. 9982 (ed. P. Groth, E. Simperl, A. Gray, et al.), 308–324. Springer International Publishing https://doi.org/10.1007/978-3-319-46547-0_29.

Medvedev, D., Shani, U., and Dori, D. (2021). Gaining insights into conceptual models: a graph-theoretic querying approach. *Applied Sciences* 11 (2): 765. https://doi.org/10.3390/app11020765.

Mordecai, Y., Fairbanks, J.P., and Crawley, E.F. (2021). Category-theoretic formulation of the model-based systems architecting cognitive-computational cycle. *Applied Sciences* 11 (4): 1945. https://doi.org/10.3390/app11041945.

Noy, N.F., Crubezy, M., Fergerson, R.W. et al. (2003). Protégé-2000: an open-source ontology-development and knowledge-acquisition environment. *AMIA Annual Symposium Proceedings* 2003: 953.

Ruttenberg, A. (2020). Basic Formal Ontology (BFO). https://basic-formal-ontology.org (accessed June 11, 2023).

Biographical Sketches

Daniel Dunbar is a PhD candidate in systems engineering at Stevens Institute of Technology. His research interests include Digital Engineering and formal knowledge representation as a means to enable better multidisciplinary collaboration. He currently works as a research assistant for the

Systems Engineering Research Center (SERC) under Dr. Mark Blackburn's leadership. Before joining Stevens, he worked in the telecommunications domain designing and implementing two-way radio systems.

Dr. Tom Hagedorn graduated from Boston University with a BS in biomedical engineering in 2010, and from the University of Massachusetts at Amherst with a PhD in 2018. His research focuses on applications of ontologies and semantic web technology to aid in various aspects of engineering design and systems engineering. This includes semantically enhanced linked data repositories, development of engineering domain ontologies, and the study of engineering methods using these tools. His doctoral work focused on semantically enabled design assistants in medical and advanced manufacturing contexts. His current work focuses on the development of digital engineering methods and frameworks built upon semantic web technologies.

Timothy D. West currently serves as the deputy director of the U.S. Air Force Research Laboratory's Operations Directorate. In this capacity, he provides leadership and oversight of the Laboratory's air and space flight test program, demonstrating combat readiness of the military's latest science and technology development programs. Prior to his current assignment, Mr. West served for nearly 30 years as a military officer in the U.S. Air Force, where he held command positions in science and technology, test and evaluation, and program management. He is a graduate of the U.S. Air Force Test Pilot School, the Air Command and Staff College, and the Air War College. Mr. West holds MS degrees in aerospace and industrial engineering from the University of Tennessee Space Institute, and a BS in mechanical engineering from the University of Kentucky. Mr. West is also a systems engineering PhD candidate at Stevens Institute of Technology.

Brian Chell, PhD, is a postdoctoral research associate with the Systems Engineering Research Center at Stevens Institute of Technology. His research interests are in distributed space systems and applying multidisciplinary design optimization techniques to mission-level analysis. Brian received a PhD in systems engineering and an ME in space systems engineering from Stevens, and a BS in aerospace engineering sciences from the University of Colorado Boulder.

John Dzielski, PhD, earned a BS in mechanical engineering from Carnegie-Mellon University in 1982, and MS and PhD degrees in mechanical engineering from the Massachusetts Institute of Technology in 1984 and 1988, respectively. He has more than 30 years of experience relating to the design, integration, and at-sea testing of unmanned undersea vehicle systems supporting various U.S. Navy missions. He has been involved in research and development of supercavitating vehicle technology since 1995. He is applying his experience in simulation-based analysis, synthesis, and validation processes to the development of formal model-based system engineering methods and processes using SysML integrated with modern analysis and design optimization tools. These methods rely heavily on semantic-web technologies and are being developed for and delivered to several different government sectors. He may be the expert on American football dynamics and aerodynamics.

Mark R. Blackburn, PhD, is a senior research scientist with Stevens Institute of Technology since 2011 and principal at KnowledgeBytes. Dr. Blackburn has been the principal investigator (PI) on 17 System Engineering Research Center (SERC) research tasks for both US Navy NAVAIR, US Army DEVCOM, and US Space Force on Digital Engineering Transformation Research Tasks. He has also been PI on a FAA NextGen and National Institute of Standards and Technology projects and has received research funding from the National Science Foundation. He develops and teaches a course on systems engineering for Cyber Physical Systems. He is a member of the SERC Research Council, OpenMBEE Leadership Team, and INCOSE Pattern Working Group focused on the Semantic Technologies for Systems Engineering initiative. Prior to joining Stevens, Dr. Blackburn worked in industry for more than 25 years. Dr. Blackburn holds a PhD from George Mason University, an MS in mathematics (emphasis in CS) from Florida Atlantic University, and a BS in mathematics (CS option) from Arizona State University.

Chapter 4

Digital Engineering Visualization Technologies and Techniques

Brian Chell, Tom Hagedorn, Roger Jones, and Mark R. Blackburn

Stevens Institute of Technology, School of Systems and Enterprises, Hoboken, NJ, USA

Problem

A major challenge in mission and systems engineering is reconciling the differences between the actual needs of the end users, and how developers perceive those needs. It is important to get these different entities on the "same page," with a shared concept of the system within its operational environment (Larson et al. 2009). It is accepted that with the maturing of DE technologies that documents are inadequate to facilitate the DE transformation. In addition, managing changes through complex toolsets and providing useful information without overloading those who do not have domain knowledge can be difficult. Visualization technologies are a good method for articulating details about the system analysis and design in a language that engineers, relevant stakeholders, and end users can all understand. The Concept of Operations (CONOPs) can be a means to bridge this gap significantly (AIAA 1993). The visualization of engineering data can take the form of simple plots and figures all the way to immersive environments enabled by gaming technology that supports virtual reality (VR) and augmented reality (AR). In the past, CONOPs have been document-based and largely static, seeing few or no updates and being of limited use throughout system development (Korfiatis et al. 2012). There is a need to articulate a CONOPs dynamically to help communicate system capabilities to the users, aid in the elicitation of requirements that may evolve over time, and validate the system of interest. This will help users and developers to realize a shared mental model and understanding of the mission, and potential solutions across a set of diverse stakeholders.

Once this shared mental model is established with a CONOPs, it is important to show how system-level changes can impact overall mission effectiveness to support decision-making relative to different operational needs. By linking enabling technologies such as visual dashboards and graphical CONOPs, we have integrated physics-based, discipline-specific, and mission-level models to support decision-making with an understanding of mission and system design trade-offs. This helps stakeholders understand the impact on the digital thread if we need to make changes to different mission objectives all the way down to different system components.

Systems Engineering for the Digital Age: Practitioner Perspectives, First Edition. Edited by Dinesh Verma.

Companion website: www.wiley.com/go/verma/systemsengineering

Background

This section provides a concise review of the literature related to operational concepts and techniques for visualization. It also sets the stage for the Armaments Interoperability and Integration Framework (IoIF), an ontology-based decision support tool and digital thread impact analysis dashboard. Chapter 3 discusses the background of IoIF in more detail. Furthermore, Figure 1.2 shown in Chapter 1 provides a simplified view of the overarching mission and systems engineering process from a more traditional point-of-view. Our collective research has formalized these notional process steps into a more integrated DE environment, with the underlying digital thread, reflected in the information model, which is now formalized through IoIF.

Operational Concepts

A CONOPs is a user-centric view of a proposed system and is used to communicate the characteristics of this system to relevant stakeholders (IEEE 1998). These characteristics include quantitative and qualitative descriptions of the system's purpose, as well as technical or operational constraints, and is best communicated in language that engages all stakeholders. As discussed in Chapter 1, this allows stakeholders and developers to reach a shared mental model of the system's required capabilities (Korfiatis et al. 2015). CONOPs development best happens early in the DE life cycle and can take place concurrently with the establishment of stakeholder needs.

Several standards exist for creating an operational concept, including those done by the Department of Defense (2000), the Institute of Electrical and Electronics Engineers (IEEE), and the American Institute of Aeronautics and Astronautics (1993). Like many systems engineering methods, developing a CONOPs has historically been achieved through a text-heavy, document-based approach. While these CONOPs documents play a large role in connecting engineers and scientists with relevant stakeholders, development is time-consuming, and generally represents a static view of the system (Cloutier et al. 2013). A systematic review of CONOPs found the process to be time-consuming and most often done only to fulfill contract requirements (Mostashari et al. 2012).

Despite the negative impact and high level of risk incurred by having an inconsistent view of the CONOPs for a system throughout its life cycle, it is difficult to create and maintain an effective one. A major challenge is that the development of a CONOPs often involves making trade-offs in a large, highly dimensional, complex trade space with stakeholders from multiple disciplines. This is a difficult task for several reasons:

1) **Value proposition:** There is generally a great deal of uncertainty in the value proposition for the system due to changing market and competitive conditions. Thus, it is challenging to translate system attributes into realizable value.
2) **Translation of concept decisions into system attributes:** Even with a common understanding of the value proposition space, it is often very difficult to translate a myriad of concept decisions into the impact on the attributes of the system that are pertinent for the creation of value.
3) **Shared mental model capability:** With a group of stakeholders from multiple disciplines, it may be very difficult to create consistent shared mental models or even to have the ability to determine if inconsistencies in the mental models exist. Unfortunately, these inconsistencies are often found much later in the system life cycle.
4) **Human dimension:** The human dimension generally manifests itself as conflicts in how the relative costs and benefits of a decision are distributed to the stakeholders. This is an area where

reward systems and organizational structures are extremely influential. Another aspect of the human dimension is the delaying effect. This may manifest itself in the form of engaging in delaying tactics as a method to avoid conflict or to stall a decision because the key stakeholders and/or decision-makers have differing opinions.

5) **Process nonlinearity:** The prevalent practice of developing CONOPs has attempted to apply a linear approach to an inherently nonlinear process.

As the complexity of systems increases, document-based CONOPs have shown to be insufficient, especially in lieu of advances in gaming engines, VR and AR. A properly maintained CONOPs should represent the system and its context throughout the design life cycle. SERC research has found a need for CONOPs development to happen more quickly and in a graphical environment (Cloutier et al. 2010). This research has led to creating more dynamically capable CONOPs that use gaming engines as an enabling technology for CONOPs development and evolution.

Gaming Engines

Advances in graphics processing units (GPUs) that support visual representations on computers that leverage gaming graphical engines have allowed for the development of a more useful, dynamic graphical CONOPs. These capabilities are a key enabling technology for a graphical CONOPs. Gaming engines allow for simulated, immersive environments, which mimic real life and can communicate difficult concepts so that stakeholders understand a system's capabilities and how it can satisfy their needs (Korfiatis et al. 2015). Early research showed that immersive environments can have the drawback of a steep learning curve (Korfiatis 2013). The rendering high-fidelity graphics creates a large computational load and can require an expensive GPUs to fully leverage the capabilities of these gaming engines. However, viable demonstrations have been created that use lower-fidelity image from a laptop without a GPU as seen in Figure 4.1.

Figure 4.1 Depiction of UAS/counter-UAS search scenario rendered in the unity gaming engine.

Early Research into Graphical CONOPs

Considering the difficulties in creating a CONOPs, and the drawbacks to a static document-based approach, researchers are formalizing methods for creating a dynamic CONOPs in an agile fashion that can communicate system capabilities to important stakeholders better. Much of the early work in developing graphical CONOPs was driven by SERC research tasks and used the Unity 3D© gaming engine, which has evolved significantly since those early projects. This included the investigation of then-current standards by cataloging CONOPs documents, finding that they were often lacking in describing mission needs and current system context. A driver for this research was to find methods for creating CONOPs more efficiently. Synergies are found when using Model-Based Systems Engineering (MBSE) tools and methods with CONOPs development (Korfiatis et al. 2015).

Another motivating factor for research into gaming engines as a method for enabling graphical CONOPs is the finding from a 2018 SERC research task that current tools cannot capture the complexity of mission scenarios, and that tools capable of modeling evolving scenarios are needed (Vesonder et al. 2018). In addition, most mission scenarios continuously evolve as complex systems are rolled out over many years and releases. Modern gaming engines such as Unity 3D© provide the necessary functionality to overcome these drawbacks and have been used in SERC research to create graphical CONOPs for several different mission scenarios (Blackburn et al. 2018). The objective of a CONOPs is to involve key stakeholders in the development process, and we have leveraged graphical CONOPs to elicit objectives and usages scenarios for research use cases from the Army sponsors.

IoIF Ontology-Based Decision Framework

The hand-created graphic, shown in Figure 4.2, comes from the Integrated Systems Engineering Decision Management (ISEDM) process (Cilli 2015). It renders multiple attributes that relate to trade-off analysis such as performance vs. cost, including information about uncertainty and long-term viability. An Assessment Flow Diagram (AFD) links each objective, the thing(s) we care about with our goal(s), to the elements of the system-of-interest's product structure (the things that we can control – our design decisions) through a set of intermediate relationships formalized as models. The AFD is a directed graph with no loops. Decision nodes are represented by boxes, assumptions represented by triangles, models represented by circles, and arrows between node

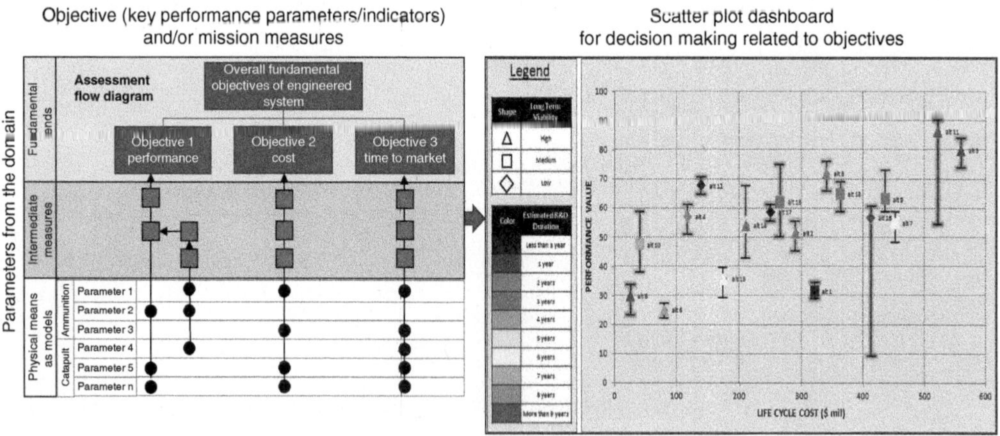

Figure 4.2 Cilli's ISEDM trade-off graphic and assessment flow diagram.

pairs to indicate influences and sequencing needed for ordering model simulations. The formalization of the AFD is discussed in Chapter 3, which characterizes the objectives' relationships between parameters of tradespace alternatives.

New Approach

The new approach to visualizing DE artifacts requires rigorous models developed with an evolving methodology on the backend, useful model diagrams or visual environments on the frontend, and a robust method for exchanging data between the two, which is enabled using IoIF as discussed in Chapter 3. SERC research has extended the semantic DE environment with visual data to help provide important information for decision-making by key stakeholders. The concept and features have been extended to computationally use IoIF to interact with graphical CONOPs, SysML models, physics-based and discipline-specific modeling and simulations.

Semantic DE Environment with Digital Thread Visualizations

The semantic triplestore repository created and updated using IoIF-enabled digital threads can serve as the backend for any number of applications. The catapult demonstration integrates tools using the specified model interface discussed in Chapter 3 to retrieve fixed information from the triplestore. By incorporating more information into the mission model via project imports or other model expansions, one can rapidly develop new data structures that are passed to applications specifically built to interact with them.

Alternatively, tools built as a front end of IoIF have the benefit of accessing information using either these predefined data structures or requesting custom data via queries to a triplestore throughout their operations. They can similarly use the direct interface discussed in Chapter 3 to update, change, or delete information as they operate, depending upon permissions.

The key methodological insight to developing either type of application is that the application need not be project-specific – rather its requests for data and update methods can be written at an appropriately abstract level. In the domain of visualization for systems engineering, applications might be written once, but then support a wide array for projects and workflows executed across many engineering domains now that they can communicate with real-time data from IoIF.

The basic workflow for any such application is straightforward. As shown Figure 3.10 in Chapter 3, the steps are:

1) Populate the model to cover the application scope (e.g., Cyber System Model and Assessment Flow Diagram such as Figure 3.13 in Chapter 3),
2) Populate IoIF from the model,
3) Compose a request or query that retrieves the needed data from IoIF (e.g., MATLAB in Figure 1.10 of Chapter 1),
4) Package the response from IoIF into the data structures used by the application (e.g., Figure 3.16 in Chapter 3)
5) Implement the desired application behaviors using those structures,
6) Render the model elements, (i.e., parameters as value properties), model simulation outputs (i.e., meta data), and requirements linked to model elements
7) Use one or more dashboards to render the data retrieved by IoIF (discussed below).

The following two sections provide two examples that illustrate how this process works.

Impact Analysis

One objective is understanding the impacts of potential changes to a system design that might be related to a change in requirements or not satisfying requirements. For example, a part has been redesigned, issues are encountered in manufacturing, or its configuration has changed. In a complex system, any change might have significant downstream impacts related to combinations of mechanical, electrical, communication or software with implications on system performance or cost. An engineer or subject-matter expert (SME) wants to know what performance parameters might be affected, what parts of the system are impacted, and the related requirements. These are the types of scenarios that have been demonstrated by the impact analysis dashboard that can render digital threads using real-time live data in an IoIF-configured triplestore repository.

The AFD has been formalized using SysML as shown in Figure 4.3 (also see Figure 3.13 in Chapter 3) as part of the system of analysis and is used to configure IoIF for purposes of trade space analysis, and impact analysis. We evolved the ontology-based Decision Framework concept for a few sponsor case studies such as the Catapult projectile launcher. We created dashboard renderings of similar information from the analyses through IoIF. The decision framework is IoIF based, including the use of the IoIF Service, which means that the decisions are directly related to the "full stack" of models as discussed in Chapter 1, and can render data in real-time as IoIF coordinates different types of analyses. The AFD and associated mission model mapping to the AFD objectives capture much of this information as shown in Figure 4.3.

The AFD captures notions of analysis sequences and of the provenance of various parameters. Other parts of the mission and system models link these parameters (i.e., SysML value properties) to things like requirements. This is ingested by IoIF through the mapping interface, and it can be accessed by third-party tools specifically aimed at visualizing, reasoning, and dynamic simulations such as graphical CONOPs to make these downstream affects apparent.

An impact analysis dashboard as shown in Figure 4.4 provides a visualization of potential effects of linked information without the need to traverse multiple SysML projects or diagrams. The dashboard is implemented using Python's Dash package, which allows the creation of interactive data dashboards accessible in web browser environments. Other packages and languages could just as

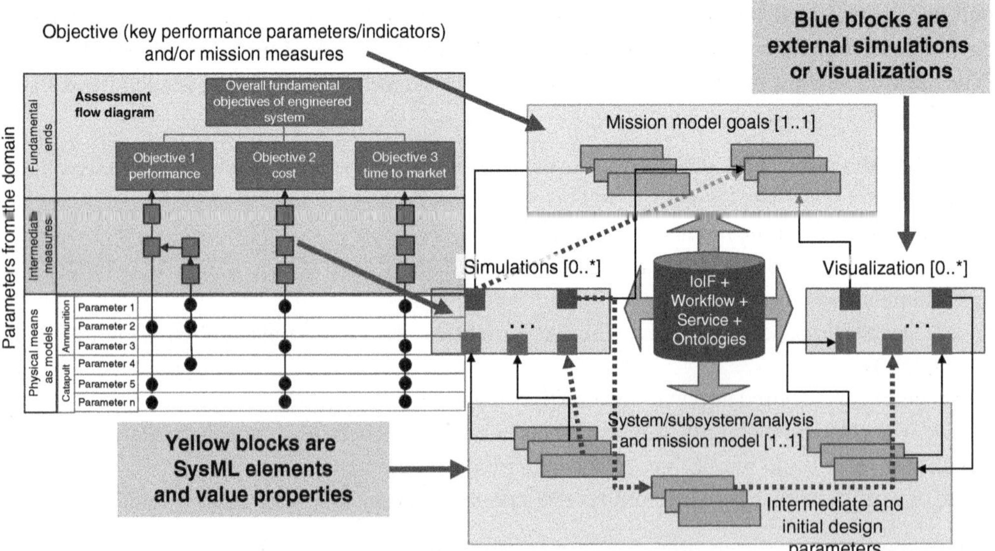

Figure 4.3 AFD formalized representation of IoIF data exchanges between analysis tools using SysML.

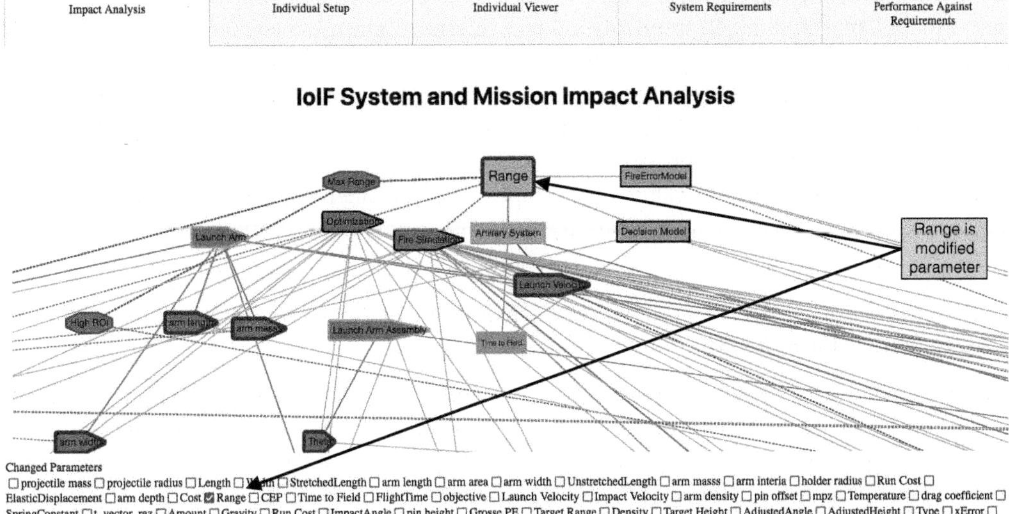

Figure 4.4 Digital thread visualization shows upstream and downstream connections to modified parameter, simulations and requirements.

easily be used, as IoIF's distributed deployment provides a language agnostic REST API to encapsulate interactions with the underlying triplestore. The baseline visualization uses a Dash implementation of the open-source Cytoscape tool, which is used to create graph visualizations. The example shows those model elements related to Range, which is a key objective for the Catapult. Upstream entities of a changed parameter (i.e., inputs to range-related simulations) have dashed line (e.g., FireSimulation). Downstream entities that may be affected by changing upstream parameters for Range (e.g., Decision Model) have solid line. Related requirements (e.g., Max Range) are in octagon shaped node.

Unlike the specified model interface discussed in Chapter 3, this visualization does not use instance-level data in the mission model, but instead on the basic structure of how it is connected both in the AFD and in other diagrams. The direct interface is used to answer a set of SPARQL queries, with the results then used to build the visualization dynamically upon launching the dashboard. The query effectively asks for data pertaining to the following questions:

- What analyses are characterized in the AFD?
- What parameters are passed to what analysis, and are they inputs or outputs?
- What system traits and components do those parameters describe?
- What requirements pertain to those traits and components?

All of this is written in a SPARQL query to the triplestore without any specific knowledge of the system, mission, or analysis, meaning that it can be used to support any model developed as described in the prior chapters. This means the impact analysis dashboard is both tool and project agnostic.

Composing the Visualization

The package to implement the dashboard provides several out the box visualization methods. The query result is thus reorganized into a data structure compatible with the desired visualization. The Cytoscape visualization is built by providing an array of elements comprising uniquely identified nodes (corresponding with unique Internationalized Resource Identifiers [IRIs] in the triplestore) and vertices connecting these (corresponding to relevant connections between parameters, models, components, and requirements). Thus, the basic workflow of the application is to transform a data payload obtained from IoIF's direct interface into these elements and to provide a means to interact and modify the resulting elements as shown in Figure 4.4.

The triplestore returns query results in the JavaScript Object Notation (JSON) format (see Figure 3.16 in Chapter 3), with field names corresponding to query variables. Since the query is static in the application, each individual result can be unpacked and assigned to visualization elements based on the variable name and value. For example, linking elements in the visualization are given directional information based upon a variable bound to an ontology classification derived from the AFD Directed Feature tags.

Passing Results Back to IoIF

Within the context of the visualization, one might want to define a new instance or update an existing instance to facilitate analysis like that seen in Chapter 3. This can be accomplished by interacting with a direct interface to IoIF. The new data can be passed back to IoIF if the data has been retrieved and packaged according to the basic workflow.

Application users are presented a simple graphical user interface (GUI) element that invokes methods that read in updated data and pass them to IoIF via its direct interface. The update consists of two SPARQL queries – one to delete existing data associated with the updated instance, and a second to write in the new data. The necessary statements are written based on user interactions, with unique nodes in the graph identified fusing the IRIs retrieved in prior queries. Thus, in this workflow a SELECT query populates the visualization interface while the DELETE/INSERT queries modify the data in IoIF.

Decision Support Dashboard

A decision support dashboard use case is helpful in illustrating how one can evolve the basic patterns developed as part of the impact analysis visualization to support highly specific data or data structures. Provided one has modeled the necessary data for a decision modeling effort (e.g., performance, cost), as notionally rendered in the hand-constructed scatter plot shown in Figure 4.2, the workflow would be precisely that of the Impact Analysis, with differences only confined to the internal workings of the application and the visualizations provided. Figure 4.5 provides a decision framework dashboard driven by IoIF, which is discussed in more detail below.

Many decision methods require very specific data types, and an application is likely to require a degree of repeatability in both the structure and content of output data. This workflow thus appears better suited to the specified model interface, which provides highly standardized outputs configured using the AFD and SysML model. This section further discusses how a SysML model can be used to configure IoIF to output data for a specific decision method.

Decision Analysis Method

The application is built to implement the ISEDM, which uses piecewise value functions and SME categorization of decision variable importance and differentiation to generate an aggregate value "score" based upon the weighted sum of individual decision variable values. The method drives the data that must be populated – at a minimum one needs the points that define the value functions, the rankings of variables, and the values of each decision variable for each alternative. Per step one of the workflows, the base system model should at least describe this data, if not populate it.

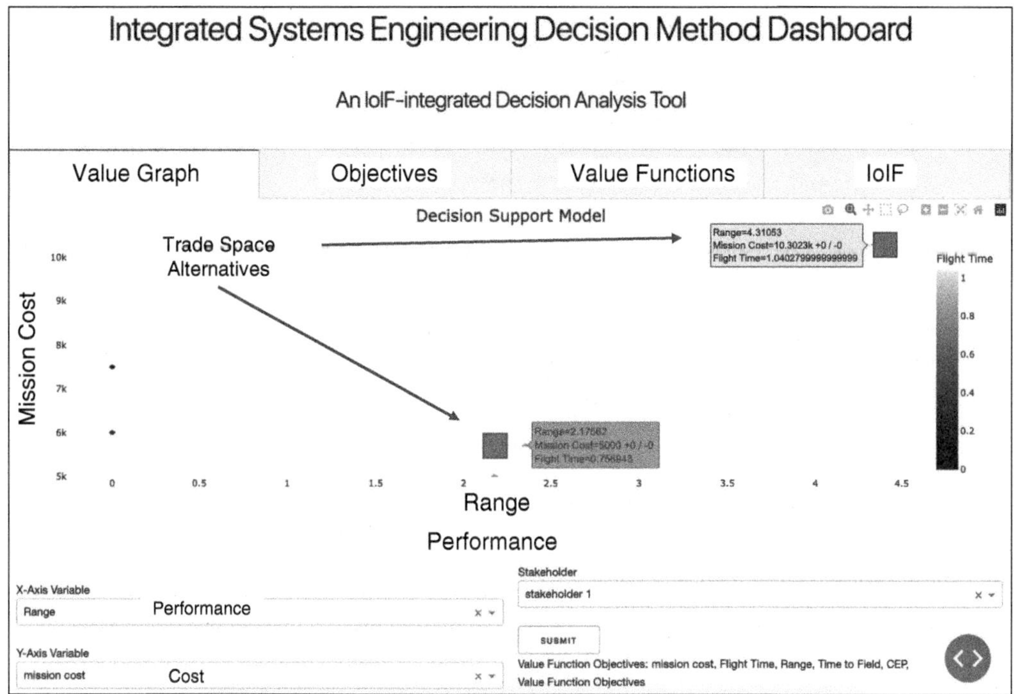

Figure 4.5 Decision framework dashboard.

Updating the System Model

In keeping with the philosophy of reusability, the decision-making SysML model is defined to be project agnostic, with its classes and ontological stereotypes kept strictly to the scope of the IoIF mid-level ontologies. The SysML's project import mechanism can import mission, system, and system of analysis models to rapidly extend an existing AFD and model to support a decision dashboard.

From a functional perspective, the decision SysML model must describe each of the various data types used in the ISEDM process, and ideally provide a means to output that data via the specified model interface. While similar results could be achieved via the direct interface, the use of the specified model interface means that decision model will not only configure the types of data that can be exchanged through IoIF but also the data structures themselves, allowing for greater development ease.

Top Level Decision Model

The decision domain comprises more than just the ISEDM process, and a general decision model was created for decision analysis, with the intent that subsequent method specific expansions could be used to implement ISEDM and other decision analysis methods. The base model describes the various entities involved in any given decision. It distinguishes between the decision itself and the process by which one analyzes preferences and ranks alternatives. The basic block diagram (Figure 4.6) uses the same stereotyping approach prescribed for integration with IoIF. In summary, decisions involve sets of alternatives. A decision analysis describes stakeholders in terms of a having sets of preferences that are translated to objectives.

This basic model of a decision is then extended with a deeper look at decision analysis domain, which is the main concern of this work (Figure 4.7). The main function is to introduce the idea of data sets that describe the preferences of some stakeholder, and the notion that a separate preference analysis might be necessary to transform that data into a mechanism by which to rank alternatives relative to one another.

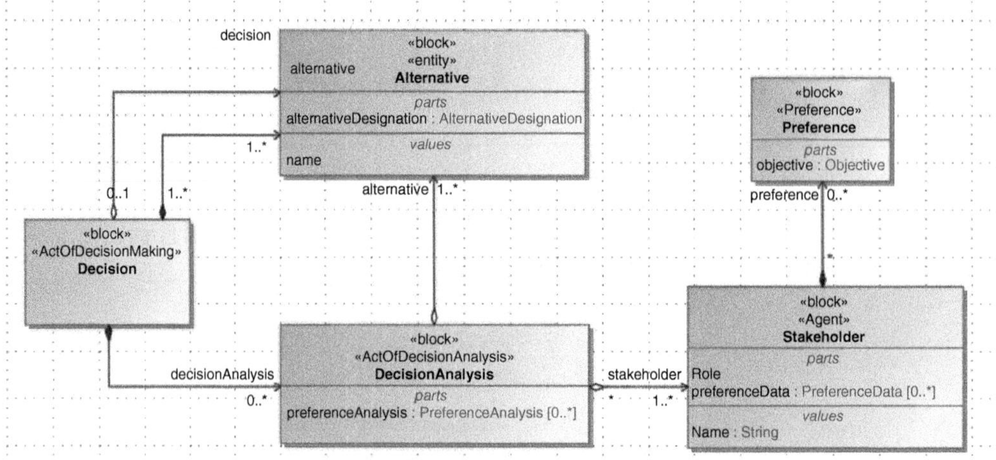

Figure 4.6 High level decision analysis block diagram.

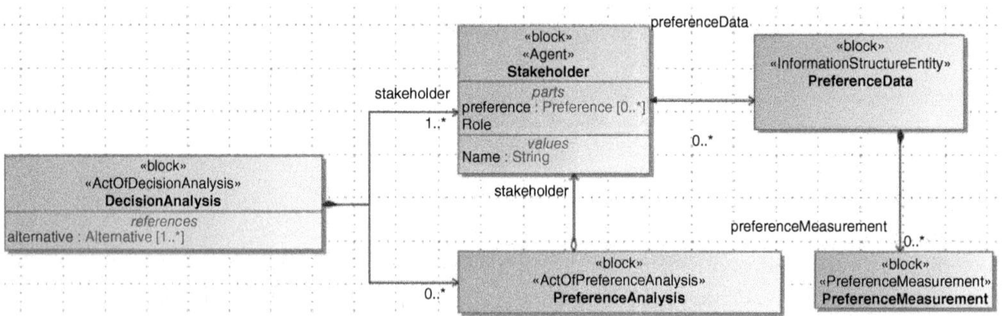

Figure 4.7 Partial view of the decision analysis block diagram.

Extending the Decision Model to a Specific Method

The high-level model is primarily a starting point from which more sophisticated descriptions of specific decision analysis methods can be formulated. Generalizations are used to relate the specific methods to the more general block diagram, and the data needed for a particular method is modeled within these more specific elements.

To implement the ISEDM model, method specific preference measurements are included to capture information relating to the method's piecewise value functions (Figure 4.8). Subtypes of objective are added to specifically add tags for cost, performance, and time objectives. The notion of attribute and SME rankings of attributes are added so that the full method may be computed by an application. Each of these serves to introduce the value properties and surrounding structure needed to unambiguously link the parameters necessary for decision-making. By instantiating the ISEDM extension in the context of another model, decision-analytic data can be captured in IoIF.

Extending the Mission Model with Decision Analysis

The decision analysis and ISEDM SysML models are imported into the mission model via a project usage. The model is then extended by generalizing intermediate classes containing the overall system analyses as shown in Figure 4.9. This step exposes all the parameters asserted or computed in the digital thread to any specified model interface.

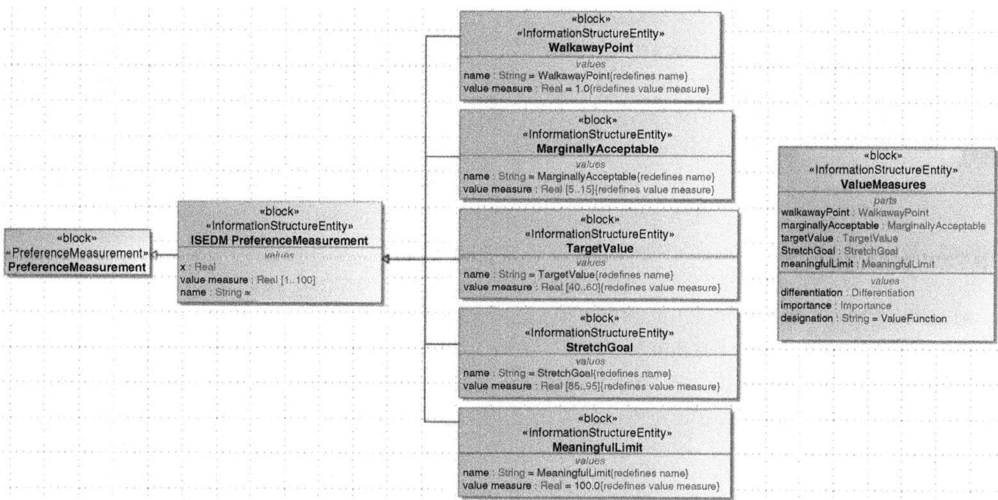

Figure 4.8 Expansion of the model to include measures used in ISEDM value functions.

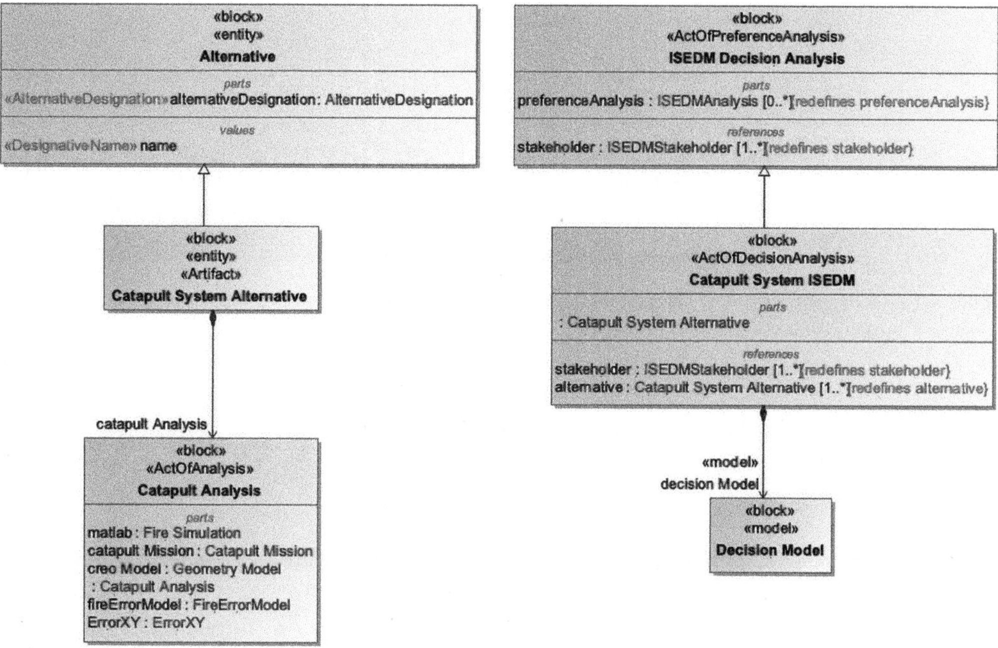

Figure 4.9 Applying the decision model to the mission model.

The second step is to add an AFD element that clarifies what parameters will be used as decision attributes (Figure 4.10). This uses the same modeling patterns and stereotyping mechanism as the rest of the AFD. Building this diagram effectively configures a data structure for the decision analysis that will be returned from IoIF when requested via the specified model interface. Finally, instances are created for the ISEDM data structures to represent the preferences of some stakeholder. This provides slots that will result in fields in the output JSON, and which can be used by an application to populate preference information relating to value functions, rankings, and weights.

Implementing the Decision Analysis Dashboard

The Decision Framework Dashboard is shown in Figure 4.5 and provides information related to the ISEDM process and has features such as Value Functions (Figure 4.11) and Priority Weighting. Following the workflow used in the implementation of the impact analysis dashboard, the next step is to form the modeled AFD into the data structures needed to implement the desired application. The AFD and applied stereotypes allow the preferences and alternative to be retrieved via the specified model interface. This means that no specialized knowledge of the model or of IoIF is needed to build the dashboard. The decision model import standardizes the structure that is received from IoIF, creating predictable fields that can be searched and interpreted with hard-coded software methods. Thus, the use of a reusable project import in the digital thread environment creates a reusable application built off IoIF's specified model interface.

As with the impact analysis case, the dashboard is built upon Python's Dash package for creating interactive data dashboards. Connecting to IoIF is accomplished using calls to the REST API of

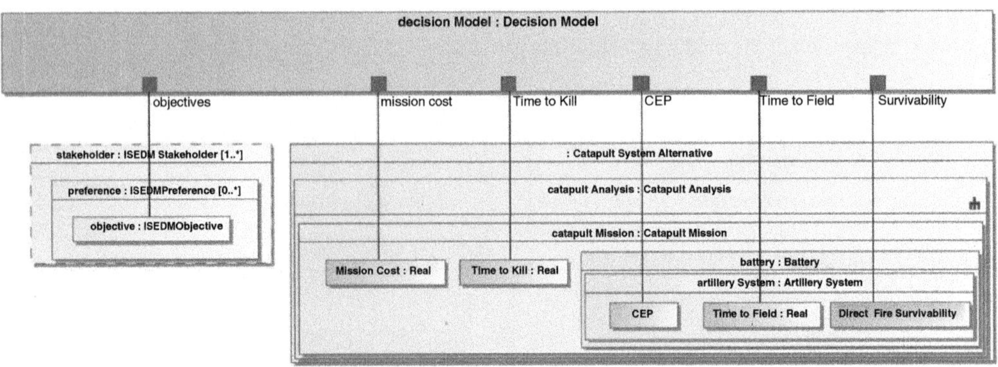

Figure 4.10 A hypothetical AFD extension to output data to run ISEDM Application.

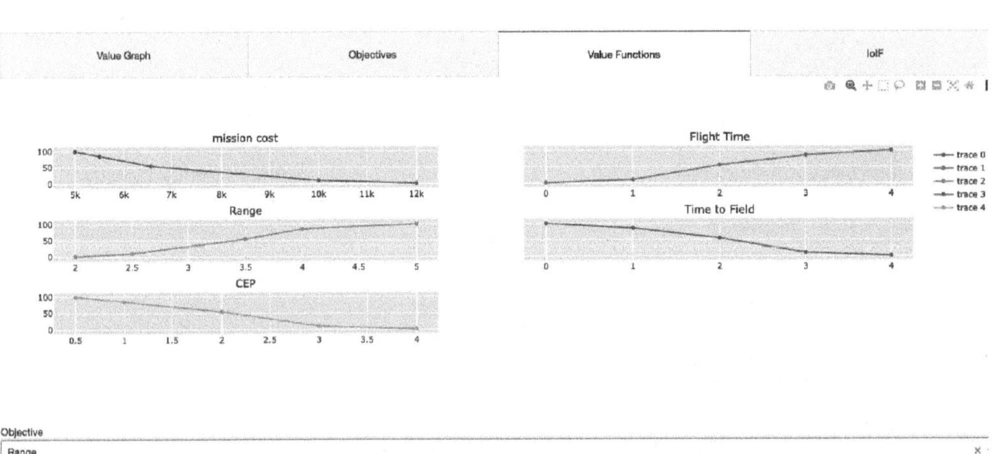

variable	WalkawayPoint	MarginallyAcceptable	TargetValue	StretchGoal	Mea
x	2	2.5	3.5	4	
value	1	10	55	85	1

Figure 4.11 Integrated systems engineering decision method dashboard (value function tab).

IoIF's service implementation. The internal functions of the dashboard interpret the IoIF-specified model interface JSON by looking for expected fields, then translates this data into structures used to implement the dashboard.

Results

Workflows

There have been several different IoIF-supported workflows developed for a few different scenarios. The Catapult workflow shown in Figure 4.12 was developed in a Jupyter Notebook and supports different analysis types, such as the baseline "Analysis as Designed" and "Analysis Requirement Changed." The typical scenario for any IoIF-enabled workflow is characterized in the AFD as discussed in Chapter 3. The typical steps associated with any workflow is to start a triplestore repository and IoIF, which connects to the triplestore. IoIF then brings in information from Teamwork Cloud (TWC) and uses the AFD to configure itself for enabling the needed communications to the various models and simulations (as discussed in Chapter 3).

Figure 4.13 is an AFD generalization. At the center is IoIF that runs with a workflow (e.g., Jupyter Notebook) and can be loaded with ontologies for different domains that are appropriate to the mission and system. Data from different types of models and visualizations use the IoIF Service or some direct capability such as the REST API (e.g., TWC) interface and other types of Linked Data. Parameters modeled as SysML value properties represent various types of simulation inputs to models (left) with outputs being mapped to Mission Measure (top) or to value properties used by Visualization and Control applications (right) that may include Dashboards, Graphical CONOPs,

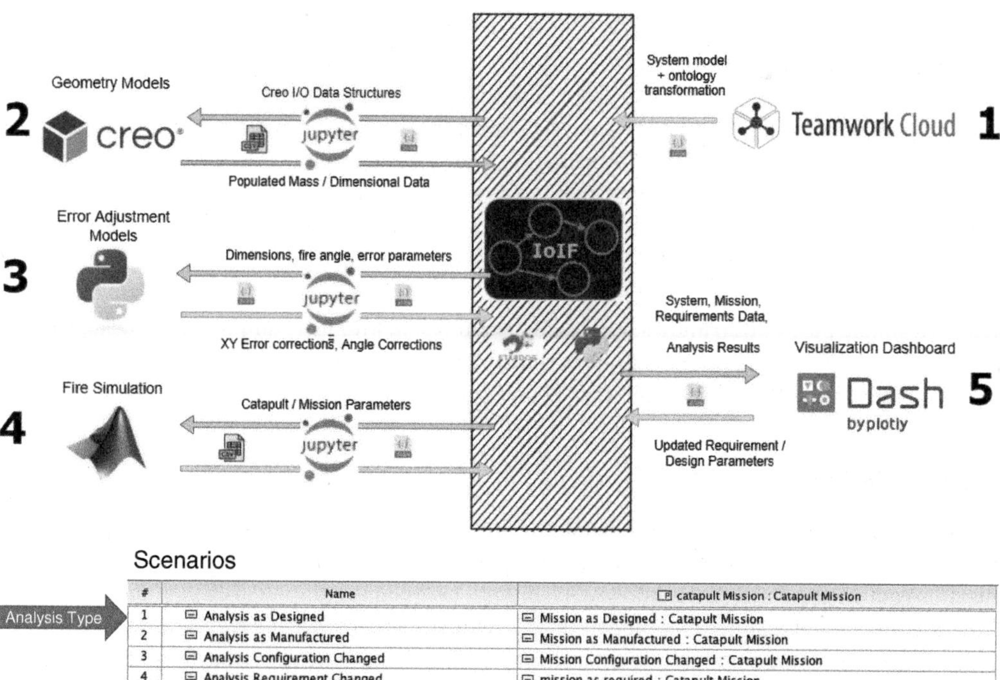

#	Name	catapult Mission : Catapult Mission
1	Analysis as Designed	Mission as Designed : Catapult Mission
2	Analysis as Manufactured	Mission as Manufactured : Catapult Mission
3	Analysis Configuration Changed	Mission Configuration Changed : Catapult Mission
4	Analysis Requirement Changed	mission as required : Catapult Mission

Figure 4.12 IoIF workflows.

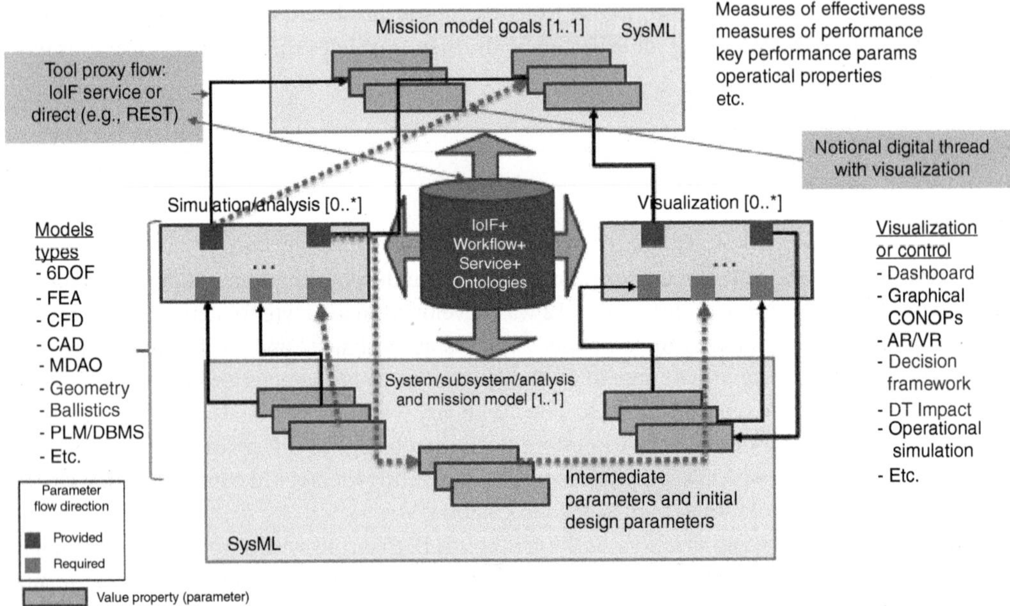

Figure 4.13 AFD generalization that can be configured for various types of analyses with digital thread.

Augmented/Virtual Reality, and Decision Framework as discussed in the New Approach section. An example of dashboard control is discussed in the following section.

IoIF-Enabled Dashboards

The New Approach Section discussed two IoIF-enabled dashboards that have been developed. Figure 4.4 showed the Digital Thread dashboard that renders the digital threads derived from the AFD connections to mission, system, analysis system, physics-based and discipline-specific models and their associated simulation. The semantic layer accessed via IoIF is also used to update the data associated with the update tab on the dashboard. The same basic principle implements the update table. A simple interactive table (Figure 4.14) can be provided in the dashboard to update instances ("Analysis Requirement Changed") and or copy data across instances, with users inputting data as necessary and the dashboard working out what parameters will be unchanged. Upon update, a simple pushbutton allows the data to be published into IoIF, and the needed steps in the workflow as shown in Figure 4.4 can be executed again. The results from those simulations can then be shown in the Decision Framework dashboard shown in Figures 4.5 and 4.11.

Graphical CONOPs

Within a DE environment, data visualization techniques, such as a graphical CONOPs, may interface with many other products, including through IoIF. These include multi-physics models, optimization routines, and MBSE workflows. The linkage between the graphical display and underlying physics-based digital engineering models is an essential part of what makes a graphical CONOPs useful. While advanced computer graphics can make any science fiction scenario seem to be feasible, using physics-based models connected to gaming engines will represent realistic scenarios and can be used to verify the system of interest for both developers and customers. For instance, in a UAS search model, adjusting the inputs to have larger rotors and a smaller battery will create a

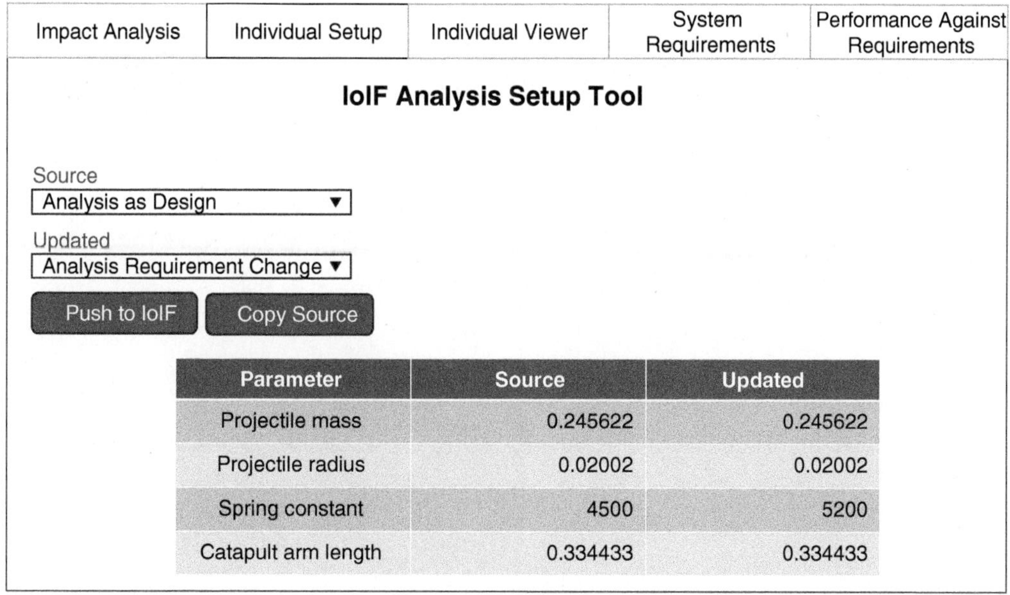

Figure 4.14 Front end GUI to update instances and modify IoIF's triplestore.

quadcopter that moves much more quickly across the environment, yet runs out of batteries in a shorter time. Making these adjustments to the design inputs and running the graphical CONOPs lets stakeholders see these design trade-offs instantly. These benefits can be realized from connecting a physics-based graphical CONOPs with other digital engineering products, for both the products and the operational concept itself.

Graphical CONOPs with Multidisciplinary Design, Analysis and Optimization

Previous SERC research investigated methods for Multidisciplinary Design, Analysis and Optimization (MDAO) (Blackburn et al. 2018; Chell et al. 2019). There were five different use cases that investigated MDAO simulation, and one was integrated to run the UAS/Counter-UAS graphical CONOPs inside of an optimization loop reflected in Figure 4.15. This required levels of automation not developed in previous graphical CONOPs research. This is due to the scenario presenting many challenges for optimization such as long run times and highly stochastic outputs. Running these manually with the visualization enabled limited the number of runs that could be executed.

The long run times of this model exacerbated computational resources required by running gaming engines. To overcome this challenge, we created a "headless" version, where no user input nor graphical outputs are required. This required coding the simulation to support batch runs and suppressing the graphical interface and CONOPs visualization. Furthermore, the underlying simulation had to be run on time scales faster than real time. With these features, it was possible to run the many (~1000) model executions, which were required for optimization. In cases where software tools do not allow headless operations, the graphical quality might have to be coarsened, or a separate model containing the same physical analyses might have to be developed if many scenario executions are necessary.

The mission scenario was wrapped into Phoenix Integration's (acquired by ANSYS) ModelCenter to automatically drive 1000s of tradespace exploration examples, both for designs of experiments and the optimization routines. This contrasts with the 10s of runs that would be possible if the mission model were run manually through the dashboard controls as shown in Figure 4.16.

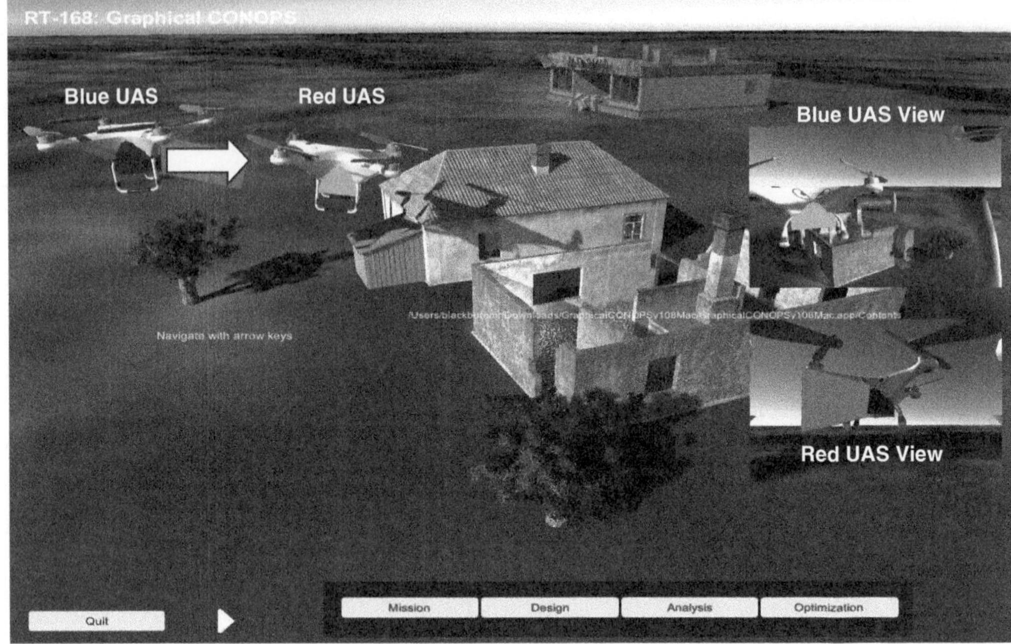

Figure 4.15 Unity gaming engine simulation of two moving UAVs with inset camera views.

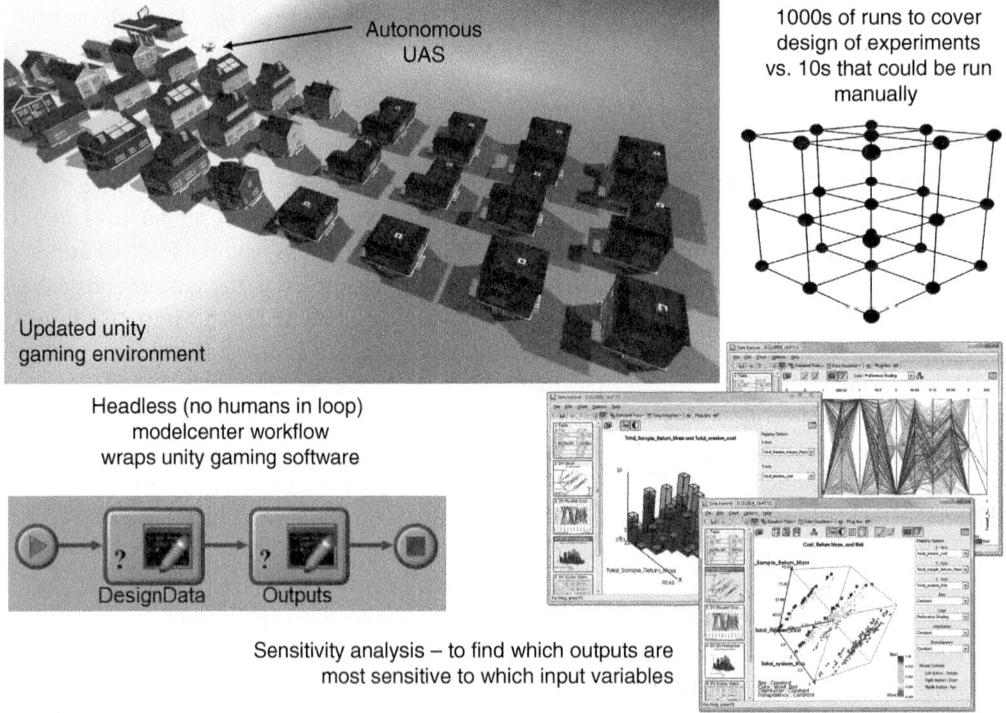

Figure 4.16 Integration of graphical CONOPs simulation with MDAO tools.

In the process of a researching methods for developing graphical CONOPs, 11 different models of an uncrewed aerial system (UAS)/counter-UAS case study were developed. These various efforts approached the graphical CONOPs problem using different technologies, both in terms of types of abstractions, level of fidelity, and humans both in- and out-of-the-loop, which has an impact on cost and value trade-offs for the simulation.

One of these UAS/counter-UAS scenarios was developed with the Unity© gaming engine. This scenario consisted of a friendly (blue) UAS searching for a target in a suburban environment, and an adversarial (red) UAS attempting to block the blue UAS from finding this target. However, in this case both quadcopters are fully autonomous and have realistic battery and flight models governing their performance. The red counterparty was observed to behave in a surprising manner that included emergent behavior. This may have occurred due to the autonomous red UAS having artificial intelligence (AI), so that it can adapt to and learn new behaviors. The simulation was fully automated, and there are no humans in the loop, except for validation of behavior from viewing the graphical CONOPs. The software communicates programmatically through file transfers, coordinated within ModelCenter, as opposed to being directed manually using the parametric. A view of this scenario is in Figure 4.15, with inset pictures of what each UAS can see, the white arrow is the search direction for the blue UAS.

The results identified a general systems framework that can be used as a backend for Graphical CONOPs in support of MDAO as well as provide inputs to other types of modeling and simulation, such as both multi-fidelity approaches to mission and system simulation. A simple user interface allows decision-makers to change the system design and see it affects mission-level performance. Users can adjust 11 design variables for the blue UAS, such as rotor and battery specifications. Then, the mission model can be run to see how well the new design performs.

Eliciting Requirements Using Graphical CONOPs

One of the best examples for the utility of a Graphical CONOPs was in eliciting user requirements from the sponsor for this research task. This can support aspect of validation by ensuring that we have the "right" requirements. Models can help the SMEs improve the elicitation and capture of requirements, using a dynamic CONOPs. We created a 2D representation of munition-related resources using Unity; this evolved over several months as shown in Figure 4.17,

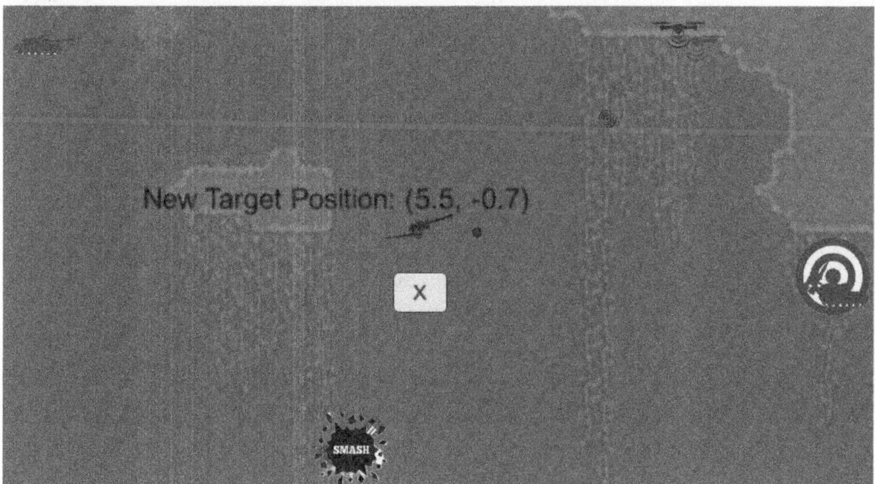

Figure 4.17 Graphical CONOPs 2D with multiple targets.

which is about the fifth version we created. In doing so, we were able to elicit requirements as we evolved the CONOPs to address the sponsor's evolving objectives. For example, one request after the initial version was to add multiple targets. In this way, the Graphical CONOPs not only support requirements elicitation but also provides a visualization of the operational capabilities.

We also used IoIF to integrate 2D graphical CONOPs with a 1D six degree of freedom (6DOF) analysis developed in MATLAB. This type of integration provides for a type of validation (or calibration) using mission-level visualization of our higher-fidelity analysis code. The objectives were not to develop the optimal munition systems, rather the research objectives were to determine how to be more systematic in developing co-simulation-based workflows so that the sponsors can assemble them with IoIF. In doing that we independently verified both 1D and 2D simulations using the same input data were able to show that both simulations produce the same outputs. This allowed us to then demonstrate using IoIF that we could drive the 2D simulation using the 1D MATLAB simulation, which could run much more quickly, but offer a visualization of the resulting runs that could be assessed by SMEs as shown in Figure 4.18.

The Unity gaming engine used for the Graphical CONOPs has precise physics simulations, and an interesting phenomenon was observed during the early versions of the CONOPs model. When the model did not include spin on the shell (projectiles are commonly spun), the simulated projectile became unstable, in a manner similar to the way a knuckle ball (no spin) in baseball will move in unpredictable directions depending on the wind. This reflects positively on the Unity software, because it reflects the actual physics on the simulated projectile.

Finally, extending the use of dissimilar models and simulations, such as the 1D and 2D that help verify that they both produce the same output, we continued this theme with 3D models and simulations as shown in Figure 4.19.

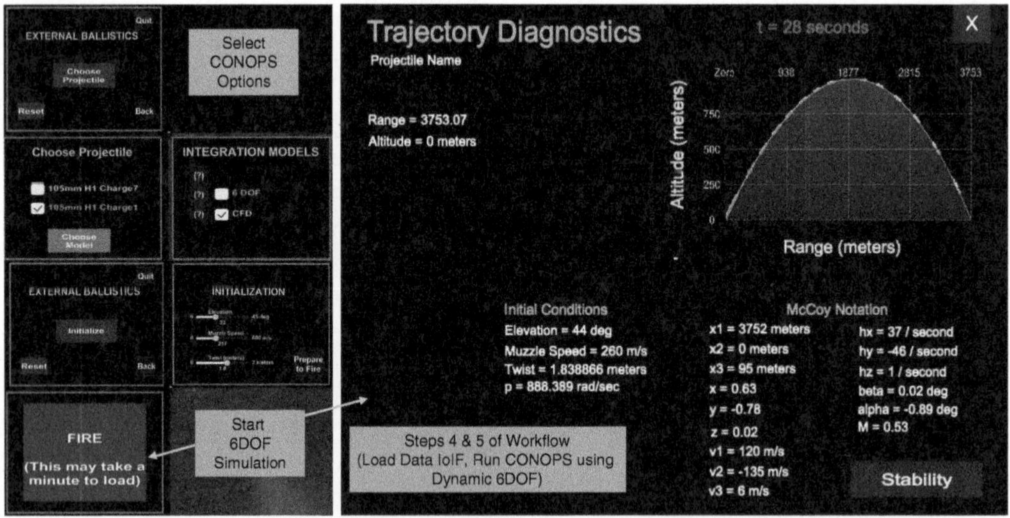

Figure 4.18 Graphical CONOPS provides visualization of 6DOF simulation.

Figure 4.19 Graphical CONOPs 3D.

Future Work

Throughout these research efforts combining visualization tools with digital engineering products, several lessons have been learned which can aid future developers with similar tasks. Product innovation can foster the use of modeling and simulation via Augmented Reality (AR) and Virtual Reality (VR) to provide quicker and more robust way of testing and demonstrating product capabilities relative to other, traditional forms of manufacturing. These traditional methods generally rely on building physical and functional prototypes in order to demonstrate how the product will operate in different existing systems. AR and VR, however, offer a promising route in that most modern development processes use a model-based approach wherein 3D models are used in lieu of traditional prototype- and document-based approaches. While some complexities exist with the direct input of these 3D models into game engines (e.g., Unity), some newer solutions have been created that help alleviate some of these challenges and even offer a more robust system that allows for broad integration with other programs and modeling engines. Thus, our proposed method, which we have named model-based systems engineering in virtual environments (MBSEVE), is a Unity-based AR testbed (early prototype shown in figure) that will take in values from other programs, such as the IoIF and integrate them with the Unity environment, which extends the Graphical CONOPs to build an ecosystem that can model various prototypes in a wide variety of environments. For example, how a new aircraft engine will fit onto an existing airframe or how different seating designs in a tank impact overall usability. The modular approach taken in this research will afford reuse of the various models and environments.

To build this MBSEVE approach, an AR operating environment is needed that can support these multi-modal feature sets and integrate seamlessly into existing workflows. We think that this approach would work best in many situations as a lot of development and engineering work is increasingly becoming virtual. To create one of these systems, a technology suite that is both robust and highly capable is needed that can work for myriad products and environments.

We plan to start investigating the use of operational AR environments and thinking through both needed hardware (e.g., head worn displays, mobile AR/webXR), input devices, tracking as well as software (e.g., Holograms, game engines, and models that could map to ontologies), as well as leveraging our IoIF approach for integrating different workflows for broader cross-domain simulations.

References

American Institute of Aeronautics and Astronautics (AIAA) (1993). *ANSI/AIAA Guide for the Preparation of Operational Concept Document, G-043-1992*. Reston, VA: ANSI/AIAA G-043-1992.

Blackburn, M., Verma, D., Dillon-Merrill, R. et al. (2018). Transforming Systems Engineering Through Model-Centric Engineering. *Final Technical Report SERC-2017-TR-110, RT-168 (ARDEC), Phase II.*

Chell, B.W., Hoffenson, S. and Blackburn, M.R. (2019). A Comparison of Multidisciplinary Design Optimization Architectures with an Aircraft Case Study. *In AIAA Scitech 2019 Forum*. p. 0700.

Cilli, M. (2015). Seeking improved defense product development success rates through innovations to trade-off analysis methods. Dissertation. Stevens Institute of Technology.

Cloutier, R., Mostashari, A., McComb, S. et al. (2010). *Investigation of a Graphical CONOPS Development Environment for Agile Systems Engineering-Phase 2*, SERC-2010-TR-007-1. Hoboken, NJ: Stevens Institute of Technology.

Cloutier, R., Hamilton, D., Zigh, T. et al. (2013). *Prototype of a Graphical CONOPs (Concept of Operations) Development Environment for Agile Systems Engineering*, SERC-2013-TR-030-2, 142. Hoboken, NJ: Stevens Institute of Technology.

Department of Defense (2000). *Operation Concept Description Standard. DI-IPSC-81430*. Department of Defense. http://everyspec.com/DATA-ITEM-DESC-DIDs/DI-IPSC/DI-IPSC-81430A_3708/.

IEEE (1998). *IEEE guide for information technology – system definition – Concept of Operations (CONOPs) document*, IEEE Std 1362-1998 (R2007), IEEE CONOPs Standard, New York, IEEE Std. 1730 (2010). IEEE Recommended Practice for Distributed Simulation Engineering and Execution Process (DSEEP).

Korfiatis, P. (2013). *Development of a Virtual Concept Engineering Process to Extend Model-Based Systems Engineering*. Hoboken, NJ: Stevens Institute of Technology.

Korfiatis, P., Cloutier, R., and Zigh, T. (2012). Graphical CONOPs development to enhance model based systems engineering. *Proceedings of the 2012 Industrial and Systems Engineering Research Conference Third International Engineering Systems Symposium, CESUN 2012* (G. Lim and J.W. Herrmann, eds.), 18–20 June 2012. Delft University of Technology. no. 18–20.

Korfiatis, P., Cloutier, R., and Zigh, T. (2015). Model-based concept of operations development using gaming simulation: preliminary findings. *Simulation and Gaming* 46 (5): 471–488.

Larson, W., Sellers, J., Kirpatrick, D. et al. (2009). *Applied Space Systems Engineering*. McGraw Hill.

Mostashari, A., McComb, S.A., Kennedy, D.M. et al. (2012). Developing a stakeholder-assisted agile CONOPs development process. *Systems Engineering* 15 (1): 1–13.

Vesonder, G., Verma, D., Hutchinson, N. et al. (2018). RT-171: Mission Engineering Competencies Technical Report. Stevens Institute of Technology. Hoboken, NJ.

Biographical Sketches

Brian Chell, PhD, is a postdoctoral research associate with the Systems Engineering Research Center at Stevens Institute of Technology. His research interests are in distributed space systems and applying multidisciplinary design optimization techniques to mission-level analysis. Brian received a PhD in systems engineering and an ME in space systems engineering from Stevens, and a BS in aerospace engineering sciences from the University of Colorado Boulder.

Dr. Tom Hagedorn graduated from Boston University with a BS in biomedical engineering in 2010, and from the University of Massachusetts at Amherst with a PhD in 2018. His research focuses on applications of ontologies and semantic web technology to aid in various aspects of engineering design and systems engineering. This includes semantically enhanced linked data repositories, development of engineering domain ontologies, and the study of engineering methods using these tools. His doctoral work focused on semantically enabled design assistants in medical and advanced manufacturing contexts. His current work focuses on the development of digital engineering methods and frameworks built upon semantic web technologies.

Roger Jones, PhD, trained in physics at Dartmouth College, Jones worked as a staff physicist at Los Alamos National Laboratory from 1979 to 1995. His primary research interests were in laser fusion and machine learning. In the early 1990s he headed projects that applied his machine learning inventions to technical problems in the private sector. In 1995 in collaboration with Citibank, Jones co-founded the Center for Adaptive Systems Applications (CASA), a company that applied neural network and adaptive technology to consumer banking. CASA was acquired by HNC Software in March 2000, at the peak of the dotcom boom. HNC Software was subsequently acquired by Fair Isaac Corporation. Much of the technology developed at CASA became part of the credit scoring offerings of Fair Isaac. Jones along with other Santa Fe scientists and entrepreneurs such as Doyne Farmer, Norman Packard, Stuart Kauffman, and David Weininger founded several other high-technology startup companies in the emerging Santa Fe technology community, dubbed by Wired Magazine as the "Info Mesa." Much of the effort of these startups focused on finance and the catastrophic reinsurance industry. By 2004 the companies Jones co-founded merged into a single company, Qforma, Inc., that focused on adaptive and predictive technologies for the pharmaceutical industry. In June 2013, Qforma merged with SkilaMederi. In 2013, Jones returned to the academic world. His current interests are in engineering and biological systems.

Mark R. Blackburn, PhD, is a senior research scientist with Stevens Institute of Technology since 2011 and principal at KnowledgeBytes. Dr. Blackburn has been the principal investigator (PI) on 17 System Engineering Research Center (SERC) research tasks for both US Navy NAVAIR, US Army DEVCOM, and US Space Force on Digital Engineering Transformation Research Tasks. He has also been PI on a FAA NextGen and National Institute of Standards and Technology projects and has received research funding from the National Science Foundation. He develops and teaches a course on Systems Engineering for Cyber Physical Systems. He is a member of the SERC Research Council, OpenMBEE Leadership Team and INCOSE Pattern Working Group focused on the Semantic Technologies for Systems Engineering initiative. Prior to joining Stevens, Dr. Blackburn worked in industry for more than 25 years. Dr. Blackburn holds a PhD from George Mason University, a MS in mathematics (emphasis in CS) from Florida Atlantic University, and a BS in mathematics (CS option) from Arizona State University.

Chapter 5

Interactive Model-Centric Systems Engineering

Donna H. Rhodes and Adam M. Ross

Massachusetts Institute of Technology, Cambridge, MA, USA

Introduction

Models are increasingly used for decision-making in systems engineering, yet while human users are essential to a model's success, research into human–model interaction has been lacking (Rhodes and Ross 2015). Models represent an abstraction of reality and can come in a variety of forms and formats. Humans use models to augment their ability to make sense of the world and anticipate future outcomes. The idea that "humans use" models highlights human interaction as a necessary factor in model-centric engineering.

Background

Motivated by the need to investigate the various aspects of humans interacting with models and model-generated data, in the context of systems engineering practice, a research program was developed under the DoD Systems Engineering Research Center (SERC). This research program, entitled *Interactive Model-Centric Systems Engineering (IMCSE),* was initiated and performed during the period 2014–2020. IMCSE research is grounded in a belief that improving human–model interaction and social dimensions of model-based environments will significantly improve the effectiveness of digital engineering practice, quality of model-supported decision-making, and cultural acceptance of a digital future.

IMCSE aimed to develop transformative results through enabling intense human–model interaction, to rapidly conceive of systems and interact with models in order to make rapid trades to decide on what is most effective given present knowledge and future uncertainties, as well as what is practical given available resources and constraints (Rhodes and Ross 2015). Future environments and practices need to leverage advancements in newer technologies, data science, visual analytics, and complex systems.

As abstractions, models are an encapsulation of reality that humans use to augment their ability to make sense of the world and anticipate future outcomes. The idea that "humans use" models, oversimplifies the relationship and neglects a key component: the human that necessarily occurs within all models. In the emerging model-centric practice, model and humans are intertwined. Models are used by humans and that interaction influences how models are conceived and used in

Systems Engineering for the Digital Age: Practitioner Perspectives, First Edition. Edited by Dinesh Verma.

Companion website: www.wiley.com/go/verma/systemsengineering

Understanding behavior of a digital engineering enterprise requires viewing human actors as endogenous constituents
• Evolving technology enables more complex and capable models but human interaction with and through models is integral to success
• Designing digital engineering environments requires understanding human capabilities and limitations
• Digital transformation requires addressing social factors in additional to technical factors
• Humans have cognitive biases, perceptual limitations, preferences, and behaviors that need to be addressed in a digital engineering context

Figure 5.1 Viewing human actors as being within the digital engineering environment.

decisions. Models and model-generated information influences decision-maker behavior in various ways. Together, humans and models exhibit dynamic behavior.

Investigation of human–model interaction and model-centric decision-making has underscored the need to view humans as endogenous constituents, that is, as actors that are inside model-centric engineering environments. This contrasts with how humans are often viewed as being "on the loop," interacting with the environment as an external agent in a supervisory role or observation role. Understanding human–model interaction is predicated on taking this endogenous point of view (Figure 5.1).

Motivation

On 20 January 2015, an invited IMCSE Pathfinder Workshop was conducted at MIT, bringing together interested stakeholders from government, industry, and academia to share knowledge and identify needs of the stakeholder community (Rhodes and Ross 2016). Participants explored the topic of interactive model-centric systems engineering and envisioned the possible future while considering both *model-centric* and *interaction-centric* viewpoints. For the purposes of the workshop, the participants were asked to view a model as an abstraction of reality to be used to inform predictions.

The *interaction-centric viewpoint* considers the stakeholders (decision-makers, analysts, system architects/engineers, involved design practitioners, end-users, etc.) who interact with models. Desired interaction may involve activities such as explore a tradespace, input data into a model, query a model, validate decisions with models, and use a model to make design choices or investment decisions. Through this, the stakeholder will then have various perceptions such as perceived accuracy of a model and whether a model works correctly. Important considerations include how to avoid cognitive biases, factors in developing trust in models, and how decisions are made on what models to use.

Taking the *model-centric viewpoint*, things to consider include how models are developed and maintained, what forms models take, how models generate information, how models accept user inputs, and how information is displayed. There are many other considerations such as examining how models are reused, deciding how sensitivity analysis is performed, and understanding how models are integrated with other models and how they are made interoperable.

Using the two perspectives of interaction-centric and model-centric to envision a desired future for interactive model-centric systems engineering resulted in the seven vignettes that follow.

Characteristics of Models

Models will have associated information that characterizes the development history of the model over time, how it has been used, and specific individuals who developed the model. There will be support provided to understand the model limitations, risks, and uncertainties. The environment will provide support to the individual that will help in understanding model limitations. Models will be designed to be intuitive for decision-making. Role-based permissions/views will provide appropriate access and abilities for the modeling development and use. Models will adapt to personal logic, evolving context, and changing questions.

Ease of Interaction

The individual interacting with a model will find it intuitive and the effort involved will be commensurate with the value the model provides. Novice users will be able to rapidly learn and benefit from use of modeling environments. Model users will find that changes to a model can easily be implemented within the modeling environment, and impacts of changes are made readily apparent. Models will be easily tailorable for levels of abstraction, for different stakeholder types, and different purposes. Modeling environments will have human error tolerance. An individual would be able to look at a given level of abstraction of a model, as well as have the ability to dig into a deeper level within a particular view. The speed of making queries will be in tune with exploration, and there will be minimal latency in query responses.

Enabling Informed Decisions

The individual will find that the model-centric environment enables more informed decision-making. Modeling environments enable decisions based on rich context information that spans the space of relevant parameters. Models will be effective in evaluating systems under alternative contexts, for example policy context. Environments will provide the capability to contrast and compare results with ease. The individual will be able to intuitively judge the models, and uncertainty in the model output will be easily apparent. There will be the means to assess and/or judge the goodness of a model-enabled decision.

Enabling Human–Human Interaction

Model-centric environments will support collaborative decision-making and design with near real-time human to human interaction. The analysis and decision behavior of each individual in the modeling process will be captured, and enable understanding of the whole system behavior in regard to meeting system objectives. Individuals will find it easy to compare their own values with those of other individuals so that they can calibrate their own values. Models will enable value-centric engineering, providing a means to make those values explicit as part of the collaborative decision process.

Guided Interaction

Interactions with models will provide guided assistance for viewing models from standpoint of other stakeholders. A model will take imprecise input from the individual and offer a guided experience prompting interaction where the individual might not even know what question to ask. The guided experience will aid individuals in dynamically interacting with the models, functioning as an "AI colleague." The individual will feel as if the environment "understands where the individual is coming from and what they are concerned about." There will be assisted capabilities and wizards for model library curation, model composability, model interrogation, and stakeholder role playing.

Model Re-Usability

The individual using an interactive model-centric environment will have some capability to retain history of changes in the models over time, including ability to track change propagation and capture usage pattern. The environment will be adaptable for the culture of the organization to enable effective reuse with confidence in the model and its appropriateness for the situation. Finding suitable models and reusing them in the individual's unique model-based environment will be easily accomplished. Models will transparently capture a user's values, communicate them clearly, and allow changes as more information is made available. An individual will easily be able to document risks and uncertainties with the models themselves, and this information will contribute to appropriate reuse. Effective approaches for model sharing will mitigate intellectual property and competitive concerns, and will not jeopardize perceived competitive advantage. Effective digital curation will enable preserving, discovering, and reusing appropriate models.

Trusted Models

The individual will trust models, with supporting evidence underlying that trust. There will be an established pedigree/heritage at variable levels of fidelity. A rich set of information (for example, test cases) will be bundled with a model. As needed for a given role, models may be "invisible" such that nonessential detailed information is hidden or have increasing levels of "transparency" for the individual who needs to examine inner workings of the model. Uncertainty in model output is apparent, as framed by specific context/use. The development environment and model libraries will foster a high trust, collaborative relationship between the model users and the model developers. Individuals will trust that sharing models will not jeopardize the need for their personal expertise. These rich characterizations of a desired future, developed during the pathfinder workshop, provided a foundation for defining and prioritizing tasks to be investigated.

Selected Highlights of IMCSE Research

Areas of inquiry explored under IMCSE included how individuals interact with models; how multiple stakeholders interact using models and model-generated information; facets of human interaction with visualizations and large data sets; how trust in models is attained; and what human roles are needed for model-centric enterprises of the future. In support of these inquiries, IMCSE research has employed many different research methods and various means to engage with the stakeholder community (Figure 5.2).

In the remainder of this chapter, we highlight selected findings of four areas of research: (1) model-centric decision-making and trust; (2) multi-stakeholder negotiation through models; (3) model choice and trade-off; and (4) model curation.

Model-Centric Decision-Making and Trust

Model-centric decision-making remains a relatively unexplored topic. Understanding decision-making in model centric enterprises is critically important, as models are increasingly used in making significant decisions throughout the system lifecycle. A recent exploratory study involving

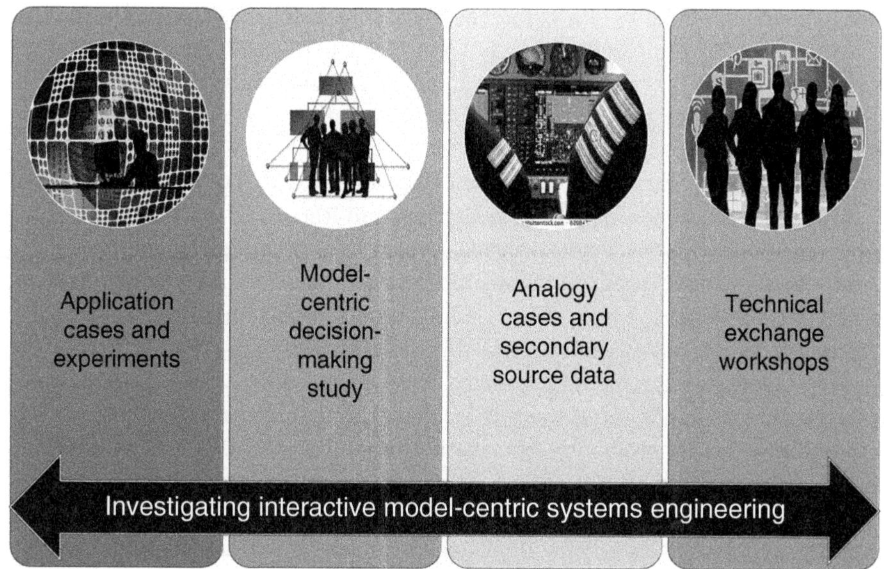

Figure 5.2 Multiple Approaches used to investigate Interactive Model-Centric Systems Engineering.

semi-structured interviews with 30 recognized experts (largely from the US defense sector) explored how various types of decision-makers and actors interact with and use models, including to what degree models are used to inform system decisions and how individuals build trust in models.

This study provided empirical insight into how human actors interact with and trust models while also providing a starting point for continued exploration into how human actors and decision-makers trust, perceive, and interact with models. Interviewees shared their opinions on how decisions are made, what factors are involved in model-centric decisions, and the enablers and barriers that are faced. Figure 5.3 shows some of the points raised by interviewees.

The initial investigation of model-centric decision-making found one major challenge in human–model interaction to be "perception of truthfulness and trust in models" as this can

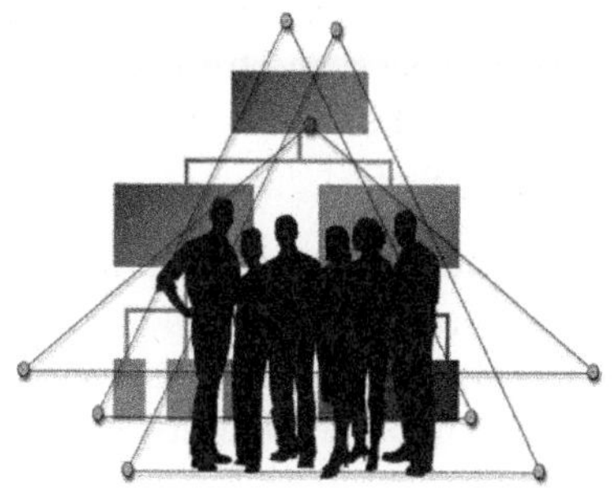

30 recognized experts

- Three actor decision flow
- Importance of intercommunication
- Understanding of assumptions and uncertainty
- Technological and social factors influencing trust
- Importance of model-related documentation
- Need for model pedigree
- Using models as primary versus supplementary
- Non-advocate role in reviews
- Transparency and trust
- Model investment bias and confirmation bias
- Factors limiting model-centric decisions
- Real-time interaction with models
- Viewing humans as endogenous

Figure 5.3 Discussion points raised by experts in model-centric decision-making study. *Source:* Adapted from Shane German and Rhodes (2018).

ultimately impact timeliness, quality, and confidence in model-based decisions (Rhodes and Ross 2015). Expert stakeholders expressed a desire not only for models to be trusted, but to be supported with underlying evidence to engender trust. Therefore, deciding to accept a model from an external source necessitates having supporting model artifacts that permit judgment of the integrity and trustfulness of the model.

The interview-based study of current model-centric decision-making revealed a pattern that we describe as a three-actor decision flow (Shane German and Rhodes 2018). The flow begins with the modelers and ends with the ultimate decision-maker. In the middle of the flow is the "through" person who is a trusted agent of the decision-maker. The data in our study suggests that as actors move further along the flow of information and have less time and ability to personally investigate a model and build their own trust in the model, their trust instead shifts more onto their people to investigate the model for them. Trust for the ultimate decision-maker is "implicitly on the models, but explicitly on the people." A model does not provide value by existing. Rather, it provides value by producing information that can be used by individuals and decision-makers to better understand a specific problem. Often, information flows from initial actors that directly interact with a model before flowing through other individuals within the decision-making process and ultimately to final decision-makers. All actors in the model-generated flow of information must have a well-calibrated understanding of the model if the information is to be effectively understood.

Model Trust as a Sociotechnical Construct

IMCSE investigation revealed the complexities of model trust, and a heuristic derived through the research is: *Model trust is a sociotechnical construct: you must examine both technological and social factors to understand how individuals develop trust in models.*

Individuals within the model-centric decision-making process rely upon various technical and social factors as they develop trust in a model. Technical factors include technical information about a model, such as its transparency, uncertainty, and input data. Social factors include the people, organizations, and relationships that shape one's trust of a model. These include factors such as credibility of the individual or organization developing the model, reliability of the relationship with the individuals recommending a model, or word-of-mouth opinions within a community concerning the model's technical performance.

Guiding Principles for Model Trust

Given both social and technical dimensions of human–model interaction, heuristics and guiding principles are desired enablers. Further investigation of the issues of models and trust revealed some of the primary reasons why we trust (or distrust) models. Based on the investigation, seven guiding principles for model trust were derived (Rhodes 2018), as follows:

1) **Transparency should always be possible, but tailorable**: Our research shows that model transparency should always be available to the user, but tailorable to needs of the individual and the situation. Individuals have been shown to desire different levels of transparency based on their role and personal preferences. Excess transparency may cause information overload that can obscure relevant information.
2) **Model-context appropriateness is a key determinant of trust:** It has been found that model trust is not as much on the model entity itself, as the model consumer's perception of the model including usefulness of the model for the situation at hand. The assumptions that are built into a model need to be acceptable for the context of use.

3) **Real-time interaction with models has upsides and downsides:** On the positive side, real-time interaction enables asking "what if" questions to gain insights and establish trust. The downside of this is that gaining results very quickly may still lead to making poor decisions.
4) **Trust may be implicitly on the models, but explicitly on people:** The research indicates that the ultimate decision-maker often relies on a trusted individual in making model-centric decisions. The decision-maker, lacking time, and ability to fully investigate and understand the model, shift trust onto their trusted agent, who acts as interpreter of information from models and modeling experts.
5) **Trust emerges from interaction between human actors, through models:** Our investigation suggests trust extends from the interaction between and across various actors (models, "interpreters," decision-makers). Trust emerges more as a result of the human interactions through models, rather than direct model interaction.
6) **Availability of model pedigree engenders trust**: Metadata and technical data are, of course, important. Our research indicates that pedigree information is also a major influence on judging trust (whereas integrity is based on technical data). Model pedigree was first described by Gass and Joel (1980) as model demographics and the term pedigree was subsequently used by Gass. Pedigree tells the decision-maker who originated the model, assumptions made, context of use, expert knowledge and investments, etc.
7) **Trust is influenced by the entangled technical and social factors:** The technical factors concerning the model itself (e.g., model fidelity, data integrity, verified algorithms) and social factors (e.g., model originator, preferences, perception of modeler expertise) cannot truly be separated. Therefore, it is important to be cognizant of how these factors interrelate and influence decisions.

Depending on both the organizational culture and the individual's preferences, different factors may play greater or lesser roles in the model trust process. Accordingly, these factors should be understood by practitioners to facilitate appropriate calibration of individual trust.

Multi-Stakeholder Negotiation Through Models

In addition to individual trust in models, trust can be engendered through the process of model-mediated activities. A research project within IMCSE sought to further explore the application of tradespace exploration (TSE) as one such model-mediated approach that previously showed promise in supporting multi-stakeholder negotiations. TSE is a design and decision-making paradigm that leverages model-based analysis of many alternatives in order to build understanding of the trade-offs between value-driving attributes and design decisions (Ross and Hastings 2005; Ross et al. 2010a). Without restricting attention to a particular implementation, generally a TSE project will follow a procedure similar to this:

1) **Problem formulation:** the structuring of the problem and scope of decision-making. This normally includes the definition of the alternative space used to enumerate potential system alternatives, the context in which those systems will operate, and the stakeholders with their value attributes used to assess the alternatives in context.
2) **Modeling/Evaluation:** the development and use of models for the purposes of evaluating the alternatives. Models can take many forms, which necessitates a selection of modeling technique(s) appropriate to the problem formulation. Creating models is itself nontrivially difficult and normally takes considerable effort without the benefit of reuse of previous models.

3) **Exploration/Analysis:** the attempt to garner knowledge and insights from the model outputs. Stakeholders and analysts are both capable of performing this step, with different strengths and weaknesses. Exploration is typically intended to generate results capable of justifying a decision to select a given alternative.

This knowledge-building process is particularly useful when applied to complex systems for which designers or analysts may not have a strong intuition of the dynamics at play. The presence of multiple cooperating or competing stakeholders is one such complexity. Early attempts to incorporate multi-stakeholder (MS) analysis into TSE simply used a value model for each stakeholder and used analysis to find alternatives that satisfied each stakeholder's model. This type of analyst-driven exploration is called "informal" MSTSE, to indicate the unlikeliness of reaching a formal agreement using the tradespace without stakeholders. Informal MSTSE has the advantage of being conducted by experts in a manner like most systems engineering activities, with the resulting lessons and insights communicated to stakeholders before they engage in the "formal" negotiation or decision-making process. However, this approach naturally risks costly iteration, as the later negotiation may raise new questions that must be sent back to the engineers responsible for tradespace analysis, thereby delaying the final decision.

Weaknesses of informal MSTSE inspired a new approach of parallel data exploration by each stakeholder, with the goal of uncovering emergent insights at the intersection of their exploration and facilitating dialogue amongst the stakeholders that could result in iterative refinement of their value models *during* exploration rather than separate from it (Ross et al. 2010b). Though effective at its intended purpose, this type of multi-stakeholder analysis was conducted entirely with the mindset, visualizations, and metrics of classic TSE. Efforts to formalize the concept of multiple stakeholders engaging with tradespace data into MSTSE sought to re-examine the latent assumptions in these methods, to confirm or reject their suitability for the multi-stakeholder problem (Fitzgerald and Ross 2014). Framing was identified as a potential key roadblock to effective MSTSE, due to the aggressively individualistic framing of traditional, single-stakeholder TSE analysis leading to misplaced reference points for decision-making.

Framing can be considered at both macro and micro levels. **Macro framing** lies outside the domain of any single-decision problem and deals with issues of writ-large beliefs and perspectives. In contrast, **micro framing** resides *within* the problem formulation, in the way information is presented and tasks are performed. Macro and micro framing can have a "weakest link" relationship in a negotiation, by which a framing trap in one may pull down the other. For example, if a stakeholder approaches MSTSE with a macro frame that is highly confrontational and individualistic, they will likely favor a micro frame, in the form of a particular visualization for example, that matches their outlook. Alternatively, if only individualistic visualizations are available, a stakeholder's macro frame may be slowly pushed into a similar aggressive, value-claiming mindset in order to reduce cognitive dissonance with their tools.

Another key element in negotiation is the default position. The **BATNA** (best alternative to a negotiated agreement) is what each stakeholder will do *on his own* if no agreement can be reached with the other stakeholders. This is an important reference point with respect to the value of any of the design alternatives under consideration as it defines the border between gains and losses. Failure to define and then leverage the BATNA during exploration reduces the situational awareness of the stakeholders.

Common BATNAs include the following:

- **Do-nothing:** if the MSTSE is strictly exploratory, inaction is likely the course of action should no agreement to proceed be made. Doing nothing typically carries zero cost and zero benefit.

- **Existing system:** for design tasks intended to improve or replace an existing system, the do-nothing alternative actually entails using the current system. This type of BATNA is one that commonly drives differences in stakeholders' bargaining leverage, as some stakeholders may be much better off with the current system than others.
- **Build preferred alternative alone:** some projects seek agreement between multiple stakeholders to reduce the cost borne by each individual. If a stakeholder is capable of affording some or all of the alternatives by themselves, those alternatives become viable BATNAs (though at a higher cost than if they could agree to share one).
- **Other opportunity:** resources that are expended on the alternatives in the tradespace represent an opportunity cost in that they cannot then be spent on other projects, which may be more valuable. This type of BATNA is the most difficult to capture, as the number of other opportunities is potentially limitless, but this fact is true for all design tasks. Usually, a small number of known viable or attractive opportunities can be considered without fear of missing better choices.

The **modeling** task itself can also propagate cooperative vs. individualistic framing implicitly into the exploration phase. When multiple stakeholders will be conducting the exploration, it is important to make sure that the modeling is satisfactory to all of them, which requires some additional management.

Joint Fact Finding (Ozawa 1991; Ehrmann and Stinson 1999) (JFF) seeks to establish credible and objective data, one of the foundations of principled negotiation (Fisher et al. 1991), to use as the foundation for evaluation of alternatives and discussion of their relative merits. If possible, all efforts should be made to convene stakeholders prior to actual exploration in order to perform JFF in support of the modeling task. JFF also helps to establish a macro frame of cooperation *before* engaging in the negotiation itself, which can help preserve positive, mutually beneficial bargaining in the face of any naturally developing competitiveness.

Not all models can be developed through JFF, as stakeholders may hold models and/or data as **private information**. If a stakeholder already possesses a model for a piece of the larger system, reusing that model can save time and effort. If they are willing to share that model (both how it works and its results) with the rest of the stakeholders as a part of a larger JFF effort, that is a valuable step in building rapport, in accordance with the principle of Full, Open, and Truthful Exchange (Raiffa 2002). However, some models' inner workings may depend on proprietary or classified information that the stakeholder is unable or unwilling to share. In this case, two approaches can be taken: the existing model can either be ignored in favor of a newly created JFF model (if possible), or "black-boxed" so that other stakeholders can only see its outputs. A black-boxed model can be fully effective if its outputs only impact the value proposition of the stakeholder who owns it. If not, other stakeholders will need to trust that the model is accurate. If a public – but presumably lower fidelity – model is available, it can be used to help validate the black-boxed model and build trust.

Entering the **exploration phase**, the dominant framing concern shifts to micro framing: the actions the participating stakeholders are asked to perform and the way the data generated by the previous steps is presented. Macro framing still has a role to play in exploration however, specifically when weighing specific alternatives as potential final agreements.

In terms of micro framing, several aspects of exploration are salient. First, alternatives must be valuated against the **BATNA as a reference point**. This provides the necessary perspective for determining the value of an alternative *as a multi-stakeholder agreement* rather than the typical, less-contextualized evaluations *in a vacuum* or *relative to other alternatives* commonly used in classic TSE activities. Second, activities should **incorporate the value statements of multiple**

stakeholders as much as possible to consistently keep each participant aware of the "group" aspect of the negotiation problem. This can prevent fixation on alternatives that are very good for one stakeholder but not for others. Third, negotiation in MSTSE exposes each stakeholder to large amounts of information that they may not have previously known, particularly the preferences of other stakeholders which are not present in classic TSE, therefore **allow stakeholders to change their mind**. New information can change subjective assessments of value (Curhan et al. 2004) and invalidate parts of the original problem formulation. Stakeholders should be encouraged to critically reassess their value statements during the negotiation. Additionally, if live value model updates are convergent in a manner leveraged by other consensus-building techniques such as the Delphi method (Golkar and Crawley 2014), these updates have the potential to open new regions of mutual value in the tradespace. Fourth, when discussing individual alternatives, effort should be made to **refer back to the macro frames** of each stakeholder. When a stakeholder refers to a design with a subjective assessment like "good," the first question should always be "why?". Each stakeholder wants a "good" design, but each has different criteria for what is "good" that includes not only their reported value model criteria but also the macro frames with which they choose to make decisions.

Table 5.1 below summarizes the MSTSE recommendations described in this section.

Model Choice and Trade-off

Systems engineers should not just try to select models to answer questions but also to better reflect upon what can be learned using different models to answer the same question. Two exploratory research studies within IMCSE have addressed the two main types of models used in early lifecycle systems engineering: evaluative models (Ross et al. 2016) and value models (Ross et al. 2015). The former tries to predict the performance and cost of potential alternative systems, while the latter tries to predict how much different levels of performance and cost *is worth* to different stakeholders. Both types of models require abstraction of reality, and both generate

Table 5.1 Summary of recommendations, with modifications for informal MSTSE.

Phase	Recommendation	Informal MSTSE
Problem formulation	Capture macro frames Create many alternatives Record key elements of problem structure Determine each stakeholder's BATNA	All of these apply except for capturing macro frames of other stakeholders. Make best estimates for stakeholders' BATNAs and value models
Modeling/Evaluation	Joint fact finding Private information	Treat modeling as normal TSE
Exploration/Analysis	Emphasize the BATNA Limit strictly individual analysis Analyze relationships Allow stakeholders to change their mind Refer back to macro frames	Continue to use BATNA-centric visualizations and analyze relationships, but limit activities related to changing stakeholder value models without their participation

Source: Adapted from Fitzgerald and Ross (2016).

essential data needed for confident decision-making. The exploratory studies have shown that using alternative model implementations simultaneously (i.e., not just picking one of them) to generate data can provide insights that can make the ultimate decisions more robust to uncertainties and deficiencies inherent in the act of modeling itself. Such robustness can be manifested in decisions that are insensitive to uncertainties or can be changed at low cost to account for new information as it unfolds.

Figure 5.4 illustrates a conceptual framework for the decision process in early system design, including the relationships between the various models that are used to support the decision-maker (Ross and Rhodes 2018). The general flow involves the creation of a design space suited to the problem in which each alternative is evaluated using a set of evaluative models to determine its performance and cost (i.e., resources required) with respect to a set of given contextual factors. Those performance and resource attributes are then fed into a value model to assess the "goodness" of each alternative, which is the key decision-making criterion. As Figure 5.4 indicates, many different specific models can play the role of evaluative or value model in a given study.

The selection of which specific model will be used is often the focus for the engineer or analyst for a given study. But the selection of the model itself can have a tremendous impact on what can be learned in the study. To explore the potential impact of value model selection, we expanded the earlier concept of interactively refining a value model (Ricci et al. 2014) into the potential use of value model trading to support the decision-making process (Ross et al. 2015). Specifically, the ability to compare the tradespace as it exists under multiple value models was shown to be a powerful means for building trust in the model results, particularly when a decision-maker may be unsure how to mathematically represent their needs early in the system life cycle.

Additionally, evaluative models come in a plethora of different forms, too many to list exhaustively. Depending on the system in question, different types of evaluative models will be appropriate and/or available. Commonly discussed characteristics of evaluative models include fidelity (or the similar concepts of accuracy and precision) and computational costs (among others such as purpose and credibility (NASA 2008)). When choosing models based on fidelity

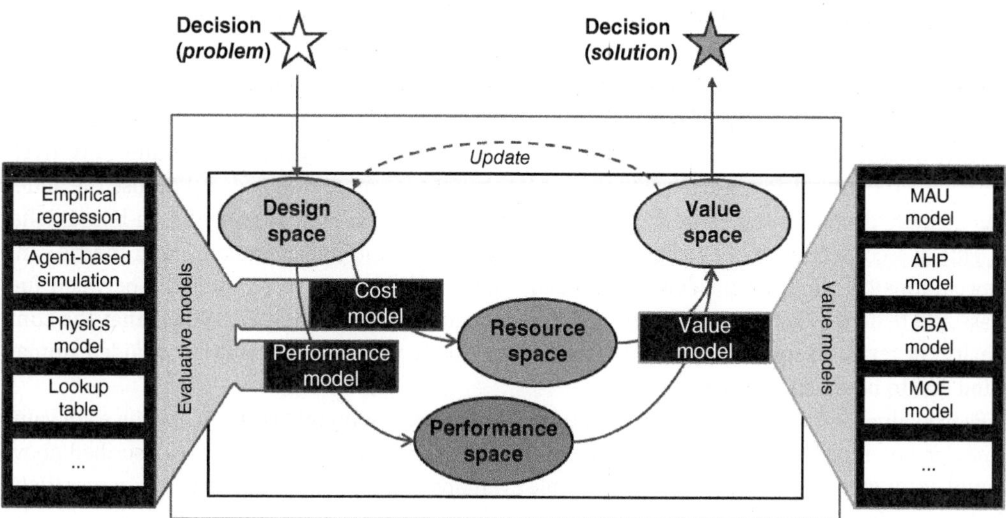

Figure 5.4 Role of key models in system decision-making, with example evaluative and value models called out.

and computational cost, there is often a trade-off, so the "best" model for a given task may be subjective. This choice may also require the consideration of the confidence the decision-maker has in each model, which may not be solely determined by its fidelity. This best-model-calculus and associated trade-offs are likely what most will think of first when hearing the phrase "evaluative model trading" (e.g., determining a model's "fit for purpose" and associated verification, validation, and accreditation (VV&A) activities (INCOSE 2015)). However, selecting the best evaluative model from a set of choices can be reframed into leveraging the use of multiple evaluative models to support the decision process. That is, instead of expending efforts to find the "right" model, what can be determined by leveraging multiple different models in order to garner potentially novel insights, especially when "fit for purpose" may be unclear early in the system life cycle? (Ross et al. 2016).

Why might engineers be interested in using multiple evaluative models? On a basic level, running multiple evaluative models and comparing their results can support cross-validation of each model and increase decision-maker confidence in their results. The use of models to support early concept decision-making may require measuring the expected performance of new or emerging technologies that have yet to be built or tested. In this case, it is possible that no evaluative models are truly validated, leading to a situation similar to that of the value model trading problem, but instead of having no "ground truth" to validate against (since value is subjective), the designers simply do not know what that ground truth is. As a result, searching for alternatives that are robust to the unknown accuracy or precision of the models is a powerful use of multiple models.

Performing model trading, rather than model selection alone, can reveal two classes of insights:

1) insights into fundamental relationships between perspectives of value and what is possible (*structural patterns for the decision problem*), and
2) insights into modeling artifacts, both in how value is captured and how evaluation is performed (*modeling artifacts*).

The first class of insights sometimes appear to emerge through the course of analysis. This may be because the relationships are buried in the interactions between factors of the problem and are not readily apparent in our mental models. For example, Figure 5.5 shows the CBA value model scatterplot for a satellite system, evaluated using evaluative model #4 (combined new material and new speed model) (Ross and Rhodes 2018). Lines show Pareto efficient points connected to similar designs with different levels of fuel. For both the electric and biprop propulsion types, there is a positive trade-off of more fuel (= more cost, x-axis) for more benefit (y-axis). But counterintuitively the small cryo propulsion designs actually get less benefit for more fuel. This is because the added fuel actually decreases the speed of those small designs in spite of increasing the on-board delta-V. This is a consequence of the confluence of physics (i.e., the rocket equation and inertia) and expectations on what is considered beneficial. The very fact that the relationship between fuel mass and benefit plays out differently in different regions of the tradespace means that the complexity (in terms of number of factors to consider) likely would overwhelm an unaided human mind due to bounded rationality.

Each evaluative model is one representation of how a system might "perform," while each value model is one representation of how a system might "be valued." The emergence described above would occur at the intersection of each possible evaluative and value model, as well as across them. Systems engineers and analysts may benefit strongly by considering not only their choice of evaluative and value model, but also how their insights might vary if they were to include more than one of each type of model.

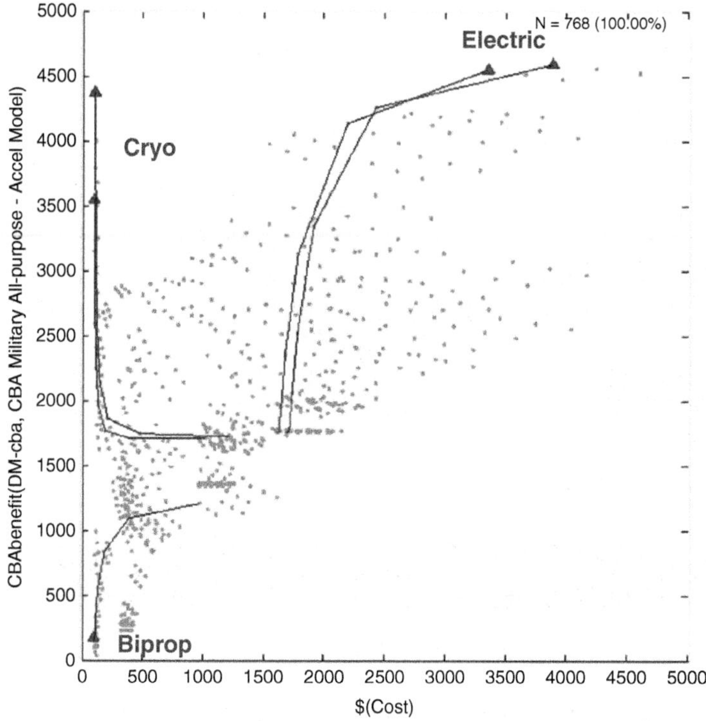

Figure 5.5 CBA value model scatterplot for evaluative model #4 (combined new material and new speed).

Model Curation

Managing multiple models, especially over a period of time, can become an increasing challenge in and of itself. Justifiably, *model curation* was discussed during the pathfinder workshop as a key future need within the systems engineering community. As digital engineering has evolved and matured, the need for curation has become even more important given that models are more often reused and shared within and across enterprises. Management of models and model portfolios is emerging as part of digital engineering practice; however, this is not typically done at an enterprise level. Model curation does not replace localized management of models, but becomes necessary as enterprises develop and acquire large collections of models. Reymondet et al. (2016) discuss a preliminary investigation into considerations for model curation in the engineering of complex sociotechnical systems, suggesting an approach for model curation for digital engineering can benefit from knowledge, practices, and experiences of the curation practice of other fields.

Curation applies to longer duration models, rather than those developed for a quick study or to simply work out a problem. As illustrated in Figure 5.6, two categories of longer duration models that necessitate model curation are (1) models used throughout the life span of a system; and (2) models to be intentionally reused for future purposes/contexts. Rouse (2015) stresses that the wealth of existing models is often not used because of a lack of knowledge of these resources and the difficulty in accessing them. Lack of access to models, mistrust of models, and perception of legitimacy of models are all barriers in model reuse and longevity. Various aspects of the future of model curation, especially the sociotechnical aspects, have been explored during the IMCSE research program.

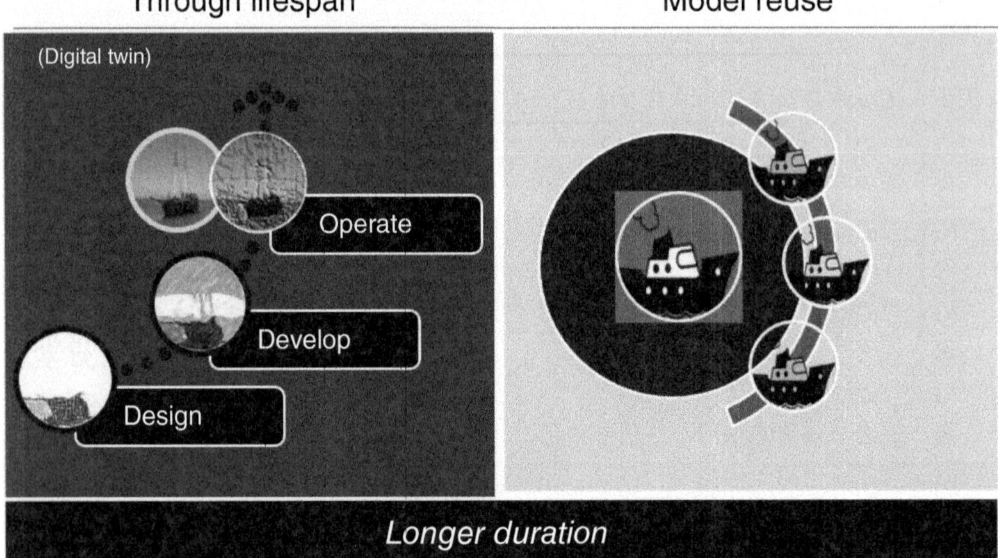

Figure 5.6 A curated model collection is comprised of longer duration models.

Model curation can be defined as: Model curation is the lifecycle management, control, preservation, and active enhancement of models and associated information to ensure value for current and future use, as well as repurposing beyond initial purpose and context. Not all models would be part of such a curated collection, but the most suitable ones will be selected for inclusion in the collection by a formal process called accession (Rhodes 2019, 2022).

A question arises as to whether a model curation function at the enterprise level could lead to more effective use of models and digital assets at all levels. Models exist at all levels of an enterprise (individual, program, business unit, enterprise) but rarely are these managed as an enterprise collection. The purpose and scope of enterprise-level model collections will vary from a collection within single enterprise for internal use only to externally governed collections (collaboratively developed) for open sharing.

Governance of an enterprise-level collection is essential but may take varied forms. Maturing an approach for model curation in the systems engineering field can leverage the work of other related curation practices, especially digital curation that offers adaptable practices. Individuals involved in model curation will need knowledge of many things, including model ontologies, model metadata, latest modeling techniques and classes of models, policies on data rights, code of ethics, and others. Skilled curators of a large model collection, similar to a museum curator role, could guide model consumers in selecting the most appropriate set of modeling assets from a repository for a given purpose (Rhodes 2019).

Curation practices promote formalism and provide for the management and control of models (and digital artifacts). Examples of curation activities include model identification, acquisition, accession, composition, evaluation, valuation, presentation, preservation, and archiving. The decision to place a model under enterprise-level curation will be driven by myriad factors such as importance of the model to a major program or project; level of importance to business strategy; need to control and protect enterprise intellectual property; need for retention of a model as objective evidence; and ability to control of multi-program access and/or multi-enterprise access.

Once a decision to place a model under curation is made, specific criteria will determine its readiness and acceptability. Our investigation identified an initial set of criteria for accepting a model into a curated enterprise model collection (accession), drawing from curation practice in other fields (Rhodes 2022). Five categories were identified as key to a making an accession decision: (1) relevance to enterprise and/or program mission; (2) economic business case; (3) completeness of metadata, data, pedigree, and documentation; (4) potential for redistribution, reuse and/or re-purposing; and (5) uniqueness of model/non-replicability.

As shown in Table 5.2, these criteria were further defined with sub-questions (adapted from digital curation best practice) to access the readiness to accept the model into the enterprise-level collection, in this case assuming a model had originated within that enterprise (Rhodes 2022). It is expected that enterprises would specify standard criteria most suitable to their enterprise mission and needs. Further validation of criteria is an area for future research.

A key benefit of model curation is that it increases the potential for effective reuse of a model on a future program, as the model and associated information (metadata, technical data, and pedigree) will be available to assess its fitness for purpose and context. Effective curation of models will have positive impact on model trust, credibility, and integrity. In addition to defined model curation implementation practices, innovative approaches and technologies may be beneficial in curation as digital engineering practices and infrastructure mature. As models become more valuable to enterprises, the question arises as to how to assess the value of a model. Potential approaches from other fields could provide consideration and approaches for valuation of digital engineering models and model collections. Significant work in the value of big data provides a foundation for work in the area, in addition to legacy approaches in valuation of institutional collections. In the future it is likely that acquisition and loan of models developed by one enterprise for use by another will increase, raising the question of how compensation for this would be determined. Further, as enterprises begin to build up substantial collections of models, the overall value of the collection itself will need to be determined (e.g., will be needed for mergers and acquisitions).

Further research is needed before the benefits of curated model collections can be realized. Three important aspects include governance, infrastructure, and the vision for model discovery in support of strategic use of the model collection.

Governance

The strategic value of the enterprise model collection will depend upon strong governance. The enterprise must appoint the enterprise authority who will be responsible for leadership and governance of the curated model collection. Some important facets of this include governance authority, model collection valuation, IP management, accreditation, and trust/credibility factors. Governance authorities will need to establish standard criteria and measures used to conduct, monitor, and control model curation practice. The enterprise must establish and enforce policies for model accreditation, model accession, model valuation, model deaccession, and model archiving.

Infrastructure

Supporting infrastructure for a model collection repository needs to be scalable and strategically managed to support the current and future needs of the enterprise. Aspects of infrastructure include security and protection, scalability of infrastructure, and enabling model sharing and remote access. Human interaction with the infrastructure will need to be designed for desired user experience, recognizing that there are likely many different types of users and varied scenarios for interaction based on the need and preferences of model consumers.

Table 5.2 Criteria for placing models under curation (criteria are adapted from work by DCC).[1]

Criteria for placing a model under curation (preliminary)	
Relevance to the enterprise and/or program mission	• Is the model relevant to the overall enterprise mission? • Is the model relevant to specific current or future program mission? • Does model (including metadata, data, model representation, documentation) fall within the model collection/repository's scope? • Are there legal requirements or guidelines that require placing the model under curation? • Is there authoritative evidence of current value to engineering field? • Is there future value in having evidence of the model's use/reuse?
Economic business case	• Does benefit of placing model under curation exceed required cost? • Has the total cost of retaining the model package over active lifespan been considered? • Has the funding source for model retention and performance of curation activities been determined and agreed upon? • Have security and safety been considered in the economic case? • Has cost of archiving model after deaccession been considered?
Completeness of metadata, data, pedigree, and documentation	• Does model documentation span the lifecycle phases during which the model was conceived, generated and used? • Is the model metadata, model pedigree information, and data pedigree complete? • Is there sufficient documentation to support sharing, access, and re-use of the model? • Is there sufficient information to judge the integrity and credibility of the model package? • Is there a sufficient set of data associated with the model to enable understanding and replication of model results?
Potential for redistribution, reuse, and/or repurposing	• Are there any IP issues, data rights issues, human subject issues or restrictions that are not addressable? • Is there evidence of model reliability and usability? • Does the model have evidence of verification and validation? • Is the model package complete (model, data, metadata, documentation, digital artifacts, etc.)? • Has the data been stored in a way that ensures its integrity has not been compromised? • Does the model meet standards and other technical criteria that allow its easy redistribution?
Uniqueness of model/ Non-replicability	• Is the model the only sole source of its content? • Can the model be easily replicated, recreated, or re-measured? • Is the cost of replicating the model financially viable? • Is there historic value and/or education value for future workforce?

Model Discovery

Enterprise model repositories are envisioned as comprised of a collection of model assets that are discoverable, retrievable, and potentially reusable. Some of the key aspects of discovering models in a model repository include approaches and technologies that support effective and efficient discovery of suitable models, searchable categorization of models, augmented model search, and decision-making. Model consumers need access to sufficient information to make decisions on making use of existing models in the collection. Model consumers must have access to information that allows them to judge the credibility of the model.

In summary, model curation is a broad topic that spans implementation of model curation practice; the roles and responsibilities of involved individuals and organizations; approaches to curate models for intended purpose and model consumer preference; and options for new technologies that enable curation and curating. This section of the paper has discussed a subset of findings from the research.

Five key insights of the research that are applicable to furthering the practice of model curation are:

1) Initial investigation of model curation has indicated the systems community will benefit from formal curation practices.
2) Lack of access to models, mistrust of models, and perception of legitimacy of models are barriers to reuse and longevity that are potentially mitigated through model curation.
3) There is potential to adapt proven practices from other fields once model curation-specific needs in the digital engineering context are clearly understood.
4) Model curation requires formally defined curation practices, as well as the resources for governance and supporting infrastructure.
5) Curation practice across the systems community will benefit from collaborative development of standard implementation enablers, such as process guidance, pedigree standards, curation templates, and many others.

Summary

In this chapter, we have shared the background and motivations for the IMCSE research program, and provided highlights of four selected studies (*model-centric decision-making and trust, multi-stakeholder negotiation through models, model choice and trade-off,* and *model curation*) from the larger research program. The research outcomes contribute to furthering the state of practice regarding various aspects of humans interacting with models and model-generated data. IMCSE aimed to develop transformative results through enabling intense human–model interaction, to rapidly conceive of systems and interact with models in order to make rapid trades to decide on what is most effective given present knowledge and future uncertainties, as well as what is practical given available resources and constraints. Ongoing and future research seeks to build on the foundations of IMCSE and transition research to practice in support of the field of digital engineering as it continues to mature.

Note

1 Digital Curation Centre, a world-leading centre of expertise in digital information curation with a focus on building capacity, capability and skills for research data management (dcc.ac.uk).

References

Curhan, J.R., Neale, M.A., and Ross, L. (2004). Dynamic valuation: preference changes in the context of face-to-face negotiation. *Journal of Experimental Social Psychology* 40 (2): 142–151.

Ehrmann, J.R. and Stinson, B.L. (1999). Joint fact-finding and the use of technical experts. In: *The Consensus Building Handbook: A Comprehensive Guide to Reaching Agreement* (ed. L.E. Susskind, S. McKearnen, and J. Thomas-Lamar). Sage Publications.

Fisher, R., Ury, W., and Patton, B. (1991). *Getting to Yes: Negotiating an Agreement without Giving in*. New York NY: Penguin Books.

Fitzgerald, M.E. and Ross, A.M. (2014). Controlling for framing effects in multi-stakeholder tradespace exploration. *Procedia Computer Science* 28: 412–421.

Fitzgerald, M.E. and Ross, A.M. (2016). Recommendations for framing multi-stakeholder tradespace exploration. In: *INCOSE International Symposium 2016*, Edinburgh, Scotland. July 2016.

Gass, S.I. and Joel, L. (1980). Concepts of model confidence. *Computers and Operations Research* 8 (4): 341–346.

Golkar, A. and Crawley, E.F. (2014). A framework for space systems architecture under stakeholder objectives ambiguity. *Systems Engineering* 17 (4): 479–502.

INCOSE (2015). Modeling and simulation, Chapter 9.1. INCOSE-TP-2003-002-04. In: *Systems Engineering Handbook: A Guide for System Life Cycle Processes and Activities*. 4e (ed. D. Walden, G. Roedler, K. Forsberg, et al.), 180–188. Wiley.

NASA (2008). Standard for models and simulations. NASA-STD-7009. July 2008.

Ozawa, C.P. (1991). *Recasting Science: Consensual Procedures in Public Policy Making*. Westview Press.

Raiffa, H. (2002). *Negotiation Analysis: The Science and Art of Collaborative Decision Making*. Belknap Press of Harvard University Press.

Reymondet, L., D.H. Rhodes, and Ross, A.M. (2016). Considerations for model curation in model-centric systems engineering. *2016 Annual IEEE Systems Conference (SysCon)*, Orlando, FL (18 April 2016). pp. 1–7. IEEE.

Rhodes, D.H. and Ross, A.M. (2016). A vision for human-model interaction in interactive model-centric systems engineering. *INCOSE International Symposium 2016*, Edinburgh, Scotland (July 2016), Vol. 26, No. 1, pp. 788–802).

Rhodes, D.H. (2018). Using human-model interaction heuristics to enable model-centric Enterprise transformation. *2018 Annual IEEE International Systems Conference (SysCon)*, Vancouver, Canada (23 April 2018). pp. 1–6. IEEE.

Rhodes, D.H. (2019). Model curation: requisite leadership and practice in digital engineering enterprises. *Procedia Computer Science* 153: 233–241.

Rhodes, D.H. (2022). Investigating model credibility within a model curation context. In: *Recent Trends and Advances in Model Based Systems Engineering*, 67–77. Cham: Springer International Publishing.

Rhodes, D.H. and Ross, A.M. (2015). *Interactive Model-Centric Systems Engineering Pathfinder Workshop Report*. Cambridge, MA: MIT.

Ricci, N., Schaffner, M.A., Ross, A.M. et al. (2014). Exploring stakeholder value models via interactive visualization. *Procedia Computer Science* 28: 294–303.

Ross, A.M. and Hastings, D.E. (2005). The tradespace exploration paradigm. *INCOSE International Symposium 2005*, Rochester, NY (July 2005), Vol. 15, No. 1, pp. 1706–1718.

Ross, A.M. and Rhodes, D.H. (2018). Interactive model trading for resilient systems decisions. In: *Disciplinary Convergence in Systems Engineering Research*, 97–112. Springer International Publishing.

Ross, A., McManus, H., Rhodes, D., and Hastings, D. (2010a). Revisiting the tradespace exploration paradigm: structuring the exploration process. *AIAA Space 2010 Conference & Exposition*, Anaheim, CA (31 August 2010). p. 8690.

Ross, A., McManus, H., Rhodes, D., and Hastings, D. (2010b). Role for interactive tradespace exploration in multi-stakeholder negotiations. *AIAA Space 2010 Conference & Exposition*, Anaheim, CA (30 August 2010). p. 8664.

Ross, A.M., Rhodes, D.H., and Fitzgerald, M.E. (2015). Interactive value model trading for resilient systems decisions. *Procedia Computer Science* 44: 639–648.

Ross, A.M., Fitzgerald, M.E., and Rhodes, D.H. (2016). Interactive evaluative model trading for resilient systems decisions. *14th Conference on Systems Engineering Research*, Huntsville, AL (March 2016). pp. 22–24.

Rouse, W.B. (2015). *Modeling and Visualization of Complex Systems and Enterprises: Explorations of Physical, Human, Economic, and Social Phenomena*. Wiley.

Shane German, E.S. and Rhodes, D.H. (2018). Model-centric decision-making: exploring decision-maker trust and perception of models. In: *Disciplinary Convergence in Systems Engineering Research*, 813–827. Springer International Publishing.

Biographical Sketches

Dr. Donna H. Rhodes is a principal research scientist in the Sociotechnical Systems Research Center at Massachusetts Institute of Technology. She is the co-founder and director of the MIT Systems Engineering Advancement Research Initiative (SEAri), a research group focused on advancing theories, methods, and practices for the engineering of complex sociotechnical systems. She is principal investigator for numerous sponsored research projects on human–model interaction, model curation, model-centric decision-making, and innovative approaches for enterprise transformation under the digital paradigm. Her research involves deep collaboration and engagement with government, industry, and other academic partners. She has been a principal investigator and collaborator in the DoD Systems Engineering Research Center (SERC). She teaches graduate courses, professional courses, and executive courses, and advises graduate students. Previously, Dr. Rhodes held systems engineering and senior leadership positions in industry. She is a past president and fellow of the International Council on Systems Engineering (INCOSE). Her contributions in the systems field have been recognized by numerous publication awards, INCOSE Founders Award, and industry awards. She received her MS and PhD in systems science from the T.J. Watson School of Engineering at Binghamton University.

Dr. Adam Ross is a research scientist in the Sociotechnical Systems Research Center at Massachusetts Institute of Technology and the President of Diakronos Solutions Inc. He is the co-founder and former lead research scientist for the MIT Systems Engineering Advancement Research Initiative (SEAri). Dr. Ross has published over 100 papers in the areas of space systems design, systems engineering, and tradespace exploration. He has received numerous awards, including the Systems Engineering 2008 Outstanding Journal Paper of the Year. He led over 15 years of research and development of novel systems engineering methods, frameworks, and techniques for evaluating and valuing system tradespaces and the "ilities" across alternative futures. He uses an approach that leverages techniques from engineering design, operations research, behavioral economics, and interactive data visualization, and is recognized as a leading expert in system tradespace exploration and change-related "ilities." He has helped both government and industry with applying analytic techniques for decision support and optimization for acquisition planning. Dr. Ross holds a dual AB in Physics and Astrophysics from Harvard University, and two SM (Aeronautics and Astronautics Engineering and Technology & Policy) as well as a PhD (Engineering Systems) from MIT.

Part II

Executing Digital Engineering

Jon Wade

Chapter 6

Systems Engineering Transformation Through Digital Engineering

Jon Wade

University of California, San Diego, CA, USA

Systems Engineering Transformation

Systems Trends

We are currently entering The Fourth Industrial Revolution (Schwab 2015), which is the ongoing automation of traditional manufacturing and industrial practices, using artificial intelligence (AI), machine learning (ML), and smart technology. The capabilities of AI and ML are increasingly blurring the lines between human and machine. Large-scale machine-to-machine communication (M2M) and the internet of things (IoT) are integrated for increased automation, improved communication and self-monitoring, and the production of smart machines that can analyze and diagnose issues without the need for human intervention. Revolutions in production technology fundamentally change society and its human development and support requirements as reflected in past industrial revolutions. The global community is calling for more attention to how systems can positively contribute to our social condition and natural environment to help advance our quality of life. A systems approach is necessary to address the social, environmental, and economic sustainable issues with our systems solutions. The following are some critical trends that are pushing current systems engineering practices into obsolescence in many critical domains and applications.

The first trend is that systems are becoming increasingly more complex in both number of components and interconnectivity. Decreasing numbers of complex systems are "stand-alone," but are instead connected to other complex systems. Not only is there an exponential growth in connectivity and interdependencies, but many of these are hidden. What once might have been a local issue is now a globally networked challenge. The examples are many. Systems of systems, and in particular Network Centric Services, have become the norm rather than the exception. Successful engineering of these systems may require understanding the technical, social, political, economic, behavioral and environmental implications to develop a system that best serves the relevant stakeholders. Quite often these are complex adaptive System of Systems (SoS) in which direct control may not be possible such that influence and the creation of a reward system to guide self-organization become the rule for the system "designer." Compounding this challenge is the need to interface with legacy systems of all shapes and forms, which is noted as the fifth trend.

Systems Engineering for the Digital Age: Practitioner Perspectives, First Edition. Edited by Dinesh Verma.

Companion website: www.wiley.com/go/verma/systemsengineering

The second trend is our increasing societal day-to-day dependency on technologically advanced and powerful, yet unpredictable SoS. The resulting impact is that these systems are no longer a matter of convenience, but instead are necessary for our daily existence. There are numerous examples of how the application of the Internet and web technology has changed our perception of how work should be accomplished – i.e., in a collaborative, distributed, evolvable manner. There is a business-to-business (B2B) web that has taken advantage of the communications medium and created extremely complex interdependent business processes and rules. There is the rise of cyber-physical systems, including everything from smart automobiles to environmental controls, to net-centric warfare. There are social networking sites and massive multi-player games. This new paradigm of interacting systems, services, and users has fundamentally changed the way systems are conceived, developed, deployed, managed, and retired. It has been the driving force behind the creation of the SoS. The notion of Systems of Systems was not created, but rather evolved organically as a technological response to the desires and needs of customers.

The third trend, which compounds the impact of the first two trends, is that a combination of customer expectations and competitive demands has greatly compressed the development and deployment lifecycles of products and services. The result is that while the complexity and criticality of new systems is increasing exponentially, the time to develop and deploy them is decreasing. It is important to differentiate between the development and the evolution of the distributed information technology (IT) network and infrastructure, and the evolution of application layer functionality. The architecture and infrastructure of the former might have a much longer life cycle than the applications which it supports.

The fourth trend is related to security. Our increasing dependence on networked systems has greatly increased their value as a target while the increased complexity and interconnectedness increases their vulnerability. Security has to be of central importance in the design and deployment of systems, particularly when they themselves are composed of systems that are unreliable and independently evolving.

The fifth trend, which relates to the first trend of complexity, is that so-called greenfield development (development that is completely new, with no legacy issues to address) is becoming the exception rather than the rule, particularly in netcentric systems that must increasingly work with legacy systems. In addition, these legacy systems are often ill-planned and unsuited for their future missions. This is not necessarily the result of poor planning, architecting, design, and execution, but rather often is the result that their future usage was unforeseen at the time of their design. The extension of a system's service life well beyond the original plan due to the significantly increasing cost and time for a replacement exacerbates the problem further. Examples of this include the now projected life of the B-52 aircraft to be 90+ years, the Aegis Combat system is expected to be 50 years instead of the original 30 years, and the existence of legacy software (e.g., Cobol programs) that is still in existence well beyond their anticipated end of life. The third trend of time compression amplifies this issue and accelerates the aging process of the legacy system infrastructure.

The final trend is found in the workforce called upon to conceive, create, and manage these complex systems – a workforce that has changed over time as much as the technological environment around it. As noted earlier, the line between human and machine is increasingly blurred, such that the workforce of the future needs to be comfortable both collaborating with other humans, and with machines with increasingly intelligent capabilities. Perhaps as a result of having grown up in a networked age with almost instantaneous feedback, our technical workforce is increasingly more interactive and experiential, while being more comfortable with change and a lack of comprehensive knowledge about a system. This new workforce often is more concerned with how things behave, rather than how the work. One result of this trend is that systems engineers and management are being trained with subject matter that may not be relevant to the challenges that they face in the field. Those who are formally trained in traditional systems engineering may find themselves in a fundamental mismatch with customers and markets because of the compressed notions of time and the newer methods, processes, and tools. Who is willing to perform, or pay for a thorough requirements analysis, when it is likely to be outdated (or at least perceived to be outdated) before it is completed? There are also the challenges of a geographically, functionally, and culturally distributed workforce.

Transformation of Systems Engineering

As noted by Snowden and Boone (2007), the move from the domain of the complicated to that of the complex requires a very different approach, in which the scientific approach of experimentation and adaptation becomes ever more important. The net result is that systems engineering needs to move from a "ballistic approach" in which the trajectory of a system's development and use is carefully calculated, to one of an intelligent device which Observes, Orients, Decides, and Acts following the cycle coined by Boyd (1976) as the OODA-loop. Boyd noted that to be successful, one must have the ability to move through this loop more quickly than their adversaries. The ramification of increasing degrees of complexity, and the other trends noted above, result in the transformation of systems engineering from a relatively open-loop discipline (SE 1.0) to a closed-loop form (SE 2.0) as noted in the table below.

Systems Engineering Gap Analysis

The following discusses the gap analysis of the capabilities that need to be provided in systems engineering methods, processes, and tools.

Systems Engineering Framework

Systems development, or just about any development, consists of the following four major activities as shown in Figure 6.1 below.

The lower activities can be classified as being associated with the problem or application space, while the upper activities are related to the solution. The left-hand activities tend to be high level and abstract, while the right-hand activities lower level and more tangible.

- **Value:** This includes understanding an environmental context, understanding the factors that influence the creation of value, and discerning how value may be created within it. Quite often,

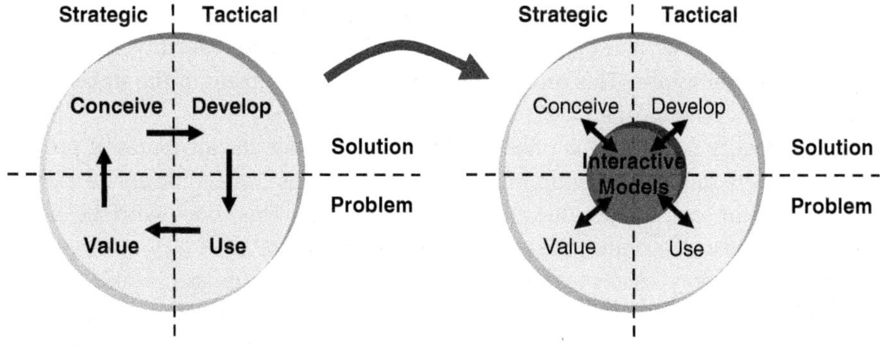

Figure 6.1 System life cycle activities. *Source:* Adapted from Wade et al. (2010).

this is the domain of executive leadership within an organization. In some organizations, this function is resident in marketing, in others it may be in sales or engineering. To be successful, whoever makes these decisions should understand the total value proposition, which includes customer needs, not just customer "wants," and how satisfying them brings value to the organization.

- **Conceive:** This includes the creation of a conceptual or abstract design that creates value. Architects and marketing may be involved in this activity.
- **Develop:** This includes the design, manufacture, and whatever else is necessary before the system can be used. This activity usually includes engineering and manufacturing.
- **Use:** This includes the distribution, deployment, use, maintenance, and eventually retirement of the system. This activity includes the entire value supply chain from sales, to manufacturing, distribution, service and, of course, the end user.

All of these activities must be considered in the system life cycle. All of these should be in the realm of the system engineer. Traditionally, these phases are executed in sequence with oral or textual documentation being used to communicate across the interfaces.

Value, Project, and Environment Attributes

In general, there are no single "best practices" independent of the value and particulars of a project and the environment in which it is being developed and deployed. It can be argued in complex systems in which the relationships between cause and effect may not be well understood, there are not even best practices (Snowden and Boone 2007). It is necessary to create a framework by which the most appropriate practices can be selected based on these criteria.

The following is a set of attributes that can be used to characterize an SE application to help determine the relevancy of an MPT. These attributes can be divided into the following three categories:

1) Stakeholder Value Attributes,
2) Project Attributes, and
3) Environment Attributes.

For an method, process, or tool (MPT) to be deemed appropriate, there should be an acceptable match in each of these categories.

Stakeholder value: The first category relates to the overall aggregated utility for the entire system. For the Intelligence Community or Military Systems, this might be the value of new capabilities delivered to the field. For commercial systems, this might be represented by some combination of return on investment (ROI), market share growth, strategic advantage, and customer satisfaction. These attributes include the value of schedule, cost, features, capacity, reliability/availability, maintainability, upgradeability, etc. This category determines the alignment of the stakeholder values and the potential strengths of an MPT.

Project: The second category relates to the project itself. These include the attributes of project size, complexity, duration, stability of requirements, etc. In some sense, this could create a set of relative risk/cost levels for each of the value attributes noted above. This, along with the alignment of the value attributes, determines the potential benefit of the MPT.

Environment: The final category relates to the environment in which the project is taking place. These attributes might relate to the ability to share knowledge between those who are responsible for the Valuation, Conception, Development, and Use of the product/service, as well as flexibility of the organization to change its operation and processes (along with its customers and supply chain), and its human and financial resources. This category can be used to expose barriers that might make a particular MPT inappropriate for a particular organization or environment.

This breakdown is similar to the one presented by Stevens (2008), which features separate contexts for Stakeholders, System, Strategic (mission and scope), and Implementation.

Attribute Impact

In this analysis, each of the attributes in the areas of Value, Project, and Environment are pushed beyond the home ground for traditional plan-driven approaches and the resulting dominant failure mechanisms are identified as the gaps that need to be addressed. An information processing paradigm will be used to represent the necessary work flow that results in the transformation of information through the four activities of Value, Conceive, Develop, and Use. If there are limited number of people involved in understanding the value proposition, creating a concept design, developing, and then using the system, the communication challenges would be minimized. However, while the use of a large number of people with a broad range of specialized skills greatly increases the scale of what can be accomplished within a given timeframe, it also creates a communication challenge. In the case of parallel computation, increasing the number of processors reduces the total time spent computing, but increases the communication overhead. At some point, the communication overhead becomes dominant and the fallacies of the mythical man month are made evident. The gaps noted below are classified as being communication or computation related. The computation related gaps are those in which the complexity of the problem is beyond a human's capabilities to successfully cope with it. Finally, there is a challenge in latency reduction, which generally drives the use of automation to reduce the amount of human effort that needs to be expended to accomplish the task in question. Thus, the methods, processes, and tools provide communication, computation (for analysis and decision-making), and automation to provide agile development and rapid fielding capability.

Value

Emergent solutions vs. project predictability: There is a clear trade-off between the desire to allow for spontaneous innovation in a program and having a predictable outcome. Clearly, plan-driven approaches are best suited to foster predictable results, but these may well not be the results that create the most value. However, it may be possible to create an environment and supply the methods and tools which reliably provide innovative results, although it is not known a priori what these results may be. In general, the greater the potential reward, the greater the project risk. This is likely to be very unsettling to those who are uncomfortable with uncertainty. There are two challenges here. The first is the ability to determine the value of the risk and reward factors. The second is the ability to successfully communicate the current status of the development and the risk to project predictability. It is necessary to satisfy both of them if one is to be able to rationally determine how to balance these opposing forces. The former is the area of greatest gap in tool availability.

Rapid value creation: An iterative approach, with timesteps on the scale of what is considered to be "rapid" and development of sufficient degree to constitute value, is clearly necessary to achieve this goal. The primary challenge is to reduce the latency through a life cycle including all the work necessary to deliver a system at the required quality level. The reduction of latency will require improved communication, analysis, and automation capability. Achieving this requires a complete tool suite supporting all three of these capabilities.

System criticality: While plan-driven approaches have traditionally been employed to reduce risk in project failure through a hierarchy of review cycles, there are many ways of achieving this goal by less human and time-intensive means. Rather than providing validation and verification at the end of a design cycle, it should be applied throughout the development process. Techniques such as Test Driven Design (TDD) and Continuous Integration are critical to achieving high-quality systems within a minimum amount of time. Analysis tools, such as code coverage tools along with determination of the potential impact in the exposed areas, are necessary to assist in making the appropriate risk vs. time to deployment decisions.

Maintainability, upgradeability, and extensibility: These attributes can generally only be achieved through the development of a suitable architecture, and support and upgrade plan. The use of a Service Oriented Architecture or Product Line infrastructure can assist in this effort such that the desired services can be deployed without having to independently create a new infrastructure. The need for such an architecture is independent of the type of developmental methods that are used to create it. Tools can assist in the process of creating such an architecture and upgrade plans, but there is no substitute for its existence.

Project

Size and complexity: Size and complexity generally determine the number of personnel that are necessary to staff on a project. This has a direct impact on the amount of communication that needs to take place in the project, which directly affects the ability for a method to scale. This is a major limiting factor for agile methods that rely on informal information exchange between capable individuals. For example, it has been noted that the method of Scrums is generally limited to teams of ten individuals or less and that there is no conclusive evidence that "Scrums of Scrums" works effectively (Carrigy et al. 2009). The major challenge is one of communication. In addition, analysis and automation capabilities can be used to reduce the effective scale of the project, reducing the number of personnel employed and thus the communication overhead.

Dynamism: % requirements change per month. There are a number of contributing factors that enable requirements churn that disables projects. The first issue is that the change in requirements needs to accurately reflect what is necessary to create the desired additional value in the deployed system. This requires that there is a means to accurately interpret the needs of the customers and transform these into a system concept along with the related requirements. Often times the customer may not know what they want until they see or experience it. The use of rapid prototyping, either virtual or physical, may be necessary to accomplish this goal.

The second issue is that changing requirements result in the need for communication and stabilization time such that everyone on the project is made aware of and understand the impact of the new requirements or direction. If the time between requirements change is less than the communication and stabilization time, then the project will cease to make forward progress. There are two ways in which this problem can be addressed. The first is to ensure that the communication and stabilization time is well understood and changes are not made more frequently than can be handled by the system. This requires appropriate policies and skills in project management. More significantly, the communication and stabilization time needs to be decreased. This can be achieved through a variety of processes and tools. For example, only those who need to be notified of the changes should be notified. This reduces churn to the absolutely minimal amount. Next the changes need to be conveyed in the most meaningful way to each person on the program so that they can quickly determine its impact and act accordingly. Again, tools can be used to accomplish this. This is an area in which model-based capabilities with automatic notification and advanced visualization are generally far superior to text-based requirements methods.

Legacy: Systems with substantial legacy constraints often impose great challenges in the area of system integration and ultimately realized system performance and value creation. These are challenges regardless of the type of development method that is being used. However, it may well be that there are rapid changes in the legacy environment outside of one's control which effectively results in dynamism in requirements. In this case, instrumentation of the legacy system may be necessary to properly characterize it to provide the information necessary to determine the requirements for the deployed system to deliver the desired value. Model-based development may be required to facilitate the translation of this information into the development process.

In addition, flexibility in the systems operational environment which includes the interfaces, application modularity, and human interfaces and governance might well dictate the rate at which the system can be changed and still achieve its desired effects. These issues tend to reduce the agility of the overall system. Some of this can be mitigated with improved user training tools and facilities. Interactive simulations of the actual use experience are generally superior to text-based documents.

Environment

Developers: The overall competency level of the developers and their understanding of the basis of project value have a large impact on their ability to work independently. Often agile methods depend on informal communications between a small number of developers, which can cause this to be a major limiter on the applicability of these methods with less capable teams. Team of teams organizational approaches (McChrystal et al. 2015) are much more amenable to agile, distributed development than traditional hierarchical organizations. Less capable teams, all else being equal, are generally less agile, but this can be partially compensated for with appropriate processes and tools. For example, while text-based documentation may not be effective to represent complex systems, model-based representations might provide the desired degree of information more intuitively to the user.

Customers: Of particular importance are dedicated, collocated CRACK (Collaborative, Representative, Authorized, Committed, Knowledgeable) performers (Boehm and Turner 2003). Understanding the value proposition for a system is critical to its success. In agile development processes, which depend upon informal communication, it is critical that these are dedicated, networked individuals such that communication can be continuous through the development process as the work is being determined on a very fine-grained basis. However, it should be possible to increase the granularity of this work specification up to the size of that actually being deployed. This will require more formalized means of communication. Again, it is critical that the intent of the changes is accurately communicated to each member of the development team and validated with the customer. While face-to-face communication is obviously best, technology (such as high-definition tele-presence systems) can be used to enhance the capabilities of remote communications. Model-based systems that can produce multi-modal, multi-sensory feedback are most useful for this. However, the competency of the customer or the proxy for the customer to convey the needs of the system is a critical factor for which no amount of methods, tools, or technology can provide a substitute. In some cases, it is best to validate the desired system with a representative set of actual customers, which may require prototype development. In any case, the understanding of customer behavior can be a limiting factor for both plan-driven or agile development methods.

Level of trust: Trust is a vital component of any high performing team. In fact, it may be seen as the foundational element (Lencioni 2002). Trust and empowerment are especially important in agile developments, which rely upon informal means of communication and personal initiative. However, one could argue that transparent processes, which enable rapid feedback, are able to increase trust as this can be based upon verifiable fact. Tools and processes can be created, which facilitate the rewards for teamwork and remove uncertainty in the state of deliverables which reduces the level of mistrust. Trust is generally the precursor to empowerment. Ultimately lack of trust and empowerment will have negative impacts on any development, whether plan-driven or agile.

Communication capability: To some extent this determines the size of the effective team and includes the ability to communicate within and between marketing, architecture, development, test, service, sales and customers. Communication depends upon trust or else it is not possible to have the necessary constructive conflict (Lencioni 2002). Many agile methods depend on informal communication, which involves voluntary communication between a variety of parties. Unless the cost of communication is low, it is unlikely to happen and the agile process will not be effective. Rather than depending upon informal communication, it is possible with processes and tools to create the means of structured communication, which is a by-product of the development process.

Culture: Does the culture thrive on chaos or order? It is generally believed that cultures that thrive on chaos are well suited to agile development and those that do not need to work within a plan-driven environment. However, one can certainly find plan-driven environments that are chaotic and agile ones that are disciplined. Agile environments in which methods and tools are employed to give personnel instant feedback on their progress produce fewer surprises and thus can be very orderly environments. Plan-driven environments that deploy a waterfall process that results in integration surprises that have been months or years in the making can be extremely gut wrenching and chaotic. Agile environments generally require that people are thoughtful, innovative, and are open to change. However, these are generally the characteristics of productive workers in any discipline. It should be noted that an organization will need to be willing to accept change if they are to take advantage of the results of a transformation of systems engineering.

Additional Areas of Opportunity

In addition to the areas mentioned above, there are other opportunities to improve the capability to rapidly deploy effective systems. Many of the areas noted above relate to the ability to communicate, analyze, and automate development activities to improve the quality of the life cycle while compressing its duration and cost. However, more can be done. For example, the flexibility of the system to adapt to future change is critical to enable the rapid addition of new features and capabilities once the system is deployed. In addition, the system could be engineered to be intelligent such that these changes can be made to be self-adaptive. Likewise, the system needs to be developed taking advantage of emerging technologies and subsystems while determining when to obsolete existing ones. It is critical to determine how best to ensure that the system provides the necessary levels of service availability and security while retaining its flexibility. Finally, it is necessary to appropriately incorporate the human element into the system ensuring that the strengths of human capabilities are leveraged and integrated with the strengths of technology.

Summary of Gap and Opportunity Areas

The following is a compilation of the gap areas described above:

- **System requirements:** creation, validation, prioritization, resolution of conflicting requirement, managing, changing, and emerging requirements and decision-making; in particular the creation of a collaborative environment that facilitates trade-off resolution and creation of a mutually understandable description of the desired system concept
- **Low-overhead communication:** the ability to provide the essential communication to keep a large organization synchronized throughout a system lifecycle with a minimum amount of overhead to provide scalable agility
- **Architectural design support:** processes and tools to support the development of an architecture, which can support the attributes of maintainability, upgradeability (flexibility), and extensibility, along with reliability, availability, security, and other emergent properties
- **Risk/opportunity management:** tools that can assist in the assessment of program risk and value creation to allow for the proper trade-offs between these competing goals based on the capabilities of the organization and the challenges of the system under development
- **Verification and validation:** an integrated set of processes and tools that can provide verification and validation throughout the lifecycle process
- **Legacy integration:** the capability to monitor and characterize the current legacy system to ensure that the addition of new applications and services have the desired capabilities, and the ability to integrate independently evolving components into a larger interoperable system
- **Human aware/self-adaptive:** the capability to optimize the use of humans in the system to take advantage of self-adaptive human capabilities

The following is a compilation of the opportunity areas described above:

- **Complexity handling capabilities:** tools and techniques that leverage technological advances in computation, visualization, information technology, and communication to provide systems engineering with the capability to manage ever-increasing system complexity thus keeping SE on the curve
- **Cycle time reduction:** a suite of processes and tools, including those noted above, which can increase the quality of the systems while compressing latency through the life cycle; these include tools that not only accelerate new development but also eliminate unnecessary work by facilitating reuse and providing correct by design construction

The digital transformation of systems engineering requires understanding the gaps between traditional MPT capabilities and the needs of future systems. However, simply incrementally evolving existing SE concepts is unlikely to dramatically increase its relevance and put it on the technology curve. True transformation requires creative thinking and significant innovation. The section titled "Transformation Through Digital Engineering" characterizes a new paradigm for SE driven by digital engineering. It presents innovation objectives and philosophy, articulates an innovation concept, and recommends critical areas of research to enable the necessary paradigm shift.

Transformation Through Digital Engineering

Foundation of Digitalization

The primary means of transforming systems engineering will be to use the same elements that are creating the systems challenges: namely technology and human capabilities. The capabilities to be leveraged are the rapid advances in computation, visualization, communication, and information technology. While the end may be in sight for Moore's Law, silicon technology capabilities have experienced an exponential rate of number of devices per chip for the past 50 years (Rotman 2020). This capability supports simulation-based systems engineering approaches in a number of areas including functional simulation, verification and validation, design analysis and synthesis, and data-mining operations. Visualization capabilities have exceeded Moore's Law rates in the recent past due to huge market demand in games and entertainment. These visualization capabilities are a critical element to enable humans to better understand hugely complex systems and make appropriate decisions regarding their capabilities, architecture, design, and implementation. Information technology, particularly the ability to analyze data and present information in a coherent way, is a necessary tool for navigating the huge amounts of data that are created during the development or use of a system, particularly one that has been instrumented for feedback.

Finally, technology is required to provide effective communication among a diverse set of stakeholders, architects, developers, and support personnel both synchronously and asynchronously. While the capabilities of individual humans are not evolving rapidly, our ability to most effectively leverage these assets with technology is critical. Garry Kasparov has noted (also known as "Kasparov's Law"), "I reached the formulation that a weak human player plus machine plus a better process is superior, not only to a very powerful machine, but most remarkably, to a strong human player plus machine plus an inferior process" (Kasparov 2017). In addition, we need to provide the means to effectively form and operate in cooperative and competitive organizations to support an effective team of teaming approach. All of this must be achieved if systems engineering is to ride the same curve that is driving the complexity of the systems that it is developing.

New Paradigm for Systems Engineering

Meeting the challenges presented by the critical system trends, addressing the gaps identified in the section titled "Systems Engineering Gap Analysis", and transforming SE into a successful, relevant and timely discipline, requires a new SE paradigm. Within this new paradigm, digitalized SE must be:

- **Agile**: Allowing for quality, timely development with an incomplete and changing set of system requirements.
- **Integrated**: Part of the main development process and not an additional set of discretionary tasks.
- **Efficient**: Providing the greatest amount of benefits with the minimal number of steps and least amount of effort.
- **Leveraged**: Enabling exponential capability growth through the leveraging of computational, visualization, communication and information technologies, and prior systems experience.
- **Extensible**: Providing the ability to expand and enhance capabilities for future growth without having to make major changes in the infrastructure.
- **Deployable**: Enabling widespread impact through workforce education and broad application.

And yet, it must be sufficiently rigorous to ensure our systems are thoroughly engineered and will work as intended, satisfying the stakeholder's needs and vision.

Some particular areas of systems science need to be developed to address the emerging systems challenges. These areas include the architecture, design, and sustainment of:

- **Dependable systems:** which includes security, availability, reliability, and resilience
- **Evolving and self-adaptive systems:** which are flexible and can efficiently and effectively be externally adapted or self-adapt to address changing environment and mission needs
- **Enterprise, systems of systems:** which include the governance support and means of influence to manage systems that cannot be directly controlled.

Innovation Objectives and Philosophy

As noted earlier in this report, in general there are no single "best practices" independent of the value and particulars of a project and the environment in which it is being developed and deployed. However, rather than develop a set of methods, processes, and tools that are tailored to a specific profile, this chapter describes areas in which digital engineering can be generally applicable to systems engineering projects of all types, but with a particular focus on addressing the aforementioned six critical systems trends for relatively critical, complex systems that are not sufficiently addressed by traditional system engineering practices.

The intention of this transformation is to do for systems engineering what Carver and Meade were able to do for integrated circuit design (Wade et al. 2020). The methodologies they developed have stood the test of time and have supported integrated circuits design for over 40 years and appear to be sufficient for the known future of digital integrated circuits. The intended digital transformation of systems engineering should be extensible over a similar period of time.

The approach taken is to focus on the innovations that provide nonlinear advances over the state of the art as incremental improvements are not believed to be adequate to address the emerging challenges. Thus, this approach focuses on leverage in two major areas:

1) the inherent capabilities of computational, visualization, communication, and information technology
2) the unique capabilities of the human mind

For the past 40–50 years, and for the foreseeable future, the technologies in the first area have been improving at an exponential rate, whereas the capabilities of the human mind have been relatively unchanging. Advances in the second, human side of the equation have been largely achieved through specialization – allowing the application of more human minds to a problem, and the development of tools, which extend human capabilities.

There are a number of differences between the capabilities of humans and technology, but the major differences are that computational technology is unsurpassed in its ability to multitask and operate with blinding speed with great precision and accuracy within well-bounded contexts. Humans are poor at multitasking and performing quick, accurate calculations, but do extremely well in loosely bounded problems making decisions based on imprecise information. Staying on the technology curve requires the leverage and integration of the underlying technologies driving systems complexity into the SE process. Given the complexity of the systems of interest, a necessary approach is to instrument them, collect the information, and use technology to enable humans to understand their emergent behavior. Computational and information technology should carry as much of the Value, Conceive, Develop, and Use load as possible, as well as the underlying

OODA-loops that support it. Effective use of technology is an essential means of tightening the feedback and reducing the delay through this loop of activities. Human interaction should be limited as much as possible to areas in which only human knowledge and capabilities provide unique value. In this space, technology should be used as a tool to assist in the human creative and decision-making process.

Rather than start with a predefined process and then focus our efforts on how tools might support these, we will look at how to optimize the combined capabilities of man and machine and develop supporting processes based on sound SE principles. These processes may then be integrated into larger existing processes or methods. It may well be that this results in disruption, but that is often the nature of technical breakthroughs. It is quite extraordinary that the current SE practices have evolved over the past half century without any significant disruptions.

In summary, we will describe an integrated, yet modular set of innovations that will enable the optimal use of technical and human capabilities to improve general SE practices while focusing on the emerging critical systems challenges. Reuse, leverage, and sustained tool development are critical elements for putting SE on the curve.

Digital Transformation Concept

The primary goal of SE is to ensure that systems are developed, which can sustainably create value. In traditional SE practices, the value proposition is developed and communicated through static written documents, if at all, usually to a separate set of people who then create a concept of operations as shown in left side of Figure 6.2. This is generally also done through a static set of written documents. This information is then passed on to another group of people who are responsible for developing the systems through the creation of an architecture, design, and implementation. Again, the communication is often document based and typically accomplished through requirements and requirements traceability. Finally, the system is deployed and used. Human interaction with the system provides a huge amount of additional complexity and uncertainty in the operation

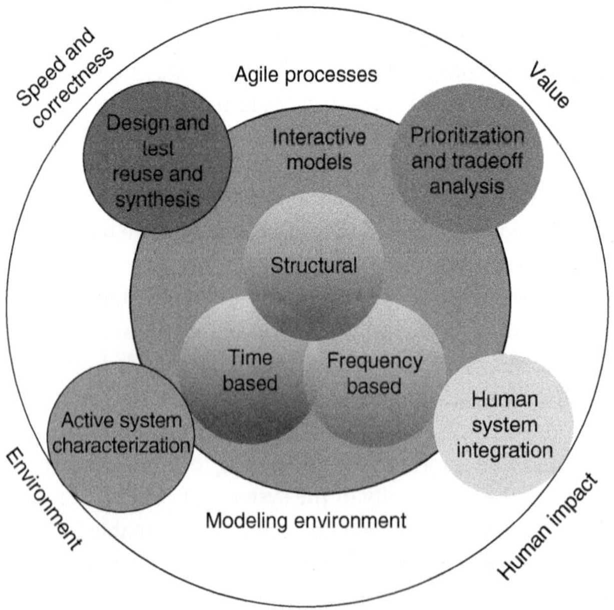

Figure 6.2 Elements of digital transformation.

of the system and its ability to create value. Determination of whether or not the appropriate level of value is created closes the loop, which impacts subsequent activities. While specialization allows many more minds to work on the system, thus amplifying human capabilities, it also results in great challenges in communication and the development of a shared mental model of what is attempting to be accomplished.

Overview

The proposed concept, as shown in right side of Figure 6.1, is one that we call Interactive Model-Centric System Engineering. In many ways, this approach is inspired by the automation and extensive use of computation, visualization, information, and communication technologies used in digital engineering. The critical attributes are that each activity of the lifecycle, and communication between these activities, are accomplished with an optimal mix of technology and human capability, through the interactive use of a consistent set of data and models, with visualization that is optimized for the particular user. The use of graphical models has the potential to provide consistent meaning to all the stakeholders and bridge these gaps.

The three key elements are:

- **Interactive**: Interactive, iterative design, execution, and re-design
- **Modeling:** The representation of information that can be processed by computation for analysis and on-the-fly simulation
- **Multi-modal/sensorial visualization:** The visual representation of information, personalized to the needs of the user

The adoption of an interactive model-centric system lifecycle model provides for the following necessary transformations necessary to move from SE 1.0 to SE 2.0 as shown in Table 6.1. In addition, tools and technology, which include models and simulation, can be used to:

- Facilitate understanding and decision making
- Improve development efficiency
- Automate processes

At the front end of the lifecycle, much of the effort is in the area of very abstract, multi-dimensional analysis and decision-making across multiple domains. This is the area in which the human mind can be productively applied if it can be given the appropriate means of viewing the

Table 6.1 The transformation of systems engineering.

Systems engineering 1.0	Systems engineering 2.0
• Systems built to last	• Systems built to evolve
• Opinion-based decision making	• Model and data-driven decision making
• Paper-based documentation	• Model-based documents
• Deeply integrated architectures	• Modularized architectures
• Hierarchical organizational model	• Ecosystem of partners
• Satisfying the requirements	• Constant experimentation and innovation
• Phase-based verification and validation	• Continuous verification and validation
• Separated development and operations	• Integrated development and operations

critical attributes and interacting with digitalized enable data analysis and simulation, per Kasparov's Law. This is likely to be too broad of a space for pure design automation. However, it is an area in which visualization, and other multi-media, multi-modal means of presenting information can greatly improve the ability of human to perform trade space analysis and decision-making. As the system representation is transformed and refined through the lifecycle into less abstract and more concrete information, increasing design efficiency becomes a major focus. Finally, as these representations start to fit into established patterns and technology, automation can be applied.

In the four activity life cycle process, much of the human contributions to add value occur at the front-end of the process, in the strategic areas of value recognition and concept creation. In these areas, tools can be used as a means to capture ideas and concepts, and translate these into representations, which can be recognized by a variety of stakeholders and contributors. As such, tools can be used to facilitate the iterative creation of shared mental models and provide a means by which to evaluate their behaviors and attributes to facilitate decision-making. Tools can be used to assist in the creation and analysis of architecture to help transform complexity into visualizations of emergent properties. Tools can provide developers with the ability to ensure that their designs can be successfully integrated to form a system that is in compliance with and supports the envisioned value proposition. These tools can be used to automate the development of system verification and validation. Finally, these tools can assist in validating the desired usage models and support the training of personnel necessary for deployment.

It is critical that the models used in this lifecycle are capable of supporting the required level of detail and fidelity for each of their users while being integrated for use through the entire life cycle. Limitations in their representational detail, views (for different users), and ability to be integrated and updated for consistency will greatly restrict their value. Thus, these are required characteristics. While all of the capabilities need not appear at once, the overall value of the tools is increased as additional tools are added to the suite.

Elements

The critical elements for innovation and research necessary to transform systems engineering to meet the emerging systems challenges are shown in Figure 6.2. These elements reduce the time required to go from idea to deployment and increase the operational effectiveness by providing closer coupling between the system solution and its usage, thereby increasing the impact of the deployed system.

At the core of this concept is the existence of interactive models. These models can be used to represent a system over the entire lifecycle, with a particular emphasis on providing the means to support a single, consistent system view to all involved stakeholders, including developers, service and support, and users. These models are intended to be both hierarchical and integrated such that various portions of the system may be modeled at different levels of precision. These will consist of state-based, capability-based and structural models. The three different types are described below.

State-based models: The first approach is to use state-based simulations to determine that the system has the correct functional behavior. This requires a set of models that can be executed sequentially to show causal, temporal relationships. This is the most common use of executable models and is the object of study for most mathematical treatments of model-based systems engineering (Wymore 2004). These models can be used to not only ensure correct functional behavior, but they can also be used to perform stress testing and fault recovery through actual or synthetic loads, and interface analysis and testing.

Capability-based models: Analysis can also be done in the "frequency" domain (capability-based), rather than the time domain (state-based) used for functional analysis. In this domain, the frequency at which certain types of activities occur is characterized and then analyzed to determine how the system responds to an ensemble of these inputs in the aggregate. In this way, time and frequency domain analysis has an analog to that used in understanding communication and signal processing systems. This type of analysis can be used to understand the performance and reliability characteristics of the system.

Structural models: Another type of models are structure based, which can be used to analyze the system with respect to its interconnectivity and aggregate behavior, independent of sequential functional behavior. For example, structural analysis can be used to better understand the dependencies of the system without the need for stimulus. An analog to this in integrated circuit design is the use of static timing analysis that looks at all potential signaling paths within the circuit independently of actual circuit stimulus, or in software the analysis of execution paths. Such analysis can be used to determine the relative flexibility of the system for change, and where and how change can most efficiently be made. A number of other system attributes can be analyzed using this approach.

For the model-based methodologies to work effectively, it is critical that the models used for each of these views are kept consistent throughout the system's life cycle. In addition, users of these models need to be informed of changes that have been made, which impact their view of the system.

Digital Capabilities

Digital transformation provides the potential for an integrated, yet modular architecture. While each of the individual areas can be pursued independently and provide incremental value, each provides additional benefit to the other areas in which the collective research provides more value than the sum of the individual parts. The relationship of these research areas to various levels of modeling, from high-level virtual abstractions to physical entities, is shown in Figure 6.3. Each of these transformation areas is described in more detail below.

Prioritization and Trade-off Analysis

While systems engineers quite often conduct prioritization-based trades, this usually is interpreted in a budgetary sense (e.g., how much thrust can we afford, how much weight can we afford), but not in the sense of how much security can we afford, or more nuanced, how much accessibility or usability will we sacrifice for the degree of security we need. So value means more than simply budgetary limitations, it also should be expected to include the overall value created for the systems stakeholders. This transformation area will determine the techniques that are available to analyze prioritization trade-offs and how they can be used to construct tools to both improve the quality of decision-making and reduce the time that is necessary to accomplish it.

Concept Engineering

Currently, the conceptualization phase of a project is done either through a laborious document driven process or in an ad hoc manner with inconsistent results. In either case, system developers and users alike often have inconsistent understandings of what the system is actually supposed to do, and how it creates value. This transformation area explores the tools and processes that provide an efficient interactive environment where multiple stakeholders can create a shared mental model during brainstorming processes – from concept of operations development throughout the life cycle.

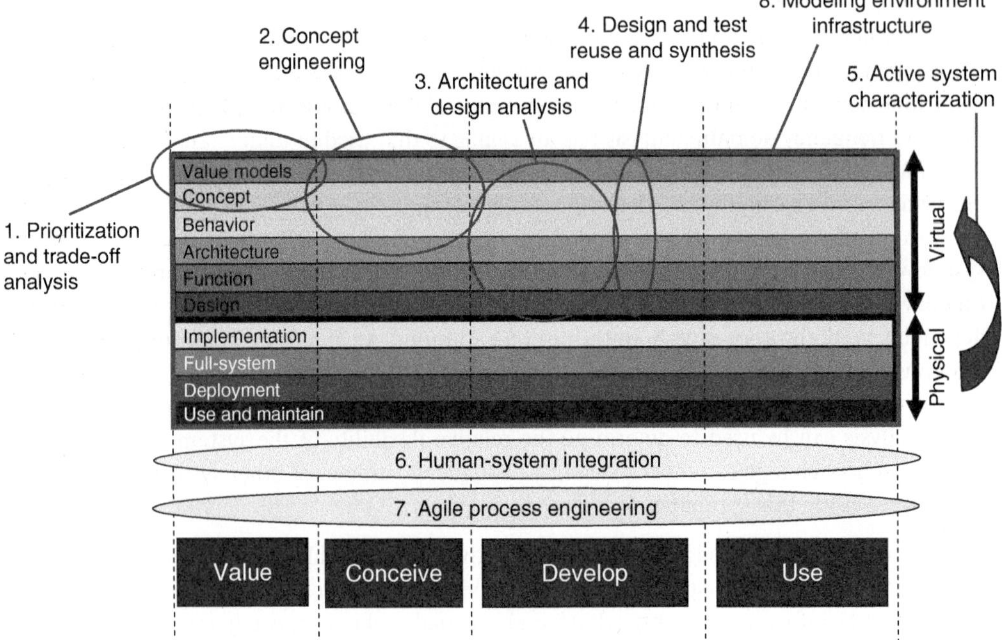

Figure 6.3 Relationship of SET research areas to modeling levels.

Architecture and Design Analysis

Architectural descriptions and designs of large systems are generally far too complex for architects and systems engineers to understand their attributes and emergent behaviors, in order to make appropriate design trade-off decisions. This transformation area focuses on the development of visualization and analysis tools to facilitate the decision-making process throughout the architectural specification and design process. Such tools may provide visualization for fragility, change propagators (areas not to touch), extensibility, modifiability, and security. Such guidance could include how and where a system should be modified with the least amount of impact to design and test.

Design and Test Reuse and Synthesis

Much time and effort may be spent in the design, development, and test of functionality that has already been implemented with similar capabilities. However, it is often difficult for a developer to know what already exists for this leverage and understand the implications of using existing technology. In addition, the developer may not have the time or inclination to make their work accessible for use by others. This transformation area explores how technology can be used to mine existing architectures, designs, and tests looking for patterns that can provide a means to categorize and catalog such work for reuse. Low-effort means of making designs and tests more accessible, and the requirements for synthesis of higher level architectures into lower level implementations using these repositories need to be explored.

Active System Characterization

It is very difficult to understand how systems actually behave based on design documents and models. In addition, systems are generally composed of legacy element and many undocumented "features". However, for new capabilities and features to be added to an existing system, it is often critical to understand its current behavior. This transformation area looks at how existing systems can be instrumented, measured, and analyzed to provide the necessary model fidelity. This applies to both system models that can aid in subsequent design efforts and also conceptual models used to validate how the system is being used and value is being created.

Human-System Integration

The need to accurately model the system is limited not just to technology but includes the human element as well. The human factor needs to be considered not only with respect to the usual human-factors issues such as ergonomics and safety, but also considering humans to be an element of the system and improving the entire system to ensure that the overall system–both technology and human components – is being optimized. Digital transformation is necessary in two areas. The first is the development of appropriate models for humans who interact with and thus are elements of a system. The second area is an exploration of the capabilities and limitations of people who are developing the system.

Agile Process Engineering

In general, there are no best practices, but rather there are practices that have been shown to be beneficial in a given environment. Agile approaches have been successful in a number of applications, but the practices may be quite brittle when the application attributes change. To realize the benefits of interactive, model-centric engineering, the supporting processes must be easily adaptable, taking full advantage of the evolving capabilities of tools and technology, within the environmental constraints. This transformation area focuses on the development of a process (or suite of process assets) and the related governance that supports the rapid development of systems in environments that may not be the natural home ground for agile development.

Modeling Environment Infrastructure

An integrated modeling environment is required to provide effective and coherent communication between the users of the collaborative modeling and design environment. Without this capability, the users would have to fall back into sequential, isolated development, which mitigates many of the advantages of modeling. This transformation area focuses on the development of such an environment that provides effective model interoperability and management. In addition, work needs to be done to understand how the environment can best present information to each user and increase design efficiency.

Epilogue

The following narrative describes a day in the life of a systems engineer in this digitally transformed world of systems engineering 2.0.

The End State

There were two more hours of testing to be completed before System-X would be released and deployed for operational use. Jane Johnson, responsible for its overall success, was confident that this would be a high-quality, on-time delivery. Her confidence was based on the fact that this was the 7th in a series of releases over the past couple of months that had continually gotten better.

They were currently releasing this product on a weekly basis, but there was talk about even more frequent releases. Jane enjoyed this talk, since the challenge to quicker releases was not in development and test now, but in fielding. It simply took too long to train users in the field how to apply the new features appropriately. Clearly more work was needed to support rapid deployment, but since the company was now focused on constant improvement, she imagined there would be effective changes soon. Of course, it had not always been like this. Critical challenges had required dramatic digital transformation a few years ago, including the emphasis on improvement.

The Old Model

Keeping an eye on the test status dashboard, Jane let her mind wander back to those pre-crisis days, when it was rarely a question of shortening cycles. In fact, there always seemed to be enough time to "do things right." Her organization was given a clear mission, which the entire team understood, at least at some level. They spent lots of time with the stakeholders building a detailed set of needs and understanding the system context. There was a whole team of SEs to translate those needs into a complete set of requirements, which became the basis for building a concept of operations, and an architecture. Requirements were decomposed into functional blocks, assigned to appropriate development organizations to build components (which were always tested extensively) and then everything was integrated into the system. Jane chuckled, remembering that integration had always been a bit of an eye-opener and the rework could cause real panic. But, because of the built in slack in the schedule, they were usually able to work things out eventually. This was the phase where the corporate heroes were often trotted out and managers and teams made or lost. Her team had been through a number of these integrations, but they were good and tended to communicate better than some of the other teams and so usually survived. They generally put their system through a series of increasingly more difficult tests and fixed things as they went. This modified "V" process was certainly better than the old Waterfall and had seemed to serve them well.

The Challenge

Jane could not put her finger on when exactly it happened, but things started to breakdown over time. It really happened across the entire process. There were all sorts of problems – things just seemed to have to happen quicker. The pace of change, losing control of the environment and technology, and the complexity of the products that made it hard to separate out functional blocks that could be worked concurrently made everything more difficult. There was not time to solicit and create a set of requirements, and when they did, the needs changed so rapidly and the requirements were useless. It was an endless task, some called it a "rock fetch" or worse, and the team often found themselves just going through the motions to satisfy the auditors. The real work was happening outside the process, because the process could not keep up. The same was true about the operational environment. They were forced by budget cuts to outsource more and use commercial off-the-shelf products whenever possible. It sure sounded great on paper, but in reality they lost control of the technology – ensuring a stable supply chain and product roadmaps for the essential system capabilities was a nightmare. In the good old days, you could usually see what you were building. Sure, there was some software thrown in, but the system decomposition was generally fairly obvious. Now software was the critical element and drove most of the complexity.

Decomposition had become a major challenge and it seemed as though everybody had to talk to everyone all the time. Why could not everyone just know what they had to develop like in the past? The tradeoff seemed to be either everyone could spend their time in meetings and know what to do, or they could do development work with the likely odds that it would not integrate. Well, at least if you got your work done you could blame someone else if it did not integrate, so most folks just plowed in to the development. Validation testing was even more challenging since the validation crews did not have a clear specification of what the system was supposed to do when it reached their shop. They did the best that they could with what they had, but ended up spending most of their time in integration work and ensuring that the system was at least not dead on arrival. The customers finally did most of the actual validation work and were none too happy about it. They were even less happy when the features and functions delivered were not the ones they needed. In fact, most new features had very little value. Nothing was going right. Jane had felt like the proverbial frog in that increasingly hot pot of water. She could not really tell when things started to get uncomfortable, but she knew that if they did not jump out of the pot the consequences would be dire. Luckily, she had made the decision to jump – bringing the entire team with her.

Critical Insights

Taking a quick hit of coffee and scanning the floor for any unusual activity, Jane sat back in her chair and actually shuddered a bit when she thought about that "call to action" offsite. Everyone knew they had to make changes, but no one seemed to know where to start. Then there was that eureka moment when suddenly they collectively seemed to realize that the "V" had it all wrong. You do not do validation at the end of the process, you do it at the beginning and you do it continually throughout the program. Validation is necessary to ensure the system provides a sound value proposition and needs to be continuous or else time and effort are wasted on low value work. But how to accomplish that when there is not a complete system in place until after it is designed? And how to get rid of the constant interruptions of sync up meetings currently plaguing their development efforts?

The only approach that seemed sound was to create a digital environment where simulation was used to model the system throughout the lifecycle. Hierarchical models could be filled in with the level of details that were needed and available at each point in development. Having a shared model in place as a central reference point would allow each member of the development, test, and deployment teams to work coherently as a team of teams without having unnecessary forced synchronizations. They could be informed of changes that impact them and disregard the rest. The intent was to create a common shared "Borg-like" state. This new environment would need to be interactive supporting iterative design, execution, and re-design; based on executable hierarchical models that provide on-the-fly simulation and can be used throughout the lifecycle; and supportive of multisensory and multimodal inputs and outputs to provide information personalized to the needs of the stakeholder or user.

Their second major insight was that the standard approaches to system analysis and design, based on decomposition and allocation of functional requirements of the system, usually generate systems that are resistant to adaptation and difficult to secure in the faces of changing threats. They needed a design approach that reflected the true objectives of the system – security, flexibility, performance, etc. – and mechanisms that balance the mission objectives with the functional capability and cost/value.

Both new system knowledge and tools would be necessary to bridge the gap and truly bring value-based system engineering to life. Analysis tools would determine both the value of flexibility and the means by which to achieve it. Other tools that could perform options analysis across tradespaces to determine value and intelligently interact with human analysts and designers were essential in the decision-making process. Everyone agreed that their current tools simply could not provide architects and designers with the information to understand the emergent behaviors of their complex systems. They also agreed that the team members were proficient at visual pattern recognition and analysis, but their capabilities of understanding these relationships from text-based information was really limited. Clearly work had to be done to support high-level value decisions involving architectural and implementation trade-offs.

The final insight was that the systems they developed needed to be adaptive if they were to keep up with the pace of change in the environment, technology, and more important, the mission. One of the most adaptive elements in any system had been largely ignored to date, and that was the human factor. Understanding the relative strengths and weaknesses of humans and incorporating them into a system appropriately was surprisingly new ground that desperately needed to be developed.

Jane could see now how the confluence of these three major insights resulted in the set of actions and activities that followed. The meeting resulted in a whirlwind of discussions, planning, research, and finally actions to transform the engineering approach.

The Change

Prioritization and Tradeoff Analysis

Jane and her team started at the front end of the process where many of the problems seemed to originate. In fact, the trouble seemed to start even before they had made any decisions. Jane realized that the critical first step in value-based decision-making was to determine the relative worth of various system attributes including flexibility, agility, resilience, sustainability, and dependability. It seemed no one had a clear understanding of how the "goodness" of solutions could be judged. Everyone knew that lower cost, higher performance, higher reliability, and shorter time to deployment were good things to achieve; but there was little agreement on the relative value of each. They decided to try models as a quantifiable means of making these determinations. At first, the models were crude, but they were refined over time and their predictions were tested and tuned based on feedback from the field. Jane had really come to appreciate the ability to show her executive management tangible and quantifiable benefits supporting the various decisions the team made.

Concept Engineering

It was clear to everyone that long drawn out requirements elicitation processes were not working and that most documentation ended up being "shelfware." In fact, the exchange of text-based documents simply could not effectively convey information to their diverse set of stakeholders. The people creating the system concepts needed a tool with which they could interact and receive timely visual feedback; something where they could quickly iterate and test their concepts. Systemigrams and other graphical tools helped at the whiteboard, but they needed more. Looking for a better solution, one of the engineers noticed her son and a group of online friends building new cities with Minecraft™ and decided to bring it in to try it out with the team. This inspiration led to the development of graphical concept of operations tools and capabilities. Over time, these capabilities evolved enabling groups of engineers, field service, and marketing people to iteratively and interactively create behavioral models of systems operation that quickly provided analysis of performance and impact. Jane was blown away when she experienced the new conceptualization process. She had not thought these people knew each other, let alone speak the same language. Yet, as it was explained to her, there they were – building and simulating concepts and comparing the results based on the prioritization models.

Architecture and Design Analysis

Although the efforts in Concept Engineering and simulation and model-aided development were paying off, there still were major gaps in the architectural specification and design of the system. In particular, it was not clear to Jane that the decisions being made maximized the value of the system. It was extremely difficult to understand the emergent attributes of the system throughout the development process. There was no longer a single "go to" expert who understood how it all worked, and how each component affected the overall operation of the system. So, they had looked at developing tools that would assist them in determining where changes could be best made in the system to avoid gratuitously breaking things. This work was quite useful in determining which parts of the system suffered from unacceptable levels of "architecture and design rot" and needed to be re-architected, redesigned, and refactored. Unexpectedly, though, it also showed which areas of the system were likely to be the least stable and thus were the greatest reliability and security risks. The tools evolved to take on more capabilities and support other nonfunctional attributes including flexibility, reliability, and performance. The development of these tools was very much an agile process, dynamically driven by what they learned along the way.

Design and Test Reuse and Synthesis

While making great progress with the efforts in the above areas, Jane recognized they still were not achieving the level of productivity that would consistently enable the development of systems at the demanded rate. One of her engineers observed that each system design was essentially a new experience; they were not really leveraging the work that had been done on previous releases. She suggested that they should focus more on finding and using patterns and on technology reuse. Additionally, they were not taking full advantage of the models created in conceptual design. There were simply too many "one offs" in each of these areas for this to be a productive use of time and effort. A task force was assigned to mine patterns from concepts, architecture, design, implementation, and tests. The findings were quite surprising as they confirmed how many different ways they architected, designed, and tested the same functionality. Much as has been done in the area of integrated circuit technology, they were able to provide consolidation with the creation of primitives, components, subsystems, and systems in a reusable library. In the case where new design was necessary, they were often able to use existing building blocks. The net result was not only in a reduction in time and effort but also an increase in quality as they were able to use well tested and characterized components.

Integration of V&V into Development

The use of simulation and modeling technology for conceptual and high-level design certainly greatly improved design productivity, but there was still the challenge of proving that the system was correct in function and operation. Already V&V was a huge effort, and it certainly was not getting any easier. It required total focus, and most importantly in Jane's company, a major culture shift. At that time in Jane's firm, the V&V staff had not been seen as the most talented and empowered people. In fact, this was where staff traditionally went when they were not selected for the architecture and design teams. The V&V team only operated late in the development pipeline, got the product late to test, always had their test time reduced, were forced to compromise their standards, and were immediately blamed for any problems that escaped into the field. In short, they felt like and often were the victims. They often worked manually and reacted to problems by throwing bodies at whatever was the crisis of the day. This was clearly not a road to success.

Jane and her team looked for a silver bullet to solve these problems, but none was found. However, the team had noticed that companies producing high-quality systems and had short release times all seemed to have similar characteristics. For example, they all had high-quality teams that focused on automating every part of the test process. Humans were removed from the process of setting up, running, and collecting test results. Instead, humans were focused on the analysis of test coverage and failure data, determining where additional tests were needed, and if so, how to automatically create them. The other major focus area was to track failures throughout the lifecycle process, provide root cause analyses, and ensure that tests were in place to find them as early as possible in the development/deployment process. To succeed with this strategy required a major commitment across the organization and without executive support it would never have been successful. Test Driven Development was included at the front-end of the process, and design and verification engineers were teamed together. No longer could someone point their finger at an individual for blame for a fault, but rather teams were held jointly accountable for schedule, functionality, and quality. They had made much progress, but cultures do not change quickly.

The conceptual and high-level design efforts provided major benefits in V&V as these hierarchical models supported the continuous development and execution of tests throughout the development process. In addition, this model-based approach was accompanied with significant levels of reuse, which provided the means by which to compose correct-by-design constructions. Finally, it was possible to quickly verify and validate incremental changes to a system to allow for rapid deployment of incremental capabilities. These approaches resulted in substantial and continuing improvement in deployed system quality. Jane noticed that in hindsight the barriers between design and test were artificial, and over time they blended into the single integrated activity of development.

Active System Characterization

Amidst all the excitement, Jane unhappily noted that while greenfield system development was the "cleanest" and resulted in an up to date set of models, she could not remember the last time they had gotten to start with a clean slate. Living with legacy was a way of life. To make matters worse, there was much that they did not know about their legacy system. It seemed to Jane that they discovered all of their system limitations by deploying new system capabilities and then waiting for user complaints. There had to be a better way. Their solution had been to develop a set of automated tools to instrument, monitor, and characterize their existing system. This was done both in the field in the form of a number of system monitors that reported real-time parametric data, as well as within the in-house system testing. Jane was pleased to know that the team made a number of new discoveries about their system and how it operated, and that these had made a significant impact on subsequent decisions.

Human-System Integration

Modeling was a major success for the organization, but it still had a major flaw: its predictions were sometimes very wrong. She saw the same issues with problems in deployed systems. When Jane asked what happened, she invariably got some excuse about operator error or the unpredictability of people. When she tried to pin down the systems engineers, they told her that they were only responsible for the technology, not for the people. After the last system failure due to "operator error," Jane had had enough. She had called in the SEs, program and product managers and read them the riot act. They were all responsible for the success of the entire system, and that included people, so they better figure out how to include people in the equation. There had been a hushed silence in the room when they realized that the blame game was over – and no one had answers. The organization had some experience with how people actually used the system in their Concept Engineering work, but they needed to take it a step further. They started by having end users interact with the product in the simulators throughout the development cycle. Later, they were able to incorporate some models of how people might interact with the system. But, where they made the most impact was in determining how to allocate system capabilities between human and machine to optimize overall system performance.

Agile Process Engineering

One of the major problems that Jane and her team encountered throughout the transition was the simple fact that it was a major transition; one that could not occur immediately and was likely to continue for many years. In fact, if successful, continual change would be a permanent aspect of their future operations. While many of the leaders in her organization talked about processes that enabled agile development, it eventually dawned on them that they needed processes that were themselves agile. There simply was not a golden process that would support the entire organization and the rest of the enterprise continuously over time. This was a breakthrough notion and was perhaps the most difficult bit of cultural change necessary support the new development paradigm. One major advantage, though, was that these processes were embedded in an integrated, highly automated environment, which kept the complexity hidden from the users. As far as Jane could tell, the intelligence of the processes guided and supported the developers to make them more effective rather than act as the traffic-cop SEs of the past.

Integrated Environment Infrastructure

While all of the tools and processes that had been created were a great help to the team, they still had some major communication problems that limited their success. Jane heard the complaints over and over again that this tool could not talk to that tool, or that data was not available in the right format, or it was impossible to find the most up-to-date information. At times the project seemed to come to a halt, while everybody synched up to the latest release. Jane called a meeting of her best and brightest to solve the problem. As it turned out, they were able to create some reasonable solutions to permit communication and data exchange between their homegrown tools, but it was a much bigger challenge for the commercial tools that they were using. Jane knew that she could not single-handedly move the market place, so she focused her efforts where they had the greatest impact. In some cases, this meant consolidating their tool use and in others it meant creating some internal interfaces and translation technology. Jane had also formed an alliance with some other systems developers to influence the CAD tool vendors, but this was going to be a long, long road. Where they had made some amazing success was in the ability for tool and environment users to find information and communicate with one another. The breakthrough idea here was to optimize the system for the user rather than the other way around. Developers were not disturbed with notifications unless there were changes to the system that impacted them. Artificial intelligence was used to make the right information available to the right people. Jane noted that while Amazon had been using the technology for years, it still surprised her when the Integrated Environment seemed to anticipate what she wanted at work. The Environment truly had become a central communication point for all of the involved stakeholders and developers.

Conclusion

Jane checked her watch and noted that two hours had gone by. She glanced at her monitor and saw that the final tests had passed and the new system was being switched over to operational status. She wondered what they might learn with this latest deployment. Whatever it was, the information would be available and used in the next release, which was already well into development. Jane realized before calling it a day that while they had accomplished much in the past few years, there was still much more that needed to be done. Continuous improvement for transformation had turned out to be just that – continuous.

References

Boehm, B. and Turner, R. (2003). *Balancing Agility and Discipline*. Addison Wesley.

Boyd, J. (1976). *Destruction and Creation (PDF)*. U.S. Army Command and General Staff College.

Carrigy, A., Colbert, E., Componation, P. et al. (2009). Evaluation of Systems Engineering Methods, Processes and Tools on Department of Defense and Intelligence Community Programs. *SERC Phase 1 Technical Report (A013)*.

Kasparov, G. (2017). On AI, Chess, and the Future of Creativity (Ep. 22). Conversations with Tyler. https://medium.com/conversations-with-tyler/garry-kasparov-tyler-cowen-chess-iq-ai-putin-3bf28baf4dba

Lencioni, P. (2002). *Five Dysfunctions of a Team: A Leadership Fable*. New York: Jossey-Bass.

McChrystal, G.S.A., Silverman, D., Collins, T., and Fussell, C. (2015). *Team of Teams*. Portfolio Penguin.

Rotman, D. (2020). We're not prepared for the end of Moore's Law. *MIT Technology Review* 123 (1).

Schwab, K. (2015). The Fourth Industrial Revolution – what it means and how to respond. *Foreign Affairs*. https://www.weforum.org/agenda/2016/01/the-fourth-industrial-revolution-what-it-means-and-how-to-respond/.

Snowden, D.J. and Boone, M.E. (2007). A leader's framework for decision making. *Harvard Business Review* 85 (11): 68–76. PMID 18159787.

Stevens, R. (2008). Profiling complex systems. *2008 2nd Annual IEEE Systems Conference*, Montreal, QC, Canada, pp. 1–6. https://doi.org/10.1109/SYSTEMS.2008.4519017.

Wade, J., Madni, A., Neill, C. et al. (2010). Development of 3-Year Roadmap to Transform the Discipline of Systems Engineering. *Final Technical Report – SERC-2009-TR-006*.

Wade, J., Madni, A., and Neill, C. (2011). An integrated, modular research architecture for the transformation of systems engineering. In *Proceedings of the 2011 Conference on Systems Engineering Research (CSER)*, Redondo Beach, CA (15–16 April), Vol. 9, Los Angeles, CA.

Wade, J., Buenfil, J., and Collopy, P. (2020). A systems engineering approach for artificial intelligence: inspired by the VLSI revolution of mead and conway. *Insight* 23 (1): 41–47.

Wymore, A. W. (2004). Contributions to the mathematical foundations of systems science and systems engineering. *Systems Movement: Autobiographical Retrospectives*. The University of Arizona, Tucson, AZ.

Biographical Sketches

Jon Wade is a professor of practice at the Jacobs School of Engineering at the University of California, San Diego, where he is the director of Convergent Systems Engineering and the executive director of the Institute for Supply Chain Excellence and Innovation. Dr. Wade's focus is on developing research and education to provide ethically sustainable solutions to critical, complex societal problems. Previously, Dr. Wade was the chief technology officer of the Systems Engineering Research Center, executive vice president of Engineering at International Game Technology, senior director of Enterprise Server Development at Sun Microsystems, and director of Advanced System Development at Thinking Machines Corporation. Dr. Wade received his SB, SM, EE, and PhD degrees in electrical engineering and computer science from the Massachusetts Institute of Technology. Dr. Wade is an INCOSE Fellow.

Chapter 7

Measuring Systems Engineering Progress Using Digital Engineering

Tom McDermott[1], *Kaitlin Henderson*[2], *Eileen Van Aken*[2], *Alejandro Salado*[3], *and Joseph Bradley*[4]

[1]*Stevens Institute of Technology, School for Systems and Enterprises, Hoboken, NJ, USA*
[2]*Virginia Tech, Blacksburg, VA, USA*
[3]*University of Arizona, Tucson, AZ, USA*
[4]*Leading Change LLC, Melbourne, FL, USA*

Motivation

Systems continue to grow in complexity, and with that growth comes greater difficulty in managing the development of these systems (DiMario et al. 2008). The utilization of models to design systems is increasingly considered a principle of good system design (Bahill and Botta 2008). MBSE was conceived as "the formalized application of modeling" to the systems engineering process (Friedenthal et al. 2007). DE and MBSE approaches are two components of enterprise digital transformation that have great promise to improve the efficiency and productivity of engineering activities, particularly for complex engineered systems. Organizations perceive and have cited many benefits of this transformation, but there has been little attention on formally measuring these benefits (Henderson and Salado 2021). Systems engineering as a discipline has long had difficulty providing quantifiable evidence of its value (Honour 2004); DE transformation provides an opportunity to better measure its value as more of the development process becomes captured in digital data, models, and tools. Transitioning from a document-based to a model-based approach is expensive, and organizations want to know if the effort and cost to adopt MBSE is worth it.

The Benefits of DE

The SERC engaged with government and industry subject matter experts (SMEs) to develop a set of metrics that should be employed to best show the value of DE and MBSE. Since there are many potential benefits, we developed a causal model based on performance measurement literature to systematically decide which metrics should be prioritized, then worked with the community of SMEs to refine that model into the set of potential measurement specifications described in the DE Measurement Framework. Our research indicated: (1) DE and MBSE have measurable benefits;

Systems Engineering for the Digital Age: Practitioner Perspectives, First Edition. Edited by Dinesh Verma.

Companion website: www.wiley.com/go/verma/systemsengineering

Table 7.1 Direct and measurable benefits of DE/MBSE.

Direct benefits	Definition	Measurable benefits
Higher level support for **Automation**	Use of tools and methods that automate previously manual tasks and decisions	Greater process automation, increased use of tools, increased efficiency, reduced cost
Early verification and validation (V&V)	Moving tasks into earlier development phases that would have required effort in later phases	More discrepancies found in reviews, earlier defect resolution, reduced rework
Strengthened testing	Using data and models to increase test coverage in any phases	Increased defect detection and resolution before fielding, reduced rework
Better accessibility of information (ASOT)	Increasing access to digital data and models to more people involved in program decisions	Increased number of users in the tools, run-time performance of the infrastructure, reduced lead time
Increased traceability	Formally linking requirements, design, test, etc. through models	Greater design correctness and completeness, better able to track product design sizing, reduced volatility and increased stability
Multiple viewpoints of model	Presentation of data and models in the language and context of those who need access	Increased number of users in the tools, greater design correctness, increased efficiency
Reusability	Reusing existing data, models, and knowledge in new development	Better data and model reuse, reduced effort, reduced lead time
Higher level of support for **Integration**	Using data and models to support both the integration of information and system integration tasks	Reduced number of product iterations before release, reduced release cycle time

(2) DE/MBSE measures can be defined and tracked, and are extensions to well-known software measures; and (3) DE/MBSE measures primarily support the systems engineering process and can provide data-driven quantitative assessment of systems engineering benefits, given an appropriate measurement framework (Henderson et al. (2021); McDermott et al. (2022); SERC-2021-TR-024 (2021)).

Previous SERC research on benefits and metrics in DE surveyed both literature and the MBSE community to broadly collect potential measures associated with benefits and adoption indicators (SERC SR-001 2020; SERC TR-002 2020). The survey results and initial DE Metrics report remain available on the SERC website: https://sercuarc.org/results-of-the-serc-incose-ndia-mbse-maturity-survey-are-in. The earlier surveys were used to narrow down a set of eight direct benefits and measurement strategies associated with DE measurement, as described in Table 7.1.

Realizing the Benefits

DE is fundamentally about increasing stakeholder involvement in the engineering development and related program management processes. There are two primary drivers of stakeholder involvement: (1) the organization has to create the environment for and measure actual use of the digital tools; and (2) the organization has to develop the associated digital project methods and processes. Many of the other benefits of a digital transformation will be lost if artifacts must be produced and activities must be conducted outside of the digital environment and related models. This leads to a measurable adoption strategy as follows:

- Methods, processes, and tools must be standardized across the project or organization
- The workforce must be trained on and work inside of the tools and associated models
- The organization must measure both the internal and external use of the tools and user experience
- Organizations need to get serious about conducting reviews "in the model"
- Organizations need to get serious about modeling and using reference architectures and reusing them across portfolios
- Tool vendors must get serious about increasing automation and inter-operability in their tools

However, adoption measures are the foundation. Real returns will accrue when the quality and timeliness of the product improve through the use of DE and MBSE. We found a set of base quality and cycle time measures that are common to most digital transformation activities and well described in the software community. These lead to a measurable product definition and development strategy as follows:

1) DE/MBSE should reduce errors throughout the system definition and design phase. Measuring **defects/errors** and resolving these earlier in the development processes is important and can be supported by the digital tools. Errors should not persist from one phase to the next.
2) DE/MBSE should reduce **cycle time** through development, integration, test, and delivery life cycle phases as well as further product support cycles. Each life cycle phase represents a team's work on the product leading to a release; a release as defined as some set of artifacts; and the work required to support, update, and then retire the product. These can all be measured in the digital environment.
3) DE/MBSE will improve **functional completeness and correctness** of the underlying description of the design. Programs and organizations need to develop the ability to quantify and analyze functions in a digital systems model. The community should put more effort in defining models associated with decomposition of capabilities, functions, and requirements, and then measuring the peer review, simulation, and test of those models.
4) Programs and organizations need to measure the **efficiency** gained in model-based review artifacts. Leading indicators of efficiency are associated with the number of review discrepancies/actions and decision signoffs or approvals at the review. To achieve return on investment (RoI), organizations must be serious about conducting technical reviews in the models, as the DoD DE Strategy states.
5) **Reduced cost**, which is the ultimate measure of return, is a lagging indicator dependent on all these other indicators. Reduced cost should be evaluated across programs in the enterprise, but more leading indicators such as defects and effort also require tracking within a program or project and must be measured to derive cost benefits. Unfortunately, this means the RoI answer is a journey, not a number.

We have found there are only a small handful of organizations and programs that have started this measurement journey. Programs must begin collecting data on these base measures before the RoI answer can be realized.

Measuring DE Activities

There are a set of activities any organization should do to begin and achieve their digital transformation. These activities should be linked to measurable outcomes. These are summarized in the following sections with a focus on the benefits of DE/MBSE as noted in Table 7.1. DE/MBSE is generally implemented as a set of methods, processes, and tools for the life cycle definition, development, and sustainment of complex engineered systems. DE/MBSE creates not only the product itself but also the digital data and models that define and then support the product over its life cycle. Because DE/MBSE processes help to define the capabilities of the eventual system, measures can serve as useful leading indicators for other product-related measures.

A challenge with measures is both ensuring that they provide information needed to support decision-making and that they are actually collected and used. A small set of measures should be tailored for each program and organization, focused on those needed for fact-based decision-making. The measures should be regularly reviewed to ensure they are being used and that the decisions made using those measures are producing the intended outcomes (PSSM 2022).

Decomposition of the DE/MBSE work activities is generally associated with models, underlying data, and the digital infrastructure supporting them. All are important concepts in the measurement approach and have related specifications.

DE/MBSE Should Reduce Errors and Cycle Time

When a system is being developed or updated, anomalies (errors, defects, omissions, etc.) are generated throughout the system definition and design phase that may not be discovered until later phases when they are more difficult or costly to correct. Measuring these anomalies and resolving these earlier in the development processes is an important benefit of digital models and can be supported by the digital tools. Anomalies that are discovered should not persist from one phase to the next. The value of DE/MBSE is first that the anomalies should be discovered earlier, and second that they should be corrected more quickly and with less effort.

Much of the work in DE/MBSE is related to what systems engineers call phases and what software engineers call iterations that happen before an engineered product hits an external user. The value of modeling is most apparent before a product hits a release point, either internally to an integration and test phase, or externally to the users. This flow is shown in Figure 7.1. In this process, there are many internal and external baselines that characterize a team's work on the overall product, and a related decision to move forward. This baseline is some set of artifacts that support or make up the overall product. In MBSE, these artifacts are related to models, in DE they are any digital artifact including data, software, etc. The value of DE/MBSE is measured as a "containment" of errors into the earlier phases, an overall reduction in the total number of errors that have to be corrected in a full release cycle, and a reduction in the cycle time required to achieve releases.

The PSM DE Measurement Framework provides a set of candidate measurement specifications for both DE Anomalies (defects and associated rework effort) and Cycle Time (PSSM 2022).

Cycle time is the elapsed time from when development work is started until the time development work has been completed and is ready for release. This time includes activities such as planning, requirements analysis, design, implementation, and testing. Cycle Time is typically targeted at measuring repeatability and predictability of team performance for well-scoped work so that

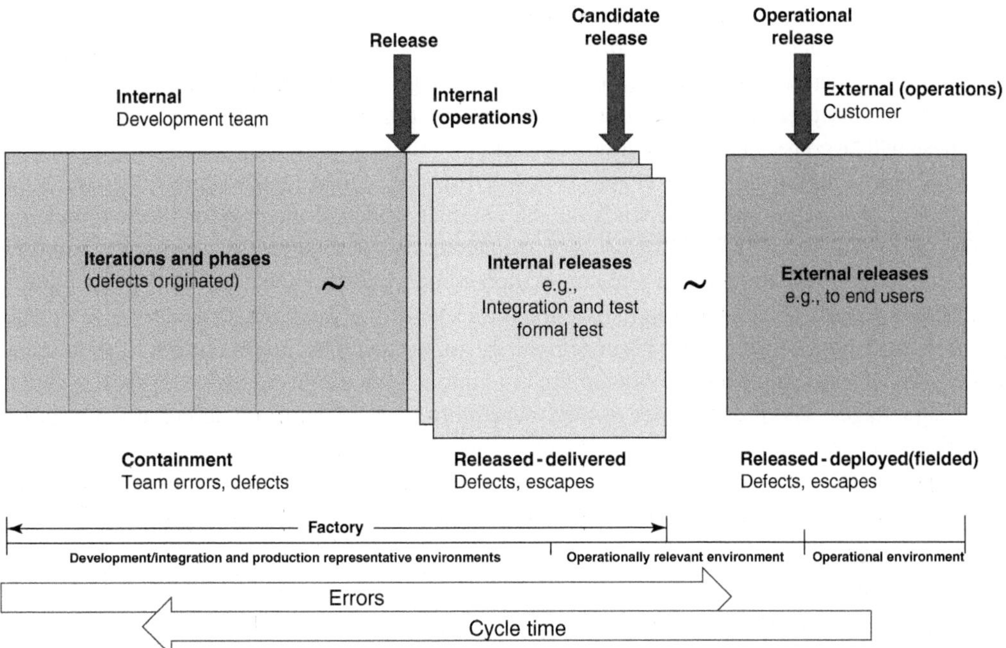

Figure 7.1 Value of DE in a development lifecycle.

results are comparable across multiple similar efforts. Cycle time for digital activities should be measured for consistency initially but organizations should see reductions in cycle times as their DE/MBSE processes mature (PSSM 2022). Cycle time measurement implies that scheduling rigor is applied to phases, iterations, and releases so that process improvement can be tracked over time and organizationally between programs.

DE/MBSE Will Improve Functional Completeness and Correctness

The use of digital models should improve the underlying description of the design. Programs and organizations need to develop the ability to quantify and analyze functions in a digital systems model in order to know that digital modeling is improving their overall product quality and usefulness over time. The community should put more effort in defining models associated with decomposition of capabilities, functions, and requirements, and then measuring the peer review, simulation, and test of those models. In the digital model one can more directly analyze completeness of models, traceability across aspects of the models, and the volatility or changes in the model as it matures.

Measures of functional completeness and correctness are typically measured within the modeling tools, generally using metrics that are calculated by the tools. These measures link to the concept of model elements. Model elements are defined as atomic (elementary) items that represent individual components, actions, states, messages, properties, relationships, and other items that describe composition, characteristics, or behavior of a system (ISO/IEC/IEEE 24641:2023 2023). A model element is an abstraction drawn from the system being modeled, representing an elementary component of a model. The number and type of model elements will be determined by the development process. There is no predefined categorization of elements – they can be defined by the underlying ontology of the system model or of the tool used to create and manage the models (PSSM 2022).

The DE measurement approach and associated measures should recognize a defined concept of a model element such that (1) the relative size of the DE effort can be measured and compared to other efforts or plans, and (2) the quality of the DE design decisions (correctness and completeness) can be measured. The project must determine the type of model elements it will measure, and these will be constrained by the tools selected.

Model Traceability is an important concept in DE/MBSE. PSSM (2022) states "the usefulness and quality of a digital model depends on the completeness and integrity of the relationships among model elements. Traceability between elements, such as requirements allocation and flow down to architectural, design, and implementation components, assures that the system solution is complete and consistent. Gaps in bidirectional traceability between the artifacts of two models might indicate where further analysis or refinement are needed. This might further apply to traceability gaps within a single model, when there is no implicit traceability between artifacts of different design stages. The prerequisites of any traceability measurement are agreed-upon, a priori guidelines and definitions, e.g., what model elements and relationships shall be traced, that apply to the specific DE model of the system." Measuring traceability is a foundational concept of DE/MBSE and many of the defects and errors found in early stage design will be discovered using traceability measures.

Model-Based Review Artifacts Will Improve Efficiency

Programs and organizations need to measure the efficiency gained in model-based review artifacts. Leading indicators of efficiency are associated with the number of review discrepancies/actions and decision signoffs or approvals at the review. To achieve RoI, an organization and/or a multi-organization program must be serious about conducting technical reviews in the models.

Reduced Cost Is a Lagging Indicator

Although cost reduction and return on investment are the metrics that program and organizational management want to know, these metrics are dependent on all the other prior discussed indicators. Reduced cost should be evaluated across programs in the enterprise, but more leading indicators such as defects and effort also require tracking within a program or project and must be measured to derive cost benefits.

Causal Analysis of DE/MBSE Benefits and Adoption Approaches

The eight direct benefits associated with DE measurement previously listed in Table 7.1 were developed from a causal measurement model (Henderson et al. 2022). These direct benefits form the roots of the causal model, considered as core activities in DE and MBSSE that lead to measurable benefit. From these direct benefits stem the secondary benefits that are the effects/result of these, including those potential metrics that are typically of interest to managers and other stakeholders such as reduced cost, improved quality, or reduced cycle time. The causal analysis informed DE metrics design. Figure 7.2 contains the full causal model.

One can visualize the complexity of the measurement challenge from this map. The eight primary benefits are in the top of the map in green boxes and font. All secondary benefits stem from these. In the map, the black nodes relate to benefits and the blue nodes relate to adoption factors. Note that the adoption factors form a causal loop beginning with "Projects/Programs use methods and processes" in the bottom left of the figure.

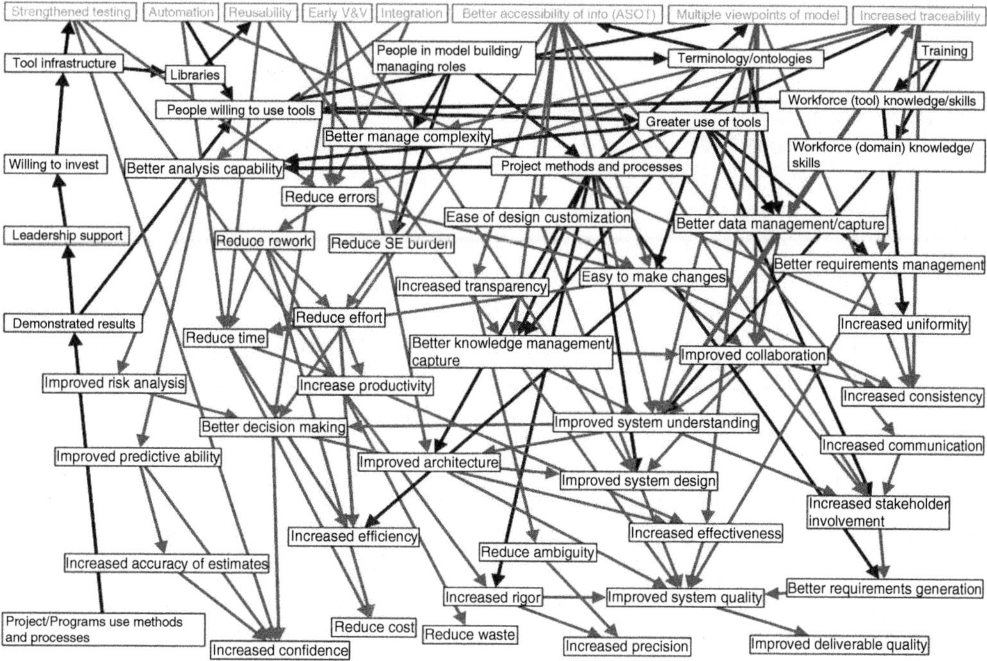

Figure 7.2 Full causal model.

There are three factors related to increased stakeholder involvement that are central to adoption. These are in blue boxes and font at the top-center of the map: greater use of tools, project methods and processes, and people willing to use the tools. Many of the other benefits of a digital transformation will be lost if artifacts must be produced and activities must be conducted outside of the digital environment and related models. Both "number of people willing to use DE/MBSSE tools" and "number or percentage of models/data sets in the ASOT" are critical leading indicators of successful DE transformation.

The causal model found a set of secondary benefits that were used to define base measures in the eventual DE Measurement Framework. These include reduced errors/defects, reduced effort, and reduced time. These are common to most digital transformation activities and well described in the software community. The causal model also shows that reduced rework and reduced cost will be important secondary benefits associated with these base measures. The causal linkages between these stand out down the center left of the map. Reduced cost, which is the ultimate measure of return, is a lagging indicator dependent on all these other indicators. Reduced cost is an important measure, particularly across programs in the enterprise, but more leading indicators such as defects and effort also require tracking within a program or project. Figure 7.3 depicts the causal flow of these adoption measures. In an enterprise measurement strategy, it is helpful to understand the causal relationships and then create a measurement strategy that links multiple measures together.

In this model, two adoption factors (maturity of tools, methods and processes, and taxonomy, ontology, and libraries) serve as enablers to the process and drive the leading indicator, "People willing to use tools." "Better accessibility of information" is the primary benefit, although it can only be measured indirectly. "Easier to make changes" measured in both effort and cycle time is the secondary benefits most appropriate to measure, and the base measures for this are effort and cycle time. Reduced cost and reduced schedule are the ultimate value indicators, but these would be lagging indicators after iterations of work complete.

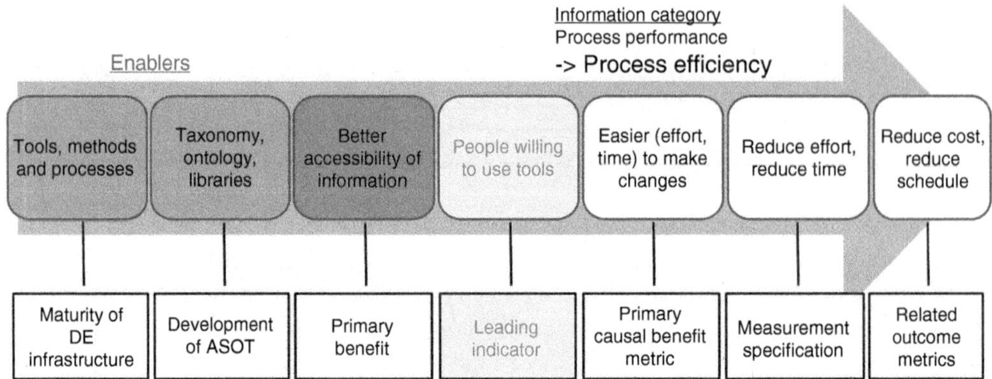

Figure 7.3 An example causal measurement strategy.

Improving functional completeness and correctness of the underlying description of the design is another measurable indicator of success. The ability to quantify and analyze functions in a digital systems model is causally related to the direct benefits in Table 7.1 of Early V&V, Strengthened Testing, Better Support for Integration, and Increased Traceability. All these again decompose to base measures of reduced errors/defects, reduced effort, and reduced rework.

Model-based review causally relates to the Better Accessibility of Information and Multiple Viewpoints of Model primary benefits. In this case, improved collaboration is the most central secondary benefit. Improved collaboration is difficult to measure directly but is an important user experience metric to assess via surveys of other means in the transformation process. Key related quantitative measures include reduced errors/defects and reduced rework, and how one tracks these measures from phase to phase of the review and decision process. The core product-related measures are defect resolution by phase and rework that must be done or traveled to later phases. Leading indicators associate with number of review discrepancies/actions and decision signoffs or approvals at the review.

Perhaps the biggest longer-term primary benefit area is Higher Level of Support for Automation. Central to automation are number of people willing to use DE/MBSSE tools (automation should increase this), easy to make changes, reduced effort, and reduced time. Reusability is a final primary benefit and is most causally related to better knowledge management/capture. Knowledge management is traditionally a difficult area to measure. In DE/MBSSE, the causal linkage is from program terminologies/ontologies and associated digital libraries, to reuse of data and models, to better knowledge management/capture.

Linking to Systems Engineering Metrics

The Systems Engineering Leading Indicators Guide, published by INCOSE, identified a set of measures to assess the effectiveness of the systems engineering process (MIT 2010). Despite the maturity of these indicators, few complete examples of actual measurement exist, primarily due to the lack of tools that can quantitatively track these measures.

Important quantitative measures supporting selected leading indicators include:

- **Requirements trends**: Model Traceability, Functional Architecture Completeness and Volatility
- **System definition change backlog trends**: Rework, Effort, Efficiency

- **Interface trends**: Model Traceability, Functional Architecture Completeness and Volatility
- **Requirements verification and validation trends**: Deployment Lead Time, Efficiency
- **Work product approval trends**: Number of Model Views/Artifacts, Deployment Lead Time
- **Review action closure trends**: Model Review Item Discrepancies
- **Defect and error trends**: Defect Detection, Defect Resolution, Rework
- **Technical measurement trends**: ASOT Frequency of Access
- **Architecture trends**: Functional Architecture Completeness and Volatility, Functional Correctness, Product Size
- **Cost and schedule pressure**: Efficiency, Rework, Deployment Lead Time

Work to update the Systems Engineering Leading Indicators Guide and link the measurement specifications in that work to the DE Measurement Framework has not yet started and so is not able to inform this work. However, one can envision how new activities like model-based reviews might change measurement of a leading indicator such as Review Action Closure Trends. Organizations should embed their DE/MBSSE measurement strategies into their systems engineering overall measurement strategy.

Summary

In today's systems, engineering creates not only the product itself but also the digital data and models that define and then support the product over its life cycle. Because DE and MBSE processes help to define the capabilities of the eventual system, these measures can serve as useful leading indicators for other product-related measures such as operational integration, evaluation, and support. DE/MBSE can also produce independent products in support of delivered data, hardware, and software products such as digital twins or other model- or simulation-based executable systems.

The benefits of DE and MBSE are associated with intangible products, defined in software, even though much of the purpose of DE/MBSE is to improve tangible products. Thus, measurement of DE and MBSE is primarily a software measurement activity. Because of this, the DE Measurement Working Group selected the PSM framework as a baseline measurement specification approach. PSM defines an information-driven measurement process focused on the technical and business goals of any organization, and allows specification of measurement goals, information, and indicators (McGarry et al. 2001). The original PSM guide to software measurement was published in 2001. In 2020, an extension to the guidance was published covering additional measurement concepts associated with software and system continuous iterative development (CID). The CID framework directly applies to DE and MBSE in the use of evolving models as the primary source of knowledge about a system and its life cycle. The DE measurement framework further extends the PSM base framework and the CID framework to cover DE/MBSE measurement concepts. Programs should use all three of these measurement frameworks in their measurement journeys.

Version 1.0 of the DE Measurement Framework was published 18 May 2022. The framework can be downloaded at https://www.psmsc.com/DEMeasurement.asp. The development of the framework was a community effort sponsored by the SERC, the Aerospace Industries Association (AIA), the International Council on Systems Engineering (INCOSE), the National Defense Industrial Association (NDIA), and PSM.

Note

1 PSMSC is an initiative sponsored by the DoD and U.S. Army to provide Project Managers with the objective information needed to successfully meet cost, schedule, and technical objectives on programs.

References

Bahill, A.T. and Botta, R. (2008). Fundamental principles of good system design. *Engineering Management Journal* 20 (4): 9–17.

DiMario, M., Cloutier, R., and Verma, D. (2008). Applying frameworks to manage SoS architecture. *Engineering Management Journal* 20 (4): 18–23.

Draft International Standard ISO/IEC/IEEE DIS 24641:2023(En) (2023). *Systems and Software engineering – Methods and tools for model-based systems and software engineering*. https://www.iso.org/standard/79111.html (accessed May 2023).

Friedenthal, S., Griego, R., and Sampson, M. (2007). INCOSE model based systems engineering (MBSE) initiative. *17th Annual International Symposium of INCOSE*, Vol. 17, no 1 (San Diego), p. 2080. https://incose.onlinelibrary.wiley.com/doi/epdf/10.1002/j.2334-5837.2007.tb02999.x (accessed February 2023).

Henderson, K. and Salado, A. (2021). Value and benefits of model-based systems engineering (MBSE): evidence from the literature. *Systems Engineering* 24 (1): 51–66.

Henderson, K., McDermott, T., Van Aken, E., and Salado, A. (2021). Measurement framework for model-based systems engineering (MBSE). In *Proceedings of the American Society for Engineering Management 2021 International Annual Conference* G. Natarajan, E.H. Ng, and P.F. Katina eds.; American Society for Engineering Management (ASEM), 979-8-9853334-0-4. https://www.proceedings.com/content/062/062095webtoc.pdf.

Henderson, K., McDermott, T., Salado, A., and Van Aken, E. (2022). Towards developing metrics to evaluate digital engineering. *Systems Engineering* 26 (1): 3–31. https://incose.onlinelibrary.wiley.com/doi/10.1002/sys.21640 (accessed February 2023).

Honour, E.C. (2004). Understanding the value of systems engineering. *Paper presented at the 14th Annual International Symposium of INCOSE*, Toulouse, France (20–24 June). International Council on Systems Engineering (INCOSE) 14 1, pp. 1207–1222. https://incose.onlinelibrary.wiley.com/doi/10.1002/j.2334-5837.2004.tb00567.x (accessed May 2023).

Massachusetts Institute of Technology, INCOSE, and PSM (2010). *Systems Engineering Leading Indicators Guide*, v. 2.0. International Council on Systems Engineering.

McDermott, T., Henderson, K., Salado, A., and Bradley, J. (2022) Digital engineering measures: research and guidance. *Insight* 25 (1): 12–18.

McGarry, J., David Card, Cheryl Jones, Beth Layman, Elizabeth Clark, Joseph Dean, Fred Hall (2001). Practical Software Measurement: Objective Information for Decision Makers. Upper Saddle River, US-NJ: Addison-Wesley Professional. www.psmsc.com.

Practical Software and Systems Measurement (2022). Digital Engineering Measurement Framework Version 1.1, PSM-2022-05-001. https://psmsc.com/DEMeasurement.asp (accessed February 2023).

SERC Technical Report SERC-2020-SR-001 (2020). *Benchmarking the Benefits and Current Maturity of Model-Based Systems Engineering across the Enterprise*. Hoboken, US-NJ: Stevens Institute of Technology.

SERC Technical Report SERC-2020-TR-002 (2020). *Digital Engineering Metrics*. Hoboken, US-NJ: Stevens Institute of Technology.

SERC Technical Report SERC-2021-TR-024 (2021). *Application of Digital Engineering Measures*. Systems Engineering Research Center.

Biographical Sketches

Tom McDermott is the chief technology officer of the Systems Engineering Research Center (SERC) and a faculty member in the School of Systems and Enterprises at Stevens Institute of Technology in Hoboken, NJ. He previously held roles as faculty and director of research at Georgia Tech Research Institute and director and integrated product team manager at Lockheed Martin. Mr. McDermott teaches system architecture, systems and critical thinking, and engineering leadership. He is a fellow of the International Council on Systems Engineering (INCOSE).

Kaitlin Henderson has a PhD in Industrial and Systems Engineering with a concentration in Management Systems from Virginia Tech in Blacksburg, Virginia. She also earned her bachelor's and master's degrees in Industrial and Systems Engineering at Virginia Tech. Kaitlin has contributed to the fields of Model-Based Systems Engineering (MBSE) and Digital Engineering (DE) with her works on MBSE adoption, MBSE domain maturity, and MBSE/DE performance measurement.

Eileen Van Aken is a professor and department head of the Grado Department of Industrial and Systems Engineering at Virginia Tech. She earned her BS, MS, and PhD degrees in industrial engineering from Virginia Tech. Dr. Van Aken served as an associate department dead and undergraduate program director from 2005 to 2016, served as interim department head from 2016 to 2018, and was appointed as department head in June 2018. Her research and teaching interests are in organizational transformation, process improvement, team-based work systems, and performance measurement system design.

Alejandro Salado has over 15 years of experience as a systems engineer, consultant, researcher, and instructor. He is currently an associate professor of systems engineering with the Department of Systems and Industrial Engineering at the University of Arizona. In addition, he provides part-time consulting in areas related to enterprise transformation, cultural change of technical teams, systems engineering, and engineering strategy.

Joseph Bradley is president of Leading Change, LLC, and an adjunct associate professor at Old Dominion University (ODU). He is currently engaged as an advisor on the Navy's efforts to build digital twins of the four naval shipyards. Since retiring as a captain in the Engineering Duty Officer community, he has served in various consulting roles, including program manager's representative for the SSGN conversion of the USS OHIO and USS MICHIGAN, as well as supporting sustainability for the COLUMBIA SSBN acquisition program. His research interests include complex system governance, digital twins, and decision-making using modeling and simulation.

Chapter 8

Digital Engineering Implications on Decision-Making Processes

Samuel Kovacic[1], Mustafa Canan[2], Jiang Li[3], and Andres Sousa-Poza[1]

[1] *Engineering Management/System Engineering, Old Dominion University, Norfolk, VA, USA*
[2] *Information Sciences Department, Naval Postgraduate School, Monterey, CA, USA*
[3] *Electrical Computer Engineering, Old Dominion University, Norfolk, VA, USA*

Introduction

Reliability-centered maintenance (RCM) is an organizational strategy that optimizes the maintenance program of a company. RCM adopts the holistic paradigm of the system engineer to ensure maintenance tasks are performed in an efficient, cost-effective, reliable, and safe manner. Organizations are actively employing digital engineering techniques to transform engineering practice within their maintenance divisions. Digital engineering (DE) provides a unique opportunity for system resilience; its probative nature providing the ability to holistically interpret a problem domain over time to aide in the maintenance of a system can be considered a critical component to pursuing RCM and value creation of the supporting decision process.

Define Problem

The same technology that is advancing organizations into the digital world has introduced a higher level of complexity that confounds decisions. Complex and fragmented processes and data leads to challenges in the decision process. In today's digitally savvy domain the decision environment must be capable of consuming digital data for analytics, visualization, and situational awareness. Digital Engineering can improve the digital process by providing increased capabilities to the decision environment.

Effective maintenance planning focuses on ensuring decision makers can rely on a machine or process providing a minimally acceptable level of operational performance, without exceeding a maximum cost. This is achieved by maximizing the use that is obtained from a device before a necessary maintenance cost is incurred. This generates several challenges. For example, there can be significant variability in predictions of the useful life of a component or device. While maximizing the use of a system component until it is closer to the point, which it will either degrade or increase the risk of experiencing a failure before it is maintained, thereby decreasing its reliability. The difficulty of maintenance planning is greatly increased as components, each with their own

Systems Engineering for the Digital Age: Practitioner Perspectives, First Edition. Edited by Dinesh Verma.

Companion website: www.wiley.com/go/verma/systemsengineering

operating characteristics and operational lives, are combined into a hierarchically nested set of machines, processes, sub-systems, and systems. The maintenance planning scenario is further complicated by social and technical considerations, availability of maintenance facilities, availability of parts, operating standards, etc.

Human decision-making is probabilistic and dynamic. The primary goal of decision science is to describe how real humans make a decision under uncertainty. Since real human decision-making typically demonstrates variations from the classically described ideal rational human, decision sciences introduced various probabilistic interpretations of human behavior. Each interpretation resulted in a different understanding of decision under uncertainty (Hertwig et al. 2019), and these new understandings of uncertainty have accentuated the limitations of the "unknown based on probability theory" (Aerts 2009). These interpretations give rise to two classifications of uncertainty that are based on the source of uncertainty (aleatory) and degree of uncertainty (epistemic) in one's knowledge. This uncertainty must be understood for an effective decision and is the premise for the tools, methods, and frameworks introduced in this chapter.

Relevant Background

The Digital Transformation Initiative has altered how we approach and engineer complex problems. Trends in digitization have been augmented by artificial intelligence (AI) and machine learning (ML) technologies capable of independent inferences that mimic human behavior. These AI/ML-based enablers gave rise to a state-of-the-art digital-physical system. This digitization trend has evolved to the Digital Twin (DT) (Grieves 2014; Jones et al. 2020) that has improved processes, decision-making, and planning for organizations to capitalize on the benefits of digital engineering. The significance of DT applications is due to the two-way information/data exchange between the physical and virtual product, which is the salient attribute of the DT capability that can augment three of the human knowledge tools: conceptualization, comparison, and collaboration. Exploring methods to understanding human compatibility in digital engineering, how conceptualization informs the decision space, and consuming digital data are necessary for the development of tools. RCM as a strategy extends predictive capabilities to proactive capabilities. Adopting a rational-based possibilistic modality over the current empirical-based probabilistic modality will improve RCM forecasting approach. The research was conducted against a maintenance backdrop based on a case study conducted on ship availability to meet mission in a complex situation in a Department of Defense (DoD) environment. The research focused on improving the decision environment, integrating digital technology with a human team member, describing a paradigm conducive to complex and emergent situations, and finally digitization and digitalization of data and processes. The work is focused on the implementation of DE in regard to a proactive RCM strategy and effective decision-making for reliability, availability, and maintainability (RAM). This research studies the activities necessary for RCM, specifically the development of a DT model and the exploration of techniques for decision space to augment the DT.

New Approaches to Decision Process

Organizations are developing RCM strategies to augment their preventive and predictive maintenance approach with a proactive maintenance approach for decreasing inefficiencies and cost. RCM has an impact both in proactively assessing part and component reliability but also aids in decision-making for prognosis and/or forecasts of system and system-of-system mission effectiveness.

A decision space must be capable of aiding the decision maker in making the best decision. There are several capabilities an organization can employ to improve the decision space: (1) develop a governing strategy to ensure the decision aligns with organizational goals, (2) adopt repeatable method for making decisions, (3) define a framework to ensure tools processes and techniques are available for the method and at the correct time, (4) identify the necessary information and structure for the decision maker, and (5) educate the organization on the decision environment. Digital engineering has a profound effect on these capabilities to provide tools and information to the decision maker.

Technology has changed the way reliability is assessed; integration of technology has created a level of complexity to the task of establishing reliability for both the technical and operational level. Greater complexity of the components means a more complex breadth of tasks and the need for advanced warnings of failure probability. Revitalized approaches to the maintenance strategy; one that will better align with today's digital evolution, sensor-based innovations, strict compliance mandates, and escalating customer demands will improve decisions. The scope of the research is to explore and exploit advanced and emerging technologies that can be integrated into the RCM strategy to improve organizations decision-making capabilities. The chapter is parsed into two categories; each distinct but dependent efforts that reflect how the technology may be most effectively used.

- The first category, digital engineering, addresses how engineering practices throughout the life cycle may exploit advanced technologies to improve predictive and proactive decision-making.
- The second category, system conceptualization decision environment, addresses how the decision-making process is improved based on increased understanding within uncertain situations. Where uncertainty is created through complexity, emergence, and/or random behaviors or interactions.

Both categories can have an enabling and disabling effect when combined and must be deliberately considered both stand-alone and integrated into complementary capabilities before implementing into an operational environment.

Augmenting the Decision Process with Digital Technology

In the context of system engineering, the whole consists of two elements. The parts and their perceived behaviors in the context of the relationships of integrating the parts. As a result, with the interaction between the physical and conceptual system and the domain, the whole can become greater than the sum of its parts; in this context, the conceptual system (domain) along with the physical system (product) complements the whole. Thus, since a stable whole is desired for decision quality, both the physical and conceptual systems should retain and maintain their individual stability. Relatedly, concepts like complexity and uncertainty that cause instability in engineered systems ensue from the interactions of the system components (Perrow 1999; Sterman 2010). Therefore, coping strategies with instability, complexity, and uncertainty that are based on data-driven relational statistical models and classical probability theory (CPT)-based probabilistic models can introduce repercussions to the decision environment (Busemeyer and Bruza 2012; Pearl 2000; Pearl and Mackenzie 2018; Busemeyer and Wang 2018a,b). To this end, one of the goals of digitalization is to take over a higher number of tasks from the human decision maker

(Wilhelm et al. 2020). In the context of DT type of advanced tools, the task can be analyzed from the information processing perspective because a DT capability can filter or process information quicker. A standard information processing model, shown in Figure 8.1, starts with observation/information gathering, analysis of this information, decision, and action; this model is a simplified linear model and does not involve any feedback and is only concerned with the input–output of each step.

To capture the real-time aspect of decision-making and information processing, two frameworks, situation awareness (SA) (Endsley 1995) and the Observe-Orient-Decide-Act loop (Osinga 2006), are commonly used. SA framework introduces three levels of SA, plus decide and act steps (Endsley 1995). The three levels of SA are interrelated ecological and mental entities. They represent the perception of the state of the environment (level 1 SA), processing the perceived information (level 2 SA), and determining the most probable belief state (level 3 SA) for decision and action, as shown in Figure 8.2.

The three levels of SA and decide and act are used to describe information processing and decision-making in various contexts. The OODA loop framework, Figure 8.3 (Osinga 2006), is like the SA framework but elaborates the subtle implicit processes in the three levels of SA. The orient phase of this framework is the most critical part of this framework because the mental processes

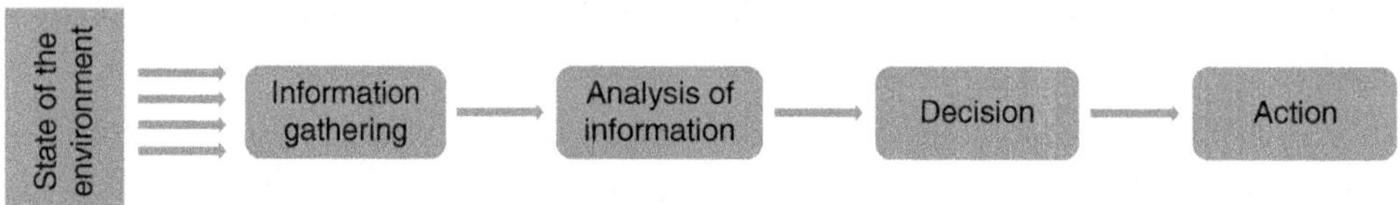

Figure 8.1 A general information processing model.

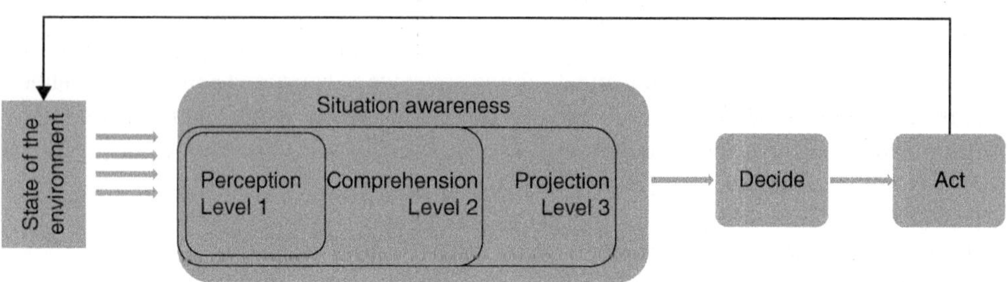

Figure 8.2 Situational Awareness Framework.

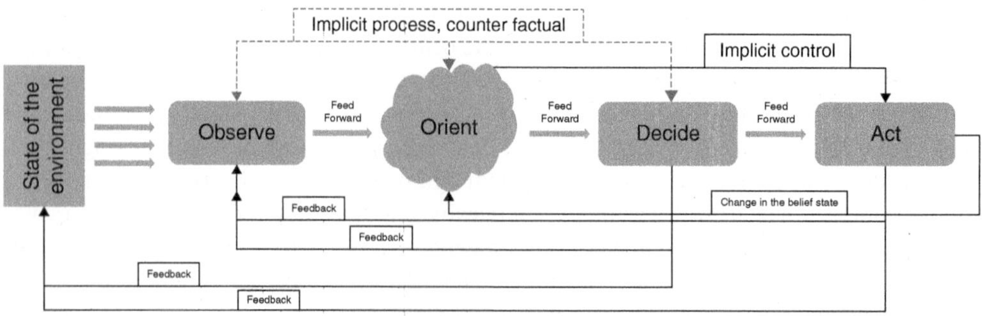

Figure 8.3 Observe, Orient, Decide, and Act (OODA) loop framework

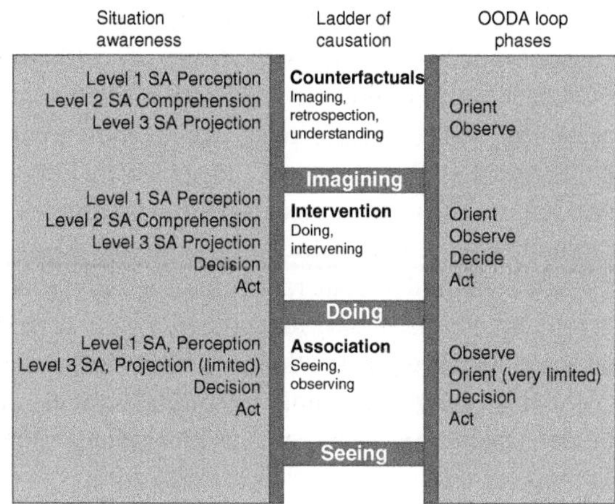

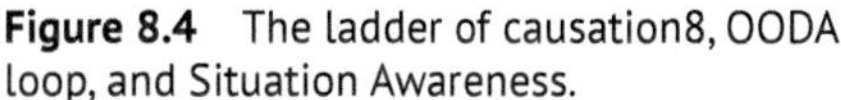
Figure 8.4 The ladder of causation8, OODA loop, and Situation Awareness.

of a decision-maker are addressed in this phase. For example, the orient phase includes counterfactual reasoning (Pearl 2000), and it is the phase in which the machines cannot outperform (Stumborg et al. 2019) the human decision-makers.

To articulate characteristics of the tasks that can be taken by a DT, the phases, and levels of these two frameworks are mapped to Pearl's ladder of causality (Pearl 2000), shown in Figure 8.4. Pearl indicates that data is dumb (Pearl and Mackenzie 2018) for both causality and conditional probabilistic descriptions such as $p(Y \mid X) > p(Y)$. Pearl and Mackenzie (2018) argue that machines are at the first step of the ladder of causation, shown in Figure 8.4. Thus, the limitation of these systems can directly or inadvertently constrain the decision and information processes in an organization. To elaborate on these limitations, SA levels and OODA loop phases were categorized concerning the three rungs of the ladder of causation, shown in Figure 8.4.

In the categorization shown in Figure 8.4, the levels of SA and phases of OODA loop frameworks are analyzed concerning their cognitive and physical characteristics. For example, the observe phase of the OODA loop and perception level of SA demonstrate both physical and cognitive characteristics. AI/ML-based state-of-the-art systems are capable of mimicking humans and, in certain situations, can be more advantageous over humans (Stumborg et al. 2019), therefore, the observe phase and perception level are categorized at the first rung (seeing) of the ladder of causation. This rung has extremely limited or no causal understanding of the phenomenon. One of the typical questions that can be answered at the seeing rung is "what is?" Or a conclusion based on the analysis at this rung would be based on conditional expectation, such as "a human bought chips"; thus, the human is likely to buy a dipping sauce. However, a human cognitive system could still glean more than AI/ML-based systems, this rung in terms of information seeking and sensor capabilities is dominated by machines. Therefore, in a decision environment that includes DT, the processes that are associated with this rung can be overtaken by DT capabilities.

The second rung (doing) of the ladder of causation is dominated by the characteristics of a human cognitive system. It involves considering the consequences of intervention, feedback from the environment, or action to the environment. One of the typical questions that can be answered at the doing rung is, "what if I do this?" Answering this type of question cannot be done by data or conditional expectation. Pearl and Mackenzie (2018) describe this rung with a do operator, $p(y|do(x),z)$, which means the probability of $Y = y$ given the intervention of $X = x$ occurs, and subsequent to this intervention $Z = z$ is observed. In the first rung, this probabilistic understanding is limited to conditional $p(y|x)$. Ergo, contrary to the first rung, the second rung (doing) can entail all levels of SA and phases of the OODA loop. Specifically, the comprehension and the orient phase are not limited as they are in the first rung. Although $p(y|do(x),z)$ can be analytically estimated by a Causal Bayesian Network (CBN) (Pearl 2000), in the case of DT augmented decision support, this could still complicate the decision processes because SA and the OODA loop entail continuous implicit and explicit feedback, as shown in Figures 8.2 and 8.3, between all the phases. More importantly, the adaptation of a DT that is based on CBN will be incomplete (Snow et al. 2022) and can induce more uncertainty. In return, this adaptation will result in simplified models of cognitive processes and impede real human decision processes. Therefore, a human-oriented DT design or integration is necessary for compatibility to occur because current DT implementations may not take over all the processes that are categorized at this rung.

The third rung (imagining) in the ladder of causation corresponds to unique characteristics of human cognitive systems, such as imagining and retrospection. One of the typical questions that can represent the characteristics of this rung is, "what if I had acted instead?". This type of question requires retroactive reasoning, such as *p(y_x|x^′,y^′)*, which means the probability of event $Y = y$ in the case of x were to be X, however, observed to be x' and Y to be *y^′*. This type of calculation requires generative theories that involve physics-based models (Pearl 2000) so that the process aspect of decision-making can be accounted for (Rescher 2000; Whitehead and Griffin 1985). Thus, this rung demonstrates the characteristics of the level 2 SA and the orient phase of the OODA loop. In addition, since human cognitive systems outperform machines (Stumborg et al. 2019) in the processes that are peculiar to level 2 SA and the orient phase of the OODA loop (Stumborg et al. 2019), this categorization demonstrates that a human-compatible DT is critical to improving decision support because the current DT implementations cannot take over the processes that are categorized at this rung.

Since DT is an AI/ML-based decision support implementation, its limits must be understood to implement human-compatible improvements. For example, without a DT, a conceptual system can interact directly with the physical system and the inseparable phases, observe, and orient can twin concomitantly with the physical system. On the other hand, introducing a DT in between the physical system and conceptual system, as shown in Figure 8.5, can hinder the required twinning. Moreover, this design can induce instability in the conceptual system and give rise to various sources of uncertainty and, in return, can annihilate the whole that both physical and conceptual system should complement.

Technical Improvements to Digital Twin

The evolution of rational theory to explain human behavior in situations that involve uncertainty reached a remarkable milestone with the introduction of the utility theory (Busemeyer and Bruza 2012; Von Neumann and Morgenstern 2007). The axiomatic basis of the utility theory

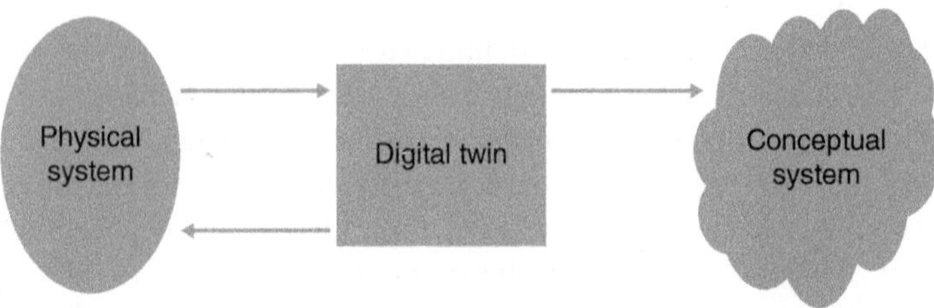

Figure 8.5 Current Digital Twin design by including the conceptual system

indicates that if preferences that are exhibited by a human satisfy the axioms of the rational action, then the choice preferences of this individual can be explicated by maximizing the expected value of a utility function. This is an important rational decision argument that means a rational agent acts to maximize expected utility, which is suitable for a decision under risk (Busemeyer and Bruza 2012). According to rational decision theories, actions follow logically deducted sequences derived from CPT, and the assignment of probabilities to events is based on the Kolmogorov axioms (Busemeyer and Bruza 2012).

Hitherto decision studies demonstrated that human behavior systemically violates the axioms of both the expected utility and subjective expected utility theories. Further research in this field resulted in descriptive theories that account for the systemic violations of the rationality axioms. A salient one of these theories is called prospect theory (Kahneman and Tversky 1979; Tversky et al. 1990), which replaces Kolmogorov probabilities with nonlinear and nonadditive decision weights. Although prospect theory accounts for some of the violations, additional paradoxes have been discovered that systemically violate the axioms of the prospect theory (Birnbaum 2008). Therefore, the lack of comprehensive descriptive decision exists (Busemeyer and Bruza 2012).

The common goal of AI research, design, and development is building rational machines (Russell 2020). Since the axiomatic foundation of this endeavor is based on CPT, the systemic violations of CPT can inadvertently control the decision processes in an engineered system. More importantly, the theories that shape the behavior of rational AI agents are for single agents acting alone; assigning probabilities for rationality for two or more becomes problematic for AI agents (Russell 2020). Since a DT implementation shares the limitations of its enablers, it may not be able to twin with the human component of the system. Thus, without apposite engineering efforts, the disparities between a human cognitive system and AI/ML-based cognitive system can inadvertently impede the engineering and management processes with an across-the-board application such as DT. In this paper, three technical areas that can improve the implementation of DT in decision support are discussed. These areas are open system decision modeling, Hilbert Space data fusion models, and quantum-like machine learning domains.

Models of Decision-Making

Depending on the theoretical foundation, a decision-making model can be a closed classical model or a closed quantum model. A closed classical model is based on a probabilistic mixture of the possible states of understanding. Typically, the time evolution of this system is expressed with the classical Markov model (Busemeyer et al. 2020; Busemeyer and Bruza 2012; Kvam et al. 2021), which results in a cumulative behavior, as shown in Figure 8.6. The state of understanding scales quickly even under the circumstance that a perturbation sustains the uncontrolled-complex environment. There are two reasons for the quick scaling. First, the classical models assume that the

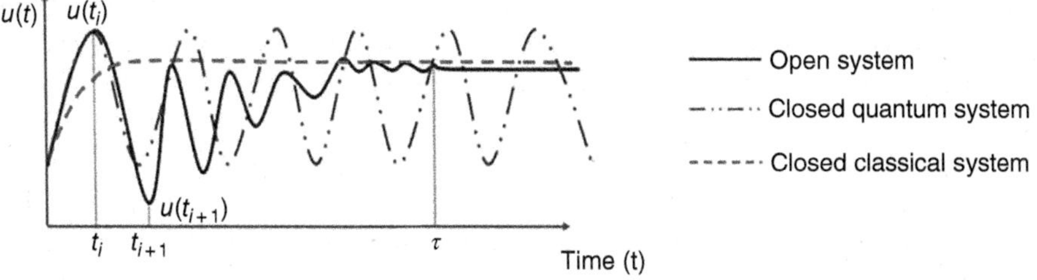

Figure 8.6 State of understanding for open system, closed quantum system, and closed classical system.

belief state of the decision-maker is always in a definite mental state. This means that the classical closed system has no ontic (internal) uncertainty. Second, the belief state of the decision-maker demonstrates an accumulated behavior and quickly scales to the most probable understanding regardless of the dynamics of the situation. Due to the accumulating nature of the model, the context-sensitive nature of the human mind cannot be captured; in other words, the variety of the mental states of the human is represented by the model.

The solid black graph shows the open system representation of the state of understanding. The closed classical model of the same system is shown by a red dashed line, and the closed quantum system is shown by a dotted dash line. Quantum models (Busemeyer and Wang 2018a) of decision-making allow a noninvasive system state evolution; the peculiar dynamics that give rise to this noninvasive evolution are the superposition of the possible decision outcomes and the unitary evolution of the system states. This unitary evolution is noninvasive and supports a temporal evolution that is implicit. This evolution is based on the primitive amplitudes, which are complex numbers, and interference term contributions can occur if the representations of the events demonstrate noncommutative relations (Busemeyer and Bruza 2012). When interference contributions to the total probability occur, the total probability can be different from 1 due to the interference term contribution (Aerts 2009; Busemeyer and Bruza 2012). These two dynamics also give rise to a measurement (ontic) uncertainty, and unless an explicit measurement (feedback, interaction) occurs, it may continue to influence the final decision.

Having a superposition of possible system states means that the entropy of the system still can be different from zero even if everything is known about the system (Von Neumann et al. 2018). This entropy can be minimized only by measuring (actualizing) the state of the system. To operationalize this concept in decision science, Busemeyer et al. (2020), Kvam et al. (2021), Kvam and Pleskac (2017), and Atmanspacher (1997, 2000, 2005) introduced a third type of uncertainty that integrates the behavioral heuristic with the quantum probability framework. This internal uncertainty is about the uncertainty of the system states that are in a superposition, which is not accounted for by the classical models. Introducing this third type of uncertainty is imperative to develop systemic strategies to cope with uncertainty (Busemeyer et al. 2020; Kvam et al. 2021).

To explicate this type of uncertainty, Kvam et al. (2021), Busemeyer et al. (2020), Martínez-Martínez and Sánchez-Burillo (2016), and Khrennikova et al. (2014) introduced the open quantum system modeling technique to decision science. By representing the state of a system with an open quantum system equation, three modeling improvements are achieved. First, the state of the systems is represented as a single probability distribution that can account for two distinct dynamics (quantum and classical Markov). Second, the time evolution of the system state in a complex and uncontrolled environment is expressed with both quantum and classical dynamics; for example, the quantum dynamics of the system state can elucidate the conceptual system's sensitivity to perturbations and information, or the classical Markov dynamics can elucidate adaptive behavior of system conceptual system that result from the interaction with the environment. Third, since the ontic type of uncertainty is about the conceptual system, a single probability distribution of the system state provides a systemic understanding of uncertainty; the quantum dynamics elucidate the ontic uncertainty, and Markov dynamics elucidate the epistemic type of uncertainty.

A Hybrid Approach: Open Quantum Model

Using an open quantum system approach to model the state of understanding means putting the mental model of the human in contact with the situation's dynamics. When a perturbation occurs to the system, it means that the system (including the conceptual system) starts engaging with an uncontrolled and complex environment. The evolution of the state of understanding can be expressed with an open system equation as ($du/dt = (1 - \gamma)$*Quantum_Dyanmics + γ*Classical_Dynamics) (Rivas and Huelga 2012). In this equation, $\gamma = 0$ represents the quantum end of the spectrum, and $\gamma = 1$ represents the classical end of the spectrum, as shown in Figure 8.7. When $\gamma = 0$, the evolution of belief states is dominated by the quantum dynamics; hence, there is high transience in the change of understanding, and acquiring more information cannot reduce the uncertainty due to the incompatible perspectives of the decision maker (Busemeyer and Bruza 2012) or information overload (Pothos et al. 2021).

Referring to the Figure 8.7, the probability of belief states highlighted is "Yes" (as one of two possible decision outcomes, the other being "No") for the six different γ values. When $\gamma = 0$, the open system model becomes like the quantum model, which demonstrates a continuous oscillation. As the γ gets closer to one, the open system starts by capturing the quantum oscillation and gradually adapting to the environment. When $\gamma = 1$, however, the system does not capture any oscillation and quickly scales to classical behavior values. The oscillation that is captured by quantum and open system models represents the internal indecisiveness of the decision-maker. The oscillation represents the ontic type of uncertainty, which is different from the epistemic type of uncertainty. Zhang et al. (2021) argue (supported by later studies (Canan et al. 2022; Snow et al. 2022)) that a driver assistant system that is designed with the open-system modeling approach can better adapt to the driver's behavior. The findings of these studies indicate that using an open quantum system model is better at capturing the internal dynamics of the decision-makers, and the same modeling technique can also be used to determine the instability in conceptual systems. For example, an open system model can capture any immediate impact of a perturbation in an environment by accounting for both physical and conceptual impacts with one single probability distribution. As shown in Figure 8.8, the open system interacts with the environment and evolves to a classical mixture at or after $t = \tau$. This means that the ontic uncertainty becomes negligible, and epistemic uncertainty governs the state of the understanding. This dissipation could take longer than the required decision time.

Relating comprehensibility and emergence to belief state of understanding via transience enables capturing contextual dynamics of complexity, uncertainty, and emergence. For example, when $\boldsymbol{\gamma} = \mathbf{0.5}$, the open system modeling approach starts with quantum oscillations. As the decision-maker adapts to the environment, the open system stabilizes around the classically most probably belief state. Using only a quantum or classical model will hinder the model from capturing the contextual effects of the decision process. The classical dynamics of the open system equation capture the epistemic uncertainty, and the quantum dynamics capture the ontic uncertainty. Transience 1 means that the change in the understanding is negligible, and transience 0 means that the change in understanding is too high so that the conceptual system experiences a high degree of uncertainty. Since the understanding of the system will demonstrate a continuous oscillation between the possible mental states, the degree of change in understanding becomes critical to cope with complexity, uncertainty, and emergence. In these types of situations, the mental model of the human is not capable of forming a coherent understanding of the situation because of not being able to form a coherent understanding of a situation humans will highly likely make an erroneous knowledge claim. After the perturbation, the belief states of the conceptual system start oscillating, and the feedback loops are needed to dissipate oscillation. In this model, the high transience in understanding is explained by the superposition of the mental states of the human. In the case of having no interaction with the environment, the superposition states maintain their vacillation due to ontic uncertainty, which is the source of the internal ambiguity or indecisiveness that human experience. What is critical to the model is that the ontic type of uncertainty cannot be observed by the others without an interruption.

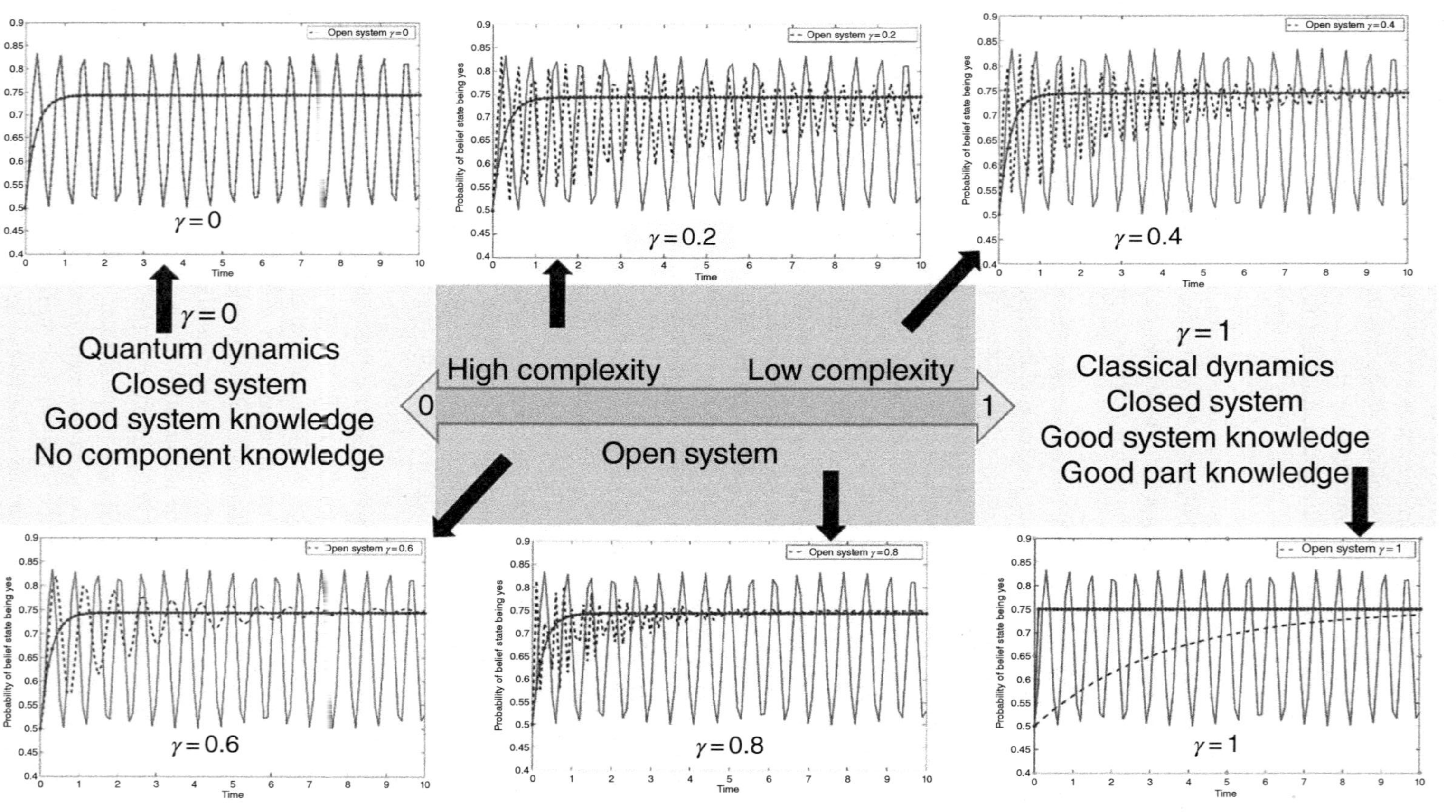

Figure 8.7 Open system parameter and complexity.

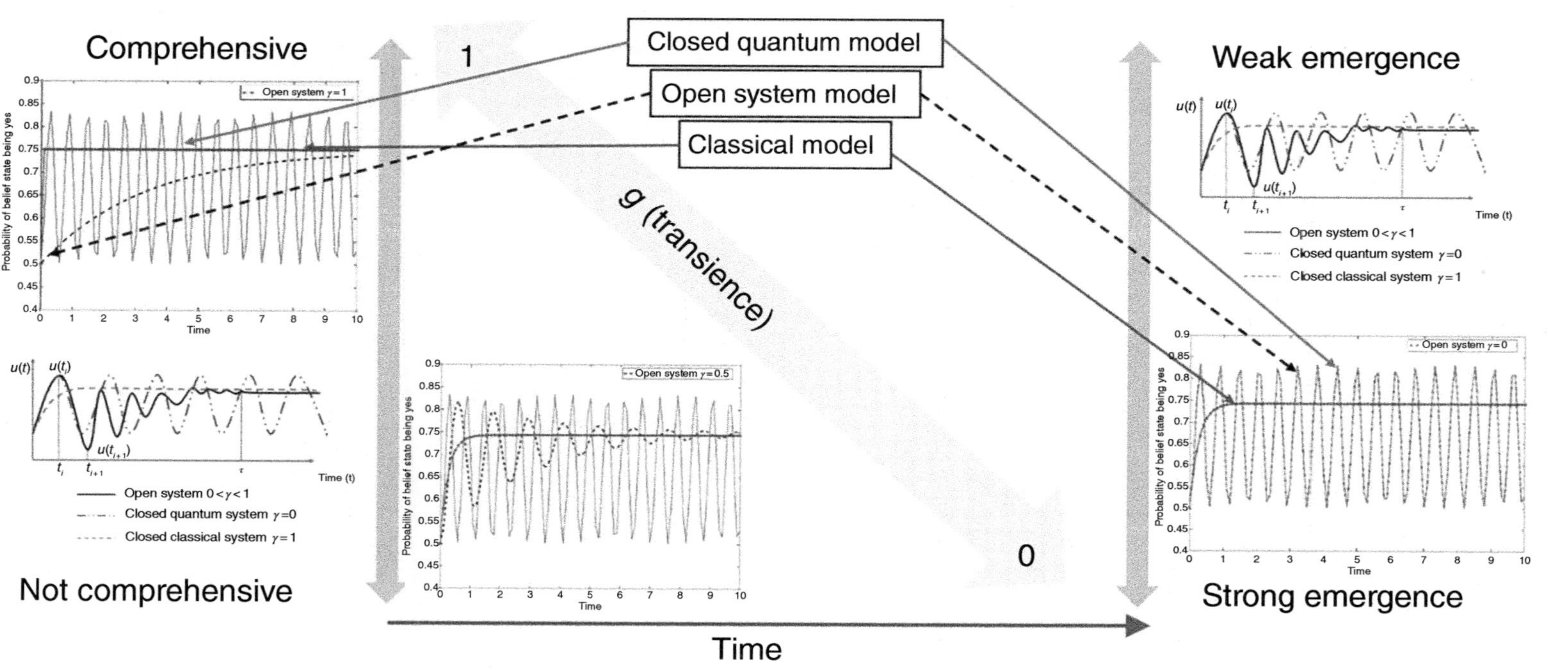

Figure 8.8 Comprehensibility and emergence to belief state.

Based on the demonstrated success of open system modeling in adapting to human behavior (Canan et al. 2022; Snow et al. 2022), improving the design of DT with open-system modeling can allow better prediction power for emergent behavior and improved twinning rate in complex situations. To improve the standard DT application shown in Figure 8.5, a human-compatible DT design can be implemented, as depicted in Figure 8.9. Using open-system modeling will augment the modeling and simulation components of DTs; specifically, using the open-system modeling will put DT in contact with the environment such that it can better adapt to the human in the loop decision environments. For example, the act phase of the OODA loop is critical to reducing the ontic type of uncertainty because it represents the physical feedback, the actualization of mental states. In the DT model shown in Figure 8.9, this physical feedback is connected to both the state of the understanding, which is based on the open-system model, and the physical system. In addition, the decision phase and controlled interaction (pre-decision) feedback are used to capture the belief state change of the conceptual system. The twinning feedback ($u(t_i)$) is the controlled feedback. If the results of the open-system model detect high transience in the change of understanding, it provides controlled feedback such that the instability in the conceptual system can be avoided.

Augmenting DTs with approaches like open-system modeling approaches provides a predictive model is built with generative theories. Using generative theories means that descriptive and predictive models are built with physics-based mathematical equations, not regression-driven models. This means that instead of introducing new predictive variables, the machine learning techniques tune their parameters to improve models. To this end, the quantum opens system modeling technique demonstrates promising results in capturing the behavior of the physical and conceptual system in one equation.

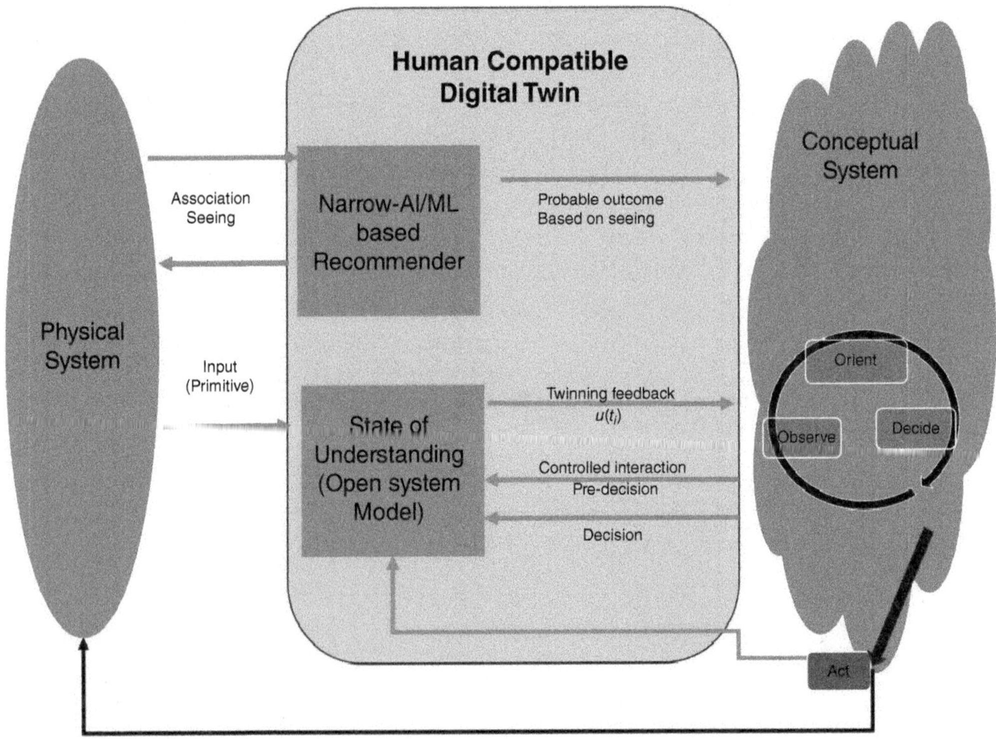

Figure 8.9 Human compatible digital twin design.

Data Fusion and Hilbert Space Modeling Contribution

An important feature of a DT is data collection from manifold instances of the system via Digital Twin Instances (DTI). DTI are the least decomposed level that you need for a coherent DT. DTI's describe the DT components of the physical product that need to be connected to allow for the necessary contextual two way information/data exchange. State-of-the-art sensors and associated technological enablers are used to capture these instances. Although the data transmission rate is considered a salient problem, the data fusion models and the creation of data tables at DTI levels are also critical for the success of the DT application. Depending on the volume of the data, DTs' data-related processes may involve collecting, summarizing, fusing, and analyzing data from various DTI contexts. Besides the state-of-the-art technological enablers, the theoretical aspect of data fusion is imperative for a human-compatible DT. The two-way connection between the physical system and DT, as shown in Figure 8.9, requires context-sensitive data summaries and fusion methods. Conceptually, the connection shown in Figure 8.9 should be acceptable for an effective DT application. However, receiving and processing data with DT and pushing it back into the system can suppress the contextuality (Busemeyer and Wang 2018b; Dzhafarov et al. 2017) and causality that can be gleaned by humans but is likely to be missed by machines.

The standard approach to handling large data sets is to summarize them by collections of contingency or cross-tabulation tables (Busemeyer and Wang 2018a,b, 2019). DTIs are the points of data collection, and all the pertinent variables at each level must be captured simultaneously for more accurate instance aggregation. Nevertheless, all the variables may not be measured at once. In this situation, the collected data would form a subset of all the variables, and each collected data can form a context m of measurement (Busemeyer and Wang 2018b; Dzhafarov et al. 2017); then, each of DTIs data set would form a collection of M different data tables. The challenge in data fusion then would be to aggregate and analyze these M different tables for Digital Twin Aggregate (DTA), the sum of all DTIs, such that the context-sensitive nature of the data maintains a coherent and interpretable representation from instances to aggregate. Busemeyer and Wang (2018b) argue that this data fusion problem arises in relational database theory and statistics. Although there are narrow solutions to this problem, e.g., Bayesian causal networks, generally, a universal relation cannot be obtained, and a sufficiently large joint distribution cannot reproduce the instances' data tables (Busemeyer and Wang 2018b, 2019). The conditions that give rise to these limitations are observed when the data tables violate consistency constraints required by classical probability theory. In return, a DTA and supporting DTIs can fail to capture causality and contextuality.

Hilbert space data fusion models (Busemeyer and Wang 2018a,b) can provide low-dimensional methods for data fusion, especially for systems that collect data from context-sensitive instances and then aim to aggregate these data sets or simulate at the edge. Hilbert space models provide powerful data fusion models for incompatible events. This is important because in a complex context-sensitive situation the variables that are used to model the complex interactions are highly likely to be incompatible, which is not recognized by the classical data fusion models. Augmented by the Hilbert space data fusion models, simulation in DT that is initiated at the DTI level can better capture the behavior of the system and can maintain the DTI level variety transitioning to DTA.

Quantum-Like Machine Learning

Quantum technology, e.g., quantum computer applications, can be an enabling technology for improved DT applications. Although the benefits of quantum computers are promising, their large-scale commercial applications are not feasible soon. Alternatively, augmenting the standard machine learning applications in classical computers with a quantum-computational framework with the quantum mathematical principles in machine learning algorithms is possible. This

Table 8.1 A summary of the difference between quantum machine learning and quantum-like machine learning.

	Quantum machine learning	Quantum-like machine learning
Environments	Quantum computer	Classical computer
Coding language	Translation of classical algorithms in the language of quantum computation	Quantum inspired information algorithms in the language of classical computer
Benefit	Computational complexity	Accuracy of the process
Invariance	Invariance under rescaling	Non-invariance under rescaling and under -tensor-copy- data pre-processing

Source: Adapted from Sergioli (2020).

approach is categorized as quantum-like machine learning. Although the focus is on the computational power of the quantum machine learning algorithms, according to Sergioli (2020) and Sergioli et al. (2019), one of the significant advantages of quantum-like machine learning algorithms is the accuracy of the quantum-like machine algorithms. Sergioli (2020) summarizes the differences, shown in Table 8.1, between quantum machine learning and quantum-like machine learning.

Since the DT concept relies on data analytics by using the data instances from DTI, along with the Hilbert space data fusion models, the quantum-like machine learning techniques can elicit various possibilities of DT applications.

Conceptual System

Causality is an investigation into the causes of the natural world around us, focused on explaining the "why" of things. Engineers often take a deterministic approach to design relying on causal relationships focused on the phenomenological nature of the domain and the systems deployed within the domain to maintain coherence through the engineering process. The ability to draw objective statements of "why," however, are hampered by the uncertainty created by complex, random, and emergent conditions. This challenges the conventional perspective of "systems" that rely on a singular perspective of the domain and the transitivity between the abstractions within the domain. The decision-makers must adopt a holistic model of the environment that allows them to explain complex relationships across the domain. Key capabilities for a prescriptive maintenance strategy in a decision process are discussed to predict the front end degradation of the holistic system and its contribution to mission. The mental model of a decision-maker and situational cues determine the possible outcomes in a decision-making situation. The situational outcomes and their associated probability values vary with the mental model of the decision maker and the cue of the information. A mental model begins establishing comprehension as soon as it starts gleaning information from cues. This process is also known as conceptual bounding, which can be represented by a domain of awareness (Canan and Sousa-Poza 2016).

The domain of awareness comprises perspectives that can represent the cues and events in a situation. Constituent perspectives of a domain of awareness can be compatible and incompatible. When perspectives are incompatible, the decision-making process can be captured by the axioms that are not supported by the classical probability theory, e.g., order effects and anomalies, unexpected events are observed. Capitalizing on theories that are comprehensive beyond the classical probability theory is critical to ameliorating uncertainty and risk management in complex situations. Specifically, since rationality is also bounded, observations and comprehension of the situation are governed by the constituents of a domain of awareness. Therefore, rationality is situational and does not need to be generalizable. This makes the rationality argument contextual as well.

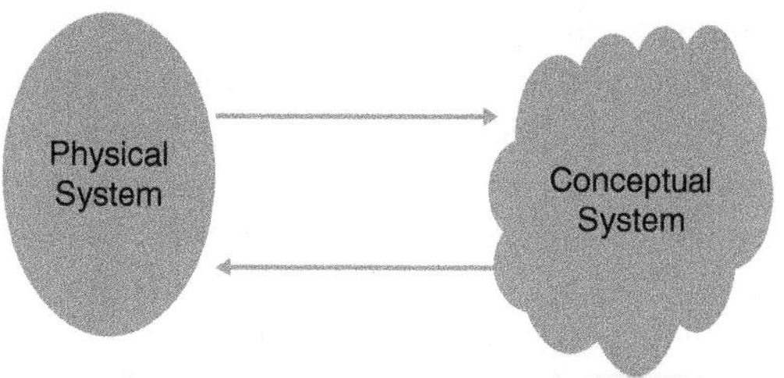

Figure 8.10 Physical, digital, and conceptual system.

To understand and model situational rationality, a participant observer dichotomy is necessary so that the epistemic states (trying to gain insight by knowing more about or setting the environment, etc.) and ontic states of the system (the representation of the world inside out, emic) and their influences on decisions can be systemically distinguished. This is important in the context of systems engineering because systems can be physical, conceptual, or a combination of these two that complement each other to support the design of an engineered system (INCOSE n.d.). System engineers maintain conceptual systems that are a model of the system in the situation and future systems (INCOSE n.d.). These conceptual models are used to project possibilities as solutions and decisions; thus, a conceptual system as a model is a comprehensive representation of the individual and the individual's understanding of the social and environmental dynamics with which the engineered system interacts, as shown in Figure 8.10.

In this case, a conceptual system interacts with the environment directly, moreover, a conceptual system forms the baseline of the process instructions to use, maintain, and retire the engineered system. A system is a perceived phenomenon by an observer and is a way of organizing a mental representation of the perceived phenomenon (INCOSE n.d.). Thus, a system does not exist independent of the human mind and cannot be understood by analyzing its physical parts alone. This analysis is always a process of abstraction and entails a conceptual bounding that includes relevant contextual variables that describe and represent the system of interest. In the context of system engineering, the whole consists of two elements. These are the parts and the perceived behaviors of the parts in the context of the relationships of integrating the parts. As a result, with the interaction between the physical and conceptual system and the domain, the whole becomes greater than the sum of its parts; however, as discussed by INCOSE (n.d.), the conceptual system (domain) along with the physical system (product) are compliments of the whole. Thus, for a stable whole, both the physical and conceptual systems should retain and maintain their individual stability. Relatedly, concepts like complexity and uncertainty that cause instability in engineered systems ensue from the interactions of the system components (Bar-Yam 2004; Perrow 1999). Therefore, coping strategies with instability, complexity, and uncertainty that are based on data-driven relational statistical models and classical probability-based probabilistic models can introduce repercussions to the decision environment (Hertwig et al. 2019; Kvam et al. 2021).

Effective maintenance planning focuses on ensuring decision-makers can rely on a machine or processes providing a minimally acceptable level of operational performance, without exceeding a maximum cost. This is achieved by maximizing the use that is obtained from a device before a necessary maintenance cost is incurred. This generates several challenges. For example: there can be significant variability in predictions of the useful life of a component or device. While maximizing the use of a system component until it is closer to the point which it will either degrade or the risk of experiencing a failure before it is maintained increases, thereby decreasing its reliability. The difficulty of maintenance planning is increased as components, each with their own operating characteristics and operational lives, are combined into a hierarchically nested set of machines, processes, sub-systems, and systems. The maintenance planning scenario is further complicated by social and technical considerations, availability of maintenance facilities, availability of parts, operating standards, etc. The overall scope and nature of uncertainty that is generated must be considered to develop an effective maintenance strategy. In the case of basic components, planners can focus on the part's failure rates. For more intricate machines and processes, Bayesian approaches are used to develop models that can identify different failure modes and the relative likelihood that they might occur.

A framework is introduced for bounding that is better suited to deal with the possibilistic nature of the problem domain. There are three specific challenges: The effect of uncertainty in reliability distributions, compounding uncertainties, and the difficulty of developing a maintenance plan within a set of interdependent maintenance plans. A set of principles are identified on which we establish the backbone of the proposed framework. The resulting framework is presented within the construct of conceptual bounding.

The separation between the conceptual system and the physical system is minimal, which means understanding is high. However, as variability within the reliability of the parts becomes significant, the approximate distance between the conceptual system and the physical system increases and confidence in the design and the product starts to deteriorate. This deterioration increases over time as shown in Figure 8.11.

Reality, Domain, Perspective (RDP) elaborates how bounding maintains the natural ties to reality in understanding dynamic and context-specific situations. The domain in the RDP construct is composed of three elements, the "observer," the "entity," and the "solution form." The complexity in this construct emanates from the disparities, or the dissonances between the individuals' domains. The domain maintains the individual's ability of understanding a problem. The degree of abstraction affects the comprehensibility and degree of bounding. The domain, *D*, can support multiple perspectives. This provides a framework that accounts for paradoxical interpretations. Since the domain supports multiple perspectives, the expected behavior becomes situationally sense making instead of being correct or incorrect. In part, a paradox emerges when two irreconcilable domains describe the same entity. The expected behaviors are both correct in their domains, however, can be perceived as incorrect in the other domains. This can happen between engineers and stake holders in the system design context in which the same notion is understood differently yet requires action.

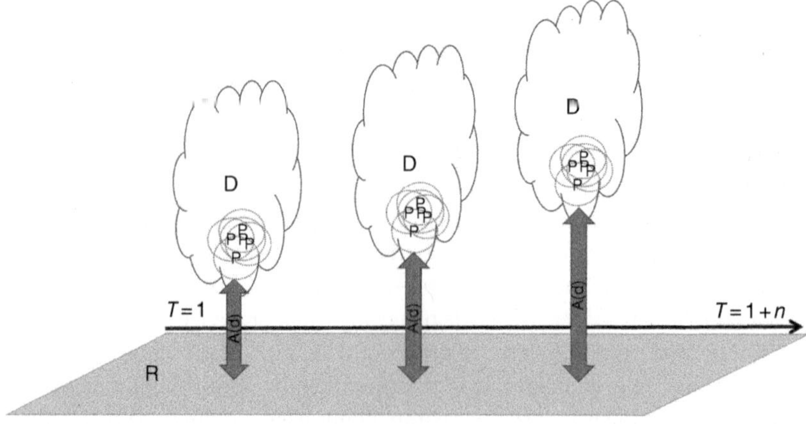

Figure 8.11 Uncertainty in R-D-P.

The distance between R and D becomes greater as more disparate perspectives are produced; in part, this generates uncertainty and can reduce the confidence that in the solution reflects reality. As contemporary systems become more technologically complicated and reliance on resources that used to be provided in house or by a handful of vendors become less likely, reliability rates for components become questionable, and has a significant impact on reliability, availability, and maintainability analysis. There is an exponential increase in randomness as each component is integrated into a system and each system into a system of system.

Confidence in reliability of the design drops as multiple paths or solutions are presented, (refer to Figure 8.12). Engineers face significant challenges when working with large-scale, socio-technical engineered systems such as a ship. Engineered systems commonly integrate existing and new components, are subjected to constant change brought on by the evolution of technology, serve multiple missions and capabilities, and integrate humans as components, with recent developments increasingly including smart machines. This result is the engineer having to deal with a significant degree of complexity, stemming from the scale of the problem, the nature of the relationships between components, and a strong dependence on the context or environment in which solutions are to be fielded. Engineers also frequently must deal with the uncertainty brought on by shifts in expectations and requirements and other changes in the environment. RCM approach is focused on determining "what failure management strategies should be applied to ensure the system achieves the desired levels of safety, reliability, environmental soundness, and operational readiness in the most cost-effective manner," which will include corrective, preventative, and proactive methods. RCM is expected to be a continuous process that uses design, operations, maintenance, logistics, engineering, and cost data across lifecycle to improve operating capability, design, and maintenance. It should make use of digital and information technologies to fully support the decision-making and planning capabilities that will be needed. To meet its effectiveness and efficiency goals, RCM must on one hand focus on single components (for example, pumps, or engines) for which traditional maintenance decision must be made. Component level decision must be weighed against their role and impact in sub-systems or multi-component engineered solutions (for example, propulsion systems) and in multi-system solutions (for example, integrated navigation) that might begin to exhibit some complex relationships and behaviors.

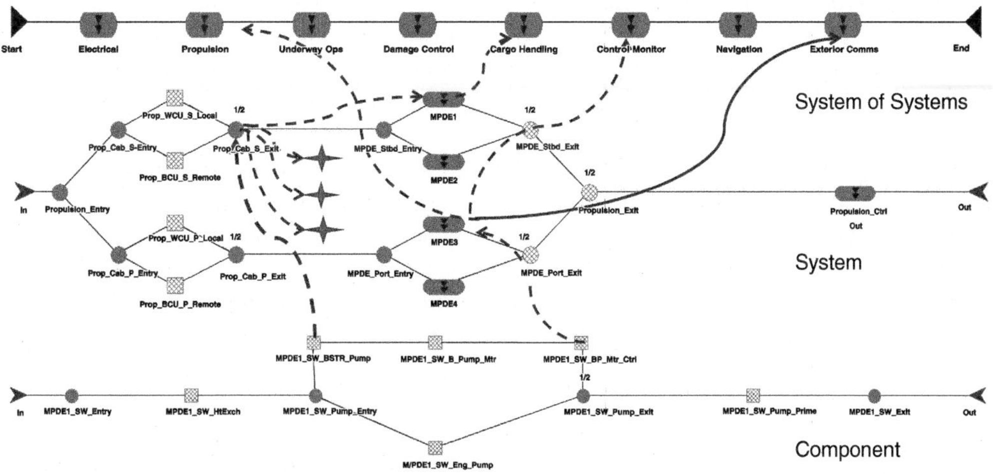

Figure 8.12 Uncertainty in reliability.

Critically, RCM can only avoid sub-optimizing total maintenance costs and not meeting assets availability goals by taking a holistic, full life cycle, view of the problem. Large scale socio-technical solutions (for example prognosis capability for a ship, or fleet readiness planning) demand extremely advanced approaches that consider the complexity of the situation. A multi-level problem is generated that is often also multi-paradigmatic. The holistic understanding of the system will be based on our understanding of the components, which results in a complex problem (intractable). For this reason, it is not possible to combine the calculated reliability of the components of a ship to readily gain an understanding of the reliability of the entire ship.

Conceptual Bounding

If the predictive capability of the current paradigm fails to resolve uncertainty in design, then a new paradigm must be adopted so that a new future state can be projected. A paradigm that must adopt principles conducive to the engineering process that allows for understanding the possibilities within the bounded domain. The intent is to illuminate the influence of a "conceptual bound"; a bounding that captures a deeper and more insightful understanding of the domain. Exploring the implications on the concept of systems in relationship to how maintenance is affected where the bounded domain must deal with a lack of a singular perspective or transitivity between abstractions. A conceptual bound is a rational construct where accuracy and precision are maintained in the shared understanding between the observers. As such, the bounding is a continuous statement that allows for the analysis, synthesis, and evaluation of a design as it evolves over time.

A Framework for Conceptual Bound

Conceptual Bounding is the process for a conceptual bound based on a possibilistic paradigm – Conceptual Bounding is axiomatic framework that depends on the interplay between its supporting principles. A conceptual bound is based on methodology, theories, and principles that improve understanding and are enacted together to inform the process and are critical for a decision process. A framework for bounding is provided to aid in the development of processes, methods, techniques, and tools in a decision space for effective decision-making in uncertain situations.

- **Reality, Domain, Perspective (RDP)**: As multiple interpretations of the system develop, whether it is between specialties or between specialties and decision-makers, the original assumptions may no longer be applicable to comprehend the situation. Multiple perspectives of the same system can generate uncertainty that requires a separation between the design and the system. This separation grows bigger over time and in return a new bounding of the domain becomes necessary to maintain understanding or reduce the separation between the concept and the system of interest.
- **Process philosophy:** According to INCOSE's (n.d.), systems can be physical, conceptual, or a combination of these two that complement each other to support an essential task in the system engineering of engineered systems. This system engineering task is to design systems that can evade ramifications when the physical system is enacted and begins interacting with the environment. A conceptual system becomes critical in this task to project possible solutions and baseline the solution for insertion into operations. A physical system exhibits observable behavior that emanates from the interaction of its constituents. However, a conceptual system is abstract and does not exhibit observable behavior. A conceptual system represents the projected future system of interest and offers various potentialities that could be actualized upon the system enacted in the environment (Lord et al. 2015). Since reality is dynamic, the engineering process must adopt an agile approach that supports exploration and exploitations for

re-alignment to so that environmental uncertainty can be obviated or minimized. Accepting a tolerance range within reliability rates suggests the need for iteration more frequently and recognizing through the iteration small incremental changes will afford a more likely or realistic result.

- **Pragmatic idealism (PI)** (Canan and Sousa-Poza 2019; Sousa-Poza and Correa-Martinez 2005). PI Introduces a probabilistic understanding of the complexity. Provides a structure for multiple, competing, or conflicting perspectives of a problem to coexist eliminating the requirement of adopting a single point of view.
- **The duality of understanding** (Brewer 2013; Sousa-Poza 2013). Provides a model of how cognitive representations of reality are generated that incorporates the phenomenological (observed) and numerological (experienced, but not observable) elements of reality. This is important because to establish an awareness, humans process the available information in the situation. This is a generative process that entails reciprocity of the perceiver (self) and the environment (other than self).
- **Complex adaptive situations methodology** (Canan et al. 2016; Canan and Sousa-Poza 2019). A holistic understanding of a situation entails the reciprocity of the elements of a situation. Therefore, it is necessary to describe adaptive behavior at the confluence of a system and human that form a situation. In doing so, a comprehensive complex adaptive situations methodology for emergent behavior can be developed.
- **Complex situation ontology** (Kovacic 2005). Taxonomical/relationship construct to draw conclusions on the nature of the solution form that the RCM will need to take
- **Situational causation**. The effects of any human action or dormant presence in a situation can only be understood through an (enduring) algedonic feedback and accounts for latent effects. In addition to utilizing the classical probability theory approaches to uncertainty, leveraging the quantum mathematical axioms will allow the bifurcation of the sources of uncertainty of the systems epistemic vs. ontic uncertainty. Representing these two sources of uncertainty allows the model to account for the situational oscillations among the perspectives of the decision-makers.

Organizations advancing its maintenance practices through the introduction of RCM focus on the holistic system in addition to its components. This elevates engineering maintenance from focusing on components to the system and simultaneously inculcating a significant degree of complexity. Conceptual Bounds provides the necessary framework for a prognosis or forecast of possible proactive maintenance solutions. To successfully achieve this, organizations will have to develop novel approaches to deal with the uncertainty that emanates from the complexity of a situation. RCM employs a prescriptive strategy used to determine what maintenance tasks are needed across the system life cycle. Like many organizations, a challenge exists to manage preventive care and keep pace with the complexity of operational assets, RCM provides a viable methodology for prognosis maintenance but must be augmented with a framework that can deal with the possibilistic nature of the domain.

Applying Conceptual Bounding

The classical view of probability conflates degrees of subjective beliefs and rules of chance. As articulated by Hacking, the probability is related to "the degree of belief warranted by evidence" (epistemic probability) and the stable relative frequencies produced by random occurrences in the environment, such as dice (aleatory probability) (Hertwig et al. 2019). These two types of probability give rise to two classifications of uncertainty that are based on the source of uncertainty (aleatory) and degree of uncertainty (epistemic) in one's knowledge. According to Kozyreva (2019), epistemic uncertainty represents the incompleteness of one's knowledge, which can be reduced by gathering more information from the environment; on the contrary, the aleatory uncertainty ensues from the statistical properties of the environment that are independent of a person's knowledge.

In decision science, it is argued that judgment under uncertainty entails contribution from both aleatory and epistemic sources of uncertainty. Recognizing this is critical because dealing with these two sources of uncertainty requires different strategies. Hence, coping strategies become critical in decision-making because using a digital twin can help reduce epistemic uncertainty resulting from an individual's epistemic constraint. Modern decision theories assume that the occurrence of every outcome of a decision or action can be assigned a probability that follows a classical distribution. The sum of the probabilities for all the possible occurrences will equal 1, even if they are ascertained by subjective means, for example, using Bayesian decision theory (Hertwig et al. 2019). This assumes that there is limited interdependence between outcomes that could result in quantum-like interference, and that the outcomes are bounded and finite. These assumptions, however, apply to a limited set of situations. More generally, the scope and nature of the problems will complicate and engender higher uncertainty due to the individual differences in a situation. In such situations, classical probability distributions fail to describe the nature of the uncertainty. Ontic uncertainty is introduced, and the outcomes take on a possibilistic nature.

For clarity, in possibilistic situations, probabilities can be assigned to outcomes, but the probabilities do not adhere to classical theory. For example, interference between factors resulting in outcomes and between subjective interpretations means that the sum of the probabilities can exceed 1. Ontologically, situations can be differentiated into a class, Realwarscheindlichkeit, that can be described using classical techniques and a class, Realmöglichkeit (Sousa-Poza and Correa-Martinez 2005), that requires techniques that account for the interference, such as quantum probability distributions.

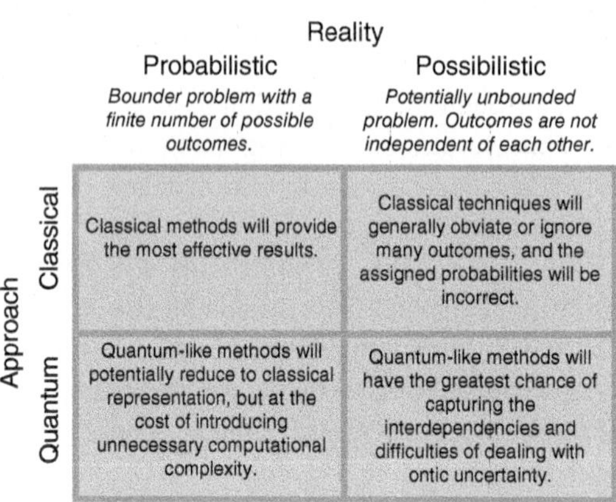

Figure 8.13 Uncertainty class and approach correspondence.

This proposed method differentiates between possibilistic and probabilistic classes of problems within a conceptual bound (refer to Figure 8.13). The environment adapts the decision-making tools and processes to maximize their effectiveness according to the conditions that are encountered. Just as critically, it will reduce the potentially negative outcomes that stem from misapplying approaches to situations for which they were not intended. The approach will inform an RCM strategy to pursue prognostic capabilities and a more informed decision-making process for fleet maintenance to complement an CBM (Condition Based Maintenance) approach.

Future Work

The Digital Transformation Initiative has expanded the solution providers' capability to solve complex problems. Digital Twin provides the capability to capture a system in contexts and can affect human interaction and decision processes. A decision space that can accommodate the domain environment so that decisions can be made is the end state for this work.

The overarching goal for a decision environment is to provide the means to execute conceptual bounding effectively. Open Systems necessitate conceptual bounding. To understand the strategic implications for proactive maintenance effective bounding of the domain is imperative, where each abstraction is as equi-probable and irreducible as the system in design. Research and development of a methodology is for an effective employment of a framework for decisions. Developing the methodology will provide for a path forward for decision-making in uncertain domains. Development of the methodology will also actualize the decision environment to understand sequencing of the process, persistence, and curation of the data, and alignment of tools and technology. Additionally, a situation diagnostic tool is required to aide in the methodology to provide situational awareness to the decision-maker while executing within the decision space. The methodology, the decision environment architecture, and situational diagnostic are all key to prognostic decision-making and organizations sojourn into proactive strategic maintenance.

A paradigmatic shift is necessary to exploit the various modeling approaches' and contributors for enabling a digital twin to interpret data within the domain and the environment and inculcate the generative process of the human to exploit this more robust understanding toward a better solution. Research at National Center for System of Systems Engineering (NCSOSE) on Complex Adaptive Situations Environment (CASE), provides the means for integrating methods, processes, techniques, and framework to effectively execute an organizational decision process. Continued research into CASE expands the current work in the following areas.

Open system modeling, the Hilbert space data fusion models, and the quantum-like machine learning techniques can elicit various possibilities of DT applications. CASE can recognize this type of uncertainty, improving the design of DT with quantum open-system modeling allowing for better prediction power for emergent behavior and improved twinning rate in complex situations. Using open system modeling will augment the modeling and simulation components of DTs such that DTs will be in contact with the environment and better adapt to the human-in-the-loop decision environment. Augmented by the Hilbert space data fusion models, a simulation in DT that is initiated at the DTI level can better capture the behavior of the system and can maintain the DTI level variety transitioning to DTA.

Integration of DT into the systems in which a DT can take over certain human tasks requires a human-centered approach enabled with CASE. Depending on the cognitive characteristic of the task, a human cannot be replaced by DT or another type of AI/ML-based system. Contextually, data fusion with the classical approaches may impede the full capability of the DT application. More importantly, knowing more does not reduce uncertainty, especially the aleatory type of uncertainty becomes critical in context sensitive situations. Therefore, the ontic type of uncertainty that can explain the indecisiveness of the decision-maker should be treated differently, CASE can recognize this type of uncertainty.

Data fusion with the classical approaches may impede the full capability of the DT application. More importantly, knowing more does not reduce uncertainty, especially the aleatory type of uncertainty becomes critical in context sensitive situations. Therefore, the ontic type of uncertainty that can explain the indecisiveness of the decision-maker should be treated differently.

Acknowledgements

This chapter is based upon research supported by the U.S. Department of Defense through the Systems Engineering Research Center (SERC) under Contracts HQ003413D0004 and W15QKN18D0040. SERC is a federally funded University Affiliated Research Center managed by

Stevens Institute of Technology. Any opinions, findings and conclusions or recommendations expressed in this material are those of the authors and do not necessarily reflect the views of the United States Department of Defense. The collaborators who supported this research are listed on the Handbook URL.

References

Aerts, D. (2009). Quantum structure in cognition. *Journal of Mathematical Psychology* 53 (5): 314–348. https://doi.org/10.1016/j.jmp.2009.04.005.

Atmanspacher, H. (1997). Cartesian cut, Heisenberg cut, and the concept of complexity. *World Futures* 49 (3–4): 333–355. https://doi.org/10.1080/02604027.1997.9972639.

Atmanspacher, H. (2000). Ontic and epistemic descriptions of chaotic systems. *AIP Conference Proceedings* 517: 465–478. https://doi.org/10.1063/1.1291283.

Atmanspacher, H. (2005). Epistemic and ontic quantum realities. *AIP Conference Proceedings* 750: 49–62. https://doi.org/10.1063/1.1874557.

Bar-Yam, Y. (2004). Multiscale variety in complex systems. *Complexity* 9 (4): 37–45. https://doi.org/10.1002/cplx.20014.

Birnbaum, M.H. (2008). New paradoxes of risky decision making. *Psychological Review* 115 (2): 463–501. https://doi.org/10.1037/0033-295X.115.2.463.

Brewer Van, E. (2013). PRISM – a philosophical foundation for complex situations. In: *Managing and Engineering in Complex Situations* (ed. S. Kovacic and A. Sousa-Poza), 45–78. Dordrecht: Springer.

Busemeyer, J.R. and Bruza, P.D. (2012). *Quantum Models of Cognition and Decision*. Cambridge University Press.

Busemeyer, J.R. and Wang, Z. (2018a). Hilbert space multidimensional theory. *Psychological Review* 125 (4): 572–591. https://doi.org/10.1037/rev0000106.

Busemeyer, J. and Wang, Z. (2018b). Data fusion using Hilbert space multi-dimensional models. *Theoretical Computer Science* 752: 41–55. https://doi.org/10.1016/j.tcs.2017.12.007.

Busemeyer, J.R. and Wang, Z. (2019). Hilbert space multidimensional modelling of continuous measurements. *Philosophical Transactions of the Royal Society A: Mathematical, Physical and Engineering Sciences* 377 (2157): 20190142. https://doi.org/10.1098/rsta.2019.0142.

Busemeyer, J., Zhang, Q., Balakrishnan, S.N., and Wang, Z. (2020). Application of quantum – Markov open system models to human cognition and decision. *Entropy* 22 (9): 990. https://doi.org/10.3390/e22090990.

Canan, M. and Sousa-Poza, A. (2016). Complex adaptive behavior: pragmatic idealism. *Procedia Computer Science* 95: 73–79. https://doi.org/10.1016/j.procs.2016.09.295.

Canan, M. and Sousa-Poza, A. (2019). Pragmatic idealism: towards a probabilistic framework of shared awareness in complex situations. In: *2019 IEEE Conference on Cognitive and Computational Aspects of Situation Management (CogSIMA)*, 114–121. https://doi.org/10.1109/COGSIMA.2019.8724208.

Canan, M., Sousa-Poza, A., and Kovacic, S.F. (2016). Semantic shift to pragmatic meaning in shared decision making: situation theory perspective. In: *WIT Transactions on State of the Art in Science and Engineering*, 1e, vol. 1 (ed. G. Rzevski and C.A. Brebbia), 93–106. WIT Press https://doi.org/10.2495/978-1-78466-155-7/008.

Canan, M., Demir, M., and Kovacic, S. (2022). A probabilistic perspective of human-machine interaction. *Proceedings of the 55th Hawaii International Conference on System Sciences*. Hawaii. 2022.

Definition of the International Council on Systems Engineering. (n.d.). International Council on System Engineering (INCOSE). https://www.incose.org/about-systems-engineering/system-and-se-definition/physical-and-conceptual (accessed 13 June 2022).

Dzhafarov, E.N., Cervantes, V.H., and Kujala, J.V. (2017). Contextuality in canonical systems of random variables. *Philosophical Transactions of the Royal Society A: Mathematical, Physical and Engineering Sciences* 375 (2106): 20160389. https://doi.org/10.1098/rsta.2016.0389.

Endsley, M.R. (1995). Toward a theory of situation awareness in dynamic systems. *Human Factors: The Journal of the Human Factors and Ergonomics Society* 37 (1): 32–64. https://doi.org/10.1518/001872095779049543.

Grieves, M. (2014). Digital twin: manufacturing excellence through virtual factory replication. *White Paper* 1 (2014): 1–7.

Hertwig, R., Pleskac, T.J., and Pachur, T. (2019). *Taming Uncertainty*. MIT Press.

Jones, D., Snider, C., Nassehi, A. et al. (2020). Characterising the digital twin: a systematic literature review. *CIRP Journal of Manufacturing Science and Technology* 29: 36–52. https://doi.org/10.1016/j.cirpj.2020.02.002.

Kahneman, D. and Tversky, A. (1979). On the interpretation of intuitive probability: a reply to Jonathan Cohen. *Cognition* 7 (4): 409–411. https://doi.org/10.1016/0010-0277(79)90024-6.

Khrennikova, P., Haven, E., and Khrennikov, A. (2014). An application of the theory of open quantum systems to model the dynamics of party governance in the US political system. *International Journal of Theoretical Physics* 53 (4): 1346–1360. https://doi.org/10.1007/s10773-013-1931-6.

Kovacic, S. (2005). General taxonomy of system [ic] approaches for analysis and design. In: *2005 IEEE International Conference on Systems, Man and Cybernetics*, vol. 3, 2738–2743. IEEE.

Kozyreva, A., Pleskac, T. J., Pachur, T, Hertwig, R. Interpreting uncertainty: a brief history of not knowing, Chapter 18 (2019). Timothy J. Pleskac, Thorsten Pachur, Ralph Hertwig Taming Uncertainty. MIT Press 343.

Kvam, P.D. and Pleskac, T.J. (2017). A quantum information architecture for cue-based heuristics. *Decision* 4 (4): 197–233. https://doi.org/10.1037/dec0000070.

Kvam, P.D., Busemeyer, J.R., and Pleskac, T.J. (2021). Temporal oscillations in preference strength provide evidence for an open system model of constructed preference. *Scientific Reports* 11 (1): 8169. https://doi.org/10.1038/s41598-021-87659-0.

Lord, R.G., Dinh, J.E., and Hoffman, E.L. (2015). A quantum approach to time and organizational change. *Academy of Management Review* 40 (2): 263–290. https://doi.org/10.5465/amr.2013.0273.

Martínez-Martínez, I. and Sánchez-Burillo, E. (2016). Quantum stochastic walks on networks for decision-making. *Scientific Reports* 6 (1): 23812. https://doi.org/10.1038/srep23812.

Osinga, F.P.B. (2006). *Science, Strategy and War: The Strategic Theory of John Boyd*. London, UK: Routledge.

Pearl, J. (2000). *Causality: Models, Reasoning, and Inference*. Cambridge University Press.

Pearl, J. and Mackenzie, D. (2018). *The Book of Why: The New Science of Cause and Effect*, 1e. Basic Books.

Perrow, C. (1999). *Normal Accidents: Living with High-Risk Technologies*. Princeton University Press.

Pothos, E.M., Lewandowsky, S., Basieva, I. et al. (2021). Information overload for (bounded) rational agents. *Proceedings of the Royal Society B: Biological Sciences* 288 (1944): 20202957. https://doi.org/10.1098/rspb.2020.2957.

Rescher, N. (2000). *Process Philosophy: A Survey of Basic Issues*. University of Pittsburgh Press.

Rivas, Á. and Huelga, S.F. (2012). Open quantum systems. an introduction. *ArXiv* 1104.5242. [Cond-Mat, Physics:Math-Ph, Physics:Physics, Physics:Quant-Ph]. https://doi.org/10.1007/978-3-642-23354-8.

Russell, S.J. (2020). *Human Compatible: Artificial Intelligence and the Problem of Control.* Penguin Books.

Sergioli, G. (2020). Quantum and quantum-like machine learning: a note on differences and similarities. *Soft Computing* 24 (14): 10247–10255. https://doi.org/10.1007/s00500-019-04429-x.

Sergioli, G., Giuntini, R., and Freytes, H. (2019). A new quantum approach to binary classification. *PLoS One* 14 (5): e0216224. https://doi.org/10.1371/journal.pone.0216224.

Snow, L., Jain, S., and Krishnamurthy, V. (2022). Lyapunov based stochastic stability of human-machine interaction: a quantum decision system approach. *ArXiv*:2204.00059 [Cs, Econ, Eess, q-Fin]. http://arxiv.org/abs/2204.00059.

Sousa-Poza, A. (2013). A narrative of [complex] situations and situations theory. In: *Managing and Engineering in Complex Situations*, 13–44. Dordrecht: Springer.

Sousa-Poza, A. and Correa-Martinez, Y. (2005). Pragmatic idealism as the basis for understanding complex domains: the trinity and SOSE. In: *2005 IEEE International Conference on Systems, Man and Cybernetics, 3*, Waikoloa, Hawaii, USA (10–12 October 2005), 2744–2750. https://doi.org/10.1109/ICSMC.2005.1571565.

Sterman, J. (2010). *Business Dynamics: Systems Thinking and Modeling for a Complex World.* Massachusetts Institute of Technology. Engineering Systems Division.

Stumborg, M., Brauner, S., Hughes, C., et al. (2019). Research and development implications for human-machine teaming in the U.S. Navy [Research Report]. CNA. *Research Memorandum DRM-2019-U-019330-1Rev.*

Tversky, A., Slovic, P., and Kahneman, D. (1990). The causes of preference reversal. *The American Economic Review* 80 (1): 204–217.

Von Neumann, J. and Morgenstern, O. (2007). *Theory of Games and Economic Behavior*, 60th anniversary ed. Princeton University Press.

Von Neumann, J., Beyer, R.T., and Wheeler, N.A. (2018). *Mathematical Foundations of Quantum Mechanics*, New ed. Princeton University Press.

Whitehead, A.N. and Griffin, D.R. (1985). *Process and Reality: An Essay in Cosmology*, Corr. ed. Free Press.

Wilhelm, J., Beinke, T., and Freitag, M. (2020). Improving human-machine interaction with a digital twin: adaptive automation in container unloading. In: *Dynamics in Logistics* (ed. M. Freitag, H.-D. Haasis, H. Kotzab, and J. Pannek), 527–540. Springer International Publishing https://doi.org/10.1007/978-3-030-44783-0_49.

Zhang, Q., Nadendla, V.S.S., Balakrishnan, S.N., and Busemeyer, J. (2021). Strategic mitigation of agent inattention in drivers with open-quantum cognition models. *arxiv*:2107.09888 [Cs, Eess]. http://arxiv.org/abs/2107.09888.

Biographical Sketches

Samuel Kovacic, PhD, is an assistant professor at Old Dominion University with over forty years' experience in industry, government, military, and academia. Dr. Kovacic research is in complexity and its implications to systems. He received his PhD from Old Dominion University in the Engineering Management System Engineering Department.

Mustafa Canan, PhD, is an Associate professor at Naval Postgraduate School (NPS). He completed his particle physics PhD (2011) at Old Dominion University and Thomas Jefferson National Accelerator and Systems Engineering (2017) at Old Dominion University. Before joining NPS, he was an NRC post-doctoral fellow and research scientist at the Air Force Research Lab, Wright-Patterson AFB. Dr. Canan's inter-disciplinary research interest includes Complex Adaptive Behavior of Systems, Dynamics of Human-Machine Teams, Quantum Information Processing, Decision Making under Uncertainty, Digital Twins, Generalized Parton Distributions, and System Engineering AI.

Jiang Li, PhD, is a professor at Old Dominion University with over twenty years' experience in industry and academia. Dr. Li's research is in applied machine learning with application in medical signal/image processing, remote sensing image analysis, neural network, and deep learning. He received his PhD from University of Texas at Arlington and completed three years of post-doctoral training at the National Institutes of Health before joining ODU.

Andres Sousa-Poza, PhD is a professor at Old Dominion University with over thirty years in Industry and Academia. Dr Sousa-Poza's research is in complex situations theory and its implications on organizations. He received his PhD from the University of Missouri, Rolla.

Chapter 9

Expedited Systems Engineering for Rapid Capability

John M. Colombi

Department of Systems Engineering and Management, Air Force Institute of Technology, OH USA

Introduction

This research examines "expedited" systems engineering (SE) as applied to the development of defense capabilities, specifically development for defense joint urgent operational needs. The life cycle of urgent need programs is driven by "time to market" as opposed to complete satisfaction of static strategic requirements, with delivery expected in months. The original hypothesis was that by identifying, finding, testing, and ultimately implementing expedited SE processes and practices, the developed solution would be more effective with a high probability of success. A potential second-order effect is that as urgent becomes the new "normal," any findings would improve systems engineering for traditional developmental programs as well.

As was noted in the SERC Systems 2020 report (Boehm 2010), SE capabilities need to be developed that achieve dramatic reduction in time that develop a fieldable first-article product, implement foreseeable classes of fielded systems changes, and rapidly adapt to unforeseeable threats. Traditional systems engineering tools, processes, and technologies poorly support rapid design changes or enhancements within acceptable schedule constraints. Perhaps an artifact of the defense acquisition process, rapid development has achieved point solutions, which make adaptation cumbersome and impractical. To increase development efficiency and ensure flexible solutions in the field, systems engineers need powerful, agile, interoperable, and scalable processes, tools, and techniques. Several defense reports have documented in-depth studies on the problems and possible solutions surrounding rapid acquisition, rapid fielding, and/or rapid prototyping (GAO 2010, 2011; DSB 2009; DoD 2010). Recommendations include acquisition process changes and introduction of new SE tools designed to handle a rapidly changing environment that challenges the current acquisition community.

First, most define "rapid" as generally delivering a capability as quickly as a few months and no longer than 24 months (DSB 2009). This definition has continued into the latest DoD Instruction 5000.02, Operation of the Adaptive Acquisition Framework (Office of the Secretary of Defense (OSD), 2020). This 2020 policy describes a flexible acquisition process with a pathway for Urgent Capability Acquisition. This research pre-dates this new pathway, but the findings can inform such efforts. The trend and strategy across the DoD, including extensive digital transformation efforts, is to speed up acquisition. Thus, "urgent" is becoming the new "normal."

Systems Engineering for the Digital Age: Practitioner Perspectives, First Edition. Edited by Dinesh Verma.

Companion website: www.wiley.com/go/verma/systemsengineering

This Systems Engineering Research Center (SERC) project on Expedited Systems Engineering examines the systems engineering and engineering management practices as applied to rapid capability and urgent needs, programs, and organizations. The successful techniques seen in rapid development and prototyping must scale to larger, more complex, supportable, and sustainable weapon systems. The SE processes and practices applied to urgent needs should provide for innovative conceptual solutions, quickly prune the design space, and identify appropriate designs that can deliver warfighting capability expeditiously.

The objective of the research was to explore and develop a scalable Expedited SE framework for hybrid programs, i.e., those exploiting rapid development, but with the intent to have traditional lifecycle considerations for deployment, maintainability, reliability, adaptability, and sustainment. Likewise, this SE framework would be applicable to more traditional acquisition programs with a desire to incorporate scaled rapid development best practices. The research evolved to consider the broader context of the people, processes, and products involved in rapid development.

A research paper (Lane et al. 2010) "Critical Success Factors for Rapid, Innovative Solutions," posed a series of questions on potential program success factors. That paper concluded that successful innovative organizations share certain characteristics:

- Driven by business value
- Take calculated risks
- Proactive management and small agile teams
- Look for solution patterns that can be re-used
- Follow concurrent engineering practices
- Provide culture and environment that supports innovation.

Research on success factors and factors categories is well documented for varied purposes in the industrial and SE communities (Hanawalt and Rouse 2009; Bullen and Rockart 1981; Boynton and Zmud 1984). Various types of success factors (Van Secter and Doolen 2011) have been documented in project planning, control, monitoring, building, project, management, organization/people, external environment, process, and technical considerations.

Critical success factors often reflect lean product development concepts (Morgan and Liker 2020) and Kanban process improvement strategies (Anderson 2010). Lean thinking is the dynamic, knowledge-driven, and customer-focused processes through which all people in a defined enterprise continuously create value and eliminate waste (Murman et al. 2002). Additional findings (Oehmen et al. 2012) have taken these lean enablers and added one additional – respect for your people. Generally, lean activities include:

- **Specify value:** Value is defined by customer in terms of specific products and services, but could include sustainability or reliability of the solution, integration with other systems (System of Systems), or schedule. Urgent operation needs must get satisfied quickly to reduce risk of harm or loss of life.
- **Value stream mapping:** Map out all end-to-end linked tasks, activities, and processes necessary for transforming inputs to outputs to identify and eliminate waste.
- **Make value flow continuously:** Having eliminated waste, make remaining value-creating steps "flow." This can be considered having the lifecycle product iterate.
- **Customers pull value:** Customer's "pull" cascades all the way back to the lowest level supplier, enabling just-in-time (JIT) production.
- **Pursue perfection:** Pursue continuous process of improvement striving for perfection.

Methodology

Grounded theory is a type of qualitative research methodology that allows theories to emerge from the collected data. It has been widely used as a method for generating theory from human responses with the intent of extracting insight from human actions and responses (Glaser and Strauss 1967). Grounded theory is a research method that starts with the researcher reading (and rereading) the subject data, coding it, analyzing and constantly comparing those codes, and then deriving a theory from that comparison. Literature is then reviewed for relevance to the proposed theory with the intent of explaining, amplifying, or disproving the discovered concept. Grounded theory accomplishes this by purposive sampling, getting and organizing data, coding it to distill the main themes, categorizing those themes, synthesizing the identified themes, and finally generating a theory.

The data for this research is based on over 30 unstructured interviews with organizations performing rapid development on U.S. military urgent needs programs. The starting point of a grounded theory research project, like any research, is to purposefully acquire data (Chun et al. 2019a). Conducting in-person site visits allowed the opportunity see facilities, meet project personnel, and visually experience how the organizations conducted business. The data are the site visit notes from the leadership of these "rapid" organizations and projects – essentially subject matter experts (SMEs) in the field. The research team developed questions that were SE focused to guide the discussions on technical and technical management processes, as well as the product or solution architecture. The responses regularly indicated the importance of people and the development team.

Coding is the core activity of grounding theory that identifies, labels, or categorizes the essence of the data. It is a shorthand that allows the researcher to filter the data and make connections from code to code. The exact codes used are unique to the researcher. Codes can be developed through inductive or deductive means. Inductive coding occurs when the researcher reads the data and assigns a label (code) to the information without any preconceived ideas. With deductive coding, the researcher has a set of predetermined codes that are then applied to the data as appropriate. Grounded theory uses a combination of inductive and deductive coding during the various stages of analysis. Inductively produced codes are applied to the data and then analyzed for themes that are filtered and assessed for relevance to the line of inquiry.

The core reoccurring effort of grounded theory is built on the continuous reexamination of discovered themes, ideas, and connections within the data (Chun et al. 2019b). Previously created codes are compared to newly created codes in a constant iterative process that seeks to discover key similarities and differences. Multiple codes are created during the initial coding process that can then be either discarded, split, or merged with other codes as they are compared. The essence of the comparison is that no code exists in a vacuum. They are each intended to return the researcher to the core inquiry of the process. The overall research approach is shown in the Figure 9.1.

There was some refinement to the original questions during the site visits with re-grouping into three categories of people, process, and product. The 37 questions were used for the remaining discussions. For example, some questions were:

- "How do you effectively incorporate/involve the end user?" (people)
- "What is the formality of engineering documentation?" (process)
- "How does your rapid development schedule drive design choices?" (product)

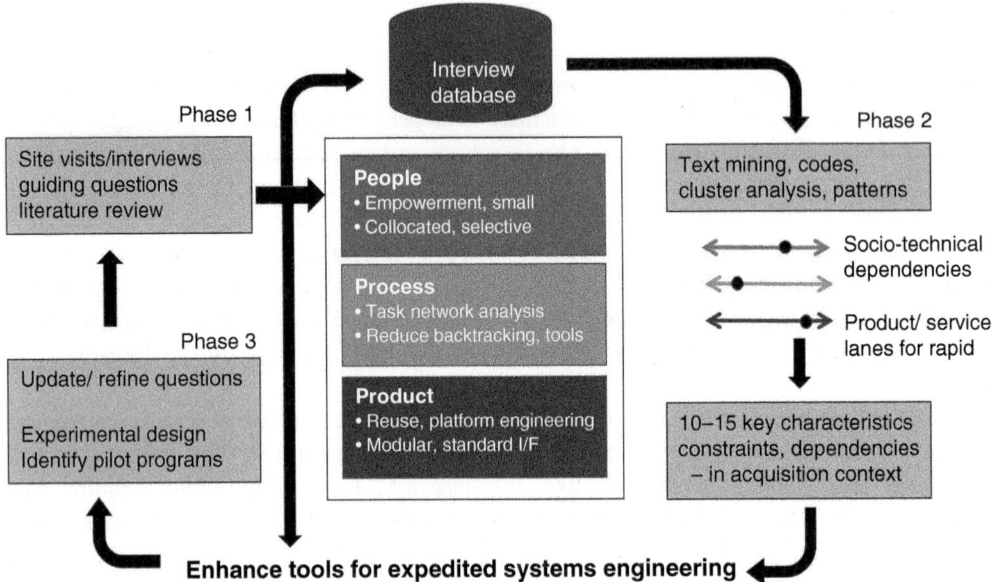

Figure 9.1 Research design and methodology.

The research targeted those organizations who have been acquiring urgent operational solutions or who had expertise in aspects of rapid nontraditional acquisition. Predominantly, the organizations were either government defense acquisition offices or select defense industries. Some of these included:

- Air Force Rapid Capabilities Office (RCO)
- U.S. Army Prototype Integration Facility (PIF)
- An aerospace industry innovation lab
- A small (agile) satellite development company
- NASA Goddard Space Flight Center
- Big Safari Program office and Program Executive Office, ISR-SOF
- Air Force Research Lab (AFRL) Center for Rapid Product Development
- SOCOM Research and Development Acquisition Center (SORDAC).

Using the interview notes, an iterative qualitative analysis was performed using ATLAS.ti™ to further explore the data for correlations and clustering. Transcribed notes from organizations were assigned to an ATLAS.ti™ hermeneutic unit. Using the observations derived through the interview process as codes, each document was coded appropriately against key words and phrases. The codes were organized into families of Product, Process, and People, which matched the clusters observed from the discussions. Originally 20 observations/findings were first identified, aggregated down to 11. This "open coding" of labels is an important part of the analysis concerned with identifying, naming, labelling, categorizing, and describing phenomena found in the discussion notes. The result was a set of consistent principles for successful DoD rapid development, with a focus on SE.

Notes were inserted into "Wordle," an online, word count software, to visualize any patterns in the data. The purpose of this qualitative tool was to give a visual representation of prominent words from the interviews. Figure 9.2 is for the full set of notes with over 23,000 words in the database. Note the prominence of "people."

Based on observations, interviews, and literature, a series of observations, or principles, begins to emerge that reflects a framework of rapid development. Originally, a stacked model was

Figure 9.2 Word Cloud from final interviews note of rapid acquisition organizations.

Table 9.1 Top 5 practices by total responses.

	Practice	Code group	Citations
1	Build and maintain trust	People	45
2	Defined set of stable requirements focused on warfighter needs	Process	44
3	Populate your team with specific skills and experience	People	40
4	Maintain high levels of motivation and expectations	People	36
5	Work to exploit maximum flexibility allowed	Process	34

proposed with increasing rigor; however, the final depiction did not infer any precedence, hierarchy, or increasing systems engineering rigor. The findings are broken into three groups:

- Group 1: Direct Responses,
- Group 2: Direct Observations, and
- Group 3: Inferred Organizational Characteristics.

These groups are discussed and decomposed in detail in the Findings section. The data presented in Table 9.1 shows the top five most common occurring practices of rapid organizations. These 5 of the 11 aggregated practices identified comprised more than two-thirds (64.2%) of the total citations made during the analysis, 199 out of 310 total codes, with three of the top five originating from the People code group.

Findings/Analysis

The Direct Responses, Direct Observations, and Inferred Characteristics, along with a parsimonious set of recommendations, are summarized in Table 9.2. Of note, there is an excitement and "vibe" to these organizations that cannot be captured in a table. These organizations appear to enjoy what they are doing, probably because they feel closer to the need, the warfighter, and their ability to

Table 9.2 Summary of findings and recommendations.

Direct responses	Recommendations
• Use Mature Technology – Focus on the state of the possible • Incremental deployment (development) is part of the product plan • Strive for a defined set of stable requirements focused on warfighter needs • Work to exploit maximum flexibility allowed • Designing out all risk takes forever – accept some risk • Keep an eye on "normalization" • Build and maintain trust • Populate your team with specific skills and experience • Maintain high levels of motivation and expectations • The government team leads the way • Right-size the program – eliminate or reduce major program oversight **Summary:** Rapid requires an integrated approach: People making judgments, Processes for task time reductions, and Product aspects focused on rapid objectives.	• Ensure that projects utilize prioritized organizational and project best practices • Train program managers, engineers, and contracting officers in organizational and cultural best practices, real-time management approaches, and the different flexibilities that exist • Encourage and enable programs to intensively share knowledge, have a risk-focused culture, and create an ambidextrous organizational structure • Share knowledge, experience, mechanisms, and lessons learned across programs and organizations • Quantitatively monitor progress in expediting SE processes, and use the measurements to improve schedule acceleration and its estimation • Use DoD rapid organizations as a testbed to introduce digitally enabled solutions • Develop the acquisition workforce using rapid programs to provide full lifecycle insight and hands-on experience
Direct observations	
Not a single Rapid, but many different flexible Rapids approaches with flexible lanes of acquisition and business practices.	
Inferred characteristics	
• Intense and efficient knowledge-sharing is used to enable stabilization and synchronization of information • Rapid organizations are characterized by a risk-tolerant culture • Rapid organizations are structured for ambidexterity with a balance between exploration and exploitation • Rapid Development organizations exercise Real-time Management	

successfully deliver technological solutions. Likewise, this excitement is contagious throughout the organizations and fosters leadership, a rapid culture, effective decision-making, and lifecycle expertise.

Based on the site interviews, the coding/tagging of the interview notes, extensive literature review, and the knowledge of the research team, the Expedited SE Framework began to develop into three groups of findings. See Figure 9.3.

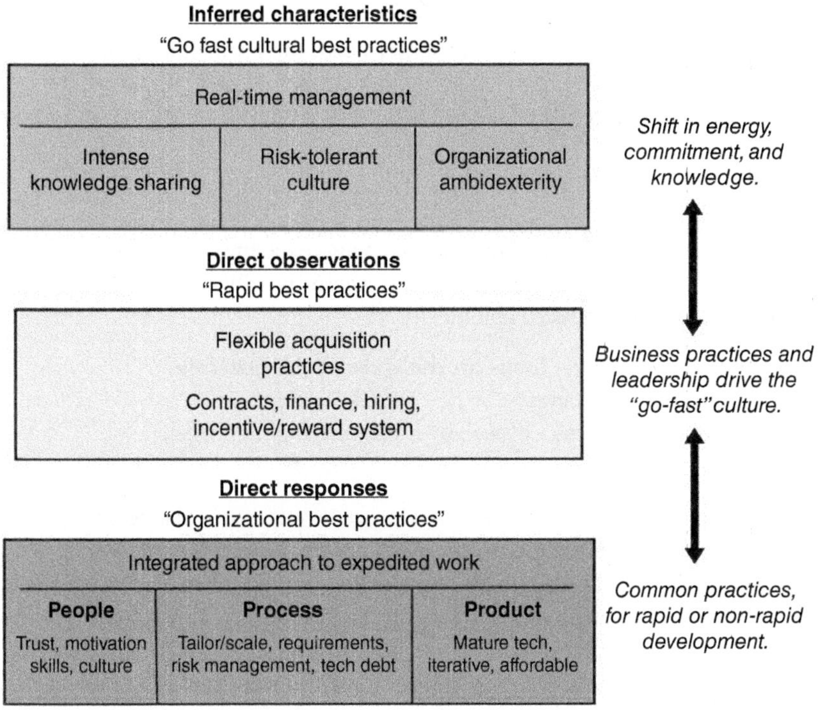

Inferred characteristics
"Go fast cultural best practices"

Real-time management
Intense knowledge sharing | Risk-tolerant culture | Organizational ambidexterity

Shift in energy, commitment, and knowledge.

Direct observations
"Rapid best practices"

Flexible acquisition practices
Contracts, finance, hiring, incentive/reward system

Business practices and leadership drive the "go-fast" culture.

Direct responses
"Organizational best practices"

Integrated approach to expedited work
People: Trust, motivation skills, culture | **Process**: Tailor/scale, requirements, risk management, tech debt | **Product**: Mature tech, iterative, affordable

Common practices, for rapid or non-rapid development.

Figure 9.3 Expedited SE framework.

- **Group 1**: Direct Responses has direct correlation to interview responses of rapid organizations and confirms best practices in project management and lean development systems.
- **Group 2**: Direct Observations captures practices of rapid organizations that "live in a rapid world," which take an integrated approach to rapid development, by leveraging people, process and product appropriately, as well as seek out and monopolize flexible acquisition practices.
- **Group 3**: Inferred Organizational Characteristics signifies a shift in energy, commitment, and knowledge in rapid organizations, demonstrating the culture of effective organizations and those seen in the agility literature.

Direct Responses for Organizational Best Practices (Group 1)

This section introduces the core findings and places those in the context of expedited acquisition and work reduction techniques. Some may read this section and see the core findings as "not new." In the current program management and lean product development communities, indeed, many of these findings have been observed in the literature (Morgan and Liker 2020; Oehmen et al. 2012). Similarly, for those who practice DoD acquisition in some niche areas (laboratory or operational prototyping, classified acquisition or platform modification shops), some of these findings may also be commonplace.

The following practices are a set of 11 consistent observations that clustered from discussion notes. Over 30 discussions were conducted from 25 site visits, including several additional informal discussions at conferences. This section describes the "Group 1" direct responses. In summary, one major finding emerged that reflects how all the direct responses combine to create an integrated approach to rapid development. There was not just one process, or one product attribute, that resulted in successful expedited systems engineering. In fact, the responses were much broader about the context of the overall environment, the warfighter need, the acquisition practices, and the combination of people, processes, and product coming together for success. Rapid requires an integrated approach: people making judgments; processes for task reduction, and product aspects focused on rapid objectives.

Product Practices

Smart decisions on the system solution can greatly lower cost and expedite product delivery. But the mindset must be to reuse and repurpose, to the maximal extent possible, mature technology, and legacy systems with an eye toward reduced complexity.

Observation 1: Use mature technology – focus on the state of the possible

- Focus on integration of mature technologies
- Reuse existing components and platforms – especially if they are flight-certified
- Reduce complexity

In rapid acquisition, untested and unproven technology poses an enormous risk to system success. Unlike most traditional acquisition programs, there is no time for technology to mature, in other words, no time for schedule slips due to immature technology struggling to develop. To avoid this pitfall, most rapid programs focus engineering efforts on the interfaces required to blend multiple existing technologies into a system capable of providing the desired set of requirements. Program teams stay abreast of emerging technology and leverage the work done by industry and other military programs. They then engineer a system-of-systems solution to meet requirements. This bounds their design space within the state of the possible. In terms of Technology Readiness Level (TRL), this means using nothing less than a TRL 6, preferably 7 or 8. This allow these teams to field quickly and generationally provide more and more capability.

Another essential characteristic of rapid product development is the reuse of existing technical capabilities. One example was a modification one organization performed to improve a small fleet of Combat Search and Rescue (CSAR) helicopters. Rather than develop a new forward looking infrared radar (FLIR) pod and lift hoist, the team examined the operational requirements requested by the user and identified existing technology currently being installed by the US Army on Department of State (DoS) and Federal Bureau of Investigation (FBI) helicopters. This reuse approach saved the program office over half of the potential contracted price (about $3M saved) and 12 months of schedule. Another significant reduction in time for this example was the time saved in flight test by using equipment that had already been flight certified. The choice of mature technology is a decision that can be approached through a balanced risk management process.

Many assume that reuse of existing components will reduce risk and deployment time directly. Rather, a different risk is realized – the risk of using a component in a potentially different application, with an unknown requirements and development (and manufacturing) history. While urgent needs may require mature technologies and nondevelopment items (NDI) or Commercial Off-The-Shelf (COTS), there may be a new set of issues for any rapid solution that may endure for a longer operational phase. An open issue that requires further research is seeking the optimal balance between short-term delivery and long-term sustainability and flexibility for future iterations.

While the term "system complexity" was not specifically mentioned during SME discussions, it can be considered a semantically equivalent concept. Complexity can be measured by the type and number of components within a solution, the degree of interconnections of those components, and the type and amount of coupling between the components. Often by reusing components, the focus becomes on quickly integrating these components with standard connections. Having well-defined interfaces allow new components to be more easily integrated later with other new components.

Observation 2: Incremental deployment (development) is part of the product plan

- "Generational development" – plan for technology maturity, advancement, and cycles
- Look for unpredicted outcomes
- Prototyping/test infrastructure

Part of the agreement for a stakeholder accepting a partial solution can include the plan for incremental development. When this concept is selected at the beginning of a development program, it enables "generational development" – an intentional plan for technology maturity, advancement, and cycles of growth. This may be done by using open architectures, and modular concepts, clearly defined system interfaces, and using industry standards. Overall, this approach will extend the system lifecycle and enhance its ability to flexibly meet the needs of an ever-changing technical and operational environment.

As rapid organizations progress through their development programs, many are constantly looking for unpredicted design outcomes. In one organization, during the latter stages of product development a specific set of questions were asked, "Who else could use this? How else could it be used? What does this enable next? and How could this be used against us?" This series of questions put this team in the right mindset to further the development and use of their products and plan beyond the immediate urgent need into a future program of record.

During site visits, many organizations had the ability and facilities to accomplish rapid prototype and test. Iteration implies incremental test and feedback. This could be at contractor facilities but was generally observed in government-owned, government-operated facilities. These rapid organizations had the ability to test or had good working relations with test organizations. Thus, they had the infrastructure to experiment on target hardware, prototype solutions, and incrementally develop, test, deliver, and get feedback.

Product/System Summary

In summary, the rapid approach keeps the system solution as simple as possible, trading out items that are not critical to success, and making maximum use of mature technologies and open interfaces within an iterative development process.

Process Practices

By first focusing on validating requirements, rapid organizations then exploited and executed their programs with the greatest flexibility allowed.

Observation 3: Strive toward a defined set of stable requirements focused on warfighter needs

- Get the requirements right, rooted in customer-derived CONOPS
- Expedite trade studies – decide and press forward
- Focus on providing the "23–80%" solution

Defining stable requirements focused on the customer needs was one of the most frequently occurring principles during the SME discussions. Not only is there not enough time to do everything a customer is asking, but customers often ask for more than they really need. It quickly became evident that every one of these organizations spends a significant amount of time up-front, face-to-face with their customer discussing requirements and operational context. They may spend more time refining a stable set of requirements than they do in actual design and production. Discussions brought to light several frustrations of the requirements development process. Customer disconnects or unrealistic expectations may emerge because customers are unaware of the state-of-the-possible. This drives concepts of operations (CONOPS) analysis, where the customer must clearly define specific needs, uses, or capabilities for the solution – in an operational context.

Equally important is an effort to keep the requirements stable. Regardless of the scope of a project, requirements creep will negatively impact the timeline of a project, delaying the delivery of operational capabilities to the warfighter. Further, requirement changes potentially weaken the scope of the project or may negate any perceived increase in baseline capability. Organizations that consistently execute rapid SE and acquisitions are rooted in high-quality stable requirements.

Rapid organizations validate requirements early and often with the customer to determine needs based on capabilities and affects. The acquiring organization must be willing to push back against unfeasible requirements, or schedule impacting requirements, in the interest of time. As one senior officer explained, "We fight hard to have the warfighter make trades" to establish requirements that are possible in the desired timeframe. Simply, focus on valid requirements that can be met by the state of the possible in a short amount of time.

Iterative development strongly impacts this observation as well. Some engineers will challenge this, saying that all requirements, especially for large acquisitions, can be defined and stabilized up front. Complexity of the technologies, system interdependencies, along with tactics, techniques, and procedures (TTPs) of the users, and a dynamic threat environment may limit the ability to stabilize requirements early. Often the total set of requirements ("100%") will change, especially if a smaller set of delivered functions is first deployed. For most complex and software systems, customers do not know precisely what they need, but instead, must iterate through a series of experiments or product releases to capture the full set of true requirements.

The short duration of rapid development projects naturally lends to more stability in requirements. Large changes in technology maturity are not often experienced in the short life cycles. There are also fewer changes in political administration (funding), leadership (rotating Colonels/Generals), and program personnel. Each change in personnel often brings a new perspectives and priorities. All these factors are less likely to change over a short period of time.

Ironically, requirements creep can become a pitfall of regular customer involvement. Several organizations emphasized the necessity to fight requirements creep once stable requirements have been established. However, stopping creep cannot be done at the expense of customer and user involvement. In this manner, an art must be developed to keep the user in the loop without allowing for spurious changes to the project once underway.

Rapid programs rarely provide the customer with 100% of what they originally ask. Interviews typically expressed the "80% solution" concept. This concept claims that most stakeholders prefer 80% of requirements delivered early, than 100% delivered (much) later. One interview commented, "50% or 23% done quickly can be very acceptable." Often eliminating or modifying certain requirements will provide the warfighter with a viable solution to a problem within an expedited, achievable timeline.

Observation 4: Work to exploit maximum flexibility allowed

- Tailor the acquisition and system engineering process to the product
- Establish a clear and short approval chain
- Document what is important and decisions made
- Use various contracting vehicles to accomplish different tasks

Because of the specialized nature of each rapid office, many have developed in-house processes adaptable to each program. This ensures each program office has a specific roadmap leading it to success, and each project lives within its own specific process and lifespan.

In the interest of time, these organizations ensure every acquisition/engineering process is highly tailored to the product. Anything not required, deemed unnecessary, or found to be non-value added is set aside. Adhering to the intent of DoDI 5000.02, they execute tasks without excess. It may appear these organizations are skipping steps in the engineering process; however, the visits indicate these steps are not skipped, but rather highly tailored. For example, a Systems Engineering Plan (SEP) may consist of few pages within some strategic document, instead of a stand-alone document.

They use formal and informal review processes, specifically performance-based and milestone-type reviews, with just the right people in attendance to make go/no-go decisions quickly. The focus is to only document important technical and programmatic information and critical decisions. In some organizations, it became evident their approval chain for reviews and program milestone approval had been shortened. Additionally, there are very few extraneous persons in the review chain that do not have some sort of approval authority, such as legal or contracting. The brevity of these approval chains often stems from a Program Management Directive (PMD) outlining the decision-making authority, typically pushing it down to a lower level within the organization. Finally, program size keeps budgets under Major Defense Acquisition Program (MDAP) thresholds and oversight.

Another practice is to combine, not skip, program level reviews. For example, test plans, Technical Readiness Review Boards (TRRB), Safety Review Boards – if deemed low risk, can be signed off at the lowest level in a single review. This is also applied to pre-milestone decision reviews as well. The approach is to shorten the approval and review process timeline by combining review processes and reducing the lull created by waiting for a review process to take place – not to diminish the quality of the product or eliminate SE analysis processes.

Another common trend is the use of various innovative contracting methods, some of which have been in place for many years, to accomplish different tasks. For example, Indefinite Delivery Indefinite Quantity (IDIQ) contracts were important to several organizations to provide as-needed support on a reoccurring basis. This approach was referred to by one organization as "creative contracting." This can only be done by contracting officers who are knowledgeable about and willing to investigate the bounds of contracting utilizing the full flexibility of the Federal Acquisition Regulation (FAR) and DoD policy.

Observation 5: Designing out all risk takes forever – accept some risk

- Creative (and implementable) solutions are allowed
- Mitigate risk through the use of mature and proven technology
- Potential for failure is accepted, providing something is better than nothing
- Determine the level of risk the customer is willing to accept

The rapid organizations operate under an uncommon risk paradigm when compared to many traditional DoD acquisition programs. In rapid, the potential for "failure" through providing only a partial or short-term solution to the field may be acceptable, as this may be preferable to delivering nothing at all. Rapid teams are made up of technical experts who cognitively assess the risks of different technical solutions throughout the design process, sometimes with formal risk assessment processes in place. This idea of risk mitigation through use of mature and proven technology led several programs to adopt the concept of demonstrations or prototyping. It often came down to the level of program or technical risk the customer is willing to accept and the limited time available.

Some of rapid deployment success hinges on expert understanding of the design space, potential technical solutions, and the ability to integrate existing technologies. Rapid programs work through a rigorous and creative design process, working to identify, eliminate, and accept risks. However, attempting to design-out all risk is a time-consuming and costly process, and not realistic if attempting to get a solution out to the customer quickly.

Observation 6: Keep an eye on "normalization"

- Track your technical debt
- Buy or maintain data rights for a build-to specification

"Normalization" is a term heard at one of the rapid acquisition offices, but the concept was reoccurring. It describes the transition of a program from a prototype, laboratory demonstration, or rapid project into a major traditional acquisition program. Most of the organizations interviewed typically work in small-rate production (a few to less than 15 units). Thus, the investments required for product implementation are minimal compared with a large aircraft program destined for a full-rate production phase and years of sustainment. However, as many rapid projects have the potential to become productized (normalized), it is advantageous for these offices to be prepared for a full-scale transition. This typically happens when an urgent need is determined to be an enduring requirement.

Technical debt, another term heard at one of the organizations, is a concept coined by Ward Cunningham in the early 1990s (Cunningham 1992) to describe the risks and compromises made in rapid development. He first applied the concept to software development.

"Shipping first time code is like going into debt. A little debt speeds development so long as it is paid back promptly with a rewrite... The danger occurs when the debt is not repaid. Every minute spent on not-quite-right code counts as interest on that debt. Entire engineering organizations can be brought to a stand-still under the debt load of an unconsolidated implementation."

Cunningham later commented that this concept has been misinterpreted and confused with the idea that you can do sloppy or poor work up front with the intention of doing a good job later (Cunningham 2011). That is not the case with the rapid organization whose primary purpose is to provide useful products to the warfighters in the field. Providing a poorly executed product to the field, however rapidly, could risk lives. It is often difficult for rapid organizations to accomplish this transition because of the time it takes to overcome the technical debts, particularly in the documentation required for a traditional program of record.

Another important concept of rapid involves data rights. Many of these organizations specifically mentioned the benefits of purchasing or maintaining some level of government owned data. The level of data required varies between programs, but the intent was consistent: Have enough data to provide the ability to modify, when necessary, maintain competition, and facilitate a transition toward normalization.

People Practices

The discussions with SMEs uncovered aspects of trust, strong team skills, empowered leadership and a unique culture with high expectations of the team.

Observation 7: Build and maintain trust

- Develop solid relationships and work to maintain them
- Empower leadership with autonomy for Program Managers/Engineers
- Consistent customer input and buy-in every step of the way

Building and maintaining trust enables empowered teams working together, being allowed to make decisions, leaders standing behind their decisions, and dealing with success or failures as they are encountered. Agile development thrives in a culture "where people feel comfortable and empowered by having many degrees of freedom." The scope of trust is an important element for expeditious behavior and extends throughout the organizations in acquisition, development, and deployment. As noted by P. Lencioni, "Five Dysfunctions of a Team," trust is the critical foundation of teamwork without which it is not possible to effectively collaborate (Lencioni 2011).

Rapid organizations repeatedly showed leadership at all levels providing "top cover" to allow teams to focus on executing the mission. When decision-making authority is placed at a low level, it shortens the process, reduces opportunity for stall time, and fosters close relationships.

Many site visits mentioned a strong and continuing relationship with the customer. From this perspective, it was vital to have the customer consistently involved in the decision-making process and to gather their feedback through the rapid life cycle. This was accomplished in many ways: short- or long-term on-site customer representatives, customer input at regular conversations/ reviews, or simply a coordination process. Regardless of how the customer was included on the team, it was clear that trust in the team's ability to deliver was vital to project success.

Trust is built through expertise, show of confidence, and record of performance. On the outside, it appears relationships exist on an organizational level. Individuals build trust with one another through demonstrated commitment and competence. A successful acquisition team must have highly skilled acquisition professionals. But it is only through the consistent application of those skills that build trust and grant individual or organizational autonomy. It is on established trust relationships with senior leadership and the customer that this autonomy allows small teams to rapidly progress toward product delivery.

Observation 8: Populate your team with specific skills and experience

- Hand pick your team. . .or grow your own
- Acquire people with the right education, experience, and personality
- Design the right team for each project

SME discussions alluded to hand picking teams and developing specific skill sets as a key aspect of success. Data indicated over 90% of the interviewed organizations handpicked their staff. Organizations identified required skills needed for each project and took necessary actions to acquire that skill set. Several methods of acquiring these skill sets were used: handpick new individuals, grow/groom current personnel (such as through mentoring, shadowing, or on-the-job experiences), hire contractor support, and reorganize teams.

Several interviews indicated that a vital trait of rapid DoD acquisition involves acute proficiency and depth concerning the application of the so-called normal acquisition process. In order to tailor the applicable rules of acquisition and engineering, team members must first understand what the rules are, and which rules or processes apply to the current situation or could be tailored. These individuals are keenly aware of the implications from omitting or tailoring a step or the challenges in executing parallel development processes.

In some cases, a person with the right attitude, personality, or motivation can make up for a lack of technical skill or experience. One organization was able to make up for a specific lack of knowledge and skill, by strategically leveraging the strengths of the personnel they had – even if that meant moving personnel around as projects progressed.

Observation 9: Maintain high levels of motivation and expectations

- Motivated, collaborative, competitive, impatient, creative, technical, independent
- Mistakes are OK, but it is not OK to repeat them
- Every member connected to the mission and vision

As the research team met with rapid organizations, a certain enthusiasm was noticed abounding in the leaders and personnel – seeming to have a culture that was traditionally military and entrepreneurial in spirit. The mindset of these individuals expressed a competitive nature and a tangible connection to helping accomplish an operational mission. They are motivated. Through discussion, this motivation appears to emanate from three primary sources. First, there is a direct connection to an operational community. Working closely with the end users creates both a connection to the operational task at hand and puts a face on the customer. For example, a team is not just rushing to develop an oxygen sensor for F-22 pilots; they are developing it for a specific named USAF pilot. Second, there is a sense of urgency; Joint Urgent Operational Needs (JUONs) by their nature are "urgent" and critically important. Providing capability to the field may very well be a matter of survival and mission success for US military members. Finally, the rapid nature of these projects provides tangible results, not typically experienced by members of the tradition acquisition and engineering community. Members of the rapid acquisition community can see the full project from concept definition through development, deployment and even operational use. This effect of seeing the result of their work directly used by operational units can be very powerful and maintain sustained levels of motivation.

A unique environmental characteristic observed in several organizations was one in which mistakes are OK, but not OK to be repeated. This concept is vital to fostering a creative, collaborative, and yet competitive environment. One specific technique observed to hone organizational skills is a "debrief culture." Originating from the operational military of reviewing a mission, focused debriefs on team performance can be extremely powerful. A debrief culture emphasizes learning from mistakes and identify root causes to improve future endeavors. The purpose of a focused debrief is to determine what went wrong and develop "lessons learned" to prevent the same errors from occurring in the next project or subsequent iterations of the current project.

Observation 10: The government team leads the way

- High level of expectations for government personnel (military and civilian)
- Focus on full use of government personnel – technical competence is expected

Rapid organizations work hard to find and hire military and government civilian experts. Government personnel are expected to run the programs, often without a prime or support contractors. Many of the rapid programs interviewed had a small support contractor footprint, if at all, compared to most major traditional acquisitions. This is not to say they did not employ or rely on contractors to provide leadership or technical support on varying scales. However, when programs did have a support contractor workforce, the expectation was still the same – the government program manager and lead engineer were expected to be the resident experts on the program.

These government teams are typically comprised of a set of functional experts as the core development team. Core functions are typically a program acquisition officer, resource/financial manager, system/lead engineer, operational expert, safety and test personnel. Technical competence is the standard, not the exception. It is expected every member of the team is technically able to run their portion of the program, with minimal personnel redundancy.

Observation 11: Right-size the program – eliminate/reduce major program oversight

- Smaller systems and budgets receive less oversight and are more stable

One benefit many of the rapid program offices enjoy is a lack of size. When you execute fast, smaller is often better. Not only do large organizations create challenges to effective management and full utilization of personnel resources, but they also tend to have larger budgets. Big programs and big budgets can easily become targets for increased oversight, longer approval chains, and funding cuts. In this sense, being big creates its own problems.

The designs and technologies selected to meet operational requirements directly impact the cost of the program. Sub-system product selection, interface complexity, sustainment considerations, and technical maturity all drive cost. Keep in mind these organizations are focusing on the 23–80% solution, are not going into mass production, and are not necessarily planning for long-term sustainment. However, these organizations intentionally took steps to reduce the overall size of their budgets. For example, the willingness to accept some types of risk buys down the cost of the design, development, and manufacturing efforts. Costs (and risk) are also reduced by using proven or mature technology. Utilizing simple or standard interfaces can also help reduce complexity and development costs.

Direct Observations for Rapid Best Practices (Group 2)

Direct Finding: Not a Single Rapid, But Many Different Flexible Rapid Lanes.

Lanes of Rapid Acquisition

Organizations indicated a number of different definitions of rapid, as well as a wide range of specific practices that were implemented to achieve results. It was observed that various organizations appeared to focus and operate within "lanes of acquisition," which could be analogous to lines of business in the commercial sector. Four lanes were observed, shown in Figure 9.4, and are defined holistically by "product" type. Thus, there is no single archetype of rapid acquisition.

The rapid lanes of acquisition observed are as follows:

Laboratory demonstration/operational prototype: This category is an activity to rapidly design, develop, and test technologies (which can be an individual technology, component, subsystem, or integrated system), typically in a laboratory or rapid prototyping environment. The intent is to demonstrate the technology first in the lab or a test environment, with eventual demonstration in the field or a demo for military utility. Evaluation can be in a realistic or simulated operational environment. Some designs may be "one-off" designs developed in the laboratory. A warfighter need may result in development of a specific technology – i.e., "technical push" – or a technology may be developed and then a warfighter need is discovered that could use the technology – i.e., a "technology pull."

Figure 9.4 Observed "Lanes of Rapid Acquisition."

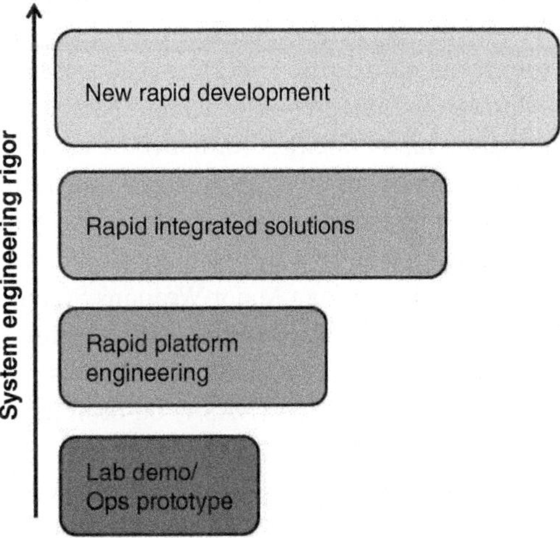

The word "prototype" was discovered in the interviews to have two possible meanings. In one case, such as in the process that DARPA typically uses, a prototype is something developed quickly in the lab, tested in the lab, and used in operations. This is slightly different from a prototype specifically intended to be used in an operational environment, i.e., as a planned test path for a program of record. An example of the latter would be a fly-off of a new fighter aircraft prototype. The lane as defined herein is meant to consider those rapid prototypes that come out of a rapid environment and are not necessarily intended to become part of a program of record, at least not at the time that the prototypes are tested.

Lastly, this category may adapt mature technologies and/or products for military use, or integration with other military systems. The process is accelerated and tailored and includes only the necessary steps to quickly develop and field a test – for example, the process may define the solution, acquire parts, build the system, test it, ship it, use it, and discard it. From the perspective of the warfighter, this would fulfill the urgent need. The people involved in laboratory programs tend to involve more researchers, scientists, and/or students.

Platform engineering: This category consists of modifying and/or integrating existing technology or technologies on top of existing platforms, with new interfaces (Muffatto and Roveda 2000). The most predominant category discovered during the site visits was some form of platform engineering, such as through replacement, upgrades, fixes, and/or enhancements to existing platforms. Platforms were typically ground vehicles or aircraft. Platform engineering also includes repurposing of existing systems, and possibly modifying them, for different missions. An example of platform engineering is the Army Prototype Integration Facility (PIF) in Huntsville, AL, whose mission is to "support Army Aviation, Missile, DoD and technology activities in the development, fabrication, integration, test/qualification of prototype tactical and ground support systems, subsystems and components."

Additionally, the PIF has capabilities that allow for the manufacture and integration of unique, difficult-to-procure, and low-rate-production items. An example of platform engineering is the Mine Resistant Ambush Protected (MRAP) vehicles, where the Army "approved the emergency funding for adding armor kits to the existing fleet of Humvees because they could be fielded more quickly" than other solutions (Lamb et al. 2009). The SE process for platform engineering is focused primarily on the interfaces and the integration of technologies and platforms. People tend to be very creative and solutions-oriented, and they have a diverse and long set of experiences with the platforms. A good commercial example of this is the popular Discovery Channel television show "Monster Garage," which takes small teams with mechanical, fabricating, or modifying expertise to transform a vehicle into a "monster machine" (Rosenburg 2003).

Integrated solutions: This category focuses on the rapid creation of a new platform or system through the integration of technologies and systems. The resulting new solution changes the original intent of either (or both) the technology or original system. The process of developing an integrated solution can occur by integrating new technology with an existing system or adding existing technologies to build a new system. The process may also integrate either new or existing systems in a new way, or on a new platform for a new mission in a new context. This category differs from Major Weapons System in the level of oversight and levels of technology risk (usually lower risk/more mature technology) and levels of cost; it is often characterized by major use of NDI or COTS components. An example would be a National Aeronautics and Space Administration (NASA) on-orbit servicing mission using existing satellite designs and ground-based robotic technology.

New rapid development: This category involves more sophisticated and complex development like traditional acquisition programs of record. Programs of record are characterized by items such as size (including dollar amount) and scope, well-defined requirements, risk level, and mission criticality. This category may also utilize new technologies for a new platform, or a new mission in a new context (such as the Integrated Solutions category above). The Air Force Next Generation Bomber (B-21) would fall into this category. Often, they include greater amounts of technology development, manufacturing, and production than the other three categories.

Other Lane Taxonomies

It is interesting to note that identification and management of "lanes of acquisition" is a common practice.

- DoD defines three categories for Major Defense Acquisition Programs (MDAPs) and the reason for its designation (DAU 2022). The Acquisition Category (ACAT) levels are generally based on dollars, with an implication that complexity and risk are synonymous with cost.
- NASA utilizes four classes of missions. These classes incorporate aspects of mission priority, risk, national interest, complexity, and cost (NASA 2021). Another set of classes used by NASA is for launch vehicles documented in their Launch Services Risk Mitigation Policy (NASA 2018). Similarly, three classes are used to assess the attributes of the mission, including cost, management, launch flight test, manufacturing audits, T&E planning, quality, risk, and safety.
- One additional example worth noting is Operationally Responsive Space (ORS) tiers (Cebrowski and Raymond 2005; Rupp 2007). ORS lanes of acquisition are focused on delivery schedule. Their three tiers were Employ (focused on existing, on-station capabilities), Deploy (including fielding of space ready technologies), and Develop (which would be technologies taking more time).

Flexibility Exists Within and Throughout All Lanes

Maybe this finding is not surprising, but there is huge flexibility both within and across these lanes of acquisition. This flexibility, in part, involved hiring and contracting processes. There was a recurring response from the interviews on the importance of people in expedited SE and urgent needs projects. For example, the right people could execute any process; but a process alone without the right people does not guarantee success. Some of the key comments from interviews include the desire for small, handpicked teams; selection of the "A" team or "a team"; and ensuring a mix of diverse experiences.

One of the first interviews discussed bringing in civilians and military into the organization. It was mentioned, "they interview 100 to hire a few." Generally, this is a unique situation for large defense programs, where personnel, both military and civilian, get assigned through normal rotations. Another organization mentioned the use of "recommendations." They, too, were highly selective and if a trusted member (such as a past employee) recommended someone, that recommendation was as good as an interview.

Another aspect that was important regarding personnel selection was personality type. In almost every case, these small rapid organizations were looking for highly motivated and competent employees. The targeted people were comfortable with uncertainty and flexibility of process. Interesting to note, not all of these rapid team members can have the same personality type. One organization mentioned they hire "Five Tiggers and one Eeyore," using archetypes from "Winnie the Pooh." In other worded, they needed to have a "naysayer" to ensure they did not run too rapid, missing critical considerations.

One organization previously had the ability to selectively choose personnel, but now receives personnel through the normal military rotation. As a result, they were no longer able to pre-select personnel according to specific skill needs or personalities. Instead, the organization instead changed its strategy to where it annually evaluated the mix of personnel, based on their competencies, skills, experiences, and desires. Then they defined the portfolio of work the organization could achieve in a given year, based on that mix of personnel. This was a very innovative way of making real-time adjustments to personnel.

Process variation was another area that was used to achieve flexibility – especially in contracting – but also in management and SE. A theme of "flexibility" arose repeatedly in the SME discussions, with organizations often detailing the contracting approach they used and how this facilitated (or hindered) the ability to execute rapidly. The application of flexibility was also dependent on the experiences of the technical personnel involved.

Flexibility and risk go hand in hand, so it is important to manage both.

For example, one organization had a several year IDIQ contract to rapidly add tasks for local on-site design and production services. These included welding, machining, cable assembly and checkout, breadboard fabrication, install, test, etc. Interestingly, these services were performed in government-owned, government-operated (GOGO) facilities. Other contracting approaches, used by this organization and several others, provided access to subject matter expertise through several sole-source contracts. This allowed immediate access to a pool of uniquely qualified and pre-identified set of individuals that could be rapidly requested to perform work as needed. Competition was also used, depending on the type of work involved (e.g., execution of a build to specification), the expertise required, and the personnel and contractors involved. Even when competition was used, the entire program understood the need to proceed quickly, such that award and delivery of weapon systems could be done in months, not years.

The choice of contracting instrument was often based on the experience, preference, and "comfort level" of the contracting officer. Contracting officers who were embedded in rapid organizations typically had the most knowledge of the flexibility available in existing DoD policy and Federal Acquisition Regulations (FARs). But most often, flexible contracting practices are not well-known, trusted, understood, or utilized because they are not the approaches used in traditional acquisition. Examples of flexible contracting mechanisms include fixed price contracts, IDIQs, sole source, commercial terms and conditions, waivers (such as for certified cost and pricing data), Federal Acquisition Regulation (FAR) Part 12 commercial tailored contracts, Other Transaction Authority (OTA), and NASA Space Act Agreements. OTAs are fixed price, milestone-based contracts, with no FARs and only the minimum statutory terms included. This is a very efficient way to rapidly get on contract, especially for accessing small businesses, entrepreneurs, or other non-traditional contractors. FAR Part 12 contracts are used to procure commercial items on a fixed price basis. NASA has used contractual flexibility with the activities involved in commercial cargo and crew for the International Space Station (ISS).

The discussions also revealed tailoring and scaling processes for systems engineering and program management. In one organization, this process was described as a "mini, compressed" equivalent of a "DoD 5000" project, scaled down to only the necessary milestones and minimal reviews appropriate for that urgent need. The focus was on execution, and the minimum amount of paperwork, meetings, and oversight. The ability to tailor in this way may depend on the organization and their use of Group 3 findings, such as intense knowledge sharing and a risk-based culture.

Inferred Organizational Characteristics for "Go Fast" Cultural Best Practices (Group 3)

In this section, additional findings are presented from the interviews and literature review that are specifically focused on expedited processes supported across an enterprise. These are referred to as

"Go Fast" cultural best practices. Here is where rapid organizations really start to differ from traditional acquisition programs. There are cultural shifts in the organization from top to bottom. From this analysis, there are four findings. The findings include knowledge sharing, a risk-focused culture, and the organization's ambidexterity (plan/execute) abilities. It also includes an ability to employ real-time management. These inferred characteristics are all cultural best practices that traverse work units, functional partitions, and project groups.

The inferred characteristics are:

1) Intense and efficient knowledge-sharing is used to enable stabilization and synchronization of information
2) Rapid organizations are characterized by a risk culture
3) Rapid organizations are structured for ambidexterity
4) Rapid DoD acquisition employs real-time management.

Observation 1. Intense and efficient knowledge-sharing is used to enable stabilization and synchronization of information

Rapid organizations repeatedly mentioned the need for intense and efficient knowledge-sharing. The knowledge-sharing had a particular purpose: synchronization and stabilization of information. Synchronization was needed because, in a rapid work environment, information is often entering the program or project quickly to different personnel. It is necessary everyone has access to the same information, otherwise, decisions will be made based on obsolete or incomplete information. Stabilization of information is also critical, particularly prior to decision-making so that the evidence for the decision is clarified, vetted, and then appropriately used.

Such continuous and intense knowledge-sharing could have the adverse effect of slowing rapid teams. However, the organizations found ways to make the intense knowledge sharing as efficient as possible. Several of the interviewees mentioned that "collocation" was necessary for fast knowledge-sharing because it allowed people to easily inform others spontaneously of new information. In fact, collocation was mentioned 30 times. This enabled everyone to have access to the project, with constant updates on all project areas, and facilitated constant integration of those different project areas. One organization would bring customers on-site during the requirements development phase to speed up the requirements definition and get a contract in "20 days vs. 1 year."

In these cases, collocation was also used to interface with the user as early and often as possible. When the warfighter was collocated with the design team, even for a partial time, the team was able to discern more quickly what the warfighter needed and do rapid trades to get to an efficient solution. Collocation, however, is impractical in those many situations in which the group is larger, temporary distributed, or involves members who travel regularly. Clearly the COVID-19 pandemic has made virtual and online team collaboration commonplace now. Since COVID-19, the use of Microsoft Teams, Zoom (ZoomGov), Jira, Confluence, Slack, and others allow for virtual collaboration, effectively changing "how" organizations rapidly shared information.

Observation 2. Rapid organizations are characterized by risk-focused culture

Consistent across the rapid organizations was a characteristic that is labeled as "a risk focused culture." Culture captures the norms that permeated the organization and its overarching beliefs and values. For rapid acquisition, this culture is "risk-focused" because it includes confronting, identifying, and understanding risk, deciding to accept it, mitigate it, or remove it, then move forward and monitor the risk. There was a constant awareness of the risks involved. The rapid organizations were not trying to eliminate all risks, since there was recognition that such an attempt would be foolhardy. Instead, they were accepting of risk.

Accepting risk meant that personnel were constantly engaged in thinking through contingency plans, identifying and exploring root causes of the risks, and monitoring the risks so mitigation actions could be taken if need.

Research on risk-focused cultures in industry has repeatedly demonstrated that they are only possible when management fosters a "climate of psychological safety" (Edmondson 2008; Edmondson and Roloff 2009). A climate of psychological safety involves leadership offering "top cover" (i.e., supporting when criticized by external parties) to their personnel as well as management expectations backed by an incentive reward structure to take appropriate risks, accept risks, and mitigate risks. In other words, the reward system must also match the risk culture. Personnel do not get fired for taking a risk or making a mistake. This culture is created when learning, storytelling, and mentoring are encouraged and practiced, and failures are shared to avoid making the same mistakes twice.

Observation 3. Rapid organizations are structured for ambidexterity

In corporate contexts, one increasingly observes the presence of "ambidextrous organizations" (Birkinsha and Gibson 2004). Ambidextrous organizations have two different structures in place: one structure focuses on exploration, and another that focuses on exploitation. The structure focused on exploitation generally has substantial routines in place, including milestones, project management practices, and specific work activities identified that should be performed in a particular order. Personnel working in exploitation structures tend to be rewarded for knowing the rules and following standard processes. The expectation conveyed by management to personnel working in an exploitation structure is take the product and process as given, with a focus on "efficiently executing to plan."

In contrast, the structure focused on exploration generally encourages entrepreneurial spirit, innovation, experimentation, learning, iteration, and risk-taking. The expectation conveyed by management to personnel working in an exploration structure is that the product or process is not given but needs to be changed in ways not initially anticipated. Personnel working in exploration structures tend to be rewarded for taking risks, generating new technologies, and creating new opportunities for the program. Such personnel tend to be explorers themselves, comfortable with ambiguity, uncertainty, and challenges that may or may not be known or resolvable.

In a corporate context, exploration structures often describe the early concept phase, while exploitation structures often describe the implementation or production phase. However, in a truly ambidextrous organization, both structures can be invoked at any point depending on the needs of the customer or new information received from outside the organization that suggests a switch of organizational structure. Several of the rapid organizations were observed exercising this ambidexterity. On the exploration side, there was a constant exploring to anticipate future need. They would not simply respond to a warfighter's statement of need, but instead would explore what the problem was and generate new ideas to help the warfighter achieve the end objective. Moreover, the rapid organizations interviewed were not simply exploring, but finding ways to explore faster. The exploitation structure existed as well, as there were project management practices, milestones/reviews, and deliverables expected. Finally, the rapid organizations were able to switch practices quickly. In one organization, an urgent need quickly converted a lab project into a fielded demonstration.

Observation 4. Rapid acquisition employs real-time management

A unique characteristic of the DoD is the opportunity that young officers are given substantial responsibility and gain experience in making decisions under uncertain conditions. The researchers found in the site discussions that the rapid organizations were managed in a manner that explicitly provided these opportunities. Personnel were empowered to learn, understand, and accept responsibility for the risks they incurred and were expected to make decisions about acquisitions, despite not having complete information. While all DoD contexts encourage empowerment at the lowest level, rapid organizations face the additional challenge that the empowerment needs to be done in real time. Empowerment requires the right people with the right skills and experiences, in the right culture and decision-making environment, to thrive.

A risk-focused culture, intense knowledge-sharing, and having both exploration and exploitation structures in place support the empowered individual; however, the research observed that these rapid practices were not the only building blocks of Group 3. A real-time management was identified and shown to have two characteristics: a forward-focus and an urgent decision process.

The researchers saw managers and engineers closely focused on future milestones that would quickly inform them when the team was diverging from the plan. One example of forward-focused thinking observed in the interviews was that progress toward test and test plans were discussed even though a team was still quite early in concept formulation discussions. Another example was that deployment specifics were being mentioned in casual conversations as new design ideas were expressed. Yet another example is "situational awareness of the user", i.e., the extent to which rapid personnel were close enough to the user group (such as a warfighter) that they knew about the latest information from the war zone or were confident that they could anticipate how the warfighter was likely to react to a new idea. Thus, empowered personnel were not simply empowered to decide, but, as they were making the decision, they needed to be able to articulate the downstream implications of that decision, and then adjust their decision-making accordingly and in real-time.

Often program personnel may not have the sole authority to make a decision; the decision needs to be made at the appropriate level or rank. Thus, empowered individuals are often in a situation in which immediate action is needed but they alone are not able to make that decision. Empowered individuals, to act quickly, must decide if action is really needed urgently, decide who has the authority to make the decision, and decide how to get the decision as quickly as possible. These experienced managers appeared to have a sufficiently detailed understanding of both the authority to which they could delegate decision-making and which people would have the right experiences and knowledge to make the decision at the lowest possible level. The interviews indicated a variety of different methods used to get a decision made, including collocating near the highest chain of command. One interviewee discussed the importance of collocation and information sharing saying, "he was 61 steps from any program manager and 123 steps from the PEO's office."

Aspects of this Go-Fast culture may be enabled by creative and effective use of collaborative information technology (IT). While not the focus of the guiding interview questions for the SMEs, current online and virtual practices, predominantly due to COVID-19, may facilitate real-time management and decision-making. However, it is important to note that the mere existence of such tools does not guarantee that team members will use them effectively, or at all. Ineffective aggregations of email(s), multiple Teams shared storage and chat, and continued local shared drives could exacerbate the information overload.

Conclusions and Future Research

This research conducted site visits on early adopters of rapid acquisition. Using a set of refined questions on people, process, and product, grounded theory was used to find clusters of responses from the interview data. The findings are broken into three groups:

- **Group 1**: Direct Responses,
- **Group 2**: Direct Observations, and
- **Group 3**: Inferred Organizational Characteristics.

The findings reflect aspects of agile and lean product development, as well as confirm many critical success factors of "time-to market" processes. Based on direct responses, observations, and inferred best practices, the following are recommendations for action.

1) Ensure that projects utilize prioritized organizational and project best practices
2) Train program managers, engineers, and contracting officers in organizational and cultural best practices, real-time management approaches, and the different flexibilities that exist.
3) Encourage programs to intensively share knowledge, have a risk-focused culture, and create an ambidextrous organizational structure
4) Share knowledge, experience, mechanisms, and lessons learned across programs and organizations
5) Quantitatively monitor progress in expediting SE processes, and use the measurements to improve both schedule acceleration and the ability to estimate it
6) Use DoD rapid organizations as a testbed to introduce digitally enabled solutions
7) Develop the acquisition workforce using rapid programs to provide full lifecycle insight and hands-on experience.

The researchers do not expect that an existing program will simply pick up and apply the framework completely. Perhaps, though, parts can be applied judiciously. While the next "big" program could be chosen as a nontraditional, rapid capability approach, the framework could also be used in a mixed (hybrid) approach. Some of the above recommendations including training and development of the workforce. The use of rapid acquisition programs, with a less than two-year lifecycle, can be an "experience developer" for young engineers. Training should continue in lean product development, agile system and software engineering, and awareness on the flexibilities inherent in hiring and contracting practices.

Personnel (from engineers to program managers to contracting officers) who participate in a rapid program see a full life cycle of acquisition from problem statement to fielding (and maybe even disposal) in a very short time. In the same timeframe for a traditional acquisition program, the personnel may see a portion of one milestone, with experiences that consist of document-driven briefings, plans, and scheduled meetings. Research on metrics that capture the true utility of these experiences should be conducted.

Related research on how to best provide these lifecycle experiences to the entire workforce should continue developing Experience Accelerators, and other game-based learning. The objective of the accelerator is to provide a safe, yet authentic environment in which participants can make mistakes without adversely affecting their careers or programs. The simulation provides the participants with the ability to view a program through the entire lifecycle, see the relationships between elements of the system, and the system developing the system and encounter the challenges faced in an expedited system development.

Given the 11 Direct Responses (Group 1 findings), the research team then put these observations in the context of transformations (functions/ tasks) needed across the life cycle. By examining the generic development process, especially for lean development, the transformations build a task network, as shown in Figure 9.5, that can be analyzed for stall, waste, missing value, lack of improvement, quality/ rework, and schedule. The major objective of rapid acquisition is to reduce the latency from start while efficiently and effectively delivering a system that meets the warfighter's urgent needs. The process of creation from start to deployment is one of transformation – be it the transformation of opportunity, to concept, to design, to finished system, and finally to deployment. Below are six techniques that can be used to reduce latency, shown in Table 9.3. These techniques reduce the total amount of work (total, new and unanticipated), increase efficiency, and increase throughput and/or decrease stall time.

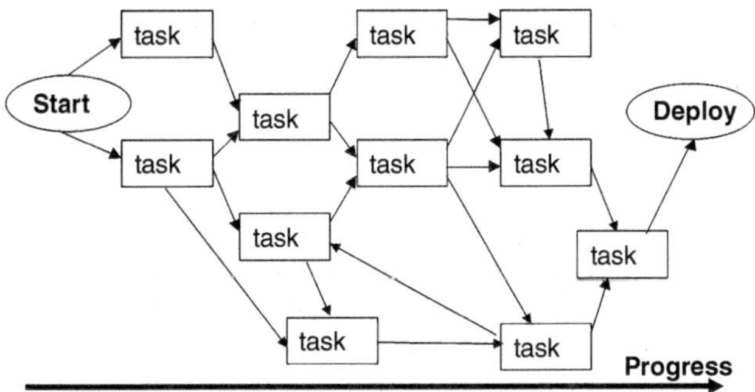

Figure 9.5 Systems engineering process as a task network, which can be minimized for schedule.

Table 9.3 Expedited program techniques.

Area	Techniques	Impact
Product	1) System simplification	Reduces total work
	2) Legacy reuse	Reduces total new work
Processes	3) Transformational efficiency	Increases efficiency
	4) Rework avoidance	Reduces rework
People	5) Parallelization	Increases throughput
	6) Improved decision making	Reduces "stall" time

1) System Simplification
2) Legacy Reuse
3) Transformational Efficiency (of each SE task)
4) Rework Avoidance (also of each SE task and across the network)
5) Parallelization of work
6) Improved Decision Making

Optimization of the task network, and implementation of these techniques should be researched.

Continuous and rapid fielding or expedited systems development play a new role in developing systems. Instead of perfecting all requirements with possible schedule impacts, modern programs must adapt themselves to have a lifecycle that is driven by a continual delivery strategy. Increasingly, schedule of delivery has become more important than cost. Organizations must fight with the "time to market" considerations. The software community has designed approaches in the last several years, such as "DevOps" and "DevSecOps" and "SecDevOps." These integrate the activities of software development, security design, and operations for continual product release.

The speed, storage, and processing of modern computing allow a more efficient model-based approached to engineering. To ensure continued U.S. superiority, the DoD in 2018 began transforming acquisition and engineering practices to use an integrated, digital, model-based approach (OSD 2018). However, this had already begun across many industries, not just defense and aerospace. This approach enables programs to prototype, experiment, and test decisions and solutions in a virtual environment before production and delivery.

Model-based systems engineering enables the development of a "Digital Twin," which allows programs to test a system in a digital environment early. This has a wide-ranging list of benefits such as being able to test or stress certain conditions on the system, as well as aiding in developing test plans for live prototypes. It can be useful to determine operating bounds/limits and estimate system performance in a variety of scenarios prior to real flight testing. Digital Engineering is being applied in the development of the Air Force's newest aircraft, weapons, and missiles (Losey 2020; Osborn 2020; Waterman 2022; Edwards 2022).

The expected benefit of digital modeling is designing, assembling, testing in the digital environment, and exploring the trade space in the virtual world early. This enhanced and expedited ability of digital assembly and testing has drastically decreased the time needed to prototype and test next-generation systems. This approach also allows for rapid response to design modifications that are all too common throughout a program's lifetime. Digital engineering and model-based systems engineering are expected to meet the defense needs of the twenty-first century, but will require new methods, processes, and tools (MPTs). With little pedigreed data to support claims, all indications are digital transformation will improve the quality and speed to support urgent operational needs, as well as large traditional programs of record. How best to scale and tailor digital engineering for small programs must be researched. All indications are the use of digital twin and digital threads may provide the Group 3 enablers. However, this digital transformation must still consider the Group 1 people, process and product findings in addition to the flexible acquisition practices of Group 2.

References

Anderson, D.J. (2010). *Kanban: Successful Evolutionary Change in Your Technology Business*. Sequim, Washington: Blue Hole Press.

Birkinsha, J. and Gibson, C. (2004). Building ambidexterity into an organization. *MIT Sloan Management Review* 45: 47–55.

Boehm, B. (2010). System 2020 – Strategic Initiative. *SERC SERC-2010-TR-009-1. Defense Technical Information Center*. https://doi.org/10.21236/ADA637406 (accessed 1 June 2023).

Boynton, A. and Zmud, R. (1984). An assessment of critical success factors. *Sloan Management Review* 25 (4): 17–27.

Bullen, C.V. and Rockart, J.F. (1981). A Primer on Critical Success Factors. MIT Libraries, Report no. 1220-81. https://dspace.mit.edu/handle/1721.1/1988.

Cebrowski, A.K. and Raymond, J.W. (2005). Operationally responsive space: a new defense business model. *Parameters* 35 (2): https://doi.org/10.55540/0031-1723.2250.

Chun, T., Ylona, M.B., and Karen, F. (2019a). Grounded theory research: a design framework for novice researchers. *SAGE Open Medicine* 7 (January): https://doi.org/10.1177/2050312118822927.

Chun, T., Ylona, M.B., and Karen, F. (2019b). Playing the game: a grounded theory of the integration of international nurses. *Collegian* 26 (4): 470–476. https://doi.org/10.1016/j.colegn.2018.12.006.

Cunningham, W. (1992). The Wycash Portfolio Management System. *Experience Report*.

Cunningham, W. (2011). Ward explains debt metaphor. http://wiki.c2.com/?WardExplainsDebtMetaphor (accessed 1 June 2023).

DAU (2022). *Acquisition Categories (ACATs) | Adaptive Acquisition Framework*. Major Capability Acquisition. https://aaf.dau.edu/aaf/mca/acat/.

DoD (2010). *Rapid Capability Fielding Toolbox Study*. Washington DC: DoD https://apps.dtic.mil/sti/citations/ADA528118.

DSB (2009). Fulfillment of Urgent Operational Needs. *Report of the Defense Science Board Task Force ADA503382*. https://dsb.cto.mil/reports/2000s/ADA503382.pdf (accessed 1 June 2023).

Edmondson, A.C. (2008). Managing the risk of learning: psychological safety in work teams. In: *International Handbook of Organizational Teamwork and Cooperative Working* (ed. M.A. West, D. Tjosvold, and K.G. Smith), 255–275. Chichester, UK: Wiley https://doi.org/10.1002/9780470696712.ch13.

Edmondson, A.C. and Roloff, K.S. (2009). Overcoming barriers to collaboration: psychological safety and learning in diverse teams. In: *Team Effectiveness in Complex Organizations: Cross-disciplinary Perspectives and Approaches* (ed. E. Salas, G.F. Goodwin, and C.S. Burke), 183–208.

Edwards, J. (2022). *Rolls-Royce, Boeing Use Digital Design Methods to Update B-52 Components.* ExecutiveBiz. March 14, 2022. https://blog.executivebiz.com/2022/03/rolls-royce-boeing-use-digital-design-methods-to-update-b-52-engines-systems/ (accessed 1 June 2023).

GAO (2010). Warfighter support: improvements to DOD's urgent needs processes would enhance oversight and expedite efforts to meet critical Warfighter needs. https://www.gao.gov/products/gao-10-460 (accessed 1 June 2023).

GAO (2011). Warfighter support: DOD's urgent needs processes need a more comprehensive approach and evaluation for potential consolidation. https://www.gao.gov/products/gao-11-273 (accessed 1 June 2023).

Glaser, B.G. and Strauss, A.L. (1967). *The Discovery of Grounded Theory: Strategies for Qualitative Research.* Aldine.

Hanawalt, E.S. and Rouse, W.B. (2009). Car wars: factors underlying the success or failure of new car programs. *Systems Engineering* 4 (13): 389–404. https://doi.org/10.1002/sys.20158.

Lamb, C.J., Schmidt, M.J., and Fitzsimmons, B.G. (2009). *MRAPs, Irregular Warfare, and Pentagon Reform*, 62. Washington, DC: National Defense University Press. https://apps.dtic.mil/sti/citations/ADA515185.

Lane, J.A., Boehm, B., Bolas, M. et al. (2010). Critical success factors for rapid, innovative solutions. In: *New Modeling Concepts for Today's Software Processes*, Lecture Notes in Computer Science (ed. J. Münch, Y. Yang, and W. Schäfer), 52–61. Berlin, Heidelberg: Springer. https://doi.org/10.1007/978-3-642-14347-2_6.

Lencioni, P.M. (2011). *The Five Dysfunctions of a Team: A Leadership Fable.* 1st ed. Jossey-Bass.

Losey, S. (2020). T-7 Red Hawk Trainer Makes Its Debut. *Defense News.* https://www.defensenews.com/air/2022/04/29/t-7-red-hawk-trainer-makes-its-debut/.

Morgan, J. and Liker, J.K. (2020). *The Toyota Product Development System: Integrating People, Process, and Technology.* New York: Productivity Press. https://doi.org/10.4324/9781482293746.

Muffatto, M. and Roveda, M. (2000). Developing product platforms: analysis of the development process. *Technovation* 20 (11): 617–630. https://doi.org/10.1016/S0166-4972(99)00178-9.

Murman, E., Allen, T., Bozdogan, K. et al. (2002). *Lean Enterprise Value: Insights from MIT's Lean Aerospace Initiative.* 2002nd ed. Houndmills, Basingstoke, Hampshire ; New York: Palgrave Macmillan.

NASA (2018). Launch Services Risk Mitigation Policy for NASA-Owned and/or NASA-Sponsored Payloads/Missions. NASA Policy Directive (NPD) 8610.7D. https://nodis3.gsfc.nasa.gov/displayDir.cfm?t=NPD&c=8610&s=7D.

NASA (2021). Risk Classification for NASA Payloads. NASA Procedural Requirements (NPR) 8705.4A. https://nodis3.gsfc.nasa.gov/displayDir.cfm?t=NPR&c=8705&s=4A.

Oehmen, J., Oppenheim, B.W., Secor D., et al. (2012). The Guide to Lean Enablers for Managing Engineering Programs. Joint MIT-PMI-INCOSE Community of Practice on Lean in Program Management. https://dspace.mit.edu/handle/1721.1/70495 (accessed 1 June 2023).

Office of the Secretary of Defense (OSD). (2020). Operation of the Adaptive Acquisition Framework. DOD INSTRUCTION 5000.02.

Osborn, K. (2020). *Air Force Flies 6th Gen Stealth Fighter - 'Super Fast' With Digital Engineering.* Warrior Maven: Center for Military Modernization. https://warriormaven.com/future-weapons/air-force-flies-6th-gen-stealth-fighter-super-fast-with-digital-engineering.

OSD (2018). *Digital Engineering Strategy.* Washington, D.C.: Office of the Deputy Assistant Secretary of Defense for Systems Engineering.

Rosenburg, D. (2003). *Inside Monster Garage: The Builds, the Skills, the Thrills.* Meredith Corporation https://www.abebooks.com/9780696218903/Monster-Garage-Builds-Skills-Thrills-0696218909/plp.

Rupp, S. (2007). *Operationally Responsive Space.* Kirtland Air Force Base. https://www.kirtland.af.mil/News/Article-Display/Article/390430/operationally-responsive-space/https%3A%2F%2Fwww.kirtland.af.mil%2FNews%2FArticle-Display%2FArticle%2F390430%2Foperationally-responsive-space%2F.

Van Secter, D. and Doolen, T.. (2011). Comparative Analysis of Critical Success Factor Research. *61st Annual IIE Conference Proceedings.*

Waterman, S. (2022). GBSD Using Digital Twinning at Every Stage of The Program Lifecycle. *Air & Space Forces Magazine.* (8 April). https://www.airforcemag.com/gbsd-using-digital-twinning-at-every-stage-of-the-program-lifecycle%EF%BF%BC/.

Biographical Sketches

John M. Colombi is a professor of systems engineering within the Department of Systems Engineering and Management at the Air Force Institute of Technology (AFIT), Wright-Patterson AFB. He has over 35 years of air force experience in developmental engineering, program management, and academia. His research interests include systems and enterprise architecture, complex adaptive systems, model-based systems engineering, digital engineering/digital twins, autonomy, and human systems integration. Dr. Colombi is a member of INCOSE and is a senior member of the IEEE. He serves as an ABET PEV for systems. He received his PhD in electrical and computer engineering and his MS with distinction in electrical engineering from the Air Force Institute of Technology. He graduated with his bachelor of science in electrical engineering (Cum Laude, AFROTC DG) from the University of Lowell.

Chapter 10

Scaling Agile Principles to an Enterprise

Michael Orosz[1], Brian Duffy[1], Craig Charlton[1], Hector Saunders[2], and Michael Shih[3]

[1] *University of Southern California Information Sciences Institute, Marina del Rey, CA, USA*
[2] *US Space Force Space Systems Command (SSC), El Segundo, CA, USA*
[3] *Booz Allen Hamilton, El Segundo, CA, USA*

Introduction

This chapter discusses the unique challenges in scaling agile and DevSecOps (development, security, and operations) to large enterprise systems. Such systems include large-scale manufacturing processes (e.g., automotive manufacturing) and service-based systems such as space-based communication systems. These systems are composed of multiple subsystems, each often developed and maintained via multiple vendors and undergo modification and upgrades on different timelines. Although many of these enterprise systems are composed of both hardware and software subsystems (e.g., space vehicle and software-based ground control), this chapter is focused primarily on software-based systems. That said, when appropriate, reference to hardware-only or hybrid hardware and software-based systems will also be noted. The targeted enterprise environment includes mid-to-large scale enterprises such as US Department of Defense acquisition programs and mission-critical systems (i.e., systems that cannot fail).

The Challenge

As noted in Chapter 37 on *Unique Challenges in Mission Engineering and Technology Integration*, enterprise systems are typically large systems of systems that involve multiple interfaces, multiple vendors/maintainers, hardware, software, human operators, end users, and are constantly evolving to meet changing customer and environmental needs and challenges. Such complex systems are normally associated with large organizations with multiple departments or divisions (e.g., human resources, executive offices, engineering, facilities, production, distribution, sales, security, etc.) and multiple vendors (e.g., suppliers (supply chain), distributors, point of sales, etc.) although they also include large information systems (e.g., financial network, etc.) and other product- or function-focused organizations (e.g., GPS enterprise consisting of space, control, and user segments). Regardless of the type of enterprise system being considered, system upgrades and new product development can apply to one, two, or more components that make up the enterprise.

Systems Engineering for the Digital Age: Practitioner Perspectives, First Edition. Edited by Dinesh Verma.

Companion website: www.wiley.com/go/verma/systemsengineering

The challenge to the acquisition professional is how to manage the development process to ensure that all components are developed as a system of systems and not as independent and isolated entities. Multiple vendors, differing timelines, delays in releases, changing requirements, the availability of reliable supply chains, and various internal and external dependencies will need to be considered.

Traditional agile development approaches that rely on one or two scrum teams – each with 7–10 team members who are focused on a small number of functions and capabilities – typically do not scale well to large enterprise system. Generally, these large systems require hundreds of developers, systems engineers, integration and testing personnel, and end-user customers focused on the development, deployment, and sustainment of the system. Scaling is not simply a matter of adding more sprint teams that operate on a key function of the enterprise system and when completed are somehow linked, integrated, tested, and deployed as a product/solution. Rather, scaling includes having an infrastructure in place that allows each sprint team to operate as a cohesive unit but also facilitate communications/collaborations between sprint teams within a project and with external system entities that interface with the program (i.e., the whole enterprise).

Another wrinkle to consider is that in many cases, particularly in US DoD acquisition programs, agile processes must be merged with traditional acquisition approaches normally associated with waterfall approaches as defined by DoDI 5000.02 (OUSD(R&E) 2020). For example, major milestones such as Milestone B and the use of earned value management (EVM) for cost and schedule performance tracking often requires that the agile approach be tailored to address these programmatic requirements.

Why Agile and Dev*Ops?

Although noted elsewhere in the handbook and discussed in the literature (Tryqa.com 2022), a major motivation for undertaking a traditional serial (or waterfall) approach to developing an enterprise system is to address the complexity of that enterprise. Due to the interconnectedness of the various systems of a system of systems enterprise, the tendency is to do all the systems engineering up front – including the detailed design. The justification for using this approach is to fix the design so that all aspects of the enterprise (i.e., interfaces, communication protocols, computations, dependencies, etc.) are known in advance to all developers involved in the development effort and to ensure everyone is operating with a common and uniform understanding of what needs to be developed.

Unfortunately, and as also noted in other chapters of this handbook and documented in the literature (Tryqa.com 2022), there are plenty of reasons why this approach is not preferred or even feasible. For example, requirements (i.e., user/customer needs) often change. In a large enterprise-based serial development program with a long development timeline, the relevancy of the original requirements and design can often become obsolete prior to the completion and delivery of the end product. Customer needs change and traditional serial approaches to systems development are typically inflexible to react to those changes.

Another challenge with the traditional waterfall method (i.e., requirements defined → initial system design – final system design → development → integration and testing → deployment) is that the integration and testing happens at the end of the development effort. Relying on integration and testing at the end of the development effort often produces two challenges. First, errors are not discovered until the end of the program. In many cases, these discrepancies often cause other problems in the system, which often results in the cascading of discrepancies. The result is a bow wave of system discrepancies accumulating at the end of the program when the original system developers are typically not available (e.g., they may have moved to another program) to address them. This situation often leads to the next challenge – the project schedule (and resulting cost) will grow as project management is forced to bring onto the project available developers and extend the project schedule to address discovered discrepancies – pushing final integration, testing, and product delivery to the right of the original schedule.

As noted in the Chapter 37, applying Agile and DevSecOps (i.e., constant integration/constant deployment) principles to the development process will help mitigate these challenges. Agile addresses the need to be flexible to incorporate changing requirements and end-user needs. DevSecOps addresses continuous integration where bugs are discovered early and often when they can be easily addressed by the development team. In addition, DevSecOps also addresses instantaneous deployment where workable/operable solutions are made available to the end-user where user feedback can easily be fed back to the development teams to allow for system design updates to reflect market needs.

Scaling Agile

The remainder of this chapter is primarily focused on software-based systems or hardware/software systems where the focus is on the software development efforts, but where hardware plays a prominent role in the development of the system (e.g., cyber-physical interface testing, etc.).

Agile and DevSecOps Review

As illustrated in Figure 10.1, the typical agile process consists of dividing work (e.g., development of a function or capability) into efforts that can typically be completed within a short period of time (e.g., 2–4 weeks) and that when completed provides value to the end user. In this chapter, these periods of time are called sprints and the works that are accomplished (i.e., the value developed) are called stories. Stories (which are decomposed from system requirements) are stored on a project backlog (i.e., a virtual queue) and assigned priorities based on when the assigned function or capability is needed (often dictated by market demand or the organization that contracted for the system to be developed). As the project moves forward and stories are completed (i.e., at the end of the sprint) these completed functions/capabilities are available to the end user for use and

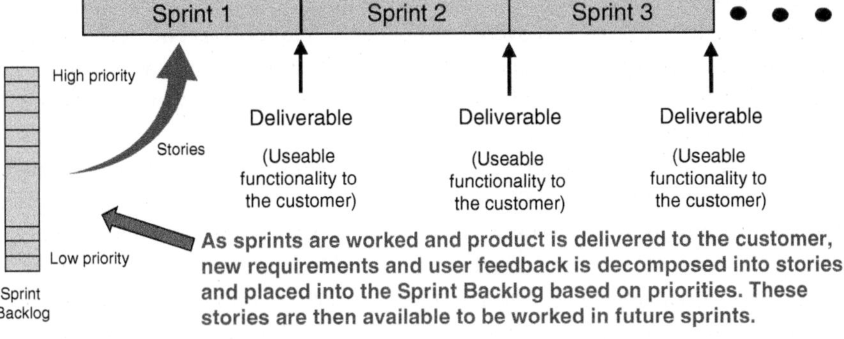

Figure 10.1 Typical agile development process (details described in text).

evaluation. Each sprint builds on the previous product resulting in a process that continuously adds value to the customer as the system is developed. Feedback from the user or changes in system requirements (e.g., a new capability is to be added) are often incorporated into the project backlog (as new stories) and addressed in the next or future sprint.

At the beginning of a sprint, decisions are made as to which story or which portion of a story is assigned to a project team member. Usually, the story is already well defined (part of the refining work from the previous sprint) and so members of the development team can quickly start working on their assigned efforts. As the development effort progresses, the story undergoes integration and testing to ensure that the new functionality works within the system it is intended to be included in. Integration and testing are part of the DevSecOps pipeline. The idea is that as new/updated software is completed and checked into Configuration Management (CM), that software is now part of a daily (or even multiple daily) build process that provides security checks and provides functional and regression testing. At the end of the sprint, the completed story is demonstrated and/or made available to the end user for evaluation and comment. Feedback from these demonstrations/deployments are provided to the sprint team for analysis and, in many cases, incorporation in the sprint (or future sprint) effort. Additional discussion on DevSecOps pipelines can be found in Chapter 37.

The advantage of using Agile and DevSecOps is two-fold. First, Agile supports the idea of fail fast, learn fast. By quickly producing a working end-product that gets into the hands of the end user, the more quickly the development team will know (from feedback) whether the product meets the needs of the customer. If the product does not meet end-user needs, that feedback will quickly be provided to the development team with information on what the end user wants (i.e., quickly learn what the customer wants or needs). The second benefit of the Agile approach – particularly when coupled with DevSecOps – is an opportunity to quickly detect system bugs (discrepancies) and other failures during development where they can be addressed and mitigated and not allowed to "flow forward" to be addressed later in the future where it will be more difficult to be addressed as the development team may already be assigned to another program.

Challenges with Scaling Agile

A major challenge with the traditional Agile/DevSecOps approach is scaling (to be addressed later) and getting end-user involvement in evaluating the newly produced product. This is particularly an issue in mission-critical environments where the product to be evaluated cannot fail or the availability of end-users for evaluation do not exist due to operational constraints. In these situations, it is imperative that as part of the integration and testing phase of the DevSecOps pipeline, a near operational environment be available (covered in detail in Chapter 37). The need for these near operational environments is especially important in large enterprise systems where it is usually not feasible for the complete enterprise to be available for updating and evaluation. This will be described in more detail in the next section.

The agile approach briefly defined in the previous section is ideal in acquisition environments where the projects are relatively small (i.e., 1000s to 10,000s of lines of code) consist of one or two sub-systems (often called Configuration Items (CIs) in US government programs) and external interfaces with external systems are usually limited to a few – including the end-user community. In many cases, however, projects are much larger in size (100,000s to millions of lines of code, multiple subsystems, and multiple interfaces and dependencies with external systems and end users). This is particularly the case with enterprise systems where multiple components of the enterprise are undergoing development on their own timelines, with different vendors and whose requirements can change impacting all other entities within the enterprise. Under these circumstances, agile and DevSecOps must be scaled.

Internal Dependencies, External Dependencies, and the Need for Some Up-Front Engineering – Understanding the Complexity of the Enterprise

As noted above, large enterprise systems typically are composed of multiple subsystems – each with internal and external interfaces and dependencies. For example, in a typical space-based enterprise acquisition program, there are three major subsystems (called segments) – space, control, and user (Wikipedia 2022). Each of these segments can be furthered decomposed into multiple subsystems. For example, in the control segment, there are multiple subsystems that make up the command-and-control segment (e.g., mission planning, missing scheduling, navigation control, monitoring, etc.). Each subsystem interfaces with other subsystems with associated dependencies. Further, within many of these subsystems, there are smaller components (e.g., sub-subsystems) with their own interfaces and processing needs. Often these smaller sub-subsystems can also be quite large (100's of lines of software code).

The key to scaling agile and DevSecOps is to understand the complexity of the system to be developed or modified. Changes within one of the enterprise systems, subsystems, or sub-subsystems can impact other components internally or externally to the overall enterprise system. Again, borrowing from the space domain, the space segment is often focused on the development and deployment of a space vehicle (SV) – a primary hardware-based system that is controlled by a software-based control segment (usually ground-based). Changes in the SV mission flight software may impact the control segment (and vice-versa). Understanding these interfaces and dependencies and having the ability to quickly react to changes is a necessary part of the development process.

Because of these dependencies, it is important that some up-front engineering be undertaken prior to the beginning of actual development. Chapter 37 discusses up-front engineering, and the reader is invited to review that chapter. That said, the following is a brief overview of the merits of undertaking up-front engineering. In the Agile approach, development is driven by the project backlog and the assigned priorities to each of the stories in that backlog. Some up-front engineering is required to (1) develop the initial project backlog and, very importantly, (2) identify dependencies between the stories in the backlog to ensure that if a story's priority is changed due to an external issue (e.g., the priority is lowered and the story is pushed to a later sprint due to the delayed delivery of an external dependency), the priority adjustment does not greatly impact other dependent stories. This is particularly important as once development is started, many of the sprint teams (as will be discussed next) may not be in constant contact with other sprint teams or other enterprise units on a continuous basis and may not be aware of dependencies if not already identified.

Sprint Teams and Structure

For large systems of systems projects that involve hundreds of thousands or more lines of source code and/or large systems of systems programs, there may be an attempt to simply keep the number of sprint teams small and increase the number of team members. As noted in the literature (PremierAgile 2022; Makadia 2021; LearNow 2022), this approach almost always results in stretching the development effort, increasing the length of the development schedule, and increasing the cost of the overall project. This is often due to the increased lines of communications that result when teams get too large. Daily scrums lengthen, information exchanged gets misunderstood, coordination between team members starts to drop, which results in stories that do not get completed within their assigned period of performance and spill over into the next development sprint. This can and almost always results in team members never being able to catch up and morale drops contributing to more development delays and high team member turnover.

A solution is to have more sprint teams with each aligned with a subsystem or sub-subsystem of the system they are developing (and another reason for some up-front engineering to understand how best to divide the project up). The challenge now is that there is a need to ensure that communications and collaborations between each sprint team is maintained and managed.

Managing Multiple Sprint Teams

Scaling Agile almost always involves establishing a hierarchy of management. The key is implementing the hierarchy without adding unnecessary bureaucracy or overhead. There are many approaches to choose from when establishing a scaled agile framework that relies on a hierarchical solution – including SAFe®, SoS, SaS, and others (Scaled Agile 2022; Spanner 2023; Scrum@ Scale 2023), but almost all of them involve establishing a single high level coordination team that monitors, coordinates and in some cases, manages the various sprint teams that make up the overall program. These oversight teams are tasked with monitoring progress of each sprint team, ensure dependencies between sprint teams are well understood and tracked and, in many cases, establish the overall objectives each team will focus on in the next few work segments.

For example, in the SAFe Scaled Agile Framework (Scaled Agile 2022), an Agile Release Train (ART) team is established to help coordinate the various sprint teams that make up the enterprise development effort. The ART team is responsible for monitoring team progress, managing the systems engineering process, and coordinating efforts between sprint teams. Often, a weekly or bi-weekly "team of teams" meeting where representatives (using the Project Owners [POs] and ScrumMaster [SMs]) from each sprint team gather to discuss overall project progress, dependency issues, and other challenge issues. The key here is that there is coordination between the various teams via frequent engagement. It is highly recommended that when scaling agile, a management team (e.g., ART team in SAFe) and frequent coordination meetings (i.e., "team of teams") be established.

Stories, Features, and Scaling of Work Tasks

In Agile, a story is typically associated with a work task that, when completed within a relatively short time (e.g., 2–4 weeks), provides value to the customer. In some cases, a single story satisfies a given requirement for the system. In other situations, a collection or group of stories are often required to satisfy a system requirement – with each story satisfying a portion of the requirement. For many projects, this structure of stories, when completed, produce value for the end user is sufficient, however, for large complex systems – particularly large enterprise systems, it may not be possible to divide requirements in work tasks that can be completed within a short period of performance (i.e., 2–4 weeks). In these situations, it may take many weeks (possibly even several months) for the development team to produce a product that provides value to the end user. The natural solution is to simply extend the period of performance (i.e., the sprint) from weeks to months to accommodate the complexity of the project. This approach is not recommended as the whole point behind agile and DevSecOps is to quickly produce value and to continuously undertake integration and testing to detect discrepancies early and often. In addition, extending a sprint

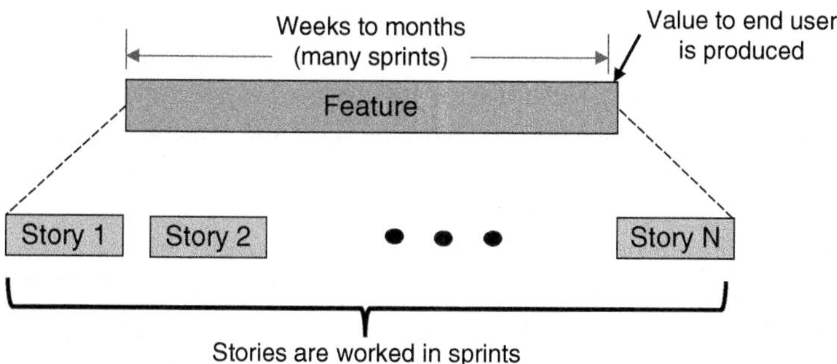

Figure 10.2 Scaling agile through the use of features and decomposed stories. Stories are worked in sprints and when all decomposed stories are completed, the feature is completed, and value is produced for the end user.

to encompass multiple months (the effects of which are described in more detail later in the chapter) can impact developer performance and morale as the need for frequent "closure" of a work package and the resulting ceremonies and time to catch a breath are less frequent. This can quickly lead to development team morale challenges that will impact the program through reduced productivity.

A solution to the above challenge is to accept that some work tasks will require multiple months to complete; however, those work tasks can be divided into smaller sub-tasks that can be completed within a 2–4-week period of performance. Often these large work tasks are called features and the sub-tasks that are decomposed from the feature are stories (Figure 10.2). In this paradigm, stories are created that when completed, add value to the feature in which they are decomposed. When all stories decomposed from a feature are completed, the feature is now complete and value to the end user is produced. More details on the concept of features, stories, sprints, and development timeboxes (called Program Increments in some scaled agile implementations (Scaled Agile 2022)) will be discussed later in the chapter.

Creation of Sprint Teams

The key to implementing scaled agile is to determine the number and types of sprint teams required to develop the product. There are two major types of sprint teams in a scaled agile environment: feature teams and enabler teams. Feature teams (sometimes called mission teams) are tasked with developing the system that will be used by the end user/customer. These teams focus on features that develop functions and capabilities that meet the needs of the end user. Examples include teams that develop scheduling and communications functionality for command-and-control systems. Enabler teams, on the other hand, build and operate the support systems (sometimes called the scaffolding) that are required to facilitate or support the feature teams in developing a product. Examples of such teams include those involved in developing software factories, systems engineering, integration and testing pipelines (both functional and regression), production, distribution, training, and other logistic functions.

Determining how many feature and enabler teams are required is heavily influenced on the size, complexity, number of subsystems, and a host of other attributes that define the overall project. Determining this information requires undertaking upfront engineering (covered earlier in the chapter) to establish the major components, interfaces, dependencies, and other attributes that define the system as well as how best to divide the overall program structure into logical components of the correct size and complexity that can be undertaken by a sprint team (7–10 team members). Determining the correct number of sprint teams is typically based on experience. However, program management should not be afraid to experiment and be supportive of possibly creating more (or possible reduce) the number of teams once the project is underway and as all teams gain experience with the system being developed and with the skills and capabilities of each team member. The key issue to remember is to keep sprint teams small (7–10 members maximum) and focused on one or two subsystems or configuration items.

Although it is recommended that sprint teams be limited to 7–10 members, there can be exceptions. In some programs, there are examples of sprint teams composed of 15–20 members. Often the members of these large sprint teams are focused on specific areas of the project that by themselves are too small to be addressed by a single sprint team, but when combined with similar "small" focused areas, define a team. For example, the management of cybersecurity, system testing, discrepancy tracking, and the management of digital engineering practices such as maintaining system models (e.g., used in MBSE) are often lumped into a single large sprint team. That said, care should be exercised in going this route as once implemented it can be difficult from a project management perspective to later devolve these large teams into smaller sprint teams. For example, cost accounts are typically created to support a sprint team. Decomposing a sprint team into multiple smaller teams can result in challenges to the cost performance tracking system used on the program. Again, it is highly recommended that adequate time be allocated up front when determining the size and number of sprint teams to avoid challenges later in the program.

Minimum Viable Products (MVPs), Minimal Marketable Products (MMPs), Sprints, and Increments

At this point, the question usually raised is what is value to the customer? In the agile world, there are two terms – minimum viable product (MVP) and minimum marketable product (MMP) – that are used to define value to the end user/customer. An MVP (Productfolio 2023) is something that is of value to the customer (i.e., some function or capability), but needs some refining to bring it to a state where the product can be deployed into the market (an MMP). MVPs are useful for allowing the customer to get an initial "look" at the product and provide feedback as to the utility of the product in meeting their needs. This feedback can then be used to inform the sprint teams as they undertake development during the next phase of development where the focus is on producing an MMP that can be "officially" deployed to the customer.

In many scaled agile frameworks (e.g., SAFe), MVPs and MMPs are produced by combining one or more features (each of which are composed of one or more stories). Often there are dependencies between these features that require one or more features to be completed prior to work being undertaken on other dependent features. In other words, these features have to be worked sequentially. In this chapter, the period of time to complete a feature is called an increment and by extension, each increment contains one or more sprints.

Each feature within an increment produces some value to the end user, but often, there is not enough value to warrant release to the end-user community. A combination of features is often required to produce a product of sufficient value to be released. Due to dependencies and available workforce, it may take multiple increments to complete all of the features required for a releasable product. An initial release – often called an MVP contains sufficient end-user value to be released to get initial feedback from the end user. In Figure 10.3, features 1–8, which span increments 1–4 are combined to produce MVP 1, features 9–2 (spanning increments 3–5) will produce MVP 2, and features 13–14 (increment 5) are part of a future MVP.

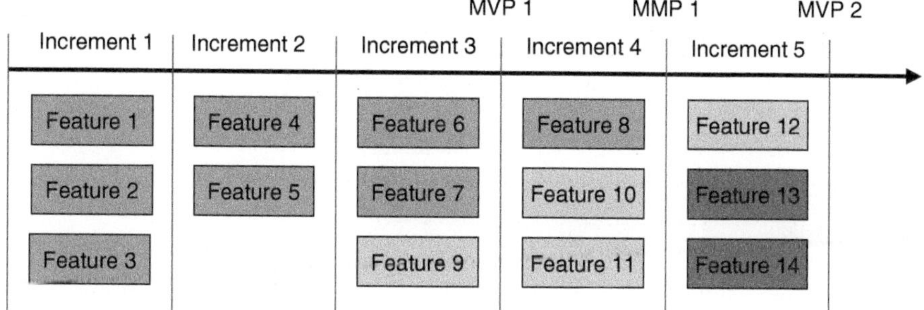

Figure 10.3 Increments where features are worked. Each increment consists of one or more sprints where stories decomposed from features are worked. Multiple features are combined to produce a minimum valuable product (MVP) or minimum marketable product (MMP).

Size of a Sprint and Increment

In the previous sections, the concept of a feature, a story, an increment, and a sprint were discussed. This section is focused on determining the size and structure of these elements. The size (i.e., number of weeks) of a sprint and the number of sprints within an increment are chosen to meet the following conditions:

- Produces something of value to the customer/market.
- Does not wear out the development team.
- Provides ample flexibility to incorporate change based on the needs of the market.
- Does not delay development due to unnecessary programmatic overhead (e.g., too many meetings, etc.)

As noted in the previous section, the output of an increment (or combination of increments) should be something of value to the end user or customer. This value can be in the form of an MVP or MMP. The output of a sprint should be something of value to the development team and also contributes to the production of an MVP or MMP. In some cases, a two-week sprint is sufficient to produce value to the team members and also contributes to the MVP/MMP at the end of one or more increments. In other cases, it may take three- or four-week-long sprints or longer (such as in hardware focused systems). One issue to monitor is how much programmatic overload team members are subjected to during a sprint. Most sprints consist of daily scrums, ceremonies, development meetings, planning meetings (for future sprints and increments), refining meetings (to modify future stories and features), and many other meetings. In some programs, these meetings and their frequency of occurrence can greatly impact (infringe) on a developer's time to develop, integrate, and test code. In such situations, it may prove useful to go with a longer sprint period of performance (e.g., 3 or 4 weeks). Program managers should not be afraid to make changes in the number of weeks a sprint covers or the number of sprints (and by extension the number of weeks) an increment covers after a project starts. Agile does not just apply to the development of the system, it also applies to how the development effort is structured. The key is to find the right balance (though experimentation) to produce value, adapt to changing requirements and needs, and not introduce fatigue and other burdens to the development team.

The Need for Rest and Retrospective

As noted in the literature (Bhaskara 2018), it is important to allow sufficient time for the development teams to decompress after spending weeks (in some cases many) developing working software at the end of a sprint or increment. Without some form of decompression, team member

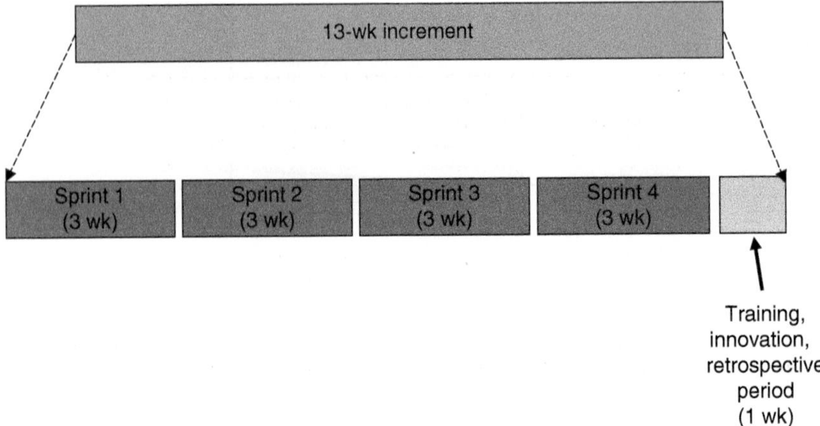

Figure 10.4 Example structure of an increment and sprint hierarchy. In this case, the increment (in which features are worked) is 13 weeks in length with four 3-week sprints and a one-week decompression/retrospective period.

fatigue will become a problem resulting in reduced productivity and in some cases, loss of key personnel (i.e., they will leave the project). In the case of an increment, it is recommended that at least a week be allocated at the end of the increment to allow for training, innovation (i.e., explore technical challenges), feature demonstrations (i.e., demonstrate value to the end-user), and retrospective (i.e., determine what worked and what did not during the increment) events to occur. In Figure 10.4, a 13-week increment consists of four 3-week sprints followed by a one-week decompression period.

Recommendations

The following are recommendations that have been shown to improve implementation of scaled agile for many large-scaled enterprise projects. Note that although these recommendations target agile/DevSecOps programs, many of them are also applicable to non-agile/DevSecOps programs such as those relying on serial/waterfall or hybrid agile/waterfall approaches. In addition, some of the recommendations are directed at the system developer while others are directed at the customer or system acquisition professional who has contracted to have a system developed using Agile and DevSecOps processes.

No Single Approach Addresses All Systems

Every project has unique characteristics and challenges. There is simply no "one size fits all" solution to scaling agile. In some cases, the solution may require combining both waterfall and scaled agile. In other cases, scrum or Kanban may be the preferred choice for undertaking agile development. It is recommended to try an approach and adapt as needed once the project is underway and the project team gains understanding of the program. Further, it is recommended that prior to starting a project, time should be invested in exploring what approaches were taken on similar projects and, select the approach that appears to best to meet the project's needs, but be open to adapting as necessary. Agile is all about nuance – the ability to adjust and refine the process as needed based on collected performance metrics (e.g., velocity, number of discrepancies detected, etc.).

Increment Planning

During increment planning, it is recommended that all stories decomposed from features assigned to the upcoming increment be sufficiently refined and assigned to the various sprints that compose the upcoming increment. Further, it is recommended that the highest priority stories be assigned to the earlier sprints, that sufficient "margin" be available in each sprint to handle unknowns and that that margin increase in size in the later sprints to accommodate potential rollover of stories from one sprint to the next. At a minimum, it is recommended that no more that 80% of the team's capacity be allocated to story point completion in a given sprint (or increment).

What does this mean? During sprint or increment planning, the key metric to track is story points. Each feature is assigned a number of story points, and each story decomposed from that feature is assigned a number of story points. Collectively, the sum of the story points assigned to all stories decomposed from a feature should equal the number of story points assigned to that feature. During planning, the scrum master will estimate the capacity of the sprint team. Capacity is the number of story points the team is capable of working during a period of time: in this case, the length of time for each sprint (i.e., 2–4 weeks). Ideally, if the complexity of the work to be done during the increment is well known and that there will be no unexpected events (e.g., no new feature added to the Increment to satisfy a recently high-priority requirement added to the project, etc.), then the total number of story points to be worked during the increment should equal the number of story points that can be worked by the project team (i.e., their capacity).

Unfortunately, the ideal is rarely achieved. Either the capacity of the team is incorrectly calculated or complexity of the work to be completed was incorrectly estimated or an unexpected high-priority feature was added to the increment. The further into an increment work progresses, the more likely these issues will present themselves. Having margin (i.e., spare story point capacity) can help absorb these unexpected events. For example, if a story cannot be completed within a sprint, it can be "rolled" into the next sprint. To do this requires having some capacity margin. In the event that an unexpected event or events do not occur, the feature team can "pull forward" stories from the next upcoming increment and get an early start on them.

Focus on MVP/MMPs

In addition to providing adequate margin when assigning stories to sprints in the next (or future) increments, it is also important for the project team to continue to focus on the upcoming MVP/MMPs. Often, due to pressures to achieve performance metrics (e.g., make a velocity threshold, etc.), project teams will focus on the low-hanging fruit – often ignoring project backlog priorities, which in many cases are linked to upcoming MVPs and MMPs. For example, in many US DoD acquisition programs, earned value management (EVM) is used to track project performance. Often, performance focuses on the number of features completed within a timebox such as an increment without regard to whether those features related to an upcoming MVP/MMP. The same problem also exists with programs that contractually provide award incentives based on the number of features completed over a specified period of performance. The solution here is to ensure that the performance metrics actually track what is important to the success of the project. In many cases, project success is not based on the number of features completed, but on the value delivered to the end user.

Up-Front Requirements Decomposition and Initial Systems Engineering

As noted previously in the chapter, agile is not an excuse for not doing up-front requirements decomposition and systems engineering. It is highly recommended that prior to the start of software development, system requirements be fully decomposed into features (and the initial set of stories – to be refined later during refinement) and that sufficient systems engineering be undertaken to (1) identify the various subsystems that compose the system, (2) identify all internal and external dependencies, and (3) gain an understanding of the complexity of the system to be developed. This is essential to ensure that a well-populated project backlog with the appropriate assigned priorities and identified dependencies be created prior to system development. Ignoring this step will almost guarantee that project schedule will be extended, and cost overruns will result as the project will have to address missing features and dependencies. The key here is to not over engineer the systems or come up with a detailed design that will change once the project starts (i.e., don't turn the agile program into a waterfall project). Rather, the goal here is to gain adequate understanding of the project to "prime the pump" to help get the project started on a good footing.

Tailoring of Tools

This recommendation applies to both the development contractor and the system acquisition organization. Often existing performance tracking and management tools such as Jira (Atlassian 2022a) may not provide the data or do so but not in sufficient detail for tracking the performance of the project. For example, developer contractors often use Jira for tracking features and stories being worked; however, it is often difficult for the customer (e.g., acquisition organization) to gain direct access to these tools. In such cases, the best that can often be achieved is for the development contractor to export the data to a file (e.g., spreadsheet). In this case, it may be up to the acquisition organization to develop the necessary tools to extract and manipulate the exported data. For example, Jira is focused on tracking what is currently being worked by the project team. Information such as how long it took to complete a story, was it blocked from being completed within a sprint (and why), and which MVP/MMP is impacted by the completion of the story are difficult to find within Jira. In such a case, it often requires an external tool to extract and track this information over a period of time.

Contract Incentives

For system acquisition organizations, one approach to keeping the project focused on agile and DevSecOps processes is to include award fee incentives in the system development contract. Often these contracts (this is particularly true for US Government contracts), the contract includes language to award fees to the development contractor if certain performance criteria are met. For agile and DevSecOps programs, such performance metrics can include meeting certain sprint or increment velocity targets, DevSecOps pipeline availability or number of high-priority features or stories are completed within a specified period of performance.

Near Operational Environments

As noted in Chapter 37, programs where an operational environment is not available to support integration and testing and customer evaluation, there is a need for a near operational environment. This is both a system engineering and scaled agile challenge. As previously noted, the challenge with large enterprise systems is that various portions of the enterprise are often developed on separate timelines. This is especially true in hybrid systems where both software and non-software subsystems are developed. Often the software subsystem is completed prior to the non-software subsystem. Integration and testing required prior to an MVP or MMP release either have to wait or can proceed if there is a near operational environment. Focus on prioritizing the development of the near operational environment (often the responsibility of an enabler team) is highly recommended. This prioritization often involves ensuring that the necessary simulators are in place prior to when they are needed by the feature teams focused on delivering a functional MVP or MMP.

Scrum vs. Kanban

There are many methods, tools, and processes that can be used to manage the day-to-day operations of a scaled agile project. Often the decision comes down to whether to use scrum or Kanban (Atlassian 2022b). For example, the sprint scrum involves implementing daily scrums (i.e., 15-minute meetings) to allow the scrum master and sprint team members to tag-up, discuss sprint progress, and address challenges. The focus here is on executing the sprint, which is usually time-boxed to be completed within weeks. Kanban Boards, on the other hand, are typically used for tracking project progress on a longer time scale (e.g., months) and often driven by priorities with time deadlines playing a secondary role (e.g., these features need to be worked now to meet an upcoming MVP release date, etc.).

It is recommended that for large-scaled agile programs, both scrums and Kanban be used. As illustrated in Figure 10.5, to track and maintain the overall project backlog, a Kanban Board is recommended. This project Kanban Board is available for viewing to all project team members. The board represents the current state of the project backlog (assigned priorities, closed features, increments features are tentatively assigned, etc.). When an increment is started features (and their decomposed stories) that are assigned to that increment are placed into their own Kanban Board (i.e., an increment Kanban). Each feature and enabler team has its own increment Kanban Board. When a sprint is executed, the assigned stories are placed and tracked in a sprint board.

Understand the Change Request Process

In addition to changing priorities (both features and stories) and the introduction of new requirements and user needs, projects often experience other types of changes and modifications that can impact project velocity. For example, a feature team may discover that a newer version of a software-based tool is required and will need to purchase a new software license. Such changes while often viewed as trivial can, in many enterprises, require following a "Change Request (CR)" process that involve many levels of review and approval. These CR processes should not be underestimated. Often, particularly with software-based tools, an organization may have a security review protocol in place to ensure that new software meets the security requirements of the organization. These reviews can require considerable time to be completed and, therefore, impact progress on completing a story or feature. In addition, some approvals may require multiple levels of review by individuals who are already burdened with other "higher-priority" tasks.

Although it is difficult to anticipate these "changes," every effort should be expanded up front prior to the start of "coding" to identify what tools and other support systems will be required for

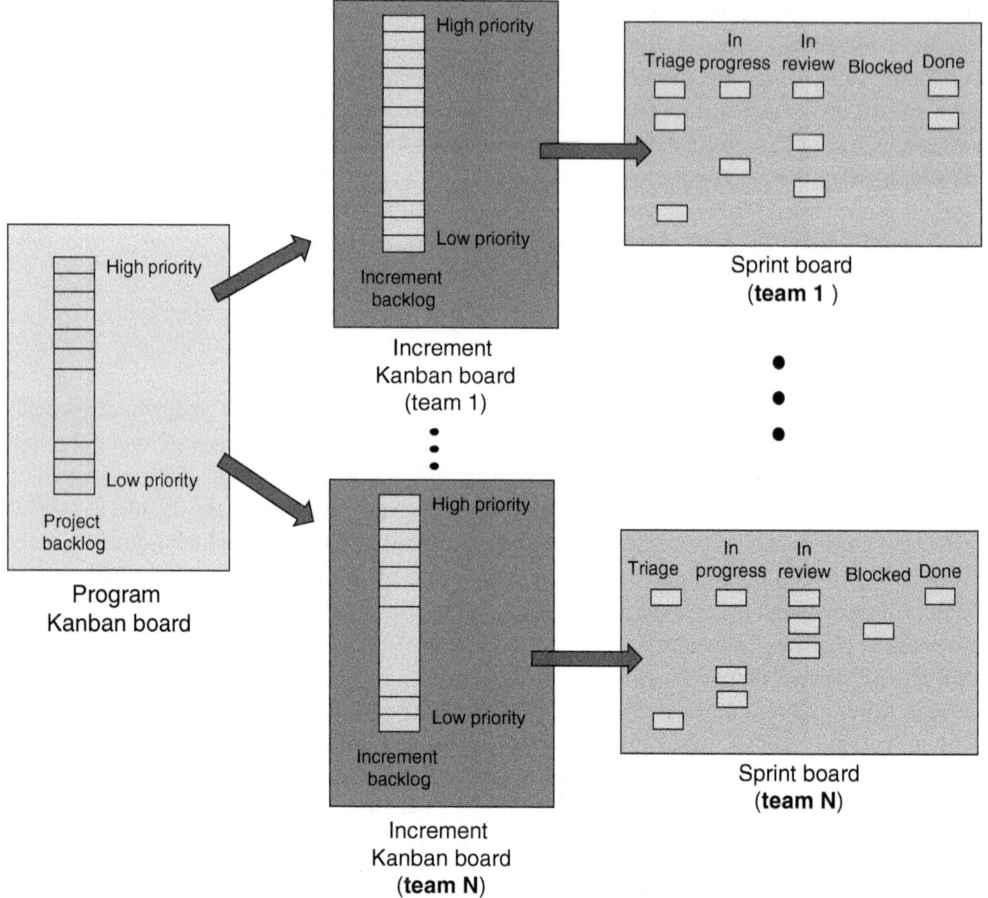

Figure 10.5 Both Kanban and sprint boards are used to manage an agile project.

the project and to be prepared when such "changes" cannot be avoided, to adapt as necessary (say bring forward another story or feature to be worked) to maintain velocity while the CR process completes.

Need for On-Boarding Training

Team member turnover (both with the development contractor and the acquisition organization – if relevant) is a fact of life for any project – particularly large enterprise systems that are continuously undergoing change to meet end-user needs. To reduce interruptions and minimize hits to project performance (i.e., velocity), it is recommended that an on-boarding training program be developed and maintained. These programs should cover the application domain of the project as well as how the project is implemented, including any nuances that differ from standard agile processes. For example, in a space-based acquisition program, a training program might include both a module that covers how space vehicles are commanded and controlled, and how agile is implemented on the project.

In many cases, on-boarding may require basic agile and DevSecOps training prior to project-specific training to accommodate team members with no prior experience or knowledge of such practices. As with some of the other recommendations in this chapter, on-boarding training is often overlooked or viewed as an unneeded expense.

Finally, and obvious to most readers, is the importance of hiring personnel that already have some familiarity with both the project domain and with agile/DevSecOps experience. This may not be possible due to high demand, but in almost every case, the time and expense invested in recruiting these skilled personnel are well worth it.

Get Licensing and IP Issues Resolved as Early as Possible

Unlike Change Requests (CR) processes that are used for unexpected events, there are also challenges when acquiring support tools and systems that are already identified for the project. For example, many organizations (both developers and acquisition) have existing software license infrastructures. In the ideal situation, when a software license is needed, the licensing organization simply grants the project the license and the project continues moving forward. In some cases, however, the license may not be in the license organization's inventory resulting in the need for the organization to negotiate with the software vendor. Not only does this negotiation process take time, but after the license is acquired, the organization may have to review (e.g., to determine if any cybersecurity vulnerabilities, etc.) and approve the use of the software (i.e., follow the CR process).

This challenge also applies to acquiring tools and systems that may have intellectual property (IP) rights. For example, a project may need access (e.g., to modify) to software that was originally developed by another vendor. Acquiring the right to gain access to this software can be a very lengthy process.

Project Owner and Communicating with Other Teams

As already noted earlier in the chapter, but re-iterated here for completeness, a major focus of the PO is to serve as the interface to the end-user (i.e., understand system requirements and end-user needs), set increment and sprint priorities, and assist the SM in addressing blockages and other resource needs. In addition, the PO also needs to interface with other project POs to ensure coordination and collaboration to avoid potential blockages and project slowdowns due to unknown dependencies and needs. When decisions are made to move stories (or features) from one sprint or increment to another to adapt to changing priorities and project needs, it is important that the PO and team members understand the impact these changes may have on other feature and enabler teams. As already noted, enterprise systems are system of systems with many interconnected interfaces and dependencies. Moving a feature from one increment to another may greatly impact progress of another team that is dependent on that feature being completed in the near future.

It is recommended that all project POs frequently (daily if possible/feasible) to discuss team progress, pending changes to the project backlog, and how to address impacts to various project teams that may result to a change. In some implementations, such meetings are called "team of teams" or "scrum of scrums" meetings. Regardless of the name, the important point is that these POs maintain coordination as the overall project is worked.

Project Team Capacity Planning

Many agile programs calculate project team capacity based on prior project velocity. Typically, the average velocity of the last two or three sprints is used for the calculating team capacity of the upcoming sprint or sprints (if planning capacity for an increment). Although vacation and personal time (e.g., sick leave, etc.) are "baked in" the velocity calculation (i.e., vacations are taken all the time and so previous productivity reflects these events), it is recommended that following be considered when using average velocity for calculating team capacity.

- Take into consideration upcoming holidays or other events that typically result in a larger amount of vacation time being taken than usual. For example, depending on the demographics of the project team, holiday breaks that are typically scheduled for a good part of December and early January are usually linked to extra vacation time being taken that is not accounted for when relying on average velocity.
- For project teams with members who may be assigned part-time to the effort, be aware that these team members may be unexpectedly pulled from the project to work another "higher-priority" effort. This happens frequently in large enterprise programs where highly skilled personnel are in great demand by multiple projects.

To reduce the impact of the last bullet (i.e., high demand on team member availability), it is recommended that project team members with a considerable breadth of experience be selected for the team. The more depth a project team has, the less impact on a project due to some team members being temporarily pulled from the project to work another task.

Working Groups

Although an agile program is designed and structured to react to requirements changes, these changes can still be disruptive. For example, if a high-priority requirements change is issued, often this requires that the project backlog be reprioritized and that features, or stories assigned to upcoming sprints, or increments may be reassigned to future sprints and/or increments. These changes in priorities and the resulting reassignment of stories and features to the future can ripple (due to dependencies) across the multiple mission and enabler teams that are working on the project – resulting on their backlogs being reprioritized and possible development delays.

In other cases, these new changes may impact modifications already being worked on the project, rendering them obsolete before they are made available to the end user. To reduce these impacts and surprises, it is recommended that the project team be engaged in various enterprise working groups (if available) chartered with developing future system requirements for the enterprise. By having a team member or members involved in these working groups, requirements changes can be anticipated and actions taken at the local level to reduce possible negative ripples (i.e., push stories and features that will change due to upcoming requirements changes further down in the project backlog).

Conclusion

There are many unique challenges in applying scaled agile to developing systems that are part of a much larger enterprise. Pitfalls include not fully identifying and understanding the system requirements that lead to populating the project backlog, not identifying internal and external dependencies that often drive determining priorities in the project backlog, not completing an adequate job up front on system architecture, and not building sufficient flexibility into the system to adapt to a changing environment.

Scaling agile to large enterprises requires considerable coordination between project teams, awareness of project team member productivity, planning of sprints and increments that allow for the unexpected, and a mindset that it is okay to make changes to the workflows as experience is gained.

Acknowledgments

This chapter is based upon research supported by the US Department of Defense through the Systems Engineering Research Center (SERC) under Contracts HQ003413D0004 and W15QKN18D0040. SERC is a federally funded University Affiliated Research Center managed by Stevens Institute of Technology. Any opinions, findings, and conclusions or recommendations expressed in this material are those of the authors and do not necessarily reflect the views of the US Department of Defense. The collaborators who supported this research are listed on the Handbook URL.

References

Atlassian (2022a). Jira software. https://www.atlassian.com/software/jira/guides/getting-started/overview (accessed 1 June 2023).

Atlassian (2022b). Kanban vs. scrum: which agile are you? https://www.atlassian.com/agile/kanban/kanban-vs-scrum (accessed 31 May 2023).

Bhaskara, A. (2018). Sprint fatigue – what is it and how to manage it?. https://medium.com/@bsaparna/sprint-fatigue-what-is-it-and-how-to-manage-it-e0745b882a2d (accessed 1 June 2023).

LearNow (2022). Challenges in scaling agile & how to overcome them? https://www.learnow.live/blog/challenges-in-scaling-agile-how-to-overcome-them (accessed 1 June 2023).

Makadia, M. (2021). Top 7 challenges in scaling agile. https://customerthink.com/top-7-challenges-in-scaling-agile/ (accessed 20 January 2023).

Office of the Under Secretary of Defense for Research and Engineering (OUSD(R&E)) (2020). Mission engineering guide. https://ac.cto.mil/wp-content/uploads/2020/12/MEG-v40_20201130_shm.pdf (accessed 1 June 2023).

PremierAgile (2022). Scaling agile challenges and how to overcome them. https://premieragile.com/scaling-agile-challenges/ (accessed 1 June 2023).

Productfolio (2023). Minimum Viable Product vs Minimum Marketable Product. https://productfolio.com/mvp-vs-mmp/ (accessed 1 June 2023).

Scaled Agile (2022). What is SAFe®? https://scaledagile.com/what-is-safe/ (accessed 1 June 2023).

Scrum@Scale (2023). How scrum scales, Scrum@Scale™ framework. https://www.scrumatscale.com/ (accessed 1 June 2023).

Spanner, C. (2023). Scrum of scrums. https://www.atlassian.com/agile/scrum/scrum-of-scrums (accessed 1 June 2023).

Tryqa.com (2022). What is Waterfall model – Examples, advantages, disadvantages & when to use it?. http://tryqa.com/what-is-waterfall-model-advantages-disadvantages-and-when-to-use-it/ (accessed 20 June 2022).

Wikipedia (2022). Ground segment. https://en.wikipedia.org/wiki/Ground_segment (accessed 14 June 2022).

Biographical Sketches

Michael Orosz directs the Decision Systems Group at the University of Southern California's Information Sciences Institute (USC/ISI) and is a research associate professor in USC's Sonny Astani Department of Civil and Environmental Engineering. Dr. Orosz has over 30 years' experience in government and commercial software development, systems engineering and acquisition, applied research and development, and project management and has developed several successful products in both the government and commercial sectors. Dr. Orosz received his BS in engineering from the Colorado School of Mines, an MS in computer science from the University of Colorado, and a PhD in computer science from UCLA.

Brian Duffy is a senior systems engineer with the University of Southern California Information Sciences Institute (USC/ISI). He conducts research and analysis to determine system engineering methods and metrics necessary to transition Major Défense Acquisition Programs from a traditional waterfall development to Agile/DevSecOps processes. Prior to USC/ISI, Mr. Duffy retired from the United States Air Force with multiple assignments related to National Security Space acquisition programs and command and control systems. Mr. Duffy holds a Masters of Aeronautical Science degree from Embry-Riddle Aeronautical University and a Bachelor of Aeronautical and Astronautical Engineering degree from the University of Washington.

Craig Charlton is a senior systems engineer at USC's Information Sciences Institute (USC-ISI) and has provided acquisition support at Space Systems Command (SSC) at the Los Angeles Air Force Base during the past 20 years on a number of leading-edge satellite systems. Prior to his position at SSC, Mr. Charlton acquired more than 25 years of experience as a software engineer and in managing software projects in the commercial world, primarily in the fields of engineering and of law enforcement. Mr. Charlton received a BA in mathematics from California State University, Long Beach.

Hector Saunders has served in the Department of the Air Force for six years. He served four years in the U.S. Air Force as a Cyber Warfare Operations Officer, and two years in the U.S. Space Force as a Program Manager. He currently leads an Integrated Product Team at Los Angeles Air Force Base to develop and modernize the GPS ground control system's software factory, which supports Agile/DevSecOps software development. Prior to his current assignment, he led cyber warfare missions and trained operators to defend the Department of the Air Forces' wide area network systems. He holds a BS in physics with minors in applied mathematics and computer science from the CUNY's The City College of New York.

Michael Shih is a senior software systems engineer with Booz Allen Hamilton, Los Angeles Office. Michael has 28 years of experience with Southern California Aerospace Industries, whose contribution covers domains such as space vehicle payload attachment FEM analysis, embedded software development within C4ISR framework, and Command & Control software development and integration of Space Ground Systems using Waterfall and/or Agile framework. Mr. Shih received his MS in computer science from USC, and his BS in mechanical, aerospace, and nuclear engineering from UCLA.

Chapter 11

System Behavior Specification Verification and Validation (V&V)

Kristin Giammarco

Department of Systems Engineering, Naval Postgraduate School, Monterey, CA, USA

Introduction

The safe and secure functioning of critical systems is dependent on system behavior, human behavior, and interactions among systems and humans. While current model-based systems engineering (MBSE) approaches and tools enable excellent capture of known and wanted behaviors, they generally fall short on ability to assist with the discovery of unknown and unwanted behaviors that have not yet been considered or documented. This is a hard problem because "unknown unknowns" seem by definition to be out of reach. Scenarios that could cause things to break down are often difficult to expose before seeing examples of those breakdowns (too often, during real operations). But a new approach and tool developed at the Naval Postgraduate School is shedding a glimmer of light on this dark corner. The Monterey Phoenix (MP) modeling environment was designed to model behaviors and interactions exhaustively to help us map out what we know we want. Students and researchers were surprised to find an additional property not deliberately designed into MP: the sets of synthetic example scenarios that MP generates also contained some unexpected yet plausible behaviors for the modeled system. That is, behaviors that no one on the project thought about before seeing them in MP-generated scenarios.

This is a highly relevant problem because those who deal with mission-critical behavior usually have an undercurrent of concern about what could go wrong that will not occur to them – or anyone – until it is too late. Those participating in mission or business processes also continue to be frustrated by excessive inefficiencies not considered when the process was initially designed. They want to reduce surprise, or at least be surprised in a synthetic environment instead of the real environment. It is critical for cybersecurity, safety, human-machine teaming, autonomy, business processes, policy, strategy, training, and many other areas to specify behaviors correctly and completely within the defined scope.

In order to find weaknesses and vulnerabilities in our systems before adversaries can exploit them, MBSE practitioners need modeling environments that not only capture what we know (or think) we want, but that also reveal assumptions we are making, requirements we are

Systems Engineering for the Digital Age: Practitioner Perspectives, First Edition. Edited by Dinesh Verma.

Companion website: www.wiley.com/go/verma/systemsengineering

overlooking, logic errors we are committing, and mistakes we are repeating. In other words, we need our modeling environments to also help us *discover what we do not know* and realize the presence of behavior we *do not want*.

This chapter is a guide for MBSE practitioners who want to develop their skills in discovering undocumented requirements, risks, and assumptions that are latent in their thinking about design, with the goal of exposing and controlling for as many of these as possible, as early as possible. The goal is to avert costly downstream design changes resulting from misunderstood requirements and missing specifications, especially constraints that were not previously realized were necessary to suppress "extra" behaviors otherwise permitted. This is not a skill learned from simply reading about it; it is best learned by practice. Readers are therefore encouraged to follow along with the example and then try the approach on their own system of interest. The approach and example described herein are excerpted from Giammarco et al. (2018), Research Task RT-176 – Verification and Validation (V&V) of System Behavior Specifications.

Background

Verification and validation (V&V) are distinct processes used for ensuring that a system meets its specifications and satisfies the user's need. *System verification* is "a set of actions used to check the *correctness* of any element, such as a system element, a system, a document, a service, a task, a requirement, etc." (SEBoK Authors 2022a). *System validation* is "a set of actions used to check the compliance of any element (a system element, a system, a document, a service, a task, a system requirement, etc.) with its purpose and functions" (SEBoK Authors 2022b). The distinction between these terms is often summarized as follows: use verification to ensure you are solving the problem right, and use validation to ensure you are solving the right problem. Verification uses the set of documented requirements as a reference point for determining correctness, whereas validation may also use value judgments derived from expectations for the system as a whole.

Modeling and simulation are frequently used to support V&V activities throughout a system's lifecycle. The fundamental purpose of V&V is usually to ascertain whether or not a modeled system will have the expected behavior when it is in operation and identify any unexpected or unwanted behaviors early so that they can be dealt with in a controlled and least costly setting (Auguston et al. 2015). Current industry standards for modeling system behavior include the systems modeling language (SysML) viewpoints for sequence, activity, use case, and state machine diagrams; system dynamics (SD) models depicting control and feedback in system processes; and agent-based models (ABM) that describe agent behaviors and interactions between agents and with the environment. SysML use case, sequence, activity, and state transition diagrams provide different views on a system's behavior, and some automated tools enable discrete event simulation of SysML activity diagrams. A current challenge with these and other diagrams is the difficulty with capturing the exhaustive set of all possible behaviors permitted by a design (Auguston et al. 2015). SD models are mathematically rigorous, yet abstract enough for a wide variety of system applications. However, SD models are best used for closed-loop systems in which component dependencies must be considered at the global level (Borshchev and Filippov 2004). Monterey Phoenix augments a typical ABM approach by adding standardization for defining agents and events using formalized event grammar and structured syntax (Ruppel 2016) and is able to generate SysML-like diagrams that represent rules for behavior (global views such as state diagrams) as well as example instances of behavior (event trace views resembling sequence diagrams). A study by Hall et al. (2022) conducted a detailed crosswalk between MP behavior diagrams and SysML behavior diagrams, laying the groundwork for the integration of MP's trace generation and model checking capability for the V&V of SysML behavior models.

The next section introduces the practitioner to MP, and the following sections provide an example application of SysML behavior specification V&V using MP.

Monterey Phoenix

Real-world MBSE practitioners working with nontrivial systems face a common challenge: training exists for how to specify a set of known wanted behaviors, but how to find out what unknown unwanted behaviors are inadvertently permitted by the specification alongside the known wanted behaviors? This is a different and harder problem. Automated support for probing the design for unwanted behaviors is beyond the scope of the most commonly used MBSE approaches and tools, and those that do offer some answer to this question often call for specialized knowledge and skills that are not common in the workforce and are therefore reserved, if used at all, for design aspects deemed the most critical. Experience shows, however, that behaviors can and frequently do emerge from interactions within and among complex systems that are not present in the models of the same. The next evolution in the advancing MBSE practitioner's journey is to learn how to make behavior models that produce not only the expected behaviors but also some unexpected behaviors that may surprise even the experts in the subject matter being modeled. Such a capability now exists and does not require specialized expertise or expensive tools.

Monterey Phoenix (Auguston 2009a,b) is a formal language, approach, and tool developed by the Navy to model and reason about *behavior* – the way in which something (e.g., a system, software, hardware, person, process) conducts activity. It is a "lightweight" (Easterbrook et al. 1998; Woodcock et al. 2009) formal method that generates sets of scenarios that are exhaustive execution traces through behavior logic rules up to a user-defined scope limit. Unlike "heavyweight" formal methods, MP does not require extensive professional preparation to generate these exhaustive sets of event traces. In support of that assertion, over one hundred college students with majors ranging from computer science, cybersecurity, and mathematics to psychology, criminal justice, and public policy have learned MP and successfully used it to produce traces that were informative to their subject matter expert mentors in the space of a five-week summer internship. Some of the MP traces lead to the identification and resolution of previously unrealized failure modes in real systems. These discoveries were made during modeling sprints conducted under the Monterey Phoenix Virtual Internship Program (MPVIP), piloted by the National Security Agency (NSA) in 2020 and 2021 and continued under the sponsorship of the Naval Research Program (NRP) in 2022 and 2023.

To prepare a behavior model that permits emergence of unexpected behaviors, we use a very simple modeling schema that employs Auguston's abstract concept of *event* (Figure 11.1). "Event" is a fundamental concept used as a building block for representing or composing activities, actions, operations, processes, states, transitions, or conditions associated with any object with behavior (Auguston 2009a). This definition of event uses abstraction to treat concepts typically modeled separately as one unified concept. MP's approach of using the abstract event concept results in simple models that give rise to complex behaviors that provoke users to think and reason about what is being modeled and question the veracity and validity of what they see in the traces that are output.

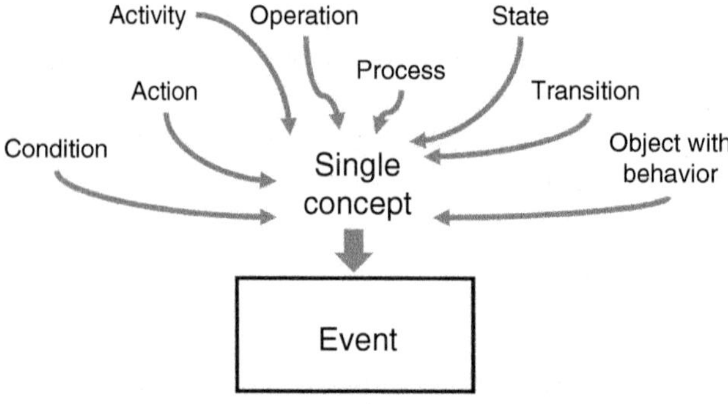

Figure 11.1 Auguston's concept of an abstract event.

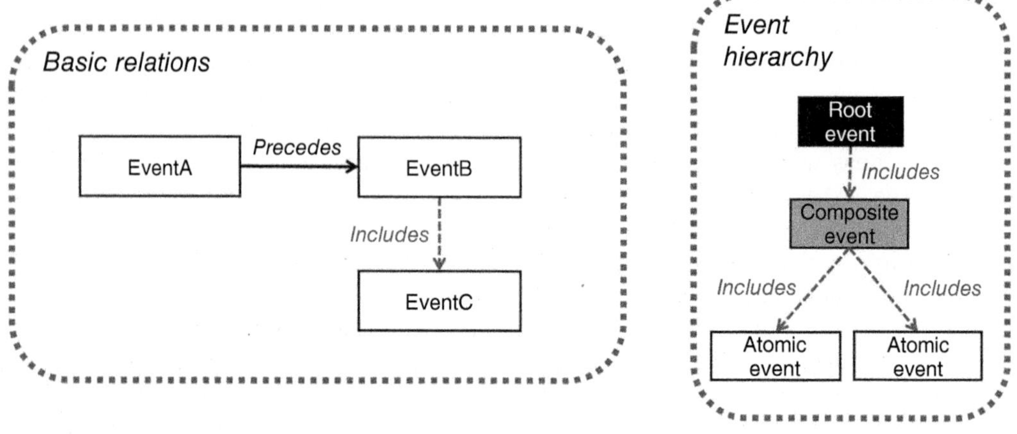

Figure 11.2 An event has two basic relations: precedence and inclusion.

Figure 11.2 shows the two basic relations of an MP event: precedence (event sequence) and inclusion (event hierarchy of root, composite, or atomic events).

MP *event grammar rules* employ the basic relations to encapsulate behaviors for modeled missions, systems, or processes. They are formal specifications for event dependencies. For example, the rule

```
ROOT Actor:        Do_this
                  Do_that;
```

means `Actor` is a root event that includes events `Do_this` and `Do_that`, with `Do_this` preceding `Do_that`. The colon denotes inclusion and separates the left- and right-hand sides of the rule. A space or new line between events denotes precedence. A semicolon denotes the end of the rule.

Grammar rules can be used to define root events as above, or composite events as follows. For example, the rule

```
Do_that: Step1 Step2;
```

means that `Do_that` is a composite event that includes events `Step1` and `Step2`, with `Step1` preceding `Step2`. The two rules together mean that `Actor` includes (`Do_this` then `Do_that`), where `Do_that` includes (`Step1` then `Step2`).

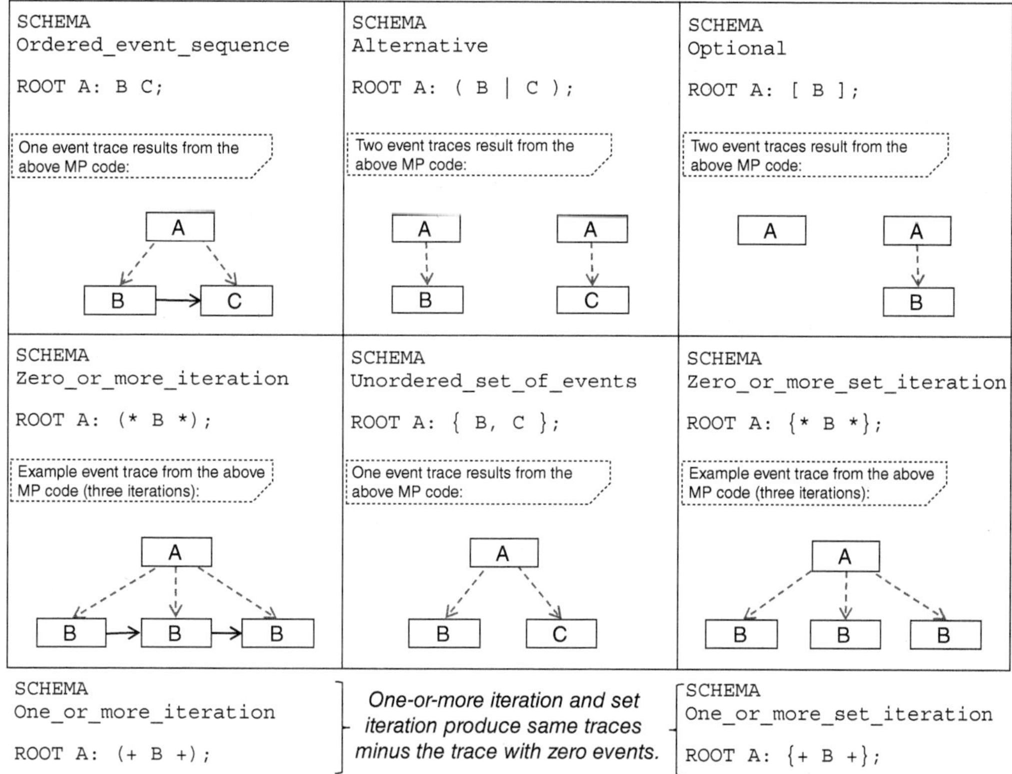

Figure 11.3 Example grammar rules with their corresponding event traces. *Source:* Adapted from Auguston (2020).

Figure 11.3 shows how these basic relations are employed along with familiar operations of concurrency, alternate, optional, and loops to compose behavior algorithms (right hand side of the rule) for separate actors or agents (left hand side of the rule) (Auguston 2020). Each path through the algorithm is visualized as an *event trace*, in which boxes represent events, solid arrows represent *precedes* relations, and dashed arrows represent *includes* relations. Each grammar rule by itself can result in one or more event traces, depending on the number and composition of logic constructs employed.

The rules shown in Figure 11.3 are example patterns that can be combined to compose the desired behavior logic. The number of traces generated is bound using patterns for zero-or-more or one-or-more iterations (Figure 11.3 middle and bottom left and right). At run time, the user sets a maximum number of event iterations called the *scope*. The scope is the upper bound limit on iteration enabling the generation of a finite and complete set of event traces. This use of multiplicity on event iteration sets MP apart from other behavior modeling approaches and tools. The set of event traces is guaranteed to be exhaustive up to the specified scope limit (Auguston 2009a, 2020). *Scope-complete* means that every valid combination of events up to the scope limit is guaranteed to be present among the event traces. If there is no iteration in the MP schema, then the execution is exhaustive, period (it is complete).

As a convention, we write a separate grammar rule for each distinct system or component to describe its behaviors independent of other systems. We then use constraints such as COORDINATE

```
SCHEMA Authentication

/*-----------------------
  USER BEHAVIORS
------------------------ */

ROOT User:   Provide_credentials
             (* CREDS_INVALID Reenter_credentials *)
             [ CREDS_VALID Access_system ];

/*-----------------------
  SYSTEM BEHAVIORS
------------------------ */

ROOT System:       Verify_credentials
                   (+ (   CREDS_INVALID Deny_access        |
                          CREDS_VALID Grant_access         ) +)
                          [ Lock_account ];

/*-------------- ---------
  CONSTRAINTS
------------------------ */

/* User and System share all instances of CREDS_VALID and CREDS_INVALID */
User, System  SHARE ALL  CREDS_VALID, CREDS_INVALID;

/* Instances of Provide_credentials precede instances of Verify_credentials */
COORDINATE   $a: Provide_credentials   FROM User,
             $b: Verify_credentials    FROM System
      DO ADD $a PRECEDES $b;  OD;

/* Instances of Deny_access precede instances of Reenter_credentials */
COORDINATE   $a: Deny_access           FROM System,
             $b: Reenter_credentials   FROM User
      DO ADD $a PRECEDES $b; OD;

/* Instances of Grant_access precede instances of Access_system */
COORDINATE   $a: Grant_access          FROM System,
             $b: Access_system         FROM User
      DO ADD $a PRECEDES $b; OD;

/* Maximum number of access denials in a single session is 3 */
ENSURE #Deny_access <= 3;

/* Lock the account if and only if access has been denied at least three times */
ENSURE #Deny_access >= 3 < -> #Lock_account == 1;

/* If the account is locked, no access is granted */
ENSURE #Lock_account == 1 -> #Grant_access == 0;
```

Figure 11.4 MP schema modeling Authentication. Root events are named for a user and a system with activities. A `SHARE ALL` statement is used to merge events in different roots, and `COORDINATE` statements add dependencies (`PRECEDES` relations in this case) between events in different roots.

statements to add dependencies like precedence, inclusion, and user-defined relations to synchronize events in different roots (Auguston 2020). `ENSURE` statements may be used to further prune the set of valid scenarios. A collection of related grammar rules and constraints is called an MP *schema*. Figure 11.4 shows an example MP schema for Authentication.

If the rules in the schema are consistent and not overly constraining, the MP schema will produce at least one valid event trace when it is run. Figure 11.5 shows the set of four traces produced by the above schema at scope 2 (two iterations).

The constraints are crafted with intention to prohibit invalid or unwanted behavior combinations from emerging in the set of generated traces. If the constraints are removed from the schema

Scope 2 Trace 1

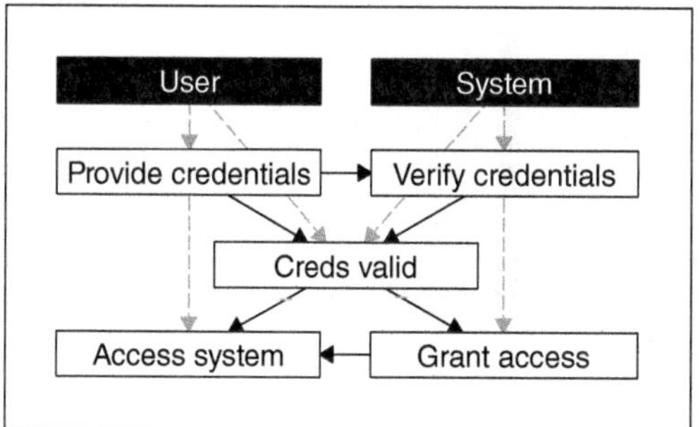

Scope 2 Trace 2

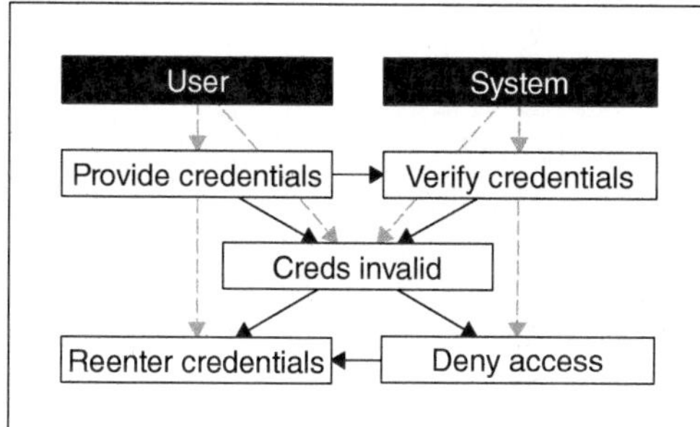

Scope 2 Trace 3

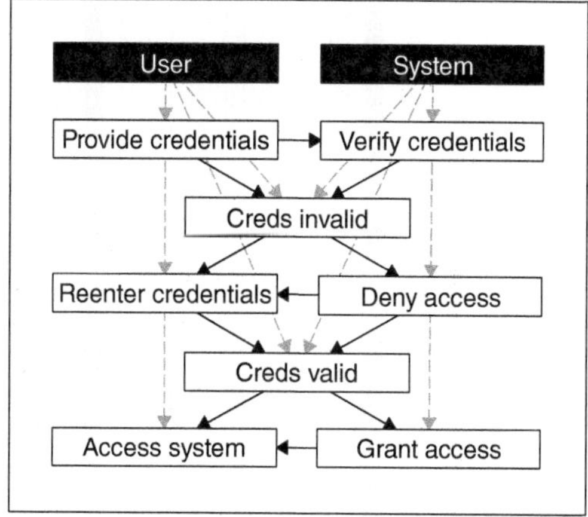

Scope 2 Trace 4

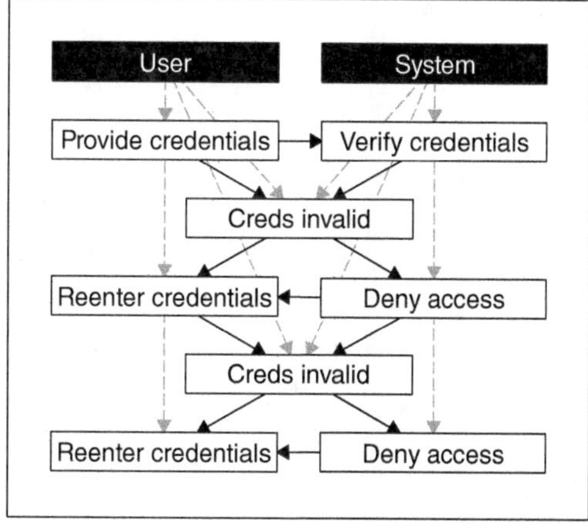

Figure 11.5 Four traces are generated at scope 2 from the schema in Figure 11.4.

in Figure 11.4, many more combinations are permitted (Figure 11.6), some of which are invalid, and some of which may also be unexpected (Figure 11.7).

MP behavior modeling can be likened to carving a statue. The raw material is the set of unconstrained traces, the chisel strikes are the constraints, and the finished statue is the final behavior model with all the desired constraints applied. Modeling provides the benefit of being able to experiment with adding and removing constraints to see how they affect the shape of the resulting traces.

Jackson (2012) states that most errors can be exposed on small counterexamples (the "small scope hypothesis"). That is, most error patterns that appear when running at a high scope (generally 4 and greater) also appear at a small scope (between 1 and 3). Experience shows that the small scope hypothesis works in practice: a user need not run the model for a high scope to expose a substantial and useful number of previously unconsidered behaviors; their shape manifests clearly enough at a low scope. For example, in Figure 11.7 (right), three iterations are sufficient to imagine an attack with repeating invalid credentials. A user need not spend the time or resources to generate 300, 30, or even 10 repetitions to see this pattern and be inspired to write a constraint to prevent

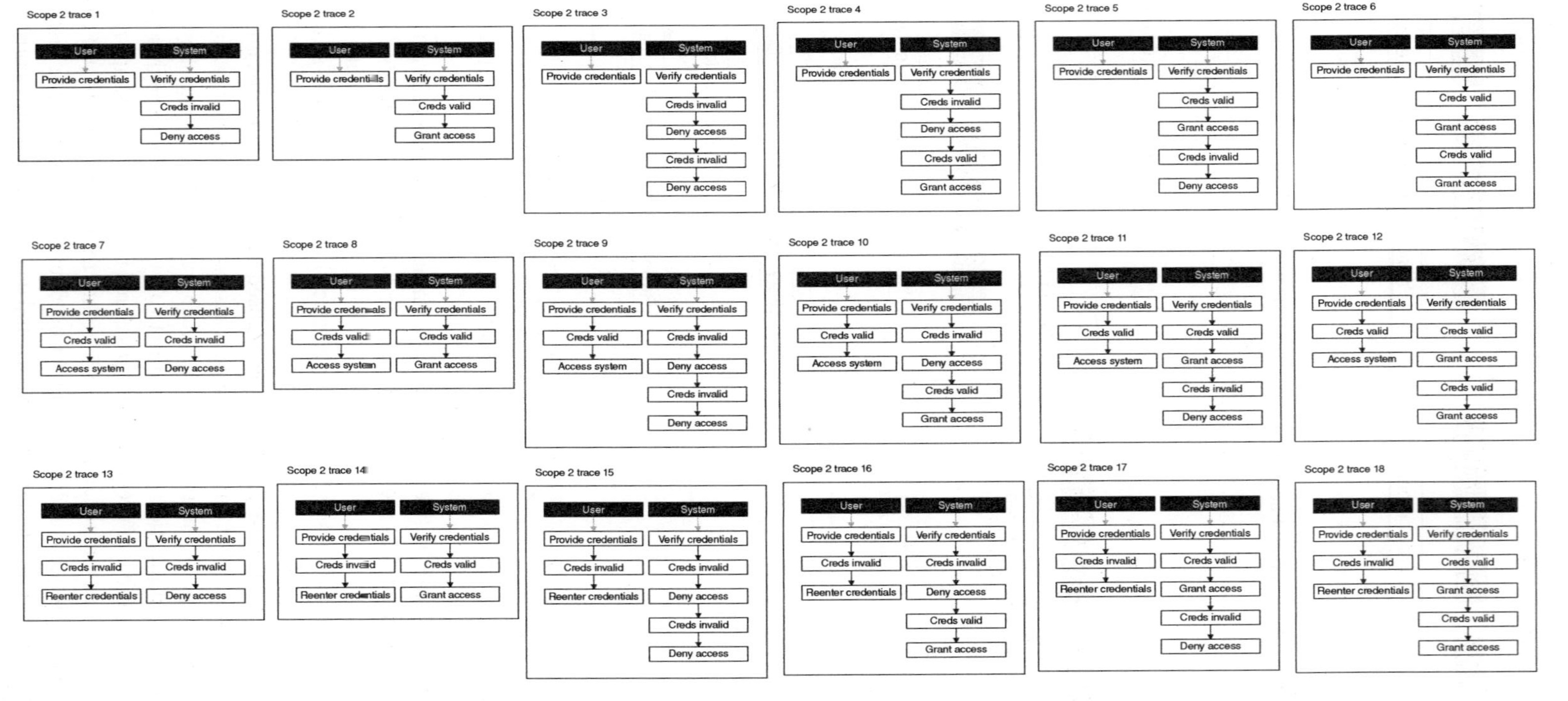

Figure 11.6 Thumbnail images of the 36 event traces generated at scope 2 when constraints in Figure 11.4 are removed.

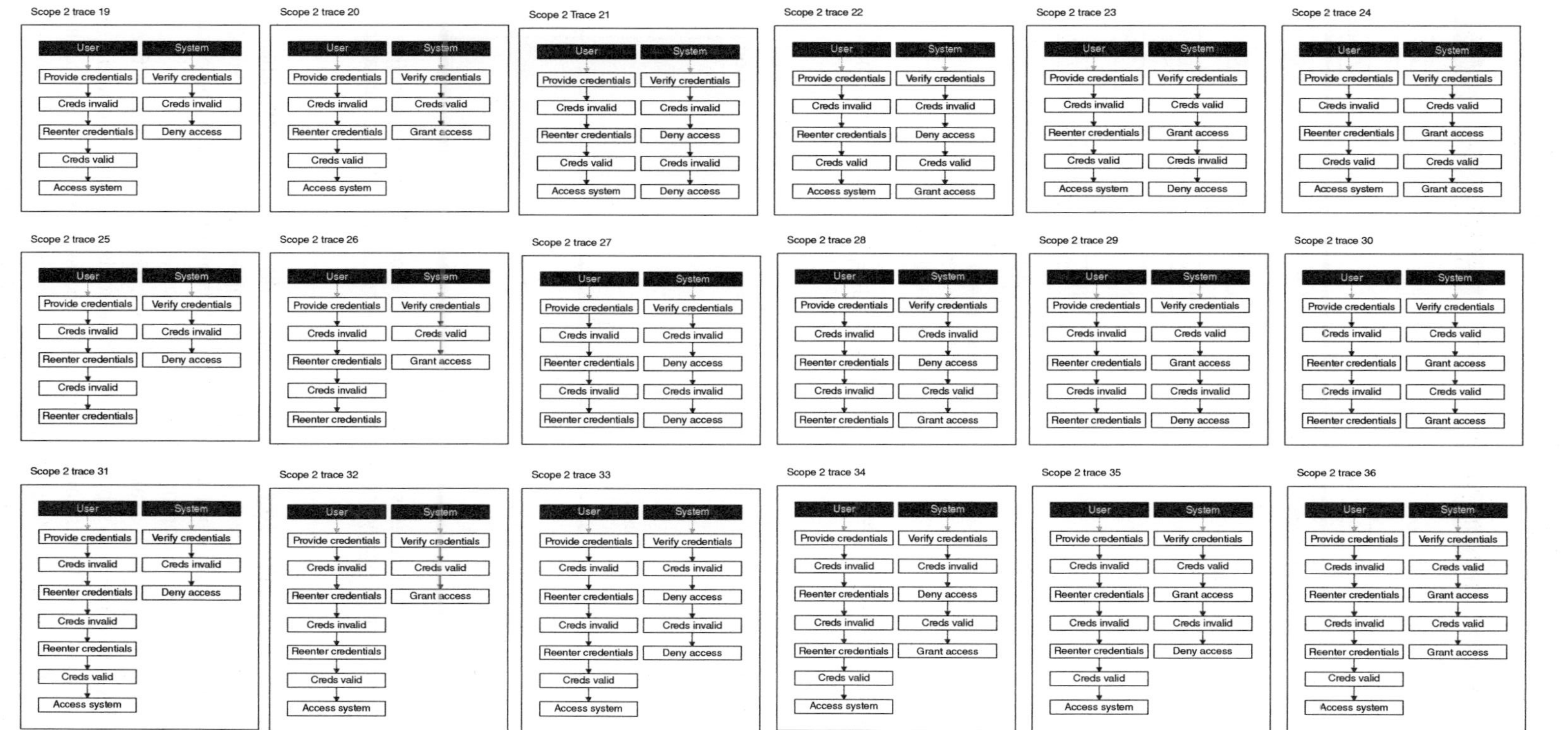

Figure 11.6 (Continued)

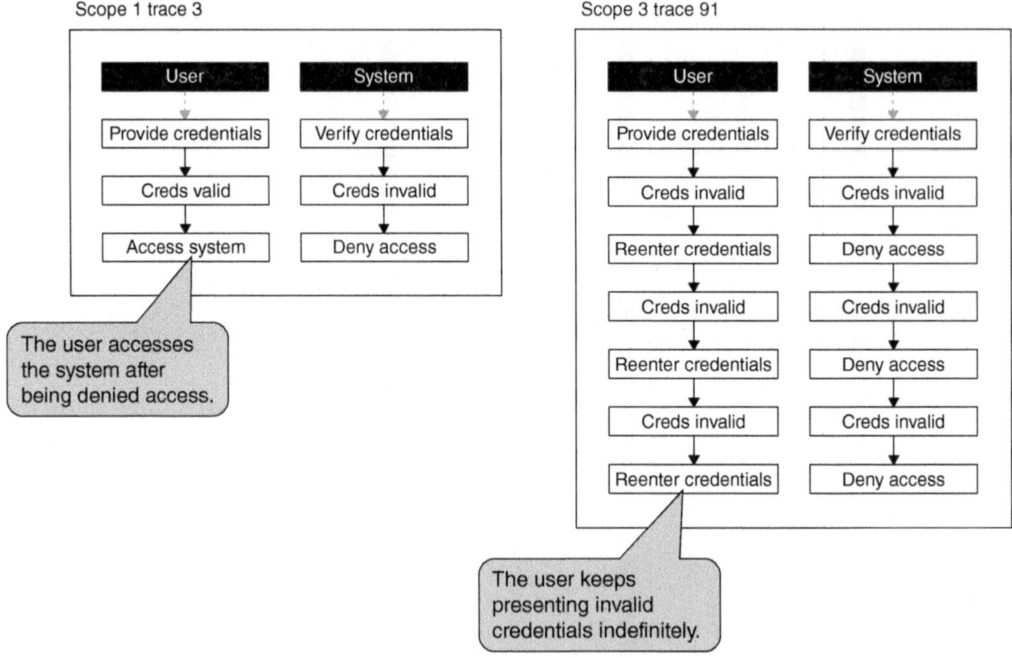

Figure 11.7 Traces 3 and 91 at scope 3 (when constraints in Figure 11.4 are removed) show some unexpected and unwanted behaviors.

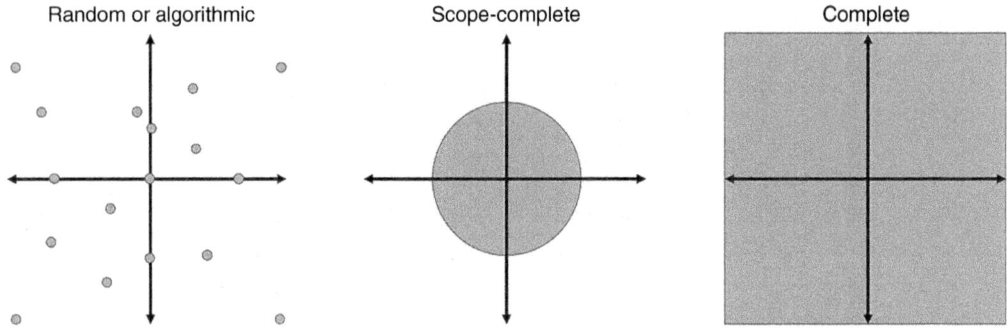

Figure 11.8 Gray regions are test cases (scenario variants, in the case of behavior modeling). MP provides a robust "sweet spot" between randomly or algorithmically selected subsets of behaviors and every possible behavior with infinite event iteration.

its permission by the system. Monterey Phoenix leverages the small-scope hypothesis to work in a "sweet spot" between an insufficient number and an impossibly high number of event traces (Figure 11.8). MP generates a finite but ample number of event traces with scope-complete formalism that allows near real-time reasoning supported by automated tools.

There are currently two MP tools available to the public: MP-Firebird (https://firebird.nps.edu) and MP-Gryphon (https://nps.edu/mp/gryphon). These tools promote education on the use of lightweight formal methods for mission, system, and process behavior modeling and provide a cost-effective tool suite for workforces charged with designing complex systems and managing their risk of exhibiting unwanted emergent behavior. MP-Firebird requires no installation, running through the web browser. MP-Gryphon is locally installable. These tools were made available to promote the development of safer, more secure, and more resilient systems in all industries and domains of application.

MP-Firebird and MP-Gryphon visualize the automatically generated event traces similarly in a sequence diagram-like format, with MP-Firebird having a swim lane option that resembles SysML activity diagrams. Each separate actor's behavior can also be visualized as an activity diagram, and global queries can be used to extract state diagrams and component diagrams from the information in the generated set of event traces. Other capabilities of MP implemented in both tools include constraint specification and support for reasoning embracing the traditional predicate calculus notation; computations for quantities such as time, cost, and other resources using event attributes; automated, rules-based trace annotations; and reports, tables, and charts for providing summaries of properties of interest at the event trace (local) or global (across all traces) levels. A collection of examples demonstrates these features and come pre-loaded with both tools available at https://nps.edu/mp/models.

Emergent Behavior Analysis

The RT-176 research was instrumental in testing a repeatable methodology (Giammarco and Giles 2017) for discovering undocumented requirements, risks, and assumptions that are latent in a design. The research led to follow on work (Giammarco 2022) that established a new area of practice called emergent behavior analysis. The goal of an emergent behavior analysis is to expose and control unwanted behaviors that a design may be inadvertently permitting as a result of undocumented assumptions, as early in the lifecycle as possible. The general approach to emergent behavior analysis with MP involves:

- uncovering strong (unexpected) emergent behaviors, postulating how they could arise, and then downgrading them to weak (expected) behaviors, and
- exposing and constraining unwanted (negative) emergent behaviors, leaving behind only acceptable (positive or neutral) emergent behaviors (Giammarco 2022).

The example in the following section will show the MBSE practitioner how emergent behavior analysis was applied on a small MP model of a SysML model.

Search and Rescue Examples

Select the Behavior Model to V&V

The main mission in the Skyzer case study features an uncrewed Air Vehicle capability for supporting search and rescue operations. The successful case scenario has been modeled as a SysML swim lanes activity diagram named "Non-Combatant Operations – Scenario 1" (shown in Figures 11.9 and 11.10 as a two-page spread). In short, this scenario describes a civilian vessel in need of medical supplies with the US Coast Guard not able to respond in time to prevent a medical emergency, but a Navy ship equipped with an Air Vehicle and the necessary medical supplies intervenes to deliver the needed medical supplies in a timely fashion. The model describes the behavior of the various actors, including the Rescuee (Victim, EPIRB), Air Force, Rescue Coordinator, Mission Commander, UAV Operator, Ground Crew, INMARSAT, GPS SAT, Ship Utilities, Control Station, Air Vehicle, and Launch and Recovery System (UCARS, RAST). This SysML diagram was selected to be the behavior model to undergo V&V with Monterey Phoenix emergent behavior analysis.

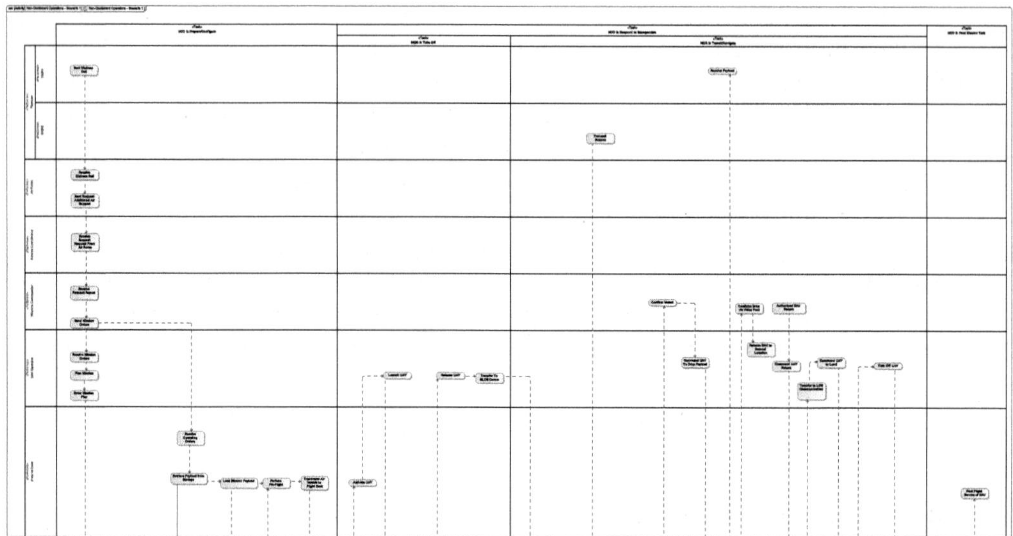

Figure 11.9 SysML diagram for "Non-Combatant Operations – Scenario 1" (upper half).

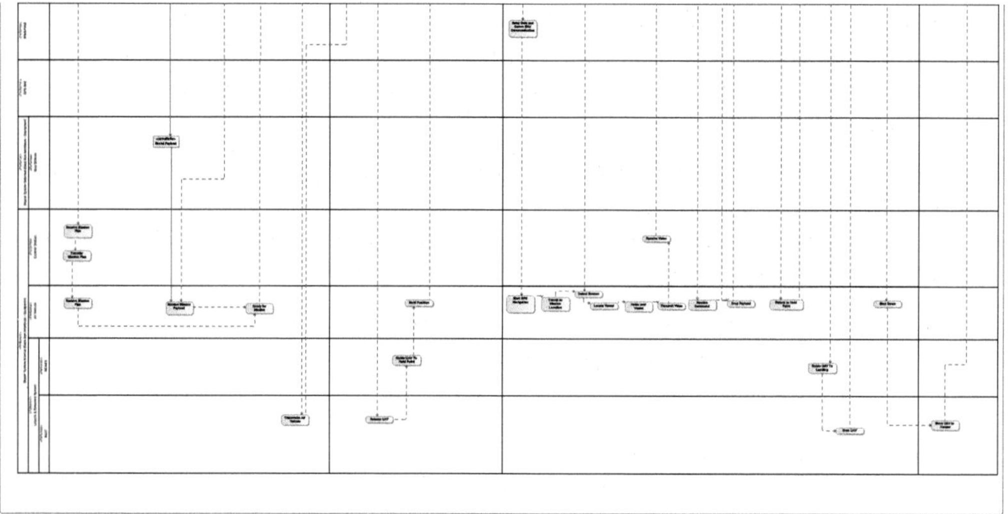

Figure 11.10 SysML diagram for "Non-Combatant Operations – Scenario 1" (lower half).

Prepare Models for V&V

The source model shown in Figures 11.9 and 11.10 is fairly large, especially for browsing as a screen capture in a document. Even browsed in the native tool, it is a lot for a human to consume in one view. The model can be segmented into four parts or phases to make it easier to review and analyze as shown in Figure 11.11 (it is not necessary to read the eyechart, just to see the phase segmentation). Reducing the complexity of a model by breaking it into time phases allows for more in-depth analyses once the model is represented in MP. Consider that your analysis may benefit from segmenting the model into phases in a similar way in whichever tool you typically use.

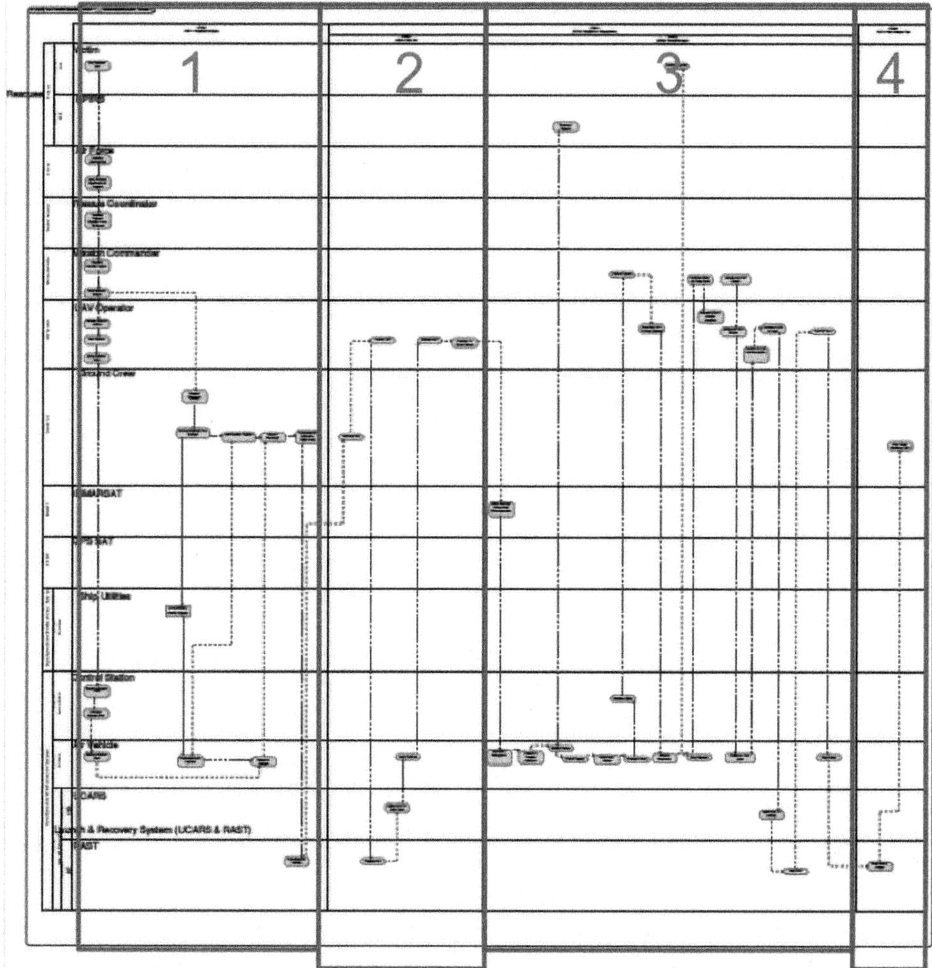

Figure 11.11 Segmentation of the model into four phases enables more room for eventual elaboration on alternative behaviors.

The four phases in the source model are called:

1) **NCO 1**: Prepare/Configure Task
2) **MOB 2**: Take Off Task
3) **MOB 3**: Transit/Navigate Task
4) **NCO 8**: Post Mission Task

Each of these phases were transcribed into the MP language (Giammarco et al. 2018). We will focus on Phase 3 for our example. See Hall et al. (2022) for a language crosswalk that supports future work in automating this transcription. If you are following along with your own model, take the two-hour self-guided tutorial available on https://nps.edu/mp to learn enough of the MP language to get started with transcribing your model into MP. You can also load and run the models by following the web addresses in the references cited in the upcoming figure captions.

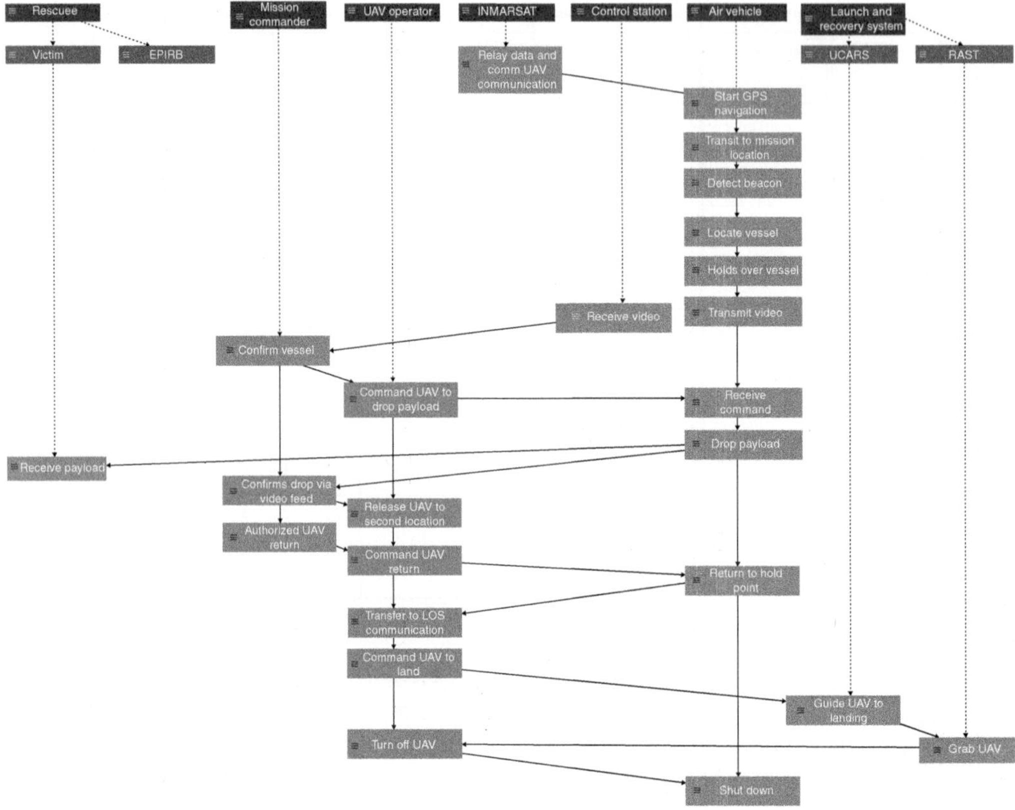

Figure 11.12 MOB 3: Transit/Navigate Task phase translated into an MP event trace. *Source:* Adapted from Giammarco and Shifflett (2018a).

After transcription, the flow of events for Phase 3 (Transit/Navigate Task) appears as a single MP event trace (Figure 11.12). This phase includes all the actions while the Air Vehicle is on station up until the Air Vehicle has returned to the ship. The same process could be repeated and followed for each of the other phases using a separate MP schema for each phase.

Expand the Model with Alternative Flows

After verifying that the event sequence in the MP model matches the activity sequence in the SysML model, the MP model can be expanded to consider other plausible alternative event flows in the system and in its environment. Alternatives can be added by stepping through the sequence and asking the question for each event: *what could occur instead of this event*? You need not come up with an alternative for every single event, but the critical decision points are identified in this way. Stepping through Figure 11.12, three critical alternatives were identified: (1) the drop could be on target, or the drop could miss the target resulting in mission failure, (2) the vessel is found or the vessel could not be found at the anticipated location, and (3) if not found, the search may continue until reaching a critically low fuel condition (bingo fuel).

With new events for "Payload Misses Target," "Vessel Not Found" and "Bingo Fuel" added to the Air Vehicle's root as alternative branches, corresponding alternatives must be added to the environment: "Payload Not Received" in the Rescuee root, and "Confirm" events acknowledging the

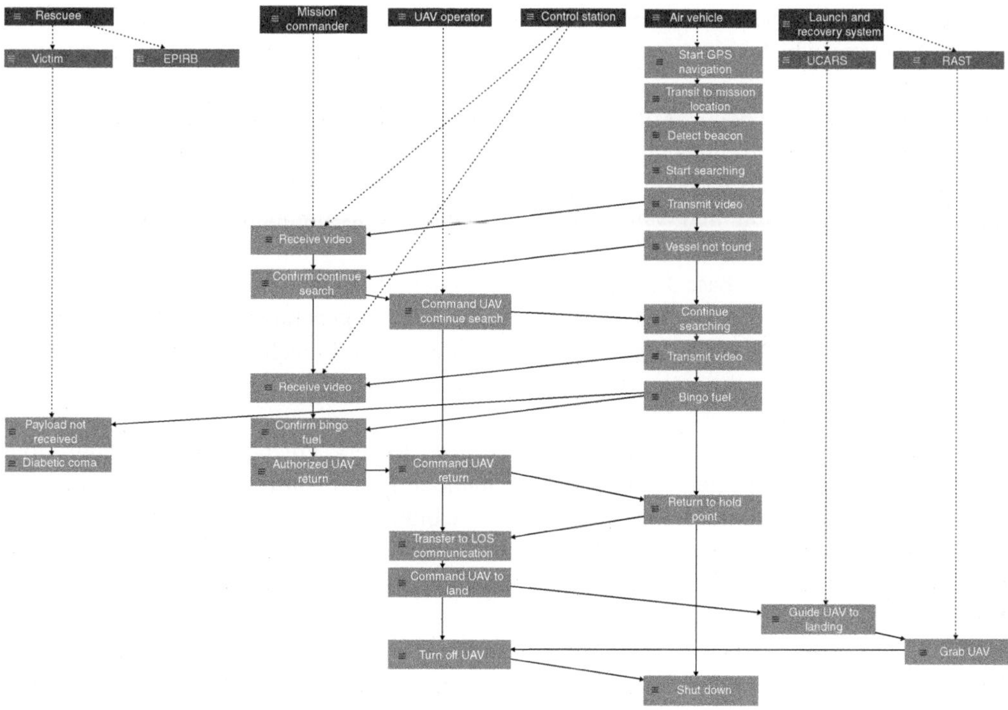

Figure 11.13 Vessel is not located and a "bingo" low fuel results in recall of Air Vehicle (scope 1, trace 8). *Source:* Adapted from Giammarco and Shifflett (2018b).

Air Vehicle's situation in the Mission Commander root. The expanded model produces eight traces at scope 1, showing the various possible combinations of events that could occur. Figure 11.13 shows one of the eight traces (the one in which a "bingo" low fuel event occurs).

If you start small and test each change to your expanded model by frequently running it and verifying the correct traces are produced, you should have a straightforward experience and be satisfied enough with the traces that you deem them ready for review and validation by the appropriate stakeholders. This discipline takes advantage of one of MP's distinguishing features – producing an abundant (and scope-complete) number of well-formed traces for V&V – but may not expose genuinely unexpected yet plausible emergent behaviors, which is the other distinguishing feature of MP. If the models are too well bound, only expected traces are allowed to emerge.

The Skyzer MP model was probed for unexpected and unwanted emergent behaviors as follows.

Probe the Model for Emergent Behaviors

Emergence occurs when global behaviors arise in the whole system from individual behaviors in parts of the system. Emergence in human-designed systems often surprises us with unexpected "extra" behaviors that are permitted due to an absence of constraints, with sometimes costly consequences. MP provides a virtual laboratory for the cultivation and discovery of this kind of emergent behavior in system designs.

If you are new to MP, or even new to any sort of behavior modeling, you are likely to make some mistakes in the logic. The presence of mistakes does not impede, and may even assist with, emergent behavior analysis. Mistakes in the grammar rules and the constraints may produce some unexpected traces, which helps you consider undocumented requirements and assumptions latent in the design.

Let us assume you are new to MP and you are learning the language at the same time you are learning the emergent behavior analysis skill, as was the main author of the Skyzer model. We will follow an early draft of the Skyzer model to show how to provoke emergence of unexpected behaviors, conduct verification and validation, and document assumptions and ideas for new requirements.

In an early draft of the Phase 3 Transit/Navigate model (presented in the previous section), no traces in which the vessel was not located were present in the generated set, despite the branch being present in the Air Vehicle's grammar rule. When entire branches of behavior are being rejected, we can infer that the grammar rule logic or constraints being applied are contradictory or too restrictive.

To see this early draft, load the model "Small_Package_Delivery_Emergent_ Behavior.mp" from the Application_examples folder on the IMPORT menu of MP-Firebird (https://firebird.nps.edu) and run it at scope 1. Both traces produced contain the branch "Locate Vessel" in the Air Vehicle, despite the option for "Vessel Not Found" in the Air Vehicle's root grammar rule. Compare this to the finished working model "Small_Package_Delivery_Creative_Expansion.mp," also in the Application_examples folder.

A first step anyone can take to verify a complex model is to break up the model and run each grammar rule alone, with no constraints, to ensure that the behavior logic for each root by themselves at least is correct. Figure 11.14 shows the Air Vehicle root grammar rule in "Small_Package_Delivery_Emergent_Behavior.mp" being run by itself, with no constraints or interactions with other roots to distract from its contents. Copy and paste the SCHEMA line and the Air Vehicle's event grammar rule into a new blank model in a separate window or deleting everything but those lines from the preloaded model if you would like to follow along.

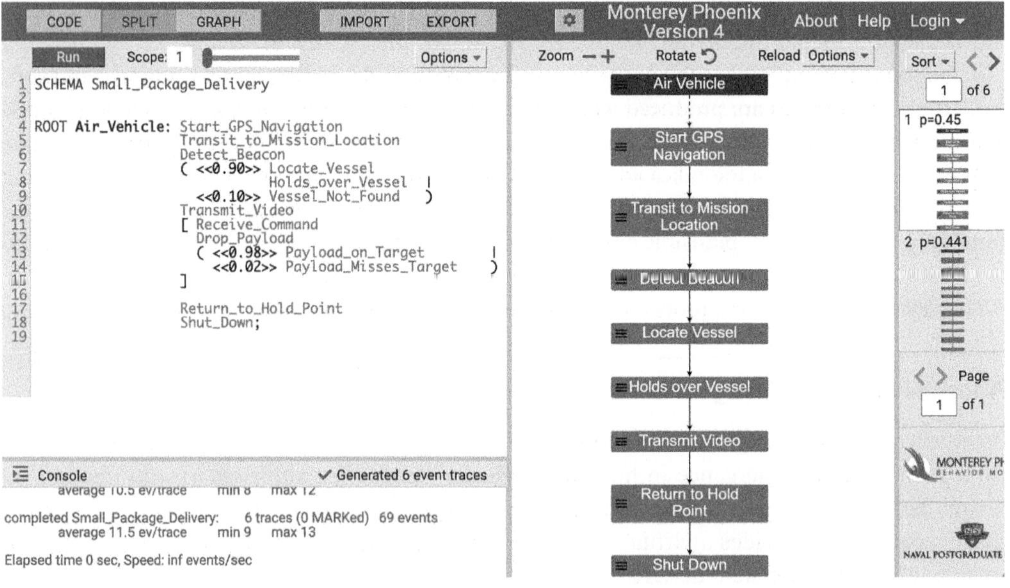

Figure 11.14 The Air Vehicle root from Giammarco and Shifflett (2018c) is run alone, producing six traces at scope 1.

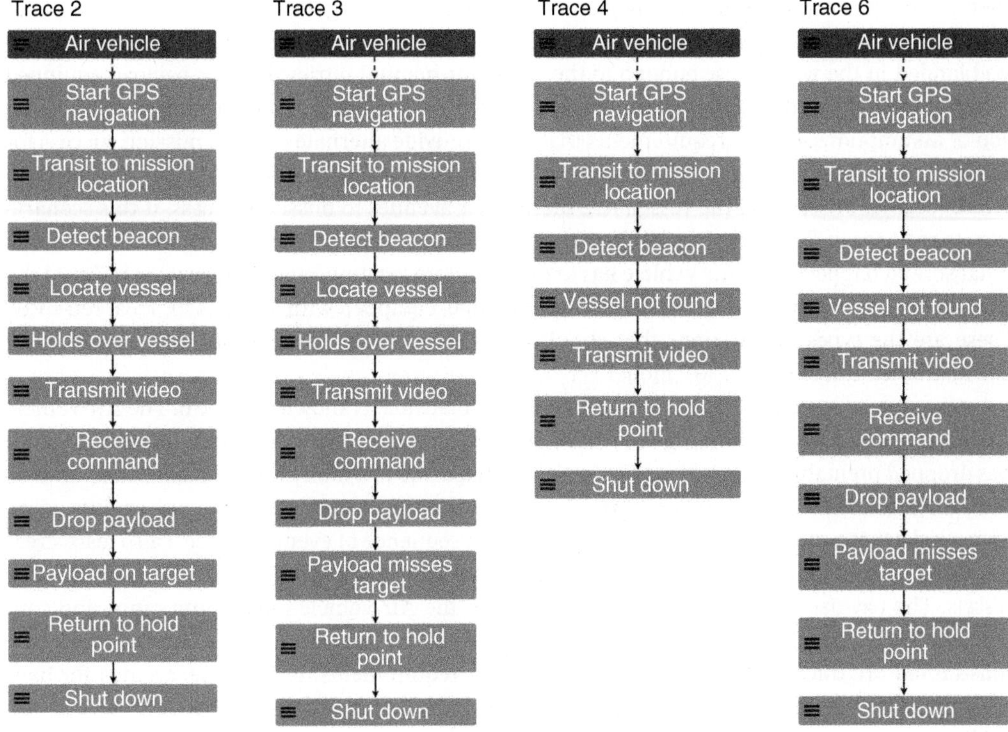

Figure 11.15 Noteworthy alternative emergent behaviors for the Air Vehicle. Far left: Baseline scenario; vessel located and payload on target (trace 2). Middle left: Vessel located but payload missed target (trace 3). Middle right: Air Vehicle needs to return before vessel is located (trace 4). Far right: Vessel not found but Air Vehicle drops payload (trace 6).

The debugging model in Figure 11.14 produces six scenarios at scope 1, four of which are shown side by side in Figure 11.15. The trace set now includes the "Vessel Not Found" branch, verifying that this behavior can express when not impeded by some constraint elsewhere in the model. But before moving on, notice that conducting a validation activity on the Air Vehicle's traces alone provide some interesting observations about other behaviors that can also express when not otherwise constrained.

In the baseline scenario (far left), the vessel was located and the payload met the target (scope 1 trace 2). These events were present in the scenario described in the SysML model. The middle left scenario (scope 1 trace 3) shows the vessel was located but the payload missed its target. This expected and plausible possibility was added during model expansion of the SysML model. The middle right scenario (scope 1 trace 4) shows the Air Vehicle needing to return to the ship before the vessel is located, another expected plausible alternative that was not explicitly modeled previously. Finally, the far right scenario (scope 1 trace 6) shows the vessel was not found and then (unexpectedly but plausibly) the Air Vehicle drops the payload.

Traces 2, 3, and 4 were sequences that were expected to be generated with the expansion of the Transit/Navigate SysML model. Their explicit presence among the generated scenarios enabled visibility and consideration to be given to them. Consider trace 3 in which the payload misses the target. As is, this trace implies that if the payload misses the target, the Air Vehicle returns to base and the mission ends, seemingly in failure. Could there be a contingency plan for making a second attempt on site, or by designing in safeguards to minimize missing the target? Or, is it incorrect to assume that this is a failure – could the payload still be retrieved by target vessel? Perhaps the payload landed in the water close enough to the vessel for manual retrieval. Is the payload equipped with flotation? Does this scenario provoke ideas for other validation-type questions, documentation of assumptions, or new requirements that could provide alternate paths to mission success for relatively little additional cost? Likewise, consider trace 4 in which the air vehicle needs to return before locating or reaching the vessel. Are there other avenues to mission success, if this scenario were to really occur? For example, could a second Air Vehicle be sent to relieve the first Air Vehicle? What should happen if the Air Vehicle has to return before locating/reaching the vessel? Could the payload be dropped at the air vehicle's maximum range, equipped with a means for local retrieval? These are the types of questions that should come up upon inspecting loosely constrained or unconstrained traces in your MP model.

An example of a scenario that was unexpected but plausible is shown in trace 6. The Air Vehicle drops the payload without finding the vessel. One possible interpretation of this scenario is that it was dropped prematurely enroute to the vessel, perhaps due to a faulty release mechanism or even a mistaken command. These are undesired behaviors that fail to meet mission requirements, identifying risks that need to be managed and tracked. But sequence of events of trace 6 also leads creative thinkers to an idea for an alternative strategy for handling package delivery to out-of-range vessels. The payload could be dropped on purpose at the Air Vehicle's maximum range and continue under its own power to make its retrieval still possible. This is an example for how minimally constrained MP traces can be used as inspiration for requirements, in this case, an idea for handling out of range vessels or air vehicles that experience a return to base condition.

All of these operational "what if" questions were exposed through MP modeling of a single SysML baseline scenario. The MP modeling of SysML behavior diagrams helped to find requirements and expose possible emergent behaviors, undocumented assumptions, and risks that may otherwise not be considered until later in the lifecycle.

For additional examples and a more formal description of the methodology for emergent behavior analysis that followed on from RT-176, please see Giammarco (2022).

Conclusions

RT-176 resulted in multiple advancements and contributions to the systems modeling community. This chapter provided a practitioner's overview of a portion of that work that provides the nearest return on investment for verifying and validating existing SysML behavior specifications. A general overview of the Monterey Phoenix (MP) approach used throughout this research task was provided, followed by an example application that practitioners can follow along with to learn the approach and apply it to a system of choice. More information about Monterey Phoenix may be found at https://nps.edu/mp (public space) and on the Naval Digital Engineering Body of Knowledge (Naval-LIFT space).

The Skyzer system model used in this chapter demonstrated the Navy-developed MP approach and tool as a candidate for integration into the Naval Enterprise Modeling Environment based on its ability to help modelers find overlooked requirements whose absence results in unwanted behaviors in the modeled design. For systems with many behavior variants, the event trace coverage and speed provided by MP exceeds that of manual trace generation while removing opportunities for inconsistencies and errors of omission. Risk management and mitigation efforts are also informed by this analysis approach for surfacing unexpected yet plausible behaviors.

Some key findings from RT-176 related to the content presented follow:

Five key modeling concepts were distilled from development of behavior models supporting this research: (1) separate behaviors and interactions, (2) model both system and environment behaviors, (3) formalize models for automatic execution, (4) properly allocate each task to a human or to a machine, and (5) use abstraction and refinement to manage large models.

Unexpected behaviors can be discovered early in models using the Monterey Phoenix approach and language. Development of a *least restrictive* MP model admits many possible behaviors, both wanted and unwanted, and some subset of those also being unexpected yet plausible. Program offices using MP should expect to become aware of and able to control for more unwanted behaviors earlier in design. Many more descriptions of emergent behaviors become available to decision-makers earlier in the system's design, so that they may be carefully planned for or removed in the design rather than haphazardly dealt with after they emerge in the actual system.

Scope-complete trace generation enables verification and validation that is more comprehensive than other methods. The small scope hypothesis (Jackson 2012) works to expose most V&V issues in behavior models at a small scope (very few event iterations, in our experience, between 1 and 3). A set of event traces that is complete up to the scope limit provides a thorough but finite set of data to test for model properties of interest, enabling MP users to guarantee the presence or absence of such properties from the system as modeled up to the specified scope.

Relaxing or restricting constraints on events and interactions steers emergent behaviors in MP models. Many behaviors that are not explicitly present nor deliberately forbidden can emerge through MP trace generation. Each grammar rule assigns a system or component of its own behaviors, any of which may eventually manifest without deliberate suppression. If an MP model is well-constrained and producing all expected traces, unexpected traces may be coaxed out of hiding by removing or temporarily commenting out constraints and inspecting the resulting traces. With MP, "positive emergence" constitutes the set of acceptable behaviors and interactions that remain after the "negative emergence" has been thoroughly exposed and pruned.

Model developers can detect, predict, classify, and control emergent behaviors with MP early in design. This research tested the first draft of a repeatable methodology for emergent behavior analysis. MP automatically *detects* all possible combinations of system behaviors and interactions permitted by the model. During inspection, each generated scenario provides a canvas for *predicting* potential future states of emergence. The emergent behaviors are *classified* as weak or strong and positive or negative. Negative emergent behaviors are *controlled* through modification of the individual behavior models or relaxation/restriction of interaction constraints.

Some recommendations pertinent to this work are as follows:

Use Monterey Phoenix for model expansion, verification, and validation: MP should be deployed with urgency on programs with critical issues, vulnerability concerns, safety hazards, and other risks that should be identified and purged sooner rather than later.

Train junior model developers how to verify and validate SysML models using MP: Junior model developers are excellent candidates for learning MP. Have them use models developed by experienced SysML modelers as source data. This will effectively teach them two skills at once: analysis of SysML models, and verification and validation of those models using MP. This approach also builds confidence in newer workforce members in applying smart ignorance (Berry 1998) to ask clarifying questions, expose tacit assumptions, and provide constructive feedback that helps to improve the quality of the SysML model and of the design it describes.

Scrub requirements with MP before writing contracts: Use MP to verify and validate high-level system requirements before they are written into contracts. This activity can help to stabilize the requirements earlier and reduce program costs by reducing the number of engineering change proposals and contract modifications later in the system's lifecycle. Record and track all V&V issues discovered in analysis using MP so that the cost of exposing and fixing each issue early can be compared to the expected value for finding and fixing the same behavior in a later lifecycle phase.

Formalize the types and definitions of emergent behavior for use in risk analysis: There is practical value to having clear criteria for classifying different types of emergent behavior for use in the conduct of risk analysis (e.g., in assigning priorities to unwanted behaviors) and for developing metrics for emergent behaviors in designs (e.g., to generate stoplight charts for behaviors of concern). Now that a collection of example models containing different types of emergent behaviors exists, the classification taxonomy (further described in Giammarco (2022)) should be more broadly tested and refined into a standard way to assign and categorize the types of emergent behavior.

Develop a graphical gateway to MP: While analysts of critical systems typically have a mathematical or formal methods background that comes with coding experience, many systems engineers have strong preferences for graphical languages and tools. Users who are turned off by the code-only interface to MP would benefit from a graphical way to create MP code, removing this barrier to the capability it delivers, which may take the form of automatic translation from SysML to MP (and the reverse, after trace generation is complete). There was some progress toward this goal (Hall et al. 2022) in crosswalking the behavior aspects of SysML with the MP language to lay the groundwork for automated transcription. Further progress was made by NPS Master's student Chris Ritter with his prototype graphical gateway tool that translates free-drawn activity models to the user's choice of MP or SysML version 2 text code, available at https://openseabird.com.

In closing, if you have read this far, you are in a position to potentially help your project or program save time, money, and reputation by finding undesired but possible behaviors that otherwise would slip by undetected until something bad happens. I understand the urgency on fixing unwanted behaviors after they occur, because before teaching and conducting research I was a practitioner immersed in the day-to-day business of urgent problem solving. I hope the work presented here helps to shift the problem-solving pattern from reactive to proactive – a "preventative maintenance" that can be conducted on the *designs* of systems. Perhaps you will find previously unrealized failure modes in which your hardware or software behaves in a way that jeopardizes the mission. Or perhaps you will discover use case variants no one previously considered, such as intentional or unintentional misuse that causes vulnerability, damage, or injury. It is entirely possible that your application of MP to real system designs saves a life or multiple lives. To prevent bad outcomes, you have to know that those outcomes are even possible in the first place, and expanding your field of awareness of these possibilities is what MP does really well. Find out about it now with MP, so that you or someone else does not have to find out about it later.

References

Auguston, M. (2009a). Software architecture built from behavior models. *ACM SIGSOFT Software Engineering Notes* 34 (5): 1–15.

Auguston, M. (2009b). Monterey phoenix, or how to make software architecture executable. In: *Proceedings of the 24th ACM SIGPLAN Conference Companion on Object Oriented Programming Systems Languages and Applications*, Orlando, FL (25–29 October 2009), 1031–1040. ACM.

Auguston, M. (2020). System and software architecture and workflow modeling language manual. https://wiki.nps.edu/display/MP/Documentation (accessed 6 June 2023).

Auguston, M., Kristin Giammarco, W., Baldwin, C., and Farah-Stapleton, M. (2015). Modeling and verifying business processes with Monterey Phoenix. *Procedia Computer Science* 44: 345–353.

Berry, D. (1999). Formal methods: the very idea, some thoughts about why they work when they work. In: *Proceedings of the 1998 ARO/ONR/NSF/ARPA Monterey Workshop on Engineering Automation for Computer Based Systems*, Electronic Notes in Theoretical Computer Science, Monterey, CA, vol. 25, 10–22. https://doi.org/10.1016/S1571-0661(04)00127-6.

Borshchev, A. and Filippov, A.. (2004). From system dynamics and discrete event to practical agent based modeling: reasons, techniques, tools. In *Proceedings of the 22nd International Conference of the System Dynamics Society*, Oxford, England (25–29 July 2004) (Vol. 22). System Dynamics Society.

Easterbrook, S., Robin Lutz, R., Covington, J. et al. (1998). Experiences using lightweight formal methods for requirements modeling. *IEEE Transactions on Software Engineering* 24 (1): 4–14.

Giammarco, K. (2022). Exposing and controlling emergent behaviors using models with human reasoning. In: *Emergent Behavior in System of Systems Engineering* (ed. L.B. Rainey and O.T. Holland), 23–61. CRC Press.

Giammarco, K. and Giles, K. (2017). Verification and validation of behavior models using lightweight formal methods. In: *Proceedings of the 15th Annual Conference on Systems Engineering Research*, Redondo Beach, CA (23–25 March 2017). Procedia Computer Science.

Giammarco, K. and Shifflett, D. (2018a). Small package delivery baseline case. https://gitlab.nps.edu/monterey-phoenix/mp-model-collection/preloaded-examples/-/blob/master/models/Application_examples/Small_Package_Delivery_Basline_Case.mp (accessed 6 June 2023).

Giammarco, K. and Shifflett, D. (2018b). Small package delivery creative expansion. https://gitlab.nps.edu/monterey-phoenix/mp-model-collection/preloaded-examples/-/blob/master/models/Application_examples/Small_Package_Delivery_Creative_Expansion.mp

Giammarco, K. and Shifflett, D. (2018c). Small package delivery emergent behavior. https://gitlab.nps.edu/monterey-phoenix/mp-model-collection/preloaded-examples/-/blob/master/models/Application_examples/Small_Package_Delivery_Emergent_Behavior.mp

Giammarco, Kristin; Ron Carlson, Mark Blackburn, et al. (2018). Verification and Validation (V&V) of System Behavior Specifications. *Final Technical Report SERC-2018-TR-116. Systems Engineering Research Center: Hoboken, NJ, USA.*

Hall, J., Le, K., Patel, K., and Savacool, M. (2022). Assessing interoperability between behavior diagrams constructed with system modeling language (SysML) and Monterey Phoenix (MP). *Master's capstone report.* Naval Postgraduate School.

Jackson, D. (2012). *Software Abstractions: Logic, Language, and Analysis* (Revised edition). MIT Press.

Ruppel, S.R. (2016). System behavior models: a survey of approaches. Master's thesis. Naval Postgraduate School.

SEBoK Authors (2022a). System verification. in *BKCASE Editorial Board. Guide to the Systems Engineering Body of Knowledge (SEBoK)*, v. 2.6 released 20 May, 2022. https://www.sebokwiki.org/wiki/System_Verification (accessed 21 February2023).

SEBoK Authors (2022b). System validation. in *BKCASE Editorial Board. Guide to the Systems Engineering Body of Knowledge (SEBoK)*, v. 2.6 released 20 May, 2022. https://www.sebokwiki.org/wiki/System_Validation

Woodcock, J., Larsen, P.G., Bicarregui, J., and Fitzgerald, J. (2009). Formal methods: Practice and experience. *ACM Computing Surveys (CSUR)* 41 (4): 19.

Biographical Sketches

Kristin Giammarco is an associate professor in the Department of Systems Engineering at the Naval Postgraduate School, where she teaches courses in system architecture and design, system integration, systems software engineering, and model-based systems engineering. She conducts research in the use and development of formal methods for mission, system, and process behavior modeling. Dr. Giammarco is a member of INCOSE and serves as the Systems Engineering PhD Program Officer and the Joint Executive Systems Engineering Management (SEM-PD21) Program Academic Associate. From NPS, Dr. Giammarco has earned a PhD in software engineering, an MS in systems engineering management, and a certificate in advanced systems engineering. She holds a BE in electrical engineering from Stevens Institute of Technology.

Chapter 12

Digital Engineering Transformation: A Case Study

Cesare Guariniello, Waterloo Tsutsui, Dalia Bekdache, and Dan DeLaurenits

Purdue University, Aeronautics & Astronautics, West Lafayette, IN, USA

Case Study of Digital Engineering from Scratch: a DoD and JPEO Journey

Digital engineering (DE) is a useful tool when it is properly utilized. In order to achieve all the advantages provided by the use of digital engineering, organizations need to perform various steps toward full implementation and a correct application of DE tools and techniques:

1) It is important that all different levels of the organization, especially the managerial side, understand the importance of DE and are ready to embrace it.
2) The organization needs to identify its own specific needs and requirements.
3) The organization needs to tailor existing products and examples to its needs.
4) The implementation of the digital transformation process must occur at all levels of the organization, with an appropriate flow of information, and keeping into account the organization-specific needs from previous steps.

In this chapter, we describe some lessons learned from a research collaboration between DoD's SERC and the Joint Program Executive Officer for Chemical, Biological, Radiological and Nuclear Defense (JPEO – CBRND), on the topic of DE and digital transformation (DT). The JPEO – CBRND, based on directives from the DoD and on the acknowledgment of some limitations of their current approach to responding to threats, initiated a process of DT "from scratch." SERC provided its expertise to support decision-making in this DT path. This collaboration highlighted some important aspects that need to be accounted for in DT at the enterprise level in such a large and complex organization.

Understanding the Needs

At the beginning of this research collaboration, JPEO and researchers from Purdue University organized a large workshop that involved various managers and executive officers from JPEO, as well as personnel more oriented toward technical applications. The workshop was structured in two parts: first, Purdue University and SERC presented examples of their studies in DE, and

Systems Engineering for the Digital Age: Practitioner Perspectives, First Edition. Edited by Dinesh Verma.

Companion website: www.wiley.com/go/verma/systemsengineering

potential applicability to JPEO needs. In particular, the team presented tools for the analysis of complex systems and System of Systems from Purdue's Analytic Workbench (Davendralingam et al. 2014). In the second part, JPEO personnel presented individual needs and requirements, as well as their expectation of which areas could be supported by DE, and what limitations or risks are perceived. This exercise already highlighted some requirements that are unique to JPEO. These include the need to share information quickly and effectively within a large organization while keeping the data safe from unwanted access or corruption, the diversity of operations performed at JPEO (which prompts the use of various different models of the same operation or system, based on the individual use that will be made of those models) and the necessity to operate DT relatively quickly, but without interrupting the current flow of operations. The workshop also evidenced the presence of some disconnection between the various offices at JPEO, typical of large organizations, which resulted in very different objectives and points of view about DE. As mentioned above, to properly implement DT and use DE, the whole organization needs to achieve some sort of mutual understanding and agreement on how the various steps will be performed and which changes will occur at each level of the organization.

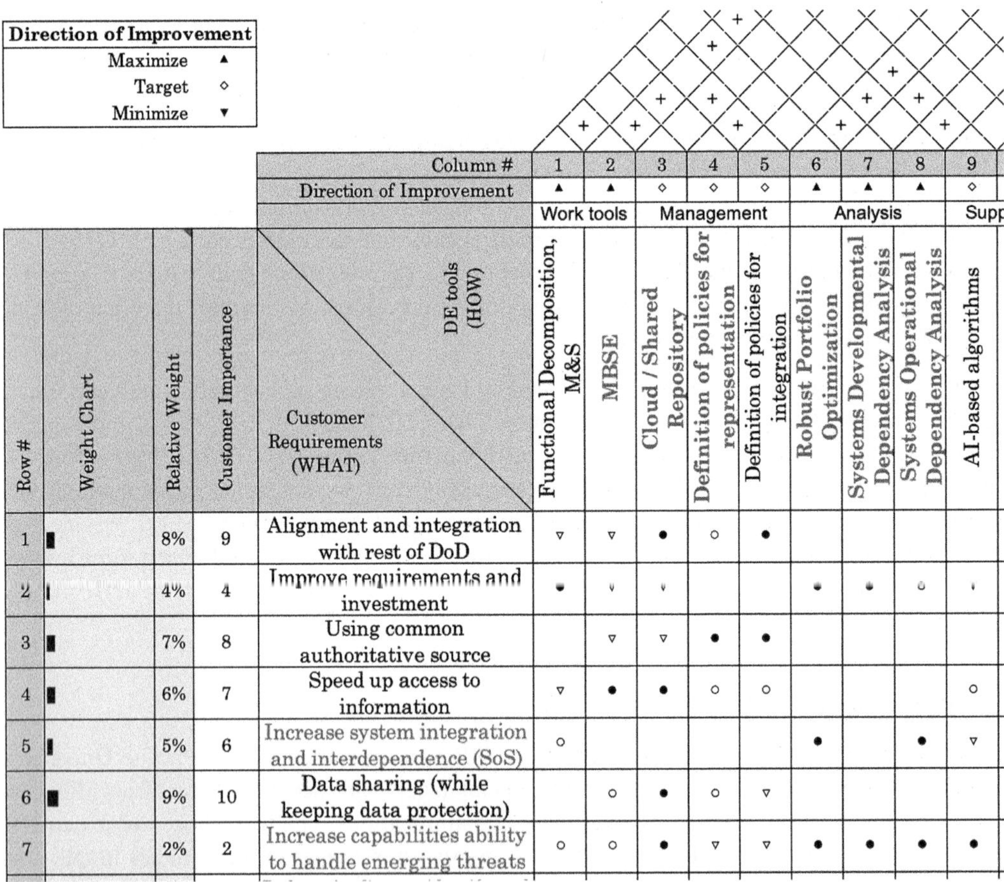

				Column #	1	2	3	4	5	6	7	8	9
				Direction of Improvement	▲	▲	◇	◇	◇	▲	▲	▲	◇
					Work tools		Management			Analysis			Supp
Row #	Weight Chart	Relative Weight	Customer Importance	Customer Requirements (WHAT) / DE tools (HOW)	Functional Decomposition, M&S	MBSE	Cloud / Shared Repository	Definition of policies for representation	Definition of policies for integration	Robust Portfolio Optimization	Systems Developmental Dependency Analysis	Systems Operational Dependency Analysis	AI-based algorithms
1		8%	9	Alignment and integration with rest of DoD	▽	▽	●	○	●				
2		4%	4	Improve requirements and investment	●	▽	▽			●	●	○	[illegible]
3		7%	8	Using common authoritative source		▽	▽	●	●				
4		6%	7	Speed up access to information	▽	●	●	○	○				○
5		5%	6	Increase system integration and interdependence (SoS)	○					●		●	▽
6		9%	10	Data sharing (while keeping data protection)		○	●	○	▽				
7		2%	2	Increase capabilities ability to handle emerging threats	○	○	●	▽	▽	●	●	●	●

Figure 12.1 Excerpt from the House of Quality representation of JPEO user needs (rows) and appropriate systems engineering and digital engineering tools (columns).

In response to the knowledge acquired at the workshop, the Purdue research group built a House of Quality (HoQ) to map the need of JPEO personnel to existing and proposed DE tools and techniques (Figure 12.1). This was a very useful tool to evaluate which steps are more important to JPEO management and to identify appropriate tools and techniques to address the various needs of the user. Both the requirements and the relative tools and techniques in the HoQ show the large variety of different aspects that need to be taken into account to introduce DE at the enterprise level.

Following this analysis, Purdue SERC researchers and JPEO needed to ensure that all involved stakeholders could properly appreciate the power of Digital Engineering. As stated above, a Digital Transformation process can be effective only if all the parties agree and collaborate on this effort. Therefore, the following step in this process was to exemplify some of the potential of DE with practical demonstrative application on small problems.

Tailoring Products for User's Needs

To demonstrate applications of DE to a problem of interest of JPEO, researchers modified two methods previously developed at Purdue University: Robust Portfolio Optimization (Davendralingam and DeLaurentis 2013) and Systems Operational Dependency Analysis (Guariniello and DeLaurentis 2017). The result was the development of a tool called Decision Support Framework (DSF), aimed at providing support to decision-making in the selection of appropriate portfolios of technologies. Following the principles of mission engineering, the user can select mission-level objectives and specify budget limits and risk aversion. The first tool, RPO, optimizes the selection of available systems in order to achieve the highest value of the mission objectives while keeping within the budget and the risk limits. The second tool, SODA, analyzes the impact of disruptions and the subsequent robustness of portfolios generated by RPO. Figure 12.2 shows an example of an input file to the DSF.

The DSF showed an example of the application of DE to a synthetic case study of interest to JPEO. This tool would help JPEO to quickly adapt to ever-changing threats and to provide a strongly objective solution to mission requirements. Furthermore, the DSF showed how a DE approach could help with modularity, with the expansion of existing problems, and with the application of different viewpoints. For example, the user can select different mission objectives with varying weights in order to accommodate different needs. Figure 12.3 shows the main window of the DSF, and Figure 12.4 shows one of the products of the DSF, which is a set of portfolios, each one corresponding to a different combination of budget and risk aversion, which maximize the performance in terms of the selected weighted mission objectives.

The DSF also contains a version of the SODA tool that has been modified to meet the needs of JPEO. This tool analyzes the impact of disruptions in systems that provide support capabilities and system-level capabilities on the mission objectives. Figure 12.5 shows the analysis of disruptions for one of the optimal portfolios of CBRND sensors.

State-of-the-Art in Digital Transformation and "Clash with Reality"

Even when demonstrating potential improvements that an enterprise can achieve thanks to DE, it is often hard to overcome the initial reticence and doubts. Studies are more and more showing and even quantifying the improvements provided by the use of DE (McDermott and Salado 2021). Organizations that reduced document-based approaches in favor of Model-based systems engineering saw multiple benefits. However, digital transformation is not an easy

A	B	C	D	E	F	G
			Support Input Requirement			
No.	System Type	System Name	Power	Transportation	Communication	Operation
-	-	-	W	Lbs	BW	#
1	Chem and Bio Detection	Aerosol Vapor Chem Agent Detector (AVCAD)	10	10	0	1
2		Analytical Lab Sys (ALS)	50	100	30	5
3		ALS Modification Work Order (MWO)	20	50	0	2
4		CBRN Dismounted Recon Sets, Kits and Outfits	10	15	0	1
5		Compact Vapor Chem Agent Detector (CVCAD)	5	2	0	1
6		Enhanced Maritime Biological Detection (EMBD)	10	15	0	1
7		US Navy Joint Biological Point Detection System	40	30	10	2
8		Joint Biological Tactical Detection System (JBTDS)	12	13	0	1
9		Joint Chemical Agent Detector (JCAD) - M4A1	8	5	0	1
10		Joint Chemical Agent Detector (JCAD) - Solid Liquid Adapter (SLA) Kit	5	12	0	0
11		Multi-Phase Chemical Agent Detection (MPCAD)	30	70	0	2
12		M256 Chem Agent Detector Kit	0	4	0	1
13		Reactive-Chemistry Orthogonal Surface and Environmental Threat Ticket Array (ROSETTA)	0	1	0	1
14	Radiological and Nuclear Detection	Joint Personal Dosimeter - Individual (JPD - IND)	5	2	0	1
15		Man-Portable Radiological Detection Sys (MRDS)	25	30	0	1
16		Radiological Detection System (RDS)	15	10	0	1
17	Multiple Detection and Integration	Capabilities to Enable Nuclear, Bio, Chem (NBC) Threat Awareness, Understanding and Response (CENTAUR)	200	300	60	10
18		CBRN Integrated Early Warning (IEW) Enhanced Capability Demo (ECD)	100	50	20	5
19		CBRN Sensors Integration on Robotic Platforms (CSIRP)	150	200	30	2
20		Non-Traditional Agent Defense	200	0	0	100
21		Nuclear, Bio, Chem Recon Vehicle (NBCRV) Sensor Suit Upgrade (SSU)	100	130	10	10
22	Support	Battery 1	0	0	0	0
23		Battery 2	0	0	0	0
24		Battery 3	0	0	0	0
25		Operators	0	0	0	0
26		Screening Obscuration Module (SOM)	0	0	0	0
27		Stryker Combat Vehicle	0	0	0	0
28		Unified Command Suite (UCS)	0	0	0	0
29		Unmanned Aerial System	0	0	0	0
30		Unmanned Ground Vehicle	0	0	0	0

1 Main Sheet | 2 SoS Capabilities | 3 Compatibility Constraints | 4 "Must Have" Systems | 5 Conditio

Figure 12.2 Excerpt of the spreadsheet modeling notional examples of CBRND sensors for use in the DSF.

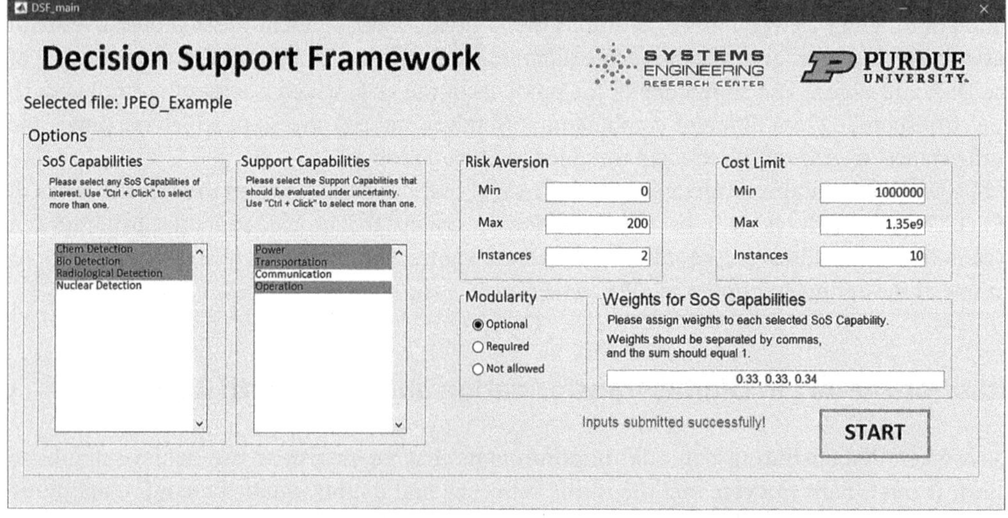

Figure 12.3 DSF main window for the CBRND sensors case study. In this specific example, three mission objectives (SoS capabilities) are highlighted, meaning that they have been selected for optimization, and three support capabilities are highlighted, meaning that they will have performance uncertainty in the run of the experiments.

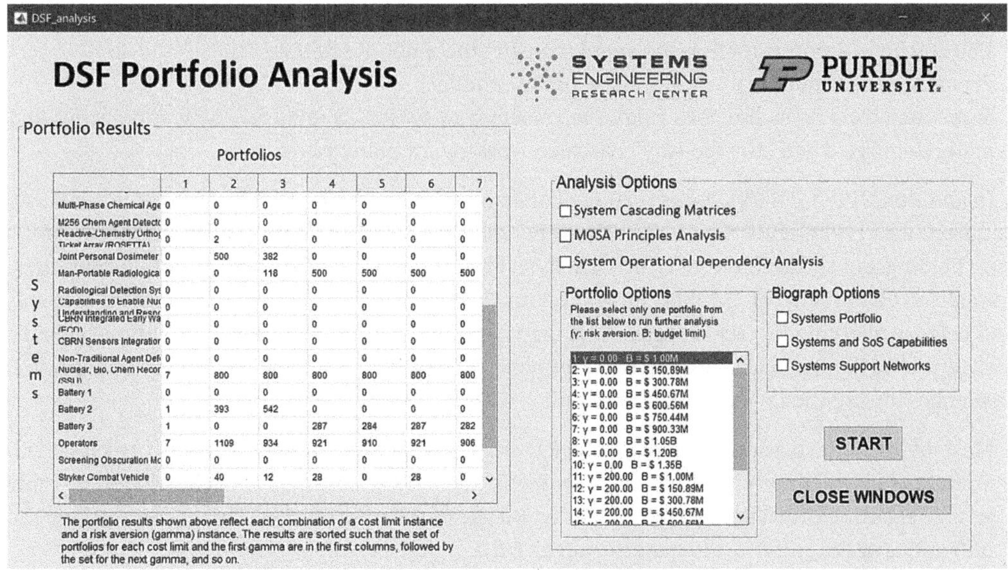

Figure 12.4 Optimal portfolios generated from a run of the CBRND scenario.

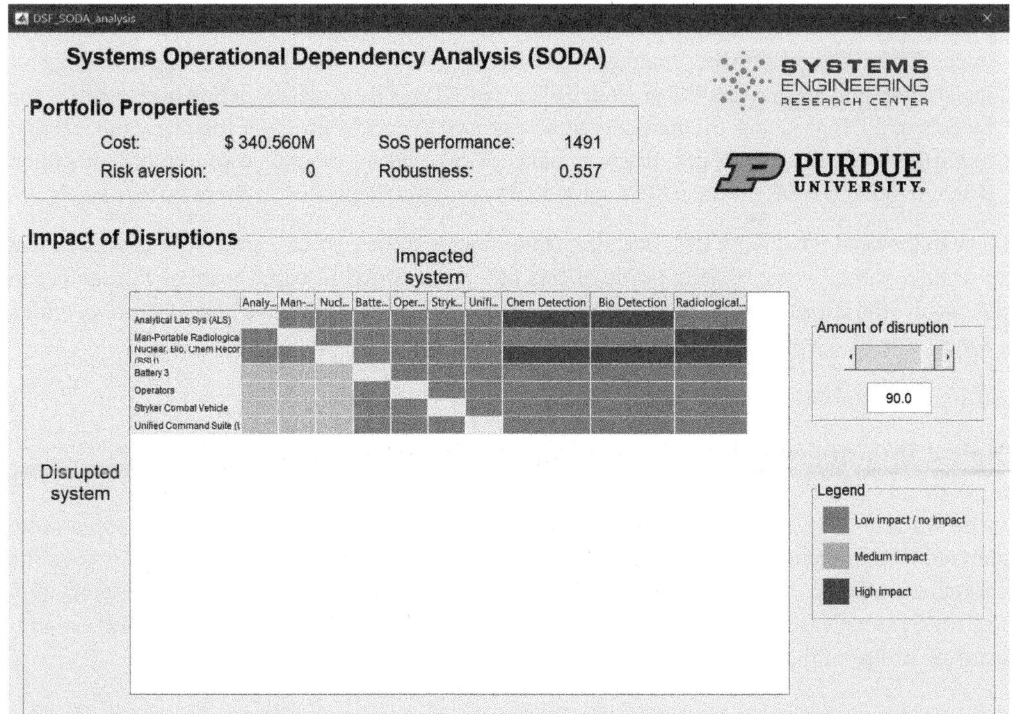

Figure 12.5 Analysis of the criticality of sensors and support systems in one of the optimal portfolios.

endeavor and requires much and widespread effort upfront. As mentioned above, proper DT needs to be implemented at all levels throughout the organization and requires "acceptance" and coordination by all involved stakeholders. In large organizations, this is not easily achievable due to multiple reasons:

- The initial investment can be expensive and time-consuming.
- The perceived improvement is too small to justify the amount of effort.
- Personnel needs to be trained in the use of DE products.
- Some managers or technicians might be reluctant to switch to relatively new approaches and methodologies when "the old way" has been working for many years.

In addition, JPEO-CBRND faces further challenges toward DT due to the nature of its work. As defined online (JPEO-CBRND 2023), "The JPEO-CBRND leads, manages, and directs the acquisition, fielding and sustainment of CBRN sensors, protective equipment, medical countermeasures, specialized equipment for U.S. Special Forces, integration and information management systems, and defense enabling biotechnologies. The organization also works closely with various government agencies that need CBRN defense equipment." The tasks of JPEO-CBRND present additional challenges toward the implementation of DE:

- Most of the information related to the technology acquired by JPEO-CBRND is classified. Therefore, every DE product needs to be implemented and maintained with the highest possible level of security for what concerns access to and use of the information. This can exacerbate the difficulties of DT and the reluctance to adopt DE.
- Furthermore, JPEO-CBRND needs to retain its capabilities while implementing DT. This makes the transition phase much more complex and intensifies the need to properly plan and execute each step of the process.
- Finally, JPEO-CBRND interacts with different branches of the government and the armed forces. This requires a further managerial decision beyond the usual technical decision regarding DE implementation in a DT process: JPEO-CBRND needs to clearly define ownership of the models and DE products, the modality of access, and to decide which of the many models that might be implemented by the different stakeholders will be the single source of truth upon which to build the DE products. The subsequent section will expand on these further needs.

Due to the need for JPEO-CBRND to start from scratch in their DT process, and to the additional constraints posed by the specific needs of this office, the SERC research team at Purdue began looking into the current implementation of DE for enterprise-level Additive Manufacturing, which is a recommended technology addition for the DoD and the US Army.

Digital Engineering Transformation and Directives (DE for AM)

In the work of SERC with JPEO, the potential additive manufacturing (AM) applications were analyzed from a technical perspective and as part of the procurement process. In this way, the researchers studied digital transformation across the enterprise of one DoD organization (i.e., JPEO) as well as across the whole DoD enterprise in a particular domain (i.e., additive manufacturing in sustainment).

Additive Manufacturing – Technical Analysis: Additive Manufacturing (AM) vs. Traditional Manufacturing (TM)

As stakeholders move forward with the AM implementation exercise, specific use case applications in the product portfolio will be analyzed to determine what applications suit AM in the acquisition process. After identifying the potential AM applications within the portfolio, the next step is understanding the cost trade-off, ease of use, flexibility, and performance associated with the AM implementation. More specifically, AM should be able to provide better performance, faster delivery/service, more flexibility in design, and/or less cost than TM. Otherwise, continuous TM use is preferred instead of converting the existing applications to AM. The stakeholders must objectively evaluate alternatives to move forward through the decision-making process. Thus, the stakeholders need to identify/define the models related to the AM. Also, the stakeholders need to understand the procurement behavior of chosen applications in the portfolio. As a result, the stakeholders will need to ask themselves questions to understand the details necessary for the AM decisions.

Examples of models
- **CAD/physical models**: performance, requirements, and ease of manufacturing
- **Operational models**: change in operations, flexibility, and mission threads
- **Supply chain models**: cost and time

Examples of questions
- What is the current or expected production volume of this product?
- What are some manufacturing details (i.e., cost, lead time, batch quantities)?
- How widely is the product being used (product penetration in the portfolio)?
- Is one manufacturing location sufficient, or should there be multiple manufacturing locations?
- Does the product have to be distributed worldwide?
- What AM capabilities does the sponsor already have?

Based on the types of models and other details obtained by answering the questions, the stakeholders can create a decision support framework to compare the AM to TM technologies (and potentially with AM/TM hybrid). As a result, the researchers compared AM to TM in the following section using the quasi-quantitative attributes.

- What scenario if AM is more suitable than TM?
- What scenario if TM is more suitable than AM?
- What scenario if hybrid AM/TM is more suitable than AM or TM alone?

AM vs. TM Comparison
In order to make a comparison between AM and TM, the researchers excluded the variables related to the mechanical, thermal, and electrical properties/performance of materials. The comparison is strictly based on the cost, production volumes, minimum lead time, and maximum part complexity. Abbreviations used in the AM vs. TM comparison are defined as follows:

Abbreviations

- **AM technology**
  3D = 3D printing
- **TM technology**
  CNC = CNC machining; Poly = Polymer casting; Rot = Rotational molding; Vac = Vacuum forming; Inj = Injection molding; Ext = Extrusion; Blow = Blow molding
- **Cost per part**
  $ = not expensive; $$ = slightly expensive; $$$ = expensive; $$$$ = very expensive
- **Production volumes**
  100 = 1; 101 = 10; 102 = 100; 103 = 1000; 104 = 10,000; 105 = 100,000

- **Lead time (minimum)**
  12 h = 12 hours; 24 h = 24 hours; 36 h = 36 hours; 1 wk = 1 week; 2 wk = 2 weeks; 3 wk = 3 weeks; 4 wk = 4 weeks; 5 wk = 5 weeks; 6 wk = 6 weeks; 7 wk = 7 weeks; 8 wk = 8 weeks or more
- **Part complexity (maximum)**
  1 = Part design is not complicated; 2 = Part design is slightly complicated; 3 = Part design is moderately complicated; 4 = Part design is complicated; 5 = Part design is very complicated

Figure 12.6 summarizes the AM vs. TM comparison by taking cost per part vs. production volume. The "3D" is the AM technology, whereas CNC, Poly, Rot, Vac, Inj, Ext, and Blow are the TM technologies. Table 12.1 summarizes the AM vs. TM comparison by taking cost per part, total production volume, lead time, and part complexity. Using this table, the stakeholder can have a rough idea of the technology for their applications. Cost per part and production volume came directly from Figure 12.6. Table 12.2 exhibits the production volume based on AM and TM technologies. Table 12.3 demonstrates the production volume based on AM and TM technologies, which is the same as Table 12.2, but with cost-per-part information. Table 12.4 conveys the lead time based on AM and TM technologies. Table 12.5 outlines the part complexity based on AM and TM technologies. Tables 12.2–12.5 follow the information depicted in Figure 12.6 and Table 12.1.

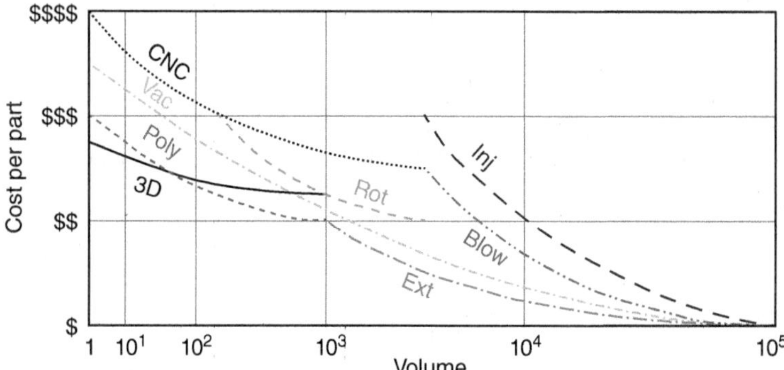

Figure 12.6 Summary plot (Shi et al. 2022), (Guide to Manufacturing Processes for Plastics n.d.). *Source:* Adapted from Patel and Gohil (2021).

Table 12.1 Summary table (Shi et al. 2022), (Guide to Manufacturing Processes for Plastics n.d.), (Patel and Gohil 2021).

		AM	TM						
Attributes		**3D**	**CNC**	**Poly**	**Rot**	**Vac**	**Inj**	**Ext**	**Blow**
Cost per part	Min	$$–$$$	$$–$$$	$$	$$	$	$	$	$
	Max	$$–$$$	$$$$	$$$	$$$	$$$–$$$$	$$$	$$	$$–$$$
Production volume	Min	1	1	1	10^2–10^3	1	10^3–10^4	10^3	10^3–10^4
	Max	10^3	10^3–10^4	10^3	10^3–10^4	10^5 or more	10^5 or more	10^5 or more	10^5 or more
Lead time	Min	12–36 h	24 h–2 wk	24 h–1 wk	4–6 wk	4–6 wk	8–10wk	2–4 wk	4–6 wk
Part complexity	Max	5	4	4	2	1	3	1	1

Table 12.2 Production volume and technologies – basic.

Production volume	AM	TM						
	3D	CNC	Poly	Rot	Vac	Inj	Ext	Blow
1	x	x	x		x			
10^1	x	x	x		x			
10^2	x	x	x	(x)	x		x	
10^3	x	x	x	x	x	(x)	x	(x)
10^4		(x)		(x)	x	X	x	x
10^5					x	X	x	x

x = production is possible.
(x) = production is likely possible, although not explicitly stated in Figure 12.6.

Table 12.3 Production volume and technologies – advanced.

Production volume	AM	TM						
	3D	CNC	Poly	Rot	Vac	Inj	Ext	Blow
1	$$-$$$	$$$$	$$$		$$$-$$$$			
10^1	$$-$$$	$$$-$$$$	$$-$$$		$$$-$$$$			
10^2	$$-$$$	$$$-$$$$	$$-$$$	($$$-$$$$)	$$-$$$		$$	
10^3	$$-$$$	$$-$$$	$$	$$-$$$	$$-$$$	($$$$)	$$-$$$	($$$$)
10^4		($$-$$$)		($$)	$-$$	$$	$$-$$$	$-$$
10^5					$	$	$	$

$, $$, $$$, or $$$$ = production is possible.
($), ($$), ($$$), or ($$$$) = production is likely possible, although not explicitly stated in Figure 12.6.

Table 12.4 Lead time and technologies.

Lead time	AM	TM						
	3D	CNC	Poly	Rot	Vac	Inj	Ext	Blow
12 h	x							
24 h	x	x	x					
36 h	x	x	x					
1 wk	(x)	x	x					
2 wk	(x)	x	(x)				x	
3 wk	(x)	(x)	(x)				x	
4 wk	(x)	(x)	(x)	x	x		x	x
5 wk	(x)	(x)	(x)	x	x		(x)	x
6 wk	(x)	(x)	(x)	x	x		(x)	x
7 wk	(x)	(x)	(x)	(x)	(x)		(x)	(x)
8 wk	(x)	(x)	(x)	(x)	(x)	x	(x)	(x)

x = production is possible.
(x) = production is likely possible, although not explicitly stated in Figure 12.6.

Table 12.5 Part complexity and technologies.

	AM	TM						
Part complexity	**3D**	**CNC**	**Poly**	**Rot**	**Vac**	**Inj**	**Ext**	**Blow**
1	(x)	(x)	(x)	(x)	x	(x)	x	x
2	(x)	(x)	(x)	x		(x)		
3	(x)	(x)	(x)			x		
4	(x)	x	x					
5	x							

x = production is possible.
(x) = production is likely possible, although not explicitly stated in Figure 12.6.

To sum up the section on AM vs. TM, the researchers have the following conclusions:

- **Cost per part**
  The AM cost per part does not alter significantly (i.e., between $$ and $$$), whereas the TM cost per part changes significantly (i.e., between $ and $$$$).
- **Total production volume**
  The range of AM's desired total production volume is smaller (i.e., between 1 and 1000) than those of TM's (i.e., between 1 and 100,000 or more).
- **Lead time (minimum)**
  The minimum lead time for AM is generally shorter (i.e., 12–36 hours) than those of TM's (i.e., between 24 hours–10 weeks).
- **Part complexity (maximum)**
  The maximum complexity can be higher in AM (i.e., 5) than in TM (i.e., 1–4).

AM vs. TM vs. Hybrid AM/TM Comparison

In addition to the simple comparison between AM and TM, comparisons can incorporate the hybrid between AM and TM (i.e., hybrid AM/TM). Thus, in this section, the researchers compare AM vs. TM vs. hybrid AM/TM using the six quasi-qualitative attributes shown in the bullet below. Table 12.6 evaluates AM, TM, and hybrid AM/TM using these quasi-qualitative attributes.

- Data protection/security
- Data repositories/access
- Agility of operation
- Agility against new threats
- Digital twin
- Economically effective production lot size

Additive Manufacturing – Management/Other JPEO-Specific Considerations

In the work of SERC with JPEO, four potential areas of research opportunity were identified for the AM technologies instead of (or in addition to) the TM technologies. In each opportunity area, the researchers conducted a literature review on the potential AM applications.

Table 12.6 Evaluation of AM, TM, and AM/TM hybrid.

Attributes	AM	TM	Hybrid AM/TM
Data protection/security	Required	Required	Required
Data repositories/access	Required	Optional (OEM only)	Required
Agility of operation	Agile	Not as agile as AM	Agile. TM is the bottleneck.
Agility against new threats	Agile	Not as agile as AM	Agile. TM is the bottleneck.
Digital twin	Required	Optional (OEM only)	Required
Economically effective production lot size	1–10^3	10^3+ (Note: TM production can be done as low as 1, although it may not be economically effective.)	1+ (Depending on application)

- **Opportunity 1**: Ergonomic face shield components and respiratory/ocular protection capability
- **Opportunity 2**: Light-weight, low-cost biological surveillance system for aerosols
- **Opportunity 3**: Wearable electronics/sensors and carriers/holders for man-portable systems
- **Opportunity 4**: Repairs in the field and service for legacy systems

Opportunity 1: Ergonomic Face Shield Components and Respiratory/Ocular Protection Capability

In the JPEO portfolio, the researchers identified ergonomic face shield components and respiratory/ocular protection capability as potential AM applications. These components may be produced using AM, rather than TM, for better ergonomics/customized fit since these components touch the end users' faces directly. To this end, engineers/developers can 3D scan the end user's face using a FARO arm (FARO n.d.), create a 3D CAD model, and print the ergonomic face shield components. The new advanced face shields and respiratory/ocular protection will protect the end-users against Chemical, Biological, Radiological, and Nuclear (CBRN) threats. Furthermore, the new technology can improve integration with individual combat equipment while minimizing or eliminating the unwanted/intrusive rubber on the face to unencumber the user while operating in a CBRN environment.

The researchers' initial idea came from the literature review on creating the orthopedic shoe insole (Cui et al. 2021), as shown in Figure 12.7. Compared to TM shoe insoles, AM shoe insoles

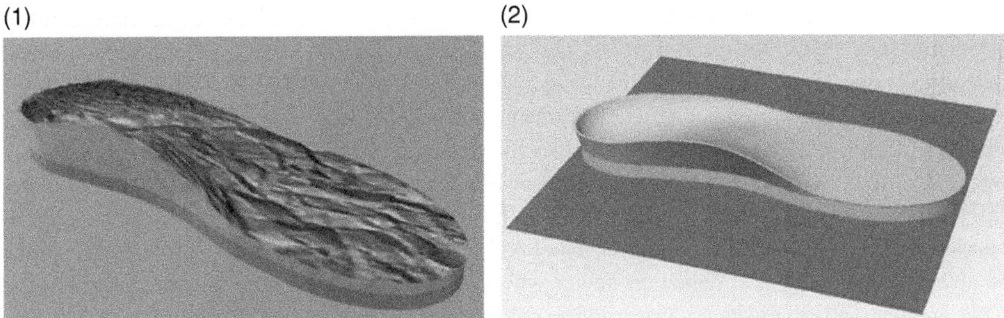

Figure 12.7 (1) 3D model used in TM, produced from the trial-and-error process, (2) 3D model used in AM, produced from the 3D scanned CAD/CAM data (Cui et al. 2021).

(i.e., "orthopedic insoles") can be made much more cost-effectively than TM shoe insoles. This is because the 3D scanning process for AM can eliminate the trial-and-error approach for product fit, thereby creating a shorter lead time and a better/customized fit for end users.

The stakeholders can apply the same design principle to respiratory and ocular protection, where a customized fit on a face can produce more protection and better ergonomics for the end users. As the researchers further conducted a literature review, it was identified that there have already been attempts to produce face shield components and respiratory/ocular protections using AM technologies (Figure 12.8). As the researchers saw the possibility of making the components of face shields using AM rather than TM, the end users in DoD can have a better-fit face shield for (1) improved ergonomics/comfort and (2) the prevention of contamination based on a better seal on/around the user's face.

An additional reason why AM is suitable for this application is that the AM process temperature is high. For instance, in fused deposition modeling (FDM), which is an extrusion-based AM technology (Figure 12.9), the material is melted for printing at high temperature (i.e., excess of 150 °C) that eliminates (or reduces) biological agents that might be on the surface. Thus, while the FDM production is taking place, the component surfaces remain sterile. However, the virus can adhere to the component surface once humans start handling the part. Thus, the AM process is inherently sterile. For example, to eliminate the COVID-19 virus (Abraham et al. 2020), the virus-containing objects must be heated for the following duration and temperature.

- 3 minutes at a temperature above 75 °C (160 °F) or
- 5 minutes at a temperature above 65 °C (149 °F) or
- 20 minutes at a temperature above 60 °C (140 °F).

Thus, the AM process is inherently sterile due to the elevated temperature during the AM printing process. On the other hand, TM typically requires a complicated assembly process, compared to AM, since TM typically cannot match the part complexity of AM.

Opportunity 2: Lightweight, Low-Cost Biological Surveillance System for Aerosols

In the JPEO portfolio, the researchers identified the Biological Warfare Agent (BWA) aerosol detection device as a potential AM application. To this end, AM will allow the tactical, lightweight/low-cost biological surveillance system that detects, collects, and identifies BWA aerosols. The researchers propose to use AM to create lightweight and low-cost outer frames for such systems. There are works of literature that deal with chemical sensing for BWA aerosols. The detection, collection, and identification of BWA aerosols can be an excellent topic for AM technology implementation if the stakeholders can create a device to achieve these goals using AM technology. To this end, fluorescence-based technology seems to be the most common technology. Thus, one possibility is to propose using AM technology to create a fluorescence-based detection system.

One example is the ultraviolet light detection and ranging (UV LIDAR) technology (Figure 12.10). The UV LIDAR system uses a laser to detect the presents of the BWA aerosols and gives early warning. The UV LIDAR technology is helpful since it can be used in the field, where soldiers can scan the area and detect the presence of BWA aerosols (Figure 12.11). Suppose the stakeholders can produce any UV LIDAR systems using AM technology. In that case, the stakeholders may be able to create a low-cost detection device with a shorter lead time. Another possibility for the BWA application is to create an air-sampling collection unit (Figure 12.12) using AM technology, as some researchers published their air-sampling unit design. Then, using the established detection technology (e.g., UV LIDAR stated above) to detect the BWA.

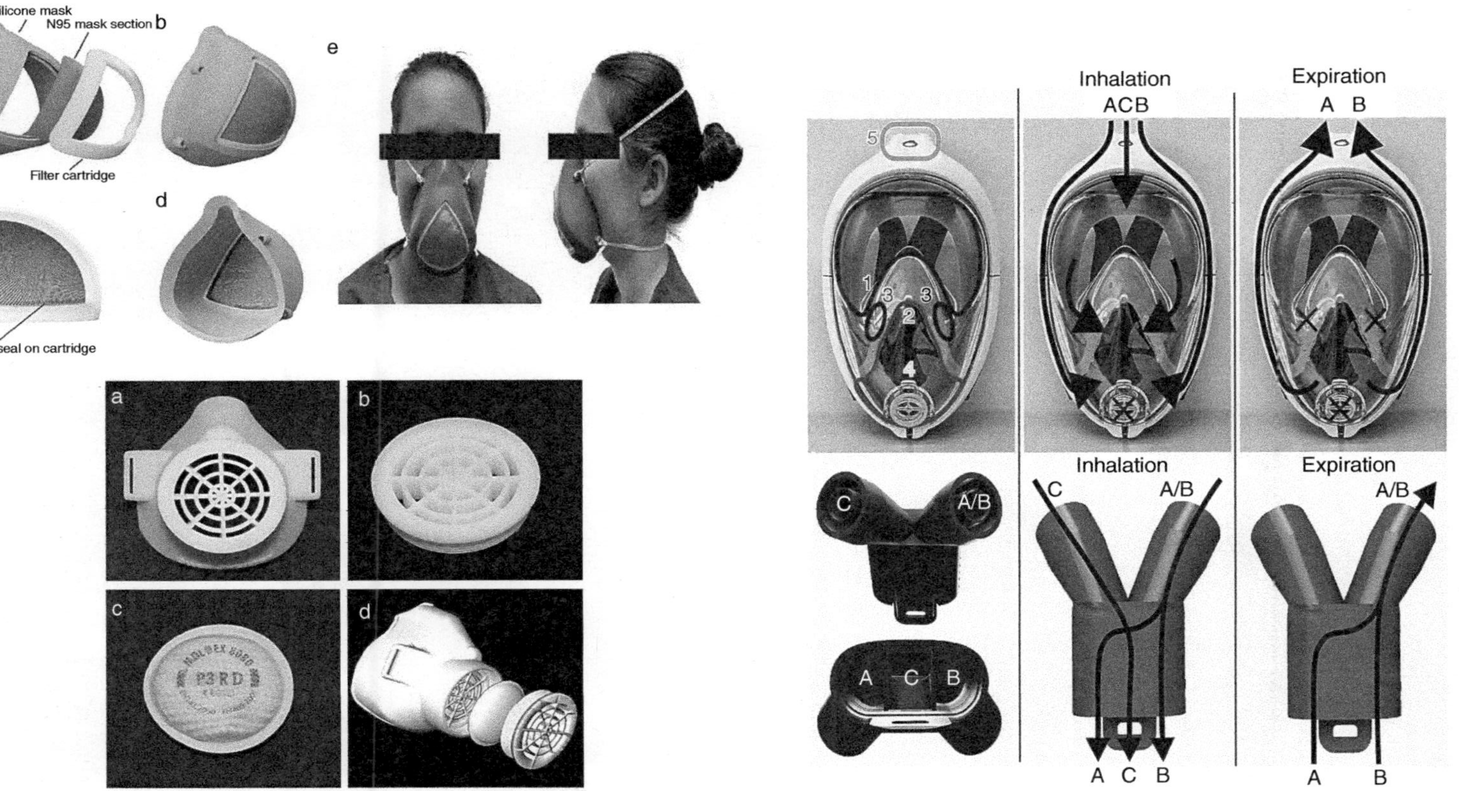

Figure 12.8 AM Masks. Top-left: AM mask CAD rendering and the use on the model. Right: inhale and exhale 3D-printed valve design. Bottom-left: AM mask pictures and CAD rendering. *Source:* Guvener et al. (2021) / Springer Nature.

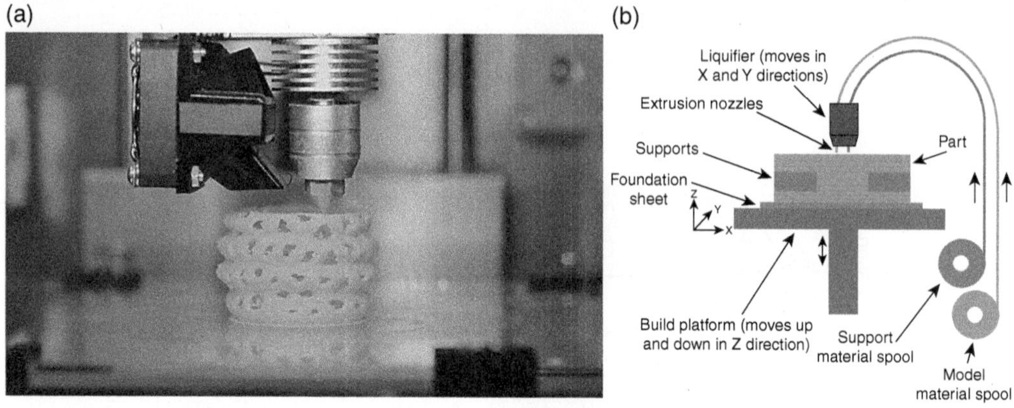

Figure 12.9 (a) FDM picture and (b) FDM schematics (The Complete Guide to Fused Deposition Modeling (FDM) in 3D Printing – 3Dnatives 2022); (Gebisa and Lemu 2018; Benesch and Redifer 2020; Abraham et al. 2020).

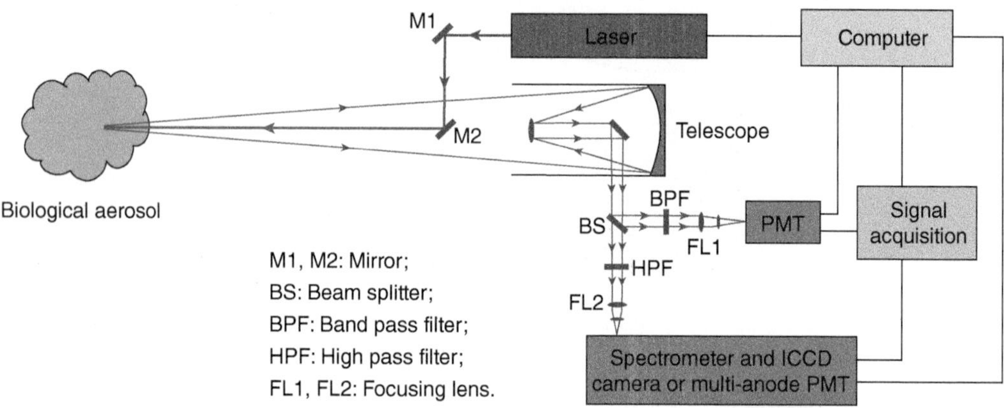

Figure 12.10 Schematics of UV LIDAR system. *Source:* Adapted from Li et al. (2019).

Figure 12.11 Example of LIF LIDAR system or detecting biological aerosols. (a) US JBSDS, (b) Canada SINBAHD, (c) Norway LIDAR, (d) UK LIDAR, and (e) Germany LIDAR systems. *Source:* Li et al. (2019)/Elsevier.

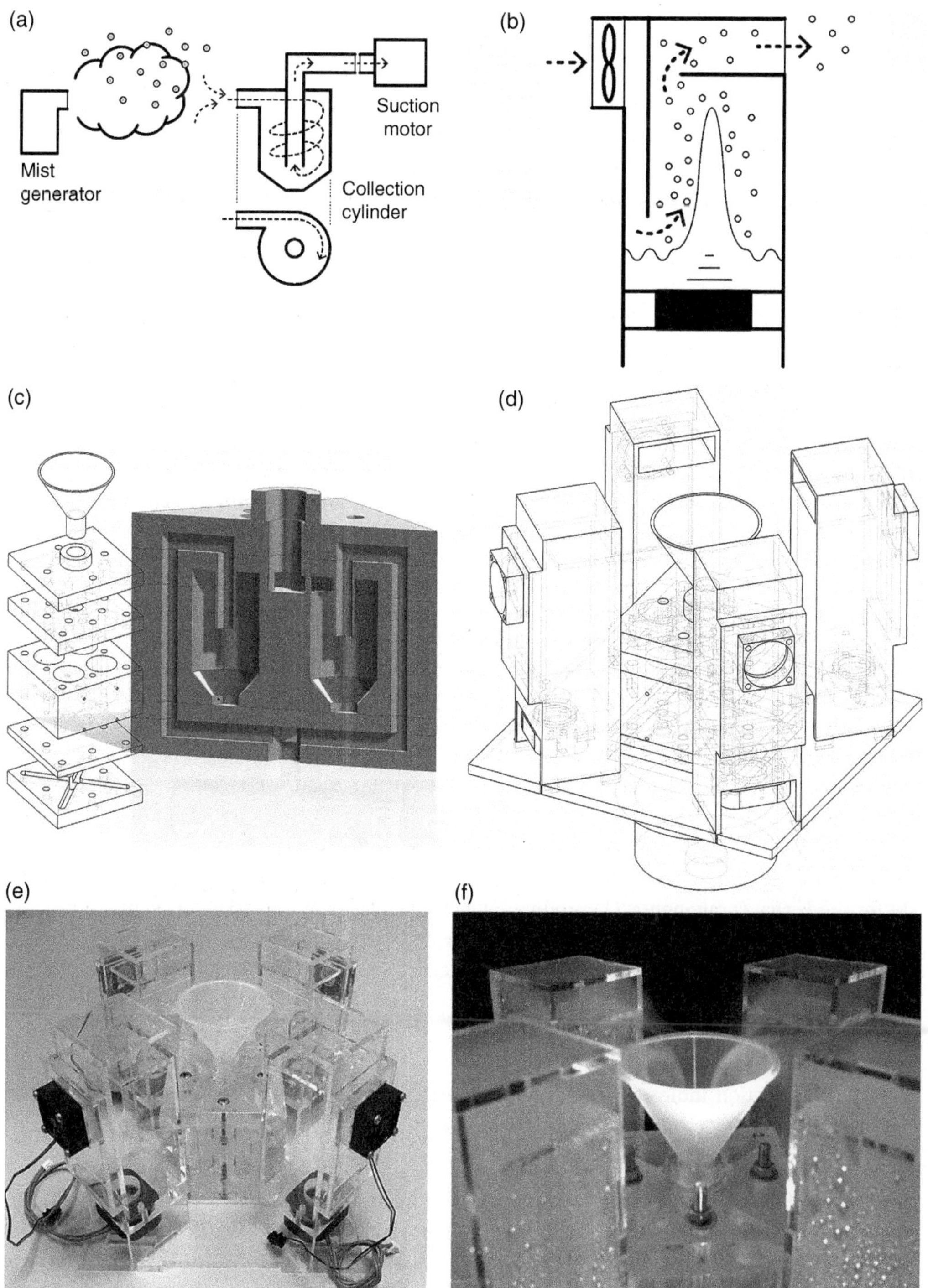

Figure 12.12 Air sampling unit. (a) air sampling scheme, (b) mist generation scheme, (c) collection container CAD rendering, (d) collection container and mist generator CAD rendering, (e) air-sampling unit, and (f) mist generation and air suction. *Source:* Saito et al. (2018) / Springer Nature / CC BY 4.0.

Opportunity 3: Carriers/Holders for Man-Portable Systems and Wearable Electronics/Sensors

In the JPEO portfolio, the researchers identified wearable electronics/sensors and carriers/holders for man-portable systems. This application uses AM to create ergonomic carriers/holders and wearable electronics/sensors for man-portable systems.

A few years ago, the Naval Surface Warfare Center – Crane, Division (NSWC-CR) created a design challenge to improve the safety of man-portable lithium-ion batteries (Lithium Battery Man Portable Hazard Containment Challenge n.d.). According to the NSWC document, individual soldiers are envisioned to carry up to 12.5 lb (5.7 kg) of battery equipment out of 50 lb (22.7 kg) of total "man-portable" weight.

Soldiers must carry large lithium-ion batteries in addition to other electronics (e.g., sensors and radio). Based on the contents of the NSWC-CR document, the batteries account for 25% of the total transported individual weight. This means that it is critical for the design of the man-potable equipment to be as ergonomically friendly as possible. To this end, ergonomically friendly carriers/holders and wearable electronics/sensors as a part of the "man-portable" equipment will significantly reduce the soldiers' stress. Therefore, the researchers propose creating AM-produced carriers/holders and wearable electronics/sensors. To this end, the researchers propose using the 3D scanners to scan each user's body shape, as discussed earlier, using the FARO arm, creating a digital model, and reverse-engineering the final product.

Opportunity 4: AM for Legacy Components and Repairs in the Field

In the JPEO portfolio, the researchers identified the spare part production in the field/remote locations as a potential AM application. Similarly, the researchers identified the spare part production for long-lead and legacy (i.e., discontinuous or obsolete) systems as a potential AM application. The common denominator for these applications is repair activity in the field.

The literature review revealed that the Defense Logistics Agency (DLA) had its legacy parts already investigated for AM applications (Parks et al. 2016). Also, there are recent pieces of literature that discuss the application of AM methodology in the legacy part (Blakey-Milner et al. 2021; Foshammer et al. 2022), in which the researchers pointed out the following advantages of using AM for the legacy components: (1) production can take place any time (i.e., on-demand production), (2) there is no need to worry about often-unpredictable maintenance of legacy production system, and (3) a manufacturer can eliminate the legacy stock from the Wearhouse.

What Can DE Do for JPEO?

The following section indicates further consideration of AM implementation at the enterprise level. More specifically, what DE can do for the sponsor (i.e., JPEO).

- **Managing product data**
  From the AM viewpoint, data management is the key to success in AM. DE allows the sponsor to manage data more efficiently. The most common product data format for AM is the STL format, which could be produced from various CAD systems (e.g., Dassault SOLIDWORKS, Dassault CATIA, and Siemens NX). The stakeholder will want to implement a secure data repository to prevent the data leak since some of the 3D-printed products in the sponsor's portfolio may be sensitive to the defense system. In addition, the repository should be able to keep track of multiple design revision levels since product assembles require multiple components. That is, each assembly has multiple components as its components and each component carries its own

revision levels. In a perfect world, everything fits; there are no assembly issues. However, in the worst case (i.e., when CAD design with incorrect revision levels is used), product assembly may become impossible due to the difference in the surface geometry, like a mating surface and tolerance.

- **Using a product lifecycle management (PLM) tool**
  DE allows the sponsor to perform PLM better. Thus, the stakeholder may wish to use a DE product lifecycle management (PLM) tool to keep correct revision of product data throughout the life cycle, starting from conceptualization, development, prototype, launch, manufacturing, service, and legacy support. A component assembly is almost always required in a product, whether AM or TM is used to create components. To this end, having fewer components using AM rather than TM will help mitigate the issue.
- **Managing personal data**
  DE allows the sponsor to manage personal data. To this end, as the sponsor starts managing data for AM, the sponsor will realize that some data are personal. For instance, scanned facial data for the ergonomic face shield components and respiratory/ocular protection results in scanned face data. To this end, the stakeholder may wish to establish the guideline to balance the following two competing requirements: (1) Ease of data access: The data must be protected with stringent control. (2) Data protection for end users: The data must be available for AM.
- **Intellectual property (IP)**
  DE allows the sponsor to manage IPs better. However, managing IP is sensitive when dealing with 3D scanners and 3D printers. That is, as the information (1) moves from physical form to digital form (i.e., reverse engineering using 3D scanners) and (2) returns from digital form to physical form (i.e., AM part printing using a 3D printer), the question of IP is not only related to the protection of IP but also the legality of who actually owns the data. For instance, when someone scans your body to design and create a lightweight wearable electronics or ergonomically friendly carrier for man-portable, what the design engineer uses during the process is to create a product using the human body as a specimen. This process requires the end-users to agree to release the 3D scanned data of their body. In addition, this process may require approval from the Institutional Review Boards (IRBs) as described by the U.S. Food & Drug Administration (Institutional Review Boards [IRBs] and Protection of Human Subjects in Clinical Trials | FDA 2022). Another example is the simple protection of IP of industrial goods. If a product in the sponsor's portfolio is scanned, printed, and distributed without proper authorization, there will be IP legal issues. Thus, the stakeholder may need to review the process with the IP legal office when considering the possibility of AM part implementation to replace the TM parts.
- **Addressing AM in different product development stages**
  DE allows the sponsor to address AM in different product development stages. The degree of AM implementation depends on the product maturity stage. Abstractly, the product maturity stages can be classified into the following four stages: introduction (i.e., proof-of-concept/prototype), growth (i.e., pre-production), maturity (i.e., mass production), and decline (i.e., legacy service). Some examples of this concept are shown below.
  - **Introduction:** Proof-of-concept/prototype
    Example: AM could be used to create a product in question to see if the brainstormed concept from the design engineering session works.
  - **Growth:** Pre-production (Small-quantity production)
    Example: AM could be used to create pre-production tools and fixtures. Also, AM could be used to develop the pre-production parts if the pre-production quantity is small, for instance, 1000 or less

  - **Maturity:** Mass production (Large-quantity production)
    Example: AM could be used to create production tools and fixtures, assuming that the surface of the AM is hard enough and withstand the wear and tear that go with the more significant production requirement. Also, AM could be used to create the production parts if the production quantity is small, for instance, 1000 or less
  - **Decline:** Legacy service
    Example: AM could be used to create service parts that are either difficult to obtain or no longer available from the manufacturer.
- **Implementing AM for various branches of DoD**
  DE allows the sponsor to implement AM in its various DoD branches (e.g., Army, Navy, Marine Corps, Air Force, Space Force, and Coast Guard). The ease of AM implementation may depend on the DoD branches. For instance, (1) a specific branch of DoD may have a very stringent policy for the product requirement so that the experimentally-made AM product may be accepted at one DoD branch but not other DoD branches; (2) a specific branch of DoD may have a group culture that does not view the new AM technology insertion favorably; and (3) a specific branch of DoD is willing to try the new AM technology. In addition, even within the same DoD branch, particular groups may be willing to accept (and pay for) the AM technology, whereas other groups in the same DoD branch may not want to do that. For instance, within the same DoD branch, special operation groups may be willing to spend time and money developing the new AM parts, whereas non-special groups may not be willing to do so.
- **Raise awareness to make a better product**
  DE allows the DoD leadership team to deal with the implementation of AM (or other new advanced manufacturing technologies) since DE raises awareness of what is coming next throughout the organization ahead of time. However, while raising awareness is essential to making changes for the better, the stakeholders may receive resistance from the people affected by the change. This may be due to human nature to resist change (people do not want to change), or this may be due to some logical reasoning. Regardless of the reason for the change, it will be crucial for the stakeholder to receive consensus from the affected individuals when making changes. For instance, if JPEO decides to work with the Army, JPEO will need to receive strong support from the Assistant Secretary of the Army – Acquisition, Logistics, & Technology (ASA(ALT)) leadership team. In this way, the AM implementation can be done in a top-down fashion.

What Does JPEO Need to Do to Prepare for DE in AM?

Through the work of SERC with JPEO, the researchers provide bullet-pointed recommendations on what the stakeholders need to do to prepare for DE in AM implementation. Since AM implementation at the enterprise-level portfolio is large and complex, DE requires the following considerations:

- The stakeholders must identify the most useful comparisons for their portfolio's acquisition decision.
- AM vs. TM
- AM vs. TM vs. hybrid AM/TM
- If the stakeholder is to pursue the AM/TM hybrid, the stakeholder may wish to clarify/define the term "AM/TM hybrid." For instance, the following questions may help.
- Is the stakeholder interested in AM and TM components used on the same part? In other words, is the stakeholder defining the product assembled with AM and TM as the "AM/TM hybrid"?

- Is the stakeholder interested in analyzing manufacturing networks that use both AM and TM? In other words, is the stakeholder defining the manufacturing networks that use both AM and TM as the "AM/TM hybrid"?
- The stakeholder may wish to identify low-production components as potential AM candidates.
- The stakeholder may want to understand what AM technology insertion in the current TM lineup makes the most sense.
- Cost is one of the most significant factors in procurement decisions.
- Does the stakeholder feel the same about the cost being the most significant factor? In other words, will the decision be strongly motivated by cost savings?
- Will the stakeholders be more motivated by agility against new threats and/or capability to respond to threats/robustness of response (than the cost savings)?
- If the stakeholders are interested in the operational agility concerning the procurement, how could they include the cost of missed opportunity?
- Can the stakeholders convert the missed opportunity (i.e., delayed mission) into a cost figure? It is often difficult to simultaneously deal with time and cost to drive the decision-making process. On the other hand, it is easy to deal with the single unit of "cost" to drive the decision-making process.
- The stakeholder may wish to investigate how AM can be used in their portfolio instead of TM (i.e., replacement of the TM portfolio by AM).
- The stakeholder may wish to investigate how AM can be used in their portfolio in addition to TM (i.e., the addition of AM to the TM portfolio).
- The stakeholder may wish to investigate the scenario in which the AM/TM hybrid can provide a faster, more effective, and long-lasting response to new threats rather than AM or TM alone.
- The stakeholder may wish to know the necessity of post-processing and unavoidable issues on the AM part surface quality (i.e., surface roughness)
- Generally, AM results in rougher surface texture compared to TM since the parts are printed layer-by-layer.
- TM and hybrid AM/TM may produce a product with smoother surfaces compared to AM alone.
- Rough surfaces are not desirable when dealing with mating surfaces, especially in the case of a hermetic seal, thereby requiring post-processing (i.e., machining, which is a part of TM).
- Rough surfaces are not an issue when dealing with stand-alone parts without assembly.
- The stakeholder may wish to confirm that the remote AM site has post-processing capabilities before implementing AM technologies.

Other vital points that the sponsor will need to pay attention to are data-related issues (i.e., the intellectual property, security of data, and definition of the owner of the models). These are sensitive topics, so decisions will need to be made case-by-case. For instance, the stakeholders need to carefully study the trade-offs of data digitalization to understand the ease of data access vs. the security of the data storage, especially when personal data (e.g., 3D scan of a face to create face masks) is involved.

Lessons Learned: How to Evaluate Digital Transformation

Digital transformation encompasses a wide range of initiatives at the enterprise level within the Department of Defense and within specific areas, such as the Joint Program Executive Office for Additive Manufacturing and Sustainability. This transformation involves incorporating digital tools and methods in mission engineering and managing the engineering life cycle for product development. One key aspect of this is the use of model-based systems engineering (MBSE), which shifts the traditional engineering approach towards utilizing models as the basis for all decisions.

Efforts to transition from document-based mission design and system engineering practices to digital-based methods have yielded promising results. One such initiative is the U.S. Navy Submarine Warfare Federated Tactical Systems (SWFTS) program, which has undergone a digital transformation over the past decade by embracing model-based systems approaches. This program has published a report that quantitatively measures the benefits of this transformation, including positive return on investment, increased efficiency compared to legacy systems, reduced efforts, and earlier identification of defects. In particular, the report found that using an MBSE approach resulted in a 9% decrease in overall problems, an 18% decrease in systems engineering hours, an 18% increase in early-stage problem detection, and a 37% decrease in reported product problems in the field (Rogers and Mitchell 2021).

The use and advancement of digital methods in mission design is an ongoing effort across various industries. One example is a process framework created by B. Williams, which links the process for Research and Engineering (OOSE) in the Office of the Secretary of Defense mission engineering process (R&D OSD ME) to an Object-Oriented Systems Engineering approach for designing and evaluating a space domain awareness mission. This framework aims to create a complete digital architecture that can be used as a reference for similar missions in the future (Williams 2022). Similarly, S. Kelly developed a digital reference architecture for the rapid design of CubeSats, utilizing previous work and experiences to establish a baseline for future mission design (Kelly 2021). There are also efforts to apply digital methodologies for specific industries, such as in Van Bossuyt et al., which focuses on using a model-based approach to design and plan offshore wind farms (Van Bossuyt et al. 2022).

From the numerous MBSE and DE applications for missions, previous projects within SERC (SERC-2021-TR-024) reported the top eight benefits of MBSE, as identified from literature surveys. These benefits include higher level support for automation, early verification and validation, reusability, increased traceability, strengthened testing, improved accessibility of information (Authoritative Source of Truth), higher level support for integration, and the ability to view models from multiple perspectives. However, despite the obvious increase in the adoption of MBSE for mission engineering, the documented benefits and the analysis constructed from measuring its benefits remain in the early stages. Many of the benefits are primarily perceived and observed, rather than quantitatively measured, due in part to the limited number of programs currently implementing a DE/MBSE process and measurement program (McDermott and Salado 2021).

The willingness to implement digital transformation in the industry may be hindered by the lack of standardized practices and tools for DE and MBSE. In addition to the ambiguity of digital mission engineering and lifecycle engineering processes, current standards lack maturity and do not sufficiently tackle integrating digital tools and programming. This complicates digital transformation, disrupting communication, and information flow instead of streamlining workflows as intended. For instance, current documentation of mission engineering is left too broad for the intention of a wide application range; however, this leaves the process with significant gaps for users to fill in and configure. There is little evidence in the literature of a well-developed DE process that is used as a standard by researchers and mission architects.

Bekdache and DeLaurentis (2023) presented an early development framework that attempts to digitally formalize the mission engineering process by leveraging DE and MBSE methods and tools with the aim of streamlining and bridging the existing gaps. The framework is outlined in Figure 12.13 and brings together elements from various sources such as the OUSD(R&E) Mission Engineering Guide (OUSD n.d.), SoS modeling and analysis defined by DeLaurentis et al. (2022), and the Space

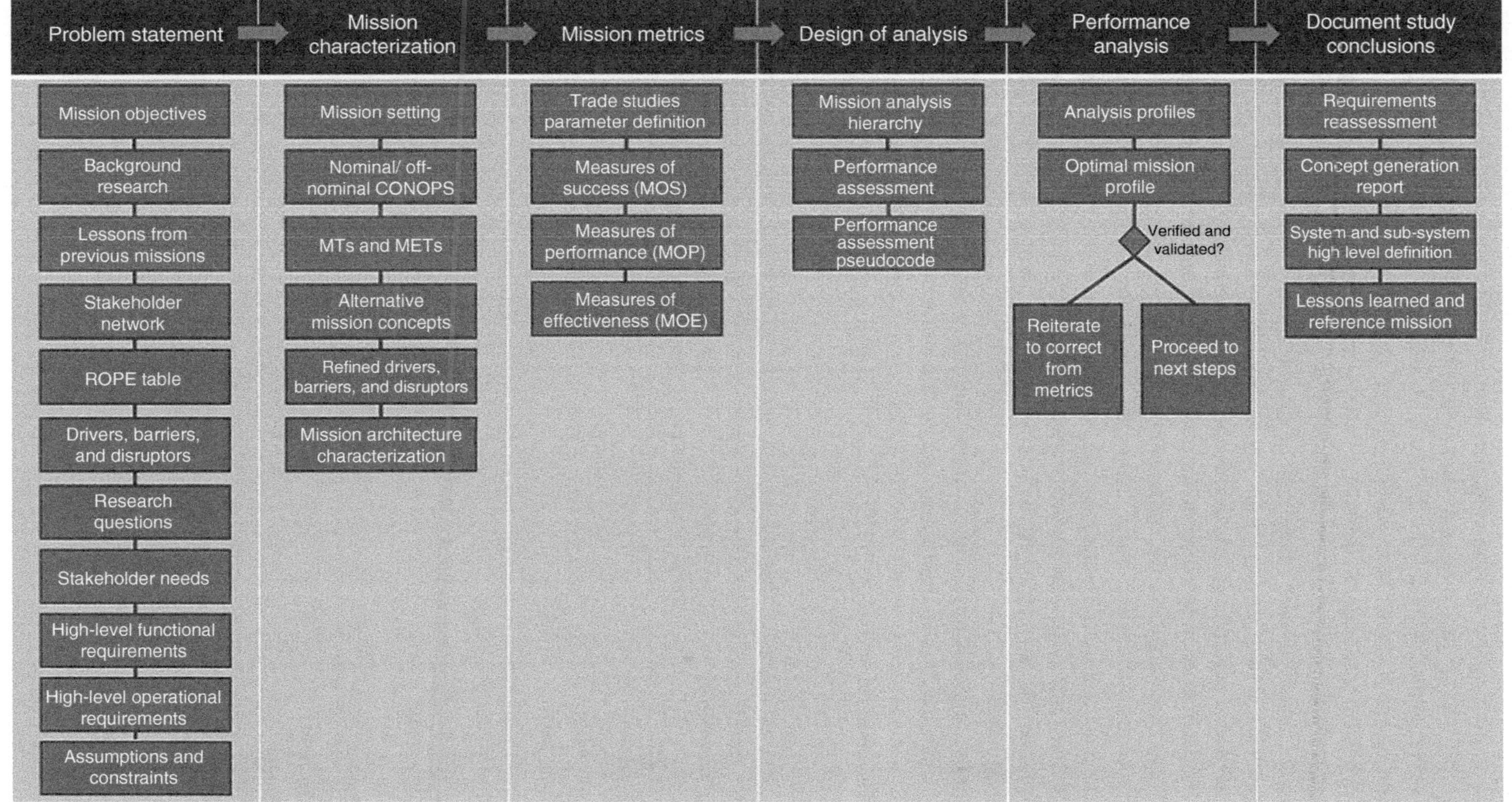

Figure 12.13 Early development of a Digital Mission Engineering Framework.

Mission Analysis and Design (SMAD) methodology (Larson and Wertz 1999). It details the sequential activities that guide the mission engineers from start to end using MBSE artifacts, such as SysML diagrams, that are linked to each other at nearly every step of the process. This digital mission engineering methodology covers Pre-Phase A and the high-level mission development of Phase A based on the NASA Project Lifecycle (NASA Systems Engineering Handbook 2020). The framework is divided into various abstraction levels, demonstrating the layers of information produced and pinpointing relevant data viewpoints for better stakeholder collaboration and communication. Some perceived and referenced qualitative metrics include increased traceability, reusability, higher level support, and improved communication and information sharing. Further studies are needed to advance the framework and develop the most suitable digital tools.

Previous SERC projects outline a set of key qualitative enterprise metrics for digital transformation that fall into five categories: quality, knowledge transfer, velocity/agility, user experience, and adoption. For each metric within these categories, it provides example inputs, outputs, and outcomes (McDermott et al. 2020). However, as digital transformation can be a prolonged process that can take months or even years, the metrics might be beneficial to use at specific times throughout the transformation. Table 12.7 suggests a classification for a list of metrics that fall under the five categories mentioned.

These metrics are mostly qualitative in nature but can be tied to some quantitative metrics that may provide useful analysis on the impact of digital transformation. These quantitative metrics may include mission construction time, total costs, efficiency, and error detection, and may be most useful during the developing and optimizing stages of digital transformation. Within the continued

Table 12.7 Suggested categorization and use of qualitative metrics based on digital transformation phase.

Metrics/Transformation period	Early-stage transformation period	Developing-stage transformation period	Optimization-stage transformation period
Increased traceability	X	X	X
Increased reusability	X	X	X
Higher-level support	X	X	X
Better communication and information sharing	X	X	X
Improved system quality		X	X
Improved consistency		X	X
Reduced time		X	X
Reduced errors and defects		X	X
Support integration		X	X
Improved system understanding		X	X
Increased willingness to adopt tool		X	X
Reduce cost			X
Manage complexity			X
Automation			X
Training and Incentives			X

search for a holistic measurement framework, more SERC projects (McDermott and Salado 2021) provide a good starting point for identifying the measurement and quantifying benefits.

For an organization to fully embrace and benefit from digital transformation, it is important to optimize various factors such as the lexicon used, digital tools and technologies, programming and modeling languages, configuration, management, user experience, and user interface. Finding the optimal combination of these elements through a portfolio discovery process can help foster a willingness to change within an organization, breaking away from traditional, manual, document-centric workflows. Within this portfolio, the investigation of human-centered design for digital environments is an important design aspect for the success of new processes and technologies. While focusing on digital engineering tools and metrics is crucial, it is also necessary to consider the human element in the long run. To ensure the success of enterprise-wide transformations, it is important to consider ways to incentivize these changes and take the organization into a new digital realm.

The Path Forward

Many times, in the field of systems engineering (as in various other fields), the reality is very different from what is learned in books and in theory. The value of digital engineering is widely acknowledged; thus, there is a large push for digital transformation. However, numerous issues can emerge, especially when the digital transformation process is applied at the enterprise level. As mentioned in Section 2, some of these issues can arise in any organization. However, the work of SERC with JPEO-CBRND also pointed out many additional difficulties that are specific to the stakeholders and the work at JPEO-CBRND. The research work presented in this chapter provided a few lessons learned that could guide future work. The most important and universal guidelines for proper DT, which have been confirmed in SERC's effort to give initial guidance to JPEO-CBRND, are the following:

- Learn the customer needs in detail
- Be sure that each stakeholder knows the DT process and is on board with the proposed approach, and at the same time, be sure to address each stakeholder's specific needs.
- Plan ahead and execute step by step, beginning with small changes, even if management will push for "quick and dirty" solutions.
- Be ready to adapt the theory to reality and address specific needs that might change throughout the DT process.
- Use literature review, state-of-the-art, and recently proposed metrics and approaches to provide guidelines to the DT process and evaluate its correct implementation.

Acknowledgments

This chapter is based upon research supported by the US Department of Defense through the Systems Engineering Research Center (SERC) under Contracts HQ003413D0004 and W15QKN18D0040. SERC is a federally funded University Affiliated Research Center managed by Stevens Institute of Technology. Any opinions, findings, and conclusions or recommendations expressed in this material are those of the authors and do not necessarily reflect the views of the US Department of Defense. The collaborators who supported this research are listed on the Handbook URL.

References

Abraham, J.P., Brian D. Plourde, and Lijing Cheng (2020). “Using heat to kill SARS-CoV-2.” *Reviews in Medical Virology*, 30(5), https://doi.org/10.1002/RMV.2115.

Bekdache, D. and Delaurentis, D.A. (2023). Towards developing a digital-enabled mission engineering framework. *2023 Conference on Systems Engineering Research (CSER),* Hoboken, NJ (March 2023).

Benesch, B. and Redifer, S. (2020). COVID-19 viral mitigation and decontamination. *DSIAC & HDIAC Technical Inquiry (TI)* response report no. DSIAC-2020-0324, 2020.

Blakey-Milner, B., P. Gradl, G. Snedden, et al. (2021). “Metal additive manufacturing in aerospace: a review.” *Materials & Design*, 209, 110008, https://doi.org/10.1016/J.MATDES.2021.110008.

Cui, W., Yiran Y., Lei D, et al. (2021). “Additive manufacturing-enabled supply chain: modeling and case studies on local, integrated production-inventory-transportation structure.” *Additive Manufacturing*, 48,102471, https://doi.org/10.1016/J.ADDMA.2021.102471.

Davendralingam, Navindran, & DeLaurentis, D. (2013). “A robust optimization framework to architecting system of systems.” *Procedia Computer Science*, 16, 255–64, https://doi.org/10.1016/J.PROCS.2013.01.027.

Davendralingam, Navindran, Daniel D, Zhemei F et al. (2014). “An analytic workbench perspective to evolution of system of systems architectures.” *Procedia Computer Science*, 28, 702–10, https://doi.org/10.1016/J.PROCS.2014.03.084.

DeLaurentis, D.A., Moolchandani, K., and Guariniello, C. (2022). *System of Systems Modeling and Analysis*. CRC Press.

FARO (n.d.). 3D Measurement, imaging & realization solutions. https://www.faro.com/en (accessed 21 August 2022).

Foshammer, J., Søberg, P.V., Helo, P., and Ituarte, I.F. (2022). Identification of aftermarket and legacy parts suitable for additive manufacturing: a knowledge management-based approach. *International Journal of Production Economics* 253: https://doi.org/10.1016/J.IJPE.2022.108573.

Gebisa, Aboma Wagari, & Hirpa G. Lemu. (2018). Investigating effects of fused-deposition modeling (FDM) processing parameters on flexural properties of ULTEM 9085 using designed experiment. *Materials*, 11(4), https://doi.org/10.3390/MA11040500.

Guariniello, C. and DeLaurentis, D. (2017). Supporting design via the system operational dependency analysis methodology. *Research in Engineering Design* 28 (1): 53–69. https://doi.org/10.1007/S00163-016-0229-0.

Guide to Manufacturing Processes for Plastics (n.d.). https://formlabs.com/blog/guide-to-manufacturing-processes-for-plastics/ (accessed 14 March 2022).

Guvener, O., Eyidogan, A., Oto, C. and Huri, P.Y. (2021). “Novel additive manufacturing applications for communicable disease prevention and control: focus on recent COVID-19 pandemic.” *Emergent Materials*, 4(1), 351–61, https://doi.org/10.1007/S42247-021-00172-Y/FIGURES/10.

Institutional Review Boards (IRBs) and Protection of Human Subjects in Clinical Trials | FDA (2022). https://www.fda.gov/about-fda/center-drug-evaluation-and-research-cder/institutional-review-boards-irbs-and-protection-human-subjects-clinical-trials (accessed 22 August 2022).

JPEO-CBRND (2023). https://www.jpeocbrnd.osd.mil (accessed 23 January 2023).

Kelly, S. (2021). A reference architecture for rapid CubeSat development. MS Thesis. Air Force Institute of Technology. March 2021.

Larson, W.J. and Wertz, J.R. (ed.) (1999). *Space Mission Analysis and Design*, 3e. Microcosm/Springer.

Li, Xin, Huang, S. and Sun, Z (2019). “Technology and equipment development in laser-induced fluorescence-based remote and field detection of biological aerosols.” *Journal of Biosafety and Biosecurity*, 1(2), 113–22, https://doi.org/10.1016/J.JOBB.2019.08.005.

Lithium Battery Man Portable Hazard Containment Challenge (n.d.). https://www.challenge.gov/challenge/lithium-battery-man-portable-containment/ (accessed 21 August 2022).

McDermott, T. and Salado, A. (2021). Application of Digital Engineering Measures. *Final Technical Report, SERC-2021-TR-024*, 2021.

McDermott Tom, E. Van Aken, N. Hutchison, M. Blackburn, M. Clifford, Y. Zhongyuan, N. Chen, A. Salado, and K. Henderson. (2020). Digital Engineering Metrics. *Technical Report SERC-2020-TR-002*.

NASA Systems Engineering Handbook (2020). https://www.nasa.gov/sites/default/files/atoms/files/nasa_systems_engineering_handbook_0.pdf (accessed 21 August 2022).

OUSD(R&E) (n.d.). Mission engineering guide. https://ac.cto.mil/mission-engineering/ (accessed 20 January 2023).

Parks, T.K., Kaplan, B.J., Pokorny, L.R., et al. (2016). Additive manufacturing: which DLA-managed legacy parts are potential AM candidates? LMI report (2016). https://apps.dtic.mil/sti/pdfs/AD1013934.pdf (accessed 21 August 2022).

Patel, Piyush, & Piyush Gohil. (2021). "Role of additive manufacturing in medical application COVID-19 scenario: India case study." *Journal of Manufacturing Systems*, 60, 811–22, https://doi.org/10.1016/J.JMSY.2020.11.006.

Rogers, E.B., & Mitchell, S.W. (2021). MBSE delivers significant return on investment in evolutionary development of complex SoS *Systems Engineering*, 24(6), 385–408, https://doi.org/10.1002/SYS.21592.

Saito, Masato, Uchida, N., Furutani, S et al. (2018). "Field-deployable rapid multiple biosensing system for detection of chemical and biological warfare agents." *Microsystems & Nanoengineering* 4(1), 1–11, https://doi.org/10.1038/micronano.2017.83.

Shi, Qian, Tsutsui, W., Bekdache, D. et al. (2022). A System-of-Systems (SoS) perspective on additive manufacturing decisions for space applications. *2022 17th Annual System of Systems Engineering Conference (SOSE)*, Rochester, NY (June 2022). IEEE, pp. 282–88, doi:https://doi.org/10.1109/SOSE55472.2022.9812665.

The Complete Guide to Fused Deposition Modeling (FDM) in 3D Printing - 3Dnatives (2022). https://www.3dnatives.com/en/fused-deposition-modeling100420174/ (accessed 21 August 2022).

Van Bossuyt, D.L., Hale, B., Arlitt, R.M. and Papakonstantinou, N. (2022). Multi-mission engineering with zero trust: a modeling methodology and application to contested offshore wind farms. *International Design Engineering Technical Conferences and Computers and Information in Engineering Conference*, St. Louis, MO (August 2022). American Society of Mechanical Engineers Digital Collection, https://doi.org/10.1115/DETC2022-90067.

Williams, B. (2022). Applied agile digital mission engineering for cislunar space domain awareness. MS Thesis. Air Force Institute of Technology, March 2022, https://scholar.afit.edu/etd/5446.

Biographical Sketches

Dr. Cesare Guariniello is a research scientist in the School of Aeronautics and Astronautics at Purdue University. He received his PhD in aeronautics and astronautics from Purdue University in 2016 and his master's degrees in astronautical Ensgineering and computer and automation engineering from the University of Rome "La Sapienza." Dr. Guariniello works as part of the Center for Integrated Systems in Aerospace led by Dr. DeLaurentis, and is currently engaged in projects funded by NASA, the DoD Systems Engineering Research Center (SERC), and the NSF. His main research interests include modeling and analysis of complex systems and SoS architectures – with particular focus on space mission architectures – aerospace technologies, and robotics. Dr. Guariniello is a senior member of IEEE and AIAA.

Dr. Waterloo Tsutsui is a senior research associate in the School of Aeronautics and Astronautics at Purdue University, IN. Before Purdue, Tsutsui practiced engineering in the automotive industry for more than ten years. Dr. Tsutsui's research interests are systems engineering, energy storage systems, multifunctional structures and materials design, and the scholarship of teaching and learning.

Dalia Bekdache is a Graduate Research Assistant in the School of Aeronautics of Astronautics at Purdue University. She is focused on aerospace systems engineering, mission engineering, and digital transformation research at the Center for Integrated Systems in Aerospace (CISA), with a minor concentration in Computational Science and Engineering. She works on the CONOPS and systems integration for urban air vehicles for NASA S2A2 project, and previously worked on Additive Manufacturing and Digital Engineering Strategy Development for the Acquisition Innovation Research Center (AIRC).

Dr. Daniel DeLaurentis is vice president for Discovery Park District (DPD) Institutes and professor of aeronautics and astronautics at Purdue University. His charge in research is via the Center for Integrated Systems in Aerospace (CISA), which he directs, working with faculty colleagues and students research problem formulation, modeling, design and system engineering methods for aerospace systems, and systems-of-systems. DeLaurentis also serves as chief scientist of the U.S. DoD's Systems Engineering Research Center (SERC) UARC, working to understand the systems engineering research needs of the defense community (primarily) and translate that to research programs that are then mapped to the nation's best researchers and students in the SERC network of 25 universities. He is fellow of the International Council on Systems Engineering (INCOSE) and the American Institute of Aeronautics and Astronautics.

Part III

Tradespace Analysis in a Digital Engineering Ecosystem – Context and Implications

Val Sitterle

Chapter 13

A Landscape of Trades: The Importance of Process, Ilities, and Practice

Valerie B. Sitterle[1] *and Gary Witus*[2*]

[1] *Electronic Systems Laboratory, Georgia Tech Research Institute, NW Atlanta, GA, USA*
[2] *College of Engineering, Wayne State University, Detroit, MI, USA*

Levels of Trades

> One-size fits all definitions . . . present hindrances to effective interdisciplinary collaboration.
> *(Boehm 2019).*

From 30 000 ft, the tradespace is about cost vs. military value. Then, it all gets complicated: which tradespace, where in the overall capability development process, which cost model in which to bin which costs, how to account for nonfunctional system qualities, etc. This chapter aims to frame the different trades that must be considered throughout the overall process of designing, selecting, and developing capabilities – especially oriented toward those for defense – and the relationships between them. Tradespaces emerge from the planning and decision trade-offs that are made across the stages of acquisition planning, programming, development, procurement, operations, and sustainment. This includes consideration of trades across different domains, levels of abstraction, and complex relationships between system-level design and operational efficacy by diverse decision-makers serving in their roles over the system acquisition and employment lifecycles.

The Defense Acquisition University (DAU) defines acquisition as "The conceptualization, initiation, design, development, test, contracting, production, deployment, integrated product support, modification, and disposal of weapons and other systems, supplies, or services (including construction) to satisfy DoD needs, intended for use in, or in support of, military missions" (Defense Acquisitions University 2023). It is through acquisition, namely the processes that support it, that the DoD acquires new capabilities necessary to meet National priorities and provide effective, affordable, and timely solutions to US Warfighters. The acquisition process is a management process that comprises three distinct systems, each of which operates independently with separate chains of command and yet together provide the overall decision support system for capability development: the Defense Acquisition System (DAS) for developing and procuring a capability, Joint Capabilities Integration and Development System (JCIDS) for identifying and validating user

* Retired.

Systems Engineering for the Digital Age: Practitioner Perspectives, First Edition. Edited by Dinesh Verma.

Companion website: www.wiley.com/go/verma/systemsengineering

requirements, and the Planning, Programming, Budgeting, and Execution (PPBE) System, for allocating resources and budgeting.

Trades occur within each distinct system – DAS, JCIDS, and PPBE – at different levels of abstraction and relevance throughout the entire process. At a high level, Smead defines the following (Smead 2015):

- A *trade* is an "attribute or characteristic (of a design, decision, etc.) with associated benefits and opportunity costs which may be exchanged in part or totality. Related term(s): parameters, input variables, objectives, constraints."
- A *tradespace* is consequently the "bounded area which considers the range of possible values (inherent or applied) for any number of attributes and characteristics, the relationships between them, and impacts on potential (design, decision, operational) outcomes."
- *Tradespace analysis* is then the "search of the bounded space to highlight the relationships between trades, their values (inherent or applied), and outcome objectives to inform decision makers. Related terms(s): multi-attribute, interdependent, decision makers."

Tradespaces are decision spaces. Trades across and within "decisions-that-need-to-be-made" to allocate current-year funding while planning for out-year funding are the foundations for practical and relevant tradespace understanding. Tradespace decisions are not (or at least should not be) a consequence of the theoretical foundations and models that underly their creation and analysis. In the broadest sense, the tradespace is about funding, contribution to military capability, and time. The remainder of this chapter will highlight the different levels of trades (i.e., decisions), challenges due to intangibles that impact effective and tractable evaluation, and the relation to real mission and operational context.

Defense Enterprise Levels of Concern

As part of the PPBE process, a Program Objective Memorandum (POM) is a "request that shows how a military department or program plans to allocate resources in response to and in line with the Defense Planning Guidance (DPG)" (Defense Acquisitions University 2023). The DPG is manifested through the combined inputs of the National Defense Strategy, the Quadrennial Defense Review which contains the NDS, the National Security Strategy (NSS), and program recommendations from the Chairman, Services, and Combatant Commands. The tradespace that emerges is considered in the context of current and projected military capability needs and costs associated with those capabilities.

To support these capability portfolio decisions, warfighting capability assessments are often conducted by their respective Service arms to capture current (i.e., baseline) and expected future warfighting capability. This latter category includes new or improved capabilities in the pipeline and expected to become available. The assessment also includes the notion of capacity and its impact on capability, whether at a system, platform, or munition level. These assessments are traditionally scoped to capture relative capability against the most stressing threat in each mission area. They are updated each POM cycle to inform decision-makers of potential impacts or improvements to warfighting capability based on funding decisions made during the POM process. Given the fluidity of the budgeting process, requests are often made for reassessments based on funding or timeline constraints, or desires to accelerate programs, and reassessments can often take days to weeks to complete.

Methods, processes, and tools (MPTs) to support these types of decisions, at this level, often draw from data and information supplied by focused system-level or mission context studies. Even so, assessments require a special lens and definition of the types of problems the process and its MPTs seek to solve. The "Applying Systems Engineering to Enterprise Systems and Portfolio Management" section of this book discusses many of the approaches to these types of problems, while a subsequent chapter in this section, "Architecting a Tradespace Analysis Framework in a Digital Engineering Environment," describes considerations for architecting the "right" framework for analyzing these and other levels of trades.

System Design and Development Analytics

> In recent years the military services have been faced by order of magnitude increases on the severity of the operational requirements imposed by their missions and in the complexity of the equipment and operations necessary to meet these requirements... Though the need for a systems approach has been recognized... there has been a lack of tools of systems analysis for linking together men and mechanisms into an integrated analytical framework. System models as analytical tools... have always played a vital role in analytical work and a sign of maturity in systems analysis will be the development of <integration across the various> physical models, abstract, and symbolic models. Models are used as the basis of evaluation tools to answer... which alternative best meets the performance requirements of the system within the imposed constraints and with the given inputs.
>
> *(Shapero and Bates 1959)*

Shapero and Bates wrote this in 1959 and yet it could read as a systems engineering abstract written today. We still face these same problems. Our technology and computational capabilities we can bring to bear on a problem space have advanced, and yet the heart of our challenge remains the same.

Tradespace exploration and analysis enables decision-makers to discover and understand relationships across the myriad of capabilities, gaps, and potential compromises that facilitate the achievement of metric objectives. Creating a tradespace for such exploration, however, requires the development and maturation of executable and scalable analytical constructs. These constructs must be implementable within the context of a larger workflow, and in tandem with each other, to guide trade space exploration and evaluate disparate concepts.

Often, these activities focus on the system-level development process, especially at its early stages (e.g., pre-Milestone A). Moreover, the analysis is frequently supported through an associated set of MPTs that focus on either optimization or parametric sweep approaches. Set-based design (SBD), reviewed by another subsequent chapter in this section, "Set-Based Design: Foundations for Practice," aims to robustly move toward optimizing a tradespace through its approach to identifying a set of potential best but different solutions that balance competing objectives.

There are a few variants to these approaches, typically differentiated through how they consider a system. Methods often "define" a system as a collection of distinct components, with system design realized through feasible combinations of various technology elements. These methods are used due to the ease of maintaining a feasible space alongside well-defined or at least easy to specify constraints and resulting attributes that are readily capturable and can be evaluated against established cost models. More recent methods include the use of genetic algorithms to discover advantageous combinations given different, competing constraints. This approach is applied to portfolio optimization as well, at levels higher than individual system analysis. The "Portfolio Optimization and Management for Systems of Systems" chapter in this volume discusses that application and a variety of methods that may be used.

Higher-level modeling and simulation (M&S) components used in early-stage evaluations may lead to refinement of which concept architectures (i.e., designs) are the most promising, or the analysis may seek instead to identify regions of design space that will yield the most operationally (or sustainably, etc.) robust systems. Some design concepts are sufficiently different that they will necessitate being represented in a computational environment as completely different model architectures or types. It may not be possible to analyze each distinct design type in an equivalent fashion. The burden to produce the information required for these analyses across entirely different system types will need fast and efficient computational methods and tools as well as require considerable planning and thought to ensure the time and cost associated with creating the data is acceptable.

The phase of development will also drive what is needed in terms of what needs to be analyzed and what information is required to support the decision needs at that point in time. Given its place in the overall DoD development process, for example, Milestone A evaluates various potential materiel solutions (i.e., technology-based systems) across a wide range of technological maturity and "best guess" performance expectations. This information is typically derived from models and simulations with varying degrees of operational fidelity for a hypothetically integrated system. The wide range of potential technologies considered may not yet have been integrated into a realized system or design, and this phase consequently focuses on identifying areas of system integration and technological risk but not on their mitigation. Approved force capability gaps presented in an Initial Capabilities Document (ICD) drive identification of acceptable capability attributes to mitigate those gaps and the associated threshold values to inform the initial drafting of the Capability Development Document (CDD). Costs are often best estimations or rough order of magnitude approximations based on prior developments. Accordingly, the Milestone A focus is on potential technological solution performance in an operational context, balanced against projected costs, specifically to address the capability gaps previously identified.

In contrast, the technological maturation and risk reduction phase leading up to Milestone B focuses on defined competing systems including commercial prototypes and/or commercial or government off-the-shelf (COTS/GOTS) systems. System integration and technological risks are more rigorously identified. As refinement of candidate solutions continues, engineering production risks are identified as well. Threshold values of the key performance parameters (KPPs) and key system attributes (KSAs) are more rigorously evaluated and matured to inform the development of a final CDD that will guide system development in subsequent phases. Cost estimates are based on engineering level data, industry proposals, and other program costing data that result in more refined projections. Milestone B trades examine operational effectiveness of specific candidate solutions (i.e., those for which there is at least a formal, integrated design) against more tailored cost projections and increasingly realistic operational requirements.

Mission and Operational Effectiveness

Many system development projects strongly focus on creating and evaluating a system's initial operational capability (IOC). By doing so, they often miss the importance of capacity in realizing an overall capability, the relation to and dependency on other systems with divergent and complementary capabilities required for success, and opportunities to make the system more cost-effectively maintainable.

In the mission and operational context, tradespace analysis involves making decisions across competing and conflicting technologies and system capabilities for warfighting effects. Three primary levels of consideration are important to distinguish strategic, operational, and tactical. Conventionally, an operation is defined as (1) a sequence of tactical actions with a common purpose or unifying theme, or (2) a military action or the carrying out of a strategic, operational, tactical, service, training, or administrative military mission; the tactical level is the level at which battles and engagements are planned and executed to achieve military objectives assigned to tactical units or task forces. A mission is defined as the task, together with the purpose, that clearly indicates the action to be taken and the reason behind it (Department of Defense 2016). An operation, therefore, supports a mission. However, doctrine states that "the strategic, operational, or tactical purpose of employment depends on the nature of the objective, mission, or task." In other words, the purpose defines the actual level of warfare and there is no clear delineation of these boundaries that would clarify or cement their use as consistent units of analysis (Harvey 2021). What is important to understand, however, is that "meeting mission requirements is the critical driver that determines how the DoD establishes operational trade-offs (e.g., size, weight, power; targeting ability; resources, etc.)" (Miller 2018).

Often when evaluating potential operational effectiveness, there is friction between approaches that incorporate significant physics and those employing high levels of abstraction such as from discrete-event simulation or agent-based methods. In some cases (e.g., electronic warfare, asynchronous communications, etc.), the collective operational performance simply cannot be well understood without underlying physics. In contrast, high-level M&S representing the operational view, especially in the form of system of systems (SoS) problems, typically contains very little detailed physics. Instead, these perspectives are rendered using agent-based modeling approaches, discrete event simulations, and event models or tables.

There are multiple, large-scale frameworks available for simulations of tactics and operations that implement higher fidelity representations. Vignettes for simulation and analysis typically take significant time to set up and modify, and entity behaviors are often programmed at such a high level of fidelity as to not be reusable in other scenarios. Even then, despite outputs that are stochastic, the results are for a single vignette and cannot scale to the sheer number of variant operational possibilities. To realize entirely different mission profiles, concepts of operations, operational environments, etc. requires that a subject matter expert create another simulation instance. A core systems engineering challenge is consequently to balance the level of fidelity required for analysis with the tradespace of vignettes necessary to make an informed decision and the time and cost to set-up, execute, and analyze the outputs.

The modern terminology for these problem types is coined "mission engineering" (ME), defined as "the deliberate planning, analyzing, organizing, and integrating of current and emerging operational and system capabilities to achieve desired warfighting mission effects. Mission engineering is an analytical and data-driven approach to decompose and analyze the constituent parts of a mission in order to identify measurable trade-offs and draw conclusions." (Deputy CTO for Mission Capabilities 2023) The key realization that serves as the basis of mission engineering analyses that was missing from many early SoS evaluations is the concept of threat.

A threat is necessary as a benchmark for evaluating effectiveness of a capability in the military sense; performance does not happen in a system-intrinsic vacuum when the entire purpose is to achieve an effect *against* something else. At its core, a threat is anything that may compromise intended functional performance. A system may be operating perfectly, but fail to achieve a desired capability due to some external, interfering influence. For example, in electronic warfare, a system's signal intended to reach a given destination may suffer from environmental interference, interference from other Blue systems, or from adversary jamming.

A threat is therefore a multidimensional concept, but the reality of using these concepts analytically is not so straightforward. Consider what description a systems engineer might receive when asking about adversarial threats a system could encounter: "Possible threats to the family of systems range from small arms (e.g., looters/rioters) to surface-to-air missiles (SAM) including man portable air defenses (e.g., terrorists), fixed-wing and rotary-wing aircraft, directed energy weapons (to include lasers and radio frequency weapons), nuclear, biological, and chemical (NBC) weapons, and information warfare" (Cellucci 2008).

While at first glance, it would seem an evaluation that comprehensive would be completely infeasible, the harsh reality is "it depends." If the system is considered critical, and all of those threats are equally possible, then investments to support analyses of those threats would be necessary. Typically, though, threats may be described as classes where each class has common characteristics; many threats may be addressed simultaneously by the same functional capability.

System use in tandem with other platforms is also critical to the concept of an operation and its effectiveness. Ideally, operational capability needs will be met by the collective operation of constituent systems as they meet their own local capability requirements. Quite often, however, these needs are not aligned. Dahmann highlighted that higher-order needs exist at a different level of abstraction than the detailed, system requirements and consider multiple alternatives to adapt constituent systems to those goals to preserve a wide range of options. Requirements driven by an operational capability need, i.e., multiple systems acting together in the context of an operation, "do not map neatly on a 1-on-1 basis with constituent system requirements" (Dahmann 2014).

Common Challenges

Some methods require a tremendous amount of work and data to create the base analytical model(s) used to evaluate a tradespace. Methods such as high-definition Design Structure Matrices, a Change Propagation Index, the Change Propagation Method, and variations on these approaches using connectivity models (i.e., matrices or networks) are very labor intensive in terms of obtaining required data and creating the component specificity required. Many cannot readily be applied to large and complex engineering systems, and new evaluation models are often required when the system design changes (Koh et al. 2013). An additional factor complicating wide use of some of these methods is that, many times, customer groups across the DoD do not have possession of or access to the original models and data used to produce the output system attributes. Often, the tradespace is the only data the customer has to analyze (Sitterle et al. 2015).

To explore a tradespace, and then to analyze it, one must first generate that tradespace. Even before a tradespace can be generated, a significant amount of related work goes into specifying the problem, potential design architectures for evaluation, design variables that will characterize those architectures, M&S components that will map the design variables to output measures on which trade-offs will be assessed, and how the overall problem specification maps to stated stakeholder requirements and distinct operational scenarios.

Regardless of the level of trades – capability portfolio selection, system design, or mission engineering – the overall process and objectives can be generalized as shown in Figure 13.1. There are three key significant areas in which the systems engineering community must continue development if we are to be cost and time effective in the practice of tradespace creation, exploration, and analysis:

1) Evolve how system engineers more effectively represent and express their problems.
2) Directly support higher-level decision-making goals through processes that are as quantifiable as possible, even in the face of great uncertainty, and are eminently traceable.
3) Create digital engineering environments that promote collaborative design, exploration, and analysis. Creating tools for use by one analyst or one analyst at a time carries significantly different considerations from creating tools intended for use by a team in active collaboration.

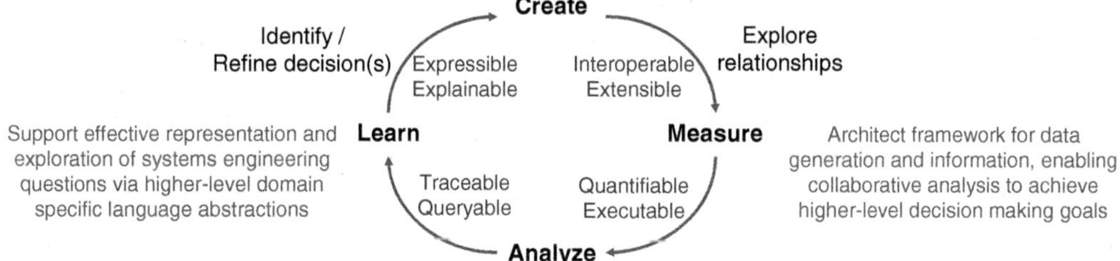

Figure 13.1 High-level tradespace needs and process.

Tradespace creation, exploration, analysis, and the associated methods, processes, and tools that support them should vary in scope, purpose, respect the level of effort required to generate a tradespace appropriate to the stage of analysis, and always be crafted to produce steps forward in terms of building on or adding to the previously existing bodies of knowledge.

System Qualities: The Nonfunctional Requirements

System quality attributes, often interchangeably called system qualities (SQs) or quality attributes (QAs), are nonfunctional requirements critical to the success of many development projects. They represent major stakeholder needs but are a major source of project overruns and failures, stakeholder value conflicts, and are poorly defined and understood (Boehm 2015). System qualities are significantly underemphasized in development efforts, perhaps in no small part because they can be so difficult to effectively describe and measure. Dr. Barry Boehm led multiple efforts aimed at developing consistent system quality ontologies and representations to mature their use in tradespace and affordability analysis. For the Systems Engineering Research Center (SERC), this occurred over seven phases of the System Qualities Ontology, Tradespace, and Affordability (SQOTA) project (Boehm 2019). His work in this area has spanned years and extends from a software engineering lens to one applicable to engineered systems in general. This latter evolution is particularly relevant to today's systems, where system capabilities and functionality are increasingly driven by data and (upgradeable) software (i.e., algorithms that ingest data, transform it, and produce output).

While the functional requirements specify what a system should do, the systems and software quality attributes describe the nonfunctional requirements (NFRs) that specify how well a system should do them. Many of these system qualities are called "ilities": reliability, availability, maintainability, usability, affordability, interoperability, and adaptability. However, the body of these system qualities includes other attributes such as security, safety, resilience, robustness, accuracy, and speed. These quality attributes are not captured in function-oriented management aids such as work breakdown structures and traceability diagrams; the NFRs generally trace to the whole system. Their requirements are often easy to specify and hard to validate (Boehm 2019).

The research in the SERC SQOTA project provides a stronger scientific basis for reasoning about quality attributes and their relationships. Resilience, for example, has significantly different definitions depending on the domain discussing it – ecology, engineering, and construction, energy development, etc. These differences are nontrivial. Further compounding the problem, most sources of quality attribute definitions limit them to single, hopefully one-size-fits all definitions for each. Boehm emphasizes throughout his work that such one-size-fits all definitions do not effectively accommodate multiple stakeholders with multiple value propositions. For example, if a system specification defines reliability as liveness to satisfy one stakeholder, it will be judged to be reliable if its liveness is very high, even though it may deliver garbled and inaccurate output and not satisfy other stakeholders whose value propositions emphasize intelligibility and accuracy.

Boehm's leadership of the seven-phase SQOTA project across multiple universities sought to address primary challenges that stem from the difficulty in defining and relating these nonfunctional requirements as identified in a 2012 Engineered Resilient Systems workshop (Boehm 2019):

- Data characterizing the SQs of tradespace options is often missing or inadequate, even though the quality of those SQs has a material impact on the relative value of the options to the stakeholders.
- The means to adequately express and analyze SQs are lacking when compared to physics-based characteristics such as weight and size.
- Needs include agreed-upon terminology to express SQs, and cross-impact models and relationships that quantitatively determine the impact of changing one SQ on the other SQs.
- The values of most SQs are scenario-dependent. Anticipated and possible future operational environments and user scenarios are usually not stated in sufficient detail to support tradespace evaluation.
- Languages and technology are lacking to cost-effectively and rapidly express vivid complex operational environments and user scenarios for a range of stakeholders.

Of special note, the SQOTA research elucidates that the system quality of affordability is often not addressed during tradespace studies. These studies instead dominantly focus on technical aspects of trades and simplistic cost assessments. They are not considering cost implications across the full life cycle, including manufacturing, operations, and maintenance. Cost is often a homogeneous bin and is not always divided according into categories in ways that provide meaningful insight to different aspects of system development. Value-based approaches offer significant potential to integrate the system qualities with other descriptive and analytical frameworks.

Ontology Foundations and Application

Boehm defines an Integrated Definition for Ontology Description Capture Method (IDEF5) for upper levels of stakeholder-based quality attributes (i.e., system qualities) alongside their contributing means. An IDEF5 ontology description is a computationally tractable representation of what exists in a given domain. It allows such representations to be purposely structured in a way that closely reflects human conceptualization. An IDEF5 further provides the means to identify primary classes of objects within a domain by isolating properties that define members of the classes and the characteristic relations that hold between domain objects (Menzel et al. 1992). Boehm first published this initial system qualities ontology in 2015 at INCOSE and then in Insight (Boehm and Kukreja, 2017), matured it for maintainability in 2016 (Boehm et al. 2016), and then revised once more as shown in Table 13.1 for the final SERC technical report.

Other members of the SERC SQOTA team, Ross and Rhodes, take a holistic view. They advance the notion of using the system qualities semantic basis as a common representation that can indicate a set of distinct, but related, metrics for measuring three aspects of a given system quality: whether it is present, the degree to which it is displayed, and the value of that system quality (Ross and Rhodes 2015, 2019). Ross and Rhodes show that a prescriptive semantic basis can be a useful structure to enable automated extraction of comprehensible system qualities statements from technical documents and, possibly, provide a computationally based means to structure and analyze such a statement (Ross and Rhodes 2019).

Table 13.1 Upper levels of stakeholder value-based system quality means-ends hierarchy.

Stakeholder value-based system quality ends	**Contributing system quality means**
Mission effectiveness	Stakeholder-satisfactory balance of physical capability, cyber capability, human usability, speed, accuracy; Impact, domain-specific objectives; Endurability, maneuverability, scalability, versatility, interoperability
Lifecycle efficiency	Development and maintenance cost, duration, key personnel, other scarce resources; Manufacturability, sustainability
Dependability	Reliability, maintainability, availability, survivability, robustness, graceful degradation, security, safety
Changeability	Maintainability, modifiability, repairability, adaptability
Composite quality attributes	
Affordability	Mission effectiveness, lifecycle efficiency
Resilience	Dependability, changeability

Source: Adapted from Boehm (2019).

Sitterle et al. take a more narrowly defined application of these findings: how these relationships, grounded in the ontology of Boehm, could help identify design alternatives most susceptible to significant, system-wide impact due to subsystem replacement (Sitterle et al. 2015). This work evaluates the sensitivity of design alternatives with respect to changes across critical design variables in a way that is readily repeatable across designs and does not assume simple relationships (changeability, adaptability, modifiability). Accordingly, a proxy picture begins to emerge with respect to how an engineering change will impact the entire system.

Synergies, Conflicts, and Architecture Analysis

Also stemming from the SERC SQOTA effort, Boehm and Kurkeja present an initial ontology of system quality attributes, defining second-level, major category system quality synergies and conflicts. Their seven-by-seven matrix includes three first-level SQs (flexibility, dependability, and resource utilization) as well as separate subsets of the complex second-level mission effectiveness SQ into separate dimensions of capability: physical capability, cyber capability, and interoperability. In the same 2017 paper that captures how improvements in one SQ result in corresponding improvements or direct conflicts in another, considering this description alongside their four-by-four matrix, reveals relationships highly germane to many efforts today.

Specifically, for flexibility, domain architecting (i.e., using domain knowledge in defining interfaces) improves reliability, modifiability, and interoperability within the domain. High-cohesion, low-coupling modules improve interoperability and reliability. Overall, this can result in reduced lifecycle costs for product line architectures. These advantages are part of the motivation behind the DOD emphasis on using a Modular Open Systems Approach (MOSA). As the SQ relationship matrices reveal, however, there are also conflicts. Flexibility achieved through high-cohesion, low-coupling modularity can result in degraded performance. Moreover, domain architecting can create multi-domain interoperability conflicts.

These concepts have motivated recent work that develops a model-based approach to evaluate architecture specifications for MOSA compliance, specifically making use of system quality attributes (Anyanhun et al. 2022). Typically, when evaluating an artifact for MOSA or other general quality attributes, it is an exercise in evaluating a system or system design for compliance with an open standard or set of requirements. However, before a system design phase, it can be beneficial to evaluate the actual architecture itself, upon which the system designs are based. This is to ensure that the architecture will foster characteristics in a system design, which collectively achieve the higher-level quality attributes and business drivers – in order to catch any architectural elements which do not foster these higher-level goals, or even work against them – and ultimately prevent system redesign.

The basis of Anyanhun et al.'s work is that, within an architecture, quality attributes serve as a traceable middle point between higher-level business objectives and lower-level architectural characteristics, which together collectively serve as a mechanism to justify that particular business objectives are achieved by specific elements within an architecture. They can serve as a versatile tool for guiding and justifying several practical and actionable uses, including but not limited to decision-making, architecture evaluation, and simply conveying priorities.

To communicate the goals of a specific architecture, a prioritization of quality attributes reflects the relative importance of each quality that the architecture intends to achieve. For various programs across different branches, missions, and stages of maturity or development, a different mixture of architectural qualities should be customized to fit each program's individual goals. For example, reusability may be important to higher levels of organization, where development and integration costs are considered across multiple platforms. Alternatively, configurability may be important to a Joint program, where a single platform is intended to be used for multiple missions and scenarios. For a robust communication of goals, where organizational priorities and architectural characteristics are holistically linked, quality attributes serve as a pivotal connective tissue within an architecture, showing that discrete elements of a particular architecture trace upward and come together to compositionally achieve an organization's business goals. Specific elements of the architecture (e.g., requirements) can be traced to a given architectural characteristic, and ultimately up to the business drivers, to convey that the architecture indeed fosters specific characteristics that will lead to the manifestation of the interoperability quality attribute in systems designed from this architecture.

Anyanhun et al.'s work provides a tangible example of leveraging quality attributes – their prioritization, value to stakeholder, and ontological basis – for the direct evaluation and analysis of architectures: trades with respect to what is manifested through an architecture. Mapping of architecture requirements to MOSA quality attribute criteria using a model-based approach establishes traceability relationships that can be queried, analyzed, and used to validate whether the architecture and resulting system will exhibit modularity and openness characteristics – all within the authoritative source of truth (i.e., the model). It also facilitates the identification and analysis with respect to how requirements changes could impact the architecture's ability to comply with required MOSA attributes.

Value and Affordability

Throughout Boehm's system quality research, he focuses on how their value is expressed. Too often, according to Boehm, system qualities are specified as single numbers, when their values actually vary by referent (primarily by stakeholder value proposition), by state (the system's internal state and its external environment state), by process (internal procedures, external operational scenarios), and by their relations with other system qualities (Boehm 2019). Using maintainability as an example, we can see these dimensions manifest quite differently:

- **Referents**: For maintainability, all stakeholders will prefer shorter above longer mean times to repair or modify, but their duration values will vary by stakeholder.
- **States**: The time and effort to effect changes or repair defects will vary by the system's internal state, such as the brittleness of its point-solution architecture, or its weak or strong modularity. The system's degree of Maintainability would also depend on aspects of its external state, such as its need to support multiple systems of systems with different stakeholders, each with different priorities and independently evolving interfaces.
- **Processes:** Maintainability values will also vary by the nature of the external processes or operational scenarios that the system needs to be involved in, such as changes across systems of systems.
- **Relations**: The seven-by-seven matrix of system quality synergies and conflicts identifies a number of relations between maintainability and other desired system qualities (Boehm and Kukreja 2017). For example, tightly coupled high-performance computing architectures and software improve computational speed but reduce maintainability. Easiest-first agile methods often improve initial usability, but make architectural commitments that often subsequently degrade scalability, safety, and/or security. On the other hand, investments in nanosensor-based monitoring systems improve both dependability but also improve maintainability. For example, aircraft autonomic logistics systems can communicate maintenance needs during flight and communicate them to the landing field, where the maintainers can prepare to turn the aircraft around in hours vs. days.

Affordability is often under-represented and over-simplified in most tradespace analysis efforts even though its inclusion can reveal numerous combinations of options for decreasing lifecycle costs and increasing lifecycle effectiveness. Lane et al. investigate the affordability tradespace with respect to how it applies to system capabilities. Specifically, they consider how expedited systems engineering activities must balance to reduce schedule and cost, encourage flexibility in architecture decisions to support future evolution of a system, and minimize technical debt that results in later rework or adversely impacts future options (Lane et al. 2013). Expedited engineering aims to deploy capabilities quickly in ways that do not overly constrain future system evolution. It seeks to expedite system development, including necessary up-front design, system prototyping or building, as well as testing and deploying a system to get capability delivered as soon as possible. This is achievable via multiple ways: lean approaches that eliminate non-value adding activities, heavy use of COTS or GOTS products, reuse or repurposing of existing systems or components, investment in product line or single-purpose architectures, etc.

Agile and lean processes are often strongly embraced as ways to provide new capabilities through integration and enhancement and rapidly respond to immediate needs. However, Lane's team reveals that this can lead to single-point solutions not flexible enough to meet the next set of needs or that could incur significant technical debt due to extensive rework and maintenance costs. Their analysis shows that valuing flexibility and investing in system and software architectures can often provide more expedient and flexible solutions than by just focusing on expedited processes, and that developers must explore a variety of options for identifying "satisficing" solutions. Evaluating these combined options can involve complex trades among affordability and other system qualities such as reliability, availability, maintainability, usability, adaptability, interoperability, scalability, and others such as safety, security, reliance, and timeliness. The choices across these trades typically relate to development processes and product architecture decisions. There is often not an optimal set of choices. Instead, the engineering team should evaluate competing stakeholder needs and make trade decisions that sufficiently balance them.

Salado and Nilchiani relate affordability directly to system requirements (Salado and Nilchiani 2015). Succinctly, they assert that "the more requirements that need to fulfilled, the more difficult is to achieve system affordability. A key aspect is that this finding does not only result from the increased requirements management effort, but from an actual reduction in the amount of systems that would possess such attribute." The set of system requirements, and the complexity that set can impart on a problem space, can fundamentally affect the affordability of the expected system.

Salado and Nilchiani explain requirements are typically produced and categorized from either (1) a design perspective, what is needed to design a system, or (2) a contractual perspective, which considers instead what is needed to procure the system and consequently changes the requirement scope. Yet neither approach effectively defines what a system must do, and so the number of requirements increases without satisfying any new stakeholder need. Additionally, a "good" set of requirements is frequently considered to be one with no conflicts between them. In some cases, however, such as software engineering, requirements may not be conflicting pairwise but are conflicting when considered simultaneously. To mitigate these challenges, Salado and Nilchiani propose a model to elicit and categorize excess-free requirement sets, identifying conflicting requirements more effectively:

- **Functional requirements (Do)**: what the system does in essence, or in other words what it accepts and what it delivers.
- **Performance requirements (Being)**: how well the system does it, which includes performance related to functions the system performs or characteristics of the system on its own, such as – ilities.
- **Resource requirements (Have)**: what the system uses to transform what it accepts in what it delivers.
- **Interaction requirements (Interact)**: where the system does it, which includes any type of operation during its life cycle.

Mission and Operational Trades in Context

Character of Mission and Operational Considerations

The big picture that drives consideration of mission and operational context for capability trades, and what systems and system configurations can supply the capability, stems from understanding what adjustments in the mixture of delivery platforms, weapon systems, and ordnance (including cost and lag times) best meet National Defense Strategy objectives. From the bottom-up perspective, the tradespace involves military capabilities of the Joint and combined arms, with quantities, endurance, performance capability, capacity, operational performance limits, and trades across performance parameters, resource needs, and contribution to military operation goals.

The tradespace balances funding planning, milestone objectives, and credibility of demonstrated evidence of cost and schedule vs. program claims. It also considers deterrence effects on adversaries by the quantity, timeliness, and technology performance, as well as the relationship between capacity at or deliverable to the point of need and the time required to re-supply or replace. All of these factors are, in turn, dependent on the tactics, techniques, and procedures associated with the operation. The tradespace decisions consequently devolve into a wide range of engineering design decisions.

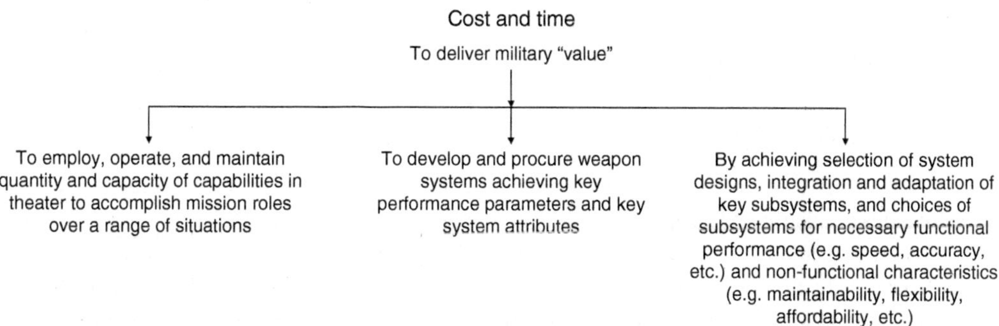

Figure 13.2 High-level description of cost and time considerations to deliver military value.

In its most simple conceptualization, the overall tradespace analysis evaluates cost as a function of military value and time required to deliver that value. This is illustrated in Figure 13.2. A more nuanced view, however, is required to understand and account for which costs and which values as well as how each is evaluated.

Whole lifecycle cost for a development program encompasses Research, Development, Test, and Evaluation (RDT&E) appropriations, production, operations and maintenance, facilities, and personnel. Choices are defined at each level. For example, in the case of unit procurements (i.e., production), are unit costs evaluated as roll-way costs of amortized over the projected useful life of the individual items? Similarly, program costs include projected costs of pre-planned and unplanned upgrades and system enhancement programs? Where will the cost impacts associated with necessary support systems (e.g., strategic and operational transportation and basing needed for a program) be accounted for?

Consider a Patriot air defense system. A Patriot system is expensive, individual rounds are expensive, and each system requires 90 personnel to operate it. How does the cost model consider the nature of its effective operation: unit cost, cost per operating hour, or cost per missile fired together with how many missile launches are required to intercept a given threat and the number and nature of those threats, etc.?

Often, the cost balance takes the form of a standard attrition model. There are two standard functional forms for kill rates characterized by a constant of proportionality in this model: (1) one proportional to the number of shooters (i.e., a target-rich environment), and (2) one proportional to the number of shooters times the number of targets (i.e., a target-poor environment). These constants of proportionality depend on weapon vs. armor characteristics and geometry as well as tactics related to range, use of cover and concealment, engagement maneuver geometry, and formation geometry. For the Patriot example, the cost-balance attrition model does not consider effects of attacks that passed through air defenses. It only considers the costs of incoming missiles and air defense missiles, though it could be extended to include resupply rates from existing stockpiles and new production and/or purchases.

For military value, there are a myriad of ways valuation can be prioritized and assessed: evaluation at the level of military value, contribution to component Service roles, achievement against Key Performance Parameter and Key System Attribute functional and nonfunctional characteristic levels, contributions to other operational objectives in the form of measures of performance (MOPs) and measures of effectiveness (MOEs) under presumed conditions, etc. This can become exceedingly challenging outside of the individual system level. On a concept of operations, for example, the order in which assets are deployed – and hence their effectiveness in their intended role – can completely change the outcome.

Role of M&S in Operational Tradespace Data Creation and Analysis

M&S can potentially address many aspects of the full spectrum of operations from rapid response to the logistics of basing and supply, providing valuable insight into a diverse set of trades. Combat evidence is rich in that it includes all aspects of operational employment and execution – arrival, forward basing delays and capacity, integration into combat operations, loiter time, ordnance capacity, out-of-action time to refuel and re-arm, etc. However, it is incomplete, uncertain, and only covers limited and specific previous situations and operations. Also, it does not directly address hypothetical potential future situations with different adversary capabilities and tactics, different ground situations, operational maneuvers, and theaters.

M&S is therefore vital for operational assessment with respect to value an individual system and its capability provides to an operation or the value certain portfolios of capabilities offer. Even so, most M&S study formulations focus on specific alternatives or parametric trade-offs, selected control variables, selected local (i.e., sub-mission) MOPs and MOEs under presumed conditions. Consequently, framing of M&S analyses can miss unanticipated, real future cases with system acquisition consequences. While M&S does not and cannot encompass all of the complexities and unknowns of real-world operations and their interactions across the spectrum of warfare, continued evolution of M&S capabilities is needed to explore hypothetical and speculative edge cases. These methods will exist outside of but based on historical combat evidence, using theoretical or statistical models to extrapolate to hypothetical potential future situations, and can be complemented with test and evaluation data as well as software-in-the-loop and hardware-in-the-loop hybrid approaches to generate the date necessary for an analysis of trades.

Different model types, at different levels of abstraction are needed to meet these challenges with the full rigor of systems engineering development and role within a defined analysis process. In many cases, especially for decision-makers to develop that intuitive understanding in the face of significant complexity, these models and their associated frameworks should provide the ability to conduct "what it" types of analysis. This is in contrast to M&S that must be changed and re-executed to discover a different result or relationship. Directly queryable model types can allow users to more effectively and quickly explore these relationships and their consequences in support of decision-making, whether for design or portfolio of capabilities definition.

Along these lines, a recent effort by Baker et al. describes the initial application of a model-based approach to a traditional view of kill chain analysis and its evolution toward a more dynamic assessment of high-level effectiveness of a multi-domain web (Baker et al. 2020). Their work applied a probabilistic graph model (PGM) approach to mission-specific kill chain analysis. As PGM nodes and their interdependencies are synthesized via the graph structure, they relationally cascade into nodes representing functions directly related to the kill chain. For example, the DoD commonly expresses kill chain assessments using the "Find-Fix-Track-Target-Engage-Assess" (F2T2EA) concept. The data model captures each of these higher-level functional capabilities. Blue and adversary capability nodes are represented consistently to ensure common components are integrated and not duplicated, providing scalability and ease of modification should conditions change.

Baker et al. also make effective use of model-based templating, developing non-platform-specific, robust sets of central relationship nodes and edges specific to a given functional area (e.g., air-based Intelligence, Surveillance, and Reconnaissance [ISR] but not specifically which air-based system or its associated capabilities). A template can then be inserted into any target set or kill chain combination and associated with specific platform capabilities. Customized templating results in a model intuitively related to the traditional, science, and technology kill chain analysis. Additional investigations show how the methods could be expanded to capture order effects associated with specific concepts of employment. The overall approach permits more complex relationships conditional on variables that could be temporal or nontemporal and enables stakeholders to evaluate how capabilities might be used together in multi-domain engagements. Moreover, the computational graph structure directly enables integration across assessments as well as flexible, queryable analyses. These data models and their associated methods transform capability design and gap discovery, leading to a better understanding of relational dependencies for senior DoD decision-makers and provide the necessary traceability, reuse of concepts, and foundations for advanced decision analysis in these problem spaces. Work continues in scalability of these methods, defining the limits and levels of abstraction at which they are most effective, and methods to reduce the cognitive burden on model developers in ways that promote consistency.

Future Systems: New Methods of Design, Evaluation, and Relation to Trades

As the systems engineering community endeavors to aid the DoD in developing and fielding new capabilities, technological advances underpinning many future systems are changing the nature of how we design systems, evaluate them, and provide assurances on their ability to function as intended:

- System capabilities and functionality are increasingly driven by data and (upgradeable) software (i.e., algorithms that ingest data, transform it, and produce output).
- Systems increasingly incorporate various levels of artificial intelligence (AI) and machine learning (ML) to achieve new capabilities.
- System capabilities are increasingly defined on data flow to and from a system and algorithms that transform that data to information, come to a decision about that information, and act based on the decision.
- Tomorrow's challenges require a different AI paradigm to help inform decisions/capabilities in environments characterized by small(er) and highly heterogeneous data, unpredictable ability to receive that data, high uncertainty, and fast response demands.
- New technology foundations endow systems with new capabilities and new characteristics that fundamentally change how people will, could, and should work with them for effective performance.

Commensurately, this creates challenges in how we create data and information for tradespace analysis. We will need new ways of dealing with data and problems instead of simply new applications of old techniques to support how we represent these systems, generate data about them to support decision-making, and analyze that data to make a decision.

For example, a common belief is that by delivering capability via software, we confer resilience via adaptability and upgradeability. Returning to Boehm's work on system qualities, he explains that tremendous discrepancy exists across the literature and domain definitions (Boehm 2019). Multiple papers over the years have sought to come to a consensus on precisely what that definition should be, but Boehm's team settles on one consistent with the INCOSE Systems Engineering Handbook's definition: "the ability to prepare and plan for, absorb or mitigate, recover from, or more successfully adapt to actual or potential adverse events" (Walden et al. 2015; Haimes 2012).

Madni and Sievers, in their subsequent chapter "Exploiting Formal Modeling in Resilient System Design: Key Concepts, Current Practice, and Innovative Approach" explain that resilience measures today are often introduced into system using *ad hoc* methods based on the best understanding of how a system is expected to be used and in what operational and environmental contexts. They develop formal resilience design methods and an approach that extends the concept of contracts commonly used in design by contract software design to accommodate design rigor and system flexibility. This work aids design, yet the new approaches to define design will be equally valuable as a basis from which to generate new types of data for trades and analysis.

Madni and Sievers note the development of resilient systems is complicated by the need to incorporate learning and adaptation capabilities to successfully endure and prevail in the face of unknowable disruptions. Systems learning in the field present a fundamental hurdle for traditional systems engineering techniques, and so Collopy et al. address challenges stemming from learning and adaptive capabilities directly (Collopy et al. 2020; Collopy and Sitterle 2019). Caught in a paradigm of specifying behavior then verifying that behavior, systems engineers have no way to test a system intended to behave in un-specified ways. The solution to this dilemma does not exist in verification testing, which tests only requirements, so Collopy's team looks to validation testing.

Traditional approaches to verification and validation testing hope to exhaustively evaluate the entire action space, however. The inputs and outputs that would be required to do so have become hopelessly impractical even for conventional complex systems (Felder and Collopy 2012). These approaches are even more obviously inadequate for systems learning in the field and formulating their own actions after development is complete. Such learning enabled systems' behavior will be as dynamic as the field data they ingest.

Collopy's team considers the spectrum of AI and ML from narrow AI/ML to the penultimate yet unattained notion of general intelligence as shown in Figure 13.3. Narrow AI\ML is typically task-specific and fixed in terms of its algorithm code and parameters once trained, exhibiting the same

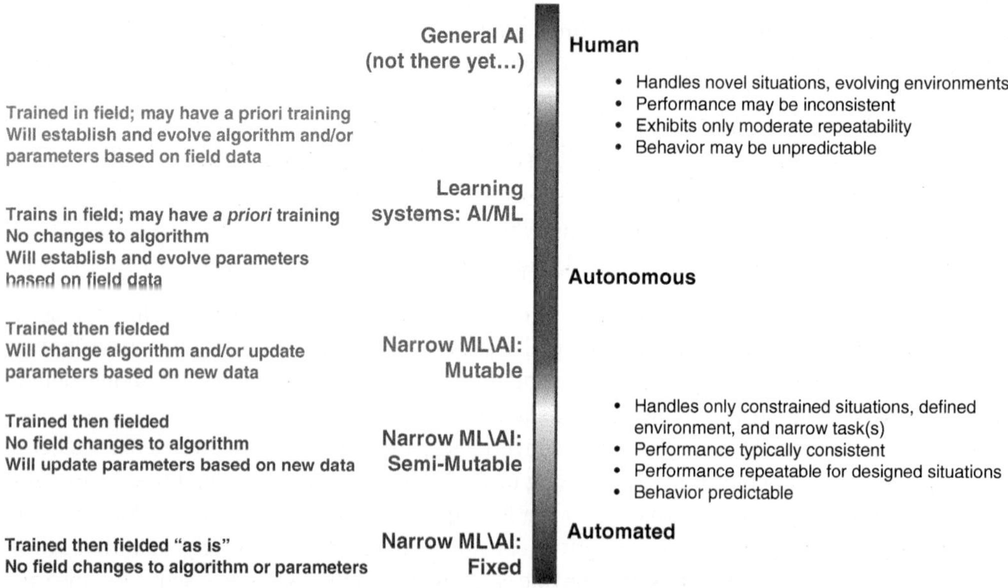

Figure 13.3 Spectrum of machine learning (ML), artificial intelligence (AI), and learning-enabled systems for validation considerations. *Source:* Adapted from Collopy and Sitterle (2019).

brittleness and systems considerations as other automated engineering systems. Momentum is building toward systems that are able to handle a greater number of operational situations with a greater degree of autonomy. The increase in potential capability manifests as increased algorithmic complexity and hence reduced understandability and predictability of system performance. It also brings potentially increased vulnerability though increased attack surfaces, especially if manned–unmanned teaming is required to achieve the operational performance.

This brings very practical considerations for the DoD. For example, when may learning enabled (LE) systems be fielded? When is training sufficient for real operation, and when must this occur prior to fielding verses when learning in the field is necessary to exploit contextual and changing data? Many current programs still use expert-curated, cleansed, and validated data sets to train ML systems, ensuring they are trained within stable, anticipated bounds. Data integrity cannot be guaranteed in the field. If a system is adapting – learning – in the field, how much of this process could or should be automated (i.e., trusting the system to learn autonomously in addition to operating with some degree of autonomy)? What are the trades we must consider to select among the options to these questions?

Especially vital for the DoD to consider is the notion of asset transferability. With ML capabilities and performance driven by data, it is critically important that data in Area X is often very different from data in Area Y. An AI\ML algorithm trained on facial or vehicle recognition in Atlanta, for example, will face a very different data environment if suddenly transplanted to Mosul. LE system performance in one theater of operations cannot be expected to be preserved in another theater of operations, at least until some initial re-learning period of time has passed. This challenges the very concept of a digital twin. Without the same data, a digital twin model could not reproduce the operations of a LE system, much less the operations of multiple instances of the same initial LE system that continued to learn in their respective Areas of Responsibility. This is, again, why new methods are needed.

As the development goals push toward increasingly autonomous, learning systems –ones more "human-like" in adaptive capabilities – the traditional culture within the intersection of the DoD and systems engineering communities still desires predictable, guaranteed performance. As this standard is beyond what can be promised even for highly trained humans, a paradigm shift in how we approach validation and develop trust in system readiness for fielding becomes necessary.

Collopy et al. pilot the development of a method for validating the spectrum of AI and learning enabled systems through the use of a formal, logical argument methodology. The concept of argument analysis is new to most of systems engineering, especially those areas focused on technical design and validation of that design. Yet when traditional deterministic analyses and decomposition approaches fail, namely when faced with uncertainty and a problem state space nearly mathematically impossible to bound, argument analysis is well-suited to support decision analysis. Using Stevens' work on replacing confirmation of static behaviors with a structured logical argument of capability success in the field as a conceptual foundation (Stevens 2017), Collopy et al. create a formal argument process specifically designed to support validation of AI and LE systems. They build on Boehm's discussion of the parallel evolution of software engineering and systems engineering away from sequential, reductionist methods toward "softer" processes emphasizing a more spiral, continuous learning approach (Boehm 2006). Boehm observes that software requirement emergence is incompatible with traditional, sequential, waterfall process models: "Fundamentally, the theory underlying software and systems engineering process models needs to evolve from purely reductionist 'modern' world views (universal, general, timeless, written) to a synthesis of these and situational 'postmodern' world views (particular, local, timely, oral) as discussed in (Toulmin and Toulmin 1992)."

Toulmin developed a softer form of deductive logic that uses a warrant construct show how evidence leads to concluding a claim is true (Toulmin 2003). The Toulmin system's strength is the warrant can employ practical reasoning and common sense. Collopy et al. found Toulmin's model particularly suits the challenge of validation for LE systems, as warrants can capture the interaction between data from various tests and other sources. A surprisingly explicit and detailed result from one test may compensate for an unfortunately ambiguous result from another test. Because Toulmin's model is formal, it organizes validation test lists and shows an explicit and reasoned connection between tests, the specific aspects, and capabilities validated by the test. Argument analysis may offer new approaches to other systems engineering problems fraught with similar complexity and uncertainty.

This work is being continued and matured in the SERC by Messmer et al. (2021), who are extending the representation of validation arguments to include alternatives and uncertainties. This makes it possible to employ the full machinery of decision analysis to evaluate organizational choices in service of systems validation, in the presence of conflicting arguments about claims and their underlying rationale.

Discussion and Conclusion

Tradespaces are decision spaces. As illustrated in Figure 13.1. tradespace analysis is an expansive process that requires the creation of a tradespace, development of metrics, analysis methods, and culminates in a decision. Decision-making, especially in the context of tradespace-type problems in a landscape increasingly moving toward a digital engineering paradigm, is also a process with distinct needs as shown in Figure 13.4. Identifying what decisions need to be made frames the entire process.

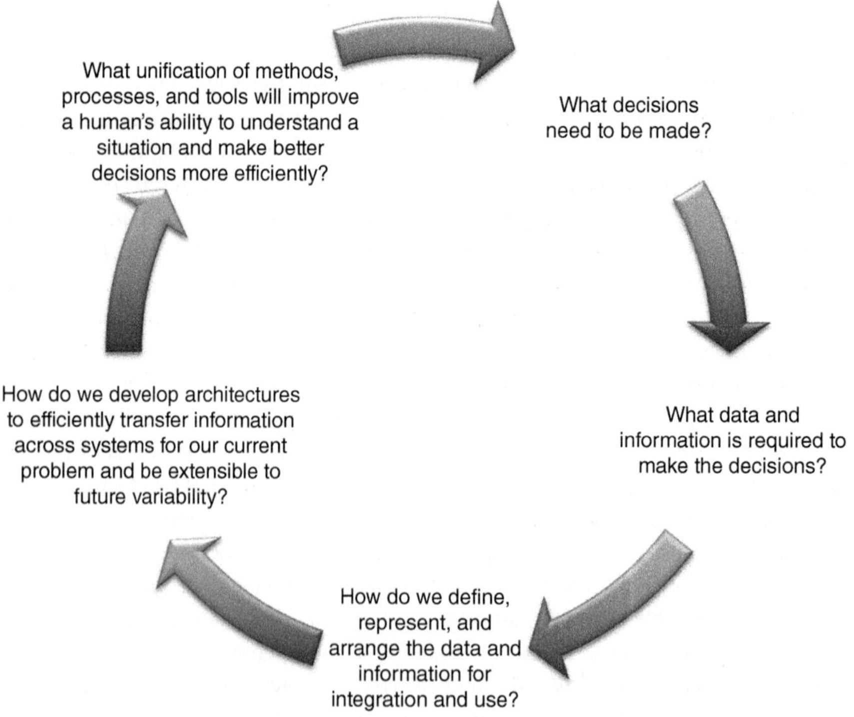
What decisions need to be made?
What data and information is required to make the decisions?
How do we define, represent, and arrange the data and information for integration and use?
How do we develop architectures to efficiently transfer information across systems for our current problem and be extensible to future variability?
What unification of methods, processes, and tools will improve a human's ability to understand a situation and make better decisions more efficiently?

Figure 13.4 Decision-making process for digital engineering environments.

Tradespace exploration and analysis is not simply for winnowing candidate selections. It should promote development of improved understanding, effectively and efficiently. A tradespace is not a two or even three-dimensional Pareto concept, but rather a hyper-dimensional problem. The challenge will be to develop tools and techniques that help analysts identify a hyper-dimensional space without losing intuitive feel for trades that may be made across these perspectives. To that end, many tradespace exploration and analysis tools are embracing dynamic views intelligently driven by how design space data maps to the systems model(s), design variables, constraints, requirements, stakeholder perspectives, etc. Frameworks that support the construction of multiple analysis types bolstered by interactive visualization can promote a more intuitive understanding for decision analysts through seeing cause and effect or at lease correlated relationships exposed via changing bounds or other key parameters of an analysis. How to architect a framework to support the type of tradespace analysis on needs is the subject of a subsequent paper in this chapter, "Architecting a Tradespace Analysis Framework in a Digital Engineering Environment."

In traditional decision analysis of tradespace data, the context of evaluation basis is on the whole of a specific instantiation of a tradespace. Many methods increasingly finding their way into practice strive for a more holistic approach toward integrating the whole tradespace analytical view with analyses targeted toward the "best" set of designs. Some sensitivity analyses may need to be global, performed across the entire tradespace to reduce a set of parameters or attributes under evaluation. In other analytical treatments, a global approach can compromise insights that may be specific to the "better" design alternatives with bias from the "poor" alternatives as well as inhibit the ability to scale both computationally and visually. Focusing some aspects of the analysis on a more tailored, decision-oriented space, a "local" data set, will enable deeper insights regarding sensitivity, uncertainty, and correlations across the "best" set of designs. In some studies, the importance of order motivates this view: whether the number of alternatives is reduced before or after applying a sensitivity analysis and incorporating the results into a decision process (Sitterle et al. 2017). It is congruent, however, with the concept of refining sets by integration (the intersection of alternative sets, each developed by a different design team) and the understanding of treatment that analysis requires. Dullen discusses these methods in a later chapter in this section, "Set-Based Design: Foundations for Practice."

When systems engineering tools focus only on the system itself, or even on sets of systems operating together yet to the exclusion of external dynamics, they fail to bring the complexities of mission engineering analyses into evaluation. This prevents attainment of a meaningful understanding of requirements and the tradespace of capabilities. This deficit also interferes with an ability to integrate mission engineering with other analyses in a flexible, scalable, relatively rapidly configurable manner.

This reveals a bidirectional problem. In current practice, individual system design processes can produce a technically feasible solution but not elucidate whether the system will be effective in an intended operational scenario. The converse is also true. Operational analysis can evaluate the effectiveness of system requirements. It cannot, however, inform a decision-maker whether or not those requirements are technically feasible. System engineering methods will need to advance effective, computationally efficient methods where mission profile requirements can be directly coupled with methods that co-evolve engineering design.

Increased integration of the operational and acquisition communities and, importantly, their models and analyses will enable us to deliver mission capabilities with greater speed and agility. One need is the capability to evaluate the trades associated with architecting a mission in a model-based scenario that includes both Blue and Red systems operating in a specific mission context. Success is then measured through two distinct pathways: (1) scenario modeling in terms of performance in the context of a mission – performance against what and to accomplish what objectives; and (2) performance modeling in specific scenarios comprising a selected mission context.

Further, the systems engineering community must continue to develop new architectural or other descriptions to relate complex relationships across operational and system-intrinsic views. Madni and Sievers present a holistic and formally grounded approach for the system quality of resilience in their chapter "Exploiting Formal Modeling in Resilient System Design: Key Concepts, Current Practice, and Innovative Approach" that follows in this section. Along these lines, we must also continue to mature effective, interpretable methods that capture how AI/ML capabilities within a system, dynamic functional reconfigurability characteristics, and human interactions will change the way we need to describe and analyze the system (1) to navigate the expanded dimensionality and still promote effective, meaningful trade space analysis; (2) to derive and analyze key performance indicators across these problem types; and (3) to identify and mitigate risk.

Other challenges for the systems engineering community include the interoperability and cross-scale integration of different M&S tools and their outputs. Pointedly, there are an increasing number of frameworks with an associated suite of M&S solutions all purported to work together from the detailed physics-level models up to a variety of higher level of abstraction models. Often, the framing of each problem space – how the problem, its elements, relationships, and outcomes are represented – is not compatible across different tools, regardless of stated interoperability at a semantic level. Additionally, just because disparate M&S components can be integrated does not mean they should. Systems engineers must take care to understand the underlying assumptions, physics when applicable, and other foundations that may make combining different sets of data meaningless or simply wrong. Validation of the scope, detail, and credibility of outputs of M&S components, especially for highly complex problems, will continue to be a challenge. Systems engineers must recognize when a model is predictive vs. when it simply confers insight into relationships, and how the nature of those outputs should be incorporated into the larger decision space for trades.

Proper levels of data abstraction for any given problem are vital to an ability to derive meaningful and actionable insight. Systems engineering needs work to mature the type and nature of abstractions necessary to more effectively represent and express challenging tradespace creation, measurement, and analysis problems, especially those dependent on the execution of multiple M&S components and the highly variable data they generate. How we specify the needed abstractions will directly impact not only how effectively the patterns may be implemented and executed in a computational environment but also how intuitively they are perceived, understood, and employed by an analyst in the development and exploration of a tradespace. As we mature our understanding in this space, we need to also provide rational, perhaps even domain-specific guidelines for when to use them.

Moreover, as a community, we need to develop a much better understanding of what types of model and analytical constructs may be reusable as well as how to specify these reusable patterns. To reason on data in any capability space, then there must be a consistent underlying "language" foundation that enables algorithms and users to do so effectively. This is especially true if there is to be any automated execution based on information derived from that data. The problem is not simply transmitting and receiving bits of data across platforms (data interoperability) but how to transfer information (derived from collected data) from separately developed systems in a form that can be accurately used for automated inference (semantic interoperability in order to reason). Semantic understanding is also highly critical when humans are in or on the decision-making loop; this may be conveyed via visualizations using any combination of methods, including those impacting other senses.

The complementary part of this challenge centers on the methods, processes, and tools to support decision-making on the basis of the available data – the human–machine cognitive creative process. More data does not always lead to better decisions. Specifically, we must develop methods to identify the critical data that will resolve a decision problem quickly, evaluate, and explore higher levels of information and courses of action derived from the data, visualizing the expanse and nature of data in hand. The overall effect of these processes is to not replace human intelligence by making the decisions automatically, but rather to enhance or augment the human analyst's intelligence through better implementation of technology. This concept is referred to as augmented intelligence, which recognizes that technology can improve human abilities to manipulate and understand information. Madni's chapter on "Augmented Intelligence: A Human Productivity and Performance Amplifier in Systems Engineering and Engineered Human-Machine Systems," later in this section, explores how augmented intelligence can evolve the practice of systems engineering as well as how it can imbue systems themselves with new capabilities that support more effective teaming with humans in the field. The corollary concept is also vital; to make best use of these approaches, certainly to aid vast and complex tradespace analyses, we will need we need models that help the AI system receive and interpret human input better.

In general, there seems to be a default mental model that human and AI system teaming, or AI–AI system teaming, is always intentional. In military operational scenarios that may arise from an emphasis on more numerous and attritable systems, this will not always be true. Multiple systems with heterogeneous capabilities, and high degrees of autonomy (even if the overall capability is simple) interact in various ways:

- Explicitly cooperate via negotiation or similar algorithmic methods,
- Cooperate in variable membership groups without explicit algorithmic direction typical of negotiation/cost function approaches, and
- Do not explicitly cooperate but collective action dynamics still create constructive or destructive dynamics.

Systems engineers will need to consider the spectrum of interactions illustrated in Figure 13.5 to evaluate our ability to deliver intended capabilities in theater. All of these relationships enable or constrain the ability to achieve mission objectives. This last category, "Human-AI Analytical Teaming," will be especially relevant for the future of tradespace analysis. How can human–AI synergy enable faster, "better" resolution of a complex problem space with high uncertainty, dynamic instability in parametric and temporal spaces, and that is computationally intractable?

Tradespace analysis and all of the activities and elements required to make it successful across the various levels and domains is a highly challenging area that will need to evolve with care and

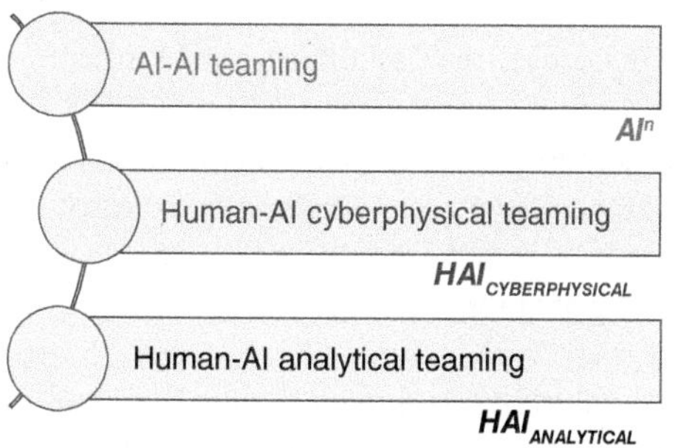
AI-AI teaming
AI^n
Human-AI cyberphysical teaming
$HAI_{CYBERPHYSICAL}$
Human-AI analytical teaming
$HAI_{ANALYTICAL}$
Interactions between and across AI-enabled systems with varying degrees of autonomy and adaptability.
Interactions between and across humans and AI-enabled systems with varying degrees of autonomy and adaptability; directly supports response to, interaction with, or creating effect in a physical environment.
Interactions between and across humans and AI-enabled systems are necessary to add context and promote abductive reasoning about complex problem spaces.

Figure 13.5 Spectrum of teaming interactions important for systems engineering.

feeding for years to come. To navigate the churn and confusion that can emanate from a myriad of new approaches and technological advances, systems engineers should focus on the outputs and not process at the expense of the outputs. How can systems engineering, and the approach a team may architect for these problems, best and purposefully help achieve these outputs?

The chapters that follow in this section highlight challenges and new opportunities for growth in the approach to creating and evaluating tradespaces as well as the processes and underlying foundations that support them. Hopefully, the foundations and needs discussed in this chapter have laid the groundwork for their understanding.

> The same behaviors that bring success to <sic> stakeholders under present conditions create a cognitive bias against future emergent innovation.
>
> *(Blair et al. 2013)*

References

Anyanhun, A., Matteson, W., Flemming, C. et al. (2022). An MBSE approach to evaluating architecture specifications for MOSA compliance. In: *25th Annual Systems and Mission Engineering Conference*, Orlando, FL (1–3 November 2022). National Defense Industrial Association.

Baker, J., Sitterle, V., Fullmer, D., and Browne, D. (2020). Application of probabilistic graph models to kill chain and multi-domain kill web analysis problems. In: *2020 Virtual Systems and Mission Engineering Conference* (10–13 November 2020). National Defense Industrial Association, (Virtual).

Blair, M.D., Nieto-Gomez, R., and Sitterle, V. (2013). Technology, Society, and the Adaptive Nature of Terrorism: Implications for Counterterror. *This report represents the views and opinions of the contributing authors. The report does not represent official USG policy or position*. UNCLASSIFIED.

Boehm, B. (2006). Some future trends and implications for systems and software engineering processes. *Systems Engineering* 9 (1): 1–19.

Boehm, B. (2015).Architecture-based quality attribute synergies and conflicts. In: *2015 IEEE/ACM 2nd International Workshop on Software Architecture and Metrics* (16 May 2015), 29–34. Washington, DC: IEEE Computer Society.

Boehm, B. and Kukreja, N. (2017). An initial ontology for system qualities. *Insight* 20 (3): 18–28.

Boehm, B., Chen, C., Srisopha, K., and Shi, L. (2016). The key roles of maintainability in an ontology for system qualities. In: *INCOSE International Symposium*, Edinburgh, Scotland (18–21 July 2016), vol. 26, No. 1, 2026–2040. San Diego, CA: International Council on Systems Engineering.

Boehm, B. (2019). System Qualities Ontology, Tradespace, and Affordability (SQOTA). *Technical Report SERC-2019-TR-012. Stevens Institute of Technology, Systems Engineering Research Center Hoboken United States.*

Cellucci, T.A. (2008). Developing Operational Requirements: A Guide to the Cost Effective and Efficient Communication of Needs. *US Department of Homeland Security, Washington, DC*. UNCLASSIFIED.

Collopy, P., and Sitterle, V. (2019). Validation of AI-Enabled and Autonomous Learning Systems. *Technical Report SERC-2019-TR-017. Stevens Institute of Technology, Systems Engineering Research Center Hoboken United States.*

Collopy, P., Sitterle, V., and Petrillo, J. (2020). Validation testing of autonomous learning systems. *Insight* 23 (1): 48–51.

Dahmann, J. (2014). 1.4. 3 System of systems pain points. In: *INCOSE International Symposium*, Las Vegas, NV (30 June–3 July 2014), vol. 24, No. 1, 108–121. San Diego, CA: International Council on Systems Engineering.

Defense Acquisitions University (2023). Department of Defense DAU glossary of defense acquisition acronyms and terms. https://www.dau.edu/glossary/Pages/Glossary.aspx (Accessed 27 January 2023).

Department of Defense (2016). DOD Dictionary of Military and Associated Terms. *DoD Dictionary of Military and Associated Terms*. JP 1-02, 8 Nov 2010 (as amended through 15 Feb 2016).

Deputy CTO for Mission Capabilities (2023). Mission engineering. https://ac.cto.mil/mission-engineering (accessed 8 Feb 2023).

Felder, W.N. and Collopy, P. (2012). The elephant in the mist: What we don't know about the design, development, test and management of complex systems. *Journal of Aerospace Operations* 1 (4): 317–327.

Haimes, Y.Y. (2012). Systems-based guiding principles for risk modeling, planning, assessment, management, and communication. *Risk Analysis: An International Journal* 32 (9): 1451–1467.

Harvey, A. S. (2021). The levels of war as levels of analysis. *Military Review*.

Koh, E.C., Caldwell, N.H., and Clarkson, P.J. (2013). A technique to assess the changeability of complex engineering systems. *Journal of Engineering Design* 24 (7): 477–498.

Lane, J.A., Koolmanojwong, S., and Boehm, B. (2013). 4.6. 3 Affordable systems: balancing the capability, schedule, flexibility, and technical debt tradespace. In: *INCOSE International Symposium*, Philadelphia, PA (24–27 June 2013), vol. 23, No. 1, 1385–1399. San Diego, CA: International Council on Systems Engineering.

Menzel, C.P., Mayer, R.J., and Painter, M.K. (1992). *IDEF5 ontology Description Capture Method: Concepts and Formal Foundations*. Texas A and M Univ College Station Knowledge Based Systems Lab.

Messmer, B., Shapiro, D., Petrillo, J. et al. (2021). Validation Framework for Assuring Adaptive and Learning-Enabled Systems. *Technical Report SERC-2021-TR-021. Stevens Institute of Technology, Systems Engineering Research Center Hoboken United States.*

Miller, E.B. (2018). DoD Principles on Mission Effectiveness and Spectrum Efficiency. Memorandum for Secretaries of the Military Departments, 3 May 2018.

Ross, A.M. and Rhodes, D.H. (2015). Towards a prescriptive semantic basis for change-type ilities. *Procedia Computer Science* 44: 443–453.

Ross, A.M. and Rhodes, D.H. (2019). Ilities semantic basis: research progress and future directions. *Procedia Computer Science* 153: 126–134.

Salado, A. and Nilchiani, R. (2015). A research on measuring and reducing problem complexity to increase system affordability: from theory to practice. *Procedia Computer Science* 44: 21–30.

Shapero, A. and Bates, C. (1959). *A Method for Performing Human Engineering Analysis of Weapon Systems* (Vol. 59, No. 784). Wright Air Development Center, Air Research and Development Command, United States Air Force.

Sitterle, V.B., Freeman, D.F., Goerger, S.R., and Ender, T.R. (2015). Systems engineering resiliency: guiding tradespace exploration within an engineered resilient systems context. *Procedia Computer Science* 44: 649–658.

Sitterle, V.B., Brimhall, E.L., Freeman, D.F. et al. (2017). Bringing operational perspectives into the analysis of engineered resilient systems. *Insight* 20 (3): 47–55.

Smead, K. (2015). *A Descriptive Guide to Trade Space Analysis*. Monterey, CA: Army TRADOC Analysis Center.

Stevens, J.S. (2017). Warranting system validity through a holistic validation framework: a research agenda. In: *INCOSE International Symposium*. (July) (Vol. 27, No. 1, pp. 654–671).

Toulmin, S. and Toulmin, S.E. (1992). *Cosmopolis: The Hidden Agenda of Modernity*. University of Chicago press.

Toulmin, S.E. (2003). *The Uses of Argument*. Cambridge University Press; Updated edition. ISBN-13: 978-0521534833.

Walden, D.D., Roedler, G.J., and Forsberg, K. (2015). INCOSE systems engineering handbook version 4: updating the reference for practitioners. In: *INCOSE International Symposium*. (October) (Vol. 25, No. 1, pp. 678–686).

Biographical Sketches

Valerie Sitterle is a principal research engineer at the Georgia Tech Research Institute (GTRI) with over 25 years of experience in engineering science, integrating engineering, natural, and physical sciences leading to the design and analysis of systems. A primary emphasis of her work integrates defense operational needs with systems sciences across multiple domains to support design and assessment of defense systems and operational and tactical concepts of employment in theater environments. She currently serves as the chief scientist for the Systems Engineering Research Division within the Electronic Systems Laboratory in GTRI, where she supports the definition and execution of R&D across the main pillars of model-based approaches and digital transformation of systems engineering to provide new capabilities and advance stakeholders' decision-making processes. She also serves as a current member of the Research Council for the Systems Engineering Research Center (SERC), a DoD UARC led by Stevens Institute of Technology. She received her BME and MS degrees in mechanical engineering from Auburn University, her MS degree in engineering science from the University of Florida, and her PhD in mechanical engineering from the Georgia Institute of Technology.

Gary Witus, is an associate professor (retired) at Wayne State University in Detroit, MI, received advanced degrees in Mathematics (University of Michigan), Industrial and Operations Engineering (University of Michigan), and Industrial and Systems Engineering (Wayne State University). He has 50 years of experience in DoD systems acquisition and military operations. He was involved in the initial operational analysis and design of the M1 Abrams tank. Later, he was a chief engineer for the M1A2 conversion from analog to digital and other system enhancements, for M1A1 System Enhancement Programs, and for other Advanced Development programs. He has been senior engineer for DARPA evaluation of advanced development projects. He founded and ran a R&D company in military robotics technologies and human–robot interactions for 16 years, followed by 12 years of research and education at the University, where he led the "Innovation with Impact and Systematic Innovation" program, and collaboration with the Systems Engineering Research Center led by Stevens Institute of Technology. He was also a direct report to General William DePuy (retired as a 4-Star, Army Vice-Chief of Staff and first Commander of the Army Training and Doctrine Command), who he honors as a teacher who said "The primary role of the peacetime army is training. I am training you into how a commander thinks." Later he said "Enough training. Now go out and accomplish something."

Chapter 14

Architecting a Tradespace Analysis Framework in a Digital Engineering Environment

Daniel Browne, Santiago Balestrini-Robinson, and David Fullmer

Georgia Tech Research Institute, Systems Engineering Research Division, Smyrna, GA, USA

Introduction

This chapter seeks to explain how to develop an architecture for conducting tradespace analysis in the context of a digital engineering environment, i.e., how should the reader develop a Digital Engineering Tradespace Analysis Framework (DETAF).

Designing and implementing a DETAF is as much an art as it is a science. It is not possible to fully describe a detailed recipe for developing such a capability, but there are general concerns that must be addressed and guidelines that can aid in the process. As such, this chapter will present these through some example applications and leverage relevant analogies when possible, instead of presenting a theory-based approach to the problem.

The chapter is organized as follows. First, an overview of the hierarchy of meaning is provided, offering the necessary context for the primary purpose behind any DETAF: traverse the hierarchy from knowledge down to data and then back up again to new knowledge. Then, a generic decomposition of a DETAF is described, identifying each of the key elements necessary for implementation. To facilitate the upfront design, key considerations are enumerated that can guide a team in ensuring the necessary use cases are satisfied by a specific DETAF implementation. Next, an example tradespace analysis workflow is reviewed, mapping each element of the workflow to the DETAF components. Finally, the authors provide lessons learned and key takeaways from their experience in building DETAFs.

The Hierarchy of Meaning

Any tradespace analysis framework is ultimately concerned with processing data to produce new information that can improve the knowledge or understanding of the stakeholder. This formulation aligns well with how researchers have conceptualized the hierarchy of meaning (Ackoff 1989), also referred to as the knowledge pyramid or the data-information-knowledge-wisdom (DIKW) hierarchy (Rowley 2007). This concept not only has a variety of names but also differs in how the levels are characterized, especially when defining the higher-level ones. Below is a description of

Systems Engineering for the Digital Age: Practitioner Perspectives, First Edition. Edited by Dinesh Verma.

Companion website: www.wiley.com/go/verma/systemsengineering

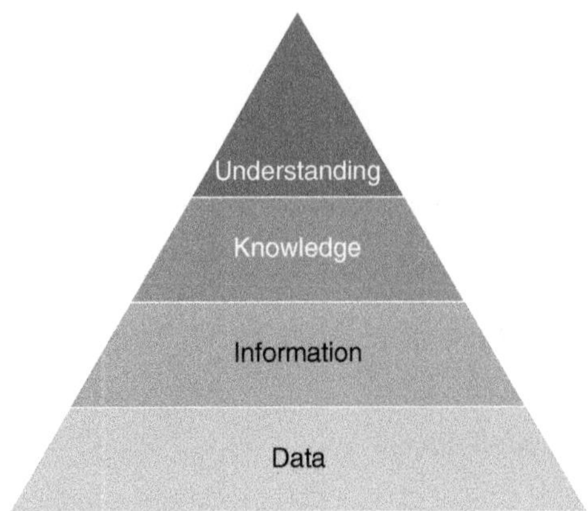

Figure 14.1 The pyramid of knowledge.
Source: Adapted from Landauer (1998).

the different levels as they best fit the concept relevant to a DETAF, with an accompanying illustration generally used to depict these concepts.

Data consists of symbols that represent objects, events, and their properties. It is a collection of raw numbers without context. For instance, a car's speedometer provides data by updating a number on your dashboard or changing the angle of a needle (Figure 14.1).

Information is data that has been made useful. It provides answers to who, what, where, when, and how many questions and helps in deciding what to do. For example, the information that you are driving at 60 mph can help you decide whether to speed up or slow down.

Knowledge is composed of instructions and know-how. It answers the "how" questions and helps decide which course of action to take. For example, your driving knowledge tells you how to control the car's speed by using the gas pedal and the brake.

Understanding consists of explanations. It answers "why" questions and enables one to evaluate which course of action to take. For instance, if the gas pedal gets stuck, you can understand why (e.g., because the gas linkage may have been twisted) and decide either to put the car in neutral and turn the ignition off, or press it further to see if it becomes unstuck. Understanding requires comprehension of causality; it offers alternative paths to achieving a goal and is an essential step to making better decisions when operating in a constrained environment. Another example of understanding is the chef that ran out of buttermilk, but realized that the purpose of the buttermilk is to provide fat and acid to the recipe, and that they could achieve the same result by mixing milk with an acid (e.g., lemon juice or vinegar).

An effective architecture for traversing this hierarchy must provide a means to travel down (i.e., elicit knowledge and understanding), formulate it in a computable form, support execution of the data in the computable form, and then facilitate development of knowledge and understanding. This process is illustrated in Figure 14.2. Ultimately, this process must be repeatable, defensible, and it must facilitate iteration, as subject matter experts and decision-makers will learn from the results produced and revise the information they contributed. The most challenging steps are to elicit and synthesize relevant understanding and knowledge, and facilitating the generation of new understanding. Nonetheless, there is still value in facilitating the traversal of the lower levels of the hierarchy.

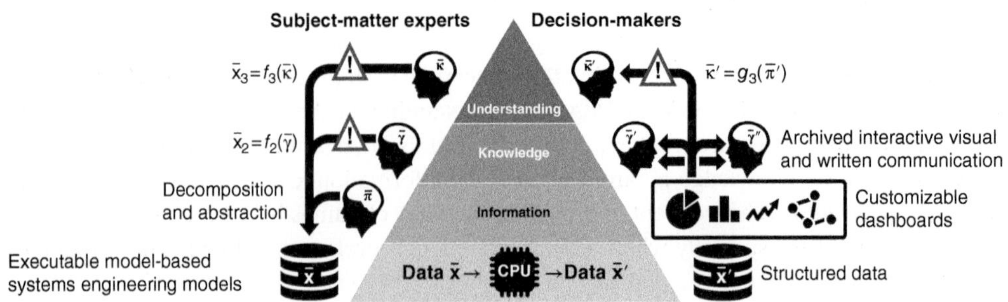

Figure 14.2 Elements required to traverse the hierarchy of meaning.

A DETAF team must first determine ways to capture the needs of the stakeholder. This can take many forms, depending on the problem at hand, and is constrained by the resources and time available. These must be balanced with the aspirational goal of developing an accurate representation of what the concept(s) being analyzed must be able to achieve (value preposition) and what constraints it must meet (costs and other limitations). The next section will delve deeper into the different approaches with examples for how to accomplish this, but in general terms this process must follow some form of modeling to make useful abstractions. This process also generally requires decomposing the relevant concepts (e.g., the stakeholders' needs, the system). Ultimately, this sub-process concludes when the relevant concepts are represented in data. If using a modern digital engineering environment (DEE), this can be done by abiding by well-established model-based system engineering (MBSE) paradigms, but it is not required to do so.

The exercise of generating knowledge ($\bar{\gamma}'$ and $\bar{\gamma}''$ in Figure 14.2) tends to be *ad hoc* or at the very least tends to be tailored to (1) the problem, (2) the organization conducting the tradespace analysis, and/or (3) the stakeholders involved. Nonetheless, in general, this process can be facilitated with low-friction tools (i.e., tools that maximize the users' ability to do value-added work) to explore the data produced. These tools should let users depict and manipulate the data quickly to help answer what-if questions on the fly, such as "what if this is the most important requirement," "what if a technology allows us to do this," "what if the labor rate is higher than expected," etc. It is also common for stakeholders to have to collaborate by comparing insights and testing hypotheses. This tends to be the most fruitful means of producing correct, dependable, and relevant knowledge. This process generally relies on similar tools but may also require understanding of statistical analysis as well. For instance, a stakeholder may need to quantitatively assess what factors drive a measure of interest. Finally, it is critical to have the ability to document the process as insights and realizations are often forgotten or become less clear over time, especially when the tools support very rapid analysis, allowing stakeholders to quickly answer questions and address hypotheses.

Finally, an often-overlooked component to this process is the practical implication to be able to efficiently produce new data from the data provided ($\bar{\mathbf{x}} \rightarrow \bar{\mathbf{x}}'$). This process can be aided through a variety of tools and capabilities. One example is to use high-performance computing (HPC) resources to be able to transform more data in less time. Another approach is to more intelligently generate the data required, e.g., by using designs of experiments, to sample the tradespace in a more efficient manner. This is particularly important when evaluating large tradespaces, which tend to grow exponentially as more factors are considered. In recent years, there has also been increasing reliance on higher-fidelity models (e.g., physics-based modeling like computational fluid dynamics or mission models like the Advanced Framework for Simulation, Integration and Modeling (AFSIM) (West and Birkmire 2020)).

The process for encoding knowledge (f_2) and understanding (f_3) is less clear and more problem and situation dependent. This often requires engaging with the subject-matter experts (SMEs) to identify the critical considerations that should be included in the model being developed. Workshops and structured interviews are a good initial step, but if a project relies on eliciting considerable knowledge from experts, it is imperative that they are consistently engaged through the entirety of the project, are presented with the findings, and are allowed to guide the evolution of the analyses and models used as they will often realize that some previously unidentified relationships or information are critical to the decisions being made. A good overview of useful approaches is presented in Elsawah et al. (2015).

Similarly, the process of generating understanding (g_3) is highly dependent on the expertise and experience of the decision-makers, the problem at hand, the time they can dedicate to exploring the data, and the time and resources available to iterate the whole process multiple times to identify high-level patterns and be able to test for causality and not simply correlation. The universal findings from developing a DETAF to solve complex decision-making problems are that (1) the process should be iterated, (2) the problem definition may need to be refined as new solutions are identified, (3) the analysis and models almost always need to be corrected or augmented, and (4) SMEs and stakeholders must be kept involved throughout the process. The most common catastrophic mistake that teams tend to make in developing a DETAF is to create it in isolation and present it to stakeholders for the first time at the end of the project.

Elements of a Digital Engineering Tradespace Analysis Framework

In general terms, a DETAF must contain five elements, as described in Table 14.1. These are not mandatory in every situation, and some are more aspirational in practice, but they provide a fairly comprehensive formulation to understand the components and explain their purpose with

Table 14.1 Elements of a digital Engineering Tradespace Analysis Framework.

Element	Description
Stakeholder needs	There must be a model that captures the stakeholders' needs, generally in the form of a hierarchy that accounts for what the system (or system-of-systems) is expected to do and what constraints it is expected to meet. There are a wide variety of approaches to elicit stakeholder needs and model requirements (Lubars et al. 1993)
Predictive models	There must be predictive quantitative models that map between system (or system-of-systems) characteristics to those that are required by the stakeholders. These are generally composed of various models that must be integrated to ensure results produced are consistent between explorations. These models may be categorized in a multitude of ways, but they are generally focused on: • Campaign or mission models to capture effectiveness, • performance models to capture system-level metrics, • cost models (e.g., acquisition cost, operational cost, life-cycle cost), and • engineering models to capture impact between size/weight/power concerns to system performance
Execution orchestrator and numerical exploration	Unless the predictive models are exceedingly simple, it is generally required to have an efficient means to execute the analysis tools, maximizing the use of available computing resources. The field of multi-disciplinary analysis and optimization (MDAO) offers a variety of methods and techniques that can be leveraged to achieve this, by not only facilitating convergence of the predictive models if necessary but also offering different approaches to optimize (e.g., numerical optimization), efficiently extract knowledge (e.g., designs of experiments), or quantify uncertainty (e.g., polynomial chaos expansion)
Visualization environment	The data produced by the Execution Orchestrator must be digestible by the modelers for verification purposes, by the SMEs for validation purposes, and more critically, by the decision-makers. This environment must support interactive visualizations to allow decision-makers to quickly answer "what if" questions, which help them build knowledge and, ideally, understanding
Reporting and archiving mechanisms	Finally, the environment must have a means to produce reports that summarize decisions, ideally with rationale from the stakeholders included, and traceability to the underlying relevant phenomena captured by the computable models. In addition, the architecture must have a means to archive and retrieve itself, so as to facilitate reproducibility in the future and maximize reuse of the capabilities developed

examples. At the very least, when architecting a tradespace analysis environment, each of these five areas needs to be considered to understand how the framework will meet the functional need provided in the description. As will be illustrated in the example workflow described later, components within a DETAF may or may not map one-to-one with these elements, based on the specifics of the implementation.

These items can be used to produce a notional morphological matrix of alternatives for the various tools and or techniques that could be included to compose the five different elements. Some are incompatible with each other, while others may be complementary. These are not intended to be the comprehensive set, as tools, methods, and techniques are constantly being developed. Nonetheless, the authors anticipate that by providing examples, the reader will better understand what is intended under each of these elements of the framework.

Considerations in the Design

There is no "one size fits all" solution for architecting a tradespace framework in a digital engineering environment. The design and implementation are dependent on the type of analysis, measures of success, anticipated user base, and other considerations. Before embarking on implementing a DETAF, the following elements need to be thoughtfully reviewed to drive the design decisions.

Questions to Be Answered

The most important step in successfully implementing tradespace analysis in a digital engineering environment is understanding upfront the purpose and goals of the DETAF. This often hinges on the type of analysis to be completed. Table 14.2 describes four common types of analyses and the types of questions that a DETAF is used to answer.

Table 14.2 Summary of types of analyses that can be supported by a DETAF.

Type of analysis	Description of problem space	Example types questions
Requirements feasibility The process of assessing the feasibility of a system's requirements to determine if they can be implemented in a practical, cost-effective manner. This should help identify and address any potential problems before the system is designed and implemented	Determine if the proposed system requirements are achievable within the given time frame, budget and other constraints. It also involves assessing the compatibility of the proposed requirements with existing systems, technology, and processes	• Can the required performance be met under reasonable costs? • What trade-offs must be made between performance measures (e.g., range vs. speed) for the given technologies available? • What level of performance is currently available today, and what is the prediction for the next advancement?

(Continued)

Table 14.2 (Continued)

Type of analysis	Description of problem space	Example types questions
Parametric analysis Analyze a system of components by gathering and analyzing numerical data. Parametric analysis allows engineers to determine how changes in one component of a system impact the other components in the system	Identify the relationships between system variables and inform decisions about system design. This type of analysis is especially useful for complex systems where multiple variables must be taken into consideration	• What parameter drives the behavior of a given metric? • What is the probability of achieving a given level of performance?
Operational analysis/ mission engineering Analyze a system's operational performance, including its ability to meet system requirements and mission objectives. It involves a comprehensive assessment of a system's operational environment, such as demand, capacity, resources, and constraints	Identify potential problems and opportunities for improvement of a system by assessing its performance, along with the performance of associated systems in the completion of a mission, operation, or campaign, under specific conditions. This can also be used to develop strategies for enhancing system performance	• Given a scenario and a friendly force, what is the likelihood of achieving mission success for a given system-of-systems configuration and capabilities? • What behaviors (e.g., TTPs, CONEMPs) are best suited for employing a new (or existing) system?
Portfolio analysis Assess a combination of discrete components to achieve an overarching capability or strategic vision. Portfolio analyses are used in acquisition planning to allocate resources under uncertain conditions and multiple objectives	This type of analysis is generally classified as a knapsack problem that requires the use of search algorithms, heuristics, and evolutionary algorithms to explore large sets of potential solutions. The analysis may include continuous variables, but is generally limited to discretized ones or ones that can be directly defined as a Boolean satisfiability problem	• What is the optimal set of components to achieve a goal given a resource constraint? • What items provide the best return on investment for a given set of objectives and budgetary, schedule, and risk constraints?

Measures of Success

Tightly related to the former consideration is an understanding of the success criteria for the DETAF. Beyond providing insight into the identified questions, one should consider what will be the return on investment from developing a DETAF? For example, with the current initiatives across the DoD to embrace and implement digital engineering in all its forms, one measure of success could be how much of the project team is embracing digital engineering and using the analytical framework to support their activities. Are the team of decision-makers leaning on the environment to be an authoritative source of truth? Is the framework embedded within the overall process perceived as a critical element to achieving the goals of a team's charter? With these types of success criteria, buy-in and involvement by the stakeholders is critical and arguably more critical than the exactness of the analysis or outputs. For the DETAF to be strongly adopted and viewed as a critical element of the larger process, it must first be trusted as an authoritative source of truth, which requires that the stakeholders have a nominal understanding of the underlying mechanics. If the DETAF is perceived to be a "magic 8-ball" by the decision-makers, they will trust it as long as the results agree with their perceived notions, but that trust is fragile. Accordingly, the level of complexity of models utilized will be heavily influenced by the experiences and knowledge of the team members, even more so than the fidelity or accuracy of the models. Again, this example is focused on success criteria of organizational development and less on the engineering challenge presented, while other efforts will focus on the rigor within the analysis, which will alter the overall approach taken. A high-level illustration of considerations necessary to develop a strong adoption dynamic of a DETAF is shown in Figure 14.3.

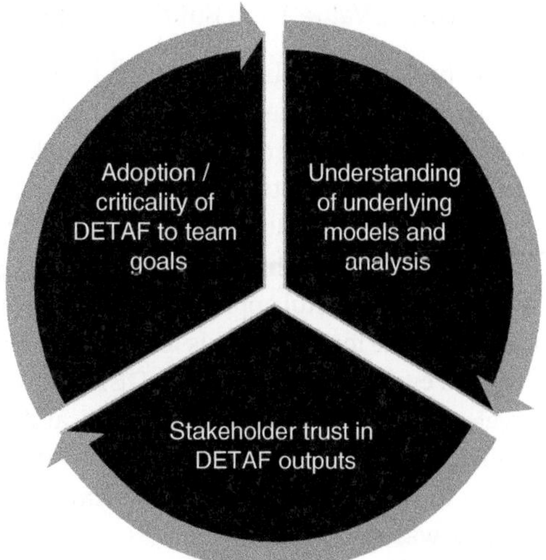

Figure 14.3 Building to strong adoption of a DETAF in a decision process often starts with the stakeholders needing a firm understanding of the underlying mechanics.

Users

A DETAF is made for a userbase. The role of the individuals in the userbase, what skills they possess, and their preferred means to analyze problems must all be primary considerations of the DETAF development team. Specifically, the development team must focus on the user interface and tailoring required to support the needed workflows and experience levels of the users. Secondly, underlying analytics need to be informed by the userbase. In some cases, complex mathematical analyses may be possible, but building trust in the DETAF outputs comes from the stakeholders understanding the innerworkings, which may drive the selected implementation approach. Similarly, different userbases will prefer to visualize information in different ways; a trade will need to be considered between educating users on new visualizations vs. utilizing familiar visualization techniques. The DETAF development team should keep in mind the tool is intended to support a decision-making exercise, and that exercise is accomplished by humans who have experience, biases, and presuppositions. Another common pitfall is pursuing novel tools (e.g., visualizations, algorithms) that do not merit use. That is not to say that DETAF teams should not be staying abreast of recent developments, but experimenting with new techniques and technologies should be balanced with the intent of the project, stakeholders' expectations, and the resources and time available.

In certain instances, a DETAF may be intended for use by a wide variety of users. In these situations, additional upfront work mapping the use cases to defined user workflows and appropriate interfaces is required. Different groups of users may require different user experiences to accomplish different elements of the overarching workflow. Furthermore, understanding the users will drive the development process with respect to software engineering rigor. If the tooling is designed to be used by the creators, then the creators are capable of debugging or resolving issues that could arise. The wider the userbase, the more rigorous the development process. This latter case may require unit testing of components, implementing checks on inputs and outputs for additional automated verification, and a means to catch and record errors that occur during use. Depending on these considerations, the software implementation approach could vary greatly. This tends to be a primary driver in the overall timeline and resources required to realize a DETAF.

Demonstrate Value Then Build in Robustness

At the beginning of the design and implementation of a DETAF, focus on outcomes. Once the three considerations described in the prior section (questions to be answered, measures of success, and users) have informed the design, seek an incremental approach in development where real value can be demonstrated early and often. A completed DETAF that comes months after the process begins is too late. More often than not, a team chartered to explore a tradespace has already moved on with something simple, perhaps a SME-based approach, to generate recommendations. A common successful approach is to identify the minimum viable product (MVP) that is necessary to create value for the analysis team, implement it, and deliver it. This generates buy-in which then creates a cycle of investing in the DETAF to enhance its offerings. It is reasonable to assume expert developer users are the first set of users, and it is wise to not bog the development down in software engineering rigor that is only valuable after the requirements and features are well understood. Trying to develop a "high-quality" framework often leads to delays, high costs, and is likely to produce wasted effort as the analysts and stakeholders improve their understanding of the problem. This understanding can only come from use of the framework. If the end goal is a widely adoptable framework, lock-in requirements through rapid prototyping and use, then use that prototype to inform a comprehensive set of detailed requirements that can then, in a parallel effort, be used by a software engineering team to develop the end product intended for long-term use and support.

Analytical Workflow with DETAF

Though all five elements highlighted in Table 14.1 are necessary to realize a DETAF, the specific manifestation of the tooling and workflow will vary from one engineering challenge to another. However, to support the reader in better understanding how these five elements are integrated together to realize a DETAF, the authors provide an example workflow here for consideration. Furthermore, Table 14.3 presents potential alternatives for each of the five steps, offering additional context to those interested in developing their own implementations.

Table 14.3 Notional matrix of alternatives for a Digital Engineering Tradespace Analysis Framework.

Element	Alternatives			
Stakeholder needs	Pairwise comparison (e.g., AHP)	Ranking (e.g., Rank ordered centroid)	Multi-attribute utility theory	Prospect theory
Predictive models	Campaign model (e.g., STORM)	Mission model (e.g., AFSIM)	Aircraft sizing and synthesis model	Component SWAP model
Execution orchestrator	Fixed point iteration	Evolutionary algorithms	Response surface methodology	Bi-level integrated systems synthesis
	Design of experiments	Artificial neural networks	Mixed-integer linear programming	
Visualization environment	Excel dashboard	JupyterLab dashboard	Custom JavaScript	Tableau
Reporting and archiving mechanisms	PowerPoint briefs	PDF reports	Git repository of models, result data, and analyses	ZIP file of digital engineering environment

Workflow Assumptions

For the workflow in this example, it is first assumed that the components in the workflow will be automatable to facilitate the generation and evaluation of large sets of design points. Thus, the mappings from design variables to capabilities and any intermediate steps must be deterministic. Other additional assumptions can be made to best meet the demands of the problem. For example, in the authors' experience, Department of Defense acquisition-focused tradespace analysis generally deals with discretized tradespaces. Meaning that, instead of identifying trades between continuous variables, a set of existing (often Government or commercial off-the-shelf – GOTS or COTS) component options offering a subset of desired functionality are used. Integrating these components together leads to variations of the new capability sought. Time to field is the primary driver for this approach; opening the tradespace to nonexistent technology would further increase a concept-to-field timeline the DoD is continually working to reduce.

Workflow Overview

Figure 14.4 offers a view of the workflow implemented by the Georgia Tech Research Institute in collaboration with different teams of Government engineers and analysts on multiple projects. Each step and component of the workflow is described in detail below.

Develop Need Statement/Define Capabilities and Increments

The first step of any design space exploration requires a well-defined overall need statement driving the search for a new solution. This component is just one part of the **Stakeholder Needs** element identified at the beginning of this chapter. Most often, authoritative documents from DoD leadership describe the need or capability gap. For consumption within the tradespace team, the need should be reiterated into a single statement or short paragraph, the goal being to offer a succinct explanation of the driving force behind all efforts without prescribing a particular solution.

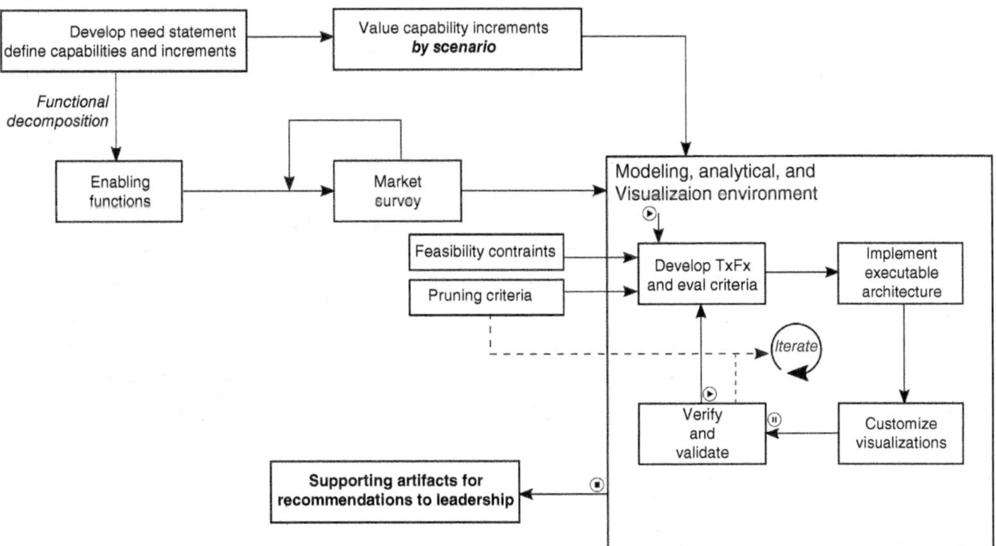

Figure 14.4 An example workflow which demonstrates the integration and interaction of the five primary components of a DETAF.

Additional detail beyond the need statement is captured in the definition of the potential capabilities the final solution may offer. Capabilities should include deterministic evaluation criteria, possibly based on a set of defined functions or on a set of captured performance parameters. Both capabilities and calculated performance parameters will be utilized as a means to compare the relative value of each alternative. Communications range or threat survivability are examples of such capabilities. Preconceived assumptions of the capabilities should be avoided in the early stages. The capabilities should capture both existing and reasonable, albeit futuristic, possibilities that create opportunities for designs to be distinguished from each other. As capability increments are valued in later steps, those that are deemed of little or no value will fall out, providing a traceable artifact of the down-select process.

Within each capability, discretized levels of capability (or capability increments) need to be defined. For example, communications range could be broken down into the following increments: local, line of sight, and over-the-horizon. Alternative problems may decompose this same capability differently to take specific electromagnetic bands or communication paths such as terrestrial verse satellite into consideration.

Any part of the workflow can and should be considered part of a living document where possible. The flexibility to update capabilities or any other element of the workflow allows the process to adopt to the problem at hand.

Value Capability Increments (*by Scenario*)

This stage continues the development of the **Stakeholder Needs** element. The value of a capability increment or combination of capability increments can vary across scenarios and in relation to another. It is likely that each capability increment will fall into one of the following categories:

a) Valued in one scenario
b) Valued across multiple scenarios
c) Not valued in any scenario

Depending on the methods used to capture this information it may also be possible to determine the interrelationships between the capabilities. . .

> . . . valued without regard to other capabilities.
> . . . valued only in combination with other capabilities.
> . . . not valued in combination with other capabilities.

Several methods exist for this process and, in general, they each call for SMEs to provide this information in a systematic way. This part of the process is a high opportunity moment in the project timeline. It is the optimal time for access with SMEs, but this time will be wasted without proper preparation. As more complex information is sought, more preparation is required. From experience, the authors' have seen how these workshops effectively capture the required information in a way that SMEs have confidence in and also promote conversations that SMEs were not previously aware of needing to have.

Functional Decomposition → Enabling Functions

Capability increments are decomposed into enabling functions. For example, the function of directing radio transmissions and the function of receiving directed transmissions would both be required for the communications line-of-sight capability. Functions provide the necessary abstraction between capabilities and the combinations of physical components that functionally perform together to provide capabilities. Therefore, this step begins to bridge the gap from the **Stakeholder Needs** element into **Predictive Models**. Functions can be defined as a mapping from components in the market survey or from performance parameters for the design.

Market Survey

A design solution is a set of component parts. The market survey provides a database of these possible components. Components have attributes such as mass, volume, power, cost, functional attributes, etc. The database is generated from research and data calls. A notional candidate solution is created from a set of components, and so an incomplete database is likely to influence the results of the process. For completeness, low Technology Readiness Level (TRL) or next-generation components should be included as entries, offering insight into future alternative capabilities as compared to today's technology options. Since the results can be significantly sensitive to the market database, it is important to understand and plan around the collection process and any associated time requirements.

The use of a market survey is specifically utilized when considering a discretized tradespace, focused on the integration of existing components to realize a new emergent capability. Alternatively, in this step, one could develop a parameterization of the functions from the previous step. This would be appropriate in a continuous tradespace environment where individual design variables can be set as opposed to selected from existing off-the-shelf solutions. Within this step, work is continuing on the **Predictive Models** element. The population of a market database or parameterization of enabling functions is providing that scaffolding that is required to develop, select, or integrate models for performance evaluation.

Modeling, Analytical, and Visualization Environment (MAVE)

Computing resources are a key enabler of implementing a DETAF. They determine how well a team can quickly generate large tradespaces of solutions, evaluate each solution against the defined capability increments, and visualize results to enable decision-makers to glean information about the impacts of different decisions. In the authors' experience, each problem requires tailoring with respect to the models, orchestration, and visualization elements. Therefore, the recommendation is to have a library of implementations available that can be manipulated and integrated for a specific problem of interest. Use of the environment is an iterative process that involves all team members including decision-makers.

Modeling, Analytical, and Visualization Environment Inputs Figure 14.4 identifies four different inputs into the modeling, analytical, and visualization environment. The dynamic nature of these inputs is a driving reason for the need to automate the environment, and this environment is composed of three of the key elements identified: **Predictive Models**, **Execution Orchestrator and Numerical Exploration**, and **Visualization Environment**.

The first input is the Capability Value Vector. This is the primary means to distinguish "goodness" between two feasible solutions. As discussed above, the relative weighting of capability increments will likely vary based on the scenario in which the in-design solution is deployed. Therefore, the analysis process needs to be executed for each set of weightings. Furthermore, as part of the verification and validation process, it is good practice to conduct sensitivity analysis on the capability increment weights.

The market survey is the second input. As is shown in the workflow, populating the market survey is an iterative process. Often there is lag in receiving replies from a data call or new component options that are identified later in the analysis process. Other times, items in the market research database are removed because they are determined to be unfit or their functionality is not represented accurately. Finally, sensitivity analysis can be conducted on the impacts of adding future technology or removing certain technology options in order to see how the "good" set of solutions changes. Again, this example assumes a discretized tradespace focused on the integration of OTS components. Therefore, in the instance of architecting a DETAF for a continuous design space, the functional parameterization and bounds on the design variables would feed into the analytical environment.

The two inputs newly introduced to the analytical framework are the Feasibility Constraints and the Pruning Criteria, both of which put limitations on performance parameters. However, they are separated based on whether they are (1) clearly determining whether a solution is *feasible* or (2) simply distinguishing *goodness* between two feasible solutions. Often, the feasibility constraints are less dynamic, being set early in the iterative analysis. It is the pruning criteria which may adjust at each iteration to further limit the tradespace into a smaller set of solution options. An example of a feasibility constraint might be the overall mass of the vehicle while a pruning criterion might be a minimum survivability metric.

Develop Transfer Functions (TxFx) and Evaluation Criteria The first step of building the MAVE is defining the logic for evaluating candidate solutions. These transfer functions can vary from low-fidelity mappings populated by SMEs of components to functions – or even directly to capabilities – up to high-fidelity physics-based models and simulations. These transfer functions are one implementation of the **Predictive Models** component of the larger DETAF. A good approach can be to start with low-fidelity mappings or simple models and through sensitivity analysis determine which aspects of the analysis have the most impact in isolating "good" solutions. For these areas of high impact, identify or build replacement higher fidelity transfer functions to improve the trust in the output data.

The evaluation criteria are generally driven directly by the valued capability increments. Generated solutions are evaluated to determine what capabilities – or rather capability increments – they offer. With the capability increments rank ordered and valued, two solutions can be compared directly by their overall *goodness*. Sometimes a base set of capability increments is required. In order to support that, pruning criteria can be introduced to ensure only solutions meeting a minimum set of capabilities are generated.

Implement Executable Architecture The executable architecture in the MAVE is synonymous with the **Execution Orchestrator and Numerical Exploration** component of the DETAF. In order to generate the large sets of designs desired in a thorough tradespace analysis, computing power needs to be properly leveraged. When developing the architecture, consider the following four characteristics listed in Table 14.4 so that the implementation offers such capability.

The authors have utilized various approaches to executable architectures. In all cases, a combination of scientific computing packages, generally available in Python modules, and web-based libraries for a user interface and visualization drawing tool have been used. There are also comprehensive commercial software solutions available which support orchestration of models and simulations across different tools to include custom made codes as well as other commercial modeling tools. Many technologies exist to support the automation of workflows and introduction of software components, such as the transfer functions and visualization generation scripts, for customization of the workflow. The authors have found that although each engineering problem requires some additional development in the implementation of the executable architecture, existing GOTS and COTS software significantly reduces the time and effort required to realize the required solution.

Table 14.4 Characteristics for an implementable executable architecture.

Standardized input formats	Select input formats (and the tools required to author those input files) that are commonly available, specifically to the design team – e.g., CSV or Excel workbooks
Automatic versioning	Ensure each execution of the analysis architecture tracks all inputs and outputs with a specific version number. It is also recommended to collect additional metadata and notes about what drove the specific iteration
Define output visualizations	Early on in the iteration cycles, select a set of output visualizations and have those automatically generated. This eases the comparison of results across executions. Additional visualizations may be added later in the design process
On- and off-ramps	Build the architecture such that there are multiple on- and off-ramps, meaning an execution can be entered at or exited at various points based on the needs of the analysis. This might be particularly useful for sensitivity analysis of a specific portion of the transfer functions

Custom Interactive Visualizations Over the past several years, large libraries of interactive visualizations have been developed, specifically for web-based applications (Bostock 2021). These are a great resource for reviewing visualization concepts and identifying a set most appropriate for the dimensions of the tradespace that need investigation. In the authors' experience, each engineering problem requires a slightly customized set of visualizations, often with a combination of static and dynamic views. This is due to the fact that traversing the hierarchy of meaning from the data generated by the executable architecture is specific to the information that needs to be conveyed, based on the stakeholders' questions. Leveraging libraries of visualization components eases the development of the needed custom visualizations, but it is an important aspect of the analytical framework to explore various alternatives and select and implement visualization strategies that distill the large amounts of information generated by the executable architecture into actionable information.

For discretized tradespaces, such as those for portfolio management problems, see the chapter 36 on "Portfolio Management and Optimization for System of Systems." In that chapter, examples of integrated visualization components demonstrate how libraries of components can be integrated to support the analysis of specific questions.

For continuous tradespaces, there are three types of interactive visualizations that have proven successful across a wide variety of projects and efforts. To describe their functionality, a simple example will be used in which there are four input (i.e., independent) variables (x_1, x_2, x_3, and x_4). These are also sometimes referred to as design or control variables as they are the ones that are under the control of the system developers. The challenge for the designers is to identify the most suitable combination of values that meet four constraints (c_1, c_2, c_3, and c_4) and maximize (or minimize) an objective (*obj*). For illustration purposes, simple linear relationships are used to define how the constraints and objective vary as a function of the input variables. In reality, these could be modeling and simulation results or a regression.

The first general-purpose interactive visualization is often referred to as a *Profiler*, with an example presented in Figure 14.5. In a profiler, the independent variables are plotted as the columns and the dependent variables as rows. This visualization generally requires having a mathematical function to plot, and is well suited for regression models. This visualization is useful for exploring the tradespace, allowing decision-makers to quickly evaluate new combinations. More importantly, this visualization can be useful for verifying that a model is correct. By evaluating the slopes depicted, SMEs can quickly assess if the model trends agree with their understanding of the model.

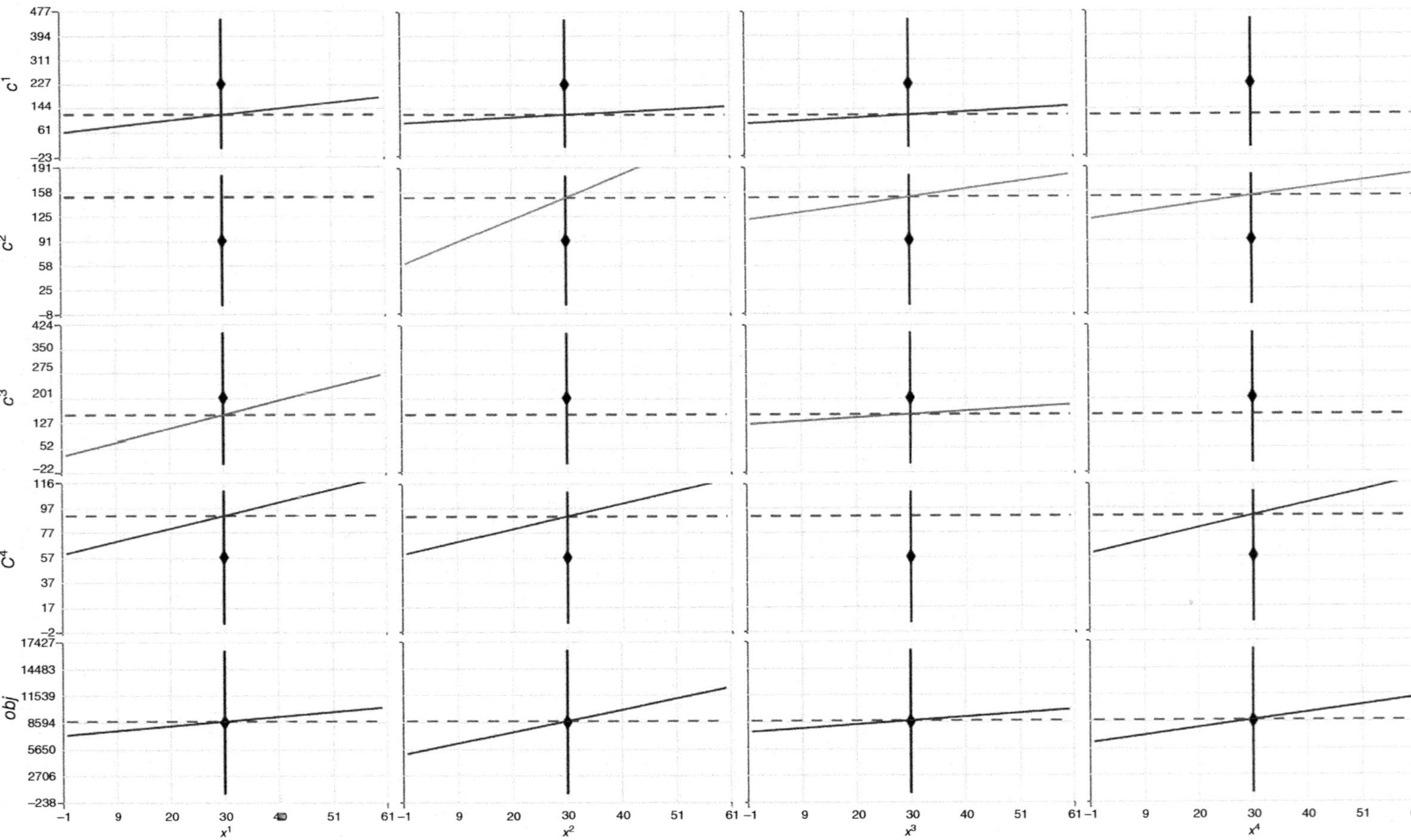

Figure 14.5 Example profiler visualization for a simple problem.

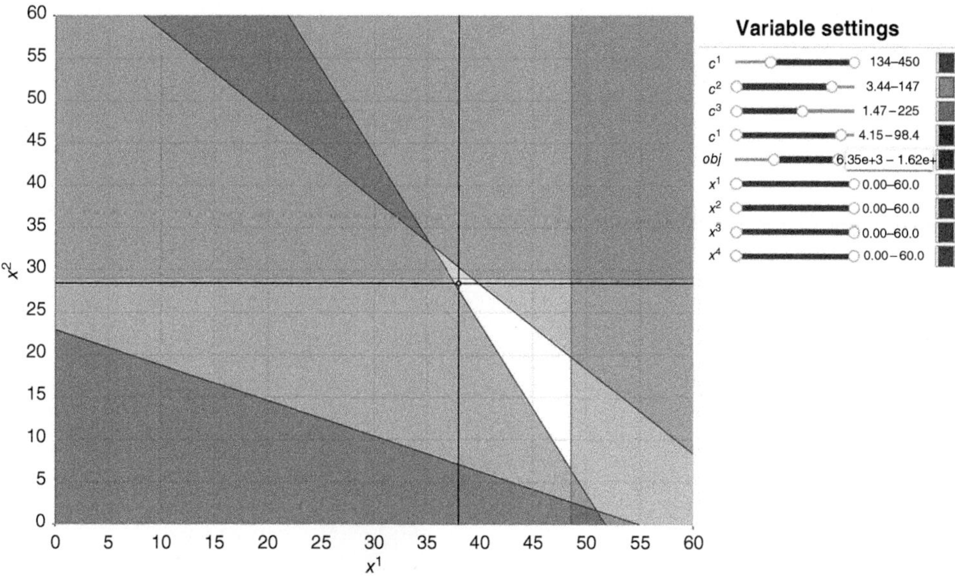

Figure 14.6 Example contour profiler visualization.

It is imperative that the visualization be interactive, as changing the value of one independent variable can change the impact that another variable has on the dependent variables (i.e., the slope in the other plots change). The detriment to this visualization is that it can be labor intensive to explore a large high-dimensional tradespace.

A second general-purpose interactive visualization technique is commonly used in aerospace engineering as it facilitates identifying feasible regions of a design space and understanding the impact of constraints. Commonly referred to as a *Contour Profiler*, an example is presented in Figure 14.6. In a contour profiler, each dependent variable is plotted as a function of two independent variables, in this case x_1 and x_2. In the example presented, the user has set a lower limit on c_1 and *obj*, and upper limits on c_2, c_3, and c_4. As with the *Profiler*, this visualization generally requires having a mathematical function to plot and is well suited for regression models. The white region represents the combinations of x_1 and x_2 that meet all the constraints. It is important to note that if x_3 or x_4 change, the shape of these constraints can change. Multiple contour plots can be integrated to assess different combinations of input variables to facilitate the exploration of the tradespace but, for high-dimensional spaces, this may not be the most effective visualization with which to do broad explorations. In those cases, a more suitable visualization may be the one presented next. Alternatively, an optimizer can be coupled with the visualization to find settings of the independent variables that maximize the unconstrained design space.

The third and final general-purpose interactive visualization is referred to as *Scatterplot Matrix*, with an example presented in Figure 14.7. In a scatterplot matrix, all dependent and independent variables are plotted against each other using scatterplots, with the histogram across the diagonal. Unlike the other two general-purpose visualizations, this visualization requires a dataset. If a parametric regression is available, a Monte Carlo Simulation can be used to generate the dataset by using uniform distributions for the input variables. The value of this visualization is that users can set limits to the inputs and outputs and identify which regions of the design space can still produce a feasible solution. This allows for top-down analysis (e.g., setting limits to effectiveness or cost)

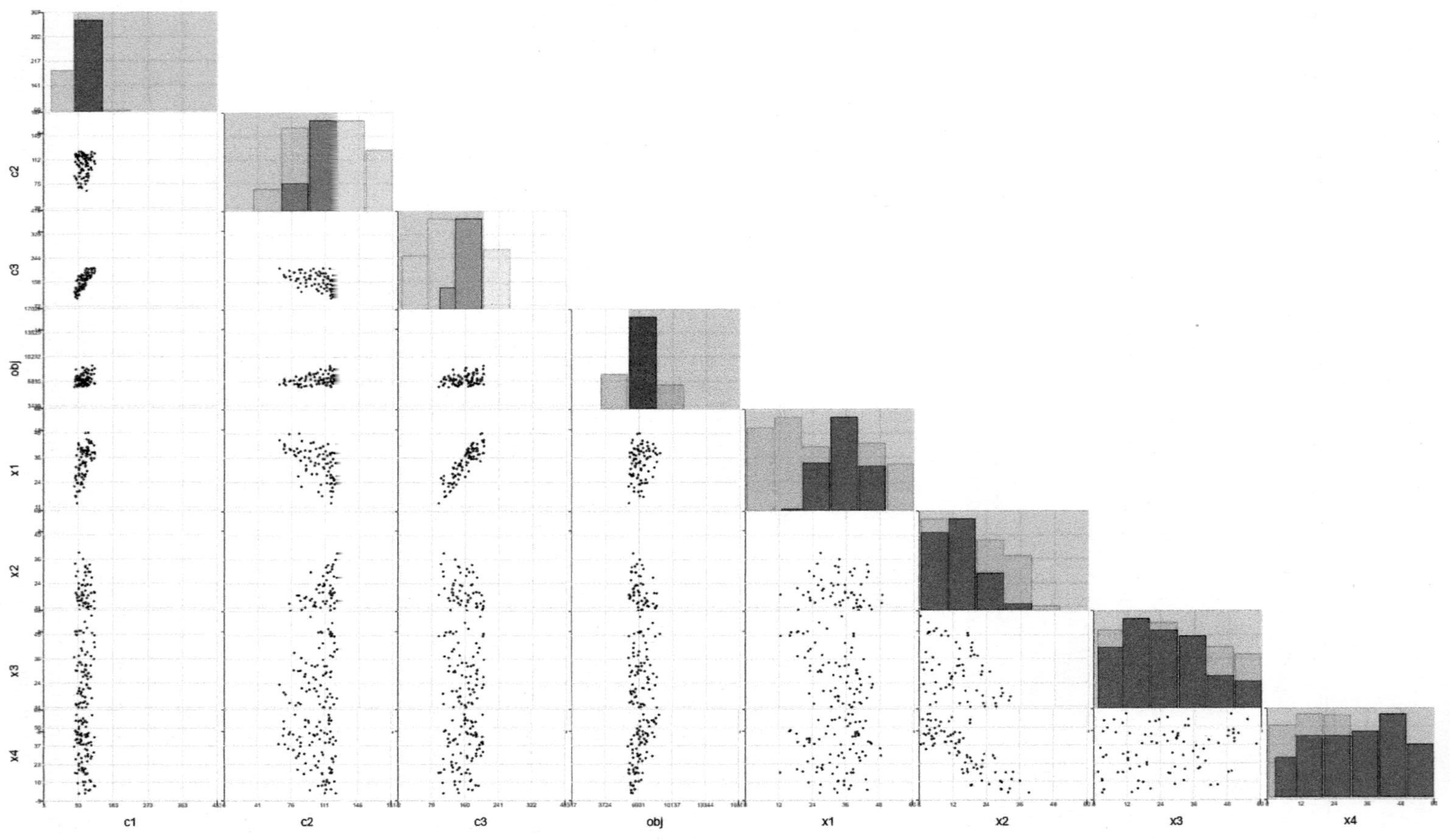

Figure 14.7 Example scatterplot matrix visualization.

and identifying the combination of input variables that can meet them. Alternatively, limits can be set to the independent variables to see what distributions and correlations are produced at the higher levels.

Verify and Validate Trust in the results from the analysis is of utmost importance. Implementing the executable architecture discussed previously is a key enabler to thorough verification and validation. For verification, where possible, automate the testing of the framework to ensure the logic is correct. Create a small set of solutions manually and ensure the automated framework offers the same results. Validation often requires the team to collaboratively review the results and execute a team "gut check." Does the data suggest a reasonable answer to the solution? Is a critical piece of functionality missing and, if so, how can we remedy our design rules to ensure we meet that need? Have certain technology options been over- or under-scored, specifically when utilizing SME mappings for transfer functions? Are we generating the proper visualizations to explore the pertinent dimensions of the design space?

Each engineering problem will require a different level of verification and validation, but the authors' experience has been that after selecting a level of fidelity for the full set of transfer functions and evaluation criteria, on the order of dozens of iterations are required before trust is built in the solution sets offered by the framework.

Iterate Iteration on the solution space is a key component of the analytical framework and tradespace analysis in general. Early iterations are required to conduct verification and validation of the transfer functions, evaluation criteria, executable architecture, and visualization output. After trust is built in the results, iterations are necessary to explore the tradespace, introduce new pruning criteria, and develop recommendations for solutions.

Supporting Artifacts for Recommendations to Leadership

The end goal of the process is to inform decision-makers. This is the **Reporting and Archiving Mechanism** component of the DETAF. An advantage of implementing the executable architecture with automation of the visualization outputs is standardization of the material presented to leadership. An additional practice that has proven successful in the past is briefing decision-makers on the look and feel of output artifacts early in the analysis process. The goal is to prepare leadership for what they will see in the future at the conclusion of the study and offer a chance to familiarize themselves with how to read and understand the results. This additional preparation allows the review of the analysis results to be more productive, with the focus being on what can be learned from the information rather than learning how to read the artifacts.

Feedback Loops

Note that for simplicity, not all feedback loops are displayed in the workflow in Figure 14.4. For example, early iterations of the analytical framework often identify missing capabilities required by the end solution. This requires a loop back to the start of the process to add a capability, additional functions, and re-evaluate the market survey results to see if additional inputs are required. Additionally, as interim results are presented to decision-makers, different aspects of the tradespace might be of specific interest to them. This could require enhancing the set of transfer functions and evaluation criteria, or building new visualizations to display different dimensions of the tradespace.

Lesson Learned

The authors share many experiences in the design, implementation, and use of different tradespace analysis environments. Across these experiences they have gathered a few key lessons which guide their continued research and development in the context of DETAFs. These lessons are highlighted specifically because of the high correlation between approach or expectation management within each context to overall project success, as seen by both the developing team and the decision-makers.

What Has Led to a Successful DETAF

In a recent effort the authors assessed 24 projects, spanning a decade, where each developed a DETAF to support tradespace analysis for a specific problem space. The goal of this study was to identify what characteristics tended to appear in the more successful vs. less successful efforts. The team assessed these projects based on quantitative and qualitative data. Quantitative data was obtained for the level of effort required, the experience of the performers, the period of performance, and the resources available (i.e., funding). Conversely, qualitative data was obtained for the considerations as captured in Table 14.5.

Table 14.5 Description of qualitative data captured in assessment of previous DETAF efforts.

Criterion	Qualitative data intended to be captured	Quantified data capture approach
Customer satisfaction	How satisfied were the customers at the conclusion of the project? Did the customers use the DETAF to help them make their decisions?	Used a 5-level qualitative scale based on formal and informal customer feedback
Insight provided	How insightful were the answers provided by the DETAF? A low score would be assigned if the results were evident without the DETAF or the DETAF did not help produce insight into the problem This could sometimes be due to externalities beyond the control of the customer and the DETAF development team, as it is not possible to predict what results will be produced a priori	Used a 5-level qualitative scale based on the recollection of the performers
Internal satisfaction	How satisfied was the DETAF development team at the conclusion of the project? This combined the satisfaction of producing a product to be proud of and burnout from overworking to meet the project's needs	Used a 5-level qualitative scale

Table 14.5 (Continued)

Criterion	Qualitative data intended to be captured	Quantified data capture approach
Technology advancement	How much did the DETAF team have to advance its software development and analytical methodology capabilities throughout the project? Did the project require the team to "think outside the box" or could they simply apply well-known and established technologies, techniques, and methods?	Used a 5-level qualitative scale
User level expertise required to use the DETAF	How broad is the userbase of the DETAF intended to be?	Used a 5-level qualitative scale ranging from "The DETAF is intended to be used primarily by the developers" to "The DETAF should be usable by competent users after completing a tutorial"
Seniority of Advocates	How senior were the advocates for the DETAF on the customer side?	Used a 10-point scale that correlated with the officer ranking of the most senior advocate, e.g., if a Navy captain was the highest ranking advocate the value would be 6.
External domain expertise required	How much did the DETAF development team have to rely on external experts to model the problem?	Used a 5-level qualitative scale. On the low end, the development team retained the necessary knowledge while on the high end the team was entirely dependent on outside expertise
Methodology complexity	How advanced were the methods and techniques used? Were they simple and traditional or did the team need to integrate multiple nontrivial technologies, techniques, and/or analysis methods?	Used a 5-level qualitative scale
Implementation complexity	How complex was the DETAF? Was it a simple Excel workbook (low end), or a full-stack web-application with complex visualizations and data structures (high end)?	Used a 5-level qualitative scale

It is important to emphasize that the data set was limited but some trends could be clearly observed. The team made every effort to ensure that the data elicited was consistent across projects by maintaining a common pool of performers that were familiar with the projects assessed. The team analyzed the data to determine which characteristics tended to produce projects with the highest customer satisfaction and that provided the most insightful results. Of all the other qualitative and quantitative factors considered, the clearest predictor for those two factors were the "User Level Expertise Required to use the DETAF" and the "Seniority of Advocates." The most successful projects developed tailored DETAFs for a well-defined userbase that had a fairly senior advocate on the customer side. Conversely, the least successful projects tended to be those that required building a DETAF for a poorly defined userbase or one that was broad with a wide range of skills. Generally, projects in which the user was a developer that manipulated the tool with the decision-making stakeholders tended to be more successful, as opposed to projects where significant effort was put forth into software engineering a robust tool. One theory as to the cause of this trend is that having developers use the DETAF with the stakeholders enabled the rapid introduction of alternative approaches, sometimes within a single working session or the next day. Increasing the rate of iteration increased the scope of the solution space explored, which resulted in better outcomes. Additionally, having an advocate at the O-6 (Navy Captain or Colonel) or higher level also proved to be common in the most successful projects. Having appropriate resources, with experienced developers and analysts proved to be important as well, but was not as clear a differentiator as the other two factors described previously.

Software Scoping

Scoping of the software engineering effort is highly dependent on the intended end user. Frameworks such as the Jupyter ecosystem (Kluyver et al. 2016) provide an interactive scripting environment that enables rapid development of capability. However, these environments do not come with guard rails to manage user workflow, comprehensive error handling, or highly tailored graphical user interfaces. In instances where the end user is a developer or comfortable in a software development environment, these drawbacks are acceptable given the advantages of rapid prototyping and analysis exploration. If the user base is broad with many different types of users, where workflow needs to be closely managed as well as input and/or error handling, sufficient time and resources need to be allocated to account for this software engineering rigor. Additionally, before making that investment in software engineering, beginning in a rapid prototyping environment will payoff greatly, refining user requirements and desired workflow through rapid iteration before transitioning to development of the product for the broader userbase.

Trade Between Model Fidelity and Cost

Ideally, a decision-maker wants the best information available. However, that best information can come at a significant cost with respect to manpower, time to develop the models, and computational resources to execute the models. Often the analytical return from the higher fidelity model is not sufficient to justify the additional resource expenditure. Model fidelity is directly related to the core questions the DETAF is intended to answer. More detail is not always better, and a fast-running, low-fidelity model may provide greater insight to decision-makers, as they are able to explore a lager tradespace in the same time. In order to fine-tune model fidelity, conduct sensitivity analysis across the set of lower fidelity integrated models. From there, determine which subset of models has the great influence on the potential outcomes, and chose to invest higher fidelity modeling time in these key areas.

Key Takeaways

This chapter provided guidance on the design and implementation of a tradespace analysis framework within a digital engineering environment. A goal of such an environment is to traverse the hierarchy of meaning, taking the knowledge from stakeholders, to distill it down into specific data to be modeled. Then in turn, those models can be leveraged to generate new data, with the framework also supporting traversing the hierarchy back up to knowledge to support decision-making. Successful architecting of such a framework requires significant involvement from stakeholders; ample iteration on the data, models, and outputs; and a focus on the outcomes of the overall process. The following subsections highlight these takeaways for a reader to guide their own practice of tradespace analysis framework architecting.

Stakeholder Involvement

The end success of a DETAF is fully dependent on the buy-in and involvement of the decision-makers and their trusted team. Involvement starts at the beginning, scoping the questions to be answered, defining measures of success, and clarifying the intended end user of the digital engineering environment and its visualization artifacts. Clarify workflow and outputs before making a significant investment in software development, and ensure the decision-makers understand the visualization artifacts such that they do effectively support the traversal of the hierarchy of meaning from data into knowledge. Keep the stakeholders engaged with intermediate deliveries of capability to ensure the capability is meeting the intended expectations.

Iteration

A DETAF requires significant iteration on the input data (based on what is available or can become available), types and detail of modeling desired, and end visualization environment. As highlighted in the Lesson Learned section, identify a minimum viable product that offers a piece of value-added capability and iterate from there. Creating value for the tradespace process is important in order to retain support from the analysis team for further investment. Success requires significant investments of time to populate the digital framework, from gathering and structuring data to conducting verification and validation exercises. Be prepared to try one approach, and then pivot based on the user experiences and decision-maker questions. Lastly, expect that delivering on expectations for a tradespace framework capability will result in new expectations. This is evidence of progress to transitioning culturally to digital engineering environments, but this can often be experienced as a moving goal post for an individual project.

Outcomes

The most important takeaway is to focus on the outcomes desired from the DETAF. Ensuring an outcome-focus is maintained is facilitated through defining measures of success upfront and reminding the development team of those throughout the process. At times, use of a DETAF is a means to expose an organization to digital transformation. In those instances, it is the adoption and use of the DETAF which measures success, even more so than the specific analytical outputs. In other instances, it is the DETAF outputs. In those cases, the creation of effective visualizations that enable decision-makers to obtain new knowledge and understanding about a problem space and its potential solutions is of paramount importance.

Similarly, understanding the userbase for the DETAF will guide the architecting process to consider outcomes. Who is the end user intended to be, and how does that inform the software development processes and end deployment environment? This further ties into the importance to demonstrate value before investing in robustness of a toolset. Moreover, leverage rapid prototyping tools to tease out the true stakeholder requirements. This is directly related to the previous takeaway to remember it is all about iteration: trying new approaches, exposing capabilities to the stakeholders, and iterating on the data, models, analytical approaches, and visualizations that compose the DETAF.

The key to a successful DETAF implementation is the usefulness it provides in better understanding the problem and solution space. Do not lose sight of the purpose for insight, in pursuit of an elegant, but potentially lacking of value-add, analysis framework.

References

Ackoff, R.L. (1989). From data to wisdom. *Journal of applied systems analysis* 16 (1): 3–9.

Bostock, M (2021). D3 Data driven documents. Library released under BSD license. Copyright 2021 Mike Bostock. https://d3js.org (accessed January 2023).

Elsawah, S., Guillaume, J.H., Filatova, T. et al. (2015). A methodology for eliciting, representing, and analysing stakeholder knowledge for decision making on complex socio-ecological systems: from cognitive maps to agent-based models. *Journal of Environmental Management* 151: 500–516. https://doi.org/10.1016/j.jenvman.2014.11.028.

Kluyver, T., Ragan-Kelley, B., Perez, F. et al. (2016). Jupyter Notebooks - a publishing format for reproducible computational workflows. In: *Positioning and Power in Academic Publishing: Players, Agents and Agendas* (ed. F. Loizides and B. Schmidt), 87–90. IOS Press.

Landauer, C. (1998). Data, information, knowledge, understanding: computing up the meaning hierarchy. In: *SMC'98 Conference Proceedings*, 2255–2260. San Diego, CA, USA: IEEE International Conference on Systems, Man, and Cybernetics (Cat. No. 98CH36218). doi:https://doi.org/10.1109/ICSMC.1998.727491.

Lubars, M., Potts, C., and Richter, C. (1993). A review of the state of the practice in requirements modeling. *Proceedings of the IEEE International Symposium on Requirements Engineering*, San Diego, CA, USA (6 January 1993). pp. 2–14. IEEE.

Rowley, J. (2007). The wisdom hierarchy: representations of the DIKW hierarchy. *Journal of Information Science* 33 (2): 163–180.

West, T.D. and Birkmire, B. (2020). AFSIM: The Air Force Research Laboratory's approach to making M&S ubiquitous in the Weapon System Concept Development process. *Cybersecurity & Information Systems Information Analysis Center (CSIAC)* 7 (3): 50–55.

Biographical Sketches

Daniel C. Browne, Georgia Tech Research Institute

Danny Browne is a principal research engineer and chief of the Systems Engineering Research (SER) Division at the Georgia Tech Research Institute. SER Division conducts research in the design and application of best practices in model-based systems engineering (MBSE) and open architectures, transition of DoD acquisition to a digital engineering approach, decision methods development, and application of strategic decision systems. Danny's current research focuses on the application of MBSE and decision systems to support DoD acquisition and portfolio tradespace analysis.

Santiago Balestrini-Robinson, PhD, Georgia Tech Research Institute

Dr. Santiago Balestrini-Robinson is a senior research engineer at the Electronic Systems Laboratory (ELSYS) in the Georgia Tech Research Institute (GTRI). His primary area of research is the development of opinion-based and simulation-based decision support systems for large-scale system architectures. Dr. Balestrini-Robinson has led teams supporting multibillion-dollar military capital equipment acquisition programs. His interests in quantitative modeling and simulation span multiple domains, ranging from large-scale military campaign level analyses to the development of "peace-time" strategic-level cyber defense models, as well as humanitarian aid and disaster relief analyses. Together with quantitative modeling and simulation-based analysis, he has also developed novel opinion-based decision support tools. These tools enable decision-makers to collaboratively explore a decision space using an end-to-end, traceable and defensible integrated and interactive visual decision support system to explore multi-level, multi-domain trade-offs that leverage single or federated sources of truth.

D. David Fullmer, Georgia Tech Research Institute
David Fullmer is a senior research engineer and head of the Applied Decision Systems Branch (ADSB) in the Systems Engineering Research Division at the Georgia Tech Research Institute. ADSB conducts DoD sponsored research in the design of decision frameworks, utilizing modern decision and analysis approaches including machine learning algorithm. He leads the Navy's Force Level Integration (FLINT) effort, which provides a structured tradespace analysis approach for the Navy's integrated budgeting process, with custom developed software hosted on SIPRNet. His research focuses on enhancing the underlying models informing such tradespace analysis tools to provide better information to decision-makers.

Chapter 15

Set-Based Design: Foundations for Practice

Shawn Dullen and Dinesh Verma

Stevens Institute of Technology, School of Systems and Enterprises, Hoboken, NJ, USA

What Is Set-Based Design?

Set-based design (SBD), so-named by Allan Ward in a study of the Toyota approach sponsored by the Air Force Office of Scientific Research, is an analysis method in which designers focus on design alternatives at both the conceptual and parametric levels (Ward et al. 1995). It differs from point-based design evaluation in that SBD preserves consideration of multiple alternatives, and does so through allowing different specialty groups in the process to each consider the design space independently. As elimination of inferior alternatives occurs, the sets gradually narrow until there is convergence upon a final solution. This avoids extensive iteration commonly found in point-based methods and enables a design team to perform a more informed, holistic assessment by preserving more design options longer in the process.

SBD, combined with concurrent engineering, forms set-based concurrent engineering (SBCE), which is the core component of the lean product development process. SBCE has been synonymous with Toyota's SBCE process, which has three principles (Sobek II et al. 1999): (1) map the design space, (2) integrate by intersection, and (3) establish feasibility before commitment. This concept is illustrated in Figure 15.1.

The first principle is to map the design space, which includes defining feasible regions, designing multiple alternatives, and communicating the sets of possibilities. Each functional team will map the design space by exploring their subsystem design space for a set of feasible solutions. In this activity, each subject-matter expert (SME) uses existing knowledge and experience to define constraints for their given functional area. When existing knowledge is not available, the SMEs rely on intuition based on experience, which could significantly limit the feasible region of the design space. SMEs generate multiple subsystem alternatives defining value ranges for each subsystem attribute. The resulting feasible region defines the design space – the tradespace – for each specialty or subsystem area of design. In order to achieve system performance under constraining conditions, each team will communicate their set of possibilities through trade-off curves, limit curves, or other means to identify the intersection of each of the design spaces.

As shown in the second column of Figure 15.1, the feasible design sets from each specialty or subsystem area are integrated, and their intersection produces an integrated set of feasible

Systems Engineering for the Digital Age: Practitioner Perspectives, First Edition. Edited by Dinesh Verma.

Companion website: www.wiley.com/go/verma/systemsengineering

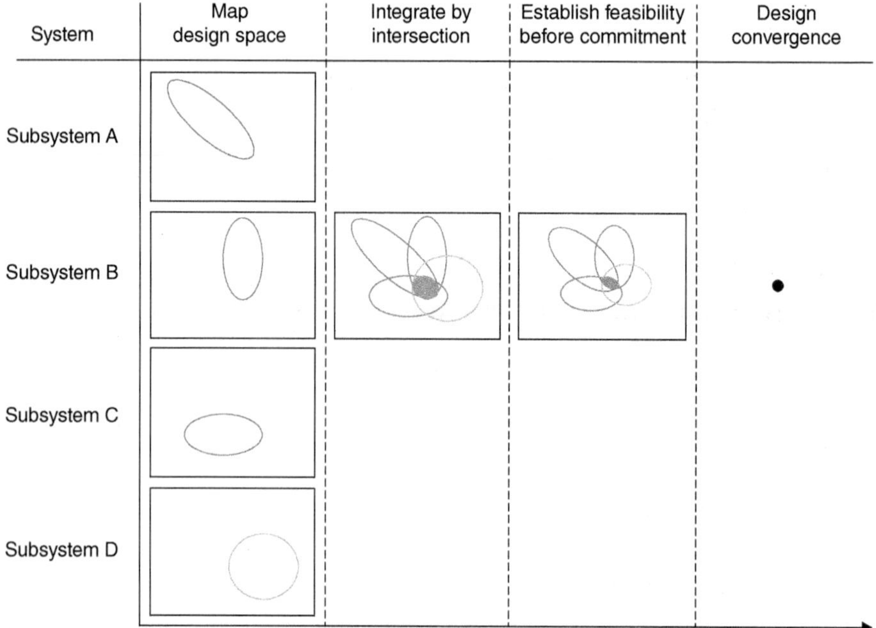

Figure 15.1 SBD Process. Each rectangle represents the entire design space for the subsystem, and each oval represents the feasible region for the subsystem. *Source:* Adapted from Dullen et al. (2021).

alternatives. This step of the process represents the second principle, integrate by intersection, and seeks to ensure conceptual robustness. Additional minimal constraints may be imposed to ensure the feasibility of achieving system performance objectives under constraining conditions.

Producing a strong integrated set of alternatives requires an equally strong approach to conveying information across the different functional areas, a process referred to as the feedback-negotiation cycle (Sobek II et al. 1999). The methods employed here to facilitate communication and the generation of feasible sets that produce a meaningful intersection across all specialty areas can have a significant role in the speed of identifying the feasible, integrated set of system alternatives as well as the nature of that set (i.e., the outcome). In some situations, the SMEs may not be able to negotiate attribute value ranges, thus creating the need for the systems engineer to resolve conflicts. The systems engineer makes the final determination of attribute value ranges and the initial integrated set of system alternatives that will move forward in the process.

The third principle is to establish feasibility before commitment by gradually narrowing sets, staying within sets once committed, and maintaining control. The teams will improve the level of detail for each design alternative as well as the level of analysis and re-evaluate the design space until a final solution is chosen. The level of detail typically starts as hand sketches and a list of parameters and will evolve into fully detailed three-dimensional (3-D) computer-aided design (CAD) models and eventually functional physical prototypes. The initial level of analysis might start from basic first principles models and mature to dynamic multi-physics models and physical testing. The multi-physics models can be considered "virtual prototyping," where information uncertainty is reduced as the model increases in fidelity, thereby improving the level of analysis. Another means of improving the level of analysis is to test a physical system prototype in an environment directly relevant to the system's intended use and performance. The design alternative sets gradually narrow as the level of detail and analysis are improved. Still, the functional teams will stay within the sets once they are committed, and the control of uncertainty will be managed through process gates. This overall process will continue to iterate until design convergence.

The Set-Based Design Difference

Point-based design (PBD) is a traditional approach where many alternative solutions converge to a single solution at the early stage of the life cycle. This single solution undergoes an iterative process of synthesis, analysis, and modification until the solution either achieves the design objectives or is considered infeasible (Ward et al. 1995). If a result of this process is deemed infeasible, the process starts over with the "next best" alternative solution.

The strength and difference of SBD is in its parallelization. In SBD, as illustrated in the first column of Figure 15.1, engineers develop a set of design alternatives in parallel at different levels of abstraction. As the design process progresses, the additional information gleaned from the various types of analyses across the area teams enables the prospective set of alternatives to be gradually refined and narrowed until the engineers are able to converge to a final solution (Sobek II et al. 1999).

What distinguishes SBD from the more widely practiced PBD is SBD's emphasis on reasoning about sets of design options (Braha et al. 2013). Ghosh and Seering developed two principles to articulate a working description of SBD, known as set-based thinking (SBT): (1) considering sets of distinct alternatives concurrently and (2) delaying convergent decision-making (Ghosh and Seering 2014). Ghosh and Seering referred to PDB as meeting neither of the two principles. To best understand SBD, however, it is important to understand the difference between similar methodologies. Creating multiple design alternatives in parallel and converging to a final solution is not a new concept. Several methods have been developed that exhibit similar traits such as the Pugh Controlled Convergence (PuCC) method (Pugh 1981), the design-build-test cycle (Wheelwright and Clark 1992), and Trade Space Exploration (Stump et al. 2004).

The PuCC method begins with engineers and designers creating a large number of design alternatives, which are then compared against a baseline design for each of the evaluation criteria using a criteria-based decision matrix called a Pugh matrix. The design alternatives with high scores are retained; those that have negative scores are discarded. Once the initial number of design alternatives has been reduced, the design team looks for opportunities to create additional design alternatives through either hybridizing some designs, modifying others, or pursuing new approaches entirely. The resulting set of design alternatives is again narrowed using the Pugh matrix evaluation process. Once again, additional design alternatives are pursued. This process of converging–diverging continues until a superior concept is selected as shown in Figure 15.2.

At first glance, the PuCC method exhibits both SBT principles. Multiple distinct design alternatives are considered concurrently, and there is delayed convergence. However, the PuCC method delays convergence but does not improve the level of information used to make a decision while the second principle of SBT requires that convergent decision-making is delayed until information is more certain. The PuCC method also does not employ the parallelization that underlies the power of SBD.

The design-build-test cycle begins by framing the problem. This activity includes clarification of stakeholder needs, defining product and manufacturing process requirements, etc. Once a clear and concise set of requirements is established, multiple design alternatives are developed across which engineers may explore the relationship between design parameters and specific customer

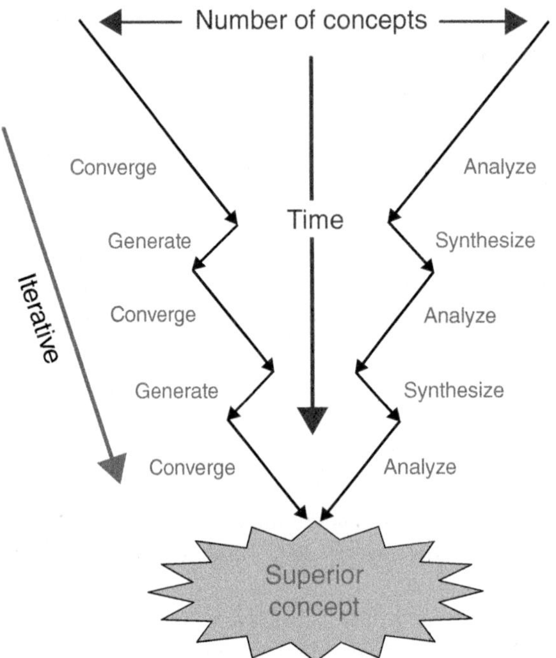

Figure 15.2 The PuCC method: converging–diverging processes.

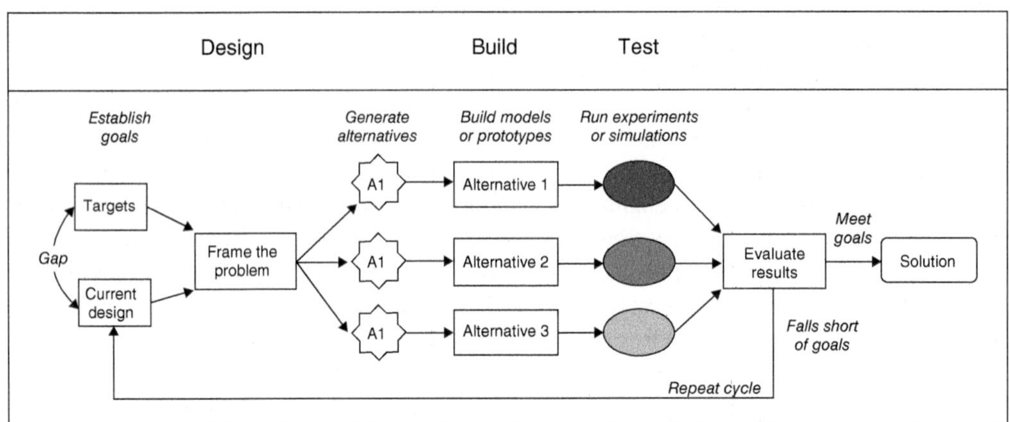

Figure 15.3 The design-build-test cycle. *Source:* Adapted from Bernstein (1998).

attributes (Wheelwright and Clark 1992). Virtual prototypes, which are simulated, or physical prototypes, which are tested, are then constructed to develop additional knowledge. The results from the simulations and tests are compared against the requirements. If the results do not meet performance requirements, engineers will search for design changes to improve performance. The entire design-build-test cycle is then repeated until all of the requirements are fulfilled, as illustrated in Figure 15.3. Compared to SBD, the design-build-test cycle meets the first principle of SBT but not the second in that it does not delay convergence of decision-making.

The final method included here for comparison to SBD is tradespace exploration, a widely practiced approach used by SMEs to help them understand the impact of their design choices and to

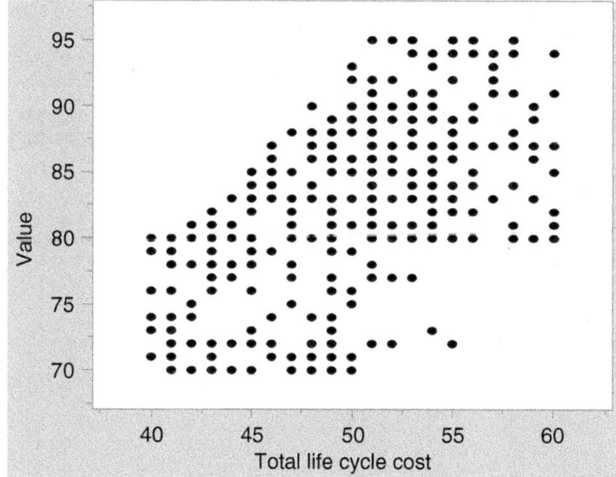

Figure 15.4 Typical tradespace representation.

define the feasible region of the design space (Miller et al. 2013). "*Tradespace exploration is a process by which a large number of alternative designs of the same system are automatically generated and graphed against two or more objectives*" (Collopy 2018). In a typical tradespace analysis process, several key steps are commonly followed:

1) SMEs build models for the system objective functions;
2) SMEs simulate these models until the feasible region of the design space is defined, which often requires thousands of runs;
3) Simulation data is recorded in database; and,
4) SMEs visualize the data and explore the tradespace (Stump et al. 2009).

This leads to the type of representation shown in Figure 15.4. Collopy emphasizes that while numerous design alternatives are generated, these are usually variant alternatives of the same basic design and not truly unique design alternatives (Collopy 2018). Consequently, this fails to meet the first SBT principle espoused by Gosh and Seering, which clearly articulates the distinction between developing sets of distinct alternatives vs. variants of a single alternative.

When to Use Set-Based Design

Set-based design is best used during the Material Solution Analysis (MSA) and the Technology Maturation Risk Reduction (TMRR) phases of the DoD life cycle, illustrated in Figure 15.5. This is roughly equivalent to the Concept stage and Development stage of the Generic life cycle described in the ISO/IEC/IEEE 15288:2015 standard (ISO 2015). During these early phases, there are high levels of information uncertainty and ambiguity. Many decision-makers will reflexively converge to a point design too early, leading to increased costs and schedule delays that occur due to the need to rework the design later in the life cycle.

Information uncertainty can result from a lack of precise knowledge about the technical problem, stakeholder needs, future threats, model accuracy, measurement error, behavior of the system in different conditions, and unanticipated interactions. Even if the problem is understood, when

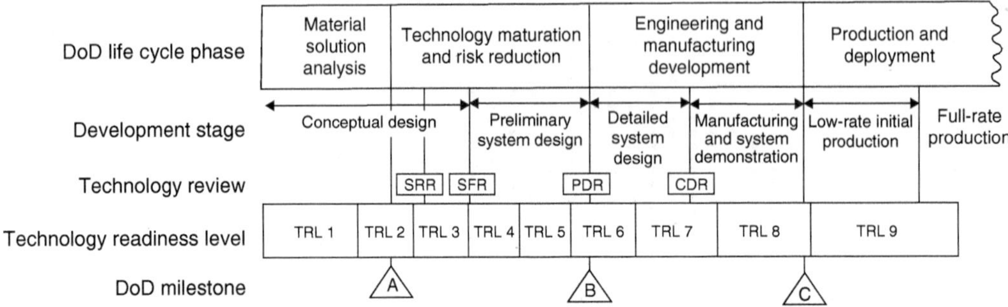

Figure 15.5 Notional mapping of the DOD life cycle, the development stages, the technical readiness levels, technical reviews, and the DOD major milestones.

the value of the other variables is unknown, the technical problem is still considered to be uncertain. If neither the variables nor the mechanism to solve the problem is known, then it is considered ambiguous (Schrader et al. 1993).

These circumstances are very common throughout the MSA and TMRR phases due to the nature of early development, which consists of conceptual design, technology design, and preliminary system design activities. These activities are associated with low technology maturation, whether from innovative concepts present in the original system design or the adaptation of known solutions to a novel application where the performance is untested and unknown. Moreover, stakeholder needs and requirements can be unclear or highly vague during these phases of development. Stakeholders will often not provide "hard" threshold requirements or, they may be convinced a specific requirement or performance measure is nonnegotiable. (Typically, however, requirements are tradeable if they are not constraints or constraining requirements.) Inevitably, new threats emerge and new operational scenarios are defined during these phases, which lead to changes in the requirements (INCOSE 2014). In any development process, the later a requirement change occurs, the greater the potential of significant rework or even new project stemming to accommodate the new needs.

SBD provides a process that is considerably more flexible and robust to changes compared to the other methods reviewed previously. In SBD, a range of potential requirements are identified during the onset of the design process. Changes, due to shifting requirements and stakeholder needs, most often fall into this range. Moreover, the parallelization of SBD means that a more diverse set of variants can be carried forward into the analysis and thereby offer a more encompassing set of alternatives for consideration by decision makers.

Knowledge Development

Relation to Development Phase

The success for any program using SBD lies in its ability to develop knowledge and reduce uncertainty in order to inform decisions with accurate and precise information. While challenging for any product development process, it is especially critical for SBD where there are multiple, potentially very different alternatives being developed and analyzed in parallel as the product progresses through the life cycle. To identify and prioritize areas that will require significant attention

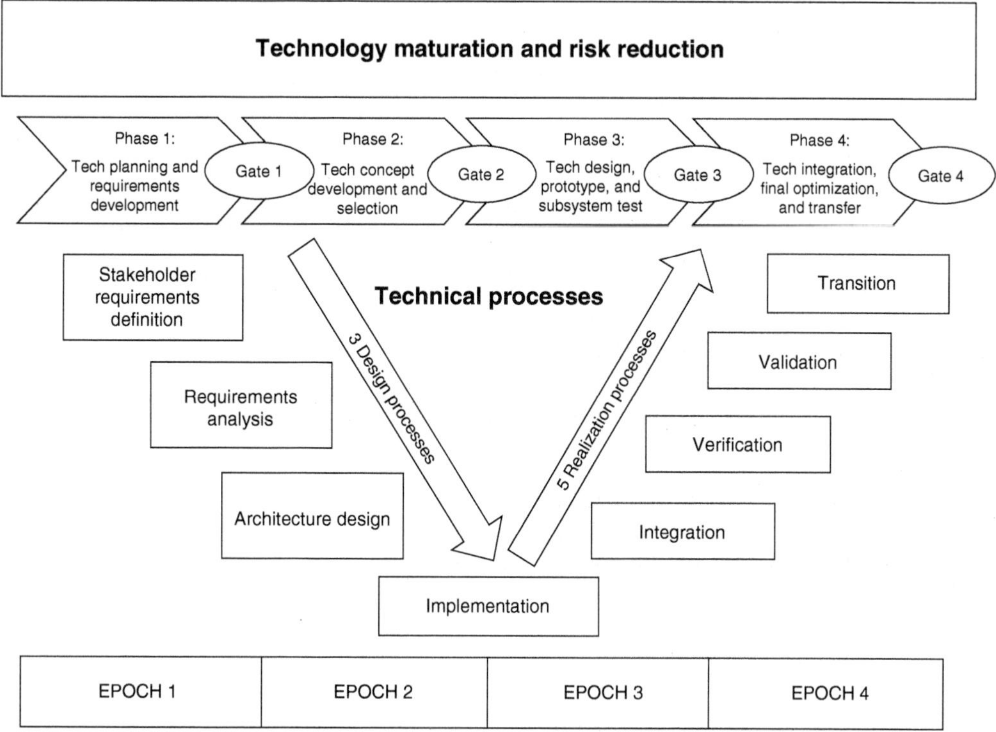

Figure 15.6 Mapping of a phase-gate to the technical processes for the TMRR phase.

for uncertainty and ambiguity reduction, the team will need to develop a knowledge point plan. This plan will need to address key areas such as requirements, user interaction, performance, interfaces, and so on. The development team then needs to align the resulting knowledge value stream, as it is called by the lean product development community, with the product value stream that captures the processes and activities used to develop a product.

Figure 15.6 shows a top-level mapping of the gates associated with successful transitioning from one phase of development to the next together with the technical processes that underly the development for the TMMR phase of the life cycle. In complement, Figure 15.7 presents an exemplar set of activities that would underly phase 1 of the process shown in Figure 15.6. At each gate, there is a major decision on how and whether the product development will proceed further, and the period of time it takes for this activity to occur across the individual phase is called an epoch. Information (i.e., from the aforementioned knowledge value stream) required to make the major decisions at the end of each epoch needs to be developed well and in a timely manner with respect to the phase activities. This will ensure that the implementation of SBD is aligned with the technical processes and their associated activities, thereby supporting effective decision-making at each gate. The knowledge point plan should be revisited over time since improvement in knowledge may lead to changes in priorities. For example, a sensitivity analysis with higher fidelity modeling may determine a new priority that necessitates a change in the knowledge point plan.

At each gate review, the knowledge point plan should be evaluated to assess its effectiveness (e.g., alignment with product value stream, likelihood of improving knowledge, priority of knowledge points, etc.) going forward into the next phase. At each phase, given the unique or changing

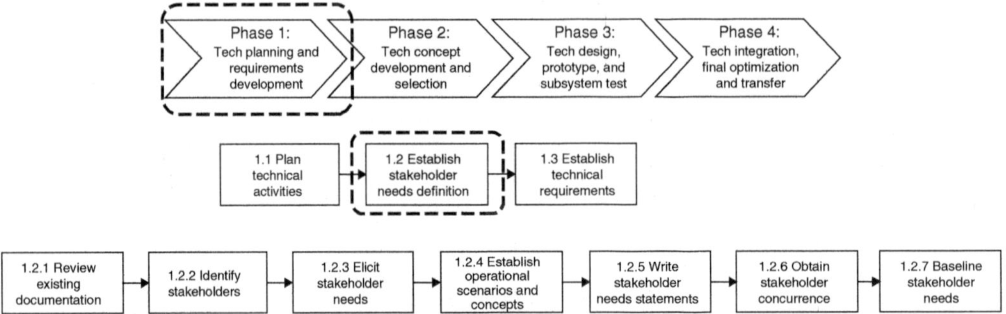

Figure 15.7 Notional sub-activities for phase 1 of TMRR phase.

level of knowledge required to make the next gate decision, the plan should include several key factors:

- an understanding of requirement tradability;
- the potential for those requirements to change or for new ones to emerge;
- the potential for the system to be used in an unintended manner;
- the potential for the system to be used in new environments and conditions outside of the current design envelope;
- validation for requirements;
- requirements that are lacking a method for verification or validation; and
- the value stream and preferences of stakeholders.

The plan should also identify a knowledge elicitation approach including methods such as shadowing or observing users, interviewing stakeholders, using mock-up prototypes or models with which users can interface, using videos or photographs of users operating similar systems to glean needs and opportunities for improvement, etc.

Stakeholder Preferences and Requirements Assessment

As mentioned previously, stakeholders will frequently provide threshold requirements that are not "hard limits" but are instead potentially tradable if not a constraint or constraining requirement. Therefore, particularly during needs elicitation, the development team should strive to identify what is truly constrained vs. what design space is open in order to effectively frame the options. In the case of a potential change in requirements due to change in operational scenarios, emerging threat characteristics, etc., the driving forces behind the potential change, the stakeholders most affected, how the changes should be expressed, and how the impact of these changes is assessed must all be accounted for and captured. Prototypes, models, and mock-ups that provide user interfaces and interaction may help develop the understanding necessary to effectively capture this information through both observation and user feedback.

Understanding and effectively accounting for stakeholder preferences are a critical part of this process as not all of their stated requirements or objectives are of equal value. A common pitfall is to collect preferences from a small set of like-minded stakeholders and ignore the fact that preferences are likely to vary across the full population of stakeholders. Specific approaches for evaluating stakeholder preferences via weighted prioritization and stakeholder valuation of preferences for multiple attributes will be discussed in the section on Design Alternative Analysis Methods and Metrics.

The knowledge point plan also includes activities to reduce information uncertainty and ambiguity associated with a lack of knowledge and to quantify inherent variation associated with any given system alternative. This can be especially challenging in projects that embody a significant amount of innovation with respect to a design. Moreover, as SBD requires multiple, distinct alternatives be developed in parallel, the ability to determine the correct information at the appropriate level of abstraction and point in time will be critical to the success of decision-making regarding design evaluation. Information precision can be accelerated by developing an overlapping strategy for virtual testing and physical testing (Tahera et al. 2017), and it is essential to identify information that is generalizable such that it can be leveraged for all or a majority of alternatives. Trade-off curves and limit curves are considered to be best practice for communicating the generalized information developed from virtual testing and physical testing. This information can be documented into an A3 report for future knowledge reuse. The plan will further need to describe how the inherent variation of a system alternative will be quantified, how measurement variation will be quantified for physical testing to include confidence intervals, how virtual testing environment will be verified and validated to include confidence intervals, how information will be documented, how information will be communicated, and how information will be reused. When presenting the knowledge point plan, a cost–benefit analysis (e.g., value of information) should accompany it.

At the beginning of a development effort, identifying areas that require additional information and priorities for gaining it can be difficult. Qualitative assessment of requirements and system functions can provide an initial evaluation that determines those most difficult to achieve. Requirements and system functions can be classified as "new, unique, and difficult" (NUD) or "easy, common, or old" (ECO). For the former, a new requirement or system function is considered totally new if no one has ever met these requirements or performed these functions before, meaning there is no experience basis at all. Unique refers to the situation where there is no experience on this particular application, but others have achieved some success on similar systems. A difficult classification means that this requirement or function has been attempted in the past and been found difficult to perform. There may be historical data that help assess the level of difficulty. A requirement or system function that is characterized as NUD typically necessitates innovation in design alternatives and is consequently associated with higher levels of uncertainty and ambiguity.

There are also situations in which each requirement or system function may individually be classified as ECO, but the set together for a system design is considered NUD. Consider the requirements expressed in Table 15.1. The first requirement has been achieved by many sports cars, so it is initially classified as ECO. The second requirement has similarly been achieved by many passenger vans, so it is too classified as ECO. Finally, the third requirement has been achieved by several compact cars, so it is classified as ECO as well. Accomplishing all three requirements at the same time, however, would be exceedingly difficult. The compilation of the three requirements, taken

Table 15.1 Example system requirements set.

Requirement statement	NUD/ECO classification
The vehicle shall go from 0 to 60 mph in less than 5 s	ECO
The vehicle shall carry a minimum of nine passengers	ECO
The vehicle shall get a minimum of 40 mi per gallon on the highway	ECO
Design a vehicle that can achieve all three requirements at the same time	NUD

together for a system, is therefore considered NUD. Tools such as a quality function deployment (QFD) correlation matrix and casual loop diagrams can help with assessment of collective sets of requirements, which can be difficult.

Increasing Knowledge Through Testing

For each requirement or system function classified as NUD, the knowledge point plan should describe how the levels of analysis and detail will evolve throughout the knowledge development process. Detail, for example, typically begins with sketches and parameter lists and will eventually mature to descriptive three-dimensional (3D) models and finally to fully functional physical prototypes. Analysis may start as basic first principles models, analogously maturing to dynamic multi-physics models and eventual physical testing. One key to success lies in the design team's ability to identify an optimal learning strategy using a mixture of virtual and physical testing, which often includes hardware-in-the-loop (HITL) testing or a hybrid of modeling and simulation together with HITL.

Figure 15.8 shows a notional progression of gradually narrowing sets of design alternatives as the level of analysis and level of detail is improved. For example, a virtual prototype could be a model that predicts the range and dispersion of a golf ball fired from a catapult. The initial level of detail could be a hand sketch of the catapult with basic parameters. The initial level of analysis could be a first principles kinematics model and first principles trajectory model. As illustrated in Figure 15.8, the initial set of possibilities is established based on evaluating the feasibility of achieving desired objectives. The level of analysis is then increased by developing 3D solid models for the set of possibilities. The level of analysis is similarly improved by using multi-body dynamics, 2D static computational fluid dynamics, and three degree of freedom trajectory models. The set of possibilities is

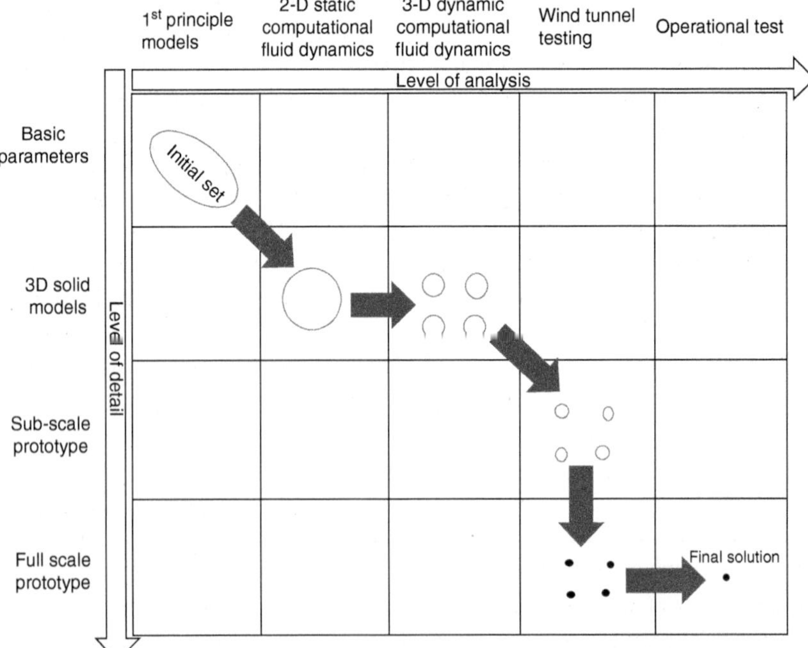

Figure 15.8 Notional narrowing of sets while increasing level of analysis and level of detail.

then re-evaluated. In the notional example, the level of analysis could be further refined through a 3D dynamic computational model and a four degree-of-freedom trajectory model. Again, the set of design possibilities will be re-evaluated and narrowed. Detail may be further augmented through rapid prototyping of the catapult and the golf ball, and analysis through data collected from physical testing. The set of design alternatives is reduced again, leading to the final stage of full-scale prototypes and operational tests that lead to a final solution decision.

Virtual prototypes, computational versions of their physical counterparts, are frequently used to evaluate form, fit, function, performance, and manufacturability. These computational models can be presented, analyzed, and tested from different product life-cycle aspects (e.g., design, manufacturing, logistics, recycling, etc.) in a manner analogous to a physical prototype of a given system (Wang 2002). For virtual testing, the plan should identify specifically what knowledge gap will be addressed through the use of a virtual prototype; the nature of the virtual model(s) and method of construction, assumptions, and limitations; the type of analysis required for the data produced; and how the virtual prototypes will be verified and validated.

Figure 15.9 illustrates the relationship between a conceptual model of a system, its realization in computational model form, and the reality of interest (i.e., that particular aspect of the problem to be measured or simulated, and therefore the basis from which the computational model and validation experiment can be constructed). The conceptual model is an abstraction of the reality of interest that describes the system based on qualitative assumptions about its elements, their interrelationships, and system boundaries.

The conceptual model is used to construct mathematical models, construct computational models, inform physical testing, and define validation requirements. It can be qualified through determination of its adequacy to address the domain of intended application to an acceptable level (Schlesinger 1979) and is validated by determining the degree to which the model is an accurate representation of the reality of interest from the perspective of the intended user (Oberkampf et al. 2002). The conceptual model can include, but is not limited to, free body diagrams, state space diagrams, circuit diagrams, logic diagrams, parametric diagrams, SysML diagrams, and UML diagrams.

The computational model is the numerical execution of a mathematical model that produces the simulation results of specified responses to include the code, type, and degree of spatial

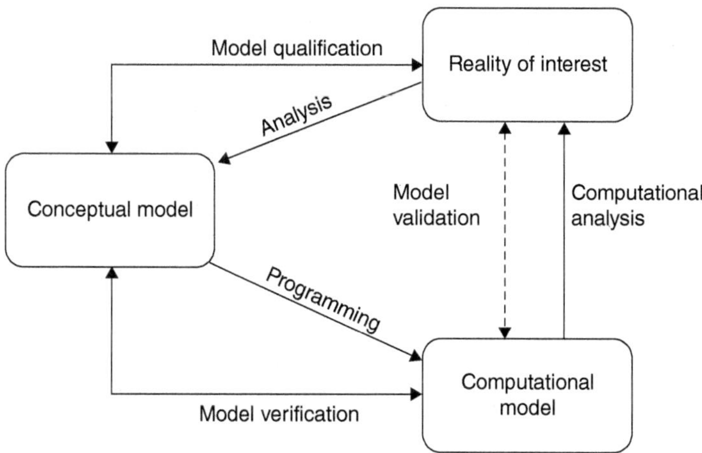

Figure 15.9 Model qualification, verification, and validation relationship. *Source:* Adapted from Schlesinger (1979).

discretization of the geometry; the temporal discretization of the governing equations; the solution algorithms to be used to solve the governing equations; and the iterative convergence criteria for the numerical solutions (ASME 2006). It is verified through determination of whether its implementation will accurately represent the conceptual description and the model solution (Oberkampf et al. 2002).

Outside of the evaluation of system design alternatives, there are trade-offs to consider in selecting the form and fidelity required of a virtual prototype: model fidelity, time, and resources required to develop and execute the model(s) and analyze its results (Gonçalves 2006); cost associated with creating, executing, and analyzing the virtual prototype; its accuracy-to-cost ratio; and how and what will be required to perform its verification and validation. Fidelity is the "*degree to which a model or simulation reproduces the state and behaviour of a real-world object or the perception of a real world object, feature, condition, or chosen standard in a measurable or perceivable manner; a measure of the realism of a model or simulation; faithfulness*" (Gross 1999). Resource considerations should include the computational burden (e.g., number of processers, storage requirements, number of licenses, etc.) and manpower, inclusive of simulation analyst, IT support, procurement, etc. to support the simulation.

Figure 15.10 depicts a potential approach to performing verification and validation (V&V) for the virtual testing phase of development. Path 1 is the virtual testing path and incorporates two verification activities: (1) code verification and (2) calculation verification. The computational model code will be verified to confirm the numerical solution algorithms are correctly programmed and that these algorithms are functioning as intended; it will also be verified for mathematical correctness and discretization errors. The assumptions, rationale, and limitations behind the methods used for verification; the results of this process; as well as descriptions of techniques used for error estimation, configuration management, code architecture, and software quality should all be included in the knowledge point plan.

In turn, the computational model will be validated by evaluating its predictive capability against the physical test results, as illustrated in path 2 of Figure 15.10. Here, the conceptual model is used to abstract the level of detail required to represent the reality of interest. A design matrix is then developed using the conceptual model and statistical methods, physical prototypes are fabricated, and a test plan of these physical models is executed. Measurement uncertainty associated with

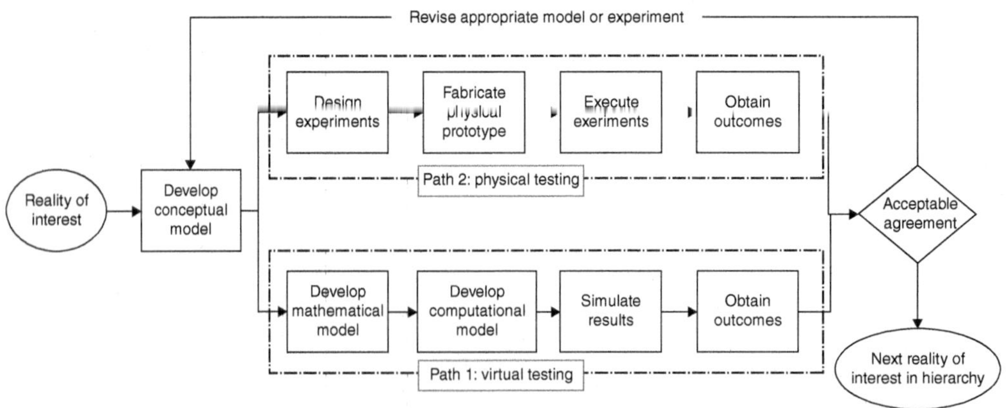

Figure 15.10 Verification and validation activities associated with virtual testing. *Source:* Adapted from ASME (2006).

providing physical testing results is an important consideration, often evaluated through measurement system analysis methods that assess the accuracy and precision across the various elements necessary to perform the testing. This may include evaluations of accuracy via bias, linearity, or stability studies and evaluations of precision via repeatability and reproducibility studies. The results of the physical testing are compared to the results of the virtual testing; acceptable agreement is defined with respect to a specified level of confidence. These methods should also be documented in the knowledge point plan.

Significant aspects to ensuring the necessary information to support decision analysis for system design alternatives includes the selection of methods and prototypes that underly the physical testing activities. Prototypes, for example, must be adjustable or variable to a degree in order to effectively account for the full range of adjustability represented across the design space under investigation. While similar enough in material and geometry characteristics as the actual product to replicate the physics, they should otherwise be made as simply, cheaply, and quickly as possible. This is especially true when destructive testing is necessary to obtain the type of data needed and, in this case, reproducibility of the prototypes is essential. Additionally, subscale testing may be necessary for effective evaluation of subsystems; this has its own associated considerations that mirror those for physical testing in general. Further, measurement methods associated with full-scale or sub-scale testing should not interfere with the functionality, dynamics, or physics-related outcomes of the design. The physical prototype should therefore account for instrumentation geometry.

Considering these aspects, the knowledge point plan should document how physical prototypes will be manufactured (inclusive of cost, schedule, level of detail required for knowledge development, quality, and risk associated with the manufacturing method), measured, and the influence the manufacturing method may have on performance. For example, in some designs, traditional methods such as computer-numerical control manufacturing of fins with unique angles may not be as cost effective or fast as additive manufacturing methods such as stereo-lithography. The end decision of which approach to use would assess the time and cost of the traditional approach against the lesser-quality surface finish and potential deflection under high-speed testing conditions produced by additive manufacturing in terms of impact on performance sensitivity and data needs.

Design Alternative Analysis Methods and Metrics

The fundamental motivation of the SBD approach is to delay decision-making that is too restrictive too early, embracing a process that supports both consideration of variant design foundations in parallel and delays final decision-making until information has more certainty. Knowledge development is central to robust, fact-based decision-making, which is why the knowledge point plan and its processes discussed in the previous section are crucial to successful SBD.

The bulk of the knowledge point plan implementation will occur during the "establish feasibility before commitment" stage of the process, when teams will improve the levels of detail and analysis for each design alternative, re-evaluating the design space until a final solution is chosen. Convergence to a final solution is not guaranteed, however. The decision-maker may still be left with a set of design alternatives, where each is capable of achieving the desired performance metrics, from which to choose the ones to go forward. To this point in the process, the design alternatives have been assessed based on feasibility, meaning whether or not a given design meets a specified set of criteria (e.g., a "go" or "no go" decision). Feasibility assessment does not consider stakeholder preferences that may favor one design over another, how well a design meets the criteria, or its robustness with respect to intended use. Consequently, additional metrics and decision-making techniques will be required to further narrow the remaining design alternatives.

Stakeholder Preferences and Valuation

As stated previously, understanding and effectively accounting for stakeholder preferences are critical parts of this process. Stakeholders may value system attributes or performance measures quite differently and with unequal prioritization. Several methods have been used in SBD to capture stakeholder preference: value models (Keeney and von Winterfeldt 2007), weighting schemes (Greco et al. 2019), well-being indicators (Canbaz et al. 2014), and fuzzy membership functions (Mendel 1995). These preferences are incorporated into different decision frameworks such as multi-attribute utility theory (MAUT), constraint satisfaction problem (CSP), fuzzy logic system (FLS) analysis, and multi-objective optimization methods (MOOM).

Multi-attribute utility theory (MAUT) is distinguished from multi-attribute value theory (MAVT) through the use of a utility function or value function, respectively. Utility functions measure preference under conditions of uncertainty or risk (Keeney and Raiffa 1993), while a value function represents a deterministic measure of preference, i.e., one under certainty (Dyer and Sarin 1979). True utility functions are more difficult to accurately assess and capture, as they require stakeholder to express preferences over lotteries that yield various attribute levels with various specified probabilities. Value functions, in contrast, are much more straightforward. Assessing a value function requires stakeholders to express preference judgments over various hypothesized outcome levels for the various attributes. Consequently, MAVT has become one of the most widespread methods in use for these types of problems as it provides the ability to capture the complex preferences of a stakeholder using a single metric such as a linear additive preference function.

There are two primary parts to multi-attribute additive value analysis: weighting the stakeholder preferences in terms of how stakeholders prioritize certain attributes or performance criteria over others, and how to assign stakeholders' assessment of value with respect to the levels of achievement for a given attribute or performance parameter based. For the former, a weighting scheme can reflect the importance of an objective measure with respect to the value of the stakeholder, and there are numerous schemes to choose from, each with advantages and disadvantages depending on the problem and how they are applied (Odu 2019; Pöyhönen and Hämäläinen 2001; Roszkowska 2013).

Considering value, stakeholders may value how well an alternative exceeds their threshold requirement or may express an objective requirement above which they place no increasing value. This information is captured in the form of value functions. Value functions may be captured via several methods, including conjoint analysis or defined as requirements-based functions scaled against objective and threshold requirement levels (Sitterle et al. 2015). This latter approach is based on a Key Performance Parameter (KPP) concept to promote comparability across analyses.

Multi-attribute additive value analysis is typically addressed via a linear additive preference function:

$$v(x) = \sum_{i=1}^{n} w_i v_i(x_i) \tag{15.1}$$

where $v(x)$ is total value of a given alternative, $i = 1, \ldots, n$ is the number of attributes being assessed and thus contributing to the total value, x_i is the alternative's score or performance measure on the

ith attribute, $v_i(x_i)$ is the single dimensional valuation of the score of x_i, and w_i is the weight of the ith attribute. In keeping with traditional utility theory, total system value is limited to the range of 0–1, and the sum of all weights equals 1. Some variants to multi-additive value methods have been developed to place a penalty on designs that fail to meet a threshold requirement (Sitterle et al. 2017), and uncertainty in the raw data score may also be taken into consideration.

Selecting Alternatives on the Basis of Value and Relation to Pareto Optimality

The linear additive preference function can be represented as an interval as illustrated in Figure 15.11. In Figure 15.11a, design alternative 1 dominates the other alternatives; there is no overlap in its range of possible total value and those of the other alternatives. This depicts a simple criterion of assessment called interval dominance, where one alternative dominates all others across an interval range (Quah and Strulovici 2009). In Figure 15.11b, however, this is not the case. There is no alternative that has no overlap with the others. When this occurs, other criteria such as the maximality criterion (Destercke 2010), E-admissibility (Levi 1975), Tukey's method for multiple comparisons (Mendenhall and Sincich 2016), elimination criterion under shared uncertainty (Rekuc and Paredis 2005), parameterized Pareto dominance (Parker and Malak 2011), or Pareto dominance (Voorneveld 2003) are often used.

Dominance criteria can aid further reduction of the set of alternatives, but situations may occur where there are several alternatives that are non-dominated. This set, called the Pareto frontier (also called a Pareto front) consists of those alternatives that are best in their valuation when another metric is held fixed. This is illustrated in Figure 15.12a, which depicts alternative value as a function of cost. Each point on the Pareto frontier possesses the greatest valuation for that cost level.

Often at this stage of analysis, decision-makers will select their own balance of preferences from the Pareto set alternatives based on personal experience or other judgments, e.g., how much capability is affordable? Other development efforts may require a bit more evaluation, however, and seek to differentiate the alternatives from one another using additional methods. One approach commonly used is to evaluate robustness, i.e., how insensitive a given alternative is to the effects of sources of variability (Taguchi et al. 2005). There are numerous approaches to measuring

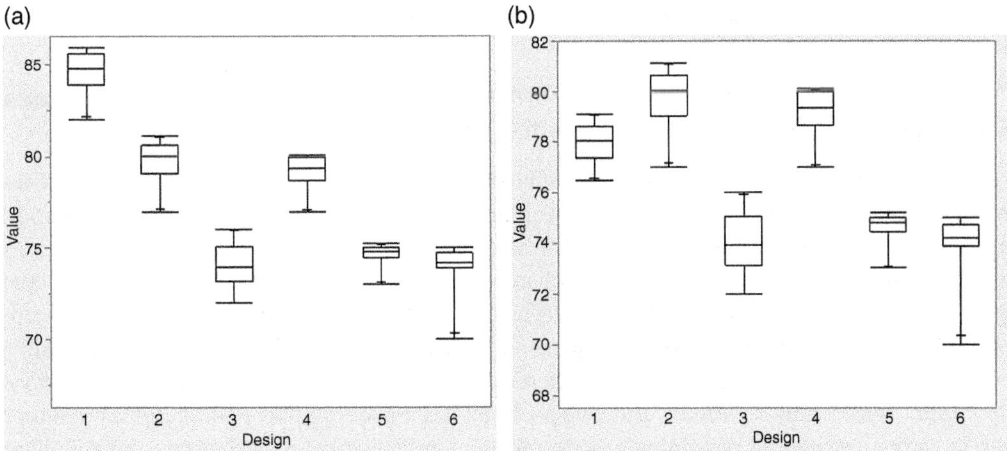

Figure 15.11 These panels illustrate scenarios requiring different dominance evaluations. (a) Illustrates a scenario with no interval overlap. (b) Illustrates a scenario with interval overlap.

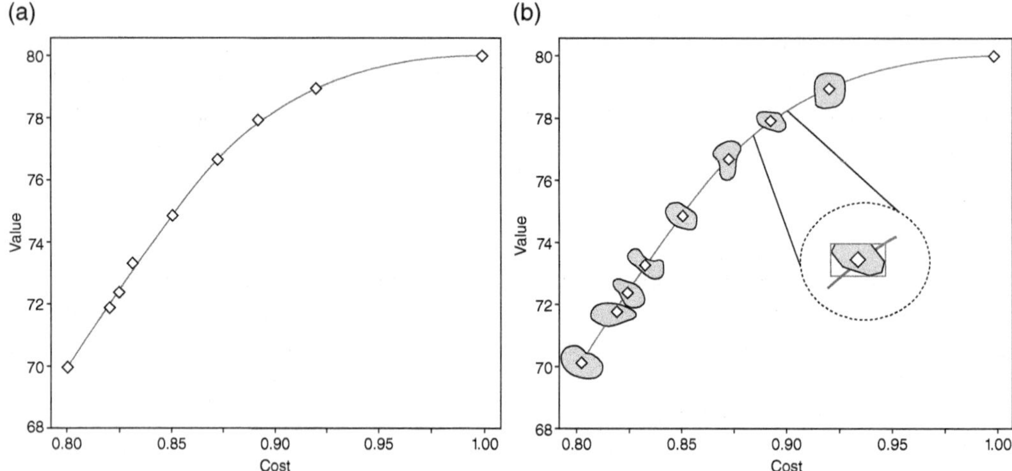

Figure 15.12 These panels illustrate two different Pareto frontier scenarios. (a) Illustrates a Pareto frontier without variation. (b) Illustrates a Pareto frontier with variation.

robustness: signal-to-noise ratios (SNRs) (Taguchi et al. 2005), sensitivity measures (Gunawan and Azarm 2005), flexibility measures (Matthews et al. 2014) or design margins (Eckert et al. 2019), and robustness test metrics (Ross 2017).

One approach is to evaluate the design alternatives under conditions of uncertainty more directly. Figure 15.12b illustrates the potential range of possible values for an alternative when uncertainty is propagated from the lowest parametric level up to the objective function (i.e., the total value). Evaluating each alternative in this manner can be time consuming and costly. To overcome these challenges, Hung and Chan developed an approach using linear combinations of objective functions to create a polygon representation of the objective variation subspace (Hung and Chan 2013a,b). In their work, an optimality influence range (OIR) quantifies the consequences of design variations on the objectives, that is, behaviors of the objective functions under uncertainty. Decision-makers then select preferences for design alternatives depending on whether the smallest influence range of objective variation (i.e., the least variation in the objective space) or the smallest influence noise (i.e., the smallest deviation of the objective variations away from the Pareto front) is most desirable.

Evaluation Based on Risk and Resilience within the SBD Process

Another common approach to further reduce the set of possible design alternatives is through risk-based measures that seek to assess the probability of failure. Among the risk-based evaluation approaches, super-quantile measures of risk are computationally attractive, scalar representations of a random variable that estimate the probability of failure at distribution tails (Buhai 2005; Rockafellar and Royset 2010). Doerry et al. developed an alternative method based on the concept of diversity and its relation to risk, employing a metric to assess the degree to which the feasible configurations within a design region are different from one another (Doerry et al. 2014). Doerry's approach is specifically oriented to the SBD problem; it does not evaluate point designs but rather sets of configurations in a remaining feasible region of design space. A feasible concept with high component diversity has many feasible configurations, implying lower risk because a common failure mode is unlikely.

Rapp et al. also developed an approach and framework specifically tailored to assessing value of a set of solutions in a SBD process, but with respect to design uncertainty and process resilience (Rapp et al. 2018). This work developed a SBD "contribution to design" (CTD) concept, which is determined stochastically and expected to produce a more resilient design process throughout the development time. Rapp et al.'s CTD combines consideration of several external factors including system performance, production cost, development time, development cost, and uncertainties associated with candidate technologies. The method centers on selecting or rejecting subsystem options at given milestones in the development process that will help achieve a "best value" CTD at the end of the development despite adapting to changes in the external factors.

Conclusion and Discussion

Design decisions during the SBD process, as opposed to a final selection at its end, do not select point solutions. The distinction, as highlighted by Ghosh and Seering (2014), lies in part in the difference between the development of sets of distinct concepts (SBD) and the development of sets of variants of a single concept (point-based design [PBD]). SBD is supported through the creation of multiple design sets, each from different teams focused on achieving different design objectives. This parallelism in set creation leads to more variation in design consideration up front, increasing confidence that a set of designs as robust as possible to identified requirements is produced by the end of the development cycle.

One of the strengths of SBD lies in its ability to contract or expand design set solutions as necessary in the face of uncertainty: stakeholder defined constraints and requirement thresholds, technological development cost and time, threat dynamics driving the need, willingness of stakeholders to relax a given requirement or not, etc. will all change over the course of development. Throughout, the SBD process strives to keep design options open for as long as possible during the design effort. This preserves the potential for discovery in the design process, reduces the likelihood of discarding innovative possibilities, and offers as much resiliency as possible to changes in these external factors that will impact the efficacy of a candidate design. Carrying these multiple design options forward, however, can incur additional costs to the development process. Ward and Sobek recognized this challenge and recommended that SBD efforts be divided according to four development categories: Tailoring, strategic breakthrough, limited innovation and reintegration, and research (Ward and Sobek 2014).

Even so, the challenge still lies in the details. SBD set creation and set reduction are critical to the success of the process, and both require better definition and formal grounding in order to provide development teams with confidence in the effectiveness, efficiency, and quality of outcomes. For many organizations, SBD is a paradigm shift from their traditional point-based design approach. The thought of carrying multiple design alternatives through the design process where the decisions to reason and narrow the design space are delayed until sufficient knowledge is generated may seem counterintuitive (Welo et al. 2019). Typically, real development processes will fall somewhere in the middle of point-based and set-based design practices. This is unsurprising as many design concept analyses require detailed engineering assessments based on precise definitions of a design. Malak and Paredis have expressed concern that these problems are "poorly suited for the set-based nature of design concepts" due to the risk of analytical uncertainty related to the imprecise nature of SBD methods (Malak Jr. and Paredis 2009).

Several research groups have consequently conducted extensive recent surveys on the current state of SBD quantitative methods and their respective application and limitations (Dullen et al. 2021; Shallcross et al. 2020; Specking et al. 2018; Toche et al. 2020). A key theme revealed across these study centers on the ongoing work to mature the SBD framework and associated processes, seeking to couple SBD more directly with mathematically grounded and tractable analytical methods. Specking et al. in particular highlighted the twin needs of uncertainty reduction and design maturation decisions, namely that design convergence required both of these activities concurrently. Identifying and isolating parameters and relationships between them in order to characterize sets of feasible designs will require extensive expertise as well as resources in the form of models, prototypes, testing, etc. Dullen et al. identified that a Markov Decision Problem (MDP) approach could be an effective method for implementing optimal learning, enabling the ability to identify efficient ways to verify and validate virtual prototypes using multi-fidelity physical prototypes.

The focus on prototypes is vital in SBD. Virtual and physical prototypes are critical to uncertainty reduction and providing data needed for parametric-based trades analyses. Extensive prototyping and testing are key to SBD in order to foster the knowledge-based environment, the institutional learning capability, and thereby accurately inform the decision-making process. This can be costly and time consuming, however, especially when not considered and planned for up front. Organizations seeking to embrace SBD for all of its advantages should invest in developing an environment (or environments) that can better facilitate large numbers of simulations, model integration, pre-processing and post-processing activities, and data storage.

Similarly, for modeling and simulation activities, the computational development effort coupled with the resources required to execute and analyze the outcomes can be extensive. If evaluation in the context of operational scenarios is required, for example, scenario development, run-time, and analysis can all take significant time and the overall complexity of the problem can make concrete inferences on system design effectiveness in the context of an operation or mission exceedingly difficult. Importantly, this statement is equally true for point-based design evaluation. It can be more of a challenge in SBD, however, when multiple design variants are still under consideration.

This highlights the importance of knowledge development and the knowledge point plan discussed earlier in this paper. Successful implementation of SBD on a development effort will be driven by how well the team develops and executes this plan and its associated activities. Providing the right information at the right time required to make a decision is key. Organizations employing SBD will need to invest in planning and infrastructure for effective and efficient knowledge development, which will include prototyping capabilities as mentioned above and training in various decision analysis methodologies used to support the process. This should be developed in the context of the nature of the project, the key decision points it requires, and the specific nature of information required to make those decisions. Relatedly, and especially important for DoD organizations, current contract language and source selection criteria may not lend itself well to effective SBD practice. New contract language and processes may be required.

Product development efforts now more than ever are in need of methodologies that can address the challenges of increased system complexities, shortening time to market, increased demands in mass customization, market instabilities, geographical barriers, improved innovation, and adaptability to emerging technologies. SBD offers many advantages over traditional approaches: its parallelism, focus on creation, and evaluation of multiple design variants rather than variants of a single design, its ability to contract and expand sets as external factors and needs change during the development process, etc. To maximize these advantages, however, organizations must plan accordingly. SBD may be an overall framework and practice to mature toward rather than an "either/or," incorporating new advances in SBD foundations and rigor as they are developed, and eventually embracing the benefits concurrent design can offer.

References

ASME V&V 10-2006 (2006). *Guide for Verification and Validation in Computational Solid Mechanics*. New York: American Society of Mechanical Engineers.

Bernstein, J.I. (1998). Design methods in the aerospace industry: looking for evidence of set-based practices. Massachusetts Institute of Technology.

Braha, D., Brown, D.C., Chakrabarti, A., et al. (2013). DTM at 25: essays on themes and future directions. *Paper presented at the ASME 2013 international design engineering technical conferences and computers and information in engineering conference*, Portland, Oregon, USA (4–7 August 2013). American Society of Mechanical Engineers.

Buhai, S. (2005). Quantile regression: overview and selected applications. *Ad Astra* 4 (2005): 1–17.

Canbaz, B., Yannou, B., and Yvars, P.-A. (2014). Improving process performance of distributed set-based design systems by controlling wellbeing indicators of design actors. *Journal of Mechanical Design* 136 (2): 021005.

Collopy, P.D. (2018). Tradespace exploration: promise and limits. In: *Disciplinary Convergence in Systems Engineering Research* (ed. A.M. Madni, B. Boehm, R.G. Ghanem, et al.), 297–307. Cham: Springer.

Destercke, S. (2010). A decision rule for imprecise probabilities based on pair-wise comparison of expectation bounds. In: *Combining Soft Computing and Statistical Methods in Data Analysis* (ed. C. Borgelt, G. González-Rodríguez, W. Trutschnig, et al.), 189–197. Springer.

Doerry, N., Earnesty, M., Weaver, C., et al. (2014). Using set-based design in concept exploration. *Paper presented at the SNAME Chesapeake Section Technical Meeting*, Arlington, Virginia, USA (25 September 2014). Society of Naval Architects & Marine Engineers.

Dullen, S., Verma, D., Blackburn, M., and Whitcomb, C. (2021). Survey on set-based design (SBD) quantitative methods. *Systems Engineering* 24 (5): 269–292.

Dyer, J.S. and Sarin, R.K. (1979). Measurable multiattribute value functions. *Operations Research* 27 (4): 810–822.

Eckert, C., Isaksson, O., and Earl, C. (2019). Design margins: a hidden issue in industry. *Design Science* 5: 1–24.

Ghosh, S. and Seering, W. (2014). Set-based thinking in the engineering design community and beyond. *Paper Presented at the Proceedings of the ASME 2014 International Design Engineering Technical Conferences & Computers and Information in Engineering Conference*, Buffalo, New York, USA (17–20 August 2014). The American Society of Mechanical Engineers.

Gonçalves, D. (2006). An approach to simulation effectiveness. *Paper Presented at the 16th Annual International Symposium of the International Council on Systems Engineering*, Orlando, Florida, USA (10–13 July 2006). Wiley.

Greco, S., Ishizaka, A., Tasiou, M., and Torrisi, G. (2019). On the methodological framework of composite indices: a review of the issues of weighting, aggregation, and robustness. *Social Indicators Research* 141 (1): 61–94.

Gross, D.C. (1999). Report from the fidelity implementation study group. *Paper Presented at the Fall Simulation Interoperability Workshop Papers*.

Gunawan, S. and Azarm, S. (2005). Multi-objective robust optimization using a sensitivity region concept. *Structural and Multidisciplinary Optimization* 29 (1): 50–60.

Hung, T.-C. and Chan, K.-Y. (2013a). Multi-objective design and tolerance allocation for single-and multi-level systems. *Journal of Intelligent Manufacturing* 24 (3): 559–573.

Hung, T.-C. and Chan, K.-Y. (2013b). Uncertainty quantifications of Pareto optima in multiobjective problems. *Journal of Intelligent Manufacturing* 24 (2): 385–395.

INCOSE (2014). *Systems Engineering Handbook: A Guide for System Life Cycle Processes and Activities, Version 4*. Wiley.

ISO/IEC/IEEE 15288 (2015). Systems and Software Engineering – System Life Cycle Processes, IEEE.

Keeney, R.L. and Raiffa, H. (1993). *Decisions with Multiple Objectives: Preferences and Value Trade-Offs*. Cambridge University Press.

Keeney, R.L. and von Winterfeldt, D. (2007). Practical value models. In: *Advances in Decision Analysis: From Foundations to Applications*, 232–252. Cambridge University Press.

Levi, I. (1975). On indeterminate probabilities. *The Journal of Philosophy* 71 (13): 391–418.

Malak, R.J. Jr. and Paredis, C.J. (2009). Modeling design concepts under risk and uncertainty using parameterized efficient sets. *SAE International Journal of Materials and Manufacturing* 1 (1): 339–352.

Matthews, J., Klatt, T., Seepersad, C.C., et al. (2014). Bayesian network classifiers and design flexibility metrics for set-based, multiscale design with materials design applications. *Paper presented at the ASME 2014 International Design Engineering Technical Conferences and Computers and Information in Engineering Conference*, Buffalo, New York, USA (17–20 August 2014). The American Society of Mechanical Engineers.

Mendel, J.M. (1995). Fuzzy logic systems for engineering: a tutorial. *Proceedings of the IEEE* 83 (3): 345–377.

Mendenhall, W.M. and Sincich, T.L. (2016). *Statistics for Engineering and the Sciences*. Chapman and Hall/CRC.

Miller, S.W., Simpson, T.W., Yukish, M.A., et al. (2013). Preference construction, sequential decision making, and trade space exploration. *Paper Presented at the ASME 2013 International design Engineering Technical Conferences and Computers and Information in Engineering Conference*, Portland, Oregon, USA (4–7 August 2013). The American Society of Mechanical Engineers.

Oberkampf, W.L. and Trucano, T.G. (2002). Verification and validation in computational fluid dynamics. *Progress in Aerospace Sciences* 38 (3): 209–272.

Odu, G. (2019). Weighting methods for multi-criteria decision making technique. *Journal of Applied Sciences and Environmental Management* 23 (8): 1449–1457.

Parker, R.R. and Malak, R.J., Jr. (2011). Technology characterization models and their use in designing complex systems. *Paper Presented at the ASME 2011 International Design Engineering Technical Conferences and Computers and Information in Engineering Conference*, Washington, DC, USA (28–31 August 2011). The American Society of Mechanical Engineers.

Pöyhönen, M. and Hämäläinen, R.P. (2001). On the convergence of multiattribute weighting methods. *European Journal of Operational Research* 129 (3): 569–585.

Pugh, S. (1981). Concept selection: a method that works. *Paper Presented at the Proceedings of the International conference on Engineering Design*, Rome, Italy (9–13 August 1981). The Design Society.

Quah, J.K.-H. and Strulovici, B. (2009). Comparative statics, informativeness, and the interval dominance order. *Econometrica* 77 (6): 1949–1992.

Rapp, S., Chinnam, R., Doerry, N. et al. (2018). Product development resilience through set-based design. *Systems Engineering* 21 (5): 490–500.

Rekuc, S., Aughenbaugh, J., Bruns, M., and Paredis, C. (2006). Eliminating design alternatives based on imprecise information. *SAE International Journal of Materials and Manufacturing*, 115 (5):208–220.

Rockafellar, R.T. and Royset, J.O. (2010). On buffered failure probability in design and optimization of structures. *Reliability Engineering & System Safety* 95 (5): 499–510.

Ross, J.E. (2017). Determining feasibility resilience: set based design iteration evaluation through permutation stability analysis.

Roszkowska, E. (2013). Rank ordering criteria weighting methods–a comparative overview. *Optimum. Studia Ekonomiczne* 5 (65): 14–33.

Schlesinger, S. (1979). Terminology for model credibility. *Simulation* 32 (3): 103–104.

Schrader, S., Riggs, W.M., and Smith, R.P. (1993). *Choice over Uncertainty and Ambiguity in Technical Problem Solving*. Elsevier.

Shallcross, N., Parnell, G.S., Pohl, E., and Specking, E. (2020). Set-based design: The state-of-practice and research opportunities. *Systems Engineering* 23 (5): 557–578.

Sitterle, V.B., Brimhall, E.L., Freeman, D.F. et al. (2017). Bringing operational perspectives into the analysis of engineered resilient systems. *Insight* 20 (3): 47–55.

Sitterle, V.B., Freeman, D.F., Goerger, S.R., and Ender, T.R. (2015). Systems engineering resiliency: guiding tradespace exploration within an engineered resilient systems context. *Procedia Computer Science* 44: 649–658.

Sobek, D.K. II, Ward, A.C., and Liker, J.K. (1999). Toyota's principles of set-based concurrent engineering. *MIT Sloan Management Review* 40 (2): 67.

Specking, E.A., Whitcomb, C., Parnell, G.S. et al. (2018). Literature review: exploring the role of set-based design in trade-off analytics. *Naval Engineers Journal* 130 (2): 51–62.

Stump, G., Lego, S., Yukish, M. et al. (2009). Visual steering commands for trade space exploration: user-guided sampling with example. *Journal of Computing and Information Science in Engineering* 9 (4).

Stump, G. M., Yukish, M., Simpson, T. W., and O'Hara, J. J. (2004). Trade space exploration of satellite datasets using a design by shopping paradigm. *Paper Presented at the 2004 IEEE Aerospace Conference Proceedings* (IEEE Cat. No. 04TH8720), Montana, USA (6–13 March 2004). Institute of Electrical and Electronics Engineers.

Taguchi, G., Chowdhury, S., and Wu, Y. (2005). *Taguchi's Quality Engineering Handbook*. Wiley.

Tahera, K., Earl, C., and Eckert, C. (2017). A method for improving overlapping of testing and design. *IEEE Transactions on Engineering Management* 64 (2): 179–192.

Toche, B., Pellerin, R., and Fortin, C. (2020). Set-based design: a review and new directions. *Design Science* 6: https://doi.org/10.1017/dsj.2020.16.

Voorneveld, M. (2003). Characterization of Pareto dominance. *Operations Research Letters* 31 (1): 7–11.

Wang, G.G. (2002). Definition and review of virtual prototyping. *Journal of Computing and Information Science in Engineering* 2 (3): 232–236.

Ward, A., Liker, J.K., Cristiano, J.J., and Sobek, D.K. (1995). The second Toyota paradox: how delaying decisions can make better cars faster. *Sloan Management Review* 36: 43–43.

Ward, A. and Sobek, D.K. (2014). *Lean Product and Process Development*. The Lean Enterprise Institute Incorporation.

Welo, T., Lycke, A., and Ringen, G. (2019). Investigating the use of set-based concurrent engineering in product manufacturing companies. *Procedia CIRP* 84: 43–48.

Wheelwright, S.C. and Clark, K.B. (1992). *Revolutionizing Product Development: Quantum Leaps in Speed, Efficiency, and Quality*. Simon and Schuster.

Biographical Sketches

Shawn Dullen is a Systems Engineering Research Center (SERC) Doctoral fellow pursuing his PhD in systems engineering from Stevens Institute of Technology. He has a BS in mechanical engineering from the University at Buffalo, an ME in mechanical engineering from Stevens Institute of Technology, and an MS in applied statistics from Rochester Institute of Technology (RIT). Mr. Dullen works for the U.S. Army Combat Capabilities Development Command Armaments

Center (CCDC AC) with 19 years of experience working different aspects of the DoD life cycle as a quality engineer and systems engineer. His current research interests are product development, set-based design, multidisciplinary optimization, optimal tolerance design under uncertainty, and Bayesian networks.

Dinesh Verma received the PhD (1994) and the MS (1991) in Industrial and Systems Engineering from Virginia Tech. He served as the founding dean of the School of Systems and Enterprises at Stevens Institute of Technology from 2007 through 2016. He currently serves as the executive director of the Systems Engineering Research Center (SERC), a US Department of Defense sponsored University Affiliated Research Center (UARC) focused on systems engineering research; along with the Acquisition Innovation Research Center (AIRC). During his 20 years at Stevens, he has successfully proposed research and academic programs exceeding $200 million in value. Prior to this role, he served as technical director at Lockheed Martin Undersea Systems, in Manassas, Virginia, in the area of adapted systems and supportability engineering. Dr. Verma has authored over 100 technical papers, book reviews, technical monographs, and co-authored three textbooks: *Maintainability: A Key to Effective Serviceability and Maintenance Management* (Wiley, 1995), *Economic Decision Analysis* (Prentice Hall, 1998), *Space Systems Engineering* (McGraw Hill, 2009). He was recognized with an honorary doctorate degree (Honoris Causa) in Technology and Design from Linnaeus University (Sweden) in January 2007; and with an honorary master of engineering degree (Honoris Causa) from Stevens Institute of Technology in September 2008.

Chapter 16

Exploiting Formal Modeling in Resilient System Design: Key Concepts, Current Practice, and Innovative Approach

Azad M. Madni[1] *and Michael Sievers*[2]

[1]*Astronautics, Aerospace, and Mechanical Engineering, University of Southern California, Los Angeles, CA 90089, USA*
[2]*Systems Architecting and Engineering, University of Southern California, Los Angeles, CA 90089, USA*

Introduction

Resilience is the ability of a system or system of systems (SoS) to continue to provide useful service despite disruptions through planning and preparation; absorbing disruption impacts; recovering service to pre-disruption performance levels; and evolving system configuration, use, and personnel training by adapting to and/or exploiting newly gained knowledge (Madni and Jackson 2009; Westrum 2007). Along with integrity, safety, reliability, fault-tolerance, and maintainability, resilience is key to dependable and available systems (Avizienis et al. 2004). It is important to note at the outset, that the resilience property is different from fault-tolerance and robustness.

Fault-tolerance, one means of achieving system dependability, comprises methods to avoid failures by detecting errors and providing means for masking or recovering from faults that are root causes of those errors. Generally speaking, fault-tolerance which is "inward" looking protects against fault conditions within a system. *Resilience*, on the other hand, allows continued trustworthy operation despite disruptive events that may have external, internal, human-triggered, or malicious origins. Fault-tolerance and resilience have somewhat overlapping goals: a resilient system is designed to protect against unknowable disruptions at design time, while fault-tolerance primarily addresses known or anticipated fault conditions. Although fault-tolerance can include some generic, high-level responses to novel disruptions, fault-tolerance does not adapt or modify system behavior based on those disruptions. Two key hallmarks of resilient systems are learning and adaptability. The properties of fault-tolerance and resilience can, and should, co-exist in systems that perform critical functions. It is important to note that resilience is not an all or nothing game. There can be levels of resilience, with different costs associated with each level. Also, resilience does not have to be introduced in all aspects of a system. In the real world, resilience is required in some aspects and not in others.

Today resilience measures are often introduced in systems using *ad hoc* methods based on the best understanding of how the system is expected to be used, its most likely contingency and disruption scenarios, and its operational environment. However, uncertainties and unknown unknowns can still result in significant safety and performance risks. The development of resilient systems is complicated by the need to incorporate *learning* and *adaptation* capabilities to successfully endure and prevail in the face of unknowable disruptions. Conventional design approaches

Systems Engineering for the Digital Age: Practitioner Perspectives, First Edition. Edited by Dinesh Verma.

Companion website: www.wiley.com/go/verma/systemsengineering

based on fault forecasting, failure mode effects and criticality analysis, and fault likelihood and impact risk assessment fall short when it comes to addressing novel disruptions and behaviors. Moreover, traditional fault injection and stress testing cannot assure protection against irregular or "unexampled" disruptions, nor can they ensure that fault-tolerance mechanisms have been correctly designed and implemented (Westrum 2007).

The risk of system failure can be reduced when provably correct, formal design approaches are used to augment or potentially replace informal and/or *ad hoc* design practices. Given the size and complexity of modern systems, proving an entire system correct is not practicable. Instead, assuring the correctness of critical core functions and resources is possible using formal design methods. While testing is not eliminated when core functions are formally designed, focusing on core functions narrows testing scope, thereby increasing the likelihood that resilience functions are correctly implemented.

It should come as no surprise that the rigor of formal design typically comes at the expense of flexibility. Since resilience requires flexibility, rigor, and resilience in system design appear to be at odds at first blush. This conclusion, in fact, is not necessarily warranted because rigor and flexibility may not be needed in the same aspects of a system. In other words, it is important to identify those aspects of a system that need flexibility and those that need design rigor. It may well be the case that flexibility and design rigor are needed in different aspects of the system.

This chapter discusses formal resilience design methods and recommends an approach that accommodates both *system flexibility* and *design rigor*. The approach extends the concept of contracts commonly used in *design by contract* (DbC) software design, a term coined by Bertrand Meyer (1992).

DbC is a formal, systematic method for creating correct software that includes a means for detecting and managing anomalous behaviors. A contract comprises pairs of assertions that define invariant assume-guarantee conditions. A guarantee is the assured post-condition when the contract assumption holds. The software use of DbC follows the pattern:

require
  precondition
do
  set of software instructions
ensure
  post-condition

The flexibility requirement for realizing resilience necessitates relaxing invariant assertions to allow for probabilistic assumptions and guarantees. Flexible contracts may be represented as Markov Decision Processes (MDPs), which are discrete-time stochastic control processes. In a MPD, decision rewards and system dynamics depend only on the current system state and the action taken in that state (Puterman 1990).

Partial observability and hidden states add uncertainty to system behavior and which actions are needed to correct undesirable situations. Partially Observable Markov Decision Process (POMDP) models are one means for modeling systems under conditions of uncertainty. A POMDP model evaluates a state probability vector based on the current state, actions undertaken, and observations that result from the action.

This chapter introduces basic resilience and machine learning terminology and then discusses the state of practice in resilience modeling. Next, a best practice approach is described based on a flexible resilience contract (FRC) approach that combines DbC and POMDP (Madni et al. 2019). This chapter concludes with an example usage of FRC followed by a chapter summary.

Problem Statement and Enabling Methods

The early stages of design tend to be fuzzy, incomplete, and based on unwarranted assumptions, poor choices, and unproven assertions. Initial requirements and design concepts get even messier when resilience requirements are imposed on the envisioned system (Goerger et al. 2014; Madni et al. 2018a). In part, the added complications are due to a need for greater specificity about what resilience means in terms of scope and level, and the types of disruptions being addressed. This chapter defines resilience as a system capability that enables continued useful system operation in the face of specific types of disruptions in uncertain environments. The concept of resilience covers fault recovery, graceful degradation, adaptive capacity and resources, and high availability. Disruptions that challenge system resilience may include knowable and unknowable faults, system misuse, cyberattacks, partial observability, and unpredictable (potentially hostile) environments. While relatively simple disruptions are easy to recognize and respond to complicated disruptions (i.e., those that involve multiple confounding conditions) and complex disruptions (i.e., those resulting from emergent behavior) pose serious challenges to restoring normal operations.

The prevailing uncertainties, complex and potentially unknowable disruptions, and difficulties in selecting suitable actions (that lead to continuation/restoration of normal operation rather than total system collapse) collectively conspire to increase the risk of system failure. Formal modeling techniques and proven resilience patterns are needed to guide the system design process. Such methods provide capabilities to learn from new evidence while also enabling trades between system model representation flexibility and design rigor in the face of uncertainties.

Disruptions, Rare Events, and High-Impact Rare Events

Disruption is a generic term that refers to events that can have a significant impact (good or bad) on the system operation. A disruption occurs when a threat succeeds in exposing an exploitable system weakness. In many situations, the weaknesses gradually appear through a chain of events in which the threats successfully evade all built-in barriers which, in some cases, involve a chain of unfortunate events, e.g., as depicted by the Swiss Cheese safety model (Reason 1990).

In the Swiss Cheese model, a system's protections are conceived as a series of semi-permeable barriers that can be visually depicted by layers of Swiss Cheese slices. The holes in the cheese at each layer represent defense weaknesses through which specific threats can potentially pass. Holes of random sizes and at various locations in each layer succeed in mostly stopping the progress of threats traversing the barriers. However, when the holes align, these threats do get through. Thus, a threat that succeeds in exploiting the gaps in defense in the set of barriers becomes an active disruption as shown in Figure 16.1.

Disruptions may have known or unknown causes, impacts, and probability of occurrences. Taleb coined the term black swan to describe low probability, high-impact events (Taleb 2010). Black swans have also been referred to as randomly occurring unknowable events with unknown or unobservable consequences. Black swans may produce positive outcomes, e.g., tablet computers and mobile phones, or negative outcomes, such as the Space Shuttle Challenger explosion.

A companion concept to a black swan is that of a gray swan. Gray swans are rare, high-impact disruptions that cannot be predicted but can be imagined. Unlike black swans, which are effectively impossible to predict (until they occur), gray swans are unlikely but possible events, sometimes designated as *unknown* events. Snowfall in tropical climates is an example of a gray swan.

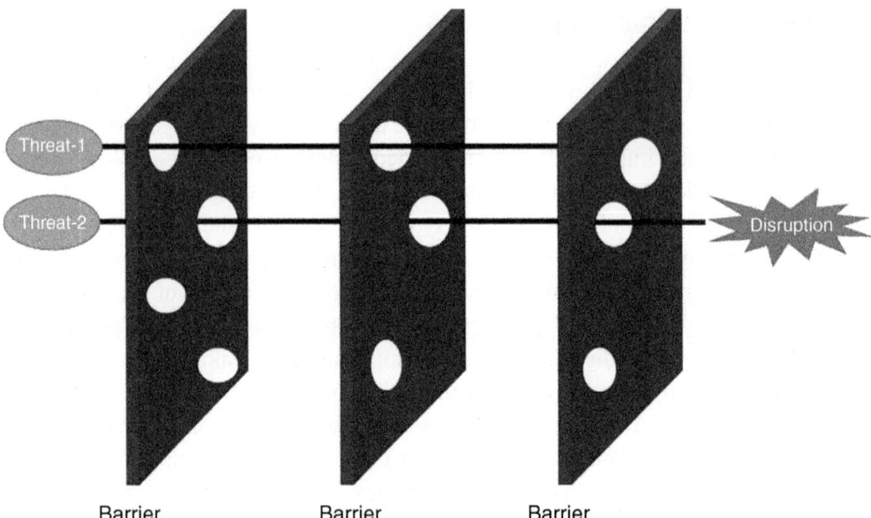

Figure 16.1 Swiss cheese model. Barriers stop Threat-1, but Threat-2 passes through and triggers a disruption.

Whether an event is unknowable or unknown depends on whether a conceptualization exists that describes the event and its consequences. To an extent, this distinction depends on the representativeness and comprehensiveness of the event probability distribution function (PDF). Three possible PDF outcomes are possible in Gholami et al. (2018):

- An event is *known* when all possible outcomes and probabilities are known for a given event space. For example, when a fair coin is flipped many times, heads occur with $p = 0.5$, and tails occur with $p = 0.5$.
- Events are *unknown* when the outcomes are known, but some or all of the probabilities of those outcomes are unknown. Although engineers did not know the likelihood that a micro-meteoroid would hit the primary mirror of the James Webb Space Telescope, they knew it was possible.
- Events are *unknowable* when the outcomes are unknown and cannot be predicted. Unknowable events become unknown once they occur and may become known if they occur sufficiently often.

Disruptions can be further classified based on their origins (Gholami et al. 2018). For example, electrical power systems category may be the result of one or more of the following: cascading technical failure, extreme natural events, cyber and physical attacks, and space weather (Gholami et al. 2018).

Design by Contract

As noted earlier, DbC, also called *programming-to-the-interface*, *contract programming*, and *programming-by-*contract, developed in the 1980s by Bertrand Meyer, is a standard approach used in modern software engineering (Meyer 1992; Enseling 2001; Cimatti and Tonetta 2012; Benvenuti et al. 1970). Meyer developed the Eiffel object-oriented programming language that includes DbC constructs, many of which are supported in other languages such as C++, Python, C#, and Java (https://www.eiffel.org/doc/, accessed January 2023; Meyer 1989). While DbC was originally created to protect against dangerous software behavior, it has found more general system use (Cimatti and Tonetta 2012).

In object-oriented software, a *computational contract* is an agreement between classes and objects that assures objects always have valid states. Software using DbC is structured as *method contracts* comprising:

- A *precondition* (requires clause) specifying what must be true when the method is used (client code).
- A *post-condition* (ensures clause) specifies what is expected of the program implemented by the method (the code within the method body)

When the preconditions are fulfilled, a method contract terminates with the post-condition satisfied and returns to the calling program. A method contract called when the preconditions are violated can choose any action, including not terminating. It is the client's responsibility to ensure that the preconditions are met when calling a method. A returned post-condition may be assumed valid when the method returns. It is the responsibility of the method implemented to ensure that post-conditions are achieved when a method returns and that the preconditions are present when the method is called.

Software that interacts with external actors (e.g., users, other software, other computer systems) is specified by *protocol contracts*. These contracts do not define a guaranteed result but determine what actors interact with it and how and when interactions are permitted.

Contracts may contain invariants, which are always true throughout the lifetime of a component. Invariants may be checked at the start or end of every use.

Assertion checks are used for evaluating contracts during runtime. For example, the Python code for integer square throws an exception if the input is <0:

```
def Int_sqroot(x):
        if x < 0:
        raise Exception ("Value of x is less than 1")
    i = 0
    k = 1
    root = 1
    while (root <= x):
        k = k+2
        i = i+1
        root = root+k
    return i
# Call square root function
j = Int_sqroot (-3)
#Output:
Exception: Value of x is less than 1
```

Contracts may be formally proven before runtime by assessing contract composition and code reachability (Balluchi et al. 2006; Büchi 1962). The former manipulates contract statements represented as mathematical expressions. The latter determines the set of all states that may be reached under dynamic operation from the initial state set.

Brute force formal proofs of correctness may be impractical for large systems, tedious to perform, and error-prone. However, complicated systems and their contracts may be decomposable into more manageable components. For example, Cimatti and Tonetta describe a property-based proof approach based on component decomposition (Cimatti and Tonetta 2012). Their concept assigns a set of contracts to each system component that is further refined by a set of contracts. Their approach integrates refinement contracts into a system architecture enabling better integration with safety-critical applications.

As elaborated in the Best Practices section, the recommended approach limits mathematical proofs to "hardcore" functions, i.e., those functions essential to implementing resilience and suitable for formal verification. The approach is based on contract refinement and deduction. All other functions use assertion checks similar to the above example for detecting runtime errors and taking actions that produce the highest discounted future reward.

Markov Process

A *Markov Process* is a stochastic model representing a sequence of possible states in which the next state depends only on the current state and probabilistic transitions between states, *T*. A Markov chain consists of a discrete set of states, and state transitions occur in discrete steps as illustrated in Figure 16.2. A finite Markov chain consists of a finite set of states. Table 16.1 shows the transition probabilities associated with Figure 16.2. Note that the rows in Table 16.1 must sum to 1.0, i.e., every state must transition to a new state, or to itself.

A *Hidden Markov Model* (HMM) is a Markov process comprising a finite number of discrete states connected by probabilistic state transitions. Each state emits an observation based on a probabilistic emission distribution. As contrasted with the Markov chain above, there is uncertainty about the system's state. Table 16.2 shows an example emission probability matrix for Figure 16.2 consisting of three possible observations at each state. If o_1 is observed, then the system could be in s_k. However, there is an equal probability that the system is in s_i. Similarly, observing o_3, the system most probably is in s_j, but there is almost an equal probability that it is in s_k. Complicating the situation further, there may be hidden states, i.e., previously unknown states with unknown transitions that are also emitting observations. Since observations cannot identify hidden states, a goal of an HMM is to learn the existence of those states and their associated emission and transition probabilities (Rabiner 1989).

A *Markov Decision Process* (MDP) is a discrete-time stochastic reinforcement learning process in which an agent determines the next action based on its current state and an action policy that maximizes a reward function (Thomas 2020). An MDP has a finite number of discrete states, probabilistic transitions, and controllable actions at each state. The next state is determined by the current state and the action taken at the current state. Actions are taken that maximize a reward function, as noted earlier.

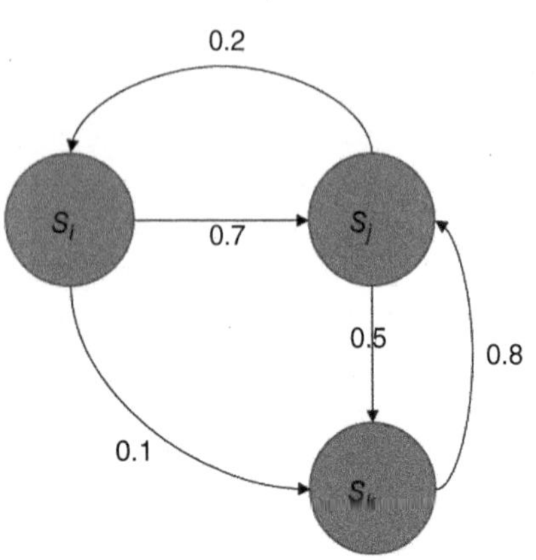

Figure 16.2 Example Markov chain.

Table 16.1 Transition matrix for Markov chain in Figure 16.2.

T	s_i	s_j	s_k
s_i	0.2	0.7	0.1
s_j	0.2	0.3	0.5
s_k	0	0.8	0.2

Table 16.2 Emission probabilities for Figure 16.2.

	s_i	s_j	s_k
o_1	0.4	0.2	0.4
o_2	0.4	0.1	0.1
o_3	0.3	0.6	0.5

MDP components comprise:

- S: the set of all possible states
- A: all possible actions an agent may take
- T: the transition dynamics, $S \times A \to S'$ where the next state, $S' \in S$, is determined by the current state and the action taken. For each state s and action a, $T(s,a) = P(s'|s,a)$ is the probability distribution over all states that a system may transition to when taking action a.
- r: the reward function, $S \times A \to \mathbb{R}$, is the value of the reward earned, $r(s,a,s')$, when taking action a when in state s and arriving in the state, s'
- $\gamma \in [0, 1]$: the discount factor applied to future rewards. When $0 < \gamma < 1$ rewards are discounted, reflecting uncertainty in the value of future actions.
- $\mu_0 \in S$ is the initial state probability distribution
- π: a policy that determines the action to take when in a state.

Figure 16.3 modifies Figure 16.2 by adding two actions, a_1 and a_2 and their transition probabilities at each state. For example, in s_i action a_1 transitions to s_j with probability 0.7 and back to itself with probability 0.3. If action a_2 is taken at s_i then there is a 0.1 probability of arriving in s_k and a 0.9 probability of staying in s_i.

A *Partially Observable Markov Decision Process* (POMDP) combines HMM and MDP. A POMDP consists of a finite number of discrete states, probabilistic state transitions with controllable actions, the next state determined by the current state and current action, an emission probability distribution at each state, and state uncertainty. A POMDP substitutes a *belief state* for an MDP's system state. A belief state, b, is a probability distribution based on an action taken (Eq. (16.1)) and the observation emitted (Smallwood and Sondik 1973; Lauri et al. 2022). The state probabilities are

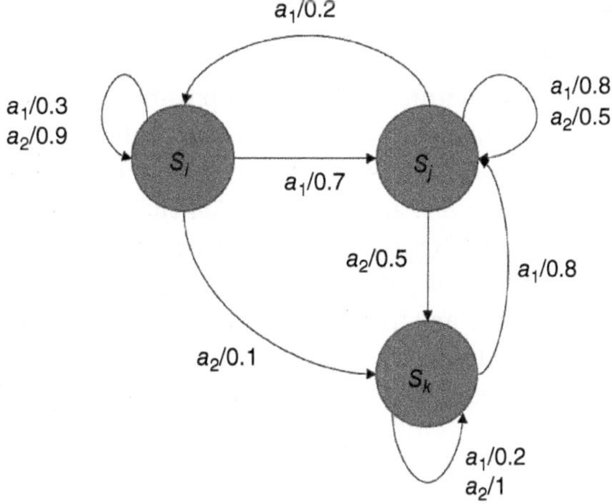

Figure 16.3 MDP showing two actions at each state.

updated as shown in Eq. (16.1), which shows the belief update, $b'(s')$, for state s' updated from the current belief state and the observation emitted, o', after taking action a.

$$b'(s') = \frac{P(o' \mid s',a)\sum_{s} P(s' \mid s,a)b(s)}{\sum_{s} P(s' \mid s,a)b(s)\sum_{s'} P(o' \mid s',a)} \quad (16.1)$$

Rewards are computed from the belief state and the instantaneous reward:

$$R(a,b) = \sum_{s \in S} r(a,s)b(s) \quad (16.2)$$

Unlike an MDP that assumes perfect state knowledge, a POMDP is potentially challenging to solve optimally. POMDP solutions such as value iteration and other heuristics provide good approximate solutions. One method developed by Smith and Simmons combines heuristic searches with piecewise linear convex value function representations (Smith and Simmons 2004). Their method makes local updates at predetermined beliefs chosen by forward-looking searches that select actions and observations.

Actions, Rewards, and Policies

Policies may be stochastic in which there is an action distribution at each state, i.e., $\pi(a \mid s)$ is the conditional probability density of a evaluated at the state, s. A deterministic policy directly maps an action to a state. A trajectory, τ, is the set of all states arrived at from a given starting point and transitioning to the next state as determined by the transition dynamics and the action selected from the policy. The reward at a given step is determined by the current state, the action taken, and the next state. The *return* associated with a trajectory is computed as the sum of the discounted rewards from the initial state to some stopping state at step $t = t_s$

$$R(\tau) = \sum_{t=0}^{t=t_s} \gamma^t r_t \quad (16.3)$$

The optimal policy maximizes the expected return:

$$\pi^* \in \operatorname{argmax}\left(E^{\pi}\left[R(\tau)\right]\right) \quad (16.4)$$

where $E\pi$ indicates evaluating all possible actions in the policy π and argmax is evaluated over all policies under consideration. An implication of Eq. (16.4) is that we want to choose a stopping state at which the discounted reward has little effect on the trajectory's return value.

The *Q-function* and *value function* enable evaluating the "goodness" of a state and action. The Q-function, $Q^{\pi}(s, a)$ assesses the expected return starting in state s, takes action a, and then chooses actions from the policy π after that. The value function, $V^{\pi}(s)$ evaluates the expected return starting in state s and then behaves according to π. Commonly, both functions are evaluated from the starting state out to infinity. But as noted previously, the evaluation need not proceed past the point at which the discounted reward has little impact on the return.

$$Q^{\pi}(s,a) = E^{\pi}\left[\sum_{t=0}^{t=T} \gamma^t r_t \mid s_0 = s,\ a_0 = a\right] \quad (16.5)$$

$$V^{\pi}(s) = E^{\pi}\left[\sum_{t=0}^{t=T} \gamma^t r_t \mid s_0 = s\right] \tag{16.6}$$

Optimal Q-function and value function are important for finding optimal policies:

$$Q^*(s,a) = \max_{\pi} Q^{\pi}(s,a) \tag{16.7}$$

$$V^*(s) = \max_{\pi} V^{\pi}(s) \tag{16.8}$$

These functions are related to each other, i.e., the Q-function can be computed from the value function:

$$Q^{\pi}(s,a) = E^{\pi}\left[r_{t+1} + \gamma V^{\pi}(s+1) \mid s_t = s, a_t = a\right] \tag{16.9}$$

Importantly, whether it is better to select an action deterministically all the time when in state s and then follow the policy or follow the policy whenever in state s and after that depends on comparing $Q^*(s, a)$ to $V^*(s)$.

Suppose $\gamma = 0.7$ and that the system has four states: A, B, C, and D. Initially, Q is the zero matrix:

$$Q = \begin{array}{c} \\ A \\ B \\ C \\ D \end{array} \begin{array}{c} \begin{array}{cccc} A & B & C & D \end{array} \\ \begin{bmatrix} 0 & 0 & 0 & 0 \\ 0 & 0 & 0 & 0 \\ 0 & 0 & 0 & 0 \\ 0 & 0 & 0 & 0 \end{bmatrix} \end{array} \tag{16.10}$$

and that the instantaneous reward matrix is:

$$R = \begin{array}{c} \\ A \\ B \\ C \\ D \end{array} \begin{array}{c} \begin{array}{cccc} A & B & C & D \end{array} \\ \begin{bmatrix} 0 & 0 & 10 & 0 \\ - & 0 & - & 10 \\ - & 2 & 0 & - \\ 0 & - & 3 & 0 \end{bmatrix} \end{array} \tag{16.11}$$

For example, looking at the top row, an action that causes the transition from state A to state C has an instantaneous reward of 10. In the third row, a transition from state C to state B has an instantaneous reward of 2.

Choosing state D as the initial state, there are three possible actions: transition to A, C, or remain in D. Randomly selecting C as the next state, $Q_{D,B} = R_{D,C} + \gamma \mathrm{Max}(Q_{C,B}, Q_{C,C})$. Since $Q_{C,B}$, $Q_{C,C} = 0$, $Q_{D,B} = 3$:

$$Q = \begin{array}{c} \\ A \\ B \\ C \\ D \end{array} \begin{array}{c} \begin{array}{cccc} A & B & C & D \end{array} \\ \begin{bmatrix} 0 & 0 & 0 & 0 \\ 0 & 0 & 0 & 0 \\ 0 & 0 & 0 & 0 \\ 0 & 3 & 0 & 0 \end{bmatrix} \end{array} \tag{16.12}$$

We are now in state C and now we can either transition to B with reward 2 or remain in C with reward 0. If we choose the action that takes us to B then we can remain in B with reward 0 or transition to D with reward 10. We now have: $Q_{C,B} = R_{C,B} + \gamma \text{Max}(Q_{B,B}, Q_{B,D}) = 2 + .7 * 10 = 9$.

$$Q = \begin{array}{c} \\ A \\ B \\ C \\ D \end{array} \begin{array}{c} \begin{array}{cccc} A & B & C & D \end{array} \\ \begin{bmatrix} 0 & 0 & 0 & 0 \\ 0 & 0 & 0 & 0 \\ 0 & 11 & 0 & 0 \\ 0 & 3 & 0 & 0 \end{bmatrix} \end{array} \tag{16.13}$$

In D we have three options: A, C, and remain in D. Choosing C we now have $Q_{D,C} = R_{D,C} + \gamma \text{ Max } (Q_{C,B}, Q_{C,C}) = 3 + \gamma \text{ Max } (Q_{C,B}, Q_{C,C}) = 3 + .7 * 11 = 10.7$

$$Q = \begin{array}{c} \\ A \\ B \\ C \\ D \end{array} \begin{array}{c} \begin{array}{cccc} A & B & C & D \end{array} \\ \begin{bmatrix} 0 & 0 & 0 & 0 \\ 0 & 0 & 0 & 0 \\ 0 & 11 & 0 & 0 \\ 0 & 3 & 10.7 & 0 \end{bmatrix} \end{array} \tag{16.14}$$

The process continues until the values in Q converge. Using the converged matrix, the optimal action for any state is the one having the highest value looking across the row for that state. For example, at this stage of the iteration, from D we want to choose the action that results in the transition to C.

For any real system, the combinations and permutations of all states and actions may be too large for practical use. However, dynamic programming (Bellman 1952; Eddy 2004), an optimality method based on the Bellman equation, breaks up a large optimization problem into a recursion of smaller, local optimization problems. Subproblems results are saved for later use rather than being recomputed in subsequent iterations. Conceptually, dynamic programming substitutes memory for execution time by saving results of many subproblems rather than taking time for recursive solutions. Dynamic programming is performed in three steps:

1) Recursion
2) Subproblem storage (memoization)[1]
3) Bottom-up

A classic example is motivated by computing the Fibonacci sequence (1,1,2,3,5,8. . .). Except for the first two values in the sequence, each successive value is calculated as the sum of the previous two values. A Python version of a brute force function that returns the n-th value in the series:

```
def fib(n):
    if n==1 or n==2:
        return (1)
    else:
        return(fib(n-2) + fib(n-1))
print (fib(5))# print the 5th number in the series
```

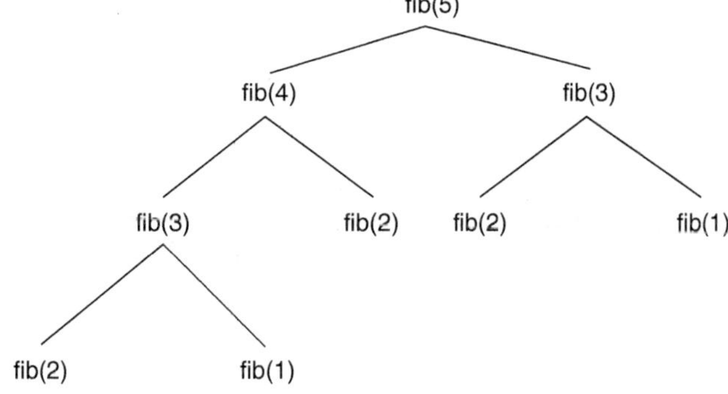

Figure 16.4 Brute force computation of the n = th value in the Fibonacci sequence executes 0 2n.

The function returns a result quickly for small values of n but is very inefficient for large n. As shown in Figure 16.4, fib (3), fib (2), and fib(1) are recomputed several times. The time required to compute the n-th value in the sequence grows on the order of 2^n.

A memoized solution stores subproblem results in an array of length $n+1$, e.g., for $n=5$, the array has length 6. Index 0 in the array is not used, index 1 stores fib (1), index 2 stores fib(2), and so forth. The array is initialized with null values and filled in as the algorithm progresses.

```
def fib(n, memo):
    if memo[n] != None:
        return memo[n]
    elif n==1 or n==2:
        return(1)
    else:
        memo[n] = (fib(n-2, memo) + fib(n-1, memo))
        return(memo[n])
memo = [None for i in range(6)]          # define a null array
memo[1] = memo[2] = 1                    # initialize memo[1] and
memo[2]
print (fib(5, memo))
```

After executing the above code, `memo = [None, 1, 1, 2, 3, 5]`.

The memoized algorithm computes the n-th Fibonacci number in time $O(2^n)$

As the memoized algorithm executes, the memo array is filled in from left to right by the recursion. In some situations, rather than recursively computing memoized values, those values may be tabulated bottom-up, as in the code below. This code determines the n-th value in the Fibonacci sequence in execution time $O(n)$.

```
def BU_fib(n):                                        # Bottom-up
algorithm
    if n == 1 or n == 2:
        return(1)
    BU = [None for i in range(n+1)]      # create a null array
    BU[1] = BU[2] = 1
    for i in range(3, n+1):
        BU[i] = BU[i-1] + BU[i-2]
    return(BU[n])
print (BU_fib(80))
```

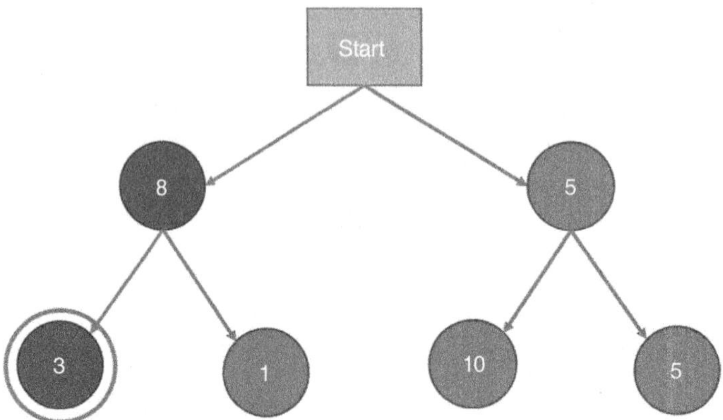

Figure 16.5 An example of the greed algorithm showing the chosen trajectory in green.

The greedy algorithm, another optimization method, generates a single solution using a bottom-up strategy. The algorithm works through a problem in stages. At each stage, a locally optimal choice is selected. The totality of all local decisions is the final solution, although that solution may not be globally optimal. Figure 16.5 shows an example greedy algorithm for selecting the trajectory with the best return. From Start, the greedy algorithm chooses the left branch over the right branch because the reward value 8 is greater than 5 in the right branch. Next, the algorithm chooses the final reward 3 over the reward 1. In this example, the actions leading to reward 8 and then reward 3 represent the optimal trajectory. By not investigating options past the first right branch, the algorithm does not find the true optimal result. Although not optimal, the suboptimal trajectory is still good enough in many circumstances.

State-of-the-Practice

This section summarizes the current practice in both resilience modeling and resilience-related machine learning. Several authors have tackled the thorny problem of how to model resilient systems, perform trade studies, evaluate hazards, and predict resilience metrics. Some of these models include sophisticated formal means for analyzing emergent behavior and uncertainty during the design process.

Resilience Metamodels

A metamodel defines the abstract entities and relationships of modeling languages used to define concrete models in a domain of interest. Metamodels consist of generalized object types, object relationships, object attributes, and compositional rules.

Bakirtzis et al. (2022) describe a cyber-physical system resilience metamodel. The metamodel combines considerations of cyber security, safety, and resilience using a formal ontology. Briefly, the metamodel apex is an entity called **Resilient Mode** that ***is managed by*** a **Function** entity. **Function** is ***recovered by*** **Resilient Mode**. **Component** ***is contained by*** **Resilient Mode** and ***performs*** **Function**. **Attack Vector** ***violates Component*** and is ***precipitated by*** **Loss Scenario**. A **Sentinel** entity ***protects against*** the **Loss Scenario** which is ***detected by monitoring*** **Function**. **Loss Scenario** ***leads to*** **Unsafe Action** that ***leads to*** **Hazard**. **Hazard** ***leads to*** **Loss** and ***elicits*** **Requirement**. A complete metamodel description, example instantiation, and discussion of its use in developing architectures and assessing operational risks is presented in Bakirtzis et al. (2022).

Post-Disruption Recovery Models

Restoring system performance post-disruption is an essential component of resilience. Wang et al. (2019) review 30 critical energy infrastructure models that cover electric power, natural gas, and fuel networks. The authors list five general model categories: optimal operation, topological network, agent-based, probabilistic, and "other" (comprising actor-based, empirical, system dynamics, and physical models).

Optimal Operation Modeling

Optimal operation models are one of the most used methods for achieving resilience in energy infrastructures. The approach adaptively optimizes infrastructure reconfiguration and repair to minimize repair time brownouts. Multi-zone microgrid architectures support high availability by providing spare capacity within a zone and zone switching when a zone fails. Repair actions and crew dispatches are planned that maximize service restoration after a disruption. The authors note that optimal operation models focus on single problems, e.g., managing resilience resources or post-disruption recovery. Adding disaster scenarios significantly increases the computational time rendering the model potentially unusable for real-time decision-making.

Topological Network Modeling

Topological network models represent power systems by undirected graphs in which nodes represent power clients and servers, and edges represent transmission lines (Lin and Bie 2018). These models are useful in evaluating power network structural vulnerabilities. Topological modeling found paths for cascading faults between interdependent power systems in one example. Another application simplified a large and complicated system by aggregating nodes into clusters that were calibrated with more detailed simulations. Although helpful in finding macro-level dependencies and potential catastrophic propagation paths, topological network models do not capture system physical properties and operational constraints. Consequently, model results may be overly optimistic. However, topological modeling versions have been used to study the impact of cyberattacks in distributed generation systems.

Agent-Based Modeling

Agents are persistent entities in a computational environment that perform autonomous functions for or with other agents or users. This form of modeling consists of rule-based, dynamic interactions between agents and is mainly used for evaluating exchanges between interdependent systems. Agents may include human interaction and may also exhibit emergent behaviors. However, computational needs for evaluating agent-based models can be significant, so most models focus on a single type of interdependence, such as physical or logical interactions.

Acheson and Dagli also discuss agent-based modeling as a means for modeling resilient SoS architectures (2016). Their model consists of three agents: SoS Manager Agent (SMA), System Agent (SA), and Threat Agent (TA). The SoS manager is responsible for developing the SoS. The SMA evolves the SoS architecture in response to Threat Agent by requesting new capabilities from constituent systems for defending against threats. The SA represents a system with the SoS. When requested by the SMA, a SA sends a message listing the capabilities it can support as determined by its willingness and ability to provide the requested capabilities. TAs represent the threats that can attack an SoS.

Probabilistic Modeling

Probabilistic models are useful in expressing uncertainties in system behavior and system disruption. Often these models comprise Monte Carlo simulations that inject randomness into the simulation for evaluating random objects and events. Probabilistic models may also be expressed by Markov chains in which transition probabilities from a state are determined by the number of times a particular path to another state is taken when an action is applied. Similarly, emission probabilities are determined by the number of times a given output is observed when arriving in a new state. Many machine learning algorithms, such as Q-Learning and Value iteration use probabilistic evaluations to learn a system's behavior, evaluate situational awareness during operation, and choose optimal actions.

Malware Attack Resilience Models

A zero-day malware attack is a previously unknown computer virus or worm. Because these attacks have not yet been seen, new anti-malware signatures are not available and existing signatures are often unsuccessful. A dynamic model developed by Tran et al. (2016) is valuable in predicting the spread and recovery from a zero-day attack. The model combines the Susceptible-Infected-Quarantined-Recovered (SIQR) model (Lai et al. 2021; Viguerie et al. 2021) currently used in forecasting coronavirus outbreaks (Tran et al. 2016; Lai et al. 2021; Viguerie et al. 2021) with the NIST SP-800-61 standard (Cichonski et al. 2012). The model comprises four nodes: **Susceptible Machines**, **Infected Machines**, **Recovered Machines**, and **Quarantined Machines**. **Susceptible Machines** impact the **Infected Machines** by the ***Incident Rate***, R_i, that is a function of the total number of computers in a system, perimeter protection effectiveness, user contact rate, quarantine contact rate, infectivity, and an incidence control function. **Infected Machines** alter the number of **Quarantined Machines** and **Recovered Machines** through a ***Quarantine Rate*** and ***Removal Rate***, respectively.

Bellini et al. developed a cyber resilience metamodel for a railway communication case study (Bellini et al. 2021). The metamodel entities comprise: Adaptive Capacity Kind, Stressor Kind, Flexibility Capacity, Dependencies, Assets, Likelihood Level, and Criticality Level. The capacity-based approach posits that resilience is an emerging system property and is not directly measurable because resilience properties can only be measured when they are used. However, resilience requires latent resources that can be activated or recombined as needed. Capacity-based approaches measure the number of latent resources as an indirect means of measuring resilience. More than just counting resources related to adaptive capacity, a resilient system also needs coping ability that assures service survivability. There is a synergy between adaptation and coping in that adaptation enables coping and coping requires adaptation. The metamodel is instantiated for radio subsystems used by rail services to show it can describe cyber threats.

Dynamic Flow Models

Goldbeck et al. discuss dynamic flow models for analyzing disaster vulnerabilities in critical, interdependent infrastructure systems (Goldbeck et al. 2019). The paper argues that traditional risk management methods are not well suited to situations with high uncertainty in risk likelihood, impact, and the effectiveness of risk control and avoidance measures. For example, the cascading effects of superstorm Sandy that struck New York City in 2012, caused power outages that hampered water removal from flooded metro tunnels. Flooding from Sandy also caused a breakdown in the liquid fuel supply chain due to damage at terminals, refineries, and pipelines combined with power outages and waterway traffic restrictions.

An infrastructure is represented as networks of nodes and directed links that interact with physical and non-physical assets. The network models infrastructure services, while assets represent system components needed for providing infrastructure services. The assets are vulnerable to disruptions that may impact infrastructure services. Assets from interdependent systems are connected by links, as are the services in one system that affect assets in another. The cross-coupling of assets and services enables the investigation of potential cascading effects.

Commonly used iterative or integrated evaluation of dynamic flow models do not consider trending and planning. As noted in Goldbeck et al. (2019), this is problematic for repairable systems because optimal repair strategies need to prioritize repair by anticipating where network capacity is most needed. Integrated models assume perfect usage predictability, which is never possible in practice.

The authors propose a rolling planning horizon that combines iterative and integrated models using a linear programming formulation. Planning is updated with each iterative step to account for newly available information. At each iteration step within a window, a planning computation is performed from that step to a time in the future.

Resilience-Related Machine Learning

Markov Decision Processes (MDP)

As explained earlier, MDP is a process in which a control agent uses the current system state and a decision rule for choosing an action to take. This form of decision rule is *Markovian* because it is independent of prior states and actions. More generally, a decision rule may be *history dependent* if it is subject to all previous actions. Decision rules may also be *randomized* if each state has a probability distribution over all acceptable actions. The decision agent makes decisions at each time epoch, as shown in Figure 16.6.

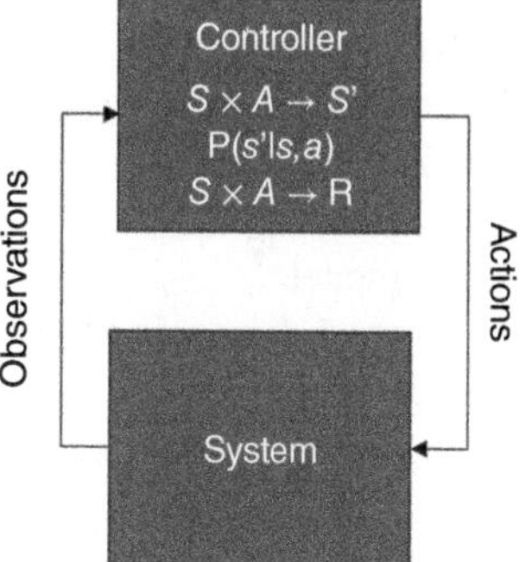

Figure 16.6 The Controller in a MDP determines the system state from system observables and actions based on maximizing a decision rule.

All decision rules are candidates for potential use in resilient systems and may be used in combination before experience in actual use. Consider a situation in which a system is believed to be in a state that has yet to be observed or analyzed. The state may have a set of candidate actions, but no information guides the selection. In this situation, a randomized decision rule might be reasonable until statistical data are available for better decisions. Cyber-resilient systems might also use randomized decision rules, making it more difficult for hackers to learn system responses.

The process by which MDP parameters are learned is straightforward. Through simulation or execution of the physical system, statistical data are collected for observations and transitions occurring for the actions taken at each state (Goldbeck et al. 2019). Parameters may then get updated during system operation as more data are collected.

Abdelmalak and Benidris (2021) describe using an MDP to enhance power system resilience during hurricanes. Their approach defines a state representing a specific system topology based on available working components. Their model assumes that failed components are not restored during the hurricane. Using a multi-objective optimization problem, the MDP looks for the trajectory resulting in minimum power loss and cost. For each state, $s_{i,t} \in S$, at time t, the minimum cost function is:

$$v_t^* = \min\left\{ v_t\left(s_{i,t}, a_{s_{i,t}}\right), a_{s_{i,t}} \in A_{s_{i,t}} \right\} \tag{16.15}$$

where $A_{s_{i,t}}$ is the set of all possible actions at state $s_{i,t}$ at time t and $v_t(s_{i,t}, A_{s_{i,t}})$ is the expected overall cost for state s_i, t when taking action $a_{s_{i,t}}$. The cost function at a given state is the sum of the instantaneous cost at time t, $C_t\left(C, a_{s_{i,t}}\right)$, and the future costs for actions taken when in $s_{i,t}$, similar to Eq. (16.4).

Wang et al. examine enhancing energy system resilience using dynamic programming for analyzing an MDP (Wang et al. 2020). A state in their model includes both failed states and states repaired:

$$s_{i,t} = \mathrm{H}_t \bigcup_{\tau=1}^{t} \tilde{F}_\tau + \tilde{R}_\tau \tag{16.16}$$

in which $\tilde{F}_\tau$ and $\tilde{R}_\tau$ are respectively, the actual component failures and the set of repaired components at time τ. As in Wang et al. (2020), their model aims to find the minimum cost determined by summing the immediate cost at $s_{i,\,t}$ and the future costs of all possible actions. The dynamic programming problem has several operational constraints that reduce the search space, but the space is still too large for a brute-force solution. Consequently, the dynamic programming approach used two simplifying approximations:

- **Post-decision state**: the state immediately after an action is taken but before arriving at the next state and next decision time due to uncertainties. Post-decision state information is used to estimate future downstream costs. By using future costs based on post-decision states, the multi-period and multi-trajectory stochastic analysis of an MDP is reduced to a one-period deterministic model.
- **Forward dynamic algorithm**: value estimates for every state are needed for the one-period deterministic model. The algorithm recursively visits each state and computes and memoizes the forward-looking values using the bottom-up method described previously.

Partially Observable Markov Decision Processes (POMDP)

A POMDP comprises a finite number of states, and like MDPs state dynamics is determined by probabilistic transitions and a finite set of controllable actions. Unlike MDPs, a POMDP depends on the emissions from the next state to determine a state probability distribution. More importantly, the states in a POMDP are only partially observable and may be difficult to solve optimally.

An experimental resilience approach combines fault-tolerance constructs for known and anticipated disruptions with POMDP and heuristics for unknown-unknowns (Sievers and Madni 2016a; Madni et al. 2018b, 2020). Fault-tolerance contracts are invariant assume-guarantee LTL statements developed from top-down (fault tree) and bottom-up (failure modes and effects analyses) that pair fault conditions to error monitors that trigger responses.

Equation (16.17) shows an example set of LTL contracts in which $G(\varphi \rightarrow X\,\psi)$ means whenever φ holds, ψ is true in the next cycle

$$\begin{aligned} &G(\text{fault} \rightarrow \mathbf{X}(\text{detection})) \\ &G(\text{detection} \rightarrow \mathbf{X}(\text{response})) \\ &G(\text{response} \rightarrow \mathbf{X}(\text{recovery})) \end{aligned} \tag{16.17}$$

Equation (16.17) states that a fault is always followed by detection, detection is followed by a response, and the response leads to recovery during the next system cycle. As shown in Figure 16.7, these contracts form the basis of a POMDP in which detections, responses, and recovery are not guaranteed, and transitions to hidden states may occur. In Figure 16.7, known state s_0 represents nominal operation, and known state s_1 represents the system when a recovery action is taken.

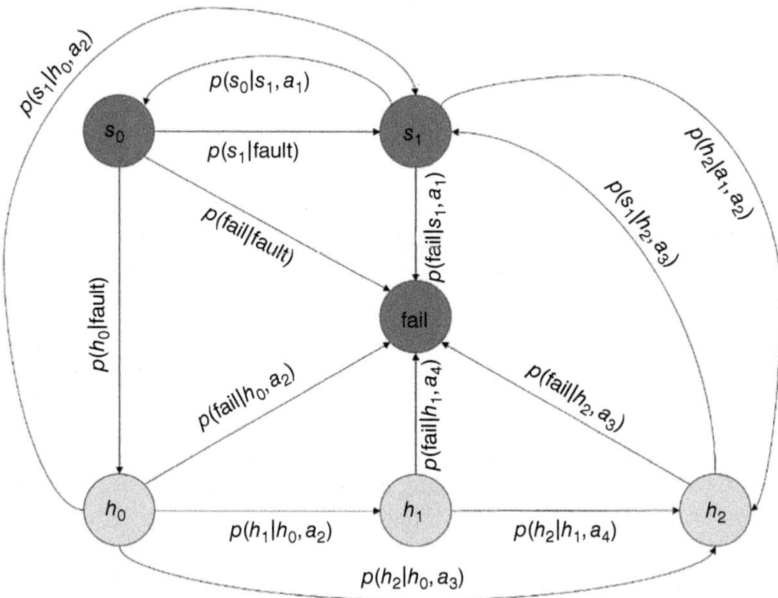

Figure 16.7 Partial POMDP representation of Eq. (16.17), not all transitions are shown.

Hidden states are shown in yellow and connected to each other and the known states. Emission and transition probabilities for the unknown states are learned during system tests and operations.

Penetration testing exercises security measures by generating and executing cyberattacks that attempt to gain control over a system. Using available knowledge, POMDP has been used for modeling cyberattack planning by creating better models of actual attack scenarios (Sarraute et al. 2012). Penetration testing increases cyber resilience by uncovering exploitable weaknesses before an actual attack occurs.

Linear Temporal Logic and Model Checking

Linear temporal logic (LTL) is an appropriate formalism for reasoning about the behavior of reactive systems. LTL is used for verifying architectures, code usage, code safety, and cyber security (Kesten et al. 1998; Clarke et al. 2009; Pnueli 1977; Vardi 2021; Gerth et al. 1995; Biggar and Zamani 2020). A system is modeled as a finite state machine, e.g., a *Büchi automaton* (Büchi 1960) that represents a system that either accepts or rejects a *ω-regular language*. An ω-regular language is an infinite-length *regular language*, and a regular language is a language (string of inputs) that is recognizable by a finite-state machine (finite automaton) (Pnueli 1977).

LTL is composed of a finite set of propositional variables, logical operations, and temporal modal operators. A propositional variable is a variable that is either **true** or **false** and used as the input to a propositional statement. In LTL system, operators describe behaviors along a single computational path.

LTL logical operations:

- $\wedge$: conjunction $\equiv\neg(\neg\varphi \vee \neg\psi)$
- $\vee$: disjunction
- $\neg$: NOT
- $\varphi \rightarrow \psi \equiv \varphi$ implies ψ
- $\varphi \leftrightarrow \psi \equiv \varphi$ is equivalent to ψ
- **true** and **false**

Unary temporal operators:

- **F** φ: Finally, φ must hold eventually in a path through the model
- **X** φ: Next: φ must hold at the next state
- **G** φ: Globally: φ must always hold the entire subsequent path

Binary temporal operators:

- ψ **U** φ: Until: ψ must hold at least until φ is true, and then φ must remain true from then on
- ψ **R** φ: Release: φ Is true until and up to the point at which ψ becomes true, and φ remains true if ψ is never true
- φ **W** $\psi \equiv$ weak release; φ remains true until ψ is true but does not require that ψ ever become true
- φ **M** $\psi \equiv$ strong release; φ remains true until ψ is true, but ψ must become true at some point

Model checking (Clarke et al. 2001) validates that a model, *M*, satisfies a property, *P* (written as $M \models P$), in which *M* is a model of computation performed by the system of interest and *P* is a logical formula describing state or trace properties. Checking involves visiting the set of reachable states from the current state in a model and ensuring that *P* holds. *P* must hold in all states for invariant assertions.

Examples of specific checks for logical formulas written with LTL primitives and operators include:

- φ: check current state properties
- **X** φ: check that φ holds at all states that succeed the current state
- ψ **U** φ: starting at the current state, do a depth-first search until φ is found or the path loops back on itself. An error is raised if ψ is false before stopping
- **F** φ: using a Büchi automaton and starting at the current state, the formula is true if no looping path is found in which φ is false in every path

As noted by Catano, cyber-resilience is measured by how well a system anticipates, withstands, adapts, and recovers from cyberattacks (Catano 2022). As with resilience in general and cyber-resilience in particular, the security community does not pay much attention to cyber issues until an attack is successful. With cyber-resilience becoming an increasing concern for military and safety-critical systems, there is an increasing need for certified software. This is software that has been proven correct and the proof delivered with the code.

Using formal methods, the approach in Catano (2022) synthesizes architectural tactics for system availability, performance, and security by extending EVENT-B models (Hoang 2017) to support LTL and cover specification of safety and liveness properties. EVENT-B models comprise observations about a system, its invariants, and its dynamic properties. A four-step process generate certified code:

- Common architecture tactics are modeled in EVENT-B including system invariants and two resilience patterns: security breach detection and recovery;
- Each architectural tactic is extended and coded in LTL and translated into EVENT-B;
- Focusing on the temporal aspects of the tactics and certified code production, program synthesis techniques generate resilient JAVA from EVENT-B models;
- Unit testing is performed for validating the certified code.

Partially Observable Stochastic Games (POSG)

Tipireddy et al. describe an approach using partially observable stochastic games (POSG) for generating automated cyber resilience policies (Welsh and Benkhelifa 2020). A POSG shares the same set of parameters found in a POMDP with the additional parameter, N, that represents the set of "players."

An attacker's policy is modeled as:

$$p\left(a_j \mid s, a_i\right) \tag{16.18}$$

in which a_j represents the attacker's action when the system is in state s and a defender takes action, a_i. The transition function from state s to s' includes both the attacker's action and the defender's action:

$$p\left(s' \mid s, a_i, a_j\right) \tag{16.19}$$

With Eqs. (16.18) and (16.19), the single defender's transition model is represented in a POMDP as:

$$p\left(s' \mid s, a_i\right) = \sum_{a_j} p\left(s' \mid s, a_i, a_j\right) p\left(a_j \mid s, a_i\right) \tag{16.20}$$

The attacker's policy model is used for computing the defender's reward as a POMDP:

$$r_i\left(s, a_i, s'\right) = \sum_{a_j} r_i\left(s, a_i, a_j, s'\right) p\left(a_j \mid s, a_i\right) \tag{16.21}$$

The defender's belief state is computed as in Eq. (16.1) using the defender's actions. The complete POMDP comprises states {normal, hacked} and defender actions {detect, no-action, restart}. A total of 27 POMDP problems were solved using Monte Carlo methods leading to 27 policy distributions. The analysis enabled examining correlations between observations, actions, beliefs, and policies over many decision support possibilities and time.

Flexible Resilience Contracts

As noted earlier, design by contract (DbC) is a formal and systematic method for creating provably correct computational systems. The semantics of DbC provide a natural means of detecting anomalous behaviors and taking corrective actions. A contract, C, is defined by a pair of assertions, $C = (A, G)$ in which A is an assumption made on the environment and G is the guarantee a system makes if the assumption is met. Assumptions are system invariants and pre-conditions, while guarantees are system post-conditions.

Flexibility for resilience necessitates relaxing invariant assertions by allowing probabilistic assumptions and guarantees. Moreover, deterministic system models are not well-suited for capturing the realities of systems that impacted by variabilities such as defects, unusual usage, environmental uncertainties, and disruptions (Li et al. 2017). The approach taken in Li et al. (2017) leverages an extension of Signal Temporal Logic called Stochastic Signal Temporal Logic (StSTL). StSTL can represent probabilistic constraints on assumptions and guarantees for modeling cyber-physical systems. The paper also describes a verification and synthesis approach that explores contract feasibility.

Xu et al. (2012) describe a probabilistic contract framework for component-based embedded systems. Similar to the method described in Li et al. (2017), the framework supports modeling components, component interactions, and uncertainty. Their method uses an Interactive Markov Chain (IMC) in which transition probabilities represent nondeterministic event outcomes. Contracts are defined by a tuple comprising a nonempty finite set of states, a set of actions, action transition relationships, transition probabilities, and the initial state. As with deterministic assertions, contract refinement is compositional, the parallel composition of implementations is satisfied by parallel composition of their contracts, and multiple contracts allocated to the same component can result in independent requirements.

Delahaye et al. (2010) discuss propositions for compositional reasoning that include a discount factor when components are combined. The discount factor gives less weight to events happening in the future than near-term events. This paper discusses system representations using Markov Decision Processes (MDPs) in which actions are taken that maximize a future reward. As noted, an MDP is representable by a Büchi automaton used for model checking (Schimpf et al. 2009).

A resilient system must accommodate known and unknown situations resulting from disruptive events. Consequently, the assumptions of full observability and comprehensive knowledge of the state space are not entirely valid. An implication is that modeling must include the possibility of hidden states, that is, states that are not known in advance but may occur during system operation. To that end, flexible contracts extend the work in Delahaye et al. (2010) by accommodating partial observability (Sievers and Madni 2016b). The framework comprises Partially Observable Markov Decision Processes (POMDPs) in which available observations are used for evaluating a belief state (Kaelbling et al. 1998). A belief state is a probability distribution over all states in a system. A system's belief state is updated as a function of the current belief state, an action taken, and the observations made after taking an action. POMDP training initially consists of supervised learning for determining the observation space associated with each known state. Transition and observation probabilities are learned through system simulations and during initial system operation in its environment. Finally, reinforcement learning during ongoing use updates POMDP parameters.

More formally, a POMDP models a decision process in which system dynamics are assumed to be a belief Markovian Decision Process (MDP), a memoryless decision process that involves transition rewards. A POMDP comprises the 4-tuple:

- β = infinite set of belief states
- α = finite set of actions
- $\rho(b, a) = \Sigma_{s \in S} b(s) R(s, s'a)$ Expected reward at b(s) on transition from s to s′ given a
- $\Lambda(b' \mid b, a) = \Sigma_{o \in O} \Lambda(b' \mid b, a, o) \Lambda(o \mid a, b)$ Transition function

in which a belief represents an understanding of a system state, $s \in S$, with uncertainty. The POMDP has a policy, π, which describes how to select actions in a belief state based on maximizing a goal defined by the reward function, ρ, within some time period, that is, $\pi : s \in S \rightarrow a \in \alpha$.

In this formulation, a resilient system comprises one or more agents responsible for evaluating and updating a set of flexible contracts associated with a system component. Component decomposition and evaluation of satisfaction use the processes described in Xu et al. (2012). Component composition and reasoning uses the approach in Delahaye et al. (2010). Agents continuously monitor their components as well as their inputs and as necessary take actions from a policy that maximizes a reward function. A key aspect of agent behavior is understanding whether an observation is a close enough match to previous observations or is sufficiently distant that it represents a new state. Agents evaluate the Mahalanobis distance (MD) between the current observation and the states associated with its component. MD is commonly used for determining whether a sample is a member of a group or an outlier:

$$\text{md} = \sqrt{\left(\overrightarrow{\text{obs}} - \vec{\mu}\right)^T \left(\text{cov}^{-1} * \left(\overrightarrow{\text{obs}} - \vec{\mu}\right)\right)}$$

in which an observation vector is compared to the mean values, $\vec{\mu}$, of observations assigned to a group. The group covariance matrix normalizes the data.

Deep Neural Networks

Deep Neural Networks have become the most powerful AI technique, leapfrogging other more established but increasingly more obsolete techniques (Kuo and Madni 2023). They are responsible for most of the current wave of successful AI and machine learning applications for image and speech recognition, natural language, big data analytics, and even deep fake videos. At the same time, over-anthropomorphized explanations invoke human notions of "learning" or "neurons" to try to explain the technology and lead to unfounded fears of synthetic intelligences running amok on our streets, in our homes, and on our battlefields. Just as systems engineers need a sufficient understanding of electrical engineering, mechanical engineering, and software engineering, they must also come to understand AI as a new engineering discipline.

Best Practice Approach

Formal correctness proofs follow one of two methods: *model checking* and *theorem proving* (Tipireddy et al. 2017). Model checking (Tipireddy et al. 2017) is well suited to verifying *control-intensive applications*, i.e., where the bulk of the code uses control structures such as if-then-else statements and operates on relatively simple data types. Tools exist today to automate model checking and generate counter examples when the checks uncover errors. Importantly, model checking used prior to physical implementation can facilitate early problem detection.

Theorem-proving involves proving system properties through deduction. It does not need to visit all system states to prove properties. Consequently, theorem proving is ideal for *data-intensive applications*. Hoare (Holzmann 2003) developed the first theorem proving concept that reasons about software using DbC-like assertions called *Hoare's Triples*. A Hoare's Triple comprises a precondition that if true prior to starting execution of a program, then a post-condition will hold when the function completes. A program can be the entire program, or a function within a program. Dijkstra extended Hoare's method with the concept of *predicate transformers*. This approach starts with the post-condition and works backward to determine the needed precondition (Hoare 1969).

The fault-tolerance domain tends to fall into the control-intensive application space. Consequently, model-checking is best suited for assuring its correctness. Conversely, resilience may involve complicated data types and heuristics that are better checked by theorem proving. While these methods employ different approaches, it is possible to combine the best of both to satisfy needs of the different system components (Dijkstra 1975). This recognition and corresponding method are one part of our best practice approach. The second part is based on the realization that not all aspects of a system need formal modeling and verification.

The POMDP modeling paradigm, a formal approach, supports both mechanistic fault-tolerance contracts and more flexible decision processes needed for managing unknown-unknowns. A POMDP model for resilience begins as an MDP in which states represent the Büchi automaton that models fault-tolerance contracts. States in the MDP are trained during test and operation using supervised learning (Ouimet 2007; Kotsiantis 2007). Supervised learning uses labeled datasets for assigning test data to specific categories. Additionally collected observations are used for determining POMDP model parameters. Observations made during subsequent operation are compared to the trained set to determine whether changes in POMDP parameters or states are needed.

The example in the next section explains this approach in greater detail.

Illustrative Example

Figure 16.8 shows the decomposition and allocation of a system resilience contracts to n subsystems. As noted in Cimatti and Tonetta (2012), a property at the system level may be proven using deduction that starts with component properties and is then iterated on with inference rules. A decomposition is deemed correct if it satisfies two conditions: (1) all correct implementations of subcomponent contracts form a correct implementation of the system, and (2) for each subcontract C', the correct implementation of all other subcontracts within a correct system operating from a correct environment for C'. A deduction tree is constructed for the system that decomposes system properties into sub-properties allocated to the components. The implementation of the components must satisfy their allocated properties. In effect, this mirrors conventional requirement decomposition and allocation, but the difference is that formal contracts are used in place of loose natural language. The following example shows how decomposition separates contracts into those that are proven by model checking and those that require theorem proving.

The illustrative example shown in Figure 16.9 is a simplified version of a quadcopter control system. A Flight Controller is the "brains" of the quadcopter that maintains its stability and quickly adjusts for gravity, and wind while following motion commands and course changes sent by an operator. The global positioning system receiver (GPSR) receives GPS transmissions and computes the quadcopter's latitude, longitude, elevation, and current time. The inertial measurement unit (IMU) contains a 3D accelerometer for measuring drone orientation relative to the Earth's surface and a 3D gyroscope that measures angular rates, i.e., six-axis gyro stabilization. Additionally, the Flight Controller receives azimuth (absolute bearing) from a Compass that determines the horizontal angle between the quadcopter direction and magnetic north. Primarily, the Flight Controller integrates acceleration with gravity estimates to calculate its current velocity. Velocity is integrated to calculate the quadcopter position that can be compared with GPS position data. Lastly the Flight Controller sends commands to Motor Speed Controllers that adjust the speed of the quadcopter propeller motors. The attacker in Figure 16.9 interferes with GPS location by sending signals that randomly spoof the quadcopter location.

For simplicity, four known states are assumed: Stable (normal flight), Unstable (unstable flight), and Lost (unknown location), and initially two unknown and untrained states, h_0 and h_1. There are five monitors: LE (GPS location), GT (GPS time), GR (gyro rate), AZ (azimuth), and MS (motor speed) from which five error observations are made and an overall system status:

- LE = |(expected location) – (computed GPS location)|
  - LEerror = *true* if LE > LEmax

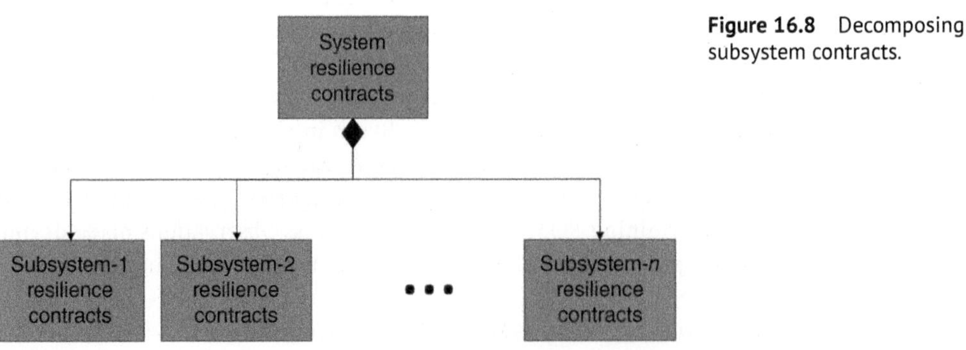

Figure 16.8 Decomposing subsystem contracts.

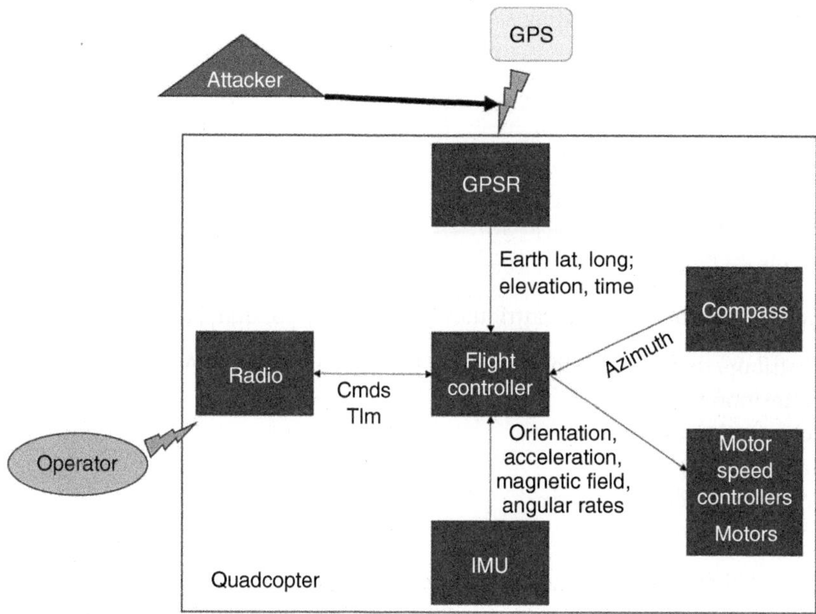

Figure 16.9 Quadcopter control system.

- GT = |(expected time) – (computed GPS time)|
  - GTerror = *true* if LE > GTmax
- GR = |(expected gyro rate) – (measured gyro rate)|
  - GRerror = *true* if GR > GRmax
- AZ = |(expected absolute bearing) – (measured absolute bearing)|
  - AZerror = *true* if AZ > AZmax
- MS = |(expected motor speed) – (measured motor speed):
  - MSerror = *true* if MS > MSerror
- SYSok = ¬ (LEerror ∨ GTerror ∨ GRerror ∨ AZerror ∨ MSerror)

Additionally, there are unknown disruptions caused by wind and the attacker. The quadcopter does not directly know wind effects and cannot know what disruptions are caused by the attacker.

The quadcopter can take one of four actions:

- a_1: follow current operator instructions
- a_2: land
- a_3: ignore a sensor
- a_4: reconfigure sensors, actuators, and control algorithms

For this example, we assume that adequate test and operating data have been collected for initial training of a POMDP model. As noted, the learning process comprises collecting observations after actions have been taken and determining the transition and emission probabilities by counting occurrences. Transition and emission tables are initially small, nonzero values for the hidden states because unknown unknowns have not occurred during test and initial operation.

This example uses Python classes for defining states:

```
class state:
    def __init__(self,name,transMtx, obsMtx, policy, cluster):
        self.name          = name
        self.transMtx      = transMtx          # P(s|state,a)
        self.obsMtx        = obsMtx            # P(o'|state,a)
        self.policy        = policy            # policy vector
        self.cluster       = cluster           # cluster matrix
```

The `__init__` method initializes the class attributes for each instantiation. State attributes are:

- `name`: a text attribute that is useful in associating the class with a POMDP state name
- `transMtx`: state transition matrix
- `obsMtx`: state observation matrix
- policy: state action policy vector
- `cluster`: state observation cluster matrix
- obs: vector of current observation

Each state has method that computes the inverse covariance of the cluster needed for the Mahalanobis distance using Python's `numpy cov` and `linalg.inv` functions. The Mahalanobis distance, md, for an observation is computed for each state from Eq. (16.22) using the Python `numpy` mean function where $\vec{\mu}$ is the mean value of each observable in a cluster.

$$\text{md} = \sqrt{\left(\overrightarrow{\text{obs}} - \vec{\mu}\right)^T \left(\text{cov}^{-1}\left(\overrightarrow{\text{obs}} - \vec{\mu}\right)\right)} \tag{16.22}$$

Tables 16.3–16.17 contain example values nominally determined by capturing data during testing initial operation. The training process comprises observing the number of times observations and transitions occur after performing an action. Reward values are determined by whether the trajectory triggered by an action is toward continuing the flight as the most important goal or toward safety if continued operation is not possible. Hidden states have small nonzero transition and emission that will be determined when observations fall outside of the values for the known states.

The correctness of state methods and table entries is confirmed through proofs of correctness. The inverse correlation function is two lines of Python:

```
covMat = numpy.cov(self.cluster, rowvar=0) # compute the
covariance matrix
self.invCov = numpy.linalg.inv(covMat)       # invert it
```

Table 16.3 $P(s' \mid \text{Stable}, a)$.

State/Action	a_1	a_2	a_3	a_4
Stable	0.978	0.698	0.978	0.548
Lost	0.01	0.2	0.01	0.25
Unstable	0.01	0.1	0.01	0.2
h_0	0.001	0.001	0.001	0.001
h_1	0.001	0.001	0.001	0.001

Table 16.4 $P(o' \mid \text{Stable}, a)$.

Observation/Action	a_1	a_2	a_3	a_4
LEerror	0.01	0.01	0.4	0.4
GTerror	0.01	0.01	0.2	0.3
GRerror	0.01	0.01	0.01	0.01
AZerror	0.01	0.01	0.3	0.1
MSerror	0.01	0.01	0.01	0.01
SYSok	0.95	0.95	0.08	0.18

Table 16.5 Action policy.

State/Reward	a_1	a_2	a_3	a_4
Stable	10	4	10	5
Lost	1	2	3	8
Unstable	−1	8	1	7
h_0	0.0001	0.0001	0.0001	0.0001
h_1	0.0001	0.0001	0.0001	0.0001

Table 16.6 Stable state cluster values.

LE	GT	GR	AZ	MS
1.2	0.1	10	0.5	2
1.5	0.2	14	0.6	2
0.5	0.05	9	0.3	1

Table 16.7 $P(s' \mid \text{Lost}, a)$.

State/Action	a_1	a_2	a_3	a_4
Stable	0.058	0.073	0.058	0.6
Lost	0.93	0.92	0.93	0.298
Unstable	0.01	0.005	0.01	0.1
h_0	0.001	0.001	0.001	0.001
h_1	0.001	0.001	0.001	0.001

Table 16.8 $P(s' \mid \text{Unstable}, a)$.

State/Action	a_1	a_2	a_3	a_4
Stable	0.1	0.675	0.06	0.4
Lost	0.03	0.03	0.2	0.3
Unstable	0.868	0.293	0.738	0.298
h_0	0.001	0.001	0.001	0.001
h_1	0.001	0.001	0.001	0.001

Table 16.9 $P(s' \mid h_0, a)$.

State/Action	a_1	a_2	a_3	a_4
Stable	0.0001	0.0001	0.0001	0.0001
Lost	0.0001	0.0001	0.0001	0.0001
Unstable	0.0001	0.0001	0.0001	0.0001
h_0	0.0001	0.0001	0.0001	0.0001
h_1	0.0001	0.0001	0.0001	0.0001

Table 16.10 $P(s' \mid h_1, a)$.

State/Action	a_1	a_2	a_3	a_4
Stable	0.0001	0.0001	0.0001	0.0001
Lost	0.0001	0.0001	0.0001	0.0001
Unstable	0.0001	0.0001	0.0001	0.0001
h_0	0.0001	0.0001	0.0001	0.0001
h_1	0.0001	0.0001	0.0001	0.0001

Table 16.11 $P(o' \mid \text{Stable}, a)$.

Observation/Action	a_1	a_2	a_3	a_4
LEerror	0.01	0.01	0.4	0.4
GTerror	0.01	0.01	0.2	0.3
GRerror	0.01	0.01	0.01	0.01
AZerror	0.01	0.01	0.3	0.1
MSerror	0.01	0.01	0.01	0.01
SYSok	0.95	0.95	0.08	0.18

Table 16.12 $P(o' \mid \text{Lost}, a)$.

Observation/Action	a_1	a_2	a_3	a_4
LEerror	0.95	0.95	0.95	0.4
GTerror	0.01	0.01	0.01	0.01
GRerror	0.01	0.01	0.01	0.01
AZerror	0.01	0.01	0.01	0.3
MSerror	0.01	0.01	0.01	0.01
SYSok	0.01	0.01	0.01	0.27

Table 16.13 $P(o' \mid \text{Unstable}, a)$.

Observation/Action	a_1	a_2	a_3	a_4
LEerror	0.095	0.01	0.095	0.2
GTerror	0.1	0.01	0.1	0.01
GRerror	0.4	0.01	0.4	0.01
AZerror	0.1	0.01	0.1	0.1
MSerror	0.3	0.01	0.3	0.3
SYSok	0.005	0.95	0.005	0.38

Table 16.14 $P(o' \mid h_0, a)$.

Observation/Action	a_1	a_2	a_3	a_4
LEerror	0.0001	0.0001	0.0001	0.0001
GTerror	0.0001	0.0001	0.0001	0.0001
GRerror	0.0001	0.0001	0.0001	0.0001
AZerror	0.0001	0.0001	0.0001	0.0001
MSerror	0.0001	0.0001	0.0001	0.0001
SYSok	0.0001	0.0001	0.0001	0.0001

Table 16.15 $P(o' \mid h_1, a)$.

Observation/Action	a_1	a_2	a_3	a_4
LE	0.0001	0.0001	0.0001	0.0001
GT	0.0001	0.0001	0.0001	0.0001
GR	0.0001	0.0001	0.0001	0.0001
AZ	0.0001	0.0001	0.0001	0.0001
MS	0.0001	0.0001	0.0001	0.0001
SYSok	0.0001	0.0001	0.0001	0.0001

Table 16.16 Initial belief values and state cluster boundaries.

State	Belief	LE	GT	GR	AZ	MS	SYSok
Stable	0.9298	0≤LE≤LEmax	0≤GT≤GTmax	0≤GR≤GRmax	0≤AZ≤AZmax	0≤MS ≤MSmax	*true*
Lost	0.04	LE≥LEmax	GT≥GTmax	0≤GR≤GRmax	AZ≥AZmax	0≤MS ≤MSmax	*false*
Unstable	0.03	0≤LE≤LEmax	0≤LE≤LEmax	GR≥GRmax	AZ≥AZmax	MS≥ MSmax	*false*
h_0	0.0001	–	–	–	–	–	*false*
h_1	0.0001	–	–	–	–	–	*false*

Table 16.17 Action policies.

State/Action	a_1	a_2	a_3	a_4
Stable	1.0	0	0	0
Lost	0	0.05	0.15	0.8
Unstable	0	0.35	0.05	0.6
h_0	0	0	0	1.0
h_1	0	0	0	1.0

Mathematically the method is rather simple, depending on two numpy functions. It is possible to look at the code for those functions or write our covariance and inversion functions and then verify the mathematics. Alternatively, it can be assumed that the code is correct based on the very large user base. However, to be sure, a few sample tests of the functions can be run and the results compared to the results from another program, e.g., MATLAB or Excel.

The Mahalanobis method, a rather simple method, once again depends on two numpy functions:

```
obs_minus_mu = obs - numpy.mean(self.cluster, axis=0)
return(numpy.dot(obs_minus_mu, self.invCov))
```

The code is a direct implementation of Eq. (16.22), and as above, it can be accepted that the two functions are correct by virtue of the user base. Of course, it is possible to write code and/or compare with other programs.

Suppose, through simulation, test, or operation, an observed monitor value is found to be distant from clusters that define Stable, Lost, and Unstable. Two options are possible: either the observation is close enough to one of these states and therefore should be added to its cluster, or it is too far from any state and therefore needs to be in its own state. In the former case, the new observation is added to the states cluster matrix and the states inverse covariance matrix method is invoked to update the values for subsequent Mahalanobis distance evaluation. There are several ways to check the significance of distances outside a cluster. One option is to evaluate p-value. P-values below 0.05 are commonly used to reject the hypothesis that an observation belongs to a cluster. Simply stated, p-value is the area under a probability distribution that is at least as extreme as an observation (Dijkstra 1975; Ouimet 2007). Another option is to re-assess the clusters from scratch.

An observation determined to be within a cluster is assigned the error observation vector associated with that cluster. For example, if the monitor values are:

LE ≥ LEmax	GT ≥ GTmax	0 ≤ GR ≤ GRmax	AZ ≥ AZmax	0 ≤ MS ≤ MSmax	*false*

Then the observation vector is:

LEerror	GTerror	GRerror	AZerror	MSerror	SYSok
true	*true*	*false*	*true*	*false*	*false*

which is an observation consistent with the Lost state. If the action taken from the state prior to Lost was expected to transition to Stable, then the policy at that prior state is revisited to reduce the reward for that action. The belief state is updated using Eq. (1), and then another action is taken.

If an observation does not belong in any existing state, the next step is to assign it to a hidden state. Our example already includes two hidden states and either can be used by assigning the observation to one of those states and updating the POMDP model attributes. Of course, with only a single appearance of the observation, the POMDP model attributes will be "sketchy" at best. However, we now have a new observation and a change in the transition and emission probabilities from the last state and the action taken, which requires updating the POMDP model attributes. As more new data is collected, it may be necessary to add more hidden states. However, at some point, there may be a sufficient number of new observations that unsupervised learning can be employed to create better clusters. Unsupervised learning looks for patterns in unlabeled data and assigns the data to clusters (Kuo and Madni 2023; Hastie et al. 2009).

The goal of supervised learning is to predict an outcome of a data sample that was not used in the training set. There are two types of supervised learning: *classification* and *regression*. Classification is used when the training data have both input and output values. Regression is applied when the training set has continuous numerical values without target labels. Because there will be both data values and known cluster assignment for this quadcopter example, a classification method is the best option. As with the covariance and matrix inversions mentioned previously, there are Python functions that provide the tools for supervised learning training and evaluation (Di Pietro 2020). Given the complexity of these functions, verification can be by virtue of the large user base. However, formal verification of custom-coded functions is performed by comparing implementations to their mathematical formulations.

Unsupervised learning uncovers previously unknown patterns within a dataset and assigns elements of the dataset to clusters on the basis of those patterns. For example, Python has k-means clustering and hierarchical clustering functions available. While either can be used for identifying patterns, k-means works best when the training data are hyperspherical, i.e., an n-dimensional sphere. On the other hand, k-means can work with large training sets while hierarchical is more limited but is less sensitive to noisy data.

The hierarchical option appears to be the best fit given the infrequency of unusual observations and the likelihood that data will be noisy. The technique begins with all data in the same cluster. At the next level in the hierarchy, the single cluster is divided into two clusters. Those clusters are divided at the next level of the hierarchy and so on. Python's `scipy` library includes `cluster.hierarchy` function that performs the needed decompositions. Verification of the `scipy` function can again depend on its high usage. As before, a custom implementation can check that the implementation matches the mathematics (Kuo and Madni 2023).

Finally, the execution of the POMDP during operation needs to be addressed to assure that it accomplishes the mission as the first priority, with safety being the second priority. For the quadcopter, a useful system property is three-axis stability, $M \mid = \textit{stable flight}$, i.e., quadcopter model, M, satisfies the property, *stable flight*. As shown in Figure 16.9, the Flight Controller receives inputs from the IMU, Compass, GPSR, and Radio.

Using the approach described in Cimatti and Tonetta (2012), an example data flow for the quadcopter is shown in Figure 16.10. The Planning Function receives a Flight Plan uploaded through the radio. The Flight Plan is a turn-by-turn and altitude list as a function of time and location. The Planning Function evaluates the Flight Plan and determines whether the plan is feasible given the quadcopter location and time constraints. It also checks the plan and location for unexpected jumps that are incompatible with safe operation. If the Plan is valid, Planning Function sends Attitude Updates to the Attitude Control Function. Attitude Control ignores Attitude Updates

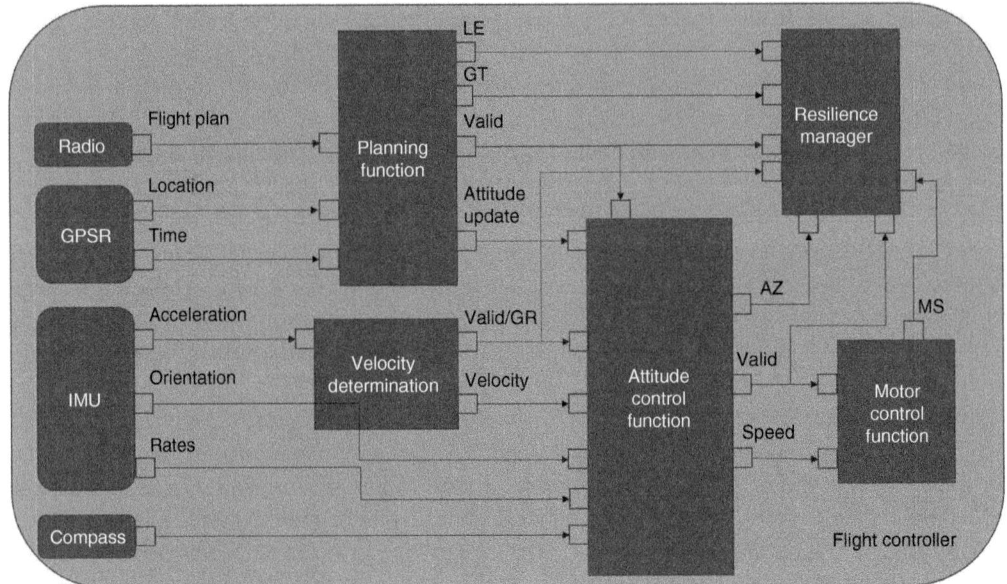

Figure 16.10 Quadcopter flight control system data flow.

when the Planning Function Valid flag is false and continues executing the previous attitude command. Attitude Control receives the Velocity vector, Orientation, and Angular Rate information and determines the needed motor Speed to adjust the quadcopter heading and altitude. Attitude Control also does a further check that determines whether the attitude updates are achievable and reasonable given the current attitude. A Valid indication from Attitude Control is an indication that the updates are possible. The Motor Control Function receives the Speed output from Attitude Control and determines the motor voltages needed for altitude and attitude adjustments. If the Valid input to the Motor Control is false, or if the required motor voltages are not feasible then the Motor Controller maintains the same voltage outputs. The Resilience Manager receives the five monitors LE, GT, GR, AZ, and MS, performs the POMDP belief state update, and commands one of the four actions (data flows not shown in figure).

As discussed in Cimatti and Tonetta (2012), contracts support compositional verification, i.e., a verification method is compositional if system properties can be deduced from component properties without using component internals. Consequently, proof of a system property is a deductive proof that use component properties by iteratively applying inference rules. Proof semantics show that a system that is built from components that satisfy component properties will satisfy the deduced property.

The system property $M \mid = \textit{stable flight}$ states that a quadcopter model must satisfy stable flight. A portion of a deduction tree for the quadcopter is shown in Figure 16.11. The figure indicates that for a model, *M*, to satisfy the property, *stable flight*, then the Planning Function must satisfy the property *valid attitude update*, Attitude Control must satisfy *valid speed*, Motor Control must satisfy *valid voltage*, and Resilience Manager must satisfy *disruption handling*. At the next level, satisfying *valid attitude update* requires that the Radio satisfy sending a *valid plan*, the GPSR satisfies sending a *valid location* and a *valid time*. The Attitude Control, Motor Control, and Resilience branches can be similarly decomposed.

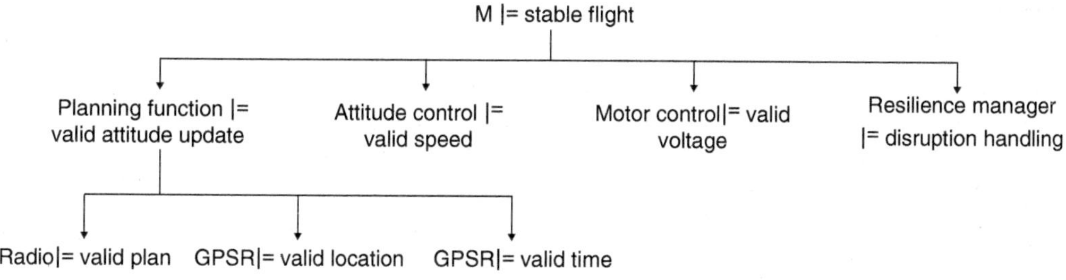

Figure 16.11 Portion of a quadcopter deduction tree.

The Planning, Attitude Control, and Motor Control functions can all be modeled by invariant contracts that lead to implementations of mathematical expressions. Proving correctness then is similar to the approach presented in the previous discussion in that a correctness proof needs to verify that the implementations match the mathematical expressions. The correctness of system components is proven by checking that every contract in the deduction tree defined contract decompositions, is correct.

The Resilience Manager implements the POMDP using the flexible contracts approach. For example, a contract might say: if $s = argmax\left(b'\right) then\ execute\, \pi_s^*$, i.e., find the maximum belief in the updated belief state *s*' and execute the action that produces the maximum reward. This contract can be decomposed into a contract that performs the belief state update, another that finds the belief state with the highest probability, then a third contract that finds and executes the action. Belief state update and finding the most probably state are performed using mathematical equations that are verified by correctness proofs. The contract that executes an action is a procedure that can be decomposed as a Büchi automaton and proven by model checking.

Chapter Summary

This chapter has addressed how formal methods can be adapted and exploited to address the challenges in engineering resilient systems. These methods are intended to replace the *ad hoc* methods being used today. After providing an in-depth review of the concept of resilience and current practice, it presents an innovative model-based approach for engineering resilient systems. A detailed illustrative example is presented to show the use of the approach. Looking to the future, resilience needs will continue to grow. For example, computer networks will continue to be threatened by internal and external cyberattacks. Protective cyber defenses may themselves fall victim to attack or become compromised by fault conditions that impact the platform and agents on which they depend. Moreover, coordinated attacks may create Byzantine situations that fool and confuse cyber defenses. The approach presented in this chapter provides a rigorous foundation to address these challenges.

Note

1 Memoization is a programming optimization method that stores the results of computationally expensive operations for later use rather than recomputing those results.

References

Abdelmalak, M. and *Benidris*, M. (2021). A Markov decision process to enhance power system operation resilience during hurricanes. *2021 IEEE Power & Energy Society General Meeting (PESGM)*. Washington, DC, USA, 2021, pp. 1–5. IEEE. https://doi.org/10.1109/PESGM46819.2021.9637871.

Acheson, A. and Dagli, C. (2016). Modeling resilience in system of systems architecture. *Procedia Computer Science* 95: 111–118.

Avizienis, A., Laprie, J.-C., Randell, B., and Landwehr, C. (2004). Basic concepts and taxonomy of dependable and secure computing. *IEEE Transactions on Dependable Computing* 1 (1): 11–33. https://doi.org/10.1109/tdsc.2004.2.

Bakirtzis, G., Sherburne, T., Adams, S. et al. (2022). An ontological metamodel for cyber-physical system safety, security, and resilience coengineering. *Software and Systems Modeling* 21: 113–117.

Balluchi, A., Casagrande, A., Collins, P., et al. (2006). Ariadne: a framework for reachability analysis of hybrid automata. *Symposium on Mathematical Theory of Networks and Systems (MTNS) (2006)* Kyoto, Japan (24–28 July 2006). Semantic Scholar. https://www.academia.edu/download/30711210/10.1.1.127.9268.pdf.

Bellini, E., Marrone, S., and Marulli, F. (2021). Cyber resilience metamodeling: the railway communication case study. *Electronics* 10 (583).

Bellman, R. (1952). On the theory of dynamic programming. *Proceedings of the National Academy of Sciences of the United States of America* 38:716–719.

Benvenuti, L., Ferrari, A., Mazzi, E., and Vincentelli, A. (1970). Contract-based design for computation and verification of a closed-loop hybrid system *International Workshop on Hybrid Systems: Computation and Control* (22 April 2008). pp. 58–71.

Biggar, O. and Zamani, M. (2020). A framework for formal verification of behavior trees with linear temporal logic. *IEEE Robotics and Automation Letters* 5 (2): 2341–2348.

Büchi, J.R. (1960). Weak second-order arithmetic and finite automata. *Mathematical Logic Quarterly* 6 (1–6): 66–92.

Büchi, J.R. (1962). On a decision method in restricted second-order arithmetic. In: *Logic Methodology and Philosophy of Science* (ed. E. Nagel, P. Suppes, and A. Tarski), 1–11. Stanford: Stanford University Press.

Catano, N. (2022). Program synthesis for cyber-resilience. *IEEE Transactions on Software Engineering*, 49 (3):962–972. https://doi.org/10.1109/TSE.2022.3168672.

Cichonski, P., Millar, T., Grance, T., and Scarfone, K. (2012). Computer Security Incident Handling Guide. NIST Computer Security Resource Center. SP 800-61 Rev. 2 https://csrc.nist.gov/publications/detail/sp/800-61/rev-2/final (accessed 2 August 2023).

Cimatti, A. and Tonetta, S. (2012). A property-based proof system for contract-based design. *38th Euromicro Conference on Software Engineering and Advanced Applications*. pp. 21–28. IEEE. https://doi.org/10.1109/SEAA.2012.68.

Clarke, E., Grumberg, O., and Peled, D. (2001). *Model Checking*. MIT Press. ISBN: 978-0-262-03270-4.

Clarke, E., Emerson, E.A., and Sifakis, J. (2009). Model checking: algorithmic verification and debugging. *Communications ACM* 22 (11): 74–84.

Delahaye, B., Caillaud, B., and Legay, A. (2010). Probabilistic contracts: a compositional reasoning methodology for the design of stochastic systems. In: *2010 10th International Conference on Application of Concurrency to System Design*, Braga, Portugal (21–25 June 2010). pp. 223–232. IEEE. https://doi.org/10.1109/ACSD.2010.13.

Di Pietro (2020). Machine Learning with Python: Classification (complete tutorial). Towards Data Science, 11 May 2020. https://towardsdatascience.com/machine-learning-with-python-classification-complete-tutorial-d2c99dc524ec (accessed January 2023).

Dijkstra, E.W. (1975). Guarded commands, nondeterminacy and formal derivation of programs. *Communications of the ACM* 18 (8): 453–457.

Eddy, S.R. (2004). What is dynamic programming? *Nature Biotechnology* 22 (7): 909–910. http://www.lmse.org/assets/learning/bioinformatics/Reading/Eddy2004NatureBiotech_D.pdf.

Enseling, O. (2001). iContract: Design by Contract in Java. InfoWorld. https://www.infoworld.com/article/2074956/icontract-design-by-contract-in-java.html#:~:text=iContract%20is%20a%20preprocessor%20for,comments%2C%20just%20like%20Javadoc%20directives (accessed 2 August 2023).

Gerth, R., Peled, D., Vardi, M.Y., and Wolper, P. (1995). Simple on-the-fly automatic verification of linear temporal logic. In: *Protocol Specification, Testing and Verification XV. PSTV 1995* (ed. P. Dembinski and M. Sredniawa). Boston, MA: IFIP Advances in Information and Communication Technology, Springer. https://doi.org/10.1007/978-0-387-34892-6_1.

Gholami, A., Shekari, T., Amirioun, M.H. et al. (2018). Toward a consensus on the definition and taxonomy of power system resilience. *IEEE Access* 6: 32035–32053. https://doi.org/10.1109/ACCESS.2018.2845378.

Goerger, S.R., Madni, A.M., and Eslinger, O.J. (2014). Engineered resilient systems: a DoD perspective. *Procedia Computer Science* 20 (28): 865–872.

Goldbeck, N., Angeloudis, P., and Ochieng, W. (2019). Resilience assessment for interdependent urban infrastructure systems using dynamic flow models. *Reliability Engineering and System Safety* 188: 62–79.

Hastie, T., Tibshirani, R., and Friedman, J. (2009). *Unsupervised Learning. The Elements of Statistical Learning*, 485–585. New York: Springer. https://link.springer.com/content/pdf/10.1007/978-0-387-84858-7_14.pdf.

Hoang, T.S. (2017). *Appendix A: An Introduction to the Event-B Modeling Method.* Berlin: Springer Verlag. https://link.springer.com/content/pdf/bbm:978-3-642-33170-1/1.pdf.

Hoare, C.A.R. (1969). An axiomatic basis for computer programming. *Communications of the ACM* 12 (10): 576–585.

Holzmann, G. (2003). Trends in software verification. *International Symposium on Formal Methods*, Pisa, Italy, Europe (8–14 September 2003). pp. 40–50. Springer.

Kaelbling, L.P., Littman, M., and Cassandra, A. (1998). Planning and acting in partially observable stochastic domains. *Artificial Intelligence* 101: 99–134.

Kesten, Y., Pnueli, A., and Raviv, L. (1998). Algorithmic Verification of Linear Temporal Logic Specifications. ICALP'98,LNCS 1443. pp. 1–16.

Kotsiantis, S.B. (2007). Supervised machine learning: a review of classification techniques. *Informatics* 31: 249–268.

Kuo, C.C.J. and Madni, A.M. (2023). Green learning: introduction, examples and outlook. *Journal of Visual Communication and Image Representation* 90.

Lai, C.C., Hsu, C.Y., Jen, H.H. et al. (2021). The Bayesian susceptible-exposed-infected-recovered model for the outbreak of COVID-19 on the diamond princess cruise ship. *Stoch Environ Res Risk Assess* 35: 1319–1333.

Lauri, M., Hsu, D., and Pajarinen, J. (2022). Partially observable Markov Decision processes in robotics: a survey. *IEEE Trans on Robotics* 1–20.

Li, J., Nuzzo, P., Sangiovanni-Vincentelli, A. et al. (2017). Stochastic contracts for cyber-physical systems under probabilistic requirements. In: *MEMOCODE '17: Proceedings on 15th ACM-IEEE*

International Conference of Formal Methods and Models for System Design, 5–14. Vienna, Austria. ACM. https://doi.org/10.1145/3127041.3127045.

Lin, Y. and Bie, Z. (2018). Tri-level optimal hardening plan for a resilient distribution system considering reconfiguration and DG islanding. *Applied Energy* 201: 1266–1279.

Madni, A.M. and Jackson, S. (2009). Towards a conceptual framework for resilience engineering. *IEEE Systems Journal* 3 (2): 181–191.

Madni, A.M., Sievers, M., Ordoukhanian, E., and Pouya, P. (2018a). Extending formal modeling for resilient systems. *2018 INCOSE International Symposium*.

Madni, A.M., Sievers, M., Madni, A. et al. (2018b). Extending formal modeling for resilient system design. *INCOSE INSIGHT* 21 (3): 34–41.

Madni, A.M., Sievers, M., and Erwin, D. (2019). *Formal and Probabilistic Modeling in the Design of Resilient Systems and System-of-Systems*. San Diego, California: AIAA Science and Technology Forum.

Madni, A.M., Erwin, D., and Sievers, M. (2020). Constructing models for system resilience: challenges, concepts, and formula models. *Systems* 8 (3).

Meyer, B. (1989). *Eiffel: An Introduction*. Interactive Software Engineering. https://se.inf.ethz.ch/~meyer/publications/eiffel/eiffel_intro.pdf

Meyer, B. (1992). Applying "design by contract". *Computer* 25 (10): 40–51.

Ouimet, M. (2007). Formal software verification: model checking and theorem proving. Embedded Systems Laboratory Technical Report ESL_TIK-00214, MIT, Cambridge, MA.

Pnueli, A. (1977). The temporal logic of programs. *Proceedings of the 18th Annual Symposium on Foundations of Computer Science*, Providence, RI (31 October–2 November 1977). pp. 46–57. IEEE.s

Puterman, M. (1990). Markov Decision processes, Chapter 8. In: *Handbooks in Operations Research and Management Science* vol. 2, 331–434.

Rabiner, L. (1989). A tutorial on hidden Markov models and selected applications in speech recognition. *Proceedings of the IEEE* 77 (2): 257–286.

Reason, J. (1990). The contribution of laten human failures to the breakdown of complex systems. *Philosophical Transactions of the Royal Society B: Biological Sciences* 327 (1241): 475–484. https://doi.org/10.1098/rstb.1990.0090.

Sarraute, C., Buffet, O., and Hoffmann, J. (2012). POMDPs make better hackers: accounting for uncertainty in penetration testing. *Proceedings of the 26 AAAI Conference on Artificial Intelligence*, Toronto, Ontario, Canada (22–26 July 2012). pp. 1816–1824.

Schimpf, A., Merz, S., and Smaus, J.G. (2009). Construction of Büchi automata for LTL model checking verified in Isabelle/HOL. *International Conference on Theorem Proving in Higher Order Logics*, Munich, Germany (17–20 August 2009). pp. 1816–1824. Springer.

Sievers, M. and Madni, A.M. (2016a). Agent-based flexible design contracts for resilient spacecraft swarms. In: *Proceedings of the AIAA Science and Technology 2016 Forum and Exposition*, SanDiego, California, USA.

Sievers, M. and Madni, A.M. (2016b). *Agent-Based Flexible Design Contracts for Resilient Space Systems*. AIAA Scitech.

Smallwood, R. and Sondik, E. (1973). The optimal control of partially observable Markov processes over a finite horizon. *Operations Research* 21 (5): 1071–1088.

Smith, T. and Simmons, R. (2004). Heuristic search value iteration for POMDPs. *Proceedings of the 20th Conference on Uncertainty in Artificial Intelligence*. pp. 520–527.

Taleb, N. (2010). *The Black Swan – The Impact of the Highly Improbably*. Random House.

Thomas, G. (2020). Markov decision processes. 6 April 2020. mdps.pdf (stanford.edu) (accessed January 2023).

Tipireddy, R., Chatterjee, S., Paulson, P., et al. (2017). Agent-centric approach for cybersecurity decision-support with partial observability. *2017 IEEE International Symposium on Technologies for Homeland Security (HST)*, Waltham, MA (25–26 April 2017). pp. 1–6. IEEE.

Tran, H., Campos-Nanez, E., Fomin, P., and Waek, J. (2016). Cyber resilience recovery model to combat zero-day malware attacks. *Computers and Security* 61: 19–31.

Vardi, M.Y. (2021). Program verification: a 70+ year history. *2021 International Symposium on Theoretical Aspects of Software Engineering (TASE)*, Shanghai, China (25–27 August 2021). pp. 1–2. IEEE. https://doi.org/10.1109/TASE52547.2021.00011.

Viguerie, A., Lorenzo, G., Auricchio, F. et al. (2021). Simulating the spread of COVID-19 via a spatially-resolved susceptible–exposed–infected–recovered–deceased (SEIRD) model with heterogeneous diffusion. *Applied Mathematics Letters* 111.

Wang, J., Zuo, W., Barbarigos-Rhode, L. et al. (2019). Literature review on modeling and simulation of energy infrastructure from a resilience perspective. *Reliability Engineering and System Safety* 183: 360–373.

Wang, C., Ju, P., Lei, S. et al. (2020). Markov decision process-based resilience enhancement for distribution systems: an approximate dynamic programming approach. *IEEE Transactions on Smart Grid* 11 (3): 2498–2510.

Welsh, T. and Benkhelifa, E. (2020). On resilience in cloud computing: a survey of techniques across the cloud domain. *ACM Computing Surveys* 53 (3).

Westrum, R. (2007). A Typology of Resilience Situations, Chapter 5. In: *Resilience Engineering Concepts and Precepts* (ed. E. Hollnagel, D. Woods, and N. Leveson). England: Ashgate Publishing Ltd.

Xu, D., Gossler, G., and Girautl, A. (2012). Probabilistic contracts for component-based design. *Formal Methods in System Design* 41: 211–231. https://doi.org/10.1007/s10703-012-0162-4.

Biographical Sketches

Azad M. Madni is a University Professor of Astronautical, Aerospace, and Mechanical Engineering and holder of the Northrop Grumman Fred O'Green Chair of Engineering. He is a member of the National Academy of Engineering, and recipient of the *2023 National Academy of Engineering's Gordon Prize* and the *2023 IEEE Simon Ramo Medal*. He is the executive director of USC's Systems Architecting and Engineering Program and founding director of USC's Distributed Autonomy and Intelligent Systems Laboratory. He is a life fellow/fellow of 10 professional science and engineering societies including IEEE, INCOSE, IISE, AAAS, AIAA, and AIMBE. His awards in systems engineering include INCOSE's Pioneer Award, Founders Award, Outstanding Service Award, and Benefactor Award. He is also the recipient of IEEE Aerospace and Electronic Systems Pioneer Award, Judith A. Resnick Excellence in Space Engineering Award, and Industrial Innovation Award, and IEEE Systems, Man, and Cybernetics Norbert Wiener Award for Outstanding Research. In addition, he is the recipient of ASME CIE Lifetime Achievement Award, and the NDIA's Ferguson Award for Excellence in Systems Engineering. He is the author of *Transdisciplinary Systems Engineering: Exploiting Convergence in a Hyperconnected World* (Springer, 2018) and co-author of *Tradeoff Decisions in System Design* (Springer, 2016). He has served as principal investigator on 97 R&D projects totaling well over $100M. His research has been sponsored by government agencies such as DARPA, NSF, NASA, DHS, DOE, NIST, AFOSR, AFRL, ARL, ONR, DOD-SERC, DOD-AIRC. His industry research sponsors include Boeing, General Motors, Northrop Grumman, and Raytheon He received his BS, MS, and PhD degrees in Engineering from UCLA with a major in engineering systems, and minors in computer methodology and AI, and engineering management.

Michael Sievers is senior systems engineer/systems at the California Institute of Technology, Jet Propulsion Laboratory (JPL), Pasadena, CA and an adjunct lecturer in Systems Architecting and Engineering Department, University of Southern California (USC), Los Angeles CA. He is responsible for the design and analysis of spacecraft avionics, software, end-to-end communication, and fault protection, as well as ground system modeling and mission concepts of operation. He has conducted research at JPL and USC in the areas of trust and resiliency and performed a DSN study that investigated dynamic trust assessments. At USC, he teaches classes in systems and systems of systems architecture, resilience, and model-based systems engineering. Dr. Sievers is an INCOSE fellow, AIAA associate fellow, and IEEE senior member.

Chapter 17

Augmented Intelligence: A Human Productivity and Performance Amplifier in Systems Engineering and Engineered Human–Machine Systems

Azad M. Madni

Astronautics, Aerospace, and Mechanical Engineering, University of Southern California, Los Angeles, CA, USA

Introduction

Systems engineering today is in the midst of a transformation inspired by the need to increase scientific rigor of systems engineering, enhance system verification and validation testing, increase MBSE coverage of the system life cycle, decrease systems engineering cycle time, and enhance performance of engineered human–machine systems (Madni and Sievers 2018). The latter, which is concerned with augmented intelligence (AugI), is the focus of this chapter.

Artificial intelligence (AI), thus far, has been successfully used in both defense and commercial sectors as a replacement for humans in performing routine tasks. However, more recently, AI is being viewed as a partner or an aid to the human to enhance joint human–machine system performance. This view is integral to the concept of AugI and adaptive human–machine teaming (HMT). The current definition of AugI is that of an AI-augmented human, where the augmentation consists of AI serving as a partner or assistant (Bird 2017). However, augmentation can and should be a two-way proposition: *AI-augmented human*, and *human-augmented AI*. In other words, while AI can augment the capabilities of the human, the human can also augment the capabilities of AI. These forms of augmentation apply both to the systems engineering process and engineered human–machine systems (Madni 2020). The fundamental change here is a shift in mindset from "humans or AI" to "humans and AI."

AugI is commonly defined as a design pattern for a human-centered partnership with AI to enhance human decision-making and learning in areas where humans have known shortcomings. AugI is concerned with keeping the human in the loop when exploiting AI technology to solve problems. In other words, AugI is a particular way of exploiting AI technology. AugI has several synonyms such as intelligence amplification, cognitive augmentation, and machine augmented intelligence, among others. Ross Ashby coined the term "amplifying intelligence" in *Introduction to Cybernetics* (Ashby 1956) to describe the means to enhance the power of selection during problem-solving, a cognitive process. The concept of AugI was revisited by J.C.R. Licklider, a psychologist and computer scientist, in his landmark book, *Man-Computer Symbiosis* (Licklider 1960). Licklider envisioned mutually interdependent, tightly coupled human brain and machine computation capabilities to capitalize on the strengths of both. Licklider viewed "man–computer symbiosis" as a special type of human–machine systems in which the partners are able to "think"

Systems Engineering for the Digital Age: Practitioner Perspectives, First Edition. Edited by Dinesh Verma.

Companion website: www.wiley.com/go/verma/systemsengineering

in ways no human has thought before and process information in ways no machine had done before. Licklider believed that greater advances can be realized through man–computer symbiosis than through AI alone.

Engelbart (1962) focused on developing computer-based means to directly manipulate information and thereby improve the performance of knowledge workers in both individual and group tasks. In particular, his work focused on increasing the scope and rate of human comprehension in specific contexts to assure effective, timely solutions. Engelbart was successful in realizing these concepts through a set of tools and natural language system (NLS).

Madni (2020) introduced AugI as a "productivity multiplier" in systems engineering and a "force multiplier" in operational military systems. Specifically, he defined a framework and methodology for identifying high payoff task regimes for AugI and provided concrete examples of AugI for both uses. The following paragraphs provide examples of the use of AugI in systems engineering and engineered human–machine systems.

AugI in Systems Engineering

Consider the process of systems architecting and design in systems engineering. The systems engineer, in collaboration with the customer, defines system objectives and constraints. Then AI refines (i.e., decomposes) the objectives, recalls known options associated with the defined objectives, assists the systems engineer in performing tradespace exploration and trade-off analysis, evaluates and ranks candidate options in terms of their degree of attainment of objectives, and presents them to the systems engineer. This illustrative allocation of functions and interaction between AI and systems engineer accounts for the strengths and limitations of both while capitalizing on the synergy between them. This is how AI can serve as a productivity multiplier for systems engineers. However, at the same time, AI has certain limitations. For example, it is difficult for AI to keep track of changing context (Madni 2020). Without full knowledge of context, it is difficult for AI to set or reset objectives. As important, it is difficult for AI to generate novel options in the face of disruptive events (e.g., budget cut) and contingency situations (e.g., loss of key person). In such circumstances, the ability to (re) set objectives, define parameter ranges, and generate novel alternatives to satisfy the new objectives relies heavily on the innate creativity and ingenuity of humans.

AugI in Engineered Systems

Consider the problem of Automatic Target Recognition (ATR) in military ground combat operations (Madni 2020). In such scenarios, it is often the case that potential threats are partially obscured from view because of natural obstructions such as intervening mountainous terrain. The ATR system needs to be able to locate and identify this vehicle. To this end, it needs to determine where to look, locate the vehicle in the visual scene, identify the type of vehicle (i.e., friend, foe, neutral), and report to the commander. In this problem, it is difficult for the AI to determine where to look (because that requires rapid contextualization, something AI is not good at). However, AI, equipped with a library of vehicles and object/pattern recognition techniques, can rapidly identify the vehicle even if partially hidden from view. This is where AugI comes in. The human, being fast in problem contextualization, can tell the AI where to look for potential threats. From that point on, the AI can locate the vehicle, identify it as a threat, neutral or friend, and notify the commander. In past work, I have described this joint capability as "shared perception." Today, it serves as an excellent example of AugI, where the AI system augments the capability of humans and the human augments the capability of the AI system. This example illustrates the concept of AugI: the human helps AI in rapid contextualization, while the AI helps the human in precise localization and threat identification. Together, their "shared perception" successfully addresses a problem that neither could successfully address without the other.

Based on the foregoing examples, AugI can be viewed as a two-way proposition, i.e., *AI-augmented human* and *human-augmented AI*. This recognition brings us to the next question: how AugI can be incorporated into systems engineering, and more specifically, into model-based systems engineering (MBSE). MBSE is the systems engineering community's response to managing complexity while supporting interdisciplinary groups in collaborating on system development using systems engineering principles and heuristics. Modeling has always been an integral part of systems engineering to support functional, performance, and other types of engineering analysis. Today, MBSE is becoming increasingly more mainstream as system modeling languages continue to mature and add new semantics needed for representing complex human–machine systems.

This chapter presents a fundamental shift in the way we think about humans and AI. Rather than view humans solely as suboptimal job performers to be replaced by AI, humans are viewed as assets to exploit when problem situations call for ingenuity and creativity (Madni 2010, 2011). This change in mindset has implications for how we architect adaptive or intelligent human–machine systems and how we think about exploiting AI in systems engineering. I describe this new mindset as "exploiting the human's capability to adapt without exceeding the human's capacity to adapt." This insight is at the heart of the AugI and HMT (Madni 2020; Bansal et al. 2019; DeChurch and Mesmer-Magnus 2010).

This chapter is organized as follows. Section 2 discusses the misperceptions of AI along with the pros and cons of AI technology today. Section 3 makes the case for AugI in both systems engineering and engineered systems. Section 4 presents a conceptual framework for exploiting AugI in complex systems. Specifically, it presents the regime of applicability of AugI and the type of activities that stand to benefit from AugI. Section 6 summarizes the key concepts in this chapter and discusses broader implications of AugI.

Separating Misperceptions of AI from Actual Pros and Cons

People have been both fascinated and uneasy about AI since its inception. This is in part due to fear of the unknown and in part due to the distorted view of AI presented by Hollywood in movies such as "2001: A Space Odyssey," "Terminator," and "The Matrix." These movies make AI systems appear sinister adding to the trepidation that some have about AI. Nevertheless, there are serious concerns about AI that cannot be ignored. The late great physicist, Stephen Hawking, cautioned that while humans and AI could co-exist, "a rogue AI" could be difficult to stop without appropriate safeguards in place. There have been other more dire pronouncements from technology leaders such as Elon Musk of SpaceX. These concerns have drawn attention to how AI is used and how to minimize the occurrence of unintended undesirable consequences. To effectively address these concerns requires a sound understanding of the limitations of AI technology.

As with any technology, there is nothing inherently good or evil about AI. It all depends on what AI is used for and the care exercised in its implementations. Like any other technology, AI can be hijacked and put to nefarious use. Such concerns, however, should not deter the pursuit of AI for societal good, and the well-being and safety of the planet. Of course, there is no denying that AI technology will profoundly impact the job market, creating several new jobs for knowledge workers, while decimating several routine or mundane jobs.

With the foregoing understanding, the pros and cons of AI as it stands today can be addressed. The pros are well-known and understood. They include the ability to offload humans in routine, structured decision-making tasks; facilitate faster decision-making; derive insights from machine learning; perform error-free processing, direct robots in performing hazardous tasks, and predict decision outcomes. These advantages of AI are discussed next.

Offload Humans in Mundane Tasks

Humans tend to get bored and suffer from lack of motivation when performing routines or mundane tasks. Also, unlike machines, they get fatigued when performing repetitive tasks over extended periods. Therefore, it only makes sense to assign such tasks to machines, and more specifically to AI if they involve decision-making in the face of disruptions and contingencies. In fact, AI has been successfully used in adaptive process automation, to exploit process flexibility and resource utilization efficiency, while allowing humans to perform cognitively demanding and creative tasks.

Faster Decisions and Actions

AI in the form of cognitive agents facilitate rapid decision-making and action execution. For example, dynamic planning and scheduling, and target detection and identification are two areas in which AI excels.

Derive Insights from Machine Learning (ML)

In today's hyperconnected world, data continues to grow exponentially. Big data, for example, has come to mean datasets in the petabytes. Human cognitive limitations preclude the ability to assimilate voluminous data. This is where AI in the form of ML, and more specifically, deep learning and green learning (Kuo and Madni 2023) can process voluminous data at blinding speed to provide timely insights for informed human decision-making.

Error-Free Processing

AI algorithms are almost infallible for performing routine tasks. AI processing can ensure near-error-free processing of data, regardless of dataset size. However, problems that call for judgment and decision-making in the face of uncertainty are best left to humans at the present time.

Hazardous Task Performance

AI-enabled robots can perform tasks that humans either cannot perform, or that pose grave risks to humans. Space exploration and explosive ordnance disposal and are two such tasks. For example, the Curiosity Mars rover, can autonomously examine Mars surface and dynamically determine the best path while continuously "learning" in its environment. An autonomous or semi-autonomous robot is ideal for disposing explosive ordnance.

Outcome Prediction

AI technologies such as computer vision can help achieve better outcomes through improved prediction. This capability is crucial for demand forecasting, oil exploration, and medical diagnosis.

AI also has its fair share of cons. To begin with, it engenders distrust as a gut response that has to be mitigated through consistent, repeatable performance. Viewed solely as a replacement for humans, AI poses economic challenges in the form of loss of low-skilled jobs, inequitable distribution of wealth that can be partly attributed to AI, and disproportionate distribution of power. Beyond the economic challenge, AI has other drawbacks such as poor judgment; lack of ethics-awareness; lack of creativity to deal with "broken plays"; inability to keep up with changing contexts, goals, and plans.

Poor "judgment" was evident in the aftermath of the shooting and hostage incident that ensued in downtown Sydney in 2014. Panicked pedestrians began calling Uber to vacate the area. Because the surge in demand was concentrated in a limited area, Uber's algorithms, which employed supply-and-demand heuristics, produced a dramatic spike in fares. It turned out that Uber's algorithms had not accounted for surge in demand resulting in such crises. While the impatient pedestrians did not care at the time, they were livid when they discovered the exorbitant ride fare that Uber charged them in that moment of crisis! This incident forced Uber to reevaluate its ride fare calculation algorithms for such emergency situations (Madni 2020).

Human intelligence, a subtle combination of knowledge, emotions, and skills, tends to be in a state of perpetual change. This flexibility enables human judgment to reflect shades of gray, even as human behavior continues to be shaped in part by real world events. Replacing adaptive and nuanced human behavior with rigid AI can potentially lead to irrational behavior within socio-technical ecosystems.

The Uber scenario described above clearly suggests that unless all contingencies are anticipated and accounted for in advance by AI algorithms, AI can reach surprisingly simplistic conclusions, and act on them to the detriment of humans and the environment. In other words, AI programs designed to pursue benign goals, may in the interest of expediency and using simplistic logic implement a highly unrealistic and unacceptable solution. One example of such a solution is "*if there is a problem with over-population in cities, an AI solution may call for reducing the population by any means available rather than develop adjacent areas and villages for some of the population to move to.*"

Today, even AI savants agree that despite dramatic advances in deep learning, we are not close to artificial general intelligence (or common-sense reasoning). In other words, we cannot build systems that can autonomously process, reason, and create solutions comparable to what the human brain can. But is that really important right now? In fact, is that the question we should be asking ourselves? Is replicating human intelligence our goal for real-world problems even though it may be a worthwhile goal for AI researchers to pursue to advance the frontiers of AI? The straightforward and simple answer to these several questions is "no." The goal today should be to amplify human intelligence and enhance joint human–AI performance through the use of AugI. AugI is focused on exploiting the synergy between human intelligence and AI to enable shared task performance when the task cannot be performed by either the human or AI alone (Madni 2020; Kamar 2016). Importantly, with AugI, the human remains the central figure in decision-making diffusing concerns of "AI taking over the world." It is important to realize that even though the underlying technologies powering AI and AugI are the same, the goals, usage, and applications are fundamentally different. AI aims to create systems that can operate autonomously, whereas AugI aims to capitalize on the synergy between human intelligence and AI to maximize human–AI performance as well as perform tasks that neither can alone. It is also important to realize that AugI is not a new technology; rather, it is a different way to think about the use of AI technology (Araya 2019; Chakraborti and Kambhampati 2018; Davis and Lessard 2018). While the pursuit of extending the frontiers of AI may continue to be an admirable goal to realize technologies that can prove invaluable in the future, society, and the planet today are better served by AugI.

Table 17.1 Exemplar uses of AI.

Software-defined vehicles	While these vehicles navigate autonomously based on autopilot, they require human intervention in certain contingency situation, e.g., car suddenly pulling into driver's lane and slamming on the brakes
Retail	AugI can suggest optimal store layout and produce placement to merchandisers based on consumer shopping patterns including foot traffic patterns; AugI can also identify where bottlenecks are occurring in retail store, and how to alleviate the problem to the human
Healthcare and medicine	AugI can provide recommendations to doctors and healthcare professionals on appropriate patient treatments based on medical data, research, and patient's medical history. For example, radiologists who used AI to assist in breast cancer detection performed better than radiologists or AI alone (Crigger et al. 2022)
Manufacturing	AugI can recommend changes to plant capacity and manufacturing lines based on projected increase/decrease in demand
Military operations	AugI can perform partially obscured threat identification for commander conducting ground-based military operations (Madni 2020)

Exemplar Uses of AugI

AugI has been used with remarkable success in several different real-world applications, ranging from autonomous vehicles and manufacturing to military operations, retail, healthcare, and medicine. Table 17.1 presents exemplar uses of AugI.

From the foregoing examples, it becomes apparent that AugI exploits social intelligence and common-sense reasoning of humans that AI does not possess at the present time. AugI is a testament to the position articulated in my book on Transdisciplinary Systems Engineering that innovation emerges at the intersection of different disciplines, human ingenuity, and technology and not from individual disciplines or technology alone. AugI is an alternative conceptualization of AI that emphasizes AI's assistive role in enhancing human performance and amplifying human capabilities. AugI is also a key enabler of human creativity and innovation, the defining characteristics of this era. Educators need to understand the impact of exploiting the synergy between human creativity and technology within a transdisciplinary engineering rubric and mindset and reflect this awareness in remaking education for the era of AugI. By reflecting this change in thinking in both curricula and courses will enable cultivation of students who value technology such as AugI as amplifiers of their capabilities and talent to undertake the challenges of the twenty-first century.

Augmented Intelligence in Systems Engineering

AI is already beginning to impact various aspects of systems engineering. These include machine learning-enabled dynamic systems modeling, adaptive process management, intelligent search and information retrieval, dynamic context management, and adaptive human-systems integration. In each case, the role of AI is in the form of AugI. This is because AugI can provide the best of AI and human intelligence while circumventing and/or compensating for their respective limitations. Figure 17.1 presents the various ways AugI can contribute to systems engineering activities using different AI technologies. As shown in Figure 17.1, AI-augmented systems engineering encompasses the system, the infrastructure support facilities, and the processes.

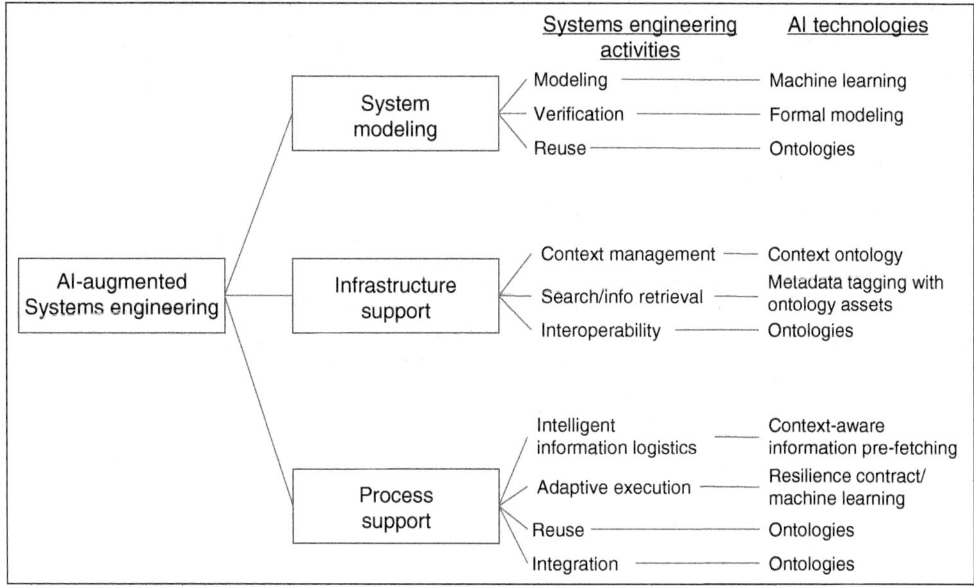

Figure 17.1 AugI can contribute to systems engineering activities using AI technologies.

At the system level, AI-augmented systems engineering can enhance system modeling, verification, and reuse. In system modeling, AugI can exploit ML in the form of reinforcement learning to collect evidence in real-time and use that to update an incomplete initial system model. I call that "closed-loop system modeling." AugI can employ formal logic in the form of contracts to ensure static model correctness. And the use of system and domain ontologies can be leveraged to identify opportunities for reuse of model fragments.

At the infrastructure support level, AugI can facilitate dynamic context management; information search and retrieval; and interoperability among subsystems and tools.

At the process level, AugI can enable adaptive interactive process management and intelligent information logistics (e.g., information pre-fetching), adaptive plan execution by exploiting flexibility afforded by AI and machine learning.

In relation to humans, AugI can take a variety of forms including memory joggers, active prompters and prodders, assistants, associates, and team orchestrators (Madni 1988). It is important to note that capabilities contribute to human-AI collaboration. Kasparov (2017) in *Deep Thinking* emphasizes the importance of human–AI collaboration rather than human replacement by AI with a cogent refrain: *"Many jobs will continue to be lost to intelligent automation. But if you're looking for a field that will be booming for many years, get into human–machine collaboration."*

Licklider (1960) was prescient in presenting his vision for cognitive assistance. He predicted that *"in not too many years, human brains and computing machines will be coupled together very tightly, and the resulting partnership will think as no human brain has thought."* To this prescient pronouncement, I offer a slight modification: *"the resulting partnership will think as no human brain or AI system in isolation could ever do."*

Thus, AugI is more than human–AI collaboration. It is exploiting the synergy between human and machine intelligence. I define AugI as the teaming of human with AI in a way that allows the team to capitalize on the strengths of each while circumventing their respective limitations. Learning to perform as a team means understanding the roles of the teammates, developing a shared awareness of mutual expectations, and being cognizant of the interdependencies with teammates. The true competitive advantage of a team resides in this knowledge of mutual interdependencies and the commitments implied by them.

However, the problem of AI–human teaming is more complicated than human–human teaming. While we know quite a bit about human teams, we know much less about AI–human teams. AI and humans have mostly different strengths and limitations. Humans are superior at rapid contextualization, considered and snap judgments, creative problem-solving, novel options generation, and responding to disruptions (known and unknown). However, humans are poor at monitoring infrequent events because they tend to experience a drop in vigilance. At the same time, they cannot keep up with rapidly occurring events, because they suffer from cognitive overload, then fatigue, and eventual drop in motivation. AI is far superior in recall of past events and cases, reaction times, rare event monitoring, and computation.

It is important to note that there are tasks that neither the human nor AI can perform by themselves. These tasks require the human and AI to work together as a team to get the job done. As noted earlier, this is the fertile regime of AugI.

Architectural Implications of AugI

AugI has several architectural implications. The most obvious implication is that since the human will always be part of the system, the system architecture needs to follow human-centered architecting principles. These principles, as defined by Madni and Jackson (2009), are the following:

- Inspectable, explainable, potentially suboptimal algorithms are preferred to opaque/black-box optimization techniques
- Flexible human-AI collaboration based on problem context and tasks
- Shared contextual awareness especially during collaborative, adaptive response
- Selective information aggregation to maximize situation and contextual awareness (context-sensitive declutter) without cognitive overload
- Mutual augmentation to maximize joint performance
- Mutual learning of priorities/preferences in both engineering and operational contexts
- Shift from human–AI function allocation to human–AI teaming in a variety of contexts

Inspectable, explainable algorithms: allow human to intervene in AI operation if needed without having to make needless assumptions. Consider route planning to a destination with specific waypoints that need to be traversed. AugI would offload the human by taking control of low-level navigation and obstacle avoidance tasks, thereby offloading the commander to concentrate on tactics. Unlike an optimization algorithm, AugI would be transparent to the commander allowing the commander to insert additional waypoints, override AugI recommendation, or take over manual control. As importantly, AI has to articulate, i.e., be able to explain its decisions and recommendations when called upon. This is an area of heated debate. What constitutes an explanation? For our purposes, we limit explanation to presenting the rules that were followed in response to triggering events.

Flexibility in human–AI collaboration interface: for AugI to aid the human at different levels in the task hierarchy, the architecture needs to make provision for human–AI interactions at multiple levels. In this regard, a context-sensitive, multi-level dashboard could ensure sustained situation awareness in the AI–human team.

Context-sensitive search and query facilitates recall compressing development cycle time. To realize this capability requires a context ontology (Madni and Madni 2004). Multi-perspective, multi-level visualization requires a domain ontology, class hierarchies for various concepts, and necessary taxonomies.

Selective information aggregation is essential for maximizing situation awareness without cognitive overload. It is achieved by aggregating those entities that the human does not interact with in specific contexts. This capability is dynamic in the sense that if context changes, certain aggregated elements may need to be disaggregated while other elements may need to be aggregated.

Shift from human–AI function allocation to exploiting synergy in human-AI teams recognizes that even in those tasks where AI is superior, there are areas that human intervention can help to maximize performance. The same is true for tasks where the human is superior, but AI can provide opportunistic useful intervention.

Conceptual Framework for Developing AugI Systems

Figure 17.2 presents a conceptual framework and guidelines for developing AugI system.

As shown in this figure, there are six partitions associated with human and AI capabilities. The square at the bottom left corner are tasks that both AI and humans don't do well. Examples of tasks that fall into this region are time-stressed decision-making with partial information, and rapid exhaustive parallel searches. This region is avoided through sound design practices. The vertical rectangle to the left is a region that humans perform exceedingly well and therefore tasks in this region are best left to humans. The horizontal rectangle at the base represents a region that is best performed by AI without human intervention. The square at the top right represents tasks that both AI and human do well. Assignment to one or the other is more a matter of their respective availability. The two trapezoids in the middle represent task regimes where both AI and human can perform, except that in the top trapezoid the human is better than AI, and in the bottom one AI is better than the human. These two regimes are a fertile ground for AugI.

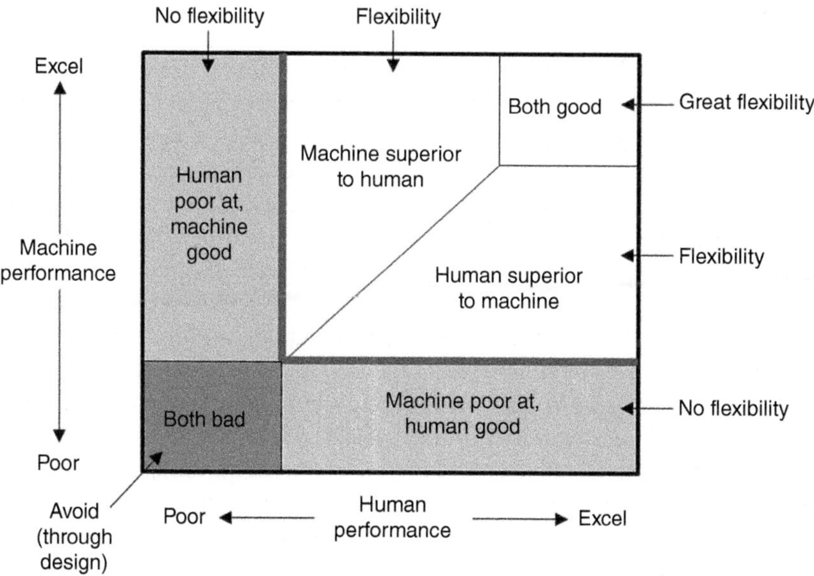

Figure 17.2 Defining the task regime that can be exploited by augmented intelligence. *Source:* Adapted from Madni and Madni (2018).

The top trapezoid represents the tasks regime where humans are generally superior to AI. However, even in this case, AI can augment the human. Examples of such augmentation are recalling past successful and unsuccessful options (e.g., designs), visually presenting multi-dimensional trade-offs, rapidly evaluating options, managing and sharing dynamic changes in context, flagging anomalous situations, and learning from both the human and data. The bottom trapezoid represents the task regime where AI is superior to humans. However, even in this case, the human can augment AI. Examples of such augmentation are goal and parameters setting for a particular mission or problem, rapidly contextualizing a previously unseen or unknown situation, generating novel options to augment the options recalled by AI using metadata tagging, making judgments in the face of uncertainty or ambiguity, and intervening at the request of AI, or when AI fails to respond in allotted time, or when the situation has not been previously experienced by AI.

Augmented Intelligence in Operational Decision-Making

With recent advances in data analytics and machine learning, some believe that machines will take over the roles of humans, offering the ability to make rapid, informed decisions based on voluminous data that no human could possibly analyze in timely fashion. They argue that with such advances, AI/ML can replace humans. This is where it makes sense to pause and review the decision-making process.

Human decision-making process begins with objective setting and structuring. This step is followed by retrieval of known options from memory, a fallible process that relies on human recall which is imperfect. What if the objective is unprecedented? This implies retrieving options from memory that might apply and more importantly, generate novel options on the fly. Humans tend to be especially good at the latter. The next step is option evaluation and prioritization. Options are evaluated, prioritized, and the best option is selected for implementation in the real-world decision-making context. Thereafter, the decision-maker monitors option execution and intervenes, if necessary.

Now let's see how augmented intelligence can improve this process, illustrated in Figure 17.3. To begin with, it is worth noting that the human decision-maker is limited because of fallible memory recall, limited computation bandwidth, and susceptibility to drop in vigilance when monitoring infrequent events. These shortcomings surface when the human decision-maker has to retrieve known options from memory; evaluate and prioritize them and monitor the execution of the selected option. All these steps can be assigned to the machine and automated. Specifically, the prioritization can be based on the aggregate weighted score of the different attributes that characterize the options.

However, the machine can do much more. The machine can employ fast-time simulation to execute each option and use the degree of attainment of the key objectives in question to prioritize the options for the human decision-maker. The machine can employ supervised learning with a training set of decision-making situations off-line to learn the preference structure of the human decision-maker in terms of the relative weights for the different attributes of options as a function of different contexts. The machine can employ AI for context monitoring/management and prompting the human when human intervention becomes necessary. The machine can also use AI to alert the human when a contingency situation or disruption is encountered. As can be seen from this example, the machine is used to automate routine tasks (i.e., recall past options), employ intelligent fast-time simulation to execute the options, and then evaluate and

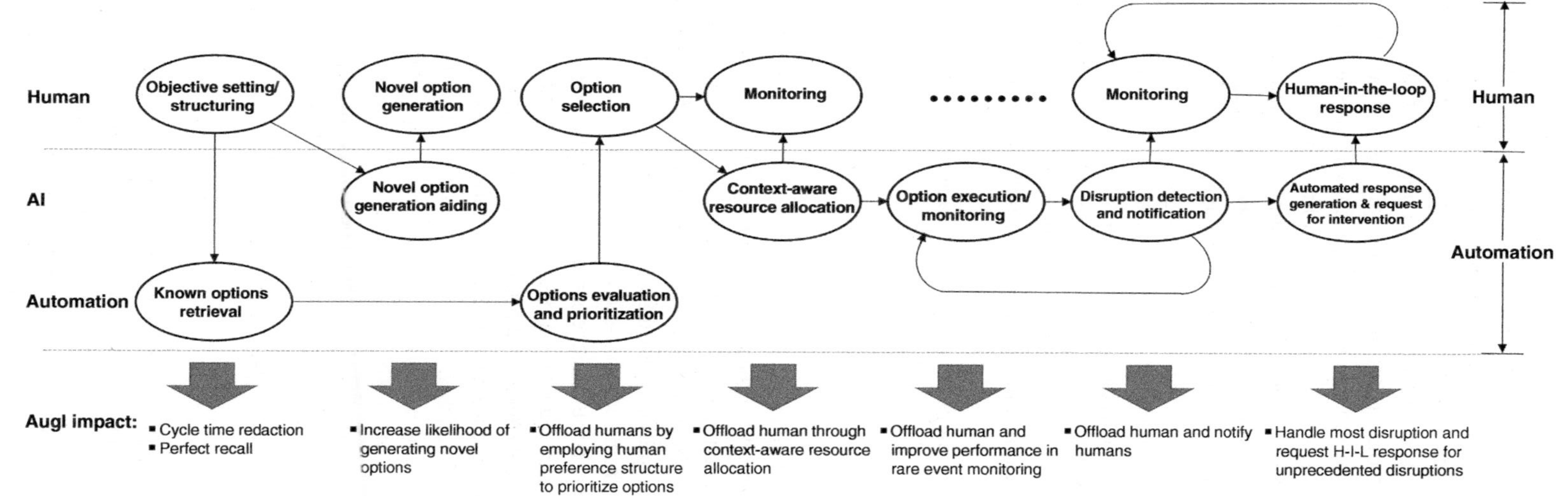

Figure 17.3 Exemplar decision-making and disruption handling process with AugI.

prioritize them using AI. Offline, supervised learning can be employed to learn the human's decision-making structure and preferences as a function of context. The result is an augmented intelligence systems that enhances joint human–machine performance in complex decision-making tasks. Importantly, this system capitalizes on the strengths of humans and machines while circumventing their respective limitations. As important, if this entire decision-making task were assigned to AI, it would result in a brittle solution. AI systems are usually tested on a specific problem or application and may exhibit superior performance on certain tasks. However, their performance can be expected to degrade significantly if the task is slightly modified to accommodate changes in the operational environment. Augmented intelligence does not have the same weaknesses of the AI system here.

The foregoing discussion highlights the fact that AI technology has quite some ways to go to claim artificial general intelligence. In particular, AI does not yet understand the physics of everyday life while machine learning does not yet support causal reasoning. These two shortcomings of AI and machine learning/deep learning provide the basis for pursuing augmented intelligence for today's complex real-world problems. By ensuring that the human is the final decision-maker who is not only context-aware but also ethics- and regulations-aware, the burden on AI is significantly alleviated. As important, the joint human–AI team is able to deliver results at speeds that only a machine can.

Augmented intelligence has significant payoff in data analysis applications as well. Three popular uses are: Data Fusion; Data Enrichment; and Search and Analysis.

Data Fusion

A key requirement in a distributed enterprise (military or civilian) is the ability to access information from an organization's relevant databases that may be co-located or distributed, internal or external, structured or unstructured. Decision-makers in these organizations need appropriately fused/aggregated information in actionable form for timely decision-making. Today accessing and searching such databases requires horizontal interoperability across heterogeneous databases. Such systems, if being built from scratch, need horizontal integration that scales with the increasing breadth of the enterprise. For existing systems, there needs to be syntactic interoperability if the databases share a common data format (e.g., XML, SQL) and communication protocols. In other cases, databases might also need semantic interoperability, i.e., ability to automatically interpret the information exchanged meaningfully and accurately to produce useful results as defined by end users of distributed databases. To achieve semantic interoperability, the content of the information exchange needs to be unambiguously defined using a common information exchange model. Very simply, this means that what is sent is the same as what is understood. There is also the concept of cross-domain interoperability, which pertains to multiple entities (e.g., social, political, legal, operational) working together toward a common goal through information exchange among the entities.

Data Enrichment

An effective augmented intelligence system exploits open source or commercial AI components and libraries to sift through data and metadata tag it to make it more easily searchable. For example, natural language processing, a central AI technology, can be used to organize and assign meaning to unstructured data such as service calls, social media posts, tweets, and product and artifact information. Sentiment analysis, which adds another AI layer, is particularly useful to service providers in that it allows them to distinguish the emotional weighting of certain data items.

Search and Analysis

An augmented intelligence system stands to benefit from a search and analysis front-end. This component is the interface between an organization's enriched data and the end user interested in gleaning insights from the data. What distinguishes an augmented intelligence solution form an AI solution is the breadth of its target users. The latter, for example, can be trained by analysts with foundational knowledge of analytics and search. End users can also be executives and decision-makers without such knowledge. For the latter, a high-level, condensed overview of the data is essential.

Augmented Intelligence in System Development and System Operation

AugI has a significant role to play in both system development and system operation. These roles are discussed in the following paragraphs.

During system development, AI can be exploited in a variety of ways to aid/facilitate the work of the systems engineer. These capabilities include human preference structure capture through supervised learning (Madni et al. 1982), context representation (Madni and Madni 2004), intelligent scenario authoring, and graphical modeling of digital twins (Madni et al. 2019), and intelligent process management. AI leverage in this mode consists of information prefetching (Madni et al. 2002), virtual prototype/digital twin state and status information during development, supervised learning algorithms, discrete event and agent-based simulation, context-aware dashboard for progress monitoring, and metrics collection. AI use in the intelligent middleware consists of ontology-enabled semantic integration, and context-based metadata tagging.

The middleware serves as brokers and mediators between the SE process layer and data sources (e.g., libraries of scenarios, digital twins, algorithms, sensors, connectors), analysis outputs, and virtual world/real-world entities. The library of modeling methods includes deterministic, stochastic, probabilistic, and parametric system modeling techniques. These include probabilistic models and reinforcement learning techniques for partially observable systems. Figure 17.4 presents the use of AugI in the systems engineering of HMS.

During system operation, AugI can take the form of context-specific search and queries, high-level commands, creative option generation during decision-making, and context-aware visualization and progress monitoring (dashboard capability). AI leverage in this mode consists of information fusion/aggregation, dynamic context management, dynamic resource allocation and redirection, reinforcement learning, fast-time simulation, question-answering, adaptive planning and decision-making, and audit trail maintenance. AI use in the intelligent middleware consists of ontology-enabled interoperability, context-based metadata tagging, integration and aggregation mechanisms, contextual filtering and context-aware information retrieval from distributed data bases, and adaptive planning and decision support. The middleware, which provides brokers and mediators between the HMS execution layer and the data sources, houses the digital thread that connects virtual world entities to real-world sensors, and real-world entities. Figure 17.5 presents the use of AugI during system operation.

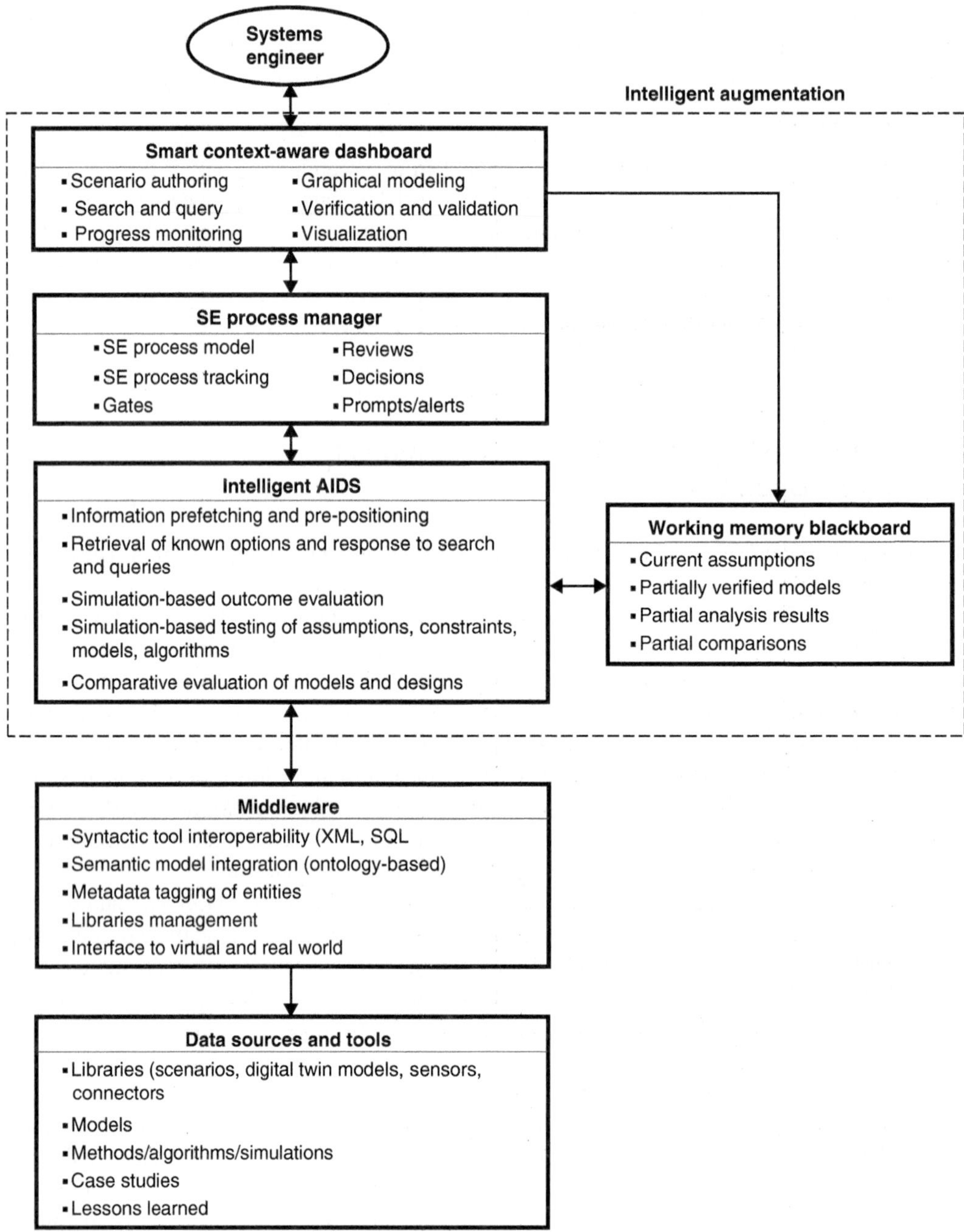

Figure 17.4 AugI in SE process.

Concluding Remarks

This chapter has presented the use of AugI in system development and system operation. The term, augmented intelligence (AugI), connotes exploitation of the symbiosis between humans and AI. AugI engenders a feeling of greater acceptance, than AI does. It does not conjure the image of intelligent robots running amuck while humans are reduced to helpless bystanders. More

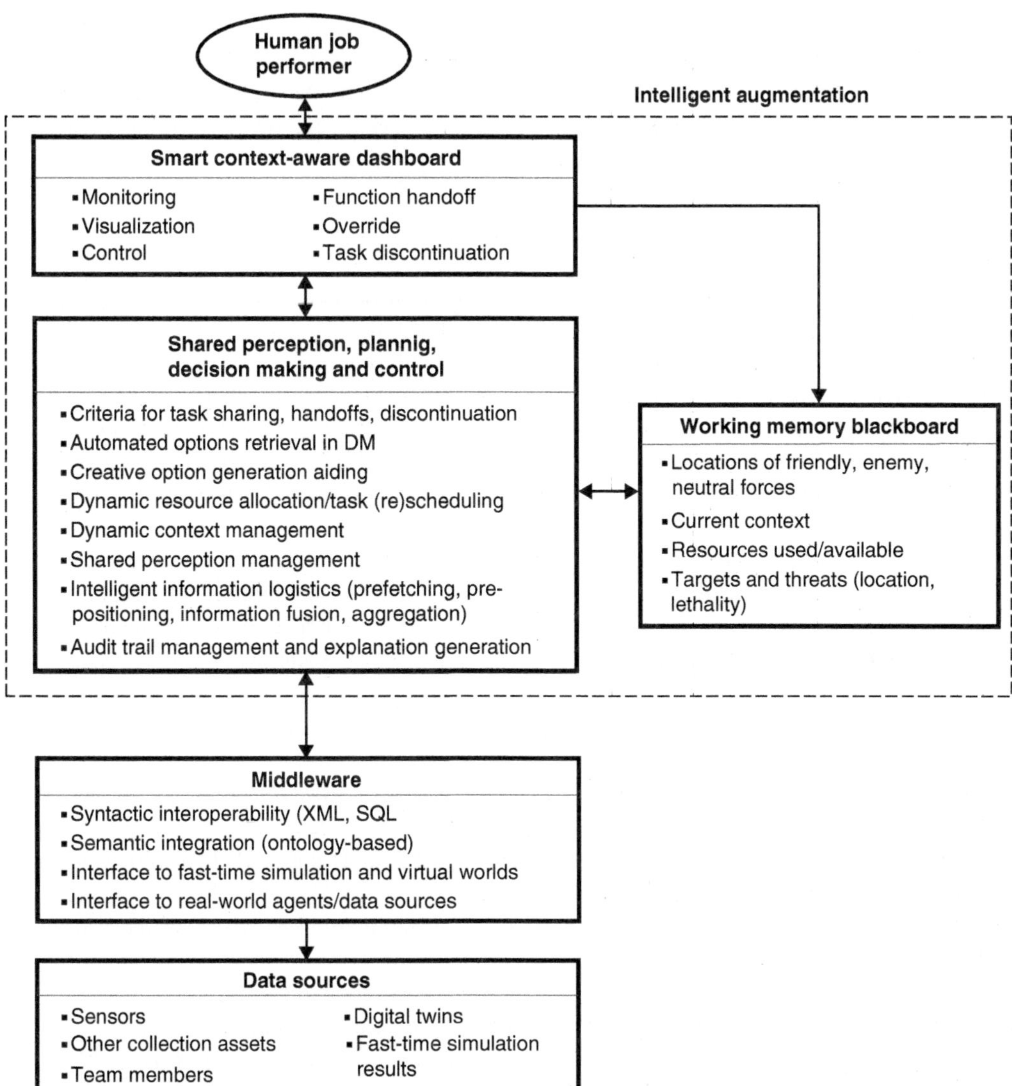

Figure 17.5 AugI in HMS operation.

importantly, AugI has crucial implications for the resulting system architecture, testing approach, and metrics. One might say that AugI is a construct that responds to the Nobel Laureate Christian Lou Lange's famous refrain: "Technology is a useful servant but a dangerous master." AugI also responds to Galbraith's (2015) caution: "We are becoming the servants in thought as in action, of the machines we have created to serve us." AugI is a means to address these legitimate concerns. It holds out hope for the future of our planet without the fear of AI taking over. As important, it dispels the misconceptions and hyperbole that surrounds AI. Ultimately, AugI is crucial to the future understanding and broader acceptance of AI (Bird 2017). Social intelligence, as it pertains to technology is a feature of AugI. AugI allows for the understanding of both the task and actions that the human is seeking to accomplish before the human does it. In general, AI systems lack social intelligence because their historical use has been autonomous operations. AugI, on the other hand, requires social intelligence because it need to explain its behavior and recommendations while allowing humans to intervene and advise or redirect its tasking.

The key concepts and framework presented in this chapter are intended to help systems engineers in incorporating AI augmentation in both systems engineering and engineered systems. A final note – in general, a human-AI team will be smarter than an AI team, because the former is capable of understanding the big picture and has common-sense reasoning which the latter lacks because of the narrow (albeit deep) view of data and information.

References

Araya, D. (2019). 3 things you need to know about augmented intelligence. *Forbes*.

Ashby, R. (1956). *An Introduction to Cybernetics*. Chapman and Hall.

Bansal, G. Nushi, B. Kamar, E., et al. (2019). Updates in human-AI teams: understanding and addressing the performance/compatibility tradeoff.*Paper Presented at the 3rd Association for the Advancement of Artificial Intelligence (AAAI) Conference on Artificial Intelligence, Proceedings of the AAAI Conference on Artificial Intelligence,* Honolulu, US-HI (27 January–1 February). Vol. 33(01), pp. 2429–2437

Bird, S. (2017). Why AI must be redefined as 'Augmented Intelligence'. https://venturebeat.com/2017/01/09/why-ai-must-be-redefined-as-augmented-intelligence (accessed 9 January 2017).

Chakraborti, T. & Kambhampati, S. (2018). "Algorithms for the greater good! On mental modeling and acceptable Symbiosis in human-agent collaboration." arXiv.1801.09854[cs.AI]. https://arxiv.org/abs/1801.09854v1.

Crigger, E., Reinbold, K., Hanson, C. et al. (2022). Trustworthy augmented intelligence in healthcare. *Journal of Medical Systems* 46: 12.

Davis, S. and Lessard, A. (2018). *The AI Revolution Begins with Augmented Intelligence*. Signafire White Paper.

DeChurch, L.A. and Mesmer-Magnus, J.R. (2010). The cognitive underpinnings of effective teamwork: a meta-analysis. *Journal of Applied Psychology* 95 (1): 32.

Engelbart, D.C. (1962). *Augmenting Human Intellect: A Conceptual Framework. SRI Summary Report AFOSR-3223*. Stanford Research Institute.

Galbraith, J.K. (2015). *The New Industrial State*, 9. Princeton University Press.

Kamar, E. (2016). Directions in hybrid intelligence: complementing AI systems with human intelligence. *Proceedings of the 25th International Joint Conference on Artificial Intelligence (IJCAI)*, New York, US-NY (9–15 July), pp. 4070-4073. AAAI Press.

Kasparov, G. (2017). *Deep Thinking: Where Machine Intelligence Ends and Human Creativity Begins*. Public Affairs.

Kuo, C.C.J. and Madni, A.M. (2023). Green learning: introduction, examples and outlook. *Journal of Visual Communication and Image Representation* 90: 103685.

Licklider, J.C.R. (1960). Man-computer Symbiosis, transactions on human factors. *Electronics* HFE-1: 4–11.

Madni, A.M. (1988). The role of human factors in expert systems design and acceptance. *Human Factors Journal* 30 (4): 395–414.

Madni, A.M. (2010). Integrating humans with software and systems: technical challenges and a research agenda. *Systems Engineering* 13 (3): 232–245, Autumn (Fall).

Madni, A.M. (2011). Integrating humans with and within software and systems: challenges and opportunities. (Invited Paper). *CrossTalk, Journal of Defense Software Engineering* "People Solutions," May/June.

Madni, A.M. (2020). Exploiting augmented intelligence in systems engineering and engineered systems. *Insight* Special Issue, Systems Engineering and AI https://doi.org/10.1002/inst. 12282.

Madni, A.M. and Jackson, S. (2009). Towards a conceptual framework for resilience engineering. *IEEE Systems Journal* 3 (2): 181–191.

Madni, A.M. and Madni, C.C. (2004). Context-driven collaboration during mobile C2 operations. *Proceedings of the Society for Modeling and Simulation International*. pp. 18–22.

Madni, A.M. and Madni, C.C. (2018). Architectural framework for exploring adaptive human-machine teaming options in simulated dynamic environments. *MDPI Systems*, special issue on "Model-Based Systems Engineering" 6 (4): 44.

Madni, A.M. and Sievers, M. (2018). Model-based systems engineering: motivation, current status, and research opportunities. *Systems Engineering*, Special 20th Anniversary Issue 21 (3): 172–190.

Madni, A.M., Samet, M.G., and Freedy, A. (1982). A trainable on-line model of the human operator in information acquisition tasks. *IEEE Transactions of Systems, Man, and Cybernetics* 12 (4): 504–511.

Madni, A.M., Madni, C.C., and Salasin, J. (2002). ProACT™: process-aware zero latency system for distributed, collaborative enterprises. *NCOSE International Symposium* 12 (1): 783–790.

Madni, A.M., Madni, C.C., and Lucero, D.S. (2019). Leveraging digital twin technology in model-based systems engineering. *MDPI Systems*, special issue on "Model-Based Systems Engineering," 7 (1): 7.

Biographical Sketches

Azad Madni is a USC President-designated university professor of astronautical engineering and holder of the Northrop Grumman Fred O'Green Chair of Engineering. He is a member of the National Academy of Engineering, and recipient of the 2023 National Academy of Engineering's Gordon Prize and the 2023 IEEE Simon Ramo Medal. He is the executive director of USC's Systems Architecting and Engineering Program and founding director of USC's Distributed Autonomy and Intelligent Systems Laboratory. He is a life fellow/fellow of 10 professional science and engineering societies including IEEE, INCOSE, IISE, AAAS, AIAA, and AIMBE. He is the recipient of INCOSE's Pioneer Award, Founders Award, Outstanding Service Award, and Benefactor Award. He is also the recipient of IEEE Aerospace and Electronic Systems Pioneer Award, Judith A. Resnick Excellence in Space Engineering Award, and Industrial Innovation Award, and IEEE Systems, Man, and Cybernetics Norbert Wiener Award for Outstanding Research. He is also the recipient of ASME CIE Lifetime Achievement Award, and the NDIA's Ferguson Award for Excellence in Systems Engineering. He is the author of Transdisciplinary Systems Engineering: Exploiting Convergence in a Hyperconnected World (Springer, 2018) and co-author of Tradeoff Decisions in System Design (Springer, 2016). He has served as Principal Investigator on 97 R&D projects totaling well over $100M. His research has been sponsored by government agencies such as DARPA, NSF, NASA, DHS, DOE, NIST, AFOSR, AFRL, ARL, ONR, DOD-SERC, DOD-AIRC. His industry research sponsors include Boeing, General Motors, Northrop Grumman, and Raytheon. He received his BS, MS, and PhD degrees in engineering from UCLA with a major in engineering systems, and minors in computer methodology and AI, and engineering management.

Part IV

Evaluating and Improving System Risk

Nicole Hutchison

Chapter 18

Complexity and Risk in Systems Engineering

Roshanak R. Nilchiani

Stevens Institute of Technology, School of Systems and Enterprises, Hoboken, NJ, USA

Prelude

The word "Complexity" is perhaps one of the most frequently used words in systems engineering and yet it is one of the least researched and studied fundamental characteristics of many engineered and natural systems. Complexity can manifest itself in the form of an amazing, sophisticated functioning order in an engineered system or the opposite, an invading entropy that disrupts, damages, or jeopardizes the function and performance of an engineered system. Risk and opportunity, on the other hand, have been viewed as an ever-existing concept that affects all phases of the lifecycle of an engineered system and is closely coupled with uncertainty. Traditionally, risk is associated with the likelihood of negative consequences of an event (Haskins et al. 2006; Simpleman et al. 2003). Complex engineered systems face various inherent as well as exogenous/environmental uncertainties, which a considerable portion of them is due to the nature of the complexity they carry. Even more simple engineered systems face various types of risks due to ambient or environmental risks that affect their life cycle. Whether the system contains a sophisticated complex order or a disorder that can propagate, both types of complexities contribute to uncertainties and emergence and as a result, a consequence of a desired or undesirable event. Therefore, the study of risk in a complex system or environment is strongly coupled with understanding the complex dynamics of a system and our ability to model aspects of complexity in a system as well as its context and environment.

Complexity

The field of complexity spans more than seven decades and connects many fields of science and engineering ranging from biology, mathematics, and chemistry to cyber-physical systems, networks, social sciences, and beyond. The term "complexity" has several definitions and various facets and characteristics in different domains of knowledge. We adopt the following definition by Willcox for our discussion in this chapter:

> Complexity is the potential of the system to exhibit unexpected behavior (Willcox et al. 2011)

Systems Engineering for the Digital Age: Practitioner Perspectives, First Edition. Edited by Dinesh Verma.

Companion website: www.wiley.com/go/verma/systemsengineering

In another word, complexity can be defined by several attributes including but not limited to multi-dependency dynamics, uncertainty, and emergence (caused by the behavior and interaction of known components) (Cotler et al. 2017; DeRosa et al. 2008; Phelan 2001; Strogatz 2003).

In 1948, Weaver published his paper named "Science and Complexity" (Weaver 1948), which over time has led to the development of complexity science. Complexity science aims at understanding the characteristics of complex systems such as emergent behavior due to reciprocities of system elements (Phelan 2001), nonlinear and dynamic interactions of elements (Cilliers 2000), and bilaterally dependent relations of elements (Strogatz) and many more topics.

Weaver (1948) described the problem of simplicity and two different kinds of complexity: organized and disorganized complexity. **Problems of simplicity** have a low number of variables that have been tackled in the past two centuries, for example, the classical Newtonian mechanics, where the motion of a body can be described with differential equations in three dimensions. In these problems, the behavior of the system is predicted by integrating equations that describe the behavior of its components (Nilchiani and Pugliese 2017; Weaver 1948). Weaver defines **organized complexity** by a substantial number of variables and "factors which are interrelated into an organic whole" (Weaver 1948). These factors all must be considered when the entire system is being analyzed. Problems of organized complexity differ from the ones pertaining to simplicity as they exceed small numbers of few variables. The type of problems of organized complexity are the ones that we often observe in biological systems as well as elegant engineered systems: they operate based on a complex order and provide a set of unique functionalities. **Disorganized complexity** is characterized by an abundance of variables in a system, each variable exhibits individual behavior, which is described as "erratic, or perhaps totally unknown" (Weaver 1948). Disorganized complexity explains the behavior of the system in form of disorder and is often characterizable by statistical techniques, which become applicable once individual behavior gives way to average behavior(s) to be assessed. Figure 18.1 shows the two types and their characteristics.

It is critical to establish the type of complexity of interest in the system before we embark on studying, modeling, and characterizing it in a system. Organized complexity is often very challenging to understand and model due to our limited cognitive ability to identify all variables and creating of the blueprint of a sophisticated functioning order and relationship of these variables in a system. In theory, we desire to design engineered systems that possess high levels of organized complexity and

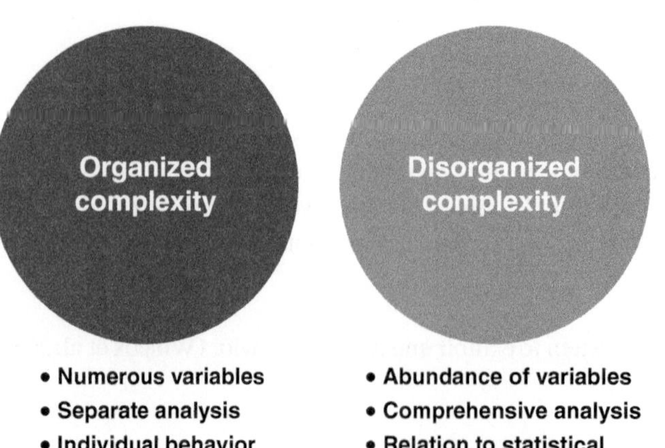

Figure 18.1 Organized and disorganized complexity as per Weaver (1948).

are able to self-organize and adapt to internal and external changes. In contrast, we would aim at identifying the variables that contribute to disorganized complexity, which increases the entropy and disorder of the system. Minimizing disorder is often very desirable since disorder has a strong relationship to the level of risk a system is exposed to. Problems of disorganized complexity and increased entropy have been studied over the past few decades and there are more advancements in quantitative measures of such disorders. By controlling and minimizing disorder in an engineered system, we reduce certain types of risks and failures in systems. Most conversations forward in this chapter as well as other research publications focus on the problem of disorganized complexity and various methodologies to understand and reduce them in engineered systems.

Complex systems exhibit the potential for unexpected behavior. The potential can manifest itself in certain situations and create an emergent behavior or it can stay hidden. Complex systems have nonlinear interactions, circular causality, and feedback loops. Logical paradoxes and strange loops are among some properties they may have. Small changes in a part of a complex system may lead to the emergence and unpredictable behavior in the system (Érdi 2008). Erdi indicates that properties of complex systems may include the following aspects:

- Circular causality, Feedback loops, logical paradoxes, strange loops
- Small changes resulting in out-of-proportion results
- Emergence/unpredictability

Complex systems are also different from **complicated systems**, and at times the two terms are used interchangeably. Complicated systems often have many parts; however, the interactions between parts and subsystems are often well known and often linear, and often not prone to emergent, nonlinear behavior. It should be noted that even a simple or complicated system can be affected by a nonlinear and complex environment and context and, therefore, show unexpected behavior when operating in a complex environment. In contrast, a complex system may possess many parts, and one or more nonlinear relationship(s) or feedback loops exists in the system that drives the emergence and unknown unknowns in the system (Nilchiani et al. 2013) (See Figure 18.2).

Complexity can also be perceived and categorized based on the point of view of the observer (Wade and Heydari 2014). Wade et al. defined three types of categories of complexity: behavioral

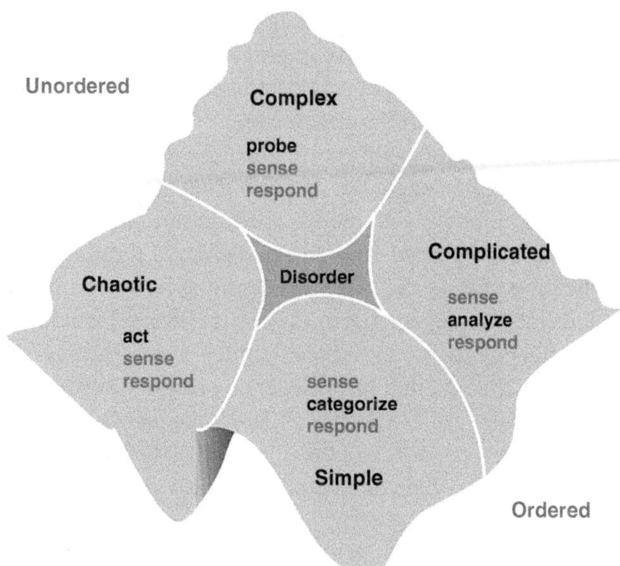

Figure 18.2 Representation of the Cynefin framework. *Source:* Adapted from Snowden and Boone (2007).

complexity, structural complexity, and constructive complexity. When the observer is external to the system and can only interact with the system as a black box, the type of complexity that can be measured is **behavioral complexity**, since it focuses on the overall behavior of the system. When the observer has access to the internal structure of the system such as blueprints and source code for engineered systems, or scientific knowledge for natural systems, then the **structural complexity** of the system is the one being measured. If the process of construction of a system is under observation, the complexity of the building process is measured, which is **constructive complexity**. This definition relates complexity to the difficulty of determining the output of the system.

Complexity and Emergence

Complex systems are often capable of demonstrating behaviors that cannot be immediately or easily explained. Most of such (emergent) behaviors are unpredictable by our current knowledge and take the stakeholders, engineers, and users of the system by surprise. Emergence is defined as "the principle that entities exhibit properties that are meaningful only when attributed to the whole, not to its parts" (Checkland 1981). In other words, an emergent phenomenon is a type of behavior at the macro level that was not hard-coded at the micro level (Page 1999) and can be described independently from the underlying phenomena that caused it (Abbott 2006).

The phenomenon of emergence is observed in both natural and engineered systems. An example of emergence in a natural system is wetness. Water molecules can be arranged in three different phases (i.e., solid, liquid, and gas) but only one of them expresses a certain type of behavior, which is high adherence to surfaces. This behavior is due to the intermolecular hydrogen bonds that affect the surface tension of water drops. These bonds are also active in the solid and liquid phases, but in those cases, they are either too strong or too weak to generate wetness. In complex systems, some properties emerge only when conditions are just right. In engineered systems, the system requirements and software specifications are supposed to be written in such a way that they are independent of their implementation. For this reason, the functions and properties they describe are emergent (Abbott 2006).

The existing literature on emergence differentiates between two types of epistemological and ontological complexity (Bedau 1997; Chalmers 2006; Kauffman 2007). Kauffman proposes two approaches to the nature of emergence. The reductionist approach, treats emergence as epistemological, meaning that the knowledge about the systems is not yet adequate to describe the emergent phenomenon; however, it can improve the explanation of the future of the system. In contrast, the ontological emergence approach assumes that we *cannot* predict the state of emergent phenomena or possible outcomes (Kauffman 2007). Ontological emergence focuses on the enormous amount of states the system could evolve into (Longo et al. 2012). We are often not only unable to predict which state will happen, but not even what are the possible states.

The definitions of emergence often do not differentiate on whether the emergent property is expected or unexpected. In engineered systems, the system engineer is in charge of the identification of the properties of the system concerning its environment, in another word, the functionalities and operation of an engineered system are desired to be fully predictable. Not only the operational environment, but also the assembly, integration, testing, and disposal environment introduce variations and changes that may trigger an emergent behavior in complex engineered systems. In the design process, it is expected to differentiate between the attributes of the system that are desired, and therefore expected, and the unexpected ones, which can be beneficial or adverse. The concept of emergence in engineered systems is also tightly coupled to systems' "ilities," or how systems respond to change.

Phase Transition and Tipping Points in Complex Systems

Phase transition is one of the most fascinating phenomena that may be observed in complex systems. Transitions between alternate states of a system have been observed and many domains of science and engineering. "When a given parameter is tuned and crosses a threshold we see a change in systems organization or dynamics" (Solé 2011). These different patterns of organizations or considered phases. It should be noted that these phenomena can present themselves as a result of interactions between multiple parts of the system. Phase transition or tipping can manifest itself in a system when it is forced outside the boundary of its original established equilibrium, which can result in a critical transition to an alternative stable state (Rietkerk et al. 2021). The tipping point can be considered as the exact situation where one or more external stressor(s) interrupts the steady state of the system's performance (Rasoulkhani and Mostafavi 2018). In a complex system, a tipping point is a qualitative change in a system (Milkoreit et al. 2018), and the exact moment when a complex system experiences an abrupt shift from one state to another, which is extremely difficult to evaluate and predict (Gerla et al. 2011; Veraart et al. 2012). The tipping point can be observed when "a small smooth change made to the parameter values of a system causes a sudden qualitative or topological change in its behavior" (Rasoulkhani and Mostafavi 2018). Natural or engineered systems can approach the tipping point when a minuscule change can amplify into an enormous shift (Scheffer 2010), which compels the system to undergo momentous changes (Kiron et al. 2012). This shift is often irreversible in nature and could lead to the creation of a new system (Huntington et al. 2012). Thus, the system must have the potential to re-achieve a stable state after it reaches the tipping point, thereby making it critical to understanding the dynamics of the system (Black and Repenning 2001; Repenning et al. 2001).

Complex systems can have multiple phase transitions and they are often thresholds or states where the entire system loses its stability, with no possibility of reverting to any prior or even the original state. Phase transitions can lead to a total system collapse (Lenton et al. 2008; Russill and Nyssa 2009). Often disturbances and changes are self-sustaining despite their small or minute origins (Russill and Nyssa 2009). Phase transition often starts when the variables causing the change are altered in a way that irrevocably causes the transition and therefore leads to the threshold (Lamberson and Page 2012).

van Nes et al. defined two different types of tipping points (van Nes et al. 2016), which they deduced from two different approaches and perspectives on tipping points. The first type is based on the concept of bifurcations as described by Scheffer et al. (2009) and is initiated by critical outside influences and variables on the complex system. These transitions can shift the state of the system into an entirely different state. The second type of tipping point originated from the domain of ecology and is related to the concept of unstable equilibria (Hodgson et al. 2015, 2016). The equilibria represent spots in the landscape of possibilities where a slope exists on each side, making them mathematic extrema (Figure 18.3). The authors describe the two different types of tipping points as *tipping due to a change in conditions* and *tipping due to a change in state.*

From the perspective of systems dynamics (SD), a tipping point or phase transition can be defined as a threshold condition that may shift the dominance of the feedback loops that control a process (Sterman 2000). Tipping points are conditions that border between two or more behavioral zones created by a dominant feedback loop (Ford and White III 2020). Even when the system starts in the desired execution mode, there is no guarantee that it will persist operating in that mode due to the presence of the possibility of a tipping point (Repenning 2001). Systems Dynamics can potentially help in revealing tipping points and their impacts on systems by specifying, formalizing, and explaining structures that create tipping points (Taylor and Ford 2006, 2008).

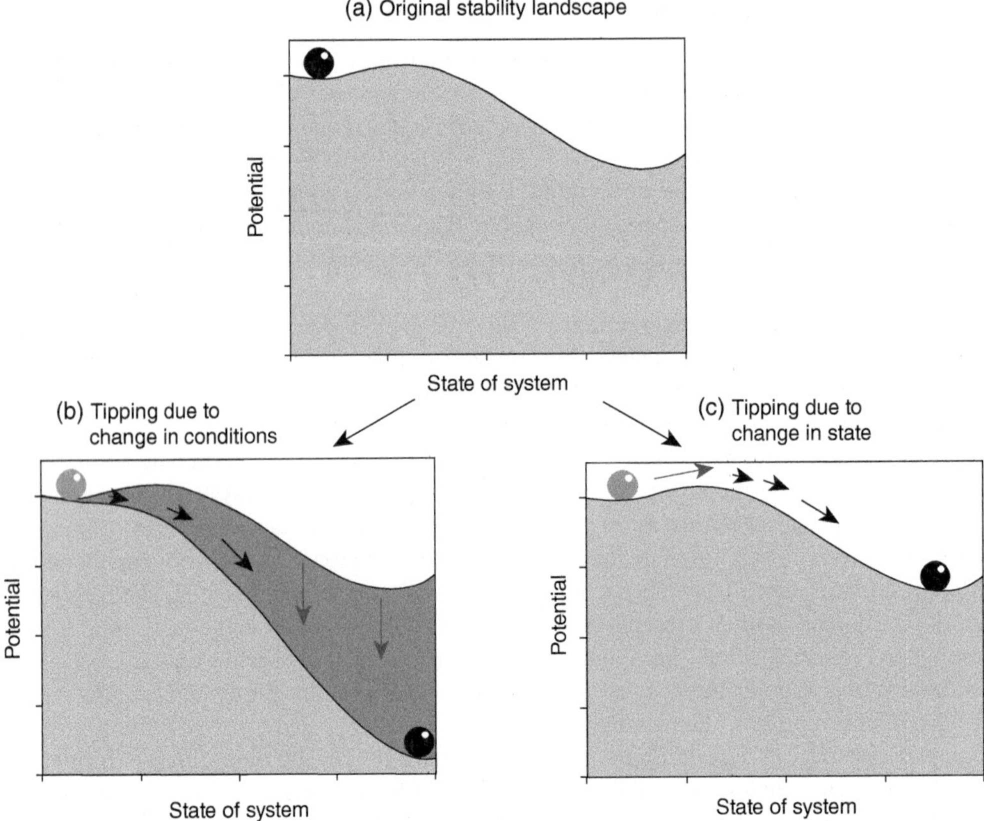

Figure 18.3 Tipping points and accompanying changes. *Source:* Adapted from van Nes et al. (2016).

The increased complexity is often associated with increased fragility and vulnerability of the system. By harboring an increased potential for unknown unknowns and emergent behavior, the probability of known interactions that lead to performance and behavior in a complex system decreases, which in turn leads to a more fragile and vulnerable system. That is, the presence of complexity in a system, even a little complexity, can swamp the behavior of the familiar, linear interactions (Nilchiani et al. 2013).

Complexity and Risk

Traditionally risk has been defined as the likelihood of an event coupled with a negative consequence of an event occurring (Conrow 2003). Conrow discussed risk as something to be avoided if possible or reduced. He also defined opportunity as the potential for the realization of the wanted and positive consequences of an event. INCOSE defines risk and opportunity management as a disciplined approach to dealing with uncertainty that is present during the entire lifecycle of a system (Haskins et al. 2006). INCOSE defines four risk categories, which include technical, cost, schedule, and programmatic risk. Technical risk focuses on if the technical requirement of a system is achieved in its life cycle. Cost risk deals with the budget overrun in the system's life cycle. Schedule risk takes a look at the possibility of failing to meet the milestones of a project.

Programmatic risk is also defined as events that are beyond the project manager's control. INCOSE also defines ambient risk or environmental risk, which is caused and created by the surrounding environment of the project (Walden et al. 2015). Risk management is a critical topic in Department of Defense acquisition programs. Defined in the *DoD Risk Management Guide* (US Department of Defense (Office of the Undersecretary of Defense Acquisition 2006)):

"Risk is a measure of future uncertainties in achieving program performance goals and objectives within defined cost, schedule, and performance constraints. Risk can be associated with all aspects of a program (e.g., threat, technology maturity, supplier capability, design maturation, performance against plan). Risk addresses the potential variation in the planned approach and its expected outcome."

Risks have the following three components:

1) A root cause, which, if understood and eliminated or corrected, would prevent a potential consequence from occurring,
2) A probability (or likelihood) assessed at present of that root cause occurring, and
3) The consequence (or effect) of that future occurrence.

A root cause is the most basic reason for the presence of a risk. Program management needs to collect and identify future root causes, likelihood, and consequences. This is often done, in the best circumstances, by assembling experts and asking them to converge on the three components. This is a group process based on the extensive practical experience of the expert panel. It is necessarily subjective, based on the memories of the panel members and their analogic reasoning.

The process of risk identification is almost always *post hoc*. One part of the *Risk Management Guide* states, "Use a proactive, structured risk assessment and analysis activity to identify and analyze root causes," and others state, "Use the results of prior event-based systems engineering technical reviews to analyze risks potentially associated with the successful completion of an upcoming review," and "During decomposition, risks can be identified based on prior experience, brainstorming, lessons learned from similar programs, and guidance contained in the program office RMP [Risk Management Plan]." (US Department of Defense (Office of the Undersecretary of Defense Acquisition 2006)).

For complex engineering development programs, often various types of risks exist that manifest themselves at different times throughout the development process. Technical and programmatic risk can result in substantial cost overruns, delays, performance issues, reduced adaptability to changing requirements, or even total cancellation of a project. One of the challenges with assessing risk using the traditional risk reporting matrices (see Figure 18.4) for engineering system development programs is that neither the likelihood nor the true consequence of risk can be objectively

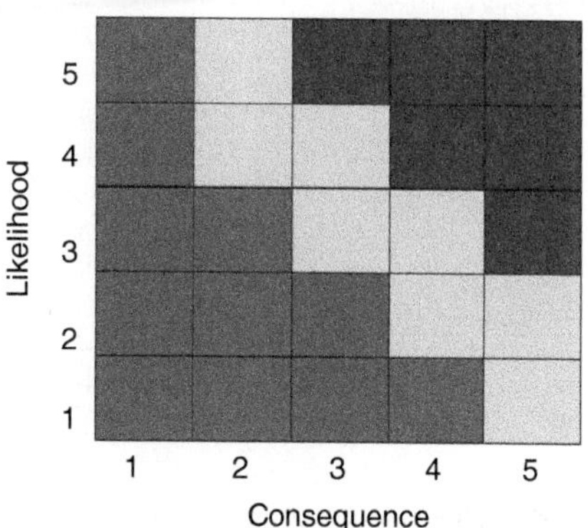

Figure 18.4 Traditional risk reporting matrix (US Department of Defense (Office of the Undersecretary of Defense)). *Source:* Adapted from Acquisition (2006).

established. For one, there is substantial uncertainty around the interactions among different components of a system as well as uncertainties about how effectively various kinds of risks can be managed across a multiplicity of interfaces.

Quantitative measurement of risk in engineering systems is an essential part of the system lifecycle management and yet a domain that is not well explored and studied. INCOSE suggests characterizing risk using cumulative probability curves with the probability of failure and consequences in a measurable and quantitative way; however, the lack of data makes it a challenging task (Walden et al. 2015). The Expected Value Model of risk suggests risk calculation as follows:

$$\text{Risk} = \text{Probability of failure}\left(\text{Pf}\right)*\text{Consequence of failure}\left(\text{Cf}\right)$$

Currently, our knowledge of the probability of failure and consequences to the system is rather limited and subjective. Without an understanding of the potential complexities hidden in a system or the environment (ambient complexities) that affect an engineered system, quantitative measures of risks are often incorrect and perhaps misleading.

Roots of Uncertainty

The roots of uncertainty are categorized in the literature as lack of knowledge, lack of definition, statistically characterized phenomena, known unknowns, and unknown unknowns (McManus and Hastings 2005). Lack of knowledge includes facts that are not known or are known only imprecisely. This type of knowledge can be gathered and the uncertainty can be reduced. Lack of definition is a type of uncertainty that exists when the elements or attributes of a system are not specified. Statistically characterized (random) variables or phenomena are things that cannot always be known precisely but can be statistically characterized, or bounded. An example of known unknowns is market uncertainty: we know the uncertainty exists, but we cannot reduce the uncertainty beforehand. Unknown unknowns may be emergent behaviors of a system; we are not even aware of their existence until they happen to the system.

These roots have different levels and depths of uncertainty. For example, the uncertainty associated with the lack of knowledge and definition is much less than the uncertainty associated with unknown unknowns (See Figure 18.5).

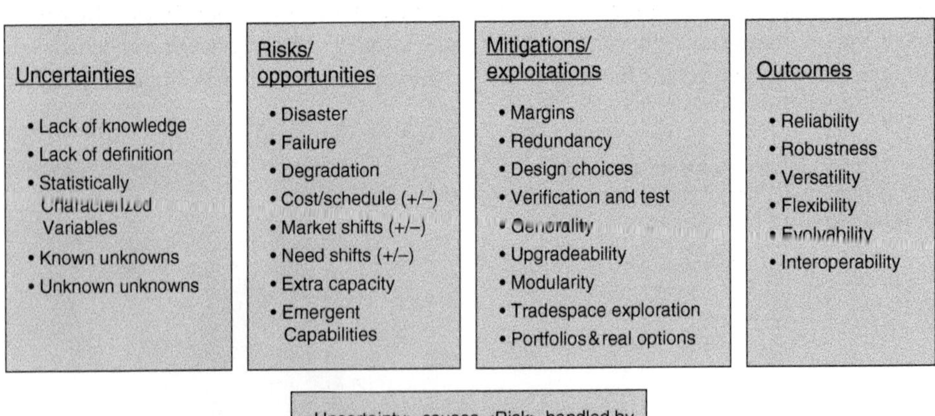

Figure 18.5 Roots of uncertainty. *Source:* Adapted from McManus and Hastings (2005).

Complexity–Risk Relationship

Complexity may be the root cause of many unforeseen uncertainties and risks. Program/project complexity can generate negative consequences that may often take the project management team by surprise. Complexity in some cases can generate positive changes and opportunities in an engineered system as well. Common and advanced methods of risk modeling, including, for example, Bayesian Networks, cannot predict the emerging risks nor predict the ripple effects of uncertainty in multiple parts of a system, other interconnected System of Systems, or even the operational environment. Complexity (organized or disorganized) is associated with the lack of knowledge about the system and its environment and often poses an intimidating effect on stakeholders, engineers, designers, and managers of these engineered systems. Without fundamentally understanding complexity, the system is at risk of being out of control, or in the control of something unknown. Often the complexity manifests in risk and risk creates more complexity. This is known as the complexity-uncertainty death spiral. Engineered system complexity can lead to unexpected or undesirable behavior, and that behavior may affect or damage part of the system and therefore move the system away from its equilibrium and release more unexpected changes to the system and its environment.

Engineered Systems and Risk

Measuring risks associated with simple and complicated systems are less challenging compared to complex systems. Problems of simplicity have a low number of variables that have been tackled in the past two centuries, and in these problems, the behavior of the system is predicted by integrating equations that describe the behavior of its components (Nilchiani and Pugliese 2017; Weaver 1948). Complicated systems may have many parts, but in general, the behavior of the system is well understood and predictable. It should be noted that the measurements of the probability of failure and consequences are often more straightforward, however, the effect of environmental or ambient complexity should be considered. Although the complicated or simple engineered system's behavior is more predictable within a certain range, the unpredictable and complex environment can introduce complexity and unforeseen risks as well as opportunities into the system (See Figure 18.6–7).

The Galileo spacecraft can be considered an example of a complicated system in interaction with environmental (ambient) complexity. The Galileo mission had originally been designed for a direct flight of about 3.5 years to Jupiter and was launched in 1989. The spacecraft was scheduled to

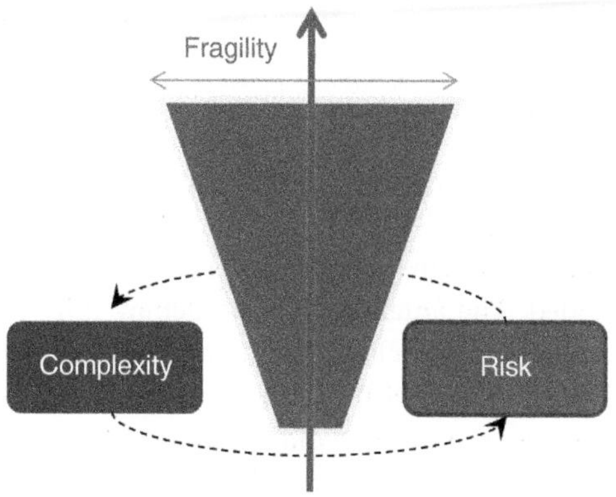

Figure 18.6 Complexity risk escalation.

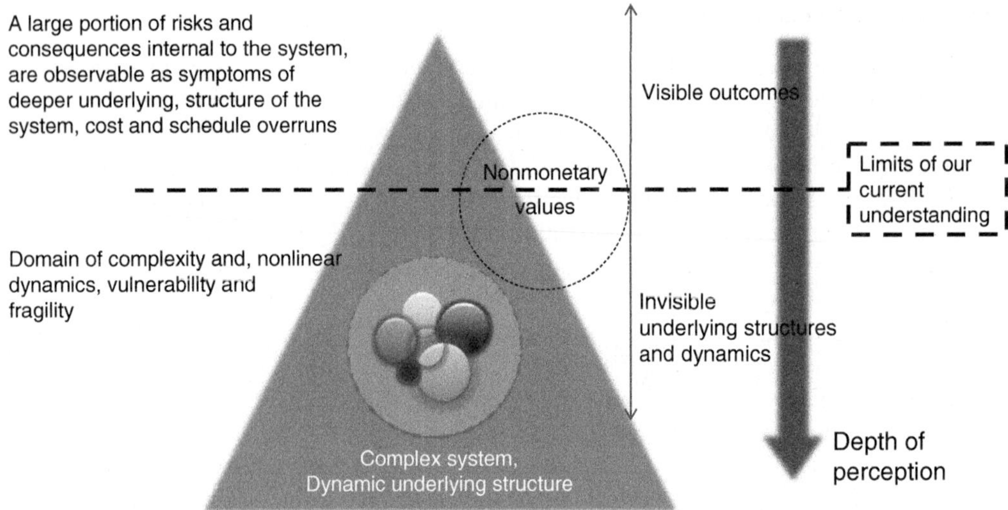

Figure 18.7 Systems adaptability and complexity cognition.

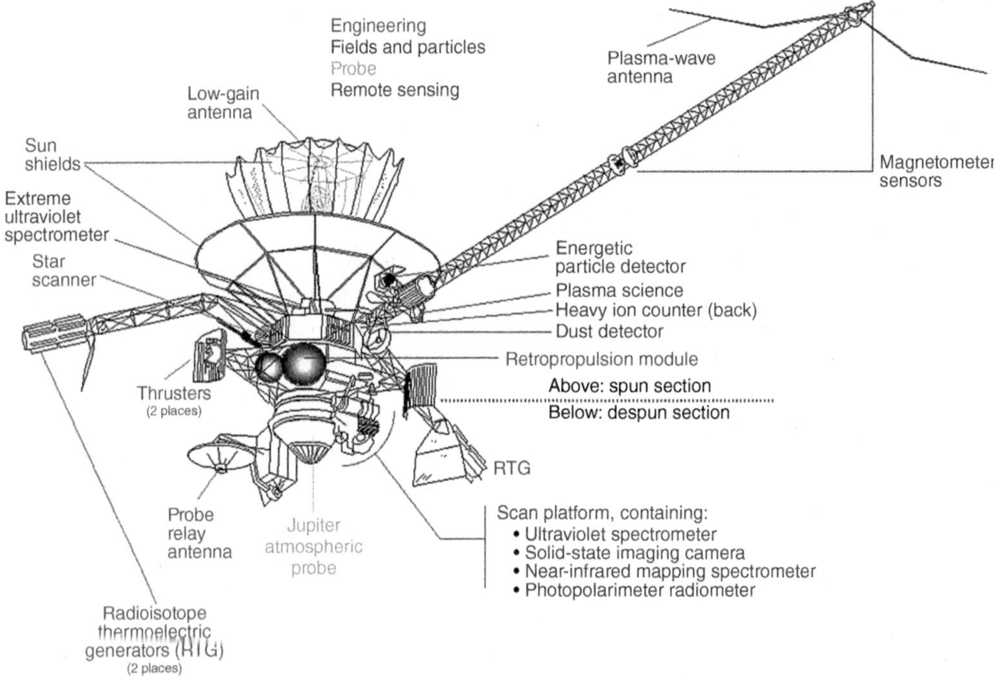

Figure 18.8 Galileo spacecraft. *Source:* Adapted from Marr (1994).

deploy its high-gain antenna in 1991 as it moved away from the Sun and the risk of overheating ended. The antenna failed to deploy fully because a few of the antenna's 18 ribs were held by friction in the closed position. The root cause of the failure to deploy was connected to the transportation of the high-gain antenna and environmental vibrations. Despite the efforts to free the ribs, the antenna did not work. From 1993 to 1996, extensive new flight and ground software was developed, and ground stations of NASA's Deep Space Network were enhanced to perform the mission using the spacecraft's low-gain antennas. To offset some of the performance loss, the spacecraft's computer was extensively reprogrammed to include new data compression and coding algorithms. An innovative combination of new, specially developed software for Galileo's on-board computer and improvements to the ground-based signal-receiving hardware in the Deep Space Network enabled the spacecraft to accomplish at least 70% of its original mission science goals using only its small, low-gain antenna, despite the failure of its high-gain antenna (Jansma 2011; Marr 1994).

Mars Rover Opportunity is another example of a complicated system in interaction with environmental (ambient) complexity and creating opportunity for lifetime extension. A series of unexpected environmental effects swept dust on the solar panel of the Mars Rover Opportunity. Opportunity landed on Mars in January 2004 and maintained communication until June 2018. It originally had more than 900 watt-hours of energy per sol, which was expected to degrade due to dust deposition on the solar panels. After a year of operation, Opportunity has regained its original power (900 watt-hours per sol). It is speculated that a part of this difference between power outputs is due to dust removal through the surface wind. Opportunity far exceeded its 90- sol primary mission and set records for life extension (5111 sols), distance traveled (45 km), and scientific discoveries for planetary rovers (Callas et al. 2019). In June 2018 a major dust storm reduced the power to the rover. In a single example, environmental uncertainties created both opportunities and risks to the same system, showcasing the power of environmental complexity and its effect on an engineered system.

Mars Polar Lander (MPL) was a part of the Mars survey program and was launched in January 1999. The spacecraft was destroyed in the process of landing. The most suspected reason for failure is the entry, deployment, and landing sequence, in which three landing legs were supposed to deploy from stowed to landing condition. Each of the legs was fitted with a Hall Effect magnetic sensor, which would generate a voltage when the legs contacted the Mars surface and shut off the descent engines after the touchdown. The touchdown sensors generated a false momentary signal at leg deployment. This was an emergent behavior in the system. The spacecraft software interpreted the signals generated at leg deployment as a valid touchdown signal. As a result, the software shut off the engines at an altitude of 40 m, and the Lander free-fell to the surface and was destroyed (Leveson 2004).

Risk and Organized Complexity

Organized complexity can be found in systems that possess a large number of interrelated variables that create an organic whole (Weaver 1948). The type of problems of organized complexity are the ones that we often observe in biological systems as well as elegant engineered systems: they operate based on a complex order and provide a set of unique functionalities. Various types of risks and uncertainties can affect the functioning of complex organized order and therefore expose the system to risk and opportunities. Some of these risks and uncertainties are including but are not limited to:

- **Environmental (ambient) risks and uncertainties**: Environmental risks can affect a complex organized functioning system and affect a subsystem, node, link, or part of the system and interrupt the order and flow of information, mass, and energy of the system and therefore the full function of the system. In an organic system, the system may have the capability of self-repair or can be repaired externally, so it will continue its complex function. If the damage is beyond repair, the organized complex system can collapse.

- **Internal risks and uncertainties**: Failure of parts, subsystems, nodes, and links in a complex system may age, fail, malfunction, or become obsolete and therefore impair the order and function of a system. If a timely repair is not possible, the organized complexity can collapse.
- **Risk of emergent behavior, phase transition, . . .**: An organized complex system under circumstances can manifest emergent behavior(s) or perhaps be exposed to an abrupt change that can push the complex system through phase transition or tipping point and therefore move the system toward a new known or unknown equilibrium. The new state of equilibrium and order may create new unforeseen risks or opportunities for the system.

Risk and Disorganized Complexity

Disorganized complexity is associated with a system with many variables, and the variable expresses erratic behavior, or perhaps totally unknown behavior (Weaver 1948). Disorganized complexity explains the behavior of the system in form of disorder and often can be characterized and measured by statistical techniques, expressing the average behavior(s) of the disorder in a system. Disorganized complexity is the most common form of complexity and the most measured type and can be detected in various levels and spectrums in all systems. One can argue that an increased level of disorganized complexity in engineered systems can lead to malfunction, failure, increased fragility, and increased entropy of a system. Increase in disorganized complexity is directly linked to increase in manifestation of various types of risk in an engineered system. For example, increase in entropy and disorder in requirement phase of the systems engineering process, may lead to more complexity in preliminary and critical design phase, manufacturing, testing, verification and validation or perhaps it may manifest itself after the system is fielded and operational. Identifying and minimizing disorganized complexity at each phase of systems engineering lifecycle is essential, as it contributes to minimizing a portion of risk and helps restoring and reorganization in a system's life cycle. One could argue that the residual disorder and entropy at each phase of lifecycle can potentially have a ripple effect through the following phases of the system's life cycle. The excess disorganized complexity and out-of-sequence changes can also contribute to rework cycles. Rework cycles will cause cost and schedule overruns (Figure 18.9).

Figure 18.10 is a conceptual illustration of the complexity level in the lifecycle of an engineering system. If an engineered system does not have the minimum required critical complexity, a system cannot perform the functions that are expected. Also, above a maximum tractable complexity level, the system development process can spiral out of control. It is the expertise, know-how, and experience of the systems engineers and management where both use standard technical management processes, such as version control, keeping dependency graphs current, keeping design changes in harmony with requirements changes, etc., that can keep the development process within the boundaries of those two and stabilize the complexity level of a system.

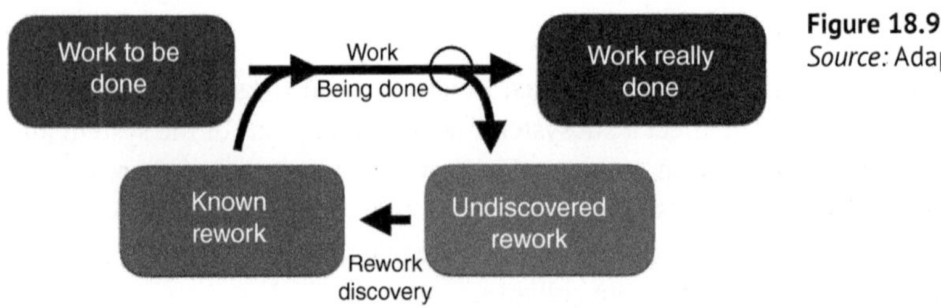

Figure 18.9 The rework cycle. *Source:* Adapted from Cooper (1994).

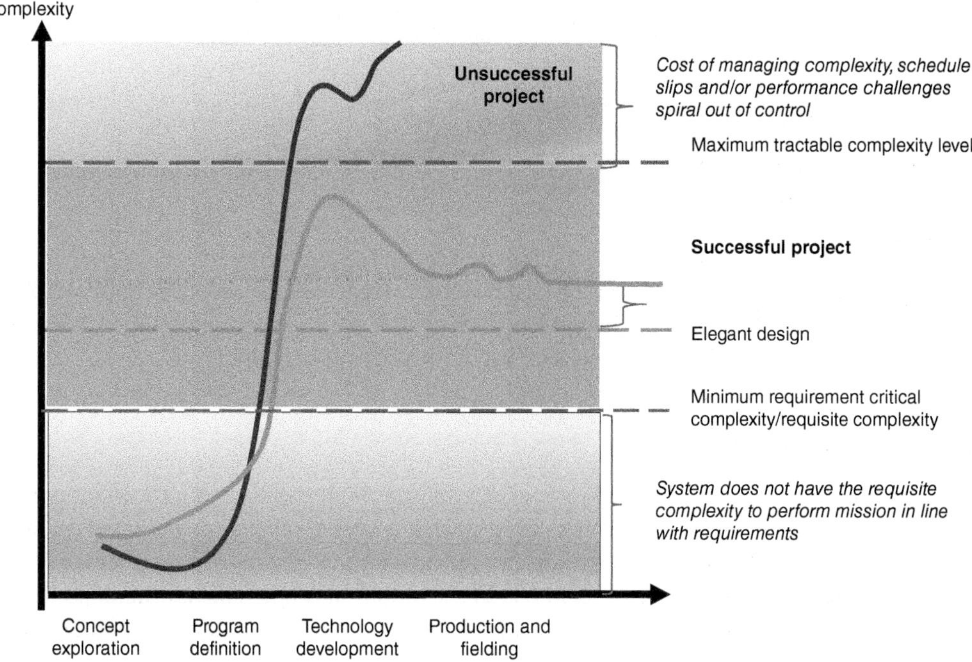

Figure 18.10 Complexity evolution throughout the systems development life cycle. *Source:* Adapted from Nilchiani et al. (2013).

Complexity Assessment and Modeling in SE

Complexity measures in academic publications have been spanning various fields of knowledge including mathematics, physics, biology, and multiple domains of engineering. One of the first and highly cited publications that provide a mathematical measure of complexity is discussed in Shannon's "Mathematical Theory of Communication," which is based on measuring the entropy of a system (Shannon 1948). Shannon explains that the entropy of a system is a set of probabilities a system has regarding its state. Another popular complexity measure is introduced by McCabe's (1976) paper in which he describes his graph-theoretic complexity measure (McCabe 1976). McCabe's complexity measure outlines the connection of graph-theory concepts and complexity and connects them to the design/structure of computer programs as well as their development. Another highly cited complexity measure is presented in Kauffman's 1996 book "At Home in the Universe" (Kauffman 1996). Kauffman describes self-organized complexity and relates it to various biological structures such as living organisms.

An overview of tens of complexity measures leads us to categorize various measures into the different fields of engineering: Signal Complexity (Electrical Engineering), Physical Complexity (Chemical, Nuclear, and Nano Engineering), Infrastructure and Network Complexity (Civil, Industrial, and Electrical Engineering), Biochemical Complexity, Design & Manufacturing Complexity (Industrial, Mechanical, and in part Electrical Engineering), Software and Code Complexity (Computer Engineering). There is significant research on measures/metrics of complexity in these fields, which have been conceptualized in Figure 18.11.

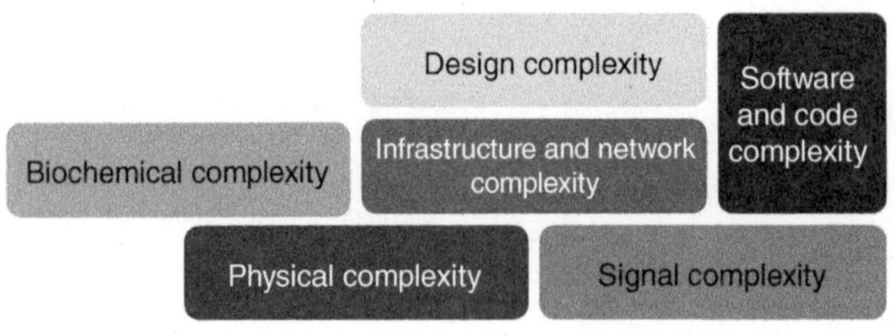

Figure 18.11 Complexity areas and scientific fields.

We will briefly discuss a few of these measures and methodologies with their applications in engineered systems in this section. There are various approaches to measuring complexity and the following section is discussing a selected brief list of measures of complexity.

Shannon Complexity Metric

Shannon's entropy is one of the classic complexity metrics built based on the theory of entropy (1948), a measurement of disorder from the physical sciences. Shannon's proposed equation measures how much information is output from a process with an entropy measurement of H. The variable p_i is the probability that character, i, is in the output message. The information content, represented by $-\log_2 p_i$, is weighted by p_i, and then summed across all the system output. The value of H is measured as the weighted average of the information content.

$$H = -\sum_i p_i \log_2 p_i$$

Cyclomatic Complexity Metric

McCabe created a complexity metric for software systems (McCabe 1976). This metric looks at the graph representation of the program and is defined as

$$v(G) = e - n + p$$

where e is the number of edges, n the number of vertices, and p is the number of connected components in the graph. This metric is named cyclomatic number and in a strongly connected graph, the Cyclomatic number is equal to the maximum number of linearly independent circuits (McCabe 1976).

Free Energy Density Rate Metric of Complexity

Chaisson proposed a metric for the evaluation of complexity based on the amount of energy of the entity under study (Chaisson 2004). He defines energy rate density, as "the amount of energy available for work while passing through a system per unit time and per unit mass" (Chaisson 2015). This metric looks at the input and output of the system, without studying its internal structure. This metric has been evaluated for multiple entities such as galaxies, stars, planets, plants, animals, societies, and technological systems, showing a rising trend in complexity (Chaisson 2014).

According to this metric, a system with a large intake of energy, and a low mass, will have a high level of complexity. From the engineering point of view, perhaps this metric may point to an inefficient system. Chiasson's metric is informative and successful based on the assumption that most of the systems under study are evolved by nature or designed with efficiency in mind. Therefore, with an assumption that there is no waste of energy or useless mass.

Propagation Cost and Clustered Cost Metrics

MacCormack presented two types of metrics for the evaluation of the complexity of software systems (MacCormack et al. 2006). The directed dependency between files in the source code is the function call. The propagation cost is the average of the visibility of modifications to dependent files, while the clustered cost considers the importance of the node scaling the relative cost accordingly.

Spectral Structural Complexity Metric

Sinha and de Weck presented a structural complexity metric based on the design structural matrix (DSM) of the system (2012). The metric is defined as

$$C(n,m,A) = \underbrace{\sum_{i=1}^{n} \alpha_i}_{C_1} + \underbrace{\left(\sum_{i=1}^{n} \sum_{j=1}^{n} \beta_{ij} A_{ij} \right)}_{C_2} \underbrace{\gamma E(A)}_{C_3}$$

where n is the number of components in the system, m is the number of interfaces, A the DSM, α_i the complexity of each component, $\beta_{ij} = f_{ij}\alpha_i\alpha_j$ the complexity of each interface, $\gamma = 1/n$ a normalization factor, and $E(A)$ the matrix energy of the DSM. C_1 is the complexity contribution of the components, C_2 is the contribution of the interfaces, and C_3 is the contribution of the topology. The application of the metric sees the evaluation of α_i through expert judgment and assumes $f_{ij} = 1$ for lack of information (Sinha and Weck 2013).

Graph Energy Metric

The metric is inspired by Hückel Molecular Orbital (HMO) Theory, which evaluates the energy of π-bonds in conjugated hydrocarbon molecules as a solution of the time-independent Schrödinger equation

$$H\psi = E\psi$$

where H is the Hamiltonian matrix, and E is the energy corresponding to the molecular orbital. The equation is an eigenvalue problem of the Hamiltonian. In 1978, Gutman defined the energy of a graph, as following:

$$E = \sum_{i=1}^{n} |\lambda_i|$$

where λ_i are the eigenvalues of the adjacency matrix representing the carbon substructure of the molecule (Gutman and Shao 2011).

A new approach introduced by Nikiforov (2007), and embraced by Sinha, evaluates the graph energy using the singular values of the matrix. This modification extends the applicability to directed graphs where the adjacency matrix is not symmetric.

Dynamic Complexity Measurement Framework

Fischi built a framework for the measurement of dynamic complexity based on the point of view of the observer (Fischi et al. 2015). The definition of complexity used in this framework is based on the system being observed, the capabilities of the observer, and the behavior that the observer is trying to predict (See Figure 18.12).

Requirement-Induced Complexity Measure

Salado proposed a problem complexity metric to measure the complexity that is induced at the requirement phase of a systems engineering process, as a function of the solution space (Salado and Nilchiani 2014). Since any systems engineering problem is defined by a set of requirements, he proposed that the problem complexity is a function of the number of requirements to be fulfilled and the level of conflict between them. Requirements often would fit one of the following categories of Functional requirements, Performance requirements, Resource requirements, and Interaction requirements. Functional requirements define the size of the problem to be solved, while the other types of requirements determine the level of conflicts a solution would need to resolve. Salado proposed a mathematical definition for problem complexity that follows the structure of COSYSMO parametric estimator (Valerdi and Boehm 2010), which uses additive factors when the variable has local effects, multiplicative factors when the effect is global, and exponential factors when the variable has global and emergent effects depending on the size of the variable.

$$C_p = K \cdot \left(\sum_{i=1}^{n} a_i \cdot r_{f_i} \right)^E \cdot \prod_{j=1}^{m} H_j^{\,b_j}$$

where C_p is the problem complexity index, K is a calibration factor that allows problem complexity to be adjusted to accurately reflect an organization's business performance. The first term represents the size of the requirement set, i.e., number of functional requirements r_f the system shall fulfill, weighted (a) to reflect the inherent difficulty of requirements and adjusted for diseconomies of scale (E). The last term represents complexity modifiers derived from the amount and types of

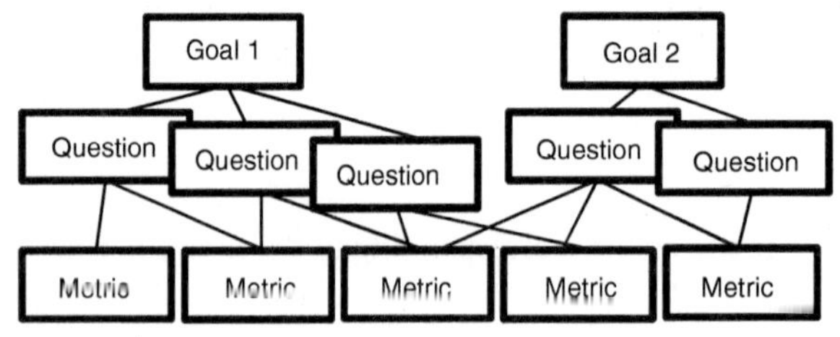

Generic dynamic complexity metric GQM model

Behavior entropy model	
Goal 1	Know and predict system behaviors
Question	What are the predicted system behaviors?
Entropy metrics	**p1** = What is the probability that the selected systems behaviors for the actual mission cannot be easily/accurately predicted/navigated in timeframe t? **p2** = What is the probability that the selected systems behaviors for the actual environment cannot be easily/accurately predicted/navigated in timeframe t?

Figure 18.12 Generic dynamic complexity metric GQM model.

conflicts (*H*), which are adjusted to reflect the influence and diseconomies of scale (*b*). Salado and Nilchiani proposed four types of conflicts, which are based on heuristics to identify conflicting requirements as follows (Salado and Nilchiani 2015):

1) A conflict may exist when two or more requirements oblige the system to operate in two or more phases of matter.
2) A conflict may exist when two or more requirements compete for the same resource.
3) A conflict may exist when two or more requirements inject opposing directions in laws of physics.
4) A conflict may exist when two or more requirements inject opposing directions in laws of society.

Spectral Structural Complexity Metrics

Spectral complexity metrics, which are based on the eigenvalues of a certain graph representation of the system were presented and studied by Pugliese (Pugliese and Nilchiani 2019). He created a set of metrics based on the adjacency matrix, the Laplacian matrix, and the normalized Laplacian matrix. or the Design Structure Matrix (DSM) or N2 matrix, it is used to represent the interfaces and their arrangement, which allows to make considerations on architectural modularity and clustering of components. The Laplacian matrix includes additional information about the degree of each component. The normalized Laplacian matrix has an interesting spectrum that is related to other graph invariants more than the spectra of the other two matrices. These three matrices are considered in their weighted variations, where edges and vertices of the graph carry different weights. He applied to two sets of random graphs, generated through Erdõs-Rényi (ER) and Barabási-Albert (BA) algorithms. The values of each metric are plotted against graph density, which is defined as

$$d = \frac{2m}{n(n-1)} \text{ undirected graphs}$$

$$d = \frac{m}{n(n-1)} \text{ directed graphs,}$$

where n is the number of nodes and m is the number of edges in the graph G.

Pugliese used graph diameter, as the maximum shortest path between all pairs of nodes in the graph. In absence of accurate information regarding the internal structure of nodes, which is usually the case in system of systems applications, where one organization cannot access data belonging to external actors, the complexity of the nodes can be approximated with the degree of the node $\alpha_i = \deg v_i$, and $\beta_{ij} = \sqrt{\alpha_i \alpha_j}$. He developed spectral metrics. The general formula for the metrics is

$$C(S) = f\left(\gamma \sum_{i=1}^{n} g\left(\lambda_i(M) - \frac{\operatorname{tr}(M)}{n}\right)\right)$$

where M is the matrix representing the system, λ_i are its eigenvalues, g and f are generic functions, and γ is a scaling coefficient that considers the size of the system. Table 18.1 shows the metrics that can be derived from this formula through combinations of these parameters. Two sets of functions, two values for the coefficient γ and three matrices, create a set of 12 metrics. The metrics are

Table 18.1 Twelve examples of spectral structural complexity metrics.

	Adjacency matrix	Laplacian matrix	Normalized Laplacian matrix
$\gamma = 1$	$E_A(G) = \sum_{i=1}^{n} \lvert \lambda_i \rvert$	$E_L(G) = \sum_{i=1}^{n} \left\lvert \mu_i - \frac{2m}{n} \right\rvert$	$E_{\mathcal{L}}(G) = \sum_{i=1}^{n} \lvert \nu_i - 1 \rvert$
	$N_A(G) = \ln\left(\sum_{i=1}^{n} e^{\lambda_i}\right)$	$N_L(G) = \ln\left(\sum_{i=1}^{n} e^{\mu_i - \frac{2m}{n}}\right)$	$N_{\mathcal{L}}(G) = \ln\left(\sum_{i=1}^{n} e^{\nu_{i-1}}\right)$
$\gamma = \frac{1}{\mathbf{n}}$	$E_{An}(G) = \frac{1}{n}\sum_{i=1}^{n} \lvert \lambda_i \rvert$	$E_{Ln}(G) = \frac{1}{n}\sum_{i=1}^{n} \left\lvert \mu_i - \frac{2m}{n} \right\rvert$	$E_{\mathcal{L}n}(G) = \frac{1}{n}\sum_{i=1}^{n} \lvert \nu_i - 1$
	$N_{An}(G) = \ln\left(\frac{1}{n}\sum_{i=1}^{n} e^{\lambda_i}\right)$	$N_{Ln}(G) = \ln\left(\frac{1}{n}\sum_{i=1}^{n} e^{\mu_i - \frac{2m}{n}}\right)$	$N_{\mathcal{L}n}(G) = \ln\left(\frac{1}{n}\sum_{i=1}^{n} e^{\nu_i - 1}\right)$

Source: Adapted from Pugliese and Nilchiani (2019).

referred to graph energy (GE), Laplacian graph energy (LGE), normalized Laplacian graph energy (NLGE), natural connectivity (NC), Laplacian natural connectivity (LNC), normalized Laplacian natural connectivity (NLNC), and where $\gamma = 1/n$, the acronym has a trailing n, such as in (GEn).

Risk and Complexity in the SE Life Cycle: Discussions

The field of systems engineering is in need of novel and applied approaches to assess disorganized complexity and risk, which will enable the stakeholders, engineers, and program managers to gain an improved assessment and insight about an engineered system risk and its manifestation at each stage of the life cycle of the system. By assessing the complexity level of an engineered system at each stage of the life cycle, we potentially can:

- Gain a more accurate and objective assessment of the risk and probability of its manifestation from very earliest stages of a development program.
- Provide the time lead and the opportunity to address and reduce excess complexity in the requirement, design, or operation phase of an engineering system. It will provide decision makers with the opportunity to reduce the risk at every stage by probing the alternatives that help in managing and reducing the excess, unwanted complexity.
- By reducing excess complexity dynamically at each stage of an engineering system development program, cost and schedule overrun that contribute to the complexity level of the development program, can be reduced dramatically.
- The optimal points, those which balance the complexity of the design with its risks, can be managed, perhaps in new and unique ways, once they have been identified and quantified. It is felt that the resulting techniques have the potential to be extended from the heavily software-related systems being studied, to a multitude of other engineering programs and projects.

Figures 18.13 and 18.14 shows the result of a previous SERC study of 31 Acquisition programs were studies based on their complexity and the amount of cost and schedule overrun based on the type of the complex system acquired. Table 18.2 also the list of these acquisition programs and their specific amount of cost and schedule overruns.

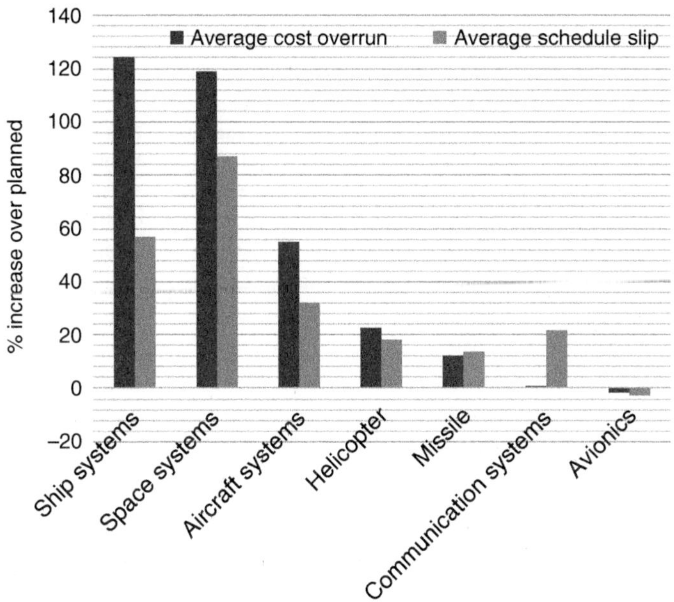

Figure 18.13 Cost overrun and schedule slips for different types of weapons systems. Most cost overruns occur for ship systems, while most schedule slips happen for aircraft. Avionic systems have had a good track record of beating both cost and schedule plans. *Source:* Adapted from Nilchiani et al. (2013).

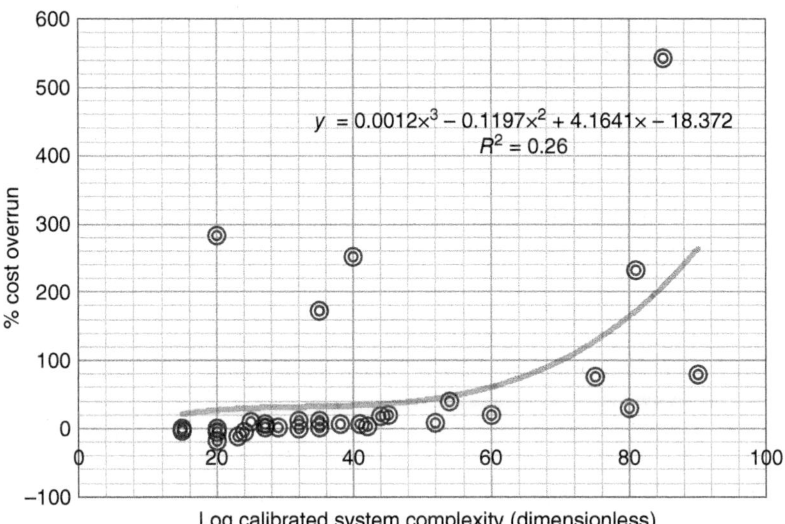

Figure 18.14 Cost overrun as a function of log calibrated architectural systems complexity. *Source:* Adapted from Nilchiani et al. (2013).

Table 18.2 A set of Department of Defense Acquisition Programs, total program cost, and percentage affect cost and schedule overruns.

Program name	Total program cost (M$)	Type of system	Primary contractor	% Cost overrun	% Schedule slip	Type of acquisitic
C-130	$6,204	Aircraft	Boeing	252	C	High TRL
E2-D Advanced Hawkeye	$17,747	Aircraft	Northrup Gruman	20.3	43.2	Medium TEL
F-35	$325,535	Aircraft	Lockheed Martin	78.2	N/A	Low TEL
FAB-T	$4,588	Aircraft	Boeing	29.1	35	Medium TEL
Global Hawk	$12,812	Aircraft	Northrup Gruman	172.2	127.3	Low TRL
Grey Eagle	$5,159	Aircraft	General Atomics	−18	N/A	High TRL
HC-130	$13,091	Aircraft	Lockheed Martin	−5.1	N/A	High TRL
MO-4C UAV	$13,052	Aircraft	Northrup Gruman	1.5	0	High TRL
P-8A Poseidon	$32,959	Aircraft	Boeing	0.1	0	High TRL
Reaper UAV	$11,919	Aircraft	General atomics	18.9	19	Medium TRL
Excalibur guided artillery	$1,781	Artillery	Raytheon	282.4	27.2	Medium TEL
IDECOM	$821	Avionic system	ITT Electronics	−0.5	−8.5	High TRL
Joint precision-approach and landing system	$25,575	Avionic system	Raytheon	−2.9	2.7	High TRL
Airborne and tactical radio System	$8,160	Communication system	Lockheed Martin	0.1	13.8	Medium TRL
Joint tactical radio system handheld	$8,358	Communication system	General dynamics	1	22.4	Medium TRL
Mobile user objective system	$6,978	Communication system	Lockheed Martin	3.8	28.9	Medium TRL
Navy multi-band terminal	$1,214	Communication system	Raytheon	−11.2	0	High TRL
Warfighter information network tactical	$5,052	Communication system	General dynamics	8.5	42	Medium TRL
Apache block IIIA	$10,737	Helicopter	Boeing	39.7	3.8	High TRL
CH-53	$22,439	Helicopter	Sikorsky	5.7	32	High TRL
AGM 88E	$1,902	Missile	ATK missile systems	10.9	22.4	High TRL
Army integrated air and missile defense	$5,529	Missile	Northrup Gruman	9.9	1.3	High TRL

Table 18.2 (Continued)

Program name	Total program cost (M$)	Type of system	Primary contractor	% Cost overrun	% Schedule slip	Type of acquisitic
Joint land attack cruise missile defense	$7,858	Missile	Raytheon	18	6.2	Medium TRL
Standard Missile RAM	$6,297	Missile	Raytheon	10.5	25.3	Medium TRL
CVN 78	$33,994	Ship	Huntington Ingalls	−4.4	13.1	High TRL
DDG 1000	$20,985	Ship	RAE systems	543	73	Low TRL
Joint highspeed vessel	$3,674	Ship	Austral USA	1	4.2	High TRL
LHA replacement assault ship	$10,095	Ship	Huntington Ingalls	5.8	13	High TRL
LCS	$32,867	Ship	Lockheed Martin	76	183	Low TEL
GPS III	$4,210	Space system	Lockheed Martin	6.8	N/A	Medium TRL
Space-based IR system (SBIRS)	$18,266	Space system	Lockheed Martin	231.2	N/A	Low TEL

Complexity (disorganized and organized) affects cost and schedule overruns (Nilchiani et al. 2013).

References

Abbott, R. (2006). Emergence explained. *arXiv* preprint cs/0602045.

Bedau, M.A. (1997). Weak emergence. *Philosophical Perspectives* 11: 375–399.

Black, L.J. and Repenning, N.P. (2001). Why firefighting is never enough: preserving high-quality product development. *System Dynamics Review* 17 (1): 33–62. https://doi.org/https://doi.org/10.1002/sdr.205.

Callas, J.L., Golombek, M.P., and Fraeman, A.A. (2019). Mars Exploration Rover Opportunity End of Mission Report. Passadina, CA: Jet Propulsion Laboratory. *JPL-CL-19-7647*.

Chaisson, E.J. (2004). Complexity: an energetics agenda: energy as the motor of evolution. *Complexity* 9 (3): 14–21.

Chaisson, E. J. (2014). The natural science underlying big history. *The Scientific World Journal, 2014*. https://doi.org/10.1155/2014/384912

Chaisson, E.J. (2015). Energy flows in low-entropy complex systems. *Entropy* 17 (12): 8007–8018.

Chalmers, D.J. (2006). Strong and weak emergence. In: *The Re-emergence of Emergence*, 244–256. Oxford, England: Oxford University Press.

Checkland, P. (1981). *Systems Thinking, Systems Practice*. New York: Wiley.

Cilliers, P. (2000). What can we learn from a theory of complexity? *Emergence*, 2(1), 23–33. https://doi.org/10.1207/S15327000EM0201_03

Conrow, E.H. (2003). *Effective risk Management: Some Keys to Success*. Aiaa.

Cooper, K.G. (1994). The $2,000 hour: How managers influence project performance through the rework cycle. *Project Management Journal* 25: 11–11.

Cotler, J., Hunter-Jones, N., Liu, J., & Yoshida, B. (2017). Chaos, complexity, and random matrices. *Journal of High Energy Physics*, (11), 48. https://doi.org/10.1007/JHEP11(2017)048

DeRosa, J.K., Grisogono, A.M., Ryan, A.J., and Norman, D.O. (2008). A research agenda for the engineering of complex systems. *2008 2nd Annual IEEE Systems Conference*, Montreal, Canada (7–10 April 2008).

Érdi, P. (2008). *Complexity Explained*. Springer.

Fischi, J., Nilchiani, R., and Wade, J. (2015). Dynamic complexity measures for use in complexity-based system design. *IEEE Systems Journal* 11 (4): 2018–2027.

Ford, D. N., & White III, R. J. (2020). Social impact bonds: the goose and the golden eggs at risk. *Systems Research and Behavioral Science*, 37(2), 333–344. https://doi.org/10.1002/sres.2632

Gerla, D. J., Mooij, W. M., & Huisman, J. (2011). Photoinhibition and the assembly of light-limited phytoplankton communities. *Oikos*, 120(3), 359–368. https://doi.org/10.1111/j.1600-0706.2010.18573.x

Gutman, I., & Shao, J.-Y. (2011). The energy change of weighted graphs. Linear Algebra and Its Applications, 435(10), 2425–2431, 15 November 2011.

Haskins, C., Forsberg, K., Krueger, M. et al. (2006). *Systems Engineering Handbook*. INCOSE.

Hodgson, D., McDonald, J. L., & Hosken, D. J. (2015). What do you mean, 'resilient'? *Trends in Ecology & Evolution*, 30(9), 503–506. https://doi.org/10.1016/j.tree.2015.06.010

Hodgson, D., McDonald, J. L., & Hosken, D. J. (2016). Resilience is complicated, but comparable: a reply to Yeung and Richardson. *Trends in Ecology and Evolution*, 31(1), 3–4. https://doi.org/10.1016/j.tree.2015.11.003

Huntington, H. P., Goodstein, E., & Euskirchen, E. (2012). Towards a tipping point in responding to change: rising costs, fewer options for arctic and global societies. *AMBIO*, 41(1), 66–74. https://doi.org/10.1007/s13280-011-0226-5

Jansma, P. (2011). Open! Open! Open! Galileo high gain antenna anomaly workarounds. *2011 Aerospace Conference*, Big Sky, MT, USA (5–12 March 2011).

Kauffman, S.A. (1996). *At Home in the Universe*. Oxford University Press.

Kauffman, S. (2007). Beyond reductionism: reinventing the sacred. *Zygon®* 42 (4): 903–914.

Kiron, D., Kruschwitz, N., Haanaes, K., and Velken, I.V.S. (2012). Sustainability nears a tipping point. *MIT Sloan Management Review* 53 (2): 69–74.

Lamberson, P. and Page, S. (2012). Tipping points. *Quarterly Journal of Political Science* 7 (2): 175–208.

Lenton, T. M., Held, H., Kriegler, E., Hall, J. W., Lucht, W., Rahmstorf, S., & Schellnhuber, H. J. (2008). Tipping elements in the Earth's climate system. *Proceedings of the National Academy of Sciences of the United States of America*, 105(6), 1786–1793. https://doi.org/10.1073/pnas.0705414105

Leveson, N. (2004). A new accident model for engineering safer systems. *Safety Science* 42 (4): 237–270.

Longo, G., Montévil, M., and Kauffman, S. (2012). No entailing laws, but enablement in the evolution of the biosphere. *Proceedings of the 14th Annual Conference Companion on Genetic and Evolutionary Computation*, Philadelphia, PA, USA (7–11 July 2012).

MacCormack, A., Rusnak, J., and Baldwin, C.Y. (2006). Exploring the structure of complex software designs: an empirical study of open source and proprietary code. *Management Science* 52 (7): 1015–1030.

Marr, J.C. (1994). Performing the Galileo mission using the S-band low-gain antenna. *Proceedings of 1994 IEEE Aerospace Applications Conference Proceedings*, Vail, CO, USA (5–12 February 1994).

McCabe, T. J. (1976). A complexity measure. *IEEE Transactions on Software Engineering*, *SE*-2(4), 308–320. https://doi.org/10.1109/TSE.1976.233837

McManus, H. and Hastings, D. (2005). 3.4.1 A framework for understanding uncertainty and its mitigation and exploitation in complex systems. *INCOSE International Symposium*, Rochester, NY, USA (13–16 June 2005).

Milkoreit, M., Hodbod, J., Baggio, J., Benessaiah, K., Calderón-Contreras, R., Donges, J. F., Mathias, J.-D., Rocha, J. C., Schoon, M., & Werners, S. E. (2018). Defining tipping points for social-ecological systems scholarship – an interdisciplinary literature review. *Environmental Research Letters*, 13(3), 033005. https://doi.org/10.1088/1748-9326/aaaa75

van Nes, E. H., Arani, B. M. S., Staal, A., van der Bolt, B., Flores, B. M., Bathiany, S., & Scheffer, M. (2016). What do you mean, 'tipping point'? *Trends in Ecology & Evolution*, 31(12), 902–904. https://doi.org/10.1016/j.tree.2016.09.011

Nikiforov, V. (2007). The energy of graphs and matrices. *Journal of Mathematical Analysis and Applications* 326 (2): 1472–1475.

Nilchiani, R.R. and Pugliese, A. (2017). A systems complexity-based assessment of risk in acquisition and development programs. *Naval Postgraduate School (NPS) Annual Acquisition Research Symposium*, Monterey, CA, USA, (26–27 April 2017).

Nilchiani, R., Rifkin, S., Mostashari, A., et al. (2013). Quantitative Risk-Phases 1 & 2. *A013-Final Technical Report SERC-2013-TR-040-3. Stevens Institute of Technology. Systems Engineering Research Center.*

Page, S.E. (1999). Computational models from A to Z. *Complexity* 5 (1): 35–41.

Phelan, S. E. (2001). What is complexity science, really? *Emergence*, 3(1), 120–136. https://doi.org/10.1207/S15327000EM0301_08

Pugliese, A. and Nilchiani, R. (2019). Developing spectral structural complexity metrics. *IEEE Systems Journal* 13 (4): 3619–3626.

Rasoulkhani, K., & Mostafavi, A. (2018). Resilience as an emergent property of human-infrastructure dynamics: a multi-agent simulation model for characterizing regime shifts and tipping point behaviors in infrastructure systems. *PLoS One*, 13(11), e0207674. https://doi.org/10.1371/journal.pone.0207674

Repenning, N. P. (2001). Understanding fire fighting in new product development. *Journal of Product Innovation Management*, 18(5), 285–300. https://doi.org/10.1111/1540-5885.1850285

Repenning, N. P., Gonçalves, P., & Black, L. J. (2001). Past the tipping point: the persistence of firefighting in product development. *California Management Review*, 43(4), 44–63. https://doi.org/10.2307/41166100

Rietkerk, M., Bastiaansen, R., Banerjee, S., van de Koppel, J., Baudena, M., & Doelman, A. (2021). Evasion of tipping in complex systems through spatial pattern formation. *Science*, 374(6564), eabj0359. https://doi.org/10.1126/science.abj0359

Russill, C., & Nyssa, Z. (2009). The tipping point trend in climate change communication. *Global Environmental Change*, 19(3), 336–344. https://doi.org/10.1016/j.gloenvcha.2009.04.001

Salado, A. and Nilchiani, R. (2014). The concept of problem complexity. *Procedia Computer Science* 28: 539–546.

Salado, A., & Nilchiani, R. (2015). A set of heuristics to support early identification of conflicting requirements. *Proceedings of the INCOSE International Symposium*, Seattle, WA, USA (13–16 July 2015).

Scheffer, M. (2010). Foreseeing tipping points. *Nature*, 467(7314), 411–412. https://doi.org/10.1038/467411a

Scheffer, M., Bascompte, J., Brock, W. A., Brovkin, V., Carpenter, S. R., Dakos, V., Held, H., van Nes, E. H., Rietkerk, M., & Sugihara, G. (2009). Early-warning signals for critical transitions. *Nature*, 461(7260), 53–59. https://doi.org/10.1038/nature08227

Shannon, C. E. (1948). A mathematical theory of communication. *The Bell System Technical Journal*, 27(3), 379–423. https://doi.org/10.1002/j.1538-7305.1948.tb01338.x

Simpleman, L., McMahon, P., Bahnmaier, B. et al. (2003). *Risk Management Guide for DOD Acquisition*, (*Version 2.0*. Fort Belvoir, VA, USA: US Department of Defense, Defense Acquisition University.

Sinha, K. and de Weck, O. (2012). Structural complexity metric for engineered complex systems and its application. In: *Gain Competitive Advantage by Managing Complexity* (ed. M. Onishi, M. Maurer, K. Eben, and U. Lindemann), 181–192. Munich, Germany: Carl Hanser Verlag GmbH & Company.

Sinha, K., and de Weck, O. L. (2013). A network-based structural complexity metric for engineered complex systems. *Proceedings of the 2013 IEEE International Systems Conference (SysCon)*, Orlando, FL, USA, (15–18 April 2013).

Snowden, D.J. and Boone, M.E. (2007). A leader's framework for decision making. *Harvard Business Review* 85 (11): 68.

Solé, R. (2011). *Phase Transitions*, vol. 3. Princeton University Press.

Sterman, J.S. (2000). *Business Dynamics: Systems Thinking and Modeling for a Complex World*. McGraw-Hill.

Strogatz, S.H. (2003). *SYNC: the Emerging Science of Spontaneous Order*. Hyperion.

Taylor, T., & Ford, D. N. (2006). Tipping point failure and robustness in single development projects. *System Dynamics Review*, 22(1), 51–71. https://doi.org/10.1002/sdr.330

Taylor, T.R.B. and Ford, D.N. (2008). Managing tipping point dynamics in complex construction projects. *Journal of Construction Engineering and Management* 134 (6): 421–431. https://doi.org/doi:10.10.1061/(ASCE)0733-9364(2008)134:6(421)1061/(ASCE)0733-9364(2008)134:6(421).

US Department of Defense (Office of the Undersecretary of Defense Acquisition, T. L.) (2006). *Risk Management Guide for DoD Acquisition, 6e, ver. 1.0)*. http://www.acq.osd.mil/se/docs/2006-RM-Guide-4Aug06-final-version.pdf.

Valerdi, R. and Boehm, B.W. (2010). COSYSMO: A Systems Engineering Cost Model. *Software Engineering*, March 2010.

Veraart, A. J., Faassen, E. J., Dakos, V., van Nes, E. H., Lürling, M., & Scheffer, M. (2012). Recovery rates reflect distance to a tipping point in a living system. *Nature*, 481(7381), 357–359. https://doi.org/10.1038/nature10723

Wade, J. and Heydari, B. (2014). Complexity: definition and reduction techniques. *Proceedings of the Poster Workshop at the 2014 Complex Systems Design and Management International Conference*, Paris, France (12 November 2014).

Walden, D.D., Roedler, G.J., and Forsberg, K. (2015). INCOSE systems engineering handbook version 4: Updating the reference for practitioners. *INCOSE International Symposium*.

Weaver, W. (1948). Science and complexity. *American Scientist* 36: 536–544.

Willcox, K., Allaire, D., Deyst, J., et al. (2011). Stochastic Process Decision Methods for Complex-Cyber-Physical Systems. Cambridge, MA: Massachusetts Institute of Technology (MIT). *AFRL-RZ-WP-TR-2011-2094*.

Chapter 19

Technical Debt in the Engineering of Complex Systems

Ye Yang[1] *and Dinesh Verma*[2]

[1] *Amazon.com Inc., New York, NY, USA*
[2] *Stevens Institute of Technology, School of Systems and Enterprises, Hoboken, NJ, USA*

Introduction

We are living in an era where modern technologies have significantly impacted every aspect of human society. In the past, technological advancements were relatively simple and straightforward, primarily focusing on solving practical problems. In recent decades, systems have become increasingly complex and sophisticated, with a large number of interconnected components collaboratively serving for a long life span and have faced enormous challenges with high stakes to safety, security, and fairness of human life and society impact.

Examples of such systems include data-intensive distributed systems, COTS-intensive cyber-physical systems, and the innovative applications of artificial intelligence (AI) and machine learning (ML) in domains across finance, defense, healthcare, transportation, etc. Such systems have the ability to play crucial roles in shaping and revolutionizing the way we live and work. The stake of design in such systems is extremely high. It is critical to consider their economic, societal, and ethical implications and to ensure such systems are used in a responsible and sustainable manner.

The metaphor of technical debt (TD) stems from and is widely adopted in the software engineering field, referring to short-term compromises in engineering decisions and artifacts in exchange for development speed or other constraints. While the metaphor is relatively new to the systems engineering field, the analogy is generally applicable. In particular, it is highly relevant to the context of engineering complex systems, concerning numerous program cancellation and obsolescence challenges due to premature decisions made in early acquisition phases. In this chapter, we adapt the metaphor, concepts, and taxonomies of TD, and discuss strategies and tactics to cope with it in the context of engineering complex systems.

What Is Technical Debt?

In agile development, rapid delivery of working software takes precedence over almost anything else. The term "technical debt" was coined by Ward Cunningham (1993), generally referring to the implied cost of future rework caused by short-term solutions for the sake of rapid feature rollout.

Systems Engineering for the Digital Age: Practitioner Perspectives, First Edition. Edited by Dinesh Verma.

Companion website: www.wiley.com/go/verma/systemsengineering

Originally, Ward used a financial analogy to justify the refactoring work to the stakeholders of a financial application (Letouzey and Whelan 2016):

> Shipping first-time code is like going into debt. A little debt speeds development so long as it is paid back promptly with refactoring. The danger occurs when the debt is not repaid. Every minute spent on code that is not quite right for the programming task of the moment counts as interest on that debt. Entire engineering organizations can be brought to a stand-still under the debt load of an unfactored implementation, object-oriented or otherwise.

Unavoidably, agile developers must make frequent design shortcuts, either unconsciously or intentionally, to meet the timebox-based development cadences. If TD is introduced unconsciously, it is typically an indicator of bad practices due to careless mistakes or lack of knowledge and experience. If TD is introduced intentionally, developers acknowledge that additional maintenance work will need to be planned later on to take care of the short-term compromises. Consequently, the TD metaphor became an effective notion to communicate, with especially nontechnical stakeholders, why additional resources are needed to be budgeted for software refactoring after the initial roll out.

In the early days, it was primarily referred to as "code debt," which requires refactoring inherent to agile development. For example, one of the early code smells is duplicated code. Duplicated code is the result of commonly practiced "Copy and Paste" behaviors, and it could be very expensive and error-prone to maintain if a future change needs to be made to the duplicated code. Over the years, the scope of "technical debt" has evolved and broadened considerably, encompassing a wide range of concepts, artifacts, and processes in the software engineering discipline. The following sections provide examples of broad software technical debt.

Software Technical Debt Landscape

Kruchten et al. proposed and evolved a widely adopted TD Landscape (Kruchten et al. 2012, 2019) to help differentiate and categorize various software TD. Kruchten et al.'s TD Landscape primarily examines TD from user-invisible software aspects that could potentially impact software systems' evolution. As illustrated in Figure 19.1, it differentiates software TD from other user-visible

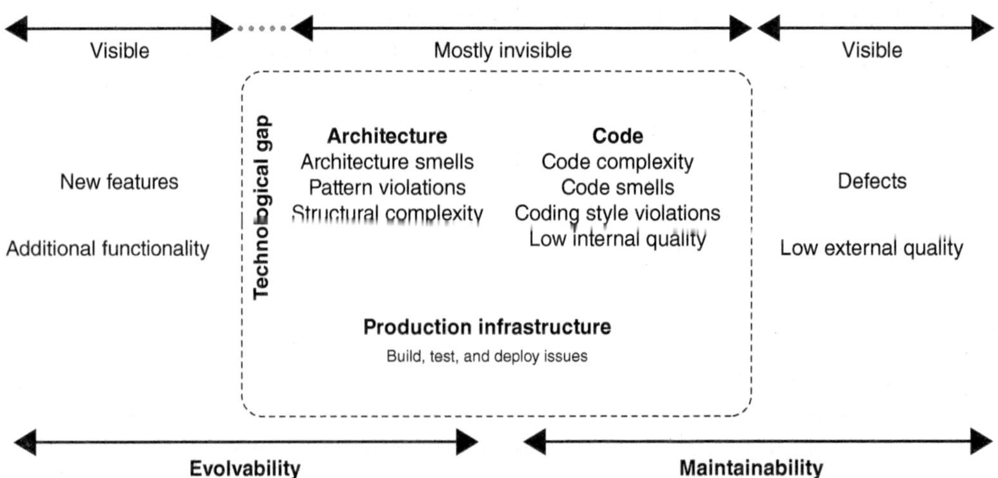

Figure 19.1 Kruchten et al.'s TD landscape. *Source:* Adapted from Kruchten et al. (2019).

elements such as new features, additional functionalities, and defects to fix. This landscape characterizes various forms of micro-level technological gaps that potentially introduce TD during development. For example, an "architecture" TD may consist of bad architecture smells, violations of known design patterns, or high structural complexity. "Code" TD may be associated with static or dynamic software code metrics such as code complexity, code smells, coding style violations, low internal quality, etc. "Production infrastructure" TD frequently refers to sub-optimal design choices or issues rooted in the underlying software build, test, and deployment systems, tools, and platforms.

Dynamism of Software Technical Debt

Analogous to monetary debt, TD accumulates over time. Think about the dynamism of TD: with respect to the accumulation of future developmental effort/cost obligations in removing or mitigating the unresolved or unaddressed technical issues. This implies, over time, a greater scope or an increased difficulty of the required rework, in various forms of upgrades, maintenance, and other sustainment activities.

TD dynamism emphasizes the financial or ownership obligations of the engineering team, so that when a design or developmental shortcut is taken, everybody on the team is aware of this newly constituted "debt," as it can create vulnerabilities or weaknesses that may be exploited. For simple, standalone systems, example technical debts such as design flaws, coding errors, or the use of to-be-outdated technologies, may be acceptable, although at the cost of a decrease in the overall reliability and stability of the system. In complex systems, TDs could accumulate at a very high rate or in unexpected manners, which would eventually make rework or maintenance very expensive or even impossible. This corresponds to cases where a TD is associated with a fundamental data model that has been baked in various ways into how things in the overall system work, or TD inherent with an underlying infrastructure component that many parts of the system have built on top of it across the system's lifetime.

Any TD actually sets up a technical context that can make a future change more costly or impossible. Over time, software practitioners learn from expensive TD rework lessons and advocate the importance of increased awareness and attention to TD. For example, at Riot Games (Clark 2018), the company which developed the PC game entitled "League of Legends," Bill Clark summarized a measurement framework following three major axes to evaluate TD: impact, fixed cost, and contagion. The impact axis measures player-facing issues (e.g., bugs, missing features, etc.), and developer-facing issues (e.g., workflow issues, random elements). The fixed cost axis measures the cost to fix and deploy a TD to production. The contagion axis measures the extent to which the technical debt will spread across the system if allowed to continue to exist. Clark's framework also highlights two types of TD that lie deep in the heart of the Riot system, those rooted in the "Foundation" of the technical stack (Foundation TD), as well as those baked in the Data models (Data TD). Empirically, Clark observed that it is sometimes hard to recognize for experienced users of a system because it's seen as "just the way it is." As time elapses, fixing the initial TD becomes extremely risky and difficult.

Technical Debt: Reality to Live With

Before proceeding to discuss the relevance of TD in the context of systems engineering, it is important to highlight several general viewpoints of TD that have evolved from software engineering practices centered on agile software development.

The presence of TD is *Inevitable* due to system complexity and uncertainty (Martini et al. 2015; Rubin 2012). If you ask anyone who has worked in or with software industry, it is highly likely that he or she has dealt with or heard a good deal about technical debt. Technical debt can occur for a variety of reasons. It may be the result of shortcuts taken during the development process in order to meet a tight deadline or to get a product to market quickly. It may also be the result of a lack of resources or expertise or a lack of understanding of the long-term consequences of certain design or implementation decisions. Examples are the unpredictable need to evolve a design over time as we gain more information and the unpreventable third-party component (e.g., COTS, open source, etc.) interface changes.

In some cases, TD is even desirable with calculated risk analysis and planning (Rubin 2012; Besker et al. 2018). Ken Rubin (2012) categorizes such TD as *Strategic* as a tool to help organizations better quantify and leverage the economics of important, often time-sensitive, decisions. Examples are strategic decisions to take shortcuts during product development to achieve an important short-term goal, e.g., meeting a time-to-market deadline. In our own experience from an artificial intelligence/machine learning (AI/ML)-based cloud service, we frequently need to act quickly to fix customer quality issues. However, it takes much longer to wait for better, long-term fixes because this would require the team to work on and train new capabilities into underlying AI/ML models. Therefore, it is often preferable to explore short-term engineering solutions to meet customer expectations.

Nonetheless, the presence of *Naïve* TD (Rubin 2012) should and can be avoided through training and process improvement. Naïve TD typically refers to irresponsible behaviors or immature practices of the people involved, e.g., engineers, business people, and so on. Example causes of naïve TD are sloppy design, poor engineering practices, and insufficient testing, as well as over-constrained project date, scope, and budget objectives. Many of these naïve TD can be surfaced and avoided if there are various quality assurance mechanisms in place, such as team design/coding standards, effective code review processes, a continuous integration and continuous deployment pipeline, retrospective meeting, and so on. For example, a thorough review of the design and code of the system can help to identify areas where technical debt may be accumulating. This can involve looking for poorly designed or implemented components, a lack of documentation or unit tests, and other issues that may make the system more difficult to maintain or enhance.

Adapting TD to the Engineering of Complex Systems

Architecture decisions are important choices made during the design and development of complex systems. Systems engineers are responsible for ensuring that systems are designed, developed, and maintained in a way that minimizes risk and maximizes efficiency and effectiveness. It is possible for systems engineering teams to learn from software engineering teams and actively manage TD and address issues as they arise in order to minimize the accumulation of issues that may result in future negative impacts.

While this involves implementing new processes or practices to identify and prevent the accumulation of technical debt, some may argue that such processes and practices could vary from domain to domain. In the remainder of this chapter, we apply the TD analogy to the engineering of COTS-intensive complex systems (Bhuta and Boehm 2007; Garlan et al. 1995; Clark and Clark 2007) as an example and introduce a TD Taxonomy for the engineering of complex systems.

This example context is chosen for convenience, as our previous work (Alelyani et al. 2019; Yang et al. 2019) demonstrated that there is a compelling need for additional metrics and measurement to enhance the understanding of, communication about, and analyzing and predicting of the life-cycle consequences incurred by the use of COTS components. This is particularly significant to mitigate the risk of early design decisions that may lead to integration and obsolescence issues later in a systems life cycle. Understanding the impact of COTS decisions through the lens of TD dynamism would potentially suggest alternative approaches to examine the life cycle ownership of many COTS-related risks and challenges, in order to more effectively plan and address potentially the many facets of potential obsolescence issues in designing, developing, maintain, and sustaining complex systems.

Impact of COTS Usage and Obsolescence Crisis

The COTS obsolescence crisis is a vivid illustration of neglected COTS TD that has not been repaid, due to the general separation of acquisition and engineering ecosystems. As acquisition life cycles can sometimes be measured in decades, engineering and program leadership decisions to rapidly acquire COTS solutions would potentially incur more future "technical debt" but most of these TD may be overlooked or considered negligible given potential ownership transitions across a system's life cycle with different organizational entities over its life span. Consequently, increased maintenance cost or unexpected cost of re-development is discovered by different engineering teams or downstream sustaining organizations. Complex system engineering teams need to ensure COTS TD visibility and accountability from its onset, as its liability is paid off in later systems maintenance or sustainment phases.

Here are some ways in which COTS-related architecture decisions can impact technical debt:

1) **Complex architecture**: A complex architecture can make it more difficult to understand and maintain a system, leading to an accumulation of technical debt over time. This is because a complex architecture can make it more difficult to identify and address issues in the system and can also make it more difficult to add new features or capabilities to the system.
2) **Lack of modularity**: A lack of modularity in the system architecture can also lead to an accumulation of technical debt. This is because a non-modular system can be more difficult to modify or update, as changes may have unintended consequences on other parts of the system.
3) **Lack of scalability**: A system that is not designed to be scalable may become more difficult to maintain and enhance over time as the system grows and changes. This can lead to an accumulation of technical debt as the system becomes increasingly complex and difficult to modify.
4) **Inefficient use of resources**: Poorly designed systems may use resources inefficiently, leading to an accumulation of technical debt as more resources are required to maintain and update the system.

For example, many U.S. Department of Defense (DoD) missions require arrays of COTS components, both hardware and software, that were not designed to be in an array. The frequent upgrading of COTS components (e.g., 18 months) is one of the root causes for many obsolescence headaches for systems whose lifetimes are typically more than 30 years. For sustainment-dominated systems, their life spans (e.g., over 20 or more years) are long enough that a significant portion of the COTS components becomes obsolete prior to the system even being deployed. It is reported that over 70% of COTS electronics become un-procurable by the time a system is fielded (Federal Acquisition Regulation 2019). As another example, the former Future Combat System had 153 relevant systems to deal with. If every COTS component was updated once a year, that would be a change every other day! In order to achieve the expected economic benefits of COTS procurement, it is critical to enable program managers to better understand obsolescence costs in order to make informed COTS commitment decisions. Clark and Clark (2007) interviewed 25 project managers on reasons why COTS-intensive systems are so difficult to maintain and concluded 11 factors have impacts on the true cost of COTS maintenance, including licensing, evaluation of new releases, defect hunting, vendor support, and so on.

It Is important to carefully consider the potential impact of architecture decisions on the technical debt of a system in order to minimize the negative impacts of technical debt over time.

These known, or perhaps unknown, costs in lower-performing organizations can be considered "organizational" debts which will need to be repaid later. A few exemplar forms for such debt repayment include planned systems upgrades, systems replacement costs, or in the worst case, defaulted systems. These unaccounted-for programmatic costs may have very negative consequences which necessitate funding shifts amongst competing systems developments, to the unforeseen detriment of other planned systems modernization programs. Conversely, a solution more collectively beneficial to the organization may be properly planned for, efficiently and effectively vetted, technology or system introduction with an outcome requiring substantially fewer resources to maintain operational reliability, availability, and maintainability over the system's life span – thus incurring less TD.

Taxonomy of COTS TD

To facilitate the discussion and communication on TD, different TD types have been described according to specific technical issues where the debt metaphor applies, e.g., architectural debt, requirements debt, and test debt, etc. Instances of such types vary considerably in nature, from suboptimal reuse of architectural components to deferred testing or the involvement of certain development communities (Bogner et al. 2021). Table 19.1 summarizes a list of COTS TD descriptions, adapted from existing work on categorizing TDs from Kruchten et al. (2012), Clark (2018),

Table 19.1 The taxonomy of TD in the engineering of complex systems.

TD category	Description	Rationale
Functionality	The degree of functionality mismatch between COTS and system needs	Hidden cost and risk due to the need to modify and extend COTS to meet specific system needs; to mask unwanted COTS features for security concerns
Performance	The degree of mismatches between COTS and system needs, w.r.t. performance properties	Hidden cost and risk due to increased need of system modeling and simulation analysis, testing to verify reliability, security, performance, etc.
Interoperability	The degree of system modularity, COTS interface complexity, and interoperability among COTS and custom components	Incurred cost and risk for later integration activities due to integration challenges. The more complex the COTS interfaces, or the higher degree of inter-dependencies between components, the more difficult and more expensive for rework
Version Conflict	Configuration management planning needs to address COTS refreshing strategy, version conflicts, as well as solution availability plan	The more COTS elements, the higher frequency of COTS new releasing, the greater risk of this TD
Documentation Support	The availability of COTS documentation and vendor support	Lack of documentation and vendor support will seriously impact on issue resolution related to obsolete COTS
System Evolution	The impact of COTS usage on long-term system evolvability to leverage innovative technologies	Requirements and design decisions imposed by COTS may place great limitation on system evolution
Organic	Policy, people, process-centric perspective of TD	Focusing on organizational decision-making, behaviors, and practices associated with personnel responsible for introductions of new technologies

and Rubin (2012). It consists of seven types, characterized according to risk indicators during early acquisition phase (e.g., identification, assessment, tailoring, etc.), that may contribute to obsolescence in later life cycles (e.g., maintenance, operation, etc.). The objectives of the COTS TD categorization is to facilitate the early identification and management of COTS TD risks and to mitigate later obsolescence risks. We will briefly introduce each of the seven types next.

1) **Functionality TD** refers to the mismatch between the desired system functional needs and what is delivered by the proposed COTS alternative, along with the excess functionality delivered by COTS that is not necessary. The former will necessitate COTS modification and custom development to bridge the functionality gap, and the latter will require additional evaluation and/or wrapper effort to disable/mask unnecessary COTS capabilities for security assurance purposes.
2) **Performance TD** refers to the mismatches between COTS capabilities and system needs with respect to quality/extra-functional properties such as reliability (e.g., mainly of hardware), safety assurance (e.g., of software and hardware), and performance in terms of bandwidth, processing capability, memory, etc. This is particularly significant for cases involving the use of COTS products developed for certain contexts but applied in new contexts with newly expected capabilities that are ultimately found to lack the desired qualities.
3) **Interoperability TD** refers to incompatible or mismatched architectural assumptions (Bhuta and Boehm 2007; Garlan et al. 1995) among multiple COTS and other components in a complex system. Each COTS comes with a set of assumptions with respect to other components or the physical environment, and insufficient COTS assessment/evaluation during the acquisition phase may lead to pre-mature COTS commitment containing undiscovered assumption mismatches that may cause COTS integration difficulty and/or obsolete components in later phases. In general, COTS interoperability risks can be categorized into three major interoperability mismatches, namely interface mismatch, internal assumption mismatch, and dependency mismatch (Bhuta and Boehm 2007).
4) **Version Conflict TD** refers to imposed requirements for upgrading COTS components with respect to its newly released versions by the commercial vendors. Commercial technology vendors typically update their products very frequently (i.e., every 6–12 months), forcing integrators to reevaluate or reengineer various aspects of the systems in order to maintain compatibility and interoperability with current technologies (Kruchten et al. 2019). Keeping up with newer versions of every COTS component is often cumbersome, expensive, and infeasible for developing and maintaining complex systems.
5) **Documentation and Support TD** refers to issues due to unavailable, insufficient, or obsolete documentation or vendor support, especially in the face of maintaining and supporting complex systems during the operation life cycle. In general, most vendors are not supportive of having their product modified and usually do not support more than two previous releases. As COTS components become the primary focus of integration efforts for development and sustainment, such systems require maintenance and support that exceeds typical COTS vendor support.
6) **System Evolution TD** refers to the inflexibility for accommodating emerging new system functionalities that are required but are out of the initial scope. System requirements imposed by COTS products may place great limitations on system evolution over time. In COTS-based systems, stakeholders may introduce some changes that may contribute to the problem of obsolescence. Candidate COTS-based solutions need to be thoroughly evaluated for imposed architectural sustainability and evolvability limitations.
7) **Organic TD** refers to any combination and degree of technological, systemic, project, and program decisions, behaviors, and practices made by the workforce, management, and/or

senior/executive leadership of the organization responsible for introductions of new technologies and systems and/or the sustainment of existing systems. This category assumes that it is possible to create a framework and leadership decision cycle to enable the capability to streamline potentially overbearing acquisition processes while focusing on core critical TD management, processes, and tools that affect systems sustainment supportability, reliability, availability, maintainability, and cost.

Managing Your TD

In the software engineering field, many researchers have looked into varying strategies, processes, factors, and tools to identify (Kruchten et al. 2012; Clark 2018; Clark and Clark 2007; Sierra et al. 2019) and manage TD in software development (Saraiva et al. 2021; Lenarduzzi et al. 2021; AlOmar et al. 2022; Li et al. 2015). Recent empirical studies also report best practices and antipatterns of TD instances in AI-based systems (Bogner et al. 2021) and ML systems (Tang et al. 2021) in facilitating long-term system usefulness. In general, TD management (TDM) corresponds to a set of activities that prevent potential TD from being incurred or deal with existing TD to keep it at an acceptable level, such as identification, measurement, prioritization, repayment, and so on (Li et al. 2015). Theoretically, managing TD involves regularly reviewing and addressing technical issues, as well as making informed decisions about trade-offs between short-term and long-term goals. By actively managing technical debt, it is possible to reduce its negative impacts and improve the overall health and stability of a software system. In Agile teams, TD has been increasingly used as a metaphor to communicate and manage design trade-offs with both technical and nontechnical stakeholders. However, in most teams, refactorings are often overlooked in prioritization and they are often triggered by development crises in a reactive fashion (Martini et al. 2015). Some of the factors are manageable, while others are external to the companies.

With respect to the engineering of complex systems, TDM activities are even more complicated, due to greater complexity, inter-dependencies among system components, and many more external factors beyond any single team's control. Many cross-cutting systems issues make it difficult to identify and manage TD due to cross-organizational boundaries. Appropriate guidelines are necessary to clarify possible confusion and highlight cross-cutting issues while identifying, classifying, measuring, and managing TD.

COTS TD Management Activities

This section provides high-level guidelines for applying the proposed TD taxonomy to identify and manage TD items at early system life cycle phases, e.g., the research or capability definition phase. Note that stakeholders from later life cycle phases, i.e., users, customers, sustaining organizations, etc., are also involved in the activities in order to facilitate shared accountability among all success-critical stakeholders. This is to ensure that the TD liability with respect to the selected solutions in the acquisition phase can be strategically paid off in later maintenance or sustainment phases.

TD Identification This activity should be conducted in alignment with and based on results from COTS assessment, modeling and simulation, prototyping, dependency analysis, etc. An example template for representing and documenting COTS TD items is provided in the following subsection. COTS TD identification may follow a bottom-up approach, starting from individual COTS candidates. There can be a many-to-many relationship between a COTS candidate and a COTS TD item, and it is possible for a COTS TD item to be associated with multiple TD types (i.e., those outlined in Table 19.1), since intensive COTS TD items in complex systems may come from the complex interdependencies among COTS hardware and software components. The COTS TD items should be frequently reviewed for updates, and synchronization across collaborative teams, in order to mitigate potential conflicts or mismatches in underlying technical assumptions.

TD Measurement This activity quantifies the benefit and cost of known COTS TD in a system through estimation techniques. This activity integrates techniques such as expert estimation, estimation models, cost categorization, solution comparison, and modeling and simulation (Li et al. 2015). More specifically, whenever a COTS TD item is identified the function, performance, and interoperability TD items need to be measured and based on intensive COTS assessment and testing results. We will provide a top-down approach for TD measurement in the following subsection.

TD Prioritization This activity ranks identified COTS TD items according to predefined rules. Sample techniques to support this activity include trade studies and cost-benefit analysis (Li et al. 2015). More specifically, this activity should be included in major milestone reviews and involve all success-critical stakeholders as early as possible. Usually, not all COTS TD can or should be repaid. Deficiencies and shortfalls almost always exist relative to needs and requirements given that resources are limited and needs are often pressing. Thus, proactive (not reactive) prioritization is essential to inform future actions (or inactions).

TD Prevention This activity aims to prevent certain COTS TD from being incurred. Process improvement, design decision support, lifecycle cost planning, early testing, and human factors analysis are techniques that support TD prevention (Li et al. 2015). More specifically, training programs, training COTS assessment teams to reduce or avoid TD items from the first three categories (i.e., functionality, performance, and interoperability), and exploring more supportive contracting options with COTS vendors to support COTS maintenance and system evolution to the greatest possible extent are ways to help prevent Organic TD.

TD Monitoring This activity watches the change of cost and benefit of unresolved COTS TD over time. Sample techniques are threshold-based, planned check, and COTS TD propagation tracking. More specifically, this is particularly critical to the Configuration Version TD type, in order to develop and employ adaptive repayment strategies (see the guidelines below); and visualization techniques may be leveraged to facilitate TD tracking.

TD Repayment This activity resolves, reduces, or mitigates COTS TD through sample strategies or activities such as COTS version upgrade, re-engineering, refactoring, incident fixing, fault tolerance, repackaging, and automation. It is important to establish COTS TD repayment strategies, with respect to particular COTS TD types. For instance, strategies for mitigating configuration version TD might include the following options: (1) skipping the new COTS version; (2) upgrading to keep up with every new COTS version; (3) upgrading COTS every other version; and (4) upgrading on a regular basis, e.g., every 18-month. As discussed under TD prioritization, not everything can (or should) be repaid.

TD Representation/Documentation This activity provides ways to represent and communicate COTS TD in a standard manner. A sample format/template for representing TD items is provided in the following subsection. It is also essential to tailor, train, and promote the usage of specific, uniform COTS TD representation to facilitate communication and management.

TD Communication This activity makes identified COTS TD visible to stakeholders through a COTS TD dashboard, backlog, dependency visualization, COTS TD propagation visualization, etc. (Li et al. 2015). Such communication should be included in major milestone reviews, and explore visualization techniques to improve awareness and visibility of COTS TD.

Among all TDM activities, the TD representation/document and prioritization are the most fundamental, while TD identification, measurement, and repayment are most important for discovering the location, significance, and developing strategies to deal with a TD item. More work is needed to further improve this TD taxonomy and for guidelines to be more adaptive for use in the complex systems context.

Representing COTS TD

From existing COTS-related literature, we identified a list of essential metrics for documenting and maintaining identified COTS TD items. This includes 15 metrics grouped into the four categories below.

- **General:** This includes six metrics:
  - **(1) ID**: A unique identifier for the COTS TD item;
  - **(2) Name**: The name of a specific COTS TD item;
  - **(3) Location**: The location of the identified TD item, e.g., the name of the COTS(s) involved;
  - **(4) Type**: The TD type(s) that the TD item is classified into (See the proposed Taxonomy in the previous section);
  - **(5) Description**: The general information on the COTS TD item; and
  - **(6) Open date/time**: The specific date/time when the COTS TD is identified.
- **Measurement:** This includes three metrics:
  - **(1) Principle**: The estimated cost of repaying the COTS TD item;
  - **(2) Interest amount**: The estimated extra cost of repaying the COTS TD item in the future rather than now, typically spent on maintenance due to quality issues of the system; and
  - **(3) Debt probability**: The probability that the COTS TD item needs to be repaid (rather than tolerated or accepted for the life of the system).
- **Contagion:** This includes two metrics:
  - **(1) Contagion**: The degree of spreading of the TD effects through interfaces to other system components, if this TD is allowed to continue to exist, e.g., using a nominal scale of 0–10; and
  - **(2) Propagation rule**: Rules specifying how the TD item impacts the related parts of the complex system, e.g., derived from the dependence graph of system components.
- **Management:** This includes four metrics:
  - **(1) Context**: The context within the system of a specific COTS TD item;
  - **(2) Mitigation strategy**: The strategies or actions for mitigating a specific COTS TD item;
  - **(3) Accountable party**: The party responsible to repay the COTS TD item, e.g., COTS vendor, integration team, program office, or other specific organization (if the TD is to be repaid at some point). It identifies the corresponding "accountable" debt-holder for an identified COTS TD liability; and
  - **(4) Intentionality**: Whether the COTS TD item intentionally or unintentionally incurred.

This template is intended for usage in early system life cycle phases, e.g., design and development of complex systems. The earlier this information is specified, e.g., at the start of a new design/development/modernization effort, the more feasible it is to appropriately assign TD monitoring and repayment responsibility within a reasonable span of authority/control. This corresponds to a possible means to offset "obsolescence" cost through more informed early design decision-making, as well as more pro-active TD management across all system lifecycle phases.

Measuring COTS TD

Technical debt can introduce additional risk into a system, as it can create significant cost overhead, vulnerabilities, or weaknesses that may be exploited, leading to a decrease in the overall affordability, reliability, and sustainability of the system. Many practitioners want tools and mechanisms to measure the TD for their projects. However, it is recognized that measuring TD isn't easy because its impact is not uniform.

In general sizing or estimating research, there are many approaches and tools, from top-down to bottom up. While it is important to follow bottom-up approaches when identifying individual COTS TD items, *the key to measure COTS TD, in our opinion, is to measure the cumulative effect of interacting COTS TD items on the lifecycle cost over time, and the ability to measure this at early acquisition phases.* To that end, we introduce a risk-based framework and mechanisms to measure COTS TD at an early phase to support the monitoring and tracking of TD dynamism over time. This framework consists of three components: the principal, the interest, and the interest probability of the TD.

Measuring TD Principal A recent study reported that the cost of managing TD in large software organizations is estimated to be, on average, 25% of the whole development cost (Martini et al. 2018). Hence, as a starting point if no historical data is available, it is reasonable to apply a simple assumption of a small percentage of the COTS acquisition cost as the TD principal. Furthermore, if a finer model is needed, we recommend an adapted, empirically based model for measuring and quantifying TD (Nugroho et al. 2011), as shown in Eq. (19.1):

$$\text{TD Principal} = \text{RF} * \text{RV} * \text{RA} \tag{19.1}$$

Where rework fraction (RF) can be assessed based on the percentage of COTS changes and/or the percentage of required work to re-qualify COTS to improve the system quality; rebuild value (RV) can be assessed based on the required effort of repairing, changing, or re-qualifying COTS to meet system expectations; and refactor adjustment (RA) can be assessed based on the engineering team's experience and use of tools to improve the productivity of system refactoring or rework.

For simplicity, each construct in Eq. (19.1) may be assessed employing the expert judgment technique. If a parametric model is needed, algorithmic cost estimating models such as COCOTS (Abts et al. 2000) can be utilized to establish a consistent framework for estimating the required effort/cost to modify and/or re-qualify COTS through tailoring, integration, and assessment in different situations or circumstances.

Measuring TD Interest In the engineering of complex systems, quantifying TD interest is more complicated since different TD items correspond with different maintenance activities. In general, the TD interest should take into consideration multiple corresponding hierarchical levels beyond the involved COTS component, as well as the severity and likelihood if the TD risk occurs. For example, ignoring certain COTS changes at the component level may require future rework in other COTS within its hosting physical unit or in other components in other physical units across the entire system. Considering a particular type of COTS TD, i.e., version conflict TD, the maintenance should involve determining COTS upgrading strategies, e.g., whether to upgrade COTS every other release, or to coordinate system component refresh every year and risk COTS obsolescence. Then, for each strategy, maintenance effort may be assessed accordingly.

Quantifying TD Interest Probability The TD interest probability is typically correlated with another TD attribute, i.e., contagion. Presumably, the more contagious the TD is, the higher its interest probability could be since the TD will accumulate at a faster pace and/or to a broader extent. However, it is very challenging and risky to come up with a one-size-fits-all metric for this. Based on our experience in parametric cost modeling, we recommend some *example* indicators, including: (1) system-level factors that measure the degree of modularity, i.e., coupling level and package density; (2) interface-level factors that define how the data or functionality flows through the system; and (3) and application-level factors that characterize the maturity of the COTS components. Additional factors such as process factors, vendor factors, and integrator factors may all play significant roles in driving the probability of TD interest.

We propose an automated technique for measuring the TD interest probability, leveraging pre-existing cost estimation inputs for budgeting purposes. Specifically, it extends the risk analysis component of COCOTS Risk Analyzer (Yang et al. 2006) to measure TD interest amount and probability, following a three-step methodology:

- Identifying cost drivers involved in the TD item(s), according to the nature of the TD item and the cost driver description. The first column in Figure 19.2 contains the size and cost drivers of COCOTS, and each driver has a self-explaining name and more specific description and rationale. This can be used to further guide the identification.
- Identifying clashing cost drivers according to the built-in risk rules COCOTS Risk Analyzer, which were obtained through expert Delphi methods and highlighted in different colors based on the probability of clashing, as shown in Figure 19.2. Each risk rule is defined as one critical combination of two COCOTS cost drivers that may cause certain undesired outcomes if they are both rated at their worst-case ratings.
- Computing TD interest probability based on the risk analysis results from the COTS risk analyzer, following Eq. (19.2) below. This is a fully-automated step, leveraging the built-in knowledge base of COCOTS Risk Analyzer. More specifically, COCOTS risk analyzer automatically examines the cost driver paired ratings and identifies risks based on 4-level risk probability assignment scheme (as shown in Figure 19.3). Each axis in Figure 19.3 is the risk potential

	Size	AAREN	ACIEP	ACIPC	AXCIP	APCON	ACPMT	ACSEW	APCPX	ACPPS	ACPTD	ACREL	AACPX	ACPER	ASPRT	APEAL	ACEAL	ACPUF	Productivity range
Size							1		1			1							
AAREN (Application Architectural Engineering)				1	1		1		1			1	1	1	1				2.09
ACIEP (COTS Integrator Experience with Product)									1										1.79
ACIPC (COTS Integrator Personnel Capability)					1	1			1		1	1	1						2.58
AXCIP (Integrator Experience with COTS Integration Processes)						1	1												1.42
APCON (Integrator Personnel Continuity)											1								2.51
ACPMT (COTS Product Maturity)										1		1							2.1
ACSEW (COTS Supplier Extension Willingness)																			1.22
APCPX (COTS Product Interface Complexity)													1						1.8
ACPPS (COTS Supplier Produce Support)																			1.48
ACPTD (COTS Supplier Provided Training and Documentation)																			1.43
ACREL (Constraints on Application System/Subsystem Reliability)																			1.48
AACPX (Application Interface Complexity)																			1.69
ACPER (Constraints on COTS Technical Performance)																			1.22
ASPRT (Application System Portability)																			1.14
APEAL (Application Evaluated Assurance Level)																	1		2
ACEAL (COTS Evaluated Assurance Level)																			2
ACPUF (Percentage of COTS' Unused Features)																			2

TD risk probability: >=50% [40%, 50%) [20%, 40%)

Figure 19.2 Built-in risk rules employed in COCOTS risk analyzer.

		Attribute 1			
		Worst case	Risk prone	Moderate	OK
Attribute 2	Worst case	Severe	Significant	General	
	Risk prone	Significant	General		
	Moderate	General			
	OK				

Figure 19.3 Assignment of risk probability levels.

ratings of a cost factor. A risk situation corresponds to an individual cell containing an identified risk probability. In this step, risk rules use the cost driver's risk potential ratings to index directly into these tables of risk probability levels. The productivity range of each cost driver represents the cost consequence of risk occurring, as summarized in the last column of Figure 19.2 (above).

$$\textbf{TD Interest Probability} = \sum_{i=1}^{15}\sum_{j=1}^{15} \textbf{Risk}_{\textbf{Probability}} * \textbf{PR}_i * \textbf{PR}_j \tag{19.2}$$

There are several reasons that we believe such an automatic, risk-based methodology is practical to enable quantitative measurement of COTS TD items. First, cost estimation models incorporate the use of cost drivers to adjust development effort and schedule calculations and reduce estimation subjectivity. As significant project factors, cost drivers can be used for identifying and quantifying project risk. Second, such cost estimation inputs are readily available during the acquisition phase. As an example, COCOTS has been integrated into several commercial estimating tools widely used in the defense domain. Third, an existing COCOTS Risk Analyzer tool can automatically obtain a COTS integration risk analysis with no additional inputs other than the set of COCOTS glue code cost drivers inputs, which were prepared to derive the COTS integration effort estimates.

If multiple COTS TD items exist, the above measurement methodology may be applied in a bottom-up approach to derive the aggregated system TD projection according to the product breakdown structures in candidate design alternatives and further aid the technical debt-aware comparison across multiple alternatives. In particular, when a system transits from the development phase to production, the TD communication offers an effective way to provide context about policy and technical decisions made in earlier development phases, as well as potential projected long-term TD concerns. In another word, it enables a systematic flow of risk information across life cycles to allow more effective program management.

Nurturing a TD-Aware Culture

The engineering of complex systems usually involves multiple acquisition, engineering, test and evaluation, and maintenance teams or organizations. It is essential to nurture the TD-aware culture across multiple collaborative systems engineering teams. "TD-aware culture" refers to the attitudes and behaviors within and across organization(s) regarding the management of technical debt. This promotes the timely synchronization, continuous transparency, and effective ownership of TD management, in order to decrease potential negative TD impact throughout the systems engineering life cycle.

First, it is important to make technical debt management a priority within the organization. The leadership and all members of the organization should internalize the importance of technical debt to the success of the system. This can involve providing training and education on the topic and highlighting the potential negative impacts of TD on the maintenance and development of systems. We also recommended setting aside dedicated time and resources for addressing technical debt, as well as making it a key focus of performance evaluations and goal setting.

Second, encouraging a culture of continuous improvement within the organization can help to ensure that technical debt is regularly reviewed and addressed. This can involve implementing agile development methods, such as Scrum, which include regular reviews of technical debt as part of the development process; implementing design and coding standards; conducting regular code reviews; or implementing processes to ensure that tests and documentation are up to date.

Third, encouraging collaboration and transparency within the organization can help to foster a culture of technical debt management. Team members who have been working on the system for an extended period of time may have valuable insights into areas of the system where technical debt may be accumulating. It can be helpful to seek input from these team members as part of the process of identifying and managing technical debt with better-aligned goals across multiple teams.

Conclusion

As systems engineering competencies and practices grow, the compelling and critical need for a systems engineering TD metaphor grows as well. Using COTS-intensive complex systems as an example context, this chapter adapts the metaphor of TD, introduces a COTS TD taxonomy, discusses how to cope with COTS TD, and establishes the TD-aware culture. This can serve as an effective mechanism to offset the ever-increasing "obsolescence" challenges in COTS-based complex systems.

Notions of COTS TD will help to increase the efficiency, effectiveness, and accountability of COTS-based complex systems design, development, maintenance, and sustainment. The use of better-informed and more-considerate COTS decision-making practices in the acquisition process will help to avoid presently unforeseen, expensive, and unaffordable obsolescence issues later in the system's life cycle, particularly in the sustainment phase.

More broadly, it would be beneficial for program and project managers to have the analysis capabilities to identify and maintain visibility of potential TD issues and better align ownership and accountability of major TD items. This will allow them to collaboratively make informed architectural and acquisition decisions and have the ability to expeditiously adjust plans across the systems lifecycle with minimal impacts to the time of new systems introductions, to new and legacy systems readiness, and to new and legacy systems sustainment costs.

References

Abts, C., Boehm, B.W., and Clark, E.B. (2000). COCOTS: A cots software integration lifecycle cost model-model overview and preliminary data collection findings. In *ESCOM-SCOPE Conference*, Orlando, FL, USA (18–21 February 2020). pp. 18–20.

Alelyani, T., Michel, R., Yang, Y. et al. (2019). A literature review on obsolescence management in cots-centric cyber physical systems. *Procedia Computer Science* 153: 135–145.

AlOmar, E.A., Christians, B., Busho, M. et al. (2022). Satdbailiff-mining and tracking self-admitted technical debt. *Science of Computer Programming* 213: 102693. https://doi.org/10.1016/j.scico.2021.102693.

Besker, T., Martini, A., Lokuge, R.E., et al. (2018). Embracing technical debt, from a startup company perspective. In *2018 IEEE International Conference on Software Maintenance and Evolution (ICSME)*, Madrid, Spain (23–29 September 2018). pp. 415–425.

Bhuta, J. and Boehm, B. (2007). Attribute-based cots product interoperability assessment. In *Proceedings of the 2007 Sixth International IEEE Conference on Commercial-off-the-Shelf (COTS)-Based Software Systems (ICCBSS'07)*, Banff, Alberta, Canada (26 February–2 March 2007). pp. 163–171.

Bogner, J., Verdecchia, R., and Gerostathopoulos, I. (2021). Characterizing technical debt and antipatterns in AI-based systems: a systematic mapping study. In *Proceedings of the 4th International Conference on Technical Debt, 2021 IEEE/ACM International Conference on Technical Debt (TechDebt)*, Virtual (19–21 May 2021). pp. 64–73.

Clark, B. (2018). A taxonomy of tech debt. Riot Games (online). https://technology.riotgames.com/news/taxonomy-tech-debt (accessed 10 April 2018).

Clark, B. and Clark, B. (2007). Added sources of costs in maintaining COTS-intensive systems. CrossTalk. *The Journal of Defense Software Engineering*. http://stsc.hill.af.mil/crosstalk/2007/06/index.html.

Cunningham, W. (1993). The WyCash portfolio management system. *ACM SIGPLAN OOPS Messenger* 4 (2): 29–30.

Fairbanks, G. (2020). Ur-technical debt. *IEEE Annals of the History of Computing* 37 (4).

Federal Acquisition Regulation (FAR), 48 C.F.R. (2019) https://www.acquisition.gov/browse/index/far

Garlan, D., Allen, R., and Ockerbloom, J. (1995). Architectural mismatch or why it's hard to build systems out of existing parts. In *Proceedings of the 17th International Conference on Software Engineering*, Seattle, WA, USA (23–30 April 1995). pp. 179–179.

Kruchten, P., Nord, R.L., and Ozkaya, I. (2012). Technical debt: From metaphor to theory and practice. *IEEE Software* 29 (6): 18–21.

Kruchten, P., Nord, R.L., and Ozkaya, I. (2019). *Managing Technical Debt: Reducing Friction in Software Development*. Addison-Wesley Professional.

Lenarduzzi, V., Besker, T., Taibi, D. et al. (2021). A systematic literature review on technical debt prioritization: Strategies, processes, factors, and tools. *Journal of Systems and Software* 171: 110827.

Letouzey, J.-L. and Whelan, D. (2016). *Introduction to the Technical Debt Concept*. Corryton, TN: Agile Alliance. https://www.agilealliance.org/wp-content/uploads/2016/05/IntroductiontotheTechnicalDebtConcept-V-02.pdf.

Li, Z., Avgeriou, P., and Liang, P. (2015). A systematic mapping study on technical debt and its management. *Journal of Systems and Software* 101: 193–220.

Martini, A., Bosch, J., and Chaudron, M. (2015). Investigating architectural technical debt accumulation and refactoring over time: a multiple-case study. *Information and Software Technology* 67: 237–253.

Martini, A., Besker, T., and Bosch, J. (2018). Technical debt tracking: current state of practice: a survey and multiple case study in 15 large organizations. *Science of Computer Programming* 163: 42–61. ISSN 0167-6423. https://doi.org/10.1016/j.scico.2018.03.007; https://www.sciencedirect.com/science/article/pii/S0167642318301035.

Nugroho, A., Visser, J., and Kuipers, T. (2011). An empirical model of technical debt and interest. In *Proceedings of the 2nd workshop on managing technical debt*, Honolulu, HI, USA (May 2011). pp. 1–8.

Rubin, K.S. (2012). *Essential Scrum: A Practical Guide to the Most Popular Agile Process*. Addison-Wesley.

da Saraiva, J. D. S., Neto, J. G., and Kulesza, U., et al. (2021). Exploring technical debt tools: a systematic mapping study. In *Proceedings of the International Conference on Enterprise Information Systems (ICEIS)*, virtual (26–28 April 2021). pp. 280–303.

Sierra, G., Shihab, E., and Kamei, Y. (2019). A survey of self-admitted technical debt. *Journal of Systems and Software* 152: 70–82.

Tang, Y., Khatchadourian, R.T., Bagherzadeh, M., Singh, R., Stewart, A. & Raja, A. (2021). An empirical study of refactorings and technical debt in machine learning systems. In *Proceedings of the IEEE/ACM 43rd International Conference on Software Engineering (ICSE)*, Madrid, Spain (22–30 May 2021). pp. 238–250.

Yang, Y., Michel, R., Wade, J. et al. (2019). Towards a taxonomy of technical debt for cots-intensive cyber physical systems. *Procedia Computer Science* 153: 108–117.

Yang, Y., Boehm, B., and Wu, D. (2006). Cocots risk analyzer. In *Proceedings of the IEEE 5th International Conference on Commercial-off-the-Shelf (COTS)-Based Software Systems (ICCBSS'05)*, Orlando, FL, USA (13–16 February 2006).

Biographical Sketches

Dr. Ye Yang is a software development manager at Amazon. Prior to joining Amazon, she was an associate professor at Stevens Institute of Technology. Her research focus lies in the interdisciplinary areas of software engineering and machine learning, and her contributions encompass innovative methods and techniques to support crowdsourced software development, mining software repositories, software process modeling, and simulation. She has also been active in professional consulting and promoting the development and transferring of research prototype toolkits in software industry. She has served as Program Co-Chairs and Program Committee member for a number of international conferences including ICSSP, ASE, ICSE, ESEM, APSEC, PROMISE, etc. She has published over 150 research papers and won three ACM/SigSoft Distinguished Paper Awards at ICSE 2019, ICSE 2020, and ASE 2021.

Dr. Dinesh Verma received the PhD (1994) and the MS (1991) in industrial and systems engineering from Virginia Tech. He served as the founding dean of the School of Systems and Enterprises at Stevens Institute of Technology from 2007 through 2016. He currently serves as the executive director of the Systems Engineering Research Center (SERC), a US Department of Defense sponsored University Affiliated Research Center (UARC) focused on systems engineering research, along with the Acquisition Innovation Research Center (AIRC). Prior to this role, he served as technical director at Lockheed Martin Undersea Systems. Before joining Lockheed Martin, Verma worked as a research scientist at Virginia Tech and managed the University's Systems Engineering Design Laboratory. He served as an Invited Lecturer from 1995 through 2000 at the University of Exeter, United Kingdom. In addition to his publications, Verma has received three patents in the areas of lifecycle costing and fuzzy logic techniques for evaluating design concepts. He was recognized with an honorary doctorate degree (*Honoris Causa*) in Technology and Design from Linnaeus University (Sweden) in January 2007; and with an honorary master of engineering degree (*Honoris Causa*) from Stevens Institute of Technology in September 2008.

Chapter 20

Risk and System Maturity: TRLs and SRLs in Risk Management

Brian Sauser

Department of Logistics and Operations Management, University of North Texas, Denton, TX, USA

System Maturity

Even with the implementation of new processes and practices within systems engineering, the challenges and the complexities systems engineers face are still significant and ever-changing. Nowhere is the need for enhanced monitoring capabilities more visible for systems than in the development of complex or large-scale systems. System development and engineering activities continue to be challenged by the formulation of larger and more complex systems or systems of systems. These emerging development paradigms are contesting many of the engineering, management, and acquisition practices that have been used for decades in the development of standalone systems. Similarly, complex system development has made the management of systems engineering activities more difficult to monitor and control due to the exponential growth of technologies and integrations being incorporated under a collective effort. Thus, there is a growing need for more systematic and systemic approaches to monitoring the development and integration of systems. This necessitates the development of new methods, processes, and tools (MPT) through best practices to govern the many unique aspects of systems development and acquisition programs and be able to compare actual progress against planned accomplishment from a technical perspective.

In life, we assess an individual's maturity – mostly subjectively – to determine their readiness to engage in a more advanced situation. We make this judgment from personal experience and observation of the individual. Then we allow the individual to enter the situation with some level of confidence (or fear) in their potential success. We rarely use well-defined MPT to aid our judgment and reduce the risk. This can have similarities to how we manage system development. *System Maturity* is a state or process of development that a system goes through regarding changes during its life cycle (i.e., conception to obsolescence/retirement), and *System Maturity Assessment* is the use of MPT to evaluate the system maturity for making lifecycle decisions. Maturation is important to understand and measure in system development because it helps build on knowledge already understood and provides information on when systems are ready to achieve certain developmental tasks or milestones. Likewise, the process of maturation does not end at deployment but continues through the life span into obsolescence or retirement. This can be described

Systems Engineering for the Digital Age: Practitioner Perspectives, First Edition. Edited by Dinesh Verma.

Companion website: www.wiley.com/go/verma/systemsengineering

by stages of maturation, but with an understanding that the milestones that define the development can be individual to the system itself.

While it is possible to subjectively assess what stage of development a simple system should be in, the difficulty increases with the size and complexity of the system. So, the questions become:

- How does one assess the maturity of a given system through its lifecycle?
- How can we use maturity assessment to manage risk and planning in systems engineering?
- How can maturity assessment be kept consistent and reproducible with less subjectivity?

The value of an assessment of system maturity using defined MPT is that it can lead to quantitative assessments that can determine whether a group of separate technology components with their associated (and demonstrated) assessments can be integrated into a system at a minimal risk to perform a required function or mission at a determined performance level. To support these challenges and the advancement of our practice of systems engineering, this chapter will describe the use of a system maturity assessment MPT and how it can be used to make informed decisions on the system's life cycle.

Maturity and Metrics

The use of maturity metrics has relied heavily on subjective assessment techniques, which then becomes the basis for making strategic acquisition decisions. These subjective assessments are labor-intensive, error-prone, and inadequate for the desired management controls. Notwithstanding the limitations of many of these metrics, any metric should not lose sight of what makes it effective and efficient in an organization (Dowling and Pardoe 2005):

1) The way the value is used should be clear.
2) The data to be collected for the metric should be easily understood and easy to collect.
3) The method of deriving the value from the data should be clear and as simple as possible.
4) Those for whom the use of the metric implies additional cost should see as much direct benefit as possible (i.e., collecting the data should not cost more than its value to the decision process).

Based on these rules we can then define metrics into two classifications: *descriptive* or *prescriptive* (Fan and Yih 1994; Tervonen and Iisakka 1996; Harjumaa et al. 2008). Descriptive metrics, sometimes referred to as hard metrics, can be objectively measured, are quantifiable, and have minimal variability when used between observers. For example, the height of an individual, proportion of telephone calls answered, or machine downtime. On the other hand, prescriptive metrics, or soft measures, are those which are qualitative, judgmental, subjective, and based on perceptual data. For example, customers' satisfaction with the speed of service or managers' assessment of staff attitude toward customers (Dowling and Pardoe 2005). With prescriptive metrics, when not used in the proper context, multiple observers can assess the same problem and yield significantly different results.

Within systems engineering, we have come to rely on prescriptive metrics for making managerial and at times engineering decisions because there is limited descriptive data. Prescriptive metrics are strongly based on human interpretation that can be influenced by personal biases and preferences. Lee and Shin (2000) found that egocentric biases and personal goals play a significant role in human beings' evaluation process. Since such cognitive bias is involved in assessment, subjectivity is inherent in our estimation and it is hard to avoid its influence (Yan et al. 2006).

Prescriptive metrics are vital in providing the richness that some descriptive metrics cannot, yet perspective metrics have been wrongfully considered as less important. While descriptive metrics consider more qualitative factors, it is possible to bridge and attempt to quantify qualities that are difficult to assess with both collectively. This chapter will describe an MPT that makes strides to moving prescriptive metrics in maturity assessment (i.e., readiness levels) closer to a descriptive state.

System Maturity Metrics

Technology Readiness Level (TRL)

In the 1990s, the National Aeronautics and Space Administration (NASA) instituted a nine-level metric called Technology Readiness Level (TRL) as a systematic metric/measurement to assess the maturity of a particular technology and to allow consistent comparison of maturity between diverse types of technologies. Given the pragmatic utility of this concept, in 1999, the Department of Defense (DoD) embraced a similar TRL concept in part due to a study by the US General Accounting Office (GAO) (now the Government Accountability Office) (see Table 20.1). This study stated that there were few metrics used within the DoD to gauge the impact of investments or the effectiveness of processes used to develop and transition technologies, and additional metrics in technology transition were needed (GAO 1999). In 2002, the GAO further articulated that the DoD needed to enable success through the demonstration of value and the credibility of new processes using metrics (GAO 2002). Additionally, the DoD made constructive changes to its approaches to acquisition that would address these issues by ensuring a weapon systems' technologies are demonstrated to an elevated level of maturity before beginning its program and using an evolutionary or phased approach to developing such systems (GAO 2006). In 2012, the GAO further defined TRL in the *Technology Readiness Assessment Guide* and discusses the continued development of readiness levels (e.g., System Readiness Level) (GAO 1999). Thus far, the TRL scale has been a key gauge of the status of maturity of a given technology and even systems within major defense and aerospace programs by monitoring capability development from concept definition through operations and support. In countless development efforts TRL has been a key indicator of progress and aided dramatically in keeping programs on track and adjudicating the perceived maturity of a technology for acquisition into a program. While the TRL has been well proven for its effectiveness in gauging individual technology maturity in research and development applications, its extrapolation to more complex systems integration (e.g., System of Systems), dictated by emerging requirements, brings about a host of issues. The GAO has best described the application of TRL in a technology readiness assessment (TRA) as a systematic, evidence-based process that evaluates the maturity of technologies (hardware, software, and processes) critical to the performance of a larger system or the fulfillment of the key objectives of an acquisition program, including cost and schedule (GAO 2020). Although, TRL was not intended to address system integration or to indicate that the technology will result in successful development of a system.

Table 20.1 Technology readiness levels.

TRL	Definition	Description
9	Actual system proven through successful mission operations	Actual application of the technology in its final form and under mission conditions, such as those encountered in operational test and evaluation. In almost all cases, this is the end of the last "bug fixing" aspects of true system development. Examples include using the system under operational mission conditions
8	Actual system completed and qualified through test and demonstration	Technology has been proven to work in its final form and under expected conditions. In almost all cases, this TRL represents the end of true system development. Examples include developmental test and evaluation of the system in its intended weapon system to determine if it meets design specifications
7	System prototype demonstration in relevant environment	Prototype near or at planned operational system. Represents a major step up from TRL 6, requiring the demonstration of an actual system prototype in an operational environment, such as in an aircraft, vehicle, or space. Examples include testing the prototype in a test bed aircraft
6	System/Subsystem model or prototype demonstration in relevant environment	Representative model or prototype system, which is well beyond the breadboard tested for TRL 5, is tested in a relevant environment. Represents a major step up in a technology's demonstrated readiness. Examples include testing a prototype in a high-fidelity laboratory environment or in a simulated operational environment
5	Component and/or breadboard validation in relevant environment	Fidelity of breadboard technology increases significantly. The basic technological components are integrated with reasonably realistic supporting elements so that the technology can be tested in a simulated environment. Examples include "high fidelity" laboratory integration of components
4	Component and/or breadboard validation in laboratory environment	Basic technological components are integrated to establish that the pieces will work together. This is relatively "low fidelity" compared to the eventual system. Examples include integration of "*ad hoc*" hardware in a laboratory
3	Analytical and experimental critical function and/or characteristic proof-of-concept	Active research and development is initiated. This includes analytical studies and laboratory studies to physically validate analytical predictions of separate elements of the technology. Examples include components that are not yet integrated or representative
2	Technology concept and/or application formulated	Invention begins. Once basic principles are observed, practical applications can be invented. The application is speculative and there is no proof or detailed analysis to support the assumption. Examples are still limited to paper studies
1	Basic principles observed and reported	Lowest level of technology readiness. Scientific research begins to be translated into applied research and development. Examples might include paper studies of a technology's basic properties

Source: Adapted from GAO (2020).

Integration Readiness Level (IRL)

When TRL is applied to components within a more complex system, the model of using individual technology maturity as a measure of readiness to integrate into system development can become confounded (Sauser et al. 2010). Similar problems also become apparent with many other technology development tools when applied in a systems context. These challenges and limitations of TRL were expressed by the Honorable Ashton Carter, then the Under Secretary of Defense for Acquisition, Technology, and Logistics (AT&L), in his Memorandum for Acquisition Professionals entitled, "Better Buying Power: Guidance for Obtaining Greater Efficiency and Productivity in Defense Spending" (2010). He states,

> The TRL review and certification process has grown well beyond the original intent and should be reoriented to an assessment of technology maturity and risk as opposed to engineering or integration risk.

Based on these fundamental conjectures, a more comprehensive set of concerns becomes relevant when considerations relating to integration, interoperability, and sustainment become equally important from a system's perspective. By looking only at the technical maturity of an individual component, TRL fails to account for the complexities involved in the integration of these components into a functional system and creates the opportunity for performance gaps to remain hidden until late in the development cycle. In other words, application of TRL to systems of technologies is not sufficient to give a holistic picture of *system* maturity since TRL is only a measure of an individual technology. Finally, multiple TRLs do not provide insight into integration between technologies or the maturity of the resulting system.

This monitoring of integration status is critical as it has been repeatedly shown that most complex system development efforts fail at the integration points. This lack of insight and the need to provide a method for monitoring integration status led to the development of a complementary concept (i.e., Integration Readiness Level or IRL) that expounds on the traditional TRL with the development of other criteria to gain a more complete perspective of system maturity. The IRL not only uses a stand-alone metric for determining readiness, but analyzes both its integration requirements and the maturity of other technologies with which it interfaces. See Table 20.2 (Sauser et al. 2010; Austin and York 2015).

IRL is a measurement of the interfacing of compatible interactions for various technologies and the consistent comparison of the maturity between integration points. For further clarification, the nine levels of IRL presented in Table 20.2 can be understood as having three stages of integration definition:

- IRLs 1-3 are considered fundamental to describing what can be defined as the three principles of integration: interface, interaction, and compatibility. These three principles are what define the subsistence of an integration effort.
- IRLs 4-7 are about assurance that an integration effort is compliant with specifications
- IRLs 8-9 relate to practical considerations and the assertion of the application of an integration effort.

System Readiness Level (SRL)

Given the utility of having a technology and integration maturity scale, the need for a system maturity scale is self-evident. In 2006, a System Readiness Level (SRL) method was developed as a descriptive model that characterizes the effects of technology and integration maturity on a systems engineering effort, particularly with respect to integrating discrete functional systems into a coherent system capability (Sauser et al. 2006). The SRL has been described using a normalized matrix of pairwise comparisons of TRLs and IRLs for any system under development, which yields a measure of system maturity (Sauser et al. 2008). The rationale behind the

Table 20.2 Integration readiness level.

IRL	Definition	Description
9	System Integration is proven through successful mission-proven operations capabilities	Fully integrated system has demonstrated operational effectiveness and suitability in its intended or a representative operational environment. Integration performance has been fully characterized and is consistent with user requirement
8	System integration completed and mission qualified through test and demonstration in an operational environment	Fully integrated system able to meet overall mission requirements in an operational environment. System interfaces qualified and functioning correctly in an operational environment
7	System prototype integration demonstration in an operational high-fidelity environment	Fully integrated prototype has been successfully demonstrated in actual or simulated operational environment. Each system/software interface tested individually under stressed and anomalous conditions. Interface, Data, and Functional Verification complete
6	Validation of interrelated functions between integrating components in a relevant end-to-end environment	End-to-end Functionality of Systems Integration has been validated. Data transmission tests completed successfully
5	Validation of interrelated functions between integrating components in a relevant environment	Individual modules tested to verify that the module components (functions) work together. External interfaces are well defined (e.g., source, data formats, structure, content, method of support, etc.)
4	Validation of interrelated functions between integrating components in a laboratory environment	Functionality of integrating technologies (modules/functions/assemblies) has been successfully demonstrated in a laboratory/synthetic environment. Data transport method(s) and specifications have been defined
3	There is **Compatibility** between technologies to orderly and efficiently integrate and interact to include all interface details	Detailed interface design has been documented. System interface diagrams have been completed. Inventory of external interfaces is completed and data engineering units are identified and documented
2	There is some level of specificity to characterize the **Interaction** (i.e. ability to influence) between technologies through their interface	Inputs/outputs for principal integration technologies/mediums are known, characterized and documented. Principal interface requirements and/or specifications for integration technologies have been defined/drafted
1	An **Interface** between technologies has been identified with sufficient detail to allow characterization of the relationship	Principal integration technologies have been identified. Top-level functional architecture and interface points have been defined. High-level concept of operations and principal use cased has been started

Source: Sauser et al. (2010) and Austin and York (2015).

SRL is that in the development life cycle, one would be interested in addressing the following considerations:

- Quantifying how a specific technology is being integrated with every other technology to develop the system.
- Providing a system-wide measurement of readiness.

Under this method, TRL evaluations for each technology and IRL evaluations of each integration are combined using matrix mathematics (explained in detail later) to produce a comprehensive

Table 20.3 System readiness levels.

SRL	Definition
9	System has achieved initial operational capability and can satisfy mission objectives
8	System interoperability should have been demonstrated in an operational environment
7	System threshold capability should have been demonstrated at operational performance level using operational interfaces
6	System component integrability should have been validated
5	System high-risk component technology development should have been complete; low-risk system components identified
4	System performance specifications and constraints should have been defined and the baseline has been allocated
3	System high-risk immature technologies should have been identified and prototyped
2	System materiel solution should have been identified
1	System alternative materiel solutions should have been considered

assessment where each technology within the system is weighted according to all its integrations and then rolled up to a system level. It is important to emphasize that the SRL is *not* a quantitatively defined rating system but is instead an analytical combination of the TRL and IRL scales. In other words, the SRL output is purely a function of the TRL and IRL inputs.

This later evolved into a collection of analytical MPT to help make informed systems engineering management decisions. The SRL and its supporting MPT have proven to be a repeatable mechanism for understanding the effects of technology and integration maturity, demonstrated utility for defining system status, and provided leading indicators for development risks. The SRL was designed to be robust, repeatable, and agile so outputs could not only be trusted and replicated, but the methodology could be easily transferred to a variety of different applications and architectures. This differs from just having a readiness scale like TRL and IRL as *maturity* is defined as the practices that support the development of a system. Therefore, SRL is more than purely a qualitative assessment. It requires the user to define the element-level contributions of the multiple technologies and integrations that make up the system. In this way, it allows managers to evaluate system development in real-time and take proactive measures by examining the status of all elements of the system simultaneously. Table 20.3 is a description of the SRL scale, which implies some assumptions – and a risk if those assumptions are invalid. Furthermore, the methodology is adaptive to use on an array of system engineering development efforts and can also be applied as a predictive tool for technology insertion trade studies and analysis.

The following sections will describe the MPT for development and use of the SRL and a System Maturity Assessment.

System Maturity Assessment

SRL Calculation

The evaluation of technology using TRL and the evaluation of each integration using IRL are combined via a set of mathematical formulas to produce an integrated assessment where each technology within the system is weighted according to all its integrations and then calculated at a system level. It is important to emphasize that the SRL is not a quantitatively defined rating system but is instead an analytical combination of the TRL and IRL scales. A fundamental assertion in the calculation of the SRL is the interpretation and use of these inputs. The TRL and IRL inputs are purely data inputs into the SRL calculation and do not assert any specific form of numerical scale, i.e., nominal, ordinal, interval, or ratio. Thus, to assert a form of scale-conversion on the inputs presupposes that the origin of the data type is known or natural. Lord (1953) describes this as, "...the numbers do not know where they came from." Since TRL and IRL are scales of nonnatural origin, their interpretations of forms of scale are also interpretable by the user of the data. Thus, $9 = 9, 8 = 8, 7 = 7$, etc. This conversion is justifiable if the conversion does not alter the scale (e.g., 8 does not become more important than 9 or 7 less important than 6) (Shah and Madden 2004, Akritas 1990). As discussed in Sauser et al. (2008) and Magnaye et al. (2010), the use of scale data in mathematically assessing progress or status without scale conversion is not without precedence, i.e., Grade Point Average (GPA), Analytical Hierarchy Process (AHP), and Failure Mode Effects and Criticality Analysis (FMECA).

The computation of the SRL is a function of two matrices:

- Matrix TRL provides a blueprint of the system's state with respect to its technologies' readiness. That is, **TRL** is defined as a vector with n entries for which the ith entry defines the TRL of the ith technology.
- Matrix **IRL** illustrates how the different technologies are integrated with each other from a system perspective. **IRL** is defined as an $n{\cdot}n$ matrix for which the element IRL_{ij} represents the maturity of integration between the ith and jth technologies.

In these matrices, the standard TRL and IRL levels corresponding to values from 1 through 9 should be normalized. Also, it has been assumed that on the one hand, a value of 0 for element IRL_{ij} defines that the ith and jth technologies are impossible to integrate. On the other hand, a value of 1 for element IRL_{ij} can be understood as one of the following with respect to the ith and jth technologies: (1) are completely compatible within the total system, (2) do not interfere with each other's functions, (3) require no modification of the individual technologies, and (4) require no integration linkage development. Also, it is important to note that IRL_{ii} may have a value lower than 1, illustrating that the technology may be a composite of different sub-technologies that are not mature.

In any system, each constituent technology is connected to at least one other technology through a bi-directional integration. How each technology is integrated with other technologies is used to formulate an equation for calculating SRL that is a function of the TRL and IRL values of the technologies and the interactions that form the system. To estimate a value of SRL from the TRL and IRL values, we recommend a normalized matrix of pairwise comparison of TRL and IRL indices. That is, for a system with n technologies, we first formulate a TRL matrix, labeled [TRL]. This matrix is a single column matrix containing the values of the TRL of each technology in the system. In this respect, [TRL] is defined in Eq. (20.1), where TRL_i is the TRL of technology i.

$$\left[\text{TRL}\right]_{n \times 1} = \begin{bmatrix} \text{TRL}_1 \\ \text{TRL}_2 \\ \dots \\ \text{TRL}_n \end{bmatrix} \tag{20.1}$$

Second, an IRL matrix is created as a symmetric square matrix (of size $n \times n$) of all integrations between any two technologies in the system. For a system with n technologies, [IRL] is defined in

Eq. (20.2), where IRL_{ij} is the IRL between technologies i and j. It is important to note that whenever two technologies are not planned for integration, the IRL value assumed for these specific technologies is the hypothetical integration of a technology i to itself; therefore, it is given the maximum level of 9 and is denoted by IRL_i

$$\left[\mathrm{IRL}\right]_{n\times n} = \begin{bmatrix} \mathrm{IRL}_{11} & \mathrm{IRL}_{12} & \dots & \mathrm{IRL}_{1n} \\ \mathrm{IRL}_{21} & \mathrm{IRL}_{22} & \dots & \mathrm{IRL}_{2n} \\ \dots & \dots & \dots & \dots \\ \mathrm{IRL}_{n1} & \mathrm{IRL}_{n2} & \dots & \mathrm{IRL}_{nn} \end{bmatrix} \tag{20.2}$$

Although the original values for both TRL and IRL can be used, the use of normalized values allows a more accurate comparison of the use of competing technologies. Thus, the values used in [TRL] and [IRL] are normalized (0,1) from the original (1,9) levels. Based on these two matrices, an SRL matrix is generated by obtaining the product of the TRL and IRL matrices, as shown in Eq. (20.3).

$$\left[\mathrm{SRL}\right]_{n\times 1} = \left[\mathrm{IRL}\right]_{n\times n} \times \left[\mathrm{TRL}\right]_{n\times 1} \tag{20.3}$$

The SRL matrix consists of one element for each of the constituent technologies and from an integration perspective, quantifies the readiness level of a specific technology with respect to every other technology in the system while also accounting for the development state of each technology through TRL. Mathematically, for a system with n technologies, [SRL] is as shown in Eq. (20.4).

$$\left[\mathrm{SRL}\right] = \begin{bmatrix} \mathrm{SRL}_1 \\ \mathrm{SRL}_2 \\ \dots \\ \mathrm{SRL}_n \end{bmatrix} = \begin{bmatrix} \mathrm{IRL}_{11}\mathrm{TRL}_1 + \mathrm{IRL}_{12}\mathrm{TRL}_2 + \ldots + \mathrm{IRL}_{1n}\mathrm{TRL}_n \\ \mathrm{IRL}_{21}\mathrm{TRL}_1 + \mathrm{IRL}_{22}\mathrm{TRL}_2 + \ldots + \mathrm{IRL}_{2n}\mathrm{TRL}_n \\ \dots \\ \mathrm{IRL}_{n1}\mathrm{TRL}_1 + \mathrm{IRL}_{n2}\mathrm{TRL}_2 + \ldots + \mathrm{IRL}_{nn}\mathrm{TRL}_n \end{bmatrix} \tag{20.4}$$

where $IRL_{ij} = IRL_{ji}$.

Each of the SRL values obtained in Eq. (20.4) would fall within the interval (0,n). For consistency, these values of SRL should be divided by n to obtain the normalized value between (0,1). Notice that [SRL] itself can be used as a decision-making tool since its elements provide a prioritization guide of the system's technologies and integrations. Thus, [SRL] can point out deficiencies in the maturation process. The SRL for the complete system is the average of all such normalized SRL values, as shown in Eq. (20.5). Equal weights are given to each technology and hence a simple average is estimated. A standard deviation can also be calculated to indicate the variation in the system maturity and parity in subsystem development.

$$\mathrm{SRL} = \frac{\left(\dfrac{\mathrm{SRL}_1}{n_1} + \dfrac{\mathrm{SRL}_2}{n_2} + \ldots + \dfrac{\mathrm{SRL}_n}{n_n}\right)}{n} \tag{20.5}$$

where n_i is the number of integrations with technology i.

Example of SRL Calculation

The following example is from the NSA *System Readiness Assessment (SRA) Engineering Handbook* and was also described by Austin and York (2015). Figure 20.1 illustrates a system architecture with 10 components and their corresponding TRLs and IRLs.

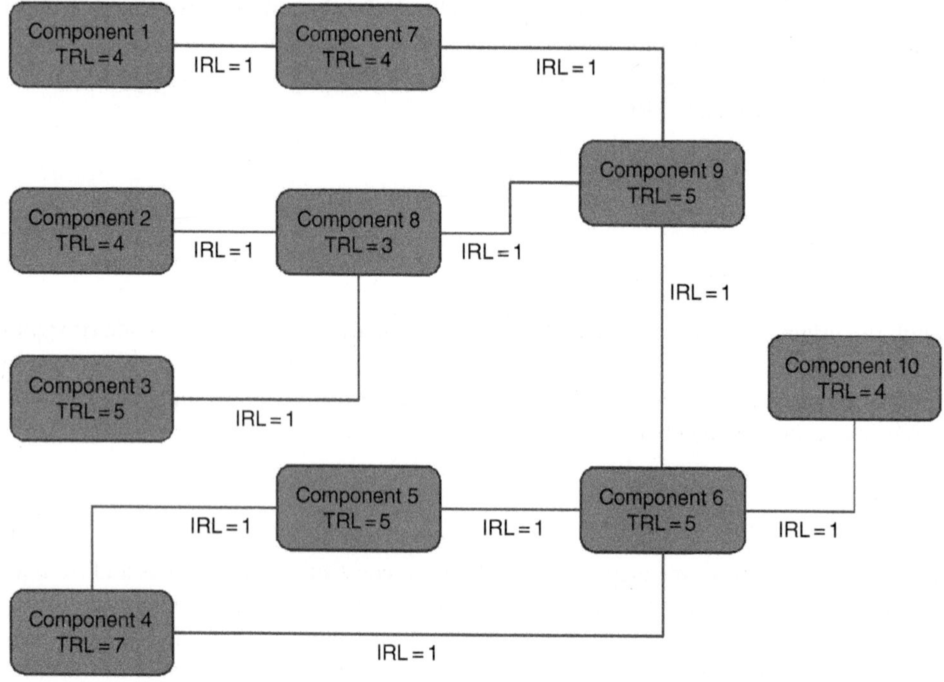

Figure 20.1 TRLs and IRLs for a 10-component system.

In the matrices represented, the TRL levels correspond to values 1 through 9 while the IRL values range from 0 to 9. Before performing the matrix math, these values are normalized by dividing by 9, the highest value. For example, an IRL of 9 has a normalized value of 1 for element IRL_{ij} and has the characteristics described in Table 20.3 with respect to the *i*th and *j*th components. Similarly, an IRL of 5 has a normalized value of $\frac{5}{9}$ or 0.556.

At a minimum, each component of a system is connected to one other. This integration is bidirectional, and it is assumed that the IRL is the same in each direction. Each component is integrated with other components in a specific way and is used to formulate and calculate the SRL.

To calculate a value of the SRL from the TRL and IRL values, an SRL matrix is generated by obtaining the product of the IRL and TRL matrices, as shown in the equation:

$$[\text{SRL}]_{n\times 1} = [\text{IRL}]_{n\times n} \times [\text{TRL}]_{n\times 1} \tag{20.6}$$

The SRL matrix consists of one element for each of the constituent components and, from an integration perspective, quantifies the readiness level of a specific component with respect to every other component in the system while also accounting for the development state of each.

Mathematically, for a system with $n = 10$ components, the SRL is as shown in the example in Figure 20.1, where TRL_i represents the individual TRLs and the IRL_{ij} are the individual IRLs between the components. SRL_i represents the readiness level of Component *i*, reflecting the readiness of *all* of its connections/interfaces. (Recall that IRL_{ij} represents the IRL only between Component *i* and Component *j*.) For the general case of a 10-component system:

The corresponding SRL_i for each component *i* is then divided by m_i, where m_i is the number of integrations of component *i* with every other component as defined by the system architecture.

This includes integration of the component with itself. The result is a normalized Component SRL value between 0 and 1.

$$\text{Component SRL}_i = \frac{SRL_i}{m_i} \tag{20.7}$$

The Composite SRL for the system is the average of the Component SRL values, as shown below, where n is the number of components:

$$\text{Composite SRL} = \frac{\left(\frac{\text{SRL}_1}{m_1}\right)+\left(\frac{\text{SRL}_2}{m_2}\right)+\left(\frac{\text{SRL}_3}{m_3}\right)+\ldots+\left(\frac{\text{SRL}_{10}}{m_{10}}\right)}{n} \tag{20.8}$$

Continuing with the example of the 10-component system illustrated in Table 20.6 where the TRLs and IRLs have the specified values shown, we generate Table 20.4. As an illustrative example, the highlighted cell in Table 20.4 indicates an IRL value of 1 for the integration readiness of the interface between Component 5 and Component 6. As indicated previously, a zero (0) is placed in the matrix where no integration is planned.

Normalizing the matrix entries, the product of the [IRL] and [TRL] matrices is given by:

$$\left[\text{IRL}\right]_{10\times 10}\times\left[\text{TRL}\right]_{10\times 1} =$$

$$\begin{pmatrix} (1)(.44)+(0)(.44)+(0)(.55)+(0)(.77)+(0)(.55)+(0)(.55)+(.11)(.44)+(0)(.33)+ \\ (0)(.55)+(0)(.44) \\ (0)(.44)+(1)(.44)+(0)(.55)+(0)(.77)+(0)(.55)+(0)(.55)+(0)(.44)+(.11)(.33)+ \\ (0)(.55)+(0)(.44) \\ (0)(.44)+(0)(.44)+(1)(.55)+(0)(.77)+(0)(.55)+(0)(.55)+(0)(.44)+(.11)(.33)+ \\ (0)(.55)+(0)(.44) \\ (0)(.44)+(0)(.44)+(0)(.55)+(1)(.77)+(.11)(.55)+(.11)(.55)+(0)(.44)+(0)(.33)+ \\ (0)(.55)+(0)(.44) \end{pmatrix} = \begin{pmatrix} 0.494 \\ 0.482 \\ 0.592 \\ 0.900 \end{pmatrix}$$

Table 20.4 TRLs and IRLs for 10 component system.

Component IDs		1	2	3	4	5	6	7	8	9	10
	TRLs	4	4	5	7	5	5	4	3	5	4
IRLs	**1**	9	0	0	0	0	0	1	0	0	0
	2	0	9	0	0	0	0	0	1	0	0
	3	0	0	9	0	0	0	0	1	0	0
	4	0	0	0	9	1	1	0	0	0	0
	5	0	0	0	1	9	***1***	0	0	0	0
	6	0	0	0	1	1	9	0	0	1	1
	7	1	0	0	0	0	0	9	0	1	0
	8	0	1	1	0	0	0	0	9	1	0
	9	0	0	0	0	0	1	1	1	9	0
	10	0	0	0	0	0	1	0	0	0	9

Table 20.5 Component SRLs for 10 component system.

Component ID	Component SRL
1	0.494/2 = 0.247
2	0.482/2 = 0.241
3	0.592/2 = 0.296
4	0.900/3 = 0.300
5	0.705/3 = 0.235
6	0.815/5 = 0.163
7	0.555/3 = 0.185
8	0.508/4 = 0.127
9	0.704/4 = 0.176
10	0.506/2 = 0.253

The resultant 10 × 1 column matrix is used to determine the Component SRLs. Each Component SRL is calculated by taking the matrix entry and dividing it by its total number of integrations including integration with itself. For example, Component 6 has five total integrations – Components 4, 5, 9, 10, and integration with itself. Thus, the Component SRL for Component 6 is 0.815/5 = 0.163. Table 20.5 shows each of the Component SRLs.

Interpreting the Results as a System Readiness Assessment (SRA)

Aside from SRL assessing overall system development, it can also be a guide in prioritizing potential areas requiring further development. That is, if considering a "systems-focused approach" to the methodology, then one cannot evaluate a system based on just a single number, such as the Composite SRL. As shown in the example, the SRL_is (technologies with their integration links considered) present a spectrum showing some subsystems whose readiness levels (i.e., SRL_i) are in different development phases other than the Composite SRL. While it could be argued that the overall SRL is only as good as the lowest SRL_i, this would also lose sight of those technologies that are potentially developing faster than the system. In understanding the value of the SRL analysis, we must understand the spectrum of SRL_i and its relationship to the Composite SRL.

Component SRLs are important, as they provide an indicator of the readiness of the individual component and its associated integrations. Examination of the individual Component SRLs identifies those components that are lagging or may be too far ahead in their "readiness." This is illustrated in the example above, where Component 8's SRL value of 0.127 – the lowest among these components – illustrates that Component 8 is lagging behind the other system components. This should be brought to the attention of the Program Manager and/or other decision-makers for a detailed risk assessment and further analysis.

The Composite SRL is determined by averaging the Component SRLs as shown below. As with any calculation involving an average, the user needs to be aware of the potential risk of masking a Component SRL that is significantly lagging or leading the average, reiterating the importance of assessing, and monitoring the individual Component SRLs.

Composite SRLs are defined on a scale from 0 to 1 with a value typically carried out to two or three decimal places. For the calculations in the example above, the Composite SRL is reported as 0.222 with 10 Component SRLs of 0.247, 0.241, 0.296, 0.300, 0.235, 0.163, 0.185, 0.127, 0.176, and

Table 20.6 Example SRL translation model.

TRL	IRL	Composite SRL_i	Midpoint between levels	Composite SRL_i range	SRL
9	9	1.000		.914–1.000	9
			0.914		
8	8	0.828		.750–.913	8
			0.750		
7	7	0.672		.601–.749	7
			0.601		
6	6	0.530		.467–.600	6
			0.467		
5	5	0.404		.349–.466	5
			0.349		
4	4	0.293		.245–.348	4
			0.245		
3	3	0.197		.157–.244	3
			0.157		
2	2	0.116		.084–.156	2
			0.084		
1	1	0.051		.000–.083	1

0.253. It could potentially be difficult to understand the difference between system readiness values that are similar (e.g. 0.247 vs. 0.241 vs. 0.296). The Table 20.6 translates composite SRL values to whole numbers consistent with TRL and IRL scaling for ease of interpretation. To translate the 0 to 1 scale to a 1 to 9 scale, the SRL Translation Model shown in Table 20.6 is used to map the decimal values to whole number values. Because the System Readiness Assessment is dependent on the system architecture configuration, a SRL Translation Model is generated for each architecture configuration when performing the SRA.

To generate the SRL Translation Model for this example architecture, a Composite SRL_i is calculated for nine system architecture configurations (each with 10 components and 10 integration links) where the TRLs for all of the components and the IRLs for all of the integration links are set equal to the same value, an integer from 1 to 9. For example, the Composite SRL of 0.051 is calculated by setting the TRL of each of the 10 components equal to 1 and the IRL of each of the 10 integration links equal to 1. The midpoints between each pair of adjacent Composite SRL_i are used as the boundaries for the corresponding Composite SRL_i Range values, as shown in Table 20.6.

The SRA shown in the example at the beginning of this section resulted in a Composite SRL of 0.222. Using the SRL Translation Model, this translates to a System Readiness Level of 3 (bolded text in Table 20.5). This indicates that immature and high-risk technologies have been identified and prototyped. Through the SRA process, areas of potential concern are identified, documented, and reported.

The SRL calculated in this example is a snapshot in time. Just as risk is assessed through the lifecycle, it is critical to measure the system readiness at multiple points along the life cycle to avoid pitfalls that can occur when readiness is only assessed once or twice. The SRA for this example was conducted early in the life cycle when all component IRLs had a value of "1." As the system development progresses, the TRLs and IRLs of the system components may mature as shown in Table 20.7.

Table 20.7 A second "snapshot" of the system TRLs and IRLs.

Component IDs		1	2	3	4	5	6	7	8	9	10
	TRLs	4	*8*	5	7	*6*	5	4	*5*	5	*7*
IRLs	**1**	9	0	0	0	0	0	*5*	0	0	0
	2	0	9	0	0	0	0	0	1	0	0
	3	0	0	9	0	0	0	0	*3*	0	0
	4	0	0	0	9	*6*	*5*	0	0	0	0
	5	0	0	0	6	9	*5*	0	0	0	0
	6	0	0	0	5	5	9	0	0	*4*	*7*
	7	5	0	0	0	0	0	9	0	1	0
	8	0	3	1	0	0	0	0	9	1	0
	9	0	0	0	0	0	4	1	1	9	0
	10	0	0	0	0	0	7	0	0	0	9

TRLs and IRLs of components that have matured are highlighted. (Since the matrix is symmetric, only the elements on one side of the diagonal are highlighted.) Note in this example that the fundamental architecture of the system has not changed, only the readiness levels of components and interfaces have matured. However, the SRA process is agile and can be used in a similar fashion should changes be made to the system architecture during the life cycle process.

The SRA for this second snapshot in time yields a Composite SRL of 0.395. Referencing the SRL Translation Model, this translates to an SRL of 5. Thus, a decision-maker can readily see improvement over the original SRL of 3. From the SRL definitions, the system's high-risk component technology development is complete and the low-risk system components have been identified.

SRAs can be performed on any size program and at any time during system development. The potential technology and integration risks will determine the frequency at which SRAs should be performed. Typically, for larger programs, a quarterly SRA is recommended while for small programs SRAs may be performed less frequently. Once the system has been defined, the system mapping completed, and the initial SRA done, subsequent or follow-on SRAs can be performed in a reasonable time.

Ideally, this type of analysis can facilitate strategic decisions about incremental technology and integration investments of limited resources. For example, in the upcoming budgetary period or fiscal year, resources may be shifted in favor of accelerating the development of the technologies and integration links that are behind and temporarily away from those that are ahead, provided such a shift is technologically and organizationally feasible. This capability can become important when a specific technology is a conduit for downstream technologies, and thus, its maturity is critical for the system to reach a certain level of maturity.

Systems Architecture and SRL

One of the current deficiencies in system maturity assessments is that they are performed independent of any systems engineering tools or supporting artifacts, which could reduce the level of subjectivity in an assessment and reliability in the results. Within the MPT of systems engineering architecting, there exists a substantial base of architectural artifacts that have the potential to significantly reduce the subjectivity and increase the reliability in a system maturity assessment. The advent of system engineering modeling tools has enabled system architects to better understand a system by depicting various views of the system and its components. For this purpose, architectural frameworks have been introduced for various domains and industries to support a common language and set of tools for developing a system.

Architectural frameworks support the need for a more structured approach to manage complexity while balancing all appropriate user perspectives. One of the widely adopted frameworks in the defense sector of the United States is the Department of Defense Architecture Framework (DoDAF). In addition, DoD subcontractors have adopted DoDAF as part of their systems engineering process, and industry consortia are currently working on adopting the DoDAF vocabulary and products to complement their standardized approaches to systems and software development.

In understanding what the SRL method means to systems engineering architecture, we first must clarify what is meant by maturity and readiness. We distinguish these two terms: the *readiness* of a system, technology, or integration implies how ready it is to be deployed on a numeric scale; and *maturity* is the characterization of the physical development that is quantified by the readiness. Thus, readiness is a scale, and maturity is the definition of each level in the scale. We also distinguish maturity within a system as having three dimensions: physical, logical, and functional. The TRL, IRL, and SRL are the maturity of the physical representation of a system or its physical architecture. Thus, the first step in applying the SRL method is to understand the physical system and its representative architecture, which shows the system design broken down into all its constituent elements (i.e., subsystems and components). Likewise, this architecture is supported by its functional and logical representations. At this point, we only focus on the physical architecture, which includes a representation of the software and hardware (i.e., products necessary to realize the concept). The physical architecture forms the basis for design definition documentation (e.g., specifications, baselines, and the work breakdown structure [WBS]). The allocation of the functional definition of the system to the physical definition completes the system's design and the definition of its components.

Conversely, a functional definition of an architecture can be mapped (or allocated) to a physical view of components. Together these views define the design (model) of the system. The system model becomes the input to the detailed design and fabrication of the physical components that comprise the system. This physical view establishes all the discrete subsystems and components that are procured or developed to produce the system. It also defines the interfaces between these subsystems and components.

The physical view also identifies the selection of technologies, including developed and commercial, that will comprise each subsystem or component. In this way, the physical architecture includes a representation of the software and hardware – components necessary to realize the system. This physical view then provides a clear definition of the system elements needed to perform the SRL assessment.

DoDAF 2.02 describes four related technical viewpoints of architecture: (DoD 2012)

1) The Capability Viewpoint (CV) identifies the requirements and delivery of the system.
2) The Operational Viewpoint (OV) identifies what needs to be accomplished and who does it.
3) The Services Viewpoint (SvcV) identifies the services and exchanges in a service-oriented architecture.
4) The Systems Viewpoint (SV) relates systems and characteristics to operational needs.

Products within this framework can be associated with the physical architecture. The SvcV and the SV are very closely related. Indeed, they could be considered two alternate views of the design/implementation of a system.

The SV represents the traditional physical design of a system, whereas the SvcV represents a more modern service-oriented architecture (SOA) design of a system. In this manner, the SV captures the information on supporting automated systems, interconnectivity, and other systems functionality in support of operating activities. As SOA becomes more predominant, it can be expected that over time, the DoD's emphasis on service-oriented environment and cloud computing may result in the elimination of the SV. Therefore, either the SV-1 or the SvcV-1 can be considered the physical design of the system depending on the architectural approach taken by the design team. These views can be considered as complementary design views.

Physical description is captured in several DoDAF products in either the SV or the SvcV model:

- SV(SvcV)-1, Systems Descriptions
- With further descriptions to be found in
- SV(SvcV)-2, Systems Flow Descriptions
- SV(SvcV)-3, Systems (Services)-Systems (Services) Matrix
- SV(SvcV)-7, Systems (Services) Measures Matrix

The target environment's elements can be established by examining the system design's physical view. If the program is using a DoDAF structure, this can be accomplished by examination of the SV-1 or SvcV-1 diagram. The component blocks on the diagram represent the elements (components) of the system. These are the candidate items to be reviewed for use in the SRL assessment. All of the connectors on the diagram represent interfaces that will be candidates for evaluation in the IRL analysis for the SRL.

To increase the understanding of all system elements, the components included in the derivative models SV(SvcV)-2, SV(SvcV)-3, and SV(SvcV)-7 can be checked to see that all required elements have, in fact, been identified. Another check can be performed by examining the program WBS and comparing the hierarchical development tasks against the elements defined for the SRL methodology. The result of this effort should be a definition of all the elements and interfaces of the system that is close to the program design and development work definition.

The remainder of this section describes the process for defining all the architectural elements to be modeled in the SRL. This can be performed in five steps:

1) Analyze the architecture to determine all the elements in the system
2) Identify the Critical Technology Elements (CTEs). These will be evaluated and scored in the SRL.
3) Identify the Non-critical Technology Elements (NTEs). These will not be evaluated and scored in the SRL. These elements will not impact the SMA Analysis.
4) Identify the Critical Technology Integrations (CTIs). These will be evaluated and scored in the SRL.
5) Identify the Non-critical Technology Integrations (NTIs). These will not be evaluated and scored in the SRL.

Note that the system under development can be at any level of decomposition. The level of decomposition is determined by the program. It should be at a level of decomposition where the program comfortably feels the major technology components can be identified. At this point, we focus on the physical model of the system. When we begin to work in the next step, we will use the functional model and, in, the allocation of the functional to the physical. The next step in the SRL architecture definition process is to identify those system components that are CTEs in the system's development. These elements will need to be evaluated, rated, and compiled in the SRL assessment. The selection of these CTE components is discussed in the next section of this paper.

Finally, DoDAF 2.02 has views and products that can be described a fit for purpose, where the emphasis is shifted to the data and artifacts rather than models. This suits the prospects of system maturity assessment using system architectures, as criteria and information harvested in a system architecture can be used to aid decision-makers when it comes to TRL and IRL.

References

Akritas, M. G. (1990). "The rank transform method in some two-factor designs." *Journal of American Statistics Association* 85, 73–78.

Austin, M. & York, D. (2015). "System readiness assessment (SRA): an illustrative example." *Procedia Computer Science* 44, 486–496.

DoD, (2012). The DoDAF Architecture Framework Version 2.02. Chief Information Officer US Department of Defense. Washington, DC, US Department of Defense.

Dowling, T. and Pardoe, T. (2005). TIMPA - Technology Insertion Metrics, CR050825, Ministry of Defense – QINETIQ, 60.

Fan, C.-F. and Yih, S. (1994). Prescriptive metrics for software quality assurance. *First Asia-Pacific Software Engineering Conference*, Tokyo, Japan (7–9 December 1994), pp. 430–438.

GAO (1999). Best Practices: Better Management of Technology Development Can Improve Weapon System Outcomes. GAO. Washington, DC, U.S. Government Accountability Office. GAO/NSIAD-99-162.

GAO (2002). DOD Faces Challenges in Implementing Best Practices. GAO. Washington, DC, U.S. Government Accountability Office. GAO-02-469T.

GAO (2006). Best Practices: Stronger Practices Needed to Improve DoD Transition Process. GAO. Washington, DC, U.S. Government Accountability Office, GAO 06-883.

GAO (2020). Technology Readiness Assessment Guide: Best Practices for Evaluating the Readiness of Technology for Use in Acquisition Programs and Projects. GAO. Washington, DC, U.S. Government Accountability Office. GAO-20-48G.

Harjumaa, L., Tervonen, I., and Salmela, S. (2008). Steering the inspection process with prescriptive metrics and process patterns. *Eighth International Conference on Quality Software*, Oxford, UK (12–13 August 2008). pp. 285–293.

Lee, M. & Shin, W. (2000). "An empirical analysis of the role of reference point in justice perception in R&D settings in Korea." *Journal of Engineering and Technology Management* 17(2), 175–191.

Lord, F. (1953). "On the statistical treatment of football numbers." *American Psychologist* 8(750–751), 19–22.

Magnaye, R. B., Sauser, B.J. & Ramirez-Marquez, J.E. (2010). "System development planning using readiness levels in a cost of development minimization model." *Systems Engineering* 13(4), 311–323.

Sauser, B., Verma, D., Ramirez-Marquez, J., and Gove, R. (2006). From TRL to SRL: the concept of systems readiness levels. *Conference on Systems Engineering Research (CSER)*, Los Angeles, CA (7 April 2006).

Sauser, B., Ramirez-Marquez, J.E., Henry, D. & DiMarzio, D. (2008). "A system maturity index for the systems engineering life cycle." *International Journal of Industrial and Systems Engineering* 3(6), 673–691.

Sauser, B., Gove, R., Forbes, E. & Ramirez-Marquez, J.E. (2010). "Integration maturity metrics: development of an integration readiness level." *Information Knowledge Systems Management* 9(1), 17–46.

Shah, D. A. & Madden, L.V. (2004). "Nonparametric analysis of ordinal data in designed factorial experiments." *Phytopathology* 94(1), 33–43.

Tervonen, I. and Iisakka, J. (1996). Monitoring software inspections with prescriptive metrics. *Proceedings of the Fifth European Conference on Software Quality*, Dublin, Ireland (16–20 September 1996). pp. 105–114.

Yan, S., Xu, Y., Yang, M., Zhang, Z., Peng, M., Yu, X., & Zhang, H. (2006). "A subjective evaluation study on human-machine Interface of marine meter based on RBF network." *Journal of Harbin Engineering University* 27, 560–567.

Biographical Sketches

Brian Sauser is a professor at the University of North Texas (UNT) in the G. Brint Ryan College of Business and College of Applied and Collaborative Studies. He currently serves as the Department Chair of the Department of Logistics and Operations Management and program coordinator of the Logistics and Supply Chain Management PhD Program. He was the degree architect of the BS in industrial distribution at UNT College of Applied and Collaborative Studies (2020–2022), founder and former Director of the Complex Logistics Systems Laboratory (2012–2022), and served as the Director of the Jim McNatt Institute for Logistics Research (2016–2019). Before joining UNT, he held positions as an assistant professor with the School of Systems and Enterprises at Stevens Institute of Technology where is he was director of the Systems Development and Maturity Laboratory; project specialist with ASRC Aerospace at NASA Kennedy Space Center; program administrator with the New Jersey – NASA Specialized Center of Research and Training at Rutgers, The State University of New Jersey; and laboratory director with G.B. Tech Engineering at NASA Johnson Space Center.

Dr. Sauser's research interest is in the engineering, management, and governance of complex systems. He teaches or has taught courses in Advanced Logistics Management, Project Management of Complex Systems, Designing and Managing the Development Enterprise, Logistics and Business Analytics, Theory of Logistics Systems, Systems Thinking, and Systems Engineering and Management. He is an Acquisition Innovation Research Center research fellow, National Aeronautics and Space Administration faculty fellow, UNT faculty leadership fellow (2018–2019), Professional Development Institute business fellow (2015–2016), IEEE senior member, associate editor of the *IEEE Systems Journal*, Editorial Board for *Systems*, past editor-in-chief of the *Systems Research Forum,* and past associate editor of the *Guide to the Systems Engineering Body of Knowledge.*

Appendix A

TRL	Criteria	DoDAF-described models		
1	**Basic principles observed and reported**			
	Do rough calculations support the concept?	CV-1,3	OV-6b	
	Do basic principles (physical, chemical, mathematical) support the concept?	CV-1	SV-3	
	Does it appear the concept can be supported by hardware?	OV-2	CV-6	
	Are the hardware requirements known in general terms?	OV-1,2		
	Do paper studies confirm basic scientific principles of new technology?	AV-1		
	Have mathematical formulations of concepts been developed?	FFP	AV-1	
	Have the basic principles of a possible algorithm been formulated?	AV-1		
	Have scientific observations been reported in peer reviewed reports?	FFP	AV-1	
	Has a sponsor or funding source been identified?	AV-1	CV-5	OV-1
	Has a scientific methodology or approach been developed?	AV-1		
2	**Technology concept and/or application formulated**			
	Has potential system or component applications been identified?	OV-1	PV-2	SV-1
	Have paper studies confirmed system or component application feasibility?	SV-1		
	Is the end user of the technology known?	AV-1	OV-1	
	Has an apparent design solution been identified?	OV-2,5a,5b	SV-1,2,6	
	Have the basic components of the technology been identified?	OV-1,2,4	SV-1,2,4,	
	Has the user interface been defined?	StdV-1,2	SV-1,2,4	
	Have technology or system components been at least partially characterized?	SV-3,4,5a,6		
	Have performance predictions been documented for each component?	SV-9		
	Has a customer expressed interest in application of technology?	CV-6	OV-5a,5b	
	Has a functional requirements generation process been initiated?	SV-3,4,5a,6	PV-2	
	Does preliminary analysis confirm basic scientific principles?	AV-1	SV-4	ScrV-4
	Have draft functional requirements been documented?	CV-1	SV-4	

(Continued)

TRL	Criteria	DoDAF-described models		
	Have experiments validating the concept been performed with synthetic data?	CV-1		
	Has a requirements tracking system been initiated?	PV-2	StdV-1	
	Are basic scientific principles confirmed with analytical studies?	AV-1	StdV-1,2	
	Have results of analytical studies been reported to scientific journals, etc.?	AV-1		
	Do all individual parts of the technology work separately? (No real attempt at integration)	AV-1	StdV-1	
	Is the hardware that the software will be hosted on available?	OV-2,5a,5b		
	Are output devices available?	OV-2,5a,5b		
3	**Analytical and experimental critical function and/or characteristic proof-of-concept**			
	Have predictions of components of technology capability been validated?	CV-2	PV-1,2	
	Have analytical studies verified performance predictions and produced algorithms?	CV-2		
	Can all science applicable to the technology be modeled or simulated?	CV-2		
	Have system performance characteristics and Measures been documented?	SV-4,7	SvcV-4,7	
	Do experiments/M&S validate performance predictions of technology capability?	SvcV-7		
	Does basic laboratory research equipment verify physical principles?	CV-2		
	Do experiments verify feasibility of application of technology?	CV-2		
	Do experiments/M&S validate performance predictions of components of technology capability?	CV-2		
	Has customer representative to work with R&D team been identified?	AV-1		
	Is customer participating in requirements generation?	CV-6	PV-1	
	Have cross-technology effects (if any) been identified?	SV-3	SV-2	
	Have design techniques been identified and/or developed?	OV-5a,5b	SV-5a,5b	
	Do paper studies indicate that technology or system components can be integrated?	FFP	CV-2	
	Has Technology Transition Agreement (TTA) including possible TRL for transition been drafted?	CV-3	SV-9	
	Are the technology/system performance metrics established?	SV-7	SvcV-7	
	Have scaling studies been started?	PV-2		
	Have technology/system performance characteristics been confirmed with representative data sets?	OV-1	PV-1	SV-1
	Do algorithms run successfully in a laboratory environment, possibly on a surrogate processor?	PV-2	SV-7	

TRL	Criteria	DoDAF-described models		
	Have current manufacturability concepts been assessed?	CV-5	PV-2,3	
	Can key components needed for breadboard be produced?	StdV-1	SV-1	
	Has analysis of alternatives been completed?	PV-2,3		
	Has scientific feasibility of proposed technology been fully demonstrated?	OV-6a	PV-2	SV-4
	Does analysis of present technologies show that proposed technology/system fills a capability gap?	CV-1,2	SV-8	
4	**Component and/or breadboard validation in laboratory environment**			
	Low fidelity hardware technology system integration and engineering completed in a lab environment	PV-2	SV-1,2,3,4,6	
	Technology demonstrates basic functionality in simplified environment	FFP	PV-1,2	
	Scaling studies have continued to next higher assembly from previous assessment	FFP	PV-1,2,3	
	BMDS mission enhancement(s) clearly defined within goals of study	CV-2,4,6	PV-1	
	Integration studies have been started	FFP	SV-1,2,3,4,6	
	Draft conceptual hardware and software designs (provide copy of documentation)	OV-4,5a,5b, 6a,6b		PV1,2
	Some software components are available	CV-3	OV 2,3	
	Piece parts and components in pre-production form exist. Provide documentation	PV-1	SV-3	
	Production and integration planning have begun. Documentation	SV-1,2,3,4,6,10a		
	Performance metrics have been established	CV-3	SV-7	SvcV-7
	Cross technology issues have been fully identified	SV-1,2,3,4,6,10a		
	Design techniques have been defined to the point where:	OV-3,4,	PV-2	
	Begin discussions/negotiations of Technology Transition Agreement	CV-3	SV-9	
5	**Component and/or breadboard validation in relevant environment**			
	High fidelity lab integration of hardware system completed and ready for testing in realistic simulated environment	SV-1,2,3,4,6,10a		
	Preliminary hardware technology engineering report completed	CV-3	PV-2	
	Detailed design drawings have been completed. Three view drawings and wiring diagrams have been submitted	CV-3	PV-2	

(Continued)

TRL	Criteria	DoDAF-described models		
	Pre-production of hardware available	CV-4,5	PV-1,2	
	Form, fit, function for application has begun to be addressed in conjunction with end user and development of staff	CV-3	PV-2	
	Cross technology effects (if any) identified and established through analysis	SV-1,2,3,4,6,10a		
	Design techniques have been defined to the point where largest problems defined	CV-4	PV-1,2	
	Scaling studies have continued to next higher assembly from previous assessment	SV-1,2,4,6,9,10a	PV-2	
	TTA has been updated to reflect data in items 1 thru 3, 5, 8	CV-4,5	PV-1,2	
6	**System/subsystem model or prototype demonstration in a relevant environment**			
	Materials, process, design, and integration methods have been employed			
	Scaling issues that remain are identified and supporting analysis is complete	SV-1,2,3,4,6,9,10a		
	Production demonstrations are complete. Production issues have been identified and major ones have been resolved	PV-1,2,3	SV-9	FFP
	Some associated "Beta" version software is available	CV-3	OV 2,3	
	Most pre-production hardware is available	CV-3	SV-6	
	Draft production planning has been reviewed by end user and developer	CV-5,6	PV-2	
	Draft design drawings are nearly complete			
	Integration demonstrations have been completed, including cross technology issue Measurement and performance characteristic validations	SV-1,2,3,4,6,9,10a		
	Have begun to establish an interface control process	SV-2,3,8	SvcV-2,3,8	
	Collection of actual maintainability, reliability, and supportability data has been started	CV-1,5	PV-2	SV-2,3,5a
	Representative model or prototype is successfully tested in a high- fidelity laboratory or simulated operational environment			
	Hardware technology "system" specification complete	SV-4,5a	SvcV-4, 5a	
	Technology Transition Agreement has been updated to reflect data in items 1 through 4, 7 through 9, 11, and 12	CV-4,5	PV-1,2	

TRL	Criteria	DoDAF-described models		
7	**System prototype demonstration in a space environment**			
	Material, processes, methods, and design techniques have been identified	SV-4	StdV-1	
	Scaling is complete	PV-2		
	Production planning is complete	CV-5,7		
	Pre-production hardware and software is available in limited quantities	PV-2	SV-3	
	Draft design drawings are complete	FFP	PV-2	
	Maintainability, reliability, and supportability data growth is above 60% of total needed data	PV-2	SV-7	SvcV-7
	Hardware technology "system" prototype successfully tested in a field environment	PV-2	CV-3	
8	**Actual system completed and qualified through test and demonstration**			
	Interface control process has been completed			
	Maintainability, reliability, and supportability data collected, and has been completed	CV-1,5	OV-4	PV-2
	Hardware technology successfully completes developmental test and evaluation	CV-1,5	OV-2,3,4	PV-2
	Hardware technology has been proven to work in its final form and under expected conditions	FFP	OV-All	SV-ALL, SvcV-All
9	**Actual system "flight proven" through successful mission operations**			
	Hardware technology successfully completes operational test and evaluation	CV-ALL	FFP	OV-All
	Training plan has been implemented	CV-1,5	OV-4	PV-2
	Supportability plan has been implemented	SV-1,2,3,4,6,10a		
	Program Protection plan has been implemented	AV-2	CV-1,2,4	PV-2
	Safety/Adverse effects issues have been identified and mitigated	SV-1,2,3,4,6,10a		
	Operation concept has been implemented successfully	OV- ALL	PV-All	SV-All

IRL	Criteria	DoDAF-described models	
1	**An interface between technologies has been identified with sufficient detail to allow characterization of the relationship**		
	Principal integration technologies have been identified	SV-1	
	Top-level functional architecture and interface points have been defined	SV-1	
	Availability of principal integration technologies is known and documented	CV-3,6	PV-2
	Integration concept/plan has been defined/drafted	FFP	
	Integration test concept/plan has been defined/drafted	FFP	
	High-level Concept of Operations and principal use cases have been defined/drafted	CV-1	
	Integration sequence approach/schedule has been defined/drafted	FFP	SV-8,9,10
	Interface control plan has been defined/drafted	OV-2	
	Principal integration and test resource requirements (facilities, hardware, software, surrogates, etc.) have been defined/identified	FFP	SV-2,3
	Integration & Test Team roles and responsibilities have been defined	SvcV-10a	
2	**There is some level of specificity to characterize the interaction (i.e. ability to influence) between technologies through their interface**		
	Principal integration technologies function as stand-alone units	SV-1	
	Inputs/outputs for principal integration technologies are known, characterized and documented	SV-2,5a	
	Principal interface requirements for integration technologies have been defined/drafted	FFP	SV-2,6
	Principal interface requirements specifications for integration technologies have been defined/drafted	SV-1	
	Principal interface risks for integration technologies have been defined/drafted	FFP	SV-1
	Integration concept/plan has been updated	FFP	
	Integration test concept/plan has been updated	FFP	
	High-level Concept of Operations and principal use cases have been updated	FFP	OV-1
	Integration sequence approach/schedule has been updated	FFP	PV-2
	Interface control plan has been updated	SV-1	
	Integration and test resource requirements (facilities, hardware, software, surrogates, etc.) have been updated	SV-2,3	
	Long lead planning/coordination of integration and test resources have been initiated	SV-6	

IRL	Criteria	DoDAF-described models		
	Integration & Test Team roles and responsibilities have been updated	OV-2,4		
	Formal integration studies have been initiated	FFP		
3	**There is compatibility between technologies to orderly and efficiently integrate and interact to include all interface details**			
	Preliminary Modeling & Simulation and/or analytical studies have been conducted to identify risks & assess compatibility of integration technologies	FFP		
	Compatibility risks and associated mitigation strategies for integration technologies have been defined (initial draft)	FFP		
	Integration test requirements have been defined (initial draft)	FFP	SV-1	
	High-level system interface diagrams have been completed	SV-1		
	Interface requirements are defined at the concept level	FFP	SV-5a	
	Inventory of external interfaces is completed	SV-3		
	Data engineering units are identified and documented	StdV-1, SV-2		
	Integration concept and other planning documents have been modified/updated based on preliminary analyses	SV-3,8		
4	**Validation of interrelated functions between integrating components in a laboratory environment**			
	Quality Assurance plan has been completed and implemented	FFP	PV-2	
	Cross technology risks have been fully identified/characterized	SV-1		
	Modeling & Simulation has been used to simulate some interfaces between components	OV-5a/b, 6a/b		
	Formal system architecture development is beginning to mature			
	Overall system requirements for end users' application are known/baselined	AV-1	CV-1	
	Systems Integration Laboratory/Software test-bed tests using available integration technologies have been completed with favorable outcomes	SV-1,4	PV-2	
	Low fidelity technology "system" integration and engineering has been completed and tested in a lab environment	SV-1,4	PV-3	
	Concept of Operations, use cases, and Integration requirements are completely defined	FFP	OV-2	SV-2
	Analysis of internal interface requirements is completed	SV-1		

(*Continued*)

IRL	Criteria	DoDAF-described models	
	Data transport method(s) and specifications have been defined	SV-2	
	A rigorous requirements inspection process has been implemented	FFP	
5	**Validation of interrelated functions between integrating components in a relevant environment**		
	An Interface Control Plan has been implemented (i.e., Inter-face Control Document created, Interface Control Working Group formed, etc.)	FFP	
	Integration risk assessments are ongoing	FFP	
	Integration risk mitigation strategies are being implemented &risks retired	FFP	
	System interface requirements specification has been drafted	FFP	SV-1
	External interfaces are well defined (e.g., source, data formats, structure, content, method of support, etc.)	SV-3	
	Functionality of integrated configuration items (modules/functions/assemblies) has been successfully demonstrated in a laboratory/synthetic environment	PV-3	SV-8
	The Systems Engineering Management Plan addresses integration and the associated interfaces	FFP	
	Integration test metrics for end-to-end testing have been defined	SV-7	
	Integration technology data has been successfully modeled and simulation	SV-5a/5b	
6	**Validation of interrelated functions between integrating components in a relevant end-to-end environment**		
	Cross technology issue measurement and performance characteristic validations completed	SV-7	
	Software components (operating system, middleware, applications) loaded onto subassemblies	SV-2	
	Individual modules tested to verify that the module components (functions) work together	PV-2	SV-8
	Interface control process and document have stabilized	FFP	
	Integrated system demonstrations have been successfully completed	FFP	PV-2
	Logistics systems are in place to support Integration	FFP	
	Test environment readiness assessment completed successfully	SV-8,9	
	Data transmission tests completed successfully	FFP	SV-2,3
7	**System prototype integration demonstration in an operational high-fidelity environment**		
	End-to-end Functionality of Systems Integration has been successfully demonstrated	SV-4	
	Each system/software interface tested individually under stressed and anomalous conditions	SV-4,8	

IRL	Criteria	DoDAF-described models	
	Fully integrated prototype demonstrated in actual or simulated operational environment	SV-8	
	Information control data content verified in system	SV-6	
	Interface, Data, and Functional Verification	SV-1,2,3,4	PV-2
	Corrective actions planned and implemented	FFP	SV-9
8	**System integration completed and mission qualified through test and demonstration in an operational environment**		
	All integrated systems able to meet overall system requirements in an operational environment	CV-6	PV-3
	System interfaces qualified and functioning correctly in an operational environment	SV-1,8	PV-2,3
	Integration testing closed out with test results, anomalies, deficiencies, and corrective actions documented	FFP	SV-9
	Components are form, fit, and function compatible with operational system	SV-1,2,3,3	
	System is form, fit, and function design for intended application and operational environment	SV-1,2,3,4	PV-2
	Interface control process has been completed/closed out	SV-1,2,3	
	Final architecture diagrams have been submitted	PV-2	
	Effectiveness of corrective actions taken to closeout principal design requirements has been demonstrated	FFP	
	Data transmission errors are known, characterized, and recorded	FPP	SV-3
	Data links are being effectively managed and process improvements have been initiated	FFP	SV-3,8
9	**System Integration is proven through successful mission-proven operations capabilities**		
	Fully integrated system has demonstrated operational effectiveness and suitability in its intended or a representative operational environment	CV-6	SV-8
	Interface failures/failure rates have been fully characterized and are consistent with user requirements	FFP	SV-1,7
	Lifecycle costs are consistent with user requirements and life-cycle cost improvement initiatives have been initiated	FFP	

Chapter 21

Managing Risk

Michael Orosz

University of Southern California, Information Sciences Institute, Marina del Rey, California, USA

Introduction

This chapter discusses approaches to managing risk during project development and sustainment with a particular focus on agile-based projects within large enterprise systems. In the *Guide to the Systems Engineering Body of Knowledge* (SEBoK), systems engineering risk is defined as:

> . . . a measure of the potential inability to achieve overall program objectives within defined cost, schedule, and technical constraints [which] has two components:
>
> 1) The probability (or likelihood) of failing to achieve a particular outcome and
> 2) The consequences (or impact) of failing to achieve that outcome (DAU 2003)"
>
> This definition is focused on identifying and managing risks to the processes used in designing, developing, and sustaining a system. The assumption is that the system being developed is targeted for an operating environment where customer needs and technical advances remain unchanged. The only concern in this development environment is ensuring that the product developed meets the original system requirements and is completed within a predefined budget and schedule. What is missing from this definition are the risks associated with failing to identify and capture evolving user, system, and technical requirements that are often driven by evolving end user or marketplace demands. Ignoring these evolving needs introduces risks, meaning you may be developing an obsolete product or one that only partially meets end-user needs.

This chapter is focused on managing the following three types of risks to systems engineering and development.

1) Project development risk
2) Technical risk
3) Obsolescence risk

Throughout the chapter, examples from various DoD space programs will be utilized.

Systems Engineering for the Digital Age: Practitioner Perspectives, First Edition. Edited by Dinesh Verma.

Companion website: www.wiley.com/go/verma/systemsengineering

Agile-Based Risk

Project Development Risk

Project development risk is focused on the failure to meet the cost and schedule constraints and technical needs associated with developing a system and/or failure to deliver it to the marketplace. This risk is closely associated with risk as referenced in SEBoK Editorial Board (2023) and is focused on the processes involved in developing the system.

Whether the development approach is a traditional waterfall or Vee model (e.g., Mooz and Forsberg 1991), agile (Agile Alliance 2023), or a hybrid of the two, this risk involves the probability of failing to accurately collect and understand systems requirements, design and develop appropriate system architecture, develop the system that meets the system requirements, and deploy and finally sustain the product.

Project risk is all about process. For example, does the project team have the necessary personnel to do the job (i.e., skillsets and experience); are the system requirements defined in sufficient detail for the project team to understand; are there adequate build, integration, and testing pipelines (e.g., DevSecOps pipeline (OWASP 2022)) available; and are sufficient verification and validation processes in place? Failure to address any of these areas will almost guarantee that the project will end up being over budget, over schedule, and/or will fail to meet end-user or market needs.

Technical Risk

Technical risk is focused on the probability of failing to incorporate new technologies and/or to recognize the evolution of external systems (e.g., within an enterprise) with which the product interacts. This risk is different from failure to initially collect and understand system requirements – both within the project and within the enterprise (which falls under project risk). Rather, this risk is focused on *addressing evolving technologies* as the project is under development or as deployed to the operating environment (i.e., the marketplace) during sustainment.

For example, in a space-based environment, there exists a technical risk that the project team responsible for the development of the ground command and control station will not be aware that there was a change in the capabilities of the space vehicle (i.e., the satellite), which is usually developed via another project and often by another vendor. This risk can also include failing to detect the performance limitations of a design. For example (and drawing from the space vehicle environment), an initial system requirement may specify a certain performance window for uploading command messages from ground control to the space vehicle. A failure to recognize that the performance window is insufficient and adapt to that realization will put the system being developed at risk of not meeting end-user/market needs.

Obsolescence Risk

Closely related to technical risk is obsolescence risk. This risk is focused on the probability of failing to adapt to changing end-user and/or market needs. This risk is present during both development and sustainment once the product or system has been deployed to the end-user community. During development, failure to incorporate evolving end-user needs will produce a system that ultimately will not meet end-user needs and requirements and will be obsolete upon deployment.

During sustainment, failure to recognize and incorporate evolving end-user needs and requirements will also lead to obsolescence. Reducing this kind of risk usually requires two actions: continuous monitoring of the marketplace and using a system architecture and sustainment process that is agile and can handle evolving end-user needs. More on this later in the chapter.

Managing Risk

Managing risk in a project or program (e.g., enterprise systems) involves focusing on system design, monitoring, and adaptation. We discuss the role of risk management of each of these areas below.

System Design

In this context, "system design" is the application of system theory to product development and includes many of the "early lifecycle" activities covered in the systems engineering Vee (Forsberg and Mooz 1994). Managing risks to a project often comes down to addressing both the design of the product (e.g., system architecture) and the processes used to develop and sustain that product. In the first case (i.e., system design), the system architecture needs to be defined for adaptability. Examples of adaptability include modular (e.g., Production and Manufacturing Management 200 and Baldwin and Clark 2000) and plug-and-play (e.g., Wang et al. 2008) architectures. Although this is simply good systems engineering, the focus is often on reducing system interfaces (for integration and testing purposes – discussed later in the chapter) and not necessarily on building systems that can adapt as technology and/or system requirements change. This is particularly a challenge when a project is extending an existing operating system that may not have a particularly modular system design. In these cases, it is often worth the cost to a project of first refactoring the system architecture (primarily software, but hardware artifacts if necessary) into a modular system prior to developing new capabilities.

Again using the space-based sector as an example, ground command and control operations typically involve the scheduling of multiple space vehicles and events. Scheduling is an active area of research, and evolved versions of schedulers frequently appear. Developing an architecture in which the scheduler is treated as a module (Figure 21.1) or a Configuration Item (CI) that can be replaced by a new or updated solution when available will help reduce technical risk and in some

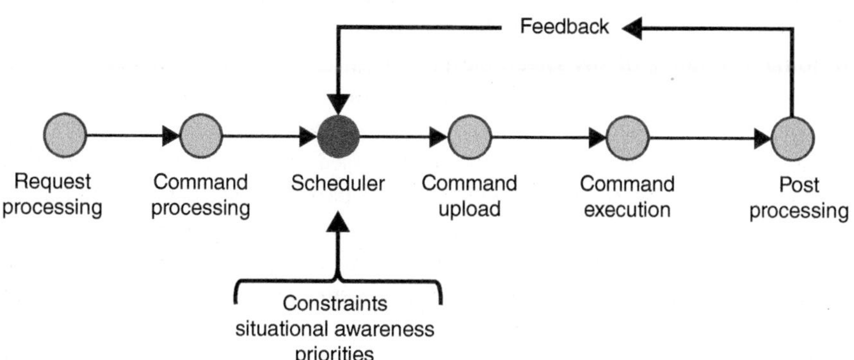

Figure 21.1 Scheduler technology is structured as a module that can be replaced with a new scheduler when available.

cases possible obsolescence risk (e.g., scheduler handles a new space vehicle design with unique operating characteristics).

For agile-based system development projects, developing a modular system architecture can often be a challenge. As noted elsewhere in this book (Chapter 31 Unique Challenges in Mission Engineering and Technology Integration and Chapter 10 Scaling Agile Principles to an Enterprise), agile-based projects often evolve the system design as the project is incrementally built and delivered to the end-user community. That said, and as noted in other chapters of the book, in many projects, particularly in hybrid agile/waterfall projects such as those often found in US DoD applications, some good upfront systems engineering is required to not only populate the project backlog with features and stories but to also lay out a modular system design (e.g., a black box system design). Often as the system is developed, the initial "black box" design may need to be modified to reflect changing system requirements and end-user needs. In such situations, it is recommended that the system design be refactored as part of the next development increment to maintain system modularity. Although this adds cost and can increase development schedules, the costs are often worth it for systems that are designed with long life cycles that include frequent updates.

Design to Reduce Integration and Testing Risk

As noted in the chapter "Unique Challenges in Mission Engineering and Technology Integration (6.4)," integration and testing can be divided into vertical and horizontal processes. Vertical integration covers integration and testing that does not involve a large number of external systems or interfaces. In other words, the integration and testing (I&T) effort is local to the module being built, integrated, and tested. For example, in Figure 21.1, the integration and testing of the scheduler module (the red filled-in circle) involves the contents of the scheduler and the interfaces between the external systems (Command Processing and Command Upload), data inputs required for the scheduler (scheduling constraints, priorities, and current situational awareness), and any execution feedback from the operational environment. In these situations, I&T can rely on simulators and "canned" data for the various inputs to the scheduler.

Horizontal I&T typically involves the complete system being developed as well as the interactions with other elements of the enterprise in which the targeted system connects. Further, these I&T efforts require testing within the actual enterprise or in a near-operational (NearOps) environment before system requirements can be signed off (i.e., system meets the system requirements). The challenge with both horizontal and vertical I&T is that the appropriate test environments may not be in place and accessible to the project team – particularly in the early stages of the project.

As noted in Chapter 31, it is highly recommended that these I&T environments be developed and in place prior to the beginning of any system development. This usually means establishing contracts with the appropriate vendors responsible for providing the test environments. Issues such as intellectual property (IP), licensing, periods of performance usage, number of users that can use the system, and other contractual issues must be addressed early in the program. Further, it is recommended that the suite of test cases that will be used as part of the verification and validation process should be developed as early as possible. This does not mean that these test cases cannot be modified later to reflect changes in the system design during development (i.e., agile processing) but the purpose of establishing an initial suite of test cases is to gauge the scope and scale of the I&T platforms.

Due to the complexity, scope, and scale of a NearOps environment (i.e., horizontal testing), in many projects there may only be one NearOps environment while there may be multiple horizontal testing environments (which typically are smaller in scope and scale). With this in mind, it is recommended (as already noted in a previous section of this chapter) that the system architecture be designed with *as few as possible modules with multiple interfaces* with other systems (both within and outside the system being developed). The lower the number of interfaces, the more likely vertical I&T can be used.

Digital Engineering

The use of digital engineering tools and processes can help reduce risk to the project. Whether starting a new project (i.e., a greenfield project) or extending an existing project (i.e., a brownfield project), it is highly recommended that model-based systems engineering (MBSE) tools be adopted and used – preferably tools that allow physics-based modeling to be included in the system model. This recommendation is particularly important for agile-based projects where new or evolved requirements and user feedback are frequently integrated into the incrementally developed system. Digital models can be used to determine upfront, before "metal is cut," the impacts of changes to the system architecture that may result when introducing new technologies into the project baseline. Physics-based modeling is important to capture the expected behavior of an operational system and predict the behavior of that system if new technology is inserted. Based on predicted behavior, decisions can be made on the appropriate actions to take to maximize "value" to the end user and the development team before incorporating a new system requirement into the baseline.

Another recommendation to consider is the use of digital twins (Wu et al. 2020). A digital twin is a digital representations of an actual operating system. As a system ages and undergoes modifications (either as part of a system-wide update or an update unique to that system), these modifications are incorporated into the digital twin of that system. These digital twins can be used to predict future performance, detect possible system failures, and predict – on an individual system basis – impacts to performance from possible system modifications.

Do Not Forget the Contracting Details

To help reduce project risk, it is essential to ensure contracting issues such as intellectual propriety (IP) ownership, licensing, usage rights, and other programmatic items are addressed early in the project contracting and development life cycle. For example, many software tools that help with project execution and management are developed and marketed by vendors outside of the United States. For many projects, this is not an issue; however, for US DoD-funded projects, there are restrictions on the purchase and use of technology manufactured and supported by vendors outside of the United States. Understanding these restrictions during contracting will reduce possible future development blockages that can increase the risk of a project not completing within budget or schedule.

Waterfall vs. Agile

There are advantages and disadvantages to using a waterfall or agile process for developing (software and non-software) systems. From a risk reduction perspective, an advantage of the agile approach is the ability to incorporate evolving end-user needs and requirements into the project baseline during development. Using this approach can help reduce the risk of producing an obsolete system, as the system evolves as the end-user needs evolve. Further, since the agile approach is designed to produce "value" to the end-user community on a frequent basis (i.e., incremental releases), the risk of delivering a non-useable system is greatly reduced in the event that project funding is reduced or a decision is made to terminate the project early. With waterfall methods, a usable system is typically produced at the end of the project, which may never be reached due to funding constraints or changing customer priorities.

That said, the risk of not producing a system or producing an obsolete system can also be minimized in waterfall-based projects. Most waterfall-based projects can be divided into segments – called Mini-Vs (Chapter 31) – in which each waterfall-based segment produces value to the end user. Another advantage of using the Mini-Vs approach is that when combined with modular contracting, each new segment can be competitively bid, allowing the customer the opportunity to get the best value possible as the overall project grows.

Finally, another approach to consider is the Spiral Model (Boehm 1986) in which the development team at each successive iteration addresses the highest risk elements solving each in successive iterations (spirals).

Situational Awareness

Fundamental to risk management is accurate and up-to-date situational awareness. Whether during system development or sustainment, understanding the environment aids decision-making and reduces risk. During system development, metrics such as the number of discrepancies (i.e., bugs) detected during testing, software lines of code developed to date (e.g., waterfall project), features attempted and completed during a program increment (e.g., an agile project), and many other metrics can provide decision-makers with a "picture" of the health of the development process. For example, a large number of "roll-over" features (i.e., features that don't get completed in a SAFe® Scaled Agile Framework (Scaled Agile 2022) – based program increment) may indicate challenges within the sprint team. In such circumstances, mitigation steps can be taken to address the challenges and help prevent the problem from cascading and introducing risk to budget and schedule.

End-User and Enterprise Engagement

Fundamental to any project – and especially agile-based projects – is end-user engagement on a frequent basis. End-user needs evolve and the sooner these needs are incorporated into the system baseline, the less likely a non-useable system will be produced. The challenge to meeting this requirement is the availability of the end-user community. This is particularly true in many US DoD projects focused on mission-critical solutions where the end-user community is typically unavailable to evaluate the most recent incremental release (i.e., minimum viable product (MVP) or minimum marketable product (MMP)). For situations where engagement with the actual end-user community is challenging, it is recommended that the project team hire subject matter experts (SMEs) to be part of the team. Although this adds a layer of expense to the overall project, the cost is typically small in the long-term as the use of SMEs can reduce future re-work to address discrepancies produced when the actual end-user community uses the final product.

In addition to frequent end-user engagement or the hiring of SMEs for the project team, it is also highly recommended that at least one member of each sprint team also participate in the various technology working groups or technical interchange meetings (TIMs) within a large enterprise in which the system being developed operates. Having one or more team members aware of pending technology and/or operational changes can help the project and individual sprint teams predict system requirement evolution and take mitigation steps to reduce project risk. For example, if an enterprise-wide working group is signaling a concept of operations (ConOps) change is pending, the sprint team can push impacted features (i.e., an agile project) further down the project backlog (i.e., lower their priorities) and work other features while waiting for the working group to officially release the new ConOps.

Continuous Integration and Testing

Finally, project risk can be reduced through continuous integration. Whether taking a waterfall, agile, or hybrid approach, continuous I&T will often detect discrepancies early, allowing time for the project team to address them as part of the development effort. Waiting until later in the project to detect these discrepancies will often result in the cascading of discrepancies – often forming a bow-wave of issues that will have to be worked off at the end of an increment (agile) or the end of the project (waterfall) where team members may not be available to quickly address them.

Anticipating and Reacting to Risk

There are some general indicators of substantial risk such as working in new domains and with new technologies, new developers, dictated schedules, and a mismatch of talent to task. A risk management process includes not only determining risk factors and risk exposure but also developing strategies to handle risks when they are realized, which may include risk avoidance or transfer of risk management to a future release rather than correcting it immediately. During this process, it is useful to review the project with respect to potential categories of risk. The SEI Taxonomy of Software Development Risks Kendall et al. (2007) is a useful aid for this purpose.

Summary

This chapter discussed approaches to managing risk during project development and sustainment with a particular focus on agile-based projects within large enterprise systems. Risk is generally defined as the measure of the potential inability to achieve overall program objectives within defined cost, schedule, and technical constraints. What is missing from this definition are risks associated with failing to identify and capture evolving user, system, and technical requirements, which are often driven by evolving end-user or marketplace demands. Ignoring these evolving needs introduces risks, meaning you may be developing an obsolete product or one that only partially meets end-user needs. This chapter focused on managing three types of risks to systems engineering and development: project development risk, technical risk, and obsolescence risk.

References

Agile Alliance (2023). *Agile 101*. https://www.agilealliance.org/agile101/. Corryton, TN: The Agile Alliance.

Baldwin, C.Y. and Clark, K.B. (2000). *Design Rules: Volume 1. The Power of Modularity*. Cambridge, MA: MIT Press.

Boehm, B. (1986). A spiral model of software development and enhancement. *ACM SIGSOFT Software Engineering Notes* 11 (4): 14–24.

DAU (2003). *Risk Management Guide for DoD Acquisition*, 5e. Belvoir, VA, USA: Defense Acquisition University (DAU)/U.S. Department of Defense, Fifth Edition, Version 2.

Forsberg, K. and Mooz, H. (1994). The relationship of system engineering to the project cycle. *Proceedings of the 12th INTERNET World Congress on Project Management*, Oslo, Norway (9–11 June 1994).

Kendall, R.P., Post, D.E., Carver, J.C. et al. (2007). *A Proposed Taxonomy for Software Development Risks for High-Performance Computing (HPC) Scientific/Engineering Applications*. CMU/SEI-2006-TN-039. Pittsburgh, PA: Software Engineering Institute (SEI), Carnegie Mellon University. https://resources.sei.cmu.edu/asset_files/technicalnote/2007_004_001_14744.pdf.

Mooz, H. and Forsberg, K. (1991). The relationship of systems engineering to the project cycle. *Joint Conference of National Council on Systems Engineering (NCOSE) and the American Society for Engineering Management (ASEM)*, Chattanooga, TN (21–23 October 1991). https://onlinelibrary.wiley.com/doi/abs/10.1002/j.2334-5837.1991.tb01484.x.

OWASP (2022). *DevSecOps Pipeline*. Wakefield, MA: Open Web Application Security Project (OWASP). https://owasp.org/www-project-devsecops-guideline/.

Scaled Agile (2022). *What is SAFe®?* https://scaledagile.com/what-is-safe/. Boulder, CO: Scaled Agile.

SEBoK Editorial Board (2023). Risk (glossary). In: *The Guide to the Systems Engineering Body of Knowledge (SEBoK)*, v. 2.8 (ed. R.J. Cloutier). Hoboken, NJ: The Trustees of the Stevens Institute of Technology. www.sebokwiki.org. BKCASE is managed and maintained by the Stevens Institute of Technology Systems Engineering Research Center, the International Council on Systems Engineering, and the Institute of Electrical and Electronics Engineers Systems Council. https://www.sebokwiki.org/wiki/Risk_(glossary)#:~:text=From%20SEBoK-,risk,failing%20to%20achieve%20that%20outcome.

Wang, S., Avrunin, G.S., and Clarke, L.A. (2008). Plug-and-play architectural design and verification. In: *Architecting Dependable Systems V* (ed. R. de Lemos). Berlin, Germany: Springer-Verlag. http://ext.math.umass.edu/~avrunin/papers/wang08-plug_play_design_and_verif.pdf.

Wu, J., Yang, Y., Cheng, X., et al. (2020). The development of digital twin technology review. *Proceedings of the 2020 Chinese Automation Congress (CAC), Shanghai, China*. https://ieeexplore.ieee.org/abstract/document/9327756/authors.

Biographical Sketches

Michael Orosz directs the Decision Systems Group at the University of Southern California's Information Sciences Institute (USC/ISI) and is a research associate professor in USC's Sonny Astani Department of Civil and Environmental Engineering. Dr. Orosz has over 30 years' experience in government and commercial software development, systems engineering and acquisition, applied research and development, and project management and has developed several successful products in both the government and commercial sectors. Dr. Orosz received his BS in engineering from the Colorado School of Mines, an MS in computer science from the University of Colorado, and a PhD in computer science from UCLA.

Part V

Model-Based Design of Safety, Security, and Resilience Systems

Tom McDermott

Chapter 22

Concepts of Trust and Resilience in Cyber-Physical Systems

Thomas McDermott[1], Megan M. Clifford[1], and Valerie B. Sitterle[2]

[1] *Stevens Institute of Technology, School for Systems and Enterprises, Hoboken, NJ, USA*
[2] *Electronic Systems Laboratory, Georgia Tech Research Institute, NW Atlanta, GA*

Trust and Resilience in Cyber-Physical Systems

Background and Introduction

Dependable and Secure Computing in Cyber-Physical Systems

Cyber-physical systems (CPS) are "engineered systems that are built from, and depend upon, the seamless integration of computational algorithms and physical components" (NSF 2016). CPS are susceptible to security threats due to the interconnected nature of their control activities, which can be vulnerable to intrusion from both local and remote adversaries. The design of CPS must address the security of these control activities and their interactions with human and machine systems. CPS are often systems of systems, increasing their attack surfaces and making it difficult to identify system boundaries. The architectural design of CPS should support their nested structure and hierarchical layers of control, with resilience as a primary architectural attribute. To address the challenges posed by CPS, traditional approaches to security, privacy, reliability, resilience, and safety may be insufficient, and the design of CPS must incorporate new and evolving methods to ensure their secure and safe functioning. Characteristics of resilience in CPS include:

- Supervisory control methods that support graceful degradation of the system in the presence of failures or malicious attacks.
- Maintenance of system availability using dynamic and potentially distributed control system elements.
- Mechanisms to ensure the integrity of control functions in the presence of failures or malicious attacks.
- Ability to address a range of human interactions across differing system performance levels.
- Ability to recognize and respond to failures that disrupt CPS elements or control functions to maintain levels of performance or to maintain safety and security.
- Ability to recognize and respond to intentional disruptions of CPS elements or control functions to maintain levels of performance or to maintain safety and security.
- Ability to protect and maintain privacy of CPS system data or control states.

Systems Engineering for the Digital Age: Practitioner Perspectives, First Edition. Edited by Dinesh Verma.

Companion website: www.wiley.com/go/verma/systemsengineering

- Ability to evaluate CPS system resilience in different environments.
- Need to determine trade-offs in system performance, safety, and security in conditions of complexity, changing threat environments, and emerging or unplanned use cases.

The concept of incorporating these characteristics into systems has a long history. However, the context in which they are being applied in CPS has evolved due to the increasing complexity of system control methods and the constantly changing threat landscape. This is a result of the integration of computers and networks in control functions, the widespread use of digital data, and the interconnected nature of cyber threats. In CPS, cyber security and physical security are intertwined and interdependent, meaning that even if each individual domain is secure, the system may still be vulnerable due to the interactions between the domains.

General Concepts of Resilience for CPS

The general framework of cyber-physical systems (CPS) resilience can be found within the engineering principles of dependable and secure computing systems. Resilience in CPS refers to the capability of the system to withstand and resist threats by utilizing specific design attributes and patterns that result in resilient outcomes. Figure 22.1 provides a general overview of the framework.

The resilience of CPS can be divided into two main categories: dependability and security attributes. Dependability and security refer to the system's ability to prevent service failures and encompasses the foundational attributes of availability, reliability, safety, integrity, confidentiality, and maintainability (Avižienis et al. 2004). These attributes must work in harmony to ensure the successful operation of the system.

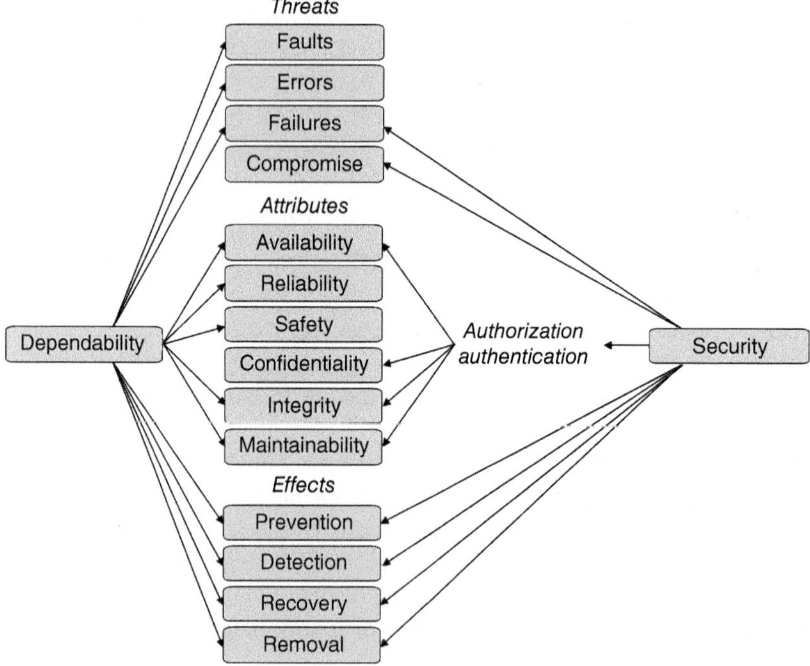

Figure 22.1 Related concerns for dependability and security in CPS. *Source:* Adapted from Avižienis et al. (2004).

It is noteworthy that there is a significant overlap between the characteristics of dependability, such as availability, reliability, safety, integrity, and maintainability, and those of security, including confidentiality, integrity, and availability. The definition of dependability in the context of resilience refers to the system's ability to avoid service failures that occur more frequently or with greater severity than is acceptable (Avižienis et al. 2004). This includes security failures.

Security, on the other hand, involves ensuring that access to the system is authorized and that access is properly authenticated, preserving the confidentiality of both system functions and data, and ensuring that the system operates as intended, without unauthorized functionality (Avižienis et al. 2004).

In the context of CPS, the relationship between dependability and trust is defined by the dependence of one system on another, and the assumption that the other system is also dependable (Avižienis et al. 2004). This dependence can be either human–machine or machine–machine in nature. The NIST framework highlights the importance of resilience in terms of trustworthiness, referring to the ability of the CPS to withstand instability, unexpected events, and return to predictable performance, even if degraded. This is a system-of-systems (SoS) concern, and CPS resilience must be considered both at the individual CPS level and in terms of relationships between CPS and other systems.

The design of CPS must take into account the trade-offs between interrelated attributes. For example, designs that enhance safety may reduce the efficacy of actions for cybersecurity, resilience, or reliability (NIST 2016). Trustworthy CPS architectures must be based on a comprehensive understanding of the physical properties and constraints of the system, the context of threats, attributes, and effects. The design process must also involve the creation and simulation of adversary models and physical and functional models of the CPS (NIST 2016).

The overall outcome of CPS resilience is reflected in the operational characteristics of the SoS. Using a definition from the U.S. Department of Defense (DoD), hardware and software components of systems should have the ability to reconfigure, optimize, self-defend, and recover with minimal human intervention. The DoD further defines three aspects of resilience in an operational context: the systems must be trustworthy, missions should be able to tolerate degradation or loss of resources, and the systems should have designs that provide means to prevail in adverse events (DoDI 2014).

In terms of knowledge and skills, a CPS systems engineer must have a comprehensive understanding of all aspects related to resilient CPS, including threats, system operation and human interactions, system vulnerabilities, approaches for resilient design, and validation of the system. Key design principles for resilient CPS (Reed 2016) include secure access control to and use of the system and its resources (system security engineering); understanding of design attributes that minimize exposure to external threats (systems security engineering and dependable computing); understanding of design patterns to produce effects that protect and preserve system functions or resources (dependable computing); approaches to monitor, detect, and respond to threats and security anomalies (cybersecurity); understanding of network operations and external security services (information systems); and approaches to maintain system availability under adverse conditions (all of the above).

Basis for a Functional Approach to Analysis of Dependability and Security

This section focuses on the importance of interdisciplinary approaches in the design and evaluation of CPS, with a particular emphasis on rigorous system modeling. The concept of cybersecurity must be understood within the larger context of resilience, with security aimed at protecting systems from malicious adversaries. The process of designing CPS typically begins by defining critical and necessary functionality, which is then decomposed into specific functional capabilities that drive system requirements. Boehm and Kukreja (2015) differentiate between functional and non-functional requirements, with the latter encompassing attributes such as security.

Traditionally, systems theory has emphasized the distinction between a system's internal structure and its external behavior, with the latter derived from the former. This approach has led to a focus on structural representations of systems in cybersecurity protection methods. However, CPS are heterarchical in nature and consist of multiple, diverse elements that interact both independently and interdependently. As a result, traditional decomposition and predictive methods are inadequate to capture the complexity of CPS. In complex systems, structure and function are intrinsically linked, and a system's structural characteristics shape its processes and behaviors.

Despite the limitations of traditional approaches, many current methods continue to focus on structural system representations. This structural bias can lead to confusion between resilience and robustness or reliability, and result in a focus on perimeter protection methods and a narrow view of threats as akin to failure modes. Cyber threats can attack both the structure and function of a system and produce effects that do not resemble typical system failures. Even graph-based methods, which are discussed in this chapter, often evaluate the resilience of complex systems based on structural properties, such as connectedness.

The literature has extensively explored the impact of node and link additions and removals on the structural properties of networks, and how networks can be made more robust against such perturbations. However, these perturbations are typically treated as external sources, not integrated into the network dynamics itself. While there has been extensive research on network dynamics, it is important to also study the dynamical processes on networks to fully capture the functional preservation of CPS.

Model-Based Assurance for CPS

The complexities of engineering design for resilient CPS are compounded by the need for system assurance. System assurance refers to the justified confidence that the system functions as intended and is free of exploitable vulnerabilities, whether intentionally or unintentionally designed or inserted at any time during the system's life cycle. A systematic set of multi-disciplinary activities is required to achieve the acceptable measures of system assurance and manage the risk of exploitable vulnerabilities.

Assurance is a set of engineering practices that serve to evaluate the design of the CPS for a reasonable set of dependability and security requirements based on the operational use of the system. In the CPS design context, the selection of technologies and processes that provide confidence the system operates as intended in the presence of both accidental and intentional vulnerabilities is the concept of "designing in" CPS dependability and security. This requires a multidisciplinary set of methods and tools, which are described in the remainder of the chapter with respect to threats, tools, modeling methods, and assurance methods.

Model-based assurance (MBA) is a process that supports system verification and validation requirements using conceptual and analytical modeling and simulation techniques. With the increasing prevalence of digital engineering and model-based systems engineering (MBSE), there is potential to transform traditional system assurance processes to more holistic and evidence-based forms using models. However, MBA faces challenges to establish the best practices and systems engineering foundations required to produce evidence to support assurance judgments. This is particularly true for CPS, which are often employed in SoS operational configurations, increasingly connected and complex, and facing diverse and sophisticated threats (NATO 2010).

MBA is the use of a model or group of models to produce evidence that a given system will perform as intended in various potential environments, operational conditions, and arrangements with other systems. To ensure that a model allows determination of whether a system design meets functional and nonfunctional requirements, the model's accuracy in representing system functional performance and characteristics must be ensured. Any final system implementation using model-based design approaches in an MBA context should lead to a system design that is safe and secure or analyzable with respect to its safety and security.

CPS create new challenges to the concept of MBA, specifically regarding what model-based approaches capture relevant and representative levels of abstraction sufficient to help validate the integrity of the system requirements and the design. To address these challenges, MBA seeks to build a model-based process in concert with existing MBSE practices to produce an evidentiary case that a system is trustworthy with respect to the properties its stakeholders legitimately rely upon within acceptable levels of risk.

The focus on analytic decomposition in many current model-based approaches needs to be augmented by approaches that enforce safe behavior, such as state-based evaluation of dynamic control. Assurance must cover any undesired or unplanned event that results in a loss and address hazards and vulnerabilities as a system state or set of conditions that, together with worst-case environmental conditions, will lead to a loss. To satisfy analysis for the purpose of assurance, a model that is an abstraction of a system's functional behavior should represent failure effects in the system, how failures propagate through the system, and observable conditions those failures manifest. A mission critical CPS must consider all classes of system failures, whether inherent or malicious, in rapidly changing external SoS contexts.

Future MBA methods, processes, and tools must go beyond traditional quality assurance scope to include the emergent dimensionality of the design space through the evolving quantification of concepts such as flexibility and resilience. They must also create a framework for describing these concepts as patterns of design that can be captured into standard practices and tool libraries. In addition, solutions can then be discussed in capabilities more dynamically. The next sections discuss these generalized patterns.

CPS Threats and Countermeasures

The unique nature of CPS means that the threats they face differ from those of traditional IT systems. Griffor has identified two specific concerns that are unique to CPS: the ability to tolerate intrusion and disruption of signal/information flow, and the need to ensure that legitimate control commands are not subject to insertion, fabrication, or replay (Griffor 2016). Table 22.1 provides a list of various types of CPS attacks, categorized according to the specific vulnerabilities that they exploit, such as software vulnerabilities or flawed network implementations, physical vulnerabilities, or a combination of both cyber and physical vulnerabilities (McDermott and Horowitz 2017).

Wan et al. further defined a set of generalized control system attack models of specific concern to CPS shown in Table 22.2 (Wan et al. 2015).

The use of attack trees and attack-graph-based visualizations provides a straightforward and widely understood approach for modeling complex coordinated attacks and assessing the impacts of attacks and countermeasures. In the context of CPS security, attack graph analysis is a commonly used method to evaluate an attacker's ability to gain access to cyber assets and exploit system functions. To facilitate this analysis, human-centered workshops can be used to define cyber assets in terms of system functions and explore the dependent functions that may be vulnerable to attack. These workshops can help identify common patterns of attack, which can then be combined in various ways to simulate different attack scenarios. When modeling CPS functionality,

Table 22.1 Taxonomy of CPS attack types.

Cyber attacks	Description
Network denial of service	Disrupt network operations to stop control signal/flow during the attack period
Malware	Installation of malicious software, or viruses, into the device
Software vulnerability exploitation	Exploitation of software vulnerabilities such as buffer overruns, numerical overflows, or backdoor applications; can be defects or intentionally created
Physical attacks	Description
Sensor spoofing	Providing false sensor data to the CPS, either by injecting false information into the sensor or into the communication paths
Signal jamming	Disrupting control/signal flow by preventing or changing the signal, primarily using external signals
Hardware vulnerability exploitation	Exploitation of hardware vulnerabilities such as network protocol errors, or side channel attacks on power, cooling, or other flows
Physical damage	Physically disrupting the CPS to disrupt control signal/flow, such as damaging interconnects or sensing surfaces
Cyber-physical attacks	Description
Insider threat	Stealing and/or modifying design or operational data for exploitation; emphasizes the importance of the CPS design environment
Identity spoofing	Providing signals or information to the CPS that appears to be legitimate, such as false network packets; used in man-in-the-middle attacks
Supply chain compromise	Introducing a flawed or vulnerable hardware or software component into a CPS when it is being manufactured or configured
Information disclosure	Monitoring CPS information to gather information needed for additional attacks or to steal private information
Social engineering	Deception or influence on the CPS engineer or operator in order to gain information for other attacks, or possible to induce poor safety or security decisions into the design
Replay attacks	Recording the control signal/flow over a period of time, in order to replace the actual signal/flow with the recorded data to confuse the system
Control system instability	Disrupting control/signal flow in order to produce control system instability

Table 22.2 Generalized control system attack models.

Attack model	Description
Interruption attack	Also called denial-of-service attack, stops the control signal/flow during the attack period
Man-in-the middle attack	Mimics the human attack behavior. When the attack happens, the control signal/flow is changed to a different manipulated signal/flow controlled by the attacker
Replay attack	Records the control signal/flow over a period of time in a vector, and when the attack starts, it replaces the actual signal/flow with the recorded data to confuse the system
Control parameter attack	Modifies the vulnerable control parameters of the system to the attacker's defined parameters. This changes the quality of control of the system
Coordinated attack	Combines two or more basic attack models, for example, combining a man-in-the-middle attack with a control parameter attack

threat parameters should be considered as additional functions to be captured in the control system design decomposition process, resulting in a graph that depicts both the control function and the associated attack.

For countermeasure patterns of specific interest to CPS was also developed by Wan et al. and is provided in Table 22.3 (Wan et al. 2015).

A functional model of a CPS can effectively combine intended control functions, threat functions, and countermeasure functions into a single model using selected patterns. This approach allows for the analysis of "reasonable assurance" by assessing the cost trade-offs associated with various countermeasure patterns. Attack graph tools already support cost/risk analyses. For instance, the Control Parameter Attack model requires moderate knowledge of the system's functional design, low attack-specific technical ability, and generally low resources from the adversary, while the countermeasure approach would have a low-to-medium implementation cost and low collateral impact to the system.

Basic countermeasure patterns can be extended to more complex attack strategies by combining them. For example, diverse redundancy is a pattern that employs redundant control systems with different hardware and/or algorithms that use voting or averaging to produce outputs. This mitigates the ability of a cyber threat to compromise all redundant elements in the system. The diversity and redundancy can be provided by components and algorithms added to the functional design of the control system. Horowitz and colleagues have collected a number of these redundancy strategies, as summarized in Table 22.4.

In the end, the result of this process is a model for a "cyber-protected system." This model comprises a collection of security design patterns, each described in terms of (1) its functional capabilities, (2) the cyber assets required to achieve those capabilities, (3) the critical cyber assets and/or functions that it will protect, and potentially (4) the specific threat functional capabilities and/or threat cyber assets that it is designed to detect or counter through direct action. These patterns, which are essentially new functions, can be incorporated into the CPS itself or into a separate monitoring device that tracks the CPS's functional behavior externally. Using a general model-based approach, different security patterns and countermeasure approaches can be implemented as updates to the CPS as new threats emerge. A variety of processes and tools are available to support this type of analysis, and the most commonly used ones are highlighted in the following section.

Tools to Evaluate Threat and Countermeasure Patterns

The design and analysis of dependable and secure CPS require new methods and tools due to their complexity and the potential for cyber-attacks. While traditional tools for fault-tolerant systems can be adapted, they are insufficient for resilient CPS. The failure analysis tools commonly used today represent the process as a set of linear causal models that sequence a series of events over time. However, they ignore nonlinear relationships, such as feedback, and tend to ignore long chains of events, which can lead to subjectivity in analysis.

A resilience analysis approach is required, which involves converting the concept of a failure in a CPS to one of availability or continuity of service. The analysis should consider vulnerable aspects, dependencies, and trustworthiness of system components. Hierarchies of control over essential CPS functions should be defined to reduce or eliminate dysfunctional interactions that can disrupt critical control functions or lead to potentially hazardous states in the controlled process.

The primary shortfall of causal event models is that they limit the analysis of countermeasure designs to linear cause-and-effect relationships. Therefore, resilience analysis should be embedded

Table 22.3 Generalized control systems countermeasure patterns.

Countermeasure	Description	Attack model countered
Isolation	Creates an isolated runtime environment (sandbox) for the critical asset that is resistant against attacks	Escalation, interruption attacks
Redundancy	Replicates the functionality of the critical asset in order to create multiple paths for high availability and fault tolerance in the case of individual function failures	Attacks that disable individual instances of critical assets and functionality
Diversification	Produces functionally equivalent variations of binaries running in software critical assets. This is an enhancement of the redundancy countermeasure	Coordinated attacks, zero-day attacks effective in identical binary copies of the critical assets
Physically unclonable function	Secures the integrity and privacy of the messages in the system using a Physical Unclonable Function (PUF) that is hard to predict and duplicate	Attacks that hijack the communication channels such as man-in-the-middle attacks
Obfuscation	Obscures the real meaning of data/signals/flows by making them difficult for an attacker to understand. It can use random sources of noise from the environment of the critical assets to increase the entropy	Attacks that require knowledge of the inner workings of the system, its functions, and its mission
Parameter assurance	Compares input data to a table of values in the system to check for large, unexpected deviations	Attacks that manipulate data files or messages that are sent to the system
Data consistency checking	Verifies the source of a parameter change	Attacks that use operator specific data entry

in a larger process that considers the functions resulting from the CPS and its external interactions and dependencies. Although failure analysis remains important, it should not attempt to purely identify all interactions and dependencies due to the complexity of the systems involved.

The design of the attack creates an additional set of control and feedback systems in the CPS, which must be countered not by tracing linear chains of events and preventing each but by creating new design patterns that serve to limit these new control behaviors from being exploited by a determined threat. The primary shortfall of causal event models is that they limit the analysis of countermeasure designs to similar linear cause-and-effect relationships (Leveson 2012).

A resilience analysis will convert the concept of a failure in a CPS to one of availability, or continuity of service. A resilience analysis will ask the questions:

- What will disrupt the control operation, what are the vulnerable aspects?
- How will that affect the availability of the control function to other parts of the system or to its human user?
- What other system components is that control function dependent upon, and to what extent?
- Can those components also be trusted?
- What is the risk (and associated cost) to ensure trust?

Linear causal analyses often omit dependencies because they are difficult to quantify, or because they are subjectively assumed to be unimportant. While failure analysis is still important, it should be part of a larger process that takes into account the functions that arise from the CPS, as well as its external interactions and dependencies. Given the complexity of these systems, the analysis should focus on defining hierarchies of control over essential CPS functions, rather than trying to identify every interaction and dependency. Resilience involves eliminating or reducing dysfunctional interactions, which are those that can disrupt the availability of critical control functions or lead to potentially hazardous states in the controlled process (Leveson 2012).

Mission and Operational Resilience Analysis

The analysis of resilience in a CPS begins with workshops that focus on the system's functions and the potential cyber threats that could exploit dependent functions. However, many resilience analysis approaches have limitations as they do not incorporate threat parameters as additional functions during the functional decomposition process that creates the CPS design. Therefore, traditional systems engineering analyses do not address cyber threat parameters as inherent to the system functional design.

To address this limitation, accepted processes for nonfunctional requirements derivation are generally human-driven, involving scenario analysis and modeling to derive lower level functions or requirements from higher level architectural models. These facilitated processes include subject-matter experts from the operational context of the CPS, CPS designers, and cyber experts with knowledge of potential attacks. The facilitator begins with questions about lost or changed functionality during operations to create a shared understanding of mission objectives and operational tasks related to the CPS. This process generates information that CPS designers and cyber threat experts can use to conduct the analysis of attack and countermeasure patterns.

One manual approach to derive these patterns is by describing "critical cyber assets" to link functional descriptions and early views of structural components. The MITRE Mission Assured Engineering (MAE) process framework is often used in the context of defense assured systems and is a starting point for the development of this approach. MITRE defines three different assessment processes, which include mission analysis, cyber threat susceptibility assessment, and cyber risk remediation assessment. Established IT guidance links cyber requirements to "critical cyber assets" or "crown jewels," while DoD guidance links cyber requirements to critical mission threads. In the CPS design, vulnerability must be assessed for both critical information assets and critical control processes against the criticality of the mission segment of interest. Three assessment processes are:

- Mission analysis or "mission-based critical asset identification." This process takes critical mission objectives or operational tasks and associates them with the critical system functions, creating an initial linking of functions to types of cyber-physical assets.
- Creation of a functional dependency graph that hierarchically links high-level mission objectives to operational tasks to information (or control) assets to sets of system assets as shown in Figure 22.2. This is a traditional mission to function to structure decomposition approach, which can be captured in a model. System assets are those structural components that mighty be exploited by a cyber attacker.
- Creation of an equivalent failure model by tracing back up the graph. The definition model is effectively a causal event chain: "if 'asset' is compromised, then function" is compromised, then "mission objective" is compromised.

Various analysis tools can be then used to assess the resulting functional and asset relationships. Three of particular interest are fault and attack tree analysis, goal-structure notation, and System-Theoretic Process Analysis – some of which will be covered in the textbook.

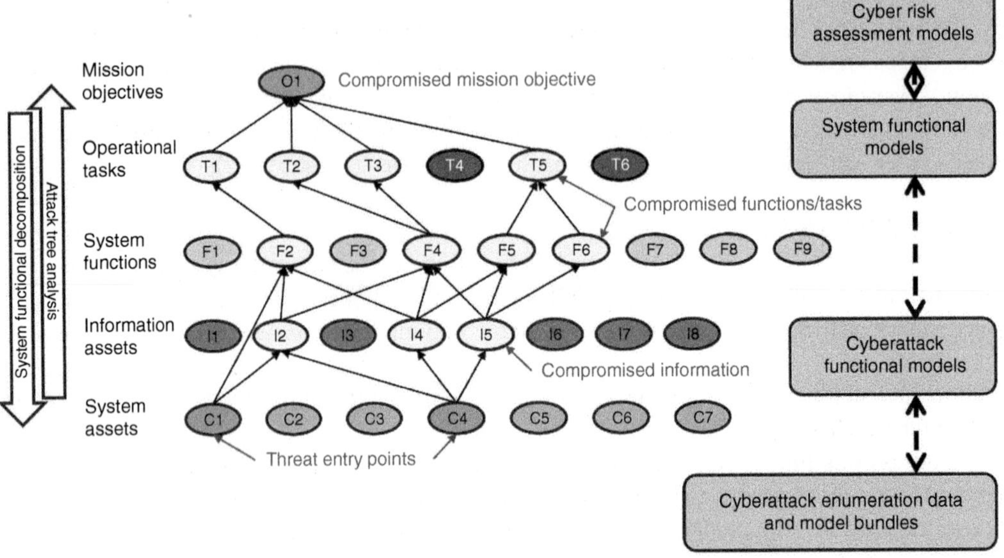

Figure 22.2 Functional modeling in cyber resilience engineering. *Source:* Adapted from Bodeau et al. (n.d.).

Table 22.4 Specific countermeasure patterns for CPS.

Countermeasure	Description
Redundancy	Checks for errors by comparing outputs from multiple components at different levels of the system. This can range from logic levels (parity or cyclic codes) to structural component levels to system levels. Checking accuracy and latency of detection is improved by lower level techniques, while higher level techniques may be more cost effective
Forward error recovery	Allows continued system operation in the presence of errors by compensating for the error with correction strategies. This provides redundancy beyond just error detection. It can also be implemented at multiple levels. From cyclic correction codes and lower levels to voting strategies at component or system levels
Triple modular redundancy	Provides forward error recovery with masking redundancy – the system has three identical modules that perform redundant operations. The output of these modules is checked by a voting module, which forwards the output based on majority vote. An error in one of the modules is masked by the majority
Backward error recovery	Tracks known good process states as recovery points. When an error is encountered the process returns to the last known good state. This is effective for temporary errors but can recur continuously (known as livelock) in the presence of permanent errors.
Dynamic redundancy	Detects errors and reconfigures the system in response. This is often referred to as "hot standby" when the reconfiguration selects a continuously running system and "cold standby" when the standby system has to be started
Diverse redundancy	Schemes like triple modular redundancy can use identical or diverse modules. Use of diverse, or non-identical, modules counters errors that may occur in common designs but also complicates the voting design
Data consistency checking	Verifies the source of a parameter change

Assurance Test and Evaluation

Assurance testing has been a longstanding discipline in the software community, with the development of assurance cases used to establish trustworthiness and assess the impact of security issues on safety regulation and cybersecurity policies. Similarly, engineering design assurance has sought to establish foundational requirements and processes for product quality assurance.

Aerospace released a document in 2009 that outlined a risk-based design assurance process flow developed by a cross-discipline, multi-company team called the Design Assurance Topic Team. In 1974, Caslake submitted a paper on the update to the quality assurance standards developed by the IEEE Nuclear Power Engineering Committee.

As interconnectivity among devices (the IoT and CPS developed), quality assurance in hardware and software and network assurance became more important. Network assurance is different from software assurance and involves quantifying risk from an IT network perspective. Formal verification is a part of the engineering process of network assurance, and there are specific tests and evaluations developed for embedded security and cyber-assurance.

However, the more holistic picture is still needed to address the challenges and opportunities in system assurance for CPS, including addressing attack vectors and ensuring compatibility among hardware, software, and documentation. To achieve system assurance, thorough model-based testing, architectures, and system synthesis of the implementation of the CPS are necessary.

New standards have been developed to enable the development of safety-critical systems, including CPS, such as DO-178C for commercial airborne software, which defines design assurance levels (DALs) with specific objectives to be satisfied and requires document traceability between certification artifacts.

Model-Based Assurance for Test and Evaluation

Hughues and Delange conducted research on model-based design and automated validation of different aircraft architectures using the Architecture Analysis and Design Language (AADL). The AADL was used to capture the system architecture, which was then processed and validated against the system requirements. However, the analysis was challenging due to the large quantity of inter-dependent results, so the team extended the analysis tool and auto-generated an assurance case from the validation results. This hierarchical notation showed the interdependencies of each requirement and provided details that were not enforced.

In 2016, researchers from Carnegie Mellon University proposed utilizing ModelPlex, a method for ensuring verification results about models, to apply to CPS implementation as a method of formal verification and validation. ModelPlex provides correctness guarantees for CPS executions at runtime by combining offline verification of CPS models with runtime validation of system executions for compliance with the model. The team also developed a technique to synthesize provably correct monitors automatically that form CPS proofs in differential dynamic logic by a correct-by-construction approach.

Sedjelmaci et al. designed and implemented a novel intrusion detection and response scheme to test, evaluate, and detect malicious anomalies that threaten unmanned aerial vehicle (UAV) networks. The team proposed detection and response techniques to monitor UAV behaviors and categorize them into appropriate lists according to the detected cyber-attack. The hierarchical scheme relied on two mechanisms: an intrusion detection mechanism running at the UAV node level and an intrusion response mechanism running at the ground station level. The proposed hierarchical intrusion detection and response scheme demonstrated a high level of security with high detection rates and low false-positive rates.

Another emerging schema for identifying failures and attacks in CPSs is the Six-Step Model (SSM), which incorporates six hierarchies of CPS: functions, structure, failures, safety countermeasures, cyber-attacks, and security countermeasures. Relationship matrices are used to identify the interrelationships between hierarchies and determine the effect of failures and cyber-attacks on CPSs. The SSM is based on two previously developed approaches: GTST-MLD and the 3-Step Model. In 2017, Sabaliauskaite proposed integrating the SSM with informational flow diagrams (IFDs) to identify possible failures and cyber-attacks and to select adequate countermeasures. IFDs are behavioral diagrams used for information flow modeling and have been used for complex CPS safety analysis. The extended IFD could be used for designing security countermeasures to achieve the required level of security, especially since the SSM Step 6 often requires the use of additional information flows. However, the applicability of this approach is only expanded upon using one example.

Formal Methods Transfer to CPS

Currently, the formal methods community and the control engineering community are engaging in a rapid integration process. One such initiative is the ERATO MMSD project, which started in 2016 and will continue until 2022, funded by Japan Science and Technology Agency (JST). The project aims to improve the quality assurance measures for CPS by utilizing a unique mathematical strategy to bridge the gap between formal methods for quality assurance in software and quality assurance in CPS. The researchers have identified several challenges in following the V-model for software requirements, such as the cost of hardware testing/simulation, correctness of designs and requirements, management of those designs and requirements, and optimization of complex systems. While formal methods have been developed to address these challenges, there are still challenges in the broader discipline of CPS.

The ERATO MMSD project seeks to provide a meta-mathematical transfer to move formal methods to heterogeneity, including coalgebraic model checking, quantitative semantics, simulation and bi-simulation notions, compositionality, and collaboration and integration with control theory and robotics. However, despite the extensive research into rigorous design of CPS over the past decade, a consistent and efficient model of integration of cyber and physical systems with the right level of fidelity for system design is still being researched. Bliudze et al. (2017) suggest that we are still far from reaching the desired degree of domain integration for the state-of-the-art CPS design, with the challenge of writing "faithful and consistent models from networks of physical components." The authors propose methodologies combining tool automation and designer ingenuity to overcome basic theoretical difficulties in CPS design.

Summary and Next Chapters

The development and operation of CPS require new approaches to address cybersecurity concerns across multiple domains. Model-based approaches that account for the complex interactions between threats, countermeasure patterns, and system behaviors are crucial for effective requirements, design, and evaluation. The primary objective is to create methods that preserve the original functional capabilities of the system while augmenting it with security design patterns. A dynamic graphical model can help explore and understand the ecosystem's structure–function relationships that produce outcomes and link security choices to a trade-based decision process that considers cost and level of security success. This approach can also visibly or quantifiably reveal the inherent diseconomies of scale that can result from overprotection.

Moreover, the same functional model building activity can be used to create a test framework that evaluates the effectiveness of countermeasure patterns. This framework involves attaching a separate functional harness to parts of the functional model, where actual test functions (fuzzers) replace specific attack patterns to evaluate system vulnerabilities. The test framework can simulate multiple attack inputs, allowing for a comprehensive evaluation of threat patterns. Current methods typically test one software component at a time, which is insufficient for addressing the system's inherent complexity. However, this foundation can help gradually build libraries of more complex system tests.

In the end, reinvigorating formal methods and verification of CPS is a challenging task due to the continuous nature. Though loss scenarios integrated with model-based assurance, and the resulting safety/security requirements and constrains, can enable designed resilience. New approaches to address resilience and trust in the development and operational use of CPS is an obvious concern; however, the following chapters provide insight into related research that addresses specific concerns, through systems engineering, with scalable solutions.

The following chapters provide a synergistic approach and include:

5.1 Introduction to Systems Theoretic Process Analysis for Security (STAP-Sec)
5.2 The "Mission Aware" Concept for Design of Cyber Resilience
5.3 The "FOREST" Concept and Meta-Model for Lifecycle Evaluation of Resilience
5.4 The Cyber Security Requirements Methodology and Meta-Model for Design of Cyber Resilience
5.5 Implementation Example: Silverfish

These chapters elucidate research efforts that explore (1) system architectures for achieving resilience; (2) system methodologies, frameworks, and analysis tools for prioritizing resilience solutions; (3) the roles and procedures for engaging operators in the real-time management of system reconfigurations that provide resilience; and (4) designing in resilience through the engineering process.

References

Avižienis, A., Laprie, J., Randell, B., and Landwehr, C. (2004). Basic concepts and taxonomy of dependable and secure computing. *IEEE Transactions on Dependa-ble and Secure Computing* 1 (1): 11–22.

Bliudze, S., Furic, S., Sifakis, J., and Viel, A. (2017). Rigorous design of cyber-physical systems. *Software & Systems Modeling* 2 (2): https://doi.org/10.1007/s10270-017-0642-5.

Bodeau, D. J., Graubart, R., Picciotto, J. et al. (n.d.). Cyber resiliency engineering framework. *DTIC*. https://apps.dtic.mil/sti/citations/AD1108457

Boehm, B. and Kukreja, N. (2015). An Initial ontology for system qualities. *INCOSE International Symposium* 25 (1): 341–356.

DoDI (2014) Department of Defense Instruction (DoDI) 8500.01, Cybersecurity. March 14, 2014.

Griffor, E. (ed.) (2016). *Handbook of System Safety and Security: Cyber Risk and Risk Management, Cyber Security, Adversary Modeling, Threat Analysis, Business of Safety, Functional Safety, Software Systems, and Cyber Physical Systems*. Syngress.

Leveson, N. (2012). *Engineering a Safer World: Systems Thinking Applied to Safety*, 13. MIT Press.

McDermott, T and Horowitz, B (2017). Human Capital Development – Resilient Cyber Physical Systems. *Systems Engineering Research Center (SERC) Technical Report SERC-2017-TR-075.*

https://sercuarc.org/publication/?id=163&pub-type=Technical-Report&publication=SERC-2017-TR-113-Hu-man+Capital+Development+%E2%80%93+Resilient+Cyber+Physical+Systems.

NATO (2010). North Atlantic Treaty Organization (NATO), Engineering for system assurance in NATO programs. Washington, DC: NATO Standardization Agency, DoD 5220.22M-NISPOM-NATO-AEP-67, February 2010.

NIST (2016). National Institute for Standards and Technology (NIST) Framework for Cyber-Physical Systems Release 1.0: Cyber Physical Systems Public Working Group (Rep.). May 2016. https://pages.nist.gov/cpspwg/ (accessed 1 June 2017).

NSF (2016). Definition from National Science Foundation, 2016, "Cyber-Physical Systems," program solicitation 16-549, NSF document number nsf16549, March 4. https://www.nsf.gov/publications/pub_summ.jsp?ods_key=nsf16549 (accessed 1 June 2017).

Reed, M. (2016). DoD Strategy for cyber resilient weapon systems. In: *Paper presented at the National Defense Industries Association, Annual Systems Engineering Conference*, Alexandria, VA (24 October 2016). NDIA. https://ndiastorage.blob.core.usgovcloudapi.net/ndia/2016/systems/18946_MelindaReed.pdf.

Wan, J., Canedo, A., and Al Faruque, M. (2015). Security-aware functional modeling of cyber-physical systems. In: *2015 IEEE 20th International Conference on Emerging Technology & Factory Automation (ETFA) 2015*, Luxembour, Luxembourg (8–11 September 2015). IEEE, 1–4. https://doi.org/10.1109/ETFA.2015.7301644.

Biographical Sketches

Tom McDermott serves as the deputy director and chief technology officer of the Systems Engineering Research Center (SERC) at Stevens Institute of Technology in Hoboken, NJ. The SERC is a University Affiliated Research Center sponsored by the Office of the Secretary of Defense for Research and Engineering. With the SERC, he develops new research strategies and is leading research on Digital Engineering transformation, education, security, and artificial intelligence applications. Mr. McDermott also teaches system architecture concepts, systems thinking and decision making, and engineering leadership. He is a lecturer in Georgia Tech's Professional Education college, where he leads a masters level course on systems engineering leadership and offers several continuing education short courses. He consults with several organizations on enterprise modeling for transformational change, and often serves as a systems engineering expert on government major program reviews. He currently serves on the INCOSE Board of Directors as Director of Strategic Integration.

Megan M. Clifford is a research associate and engineer at Stevens Institute of Technology. She works on various research projects with a specific interest in systems assurance and cyber-physical systems. She previously worked on the Systems Engineering Research Center (SERC) leadership team as the Chief of Staff and Program Operations, was the Director of Industry and Government Relations to the Center for Complex Systems and Enterprises (CCSE), and held several different positions in industry, including Systems Engineer at Mosto Technologies while working on the New York City steam distribution system.

Valerie Sitterle is a principal research engineer at the Georgia Tech Research Institute with over 25 years of experience in engineering science, integrating engineering, natural, and physical sciences leading to the design and analysis of systems. A primary emphasis of her work integrates defense operational needs with systems sciences across multiple domains to support design and assessment of defense systems and operational and tactical concepts of employment in theater environments. She currently serves as the chief scientist for the Systems Engineering Research Division within the Electronic Systems Laboratory in GTRI, where she supports the definition and execution of R&D across the main pillars of model-based approaches and digital transformation of systems engineering to provide new capabilities and advance stakeholders' decision-making processes. She also serves as a current member of the Research Council for the Systems Engineering Research Center (SERC), a DoD UARC led by Stevens Institute of Technology. She received her BME and MS degrees in mechanical engineering from Auburn University, her MS degree in engineering science from the University of Florida, and her PhD in mechanical engineering from the Georgia Institute of Technology.

Chapter 23

Introduction to STPA-Sec

Cody Fleming

Iowa State University, Mechanical Engineering, Ames, IA, USA

System Theoretic Process Analysis for Security (STPA-Sec)

Introduction and Background

Cybersecurity generally follows a software-oriented perspective, and the legacy of cybersecurity is one that focuses on pure IT software systems rather than those that interact with and potentially change the physical world. Meanwhile, it has been increasingly recognized that software assurance methods must focus on integrating security earlier in the acquisition and development cycle of software (Mead and Woody 2016). For example, certain approaches use existing malware to inform the development of security requirements in the early stages of the software life cycle (Mead et al. 2015), which seeks a similar approach to the one presented in this chapter. However, it follows the standard, bottom-up approach of identifying threats and generates solutions based on those threats. These techniques work well for IT software systems yet are insufficient for cyber-physical systems. Many cybersecurity approaches are not effective for cyber-physical systems in part because an attack on a physical system is not necessarily detectable or counteracted by cyber systems (Hu 2013). Recent developments in threat modeling are geared specifically for cyber-physical systems and consider the physical component that many other methods do not (Burmester et al. 2012). However, this approach still relies on historical threats to identify vulnerabilities. STPA-Sec (Young and Leveson 2013), however, aims to reverse the tactics-based bottom-up approach of other cybersecurity methodologies and furthermore develop approaches that consider hardware, software, and the environment in with which these systems interact and operate.

This chapter builds on STPA-Sec, presenting a method that is informed explicitly by mission and system stakeholders via the concept of the War Room. This aids the STPA-Sec analysis by minimizing the chance of outputs not matching the perspectives and experiences of the stakeholders. Second, this chapter describes a mission-aware viewpoint to augment the STPA-Sec analysis. That is, one could have the exact same system in a completely different mission context and would potentially want to choose different security solutions. The incorporation of the mission into the analysis scopes the security problem above the cyber-physical system level, which both opens possibilities for potential vulnerability solutions and motivates the choice of security or resiliency-based solutions.

Systems Engineering for the Digital Age: Practitioner Perspectives, First Edition. Edited by Dinesh Verma.

Companion website: www.wiley.com/go/verma/systemsengineering

Furthermore, the STPA-Sec approach frames the cybersecurity (or cyber resilience) problem first as a *safety problem*. In addition to preventing traditional notions of accidents, STPA-Sec allows for reasoning about mission success (or failure) in the context of system safety. As such, this chapter is less about cybersecurity techniques and more about system safety techniques, which motivates the need for a brief introduction into system safety techniques.

Chain-of-Events Accident Causality Models

In modern, complex engineered systems, there is an abundance of software and human–computer interaction, and thus the nature of accident causation has changed relative to the purely mechanical systems of the past. It is therefore important to review the accident causality models upon which hazard analysis techniques are based, which motivates the need for STPA (a general hazard analysis technique) and its security variant, STPA-Sec.

A major difference between Systems Theoretic Process Analysis (STPA) and traditional hazard analysis techniques is the underlying model of accident causality. Most current hazard analysis and safety assessment techniques are based on a chain-of-events model of causality, where the events represent component failures. Each failure event leads to the next one in the chain with a direct relationship between the two. A popular chain-of-events model is called the Swiss Cheese Model (Reason 1990).

The Swiss cheese model argues that accidents are caused by failures in four stages: organizational influences, unsafe supervision, preconditions for unsafe acts, and unsafe acts (Reason 1990). Each stage can be represented by a slice of Swiss cheese and the holes in the cheese represent a failed or absent defense in that layer. Figure 23.1 depicts this model, which is popular in aviation, defense, and several other sectors (Nance 2005).

The model assumes that the holes representing individual weaknesses are randomly varying in location and size. Eventually the holes come into alignment so that a path is possible through all slices, representing a sequence of failures throughout several layers of defense. Failures then

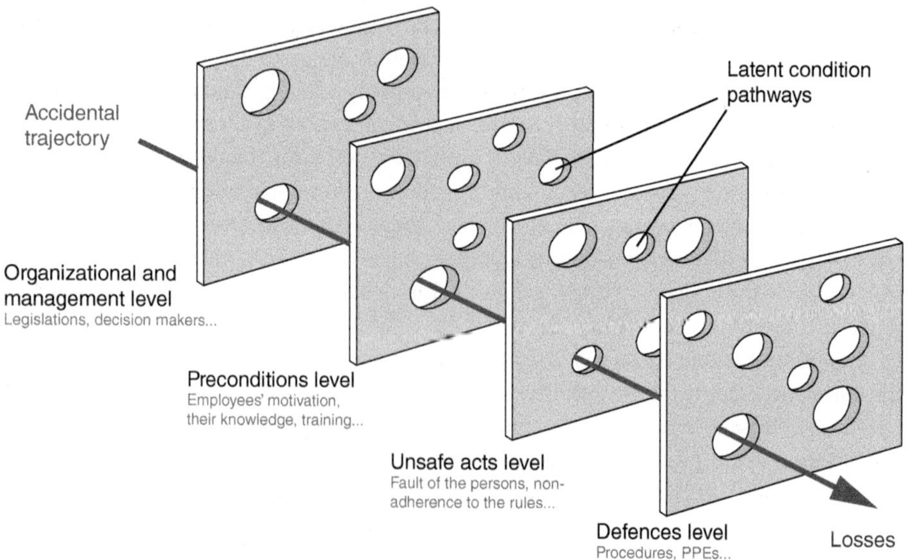

Figure 23.1 Swiss Cheese model of accident causality. *Source:* Adapted from Reason (1990).

propagate along the path through each defense barrier and cause an accident. The Swiss cheese model not only assumes a linear chain of events structure but also assumes random behavior of components and independence between failures in each layer. Critics argue that these assumptions do not hold in practice, especially for safety-critical software and human behavior (Hollnagel et al. 2007; Dekker 2013).

Many of the hazard analysis techniques used for development and certification of modern engineered systems assume a linear, chain-of-events causality model. For example, "each (event sequence diagram) begins with a triggering event (primary system failure) and proceeds through a logic diagram to show how backup systems and procedures are used either in series or parallel to prevent the ultimate bad outcome from occurring" (JPDO 2012).

Techniques Based on Chain-of-Events Model

Fault tree analysis (FTA) was developed in the 1960s at Bell Laboratories (Watson 1961) and expanded in the 1980s to formalize and standardize the rules of application (Vesely et al. 1981). FTA is the primary technique for scenario modeling in probabilistic risk assessment (PRA) (Modarres 2006) and is perhaps the most popular hazard analysis technique. For aeronautics applications, FTA is recommended to be applied in between "Preliminary Design" and "Detailed Design" (as well as later phases) (SAE 1996). FTA is most effective when more design details are available than what is typically available during early phases of systems engineering, and it is therefore rarely used during concept development, system architecting, or requirements definition (Fleming and Leveson 2015a,b).

Failure modes and effect analysis (FMEA) was developed to systemically evaluate the effect of individual component failures on system performance. The United States Department of Defense developed FMEA in 1949 as part of its weapon systems program (USDoD 1949), and the technique has been applied in many domains including aerospace (SAE 1996). Like FTA, FMEA is based on a chain-of-events model of accident causation. However, unlike FTA, FMEA is an inductive process that starts with a basic component failure or fault, and then the analyst reasons about this failure's effect on the overall system behavior. FMEA may be described as a "bottom-up" approach, while FTA is a "top-down" approach that begins with system-level events rather than component-level events.

Event tree analysis (ETA) was developed in 1974 during a nuclear power plan study (Rasmussen 1975) and has been adopted in the aerospace domain (RTCA 2008). Like FTA and FMEA, ETA is based on the chain of events accident model, and in fact event trees were originally intended to be combined with fault trees as part of an overall probabilistic risk assessment (Leveson 2012). The first step in ETA is to identify an initiating failure event such as a structural failure or loss of fuel. The next step is to list, in the anticipated sequence of operation, the set of barriers or protective functions intended to prevent the initiating event from leading to an accident. Last, a logical tree is constructed by tracing forward in time from the initiating event and inserting a binary branch at each barrier to represent the possible success or failure of that barrier.

ETA is similar to FMEA in that it is forward-searching, or "bottom up," and traces forward from a failure condition to a potential hazardous state. Alternatively, ETA is similar to FTA in its logical decomposition of conditions, and event trees are effective for quantitative analysis if the probabilities of each barrier condition are known. Another similarity between ETA and FTA is that each barrier is assumed to operate independently, which means that the probability of each end state is simply the multiplication of each of the probabilities along the path to its end state.

The Need for a Different Accident Model and a New Approach

Based on some chain-of-events structure such as the Swiss cheese Model, traditional safety engineering techniques then focus on preventing or reducing the probability of component failure to prevent accidents. FTA and FMEA, at least as commonly used, also assume that most of the component failure modes are independent. Human operators and software are treated as if they fail like mechanical hardware, and likelihood of error is assigned to them (often with either a probability of 0.0 or 1.0). Given the critical role that software and human decision-making play in modern complex systems, these assumptions are unhelpful and/or unrealistic.

Accidents often arise due to unanticipated failures or due to unsafe interactions among components that have not failed. Starting a hazard analysis from failures puts the analysis at risk of identifying only a subset of the possible causes, as opposed to beginning with hazards and identifying the interactions that could possibly lead to hazardous states, including those not involving component failure (see Figure 23.2).

The systems approach espoused by STPA recognizes that software does not "fail," but merely performs the way it was designed: it can therefore be hazardous due to flawed requirements (or implementation) or unsafe interactions with the rest of the system. Most software-related accidents arise due to flaws in the software requirements (Leveson 2012); therefore, obtaining a complete and correct set of safety-related requirements and constraints on software behavior is key to preventing software-related accidents.

Human operators also do not fail in the sense that hardware does and most of their errors are not random. Instead, humans are influenced by the design and operation of the overall system, as well as the operational context, and can thus make unsafe decisions. These decisions may be due to the factors in Figure 23.4, such as incorrect mental models of the process they are controlling, possibly due to missing or incorrect feedback.

The human error identification process in these traditional methods is incomplete, but the problems involve more than just incompleteness. In methods such as FTA or FMEA, human error is often treated in the same way as a physical failure, that is, as a deviation from a predefined behavior or procedure. Unfortunately, this treatment of human error oversimplifies it as a binary decision between right and wrong. Many of the most important situations involved in accidents are

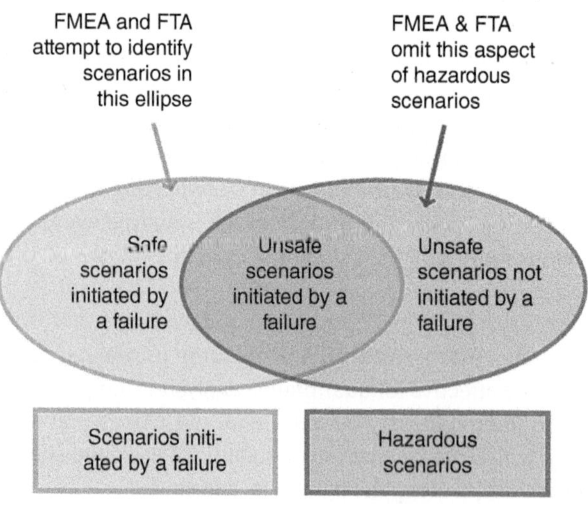

Figure 23.2 Consequences of equating safety and reliability. *Source:* Adapted from Fleming (2015).

overlooked because they are difficult or impossible to model in this way, including when (Fleming et al. 2013):

- The correct behavior is not predefined or not clear.
- The prescribed behavior is thought to be incorrect by the person responsible for following it.
- Procedures conflict with each other, or it is not clear which procedure applies.
- The person has multiple responsibilities or goals that may conflict.
- The information necessary to carry out a procedure is not available or is incorrect.
- Past experiences and current knowledge conflict with a procedure.
- The procedure is misunderstood or the responsibility for the procedure is unclear.
- The procedure is incorrect.

Although some accidents still occur today due to failures or combinations of failures, many of them arise due to a complex combination of factors. As such, identifying a failure and its predicted effect is insufficient. Most accidents happen due to the interaction of components and dynamic behavior of the agents – which include human operators and software – and only sometimes is a failure even necessary. Examples of these complex factors abound in aviation, for example Air France 447 (Wise et al. 2011), Asiana 214 (Sherry and Mauro 2014), and several runway overshoot accidents (CIAIAC 2006; Hawkins et al. 2013).

STAMP

System-Theoretic Accident Model and Processes (STAMP) was created to capture more types of accident causal factors including social and organizational structures, new kinds of human error, design and requirements flaws, and dysfunctional interactions among non-failed components (Leveson 2012). Rather than treating safety as a failure problem or simplifying accidents to a linear chain of events, STAMP treats safety as a control problem in which *accidents arise from complex dynamic processes that may operate concurrently and interact to create unsafe situations.*

Accidents can then be prevented by identifying and enforcing constraints on component interactions. This model captures accidents due to component failure, but also explains increasingly common component interaction accidents that occur in complex systems without any component failures. For example, software can create unsafe situations by behaving exactly as instructed or operators and automated controllers can individually perform as intended but together, they may create unexpected or dangerous conditions.

STAMP is based on systems theory and control theory. In systems theory, emergent properties are those system properties that arise from the interactions among components, and safety is a type of emergent property. The emergent properties associated with a set of components are related to constraints upon the degrees of freedom of those components' behavior (Checkland 1999). There are always constraints or controls that exist on the interactions among components in any complex system. These behavioral controls may include physical laws, designed fail-safe mechanisms to handle component failures, policies, and procedures. Such controls must be designed such that the safety constraints are enforced on the potential interactions between the system components. In air traffic control, for example, the system is designed to prevent loss of separation among aircraft.

System safety can then be reformulated as a system control problem rather than a component reliability problem – accidents occur when component failures, external disturbances, and/or potentially unsafe interactions among system components are not handled adequately or controlled, leading to the violation of required safety constraints on component behavior (such as maintaining minimum separation). System controls may be managerial, organizational, physical, operational, or

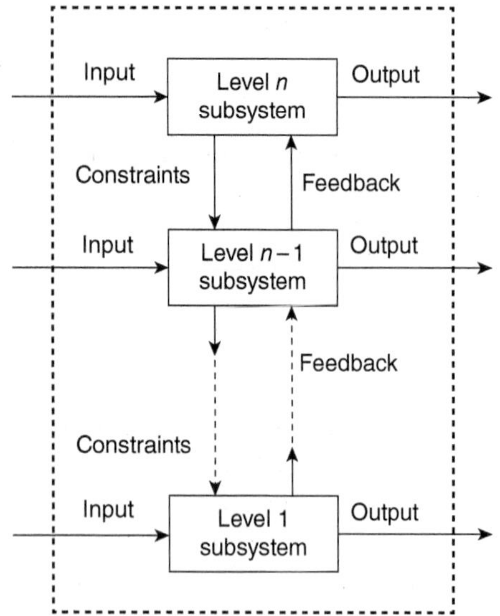

Figure 23.3 Basic features of a hierarchical system.

in manufacturing. In STAMP, the safety controls in a system are embodied in the hierarchical safety control structure. For deeper reading on hierarchy theory, see Checkland (1999) and Ahl and Allen (1996).

Control processes operate throughout the hierarchy whereby commands or control actions are issued from higher levels to lower levels and feedback is provided from lower levels to higher levels (see Figure 23.3). Accidents arise from inadequate enforcement of safety constraints, for example due to missing or incorrect feedback, inadequate control actions, component failure, uncontrolled disturbances, or other flaws. STAMP defines four types of unsafe control actions that must be eliminated or controlled to prevent accidents:

1) A control action required for safety is not provided or is not followed
2) An unsafe control action is provided that leads to a hazard
3) A potentially safe control action is provided too late, too early, or out of sequence
4) A safe control action is stopped too soon or applied too long

One potential cause of a hazardous control action in STAMP is an inadequate process model used by human or automated controllers. A process model contains the controller's understanding of (1) the current state of the controlled process, (2) the desired state of the controlled process, and (3) the ways the process can change state. This model is used by the controller to determine what control actions are needed. In software, the process model is usually implemented in variables and embedded in the program algorithms. For humans, the process model is often called the "mental model" (minghui). Software and human errors frequently result from incorrect process models. Accidents like this can occur when an incorrect or incomplete process model causes a controller to provide control actions that are hazardous. While process model flaws are not the only cause of accidents in STAMP, it is a major contributor.

The generic control loop in Figure 23.3 shows other factors that may cause unsafe control actions. Consider an unsafe control action for an air traffic controller: a flight crew is instructed to increase altitude while another aircraft is flying through that new altitude. The control loop in Figure 23.3 would show that one potential cause of that action is an incorrect belief that the airspace above the aircraft is clear (an incorrect process model). The incorrect process model, in turn, may be the result of inadequate feedback provided by a failed sensor, the feedback may be delayed, the data may have been corrupted, etc. Alternatively, the system may have operated exactly as designed but the designers may have omitted a feedback signal or the feedback requirements may be insufficient.

STPA

Systems-theoretic process analysis (STPA) is a hazard analysis technique that is based on the STAMP accident causality model. STPA is more powerful than failure-based techniques in the ability to capture a wider array of hazardous behaviors, including organizational aspects, requirements flaws, design errors, complex human behavior, and component failures (Levesonsafterworld).

While many hazard analysis techniques stop once a sequence of events or failures has been identified, STPA helps explain the complex reasons why a sequence of events might occur, including underlying processes and control flaws that may exist without any component failure.

Although STPA is relatively new compared to traditional methods, it has been demonstrated successfully on a wide range of systems including aviation (Fleming 2013), spacecraft (Ishimatsu et al. 2010, 2011), Fleming et al. (2012), missile defense systems (Pereira et al. 2006), civil infrastructure (Dong 2012), and others, as well as a broad range of system engineering activities including requirements development and system architecting (Fleming 2017; Carter et al. 2018; Bakirtzis et al. 2018).

STPA starts from fundamental system engineering activities, including the identification of losses or accidents to be avoided, the hazardous behavior that could lead to these losses, safety requirements and constraints, and the basic system control structure used to avoid these losses. The primary goal of STPA is to generate detailed safety requirements and constraints that must be implemented in the design to prevent the identified unacceptable losses. It achieves this goal by identifying unsafe control behavior and the scenarios that can lead to this behavior, including component failure scenarios.

The goal of STPA is the same as any safety engineering activity, but STPA includes a broader set of potential scenarios including those in which no failures occur but the problems arise due to unsafe and unintended interactions among the system components. STPA also provides more guidance to the analysts than fault tree analyses mentioned above. Functional control diagrams and a set of generic causal factors are used to guide the analysis. Figure 23.4 shows the basic causal factors used in an STPA analysis of a functional control diagram.

STPA-Sec

Traditionally, there is no "science" to applying security as a structured assistant to mission success. Indeed, it is often true that the procedure of securing mission-critical systems is based upon an unstructured and ultimately random security assessment that might or might not lead to mission

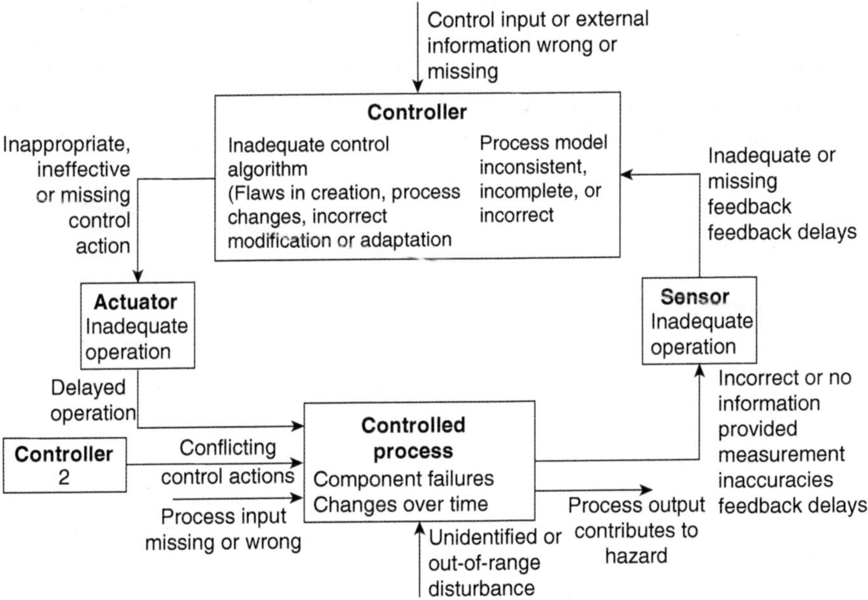

Figure 23.4 STPA Control loop with causal factors.

degradation (see, for example, policy lists). This is problematic because security should not be exercised for the sake of security but, in general, should be used as a tool to avoid possible transitioning to states that violate the system's expected service. Avoiding this transitioning to hazardous states is the raison d'être of security. Following that definition, any security measure that goes beyond providing assurance of safe behavior or any measure that does not adequately assure the safe behavior of the system during a mission is a loss of resources and, hence, can inadvertently be a hindrance in the command and control of military systems. STPA-Sec (Young and Leveson 2013) was created with the goal of creating just such a science, and what follows is a development of STPA-Sec that explicitly considers achieving mission objectives in the face of security threats and vulnerabilities.

The STPA-Sec approach seeks to identify how security issues can lead to accidents or unacceptable losses. In particular, the method begins by first describing, in a systems- and control-theoretic way, the behavior of the system under control and other actors within the overall mission and how that behavior can become unsafe. The general process for conducting STPA-Sec is outlined in Figure 23.5.

The first step of building the model is identifying the mission to be performed, which is to be identified by the relevant stakeholders of any system and/or its mission. This statement takes the form of, "The mission is to perform a task, which contributes to higher-level missions or other purposes." The specific language used here serves to outline the general purpose and function of the mission succinctly and precisely.

After defining the mission, the next step is to define the unacceptable outcomes or losses associated with the mission. For example, an armed UAV conducting a strike mission may have an unacceptable loss defined as any friendly casualties occurring. These losses or outcomes were either explicitly or implicitly identified and prioritized by the War Room stakeholders. For example, the failure to destroy a target may be less important to mission commanders than inflicting friendly casualties.

After defining the unacceptable losses, one defines a set of hazardous scenarios that could potentially result in an unacceptable outcome. Some of these scenarios may have been described in the so-called War Room; however, it is likely that many will be defined by the analysts on their own. For example, in the UAV mission and unacceptable loss described above, a hazardous scenario might involve friendly forces within the targeting area. This on its own does not necessarily lead to the unacceptable loss of friendly casualties, but such an outcome is certainly a possibility if the munition is launched. The set of hazardous scenarios does not have to be an exhaustive list; however, the analysts should strive to define a set of hazards that have a reasonable chance of occurring

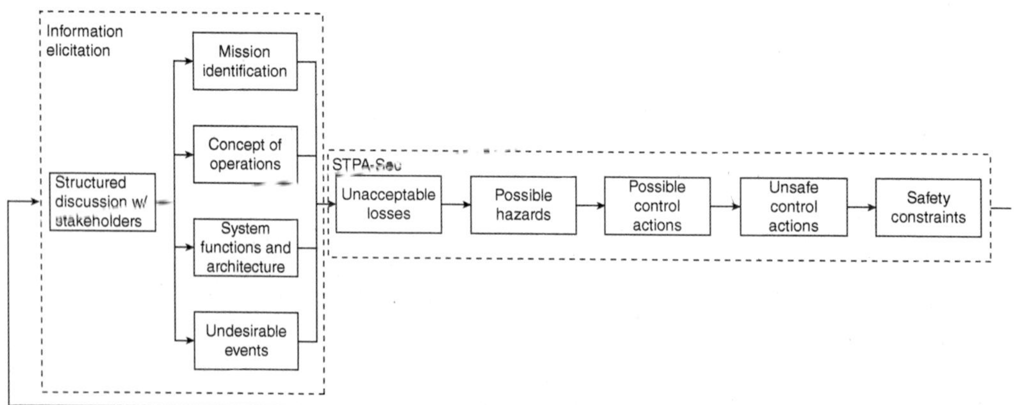

Figure 23.5 Conceptual overview of STPA-Sec Process.

during a mission and might be provoked by cyberattacks that corrupt relevant system software. The following list and table describe losses and hazards for a UAV reconnaissance mission.

L1. Loss of resources (human, materiel, etc.) due to inaccurate, wrong, or absent information
L2. Loss of classified or otherwise sensitive technology, knowledge, or system/s
L3. Loss of strategically valuable materiel or personnel/civilians due to loss of control of system

After defining the set of hazards, the analysts then develops a functional hierarchy and the control actions that can be taken at each level during the mission. For example, in a typical mission, there might be three functional levels or actors: the mission planner, the system operator, and the physical system. Obviously, the defined functional levels depend on how the analysts define them and can vary depending on the system in question; yet this step is necessary as it allows us to scope the model to a reasonable degree of granularity. Figure 23.6 depicts the hierarchical functions of

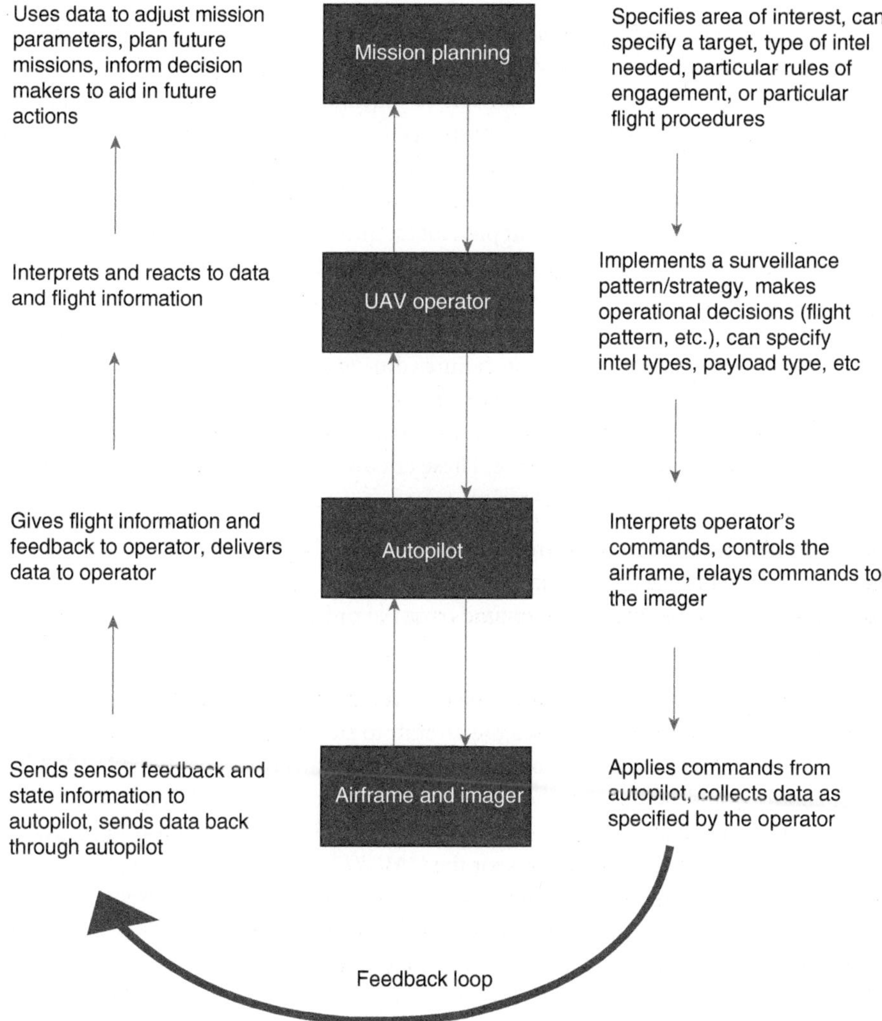

Figure 23.6 Hierarchical model defining expected service and mission context of a UAV. Each level is defined by a generic control structure. Inadequate control in each level can cause an adversarial action to degrade the expected service and produce a hazardous state in Table 23.1.

Table 23.1 Hazards for a UAV reconnaissance mission.

Hazard	Worst-case environment	Associated losses
H1-Absence of information	Imminent threat goes undetected	Manpower, materiel, territory, etc. **L1**
H2-Wrong or inaccurate information	Threat is incorrectly identified/characterized	Same as above **L1**
H3-Loss of control in unacceptable area	UAV is lost in enemy territory and suffers minimal damage in crash/landing	Compromise of critical systems, intelligence, and/or other potentially classified information or technology **L2, L3**

the control structure, an abstract description of each function's control responsibilities, and information or feedback received from the lower levels.

Next, the analysts define the control actions that can be taken at each level. A generic control loop is presented in Figure 23.4. In general, the control action at one functional level enacts a change onto a controlled process at a lower level via an actuator and then the controller receives feedback from the controlled process via a sensor. For example, a control action in the mission planning functional level could be defining a flight plan for an unmanned reconnaissance mission, and a control action at the operator level in the same mission might be commanding the vehicle to make a 30° turn to the north. The control actions and functional levels should be represented in a flow diagram that represents the planned order with respect to the mission. This will help analysts establish the "baseline" order of operations and procedures during the mission, which can be used later to analyze deviations from standard operating procedure.

After defining the control actions within a mission, the next step is to define the circumstances in which a particular control action could be unsafe. These circumstances can generally be defined as being a part of one of the four following categories:

- not providing a control action causes a hazard;
- providing the control action causes a hazard;
- the control action is performed at the incorrect time or out of order;
- the control action is applied for too long or stopped too soon.

For example, in the reconnaissance mission described above, the turning control action could be unsafe when it is applied for too long and causes the aircraft to stall out. For each control action, the analysts should identify a circumstance for each of the four categories mentioned above in which the control action would be unsafe. These unsafe control actions are placed into a table, which then allows us to easily access information when needed during the next phase of analysis. Table 23.2 contains a set of unsafe control actions for the UAV reconnaissance mission.

To interpret the table, the relevant action is in the leftmost column (and in this case, an assignment of the action to a specific level in the hierarchy). The four columns to the right represent scenarios where the control action in the same row may become hazardous. For example. "Providing" (the 3rd column in the table) the "designate area of interest" control action leads to hazard H1 and/or H2 if the location of the area is incorrect, which could have downstream implications on mission success.

Now that we have identified the control actions available in the system and the conditions under which they create hazardous scenarios, we can identify a set of constraints that can be

Table 23.2 Hazardous or potentially unsafe control actions for UAV mission.

Control action	Not providing causes hazard	Providing causes hazard	Incorrect timing or order	Stopped too soon or applied too long
CA 1.1-Designate area of interest		Area is wrong or will not provide needed information H1, H2	Area designated no longer of use H1, H2	Area would be useful at another time H1, H2
CA 1.2-Specify target	Surveillance is not focused enough H1, H2	Target is wrong or does not provide needed information H1, H2	Target no longer of interest or does not provide needed information H1, H2	Needed information occurs before or after surveillance H1
CA 1.3-Indicate intel type	Gather too much or too little data to be useful H1, H2	Intel type is not appropriate for what is needed H1, H2	Type of intel collected at wrong time (i.e. SIGINT during time with no signals) H1, H2	Miss desired type of intel H1
CA 1.4-Create rules of flight or engagement	UAV strays into inappropriate area H3	UAV cannot collect needed information H1, H2	Same as left H1, H2	Same as left H1, H2
CA 2.1-Designate surveillance strategy	Surveillance is ineffective, does not provide needed information H1, H2	Surveillance strategy is inappropriate, does not provide needed information H1, H2	Similar to left H1, H2	Similar to left H1, H2
CA 2.2-Set flight parameters	UAV strays into inappropriate area H3	UAV has inappropriate field of view H1, H2	Similar to left H1, H2	Similar to left H1, H2
CA 2.3-Start process (begin surveillance)	Information not collected H1	Inappropriate information collected H1, H2	Needed information not collected H1, H2	Same as left H1, H2
CA 2.4-Make maneuver command	UAV strays into inappropriate area, field of view not adjusted appropriately H1, H2, H3	Same as left H1, H2, H3	Same as left H1, H2, H3	Same as left H1, H2, H3
CA 2.5-End process	?	Needed information not collected H1, H2	Same as left H1, H2	Same as left H1, H2
CA 2.6-Make data collection command	Needed information not collected H1, H2	Same as left H1, H2	Same as left H1, H2	Same as left H1, H2

(Continued)

Table 23.2 (Continued)

Control action	Not providing causes hazard	Providing causes hazard	Incorrect timing or order	Stopped too soon or applied too long
CA 3.1-Compute, translate, or interpret command	Stable flight not achievable, field of view not appropriately adjusted H1, H2, H3	Stray into inappropriate area, field of view not appropriate H1, H2, H3	Same as left H1, H2, H3	Same as left H1, H2, H3
CA 3.2-Send signal	Same as above H1, H2, H3	Same as left H1, H2, H3	Same as left H1, H2, H3	Same as left H1, H2, H3
CA 3.3-Interpret feedback	Same as above H1, H2, H3	Same as left H1, H2, H3	Same as left H1, H2, H3	Same as left H1, H2, H3
CA 3.4-Determine orientation and location	Same as above H1, H2, H3	Same as left H1, H2, H3	Same as left H1, H2, H3	Same as left H1, H2, H3
CA 3.5-Report information	Operator does not get information, same as above H1, H2, H3	?	Cannot fly properly H3	Same as left H3
CA 4.1-Move control surface	Stray into inappropriate area, field of view not adjusted properly H1, H2, H3	Same as left H1, H2, H3	Same as left H1, H2, H3	Same as left H1, H2, H3
CA 4.2-Take picture or collect data	Needed information not collected H1, H2	?	Needed information not collected H1, H2	Same as left H1, H2
CA 4.3-Send data	Same as above H1, H2	?	Same as above H1, H2	Same as left H1, H2
CA 4.4-Send feedback	Same as 4.1 & 4.2 H1, H2, H3	Same as left H1, H2, H3	Same as left H1, H2, H3	Same as left H1, H2, H3

applied to the behavior of the system to limit the possibility of a hazardous scenario leading to an unacceptable loss. The constraints defined for the control actions outlined in Table 23.2 is presented in Table 23.3.

In addition to the constraints that should be applied on the system, analysis of the STPA-Sec model identifies areas that should receive the most attention in order to increase security and resiliency against cyberattacks that can produce unacceptable mission outcomes. For the UAV reconnaissance mission identified in this example, the most pressing unacceptable outcome relates to military commanders not receiving vital information about potential enemy activity within an area of interest. In this case, the integrity of the video feed coming from the UAV should receive top priority. Developing and evaluating measures for ensuring integrity of the video feed (or assuring

Table 23.3 Safety constraints for a subset of control actions.

Control action	Safety constraint
CA 1.1-Designate area of interest	The mission planner shall always clearly define the area of interest to align with any future mission that the for which the reconnaissance is needed
CA 1.2-Specify surveillance target	The mission planner shall indicate as specific a target as possible for the reconnaissance
CA 1.3-Indicate type of intelligence needed	The mission planner shall designate a specific type of intelligence that the mission is going to collect
CA 1.4-Create rules of flight or engagement	The mission planner shall indicate a specific set of rules of engagement to prevent confusion
CA 4.1-Move control surface	Control surfaces shall only move upon receiving authentic commands from the flight control system
CA 4.2-Take picture or collect data	Data collection shall only occur upon authentic command from the operator
CA 4.3-Send data/ feedback	The component shall relay collected data or send feedback to the appropriate monitors at regular intervals

that the system can identify when integrity has been lost) is outside of the scope of the current project.

This analysis also assists in deciding which areas of the system to devote further time and energy in terms of identifying detailed causal scenarios according to Figure 23.4. See Thomas and Leveson (2018) for more guidance and examples on how to identify these scenarios, which can also provide more refinement and targeted efforts to identify security flaws and mitigations, which will be further explored momentarily.

Outcome of the STPA-Sec Process

At the beginning of this process, several unacceptable outcomes of the mission are identified and prioritized. The next step is to identify (1) how those losses may occur and then (2) how they can be avoided or mitigated. In the previous step, the circumstances under which control actions would become unsafe were identified. Now, those circumstances are used to derive security requirements and constraints on the behavior of the system. For example, a constraint in the UAV strike mission mentioned previously could be "no fire munition command shall be issued when friendly forces are in the targeting area." These constraints are presumably already reflected in the operational procedures of the system.

Finally, we can identify causal scenarios using all the previously defined information to determine how an unacceptable loss may occur. Using the UAV strike mission as an example, an unacceptable loss can occur when the operator issues a fire munition command when there are friendly forces within the target area because his or her sensors indicated otherwise. Such a scenario could feasibly be the result of a precedented Denial-of-Service Attack on the operator's sensor. By creating these causal scenarios, we seek to determine the most likely or most damaging pathways for potential security breaches. Furthermore, creating the STPA-Sec model helps identify the most critical components, features, or functionality in a system with respect to mission success. This information can then be used to guide which cyber-security or cyber-resiliency measures are implemented in the future.

Possible Future Directions and Existing Extensions

This chapter presented an approach to augmenting security analysis, particularly for cyber-physical systems, that is based on the STAMP accident causality model and associated STPA hazard analysis methodology. This framework is based on a top-to-bottom identification of unacceptable losses or outcomes to a particular mission that the system is expected to perform. The method, called STPA-Sec, examines how the paths to those outcomes can occur and can then in turn be avoided. This chapter presents an application of this approach to a hypothetical tactical reconnaissance mission using a small UAV and generated a set of constraints that should be present in the behavior of this example system to avoid pathways to unacceptable outcomes. In addition, this approach identifies the areas most critical to mission success as starting points for future implementations of security or resiliency solutions.

A future direction based on the findings of this work includes implementing the identified system constraints on a model and formally checking that they can avoid unacceptable losses to the mission. Additionally, this work could be extended by closing the loop and testing security or resiliency solutions' effects on the behavior of the system in its mission. This would allow security and resiliency solutions to be evaluated based on their cost, complexity of implementation, and effectiveness at preventing unacceptable mission outcomes.

Although this chapter is intended to provide a motivation for, as well as a description of STPA-Sec, there is obviously more to assuring that systems are resilient in the face of cyberattacks. Turning the potentially hazardous scenarios into design solutions requires trade studies and other system engineering techniques as well as knowledge of how vulnerabilities and attacks may propagate through the system (note that STPA-Sec is relatively agnostic to how an attack may occur or whether vulnerabilities exist but rather focuses on how a system can behave unsafely).

A major challenge in engineering for the cyber security (or resilience) of complex systems is the assessment of the system's security posture at the early stages of its life cycle. In the defense community, it has been estimated that 70–80% of the decisions affecting safety and security are made in the early concept development stages of a project (Horowitz et al. 2017). Therefore, it is advantageous for this assessment to take place before lines of code are written and designs are finalized. To allow for security analysis at the design phase, a system model has to be constructed, and that model must reasonably characterize a system as well as be sufficiently detailed to enable matching of attack vectors mined from databases. Matching possible attack vectors to the system model facilitates detection of possible security vulnerabilities in timely fashion. One can then design systems that are secure by design instead of potentially having to add bolt-on security features later in the process, an approach that can be prohibitively expensive and limited in its mitigation options. Consequently, employing a model reduces costs and highlights the importance of security as part of the design process of CPS.

In this vein, several extensions of STPA-Sec or "add-ons" have been developed. First, several attempts have been made at integrating STPA with more formal model-based engineering tools such as SysML. One example augments STPA-Sec models with an underlying set of semantics related to cyberattacks and vulnerabilities that enables traceability between system architecture information, mission context, component-level vulnerabilities, and the potential mission effects of the vulnerability (Carter et al. 2019). Second, a set of algorithmic tools to apply attack patterns and/or identify vulnerabilities to these augmented STPA-Sec models has shown promise in automating the cyber security analysis, particularly at the early stages of systems engineering (Bakirtzis et al. 2020).

References

Ahl, V. and Allen, T.F.H. (1996). *Hierarchy Theory: A Vision, Vocabulary, and Epistemology*. Columbia University Press.

Bakirtzis, G., Carter, B.T., Elks, C.R., and Fleming, C.H. (2018). A model-based approach to security analysis for cyber-physical systems. In: *2018 Annual IEEE International Systems Conference (SysCon)* (23–26 April 2018), 1–8. New York: IEEE.

Bakirtzis, G., Ward, G., Deloglos, C. et al. (2020). Fundamental challenges of cyber-physical systems security modeling. In: *2020 50th Annual IEEE-IFIP International Conference on Dependable Systems and Networks-Supplemental* Volume (DSN-S)(29 June–2 July 2020), 33–36. New York: IEEE.

Burmester, M., Magkos, E., and Chrissikopoulos, V. (2012). Modeling security in cyber–physical systems. *International Journal of Critical Infrastructure Protection* 5 (3): 118–126.

Carter, B.T., Bakirtzis, G., Elks, C.R., and Fleming, C.H. (2018). A systems approach for eliciting mission-centric security requirements. In: *2018 Annual IEEE International Systems Conference (SysCon)* (23–26 April 2018), 1–8. New York: IEEE.

Carter, B.T., Fleming, C.H., Elks, C.R., and Bakirtzis, G. (2019). Cyber-physical systems modeling for security using SysML. In: *Systems Engineering in Context*, 665–675. Cham: Springer.

Checkland, P. (1999). *Systems Thinking, Systems Practice: Includes a 30-Year Retrospective*. Wiley.

CIAIAC (2006). Failure of Braking, Accident Occurred on 21 May 1998 to Aircraft Airbus A-320-212 Registration G-UKLL at Ibiza Airport, Balearic Islands. *Technical Report, Ministerio de Fomento Subsecreteria, Spain.*

Dekker, S. (2013). *The Field Guide to Understanding Human Error: Second Edition*. Ashgate Publishing, Ltd.

Dong, A. (2012). Application of CAST and STPA to railroad safety in China. Master's thesis. Massachusetts Institute of Technology.

Fleming, C.H., Leveson, N.G., and Placke M.S. (2013). Assuring Safety of NextGen Procedures. In Proceedings of the 10th USA/Europe Air Traffic Management Research and Development Seminar (ATM2013), Chicago, Illinois (10-13 June 2011).

Fleming, C.H. (2015). Safety-driven early concept analysis and development. PhD dissertations. Massachusetts Institute of Technology.

Fleming, C.H. (2017). Systems theory and a drive towards model-based safety analysis. In: *2017 Annual IEEE International Systems Conference (SysCon)* (24–27 April 2017), 1–5. New York: IEEE.

Fleming, C.H. and Leveson, N.G. (2015a). Integrating systems safety into systems engineering during concept development. *INCOSE International Symposium* 25 (1): 989–1003.

Fleming, C.H., and Leveson, N.G. (2015b). Including safety during early development phases of future air traffic management concepts. In Proc. 11th USA/European Air Traffic Management Seminar, Lisbon, Portugal (23–26 June 2015).

Fleming, C., Ishimatsu, T., Miyamoto, Y., et al. (2012). Safety guided spacecraft design using model based specifications. In Proceedings of the 5th IAASS Conference, Versaille, France (17–19 October 2011). Noordwijk, The Netherlands: European Space Agency Communications; ESTEC.

Fleming, C.H., Spencer, M., Thomas, J. et al. (2013). Safety assurance in nextgen and complex transportation systems. *Safety Science* 55: 173–187.

Hawkins, R., Habli, I., and Kelly, T. (2013). The principles of software safety assurance. In: *31st International System Safety Conference* (12–16 August 2013). Boston, Massachusetts USA: The International System Safety Society.

Hollnagel, E., Woods, D.D., and Leveson, N. (2007). *Resilience Engineering: Concepts and Precepts.* Ashgate Publishing, Ltd.

Horowitz, B., Beling, P., Fleming, C. et al. (2017). *Security Engineering FY17 Systems Aware Cybersecurity*. Stevens Institute of Technology Hoboken United States.

Hu, F. (2013). *Cyber-Physical Systems: Integrated Computing and ENGINEERING DEsign*. CRC Press.

Ishimatsu, T., Leveson, N., Thomas, J. et al. (2010). Modeling and hazard analysis using STPA. In: *Conference of the International Association for the Advancement of Space Safety*, Huntsville, Alabama (19–21 May 2010). Noordwijk, The Netherlands: European Space Agency Communications; ESTEC.

Ishimatsu, T., Leveson, N., Fleming, C. et al. (2011). Multiple controller contributions to hazards. In: *5th IAASS Conference* Versaille, France (17–19 October 2011). Noordwijk, The Netherlands: European Space Agency Communications; ESTEC.

JPDO (2012). Capability Safety Assessment of Trajectory Based Operations v1.1. *Technical Report, Joint Planning and Development Office Capability Safety Assessment Team.*

Leveson, N.G. (2012). *Engineering a Safer World*. MIT Press.

Mead, N.R. and Woody, C. (2016). *Cyber Security Engineering: A Practical Approach for Systems and Software Assurance*. Addison-Wesley Professional.

Mead, N.R., Morales, J.A., and Alice, G.P. (2015). A method and case study for using malware analysis to improve security requirements. *International Journal of Secure Software Engineering (IJSSE)* 6 (1): 1–23.

Modarres, M. (2006). *Risk Analysis in Engineering: Techniques, Tools, and Trends*. CRC Press.

Nance, J.J. (2005). Just how secure is airline security?: The Swiss cheese model and what we've really accomplished since 9/11. *ABC News* 4: 12.

Pereira, S. J., Lee, G., and Howard, J. (2006). A System-Theoretic Hazard Analysis Methodology for a Non-Advocate Safety Assessment of the Ballistic Missile Defense System. *Technical Report, DTIC Document.*

Rasmussen, N.C. (1975). *Reactor Safety Study: An Assessment of Accident Risks in US Commercial Nuclear Power Plants*, vol. 4. National Technical Information Service.

Reason, J. (1990). *Human Error*. Cambridge University Press.

RTCA (2008). Safety, performance and interoperability requirements document for the in-trail procedure in the oceanic airspace (ATSA-ITP) application. DO-312, Washington, DC.

SAE (1996). *ARP-4761A, Guidelines and Methods for Conducting the Safety Assessment Process on Civil Airborne Systems and Equipment*. Society of Automotive Engineers. S-18.

Sherry, L. and Mauro, R. (2014). Controlled flight into stall (CFIS): Functional complexity failures and automation surprises. In: *Integrated Communications, Navigation and Surveillance Conference (ICNS)* (8–10 April 2014), D1–1. New York: IEEE.

Thomas, John, and Leveson, Nancy. (2018). STPA Handbook.

USDoD (1949). *MIL-P-1629 – Procedures for Performing a Failure Mode Eject and Critical Analysis.* United States Department of Defense.

Vesely, W. E., Goldberg, F. F., Roberts, N. H., and Haasl, D. F. (1981). Fault Tree Handbook. *Technical Report, DTIC Document.*

Watson, H. (1961). *Launch Control Safety Study*. Murray Hill, NJ, USA: Bell Telephone Laboratories.

Wise, J., Rio, A., and Fedouach, M. (2011). What really happened aboard Air France 447. *Popular Mechanics* 6: 35–36.

Young, W. and Leveson, N. (2013). Systems thinking for safety and security. In: *Proceedings of the 29th Annual Computer Security Applications Conference*, New Orleans, LA, USA (9–13 December 2013), 1–8. New York: Association for Computing Machinery.

Biographical Sketches

Cody Fleming is an associate professor in mechanical engineering at Iowa State University and a core member of both the ISU's Visualize, Reason, Analyze, Collaborate (VRAC) Center and the ISU Translational AI Center (TRAC). He has been at ISU since 2020, was a member of the University of Virginia faculty from 2015 to 2020 and received his PhD in aeronautics and astronautics from MIT in 2015. Within systems engineering and in particular system safety, his research involves theory and methods for control systems and autonomous systems, including analytical and data-driven methods for prediction, decision-making, and control.

Chapter 24

The "Mission Aware" Concept for Design of Cyber-Resilience

Peter A. Beling [1], Megan M. Clifford [2], Tim Sherburne [1], Tom McDermott [2], and Barry M. Horowitz [3]

[1] *Virginia Polytechnic Institute and State University, Hume Center for National Security and Technology, Blacksburg, VA, USA*
[2] *Stevens Institute of Technology, School for Systems and Enterprises, Hoboken, NJ, USA*
[3] *University of Virginia, Charlottesville, VA, USA*

Mission Aware Concept for Cyber Resilience

For several years, the main focus of the Trusted Systems research thrust within SERC has been the creation of tools and methods that enhance the design of cyber-resilient cyber-physical systems. Researchers from various institutions have played a key role in this effort and have developed the Mission Aware (MA) framework, which integrates and aligns cyber engineering requirements with the development life cycle and systems engineering processes (Horowitz et al. 2014, 2015a,b, 2017, 2018; Beling et al. 2019, 2021; Carter et al. 2019; Fleming et al. 2021; Bakirtzis et al. 2022).

Cyber-physical systems (CPS) and high-integrity systems often incorporate a range of hardware and software components to offer crucial capabilities, including advanced computing platforms, control systems, sensors, and communication networks. These components monitor system operational conditions and control assets as needed for their designated mission. Examples of CPS include energy operation centers, wearable devices, connected vehicles, defense and military operations, and homeland security monitoring centers.

The critical nature of these systems demands comprehensive, effective, and cost-efficient security analysis throughout their lifecycle, especially during the early stages of concept development. This is emphasized in the defense community, where it is believed that 70–80% of decisions affecting safety and security are made during the early stages of a project (Department of Defense 2014). Based on these insights, we identify two distinct needs in CPS security: (1) the need to derive resilience from both the system and mission context, and (2) the need to establish systematic risk-based security analysis early in the system's life cycle that goes beyond simple security compliance checklists.

Recent reports on system vulnerabilities, which not only impact infrastructure but also potentially result in loss of life, highlight these needs. One such example is the Boeing 757, which has been shown to have significant exploitable vulnerabilities in a real-world setting (Biesecker 2017). Perimeter-based security measures have shown some success in protecting CPS, but they are prescriptive (e.g., use a firewall, encrypt communication channels) without taking into account the

Systems Engineering for the Digital Age: Practitioner Perspectives, First Edition. Edited by Dinesh Verma.

Companion website: www.wiley.com/go/verma/systemsengineering

nature and purpose of the system and its mission. In contrast, mission-centric cybersecurity acknowledges the specific expected service that needs protection, providing a basis for potential defenses and focusing the system designer's resources and efforts on mitigating potential vulnerabilities (Horowitz et al. 2018, 2021).

To tackle these challenges, Mission Aware was developed as a proactive, model-based, and strategic cybersecurity approach. It starts by understanding the system's mission, expected behavior, and potential hazards, and who it serves and why. By answering these questions, the mission requirements, admissible functional behaviors, and potential system structure can be modeled from top to bottom. This leads to systematic analysis by combining all three domains in a well-defined mission specification. Finally, public vulnerability repositories (e.g., CAPEC, CWE, and CVE from MITRE) and NIST security frameworks may be used to assess the security posture of critical subsystems that may impact mission degradation (Adams et al. 2018) (Figure 24.1).

The MA approach incorporates methods for evaluating cyber-physical system threats and attacks, a framework for formulating requirements and design concepts for cyber resilience, and model-based tools for the selection of resilient architectures. MA was created in response to the need for systematic methods and tools that can be used throughout the systems engineering process, from mission engineering to developmental and operational testing. Formal assurance methods for systems and software require that the design information be at its final level of detail, which may not always be feasible. Therefore, formal assurance language and mathematical methods, combined with operational analyses, make assumptions about the design. However, it is crucial to identify mission-critical vulnerabilities during the development stages.

Figure 24.2 illustrates the assurance methodologies at different stages of the system life cycle. Methodologies applied early in the life cycle concentrate on loss causation and resilience, while those applied later focus on risk management and assurance. There should, however, be ongoing evaluations of assurance-related quality attributes. In previous projects, the SERC research team found that the System Theoretic Accident Model and Processes (STAMP) and System Theoretic Processes Analysis Security (STPA-Sec) were effective at the conceptual stage for the process model (Beling et al. 2021; Fleming et al. 2021; Bakirtzis et al. 2022). The research team also developed the Cyber Security Requirements Methodology (CSRM), which incorporates security requirements into a model-based systems engineering (MBSE) meta-model (Horowitz et al. 2018; Beling et al. 2019). Finally, the team developed the Framework for Operational Resilience in Engineering and System Test (FOREST), which provides a structured approach to the derivation of testable requirements for cyber resilience (McDermott et al. 2022). STPA-Sec, FOREST, and CSRM are covered in Chapters 5.2, 5.4, and 5.5, respectively.

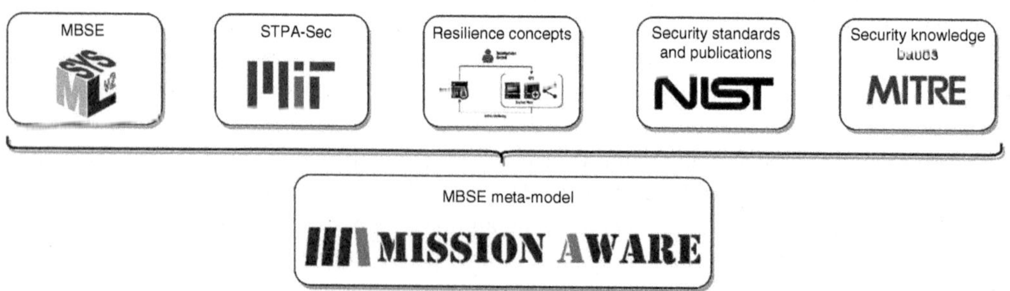

Figure 24.1 Mission aware formation of the MA meta-model. *Source:* Adapted from Horowitz (2018).

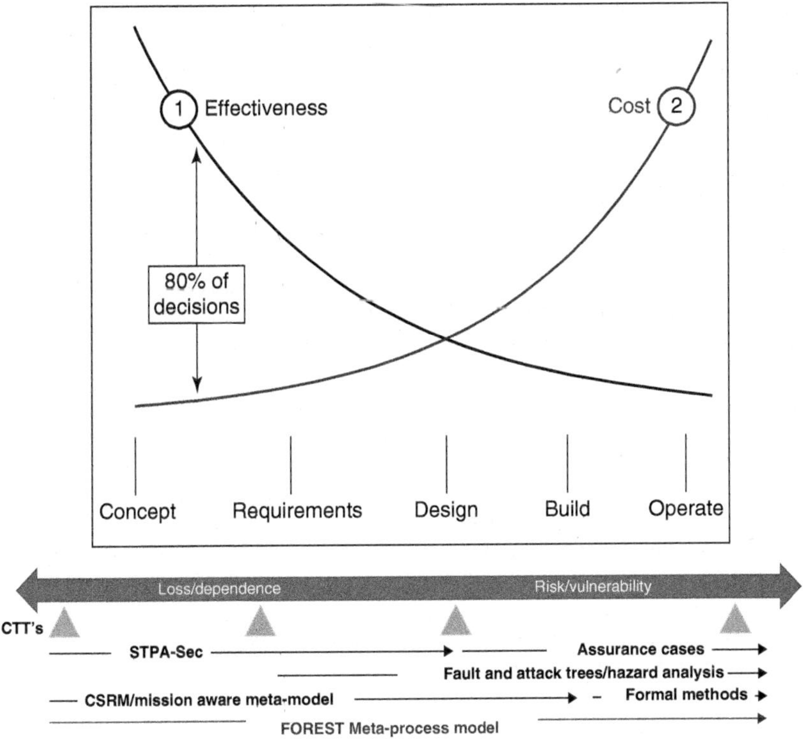

Figure 24.2 Assurance methodologies at various stages of the life cycle.

Resilience Mechanisms

Mission Aware aims to provide a framework for designing systems that are resilient against cyberattacks. It focuses on using monitoring processes – which we term *sentinels* – to detect symptoms of cyberattacks and a response mechanism that reconfigures the system into a *resilience mode* of operation. Resilience modes and detection patterns are a finite and manageable set of strategies that can be specified in an MBSE environment. This approach is primarily applied in cyber-physical systems, such as vehicles and weapons systems, rather than pure cyber and networking systems.

During the early conceptual and requirements phases, patterns can be specified and used as a basis for requirements, architecting, design, and provisioning for verification and validation activities. Table 24.1 provides a library of resilience modes and detection patterns. The mode refers to the application of the pattern in the system, with a description of how the attack model can be countered and implemented. Table 24.1 is based on research conducted by the University of Virginia, Siemens, and the Stevens Institute of Technology (McDermott et al. 2020).

Figure 24.3 illustrates the basic architecture of a sentinel-based resilience mechanism. The sentinel monitors critical system functions and assesses their performance using rule-based or statistical methods. If the performance is anomalous or unacceptable, the sentinel initiates a reconfiguration to engage a resilient mode of operation that uses alternative hardware and software. The resilient modes of operation are designed using components from different manufacturers to minimize the risk of idiosyncratic attacks.

Table 24.1 Library of resilience modes and detection patterns.

Mode/Pattern	Description	Attack model countered
Trusted Kernel or Guard	Creates a small control system within the CPS that independently monitors and/or manages all resource access	Escalation, interruption attacks
Isolation	Creates an isolated runtime environment (sandbox) for the critical asset that is resistant against attacks	Escalation, interruption attacks
Redundancy	Replicates the functionality of the critical asset in order to create multiple paths for high availability and fault tolerance in the case of individual function failures	Attacks that disable individual instances of critical assets and functionality
Diversification	Produces functionally equivalent variations of binaries running in software critical assets. This is an enhancement of the redundancy countermeasure	Coordinated attacks, zero-day attacks effective in identical binary copies of the critical assets
Physically unclonable function	Secures the integrity and privacy of the messages in the system using a Physical Unclonable Function (PUF) that is hard to predict and duplicate	Attacks that hijack the communication channels such as man-in-the-middle attacks
Obfuscation	Obscures the real meaning of data/signals/flows by making them difficult for an attacker to understand. It can use random sources of noise from the environment of the critical assets to increase the entropy	Attacks that require knowledge of the inner workings of the system, its functions, and its mission
Parameter assurance	Compares input data to a table of values in the system to check for large, unexpected deviations	Attacks that manipulate data files or messages that are sent to the system
Data consistency checking	Verifies the source of a parameter change	Attacks that use operator specific data entry
Limiting circuits	Limits resource use (power, memory) to prevent overload	Power system attack

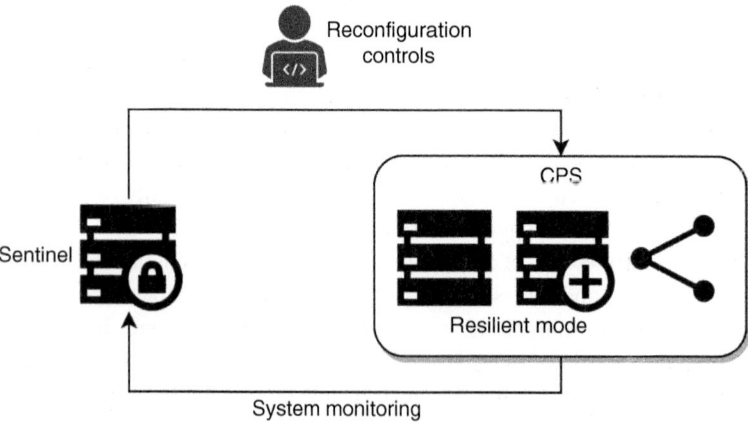

Figure 24.3 Sentinel-based system for resilience against cyberattack.

Sentinels are implemented by adding dedicated software and hardware for sensing, computing, and communication. However, as they may represent potential targets for attack and pathways for accessing other subsystems, sentinels should be designed to be more secure than their host system. The detection process is automated, but the level of reconfiguration automation may vary.

The following sections describe the process of attack detection by sentinels, their response to detection, and the selection of their placement. The discussion includes examples of specific engineering patterns for sentinels and resilient modes of operation.

Figure 24.4 provides a detailed illustration of sentinel functions. As shown in the figure, the sentinel receives data to support its monitoring role through interfaces connected to the system functions or sensors monitoring those functions. These interfaces can be either wired or wireless. The sentinel then processes the collected data to ensure it can be integrated and used for analysis and decision-making. This process includes setting data rates, formats, and communication protocols for internal use within the sentinel. After the data is processed, the sentinel analyzes it to detect and locate cyberattacks. A range of analytical methods may be used, including simple threshold mechanisms, statistical methods for detecting anomalies in multivariate data based on deviation from the mean, and machine learning methods.

Upon detecting an attack, the sentinel alerts system users and provides information on the attack and the steps required for reconfiguration to ensure continued operation. The level of automation in the resilience response will depend on the system design, which can range from totally automated, where the sentinel determines the reconfiguration and either engages directly with control elements or informs operators who will execute the reconfiguration, to semi-automated, where system operators receive automated attack alerts and reconfiguration recommendations and make decisions based on that input and their understanding of the system context, to manual, where system operators receive automated attack alerts and make decisions using their own understanding of the system and context. The sentinel must also prepare and disseminate its results for users in more strategic roles, such as forensics and systems adaptation for managing resilience over longer time cycles.

The hardware/software design of sentinels will vary depending on the system being supported and can be implemented through a single computing node or a highly distributed set of nodes. The

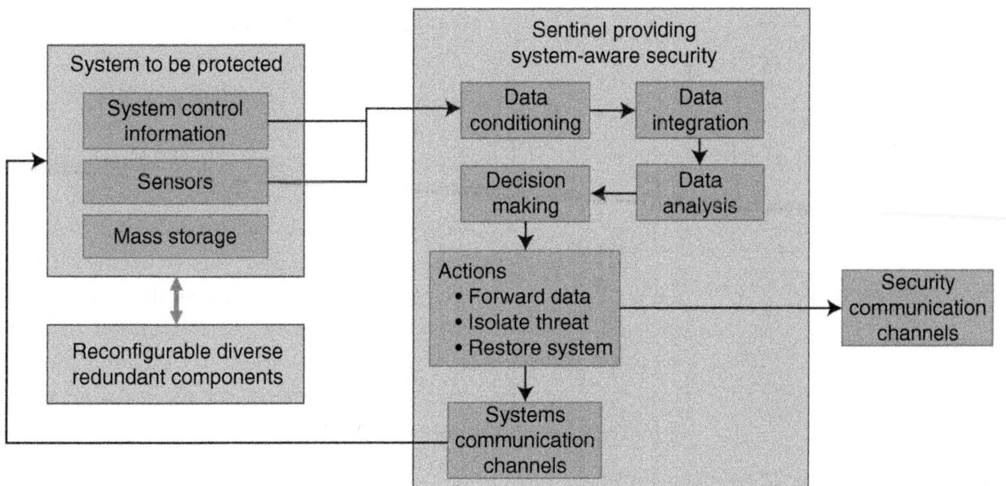

Figure 24.4 Detailed representation of sentinel functions. *Source:* Adapted from Horowitz (2020).

design should be based on reusable design patterns for optimal quality and cost advantages. Examples of patterns for detecting cyberattacks include:

- discovery of data inconsistencies;
- detection of unauthorized changes in system operational parameters;
- recognition of significant communication incompatibilities; and
- various methods for anomaly detection.

The implementation of specific solutions will vary across different systems and the consequences of a cyberattack can also vary, so risk-based decisions are necessary in terms of which design patterns reduce attack risks most effectively. SERC research has included development of prototype designs using some of these patterns for detecting cyberattacks in UAVs, police cars, 3D printers, and military systems.

The following sections describe how sentinels detect attacks, their response after detection, their placement, and how to test them. The chapter concludes with a case study of a hypothetical weapons system.

Sentinel Design Patterns

This section provides examples of several patterns for sentinels design with the goal of detecting anomalous system behavior as an indicator of possible cyberattack. A logical architecture for each pattern is presented, along with a message flow and discussion of system architecture that may be appropriate for the application of the sentinel pattern.

The *changing control input pattern*, as depicted in Figure 24.5, provides a logical architecture for monitoring control paths in hierarchical cyber-physical systems. The sentinel checks the control path through a series of controllers to maintain the consistency of control actions, for instance, ensuring that Control Action B is a logical outcome of Control Action A. This pattern is beneficial for detecting any cyber tampering in Controller B or the control path for Control Action B. Cyber physical systems with a hierarchical control structure can derive value from this detection pattern, as it provides a means of ensuring the integrity of control actions.

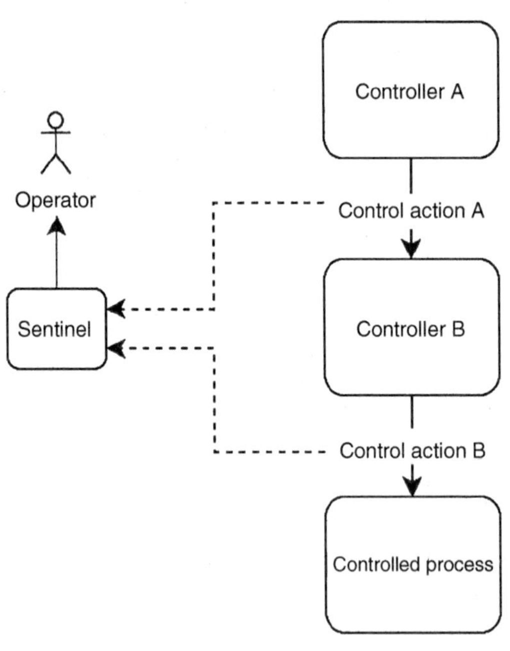

Figure 24.5 Changing control input pattern for detection of cyberattacks. *Source:* Adapted from Beling et al. (2023).

The resource introspection pattern, as illustrated in Figure 24.6, offers a logical architecture for monitoring resource utilization in cyber-physical systems. The sentinel keeps an eye on the CPU, memory, link, and other resources utilized by the controller or controlled process to ensure they align with the current operating state and mode of the system. For instance, it checks if the throughput of feedback messages matches the CPU utilization of the controlled process, which is the expected source of these messages. Cyber-physical systems for which there is a clear understanding of resource utilization semantics are an appropriate application of this pattern.

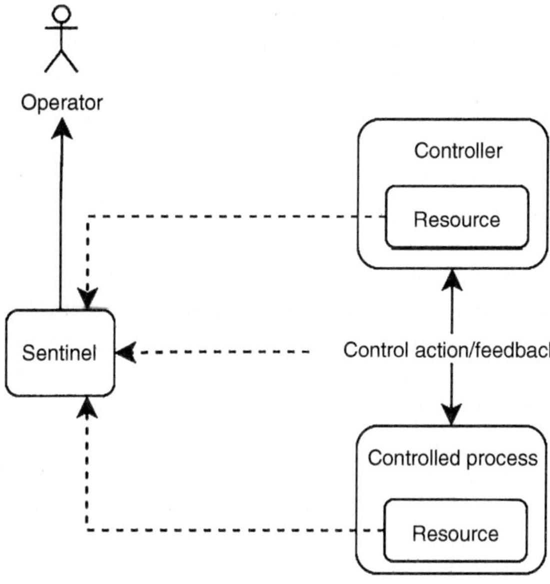

Figure 24.6 Resource introspection pattern for detection of cyberattacks. *Source:* Adapted from Beling et al. (2023).

Attestation Using TPM

The *attestation using a Trusted Platform Module (TPM) pattern*, as shown in Figure 24.7, provides a logical architecture for verifying the integrity of software and configuration in cyber-physical systems. During the boot process of the controller, secure hashes (SHA256) of software and configuration partitions are generated and recorded in the platform configuration registers (PCR) of a TPM. The firmware that performs the initial partition hash typically comes from a write-once memory location. After the boot sequence is completed, if all PCR values hold the correct SHA256 values, a shared secret is unlocked within the TPM, allowing for the calculation of a time-based one-time password (TOTP). The TOTP is reported to the Sentinel, which verifies the integrity of all software and configuration partitions by comparing it to prior knowledge of the controller's shared secret. Cyber-physical systems undergoing regular deployment or maintenance phases are susceptible to tampering of software or configuration and can benefit from this detection pattern, as it provides a means of verifying the integrity of these components.

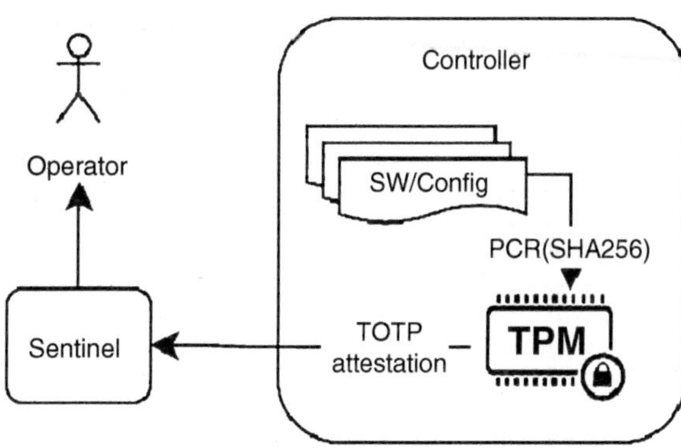

Figure 24.7 Attestation using a trusted platform module (TPM) pattern for detection of cyberattacks. *Source:* Adapted from Beling et al. (2023).

Resilience Mode Design Patterns

This section showcases several patterns for resilient modes of system operation. In the event of detecting abnormal system behavior, a sentinel will suggest a resilient mode of operation to address the issue. The patterns are presented with a logical architecture, a description of the message flow, and a discussion of the system architecture that is suitable for implementing the resilience pattern.

The *diverse redundant controller resilience pattern*, as illustrated in Figure 24.8, provides a logical architecture for cyber-physical system controllers that support mission-critical functions. The sentinel or the operator can trigger the switch of the active controller based on mission requirements. The diversity of implementation and supplier reduces the risk of abnormal system behavior being transferred to the redundant controller. Cyber-physical system controllers that are crucial to mission success can take advantage of this resilience pattern to ensure continued operation in the face of abnormal system behavior.

The *path diversity resilience pattern*, as depicted in Figure 24.9, provides a logical architecture for cyber physical system control paths that support mission-critical messaging. The sentinel or the operator can trigger the switch of the active path based on mission requirements. The diversity of path technology reduces the risk of abnormal system behavior being transferred to the redundant

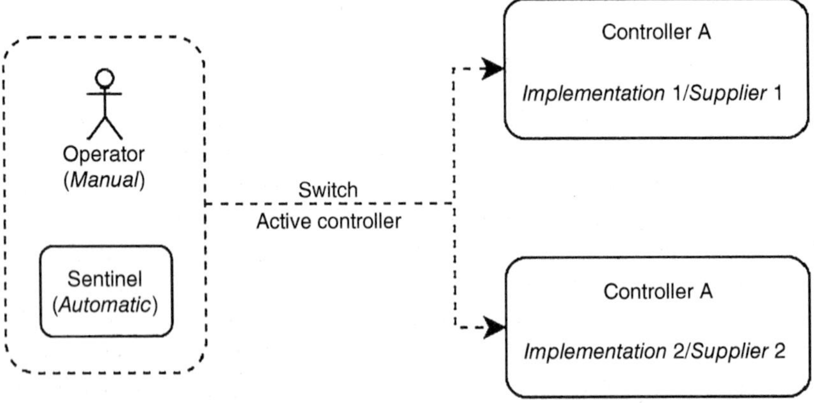

Figure 24.8 Diverse redundant controller resilience pattern. *Source:* Adapted from Beling et al. (2023).

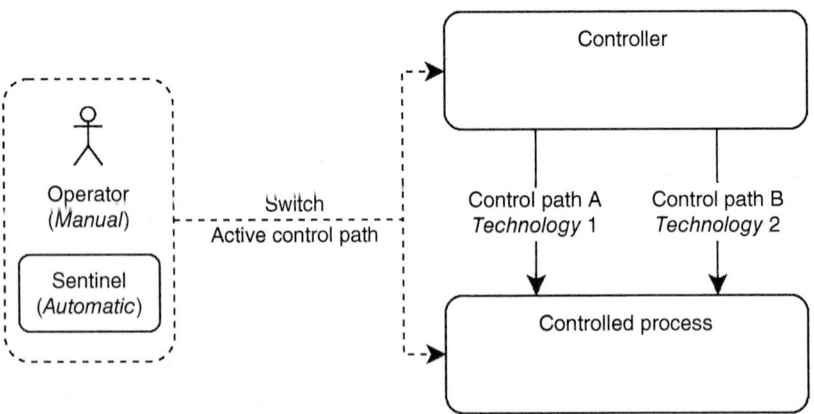

Figure 24.9 Path diversity resilience pattern. *Source:* Adapted from Beling et al. (2023).

path. Cyber-physical system control paths that play a crucial role in mission-critical messaging can take advantage of this resilience pattern to ensure continued operation in the face of abnormal system behavior.

The *protected restore resilience pattern*, as illustrated in Figure 24.10, offers a logical architecture for cyber physical systems undergoing regular deployment or maintenance phases. The sentinel or the operator can initiate the restore of software or configuration for a controller based on mission requirements. Restoring a protected backup can disrupt a cyberattacker's access and return the controller to a known state of operation. Cyber-physical systems that are susceptible to tampering of software or configuration during deployment or maintenance can take advantage of this resilience pattern to maintain the integrity of their systems in the face of potential cyber threats.

Earlier, we discussed reusable design patterns for detecting cyberattacks and designs that utilize diverse redundancy for continued operation are also reusable. However, the implementation and risk reduction value of these designs depend on the specific system being protected. Diverse solutions generally do not perform as well as the normal mode of system operation, but they may still be acceptable for continued operation. Some examples of opportunities for diverse redundancy include:

- The use of various sensors for providing situational awareness information, such as radar, infrared, audio, video, and many other technologies that could be used for surveillance subsystems.
- The use of different navigation subsystems, such as GPS and inertial navigation.
- The use of commonly used subsystems that are designed and produced with different hardware and software by different manufacturers, such as different operating systems, application software, microelectronics components, and communications switches.

The implementation of diverse redundancy creates a challenge for cyber attackers, as they must understand the system design and develop multiple attacks on the diversely redundant subsystems to effectively disrupt the targeted system. This should increase the cost, time, technical complexity, and risk for creating successful cyberattacks and deter attackers from attempting to disrupt the system. However, for this solution to be effective, it must be low cost, timely, low risk, and effective.

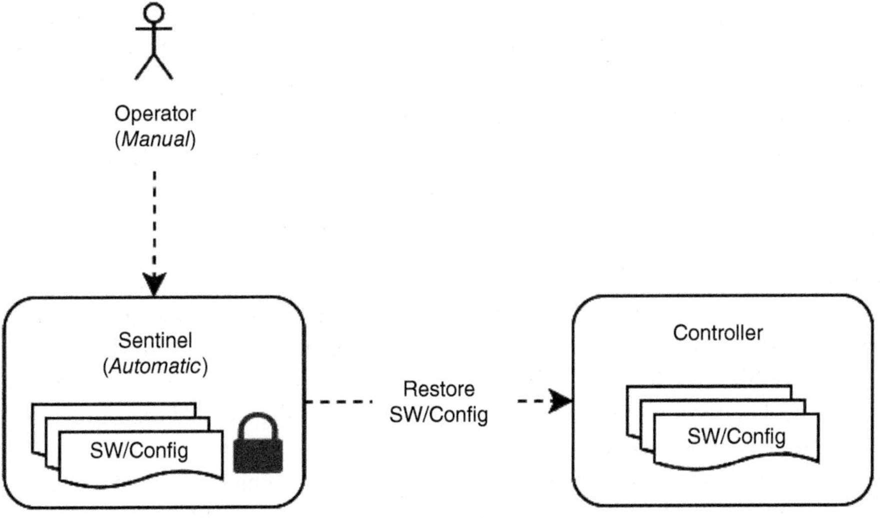

Figure 24.10 Protected restore resilience pattern.

Designers of resilient systems must consider the potential losses in performance that may result from reconfiguring the protected system, as well as the operational acceptability of such losses. Resilience can also be achieved by integrating multiple approaches for achieving diversity, serving both the detection of attacks and reconfiguration responses.

For example, one of the design patterns derived from the authors' research efforts is referred to as configuration hopping with voting (Horowitz 2020). An experimental application of this design pattern, utilizing multivariant programming via the use of three diversely manufactured communication switches and through comparison of message content going into and coming out of the switches, could determine if there was an inappropriately performing switch. If so, the improperly performing switch could be taken out of service while continuing system operation (Horowitz 2020).

In addition, to make matters more complex for a cyberattacker intent on changing message content, the design pattern included the use of a moving target technique, dynamically changing which switch is to be operationally employed once every few seconds, with the use of randomly selected times for moving the potential targets (Horowitz 2020). Since the diversely implemented switches were not closely synchronized in terms of order of messages and their timing, use of moving target defense brought with it the potential to create problems due to the timing of message processing within the diverse switches. To address this problem, message content comparisons were done in a batched manner at sufficiently spaced intervals to reduce the percentage of deviating messages due to timing. The sentinel detection algorithms were designed to permit missing messages as a normal situation when the deviations occurred close to the switching times, and the operational system depended on its existing communication protocols to assure that missing messages due to dynamic changing of the switch in operation were either resent in a timely manner or were acceptable for loss at low rates. Operational prototype-based experiments related to control of a ship's propulsion system were conducted to measure message loss rates. Results indicated that the number of lost messages due to a 20-hop/s resilience design was acceptably low (Horowitz 2020).

As an example, configuration hopping with voting uses multivariant programming with three diverse communication switches. By comparing message content going in and out of the switches, the sentinel can detect an improperly performing switch and remove it while continuing system operation. The pattern also employs a moving target technique, dynamically changing the switch in use every few seconds at random times, making it more difficult for a cyberattacker to change message content. To address potential timing issues, message content comparisons are done in batches at intervals to reduce deviations due to timing. The sentinel algorithms allow for missing messages close to switching times and the system relies on communication protocols to resend or accept low loss rates of missing messages. Prototype experiments in controlling a ship's propulsion system showed acceptably low message loss rates with a 20-hop/s resilience design (Babineau et al. 2012).

STPA and STPA-Sec

Systems-Theoretic Accident and Processes (STAMP) is a safety analysis method that is based on the concept of causation (Leveson and Thomas 2018). It models each level of a system as a control process and assumes that accidents can be caused by unsafe interactions of system components, in addition to component failures. This layered approach to safety provides a notion of consequence within the safety model and asserts that emergent properties, such as safety and security, cannot be assured by examining subsystems in isolation.

STPA (System-Theoretic Process Analysis) is a type of STAMP modeling that is mainly used to identify hazardous conditions and states proactively (Leveson and Thomas 2018). STPA-Sec extends STPA to bring the benefits of loss-oriented safety assessment to security (Young and Leveson 2013;

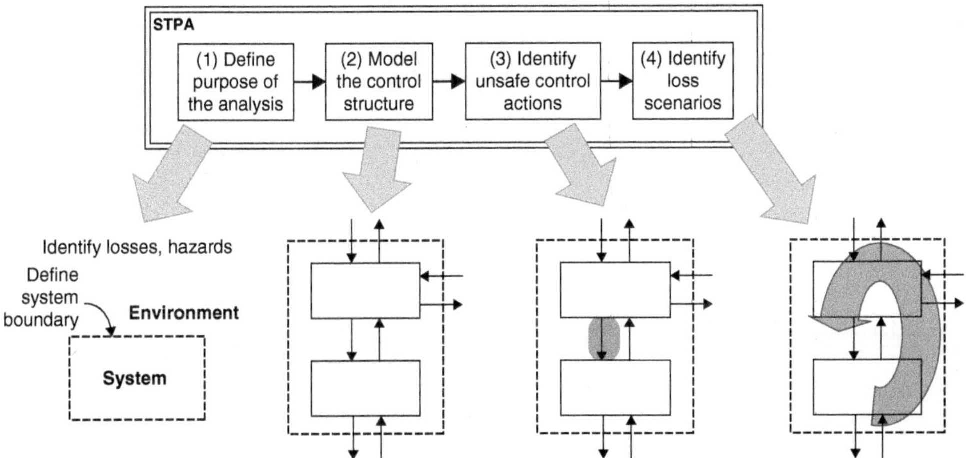

Figure 24.11 System theoretic process analysis (STPA) overview (STPA Handbook). *Source:* Adapted from Leveson (2018).

Leveson and Young 2014; Young and Porada 2017). The hierarchical control concept in STAMP is consistent with multiple model-based systems engineering (MBSE) block diagrams, such as architectural or behavioral diagrams, as they can be enhanced to model unsafe control actions in addition to the control system that defines the behavior and architecture of the system. MBSE is also based on a hierarchical notion that systems can be modeled through different views at different levels of abstraction. As a result, STAMP entities are an important view of safety and security (Figure 24.11).

The first step in STPA involves defining the system boundary, purpose, and prioritized losses and hazards. This information drives the system architecture tradespace analysis. The next step involves modeling the hierarchical control structure for key system behaviors. The analysis then considers how unsafe or hazardous control actions could lead to hazardous states and system losses. Finally, scenarios that could result in unsafe or hazardous actions, including intentional cyber security attacks, are considered.

The scenarios are then evaluated to determine remediation mechanisms using sentinel detection patterns and resilience architecture patterns. The architectural tradespace includes the set of sentinels and resilience modes that effectively mitigate the most likely cyberattacks and minimize the highest priority mission losses within the constraints of development time and budget.

Additional details on STPA-sec and its use within the Mission Aware framework can be found in the chapters in this cluster on STPA and CSRM.

Mission Aware MBSE Meta-Model

The MBSE element of Mission Aware is based on the Vitech Corporation's successful meta-model (Scott and Long 2018), which is publicly available and compliant with the SysMLv2 requirement for precise systems engineering semantics. The Vitech metamodel covers critical systems engineering concepts and their interconnections, including requirements, behavior, architecture, and testing (Figure 24.12). Chapter 5.5 presents the full Vitech meta-model and the key extensions to it that are made to capture the MA framework and the FOREST methodology (Chapter 5.4). The extended meta-model supports an incremental, layered approach through consistent relationships across the requirements, behavior, architecture, and test domains. Using a metamodel provides

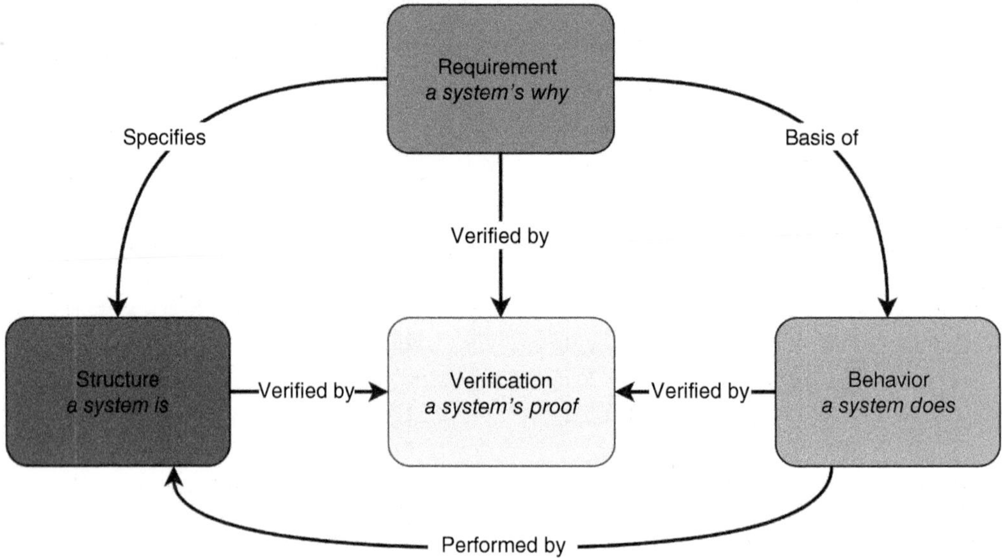

Figure 24.12 General system meta-model from Vitech Corporations Genesis Product. Chapter 6.5 provides a detailed extension of this model to include MA concepts. *Source:* Adapted from Scott and Long (2018).

three main benefits: aligning the documentation methods, reducing workload and preventing errors; adhering to systems engineering best practices; and connecting performance metrics such as safety, security, and resilience with the system design process.

Summary

Mission Aware aims to provide a framework for designing systems that are resilient against cyber-attacks. The framework focuses on the use of engineered mechanisms for detecting and responding to potential cyber-attacks. During the early conceptual and requirements phases, patterns for these engineered mechanisms can be specified using model-based systems engineering (MBSE) techniques and used as a basis for requirements, architecting, design, and provisioning for verification and validation activities. This approach is primarily applied in cyber-physical systems, such as vehicles and weapons systems, rather than pure cyber and networking systems.

The Mission Aware MBSE Meta-Model will be delved into further in for the design of cyber resilience in Chapter 5.5, and demonstrated through Silverfish, a weapon systems use case, in Chapter 5.6. The efforts show significance in the lifecycle evaluation of resilience, with specific affect to the concept and design of cyber resilience. Again, to be effective in resilience engineering, we must be able to reason about system functions, operational tasks, and missions.

References

Adams, S., Carter, B., Fleming, C., and Beling, P.A. (2018). Selecting system specific cybersecurity attack patterns using topic modeling. In: *2018 17th IEEE International Conference on Trust, Security and Privacy in Computing and Communications/12th IEEE International Conference on Big Data Science and Engineering (TrustCom/BigDataSE)*, New York (1-3 August 2018), 490–497. IEEE.

Babineau, G.L., Jones, R.A., and Horowitz, B. (2012). A system-aware cyber security method for shipboard control systems with a method described to evaluate cyber security solutions. In: *2012 IEEE Conference on Technologies for Homeland Security*, Waltham, Massachusetts, USA (13–15 November 2012), 99–104. IEEE.

Bakirtzis, G., Sherburne, T., Adams, S. et al. (2022). An ontological metamodel for cyber-physical system safety, security, and resilience coengineering. *Software and Systems Modeling* 21 (1): 113–137.

Beling, P., Horowitz, B., Fleming, C., et al. (2019). Model-Based Engineering for Functional Risk Assessment and Design of Cyber Resilient Systems. *University of Virginia Charlottesville United States, Technical Report.*

Beling, P., McDermott, T., Sherburne, T., et al. (2021). Developmental Test and Evaluation and Cyberattack Resilient Systems. *Systems Engineering Research Center, Technical Report.*

Beling, P., Sherburne, T., and Horowitz, B. (2023). Sentinels for cyber resilience, in autonomous intelligent agents for cyber defense. *Springer* 87: 425–444. [forthcoming].

Biesecker, C. (2017). Boeing 757 testing shows airplanes vulnerable to hacking, DHS says. http://www.aviationtoday.com/2017/11/08/boeing-757-testing-shows-airplanes-vulnerable-hacking-dhs-says/ (accessed 13 November 2022).

Carter, B., Adams, S., Bakirtzis, G. et al. (2019). A preliminary design-phase security methodology for cyber–physical systems. *Systems* 7 (2): 21.

Department of Defense (2014). Cyber security. http://www.dtic.mil/whs/directives/corres/pdf/850001_2014.pdf (accessed 14 November 2022).

Fleming, C.H., Elks, C., Bakirtzis, G. et al. (2021). Cyber-physical security through resiliency: a systems-centric approach. *Computer* 54 (6): 36–45.

Horowitz, B.M. (2020). Cyberattack-resilient cyberphysical systems. *IEEE Security & Privacy* 18 (1): 55–60.

Horowitz, B., Beling, P., Skadron, K., et al. (2014). Security Engineering Project-System Aware Cyber Security for an Autonomous Surveillance System on Board an Unmanned Aerial Vehicle. *Systems Engineering Research Center, Hoboken, NJ, Technical Report.*

Horowitz, B., Beling, P., Humphrey, M., and Gay, C. (2015a). System Aware Cybersecurity: A Multi-Sentinel Scheme to Protect a Weapons Research Lab. *Stevens Institute of Technology, Hoboken, NJ, Technical Report.*

Horowitz, B., Beling, P., Skadron, K., et al. (2015b). Security Engineering Project. *Systems Engineering Research Center, Hoboken, NJ, Technical Report.*

Horowitz, B., Beling, P., Fleming, C., et al. (2017). Security Engineering FY17 Systems Aware Cybersecurity. *Stevens Institute of Technology, Hoboken, United States, Technical Report.*

Horowitz, B., Beling, P., Fleming, C., et al. (2018). Cyber Security Requirements Methodology. *Stevens Institute of Technology, Hoboken, United States, Technical Report.*

Horowitz, B., Beling, P., Clifford, M., and Sherburne, T. (2021). Developmental Test and Evaluation (DTE&A) and Cyber Attack Resilient Systems - Measures and Metrics Source Tables. *Systems Engineering Research Center, Technical Report.*

Leveson, N.G. and Thomas, J. (2018). STPA handbook. https://psas.scripts.mit.edu/home/get_file.php?name=STPA_handbook.pdf (accessed 12 November 2022).

Leveson, N.G. and Young, W. (2014). An integrated approach to safety and security based on systems theory. *Communications of the ACM* 57 (2): 31–35.

McDermott, T., Fleming, C., Clifford, M.M. et al. (2020). *Methods to Evaluate Cost/Technical Risk and Opportunity Decisions for Security Assurance in Design.* Stevens Institute of Technology, Systems Engineering Research Center Hoboken United States.

McDermott, T., Clifford, M.M., Sherburne, T. et al. (2022). Framework for operational resilience in engineering and system test (FOREST) Part I: methodology–responding to "security as a functional requirement". *Insight* 25 (2): 30–37.

Scott, Z. and Long, D. (2018). One model, many interests, many views. Vitech Corporation, *Technical Report*. http://www.vitechcorp.com/resources/white_papers/onemodel.pdf

Young, W. and Leveson, N.G. (2013). Systems thinking for safety and security. In *Proceedings of the Annual Computer Security Applications Conference* (ACSAC 2013), New Orleans, LA (9 December 2013). ACM.

Young, W. and Porada, R. (2017). System-theoretic process analysis for security (STPA-Sec): cyber security and STPA. In *2017 STAMP Conference* (27–30 March 2017). MIT Press.

Biographical Sketches

Peter A. Beling is a professor in the Grado Department of Industrial and Systems Engineering and associate director of the Intelligent Systems Division in the Virginia Tech National Security Institute. Dr. Beling's research interests lie at the intersections of systems engineering and artificial intelligence (AI) and include AI adoption, reinforcement learning, transfer learning, and digital engineering. He has contributed extensively to the development of methodologies and tools in support of cyber resilience in military systems. He serves on the Research Council of the Systems Engineering Research Center (SERC), a University Affiliated Research Center for the Department of Defense.

Megan M. Clifford is a research associate and engineer at Stevens Institute of Technology. She works on various research projects with a specific interest in systems assurance and cyber-physical systems. She previously worked on the Systems Engineering Research Center (SERC) leadership team as the Chief of Staff and Program Operations, was the Director of Industry and Government Relations to the Center for Complex Systems and Enterprises (CCSE), and held several different positions in industry, including Systems Engineer at Mosto Technologies while working on the New York City steam distribution system.

Barry M. Horowitz held the Munster Professorship in Systems Engineering at the University of Virginia, prior to his retirement in May 2021. His research interests include system architecture and design.

Tom McDermott serves as the deputy director and chief technology officer of the Systems Engineering Research Center (SERC) at Stevens Institute of Technology in Hoboken, NJ. The SERC is a University Affiliated Research Center sponsored by the Office of the Secretary of Defense for Research and Engineering. With the SERC, he develops new research strategies and is leading research on Digital Engineering transformation, education, security, and artificial intelligence applications. Mr. McDermott also teaches system architecture concepts, systems thinking and decision making, and engineering leadership. He is a lecturer in Georgia Tech's Professional Education college, where he leads a master's level course on systems engineering leadership and offers several continuing education short courses. He consults with several organizations on enterprise modeling for transformational change, and often serves as a systems engineering expert on government major program reviews. He currently serves on the INCOSE Board of Directors as Director of Strategic Integration.

Tim Sherburne is a research associate in the Intelligent System Division of the Virginia Tech National Security Institute. Sherburne was previously a member of the systems engineering staff at the University of Virginia supporting Mission Aware research through rapid prototyping of cyber-resilient solutions and model-based systems engineering (MBSE) specifications. Prior to joining the University of Virginia, he worked at Motorola Solutions in various Software Development and Systems Engineering roles defining and building mission critical public safety communications systems.

Chapter 25

The "FOREST" Concept and Meta-Model for Lifecycle Evaluation of Resilience

Tim Sherburne[1], Megan M. Clifford[2], Barry M. Horowitz[3], Tom McDermott[2], and Peter A. Beling[1]

[1] *Virginia Polytechnic Institute and State University, Hume Center for National Security and Technology, Blacksburg, VA, USA*
[2] *Stevens Institute of Technology, School for Systems and Enterprises, Hoboken, NJ, USA*
[3] *University of Virginia, Charlottesville, VA, USA*

The Framework for Operational Resilience in Engineering and System Test (FOREST)

Introduction and Background

The Department of Defense (DoD) is increasing efforts to address the operational risks posed by cyber-attacks and insider threats by utilizing system assurance and system resilience. System assurance refers to the confidence that a system is free of exploitable vulnerabilities and operates as intended, while system resilience refers to a system's ability to recover from a loss of function. The system resilience approach allows for early consideration of cyber risks in the systems engineering process and can be analyzed in terms of system functions prior to the design stage (Reed 2016; Ross 2022).

Digital engineering and model-based systems engineering (MBSE) are widely used in the conception, design, integration, and validation of mission-critical systems, but there is still a lack of integrated modeling and dynamic simulation support for systems engineering in terms of operational and cyber resilience from concept to design. The development of common standards, methods, processes, and tools is necessary, along with new metrics, methods, and tools for hazard mitigation. New approaches must also be found to support modeling system functions in the pre-design stages in the context of MBSE (DoD 2014; Beling et al. 2019; Horowitz et al. 2018a).

Testing and evaluating cyber resilience poses unique challenges as resilience behaviors are only exhibited when critical functions are lost. This requires the creation and emulation of functional models and reasoning about the internal system states resulting from disruptions caused by system failures or successful attacks. The Systems Engineering Research Center (SERC) has focused on developing methods and tools for cyber resilience in cyber-physical systems, with the objective of creating an end-to-end systems engineering methodology that connects mission-level resilience analysis and system development activities. This methodology includes a meta-process model

Systems Engineering for the Digital Age: Practitioner Perspectives, First Edition. Edited by Dinesh Verma.

Companion website: www.wiley.com/go/verma/systemsengineering

called the Framework for Operational Resilience in Engineering and Systems Test (FOREST) and a reference architecture meta-model called Mission Aware. These models are utilized to make security and related resilience decisions in capability development using a standard, risk-based approach and to derive measures and metrics that can serve as the basis for test and evaluation in a rigorous systems engineering process. In turn, the framework provides developers with insights that support the formation of testable requirements for operational resilience and the design of systems that are immune to unknown vulnerabilities and threat tactics (Horowitz et al. 2021; Beling et al. 2021).

FOREST

The FOREST model is a process framework designed to break down operational resilience into its key components, including options, information flows, and decisions that occur during system attacks and resilience responses. The model is made up of eight Testable Resilience Efficacy Elements (TREEs), which are complex and overlapping, addressing the intersections of technology, doctrine, and people. These TREEs provide a comprehensive view of resilience that can aid in the development of test plans and metrics for both technological and operational aspects of the system. The FOREST model is designed to be leveled, iterative, and cyclical, allowing for continuous evaluation of the system design at various stages of the life cycle (Figure 25.1).

The eight TREEs are represented in Figure 25.2 and are explained in detail in this chapter. The first three TREEs are aligned with the Mission Aware reference architecture, emphasizing the importance of active sensing in detecting system anomalies. The framework then focuses on incident isolation and diagnostic information used in the selection of resilience mode responses. It also covers operator response and supporting technology for resilience solutions and the ability to run tests for confidence in resilience modes. The model includes augmented intelligence support, decision support, and archiving for post-event analysis and adaptation (Beling et al. 2021).

The FOREST model is integrated into the design model, providing a way to evaluate and test at every decision stage. However, it does not provide explicit guidance on the requirements-setting, architecting, design, or engineering tasks that may be associated with each of the eight TREEs. Tools and methods are needed to help with tasks such as designing system capabilities for attack sensing, isolation, and response. While there are libraries of patterns for these resilience functions, choosing the appropriate patterns for a complex system considering budgetary constraints and operational requirements can be a challenging decision.

FOREST aims to assist system owners and engineers in understanding the problem of what to protect and why, which decision patterns to use where, and when decision support for operators may be necessary. It is intended to be used in conjunction with risk assessment and design methodologies, and this chapter illustrates how the methodology for resilience architecting and preliminary design can be adapted to work with FOREST.

Attack Sensing

This aspect of resilience focuses on identifying successful cyberattacks and notifying system operators. Attack detection methods can range from manual recognition of an attack, such as a denial-of-service attack that crashes a system, to using sensors that detect symptoms indicating a persistent cyberattack, such as inputs and outputs that significantly deviate from the system design. The design must also consider resilience against specific threats, such as buffer overflow attacks.

FOREST requires documentation on the attack detection techniques used in the resilience design, the reasoning behind their selection, and the methods considered for testing and evaluating the implementation. This information should be easily traceable from the specific system-level risks posed by cyberattacks to the resilience solutions included in the design. Model-based system

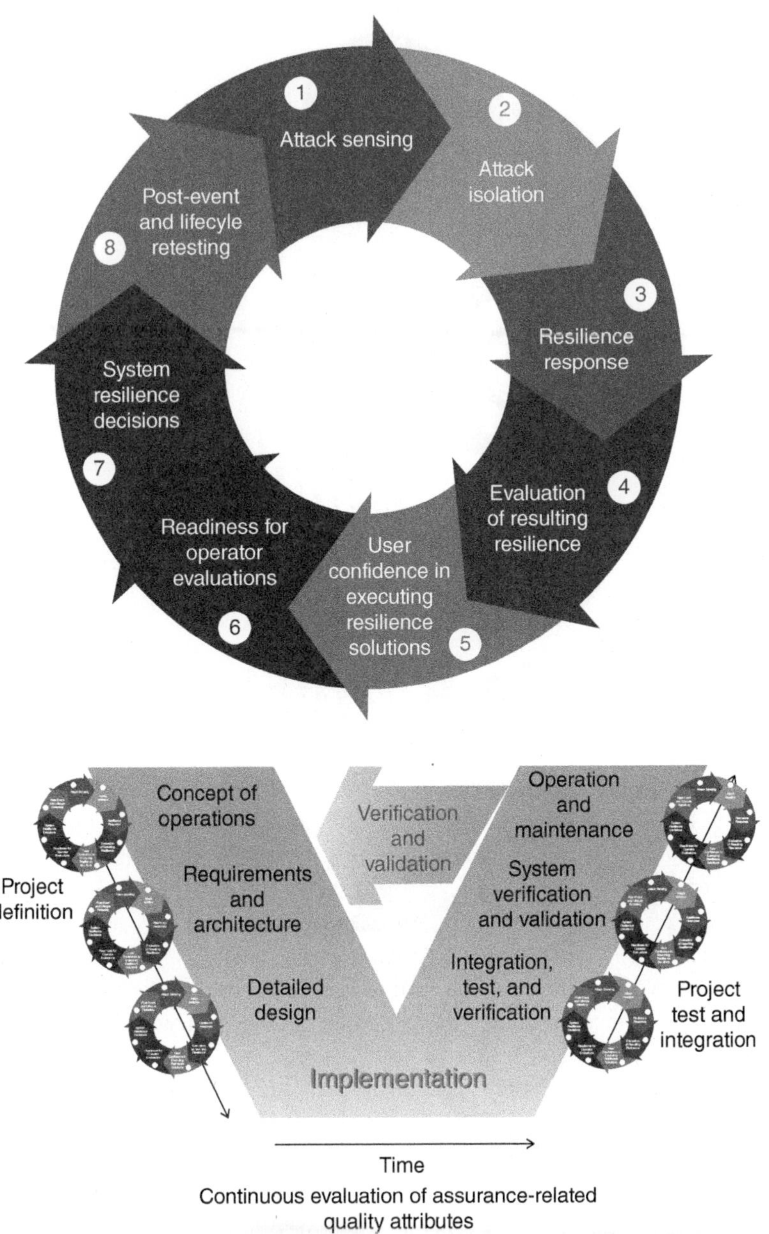

Figure 25.1 The framework for operational resilience in engineering and system test (Horowitz et al. 2017).

engineering tools can be helpful in fulfilling this aspect of the framework, as illustrated in the Silverfish case study example in this cluster of chapters.

Test and Evaluation for Attack Sensing

The effectiveness of cyberattack detection methods can vary, leading to different test designs and evaluation metrics. For instance, solutions that monitor deviations in software execution to detect attacks may not be able to determine the attack's consequences. Conversely, solutions that monitor inputs and outputs to detect system malfunctions likely caused by an attack may not identify the

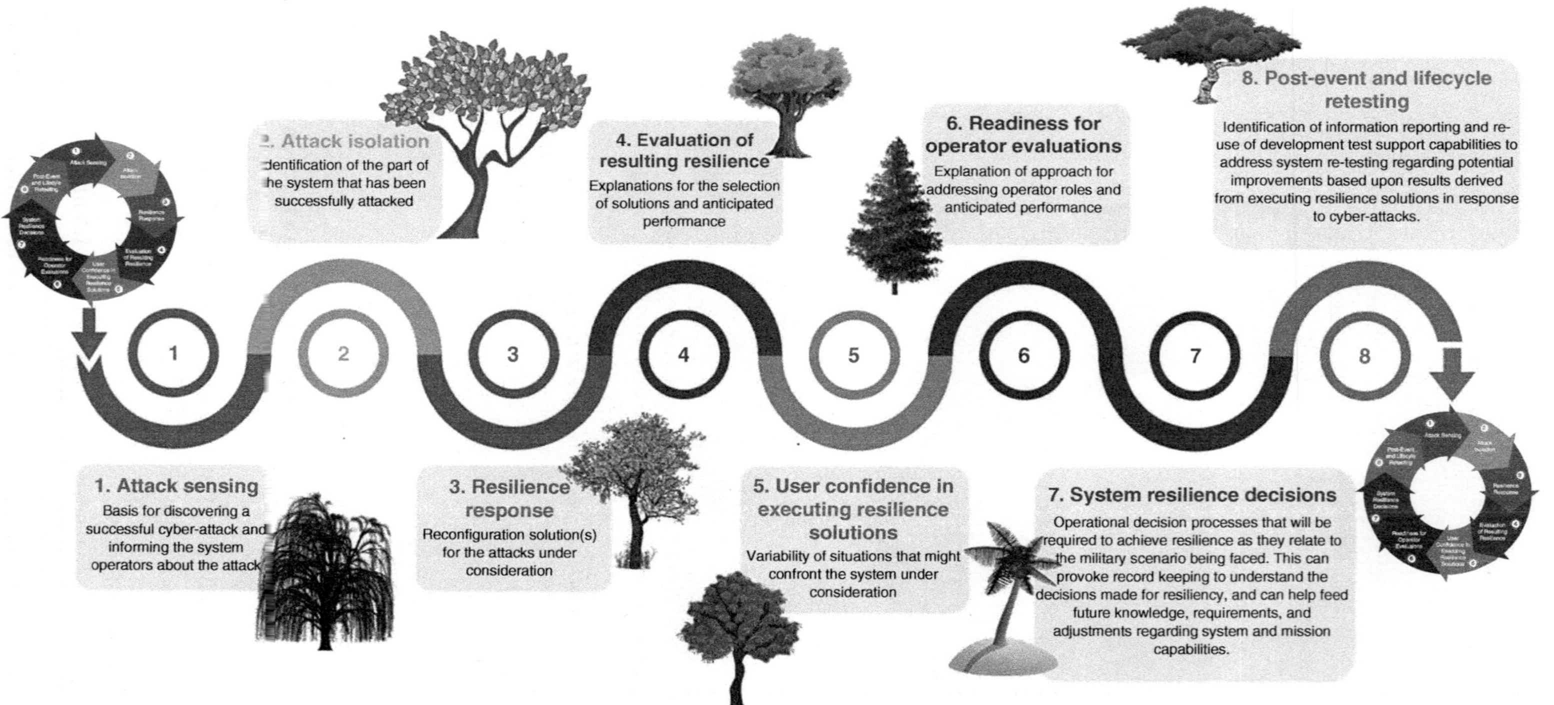

1
2
3
4
5
6
7
8
1. Attack sensing
Basis for discovering a successful cyber-attack and informing the system operators about the attack
2. Attack isolation
Identification of the part of the system that has been successfully attacked
3. Resilience response
Reconfiguration solution(s) for the attacks under consideration
4. Evaluation of resulting resilience
Explanations for the selection of solutions and anticipated performance
5. User confidence in executing resilience solutions
Variability of situations that might confront the system under consideration
6. Readiness for operator evaluations
Explanation of approach for addressing operator roles and anticipated performance
7. System resilience decisions
Operational decision processes that will be required to achieve resilience as they relate to the military scenario being faced. This can provoke record keeping to understand the decisions made for resiliency, and can help feed future knowledge, requirements, and adjustments regarding system and mission capabilities.
8. Post-event and lifecycle retesting
Identification of information reporting and re-use of development test support capabilities to address system re-testing regarding potential improvements based upon results derived from executing resilience solutions in response to cyber-attacks.

Figure 25.2 The FOREST and the eight TREEs.

specific type of attack. Therefore, some solutions may use a combination of methods to understand both the attack technique and its impact on the system.

Testing solutions that detect issues in software execution involves using attack software to initiate attacks and measure the likelihood of detection, false detections, and detection delays. It is also crucial to verify that the attack information is properly conveyed to system operators. Conversely, tests for solutions that recognize functional issues without attacking the system require emulating deviations caused by potential cyberattacks and evaluating the system's ability to respond and recover. In this case, metrics such as the percentage of normal system functionality sustained and the time it takes for recovery are essential. By using these test designs and evaluation metrics, engineers can effectively evaluate and improve the resilience of the system to cyberattacks (Adams et al. 2018; Fleming et al. 2021; Horowitz et al. 2014).

Attack Isolation

This aspect of resilience is crucial for identifying the part of the system that has been successfully compromised by a cyberattack. The level of attack isolation specificity can vary depending on the detection method used and the physical system's design. This can range from identifying a broad component of the system to pinpointing a specific computing element in a distributed system. The information gathered during this step, combined with the attack detection techniques, forms the basis for responding to a successful cyberattack.

While FOREST acknowledges that some elements of attack isolation can be automated, it recognizes the importance of manual verification through responder-initiated diagnostics. This helps improve the understanding of the situation and refine the technical and operational response. Additionally, the information collected during the attack isolation process should be saved and shared with the wider cybersecurity community. This can help enhance existing solutions and evaluate adversary attack capabilities.

The framework requires documentation on the methods used to determine the effectiveness and accuracy of the attack isolation process. It also requires documentation on the strategies considered for testing and evaluating the implemented design. By following these documentation requirements, engineers can better understand the effectiveness of their attack isolation process and refine it to enhance the system's resilience to cyberattacks.

Test and Evaluation for Attack Isolation

The success of a cyber-resilience response to a cyberattack relies on the ability to identify which parts of the system have been targeted. The design of the solution must consider both the technical and operational aspects to provide a clear understanding for the response team. This includes documenting the situational awareness concepts and protocols for collecting and utilizing information from follow-up diagnostics. Testing should validate that the isolation and diagnostic processes provide sufficient information to support the desired resilience response. The metrics for evaluating the isolation should consider the accuracy of the automated processes and the value added by the follow-up diagnostics, compared to any delays they may introduce.

Resilience Options

This TREE in the framework deals with the reconfiguration of the system and containment of safety consequences in the aftermath of a successful cyberattack. The objective of a cyber-resilience solution is to keep critical system functions running and to control the immediate operational and safety impacts of the detected attack. The response to an attack can range from manual actions by operators, to restarting the system, to real-time reconfiguration using redundant components designed for resilience. It is important to note that the reconfigured system may not operate at full capacity and that operators may need to play extended roles during recovery. The time needed for reconfiguration is a crucial factor to consider in response planning. The immediate consequences of an attack can vary, with some being fully resolved and others not contained at all. The framework requires documentation of the resilience measures, their technical design, operator functions, and evaluation metrics, including time to reconfigure and expected performance losses. The methods for testing and evaluation should also be documented.

Test and Evaluation for Resilience Options

Researchers in the field of cyberattack resilience have developed a diverse range of resilience solutions. As previously outlined, two main design opportunities have arisen based on the detection of software execution disruptions and system functional errors. These resilient responses can include shifting from a corrupted sensor to an alternate navigation system, providing alternate control methods in case of corruption of control inputs, and switching between communications techniques in case of corruption of data communications between sub-systems. It is important for resilience designers to document their design approach and the related test and evaluation activities. Where feasible, quantitative metrics such as delay times in reconfiguring the system and reductions in system performance should be developed to measure the effectiveness of the resilience solutions. For design features that cannot be quantitatively confirmed, demonstrating their performance in various circumstances and with different operators should be sufficient to prove their resilience.

Evaluation of Resilience Options

Based upon information provided in TREEs 1–3 above, this part of the framework calls for documentation that provides explanations for the selection of solutions, the anticipated performance of the reconfigured system (including time to reconfigure), and the basis for deciding that the resulting operational capabilities are satisfactory. Some of the results may be quantitative in nature, and directly tested, while some of the evaluation may be qualitative in nature, in which case the basis for justification should be explained.

Test and Evaluation for Resilience Options

Continuous evaluation of resilience options is crucial to the operational resilience of a system, given the vast number of options available. However, an aspect that is often overlooked in development is the test and evaluation of resilience options. Documentation, whether in models or written reports, should be easily traceable to the tests and the reasoning behind the chosen options. For example, technical availability of resilience modes is necessary for the mission and is a system limitation. This can be partly measured quantitatively. Another example is the test and evaluation of operator judgment regarding the usability and failure transparency of resilient modes. This can be measured qualitatively through operator ratings to better understand the usability and transparency of resilient modes. By conducting thorough and well-documented test and evaluation of resilience options, engineers can ensure that the selected options will effectively enhance the system's resilience. It also provides a basis for continuous improvement and refinement of the system's resilience in response to new threats and challenges.

Operational Confidence in Executing Resilience Solutions

This element of the framework addresses the various situations that a system may encounter, including the possibility of cyberattacks that can disable or corrupt the intended solutions. Designs may include the capability for rapid testing to verify that technical and operational adjustments perform as intended when inserted into use. These tests may be fully predefined or left to the operators to define. For example, if a UAV auto-pilot is replaced, the pilot may want to control the aircraft for a certain period to ensure that nothing unexpected happens. The framework requires documentation of the basis for achieving a high level of confidence and the related test and evaluation methods. By documenting the methods used to achieve high confidence in technical and operational adjustments, engineers can ensure that the system's resilience is maintained. This documentation also allows for continuous improvement and refinement of the system's resilience as new situations and challenges arise.

Requirements Specification Using FOREST

A key concern of any systems engineering model is an understanding of the system's architecture, including its components and physical links that connect them. Components may include hardware elements, software elements, external systems, and/or humans. Of equal concern is an understating of the expected behavior of the system being modeled. Behavior elements include functions, their input and output items as well as any resources provided or consumed. The call structure provides an understanding of behavior control flow including looping, parallel execution, path selection with exit choices, etc. Components perform functions, thereby linking the physical architecture with the behavior model. These standard system modeling entities define the engineering process itself and provide structure to the essential design artifact of the system under design.

As mentioned above, MBSE entities and relationships do not address "-ilities" necessary for the design of cyber-physical systems (CPS). Additional entities for safety, security, and resilience that are specifically related to CPS must be added to provide evidence for the correct behavior of CPS. Such performance metrics are defined in the augmentation of the CPS meta-model and related to already standardized MBSE entities with properly defined relationships. This is an important addition to the standard meta-model provided by Vitech. By adding structure to performance metrics systems engineers can design CPS that provide operational assurance in the face of hazards or security violations.

Traditional system performance metrics are captured as parameters of links, components, and/or functions with a constraint definition defining the equations and relationships between individual entities. Consideration of safety, security, and resilience performance metrics require augmentation of the standard MBSE meta-model with additional concepts to capture both an operational risk perspective and an adversarial attacker perspective (Table 25.1).

Safety and security often require specification of system behavior as a set of feedback control loops. As such, specializations of control action and feedback are provided as subtypes of the standard function input output item. While this phraseology is borrowed from STAMP, it applies to a large number of safety and security methods. STAMP in some sense distills any general framework for "-ilities" at a higher abstraction level – by leveraging notions of uncontrolled actions and control hierarchy – that is suited for use in a meta-model. Specifically, losses, hazards, and hazardous actions are captured and related (by means of leads to) as part of a methodical operational risk assessment process. Additionally, explicit associations are captured to understand an unsafe action as a variation of a specific control action with the process model system state that provides the context for the control action to become unsafe, which is borrowed from the domain of control theory and governs all CPS to some extent.

Table 25.1 Testable resilience efficacy element (TREE).

Testable resilience efficacy element (TREE)	TREE number	Description
Attack **Sensing**	T.1	This element of resilience provides the basis for discovering a successful cyber-attack and informing the system operators about the attack
Attack **Isolation**	T.2	This element of resilience solutions addresses identification of the part of the system that has been successfully attacked
Resilience **Options**	T.3	This element of resilience solutions addresses the reconfiguration solution(s) for the attacks under consideration as well as the immediate containment of safety-related consequences
Evaluation of resilience options	T.4	This part of the framework calls for documentation that provides explanations for the selection of solutions, the anticipated performance of the reconfigured system (including time to reconfigure), and the basis for deciding that the resulting operational capabilities are satisfactory
Operational **Confidence** in executing resilience solutions	T.5	The framework calls for documentation of the basis for achieving high enough confidence and the related test and evaluation methods
Readiness for operational execution (real-time mission context)	T.6	The framework will expect explanation of the basis for the system design approach regarding test support for addressing operator roles and anticipated performance
System resilience decision and **Execution**	T.7	The framework will look for the rationale for who decides on what, and the training and tech support required for decision-makers
Post-event and lifecycle test responses	T.8	This portion of the framework addresses identification of information reporting and reuse of development test support capabilities to address system retesting regarding potential improvements based upon actual results derived from executing resilience solutions in response to cyberattacks

Mission Aware Elicited Requirements for Resilience

An important step in assessing any performance metric is to first identify loss scenarios that can lead to unsafe actions. These loss scenarios are the complement of the stakeholder requirements or otherwise define the mission of the system. In the domain of CPS, unsafe behavior and security violations are intertwined, meaning that an attacker could transition the system to a hazardous state. To augment the safety loss scenarios, databases, for example MITRE CAPEC (CAPEC - Common Attack Pattern Enumeration and Classification (CAPEC^{TM}) n.d.), which contain attack patterns, are consulted. The meta-model relates the notion of loss scenario with the notion of recovery and resilience by identifying how a sentinel, which is a type of remediation, could protect against the loss by first indicating how it can be detected by monitoring a link, resource, or function and then how the system reconfigures using a specific resilient mode.

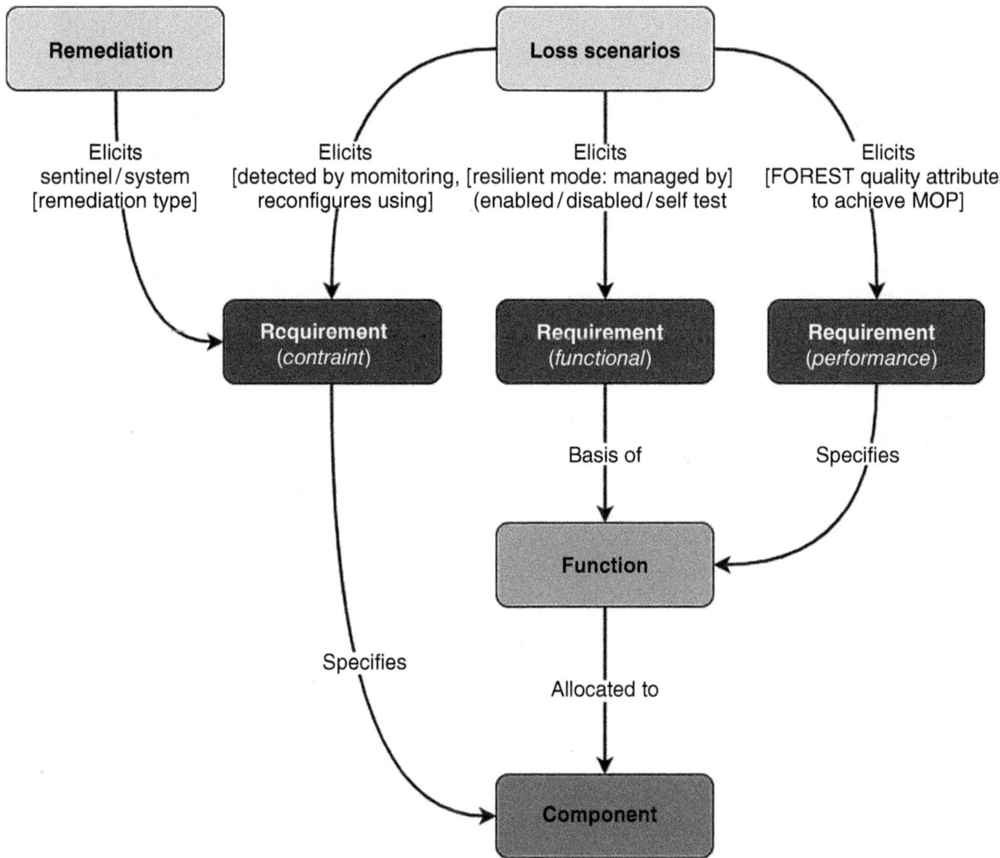

Figure 25.3 Identification of loss scenarios and remediations elicitation. *Source:* Adapted from Beling et al. (2021).

The identification of Loss Scenarios and Remediations enables elicitation (Figure 25.3) of various types of System Requirements:

- **Constraints**
  - that provide Sentinel functions
  - that enable System Monitoring by a Sentinel
  - that provide System Resilient Modes
- **Functions** – that enable System Management (enable/disable/self-test) of Resilient Modes
- **Performance** – that bound FOREST quality attributes that achieve Mission MOPs

A set of System Quality Attributes (-ilities) are provided for the FOREST TREE steps (Table 25.2). The quality attributes are used as an instrument to evaluate cyber-resilience system design choices and as validation criteria during system test. As noted, some quality attributes are directly *measurable* by the system, some are *rated* by the operators of the system, and others are *considerations* for system development teams by illustrating the system limitations.

The quality attribute definitions can be found in the below table (Table 25.3).

Every FOREST step is associated with the appropriate SE artifact and quality attributes (Figure 25.4).

Table 25.2 Relationships between system quality attributes and FOREST TREEs

Quality attribute	T.2: Sense	T.2: Isolate	T.3: Options	T.4: Evaluate	T.5: Confidence	T.6: Readiness	T.7: Execution	T.8: PostEvent
accuracy	✓	✓						
adaptability						●		
affordability								☑
availability				✓		●		
composability			●					☑
extensibility								☑
failure transparency				●				
learnability					●			
predictability					●			
recoverability				●				
repeatability					●			
safety			✓					
stability							✓	
survivability						✓		
testability	✓						✓	☑
timeliness	✓	✓		✓	✓	✓	✓	☑
usability				●				

✓ System Measure ● Operator Rating ☑ Development Consideration.

Table 25.3 Quality attribute definitions.

Quality attribute	Definition
accuracy	Closeness of measurements to a specific value
adaptability	Refers to a process where a system adapts its behavior to individual users based on information acquired about its user(s) and its environment (mission)
affordability	Program cost
availability	The degree to which a system, subsystem or equipment Is in a specified operable and committable state at the start of a mission, when the mission is called for at an unknown, i.e. a random, time The probability that an item will operate satisfactorily at a given point in time when used under stated conditions in an ideal support environment
composability	Composability is a system design principle that deals with the interrelationships of components. A highly composable system provides components that can be selected and assembled in various combinations to satisfy specific mission requirements
extensibility	The ability to extend a system and the level of effort required to implement the extension. Extensions can be through the addition of new functionality or through modification of existing functionality. The principle provides for enhancements without impairing existing system functions
failure transparency	In a distributed system, failure transparency refers to the extent to which errors and subsequent recoveries of hosts and services within the system are invisible to users and applications. For example, if a server fails, but users are automatically redirected to another server and never notice the failure, the system is said to exhibit high failure transparency
learnability	A quality of products and interfaces that allows users to quickly become familiar with them and able to make good use of all their features and capabilities
predictability	The degree to which a correct prediction or forecast of a system's state can be made either qualitatively or quantitatively
recoverability	The property of being able to return to a normal state of health for the system, subsystem, equipment, process, or procedure
repeatability	The closeness of the agreement between the results of successive measurements of the same measure, when carried out under the same conditions of measurement
stability	The state of system, subsystem, equipment, process, or procedure remaining stable over time and not needing changes
survivability	The ability of a system, subsystem, equipment, process, or procedure to continue to function during and after a natural or man-made disturbance
safety	Safety is the state of being "safe," the condition of being protected from harm or other non-desirable outcomes Safety can also refer to the control of recognized hazards in order to achieve an acceptable level of risk
testability	The degree to which a software artifact (i.e. a software system, software module, requirements- or design document) supports testing in a given test context. If the testability of the software artifact is high, then finding faults in the system (if it has any) by means of testing is easier
timeliness	The characteristic of being able to complete a required task or fulfill an obligation before or at a previously designated time
trustworthiness	The degree to which a system can be expected to preserve the confidentiality. Integrity, and availability of the information being processed, stored, or transmitted by the system across the full range of threats
usability	The capacity of a system to provide a condition for its users to perform the tasks safely, effectively, and efficiently while enjoying the experience

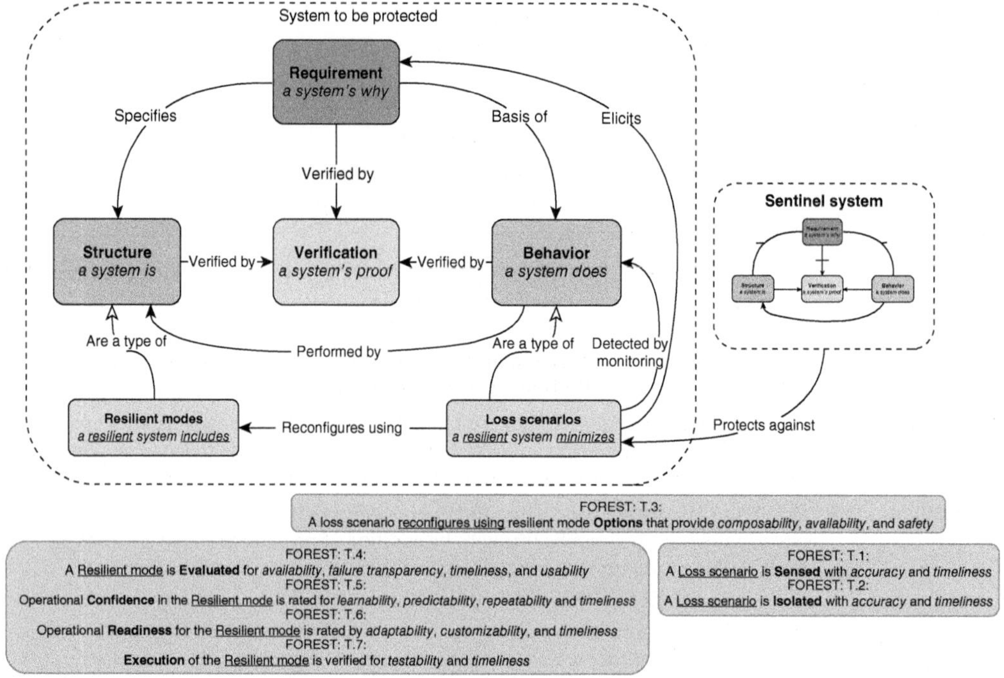

Figure 25.4 Mission Aware MBSE meta-model with FOREST quality attributes.

In the Silverfish Case Study, the FOREST and TREEs will be reviewed. Mission Aware has also been adapted to support FOREST. In MA and CSRM, the sentinel is a pattern that monitors the system for unplanned behaviors and employs or recommends resilience modes of operation in response to behavioral deviations. For systems models to be useful for design and evaluation, experience has shown that the modeling must be done from both a control and structural perspective. The STPA framework supplies concepts of controllers that provide information to one another at each level. Those control functions can be the target of behavior monitoring using sentinel patterns that provide for automated detection of unplanned behaviors and application (or switching in) of these resilient modes with various levels of automation (Leveson and Young 2014):

- **Totally automated**: Sentinel determines what to do and informs appropriately trained system operators regarding automated execution.
- **Semi-automated**: System operators receive automated recommendation(s) from Sentinel and, accounting for both battle context and a broader set of information available to them, decide on what to do.
- **Manual**: Operators, or higher levels in the command hierarchy, determine what to do.

FOREST Requirement Templates

A set of requirement templates are provided in the table below as an aid to defining a set of testable requirements for each identified loss scenario and resilient mode, for the system of interest, aligned to the FOREST TREES (Table 25.4).

The research enabled the team to develop a larger set of requirements that also allowed traceability to promote a more orchestrated effort on designing in resilience to the system.

Table 25.4 FOREST requirement templates.

#	Name	Text
1	T.1.1 TREE.Sense –Monitor	The system shall sense <id:name> Loss Scenario by monitoring <id:name> (Link/ esource/ unction)
2	T.1.2 TREE.Sense – Abnormal Behavior	The <abnormal system behavior spec.> for <id:name> (Link/Resource/Function) shall trigger sensing of <id:name> Loss Scenario
3	T.1.3 TREE.Sense – Logged	Abnormal system behavior sensed for <id:name> Loss Scenario shell be logged for post event analysis
4	T.1.4 TREE.Sense – Alert	The system shall alert users via <alert mechanism to a triggered <id:name> Loss Scenario
5	T.1.5 TREE.Sense – Time Spec	The system shall alert of a triggered <id:name> Loss Scenario within <time spec>
6	T.1.6 TREE.Sense – Accuracy Spec	The system shall alert of a triggered <id:name> Loss Scenario with accuracy of <accuracy spec>
7	T.1.7 TREE.Sense – Injection	A test support system shall provide injection controls for emulation of <id:name> Loss Scenario
8	T.1.8 TREE.Sense – Test Coverage Measure	A test support system shall measure test coverage of <ld:name> Loss Scenario
9	T.2.1 TREE.Isolate - Source	The system shall isolate the (Component/Link)that is the source of the abnormal behavior associated with <id:name> Loss Scenario
10	T.2.2 TREE.Isolate – Alert	The system Shall alert users via <alert mechanism> to the isolated <id:name> (Component/ Link) as the source of the abnormal system behavior associated with <id:name> Loss Scenario
11	T.2.3 TREE.Isolate – Time Spec	The system shall alert of isolated <id:name> (Component/ Link) within <time spec>
12	T.2.4 TREE.Isolate – Accuracy Spec	The system shall alert of isolated <id:name> (Component/ Link) with accuracy of <accuracy spec>
13	T.3.1 TREE.Option – Resilient Mode	The system shall provide <id:name> Resilient Mode as a reconfiguration option for <id:name> Loss Scenario
14	T.3.2 TREE.Option – Abort Unsafe	The system shall abort unsafe <id:name> Function that is triggered by <id:name> Loss Scenario
15	T.3.3 TREE.Option – Composability rating	The test support system shall provide operator composability rating mechanism for <id:name> Resilient Mode option
16	T.4.1 TREE.Evaluate - Failure Transparency Rating	The test support system shall provide mechanisms for operator evaluation of system failure transparency for <id:name> Resilient Mode
17	T.4.2 TREE. Evaluate – Recoverability Rating	The test support system shall provide mechanisms for operator evaluation of system recoverability for <id:name> Resilient Mode
18	T.4.3 TREE. Evaluate – Useability Rating	The test support system shall provide mechanisms for operator evaluation of usability for <id:name> Resilient Mode
19	T.4.4 TREE. Evaluate – Time Spec	The test support system shall measure timeliness of operator evaluation of <id:name> Resilient Mode

(*Continued*)

Table 25.4 (Continued)

#	Name	Text
20	T.4.5 TAEE. Evaluate – Availability Spec	The system shall ensure technical availability of <id:name> Resilient Mode within <availability spec>
21	T.5.1 TREE Confidence – Management Controls	The system shall provide management control mechanisms for operators to pain confidence in <id:name> Resilient Mode
22	T.5.2 TREE. Confidence – Learnability Rating	The test support system shall provide mechanisms for rate operator confidence in learnability of <id:name> Resilient Mode management controls
23	T.5.3 TREE. Confidence – Predictability Rating	The test support system shall provide mechanisms to rate operator confidence in predictability of <id:name> Resilient Mode management controls
24	T.5.4 TREE. Confidence – Repeatability Rating	The test support system shall provide mechanisms to rate operator confidence in repeatability of <id:name> Resilient Mode management controls
25	T.5.5 TREE Confidence – Time Spec	The test support system shall measure timeliness of management control mechanisms for <id:name> Resilient Mode
26	T.6.1 TREE.Readiness – Availability Rating	The test support system shall provide mechanisms for rating operational availability for <id:name> Resilient Mode option
27	T.6.2 TREE.Readiness – Adaptability Rating	The test support system shall provide mechanisms for rating operator adaptability incorporating <id:name> Resilient Mode into the current mission
28	T.6.3 TREE.Readiness – Time Measure	The test support system shall measure timeliness of operator readiness decisions for <id:name> Resilient Mode
29	T.6.4 TREE.Readiness – Survivability Measure	The system shall measure mission function survivability for <id:name> Resilient Mode
30	T.7.1 TREE.Execution – Resilient Mode	The system shall provide management controls for execution of <id:name> Resilient Mode option
31	T.7.2 TREE.Execution – Stability Measure	The system shall measure mission function stability for <id:name> Resilient Mode
32	T.7.3 TREE Execution – Time Spec	The system shall ensure execution of <id:name> Resilient Mode is whin <time spec>
33	T.7.4 TREE.Execution - Test Coverage Measure	The tech support system shall measure test coverage of <id:name> Resilient Mode
34	T.8.1 TREE PostEvent – Loss Scenario Incidents	The system shall support post event analysis of <id:name> Loss Scenario incidents
35	T.8.2 TREE.PostEvent – Resilient Mode Utilization	The system shall support post event analysis of <id:name> Resilient Mode utilization

Summary

In conclusion, the FOREST framework offers a comprehensive approach to enhancing cyber resilience during system development, applicable to any system-level resilience concerns beyond cybersecurity. The methodology systematically evaluates a system's components under attack or disruption, leading to the development of functional requirements and views expressed in a model-based systems engineering tool. The FOREST meta-process model and Mission Aware reference architecture meta-model form the core of the methodology, guiding decision-making for security and related resilience in capability development through a standard risk-based approach. This methodology, described in detail in this chapter, can help engineers improve the resilience of their systems and their ability to withstand cyberattacks and disruptions. With its potential to enhance system functionality, FOREST represents an important contribution to the field of systems engineering and cyber resilience.

References

Adams, S., Carter, B., Fleming, C., and Beling, P.A. (2018). Selecting system specific cybersecurity attack patterns using topic modeling. In: *2018 17th IEEE International Conference on Trust, Security and Privacy in Computing and Communications/12th IEEE International Conference on Big Data Science and Engineering (TrustCom/BigDataSE)* (1–3 August 2018), 490–497. New York: IEEE.

Beling, P., Horowitz, B., Fleming, C., et al. (2019). Model-Based Engineering for Functional Risk Assessment and Design of Cyber Resilient Systems. *University of Virginia Charlottesville United States, Technical Report.*

Beling, P., Horowitz, B., Beling, P., et al. (2021). An Agile Engineering Framework to Support the Development of Sustainable and Resilient DoD Systems, *Technical Report SERC-2021-TR-015 V2.* https://sercproddata.s3.us-east-2.amazonaws.com/technical_reports/reports/1639493452.A013_SERC%20WRT%201022_Technical%20Report%20SERC-2021-TR-015_V2.pdf (accessed 20 November 2022).

CAPEC - Common Attack Pattern Enumeration and Classification (CAPECTM) (n.d.). MITRE. Boca Raton, FL: IEEE. https://capec.mitre.org/ (accessed 29 June 2023).

Department of Defense (2014). Cyber security. http://www.dtic.mil/whs/directives/corres/pdf/850001_2014.pdf (accessed 14 January 2023).

Fleming, C.H., Elks, C., Bakirtzis, G. et al. (2021). Cyber-physical security through resiliency: a systems-centric approach. *Computer* 54 (6): 36–45.

Horowitz, B., Beling, P., Skadron, K., et al. (2014). Security Engineering Project-System Aware Cyber Security for an Autonomous Surveillance System on Board an Unmanned Aerial Vehicle. *Systems Engineering Research Center Hoboken NJ, Technical Report.*

Horowitz, B., Beling, P., Fleming, C., (2017). Security Engineering FY17 Systems Aware Cybersecurity. *Stevens Institute of Technology Hoboken United States, Technical Report.*

Horowitz, B., Beling, P., Fleming, C., et al. (2018a). Cyber Security Requirements Methodology. *Stevens Institute of Technology Hoboken United States, Technical Report.*

Horowitz, B., Beling, P., Fleming, C., et al. (2018b). Cyber-Security Requirements Methodology. *Systems Engineering Research Center, Technical Report.*

Horowitz, B., Beling, P., Clifford, M., and Sherburne, T. (2021). Developmental Test and Evaluation (DTE&A) and Cyber Attack Resilient Systems - Measures And Metrics Source Tables. *Systems Engineering Research Center, Technical Report.*

Leveson, N.G. and Young, W. (2014). An integrated approach to safety and security based on systems theory. *Communications of the ACM* 57 (2): 31–35.

Reed M (2016). DoD strategy for cyber resilient weapon systems. *Paper presented at the National Defense Industries Association, Annual Systems Engineering Conference*, Alexandria VA, October 2016.

Ross, R. S. (2022). *Engineering Trustworthy Secure Systems.* https://doi.org/10.6028/nist.sp.800-160v1r1 (accessed 29 June 2023).

Biographical Sketches

Peter A. Beling is a professor in the Grado Department of Industrial and Systems Engineering and associate director of the Intelligent Systems Division in the Virginia Tech National Security Institute. Dr. Beling's research interests lie at the intersections of systems engineering and artificial intelligence (AI) and include AI adoption, reinforcement learning, transfer learning, and digital engineering. He has contributed extensively to the development of methodologies and tools in support of cyber resilience in military systems. He serves on the Research Council of the Systems Engineering Research Center (SERC), a University Affiliated Research Center for the Department of Defense.

Megan M. Clifford is a research associate and engineer at Stevens Institute of Technology. She works on various research projects with a specific interest in systems assurance and cyber-physical systems. She previously worked on the Systems Engineering Research Center (SERC) leadership team as the Chief of Staff and Program Operations, was the Director of Industry and Government Relations to the Center for Complex Systems and Enterprises (CCSE), and held several different positions in industry, including systems engineer at Mosto Technologies while working on the New York City steam distribution system.

Barry M. Horowitz held the Munster Professorship in Systems Engineering at the University of Virginia, prior to his retirement in May 2021. His research interests include system architecture and design.

Tom McDermott serves as the deputy director and chief technology officer of the Systems Engineering Research Center (SERC) at Stevens Institute of Technology in Hoboken, NJ. The SERC is a University Affiliated Research Center sponsored by the Office of the Secretary of Defense for Research and Engineering. With the SERC, he develops new research strategies and is leading research on Digital Engineering transformation, education, security, and artificial intelligence applications. Mr. McDermott also teaches system architecture concepts, systems thinking and decision-making, and engineering leadership. He is a lecturer in Georgia Tech's Professional Education college, where he leads a masters level course on systems engineering leadership and offers several continuing education short courses. He consults with several organizations on enterprise modeling for transformational change, and often serves as a systems engineering expert on government major program reviews. He currently serves on the INCOSE Board of Directors as director of Strategic Integration.

Tim Sherburne is a research associate in the Intelligent System Division of the Virginia Tech National Security Institute. Sherburne was previously a member of the systems engineering staff at the University of Virginia supporting Mission Aware research through rapid prototyping of cyber-resilient solutions and model-based systems engineering (MBSE) specifications. Prior to joining the University of Virginia, he worked at Motorola Solutions in various Software Development and Systems Engineering roles defining and building mission critical public safety communications systems.

Chapter 26

The Cyber Security Requirements Methodology and Meta-Model for Design of Cyber-Resilience

Tim Sherburne[1], Megan M. Clifford[2], Barry M. Horowitz[3], and Peter A. Beling[1]

[1] *Virginia Polytechnic Institute and State University, Hume Center for National Security and Technology, Blacksburg, VA, USA*
[2] *Stevens Institute of Technology, School for Systems and Enterprises, Hoboken, NJ, USA*
[3] *University of Virginia, Charlottesville, VA, USA*

Introduction and Background

This chapter continues the theme of the cluster, namely methods for achieving cyber resilience in cyber-physical systems. Resilience may be defined as the ability of systems to resist, absorb, and recover from or adapt to an adverse occurrence during operation that may cause harm, destruction, or loss of ability to perform mission-related functions. Cyber resilience aims to deal specifically with attacks that can arise through compromise of the cyber elements of a system. In most cases, resilience is a property that must be engineered into the system.

Mission Aware (MA), covered in detail in a previous chapter, is a reference architecture for operational resilience of cyber-physical systems that was developed under prior SERC research efforts. As illustrated in Figure 26.1, the primary feature of the MA architecture is a *sentinel* that monitors the system or mission being protected, detects abnormal behavior or other signs of loss of function, alerts system users or mission owners to detected loss of function, and has the capability to switch the system or mission to a *resilient mode of operation*, which is a distinct and separate method of operation for a component, device, or system based upon a diverse redundancy or other design pattern. Resilient modes of operation are designed so that the system can still meet the primary objectives of the mission, though with possible loss of operational performance. A sentinel should be designed with simplicity in mind so that it is more easily secured. For example, a sentinel with a few hundred lines of code is easier to check for malicious code injections than a complex system (Beling et al. 2019; Horowitz et al. 2014, 2017a,b; Carter et al. 2019).

Many engineering patterns exist for resilience mechanisms, and each has associated implementation and operational costs or performance burdens. These costs and performance burdens make it evident that the sentinel-based resilience architecture from Figure 26.1, or indeed any other engineered method for resilience, cannot be used to protect all system functions. It is necessary to pick and choose not only the mechanisms to use but which functions to monitor. Therefore, a systematic, tractable, and rigorous method is necessary to support decision-making for implementing resilience solutions in cyber-physical systems (Bakirtzis et al. 2017, 2018).

Systems Engineering for the Digital Age: Practitioner Perspectives, First Edition. Edited by Dinesh Verma.

Companion website: www.wiley.com/go/verma/systemsengineering

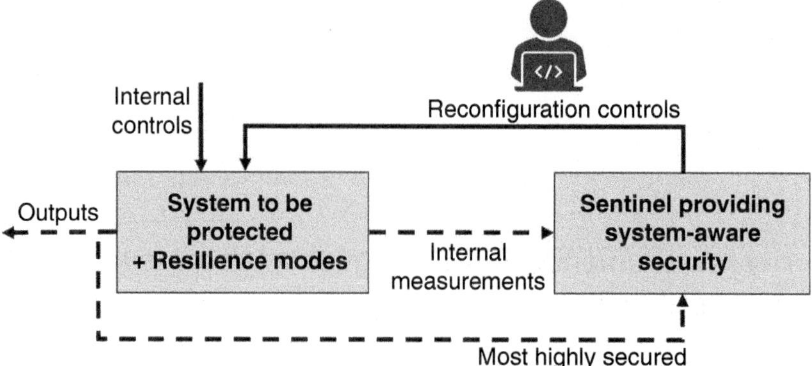

Figure 26.1 Sentinel-based resilience architecture.

In the MA methodology (Chapter 24), an MBSE Meta-Model integrates systems models, resilience architectures, and a risk assessment derived from the Systems Theoretic Process Assessment (STPA) and its security-focused derivative, STPA-Sec (Chapter 23). This chapter introduces a companion methodology called the Cyber Security Requirements Methodology (CSRM). CSRM describes how MA and STPA-Sec can be used in the context of deriving an architecture for resilience that satisfies stakeholder needs. It should be noted that CSRM can also be used in conjunction with the Framework for Operational Resilience in Engineering and System Test (FOREST) to elicit testable requirements for resilience.

Cyber Security Requirements Methodology Overview

The Cyber Security Requirements Methodology (CSRM) for cyber-physical systems presented in this research activity is risk-based, involving the assessment of potential consequences of specific attack scenarios and their likelihood of occurring. The consequences may range from human harm, loss of control, corruption, or delay of situation awareness information, to the denial of system operations. CSRM acknowledges that the system's owners, operators, and users are in the best position to prioritize the consequences that need to be avoided. However, the likelihood of attacks is determined by the adversaries who prioritize and execute them. Cybersecurity solutions aim to minimize the likelihood of attacks, while cyberattack resilience solutions address the consequences of detected attacks. The six-step CSRM is structured to address this risk division and is executed by four teams, each with the relevant expertise to carry out the steps.

The purpose of the CSRM is to improve the current preliminary design processes for new cyber-physical systems by offering an efficient and timely process to meet their cybersecurity requirements. Cybersecurity elements such as defense and resilience are complementary and best approached collectively during the design phase of the system. This stage is critical, as significant initial decisions can be made regarding system architecture, including cybersecurity measures. Analysis and rapid prototyping/simulation tools, such as SysML, can also be used to aid in decision-making regarding cybersecurity requirements (Horowitz et al. 2018).

The elements required for achieving cyber security are interdependent, during the phase of system design, important initial decisions can be made regarding the system architecture, including:

- Separation and isolation of hardware and software supporting different system functions,
- Use and selection of off-the-shelf products, accounting for historical cyberattacks,

- Dependence on defense capabilities, with specific solutions to be selected when design is sufficiently mature,
- Where within the new system's development process to focus the most emphasis and corresponding resources regarding SW development processes (quality assurance tools, testing, developer skills, life cycle support, etc.),
- Design and performance requirements for resilience-related capabilities both for immediate implementation and to facilitate simpler addition in preparation for higher likelihood requirements over the life cycle,
- Addressing the operator-related aspects of resiliency through rapid prototyping experiments and exercise-related support tools.

The four teams can be diluted and best designated as an SE Team, a System Operator (Blue) Team, a Security Analyst (Red) Team, and a Test Team.

The Systems Engineering Team is comprised of members with an extensive skillset – ranging from technical and operational experience with strong analytical skills and the ability to utilize system description and assessment tools. The SE Team is required to develop an initial high-level system design, or at least provide the overall cyber-physical system project's SE team, without the cyberattack-related resilience features. Then, based upon the Blue Team's prioritized consequence avoidance assessment, the SE Team would derive potential resilience features and the architecture for their implementation.

The SE Team would develop integrated solution alternatives that consider the complete risk analysis, including both Blue and Red Team assessments, after receiving the Red Team's prioritized solution evaluations. In addition, the SE Team would manage the methodology process and update system descriptions to incorporate new solutions resulting from the CSRM process (Figure 26.2).

The Blue Team should consist of individuals experienced in managing systems under duress, such as electronic warfare attacks or weapon-fire attacks, in addition to cyberattacks. Familiarity with the operational practices of legacy systems related to the system under development would be beneficial. The team prioritizes the consequences to be avoided for various system functions, such as denial of service or information corruption to operators, focusing on the Consequence

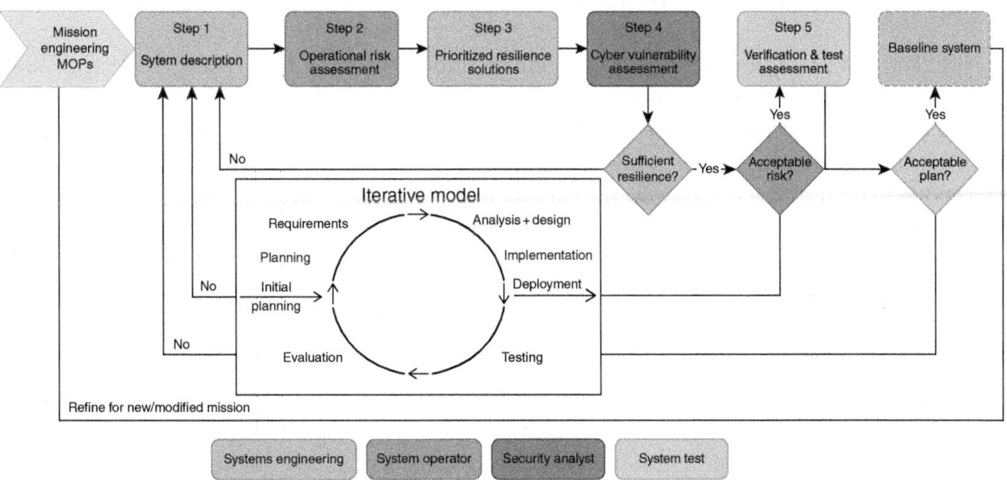

Figure 26.2 Cyber security requirements methodology, iterative. Source: Adapted from Horowitz et al. (2018).

component of risk. The SE Team should support the Blue Team in interpreting the tool-based representation of the system, as needed. Notably, CSRM does not require input from cyber security experts for Consequence analysis.

The Red Team evaluates the probability of potential cyberattacks with and without the application of potential solutions to the overall system design. Their assessment should prioritize software quality, defense, and resiliency solutions while considering past cyberattacks and software vulnerabilities. The team then suggests alternative solutions and assess their impact on related cyberattack likelihoods. A key feature of CSRM is the Red Team's combination of cyberattack expertise and cyber security expertise working together to develop an iterative assessment that links solution selection with influencing attack likelihoods.

The Test Team evaluates the different points in which the system can be tested during developmental and operational phases of the system. In conjunction with the Framework for Operational Resilience in Engineering and System Test (FOREST), there are eight Testable Requirements Elicitation Elements (TREEs). These, paired with the CSRM, enable identification of quality attributes that can be measured by the system, operator, and/or system limitation/development consideration (quantitative for the system measurements and qualitative for the operator). This team also helps elicit the best input of resilient modes by way of creating a meta-model for evaluation (Figure 26.3).

To enable team members to collaborate and complete tasks at each step of the process, the CSRM used a small set of support tools. The CSRM is tool-agnostic; however, there are specific needs at each step. It can be up to the program to decide whichever tool to use to satisfy the need. For instance, one tool used in the research includes No Magic's MagicDraw software, which is used to create and modify system models in the Systems Modeling Language (SysML). Another is the Systems Theoretic Accident Model and Process (STAMP) hazard analysis methodology is used to identify key requirements, critical functions, and to organize modeling efforts. Finally, the Red Team uses the Cyber Body of Knowledge (CYBOK) tool, developed by the research universities, to identify and quantify the likelihood of attacks (Leveson 2011; CYBOK 2017).

CSRM Steps and Application

The steps of CSRM are:

1) The SE team produces a system description that includes the system architecture and functional descriptions of the system components that satisfy the Mission MOPs.
2) The Blue Team performs an operational risk assessment of the current design and produces a prioritized list of undesirable outcomes.
3) The SE Team uses the system description and the risk assessment to produce a list of possible resilient solutions.
4) The Red Team prioritizes cyber defense and resilient solutions using their prior knowledge of attack likelihood.
5) The Test Team produces a verification and test plan that addresses defined resilience modes for the system.
6) The process iterates until an acceptable baseline system is defined that is deemed acceptable to all teams:
   a) As necessary, the SE Team updates the system description to include the red team recommendations.

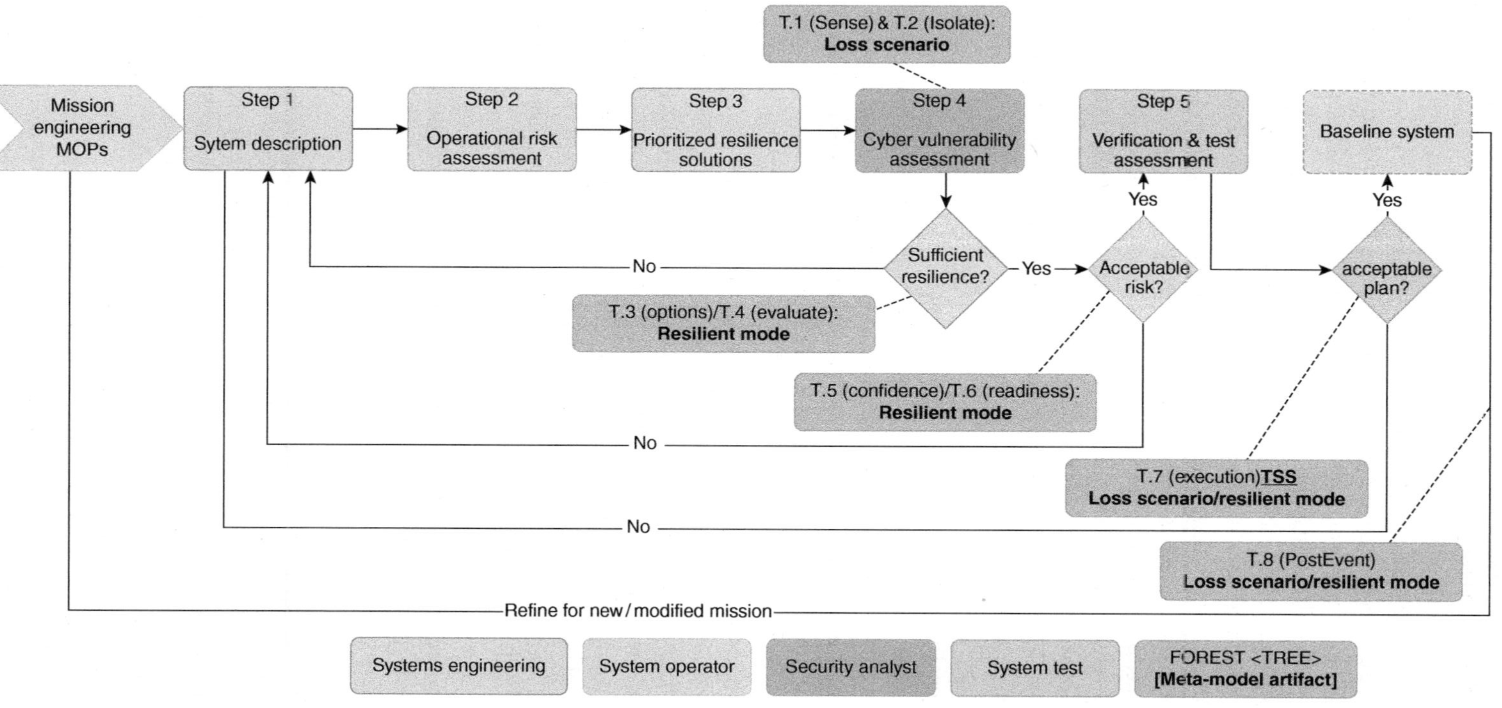

Figure 26.3 Cyber security requirements methodology (with FOREST overlay). Source: From Horowitz et al. (2021).

b) As necessary, the Blue revises the operational risk assessment to account for the new system design and Red Team assessment.

c) As necessary, the Test team revises the verification and test plan to account for new system design and red team assessment.

The Silverfish use case is expanded upon in Chapter 27; however, the remainder of this chapter presents the development and application of rapid prototyping and analytical tools for establishing system cybersecurity requirements with the CSRM.

The rapid prototyping and simulation provides a mechanism to explore system resiliency design alternatives and associated user experience impacts early in the development lifecycle prior to specifying a particular design and implementation. In numerous examples, the rapid prototyping demonstrated human factors-related issues that could emerge from suggested requirements. Since most decision relate to trade-offs between a system specification (or an -ility) and cyber security, the rapid prototyping and simulation, along with the constructed steps of the CSRM, helps design in engineered resilience in a proactive method.

As mentioned, while the SE Team oversaw the entire process, the activities associated with the individual steps were managed by different teams.

CSRM Step 1: System Description

The Mission Engineering measures of performance(s) feeds into this first step of the CSRM process.

During Step 1, which is managed by the SE Team, an initial SysML description of the Silverfish architecture was created. Details on the SysML description of the Silverfish architecture are in other chapters and can also be found online at sercuarc.org. Cyber security considerations were intentionally excluded from this stage, with the SE Team focusing on the technical design of the system while relying on the Blue Team for input on operational aspects.

At this early stage of development, the SysML model primarily serves to document the core functionality and basic architecture of the system. Discussions with the Blue Team results in incremental updates to the model as the baseline architecture was defined and agreed upon. Any updates are reflected in the internal block diagram of the system composition.

The SysML model established during Step 1 served as a reliable point of reference for both the SE Team and the Blue Team. At this stage, a basic model of the system lays the foundation for the SE Team's loss-based engineering hazard analysis (the research used a STAMP-based hazard analysis).

In addition to creating the SysML model, the SE Team also developed the rapid prototype/simulation vehicle described in the use case during Step 1. This vehicle facilitated the evaluation of the system's performance.

CSRM Step 2: Operational Risk Assessment

The Blue Team manages Step 2, which involved identifying and prioritizing unintended operational consequences that should be avoided in the application of the system. The team analyzed the SysML description and mission engineering measures provided by the SE Team and identified a set of consequences, such as the manipulation of operator weapon control commands by an adversary. The SE Team provided clarifying information on the technical design choices, and the Blue Team prioritized the use consequences to be avoided.

After the Blue Team exercise, the SE team updates the SysML model by adding the consequence prioritization to a requirements diagram. They convert the consequences into requirements language and arrange them hierarchically to reflect the rankings provided by the Blue Team.

Additionally, the SE team generates more requirements by utilizing STAMP-derived formulations based on the Silverfish system's purpose and functionality. The figure below displays the requirements diagram that was updated.

The Blue Team consequences are given names such as "Blue 2.1" that correspond to their ranking in the exercise. The additional STAMP-derived requirements are located near the top of the diagram and serve as root nodes from which the other requirements can be traced.

The following Section describes how this tabular result is utilized by the SE Team in Step 3 of CSRM.

CSRM Step 3: Prioritized Resilience Solutions

In Step 3 of the CSRM, the SE Team is tasked with devising potential resilience solutions in response to the prioritized consequences identified by the Blue Team. An example of this is from one of the research efforts. In the effort, three distinct options were created to offer cyber security resilience, with the understanding that any combination of these options may be utilized to satisfy the resilience requirements. The tree areas for resilience selected by the SE Team were:

- Resilient weapon control capabilities (including data consistency checking design pattern, and diverse redundant HW/SW) implementation for the operator's vehicle-mounted computer
- Diverse redundant communications sub-systems,
- Resilient situation awareness capabilities (including diverse redundant sensor voting and situation awareness introspection design patterns)

The SE Team considered the results of the rapid prototyping/simulation efforts from Step 1 of the CSRM when selecting potential resilience requirements. The three resilience options developed included the "Real-Time Resilience Confidence Testing" design pattern, which allows operators to initiate pre-designed tests of the diverse mode of operation to achieve resilience. However, this testing comes at the cost of time, which could be operationally critical depending on the attacker's intent. The operator must decide whether to trade-off time for confidence in achieving the desired resilience outcome based on the mission context. Another newly developed design pattern is "Situation Awareness Introspection," which involves comparing the level of operator's situation awareness activity with the associated machine utilizations related to those displays. This design pattern can detect adversarial activities even when the operator's situation awareness display shows no adversarial information.

CSRM Step 4: Cyber Vulnerability Assessment

In Step 4 of the CSRM, the Red Team made assessments based on the SysML descriptions provided by the SE Team on the software engineering, cyber defense, and cyberattack resilience requirements. While the conclusions drawn were reached by experience and knowledge, historical cyberattack data was also reviewed using the CYBOK tool. It is worth noting that the Red Team did not have access to the results from Step 2 of the CSRM to ensure that the assessments in Step 4 were primarily based on technical factors rather than operational factors derived from the Blue Team.

To provide insight into the process, here were the list of conclusions drawn by the Red Team in one example from the research:

1) Encryption was deemed a desirable security requirement based on the weapon control integrity requirements provided in the Silverfish functional description and assumed by the SE team in their SysML descriptions.

2) The design should require separation of the weapon control system hardware and software from the situation awareness hardware and software, including separate operator displays, based on the projected relative complexity of the situation awareness software compared to the control software needed for weapon control from a software engineering perspective.
3) Utilization of a comprehensive set of software quality tools, including static and dynamic testing tools, extensive use of end-to-end testing, and assembly of a high-end team of software designers and developers focused on weapon control software development, should be required for the highly focused weapon control subsystem functionality.
4) Assuming the adoption of isolation and proposed development practices for weapon control software, a diverse redundancy resilience requirement for weapon control was suggested to be of low priority, meaning resilience was less critical if the hardware/software implementation made attacks aimed at the weapon control function challenging enough.
5) To avoid potential attacks through the C2 system, adoption of a voice-only military communication system was suggested to be extended to higher levels of command.
6) The communication subsystem was considered the highest priority for resilience, with diverse redundancy being used to address attacks resulting in denial of service and message delays.
7) The resiliency design for situation awareness was suggested to be the second highest priority for resilience, including diverse redundant IR sensors as a basis for addressing both reliability and cyberattack resilience requirements.
8) If potential scenarios require interactions across separately protected, closely located protected areas, an operator authentication design requirement should be considered.

The proposed set of architectural and system design requirements may lead to a potential conflict between the priorities developed by the Blue Team related to consequences and those produced by the Red Team for cyber security. The Red Team's suggestion of lower resilience priority for weapon control functions, although balanced by the implementation of isolation and enhanced software engineering requirements, is one of the most notable examples of such contention. Overall, the emphasis on high-priority software engineering costs may impact the affordability of cyber defense and cyberattack resilience opportunities.

In the end, engineering trade-offs are elucidated further so that architectural and design requirements can be updated in the system description if there is not sufficient resilience.

CSRM Step 5: Verification and Test Assessment

In Step 5 of the CSRM, the SE Team is responsible for incorporating the Red Team's recommendations into a corresponding set of SysML representations, which is evaluated by the Blue Team in preparation for management decisions on the architecture and preliminary design. For example, The SE Team can recognize that the disparity in resilience prioritization for the weapon control subsystem is a critical issue, and as a result, propose an unanticipated system architecture-related suggestion: to consider providing cyberattack detection capability for the isolated weapon control system, in the event that the resilience requirement for that part of the system was dropped. This would involve designing the system to detect cyberattacks and prevent their immediate consequences, leaving operational commanders to decide on further steps for continuing operations. SysML representations were prepared for the Step 6 to present the Red Team's findings to the Blue Team.

Furthermore, Step 5 requires the SE Team to initiate cost analyses to inform management decisions on design alternatives.

CSRM Step 6: Process Iterates Until Acceptable Baseline System Is Defined and Deemed Acceptable to All Teams

In Step 6, the SE Team received feedback from the Blue Team on the Red Team recommendations from Step 4 and the SE Team responses from Step 5. As an example, the Blue Team can support the recommendation on the separation of situation awareness and weapon control functions, convinced the operators could manage two separate displays. Another example is the Blue Team can also agree with the Red Team's proposal to limit communication to voice-only for higher-level command. It is understanding and knowing what is viably possible operationally for the system.

Another example is that the Blue Team could also support an SE Team suggestion to include detection-only capability for the separated weapon control function and to provide situation awareness resilience even if weapon control was not resilient. The Blue Team would provide their rationale on knowing that resilient awareness would ensure operator safety and provide higher-level command with the ability to take resilience-related actions that did not rely on the use of the weapon control system functions.

The Blue Team could also decide to explore a suggestion further – whether through MBSE efforts or another. An example would be the Blue Team wanting to explore the likelihood of closely located deployments that prompted a Red Team's proposal to include suggestions for technology-based authentication of operators.

CSRM Process Conclusion

The process provides valuable insight into engineering and designing in resilience to systems. For instance, with the research, it showed that in the early stages of a system design, the steps propelled prototype demonstrations that raised previously unknown concerns. It also aided in further exploration of design to better equip users and future modifications. The methodology also aided in the buildout of Sentinels and use cases for substantiation of specific requirements (e.g., timing requirements). A significant timing requirement emerged during a simulation scenario where a cyberattack modified the fire command to the operator in a research example. The crucial explication provided the team(s) with the necessary knowledge of Sentinel placement and design. Undesired by-products of the system were also shown through the steps as well. The use of SysML representations for the various resilience options helped progress engineering practices, and the conversations and traceability that surround them.

The risk assessment conducted via CSRM has a clearer basis for prioritization, namely human safety and weapon effectiveness. System resilience is a system design topic, and requirements for resilience depend upon related cyber defense, software, and hardware engineering requirements. As a result, all the inter-related areas of cyber security should be approached concurrently.

CSRM with the Framework for Operational Resilience in Engineering and System Test

As mentioned, the CSRM in conjunction with the Framework for Operational Resilience in Engineering and System Test (FOREST), enable identification of quality attributes that can be measured by the system, operator, and/or system limitation/development consideration (quantitative for the system measurements and qualitative for the operator). The eight Testable Requirements Elicitation Elements (TREEs) are easily traced through the CSRM. The pairing of framework and methodology also helps elicit the best input of resilient modes by way of creating a Meta-Model for evaluation.

The associated TREEs are connected to the relevant CSRM step and provide guidance for judging the completeness of the system description, from a resilience perspective, and for driving the iterative loops of CSRM. The TREE connections are:

- During the cyber vulnerability assessment (Step 4) TREE.1 and TREE.2 ensure that appropriate interfaces and mechanisms are defined to enable *Sensing* and *Isolation* of each Loss Scenario.
- The "sufficient resilience" loop is driven by TREE.3 and TREE.4 that ensure each Loss Scenario has one or more Resilient Mode reconfiguration *Options* and that appropriate interfaces and mechanisms are defined to enable technical *Evaluation* of each Resilient Mode.
- The "acceptable risk" loop is driven by TREE.5 and TREE.6 that ensure appropriate interfaces and mechanisms are defined to enable operational judgement of *Confidence* and *Readiness* for usage of each Resilient Mode.
- The "acceptable test" loop is driven by TREE.7 that ensures all Loss Scenarios and Resilient Modes have appropriate Test Support interfaces and mechanisms to drive and measure *Execution*.
- TREE.8 ensures that appropriate *Post Event* analysis mechanisms exist to adapt the Baseline System for new or refined Missions.

Chapter 25 illustrated the FOREST requirements template, and the CSRM can continue to drive the TREES to the degree that requirements can be extracted and then injected into the engineering of the system. As stated, the process also helps elicit the best input of resilient modes by way of creating a Meta-Model for evaluation. The MBSE enables digital engineering through shared data and models that can be tested and developed to originate resiliency.

Mission Aware MBSE Meta-Model

Mission Aware is built on the proven Vitech Corporation Meta-Model, which has been successfully used in various industries (Long and Scott 2011). This Meta-Model, compliant with SysMLv2 standards for precise systems engineering, is available to the public. To address the safety, security, and resilience needs of cyber-physical systems, the Meta-Model has been enhanced with additional entities and interactions.

The Mission Aware Meta-Model encompasses essential systems engineering concepts and their connections, such as requirements, behavior, architecture, and testing (as shown in Figure 26.4). This integrated model not only defines the final specifications of a system but also tracks the progress toward that specification, including concerns raised and addressed, and risks identified and managed. The Meta-Model supports an incremental, step-by-step approach by establishing consistent relationships between the different domains of requirements, behavior, architecture, and testing.

The use of a Meta-Model brings several benefits, including the alignment of various documentation methods, which reduces workload and minimizes the possibility of errors. The Meta-Model also ensures compliance with best practices in systems engineering and links performance metrics such as safety, security, and resilience to the system design process. By utilizing the Mission Aware Meta-Model, engineers can improve the consistency and effectiveness of their systems engineering processes and enhance the resilience of the system to cyberattacks.

The MA MBSE Meta-Model (Figure 26.4) extends the Vitech model to include concepts from the MA approach to resilient system design. Specifically, the extended Meta-Model includes resilient modes and extends the behavior with consideration for loss scenarios. Tables 26.1 and 26.2 provides a detailed description of the extensions.

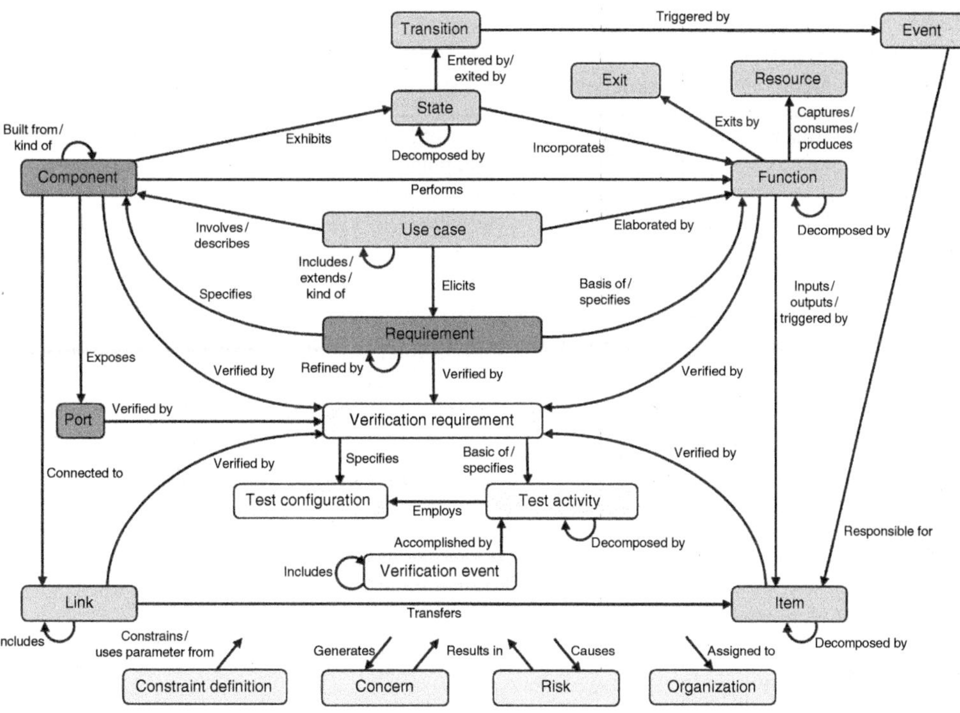

Figure 26.4 The Vitech systems Meta-Model. *Source:* Adapted from Long and Scott (2011).

Table 26.1 MBSE Meta-Model augmentations for Mission Aware.

Element	Entity	Description
Control structure	Control action	A controller provides control actions to control some process and to enforce constraints on the behavior of the controlled process
	Feedback	Process models may be updated in part by feedback used to observe the controlled process
	Context	The set of process model variables and values
Risk	Loss	A loss involves something of value to stakeholders. Losses may include a loss of human life or human injury, property damage, environmental pollution, loss of mission, loss of reputation, loss or leak of sensitive information, or any other loss that is unacceptable to the stakeholders
	Hazard	A hazard is a system state or set of conditions that, together with a particular set of worst-case environmental conditions, will lead to a loss
	Hazardous action	A hazardous action is a control action that, in a particular context and worst-case environment, will lead to a hazard
Vulnerability	Loss scenario	A loss scenario describes the causal factors that can lead to the unsafe control and to hazards. Two types of loss scenarios must be considered: (1) Why would unsafe control actions occur? (2) Why would control actions be improperly executed or not executed, leading to hazards?
	Remediation	The hygiene practice or resilience mechanism to protect against a loss, hazard, loss scenario, or attack vector

(*Continued*)

Table 26.1 (Continued)

Element	Entity	Description
Mission Aware	Attack pattern	An inventory (check list) of potential paths or means by which a hacker can gain access to a computer or network server in order to deliver a payload or malicious outcome. Attack patterns enable hackers to exploit system vulnerabilities, including the human element
	Attack vector	3-way associative class between attack pattern, component / link, and remediation which tracks likelihood and severity of the attack pattern after remediation
	Hygiene practice	A routine practice (check list) of basic security capabilities to reduce cyber risks due to common or pervasive threats
	Resilient mode	A configuration of a target system that remediates one or more loss scenarios
	Sentinel	A highly secure subsystem responsible for monitoring and reconfiguration of resilient modes for a target system

Source: Adapted from Horowitz et al. (2018).

Table 26.2 MBSE Meta-Model entities.

Element	Entity	Description
Requirement	Requirement	A requirement is either an originating requirement extracted from source documentation for a system, a refinement of a higher-level requirement, a derived characteristic of the system or one of its subcomponents, or a design decision
Physical	Component	A component is an abstract term that represents the physical or logical entity that performs a specific function or functions
Interface	Link	A link is the physical implementation of an interface.
	Item	An item represent flows within and between functions. An item is an input to or an output from a function
Functional	Function	A function is a transformation that accepts one or more inputs (items) and transforms them into outputs (items)
	Call structure item	Recursive call structure, for example, select, parallel, loop, for each function
	Exit	An exit identifies a possible path to follow when a processing unit completes
	Resource	A resource is an element, for example, power, MIPS, interceptors, that the system uses, captures, or generates while it is operating
Miscellaneous	Constraint definition	A constraint definition captures a parametric constraint as an expression, identifying the independent variable(s), with associations to the system parameters

The high-level SE artifacts for system design are requirements, structure, behavior, and verification. A resilient system design extends the structure with Resilient Modes and extends the behavior with a consideration for Loss Scenarios. The detailed MA MBSE Model assists the design process with maintaining the rigor of the design. Among others, there are three primary benefits to using a Meta-Model (Figure 26.5):

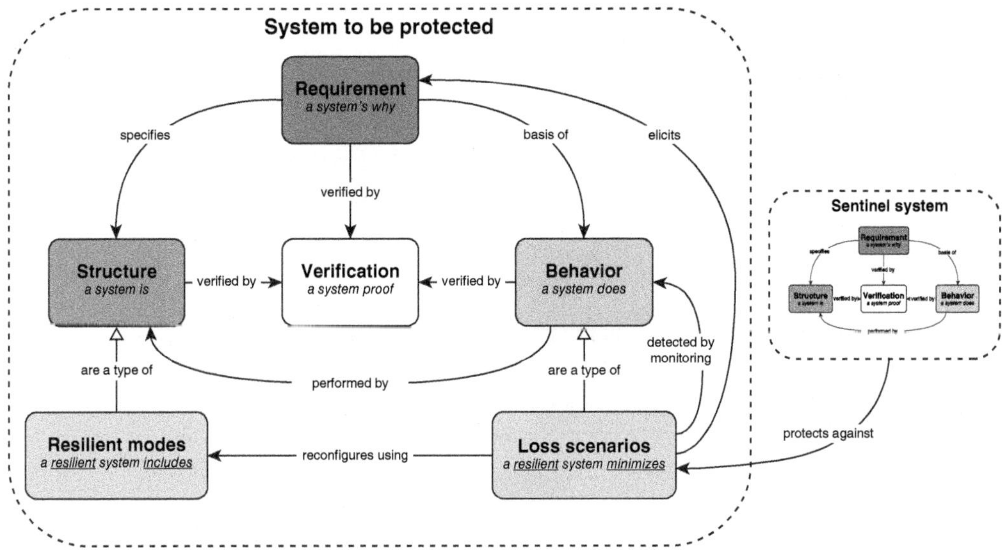

Figure 26.5 MBSE Meta-Model augmentations for Mission Aware (Horowitz et al. 2021).

1) Streamlining the methods of documentation to ensure consistency and reduce the workload associated with maintaining the documents, while preventing errors.
2) Enforcing a set structure to remain in compliance with systems engineering best practices.
3) Connecting performance metrics such as safety, security, and resilience with the system design process.

A fundamental concern of any systems engineering model is an understanding of the system's architecture, including its `components` and physical `links` which connect them. Components may include hardware elements, software elements, external systems, and/or humans. Of equal concern is an understating of the expected behavior of the system being modeled. Behavior elements include `functions`, their input and output `items`, as well as any resources provided or consumed. The `call structure` provides an understanding of behavior control flow including looping, parallel execution, path selection with exit choices, etc. Components perform functions, thereby linking the physical architecture with the behavior model. These standard system modeling entities define the engineering process itself and provide structure to the essential design `artifact` of the system under design.

However, MBSE entities and relationships do not address the -ilities necessary for the design of CPS. Additional entities or safety, security, and resilience that are specifically related to CPS must be added to provide evidence or the correct behavior of CPS. Such performance metrics are defined in the augmentation of the CPS Meta-Model and related to already standardized MBSE entities with properly defined relationships. This is a critical addition to the standard Meta-Model provided by Vitech. By adding structure to performance metrics systems engineers can design CPS that provide operational assurance in the face of hazards or security violations.

The MA Meta-Model detailed view illustrates further the CSRM steps and associated Meta-Model entities. It also displays the specific elements within the Meta-Model and its corresponding CSRM step (Figure 26.6).

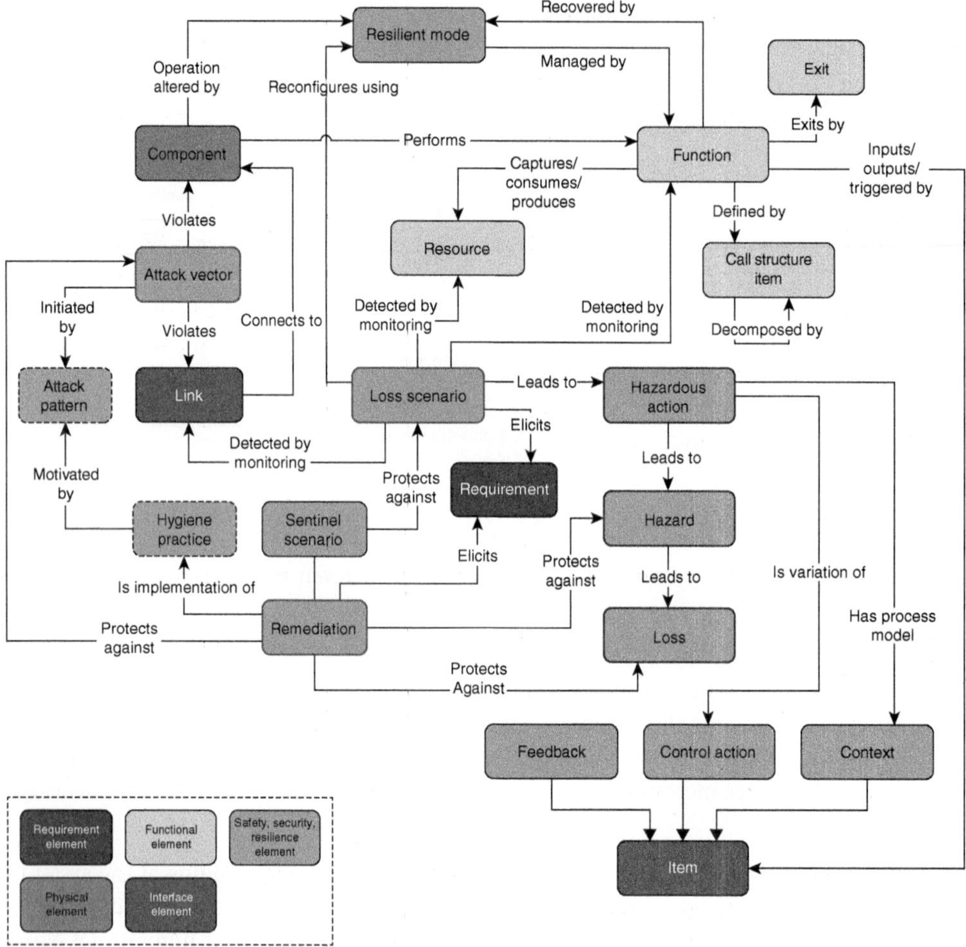

Figure 26.6 The Mission-Aware Meta-Model. *Source:* Adapted from Horowitz et al. (2018, 2021).

CSRM.1. System Description (Mission, Architecture, Behavior)
- Use Case/Requirement
- Component, Link
- Function, Exit, Resource, Control-Action, Feedback, Context, Call Structure Item

CSRM.2. Operational Risk Assessment
- Loss, Hazard, Hazardous Action

CSRM.3. Prioritized Resilience Solutions
- Resilient Mode

CSRM.4. Cyber Vulnerabilities Assessment
- Loss-Scenario, Remediation, Elicited Requirements

CSRM.5. Cyber Resilience Test Activity/Verification and Test Assessment
- Verification of Loss Scenario abnormal behavior
- TREE (Sense, Isolate, Option, etc.)
- Test configuration of Resilient Modes and Sentinels

The traditional system performance metrics are captured as parameters of links, components, and functions, with `constraint definitions` outlining the equations and relationships between entities, individually. To properly address the considerations of safety, security, and resilience in performance metrics, the standard MBSE Meta-Model must be augmented with additional concepts to account for both operational risk and adversarial attacker perspectives (as shown in Table 26.2).

Safety and security often require specification of system behavior as a set of feedback control loops. As such, specializations of `control action` and feedback are provided as sub-types of the standard function input–output `item`. While this phraseology is borrowed from STAMP, it applies to a large number of safety and security methods. STAMP in some sense distills any general framework for "-ilities" at a higher abstraction level – by leveraging notions of uncontrolled actions and control hierarchy – that is suited for use in a Meta-Model. Specifically, `losses`, `hazards`, and `hazardous actions` are captured and related (by means of leads to) as part of a methodical operational risk assessment process. Additionally, explicit associations are captured to understand an unsafe action as a `variation` of a specific control action with the `process model` system state that provides the `context` for the control action to become unsafe, which is borrowed from the domain of control theory and governs all CPS to some extent.

A crucial step in evaluating any performance metric is first identifying `loss scenarios` that could lead to unsafe actions, which are complementary to the stakeholders' requirements or define the system's mission. In CPS, unsafe behavior and security violations are intertwined, meaning an attacker could transition the system to a hazardous state. To augment safety loss scenarios, databases such as MITRE CAPEC (CAPEC), containing `attack patterns`, can be consulted. The Meta-Model links loss scenarios to recovery and resilience by identifying how a `sentinel`, a type of `remediation`, could protect against the loss by detecting it through `monitoring` a link, resource, or function, and then how the system reconfigures using a specific `resilient mode`.

Building Blocks for Cyber Resilient Systems

The synergistic capabilities of MBSE, STPA-Sec, Resilience Concepts, Mission Aware, FOREST, CSRM, Security Standards and Knowledge Bases enable bypassing traditional vulnerability assessments and risk/consequence models for design of cyber resilience and places emphasis on desired results instead of simply available solutions. The adjoining enables clear articulation of desired results with better solution alignment that allows for innovation. The intrinsic iterative component permits modularity as well. Moving more towards functional requirements, the importance of qualitative descriptions of an *activity to perform* (activity) or *purpose to achieve* (result) usher in better comprehensive models. In the next chapter, the Silverfish use case begins to demonstrate the efforts for designing in cyber resilience with engineering.

References

Bakirtzis, G., Carter, B.T., Fleming, C.H., and Elks, C.R. (2017). Mission aware: evidence-based, mission-centric cybersecurity analysis. *ArXiv-Eprints* .

Bakirtzis, G., Carter, B.T., Fleming, C.H. and Elks, C.R. (2018). A model-based approach to security analysis for cyber-physical systems. *IEEE International Systems Conference*, Vancouver, Canada (23–26 April 2018) IEEE.

Beling, P., Horowitz, B., Fleming, C., et al. (2019). Model-Based Engineering for Functional Risk Assessment and Design of Cyber Resilient Systems. *University of Virginia, Charlottesville, United States, Technical Report.*

Carter, B., Adams, S., Bakirtzis, G. et al. (2019). A preliminary design-phase security methodology for cyber-physical systems. *Systems* 7 (2): 21.

Cyber Security Body of Knowledge (CyBOK). (2017). https://www.cybok.org/

Horowitz, B., Beling, P., Skadron, K., et al. (2014). Security Engineering Project-System Aware Cyber Security for an Autonomous Surveillance System on Board an Unmanned Aerial Vehicle. *Systems Engineering Research Center, Hoboken NJ, Technical Report.*

Horowitz, B., Beling, P., Fleming, C., et al. (2017a). Security Engineering FY17 Systems Aware Cybersecurity. *Stevens Institute of Technology, Hoboken, United States, Technical Report.*

Horowitz, B., Beling, P., Fleming, C., et al. (2017b). Security Engineering – FY17 Systems Aware Cybersecurity. *Systems Engineering Research Center, Technical Report SERC-2017-TR-114.*

Horowitz, B., Beling, P., Fleming, C., et al. (2018). Cyber Security Requirements Methodology. *Stevens Institute of Technology, Hoboken, United States, Technical Report.*

Leveson, N. (2011). *Engineering a Safer World: Systems Thinking Applied to Safety.* MIT Press.

Long, D., & Scott, Z. (2011). *A Primer for Model-Based Systems Engineering.* Lulu.com.

Biographical Sketches

Peter A. Beling is a professor in the Grado Department of Industrial and Systems Engineering and associate director of the Intelligent Systems Division in the Virginia Tech National Security Institute. Dr. Beling's research interests lie at the intersections of systems engineering and artificial intelligence (AI) and include AI adoption, reinforcement learning, transfer learning, and digital engineering. He has contributed extensively to the development of methodologies and tools in support of cyber resilience in military systems. He serves on the Research Council of the Systems Engineering Research Center (SERC), a University Affiliated Research Center for the Department of Defense.

Megan M. Clifford is a research associate and engineer at Stevens Institute of Technology. She works on various research projects with a specific interest in systems assurance and cyber-physical systems. She previously worked on the Systems Engineering Research Center (SERC) leadership team as the Chief of Staff and Program Operations, was the Director of Industry and Government Relations to the Center for Complex Systems and Enterprises (CCSE), and held several different positions in industry, including systems engineer at Mosto Technologies while working on the New York City steam distribution system.

Barry M. Horowitz held the Munster Professorship in Systems Engineering at the University of Virginia, prior to his retirement in May 2021. His research interests include system architecture and design.

Tim J. Sherburne is a research associate in the Intelligent System Division of the Virginia Tech National Security Institute. Sherburne was previously a member of the systems engineering staff at the University of Virginia supporting Mission Aware research through rapid prototyping of cyber-resilient solutions and model-based systems engineering (MBSE) specifications. Prior to joining the University of Virginia, he worked at Motorola Solutions in various Software Development and Systems Engineering roles defining and building mission critical public safety communications systems.

Chapter 27

Implementation Example: Silverfish

Tim Sherburne[1], *Megan M. Clifford*[2], *and Peter A. Beling*[1]

[1] *Virginia Polytechnic Institute and State University, Hume Center for National Security and Technology, Blacksburg, VA, USA*
[2] *Stevens Institute of Technology, School for Systems and Enterprises, Hoboken, NJ, USA*

Implementation Example: Silverfish

Introduction and Background

Silverfish is a hypothetical system but was deemed by the ARDEC team on the System Engineering Research Center (SERC) Research Task (RT)-191 as sufficient for cyber use case development. Initially meant for the Cyber Security Requirements Methodology, the Silverfish use case quickly gained attention in several cyber research efforts. To best serve the field manual, this chapter is written in chronological and practical order (Horowitz et al. 2018b).

Concept of Operations (CONOPS)

The Silverfish System (Horowitz et al. 2015b) is a rapidly deployable set of fifty (50) individual ground-based weapon platforms (referred to as obstacles) controlled by a single operator. The purpose of the system is to deter and prevent adversaries from trespassing into a designated geographic area that is located near a strategically sensitive location. The system includes a variety of sensors to locate and classify potential trespassers as either personnel or vehicles (Figure 27.1).

An internal wireless communication system is used to support communication between the sensors and the operator and supports fire control communications between the operator and the obstacles. The sensors include obstacle-based seismic and acoustic sensors, infrared sensors, and an unmanned aerial vehicle-based surveillance system to provide warning of potential adversaries approaching the protected area. The operator is located in a vehicle and operates within visual range of the protected area. The operator is in communication with a higher-level command and control (C2) system for exchange of doctrinal-related and situation awareness information. A more detailed functional description of the system is presented below.

- **Purpose**: Deter and prevent, when and where necessary, via the use of rapidly deployable obstacles, adversarial tracked vehicles (assumed maximum speed – 10 mph) or individuals from trespassing into geographic areas that are close to strategically sensitive locations.

Systems Engineering for the Digital Age: Practitioner Perspectives, First Edition. Edited by Dinesh Verma.

Companion website: www.wiley.com/go/verma/systemsengineering

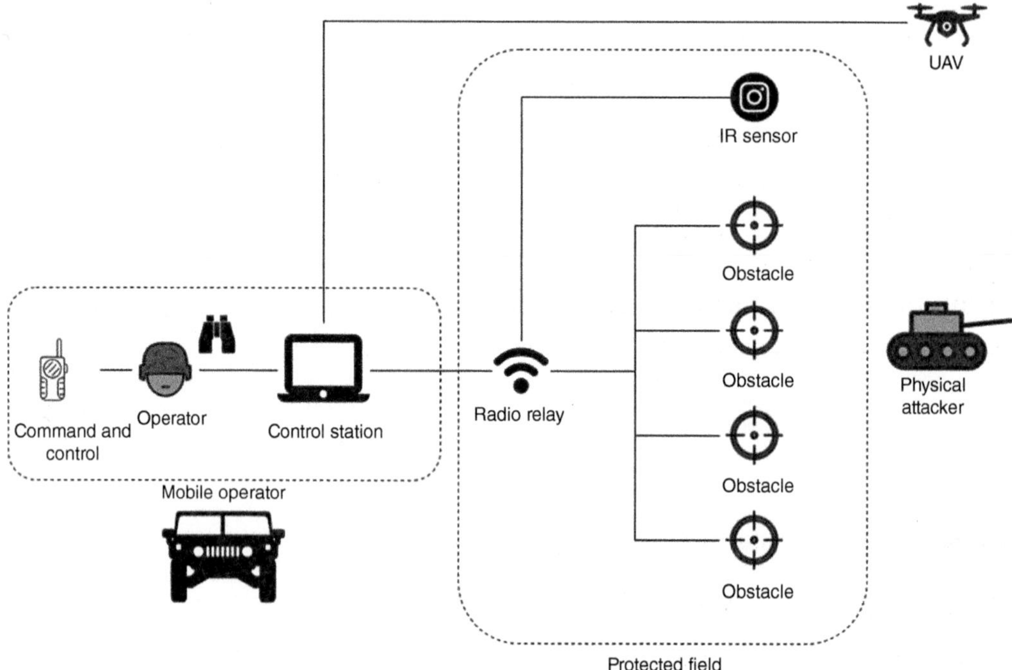

Figure 27.1 Silverfish reference architecture / overview. Source: Adapted from Horowitz et al. (2014).

- **Prohibited area**: 100 acres of open field space (100 acres, approximately 0.16 square miles = 0.4-mile × 0.4 mile area). At maximum speed, a vehicle would take about three minutes to cross the prohibited area.
- **Obstacle deployment**: About 50 obstacles are available to be distributed over the 100-acre protected area (each obstacle is designed to protect a 300 × 300 foot area). Two types of obstacles can be deployed. One type of obstacle addresses anti-personnel requirements. It contains six (6) short-range sub-munitions, each covering a 60-degree portion of a circular area to be protected. The second type of obstacle contains a single munition capable of impacting a tracked vehicle.
- **Operation**: The operator, located in a vehicle that is operated close to the prohibited area (~150 m away), remotely controls individual obstacles and their sub-munitions, based upon sensor-based and operator visual surveillance of the prohibited area.
- **Prohibited area surveillance**: The operator is supported by obstacle-based acoustic and seismic sensors (geophones and accelerometers) that can detect and distinguish between vehicles and people, redundant infrared sensors that can detect and track the movement of people and vehicles, and real-time Video/IR derived early warning information regarding people and vehicles approaching the prohibited area provided by a UAV managed by the operator. The UAV is used to provide warning information. The operator can relocate his or her vehicle for improved visual observation.
- **Obstacle design features**: The obstacle-based sensors provide regular operator situation awareness reports (seconds apart) when they detect a trespasser. They provide, at a lower data rate (e.g., a minute apart), general health related information, including reports on their location (GPS-based), their on-off status, and their remaining battery life. Should a weapon be fired, the obstacle

confirms the acceptance of commands and the actual firing events. To address potential tampering risks, obstacle-based software can only be modified by electrically disconnecting their platform-based computer from the obstacle, and removal results in self-destruction of that computer.

- **Infrared sensor configuration**: A single pole-mounted IR sensor is assumed to be capable of providing surveillance of the entire protected area. A second sensor is provided for redundancy and can be used to provide surveillance of areas that the single sensor is not able to observe. The IR sensors provide the same type of operator situation awareness data at the same rates as the obstacle-based sensors, but in addition provide tracking information to enable the operator to project future locations of moving vehicles or people.
- **Requirements for avoiding errors**: Concerns exist regarding detonating sub-munitions in cases where non-adversarial vehicles or people, by chance, enter the prohibited area. Concerns also exist about failing to fire munitions when an adversary is approaching a strategically sensitive location via the prohibited area. The operator, when possible, can use visual observations to increase confidence regarding fire control.
- **Operator functions**: The operator can set the obstacles into either on or off modes and can cause individual or designated groups of obstacles/sub-munitions to detonate when in on mode. Obstacles can be commanded to self-destroy designated critical information in order to prevent adversaries from collecting such information for their own purposes. The operator also can launch a quad-copter drone (UAV) to provide video/IR-based early warning information regarding potential trespassers of the protected area (~5 minute warning for vehicles approaching at a 10 mph speed).
- **Communications systems**: The Operator, the higher level C2 System, and UAV operate on a shared radio system that is integrated to a relay node(s) that couples into the Silverfish system's integrated wireless communication network. The communication system includes digital interfaces that support formatted data transfers between the operator's system, the UAV subsystem, the individual obstacles, the IR subsystem, and the C2 Center. The communication system also supports short message text and voice communications between operator and C2 system.
- **Operator control station**: The operator is provided with a vehicle-mounted computer(s) subsystem that provides situation awareness information including individual obstacle status, and sensor-based situation awareness information. The subsystem also provides computer-based entry and corresponding weapon system feedback for fire control related inputs from the operator. The control station also supports required digital situation awareness-related reporting to the C2 center, as well as support for UAV control.
- **Command center controls**: The C2 center digitally provides weapon control information for the operator (determines weapon system on/off periods, provides warning of periods of higher likelihood of attack, provides forecasts of possible approach direction to the prohibited area, enables operation with/without UAV support, etc.). As determined by either the operator or the C2 center, out of norm situations can be supported through rapid message communications between the C2 center and the operator.
- **Forensics**: All subsystems collect and store forensic information for required post-mission analysis purposes.
- **Rapid deployment support**: All subsystems enable rapid deployment testing to confirm readiness for operational use.

The Silverfish user interface shows the grid layout (Figure 27.2).

Silverfish grid layout

- Prohibited area:
  - ~100 acres ≈ .16 sq. miles (.4 × .4)
- Obstacle deployment:
  - ~50
  - 7×7 grid (A1-G7)
  - Aligned to compass coordinates
    - Operator observation point
- Cell grid
  - ≈ 300 ft. × 300 ft.
  - 6 Munitions per cell (ready/fired state)
- Vehicle traversal:
  - Max Speed = 10 mph ≈ 15 ft./sec.
  - 20 seconds/grid
  - 2.3 minutes/protected area

NW					N					NE
		A	B	C	D	E	F	G		
	1								1	
	2								2	
	3								3	
W	4								4	E
	5								5	
	6								6	
	7								7	
		A	B	C	D	E	F	G		
SW										SE

Figure 27.2 Silverfish user interface. Source: Adapted from Horowitz et al. (2018a).

Requirements, Architecture, and Preliminary Design

CSRM: Mission Engineering MOPs

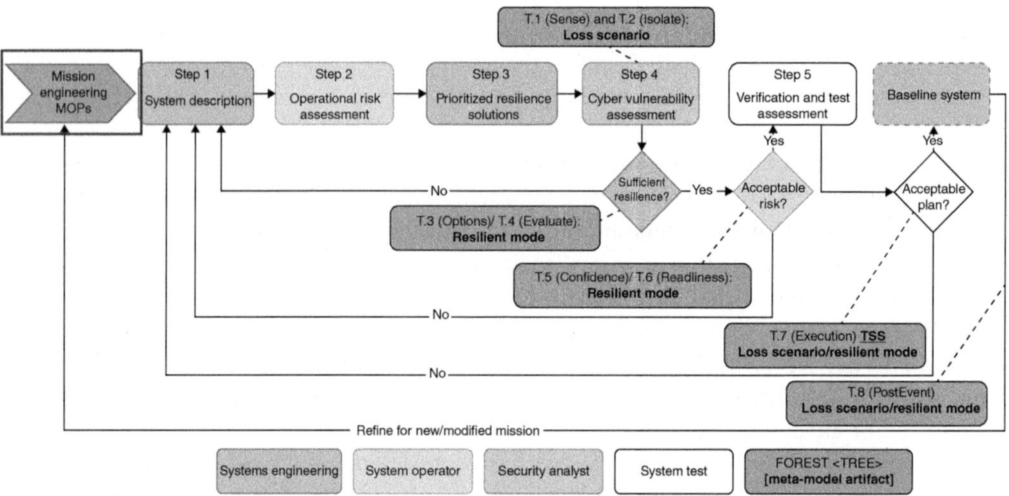

Two example missions (Table 27.1) are defined for Silverfish with derived measures of performance (MOP) requirements affecting its resilience operation. For example, the "isolated area" mission expects faster moving vehicle-based physical attackers, thereby deriving shorter resilience recovery times for the Fire Control subsystem than an "urban area" mission, where slower moving person-based attackers are more likely.

System Losses (Table 27.2) are identified by the mission operators and prioritized for importance. For Silverfish, the highest priority unacceptable loss is serious injury or loss of life for both military personnel and civilians.

Table 27.1 Mission engineering requirements.

Mission requirement	Type	Origin
ME.100.1: Silverfish shall support an isolated area mission	Composite	Originating
ME.100.1.1: A Silverfish isolated area mission shall expect vehicle-based physical attackers	Constraint	Derived
ME.100.1.1.1: A Silverfish isolated area mission shall support fire control resilience recovery within x seconds	Performance	Derived
ME.100.1.2: A Silverfish isolated area mission shall prioritize sensor confirmation of physical attackers	Constraint	Derived
ME.100.1.2.1: A Silverfish isolated area mission shall support redundant sensor availability of xx.xx %	Performance	Derived
ME.100.2: Silverfish shall support an urban area mission.	Composite	Originating
ME.100.2.1: A Silverfish urban area mission shall expect person-based physical attackers	Constraint	Derived
ME.100.2.1.1: A Silverfish urban area mission shall support fire control resilience recovery within y seconds	Performance	Derived
ME.100.2.2: A Silverfish urban area mission shall prioritize operator confirmation of physical attackers	Constraint	Derived
ME.100.2.2.1: A Silverfish urban area mission shall support dual operator availability of yy.yy %	Performance	Derived

Table 27.2 Silverfish system requirements.

System requirement	Type	Origin
SF.100.1: Silverfish shall provide a field deployable distributed network of munitions	Functional	Originating
SF.100.2: Silverfish shall provide a field deployable distributed network of situational reporting sensors	Functional	Originating
SF.100.3: Silverfish shall support a command and control interface	Functional	Originating
SF.200.1: Silverfish shall support near real-time situational reports	Performance	Originating
SF.200.2: Silverfish shall be deployed, on average, within 4 h by 4 deployment technicians	Performance	Originating
SF.300.1: Silverfish shall protect a field of up to 100 acres	Performance	Originating
SF.300.2: Silverfish shall support up to 50 obstacles	Performance	Originating
SF.300.3: Silverfish batteries shall support 7 d of continuous operation	Performance	Originating
SF.500: Silverfish shall support cyber hygiene practices	Composite	Design Decision
SF.502.1: Silverfish shall support network encryption	Composite	Design Decision
SF.502.1.1: Silverfish shall encrypt wired LAN traffic	Functional	Derived
SF.502.1.2: Silverfish shall encrypt wireless radio relay traffic	Functional	Derived
SF.502.2: Silverfish shall support at-rest encryption	Functional	Design Decision
SF.502.3: Silverfish shall support user authorization	Functional	Design Decision
SF.505.1: Silverfish shall support a log management solution	Functional	Design Decision
SF.506.1: Silverfish deployed components shall support Linux OS	Constraint	Design Decision
SF.506.1.1: Silverfish shall support AppArmor Linux security module	Constraint	Design Decision

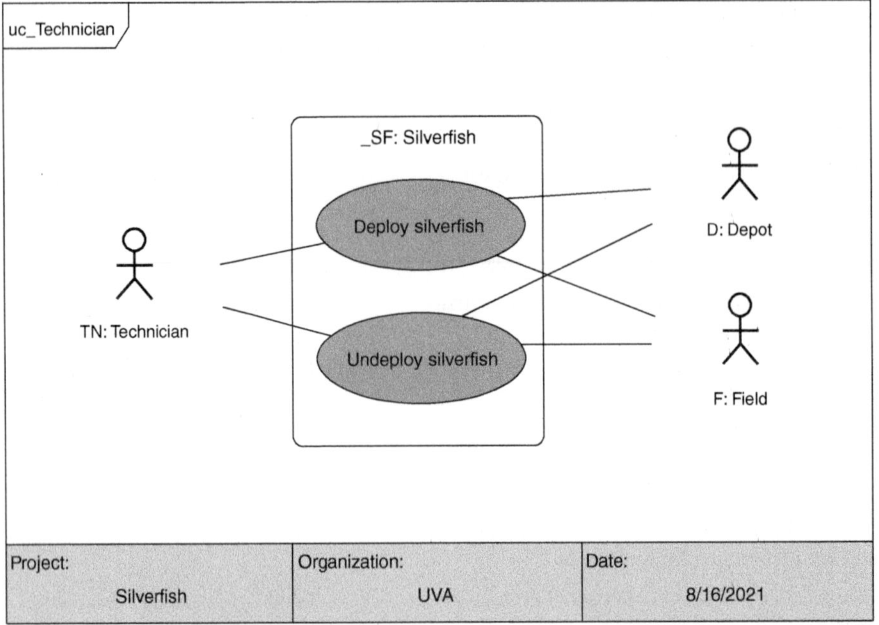

Figure 27.3 Silverfish: technician: use case.

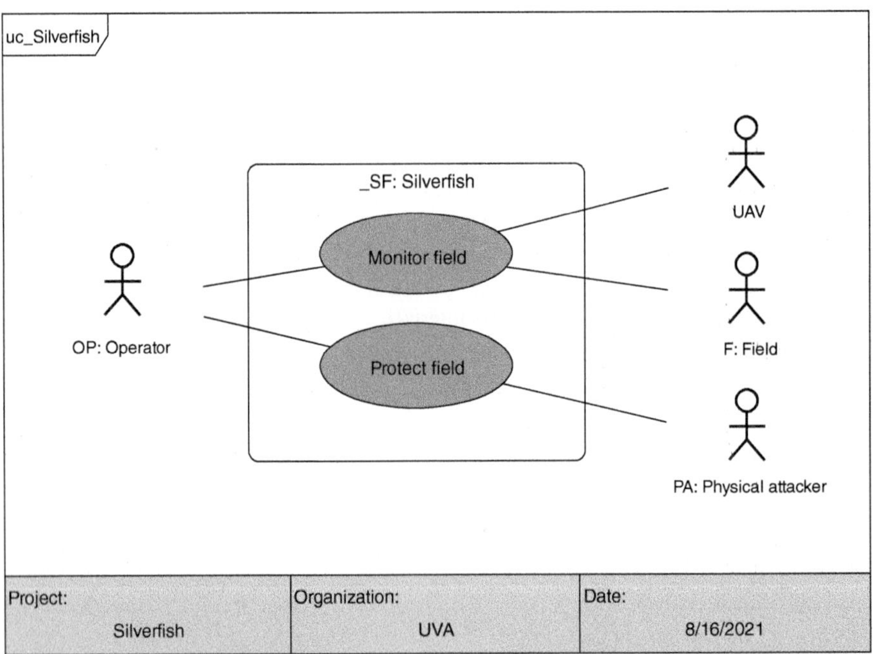

Figure 27.4 Silverfish: operation use case.

CSRM.1: MBSE System Description

The Technician use cases are presented in Figure 27.3. Silverfish: Technician: Use Case and include Deployment from Depot to Field and UnDeployment from Field back to Depot.

The Operator use cases are presented in Figure 27.4 and includes Monitoring of the protected field with the Situational Aware Control Station using Sensors and Protecting the field with the Fire Control Station using Obstacle munitions.

The Tester use cases are presented in Figure 27.5. Silverfish: Test Support Use Case and include the typical functional and performance verification. Additionally, verification of Cyber Resilience is included describing injection of Loss Scenarios and remediation via Resilient Modes.

In addition, there is a Mission Aware use case to protect Silverfish with the operator, technician, Silverfish, and cyberattacker as actors (Figure 27.6).

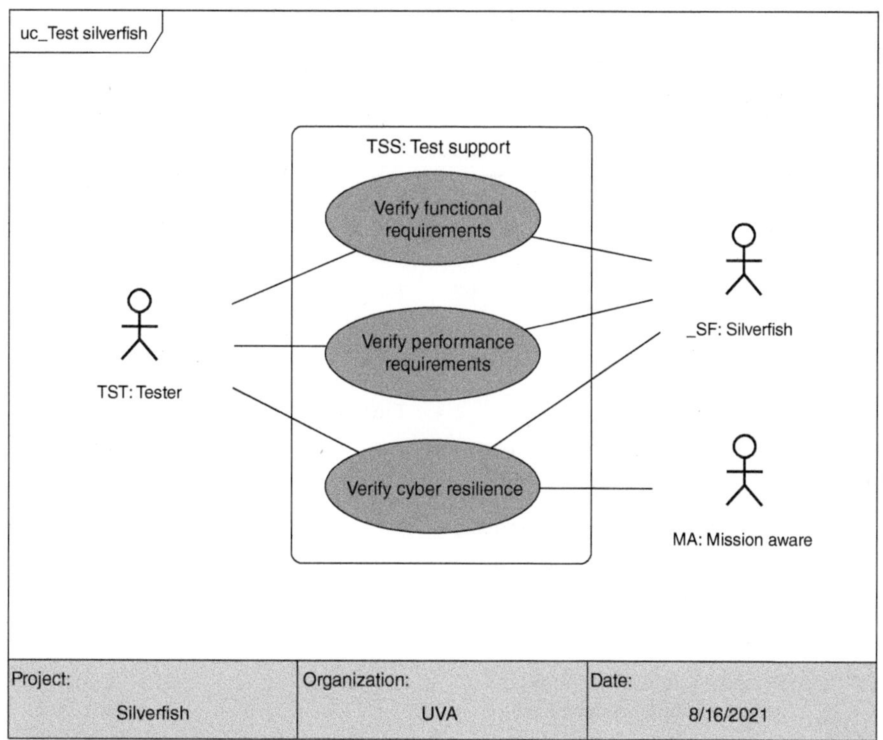

Figure 27.5 Silverfish: test support use case.

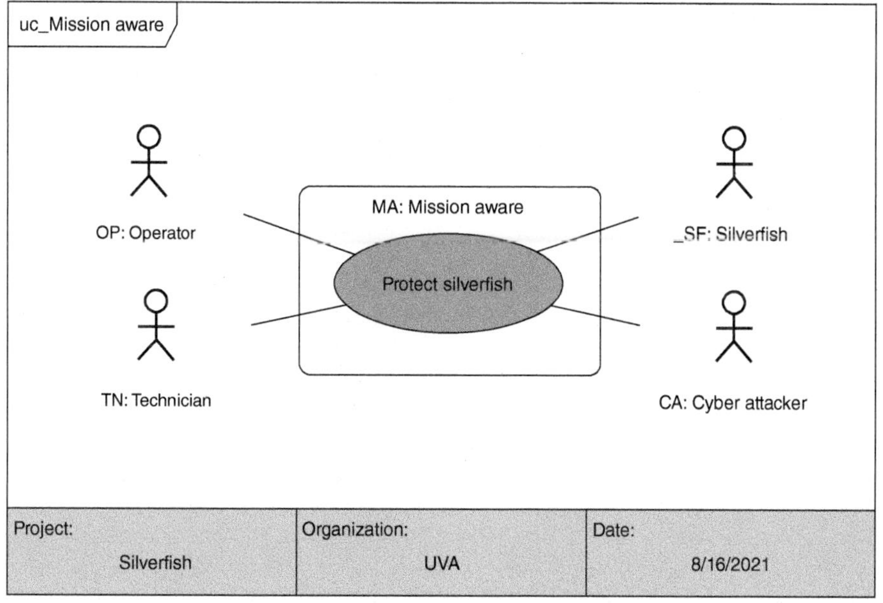

Figure 27.6 Silverfish: Mission Aware use case (Horowitz et al. 2021).

A representative set of Silverfish system requirements are captured (Table 27.3. Silverfish System Requirements) covering functionality, performance, capacity, and design constraints. These system requirements will be further expanded as part of the Resilient Mode Analysis and Loss Scenario Identification with focus on the FOREST quality attributes.

Table 27.3 Silverfish system functions (behavior).

System function	Description	Decomposed by: function	Triggered by: control action
F.4.10:SF: Fire	Select and fire one or more munitions for one or more obstacles	F.4.10.1:CS: Input Fire Munition Command F.4.10.2:RR: Transfer Fire Munition Command F.4.10.3:OBS: Initiate Fire Munition	OP.1.1:OP: CA: L1-Fire
F.4.10.1:CS: Input Fire Munition Command	Process operator input to fire one or more munitions for one or more obstacles, manage munition fire state, and wireless transmit fire command to selected munitions		OP.1.1.1:CS: L2-Operator Fire Control Action
F.4.10.2:RR: Transfer Fire Munition Command	Wirelessly transfer munition fire commands from control station to obstacles		OP.1.1.2:RR: L2-Transfer Fire Control Action
F.4.10.3.OBS: Initiate Fire Munition	Detonate selected mentions and update munition state to fired		OP.1.1.3.OBS: L2-Initiate Fire Control Action
F.4.13:SF: Monitor Field	Monitor field for physical attackers (human or vehicle) by fusing UAV, IR, acoustic and seismic sensor analytics	F.4.13.1:UAV: Report UAV Analytics F.4.13.2:LAN: Transfer UAN Analytics F.4.13.3:IR: Report IR Analytics F.4.13.4:OBS: Report Acoustic Seismic Analytics F.4.13.5:RR: Transfer Acoustic Seismic IR Analytics F.4.13.6:C5: Perform Situational Fusion	F.1.1:F:FB: L1-Sensor Signature
F.4.13.1:UAV: Report UAV Analytics	Periodically report UAV sensor analytics		F.1.1.6:UAV: Sensor Feedback

Table 27.3 (Continued)

System function	Description	Decomposed by: function	Triggered by: control action
F.4.13.2:LAN: Transfer UAV Analytics	In vehicle transfer of sensor data		F.1.1.3:LAN: Sensor Transfer Feedback
F.4.13.3.IR: Report IR Analytics	Periodically report IR sensor analytics		F.1.1.2:IR: Sensor Feedback
F.4.13.4:OBS: Report Acoustic Seismic Analytics	Periodically report Obstacle sensor analytics		F.1.1.4:OBS: Sensor Feedback
F.4.13.5:RR: Transfer Acous tic: Seismic IR Analytics	Wirelessly transfer sensor data		F.1.1.5:RR: Sensor Transfer Feedback
F.4.13.6.CS: Perform Situational Fusion	Fuse sensor data into an integrated situational aware user interface of the protected field		F.1.1.1:CS: Sensor Feedback

The Silverfish system architecture and initial representations were also needed to be created prior Step 2, Operational Risk Assessment. These initial system descriptions define the basic composition, architecture, and concept of standard operation for the Silverfish system. The major components of the baseline system are defined in a platform-based design along with basic functional descriptions of the information exchange between each component. The system context incudes the full lifecycle from research and development to depot to field deployment to study cyberattackers throughout the life cycle. The system context defines the human interfaces to the Silverfish System including Command and Control, Deployment Technicians, and Mission Operators. The system-of-system interfaces include the Unmanned Ariel Vehicle (UAV) and Mission Aware Sentinel.

The Silverfish system architecture shows the wired and wireless interfaces between Control Station, Obstacle, and IR sensor subsystems. The blue border components show the external interfaces to the Silverfish subsystems (Figures 27.7 and 27.8).

Then, a representative set of top-level system functions is defined and decomposed into subfunctions with a triggering control action defined for each. The primary system functions for Silverfish are to Fire the Obstacle munitions and to Monitor the Protected Field for Physical Intruders. The system behavior, functions, are looks at vs. the control structure. System function examples include graphical control structure vs. tabular view, decomposition of functions, and triggered by control actions/feedback (Figure 27.9).

The Silverfish Components are described in the table below with the list of Functions *performed* by each Component (Table 27.4).

The Links between Silverfish Components are described in the below table with the list of Items (Control Actions/Feedback) transferred by each Link (Table 27.5).

The function/item relationships are important considerations for Mission Aware/STPA analysis of Loss Scenarios and Resilient Modes.

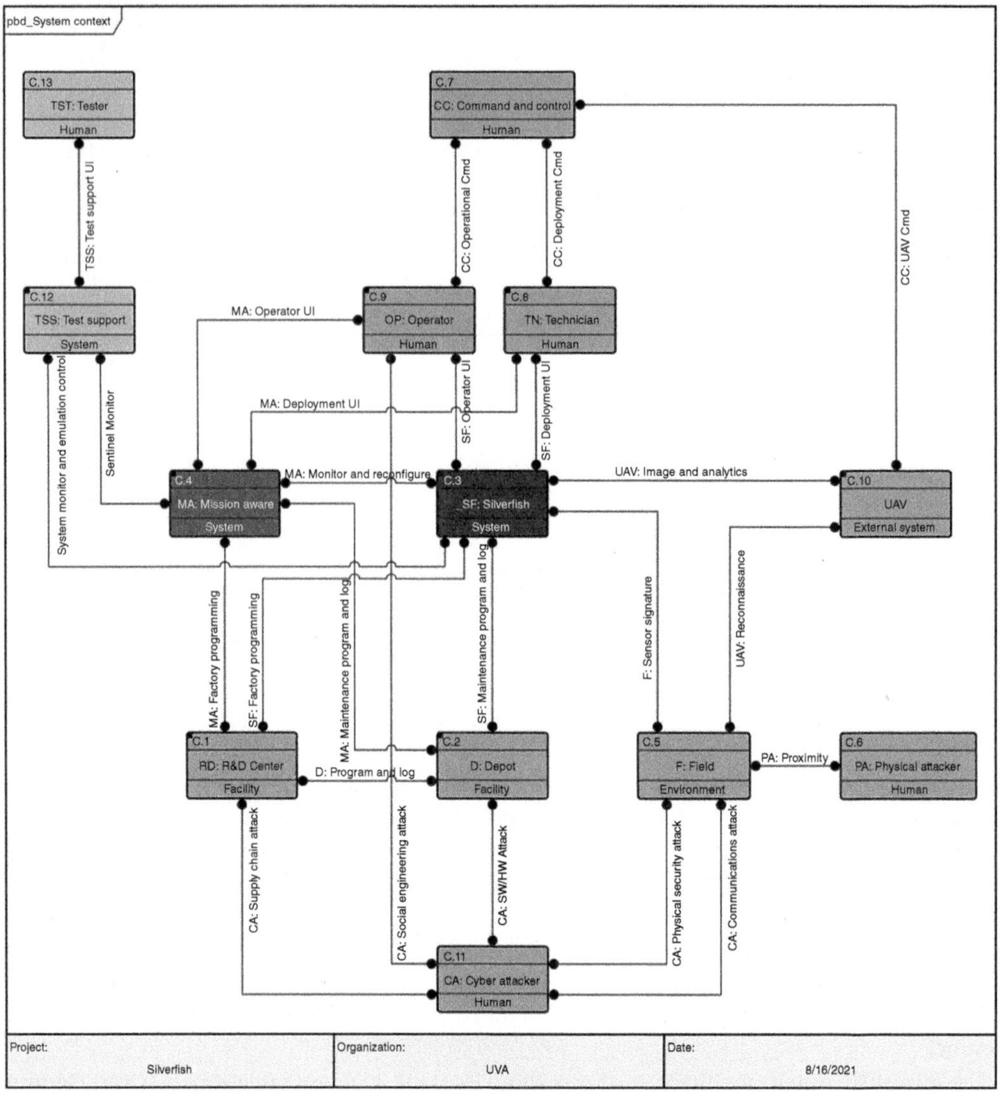

pbd_System context
C.13
TST: Tester
Human
TSS: Test support UI
C.12
TSS: Test support
System
C.7
CC: Command and control
Human
CC: Operational Cmd
CC: Deployment Cmd
CC: UAV Cmd
C.9
OP: Operator
Human
C.8
TN: Technician
Human
MA: Operator UI
MA: Deployment UI
SF: Operator UI
SF: Deployment UI
System monitor and emulation control
Sentinel Monitor
C.4
MA: Mission aware
System
MA: Monitor and reconfigure
C.3
SF: Silverfish
System
UAV: Image and analytics
C.10
UAV
External system
MA: Factory programming
SF: Factory programming
MA: Maintenance program and log
SF: Maintenance program and log
F: Sensor signature
UAV: Reconnaissance
C.1
RD: R&D Center
Facility
D: Program and log
C.2
D: Depot
Facility
C.5
F: Field
Environment
PA: Proximity
C.6
PA: Physical attacker
Human
CA: Supply chain attack
CA: Social engineering attack
CA: SW/HW Attack
CA: Physical security attack
CA: Communications attack
C.11
CA: Cyber attacker
Human
Project:
Silverfish
Organization:
UVA
Date:
8/16/2021

Figure 27.7 Silverfish: system context. Source: Adapted from Horowitz et al. (2021).

CSRM.2: Operational Risk Assessment

T.1 (Sense) and T.2 (Isolate):
Loss scenario
Mission engineering MOPs
Step 1
System description
Step 2
Operational risk assessment
Step 3
Prioritized resilience solutions
Step 4
Cyber vulnerability assessment
Step 5
Verification and test assessment
Baseline system
No
Sufficient resilience?
Yes
Acceptable risk?
Acceptable plan?
T.3 (Options)/ T.4 (Evaluate):
Resilient mode
T.5 (Confidence)/ T.6 (Readliness):
Resilient mode
No
T.7 (Execution) TSS
Loss scenario/resilient mode
No
T.8 (PostEvent)
Loss scenario/resilient mode
Refine for new/modified mission
Systems engineering
System operator
Security analyst
System test
FOREST <TREE>
[meta-model artifact]

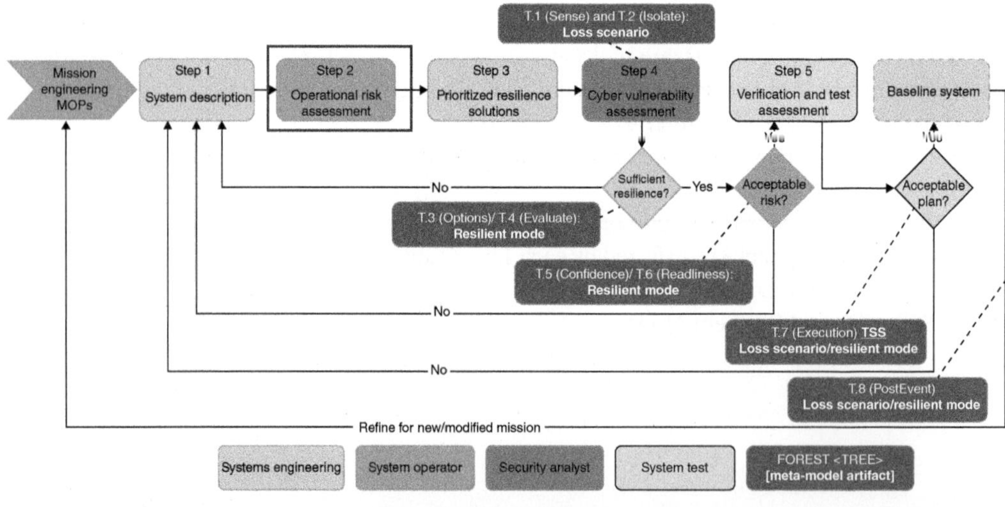

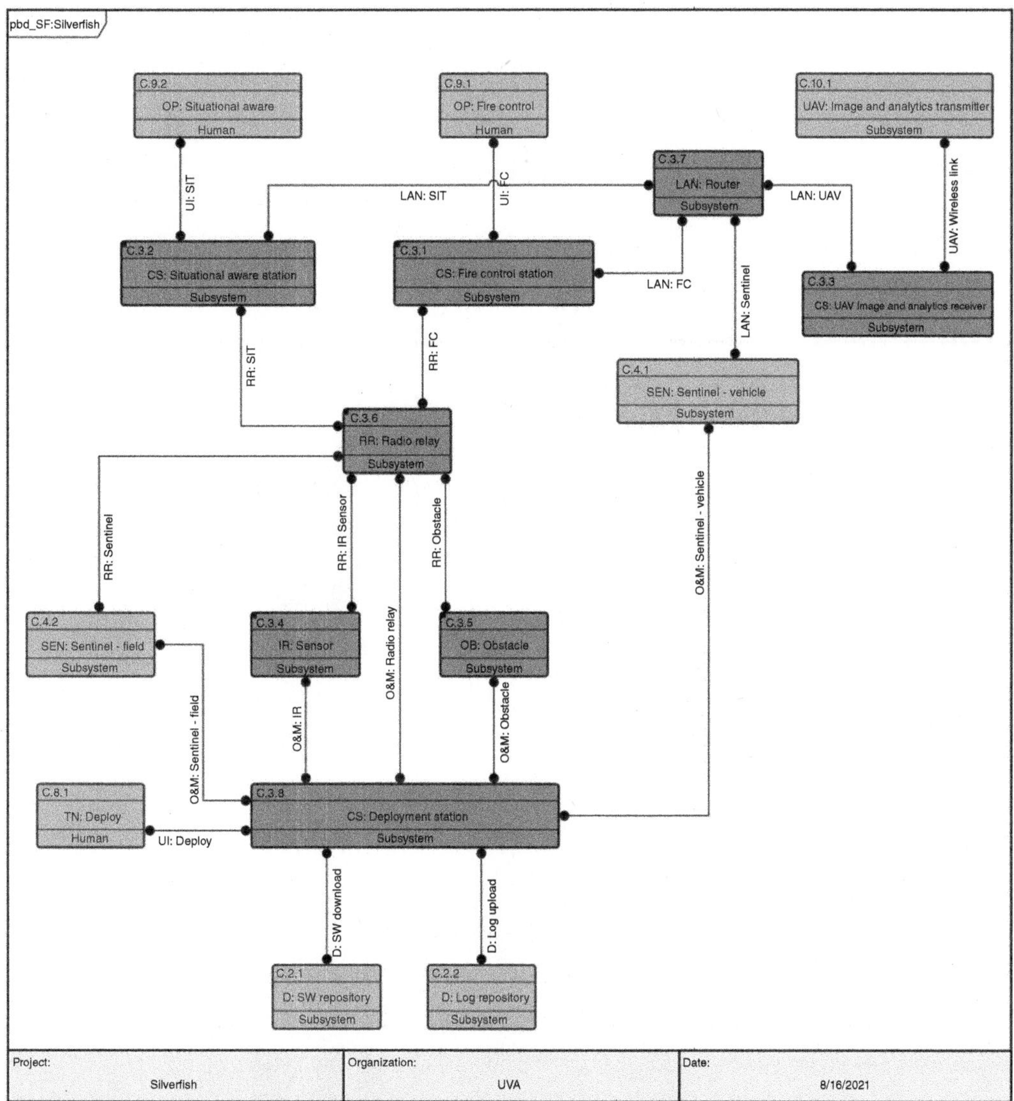

Figure 27.8 Silverfish system. Source: Adapted from Horowitz et al. (2021).

The Silverfish operational risk assessment is presented in three tables below using the STPA (Leveson and Thomas) structured elicitation process. Each loss has an associated priority, defined by the mission operator, to align system capabilities within programmatic constraints. Each table shows the traceability between STPA steps (`leads to / is caused by / is variation of` associations) (Tables 27.6, 27.7, and 27.8).

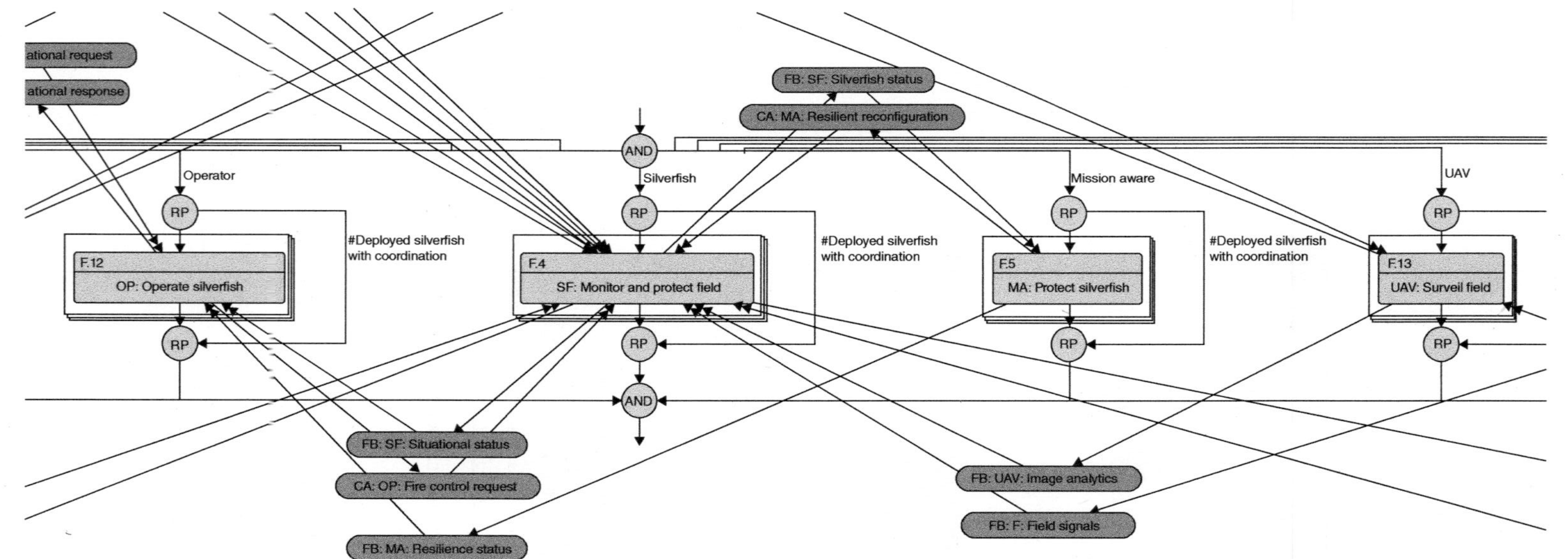

Figure 27.9 Silverfish System Control Structure. Source: Adapted from Horowitz et al. (2021).

Table 27.4 Silverfish system components.

System component	Description	Performs: function
C.3.1:CS: Fire Control Station	The Fire Control (FC) Station maintains the status of all Obstacles and provides the Operator interface for munition firing	F.4.10.1:CS: Input Fire Munition Command
C.3.2:CS: Situational Aware Station	The Situational Aware (SIT) Station maintains sensor status and provides the Operator user interface for field monitoring	F.4.13.6:CS: Perform Situational Fusion
C.3.3:CS: UAV Image and Analytics Receiver	A receiver for UAV images and analytics that feed into the situational aware control station providing enhanced field awareness	F.4.13.1:UAV: Report UAV Analytics
C.3.4:IR: Sensor	A full field Infrared Sensor that provides field awareness day and night	F.4.13.3:IR: Report IR Analytics
C.3.5:OB: Obstacle	An Obstacle provides both directional munitions and seismic/acoustic sensor for situational awareness	F.4.10.3:OBS: Initiate Fire Munition F.4.13.4:OBS: Report Acoustic and Seismic Analytics
C.3.6:RR: Radio Relay	The Radio Relay provides wireless connectivity between vehicle mounted control stations and field deployed obstacles and sensors	F.4.10.2:RR: Transfer Fire Munition Command F.4.13.5:RR: Transfer Acoustic and Seismic and IR Analytics
C.3.7:LAN: Router	In vehicle LAN	F.4.13.2:LAN: Transfer UAV Analytics

Source: Adapted from Horowitz et al. (2018b).

Table 27.5 Silverfish system links.

System link	Connects to: Component	Transfers: item
LAN.2:LAN: SIT	C.3.2:CS: Situational Aware Station C.3.7:LAN: Router	F.1.1.3:LAN: Sensor Transfer Feedback
LAN.3:LAN: UAV	C.3.3:CS: UAV Image and Analytics Receiver C.3.7:LAN: Router	F.1.1.3:LAN: Sensor Transfer Feedback
RR.1:RR: FC	C.3.1:CS: Fire Control Station C.3.6:RR: Radio Relay	OP.1.1.2:RR: L2-Transfer Fire Control Action
RR.2:RR: SIT	C.3.2:CS: Situational Aware Station C.3.6:RR: Radio Relay	F.1.1.5:RR: Sensor Transfer Feedback
RR.3:RR: Obstacle	C.3.5:OB: Obstacle C.3.6:RR: Radio Relay	F.1.1.5:RR: Sensor Transfer Feedback OP.1.1.3:OBS: L2-Initiate Fire Control Action
RR.4:RR: IR Sensor	C.3.4:IR: Sensor C.3.6:RR: Radio Relay	F.1.1.5:RR: Sensor Transfer Feedback

Source: Adapted from Horowitz et al. (2018b).

Table 27.6 Silverfish STPA losses.

Loss	Priority	Is caused by: Hazard
L.1: Loss of life or serious injury to military	1	H.1: Weapon mis-fire H.2: Excessive time and/or personnel to deploy H.3: Excessive time and/or personnel to un-deploy
L.2: Loss of life or serious injury to civilian	1	H.1: Weapon mis-fire
L.3: Loss of protected area assets	2	H.1: Weapon mis-fire H.2: Excessive time and/or personnel to deploy
L.4: Loss of classified mission HW / SW	3	H.3: Excessive time and/or personnel to un-deploy

Source: Horowitz et al. (2021).

Table 27.7 Silverfish STPA hazards.

Hazard	Leads to: Loss	Is caused by: Hazardous Action
H.1: Weapon mis-fire	L.1: Loss of life or serious injury to military L.2: Loss of life or serious injury to civilian L.3:Loss of protected area assets	HCA.1: Incorrect Fire HCA.2: No Fire
H.2: Excessive time and/ or personnel to deploy	L.1: Loss of life or serious injury to military L.3: Loss of protected area assets	HCA.3: Unable to set Location
H.3: Excessive time and/ or personnel to un-deploy	L.1: Loss of life or serious injury to military L.4: Loss of classified mission HW / SW	

Source: Horowitz et al. (2021).

Table 27.8 Silverfish hazardous actions.

Hazardous action	Leads to: hazard	Is caused by: loss scenario	Variation of: control action
HCA.1: Incorrect Fire	H.1: Weapon mis-fire	LS.1: Manipulated Fire Command	SF.1.1: SF: L1-Fire Munition
HCA.2: No Fire	H.1: Weapon mis-fire	LS.2: Situational Injection LS.3: Situational Delay LS.5: Delayed Fire Command	OP.1.OP: CA: L1-Fire
HCA.3: Unable to set Location	H.2: Excessive time and/or personnel to deploy	LS.4: Tampered Deployment	TN.2:TN: L1-Set Location

Source: Horowitz et al. (2021).

CSRM.3: Resilient Modes

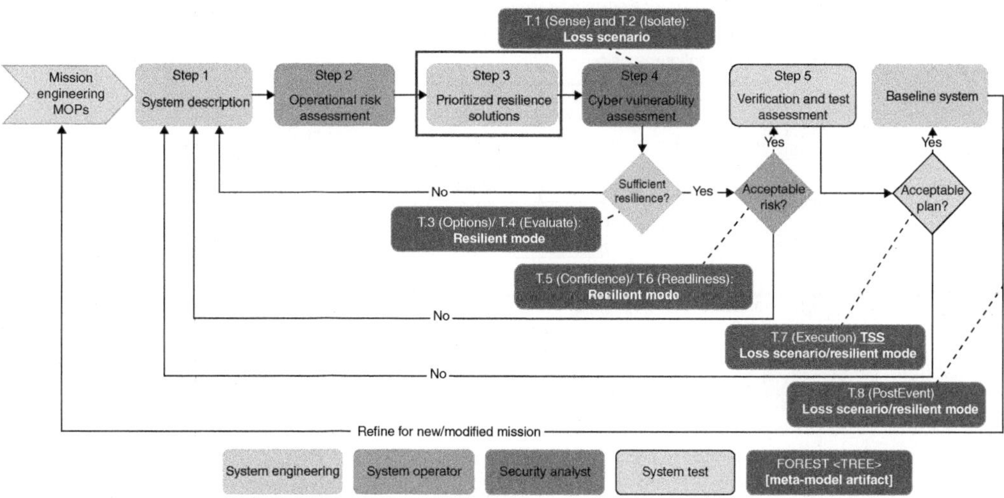

The Silverfish resilient modes are defined in the below. Each resilient mode identifies the associated loss scenarios that it provides reconfiguration for and affected components that it provides alternative operation for (diverse redundancy) (Figure 27.10, Table 27.9).

In the Mission Aware effort for Silverfish, Resilient Modes are identified for the system and are defined relationships by Operation Altered by, Managed by, and Recovered by.

CSRM.4: Cyber Vulnerability Assessment

The Silverfish vulnerability assessment is captured in the three tables below. Each loss scenario identifies the resilient mode used to reconfigure the system to maintain mission function.

If applicable, each remediation identifies the implemented hygiene practice and the attack vector and loss scenarios that it `protects against` (Figure 27.11).

The Cyber Vulnerability Assessment through Mission Aware identifies the *Loss Scenarios* for the system, defined by relationships `reconfigures using` and `leads to`. It identified *Remediations* through the defined relationships of `protects against` and `is implementation of`. It also enables a checklist of hygiene practice (MITRE) and Attack Pattern (CAPEC) (Tables 27.10–27.12).

The Loss Scenarios elicit system requirements that constrain the system structure to provide the identified monitoring mechanisms and related resilient modes. Additionally, the system requirements are elicited that refine the system behavior to enable management (enable/disable/self-test, etc.) of the related resilient modes. A sample set of system requirements are elicited that specify the performance for the FOREST quality attributes (which can be found in previous chapters) that achieve the Mission MOPs.

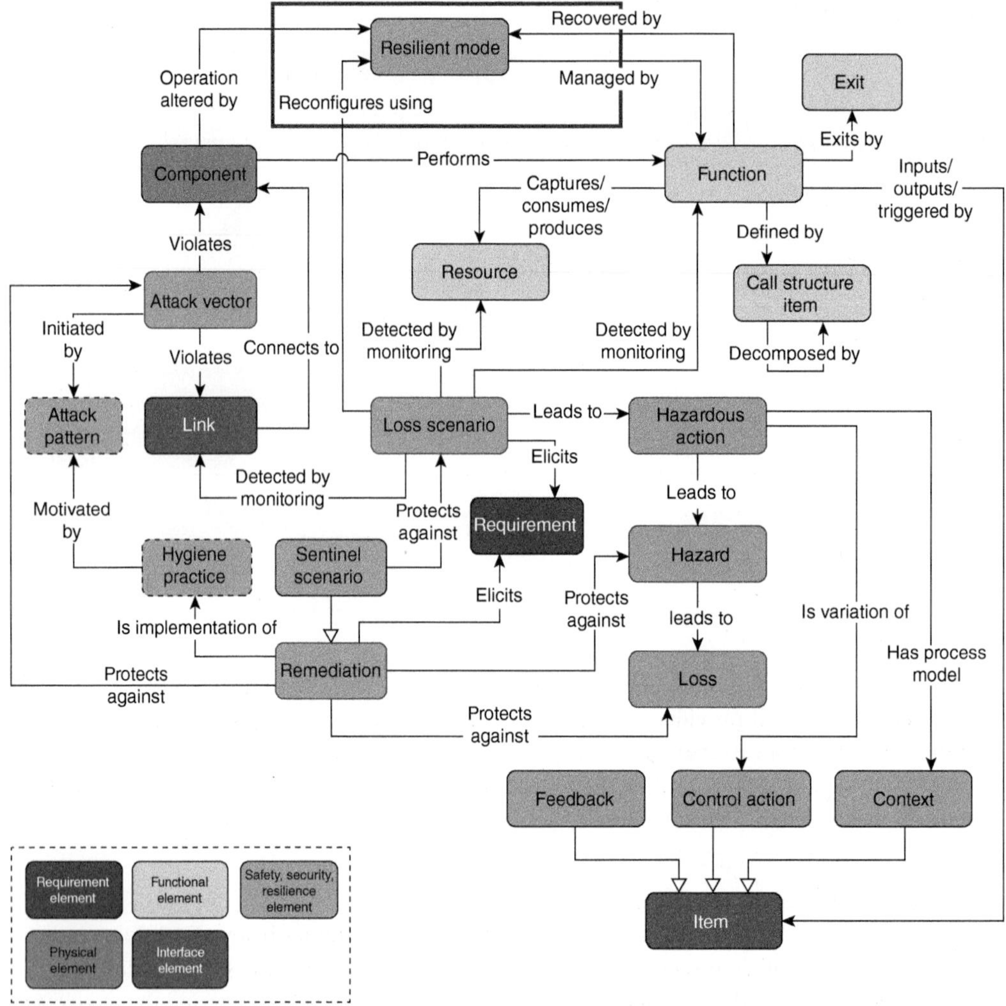

Figure 27.10 Resilient modes. Source: Adapted from Horowitz et al. (2021).

Table 27.9 Silverfish CSRM resilient modes.

Resilient mode	Provides reconfiguration for: loss scenario	Provides alternate operation for: component
RM.1: Diverse Redundant Radio Relay	LS.2: Situational Injection|| LS.3: Situational Delay	C.3.1: CS: Fire Control Station|| C.3.4: IR: Sensor||C.3.6: RR: Radio Relay
RM.2: Diverse Redundant Fire Control	LS.1: Manipulated Fire Command	C.3.1: CS: Fire Control Station
RM.3: Diverse Redundant IR Sensors	LS.3: Situational Delay	C.3.4: IR: Sensor
RM.4: Obstacle Restore	LS.4: Tampered Deployment	C.3.5: OB: Obstacle
RM.5: Operator Reposition	LS.3: Situational Delay	C.9: OP: Operator

Source: Adapted from Horowitz et al. (2018b).

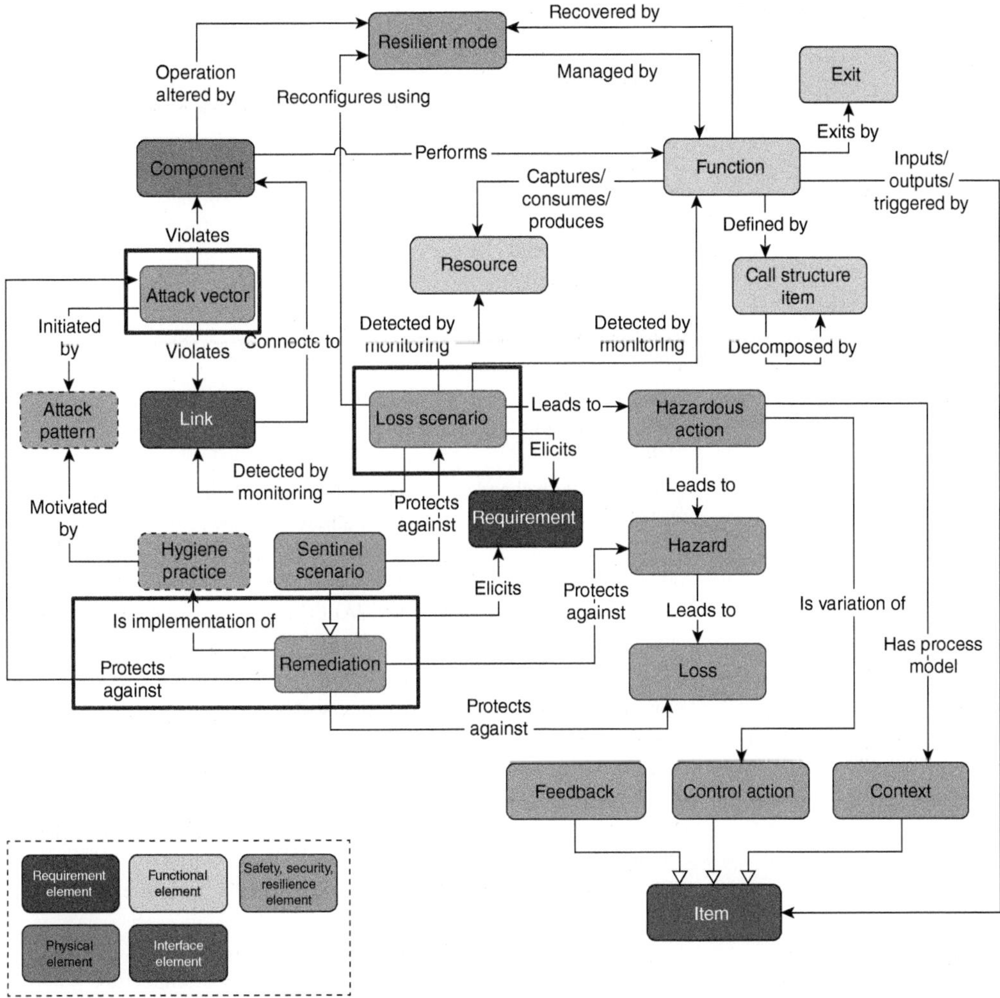

Figure 27.11 Cyber vulnerability assessment. Source: Adapted from Horowitz et al. (2021).

Table 27.10 Silverfish STPA loss scenarios.

Loss scenario	LEADS TO: HAZARDOUS ACTION	Reconfigures using: resilient mode
LS.1: Manipulated Fire Command	HCA.1: Incorrect Fire	RM.2: Diverse Redundant Fire Control
LS.2: Situational Injection	HCA.2: No Fire	RM.1: Diverse Redundant Radio Relay
LS.3: Situational Delay	HCA.2: No Fire	RM.1: Diverse Redundant Radio Relay RM.3: Diverse Redundant IR Sensors RM.5: Operator Reposition
LS.4: Tampered Deployment	HCA.3: Unable to set Location	RM.4: Obstacle Restore
LS.5: Delayed Fire Command	HCA.2: No Fire	

Source: Horowitz et al. (2021).

Table 27.11 Silverfish CSRM remediations.

Remediation	Is implementation of: hygiene practice	Protects against: attack vector
REM.CH.PRO.1: Deployment Account	CPRAC.1: Eliminate Default Access CPP.AC.2: Physical or Procedural Access CPP.AC.3: Require Authentication CPP.AD.1: Minimize administrative privileges CPRUI.1: Unique Identifiers	SF.CAPEC.122: Privilege Abuse
REM.RES.DEF.1: Link encryption	CPP.BD.1: Control and protect information	LS.1: Manipulated Fire Command LS.2: Situational Injection RR.CAPEC.94: Radio Relay Man in the Middle RR.CAPEC.117: Radio Relay Interception
REM.RES.DEF.2: Voice only command and control		CC.CAPEC.607: Command and Control Jamming
REM.RES.DEF.3: Sentinel: Field – OBS: Measured Boot	CPP.CM.1: Manage configurations CPRCM.3: Constrain installation CPP.S1.1: Inventory software CPP.VU.1: Vulnerability detection	LS.4: Tampered Deployment OBS.CAPEC.439.CONFIG: Obstacle Configuration Modification during Distribution OBS.CAPEC.439.MALWARE: Obstacle Mal-ware during Distribution OBS.CAPEC.439.SW: Obstacle Software
REM. RES.DR.1:Sentinel: Vehicle – Weapon Mis-Fire		FC.CAPEC.438: Fire Control Modification during Manufacture LS.1: Manipulated Fire Command
REM.RES.DR.2:Sentinel: Vehicle – Weapon Delay Fire		FC.CAPEC.438: Fire Control Modification during Manufacture LS.5: Delayed Fire Command
REM.RES.DR.3:Sentinel: Field – Situational Delay		IR.CAPEC.438: IR Modification during Manu facture LS.3: Situational Delay
REM.RES.DR.4:Sentinel: Field – Situational Injection		LS.2: Situational Injection RR.CAPEC.594: Radio Relay Injection
REM.RES.HARD.1: Isolate fire control and situational awareness		H.1: Weapon mis-fire
REM.RES.HARD.2: Hardened fire control software		H.1:Weapon mis-fire LS.1:Manipulated Fire Command
REM.RES.HARD.3: Tamper resistant hardware	CPP.RM.2: Physical blocks	L.4: Loss of classified mission HW/SW OBS.CAPEC.507: Obstacle Theft

Source: Horowitz et al. (2021).

Table 27.12 Silverfish sentinel-based remediations.

Remediation	Protects against LS/AID	Elicits: requirement
REM.RES.DR.1:Sentinel: Vehicle – Weapon Mis-Fire	FC.CAPEC.438: Fire Control Modification during Manufacture LS.1: Manipulated Fire Command	MA.100.1.1: The vehicle Sentinel shall protect against manipulated fire commands
REM.RES.DR.2:Sentinel: Vehicle – Weapon Delay Fire	FC.CAPEC.438: Fire Control Modification during Manufacture LS.5: Delayed Fire Command	MA.100.1.2: The vehicle Sentinel shall protect against delayed fire
REM.RES.DR.3:Sentinel: Field – Situational Delay	IR.CAPEC.438: IR Modification during Manufacture LS.3: Situational Delay	MA.100.2.2: The field Sentinel shall protect against situational delay
REM.RES.DR.4:Sentinel: Field – Situational Injection	LS.2: Situational Injection RR.CAPEC.594: Radio Relay Injection	MA.100.2.1: The field Sentinel shall protect against situational injection

Source: Horowitz et al. (2021).

CSRM.5: Verification and Test

This section highlights aspects of system test relevant to Cyber Resiliency verification.

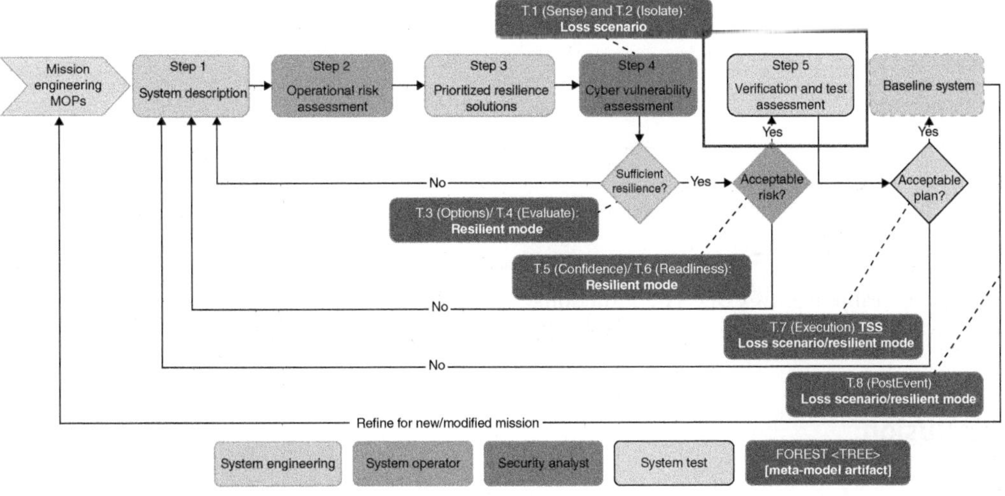

A top-level set of verification requirements are described in Table 27.13. Silverfish TE Cyber Resilience Verification Requirements. Each requirement drives verification of a Loss Scenario and associated Resilient Modes with specification of the appropriate Test Configuration and Test Activity to achieve the test (Figure 27.12).

A set of Test Configurations are described in the table below. The Components that form the configuration include relevant Resilient Modes and Sentinels. Additionally, the set of Test Activities that employ the configuration are listed (Table 27.14).

Table 27.13 Silverfish TE cyber resilience verification requirements.

Verification Requirement	Verifies: Requirement	Specifies: Test Activity / Configuration
TE.1: TE shall verify resilience metrics (parameters) of manipulated fire commands loss scenario and associated resilient modes	MA.100.1.1: The vehicle Sentinel shall protect against manipulated fire commands	TA. 1:TE: Cyber Resilience TC.1: Fire Control Subsystems
TE.2: TE shall verify resilience metrics (parameters) of situational injection loss scenario and associated resilient modes	MA.100.2.1: The field Sentinel shall protect against situational injection	TA.1:TE: Cyber Resilience TC.2: Situational Aware Subsystems
TE.3: TE shall verify resilience metrics (parameters) of situational delay loss scenario and associated resilient modes	MA.100.2.2: The field Sentinel shall protect against situational delay	TA.1:TE: Cyber Resilience TC. 2: Situational Aware Subsystems
TE.4: TE shall verify resilience metrics (parameters) of tampered deployment loss scenario and associated resilient modes	MA.100.2.3: The field Sentinel shall protect against tampered deployment	TA.1:TE: Cyber Resilience TC.3: Deployment Subsystems
TE.5: TE shall verify resilience metrics (parameters) of delayed fire loss scenario and associated resilient modes	MA.100.1.2: The vehicle Sentinel shall protect against delayed fire	TA.1:TE: Cyber Resilience TC.1: Fire Control Subsystems

A general Test Activity for Cyber Resilience Verification is described in Figure 27.13. Silverfish: TE: Cyber Resilience: EFFBD. The Functional Flow Block Diagram describes a generic loop for each Loss Scenario and each associated Resilient Mode. The steps of the loop include:

Conclusion

In conclusion, this use case highlights a practical and innovative application of model-based system assurance methodologies. The approach evaluates the quality of requirements and design solutions based on safety and security risks in the presence of a cyberattack. The methodology considers trade-offs between hazard/risk, cost, and threat adversary properties while also recognizing the interdependencies between cybersecurity and system safety. The traditional assurance processes in the cybersecurity domain are inadequate, and this approach calls for further development of new metrics, methods, and tools for hazard mitigation. In the context of digital engineering, finding the right approach to support systems engineering activities for end-to-end solutions

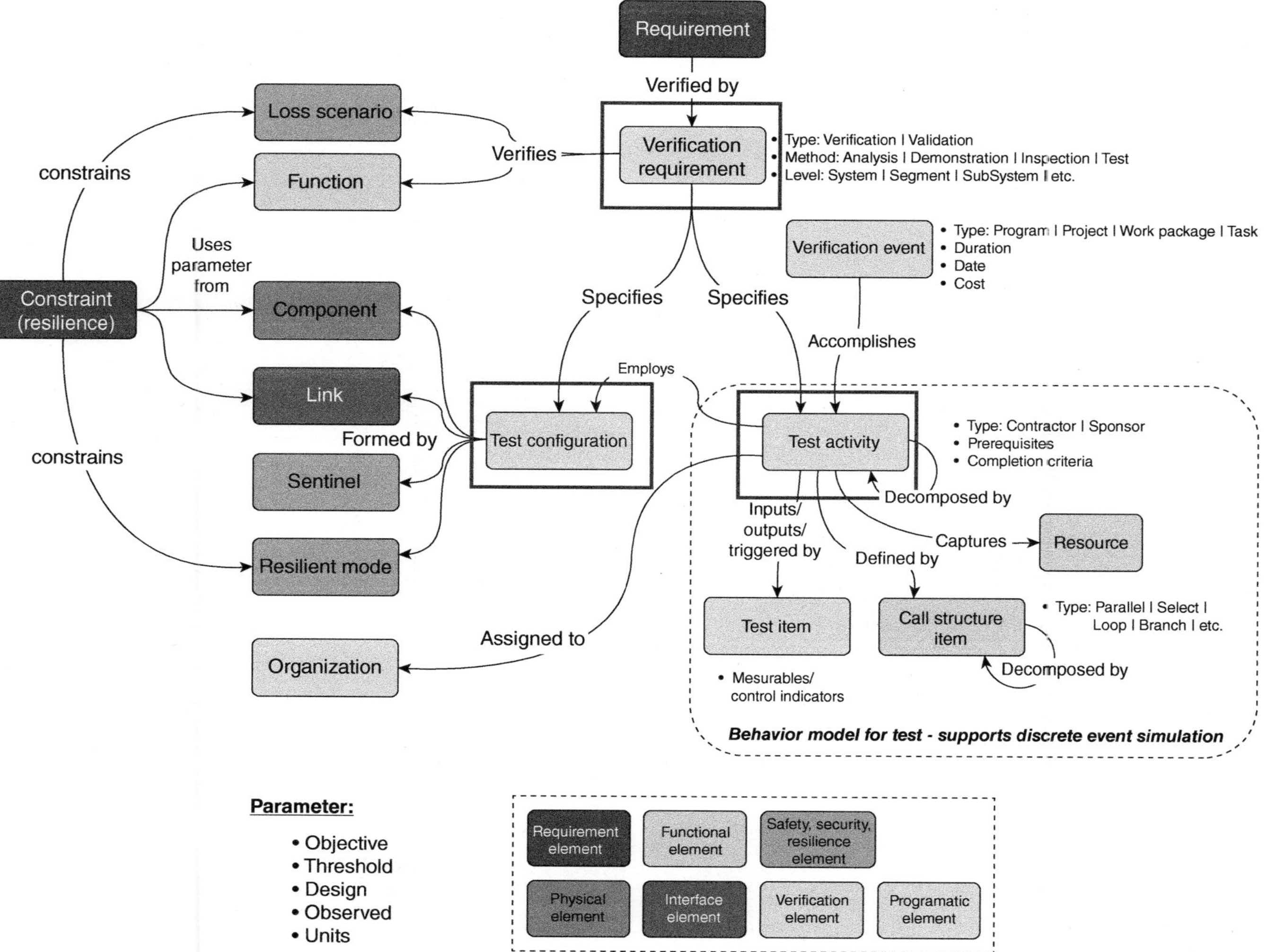

Figure 27.12 CSRM.5 Mission Aware Verification and Test Assessment. Source: Adapted from Horowitz et al. (2021).

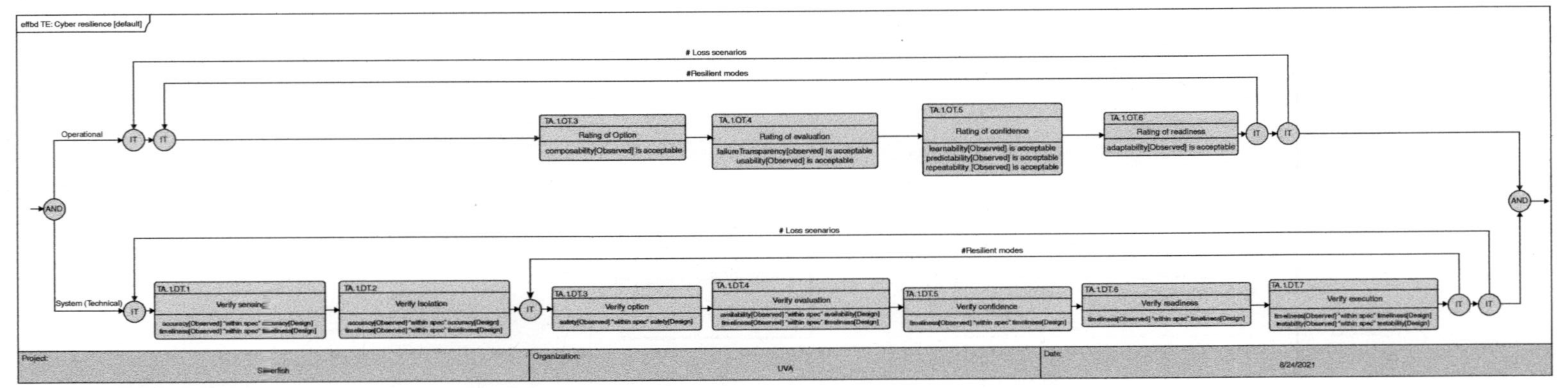

Figure 27.13 Silverfish: TE: Cyber Resilience: EFFBD.

Table 27.14 Silverfish test configurations.

Test configuration	Formed by: component/resilient mode	Employed by: test activity
TC.1: Fire Control Subsystems	C.3.1:CS: Fire Control Station C.3.5:08: Obstacle C.3.6:RR: Radio Relay C.4.1:SEN: Sentinel – Vehicle RM.2: Diverse Redundant Fire Control	TA.1:TE: Cyber Resilience
TC.2: Situational Aware Subsystems	C.3.2:CS: Situational Aware Station C.3.4:IR: Sensor C.3.5.1:08: Munition C.3.6:RR: Radio Relay C.4.2:SEN: Sentinel - Field C.10: UAV RM.1: Diverse Redundant Radio Relay RM.3: Diverse Redundant IR Sensors RM.5: Operator Reposition	TA.1:TE: Cyber Resilience
TC.3: Deployment Subsystems	C.3.5:OB: Obstacle C.3.6:RR: Radio Relay C.3.8:CS: Deployment Station C.4.2:SEN: Sentinel – Field RM.4:Obstacle Restore	TA.1:TE: Cyber Resilience

in the context of the mission is crucial. The use case emphasizes how cyber resilience can be designed in by considering associated requirements, engineering trade-offs, and quality attributes holistically. The use case follows the CSRM steps and FOREST methodology, demonstrating their effectiveness in enhancing cyber resilience.

In all, the work will continue to explore (1) system architectures for achieving resilience, (2) system methodologies, frameworks, and analysis tools for prioritizing resilience solutions, (3) the roles and procedures for engaging operators in the real-time management of system reconfigurations that provide resilience, and (4) designing in resilience through the engineering process.

References

Adams, S., Carter, B., Fleming, C., and Beling, P.A. (2018). Selecting system specific cybersecurity attack patterns using topic modeling. In: *2018 17th IEEE International Conference on Trust, Security and Privacy in Computing and Communications/12th IEEE International Conference on Big Data Science and Engineering (TrustCom/BigDataSE)*, New York, NY, USA (1–3 August 2018), 490–497. IEEE.

Beling, P., Horowitz, B., Fleming, C., et al. (2019). Model-Based Engineering for Functional Risk Assessment and Design of Cyber Resilient Systems. *University of Virginia, Charlottesville, United States, Technical Report.*

Berman, M., Adams, S., Sherburne, T. et al. (2019). Active learning to improve static analysis. In: *2019 18th IEEE International Conference on Machine Learning and Applications (ICMLA)*, Boca Raton, FL, USA (16–19 December 2019), 1322–1327. IEEE.

Bodeau, D., Graubart, R., and Laderman, E. (2019). Relationships Between Cyber Resiliency Constructs and Cyber Survivability Attributes. *MITRE. Technical Report.*

Carter, B., Adams, S., Bakirtzis, G. et al. (2019). A preliminary design-phase security methodology for cyber-physical systems. *Systems* 7 (2): 21.

Department of Defense (2014). Cyber security. http://www.dtic.mil/whs/directives/corres/pdf/850001_2014.pd (accessed 12 November 2022).

Department of Defense (2018). Manual for the operation of the joint capabilities integration and development system (JCIDS). https://www.dau.mil/cop/rqmt/DAU%20Sponsored%20Documents/Manual%20-%20JCIDS,%2031%20Aug%202018.pdf (accessed 13 November 2022).

Fleming, C.H., Elks, C., Bakirtzis, G. et al. (2021). Cyber-physical security through resiliency: a systems-centric approach. *Computer* 54 (6): 36–45.

Horowitz, B., Beling, P., Skadron, K., et al. (2014). Security Engineering Project-System Aware Cyber Security for an Autonomous Surveillance System on Board an Unmanned Aerial Vehicle. *Systems Engineering Research Center Hoboken NJ, Technical Report.*

Horowitz, B., Beling, P., Humphrey, M., and Gay, C. (2015a). System Aware Cybersecurity: A Multi-Sentinel Scheme to Protect a Weapons Research Lab. *Stevens Institute of Technology, Hoboken, NJ, Technical Report.*

Horowitz, B., Beling, P., Skadron, K., et al. (2015b). Security Engineering Project. *Systems Engineering Research Center Hoboken NJ, Technical Report.*

Horowitz, B., Beling, P., Fleming, C., et al. (2017). Security Engineering FY17 Systems Aware Cybersecurity. *Stevens Institute of Technology, Hoboken, United States, Technical Report.*

Horowitz, B., Beling, P., Fleming, C., et al. (2018a). Cyber Security Requirements Methodology. *Stevens Institute of Technology, Hoboken, United States, Technical Report.*

Horowitz, B., Beling, P., Fleming, C., et al. (2018b). Cyber-Security Requirements Methodology. *Systems Engineering Research Center, Technical Report.*

Horowitz, B., Beling, P., Clifford, M., and Sherburne, T. (2021). Developmental Test and Evaluation (DTE&A) and Cyberattack Resilient Systems - Measures and Metrics Source Tables. *Systems Engineering Research Center, Technical Report.*

Leveson, N. G. and Thomas, J. P. (2018). STPA handbook. https://psas.scripts.mit.edu/home/get_file.php?name=STPA_handbook.pdf (accessed 13 November 2022).

Pitcher, S. (2018). New DOD approaches on the cyber survivability of weapon systems. https://www.itea.org/wp-content/uploads/2019/03/Pitcher-Steve.pdf (accessed 25 March 2019)

Scott, Z. and Long, D. (2018). One Model, Many Interests, Many Views. *Vitech Corporation. Technical Report.* http://www.vitechcorp.com/resources/white_papers/onemodel.pdf (accessed 20 June 2023).

The MITRE Corporation. (2007). Common attack pattern enumeration and classification (CAPEC). https://capec.mitre.org/ (accessed 23 November 2022).

Young, W. and Leveson, N.G. (2013). Systems thinking for safety and security. In: *Proceedings of the Annual Computer Security Applications Conference (ACSAC 2013)*, New York, NY (1 December 2013), 357–366. ACM.

Young, W. and Leveson, N.G. (2014). An integrated approach to safety and security based on systems theory. *Communications of the ACM* 57 (2): 68–75.

Young, W. and Porada, R. (2017). System-theoretic process analysis for security (STPA-Sec): Cyber security and STPA. In *2017 STAMP Conference*, Cambridge, MA (27–30 March).

Biographical Sketches

Peter A. Beling is a professor in the Grado Department of Industrial and Systems Engineering and associate director of the Intelligent Systems Division in the Virginia Tech National Security Institute. Dr. Beling's research interests lie at the intersections of systems engineering and artificial intelligence (AI) and include AI adoption, reinforcement learning, transfer learning, and digital engineering. He has contributed extensively to the development of methodologies and tools in support of cyber resilience in military systems. He serves on the Research Council of the Systems Engineering Research Center (SERC), a University Affiliated Research Center for the Department of Defense.

Megan M. Clifford is a research associate and engineer at Stevens Institute of Technology. She works on various research projects with a specific interest in systems assurance and cyber-physical systems. She previously worked on the Systems Engineering Research Center (SERC) leadership team as the Chief of Staff and Program Operations, was the Director of Industry and Government Relations to the Center for Complex Systems and Enterprises (CCSE), and held several different positions in industry, including systems engineer at Mosto Technologies while working on the New York City steam distribution system.

Tim Sherburne is a research associate in the Intelligent System Division of the Virginia Tech National Security Institute. Sherburne was previously a member of the systems engineering staff at the University of Virginia supporting Mission Aware research through rapid prototyping of cyber resilient solutions and model-based systems engineering (MBSE) specifications. Prior to joining the University of Virginia, he worked at Motorola Solutions in various Software Development and Systems Engineering roles defining and building mission critical public safety communications systems.

Part VI

Analytic Methods for Design and Analysis of Missions and Systems-of-Systems

Dan DeLaurenits

Chapter 28

Unique Challenges in System of Systems Analysis, Architecting, and Engineering

Judith Dahmann[1] *and Dan DeLaurenits*[2]

[1] *The MITRE Corporation, MITRE Labs, McLean, VA, USA*
[2] *Purdue University, Aeronautics & Astronautics, West Lafayette, IN, USA*

What Are "Systems of Systems" and What Makes Them Different from Other Systems?

Simply put, systems of systems (SoS) are systems comprised of multiple constituent systems that are independent and useful on their own, but when combined into an SoS provide value beyond any one of the constituent systems. These constituent systems are typically independent from the SoS and the other constituent systems, with their own management, funding, stakeholders, requirements, and development processes.

The widely accepted characterization of SoS (Maier 1998) identifies two features that distinguish SoS:

- **Operational independence of constituent systems**
  In an SoS, constituent systems operate independently of the SoS and other systems. Most often these systems existed prior to the formation of the SoS and in many cases these systems are deployed and in use when called upon to support a new capability.
- **Managerial independence of constituent systems**
  The systems in an SoS are managed independently and their owner/managers may be evolving the systems to meet their own needs.

Maier's earlier work (and that of others) also put forth characteristics that, while not unique to SoS, are often present in them, such as:

- **Geographical distribution**
  In many cases, constituent systems in an SoS are geographically distributed, although many view this as a less significant or secondary characteristic of SoS.
- **Evolutionary development processes**
  SoS development is based on developments in the constituent systems. These developments may take place asynchronously based on the independent development processes of the constituent systems. This means that the SoS will evolve incrementally rather than be "delivered" as normally envisioned in a single system development or acquisition.

Systems Engineering for the Digital Age: Practitioner Perspectives, First Edition. Edited by Dinesh Verma.

Companion website: www.wiley.com/go/verma/systemsengineering

- **Emergent behavior**
  Emergence is described as

> Emergent system behaviour can be viewed as a consequence of the interactions and relationships between system elements rather than the behaviour of individual elements. It emerges from a combination of the behaviour and properties of the system elements and the systems structure or allowable interactions between the elements, and may be triggered or influenced by a stimulus from the systems environment.
>
> *(SEBoK Contributors 2021c)*

In many ways, emergence is the objective of an SoS where multiple systems are brought together to generate capability, which results from the interaction of the constituent systems. However, unanticipated, and undesirable emergent behaviour is a risk of SoS.

The differences between systems and SoS as they relate to systems engineering ("INCOSE" 2018) are shown in Table 28.1. As the table indicates, these differences are not cut and dry, especially as monolithic systems themselves have become more heterogeneous and hence complex.

Since the term SoS began to be used in the 1950s there has been considerable literature on definitions of SoS (Jamshidi 2005). However, by 2019, the community had evolved to the point that several SoS standards had been adopted ("INCOSE" 2020) by the International Standards Association along with a definition of SoS.

ISO/IEC/IEEE 21839 ("ISO" 2019) provides a definition of SoS and constituent system:

System of Systems (SoS): Set of systems or system elements that interact to provide a unique capability that none of the constituent systems can accomplish on its own. Note: Systems elements can be necessary to facilitate the interaction of the constituent systems in the system of systems.

Constituent systems: Constituent systems can be part of one or more SoS. Note: Each constituent is a useful system by itself, having its own development, management goals and resources, but interacts within the SoS to provide the unique capability of the SoS.

While many SoS exist and are used in everyday life, not many were recognized and designed as such. As a result, they develop and evolve without benefit of SE. When SoS are recognized and treated as an SoS, they can be categorized as one of four SoS types, based on the authority relationships between the

Table 28.1 Comparing SoS with systems.

Systems tend to …	Systems of systems tend to …
Have a clear set of stakeholders	Have multiple levels of stakeholders with mixed and possibly competing interests
Have clear objectives and purpose	Have multiple, and possibly contradictory, objectives and purpose
Have clear operational priorities, with escalation to resolve priorities	Have multiple, and sometimes different, operational priorities with no clear escalation routes
Have a single lifecycle	Have multiple lifecycles with elements being implemented asynchronously
Have clear ownership with the ability to move resources between elements	Have multiple owners making independent resourcing decisions

SoS and the constituent systems. This is not the only way to characterize SoS types but given the importance of the independence of the constituent systems in an SoS on SE, these types have been found to be useful in describing SoS. In reality, most SoS are fluid and operate under a combination of these types. Understanding the types provides a useful framework for understanding SoS.

Table 28.2 displays these types that were originally based on work done by Maier (1998) and expanded as part of the development of guidance for SoS systems engineering in the US DoD (Dahmann and Baldwin 2008).

Directed SoS are integrated SoS built and managed to fulfil specific purposes. Directed SoS are centrally managed and evolved. While the constituent systems in a direct SoS systems maintain the ability to operate independently, their normal mode of operations is subordinated to the central purpose of the SoS.

At the other end of the SoS spectrum are **Virtual SoS**. In this type of SoS, there is no a central management authority and no commonly agreed purpose for the SoS. Virtual SoS exhibit emergent behaviors that rely upon relativity invisible mechanisms to maintain the SoS. The best example of a Virtual SoS is the Internet.

In **Collaborative SoS**, the constituent systems interact voluntarily to fulfil agreed purposes and the systems themselves collectively decide how to interoperate, by enforcing and maintaining standards. There is no central authority, rather a Collaborative SoS is based on the agreements among the systems alone.

Finally, **Acknowledged SoS** essentially fall between directed and collaborative, with recognized objectives, a designated manager, and resources for the SoS. In parallel the constituent systems retain their independent ownership, management, and resources. This acknowledged case tends to be the most common in defense, with top level mission objectives balanced with the objectives of the owners of the systems that support the SoS.

Typically, Acknowledged SoS are not new developments. They usually arise to address a new capability need by leveraging available systems. As a result they tend to take the form of an overlay to an ensemble of existing systems with the objective of improving the way the systems work together to meet a new user need. Under these circumstances, the SoS manager, when designated,

Table 28.2 System of systems types.

Type	Definition
Directed	Directed SoS are those in which the SoS is engineered and managed to fulfill specific purposes. It is centrally managed during long-term operation to continue to fulfill those purposes as well as any new ones the system owners might wish to address. The component systems maintain an ability to operate independently, but their normal operational mode is subordinated to the centrally managed purpose
Acknowledged	Acknowledged SoS have recognized objectives, a designated manager, and resources for the SoS; however, the constituent systems retain their independent ownership, objectives, funding, development, and sustainment approaches. Changes in the systems are based on cooperative agreements between the SoS and the system
Collaborative	In collaborative SoS, the component systems interact more or less voluntarily to fulfill agreed-upon central purposes
Virtual	Virtual SoS lacks a central management authority and a centrally agreed-upon purpose for the system of systems. Large-scale behavior emerges – and may be desirable – but this type of SoS relies upon relatively invisible, self-organizing mechanisms to maintain it

Source: Defense Acquisition University (DAU) (2023).

typically does not control the constituent systems in the SoS and consequently is in a position of influencing rather than directing constituents to meet SoS need.

In sum, SoS can be categorized into four types based on the authority relationships between the SoS and the systems. These types – Directed, Acknowledged, Collaborative, and Virtual – are one way to conceptualize SoS. In an actual SoS, very often the SoS is comprised of elements that exhibit characteristics of the different types across the SoS. In some cases, the SoS owner may also have authority over some of the constituent systems, while others maintain their independence. Also, particularly in non-defense applications, the communications infrastructure supporting the information exchange may itself be an SoS, possibly a Virtual SoS (e.g., the internet). These complex "management" relationships and their implications are an important source of the uniques challenges SoS pose for analysis and engineering.

Why Should We Care? Because SoS Are Increasingly Ubiquitous

While the idea of SoS was seen as novel when first introduced, the situation today is such that we rarely have a system with is not part of a larger systems of systems. Increased networking and interconnectedness of systems today contributes to growth in the number and domains where SoS are becoming the norm, particularly with the considerable convergence among systems of systems, cyber-physical systems, and the internet of things (de C Henshaw 2016).

Application of SoSE is broad and is expanding into almost all walks of life. Originally identified in the defense environment, SoSE application is now much broader and still expanding. The early work in the defense sector has provided the initial basis for SoSE, including its intellectual foundation, technical approaches, and practical experience. In addition, parallel developments in information services and rail have helped to develop SoSE practice (Kemp and Daw 2014). Now, SoSE concepts and principles apply across other governmental, civil, and commercial domains.

As noted in the SoS knowledge area of the SE Body of Knowledge (SEBoK) (*SEBoK Contributors* 2021b), SoS proliferate across domains today as illustrated in Table 28.3, which has been adapted from SEBoK:

Beyond the domains, SoS can range in complexity and scope as shown in Figure 28.1.

On one end of the spectrum is an SoS focused almost entirely on technical integration as shown in Figure 28.1 with an example from an European Commission (EC) research project

Table 28.3 SoS across domains.

Transportation	Aviation and air traffic management, rail network, transports, highway management, and space systems
Energy	Smart grid, smart houses, and integrated production/consumption
Health care	Regional facilities management, emergency services, and telemedicine
Defense	Military missions such as integrated air/missile defense, multi domain battle
Natural resource management	Global environment, regional water resources, forestry, and recreational resources
Disaster response	Responses to disaster events including forest fires, hurricane, floods, and terrorist attacks
Consumer products	Integrated entertainment and household product integration
Business	Banking and finance, supply chain
Media	Film, radio, and television

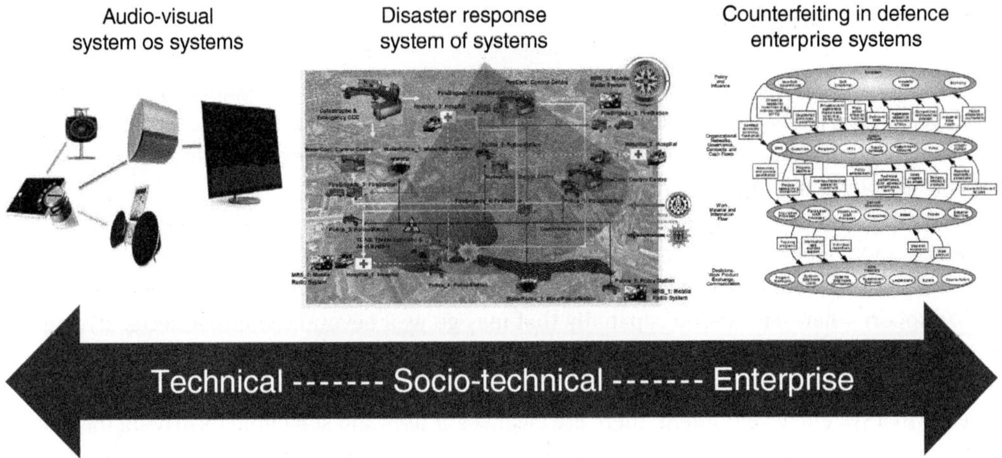

Figure 28.1 Scale and scope of SoS.

COMPASS (Fitzgerald 2013), which used as a case study in SoS the integration of the components of a consumer audio-visual "system." Here the focus was on maintaining the quality of the user experience across various combinations of constituent system combinations with a clear focus on the technical integration of the constituent systems. Components of the audio-visual SoS – the constituents – are developed independently from the SoS posing the challenge for the SoS of enabling SoS behavior – quality of the set of components – without control over the constituents.

Other examples of systems of systems move into the sociotechnical domain where not only are the systems integral to the SoS capability but also the organizations and their processes. This is shown in the figure by the example of a disaster response SoS, another SoS case example developed by the DANSE project (Engel and Browning 2008), another EC SoS research activity. In disaster response, the coordinated activities of the various responder organizations are as important to the success of the SoS as is the integration of the supporting technical systems. Again, the lack of control over the behavior of constituents, given their different objectives and dynamic responses to the changing environment, poses challenges for the SoS.

Finally, SoS can be broader in their scale and scope and address enterprise level concerns. In the figure, this is shown by work on fighting the problem of counterfeiting as an enterprise (Bodner 2014) comprised of a wide variety of systems, organizations, policies, and competing efforts. In this "enterprise" SoS, the expanded breath of the SoS to include development, management, and operation of defense systems and the legal, political, and economic factors affecting the independent organization elements of the SoS dominate the complexity beyond the technical issues.

So, in answer to the question "why do we care about SoS?" – the prevalence of SoS today means that whether our concern is analysis and engineering of technical systems, socio-technical systems or enterprises, SoS dominate most areas, making consideration of SoS critical.

What Are the Implications of SoS Characteristics on SoS Analysis and Architecting?

So, given the distinguishing characteristics of SoS, what make SoS analysis and an engineering hard?

Management and Oversight

Most SoS are comprised of systems that have their own users, funding, management structures, and development plans. This means that SoS development must accommodate the constraints of these independent systems in managing changes to meet SoS objectives. The resulting political and cost considerations all impact technical engineering activities.

In a typical system development, the engineering focus is on the physical system design and implementation. However, as discussed above in SoS scale and scope, the SoS systems engineer is often integrating not only the technical systems but also the organizational operations across the SoS to support a new end-to-end capability that may go well beyond the initial needs driving the development and engineering of the constituent systems. Challenges for modeling and analysis due to this situation are also stressing.

Ideally in a system development, there is a clear set of user and stakeholders driving the system requirements. In an SoS, however, there are multiple levels of users and stakeholders. Stakeholders for the constituent systems have their own perspectives and needs, which may not align with those of the SoS.

Finally, in a system development, typically there is a program manager and organic systems engineering with aligned funding and management responsibility for the system. In an SoS, each constituent system has its own manager, systems engineer and resources, and in all but the directed case these are managerially independent of the SoS.

As discussed in the US DoD Guide to SoS SE:

"SoS governance is complex. It includes the set of institutions, structures of authority, and the collaboration needed to allocate resources and coordinate or control activity. Effective SoS governance is critical to the integration of efforts across multiple independent programs and systems in an SoS. While the SoS will have a manager and resources devoted to the SoS objectives, the systems in the SoS typically also have their own PMs, sponsors, funding, systems engineers, and independent development programs. Some systems may be legacy systems with no active development underway. In addition, some systems will participate in multiple SoS. Consequently, the governance of the SoS SE process will necessarily take on a collaborative nature" ("DoD" 2018). These considerations are not unique to defense, as Keating (2014), Gorod et al. (2008), and others have written.

Operational Focus and Goals

In system development, the objective of the system will drive the systems requirements. In SoS, the objectives of the SoS may not align with those of the constituent systems. These systems very often continue to support their original mission functions as well as new SoS mission functions. These multiple mission objectives can lead to issues of competing management and technical authority

Again, as described in the US DoD SoS SE guide:

> For a single system within an operational environment, the mission objectives are established based on a structured requirements or capability development process along with defined concepts of operation and priorities for development. ... There is a strong emphasis on maintaining a specific, well-defined operational focus and deferring changes until completion of an increment of delivery. SE inherits these qualities in an individual system development. On the other hand, SoS SE is conducted to create operational capability

beyond that which the systems can provide independently. This may make new demands on the systems for functionality or information sharing which had not been considered in their individual designs. In some cases, these new demands may not be commensurate with the original objectives of the individual systems.

Implementation

Systems engineering is typically implemented as part of an established systems development and acquisition process with clear decision points and a well-established set of engineering activities and reviews aligned with those decision points. In contrast, in an SoS, systems engineering addresses the set of systems which contribute to SoS capability objectives, including legacy systems, new systems or systems still in development. The challenge is to evolve the SoS capabilities by leveraging the asynchronous developments of the constituent systems.

Again, an extract from the US DoD SoS SE Guide provides a good description of this:

> Typically, SoS involve multiple systems that may be at different stages of development, including sustainment. SoS may comprise legacy systems, developmental systems in acquisition programs, technology insertion, life extension programs, and systems related to other initiatives. The SoS manager and systems engineer need to accept the challenge to expand or redefine existing SE processes to accommodate the unique considerations of individual systems to address the overall SoS needs. It is the role of the SoS systems engineer to instill technical discipline in this process. The development or evolution of SoS capability generally will not be driven solely by a single organization but will most likely involve multiple ... program managers ... and operational and support communities. This complicates the task of the SoS systems engineer who must navigate the evolving plans and development priorities of the SoS constituent systems, along with their asynchronous development schedules, to plan and orchestrate evolution of the SoS toward SoS objectives. Beyond these development challenges, depending on the complexity and distribution of the constituent systems, it may be infeasible or very difficult to completely test and evaluate SoS capabilities.

Engineering and Design Considerations

Engineering of individual systems focus on establishing system boundaries, defining interfaces, developing approaches to ensure system performance and behavior, using establish metrics to assess system development progress and performance.

> In an SoS, things are often more complex. Again, the discussion in the DoD SoS SE guide describes this well:
>
> In an SoS, it is important to identify the critical set of systems that affect the SoS capability objectives and understand their interrelationships. It can be difficult to establish the boundaries of an SoS since the constituent systems of the SoS typically will have different owners and supporting organizational structures beyond the SoS management.
>
> Further, an SoS can place demands on constituent systems that are not supported by those systems' designs. Combinations of systems operating together within the SoS contribute to the overall capabilities. Combining systems may lead to emergent behaviors more than is

usually seen in single systems. As with emergent behaviors of single systems, these behaviors may either improve performance or degrade it.

In addition, beyond the ability of the systems to support the functionality and performance called for by the SoS, there can be differences among the systems in characteristics that contribute to SoS "suitability" such as reliability, supportability, maintainability, assurance, and safety. ... The challenge of design in an SoS is to leverage the functional and performance capabilities of the constituent systems to achieve the desired SoS capability as well as the crosscutting characteristics of the SoS to ensure the meets the broader user needs.

What Does This Mean in Terms of Persistent SoS Challenges?

Given the characteristics of SoS, the types of SoS and the differences in systems engineering of systems and SoS, what then are the challenges systems engineers face when it comes to applying SE to SoS? Drawing on work done under the auspices of the Internal Council on Systems Engineering (INCOSE) SoS working group (SoSWG), this section describes what have been termed "SoS Pain Points," areas where systems engineers face challenge when applying SE to SoS.

One of the initial activities of the INCOSE SoS Working Group was to understand the issues of importance or "pain points" in SoS as the basis for planning working group initiatives. A "SoS Pain Point" survey was constructed asking respondents to identify their priority SoS areas of concern. The results of the survey were reviewed and sorted into major challenge areas. The key areas and issues were summarized in a white paper which was presented to the SoSWG in June 2012 for review and comment. Based on initial feedback, short descriptions of the key pain points were drafted and posted for additional feedback. The pain points were subsequently updated and circulated for discussion at the INCOSE International Symposium in June 2013, and the final results were presented in a paper to the INCOSE International Symposium in July 2014.

The results identified seven areas of particular concern to systems engineers when applying SE to SoS. These are depicted in Figure 28.2 and summarized in the INCOSE Handbook (2015):

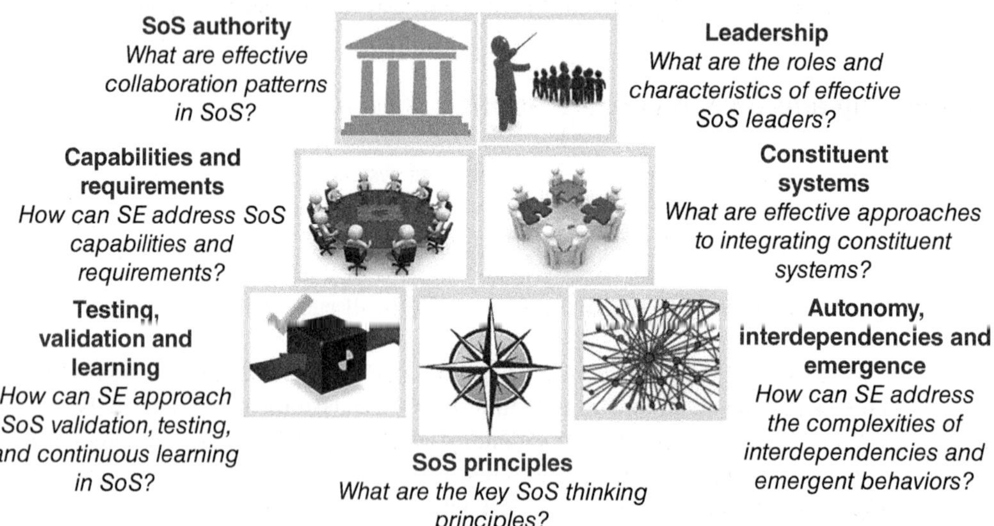

SoS authority
What are effective collaboration patterns in SoS?
Leadership
What are the roles and characteristics of effective SoS leaders?
Capabilities and requirements
How can SE address SoS capabilities and requirements?
Constituent systems
What are effective approaches to integrating constituent systems?
Testing, validation and learning
How can SE approach SoS validation, testing, and continuous learning in SoS?
Autonomy, interdependencies and emergence
How can SE address the complexities of interdependencies and emergent behaviors?
SoS principles
What are the key SoS thinking principles?

Figure 28.2 SoS pain points. *Source:* Adapted from Dahmann (2015).

SoS authorities: In an SoS, each constituent system has its own local "owner" with its stakeholders, users, business processes and development approach. As a result, the type of organizational structure assumed for most traditional systems engineering under a single authority responsible for the entire system is absent from most SoS. In a SoS, SE relies on cross-cutting analysis and on composition and integration of constituent systems which, in turn, depend on an agreed common purpose and motivation for these systems to work together toward collective objectives that may or may not coincide with those of the individual constituent systems.

Leadership: Recognizing that the lack of common authorities and funding pose challenges for SoS, a related issue is the challenge of leadership in the multiple organizational environments of an SoS. This question of leadership is experienced where a lack of structured control normally present in SE of systems requires alternatives to provide coherence and direction, such as influence and incentives.

Constituent systems' perspectives: Systems of systems are typically comprised, at least in part, of in-service systems, which were often developed for other purposes and are now being leveraged to meet a new or different application with new objectives. This is the basis for a major issue facing SoS SE; that is, how to technically address issues that arise from the fact that the systems identified for the SoS may be limited in the degree to which they can support the SoS. These limitations may affect the initial efforts at incorporating a system into an SoS, and systems "commitments" to other users may mean that they may not be compatible with the SoS over time. Further, because the systems were developed and operate in different situations, there is a risk that there could be a mismatch in understanding the services or data provided by one system to the SoS if the particular system's context differs from that of the SoS.

Capabilities and requirements: Traditionally (and ideally), the SE process begins with a clear, complete set of user requirements and provides a disciplined approach to develop a system to meet these requirements. Typically, SoS are comprised of multiple independent systems with their own requirements, working towards broader capability objectives. In the best case, the SoS capability needs are met by the constituent systems as they meet their own local requirements. However, in many cases, the SoS needs may not be consistent with the requirements for the constituent systems. In these cases, the SoS SE needs to identify alternative approaches to meeting those needs through changes to the constituent systems or additions of other systems to the SoS. In effect this is asking the systems to take on new requirements with the SoS acting as the "user."

Autonomy, interdependencies, and emergence: The independence of constituent systems in an SoS is the source of a number of technical issues facing SE of SoS. The fact that a constituent system may continue to change independently of the SoS, along with interdependencies between that constituent system and other constituent systems, add to the complexity of the SoS and further challenges SE at the SoS level. In particular these dynamics can lead to unanticipated effects at the SoS level leading to unexpected or unpredictable behavior in an SoS even if the behavior of constituent systems is well understood.

Testing, validation, and learning: The fact that SoS are typically composed of constituent systems which are independent of the SoS poses challenges in conducting end-to-end SoS testing as is typically done with systems. Firstly, unless there is a clear understanding of the SoS-level expectations and measures of these expectations, it can be very difficult to assess level of performance as the basis for determining areas that need attention or to assure users of the capabilities and limitations of the SoS. Even when there is a clear understanding of SoS objectives and metrics, testing in a traditional sense can be difficult. Depending on the SoS context, there may not be funding or authority for SoS testing. Often the development cycles of the constituent systems are tied to the needs of their owners and original ongoing user base. With multiple constituent systems subject to asynchronous development cycles, finding ways to conduct traditional end-to-end testing across the SoS can be difficult if not impossible. In addition, many SoS are large and diverse making traditional full end-to-end testing with every change in a constituent system prohibitively costly. Often the only way to get a good measure of SoS performance is from data collected from actual operations or through estimates based on modeling, simulation, and analysis. Nonetheless the SoS SE team needs to enable continuity of operation and performance of the SoS despite these challenges.

SoS principles: SoS is a relatively new area, with the result that there has been limited attention given to ways to extend systems thinking to the issues particular to SoS. Work is needed to identify and articulate the cross-cutting principles that apply to SoS in general, and to developing working examples of the application of these principles. There is a major learning curve for the average systems engineer moving to an SoS environment, and a problem with SoS knowledge transfer within or across organizations.

How Are SoS Challenged by Complexity?

Finally, understanding how the characteristics of SoS make this class of systems particularly susceptible to complexity provides another perspective on the unique challenges SoS pose for analysis and engineering.

Based on the SEBOK: "complexity is a measure of how difficult it is to understand how a system will behave or to predict the consequences of changing it. It occurs when there is no simple relationship between what an individual element does and what the system as a whole will do, and when the system includes some element of adaptation or problem solving to achieve its goals in different situations. It can be affected by objective attributes of a system such as by the number, types of and diversity of system elements and relationships, or by the subjective perceptions of system observers due to their experience, knowledge, training, or other sociopolitical considerations" (*SEBoK Contributors* 2021a).

Complexity in SoS is included in the Maier characteristic of SoS of emergence (Maier 1998). It is the core of the SoS pain point on autonomy, interdependencies, and emergence (Dahmann 2015). It is inherent in the management and oversight implications of systems. Understanding how SoS are particularly susceptible to complexity provides insight into SoS challenges. Using work on complexity by INCOSE (Watson et al. 2019) as the frame of reference, the following section reviews key dimensions of complexity and how these apply to SoS.

Diversity: The structural, behavior, and system state varieties that characterize a system and/or its environments.

How SoS exhibit diversity: SoS can exhibit tremendous diversity across the various constituent systems, which provide a range of different behaviors, functionality, and technical approaches.

Why? By definition, SoS are comprised of multiple independent systems with their own users, management structures, requirements, etc. often developed prior to their membership in an SoS, increasing the likelihood that there will be differences among the constituents of an SoS. ***Challenge:*** *SoS Authority; Management/Governance.*

Connectivity: The connection of the system between its functions and the environment. This connectivity is characterized by the number of nodes, diversity of node types, number of links, and diversity in link characteristics. Complex systems have multiple layers of connections within the system structure.

How SoS exhibit connectivity: SoS include connectivity within each constituent system, among constituents in the SoS and between the SoS and its environment. SoS are comprised of "connected" constituent systems, so in addition to the connectivity within each constituent, an SoS by its nature is characterized by additional connectivity among constituents.

Why? SoS typically have large numbers of nodes, a diversity of node types, a large number of links, and diversity in link characteristics, as well as multiple layers of connections within the system structure. Discontinuities (breaks in a pattern of connectivity at one or more layers) are often found in SoS. ***Challenge:*** *SoS Autonomy, Interdependence, and Emergence.*

Interactivity: The behavior stimulus and response between different parts of a system and the system with its environment. Complex systems have many diverse sources of stimulus and diverse types of responses. The correlation between stimulus and response can be both direct and indirect (perhaps separated by many layers of system connectivity). The types of stimuli and responses vary greatly. The levels of stimuli and responses can range from very subtle to very pronounced. The timeframe for system responses can vary hugely.

How SoS exhibit interactivity: In an SoS, each constituent system interacts with its users, and each constituent system in the SoS interacts with one or more other constituent systems.

Why? An SoS is characterized by interaction among multiple independent systems (i.e., the constituent systems), which produces the emergent capability which is not present in the constituent systems alone. SoS have many diverse sources of stimulus and diverse types of responses. The correlation between stimulus and response can be both direct and indirect and may be spread across multiple systems in the SoS. ***Challenge:*** *SoS Authority; SoS Autonomy, Interdependence, Emergence.*

Adaptability: Complex systems proactively and/or reactively change function, relationships, and behavior to balance changes in environment and application to achieve system goals.

How SoS exhibit adaptability: To the degree users are considered part of an SoS, with multiple users across the SoS, each user may respond to changes in their system, which can occur as a result of interaction with other systems in the SoS.

Why: SoS are composed of operationally independent constituent systems; hence the operators of each of these systems may have their own rules of engagement and may react or adapt to changes in different ways based on their local objectives. In addition, systems that are autonomically adaptive may react differently from other systems in the SoS. ***Challenge:*** *SoS Management and Governance; SoS Autonomy, Interdependence, Emergence.*

Multiscale: Behavior, Relationships, and Structure exist on many scales, are ambiguously coupled across multiple scales, and are not reducible to only one level.

How SoS exhibit multiscale: SoS may contain constituent systems of varying scales to some extent, but only over a small number of orders of magnitude.

Why? Because SoS are composed of existing systems, these may be very different and at different scales. ***Challenge:*** *Constituent System Diversity; SoS Autonomy, Interdependence, Emergence.*

Multi-perspective: Multiple perspectives, some of which are orthogonal, are required to comprehend the complex system.

How SoS exhibit multi-perspective: Each constituent system brings its own perspective to the SoS and the SoS environment.

Why? SoS are typically comprised of multiple independent systems that were developed and operated prior to the existence of the SoS; hence they each bring with them their own perspectives, which may or may not be aligned with the SoS. A complete understanding of the SoS requires understanding all of these different perspectives. ***Challenge:*** *Constituent System Diversity; SoS Autonomy, Interdependence, Emergence.*

Behavior (not describable as a response system): Complex system behavior cannot be described fully as a response system. Complex system behavior includes nonlinearities. Optimizing system behavior cannot often be done focusing on properties solely within the system.

How SoS exhibit behavior: While the behaviors of the individual systems may be predictable, particularly when the numbers of systems are large and they have multiple internal behaviors, the SoS behavior can become unpredictable.

Why? Each constituent system has been designed to be operated independently and safely within its own context, without regard to the potential impact on the behavior of itself, on other systems and on overall SoS behavior. ***Challenge:*** *SoS Autonomy, Interdependence, Emergence.*

Dynamics: Complex systems may have equilibrium states or may have no equilibrium state. Complex system dynamics have multiple scales or loops. Complex systems can stay within the dynamical system or generate new system states or state transitions due to internal system changes, external environment changes, or both. Correlation of changes in complex systems to events or conditions in the system dynamics may be ambiguous.

How SoS exhibit dynamics: Following the discussion of behavior, these effects can be dynamic and impact other systems or have feedback loops leading to dynamic complexity.

Why? Notably constituent systems may not only be managed independently, they may operate independently, increasing the prospects of dynamic complexity. Indeed, the presence of a constituent system is uncertain in an SoS. ***Challenge:*** *Constituent System Diversity; SoS Autonomy, Interdependence, Emergence.*

Representation: Representations of complex systems can be difficult to properly construct with any depth. It is often impossible to predict future configurations, structures, or behaviors of a complex system, given finite resources. Causal and influence networks create a challenge in developing "requisite" conceptual models within these time and information resource constraints.

How SoS exhibit representation: One feature of SoS is that boundary conditions can be hard to define, which includes not only which constituent systems should be included in representation of an SoS but also which behaviors of a constituent system play a role in the SoS making representation of an SoS a challenge.

Why? SoS can best be defined as the set of systems which contribute to a capability, noting that it is often not until you observe an SoS in operation that you may be able to identify the contributors, and as constituents adapt these may change. ***Challenge:*** *Constituent System Diversity; SoS Autonomy, Interdependence, Emergence.*

Evolution: Changes over time in complex system states and structures (physical and behavioral) can result from various causes. Complex system states and structures are likely to change as a result of interactions within the complex system, with the environment, or in application. A complex system can have disequilibrium (i.e., non-steady) states and continue to function.

Complex system states and structures can change in an unplanned manner and can be difficult to discern as they occur. The changes in the states and structure of a complex system are a natural function of (is often present in) the complex system dynamics. Changes can occur without centralized control, due to localized responses to external and/or internal influences.

How SoS exhibit evolution: Rarely do we ever "develop and field" an SoS, rather SoS are typically composed of systems that existed before the SoS; changes in an SoS result from changes in one or more of the constituent systems (or in the SoS environment), making SoS development an evolutionary process.

Why? One of the Maier characteristics of SoS is evolutionary development – since constituent systems in an SoS are independent, changes are typically made in the constituent systems asynchronously, leading to incremental or evolutionary development. ***Challenge:*** *SoS Autonomy, Interdependence, Emergence.*

System emergence not predictable behavior: Unexpected Emergence (Complex) – Emergent properties of the holistic system unexpected (whether predictable or unpredictable) in the system functionality/response. Unpredictable given finite resources. Behavior not describable as a response system.

How SoS exhibit emergence: Because each system may operate independently, interactions among systems may lead to unexpected effects.

Why? By definition, SoS are comprised of multiple independent systems, which means that changes in operation of one system could lead to new behavior in another, leading to unpredictable results. ***Challenge:*** *SoS Autonomy, Interdependence, Emergence.*

Disproportionate effects: Details seen at the fine scales can influence large-scale behavior. Small-scale modifications can result in radical changes of behavior. Scale can be in terms of magnitude of effect or aggregate amount of change. Weak ties can have disproportionate effects.

How SoS exhibit disproportionate effects: A risk in SoS is that the change in one constituent system may have cascading effects on others and the SoS as a whole – a good example is large area power outages resulting from a squirrel chewing through a single power line.

Why? Again, the independence of the constituent systems means that the full range of interactions and results may not be understood. ***Challenge:*** *SoS Autonomy, Interdependence, Emergence.*

Indeterminate boundaries: Complex system boundaries are intricately woven with their environment and other interacting systems. Their boundaries can be nondeterministic. The boundary cannot be distinguished based solely on processes inside the system.

How SoS exhibit indeterminate boundaries: One feature of SoS is that boundary conditions can be hard to define, which includes not only which constituent systems should be included in representation of an SoS but also which behaviors of a constituent system play a role in the SoS making representation of an SoS a challenge.

Why? SoS are the set of systems which contribute to achieving a desired capability. Particularly in large socio technical SoS, it may be difficult to determine the factors that are contributors, and these can change over time. ***Challenge:*** *Constituent System Diversity; SoS Autonomy, Interdependence, Emergence.*

Contextual influences: All systems reside in natural and social environments and relate to these. In the relationship between the system and the natural and social environments, there can be complexity. This complex interaction depends on the social application of the system. Social systems often strive to achieve multiple, sometimes incompatible, objectives with the application of the same system.

How SoS exhibit contextual influences: The wider external environment can have an impact on both the constituent systems and their interactions and consequently on the overall SoS capability.
Why? The constituent systems themselves interact with the environment independently and these interactions can affect both the constituent behavior and the way they affect other constituents and as a result the SoS outcomes. ***Challenge:*** *Constituent System Diversity; SoS Autonomy, Interdependencies, Emergence.*

Mission Engineering Illustrates SoS Challenges

Recently, there has been increased interest in the explicit application of SoS engineering to achieve mission objectives, referred to a "Mission Engineering." This recent interest is based on US Defense recognition that while engineering effective defense systems is critical, it is equally important that defense systems work together effectively as an SoS and, importantly, that when deployed in an operational environment effectively achieve mission outcomes.

Legislation ("NDAA" 2017) called for defense Mission Integration Management (MIM). Mission (*DoD Joint Chiefs of Staff* 2023) is defined as the "task, together with the purpose, that clearly indicates the action to be taken and the reason thereby. More simply, a mission is a duty assigned to an individual or unit." MIM is the "synchronization, management, and coordination of concepts, activities, technologies, requirements, programs, and budget plans to guide key decisions focused on the end-to-end mission" (DoD 2020). Mission engineering is the "technical sub-element of MIM as a means to provide engineered mission-based outputs to the requirements process, guide prototypes, provide design options, and inform investment decisions." (DoD 2020) Making mission engineering "the deliberate planning, analyzing, organizing, and integrating of current and emerging operational and system capabilities to achieve desired operational mission effects" (Department of Defense 2013).

As shown in Figure 28.3, mission engineering treats the end-to-end mission as the "system."

Individual systems, including organizations and other non-material elements, are components of the larger mission "system" or system of systems. System engineering is applied to the systems-of-systems supporting operational mission outcomes, through the use of "mission threads" or the set of key actions required to successfully execute the mission. Mission engineering goes beyond data exchange among systems to address cross cutting functions, end-to-end control and trades across systems. Technical trades exist at multiple levels not just within individual systems or components. The composition of an SoS to support a mission will be tailored to the specific scenario and mission context – in effect SoS are composed to meet specific mission needs adding another dimension of complexity to SoS. Well-engineered composable mission architectures foster resilience, adaptability, and rapid insertion of new technologies.

The US Defense Department has developed guidance on mission engineering (DoD 2020) and there is a growing body of mission engineering practice (*SEBoK Contributors* 2021d) in mission engineering competencies (Hutchison et al. 2018), education (Van Bossuyt et al. 2019), and methods (Beam 2015; Beery and Paulo 2019; Dahmann 2019). While mission engineering has focused on Defense missions, the practice of mission engineer can apply to not defense areas as well – some of the examples in the previous section – disaster response, for example, provide opportunities for broader application of mission engineering.

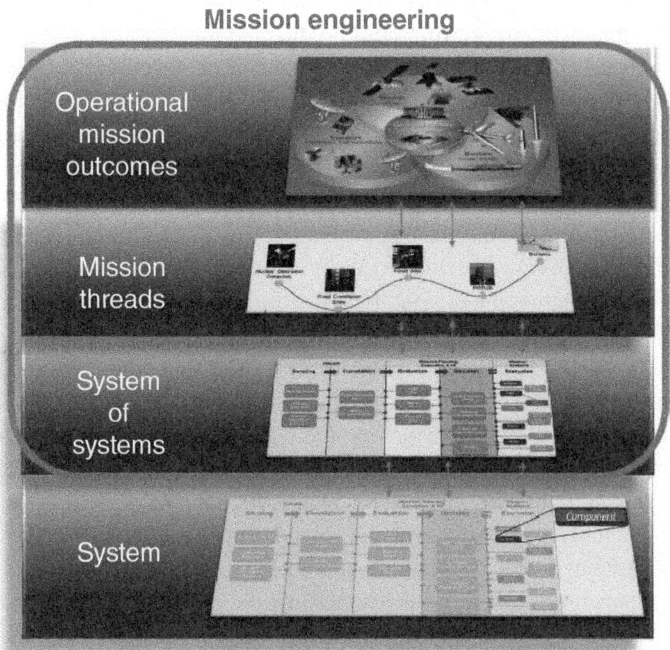

Figure 28.3 Mission engineering links systems of systems with mission outcomes.

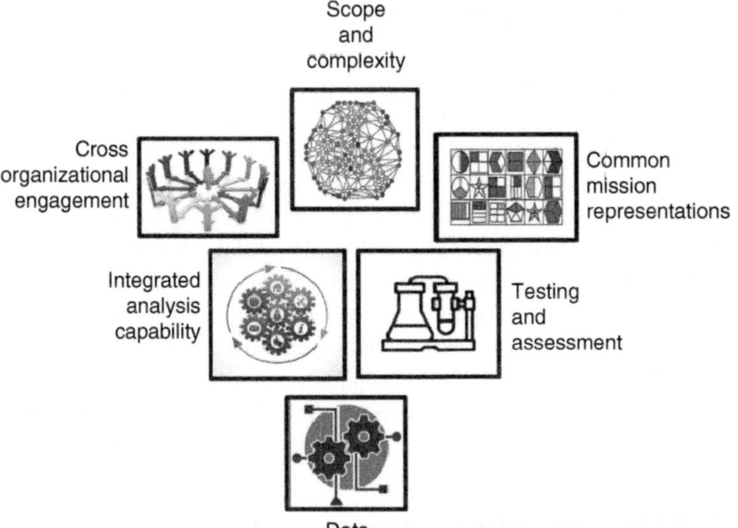

Figure 28.4 Mission engineering challenges. *Source:* Adapted from Zimmerman and Dahmann (2018).

As practice of ME grows, the experience illustrates many of the challenges that face SoS more generally. In particular, Zimmerman and Dahmann (2018) reviewed the challenges that face mission engineering, as shown in Figure 28.4.

Scope and complexity: Most missions in defense are broad in scope and complex. Missions span multiple systems, organizations, and scenarios. The execution of the ballistic missile defense mission, for example, crosses multiple organizations and a wide range of types of systems multiple system interdependencies. Stakeholders exist at both the system and mission levels with competing interests and priorities and no directed interest in mission engineering. Multiple stakeholders with their own interests, motivations, and perspectives, often participate with their own models and analysis tools at the system, components, and mission function multiplying the challenges facing SoS in general. To address the mission in a coherent way requires methods, processes, and tools, which can provide a shared view of key elements of the mission.

Cross organizational engagement: Missions depend on effective interaction among systems owned, developed, managed, and operated by different organizations. Each organization has its own systems engineering processes and tools to support the needs of its organization. To effectively, engineer across the mission requires the same type of effective interaction across engineering as is needed across systems in any SoS.

Integrated analysis capability: One thing that differentiates ME from SoSE is that with ME, you not only need to ensure that the SoS is well engineered (the domain of SoS), but with ME, the focus is on whether the SoS when implemented in the operational context actually achieves the mission outcomes. It has been said that SoSE ensures the SoS is built right, but ME ensure the SoS incorporates the right capabilities. This means that ME faces challenges of developing integrated analysis capabilities that bridge engineering and mission effects, facing the difficulty of linking changes in systems or SoS engineering models with impacts on missions in operational or mission simulations.

Common mission representations: Effective ME requires a common view of the mission – CONOPs, systems capabilities, threats – to provide a shared framework across the mission, which can be used as context for more detailed views of specific issues related some elements of the mission, Component mission perspectives, and the view of the systems, providing a perspective that addresses the concerns of the various stakeholders but in the larger common mission context.

Testing and assessment: One factor that leads to the complexity of ME is that, like in many SoS, the systems are independent, are at different stages of their lifecycles and their development cycles are geared toward their system users, which means the mission-level engineering has a limited ability to synchronize and validate impacts of system changes on the missions. This leads to the broader challenge of how to test capability across multiple system lifecycles: legacy systems, systems under development, emerging solutions, and technology insertion in a mission context.

Data: Data is critical to effective engineering at any level and common data shared across models and analyses is key to successful mission engineering. In the absence of ME, typically each organization invests considerable resources to develop data which is often not known or shared across a mission. The need for data on missions, systems, interfaces, interactions, and interdependencies faces the challenges that this data is typically very distributed, maintained in various forms by different organizations, it often focuses on specific system needs and do not address interdependencies and interactions. Even when available, the needed data can be hard to locate or access.

What Can We Conclude About the Unique Challenges...?

The unique challenges posed by SoS are many ... and this chapter could have been much longer! However, there has been much progress in recent years addressing SoS-unique challenges across a wide variety of application domains. The remaining chapters in this section provide a sampling of progress emanating from relevant SERC activities.

References

Beam, D.F. (2015). Systems Engineering and Integration as a Foundation for Mission Engineering. *Naval Postgraduate School.*

Beery, P. and Paulo, E. (2019). Application of model-based systems engineering concepts to support mission engineering. *Systems* 7 (3): 44.

Bodner, D. (2014). Enterprise modeling framework for counterfeit parts in defense systems. *Procedia Computer Science.* 36. https://doi.org/10.1016/j.procs.2014.09.016.

de C Henshaw, M.J. (2016). Physical systems, the internet-of-things. . . whatever next? *Insight* 19 (3): 51–54. https://doi.org/10.1002/inst.12109.

Dahmann, J. (2015). Systems of systems pain points. *INCOSE International Symposium*, Seattle, WA (16 July 2015).

Dahmann, J. (2019). Keynote Address: mission engineering: system of systems engineering in context. *Proceedings of the IEEE System of Systems Engineering Conference*, Anchorage, Alaska (19–20 May).

Dahmann, J.S. and Baldwin, K.J. (2008). Understanding the current state of US defense systems of systems and the implications for systems engineering. *2008 2nd Annual IEEE Systems Conference*, Montreal, Canada (7–10 April 2008).

Defense Acquisition University (DAU) (2023). Defense Acquisition Guidebooks. Acquisition Guidebooks | Adaptive Acquisition Framework (dau.edu) (accessed 21 August 2023).

Department of Defense. (2013). Defense acquisition guidebook. https://at.dod.mil/sites/default/files/documents/DefenseAcquisitionGuidebook.pdf (accessed 26 June 2023).

DoD (2018). *Systems Engineering Guide for Systems of Systems.* Arlington, VA: *U.S. Department of Defense, Director, Systems and Software Engineering Deputy Under Secretary of Defense (Acquisition and Technology) Office of the Under Secretary of Defense (Acquisition,* Technology and Logistics).

DoD (2020). Mission Engineering Guide. US Department of Defense. Mission Engineering Guide. November 2020.

DoD Joint Chiefs of Staff (2023). Joint Operations (3.0). https://Joint Chiefs of Staff > Doctrine > Joint Doctrine Pubs > 3-0 Operations Series (jcs.mil) (accessed 21 August 2023).

Engel, A. and Browning, T. (2008). 2.1.2 Designing systems for adaptability by means of architecture options. *Systems Engineering.* 11: 125–146. https://doi.org/10.1002/sys.20090.

Fitzgerald, J.S. "Comprehensive Modelling for Advanced Systems of Systems Model-based Engineering for Systems of Systems: the COMPASS Manifesto." (2013).

Gorod, A., Sauser, B., and Boardman, J. (2008). System-of-systems engineering management: a review of modern history and a path forward. *IEEE Systems Journal* 2 (4). https://doi.org/10.1109/JSYST.2008.2007163.

Hutchison, N. A., Luna, S., Miller, W. D., et al. (2018). Mission engineering competencies. *2018 ASEE Annual Conference & Exposition*, Salt Lake City, UT (24–27 June 2018). American Society for Engineering Education.

INCOSE SE Handbook. (2015).

INCOSE (2018). In Systems of Systems Primer. *INCOSE-TP-2018-003-01.0.*

INCOSE (2020). In Systems of Systems Standards Quick reference Guide. *INCOSE 2020.*

ISO. (2019). *International Organization for Standardization (ISO)/IEC/IEEE 21839 —Systems and Software Engineering—System of Systems Considerations in Life Cycle Stages of a System.*

Jamshidi, M. (2005). System-of-systems engineering – a definition. *Proceedings of the IEEE International Conference on Systems, Man and Cybernetics*, Waikoloa, Hawaii, USA (10–12 October 2005). IEEE.

Joint Publication (JP) 3-0. Doctrine for Unified and Joint Operations (Washington, DC: The Joint Staff, September 2001).

Keating, C. B. (2014). Governance implications for meeting challenges in the system of systems engineering field. *9th International Conference on System of Systems Engineering, SoSE 2014,* Glenelg, Australia (9–13 June 2014). https://doi.org/10.1109/SYSOSE.2014.6892480

Kemp D., Daw A. 2014, INCOSE UK Capability Systems Engineering Guide

Maier, M.W. (1998). Architecting principles for systems-of-systems. *Systems Engineering: The Journal of the International Council on Systems Engineering* 1 (4): 267–284.

NDAA. (2017). National Defense Authorization Act (NDAA) for Fiscal Year 2017, Section 855.

SEBoK Contributors (2021a). Complexity. *SEBoK*. https://sebokwiki.org/w/index.php?title=Complexity&oldid=67074 (accessed 21 August 2023).

SEBoK Contributors (2021b). Applications of Systems Engineering. SEBoK. https://sebokwiki.org/w/index.php?title=Applications_of_Systems_Engineering&oldid=67057 (accessed 21 August 2023).

SEBoK Contributors (2021c). Emergence. *SEBoK*. https://sebokwiki.org/w/index.php?title=Emergence&oldid=67188 (accessed 21 August 2023).

SEBoK Contributors (2021d). Mission Engineering. *SEBoK*. https://sebokwiki.org/w/index.php?title=Mission_Engineering&oldid=67286 (accessed 21 August 2023).

Van Bossuyt, D.L., Paulo, E., Beery, P. et al. (2019). The naval postgraduate school's department of systems engineering approach to mission engineering education through capstone projects. *Systems* 7 (3): 38. https://doi.org/10.3390/SYSTEMS7030038.

Watson, M., Anway, R., McKinney, D., et al. (2019). Appreciative methods applied to the assessment of complex systems. *INCOSE International Symposium*, Orlando, FL (20–25 July 2019). 29(1), 448–477.

Zimmerman, P., and Dahmann, J. (2018). Digital engineering support to mission engineering. *21st Annual National Defense Industrial Association Systems and Mission Engineering Conference*, Tampa, FL (22–25 October 2018).

Biographical Sketches

Dr. Judith Dahmann is a technical fellow at the MITRE Corporation and the Systems of Systems Engineering lead in the MITRE Systems Engineering Innovation Center. She is the MITRE project leader for Systems and Mission Engineering Technical Support activities in the US DOD Office of the Under Secretary of Defense for Research and Engineering supporting mission engineering activities for selected priority Defense missions and the application of digital engineering to mission engineering. She was the technical lead for development of the DoD guide for systems engineering of systems of systems (SoS) and was the project lead for International Standards Organization (ISO) 21839, the first ISO international standard on "SoS Considerations for Systems Throughout their Life Cycle." Dr. Dahmann is also part of the MITRE team supporting the Nuclear Command, Control, and Communications (NC3) Enterprise Center, providing leadership in the application of systems of systems engineering and digital engineering to NC3 modernization. Prior to this, Dr. Dahmann was the chief scientist for the Defense Modeling and Simulation Office for the US Director of Defense Research and Engineering (1995–2000) where she led the development of the High Level Architecture, a general-purpose distributed software architecture for simulations, now an IEEE Standard (IEEE 1516). Dr. Dahmann is a fellow of the International Council on Systems Engineering (INCOSE) and the cochair of the INCOSE Systems of Systems Working Group and cochair of the National Defense Industry Association SE Division SoS SE Committee.

Dr. Daniel DeLaurentis is vice president for Discovery Park District (DPD) Institutes and professor of aeronautics and astronautics at Purdue University. His charge in research is via the Center for Integrated Systems in Aerospace (CISA), which he directs, working with faculty colleagues and students research problem formulation, modeling, design and system engineering methods for aerospace systems and systems-of-systems. DeLaurentis also serves as chief scientist of the U.S. DoD's Systems Engineering Research Center (SERC) UARC, working to understand the systems engineering research needs of the defense community (primarily) and translate that to research programs that are then mapped to the nation's best researchers and students in the SERCs network of 25 universities. He is fellow of the International Council on Systems Engineering (INCOSE) and the American Institute of Aeronautics and Astronautics.

Chapter 29

System of Systems Analytic Workbench

Cesare Guariniello[1], Payuna Uday[2], Waterloo Tsutsui[1] and Karen Marais[1]

[1] *School of Aeronautics and Astronautics, Purdue University, West Lafayette, IN, USA*
[2] *Stevens Institute of Technology, Systems Engineering Research Center, Hoboken, NJ, USA*

Genesis and Description of the Analytic Workbench

Evolving and refining large groups of interdependently operating systems, or systems of systems (SoS), present significant technical, operational, and programmatic challenges. SoS generally involve integrating multiple independently managed systems to achieve a unique capability, therefore requiring collaboration and negotiation as well as control. In such complex systems, human behavioral and social phenomena in collaboration are critical to proper functionality, as are cascading impacts from interdependencies; altogether, emergent outcomes are the norm. Handling situations characterized by the presence of cascading impacts and emergent behavior goes well beyond the immediate mental faculties of decision-makers and even capabilities of existing system-level decision-support tools. The current cutting edge in analysis for SoS seeks a collection of methods, processes, and tools that provides SoS practitioners with meaningful quantitative insights into projected SoS behavior and the possibilities for evolving the SoS. Policies set forth in the last few years in acquisition guidance documents, emerging SoS standards, and informal guidance, such as the US Department of Defense (DoD) Systems Engineering Guide for Systems of Systems (US DoD 2008a) and Acquisition Guidebooks and References (US DoD 2008b), provide useful guidance but are in need of a supporting analytic perspective to facilitate more informed decision-making. Several research groups are working on advancements in this important area. Ongoing research focuses on "situational awareness" products for both SoS and constituent system-level decision-support as well as strategic approaches for modeling SoS architectures and their ability to restructure quickly to respond to failures, new needs, and new missions. One area of applicability of SoS modeling and analysis is civil and commercial air transportation, especially relevant with the emergence of "smart, connected" cyber-physics system networks, Internet-of-things, and more.

This chapter discusses a multidisciplinary effort undertaken by the SERC over the last decade to create an Analytic Workbench (AWB) of computational tools to facilitate better-informed decision-making about SoS architectures. The work is motivated by the idea that SoS practitioners typically have information and archetypal technical queries that can be mapped to appropriate analysis methods best suited to provide outputs and insights directly relevant to the posed questions. These archetypal, technically driven queries are mapped to relevant methods that can provide analytical outputs to directly support SoS acquisition and architectural decisions.

Systems Engineering for the Digital Age: Practitioner Perspectives, First Edition. Edited by Dinesh Verma.

Companion website: www.wiley.com/go/verma/systemsengineering

SERC and DoD Needs

Acknowledging the need for more research in the fields of Systems Engineering and Systems of Systems, the DoD initiated various projects with Purdue University, through the SERC, aimed at modeling and simulation of the impact of interdependencies between systems. The initial analysis of SoS structure, interdependencies, and evolution was based on the wave model (Figure 29.1, from Dahmann et al. 2011), the SoS equivalent of the V-model for systems engineering. Besides the existing guidelines by the DoD and the description of basic needs of the sponsor, we faced an open-ended problem, where we had to both lay the theoretical foundation for this type of SoS analysis of interdependencies as well as demonstrate the usefulness of this work through the implementation and demonstration of appropriate tools on various applications. We describe here the series of initial choices that, within the boundaries of DoD needs and requirements, characterized the beginning of the AWB development and shaped the path forward.

Rationale for Choices

In particular, the AWB team made three major initial choices:

- **Developing an Analytic Workbench of tools**, each of which addresses a specific aspect of interdependencies between systems in an SoS, rather than "one tool to solve them all." This approach guarantees that each tool is designed to give answers about the specific aspect required by the user. When multiple aspects of an SoS are of interest, different tools of the AWB can be applied to the same problem.
- **Using a network representation of SoS**: This network representation directly models interdependencies between systems, with a graphical representation that facilitates the use and understanding of the tools; and the network-based representation is a common representation for the tools, which facilitates the use of different tools on the same problem.
- **The theoretical formulation and basic implementation of the tools are domain-agnostic**, providing a solid theoretical foundation that allows the tools to be applied to many different fields.

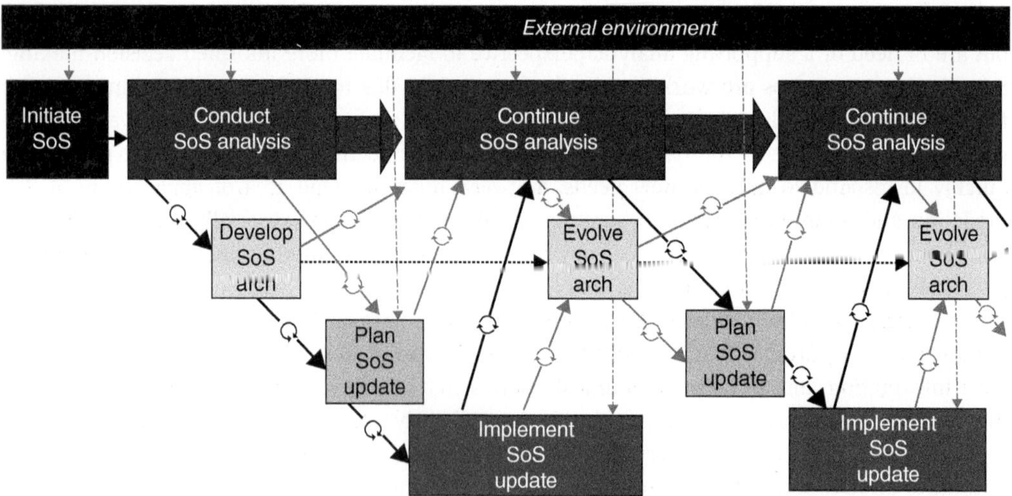

Figure 29.1 Wave model of SoS analysis and update.

Initial Tools in the Analytic Workbench

This section provides a brief description of the five tools created and implemented during the first development of the AWB. Each of the tools underwent improvements and enhancements over the years.

Robust Portfolio Optimization

The robust portfolio method adopts an "investment-like" perspective in the system of systems engineering, where the objective is to balance the risks in holding financial assets against the expected return on investment (Davendralingam and DeLaurentis 2015). RPO treats SoS as a portfolio of systems that can be acquired or connected following feasible rules. In the context of an SoS, the expected returns correspond to an expected capability due to investing in a system, and the risks are the developmental or operational risks of the individual system. The method seeks to maximize the overall SoS performance measured using a predetermined capability index and to maintain cost and risk (e.g., developmental and cost risk) within acceptable levels while accounting for the impact of data uncertainty.

RPO models an SoS as a network of discrete nodes, each with a predefined set of features. Each node has a set of requirements (inputs) and capabilities (outputs). The connection between compatible nodes, based on rules involving the inputs and outputs, allows for feasible architectures to be developed (Figure 29.2). Each system provides capabilities at the system level, and a combination of these system-level capabilities provides SoS-level capabilities. Several systems provide support capabilities necessary for the adequate operation of all the systems in the selected portfolio.

Using RPO for a particular SoS design problem yields a set of Pareto optimal solutions (each solution is a portfolio of systems) corresponding to a user-defined risk aversion factor. The stakeholders of the complex SoS problem can then explore the design space of available options resulting from the RPO analysis based on dependencies between various systems.

System Operational Dependency Analysis

Systems Operational Dependency Analysis (SODA, Guariniello and DeLaurentis 2017) is a method to model and analyze systems' input/output behavior in the operational domain. SODA analysis computes the operability of each system as a function of its own internal status and of the operability of the other systems in the network, based on the topology and the features of the dependency. The model can therefore evaluate the impact and propagation of failures and disruptions from the nominal operability, and the effect of architectural design decisions on global

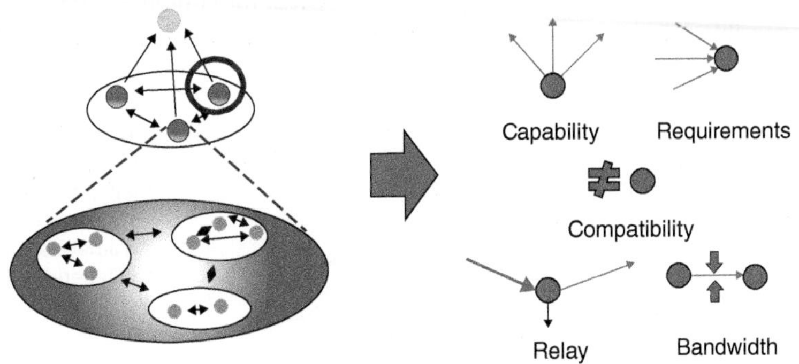

Figure 29.2 Dependencies between systems are modeled and treated as constraints in the Robust Portfolio Optimization methodology.

behavior. Based on these results, the user can quantify various metrics of interest, for example, robustness and resilience. Using SODA, designers and decision-makers can quickly analyze the operational behavior of SoS and evaluate different architectures under several working conditions. The user can compare architectures based on metrics of interest, the trade-off between competing desired features, identify the most promising architectures, as well as the causes of the observed behavior, and discard architectures that lack the requested features. This way, in the early design process, promising architectures can be taken into consideration and improved based on the information given by the model parameters and the observed behavior, thus supporting the process of concept selection.

SODA is a three-parameter piecewise linear model suitable for analyzing system dependencies, including partial dependencies and their effect on system behavior. Each system is characterized by its internal health status. Each interdependency is modeled with three parameters, characterizing the strength, criticality, and impact of the specific one-to-one dependency. SODA computes the impact of cascading disruptions on the behavior of the entire SoS. Figure 29.3 shows a simple three-node SODA network.

SODA analysis can be a deterministic evaluation of the status of a single instance of the system, using a single set of the internal status of each system, or a stochastic evaluation of the overall behavior, using a probability distribution for the internal status of each system, which results in stochastic distributions of operability. The operability of the nodes of interest is used to analyze and evaluate properties of the overall system, such as robustness, resilience, and risk.

Systems Developmental Dependency Analysis

Systems Developmental Dependency Analysis (SDDA, Guariniello and DeLaurentis 2013) is used to assess the impact of partial developmental dependencies between components in an SoS. A parametric model of developmental dependencies outputs a schedule of the development of the SoS, accounting for partial dependencies and for the cascading impact of delays. SDDA evaluates the development schedule of an SoS based on the current and expected performance of each system (in terms of development time and delays), the model of the dependencies, and the amount of accepted risk. SDDA supports educated decision-making in the development and revision phases of systems architecture. In particular, throughout the whole development phase, the information produced by SDDA can be used to identify criticalities and bottlenecks, quantify possible partial delay absorption, and assess the best time to begin the development of each system, accounting for development cost, stakeholder decisions, and risk. More recent applications of SDDA support decision-making for technology prioritization (Guariniello et al. 2021).

The outcome of SDDA analysis is the beginning time and the completion time of the development of each system, as well as an assessment of the combined effect of multiple dependencies and possible delays in the development of predecessors. The lead time (i.e., the time by which a system

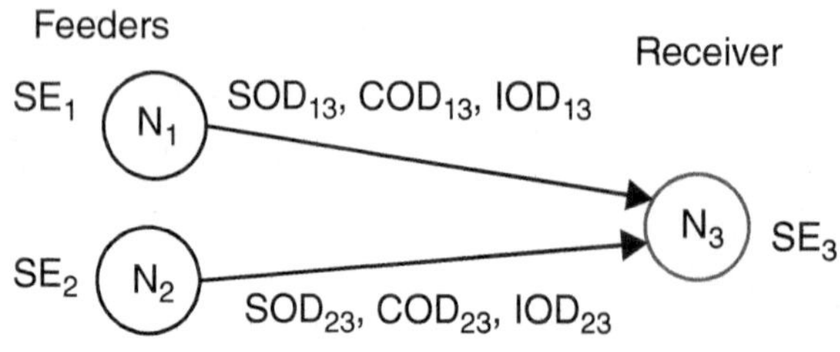

Figure 29.3 SODA network with three nodes, each with its own internal status (Self-Effectiveness, SE) and two dependencies, each characterized by three parameters (Strength, Criticality, and Impact Of Dependency).

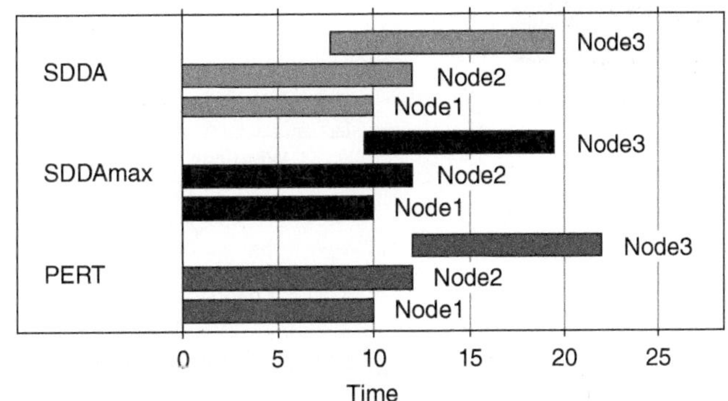

Figure 29.4 Development schedule of a three-node SoS as shown in Figure 29.3, according to SDDA, a more risk-averse version of SDDA (SDDAmax), and PERT.

can begin to be developed before a predecessor is fully developed) is calculated based on the parameters of the dependencies and the performance of the predecessors. SDDA allows for deterministic or stochastic analysis. When using deterministic analysis, SDDA evaluates the impact of a single instance (i.e., one given amount of delay in each system), resulting in one beginning time and completion time for each system. When using stochastic analysis, the amount of delay in each system follows a given probability density function. Consequently, the beginning and completion time of each system will also be a probability density function. Results from the analysis are used to compare different architectures in terms of development time, ability to absorb delays, and flexibility. Figure 29.4 shows an example of a development schedule resulting from deterministic SDDA analysis and compared with PERT (Malcolm et al. 1959).

System Importance Measures (SIMs)

The Systems Importance Measures Resilience Design methodology is a four-phase method that highlights the relative importance of different systems to overall SoS resilience via a prescribed set of measures and provides design guidance on how to improve the overall SoS resilience (Uday and Marais 2014). The four phases of the process (Figure 29.5) are:

- **Phase 1**: Identify potential disruptions ("What can go wrong?")
- **Phase 2**: Determine impacts of disruptions ("What are consequences of unmitigated disruptions?")
- **Phase 3**: Determine current SoS resilience ("How well is the SoS able to handle the disruptions?")
- **Phase 4**: Improve SoS resilience using Design Principles ("What can be done to improve SoS resilience?")

The method provides a platform for multiple analysts and decision-makers to study, modify, discuss, and document options for implementing SoS resilience in a fashion that scales with the SoS size. The visual nature of the resilience map provides a useful, highly intuitive, and immediate way for summarizing key points of concern in iteratively building resilience into an SoS.

Given the importance of SoS, managing their resilience is vital to national security, global economies, and in many cases, public health and safety. There are many reasons why an SoS may not be resilient: design flaws, unanticipated disruptive events, emergent behavior of operational evolution, poor contingency planning and execution, or limitations at the organizational level. Given the diversity and often wide geographic distribution of SoS constituent systems, inclusion of backup systems for a SoS is often impractical and costly. Additionally, high levels of

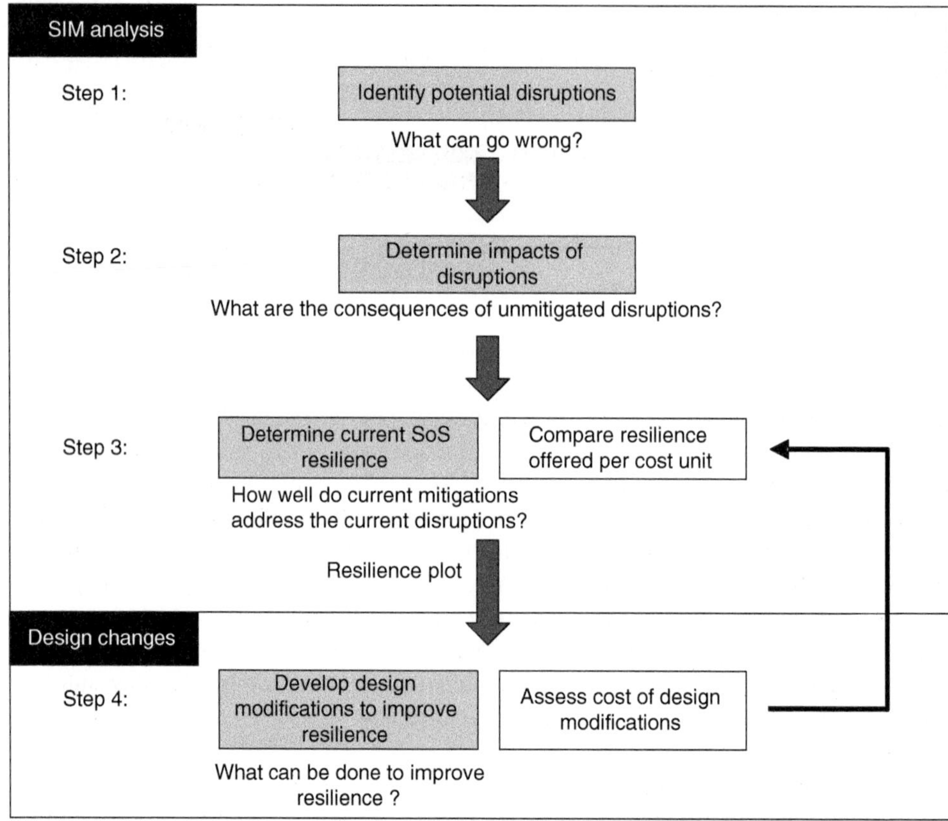

Figure 29.5 Steps of SIMs methodology.

interdependency between the systems increase the risk of failures cascading throughout the SoS. At the same time, SoS features giving rise to these hurdles also offer the opportunity to improve the resilience of the overarching system through unconventional means.

Measuring resilience is a pre-requisite for designing resilience and more generally of incorporating resilience into cost-benefit analyses and other trade-offs (Francis and Bekera 2014). SIMs is a method to quantify resilience in SoS. Similar to the component importance measures used in reliability engineering (Elsayed 1996; Van der Borst and Schoonakker 2001; Rausand and Høyland 2004; Ramirez-Marquez and Coit 2007), the SIMs help identify and rank components or systems according to their impact on resilience.

The approach followed in SIMs allows for the explicit identification of systems that affect resilience, either by being vulnerable to disruptions or by being important to disruption mitigation. SIMs output three resilience metrics that rank the constituent systems based on their impact on the overall SoS performance by answering the following three questions:

1) What is the SoS-level impact of a particular disruption if no mitigations are available?
2) What is the SoS-level impact of a particular disruption using existing mitigations?
3) How effective is particular mitigation in responding to a disruption?

These three System Importance Measures (SIMs) rank or prioritize the constituent systems of an SoS based on their resilience significance.

SIMs allow designers to identify how each system or component contributes to, or is vulnerable to, disruption. The SIMs can be used to assess and improve the resilience of both existing (fielded) SoS and new (undeployed) SoS by highlighting how well or how poorly current SoS structures can handle disruptions. The SIM analysis points to inadequacies that need to be addressed and helps designers evaluate the strengths and weaknesses of potential SoS architectures.

Multi-stakeholder Dynamic Optimization (MUSTDO)

The dynamic nature of SoS evolutions and decoupled nature of decision-making due to localized authority within an SoS (such as seen in an "acknowledged" SoS) means that the interplay of tactical and strategic decisions can result in increased risks in an SoS evolution. This temporally coupled nature of decision-making, combined with ubiquitous uncertainty, further exacerbates the already existing complexities in SoS architectural decision-making.

A top manager with partial authority is more common than a centralized authority; therefore, mechanisms are required that can influence the behavior of the participating stakeholders. It is also crucial to include both the near-term and long-term objectives in the early design phase, to improve the capability or reduce the cost in the long run. The Multi-Stakeholder Dynamic Optimization (MUSTDO, Fang and DeLaurentis 2015; Fang et al. 2018) methodology supports the coordination of conflicts via the integration of a transfer contract mechanism (to enable decentralized, multi-stage decision-making between stakeholders) and Approximate Dynamic Programming (ADP). The idea here is to relegate the quantitative complexities of decision-making coordination to the algorithm, while delegating the decision-making and trade-off assessment to the SoS decision-maker, within a coordinated quantitative framework.

A transfer contract is defined as the compensation that each participating stakeholder needs to pay to other stakeholders for consuming the shared resources provided by the top-level manager. The transfer contract can influence the decisions that each participating stakeholder makes to achieve the best use of the limited resources. Meanwhile, the stakeholders can retain part of their private information during the negotiation process. When extending the problem to a long-term horizon (Figure 29.6), ADP (Bertsekas and Tsitsiklis 1995; Powell 2007) is a well-recognized method for addressing multi-stage decision-making problems under uncertainty in Operations Research (OR). The combination of the transfer contract coordination mechanism and ADP constitutes the MUSTDO method.

This method is valuable in (1) providing a method to address the multi-stage composition decisions, which reduces the complexity due to the increasing number of systems and amount of uncertainty, and (2) providing a new perspective to coordinate conflict in an acknowledged SoS problem, which enables the efficient use of resources.

Story of Success: How We Helped Users and Learned from Users

The AWB has been one of the most successful products of research at SERC. This is due to multiple reasons: first, the AWB addresses areas of important and well-acknowledge needs in the SoS community; second, the AWB provides practical methodologies and tools to achieve preliminary assessment and evaluation of SoS in the early stage of design or modification; third, since the first development of the AWB tools, the researchers followed a progressive approach of more and more complex applications, which allowed them to test, verify, and improve the tools over the years. This section will describe the first steps in this process of initial applications and of gathering knowledge from the first users of the AWB tools.

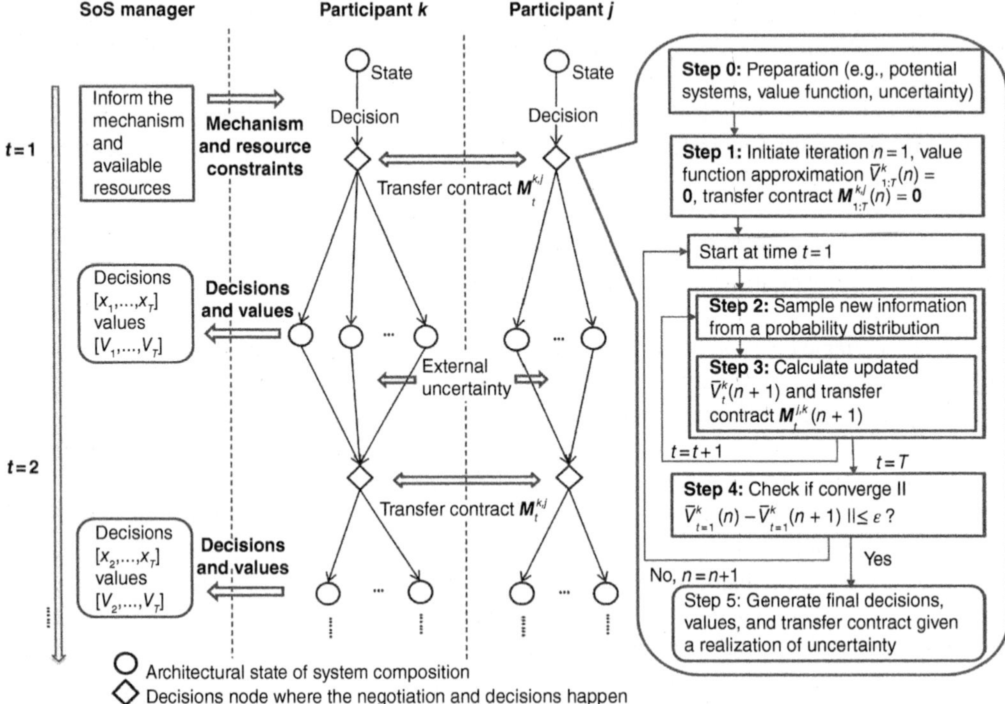

Figure 29.6 Workflow of MUSTDO.

Initial Implementation and Naval Warfare Scenario

The first demonstration of the AWB used a synthetic Naval Warfare Scenario (NWS), based on the concept of Littoral Combat Ships (LCS, Abbott 2008). As shown in Figure 29.7, the NWS is an SoS where various systems, including LCS, Unmanned Aerial Vehicles (UAV), combat ships, Remote Controlled Vehicles (RCV), Underwater Vehicles (UWV), helicopters, and satellites form a network of communication and operational dependencies that provides Anti-Submarine Warfare (ASW) capabilities, Mine Countermeasures (MCM) capabilities, and Surface Warfare (SUW) capabilities.

Using the same Concept of Operations for the NWS, the problem was specialized for each individual tool, by using literature review, Agent-Based Modeling (ABM), and expert judgment to identify the necessary information to run the tools on the NWS problem.

Systems Developmental Dependency Analysis was used to identify the schedule of development of the NWS SoS according to different development architectures (Guariniello 2016). Figure 29.8 shows the developmental dependencies of various NWS subsystems in one of the developmental architectures used to run SDDA analysis. Figure 29.9 shows the corresponding schedule of development of the different NWS architectures, when no delays occur. More SDDA analysis was performed to show the stochastic impact of expected delays.

The same systems in the three NWS developmental architectures have been used to create SODA networks. These SODA models, providing information on the operational dependencies and on the mission-level capabilities of detection and engagement of the adversary, were used in conjunction with the SDDA model to assess the development of partial and final capabilities

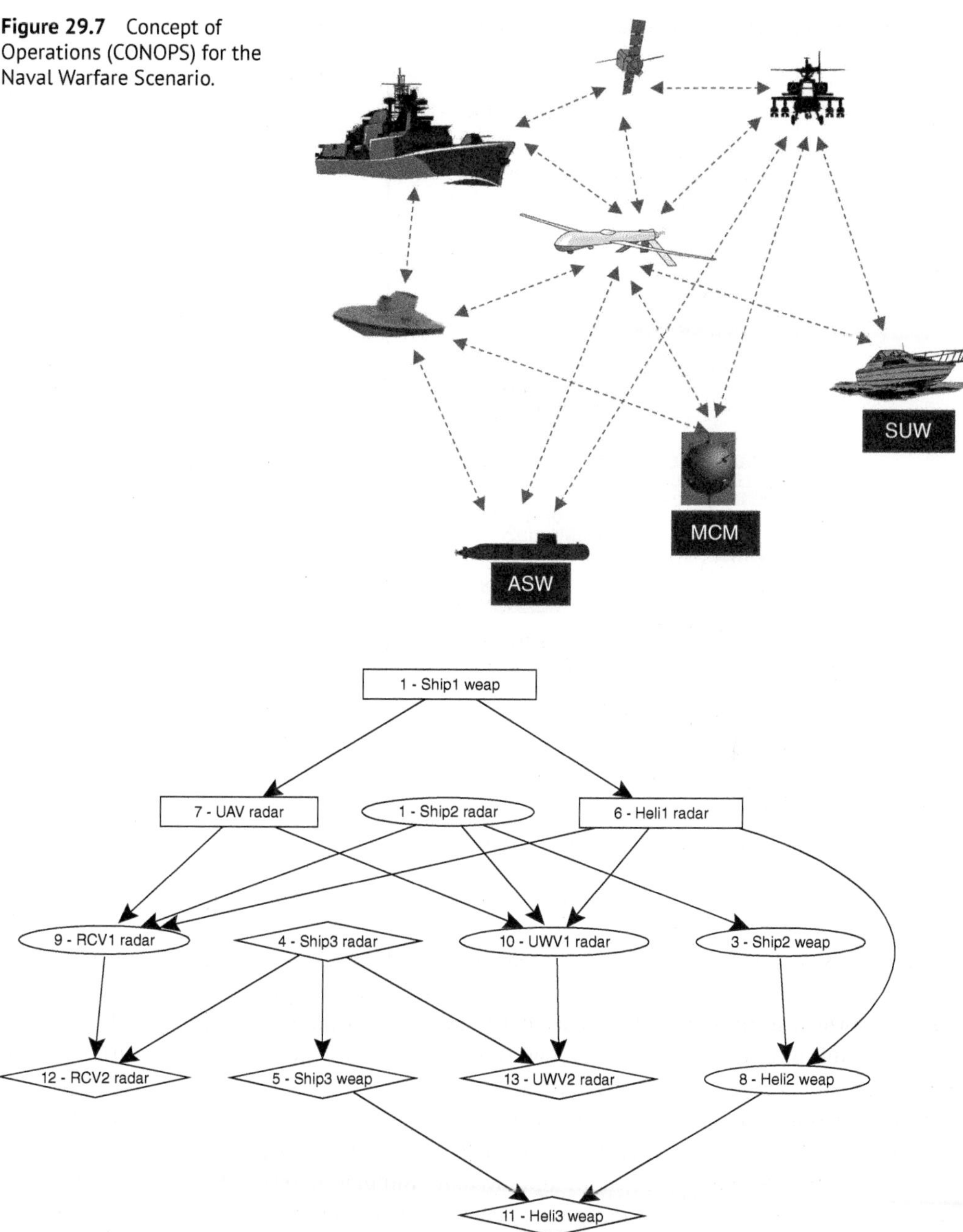

Figure 29.7 Concept of Operations (CONOPS) for the Naval Warfare Scenario.

Figure 29.8 Developmental dependencies for architecture A of the NWS. *Source:* Adapted from Guariniello (2016).

over time, when developing the NWS according to three different architectures, as shown in Figure 29.10.

Robust Portfolio Optimization was used to evaluated portfolios when different types of NWS systems are available, each characterized by its own capabilities, uncertainty, and cost. Figure 29.11 shows different portfolio selections corresponding to different levels of conservatism. Higher

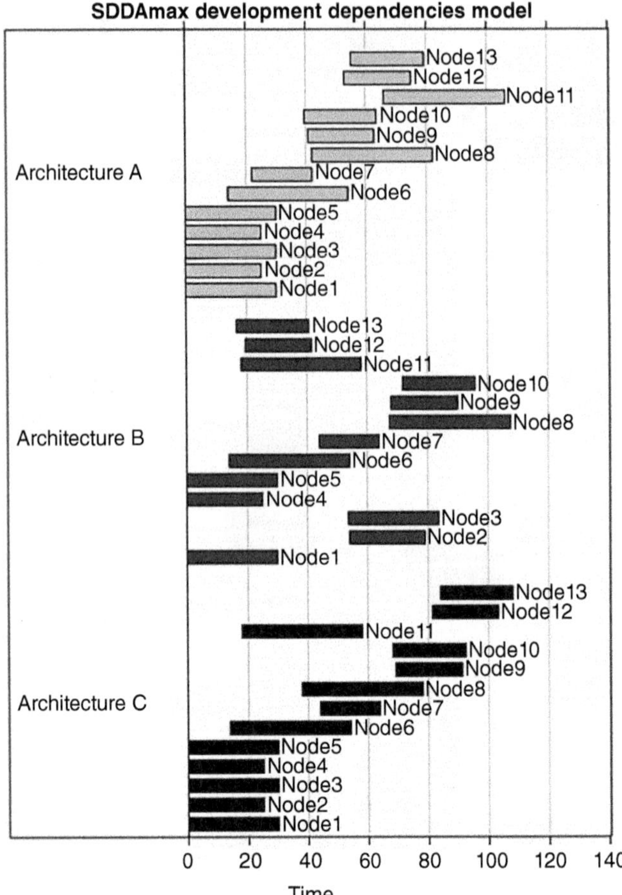

Figure 29.9 Schedule of development of the three different NWS architectures, according to SDDA model. Node number as in Figure 29.8.

conservatism will result in preferred selection of systems with lower uncertainty. The different portfolio result in different SoS-level capabilities, as shown in Figure 29.12, and can be used as input architectures for the other tools in the AWB.

System Importance Measures were also calculated for the baseline NWS network, where different disruptions were imposed on the satellite, UAV, helicopter, and LCS. Some disruption affected only one system; other disruption affected multiple systems. Simulation resulted in the computation of the importance of each of these disruptions. The use can not only assess disruption in different combination of systems, but also evaluate which disruption will result in acceptable vs. unacceptable performance, based on level of acceptance set up by the user. Figure 29.13 shows examples of System Disruption Importance (SDI) in the NWS.

This tool has also been used to assess the impact of mitigation measures on the NWS. For each of the disruptions shown in Figure 29.13, four different mitigation measures have been implemented. The result of the combination of disruption and mitigations is quantified in the System Disruption Conditional Importance (SDCI) measure. As shown in Figure 29.14, some mitigations may have a major effect on certain disruptions, bringing the SoS back within acceptable performance range, but at the same time they might have only a marginal effect against

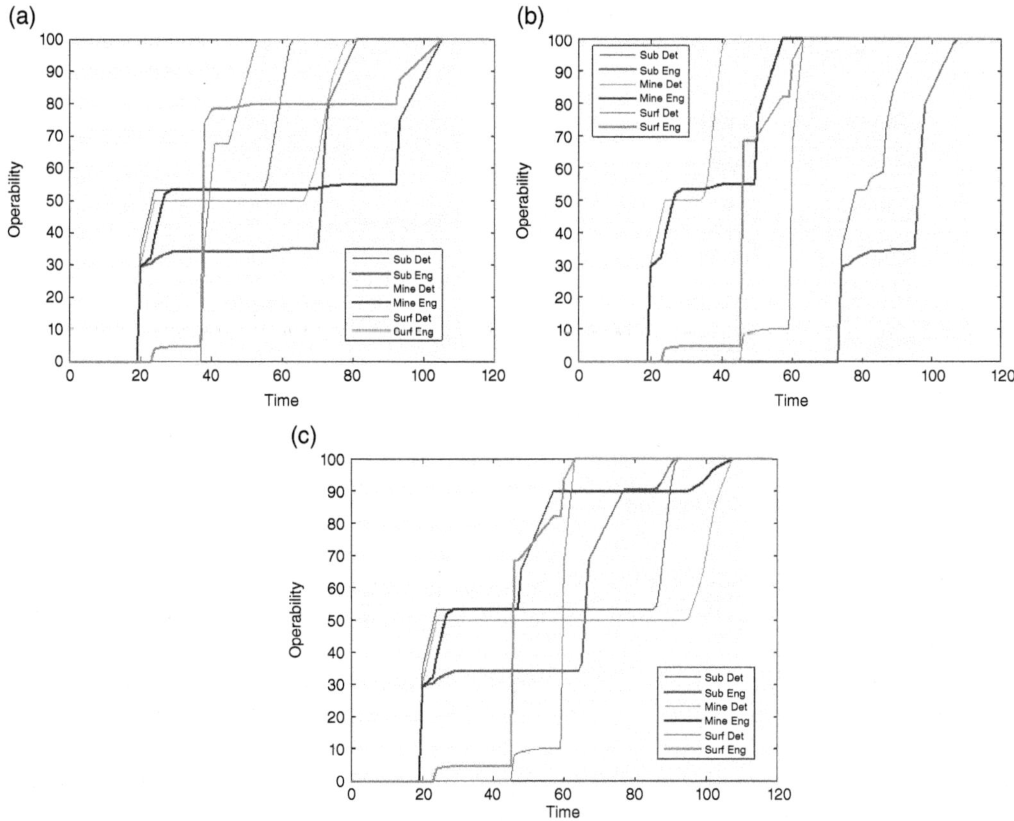

Figure 29.10 Development of partial capabilities of submarine detection and engagement, mine detection and engagement, and surface vehicles detection and engagement in three different NWS development architectures. (a) Architecture A. (b) Architecture B. (c) Architecture C.

Figure 29.11 RPO selection of systems for the NWS, under different values for conservatism (risk aversion).

Systems	Available system packages	Gamma (level of conservatism)			
		0.01	0.21	0.41	0.61
ASW	Variable depth	-	-	-	-
	Multi Fcn Tow	x	x	x	x
	Lightweight Tow	-	-	-	-
MCN	RAMCS II	-	-	-	-
	ALMDS (MH-60)	x	x	-	-
SUW	N-LOS missiles	x	x	x	-
	Griffin missiles	-	-	-	x
Seaframe	Package 1	-	-	-	-
	Package 2	-	-	-	-
	Package 3	x	x	x	x
Comm.	System 1	-	-	-	-
	System 2	x	x	x	x
	System 3	-	-	-	-
	System 4	-	-	-	x
	System 5	-	-	-	-
	System 6	x	x	x	-

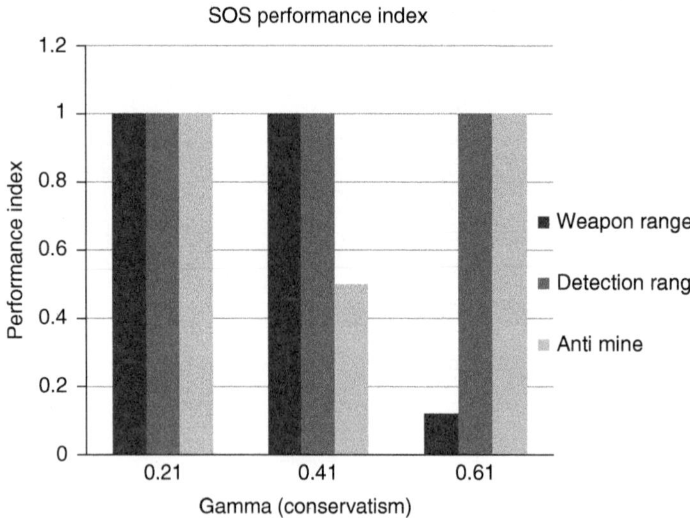

Figure 29.12 SoS capabilities provided from the various portfolios.

other disruptions. This demonstration of the use of SIMs on the NWS provided good insight into possible uses of SIMs to evaluate choices not only while developing a SoS but also when operating it.

Finally, we used the *Multi-Stakeholder Dynamic Optimization* tool to show how the NWS can be developed even when the SoS manager does not have authority to impose contractual choices. Decision based on transfer contract and dynamic optimization result in stakeholder decisions similar to those that a central authority would have made. Figure 29.15 shows an example of application of MUSTDO at one step of the NWS SoS acquisition.

NanoHub GUI

After the initial implementation and demonstration of the individual tools, SERC researchers at Purdue wanted to reach out more users to obtain further use and evaluation of the AWB. Therefore, graphic user interfaces (GUI) were implemented and made available on Purdue University's Nanohub website (DeLaurentis et al. 2017). Figure 29.16 shows one of the GUIs available on Nanohub. Dozens of users from many different countries ran analyses with the GUIs on Nanohub, and some of them provided valuable feedback to make the tools more user-friendly and to address necessary changes in the explanation of some of the fundamental theoretical concepts.

Inclusion in Academic Curriculum

Finally, the initial diffusion of both theoretical foundation and practical application of the AWB tools came from its inclusion in the curriculum of the System-of-Systems Modeling and Analysis course at Purdue University (AAE 560), a graduate-level class in the School of Aeronautics and Astronautics. This course saw a mix of students and professionals use tools in the AWB on a wide variety of application, including space architecture, water distribution systems, smart power grids, and multidomain battle scenarios. This last step of initial demonstration, which produced various publications, posed the basis for collaboration of the researchers who implemented the AWB with sponsors and partners in various fields.

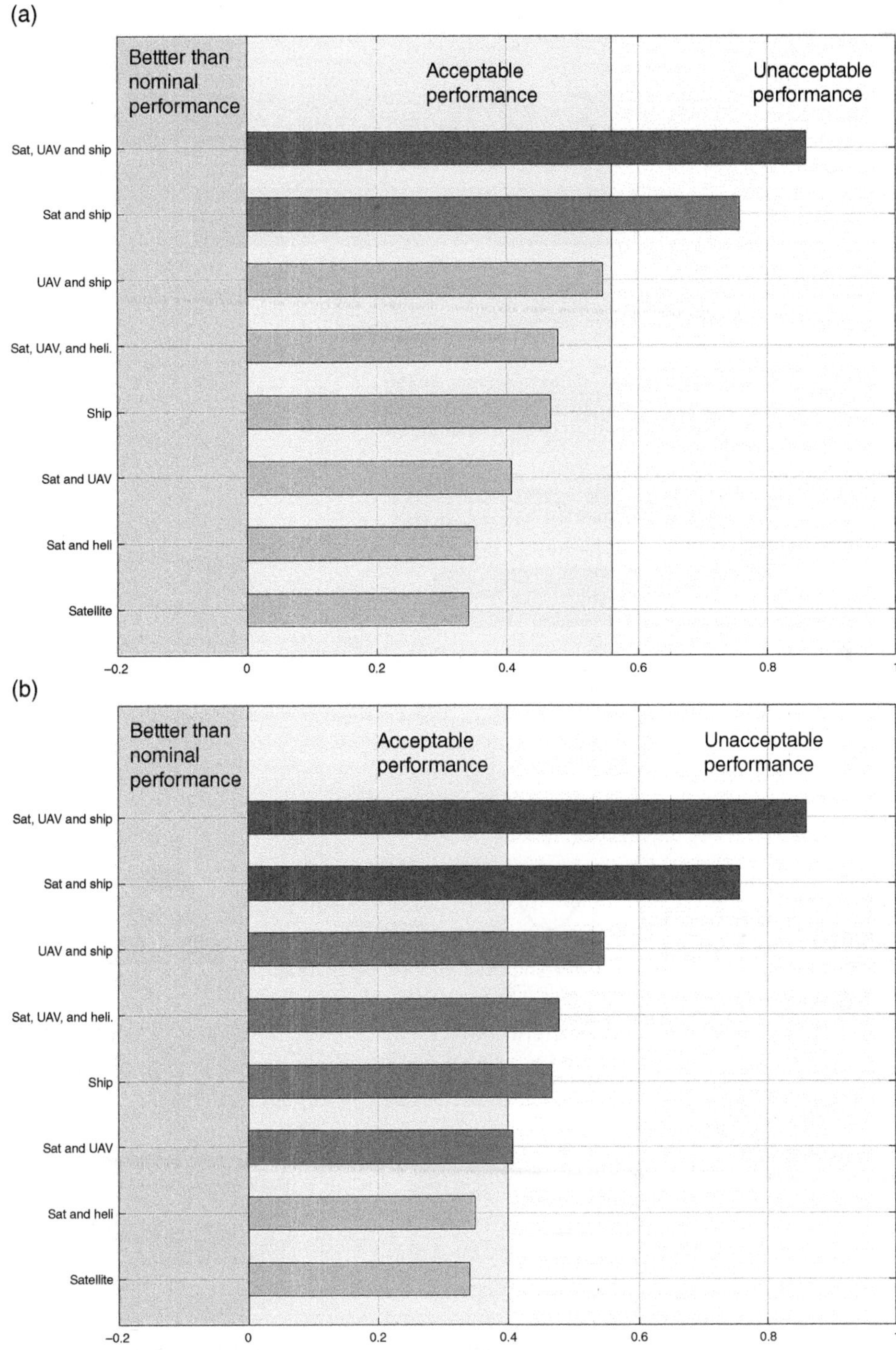

Figure 29.13 Systems Disruption Importance. Given the disruption in the systems on the *x*-axis, the *y*-axis shows the loss in operability, that is the System Disruption Importance. The higher the loss, the worse is the impact of the disruption. (a) High tolerance of disruption (performance is acceptable with SDI up to 0.56). (b) Low tolerance of disruption (performance is acceptable with SDI up to 0.4).

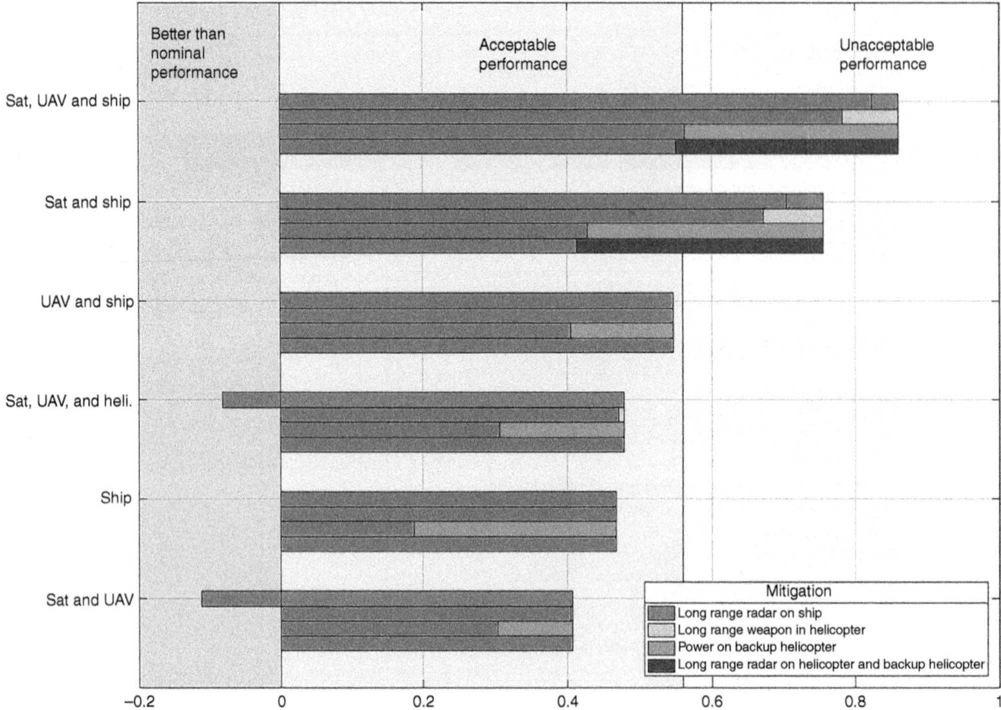

Figure 29.14 System Disruption Conditional Importance (SDCI) for various types of mitigation measures.

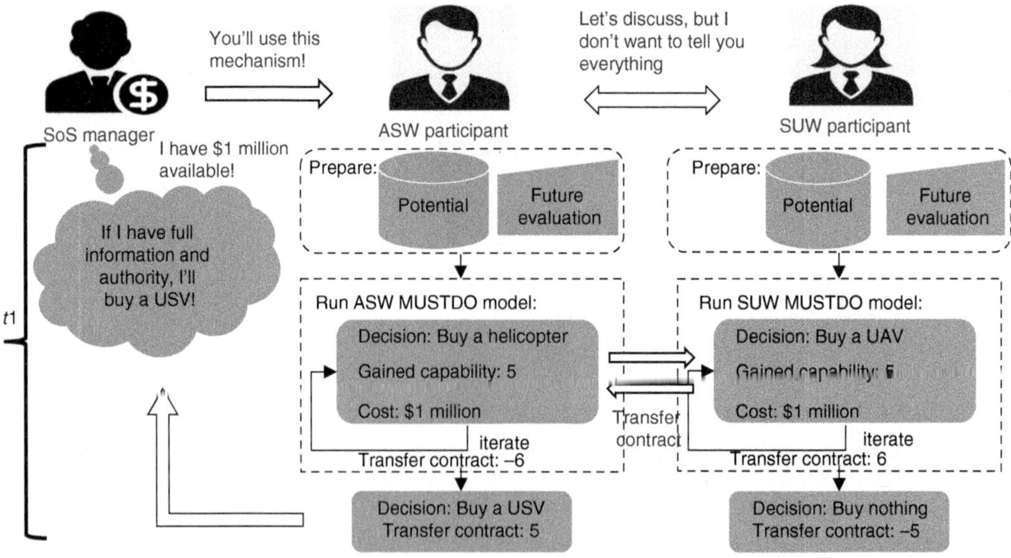

Figure 29.15 Use of transfer contract at each step of a dynamic optimization problem to develop the NWS SoS.

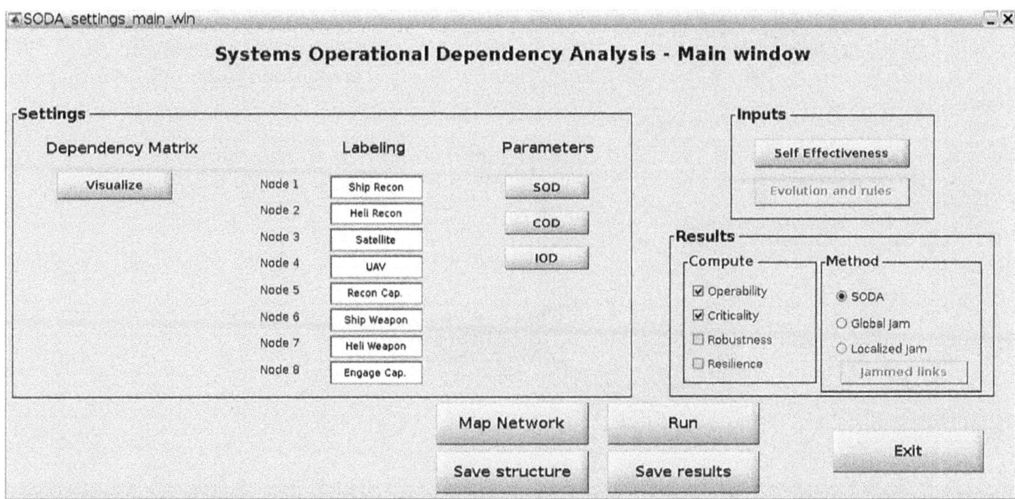

Figure 29.16 Graphic user interface (GUI) window for data input in SODA, as available on Nanohub.

Further Applications of the Analytic Workbench

The series of applications of the AWB that followed the initial demonstration stage was characterized not only by diffusion into other fields but also by the development of new ideas and concepts, derived from the variety of different perspectives provided by users. For example, the AWB tools were modified to address specific user needs and to support systems engineering decision-making. Further evolution of the AWB allowed the researchers to use the AWB with concepts from machine learning (ML) and uncertainty quantification (UQ). At the same time, part of the effort focused on combined application of multiple tools and on more and more advanced GUI.

Moon and Mars Exploration Architectures: Propulsion and Habitat Studies for NASA

The first studies for NASA Marshall Space Flight Center (MSFC) were directly derived from the analysis of SoS architectures for crewed exploration of Mars (Guariniello 2016). The multi-year work of Purdue researchers with NASA mostly used SODA and SDDA. The research sponsors directed the development of analysis of a hierarchical architecture, with each large, high-level system (spacecraft, communication satellites, habitats, and so on) expanded into low-level operational networks of subsystems. Then, the focus was put on comparison of different interplanetary propulsion systems: chemical, nuclear thermal, and solar electric. Intensive literature review and collaboration with subject-matter experts (SMEs) provided the necessary input data to the tools. Figure 29.17 shows one of the operational dependency networks developed to run SODA analysis and identify the most critical subsystems in the propulsion systems.

Figure 29.18 shows an example of the results of SODA analysis, identifying the most critical subsystems in each of the different propulsion systems and quantifying the impact of different amounts of disruption in each subsystem.

When NASA put priority on Moon exploration as a step toward Mars, the AWB was applied to the analysis of systems and subsystems in the habitat of the Lunar Gateway (Chavers et al. 2019). Figure 29.19 shows the functional decomposition of the habitat, which has been used as the baseline to develop the operational dependency networks to run SODA analysis (DeLaurentis et al. 2022).

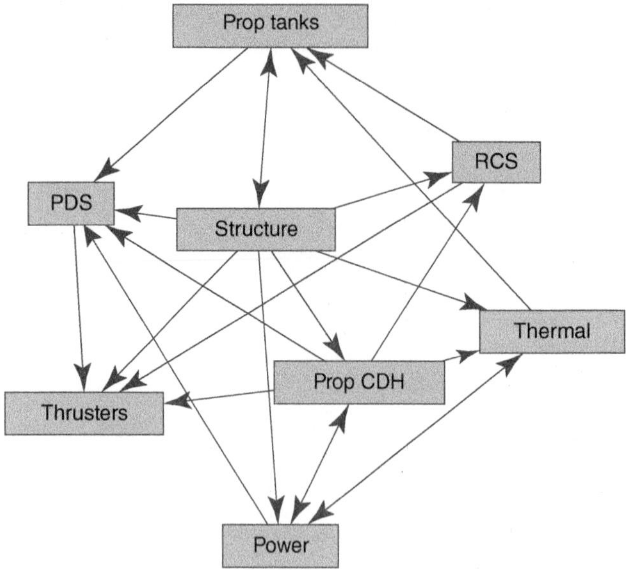

Figure 29.17 SODA network of subsystems of a chemical propulsion system.

The sponsors, in this case, expressed interest in ways to obtain a quick look at the big picture and assess the criticality of multiple subsystems at once. Therefore, a Disruption Impact Matrix user interface was developed, as shown in Figure 29.20.

Cryogenic Fluid Management and Technology Prioritization

The next application developed with NASA used the Systems Developmental Dependency Analysis tool. This is one of the cases where a combination of the user's needs and of the desire of the researcher to explore new directions resulted in enormous improvements in the tools and their practical application. When NASA MSFC requested a study of cryogenic fluid management (CFM) technologies, the research project built on the SDDA tool to include considerations on the yearly budget and prioritization of technologies based on mission requirements and stakeholder preferences (Guariniello et al. 2021).

The optimal schedule of development is based on the available budget, cost of development of technologies, and developmental dependencies, as modeled in SDDA. Figure 29.21 shows an example of a different development schedule resulting from varying the available budget. The enhanced tool can also provide a sand chart of the distribution of budget over the years for the development of CFM technologies, as shown in Figure 29.22.

Finally, the new version of SDDA developed within this practical application can provide an optimized development schedule, keeping into account stakeholder preferences, while still satisfying the constraints imposed by developmental dependencies according to the SDDA model. Figure 29.23 shows the development schedule of CFM technologies when stakeholders express their preference that some technologies are developed as soon as possible, given the constraints of budget and developmental dependencies.

Use of the Analytic Workbench in the Context of Artificial Intelligence

While the tools in the AWB were developed keeping in mind principles of systems engineering and with capabilities of interaction with model-based systems engineering (MBSE, Guariniello

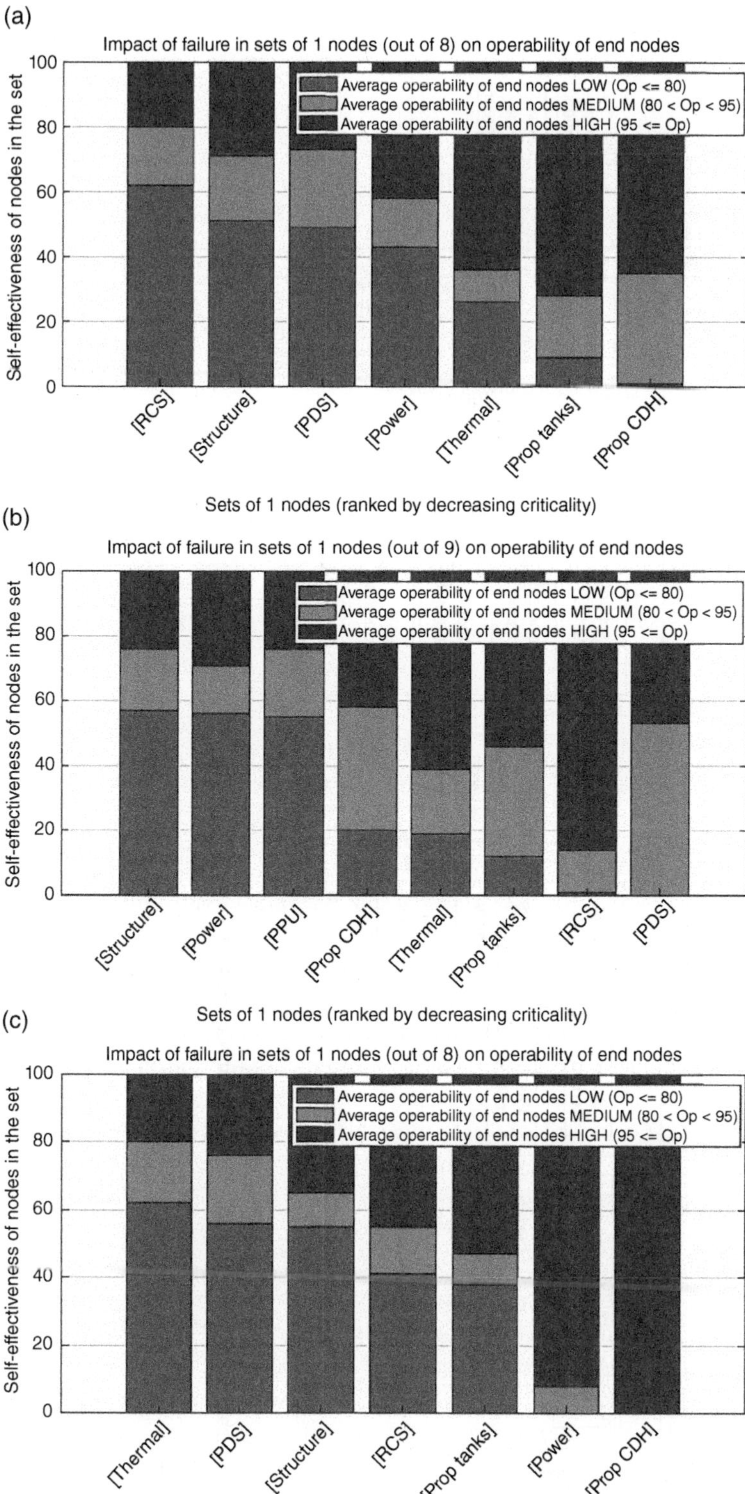

Figure 29.18 Impact of disruptions in the propulsion subsystems. Shades indicate the nominal, sub-nominal, and critical status of the subsystems of interest when the system indicated at the bottom of the bar experiences increasing disruptions. (a) Chemical. (b) Nuclear thermal. (c) Solar electric.

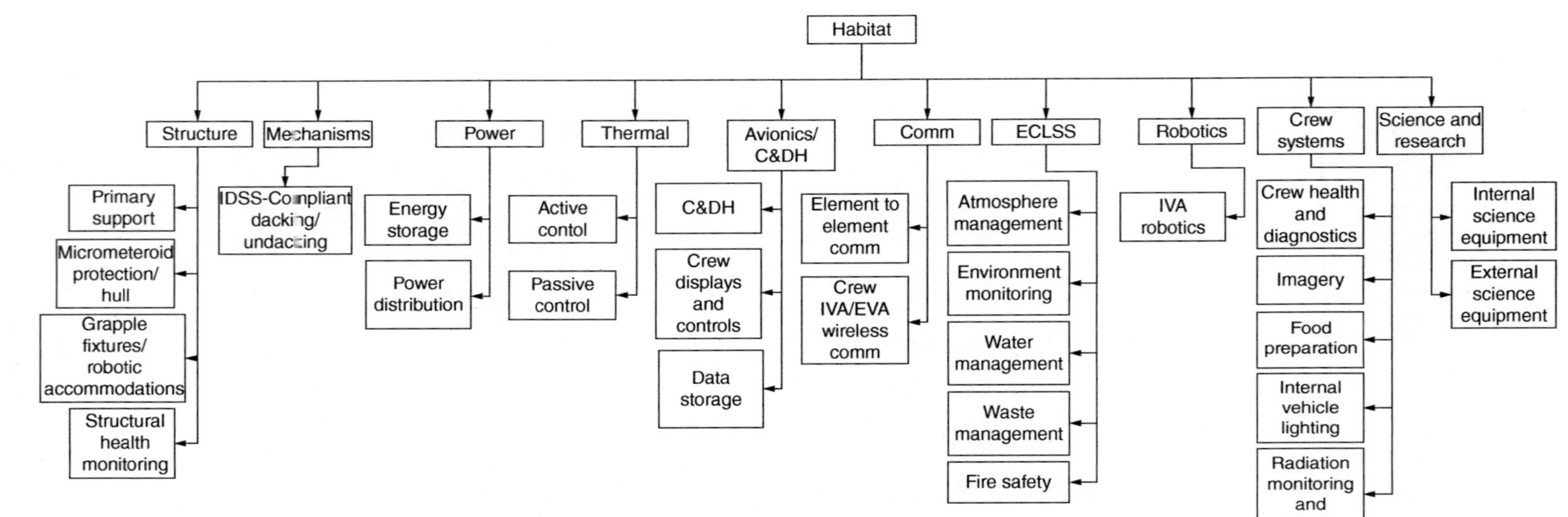

Figure 29.19 Systems and subsystems of the Lunar Gateway Habitat.

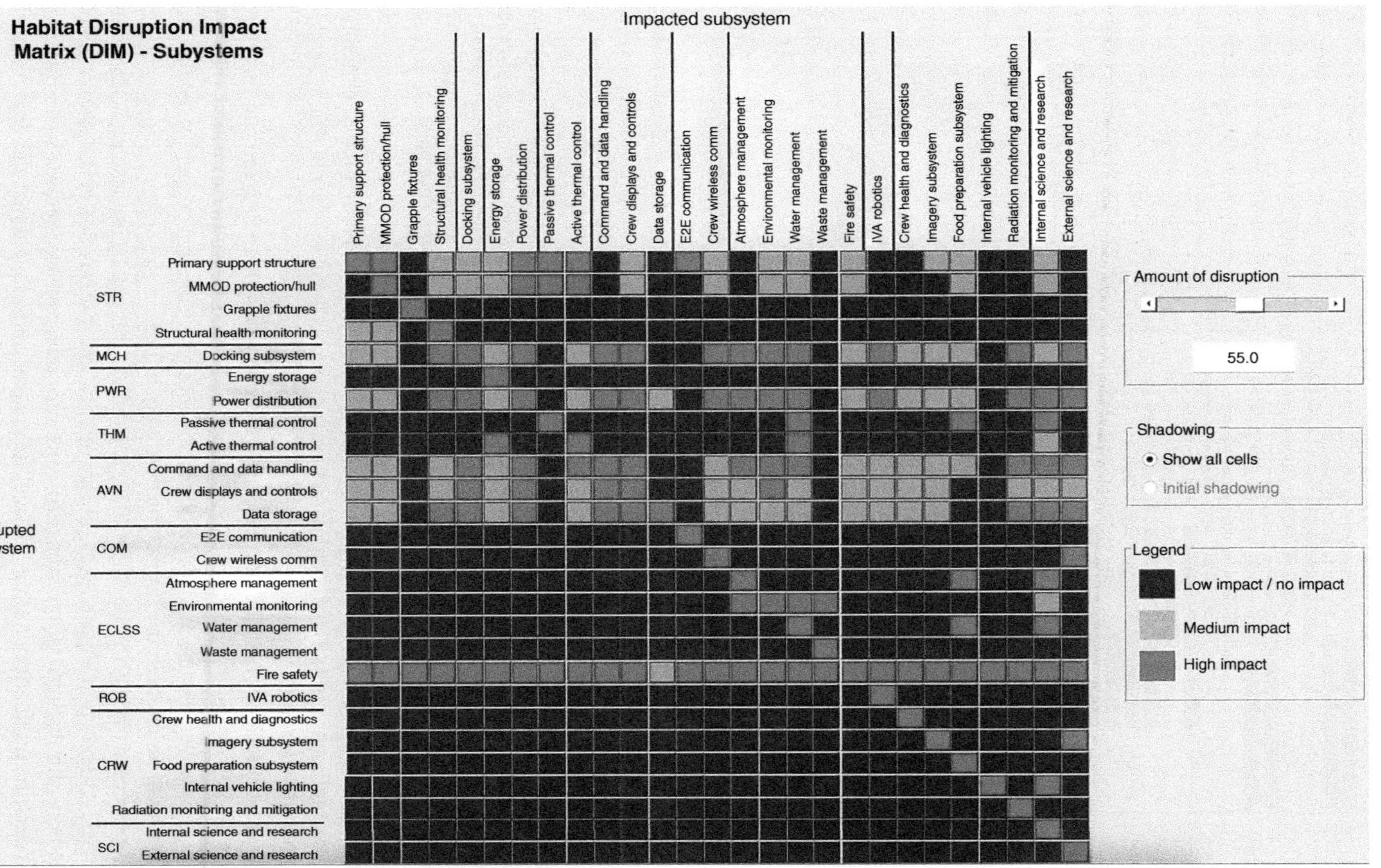

Figure 29.20 Disruption Impact Matrix for the Lunar Gateway habitat. Shades indicate low, medium, and high impact of disruption of each subsystem on the other subsystems, according to the legend in the figure.

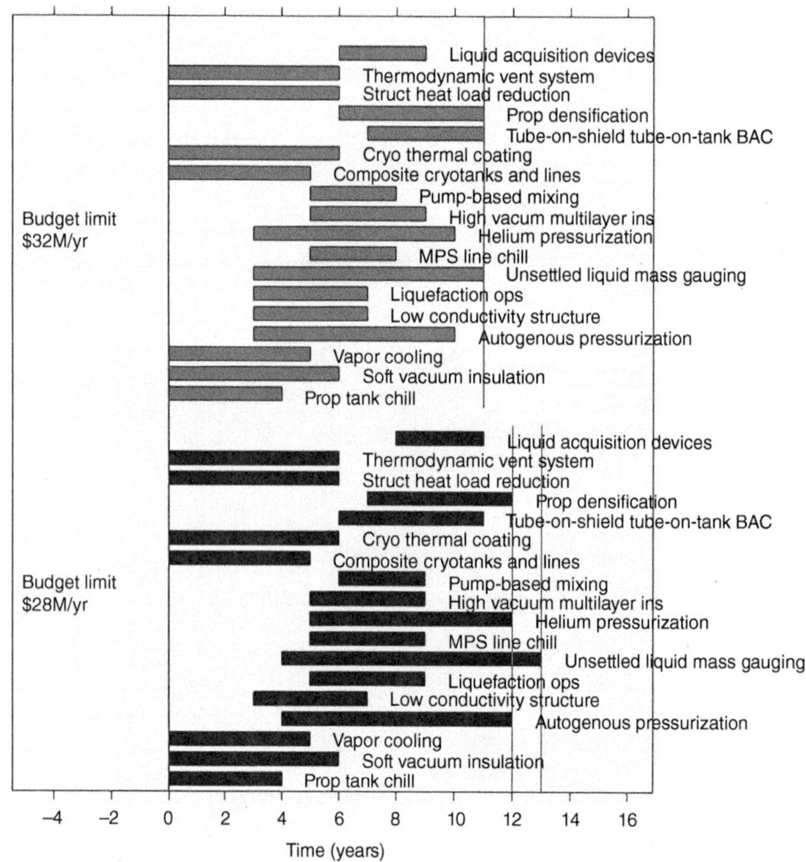

Figure 29.21 Development schedule of CFM technologies with different budget available. On the top part of the diagram, the vertical line identifies the time when all CFM technologies for Moon and Mars exploration have been developed. On the bottom part of the diagram, there are two vertical lines: the line on the left indicates the time when all CFM technologies for Moon exploration have been developed, and the line on the right indicates the time when all CFM technologies for Mars exploration have been developed.

et al. 2018), a further step for better practical application of the tools was their use in combination with artificial intelligence (AI). In particular, an application was developed for SODA to be used in combination with ML and UQ.

SODA provides information about the criticality of each system in an SoS and can therefore be used to assess which alternative SoS architecture have better qualities of performance, robustness, and resilience. However, the use of SODA alone is not enough to obtain a deeper understanding of how the features of an architecture impact the desired qualities.

The work presented in Guariniello et al. (2019, 2020) filled this gap by interfacing SODA with AI methodologies. As shown in Figure 29.24, in this application, various SoS architectures (satellite constellations in the demonstration scenario) were characterized by features including orbital semi-major axis and inclination, number of satellite systems and subsystems, and types of satellite attitude control systems, guidance and navigation systems, and power systems. The architectures were analyzed with SODA and sorted against metrics, including mass, performance, serviceability, resilience, and robustness. Machine learning methodology, in particular artificial neural networks, was used to identify relationships between architectural features and desired metrics.

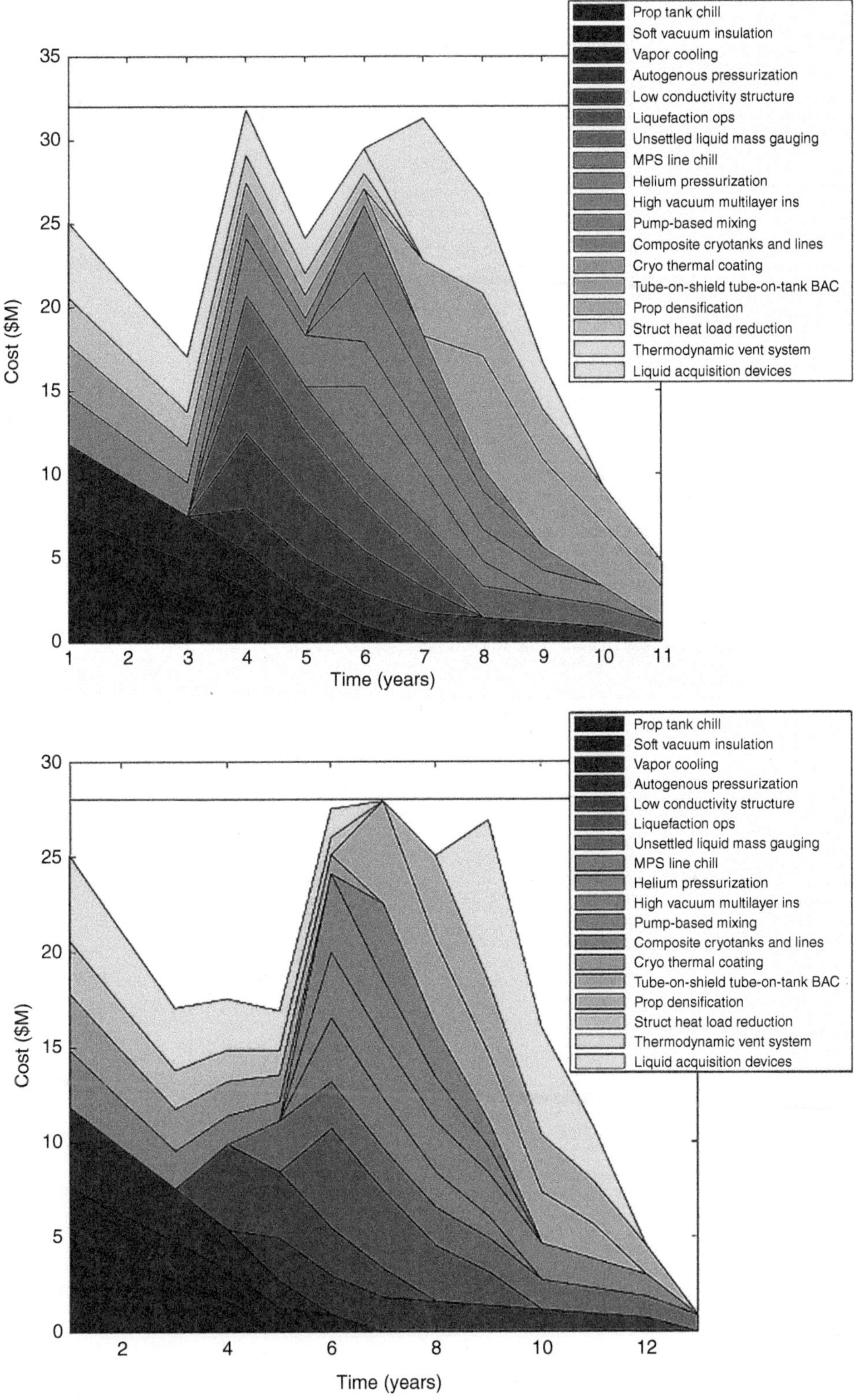

Figure 29.22 Distribution of budget for the development of CFM technologies. Left: $32 million yearly limit. Right: $28 million yearly limit.

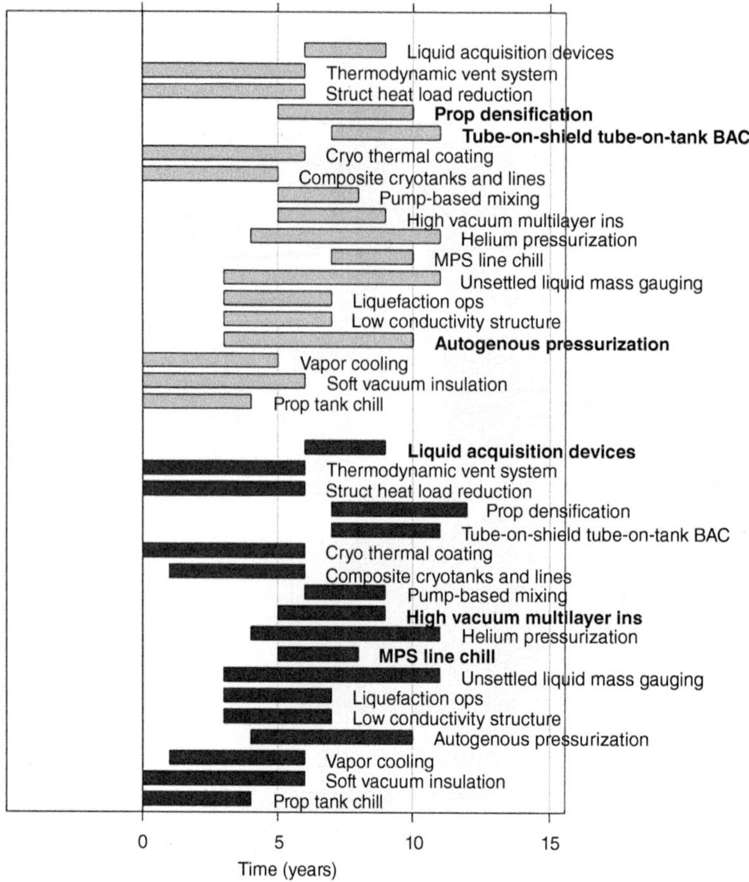

Figure 29.23 CFM development schedule when different technologies (marked in bold) are preferred.

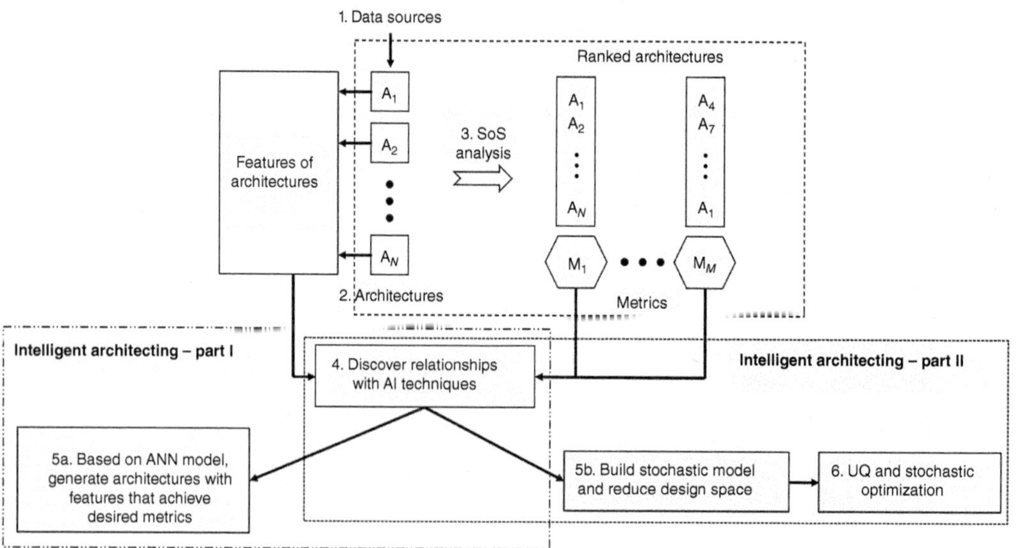

Figure 29.24 Flowchart of the application of AI to enhance the AWB.

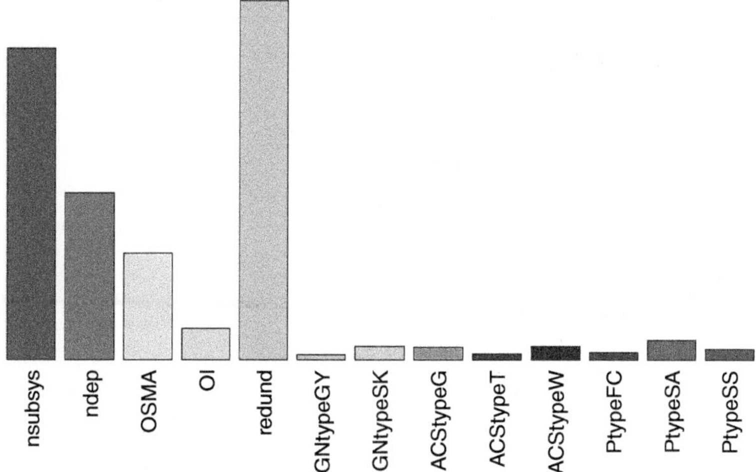

Figure 29.25 Relevance of various architectural features on the SoS-level metric of resilience. The redundancy, number of on-board subsystems, and their dependencies are the most important characteristics driving resilience, followed by orbital semi-major axis (OSMA) and orbital inclination (OI). The other features, including various types of Guidance and Navigation systems, Attitude Control systems, and Propulsion types, have little impact on resilience.

To reduce the design space, UQ and stochastic optimization have been used to identify what architectural features have the highest amount of impact on metrics of interest. For example, Figure 29.25 shows the relevance of the SoS architectural features on resilience. This approach allows designers to focus on a reduced design space, modifying the features that will provide the desired values of metrics of interest.

Multi-domain Battle Scenario and Development of the Decision Support Framework

After various demonstrations and applications that showed the power of the conceptual foundation of the AWB tools, the next step undertaken by researchers has been the development and implementation of better interfaces, and the enhancement of usability of multiple tools on the same problem. These steps produced a tool called the Decision Support Framework (DSF), where hierarchical SoS architectures are produced with the use of RPO and later further analyzed with SODA. The initial demonstration has been executed on a multi-domain battle scenario, using requirements from mission engineering. Figure 29.26 shows the main GUI window for the DSF.

Continuous Development

As evident from the long history of development and use of the AWB in the previous SERC projects, the tools are continuously being improved, with more features added to increase user-friendliness and to address new needs and requirements brought to the attention of the researchers. To avoid unnecessary complexity in further developing the tools, we are taking steps to evolve the subset of AWB tools contained in the DSF, that is RPO and SODA. In an ongoing research project on data-driven capability portfolio management, we adapted previously developed tools to create

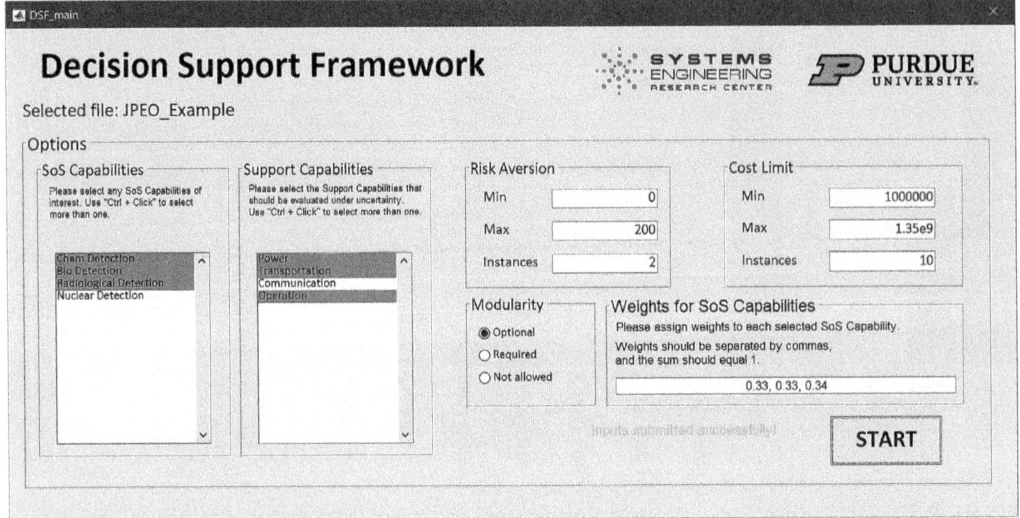

Figure 29.26 GUI for the Decision Support Framework.

a decision-support prototype tool for complex mission threads. The intention of the prototype tool development is for the stakeholders to use the prototype tool to inform decisions in Integrated Acquisition Portfolio Reviews (IAPRs).

Continuous Improvement of AWB Tools, Inclusion of External Tools, and New Graphic User Interfaces (GUIs)

As described in the section titled "Multi-domain Battle Scenario and Development of the Decision Support Framework", in the DSF (Figure 29.27), RPO and SODA provide supportive information for the SoS decision-making process. Furthermore, the DSF also incorporates assessment of adherence to the principle of Modular Open Systems Approach (MOSA) that employs modular architectures using widely accepted standards (Dai et al. 2022) and balances modularization and cohesion in system architecture (Davendralingam et al. 2019).

Figure 29.28 shows the relationship between the AWB and the DSF. The DSF includes only a subset of tools from the AWB, and these tools are used sequentially, with inputs to SODA based on a rough estimate of operational dependencies, resulting from RPO analysis. In ongoing research, we decided to start from the DSF, to avoid overcomplicating the problem by considering all the tools in the AWB for combined use.

Typically, systems engineering tools focus on the system at the level that the stakeholders are focusing on without looking at the portfolio of systems. That is, the systems engineering tools may not translate the complexities of mission engineering analysis into the evaluation in a way that is both (1) meaningful to the requirements within the trade space of capabilities and (2) flexible, scalable, and configurable to integrate with other analyses. To this end, recommendations from the DoD advisory panel suggested that the DoD approach should take a more holistic and portfolio-centric method for acquisitions rather than the current program-centric approach. Thus, the researchers developed system-of-systems engineering tools instead of dealing with systems engineering tools. This way, systems and technologies are evaluated within an overall portfolio, exposing how each component plays a role in the realized capability while connecting the mission needs of warfighters with acquisition decisions.

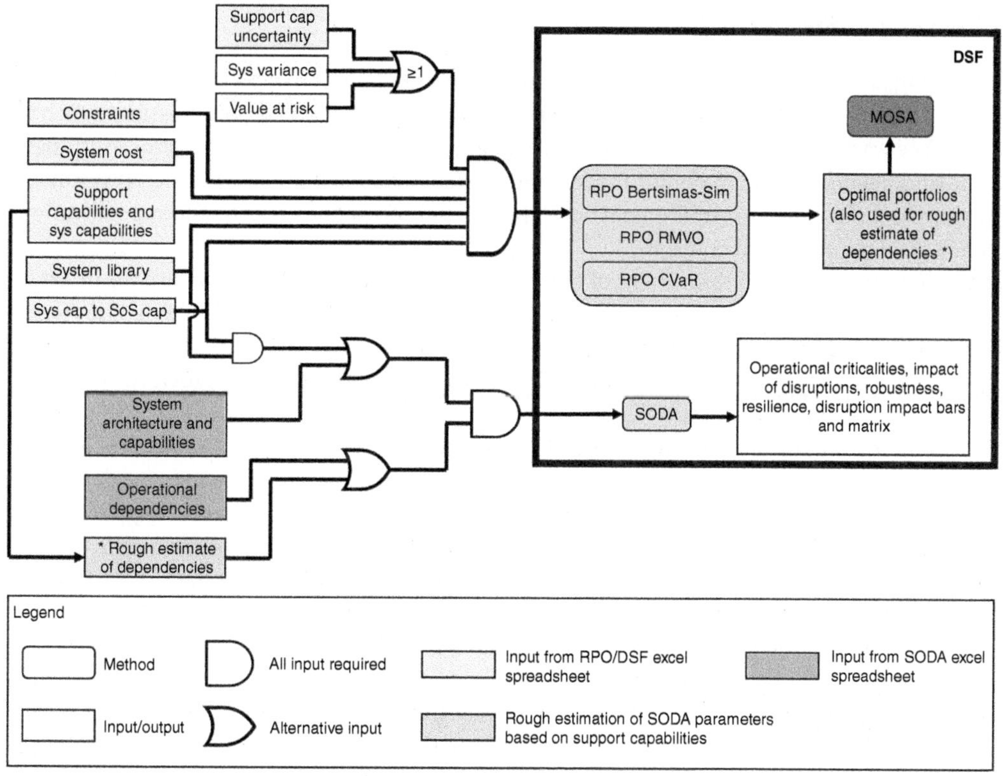

Figure 29.27 Schematics of DSF.

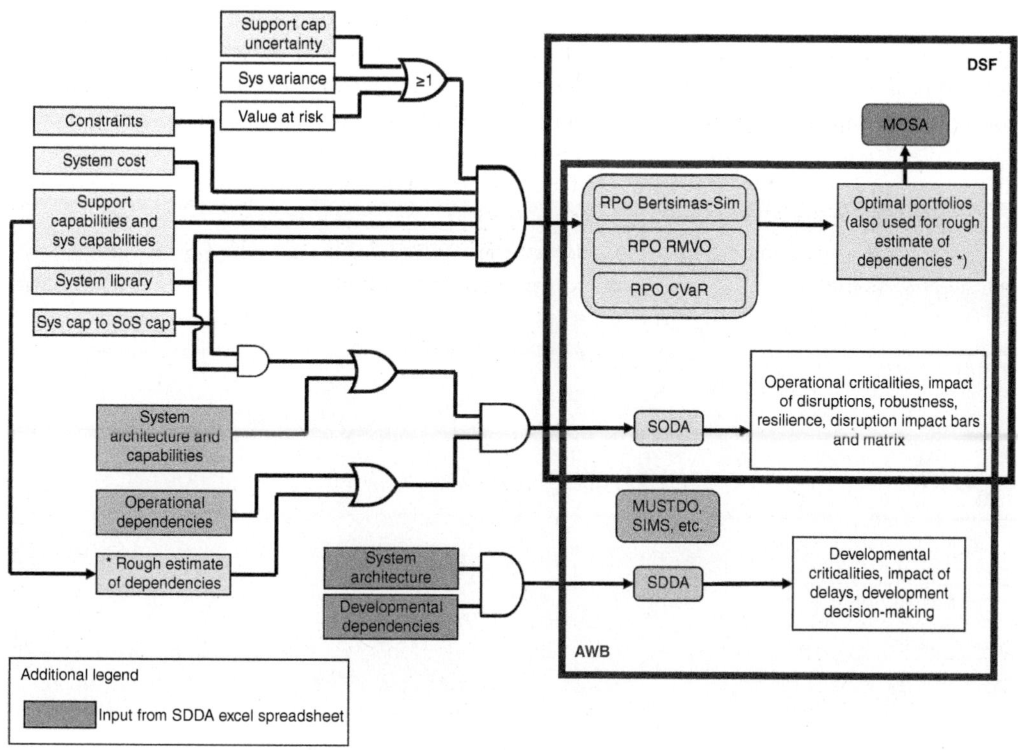

Figure 29.28 Relationship between the AWB and the Decision Support Framework, which is a specific subset of AWB.

In an ongoing research effort, the DoD's need for efficient mission engineering analysis drove the prototype SoS tool refinement and external tool insertion for complex missions, such as anti-surface warfare (ASuW) missions, as shown in Figure 29.29. Thus, mission engineering analysis and architecture development for modernization decisions must include detailed investments and prioritization related to requirement development and capabilities selection to support various technological improvement concepts.

This effort saw not only a refinement of the previously existing AWB tools, but also the insertion of external tools in the AWB. Figure 29.30 demonstrates the current status of the AWB. Prior to this effort, a set of MATLAB-based scripts was used to implement RPO (Shah et al. 2015). Researchers converted these scripts into a Python-only library. The Python-based RPO provided better and easier integration and validation.

Furthermore, the researchers implemented GUIs based on Jupyter Notebook to define problem data and visualize RPO and SODA outputs. This way, all tools can rely on open-source dependencies and have better portability, which is of fundamental importance for practical application of the AWB. As a result, all prototype toolsets were integrated into version-controlled and installable Python products. The process included adding unit and integration testing, static code analysis, and implementation of Continuous Integration and Continuous Delivery (CI/CD). Moreover, the researchers implemented new mission engineering applications and scenarios for RPO, as new mission threads necessitated new requirements for representation of capabilities. Finally, the researchers worked on automation of the data transfer from RPO to SODA.

In addition to refining the AWB tools, the researchers added external inputs and tools to the AWB toolset. Figure 29.30 shows these additions. First, the figure (or Figure 29.30, if needed, though the repetition is inelegant) shows a block marked "new technology injection", which is a framework that allowed to create models of Earth observation using Synthetic Aperture Radar (SAR) satellites. The injection of new systems in the input to AWB tools was conducted with Tradespace Analysis Tool for Designing Constellations (TAT-C) (Chell et al. 2022). The TAT-C is a design-phase mission-analysis tool. The tool was built for testing new earth science mission concepts and implemented in Jupyter Notebook. The TAT-C provided the spatial and temporal

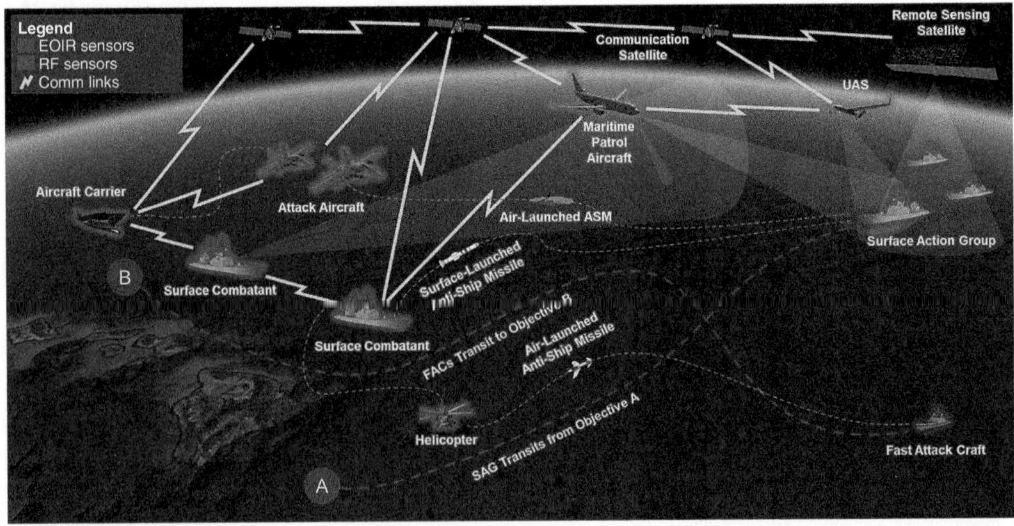

Figure 29.29 An ASuW example problem: Notional ASuW Scenario OV-1 (provided by GTRI).

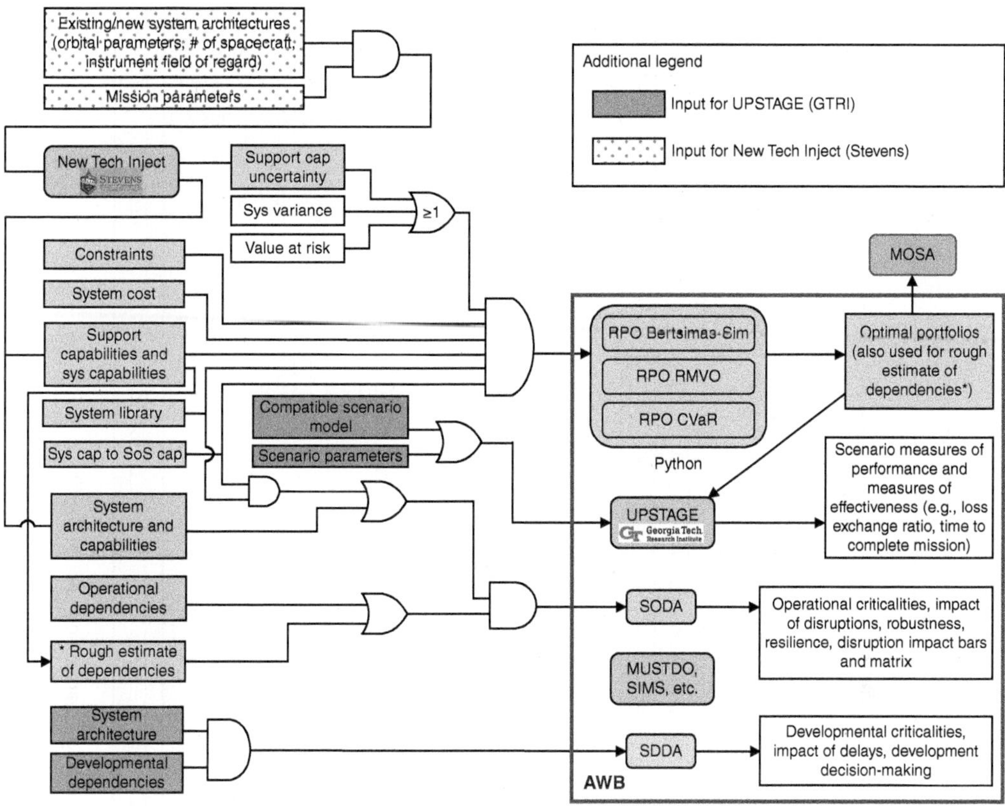

Figure 29.30 Addition of tools developed by the AIRC WRT-1049.5 researchers.

earth coverage metrics output. Thus, this new tool was incorporated into the decision tool for critical missions, such as ASuW threads.

Furthermore, researchers in this project incorporated in the AWB a new tool, called the Universal Platform for Simulating Tasks and Actors with Graphs and Events (UPSTAGE). The UPSTAGE is a multi-solution, hybrid simulation, and wargaming framework to support quick-turn operational scenario analysis against evolving and competitive adversaries.

In the version of RPO created for DSF, the input was given in a spreadsheet format (Figure 29.31). This format is convenient for data entry for a complex mission. JSON files have also been introduced to simplify automation of data processing by machines.

Figure 29.32 shows a screenshot of the SoS-AWB GUI developed as a part of this project. The researchers needed to create a better GUI since using complex mission threads, like those of the ASuW problem described above, with multiple analysis tools necessitated the need for usability enhancement.

Figure 29.33 depicts the RPO results of the ASuW example problem with various amount of conservatism. Conservatism provides a degree of protection from uncertainty (Davendralingam et al. 2018; Guariniello et al. 2018). Using the RPO result, we can observe the SoS performance index as a function of cost. For example, the relationship is linear at a lower cost, meaning that the higher investment results in higher performance at this stage. On the other hand, when the cost is higher, the relationship is no longer linear. This observation is interpreted as solutions that are approaching the point of maximum yield in terms of SoS performance. From the result, the

No.	System Type	System Name	Support Input Requirement													System Capabilities (Outputs)					Cost
			VLS Cell	Harpoon Launcher	NSM Launcher	Large Hardpoint	Small Hardpoint	Hellfire Hardpoint	NextGen Hardpoint	Officers Personnel	Enlisted Personnel	Flight Personnel	Cueing	SATCOM Bandwidth	Area Air Defense	SC 1 = Maritime Surveillance	SC 2 = Identify Surface Contacts	SC 3 = Jam Ship Radars	SC 4 = Standoff Range	SC 5 = Disable Surface Combatant	
-	-	-	#	#	#	#	#	#	#	#	#	#	Non-consumable	Notional BW: 1/3/9 - L/M/H	Area Air Defense	Notional Area Surveillance Capability (1/3/9 - L/M/H)	Notional ID Capability (1/3/9 - L/M/H)	Notional Jamming Capability (1/3/9 - L/M/H)	Weapon Range (nm)	P_hit SC	($MUSD)
1	Space Assets	Legacy SAR Satellite								5	12			9		9					$1.923
2		Small SAR Satellite								1	2			1		9					$0.577
3		EO/IR Satellite								1	5		1	3		3					$4.808
4		Comm Satellite								3	12										$1.923
5	Aviation Assets	MQ-4C								1	2		1	3		3	9	1			$0.952
6		P-8A								2	5	2	1	1		3	9	3			$1.827
7		EA-18G									7	3		1		1	3	9			$0.888
8		F/A-18E/F										1	1	1			1				$0.643
9		MH-60S									2	2	1	1			1	1			$0.359
10		F-35B										1	1	1			3	3			$0.971
11		F-35C										1	1	1			3	3			$0.908
12	Surface Assets	INDEPENDENCE (LCS-2)								11	67			3		1					$2.115
13		FREEDOM (LCS-1)								11	87			3		1					$2.115
14		ARLEIGH BURKE (DDG-51)								25	278			3		1					$1.538
15		MAHAN (DDG-72)								25	278			9		1					$1.827
16		OSCAR AUSTIN (DDG-79)								28	295			9		1					$2.308
17		JACK LUCAS (DDG-125)								28	295			9		1					$4.231
18		ZUMWALT (DDG-1000)								10	165			3		3					$7.692
19		TICONDEROGA (CG-47)								30	300			3		1					$2.885
20		BUNKER HILL (CG-52)								30	300			9		1					$3.077
21		WASP (LHD-1)								40	600			9	2						$3.692
22		AMERICA (LHA-6)								38	550			9	2						$6.538
23		FORD (CVN-78)								300	2300			9	3						$19.997
24	Munitions	AGM-84H/K				1													150	10%	$3.300
25		BGM-109 Blk V	1											1					1350	2%	$1.409
26		RIM-174	1																130	85%	$4.318
27		AGM-158D JASSM-XR							1					1					970	95%	$1.500
28		AGM-158C LRASM				1								1					300	99%	$3.960
29		AGM-84D				1													50	5%	$0.500
30		AGM-84F				1													170	5%	$0.600
31		RGM-84F		1															150	5%	$0.600
32		AGM-119					1												100	10%	$0.800
33		RGM-184A (NSM)			1									1					100	80%	$2.194
34		AGM-114L						1											6	10%	$0.150
35	Personnel	Navy Officer Personnel																			$0.006
36		Navy Enlisted Personnel																			$0.004
37		Navy Flight Personnel																			$0.005

Figure 29.31 Screenshot of RPO input file for the ASuW example problem (showing a partial segment of the file).

Analytic Workbench
SYSTEMS ENGINEERING RESEARCH CENTER
AIRC ACQUISITION INNOVATION RESEARCH CENTER
Run Analysis
Start
Log
Analysis
RPO - Robust Portfolio Optimization
SODA - System Operational Dependency Analysis
SDDA - System Developmental Dependency Analysis
GTRI - Notional GTRI Tool
SI - Notional Stevens Institute Tool
VT - Notional Virginia Tech Tool
Scenario
Scenarios are currently analysis-specific. Please select an analysis tool first.
Use built-in example
Build a new scenario - coming soon
Import from file:
File
Open (0)
Append to output
Next

Figure 29.32 Screenshot of AWB GUI.

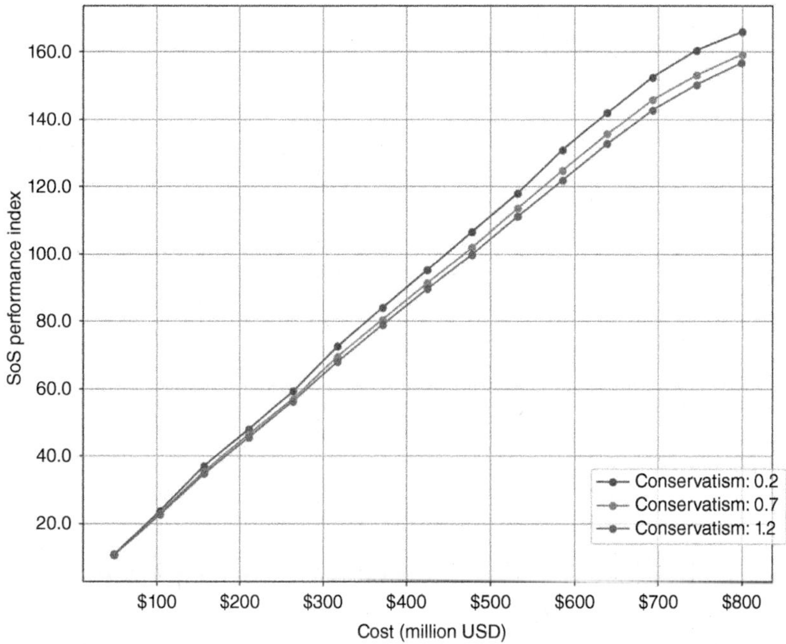

Figure 29.33 Screenshot of RPO results of the ASuW example problem with various degrees of conservatism.

stakeholders can visualize the cost (i.e., investment) and SoS performance index (i.e., return), so they can make an informed decision based on the model-based analysis.

Figure 29.34 presents the results of RPO analysis in an ASuW example with possible alternatives for the portfolio. The potential cost of alternatives ranges from US\$50M to US\$800M. Using this result, the stakeholders can understand precisely what combinations of assets are optimal for the

	Low-cost alternatives							High-cost alternatives						
Alternative	0	1	2	3	4	5	6	38	39	40	41	42	43	44
Objective value	10.7	10.7	10.7	22.6	22.2	21.8	35.4	146.1	159.6	157.6	156.5	169.5	166.7	164.6
Cost	$ 50.00	$ 49.94	$ 49.96	$ 103.53	$ 103.53	$ 103.57	$ 157.14	$ 692.86	$ 746.40	$ 746.36	$ 746.42	$ 800.00	$ 800.00	$ 800.00
Max conservatism	0.2	0.7	1.2	0.2	0.7	1.2	0.2	1.2	0.2	0.7	1.2	0.2	0.7	1.2
Legacy SAR satellite	0	0	0	0	0	0	0	0	1	1	0	0	0	0
Small SAR satellite	1	1	1	2	2	2	3	6	5	5	8	8	8	8
EO/IR satellite	0	0	0	0	0	0	0	0	0	0	0	0	0	0
Comm satellite	2	2	2	4	4	4	6	19	20	20	20	20	20	20
MQ-4C	12	12	12	15	15	20	20	20	20	20	20	7	1	0
P-8A	4	4	4	3	4	4	4	4	4	4	4	4	4	4
EA-18G	0	0	0	0	0	0	0	0	0	0	3	1	1	1
F/A-18E/F	0	0	0	0	0	0	0	61	83	114	107	92	88	87
MH-60S	0	0	0	0	0	1	1	0	1	1	0	0	1	0
F-35B	0	0	0	8	8	0	20	0	0	0	0	0	0	0
F-35C	0	0	0	0	0	0	0	19	77	46	50	67	71	72
INDEPENDENCE (LCS-2)	0	1	1	0	0	1	0	0	0	0	0	0	0	0
FREEDOM (LCS-1)	1	0	0	0	0	0	0	0	0	0	0	0	0	0
ARLEIGH BURKE (DDG-51)	1	1	1	1	1	2	1	5	3	2	2	3	3	3
MAHAN (DDG-72)	0	0	0	0	0	0	0	0	0	0	0	0	0	0
OSCAR AUSTIN (DDG-79)	0	0	0	0	0	0	0	0	1	0	0	1	0	0
JACK LUCAS (DDG-125)	0	0	0	0	0	0	0	0	0	0	0	0	0	0
ZUMWALT (DDG-1000)	0	0	0	0	0	0	0	0	0	0	0	0	0	0
TICON DEROGA (CG-47)	0	0	0	0	0	0	0	0	0	0	0	0	0	0
BUNKER HILL (CG-52)	0	0	0	1	1	0	1	1	2	3	3	2	3	3
WASP (LHD-1)	0	0	0	1	1	0	1	0	0	0	0	0	0	0
AMERICA (LHA-6)	0	0	0	0	0	0	0	0	0	0	0	0	0	0
FORD (CVN-78)	0	0	0	0	0	0	0	1	2	2	2	2	2	2
AGM-84H/K	0	0	0	0	0	0	0	0	0	0	0	0	0	0
BGM-109 Blk V	5	5	5	0	0	28	2	263	145	174	179	211	228	230
RIM-174	0	0	0	0	0	0	0	0	0	0	0	0	0	0
AGM-158DJASSM-XR	0	0	0	27	25	0	45	36	99	61	60	77	66	61
AGM-158C LRASM	0	0	0	0	0	0	0	0	0	0	0	0	0	0
AGM-84D	6	6	6	6	7	7	8	125	174	236	221	191	183	113
AGM-84F	2	2	2	0	1	1	0	5	0	0	1	1	1	69
RGM-84F	8	8	8	16	16	16	15	48	48	40	40	48	48	48
AGM-119	0	0	0	0	0	0	0	0	0	0	0	0	0	0
RGM-184A (NSM)	1	1	1	0	0	1	0	0	0	0	0	0	0	0
AGM-114L	19	19	19	0	0	26	0	0	0	0	0	0	0	0
Navy officer personnel	64	66	67	133	137	109	145	577	864	847	863	849	851	854
Navy enlisted personnel	439	419	420	1277	1285	747	1334	4315	6668	6395	6409	6630	6655	6633
Navy flight personnel	9	11	12	15	19	15	31	92	171	173	180	183	175	174

Altenatives 7–37 were omitted from this figure.

Figure 29.34 Screenshot of RPO results of ASuW example problem, shown with low-cost (Alternatives 0 to 6) and low-cost alternatives (Alternatives 38 to 44).

specific mission thread, how much the operation costs, and how much SoS performance (i.e., objective values) they can obtain. In some cases, the stakeholders can see that the objective values and costs may be similar. However, the allocations of assets can be very different, due to consideration of uncertainty, constraints, and required support inputs for each system in the portfolio.

Future of the Analytic Workbench

The story of the initial development of the AWB and its subsequent evolution and application to a variety of problems highlighted the importance of a gradual approach to the development of the current product. It also showed how the concurrent development of tools and application to practical problems can suggest the best directions of evolution of the tools. Various options are currently being considered for the future of the AWB, including implementation improvements, conceptual changes, and development of better user interfaces. One of the most important areas of improvement that will touch multiple aspects of the AWB concerns the "independence and compatibility" of the tools. In the current configuration, most of the applications that make use of more than one AWB tool use a waterfall or sequential approach (Figure 29.35). In this approach, the data flow is unidirectional, and the direction of the data flow is predetermined. Each tool analysis is treated as a distinct event. Moreover, each analysis is completed within the tool before moving on to the subsequent tool for further analysis. The advantage of this approach is its simplicity. However, this approach has two shortcomings due to its inherent rigidity: first, the rigidity limits the usability of the tools since there is no feedback among them. The waterfall approach assumes that interactions with the stakeholders are unnecessary during the analysis and that the requirements do not change throughout the process. Second, the sequential use of the tools does not guarantee that the capabilities of each tool will be used at their fullest. For example, running the analysis of operational criticalities only on portfolios resulting from RPO can potentially prevent the user from identifying solutions with very high robustness according to SODA analysis, but which are discarded by RPO as being suboptimal in terms of nominal performance.

Complex SoS problems are dynamic in nature due to the multiple interactions among systems. In addition, they present multiple characteristics, each of whom is analyzed by a subset of the AWB tools. An agile approach (Figure 29.36) can address the difficulty associated with dynamic problems. In this approach, there is a multi-directional data flow among tools. Data propagation can occur among tools so that the tools can provide feedback to each other. Thus, the agile approach is suitable for dynamic problems.

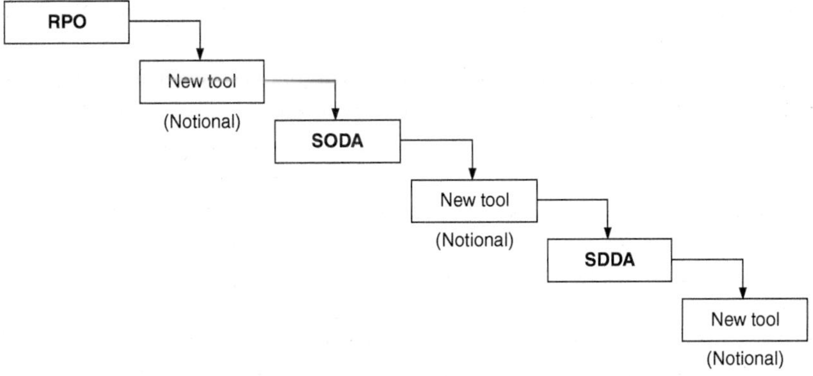

Figure 29.35 Waterfall approach in AWB.

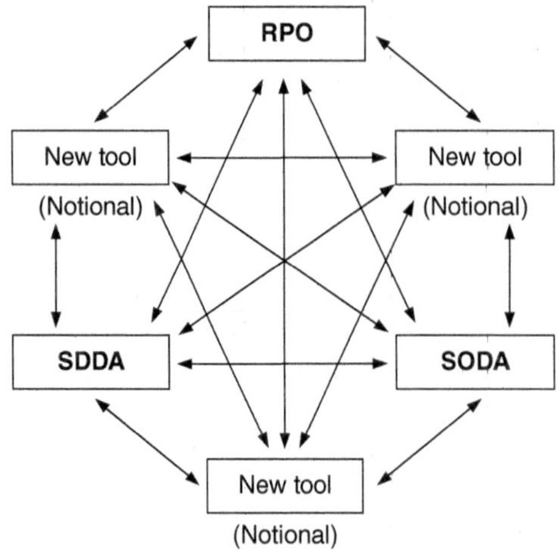

Figure 29.36 Agile approach in AWB.

An additional advantage of the agile approach is the integration of AWB with external tools. In complex SoS problems, the AWB must interact with additional tools. If, for instance, researchers work on a hypersonic vehicle design problem as an SoS problem, the AWB tools must accommodate physics-based tools incorporation with the AWB tool suite. Physics-based tools can include aerodynamics-modeling tools, ablation-modeling tools, and trajectory-modeling tools. When the researchers use the AWB tools in conjunction with the physics-based tools, the conglomerate of tools can collectively analyze the problem, thereby developing high-fidelity final solutions. The difficulty with the agile approach is to define appropriate representations of the systems that support the multiple flows of data, as well as a framework to determine how to properly drive the propagation of data among the tools.

Conclusions

This multi-year, multi-researcher effort has been educational in many ways. We learned a lot about how to model and analyze complex SoS. While the theoretical development of the different tools posed interesting intellectual challenges, the practical aspects of making the workbench work posed their own challenges. Academic researchers often only develop tools to the point where they can demonstrate theories and methods they developed. In this case, we wanted to create a practical workbench that others could use, and in addition this workbench would consist of multiple different tools. Our experience should be a helpful example to other researchers looking to take academic theories to practical application.

This chapter showed how the AWB was developed with both the academic perspective, including innovative and groundbreaking research, addressing identified gaps in the domain of SoS analysis, and the practical perspective of how to quickly provide support to users and sponsors. The commonly occurring discrepancy between academic research and practical application is due to factors on both ends of the relationship: from the academic side, sometimes the general attitude is unique, especially among those who work on fundamental research. Some researchers communicate with other researchers within their discipline with a language that is so specialized that can appear esoteric to an outsider. If the conversation is incomprehensible, external users who have practical problems to solve might perceive this not only as a complication, but as a barrier that academic researchers are building.

From the side of practical applications, some users often exhibit a preconception that academic research has little relevance to real-life applications. An old article published in Nature in 1892 stated, "if universities do not study useless subjects, who will?" (Fitzgerald 1892). While a gap actually exists between academic research, especially fundamental research, and practical applications, our effort with the AWB highlighted how an approach is possible where academic research can be supported by practical applications, in a combined effort. This effort requires researchers to be able to explain how the fundamental concepts that they developed can be applied in practice and requires users to be able to understand at least the basics of these concepts, so as to be able to provide guidance to the evolution of the tools. The clear advantage in this ongoing effort is the development of tools that provide fundamental research and reduction of important gaps identified in the field of SoS, as well as usable and useful support to decision-making in practical problems.

References

Abbott, B.P. (2008). *Littoral Combat Ship (LCS) Mission Packages: Determining the Best Mix*. Monterey, CA: *Naval Postgraduate School*.

Bertsekas, D. P. and Tsitsiklis, J. N. (1995). Neuro-dynamic programming: an overview. *34th IEEE Conference on Decision and Control*. New Orleans, LA.

Chavers, G., Suzuki, N., Smith, M. et al. (2019). *NASA's human lunar landing strategy*. Washington, DC: International Astronautical Congress.

Chell, B., LeVine, M.J., Capra, L. et al. (2022). Conceptual design of space missions integrated with real-time, in situ sensors. In: *Transdisciplinarity and the Future of Engineering* (ed. B.R. Moser et al.), 350–359. IOS Press.

Dahmann, J., Rebovich, G., Lane, J. et al. (2011). An implementers' view of systems engineering for systems of systems. In: *2011 IEEE International Systems Conference*, 212–217. Montreal, Canada.

Dai, M., Guariniello, C., and DeLaurentis, D. (2022). Implementing a MOSA decision support tool in a model-based environment. In: *Recent Trends and Advances in Model Based Systems Engineering* (ed. A.M. Madni, B. Boehm, D. Erwin, et al.), 257–268. Springer International Publishing.

Davendralingam, N. and Delaurentis, D.A. (2015). A robust portfolio optimization approach to system of system architectures. *Systems Engineering* 18 (3): 269–283.

Davendralingam, N., Guariniello, C., and Delaurentis, D. (2018). A robust portfolio optimization approach using parametric piecewise linear models of system dependencies. In: *Disciplinary Convergence in Systems Engineering Research* (ed. A.M. Madni, B. Boehm, R.G. Ghanem, et al.), 83–96. Springer International Publishing.

Davendralingam, N., Guariniello, C., Tamaskar, S. et al. (2019). Modularity research to guide MOSA implementation. *The Journal of Defense Modeling and Simulation* 16 (4): 389–401.

Delaurentis, D.A., Marais, K., Davendralingam, N., et al. (2017). System of systems analytic workbench toolset. *Tools Available on Nanohub, Version 1.5*. https://nanohub.org/resources/plottool (accessed 24 June 2023).

DeLaurentis, D.A., Moolchandani, K., and Guariniello, C. (2022). *System of Systems Modeling and Analysis*. Boca Raton, FL: CRC Press.

Elsayed, E. (1996). *Reliability Engineering*. Addison Wesley Longman Inc.

Fang, Z., Davendralingam, N., and DeLaurentis, D. (2018). Multistakeholder dynamic optimization for acknowledged system-of-systems architecture selection. *IEEE Systems Journal* 12 (4): 3565–3576.

Fang, Z. and DeLaurentis, D. (2015). Multi-stakeholder dynamic planning of system of systems development and evolution. *Conference on Systems Engineering Research (CSER)*, Hoboken, NJ (17–19 March).

Fitzgerald, G.F. (1892). The value of useless studies. *Nature* 45 (1165): 392.

Francis, R. and Bekera, B. (2014). A metric and frameworks for resilience analysis of engineered and infrastructure systems. *Reliability Engineering and System Safety* 121: 90–103.

Guariniello, C. (2016). *Supporting Space Systems Design via Systems Dependency Analysis Methodology*. Purdue University.

Guariniello, C. and DeLaurentis, D. (2013). Dependency analysis of system-of-systems operational and development networks. *Procedia Computer Science* 16: 265–274.

Guariniello, C. and DeLaurentis, D. (2017). Supporting design via the system operational dependency analysis methodology. *Research in Engineering Design* 28 (1): 53–69.

Guariniello, C., Fang, Z., Davendralingam, N. et al. (2018). Tool suite to support model based systems engineering-enabled system-of-systems analysis. In: *IEEE Aerospace Conference*, 1–16. Big Sky, MT.

Guariniello, C., Marsh, T.B., Diggelmann, T., and DeLaurentis, D.A. (2021). System-of-systems methods for technology assessment and prioritization for space architectures. In: *IEEE Aerospace Conference (50100)*, 1–13. Big Sky, MT.

Guariniello, C., Mockus, L., Raz, A.K., and DeLaurentis, D.A. (2020). Towards intelligent architecting of aerospace system-of-systems: Part II. In: *IEEE Aerospace Conference*, 1–9. Big Sky, MT. https://doi.org/10.1109/AERO47225.2020.9172585.

Guariniello, C., Mockus, L., Raz, A.K., and DeLaurentis, D.A. (2019). Towards intelligent architecting of aerospace system-of-systems. In: *IEEE Aerospace Conference*, 1–11. Big Sky, MT.

Malcolm, D.G., Roseboom, J.H., Clark, C.E., and Fazar, W. (1959). Application of a technique for research and development program evaluation. *Operations Research* 7 (5): 646–669.

Powell, W.B. (2007). *Approximate Dynamic Programming: Solving the Curses of Dimensionality*. Hoboken, NJ: Wiley.

Ramirez-Marquez, J.E. and Coit, D.W. (2007). Multi-state component criticality analysis for reliability improvement in multi-state systems. *Reliability Engineering and System Safety* 92 (12): 1608–1619.

Rausand, M. and Høyland, A. (2004). *System Reliability Theory: Models, Statistical Methods, and Applications*. Hoboken, NJ: Wiley-Interscience.

Shah, P., Davendralingam, N., and DeLaurentis, D.A. (2015). A conditional value-at-risk approach to risk management in system-of-systems architectures. In: *System of Systems Engineering Conference (SoSE)*, 457–462. San Antonio, TX.

Uday, P. and Marais, K.B. (2014). Resilience-based system importance measures for system-of-systems. *Conference on Systems Engineering Research (CSER)*, Redondo Beach, CA (21–22 March).

US DoD (Department of Defense). (2008a). Systems engineering guide for system-of-systems. https://acqnotes.com/wp-content/uploads/2014/09/DoD-Systems-Engineering-Guide-for-Systems-of-Systems-Aug-2008.pdf (accessed 24 June 2023).

US DoD (Department of Defense). (2008b). Acquisition Guidebooks and References. https://aaf.dau.edu/guidebooks/ (accessed 24 June 2023).

Van der Borst, M. and Schoonakker, H. (2001). An overview of PSA importance measures. *Reliability Engineering and System Safety* 72 (3): 241–245.

Biographical Sketches

Dr. Cesare Guariniello is a research scientist at Purdue's School of Aeronautics and Astronautics. He received his PhD in aeronautics and astronautics from Purdue University in 2016 and his master's degrees in Astronautical Engineering and Computer and Automation Engineering from the University of Rome "La Sapienza." Dr. Guariniello works as part of the Center for Integrated Systems in Aerospace led by Dr. DeLaurentis and is currently engaged in projects funded by NASA, the DoD Systems Engineering Research Center (SERC), and the NSF. His main research interests include modeling and analysis of complex systems and SoS architectures – with particular focus on space mission architectures – aerospace technologies, and robotics. Dr. Guariniello is a senior member of IEEE and AIAA and a member of INCOSE.

Dr. Payuna Uday is a research scientist at the Systems Engineering Research Center (SERC). Her research in systems engineering aims to improve decision-making within the context of designing and operating large-scale, complex systems. She is particularly interested in exploring interdisciplinary approaches to develop tools and metrics that better inform strategies and policies. She received her PhD and master's degrees from the School of Aeronautics and Astronautics at Purdue University. She also holds BTech in electronics and communication engineering from the National Institute of Technology in Trichy, India. Prior to her current position, she was a senior consultant at Landrum & Brown working on air transportation systems, specifically airfield and airspace planning for major US airports.

Dr. Waterloo Tsutsui is a senior research associate in the School of Aeronautics and Astronautics at Purdue University in West Lafayette, IN. Before Purdue, Tsutsui practiced engineering in the automotive industry for 10+ years. Tsutsui's research interests are systems engineering, energy storage systems, multifunctional structures and materials design, and the scholarship of teaching and learning.

Dr. Karen Marais' research focuses on developing ways to deliver value through reliability, safety, and sustainability. On the theme of reliability, she is interested in how decisions about reliability and maintenance affect the value delivered by systems. She has shown, for example, that a focus on cost can result in a decrease in value. On the theme of safety, she is interested in how we can design and operate complex systems to work with human nature, rather than attempting to force people to be constantly vigilant and never make mistakes. She is currently extending this safety work to the prevention of other project failures too. Dr. Marais also works on data-driven approaches to improving aviation safety, in particular, on developing ways to use smartphone and other device data to improve General Aviation safety. On sustainability, Dr. Marais' primary focus is on the environmental impacts of aviation. Here, she investigates operational improvements that can reduce aviation emissions. She is the author or co-author of several technical publications, including 21 journal papers, 2 reports for the National Academies, and 2 book chapters. She received an NSF CAREER award in 2014. Dr. Karen Marais has worked in engineering for two decades. Dr. Marais holds a B Eng in electrical and electronic engineering from the University of Stellenbosch and a BSc in mathematics from the University of South Africa. She also holds a master's degree in space-based radar from MIT. She received her PhD from the Department of Aeronautics and Astronautics at MIT in 2005. Prior to graduate school, she worked in South Africa as an electronic engineer.

Chapter 30

Computational Intelligence Approach to SoS Architecting and Analysis

Cihan Dagli, Richard Threlkeld, and Lirim Ashiku

Missouri University of Science and Technology, Engineering Management and Systems Engineering Department Rolla, MI, USA

Introduction to SoS Architecting and Analysis

"System-of-Systems" is an emerging and essential multidisciplinary area of systems engineering. The SoS definition is as follows: a set or arrangement of systems that results when independent and valuable systems are integrated into a more extensive system that delivers unique capabilities greater than the sum of the capabilities of the constituent parts. The individual systems alone cannot independently achieve the overall goal of the SoS and are dependent upon each other. SoSs consists of several complex adaptive systems that behave autonomously but cooperatively (Dahmann et al. 2008). The constant interaction between the systems and the interdependencies produces emergent properties that cannot be fully accounted for, analyzed, and optimized by the "normal" systems engineering practices and tools. SoS engineering, an emerging discipline in systems engineering, attempts to form a methodology for approaching SoS problems (Luzeaux et al. 2013).

Complex systems in the future will contain complicated logic and reasoning in intricate arrangements. The architecture organization of these systems involves lots of web-like connections and demonstrates the ability of individualized adaptability. The complex systems are designed for autonomy and may exhibit single or multiple emergent behaviors that can be illustrated. The challenge faced with Complex Adaptive Systems (CAS) design is to create an organized complexity that will allow a system or systems to complete its goals. The Complex Adaptive System-of-Systems (CASoS) approach analyzes this considerable uncertainty in socio-technical systems.

Dahmann et al. (2011) address the four categories of SoS: Directed, Collaborated, Acknowledged, and Virtual, these SoSs with various configurations can address managerial challenges and provide optimal results to the SoS manager.These four types of SoS differ based on their managerial control over the participating systems in the SoS and their structural complexity. The spectrum of SoS ranges from Virtual SoSs that are complex systems to directed SoS that represent multiple complicated systems.

Since SoSs grow in complexity and scale with time, it requires that architectures are flexible for understanding and governance for proper management and control. Systems architecting can be defined as specifying the structure and behavior of an envisioned system. Classical system architecting deals with static systems, whereas SoS architecting processes must be completed at a

Systems Engineering for the Digital Age: Practitioner Perspectives, First Edition. Edited by Dinesh Verma.

Companion website: www.wiley.com/go/verma/systemsengineering

meta-level. The architecture achieved at a meta-level is known as the meta-architecture. The meta-architecture sets the tone of the architectural focus (Malan and Bredemeyer 2001). It narrows the scope of the reasonably large domain space and boundary. The architecture is still not fixed, but meta-architecture provides multiple alternatives for the final architecture that the SoS will utilize. Thus, architecting can be referred to as filtering the meta-architectures to finally arrive at the desired architecture for the SoS. SoS architecting involves integrating multiple systems architectures to produce a large-scale system meta-architecture for a specifically designated mission (Dagli and Kilicay-Ergin 2008). SoS achieves the required goal or goals by introducing collaboration between existing system capabilities that are required to create a more extensive capability based on the meta-architecture selected for the SoS. The degree of influence on individual systems architecture through the guidance of the SoS manager in implementing SoS meta-architecture can be classified as directed, acknowledged, collaborative, and virtual. Acknowledged SoSs have documented objectives, an elected manager, and defined resources for the SoS. Nonetheless, the constituent systems retain independent ownership, objectives, capital, development, and sustainment approaches.

The four general types of SoS that are described below:

Acknowledged

- Elected Manager and known resources for the SoS
- Documented objectives for the SoS

Virtual

- Virtual SoS lacks a centrally agreed upon purpose and management authority for the SoS.
- Large-scale behavior emerges and must rely upon relatively invisible mechanisms to maintain it.

Collaborative

- In collaborative SoS, the component systems generally interact voluntarily to fulfill agreed-upon central purposes.
- Acknowledged (FILA-SoS integrated model is based on Acknowledged SoS).
- Acknowledged SoS have resources, a designated manager, and recognized objectives for the SoS; however, the systems maintain their own independent ownership, objectives, funding, and development and sustainment approaches.
- Changes in the systems are based on collaborative interactions between the SoS and the system.

Directed

- Directed SoSs are the SoSs that are an integrated system-of-systems built and managed to fulfill specific purposes.
- The directed SoS is centrally managed during operations to continue to fulfill those purposes and goals, including any changes the system owners might wish to address.
- The sub-component systems maintain an ability to operate independently, but their standard operational mode is subordinated to the central SoS managed purpose.

The dynamic planning for a SoS is a problematic endeavor. Department of Defense (DoD) programs have to constantly face challenges in the acquisition process to integrate new systems and upgrade existing systems over time under constraining threats, reduced budgets, and resource uncertainty over time. The DoD must analyze and project future scenarios and assess the impact of technology on the DoD and stakeholders' changes to the SoS. The DoD is looking for options that assist in affordable acquisition selections and decrease the cycle time for early acquisition and other technology add-ons. FILA-SoS provides decision aiding for stakeholders and the SoS manager.

Acknowledged SoS lie in between this spectrum. This particular SoS is the focal point of our research endeavor. Acknowledged SoS and Directed SoS share some similarities, as both have SoS objectives, management, funding, and authority (Dahmann and Baldwin 2008). Nevertheless, unlike Directed SoS, Acknowledged SoS systems are not subordinated to SoS. However, Acknowledged SoS systems retain their management, funding, and authority in parallel with the SoS. Collaborative SoS are similar to Acknowledged SoS systems in that systems voluntarily work together to address shared or common interests.

Flexible and Intelligent Learning Architectures for SoS (FILA-SoS) integrated model is an approach to address pain points and optimize SoSs. FILA-SoS is based upon the wave model and assists in decision-making for the SoS manager. The model developed called the FILA-SoS utilizes straightforward system definitions methodology and an analysis framework that supports the analysis and understanding of the key trade-offs and requirements by various SoS stakeholders and decision-makers in a short time. FILA-SoS and the Wave Process examine four of the most challenging parts of systems architecting:

1) Examining uncertainty and variability of the capabilities of the SoS and availability of potential component systems within the SoS
2) Resourcing the evolution of the SoS resources, environment, and needs over time
3) Accounting for different approaches and motivations of the stakeholders and autonomous component system managers
4) Optimizing fixed budget and resources for SoS characteristics in an uncertain and dynamic environment

FILA-SoS has many unique capabilities, including integrating models for modeling and simulating SoS systems. FILA-SoS can input multiple SoS systems and can be integrated, optimizing multiple waves. The tool and methodology also have modularity because the model runs models independently and integrates with other models utilized for analysis and optimization. Several models for determining SoS behavior and architecture generation for different complex systems exist. FILA-SoS provides value by aiding the SoS manager in real-time and future decision-making. It also assists the SoS manager in understanding the emergent behaviors of SoSs in an acquisition environment and their impact on SoS architecture goals and quality. FILA-SoS serves as a methodical approach to studying different types of SoS and systems of semi-cooperative, cooperative, and noncooperative. It enables the identifications of intra and interdependencies among SoS elements and the acquisition environment. FILA-SoS provides a "What-if" Analysis depending on variables and resources for the SoS that can be changed during the acquisition process and through wave cycles. It can simulate any architecture utilizing colored Petri nets. FILA-SoS can also simulate rules of engagement and behavior settings: all systems are noncooperative, all systems are semi-cooperative, and all systems are cooperative or a combination. The methodology and tool function as a testbed for decision-makers and stakeholders to evaluate operational guidelines for managing various complex acquisition environment scenarios.

In the real world, systems are complex, nondeterministic, evolving, and have human-centric capabilities. The connections of all complex systems are nonlinear, globally distributed, and evolve in space and time. Because of nonlinear properties, system connections create emergent behavior. Developing an approach to understanding and optimizing large-scale and complex systems is imperative to the architecting of SoSs. The approach and goal are not to constrain the system's design but to design the system such that it controls and adapts itself to the environment quickly, robustly, and dynamically. These complex entities change rapidly and dynamically, including socioeconomic and physical systems. SoSs exist in many areas of the economy and include transportation, health, energy, cyber-physical systems, economic institutions, and communication infrastructures.

System of System Challenges

CASs are at the edge of chaos as they maintain stability through constant changes and evolution. Order and chaos are two complementary states of our world and environment. A dynamic balance exists between these states, and there are constant trades to determine the balance. The order ensures consistency and predictability of systems possible. When there is too much order, rigidity, and suppression of creativity, chaos changes the environment creating disorder and instability, but it can also lead to emergent behaviors being created and allows novelty and creativity. A sufficient balance of order and chaos is necessary for a system to maintain an ongoing identity and enough chaos to ensure growth and innovative development. One of the challenges in Complex Adaptive Systems design is to design an organized complexity that will allow a system to achieve its goals. SoSs are naturally complex systems due to the characteristics that are component systems, operationally independent elements, and are also managerially independent of each other. The component systems preserve existing operations independent of the SoS. SoS has an evolutionary development with time and shows one or many emergent behaviors due to the large-scale complex structure which. Emergence means the SoS can perform functions that do not reside in any single subcomponent system.

INCOSE SoS working group survey identified seven "pain points" presented as questions for systems engineering of SoS, which are listed in Table 30.1 (Dahmann 2012).

The importance and impact on systems engineering of each pain point is illustrated below:

- **Lack of SoS Authorities and Funding and Leadership** pose several management and governance issues for SoS. These conditions significantly impact the ability to implement systems engineering (SE) utilizing classical SE methodologies. Also, this problem affects the modeling and simulation activities of the SoS to optimize the systems accurately.

Table 30.1 SoS and enterprise architecture activity pain points.

SoS and enterprise architecture activity pain points	Question
Lack of SoS authorities and funding	What are effective collaboration patterns in systems of systems?
Leadership	What are the roles and characteristics of effective SoS leadership?
Constituent systems	What are effective approaches to integrating constituent systems into a SoS?
Capabilities and requirements	How can SE address SoS capabilities and requirements?
Autonomy, interdependencies, and emergence	How can SE provide methods and tools for addressing the complexities of SoS interdependencies and emergent behaviors?
Testing, validation, and learning	How can SE approach the challenges of SoS testing, including incremental validation and continuous learning in SoS?
SoS principles	What are the key SoS thinking principles, skills, and supporting examples?

Source: Adapted from Dahmann (2012).

- **Constituent Systems** are essential roles in the SoS. These systems have different interests and ambitions, which may or may not be aligned with the SoS. Data, models, and simulations for these systems will naturally have to be altered to address the systems' specific needs. The altering may not support the SoS analysis or engineering design.
- **Autonomy, Interdependencies, and Emergence** are ramifications of the constituent systems' varied behaviors and interdependencies, describing them as complex adaptive systems. Unpredictable emergence comes naturally in such a state. While modeling and simulation is a valuable tool for representing and measuring these complexities, it is often hard to accurately achieve real-life emergence. This is typically due to a limited understanding of validation's fundamental issues and consequences.
- **Capability of the SoS** and systems individual capability needs may be abstract and high. The need definition and capabilities need to align with the requirements of the SoS mission. The SoS mission depends upon the constituent systems for success, which may or may not be able or willing to address them.
- **Testing, Validation and Learning** becomes extremely difficult because the systems continuously keep adapting and evolving in addition to the SoS environment, which includes stakeholders, governments, etc. Creating a practical testbed for modeling and simulating the large dynamic SoS is challenging due to this constant change. Modeling and simulation can solve part of this problem with live enhancement testing and addressing risks in the SoS when testing is not feasible; however, this requires an accurate representation of the SoS, which can be difficult to be accurate, as discussed in earlier points.

SoS Principles are still being developed, understood, and implemented. The effectiveness is still being determined from these principles, which puts pressure on the continuous development of SoS engineering. Similarly, there is an absence of a well-established agreeable space of SoS principles to drive development and knowledge. This constricts the effective use of potentially powerful tools.

The environment and the systems are continuously changing. Let there be an initial environment model representing the SoS acquisition environment. As the SoS acquisition progresses, these variables are updated by the SoS Acquisition Manager to reflect the current acquisition environment. Thus, the new environment model at a new time has different demands. To fulfill the demands of the mission, a methodology is needed to assess the overall performance of the SoS in this dynamic situation. The motivation for evolution are the changes in the SoS environment (Chattopadhyay et al. 2008). The environmental changes consist of:

- SoS Stakeholder Preferences for key performance attributes
- Interoperability conditions between new and legacy systems
- Additional mission responsibilities to be accommodated
- Evolution of individual systems within the SoS

Architecture evaluation is an additional SoS challenge area as it lends itself to a fuzzy approach because the criteria are frequently non-quantitative, subjective (Pape and Dagli 2013), or based on difficult to define or even unpredictable future conditions, such as "robustness." Individual attributes may be ill-defined, mathematically precise, and in linear functional form from worst to best. Some attributes might offset others depending on how good or bad they affect the SoS performance. Several average attributes coupled with one very poor attribute may be better than an architecture with all marginally good attributes, or vice-versa. A fuzzy approach allows many of these considerations to be handled using a reasonably simple set of rules and to include non-linear characteristics in the fitness measure. The simple rule set allows for minor adjustments to be made to the model to see how small changes affect the outcome of the chosen architecture.

Overview of the FILA-SoS Model Approach

FILA-SoS utilizes a straightforward system definitions methodology and an analysis framework that supports the exploration and understanding of the key trade-offs and requirements by a broad range of system-of-system stakeholders and decision-makers. FILA-SoS and the Wave Process address the most challenging aspects and pain points of system-of-system architecting:

- Integrating uncertainty and variability of the capabilities and availability of potential component subsystems.
- Resourcing the evolution of the SoS needs, resources, and environment over time.
- Accounting for the different approaches and motivations of the autonomous component system managers.
- Optimizing SoS characteristics in an uncertain and dynamic environment with constrained budget and resources (Figure 30.1)

FILA-SoS follows the Dahmann's proposed SoS Wave Model process closely for architecture development of the DoD acquisition process. The overall idea is to select a set of systems and interfaces based on the needs of the architecture in an entire cycle called the wave. Within the wave,

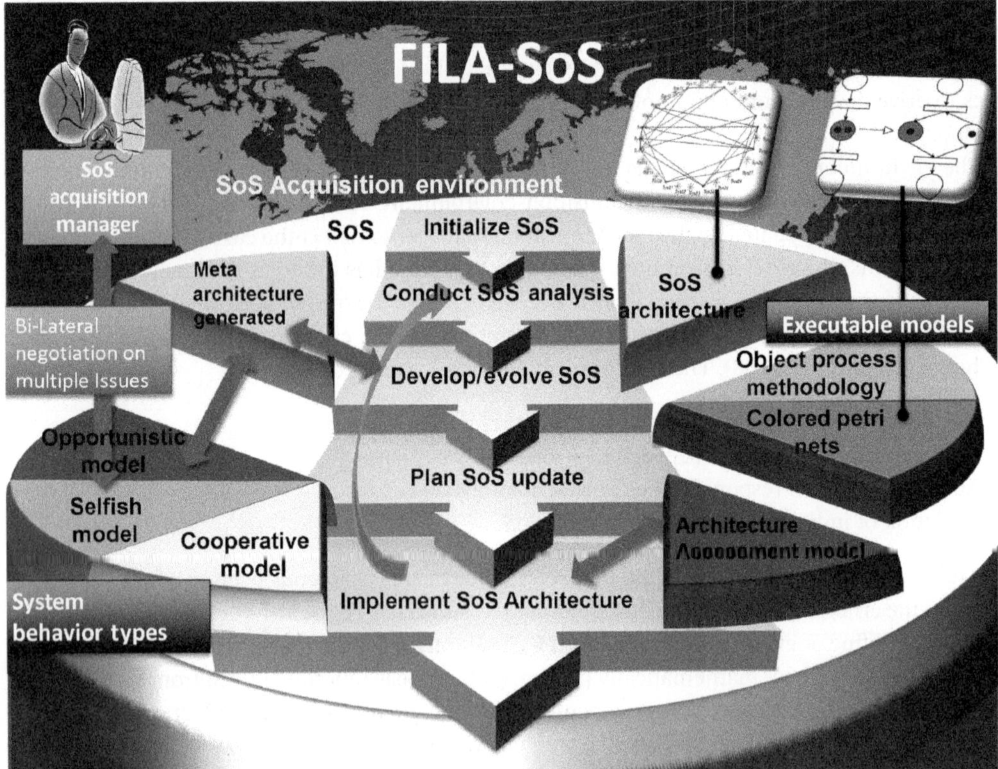

Figure 30.1 FILA-SoS overview.

there may be several negotiation rounds, which are called epochs. After each wave, the systems selected during negotiation in the previous wave remain part of the meta-architecture, while new systems could replace those left out.

The value added by FILA-SoS to systems engineering is that it aids the SoS manager in future decision-making and can help understand the emergent behavior of systems in the acquisition environment. FILA-SoS also facilitate a "What-if" Analysis using variables such as SoS funding and capability priority that can be changed as the acquisition progresses through wave cycles. The parameter setting for all negotiation models can be changed, and rules of engagement can be simulated for different combinations of systems behaviors from various SoS.

Meta-Architecture Generation

In practical application, multiple agents interact with each other to determine their strategies as one's strategy affects the others' outcome. Modeling agents' decision-making processes when interacting with each other can be challenging. There are two categories decision-making sequences: simultaneous decision-making, i.e., when all the agents determine their strategies simultaneously; and hierarchical decision-making, i.e., an order exists for the agents to announce their decisions. In the former category, game theory approaches are used to model the decision-making processes. In the latter case, the inherent hierarchical decision-making process can be modeled using a multi-level optimization approach.

Multi-level optimization models are observed when a set of agents has the authority to affect the decision-making process of the other agents (Anandalingam and Friesz 1992). Different optimization problems exist at different levels in multi-level optimization problems, corresponding to the agents' decision-making in different hierarchical order. The optimization problem of an agent with lower hierarchical order is integrated as a constraint within the optimization problem of an agent with higher hierarchical order. The agent with the higher hierarchical order can inform the following decision makers' optimum strategies while determining their strategy.

Multi-level optimization problems are considered to be one of the most challenging classes of optimization problems. SoS architecting is a multi-objective optimization problem, and considering the possibility of integrating several individual systems results in a multi-objective multi-level optimization problem. Due to the complexity of such problems in large SoSs, the theoretical properties of such problems are limited, and generally, heuristics methods are utilized to provide reasonable quality solutions.

Membership Functions

Membership functions (MF) map the fuzzy values to the real-world values and show the fuzziness between the grades within each attribute. The Matlab Fuzzy Toolbox has built-in shapes and mathematical representations for membership functions. Trapezoidal, Triangular, and the Gaussian smoothed corners of trapezoidal shapes are some of the available options. In real space, further scaling is required for the individual variables of the MFs. The MFs cross at the 50% level between the numbers in the granularity scale from 1 to 4. As an example of an FIS, 1 = Unacceptable, 2 = Marginal, 3 = Acceptable, and 4 = Exceeds (expectations) for each attribute: Performance, Affordability, (Developmental) Flexibility, and Robustness. There is no requirement that the scaling is the same for different attributes. Translating real values to fuzzy values is called fuzzification or fuzzifying. Multiple criteria are combined through the rules in fuzzy space, and the fuzzy output value is de-fuzzified to a crisp value for the SoS assessment. Examples are shown in the use cases.

Nonlinear Trades in Multiple Objectives of SoS

Fuzzy logic can fit highly nonlinear surfaces even with a few rules. The combination of MF shapes and combining rules to fit quite nonlinear surfaces can meet the several required dimensions of complex problems. The actual scaling can be conducted through linguistic variables discovered through the interactions of the stakeholders. The linguistic variables are typically terms such as "very bad," "good," "excellent," etc., but are typically mapped to a quantitative value. For instance, "excellent affordability is a cost between \$10M and \$14M," or "acceptable affordability is a cost between \$16M and \$18M." Suppose the attribute evaluation elements can be categorized in such fuzzy terms as this; relatively simple rules for combining them can result in a straightforward overall SoS evaluation from the resultant FIS or fuzzy rule-based system.

Combining SoS Attribute Values into an Overall SoS Measure

A Mamdani FIS allows the combination of as many input attributes as desired (Fogel 2006). Each attribute is correlated to an objective or dimension in a multi-objective optimization problem. This concept is expanded to include networks of fuzzy systems to cover deep and complicated problems with many dimensions (Gegov 2010), and uncertainties are extended to Type II fuzzy sets. If rules of the form are combined with rules of the form "if attribute one and attribute four are excellent, but attribute five is marginal, then the SoS is better-than-average," etc., which allows for asymmetry or nonuniform weighting among attributes, then very complex evaluation criteria may be described for the SoS.

A Mamdani Type I fuzzy rule set is also known as a Fuzzy Associative Memory (FAM) to combine the attribute values into the total SoS fitness score. Attribute measures are calculated from fuzzy variables from the mappings, and the rules are followed to form a fuzzy measure for the SoS architecture. The SoS architecture is represented as mathematical chromosomes. That measure may be de-fuzzified back to a crisp value for final comparison in the GA through an equivalent mapping in the output space of the overall value. The rules are purposely simple because it is easier for the analyst to understand and explain them to the stakeholders. Also, a few rules within the fuzzy logic system can be advantageous in defining the shape of the resulting surface. Some sensitivity analysis can be completed on the rule sets, and minor changes in the rules may be displayed for comparison. A few examples of the rules are typically in the form: "if all attributes are excellent, then the SoS is superb," "if all attributes except two are superb, then the SoS is still superb," "if any attribute is completely unacceptable, then the SoS is unacceptable." Ten or more rules can give an excellent estimate of the stakeholders' intentions, including significant nonlinearities and complexity (Gegov 2010). The Mamdani FIS allows satisfying two contradictory rules simultaneously by simply including them both in calculating the resultant output value.

The analyst and stakeholders easily understand the linguistic form of these rules. For example, "if any attribute is unacceptable, then the SoS is unacceptable" can be expressed linguistically as a single sentence. However, mathematically a separate rule for each attribute is tested individually to implicate the unacceptability of the SoS. The rules typically come from stakeholder interviews and can be edited by the facilitator and stakeholders working together. If consensus cannot be reached on a rule statement among the stakeholders, both rules can be added to the rule set for evaluation of the architectures. This approach can also help explain the issues to the stakeholders.

Exploring the SoS Architecture Space with Genetic Algorithms Combining the Fuzzy Approach with the GA Approach

One class of evolutionary algorithms that FILA-SoS utilize is the genetic algorithm (GA). GA's key featurc is to evaluate the overall fitness of a series of chromosomes in a "population." Then the algorithm sorts the chromosomes by their fitness and proceeds to the next generation through mutations or crossovers of a fraction of the better fitness chromosomes in that generation. Special rules, mutation rates, and crossover points for certain sections of the chromosome (genes), or deciding which parents are combined, can all be changed as part of the GA approach.

The GA first generation begins with a population of random arrangements of chromosomes built from the meta-architecture, which spans the search space, then sorts them by fitness. A fraction of the better-performing chromosomes is selected for propagation to the next generation through mutation and transposition. A few poorly performing chromosomes may also be included for the next generation to avoid the danger of becoming stuck on a purely local optimum. However, proper selection of transposition and mutation processes can also help avoid this problem.

Combining the Fuzzy Approach with the GA Approach

One class of evolutionary algorithms that FILA-SoS utilize is the genetic algorithm (GA). GA's key feature is to evaluate the overall fitness of a series of chromosomes in a "population." Then the algorithm sorts the chromosomes by their fitness and proceeds to the next generation through mutations or crossovers of a fraction of the better fitness chromosomes in that generation. Special rules, Mutation rates, and crossover points for certain sections of the chromosome (genes), or deciding which parents are combined, can all be changed as part of the GA approach.

The GA first generation begins with a population of random arrangements of chromosomes built from the meta-architecture, which spans the search space, then sorts them by fitness. A fraction of the better-performing chromosomes is selected for propagation to the next generation through mutation and transposition. A few poorly performing chromosomes may also be included for the next generation to avoid the danger of becoming stuck on a purely local optimum. However, proper selection of transposition and mutation processes can also help avoid this problem.

When the tradable values of the attributes are determined from the SoS stakeholder interviews, the range of values of each attribute is decided upon. In each dimension, the entire range is mapped with the stakeholders and contiguously to the granularity described by the membership functions. It is possible that there is not an acceptable arrangement of systems and interfaces. By iterating adjustments of the attribute membership function's edges against a population of randomly generated populations of chromosomes, an acceptable analysis of the SoS behavior could be determined. The exploration phase allows the setting of different membership function edges to take advantage of the variability and changes in the evaluation functions to adapt the GA search toward regions that are more likely to produce a compromise from among the competing attributes.

Use Cases and Applications

There are many use cases in several fields of utilizing meta-architecting systems of systems. Two use cases highlighted in this section are cybersecurity and healthcare. Two very different sectors contain the application of meta-architecting and utilize the FILA-SOS methodology and tools (Ashiku et al. 2022; Dagli et al. 2009, 2013; Dahmann et al. 2009; Threlkeld et al. 2022).

Healthcare Use Case Introduction

According to the US Renal Data System, approximately 750,000 are affected by the end-stage renal disease (ESRD) (Ceren Ersoy et al. 2021). Two primary treatments for ESRD are dialysis and kidney transplant. While both treatments are required to meet the demand for ERSD, transplantation improves quality of life and better survival outcomes on average (Hart et al. 2021). The demand for kidney transplants far outpaces the supply. In the United States, over 115,000 people are on the waiting list for kidney transplants, but in 2019 only 24,273 kidneys were transplanted (Mohan et al. 2018).

As seen in Figure 30.2, the number of kidney transplant waitlist registrations far outweighs the supply. The difference between the number of people on the waiting list and the number of kidney transplants conducted has been increasing at an average rate of 6% from 2010 to 2020. The demand for kidneys is expected to increase continually based on the observed trend from 2010 to 2020. Even with the great demand for kidney transplants, approximately 20% of procured deceased donors' kidneys are discarded in current practice. These lower-quality kidneys (KDPI > 85) have a high probability of being discarded (Axelrod et al. 2018). Although some high-risk kidney discards are unavoidable, even lower-quality kidneys have been life-extending and cost-effective for the appropriate candidates (Alhamad et al. 2019). The discard rate of kidneys rises with the Kidney Donor Profile Index, a regressor of several donor characteristics into a single value denoting graft failure after transplant (Aubert et al. 2019). The high discard rate is linked to high KDPI scores; a

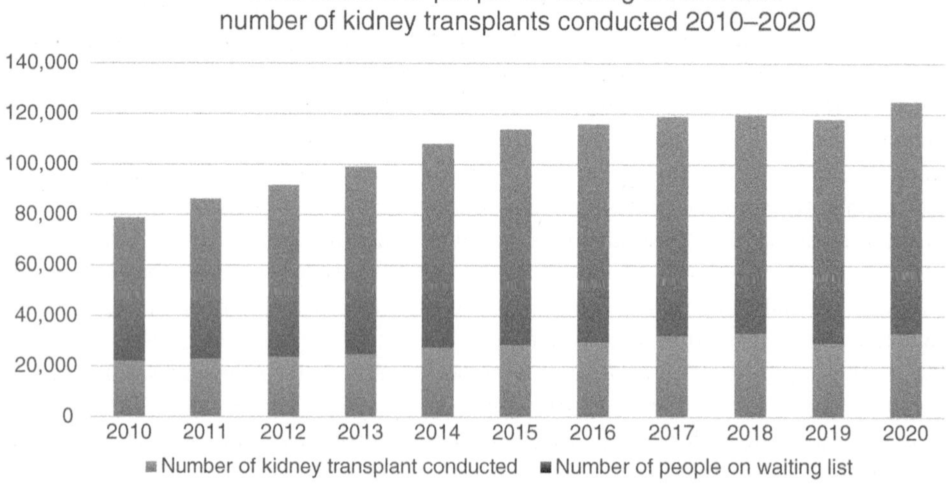

Figure 30.2 Kidney transplants and number of waitlisted candidates. *Source:* Adapted from Ceren Ersoy et al. (2021).

high-KDPI deceased donor kidney is the potential opportunity for increased kidney utilization addressed by optimizing the kidney transplant SoS. Architecture is a fundamental organization of a system, embodied in its components, their relationships to each other and the environment, and the principles governing its design. A meta-architecture is an architecture of the SoS that visualize the interfaces and interactions between each system (IEEE Recommended Practice for Architectural Description for Software-Intensive Systems. 52010-2011 n.d.). A brief analysis of the literature could assist in developing an optimized meta-architecture with a well-defined transplant SoS. The literature suggests multiple approaches that have been utilized to optimize SoS, even in the organ transplant domain. Beliën et al. developed a mixed-integer linear programming long-term decision model to optimize the location of organ transplant centers (Beliën et al. 2013). The Lagrange theory related to the model is studied and presented as a mathematical model based on networks, addressing cost optimization (Caruso and Daniele 2018). A bi-objective mathematical programming model is discussed by Ersoy et al. that optimizes cost and time while considering organ priorities (Ceren Ersoy et al. 2021). While these models optimize certain parts of the organ transplant SoS, they do not attempt to optimize the entirety of the SoS and identify an optimal meta-architecture. Haris and Dagli utilize architecture alternatives using genetic algorithms with fuzzy logic to optimize multi-objective key performance attributes with rule changes and constraints (Haris and Dagli 2011). The SoS Explorer following methods defined by Pape et al. (2013) utilize genetic algorithms of non-gradient descent using fuzzy inference rules to support selecting the optimal meta-architecture for KPAs. This approach has been used in other domains with complex systems, such as cybersecurity (Ashiku and Dagli 2019a, 2019b), inspection systems (Karim and Dagli 2020), and the Department of Defense (Lesinski et al. 2016), but not in the healthcare space. To increase understanding of the architecture's potential for implementation and development, the stakeholders, users, domain experts, and a transdisciplinary research team were created to address the issues with the kidney transplantation SoS (Fritz et al. 2019).

Methodology

Workshops and Interviews with Stakeholders

The first step was understanding and analyzing the stakeholders, work processes, impactful capabilities, metrics, and current functions. The research team conducted three workshops to host tens of stakeholders from the kidney transplant community. The workshops allowed the research team to develop and understand the architecture and the stakeholder's pain points and interactions. The spiral development continuously developed the architecture, and the workshops functioned as an initial validation and verification of the methodology for developing the meta-architecture. The workshops were two hours meetings with significant representation from the stakeholders. We used the first workshop to define the systems and validate the initial workflows of the kidney transplant system based on tens of interviews conducted with the stakeholders. The second workshop adjusted the workflows and introduced decision support systems (Threlkeld et al. 2021). In the last workshop, we proposed metrics summarizing the kidney procurement system for validation.

The stakeholders involved in the kidney transplant system are Organ Procurement Organizations (OPOs), Transplant Centers (TCs), and patients. The OPO is the system that owns the capability of providing kidney offers to transplant centers. The OPO consisted of several key personnel involved in the workshops, including directors, managers, and kidney coordinators. The OPO wants to maximize the number of kidneys transplanted and sends the offers to various TCs in the designated area depicted by the United Network for Organ Sharing (UNOS). UNOS is the governing body of the entire kidney SoS. It provides policy and the patient merit list for the order in which patients are offered kidney transplants. The transplant center's primary capability is to perform a kidney transplant that maximizes life-years gained. The direct personnel involved are the kidney coordinator and the transplant surgeon. All these stakeholders are systems within themselves and have a diverging interest in the kidney SoS. These stakeholders and personnel formed the basis for developing the KPAs and the systems' capabilities.

Key Performance Attributes

The KPAs were developed and defined from the workshops and the interviews with stakeholders of the kidney transplant system. KPAs are calculated by Eqs. (1)–(6), defined as systems, $S(X,i) = 1$ if the i system participates in chromosome X of the meta-architecture and 0 otherwise. Interfaces, $I(X,i,j) = 1$ if the i and j systems have an interface connection in chromosome X, and 0 otherwise. The C_{xxx} represents system characteristics, i value of performance where "*xxx, i*" indicates the characteristic value for the system *I*. The δ variable represents the augmenting factor of the interfaces between active systems. The *Performance* KPA calculates the overall kidneys transplanted for the given participating systems. The *Discard Rate* KPA is the number of kidneys discarded by the SoS. The *OoverE* is the observed kidney transplanted divided by the expected number of kidneys provided by UNOS. The *Credibility* KPA is the F_{score} of the decision support systems. The *Affordability* KPA is a function of each participating system's initial and upkeep costs and the costs of the interacting interfaces. The *Acceptability* KPA is calculated by the feedback score given by each participating system and stakeholder. The optimal meta-architecture is computed using a Simple SOGA genetic algorithm with 10,000 maximum evaluations. Figure 30.3 gives a summary of these equations.

Systems of Systems Explorer

SoS Explorer is a software tool written by the Engineering Management and Systems Engineering Department at Missouri S&T (Missouri University of Science and Technology n.d.). It provides a framework for defining SoS problems and computationally generating optimal architecture. The objectives determine the overall performance of the architecture. These objectives may be defined using Python, MATLAB, or F#. We can then use the selected optimizer to generate optimum

$$\text{Performance} = \sum_i^{Ns} S(X,i) * C_{\text{kidneytransplanted},i} \prod_j^{Ns} [1 + \delta S(X,j) I(X,i,j)] \quad (1)$$

$$\text{DiscardRate} = \sum_i^{Ns} S(X,i) * (C_{\text{Discardrate},i} \prod_j^{Ns} [1 + \delta S(X,j) I(X,i,j)] \quad (2)$$

$$\text{OoverE} = \sum_i^{Ns} S(X,i) * (C_{\text{ExpectedvdObserved},i} \prod_j^{Ns} [1 + \delta S(X,j) I(X,i,j)] \quad (3)$$

$$\text{Credibility} = \sum_i^{Ns} S(X,i) * C_{\text{Fscore},i} \prod_j^{Ns} [1 + \delta S(X,j) I(X,i,j)] \quad (4)$$

$$\text{Affordability} = 1 - \sum_{i=1}^{Ns} S(X,i) * [C_{\text{Operating},i} + \sum_{j=1 \neq i}^{Ns} I(X,i,j)(C_{\text{Inital},j})] \quad (5)$$

$$\text{Acceptability} = \sum_i^{Ns} S(X,i) * (C_{\text{Usersurvey},i} \prod_j^{Ns} [1 + \delta S\,(X,j) I(X,i,j)] \quad (6)$$

Figure 30.3 KPA equations.

architectures, which are displayed in the graphical user interface (GUI) and may be interacted with by the user. Solutions may also be stored as Excel Open XML files (XLSX) or graphically as Portable Network Graphics (PNG) images. For evaluation, the objectives require an architecture that is a set of systems and interfaces in this framework. System information is represented in its characteristics, capabilities, and feasible interfaces. The objectives are evaluated by an optimizer, of which three evolutionary algorithms are included: NSGA-III, MaOEA-DM, and Simple SOGA. Both single and multiple objective optimizations are supported. Furthermore, constrained optimization is supported, and constraints may be added using Python, MATLAB, or F#.

Membership Functions and Fuzzy Assessor

To quantify the ambiguity of the KPAs, a fuzzy assessor and membership functions were developed using MATLAB's fuzzy designer. The granularity of the membership functions is chosen based on the workshops and interviews with stakeholders. In this case, we considered a granularity of five for both input (KPAs) and output (overall objective). However, the granularity can be varied with each KPA. We regarded membership functions based on stakeholders' definitions of *Low, Below Average, Average, Above Average,* and *High* for the KPAs. The process is repeated with each KPA with a group of stakeholders to quantify the KPA.

The output example is a similar exercise with the stakeholder group. The overall objective is evaluated using rules applied to the membership function of KPAs. The membership functions such as *very unacceptable, unacceptable, tolerable, desirable,* and *very desirable* represent the overall assessment of the meta-architecture chosen by the SoS explorer. These membership functions have set rules such as if *Performance, Discard Rate,* and *OoverE* are low and *Credibility, Affordability,* and *Acceptability* are below average, then the *overall output* of the system is *unacceptable.* These rules are created to determine the output desirability of the overall SoS.

Rules

A set of rules is generated to create a fuzzy assessor to evaluate the multiple objectives in assessing the stakeholders' nonlinear trade-offs. Hence, it allowed us to evaluate a six-dimensional nonlinearity of each architecture selected through SoS Explorer. For instance, if performance is low, discard rate is low, OoverE is low, credibility is low, affordability is below average, and acceptability is low, then the overall assessment is that the meta-architecture is very unacceptable. Another side of the spectrum example is if performance is high, discard rate is high, OoverE is high, credibility is above average, affordability is above average, acceptability is high, then the meta-architecture is very desirable. These rules create a KPA dimensional surface to determine the objective function of the meta-architecture of the SoS.

Results

We will examine two use cases: the first is where there is only one donor kidney, and every system has two options to choose from, and the optimal meta-architecture is calculated. There are three kidneys and three systems to select in the second use case, and the optimal meta-architecture is selected. The general methodology described previously is applied to both use cases (Figure 30.4).

The fuzzy Gaussian membership function plot shows the low, below average, average, above average, and high fuzzy membership functions. The performance is calculated using the

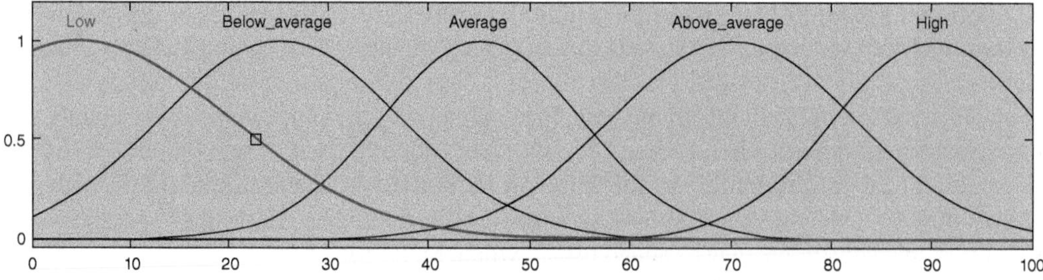

Figure 30.4 The Gaussian membership function plot for the performance KPA.

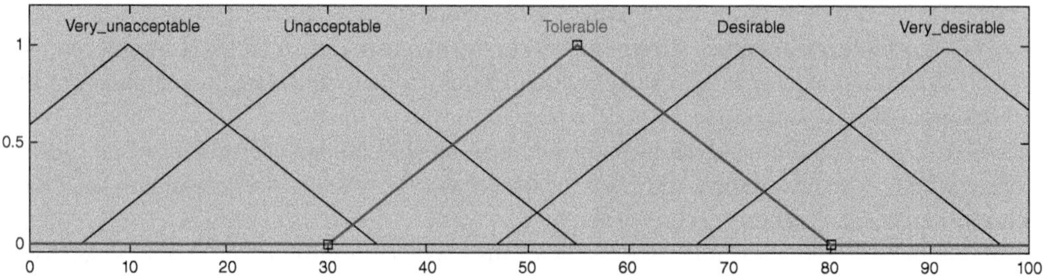

Figure 30.5 Membership functions for the performance key performance attribute.

previously discussed equations and rated based on the fuzzy membership functions. The remaining KPAs were developed using a similar methodology. The objective function is illustrated in Figure 30.5 using triangular fuzzy membership functions. The overall desirability is either very unacceptable, unacceptable, tolerable, desirable, or very desirable based upon the rules and fuzzy memberships of the KPAs.

Both the KPA membership functions, and the overall desirability of the meta-architecture were used in both use case 1 and use case 2.

Use case 1 utilizes a SoS consisting of one kidney donor, two OPOs, two T.C.s, two Organ Procurement Hospitals, two ground transportations, two air transportations, and two decision support systems. Figure 30.6 shows each system, its interface, and its interactions. Each line shows a potential connection and interface of each of the systems. The systems that participate in the systems of systems were chosen to advance each of the capabilities required for the meta-architecture. Each circle within Figure 30.6 represents potential systems that can participate in the meta-architecture.

Each circle that is not shaded is not participating in the meta-architecture. Figure 30.6 shows which systems advance the capabilities required by the SoS. UNOS provides the policy to attempt to optimize through constraints of the behaviors of the systems. OPO 2 provides the kidney offer to TC 2. These capabilities are represented as True in the matrix. Figure 30.7 shows a map representation of Figure 30.6. The potential systems, connections, and interfaces are similarly represented in Figure 30.7.

The meta-architecture for use case 1 was calculated at an overall value of 75.5 and is a desirable overall architecture based upon the fuzzy membership functions. Use case two expands on the flexibility of the meta-architecture, demonstrating three donors and three of each of the systems. The meta-architecture for use case 2 was optimized with a value of 73.8 based on the KPAs. The optimized meta-architecture is deemed desirable due to its 73.8 overall desirability value.

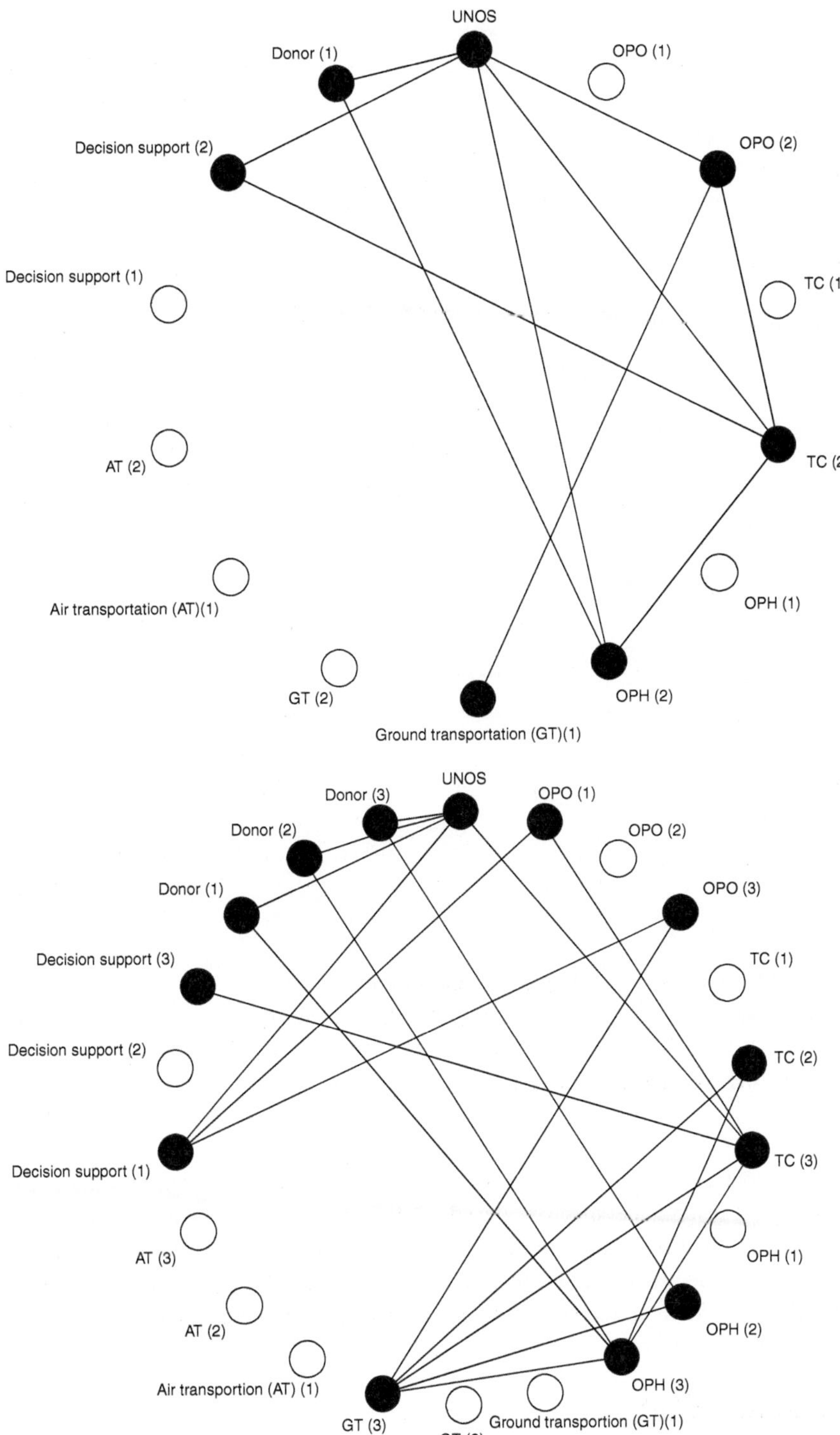

Figure 30.6 Meta-architecture for use case 1 and 2.

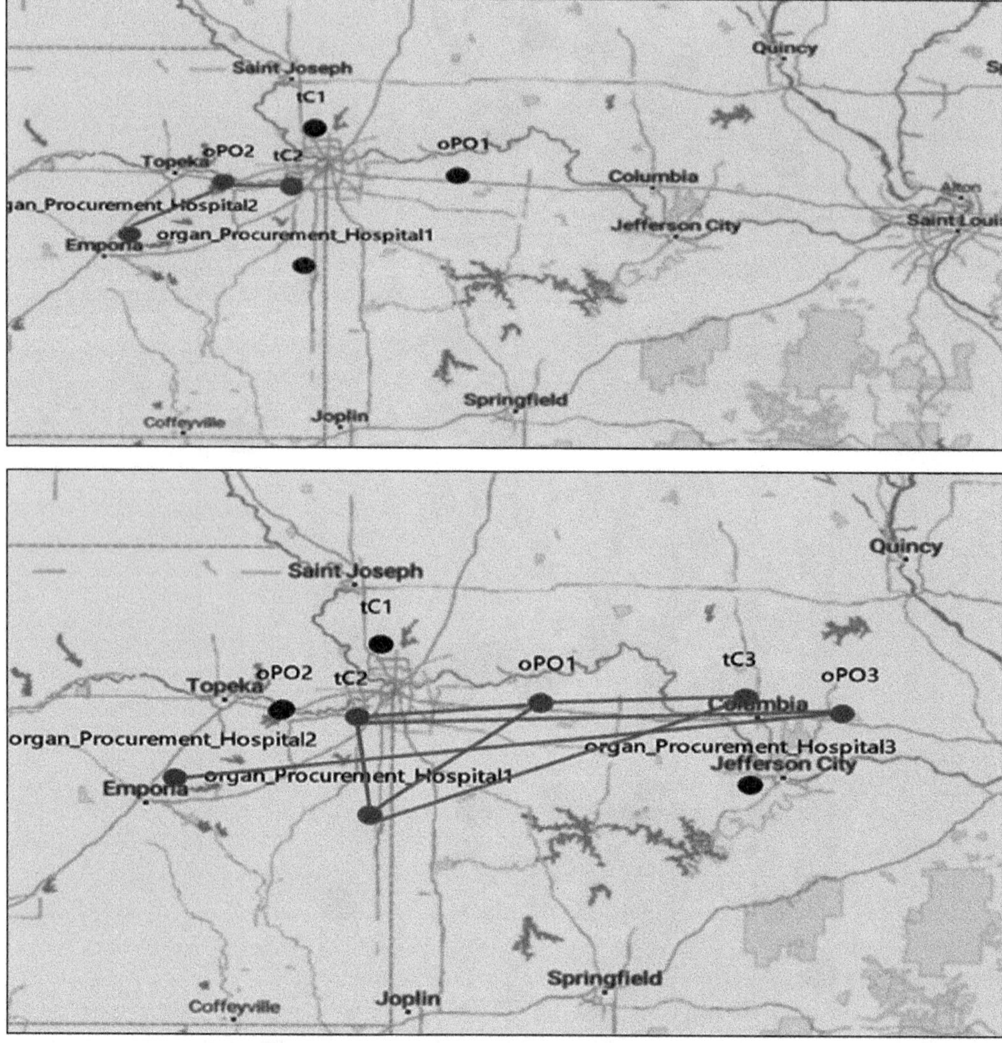

Figure 30.7 Map of meta-architectures for use cases 1 and 2.

The methodology and tool can assist UNOS, OPOs, and TCs in optimizing the donor kidneys for patients from a system of a systems perspective. Based on different regions and stakeholders, the tool can customize KPAs and weighting towards the ultimate objective. This optimization can focus on reducing discard while also considering other KPAs. Future work is still in development for validation and verification with all the stakeholders of the kidney transplant SoS.

Conclusion and Future Work

A methodology was created for the organ procurement SoS to optimize a meta-architecture for selecting the best participating systems for a given set of donor kidneys. Utilizing the KPAs, fuzzy membership functions, rules, and genetic algorithms, two use cases were optimized for the most objective value to the systems. The methodology allows organ procurement stakeholders to develop further or create new KPAs and fuzzy membership functions to optimize their SoS. Future work will integrate the meta-architecture methodology into the AnyLogic platform to perform predictive simulation as an agent-based simulation. The team will also meet with the stakeholders again to conduct additional validation and verification of the method. Improvements will be made to the methodology and make it more accepted by the kidney transplant community. The agent-based simulation can be changed with an expanding list of stakeholders in the organ transplant community. The simulation will enable a real-time analysis to create an optimal meta-architecture for numerous kidneys to reduce kidney discard.

A Use Case of FILA-SoS in Cybersecurity

Introduction

Similar to technological advancements producing a digital society, adversaries are also exploring new techniques and opportunities to advance cybercrime. Reference (Huang et al. 2018) indicates that the estimated costs of cybercrime are expected to reach $6 trillion by 2021 as opposed to $3 trillion in 2015. Securing against imminent cyber-attacks is a crucial concern for digital technology, particularly for organizations that utilize service applications for which there is little knowledge about how they are developed or what is involved in its execution. These applications may be vulnerable to trap doors used as entry points to allow debugging and testing the applications without the need to undergo usual security access procedures (Acheson et al. 2013). The norm of using e-banking or e-payment systems has been scrutinized as easy targets by adversaries manipulating lazy users into performing actions that enable access to confidential information; usually performed virtually via phone, sms, or email. The most common types of social engineering affecting banking organizations include vishing or phishing. Malicious software that invades, damages, or disables computer systems by taking partial or complete control of system operations to infect, steal, or block information for financial or organizational gains is known as malware (ISO/PAS 19450:2015(en) n.d.); the most widespread malware in banking are Trojans and ransomware.

Research has introduced various methods in generating a meta-architecture related to the problem solution space based on the class of SoS and the leverage the SoS manager has over its constituents. A contract negotiation model for constituent participation in the acknowledged SoS is emphasized by Qin, Dagli, and Amaeshi (Haris and Dagli 2011), where the SoS manager attempts to reach conformity through bargaining with selected participants to achieve the SoS's overall goal. An agent-based model for acknowledged SoS in (Cybersecurity n.d.) implemented in software as an abstract class preserving the independence of constituents while representing real world situation, emphasizes the need of system-to-system interaction in achieving the overall objective; participation and capabilities is not only negotiated with SoS manager but also negotiated from system to system. In Haris and Dagli (2011), use of a genetic algorithm with fuzzy logic to determine best fit of multi-dimensional KPAs demonstrated that fuzzy rule changes produced results mimicking environmental effects. Coupling of genetic algorithm with fuzzy logic, type-1, allows attribute evaluation on overlapping regions caused by the fuzziness of the attributes and presented with membership functions at some level of granularity is emphasized in Pape et al. (2013) and Lesinski et al. (2016). Agarwal et al. (2015) emphasizes on Flexible and Intelligent Learning Architectures (FILA) for SoS to generate an opening meta-architecture or a baseline, then negotiate with participating constituents to generate the next meta-architectures where SoS manager negotiates performance, funding, and deadlines for participation. In our model, we will combine methods defined by Pape et al. (2013) and Lesinski et al. (2016), coupling of genetic algorithm with the fuzzy logic and will complement it with Agarwal et al. (2015) to account for evolving cyber threats and defense mechanisms.

Approach

Optimization will be demonstrated through use of SoS Explorer architecting tool connecting genetic algorithm with FIS to provide an intuitive and interactive visualization of meta-architectures. We perform a single optimization using simple Self-Organizing Genetic Algorithm (SOGA) coupled with type-I Fuzzy Logic Toolbox in MATLAB© to handle evaluation of overlapping membership functions to serve as new fitness from which survivors for next iteration are chosen; crossover and mutation is followed by evaluating KPA fitness with process reiteration until termination cycles have been exhausted or convergence has occurred (Pape et al. 2013; Agarwal et al. 2015; Missouri University of Science and Technology n.d.). SoS Explorer includes potential participating constituents as systems; for each system there are related characteristics defining the system and related capabilities that the system will exhibit to demonstrate the ability to contribute to the overall objective of cybersecurity. Table 30.2 presents potential systems and associated capabilities represented in binary form of either 1 or 0. In this case, 1 indicates that the system advances that particular capability, whereas a 0 reflects otherwise. In our example, we have 24 potential participating systems and 10 capabilities. Two of the systems are expanded into constituents to incorporate defense mechanisms positioned in different settings, which collaborate to generate a higher performance value, meaning that their ability to secure from various types of anomalies that fall under same category improves with constituent interaction. Yet, in this example, not all systems have the ability to interface with all other systems; although this statement may be contested in other settings, in this case research and experience shows that adding an interface between unrelated (noncommunicating) systems will not contribute to an increase in emergent behavior value. Hence our 24×24 matrix is a mixture of 1s and 0s, reflecting the potential to interface with 1 and 0 otherwise. On the other hand, system characteristics are defined in a universe of discourse of 0 through 10 denoting lower and upper bound on level of performance. System characteristics related to assessing cybersecurity attributes include the ability to defend cyber space from attacks (protect), time performance of detecting an attack (detect), time performance of responding to an attack (respond) and estimated time to full recovery (recover). Each of the systems is associated with distinct characteristics' values which may change based on threat dynamics, advances in system capabilities and performance, or system integrations that may be able to handle additional threats with different level of performance. In addition to the said characteristics, we also included system costs and defined them into two categories considering system development and operations. In this case, we extrapolated the development costs over the lifecycle of the system, and defined it as the interface development cost (IF_Dev_cost) and operational cost as (Oper_cost). KPAs, depicted in Figure 30.8, are assessed in terms of the system characteristics measuring the performance of securing from threats or attacks; the assessed values indicate the effectiveness of the meta-architecture for a given KPA.

They are presented in a normalized universe of discourse of 0 to 100, denoting lower and upper bound of performance. Such values will serve as inputs to the FIS and integrated into the fitness function of the tool's genetic algorithm (GA), Figure 30.9, to facilitate in selecting optimum meta-architecture.

Table 30.2 Potential systems and their capabilities.

	Capabilities									
Systems	Malware	Social engineering	Password attack	DDoS	MiTM	Driveby downloads	Sniffers	Malicious insiders	Trap doors	Negligent employee
Anti-malware	1	0	1	0	1	0	1	0	0	0
IDPS-host	1	0	0	1	1	0	1	0	1	0
IDPS-network	0	0	0	1	1	0	1	0	0	0
IDPS-signature	0	0	0	1	1	0	1	0	0	0
IDPS-behavior	0	0	0	1	1	0	1	0	0	0
IDPS-rule	0	0	0	1	1	0	1	0	0	0
IDPS-anomaly	0	0	0	1	1	0	1	0	0	0
Cont. planning	0	1	1	0	0	0	0	1	1	1
Pen. testing	0	1	1	0	0	1	1	1	0	0
UEBA	0	1	1	0	0	0	0	1	1	1
SAT	0	1	0	1	0	0	0	1	1	1
Policy	0	1	0	1	0	0	0	1	1	1
FW-packet	0	0	1	0	1	1	0	0	0	0
FW-circuit	0	0	1	0	1	1	0	0	0	0
FW-stateful	0	0	1	0	1	1	1	0	0	0
FW-proxy	0	0	1	0	1	1	1	0	0	0
FW-nextG	0	0	1	0	1	1	1	0	0	0
Cryptography	0	0	1	0	1	0	0	1	1	1
MLS	0	0	1	0	1	0	0	1	1	1
Whitelisting	0	0	1	0	0	1	0	1	1	1
UAC	0	1	1	0	0	0	0	1	1	1
Auto-backup	0	0	0	0	0	1	0	1	1	0
Patching	0	0	0	0	1	0	0	0	1	0
User education	0	1	0	0	1	1	0	1	1	1

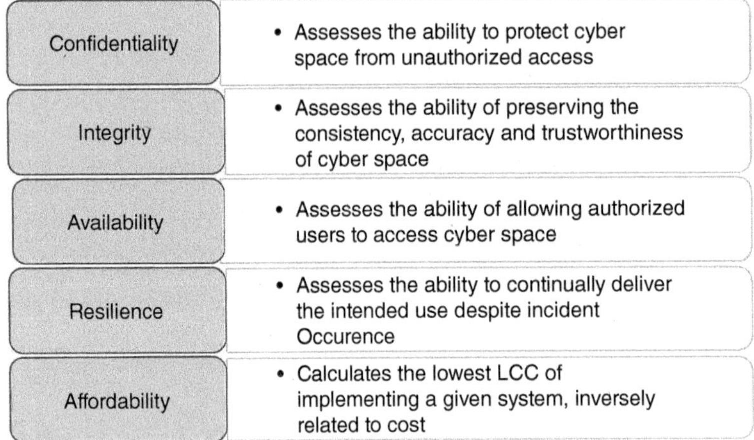

Figure 30.8 Cybersecurity key performance attributes.

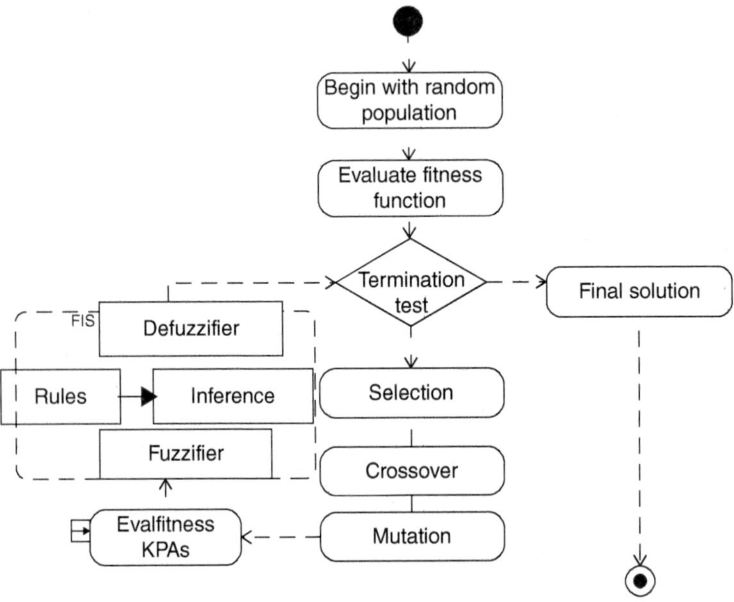

Figure 30.9 Genetic algorithm coupling with fuzzy inference system.

FIS rules in the form of IF-THEN statements using AND or OR operators in the antecedents will represent the relationship of multi-inputs to single-output space, offering flexibility in defining a consequent to be true to a certain degree. Base rules were generated by initially mapping input membership functions to similar (same granularity if level of granularity is equal) output membership function. From there, we summoned rules by observing 3D surfaces generated as we considered the relationship of two KPAs to the output.

Figure 30.10 illustrates the 3D surfaces that we studied in refining our rules. Based on rules of the overlapping membership functions, the aggregated result will undergo centroid defuzzification to achieve a crisp value. The meta-architecture will be presented in SoS Explorer as a set of active systems and interfaces (known as a chromosome) comprising the meta-architecture; partial representation is depicted in Figure 30.11.

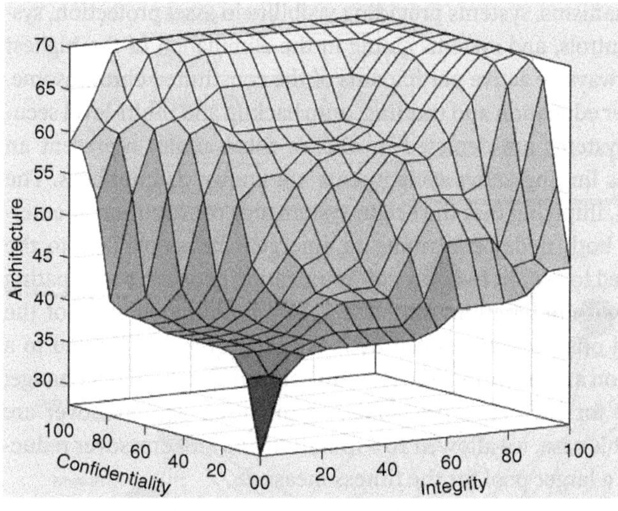

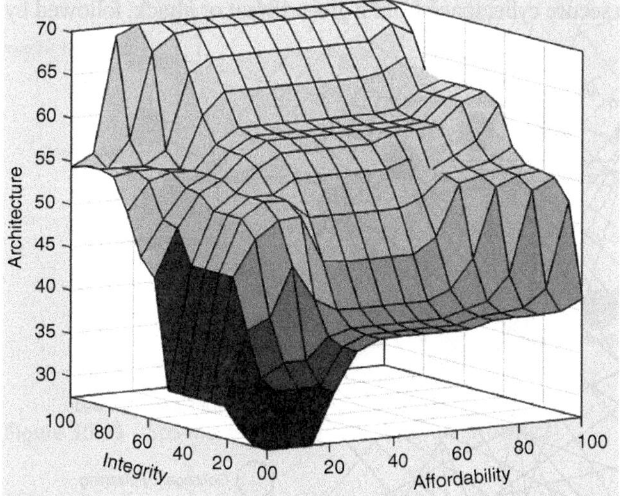

Figure 30.10 3D surface generated from rules of 2 KPA inputs to output.

Simple SOGA #1
$\mu = 40, \lambda = 40$
$P_m = 0.1, P_c = 0.85$
Iterations = 10,000
Negotiation: Optimal

Figure 30.11 Simple SOGA parameters.

Although Table 30.2 emphasizes systems capabilities of securing from the listed types of attacks, that the system characteristics will assess how well a specific system will defend from a given threat or attack. Various types of intrusion detection prevention systems and firewalls will be utilized to secure cyber space at distinct network layers or secure from various types of known and unknown anomalies.

KPAs will be calculated in terms of the participating systems and their interfaces; then, they are normalized by using the total feasibility space of SoS. Developed systems' interfaces producing emergent behavior that increase the assessment value will be rewarded for participation, whereas others causing a decrease in value will be penalized; hence, we will use an augmenting factor for interfaces between active systems to the objective function. Additionally, we introduce variables to correlate the trade-off weight of system characteristics onto the competing objectives.

Results

Meta-architectures generated by the SoS Explorer, resulting from the genetic algorithm optimization, will reflect cybersecurity measures that best fit the dynamic threat/attack management. In

modeled in terms of probability of success (Qin et al. 2017). In our case, we altered system characteristics values at different time intervals while penalizing a system for capabilities that do not exhibit added emergent behavior, thus generating the meta-architecture.

Cybersecurity Use Case Conclusion

The paper addressed cybersecurity for financial institutions, primarily in the banking industry, emphasizing both traditional and e-banking that require access to cyberspace. We presented the use of SoS Explorer as a tool to solve many-objective optimization characterizing participating systems' influence on the meta-architecture in terms of a single overall objective of cybersecurity. To determine the leverage systems had on the meta-architecture, we manually activated or deactivated systems and interfaces to observe changes in the overall value. Additionally, we adjusted the level of performance at different time intervals to characterize technology capability changes and the need for constituent participation to offset, if any, performance degradation. Future work calls for neural network structure to monitor a shift in architecture in anomaly protection, detection, and recovery. Machine learning for anomaly detection will amend cybersecurity architectures to target events outside the subset of anomalous activities, thus turning defense mechanisms into proactive systems. Paper limitations include the lack of cyberattack data on financial institutions, the effect such attacks had on organizations, and perhaps the lack of consolidated data sets associated with the types of attacks.

Conclusion

Developing models of acknowledged SoS architectures can assist in discovering and defining issues, satisfying problems with stakeholder needs, and analyzing the impact of policies through the rules on architecture selection. Key performance attributes that depend on architecture selection can be discovered through facilitated interactions with stakeholders and SMEs. Relatively simple fuzzy rule-based systems can be created and combined with the KPA evaluations for an overall SoS assessment. A fuzzy genetic approach is utilized for finding solutions to several SoS architecting problems integrated with a restrictive meta-model of undirected network graphs representing the system interfaces. Defining the boundaries of the membership functions and changing them independently is an excellent way to get answers quickly about the problem. The variable changing can show the mapping between fuzzy and real-world variables. Following the FILA-SoS makes it relatively easy to switch back to the real values to perform a "what-if analysis." Exploring the additional resources gives real-life examples of the FILA-SoS methodology and presentations on the topics. The process of finding "good" suggested architectures through applying the fuzzy genetic approach appears to help propose architectures and evaluation steps following the wave model of the evolution of an acknowledged SoS.

More Use Cases in Published Papers and Resources of Meta-Architecting

1 Qin, R., Dagli, C.H., and Amaeshi, N. (2016). A contract negotiation model for constituent systems in the acquisition of acknowledged system of systems. *IEEE Transactions on Systems, Man, and Cybernetics: Systems* 47 (11): 3050–3062.

2 Konur, D., Farhangi, H., and Dagli, C.H. (2016). A multi-objective military system of systems architecting problem with inflexible and flexible systems: formulation and solution methods. *OR Spectrum* 38 (4): 967–1006.

3 Konur, D. and Dagli, C.H. (2015). Military SoS architecting with individual system contracts. *Optimization Letters* 9 (8): 1749–1767. http://link.springer.com/article/10.1007/s11590-014-0821-z.
4 Kilicay-Ergin, N. and Dagli, C. (2015). Incentive-based negotiation model for SoS acquisition. *Systems Engineering* 18: 310–321. http://onlinelibrary.wiley.com/doi/10.1002/sys.21305/full.
5 Acheson, P., Dagli, C., and Kilicay-Ergin, N. (2013). Fuzzy decision analysis in negotiation between the SoS agent and the system agent in an agent-based model. *International Journal of Soft Computing and Software Engineering* 3 (3): 25–29. www.jscse.com. ISSN: 2251-7545.

References

Acheson, P., Dagli, C., and Kilicay-Ergin, N. (2013). Model based systems engineering for system of systems using agent-based modeling. *Procedia Computer Science* 16: 11–19. https://doi.org/10.1016/j.procs.2013.01.002.

Agarwal, S., Pape, L.E., Dagli, C.H. et al. (2015). Flexible and Intelligent Learning Architectures for SoS (FILA-SoS): architectural evolution in systems-of-systems. *Procedia Computer Science* 44: 76–85. https://doi.org/10.1016/j.procs.2015.03.005.

Alhamad, T., Axelrod, D., and Lentine, K.L. (2019). The epidemiology, outcomes, and costs of contemporary kidney transplantation. In: *Chronic Kidney Disease, Dialysis, and Transplantation*, 4e (ed. J. Himmelfarb and T.A. Ikizler), 539–554.e5. Elsevier. https://doi.org/10.1016/B978-0-323-52978-5.00034-3.

Anandalingam, G. and Friesz, T.L. (1992). Hierarchical optimization: an introduction. *Annals of Operations Research* 34 (1): 1–11.

Ashiku, L. and Dagli, C. (2019a). Cybersecurity as a centralized directed system of systems using SoS explorer as a tool. In: *2019 14th Annual Conference System of Systems Engineering (SoSE)*, 140–145. https://doi.org/10.1109/SYSOSE.2019.8753872. IEEE.

Ashiku, L. and Dagli, C.H. (2019b). System of systems (SoS) architecture for digital manufacturing cybersecurity. *Procedia Manufacturing* 39: 132–140. https://doi.org/10.1016/j.promfg.2020.01.248.

Ashiku, L., Threlkeld, R., Canfield, C., and Dagli, C. (2022). Identifying AI opportunities in donor kidney acceptance: incremental hierarchical systems engineering approach. In: *2022 IEEE International Systems Conference (SysCon)*, 1–8. IEEE.

Aubert, O., Reese, P.P., Audry, B. et al. (2019). Disparities in acceptance of deceased donor kidneys between the United States and France and estimated effects of increased US acceptance. *JAMA Internal Medicine* 179 (10): 1365–1374. https://doi.org/10.1001/jamainternmed.2019.2322.

Axelrod, D.A., Schnitzler, M.A., Xiao, H. et al. (2018). An economic assessment of contemporary kidney transplant practice. *American Journal of Transplantation* 18 (5): 1168–1176. https://doi.org/10.1111/ajt.14702.

Beliën, J., De Boeck, L., Colpaert, J. et al. (2013). Optimizing the facility location design of organ transplant centers. *Decision Support Systems* 54 (4): 1568–1579. https://doi.org/10.1016/j.dss.2012.05.059.

Caruso, V. and Daniele, P. (2018). A network model for minimizing the total organ transplant costs. *European Journal of Operational Research* 266 (2): 652–662. https://doi.org/10.1016/j.ejor.2017.09.040.

Ceren Ersoy, O., Gupta, D., and Pruett, T. (2021). A critical look at the U.S. deceased-donor organ procurement and utilization system. *Naval Research Logistics (NRL)* 68 (1): 3–29. https://doi.org/10.1002/nav.21924.

Cybersecurity (n.d.). https://www.pwc.com/us/en/services/consulting/managed-services/cybersecurity-operations-managed-services.html.

Chattopadhyay, D., Ross, A.M., and Rhodes, D.H. 2008. A framework for trade-space exploration of systems of systems. *6th Conference on Systems Engineering Research*, Los Angeles, CA (April 2008). IEEE.

Dagli, C.H. and Kilicay-Ergin, N. (2008). System of systems architecting. In: *System of Systems Engineering: Innovations for the Twenty-first Century*, vol. 58 (ed. M. Jamshidi), 77–100. Wiley.

Dagli, C., Ergin, N., Enke, D., et al. (2013). An Advanced Computational Approach to System of Systems Analysis & Architecting Using Agent-Based Behavioral Model (No. SERC-2013-TR-021-2). Missouri University of Science and Technology, Rolla.

Dagli, C.H., Singh, A., Dauby, J.P., and Wang, R. (2009). Smart systems architecting: computational intelligence applied to trade space exploration and system design. *Systems Research Forum* 3 (02): 101–119. World Scientific Publishing Company.

Dahmann, J., Lane, J., Rebovich, G., and Baldwin, K. (2008). A model of systems engineering in a system of systems context. *Proceedings of the Conference on Systems Engineering Research*, Los Angeles, CA, USA (April 2008). IEEE.

Dahmann, J., Baldwin, K.J., and Rebovich Jr, G. 2009 Systems of systems and net-centric enterprise systems. *7th Annual Conference on Systems Engineering Research*, Loughborough, England (April 2009). IEEE.

Dahmann, J., Rebovich, G., Lowry, R. et al. (2011). An implementers' view of systems engineering for systems of systems. In: *2011 IEEE International Systems Conference (SysCon)* (April 2008), 212–217. IEEE.

Dahmann, J.S. and Baldwin, K.J. (2008). Understanding the current state of US defense systems of systems and the implications for systems engineering. In: *2008 2nd Annual IEEE Systems Conference*, 1–7. IEEE.

Dahmann, J. (2012). INCOSE SoS Working Group Pain Points. *Proc TTCP-JSA-TP4 Meeting.*

Fogel, D.B. (2006). *Evolutionary Computation: Toward a New Philosophy of Machine Intelligence*, vol. 1. Wiley.

Fritz, L., Schilling, T., and Binder, C.R. (2019). Participation-effect pathways in transdisciplinary sustainability research: an empirical analysis of researchers' and practitioners' perceptions using a systems approach. *Environmental Science & Policy* 102: 65–77. https://doi.org/10.1016/j.envsci.2019.08.010.

Gegov, A. (2010). *Fuzzy Networks for Complex Systems: A Modular Rule Base Approach.* Berlin Heidelberg: Springer-Verlag.

Haris, K. and Dagli, C.H. (2011). Adaptive reconfiguration of complex system architecture. *Procedia Computer Science* 6: 147–152. https://doi.org/10.1016/j.procs.2011.08.029.

Hart, A., Lentine, K.L., Smith, J.M. et al. (2021). OPTN/SRTR 2019 annual data report: kidney. *American Journal of Transplantation* 21 (S2): 21–137. https://doi.org/10.1111/AJT.16502.

Huang, K., Siegel, M., and Madnick, S. (2018). Systematically understanding the cyber attack business: a survey. *ACM Computing Surveys* 51 (4). https://doi.org/10.1145/3199674.

IEEE Architecture Working Group. (2000). IEEE Recommended Practice for Architectural Description of Software-Intensive Systems. IEEE std 1471.

ISO/PAS 19450:2015(en) (n.d.). Automation systems and integration – object-process methodology. https://www.iso.org/obp/ui/#iso:std:iso:pas:19450:ed-1:v1:en.

Karim, M.M. and Dagli, C.H. (2020). SoS meta-architecture selection for infrastructure inspection system using aerial drones. In: *2020 IEEE 15th International Conference of System of Systems Engineering (SoSE)*, 23–28. https://doi.org/10.1109/SoSE50414.2020.9130538.

Lesinski, G., Corns, S.M., and Dagli, C.H. (2016). A fuzzy genetic algorithm approach to generate and assess meta-architectures for non-line of site fires battlefield capability. In: *2016 IEEE Congress on Evolutionary Computation (CEC)*, 2395–2401. https://doi.org/10.1109/CEC.2016.7744085.

Luzeaux, D., Ruault, J.-R., and Wippler, J.-L. (ed.) (2013). *Large-scale Complex System and Systems of Systems*. Wiley.

Malan, R. and Bredemeyer, D. (2001). Architecture Resources. Defining Non-Functional Requirements.

Missouri University of Science and Technology (n.d.). *SOS Explorer*. https://emse.mst.edu/sos-explorer/ (accessed 13 December 2022).

Mohan, S., Chiles, M.C., Patzer, R.E. et al. (2018). Factors leading to the discard of deceased donor kidneys in the United States. *Kidney International* 94 (1): 187–198. https://doi.org/10.1016/J.KINT.2018.02.016.

Pape, L. and Dagli, C. (2013). Assessing robustness in systems of systems meta-architectures. *Procedia Computer Science* 20: 262–269.

Pape, L., Giammarco, K., Colombi, J. et al. (2013). A fuzzy evaluation method for system of systems meta-architectures. *Procedia Computer Science* 16: 245–254. https://doi.org/10.1016/j.procs.2013.01.026.

Qin, R., Dagli, C.H., and Amaeshi, N. (2017). A contract negotiation model for constituent systems in the acquisition of acknowledged system of systems. *IEEE Transactions on Systems, Man, and Cybernetics: Systems* 47 (11): 3050–3062. https://doi.org/10.1109/TSMC.2016.2560520.

Threlkeld, R., Ashiku, L., Canfield, C. et al. (2021). Reducing kidney discard with artificial intelligence decision support: the need for a transdisciplinary systems approach. *Current Transplantation Reports* 8 (4): 263–271. https://doi.org/10.1007/s40472-021-00351-0.

Threlkeld, R., Ashiku, L., and Dagli, C. (2022). Complex system methodology for meta architecture optimization of the kidney transplant system of systems. In: *2022 17th Annual System of Systems Engineering Conference (SOSE)*, 304–309. IEEE.

Biographical Sketches

Cihan H. Dagli (University of Missouri Science and Technology) is a fellow of INCOSE and IISE; life member of IEEE and a member of NDIA; and a fellow of International Foundation for Production Research. He is professor of engineering management and systems engineering at the Missouri University of Science and Technology. Dr. Dagli is the founder and director of the Missouri S&T's System Engineering graduate program. He is the director of Smart Engineering Systems Laboratory and a senior investigator in DoD Systems Engineering Research Center-URAC. His current research interests are in the areas of Architecting Cyber Physical Systems, Complex Adaptive Systems, System of Systems, Data Analytics and Machine Learning. He is currently working on systems architectures and meta-architectures for self-organizing systems ensembles and deep neural networks in creating adaptive behavior. This research builds on Flexible and Intelligent Learning Architectures for SoS (FILA-SoS) research completed as a part of research work within SERC for DoD. He has published more than 483 papers in refereed journals and proceedings, 35 edited books, and cited 5635 times based on google scholar. He has consulted with various companies and international organizations including The Boeing Company, AT&T, John Deere, Motorola, US Army, UNIDO, and OECD.

Richard Threlkeld (University of Missouri Science and Technology) has over a decade of experience with research and development in operations, academics, defense contractors, and commercial entities. He has worked for Georgia Tech Research Institute, Lockheed Martin Missiles and Fire Control, and General Electric Research. He currently serves and the cofounder of two small businesses Axiom AI and Fulcra Robotics. He served as a captain in both active and reserve components of the United States Army. After active duty, he continued to serve the Department of Defense in academia working as a principal investigator for multiple defense projects with DTRA, JPEO, SCO, and the Army. He also has experience as a principal investigator and chief architect working for traditional defense contractors on multiple DoD and digital enterprise projects. In addition, he also has led commercial and government research covering a wide range of technologies including digital twins, model-based engineering, position, navigation, and timing, power electronics, robotics, autonomy, quantum, artificial intelligence, and harsh environment electronics. He received his undergraduate degree in chemical and biomolecular engineering from the Georgia Institute of Technology, as well as a masters in systems engineering from Georgia Tech. He also received a master of science in environmental management form Webster University. He is currently a PhD student at Missouri University of Science and Technology to follow his passion for serving academia.

Lirim Ashiku (University of Missouri Science and Technology) has been a software engineer and consultant for over 15 years. His undergraduate studies were done at the University of Wisconsin – Madison in the field of Electrical and Computer Engineering. He acquired a master's degree in information technology focused on software engineering from Walden University. He is currently a PhD student at Missouri University of Science and Technology to follow his passion for serving academia. His main interests include machine learning, deep learning, many-objective optimization of deep neural architecture, and human-machine integration. He has published 10 publications in the field of network security, organ allocation, and machine learning.

Chapter 31

Unique Challenges in Mission Engineering and Technology Integration

Michael Orosz[1], Brian Duffy[1], Craig Charlton[1], Hector Saunders[2], and Ellins Thomas[3]

[1] *University of Southern California Information Sciences Institute, Marina del Rey, CA, USA*
[2] *US Space Force Space Systems Command (SSC), El Segundo, CA, USA*
[3] *Booz Allen Hamilton, El Segundo, CA, USA*

Introduction

The objective of this chapter is to introduce the unique challenges in mission engineering and technology integration in developing systems that are part of a much larger enterprise. Such systems include large-scale manufacturing processes (e.g., automotive manufacturing) and operational service-based systems such as space-based communication systems where there are multiple subsystems, each often developed and maintained via multiple vendors and undergo modification and upgrades on different timelines. Although many of these enterprise systems are composed of both hardware and software subsystems (e.g., space vehicle and software-based ground control), this chapter is focused primarily on software-based systems. That said, when appropriate, any reference to hardware-only or hybrid hardware and software-based systems will also be noted. The targeted enterprise environment includes mid-to-large scale enterprises such as US Department of Defense acquisition programs and mission-critical systems (i.e., systems that cannot fail).

In this chapter, mission engineering refers to applying systems engineering processes and principles to the complete product life cycle – requirements analysis, design, development, integration, testing, deployment, and sustainment of a complex systems of systems project. Such processes and principles include DevSecOps, digital engineering, model-based systems engineering (MBSE), Agile, and other systems design and development processes. Technology integration refers to the processes and principles of inserting technology into both the engineering and development processes of a systems acquisition program.

As discussed later in the chapter, what makes mission engineering and technology integration difficult with mid-to-large scale enterprise systems are the large number of inter-system dependencies, the multiple vendors that maintain and upgrade these subsystems, the various timelines used to update and sustain these subsystems. All of this occurs in an environment where end-user needs and the operating and development environments themselves are continuing to undergo technical change. For example, in a space-based communications enterprise, often the ground-based command and control system (C2) is completed prior to the completion of the hardware-centric space vehicles that the C2 system will command. Issues such as systems integration and

Systems Engineering for the Digital Age: Practitioner Perspectives, First Edition. Edited by Dinesh Verma.

Companion website: www.wiley.com/go/verma/systemsengineering

testing can be difficult to undertake when major portions of the enterprise are not available, or delivery is delayed for a variety of reasons – including supply chain challenges or evolving end-user requirements and needs.

Traditional approaches, such as a serial or waterfall processes where each phase of a product life cycle is undertaken and completed before the next phase is started are not well suited to developing subsystems of a large enterprise due to the challenges already mentioned (e.g., inability to absorb changing requirements, schedule slippages due to delayed delivery of external dependencies, etc.). As noted later in this chapter, there are approaches for addressing the challenges of the waterfall method. However, even these solutions do not fully address the challenges which programs may experience when a lack of continuous integration and testing results in late discrepancy detection, and a subsequent accumulation of a bow-wave of issues that must be addressed at the end of the program. Addressing this bow-wave of discrepancies at the end of product development often comes at the expense of increased project schedule and cost. A pure Agile (The Manifesto Authors 2001) approach – which has been applied successfully in many software-focused organizations (e.g., Amazon, Google, etc.) where unfettered market demands dictate what is developed – are also difficult to apply to large enterprise systems where the product being developed targets a specific customer-defined need (i.e., the customer is funding the specific development effort, not the development organization). Under these circumstances, there are often deliverables with specific functions that must be delivered at specific milestones that can impact an Agile approach.

Despite the challenges noted above regarding Agile, the focus of this chapter is on introducing and adapting Agile/DevSecOps processes and principles to the mission engineering and management of the overall project effort. Whether a software, hardware, or a hybrid hardware/software program, the ability to quickly respond to changing customer requirements and to adapt to evolving technology is an important ingredient in supporting mission engineering and technology integration. It is the combining of both Agile and DevSecOps processes and principles that provides the most value to the mission engineering and technology integration process. The ability to adapt or absorb changing system requirements on a short-time interval and use continuous integration to detect discrepancies early in the development effort will help keep system acquisition programs on schedule, on budget, and relevant to the customer/user community.

The remainder of this chapter discusses in more detail the challenges presented by mission engineering and technology integration in a large-scale enterprise system, the adoption of modern mission engineering and systems development practices to address various challenges, a description of a space-based acquisition program that has adopted many of the recommendations listed in this chapter, and finally, a summary of recommendations to help the systems engineering practitioner.

Mission Engineering and Technology Integration

There are many definitions of mission engineering (OUSD(R&E) 2022) and technology integration (Wiegers and Beatty 2013); however, in this chapter, mission engineering refers to applying systems engineering processes and principles to the design, development, integration, testing, deployment and sustainment of a complex systems of systems project. Such processes and principles include DevSecOps, digital engineering, model-based systems engineering (MBSE), Agile, and other systems building processes.

Technology integration refers to the processes and principles of inserting technology into both engineering and development processes. Technology integration often occurs when a new system is in the process of development and system requirements are still evolving, or during sustainment of an existing operational system. An example of technology integration into an existing system is the insertion of a new communications protocol (i.e., a requirement change to address security vulnerabilities) into an existing space-based communication system's baseline. In this example, there will be updates made to the command and control (C2) system as well as modifications to the corresponding space vehicles (which in many cases have yet to be built and delivered). Issues such as maintaining backward compatibility and the flexibility to absorb future changes into the system are within the domain of technology integration.

Technology integration also involves the evolution of the actual development process itself. For example, many US Department of Defense acquisition programs are transitioning from the traditional DoDI 5000.02 waterfall methods (OUSD(A&S) 2020) to a more combined Agile and Dev*Ops (Tanner 2023) approach. For these programs, security often takes the place of the asterisk in Dev*Ops (Figure 31.1), which often requires the insertion of new technologies (e.g., Kanban board, security scanning technology, performance tracking and methods) into an existing development platform.

The Challenge

As noted in the introduction, the focus of this chapter is mainly on software-based systems, however, where appropriate, references to hardware and/or combined hardware and software systems will also be covered. For example, both hardware and software development along with their dependencies are commonly found in large enterprise systems.

Enterprise Systems

Although this chapter is focused on large enterprise systems, many of the challenges discussed are also applicable to smaller projects. For example, the need to do some aspect of up-front

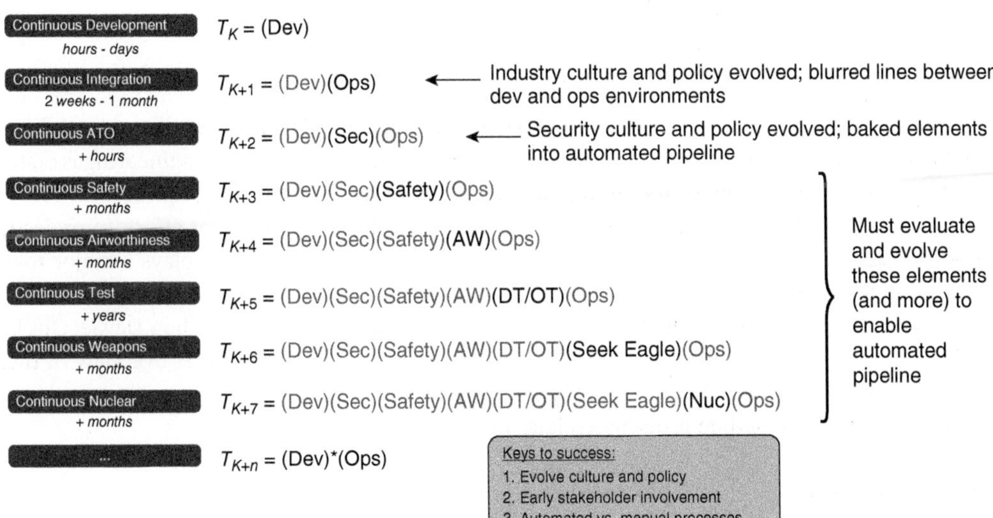

Figure 31.1 Dev*Ops. DevOps (i.e., nothing takes the place of the asterisk) is typically found in many development programs. In recent years, due to cyber vulnerabilities (many software-focused programs), security has been a key focus resulting in DevSecOps (i.e., Security takes the place of the asterisk).

engineering for an Agile-based project applies whether the project involves a component of a large enterprise or is a stand-alone application.

Enterprise systems are typically large systems of systems that involve multiple interfaces, multiple vendors/maintainers, hardware, software, human operators, end users, and are constantly evolving. Such complex systems are normally associated with large organizations having multiple departments or divisions (e.g., human resources, executive offices, engineering, facilities, production, distribution, sales, security, etc.), and multiple vendors (e.g., suppliers [supply chain], distributors, point of sales, etc.) although they may also include large information systems (e.g., financial network, etc.) and other product- or functional-focused organizations (e.g., GPS enterprise consisting of space, control, and user segments). Regardless of the type of enterprise system being considered, system upgrades and new product development can apply to a one or more components or to subsystems that compose the enterprise. The challenge to the acquisition professional is how to manage the engineering process to ensure that all components and the overall system in general are well engineered to support the development and sustainability of the system. Multiple vendors, differing deliverable timelines, delays in releases, changing requirements, the availability of reliable supply chains, and various inter and external dependencies need to be considered and built into the engineering process. For example, in a typical space-based enterprise system, software development often precedes hardware development (i.e., a software-based command and control system is completed prior to the availability of the hardware-based space vehicle(s)). In this example environment, the full integration and testing of the enterprise will not occur until the space vehicle(s) are delivered and operational. To reduce the discovery of system discrepancies (e.g., system performance does not meet the customer's requirements), it is very important to have a host hardware (or simulated hardware) solution available that represents the end-use system as closely as possible (a near-operational environment). The goal of this representative environment is to enable integration and testing during software development that captures and addresses discrepancies during development when engineering and development teams are available. Near-operational environments will be discussed later in this chapter.

Traditional System Development Methods and Their Challenges

Systems engineering is closely linked to the type of system being developed and the methodology used to develop the system. In waterfall/serial projects, the bulk of the systems engineering is often undertaken at the beginning of the project while systems integration and testing (used to verify and validate the system) are usually undertaken at the end of the development effort. In projects using the Agile method (discussed later in the chapter), systems engineering plays a major role throughout the development process. In such environments, the system architecture evolves throughout the development, integration, and testing processes – particularly when DevSecOps is involved – is undertaken continuously throughout the project. With that, a brief overview of the waterfall/serial and related methods – along with systems engineering challenges – is presented. The Agile/DevSecOps method is discussed later in the chapter.

Waterfall/Serial Method

In the past several decades, the traditional approach to mission engineering was to complete all the system engineering up front (e.g., functional analysis and architecture development), followed by the primary design, detailed design, development, integration, testing, and a final phase of deployment (Figure 31.2).

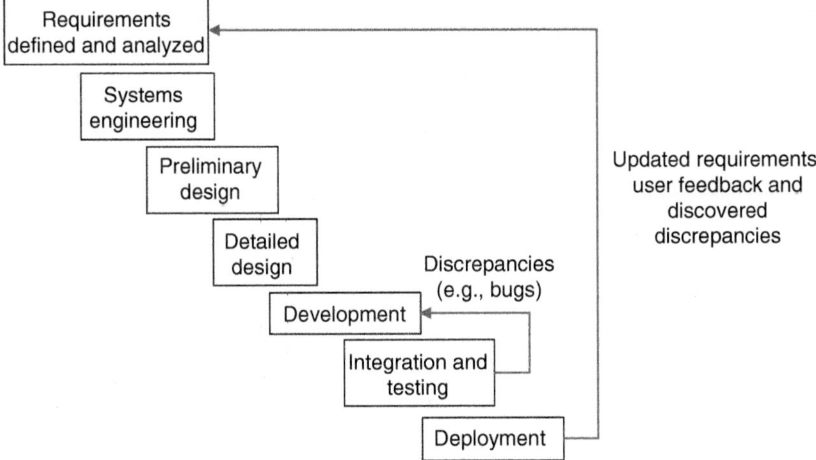

Figure 31.2 Development waterfall/serial process flow used in systems development. Discrepancies discovered during integration and testing can lead to extended schedules as the discrepancies are worked off. Undertaking complete systems engineering at the beginning of the development effort often leads to rigid and inflexible systems.

The waterfall or serial approach to product development (heavily used in the US DoD OUSD(A&S) (2020) and in many parts of industry, particularly in large enterprise systems) assumes that the system requirements and operational needs of the customer are well understood and remain static (and relevant) as the product is developed and fielded. In some situations, this assumption may very well be accurate, and the waterfall/serial method is appropriate. One such example is the updating of a circuit board for an on-board automotive application. However, for a vast majority of programs, the assumption that system requirements, system design and end-user needs do not change over the life of a project typically do not align with reality, often resulting in systems being developed and fielded that do not meet customer needs. In some cases, due to the length of the project, the system requirements and user needs became obsolete – often surpassed by the competition. For example, by placing integration and testing towards the end of the development timeline, discrepancies (e.g., software bugs, hardware faults, etc.) are not detected until late in the development cycle. In many cases, discrepancies will build (i.e., creating a "bow wave" of discrepancies) resulting in the stretching out the development schedule (and increased cost of the overall project) to allow the development team time to address the discrepancies and "catch-up" with integration and testing. Meanwhile, as the development schedule is stretched, new technologies and evolving customer needs are not integrated into the developing system as the original system design is "cast in stone."

"Mini-Vs"

One approach to addressing the challenges of the traditional waterfall/serial method is to divide the serial development process into what might be referred to as "mini-V" intervals. As documented in Figure 31.3 and discussed in (Wikipedia 2023a), the traditional waterfall method can be viewed as a V-shaped effort that starts with requirements analysis, followed by design, development, integration and testing, and finally deployment. A "mini-V" interval approach to system development consists of dividing up the overall project into a series of "mini-Vs" with each "mini-V" producing a product of value to the end-user. At the end of each "mini-V" increment, a deliverable is produced, and the next "mini-V" increment development effort will build on the previous

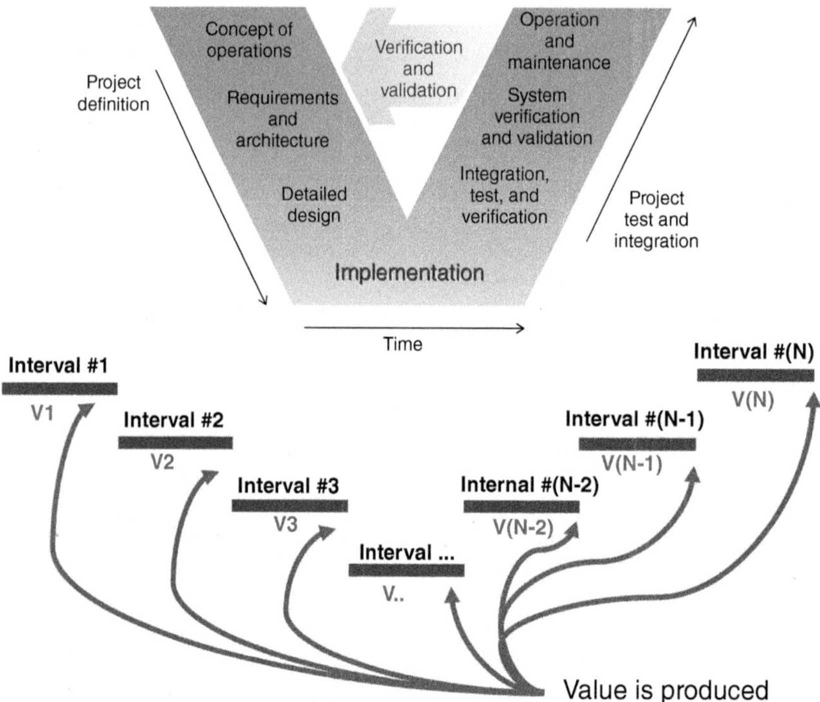

Figure 31.3 Upper graphic (adopted from Wikipedia 2023a) shows the "V" that a traditional waterfall/serial development process follows (and as outlined in Figure 31.2). At the end of the "V" process, a product is produced. A method for addressing product obsolescence and to better respond to changing requirements and customer needs is to divide the overall project into a series of "mini-Vs" where each "mini-V" follows a waterfall/serial approach, where value to the customer is produced at the end of the effort, but the timeline is much shorter in duration (weeks or months instead of years). As one "mini-V" completes, updated requirements and customer needs are introduced into the next "mini-V".

"mini-V" effort. The advantage of this approach is that the customer will receive frequent deliverables over time (instead of at the end of a lengthy development period) and, at least in theory, new requirements and user feedback can be incorporated into the design of the next "mini-V" increment. In practice this rarely happens because within each "mini-V" increment, integration and testing is still pushed to the end of the development cycle within the "mini-V." Discrepancies discovered at the end of the shortened development cycle will often accumulate resulting in a bow wave of discrepancies that will spill over into the next "mini-V" increment. This pattern will often repeat resulting in extended schedules and project cost overruns.

Technology Integration Challenges

Whether using a traditional waterfall/serial development approach or a more flexible/adaptable agile approach (discussed later), the introduction of technology can be a challenging problem. As previously noted, in the waterfall/serial approach, the systems engineering and design are undertaken up front and are assumed to remain static throughout the development life cycle. New requirements (often involving the introduction of new technology) are difficult to insert into the baseline due to several factors. Factors such as a rigid system design that was not developed to incorporate change, lack of models to fully (and quickly) understand the impact that the insertion of new technology will have on the project and, in the case of enterprise systems, the full impact that a new technology integration will have on other systems. In such "static" systems, technology integration often means development must stop to allow for a complete system engineering review and update resulting in slipped schedules and project cost overruns.

Even for systems that employ a more agile or flexible approach to development, technology integration can still present a problem. Although the agile approach is designed to adapt quickly to changing requirements and customer needs, if the current system is not designed to evolve then the same issues that plague technology integration in the waterfall/serial approach (i.e., delayed schedules, reworked system designs, etc.) will result. This is particularly a problem in brownfield projects (i.e., extending an existing product) with a large number of internal and external dependencies. In such cases, impacts from the insertion of new functionality involving one or more of these dependencies will often ripple throughout the interconnected system of systems. Modular system designs with minimum interdependencies can help address this challenge and are discussed later in the chapter.

Model Based Systems Engineering (MBSE)

In an attempt to address the challenges of making engineering systems less rigid and more flexible, model-based systems engineering (MBSE) methods have been applied. MBSE applications, such as Cameo Systems Modeler (Dassault Systems CATIA 2023a) and MagicDraw (Dassault Systems CATIA 2023b), involve the development of one or more (usually connected) models that exhibit both the system architecture and behavior (e.g., performance, supported functionality, end-user interactions, etc.) of a system within a simulated operational environment. The advantage of using such models is that modifications and technology insertions to the baseline system can first be applied to the model(s) to determine any potential impacts without the need for writing any software or the "cutting of metal" for a hardware-based product. MBSE models allow "what-if" analysis to explore impacts and develop possible courses of action to address changing requirements.

Closely related to MBSE models is the concept of a digital twin. Digital twins represent (almost always implemented in software) an exact replica of an existing operational system. For example, a digital twin of an aircraft engine is a digital model of that specific engine. In this example, the digital twin is used to predict future system behavior (as the engine is used and ages) as well as predicting any impacts to the engine due to changing conditions (e.g., different fuel supply or the addition of a new oiling system).

The challenge here is that these models may not be complete or may not be consistently updated to reflect the true system as the system evolves during development and sustainment. This is discussed in more detail in the recommendations section of this chapter and often requires instilling a disciplined workflow process to ensure that digital model updates are a part of the development effort, and that work tasks are not officially signed off until the model updates are completed.

Integration and Testing

Regardless of the system development effort chosen (e.g., waterfall, Agile, hybrid, etc.), all methods require integration and testing to verify and validate the system. Integration and testing usually consists of two types of integration processes – either vertically or horizontally focused. Vertically focused integration assumes that a given subsystem can be successfully integrated and tested with few if any external dependencies (i.e., minimal interfaces with other subsystems, etc.).

In situations where a few external dependencies exist, or interfaces are predictable, simulators can often be used as proxies for these components.

Horizontal integration and testing, however, is focused on integrating the target subsystem with other dependent subsystems. In other words, the complete system – possibly even the complete enterprise – is integrated and tested. For projects where some of these dependent subsystems do not yet exist, or are not available due to higher priority demands, a suitable substitute is required. In many cases, however, a suitable substitute may not be available or may be delayed. In such cases, integration and testing delays, and corresponding late discrepancy discovery, will result in schedule and project cost overruns. As noted later in this chapter, a possible engineering solution to this problem is the development of a near-operational environment that is used for horizontal integration and testing.

Recommendations

The Agile/DevSecOps Approach

Based on the waterfall/serial challenges previously discussed, it is highly recommended that when starting a new project – whether developing a new product or updating an existing project – a combined agile and DevSecOps process should be used for the development and sustainment of the project. As noted elsewhere in this field guide (e.g., chapter on *Scaling Agile Principles to an Enterprise*), agile-based systems are designed to adapt to changing requirements (e.g., Request for Change (RFC) orders), to incorporate user feedback (from using/evaluating previously releases of the evolving product) during the development process and to provide continuous value to the user (Figure 31.4).

For example, most agile-based projects produce a product at the end of a pre-defined interval (a sprint, increment, etc.), also referred to as timeboxing (Wikipedia 2023b). Subsequent development intervals add value (new or updated functionality) to what has already been produced. If a lack of funding or some other reason results in the project being prematurely terminated, value has still been produced. The most recently completed "product" from the most recent completed

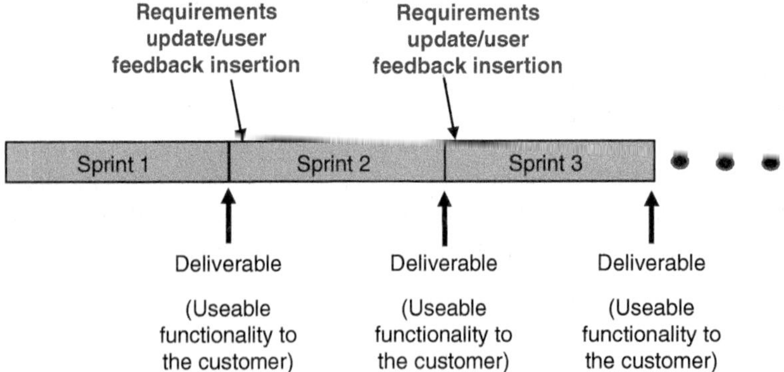

Figure 31.4 At the end of each sprint, a deliverable product is produced that contains functionalities that are useable to the customer. Each sprint builds on the previous product resulting in a process that continuously adds value to the customer as the system is developed. In addition, new/updated requirements and customer feedback (obtained from using previous deliverables) are incorporated into the next sprint development effort.

Table 31.1 Agile benefits compared to waterfall/serial approaches.

Waterfall	Agile
• Customer may get anywhere between 0% and 100% of the functionalities • Customer has to wait until the end of development to get any functionality (e.g., 4+ years) • No guarantee that it provides any functionality for the end user • Little time for contractor to recover when critical issues are discovered during Critical Design Review (CDR) and Final Qualification Test (FQT) phase • No ability to react to changing needs and priorities	• Throughout development, customer gets incremental functionality (e.g., 2–10 wk) • Functionality that can be used by the end user • Frequent, early integration and testing reduces the bow wave of DRs typically found at the end of a Waterfall effort • More robust and relevant software through frequent demonstrations and (ideally) end-user engagement • Adapts to changing needs and priorities • Agile team consist of 5~9 team members with the goal of building working product

development interval has value – it provides a level of functionality that is useful to the customer community. This greatly differs from waterfall/serial processes where a useful product is produced only at the end of the product development cycle. In a waterfall/serial process, if a lack of funding or some other reason results in the premature termination of the project, there is typically no useful product produced. In many such software development cases, the only product produced is source code that is of little use to the customer. Table 31.1 lists advantages that an agile process offers over that of the traditional waterfall/serial approach.

One of the major advantages often cited for using agile is the concept of "fail early, fail often, but always fail forward" attributed to John C. Maxwell (Aparicio 2023). The objective is not to fail for the sake of failing, the objective is to learn from failures during development while minimizing the resource impact of finding and fixing issues (i.e., "Build a little, test a little, deploy a little"). For many products and their domains, such a view should and needs to be embraced, however, this view can be a challenge when working in domains where failure simply cannot be tolerated due to the nature of the domain. Such systems include those developed for mission-critical operations (e.g., flight control, etc.). As discussed later in this chapter, systems created for mission-critical operations require a unique development environment where near-operational environments need to be part of the overall system architecture.

Continuous Integration and Deployment

A major part of mission engineering is having the appropriate development platform. For software-based systems, such a platform usually consists of a software factory (more details below) that promotes both development, deployment, and sustainability. Critical to the software factory is the concept of continuous integration and continuous deployment (InfoWorld 2022). CI/CD is applicable to both the waterfall/serial and agile approach to systems development.

A small note regarding CI/CD and the difference between delivery and deployment (i.e., the D in CI/CD). Delivery refers to producing a product (whether at the end of a sprint, an increment or the end of a waterfall effort) that exhibits value to the customer but may not be ready to be deployed into the operational environment. For example, in the agile community, the notion of an MVP (minimum viable product) is often treated as a deliverable – a product that contains value to the customer but may need some refinement before being deployed. Often this refinement comes from being used by a small subset of the user community. In some development communities, the MVP can be thought of as a "beta" release of a product. Deployment on the other hand refers to a product that is ready to be deployed into the operational environment. In the agile community, the notion of an MMP (minimum marketable product) is often treated as a deployable product. More information on MVPs and MMPs can be found in Productfolio (2020). Please note the exception to deployable MMPs in the next paragraph.

The objective of any product development is the deployment of a working and useful system that meets the needs of the customer. Unfortunately, and as noted earlier in the chapter, it is often difficult to time the deployment of a product that coincides with the delivery of all other components of the enterprise system. For example, in the space-domain environment, command and control systems are often developed and ready to be deployed far in advance of availability of the space vehicles (SVs), which the C2 systems will command. In these circumstances, the best that can be achieved is continuous delivery of a working product (i.e., an MVP) that may be put on the virtual shelf for later use. Delivery indicates a product that has been fully tested against a near-operational environment (discussed later); however, that product has yet to be tested and deployed with the *actual* operational environment. Under these circumstances, the delivered MVP (as opposed to a deployed MMP) will not be fully integrated, tested, and ready for deployment until the remaining components of the operating enterprise are available.

The Software Factory

The typical software factory supports the DevSecOps concept by implementing and supporting development (Dev in DevSecOps), integration and testing (covers both Dev and Sec in DevSecOps), and delivery/deployment (covers Ops in DevSecOps) components. As depicted in Figure 31.5, DevSecOps is a control/feedback system where development is influenced by integration and testing (discovered discrepancies are addressed) and operations (discovered discrepancies and customer feedback).

Software factories can be difficult to build or adapt from a predecessor project. Creating a software factory involves integrating various open source and purchased programs onto a bare-metal system. Licensing, purchase orders, approval cycles, version compatibilities, security protocols, and a host of other factors can often delay the deployment of an operational software factory. In addition, issues such as maintenance and security patch management as well as system evolution to address technology change also need to be considered. Projects should plan lead time accordingly to avoid software development delays.

Near-Operational Environment

As previously noted, integration and testing come in two flavors: vertical and horizontal. Vertical integration and testing focus on portions of the overall system that can be integrated and tested in isolation from the rest of the system, or with some simple simulated interfaces. Horizontal integration and testing focus on situations where the complete system (i.e., all subsystems – both internal and external) is required due to dependencies. Ideally, horizontal integration and testing is undertaken in the actual environment. As previously noted, having access to the operational environment and various external systems may not be possible due to the mission-critical nature of the domain (e.g., command and control of a space-based system). Under these circumstances, it is critical that a near-operational environment be available that simulates the actual operational environment as accurately as possible. This often requires not only the use of simulators, but that they are high-fidelity simulators, which exhibit the actual physics of the targeted domain. Ensuring that a near-operational environment is available is part of mission engineering. In other words, mission engineering is not only responsible for understanding the

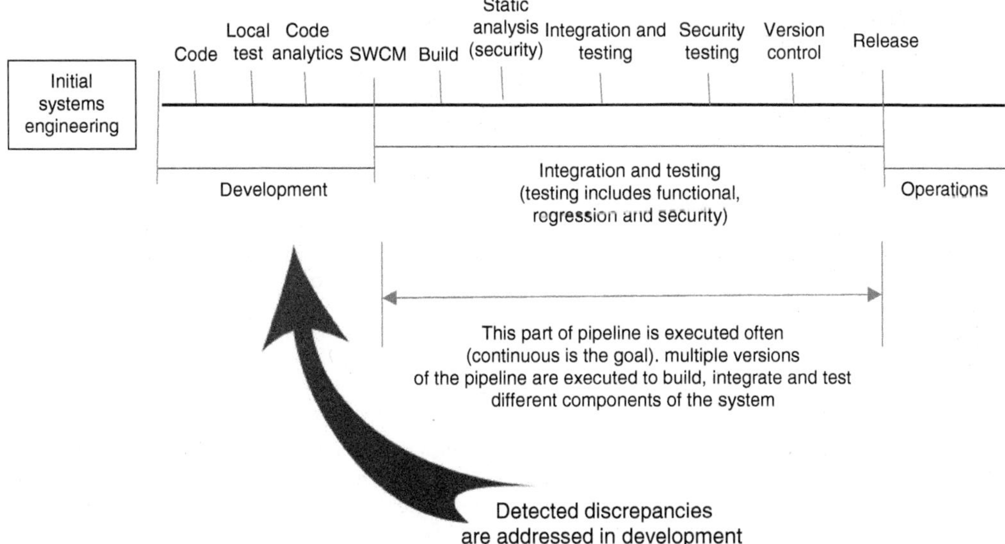

Figure 31.5 A typical DevSecOps pipeline. As discussed in the text, there is a need for initial systems engineering to design the initial systems architecture and initially populate the development project backlog. The DevSecOps pipeline starts with development – code is developed, tested locally, and then checked into a software configuration management (SWCM) system. This code is now available for the integration and testing portion of the pipeline. A build is initiated that creates a standalone component of the overall system. At this point static analysis is undertaken to check for security vulnerabilities. After the static analysis phase, the newly built subsystem component is integrated with other built subsystem components (including simulators where subsystems do not exist). This integrated system then undergoes functional, regression and dynamic security testing. Functional testing tests newly added functionality. Regression testing ensures previous functionality still works. Once the overall system is fully tested, the functional test for the newly created function ("value") then becomes part of the suite of regression tests. Dynamic security checking is used to check for security vulnerabilities of the operating system. A completed system build is then assigned a build version and is available for release. Multiple versions of the integration and testing portion of the pipeline can exist to maximize system development performance (see text).

requirements and needs of the product being developed but also needs to understand the various external systems that are part of the overall enterprise in order to ensure that the simulator is as accurate as possible.

Modular and Isolated Systems

Often, due to the complexity of a near-operational environment, the availability of such an environment may be delayed or once available, the system may have such high demand that its use will be rationed. Under this circumstance, the fallback approach is to undertake integration and testing using the vertical build environments. This is where mission engineering plays a vital role in the design of the overall system environment. Not only is minimizing interfaces between multiple subsystems (Figure 31.6) a good software engineering practice but it is also a good mission engineering practice. The goal of mission engineering in this situation is to develop a system of systems such that a majority of the subsystems can operate or at least be tested (via vertical testing) in isolation or, if connected to other subsystems, the interfaces can easily be simulated.

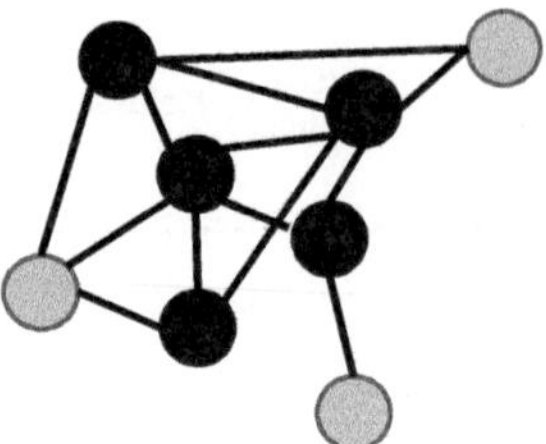

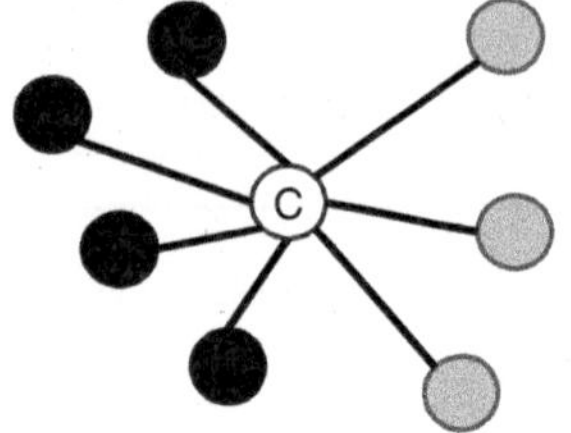

System A: Many internal and external interfaces

System B: Minimal internal and external interfaces

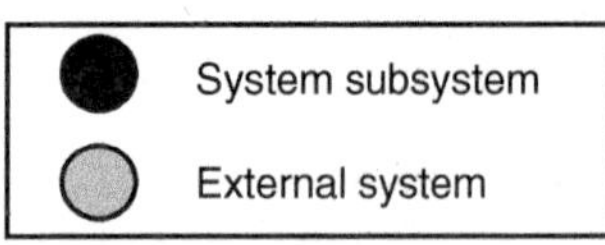

Figure 31.6 The goal of good mission engineering is to reduce the number of internal and external interfaces in the system design. System A has many internal and external subsystem interfaces reducing the opportunity to rely on vertical integration and testing. System B contains one critical subsystem (labeled "**C**") with many internal and external interfaces. In the case of System B, a simulator can be used for the interface with the "**C**" subsystem component allowing vertical integration and testing. Horizontal integration and testing are only required for the "**C**" subsystem.

Functional, Regression, and Security Testing

It is recommended that a DevSecOps operation is able to support functional, regression, and security testing. Functional testing is where the system is tested to ensure that a new function being added to the baseline is fully implemented and meets the requirements of the system. Regression testing is used to ensure that previous functionality of the system is not adversely impacted due to the insertion of new technology into the baseline. For new systems, initial testing is strictly functional since there is no previous functionality to test. As the new system is developed and new functions are inserted, previously developed functional tests now become part of the regression test suite. Finally, in support of the security aspect of the DevSecOps, security testing is used to ensure that system vulnerabilities (e.g., those that are prone to cyberattacks, etc.) are detected and addressed. Ideally, all three testing elements are used during the DevSecOps process.

It is recommended that security testing involve the establishment of coding standards (for the developers), static testing, and dynamic testing. Coding standards (such as memory safety) are typically verified during code walkthroughs. For customers who are actively engaged in the development process (e.g., US DoD acquisition personnel), it is recommended that members of the customer team participate frequently (as an observer) in these code walkthroughs. Static scanning involves applying cyber vulnerability detection tools (e.g., Fortify by MicroFocus (2023)) against the as-built code in a non-running environment. These tools explore the as-built software looking for embedded cyber vulnerabilities – based on a vendor provided rule pack – and typically generate a report listing any detected issues that may be system discrepancies that need to be addressed by the developers. Dynamic scanning involves applying cyber vulnerability detection tools (e.g., WebInspect also by MicroFocus) to a running system. These tools explore the running system, often applying various penetration tests to test for cyberattack vulnerabilities. Vulnerabilities are reported and should also be treated as discrepancies to be addressed by the development team.

Multiple Pipelines

As previously noted, one of the advantages of continuous integration (CI) in the CI/CD pipeline is the early detection of discrepancies. Early detection enables discrepancies to be quickly addressed and not allowed to cascade into building a "bow wave" of problems that must be addressed later in the development cycle, where the engineering and development teams may not be available. For large enterprise efforts, it is recommended that multiple pipeline runs be available. The first type of pipeline is focused on the development effort where the engineer/developer is developing a particular unit of code and needs to undertake integration and testing to complete the development effort before checking the source code back into the baseline repository. These pipelines are often customizable to allow the developer to define the type of build and testing (e.g., unit, horizontal or vertical) that is desired. For example, security scanning may be selectively "turned off" to speed up debugging.

A second type of pipeline is focused on integrating and testing the baseline – along with any newly added functionality – to detect discrepancies early. These pipelines can also be customizable (e.g., vertical or horizontal), but they generally are run automatically (daily, nightly, or weekly) to ensure that the baseline system is continuously integrated and tested. These pipelines can be configured to undertake functional, regression, and/or security testing as part of their configuration.

Requirements Decomposition

A critical step in mission engineering is understanding systems requirements. System requirements define how the system will be used, the functionality that will be supported, and the operating environment in which the system will operate (e.g., stand alone, part of an enterprise, etc.). Internal and external interfaces, system architecture, performance parameters, and user interfaces are all driven by the system requirements. Although the agile framework is designed to set goals or objectives, then "do a little coding, test, learn and adapt," there is still a need to complete good systems engineering up front. That often means taking the time to conduct a requirements decomposition.

In the agile environment, requirements decomposition produces two results: an understanding of the initial systems architecture and an initial project backlog containing work tasks (with initial assigned priorities) that produce some value toward a deliverable to the customer. The key word here is "initial." The system architecture and the project backlog will change (i.e., be refined) as the system is developed. However, having an initial architecture and project backlog help with determining gaps in the overall system design, provide an understanding of the system interfaces, identify performance parameters, identify internal and external dependencies, and provide an estimate of the resources and critical skills required for the overall project. Ideally, a pre-project system study, or planning phase, can benefit a project by allowing time for the initial systems engineering and architecture (e.g., lay out the skeleton of the system) to be completed before the detailed development of the actual solution dominates the day-to-day operations of the project.

Additional Recommendations

The following are additional recommendations that apply to systems engineering in general.

Model-Based Systems Engineering (MBSE)

As noted above, the key to absorbing change in a system design (no matter what development process is adopted) is to have a representative model of the system that can be manipulated to understand the impact of any change to the system. The key issue with MBSE is how much detail should be included in the model? For waterfall/serial development approaches, a very accurate model needs to be developed to ensure the systems design will support the system requirements. This usually requires extensive up-front development with full knowledge of the enterprise system, all dependencies (internal and external) and building in the necessary "hooks" in anticipation of new requirements being added to the project during development. In addition, since these models typically exhibit the behavior of the system, the models themselves will undergo change to reflect changes to the system design as the system is built.

For agile-based systems, the same challenges exist. However, the model itself does not initially need to be as detailed as that of the waterfall/serial effort and should favor breadth rather than depth. In agile, it is recommended that the model contain sufficient detail to help guide the development team with an initial system architecture and project backlog. Often, all that is needed is a black box representation of the system where the key functions/operations are represented by a black box (along with an initial description of what the box does) along with all the interfaces to and from that black box/function. As the project is developed, the model is updated with additional information to reflect the actual system. When development of a function is started, it is recommended that the details of the black box be defined prior to the start of coding (i.e., design drives the development). As the function is developed and tested, the model is updated to reflect reality. Although this process appears to be time-consuming, the benefits far outweigh the disadvantages as a functioning model now exists of the system. This model can be modified to determine the differences that new requirements may have on the system, prior to developing the function which supports the requirement. Additionally, as model interfaces evolve to become interoperable, the project's model can be integrated with other enterprise(s) models to demonstrate how complex systems of systems interact; without the use of resource intensive physical-world demonstrations.

Finally, it is highly recommended that the system requirements and test cases/procedures used to verify and validate the completed system be included in the model (i.e., have links between system requirements, decomposed features/stories, test cases, and test results). This is particularly important for large enterprise systems where many requirements and test cases exist that may need to be reviewed much later in the project to ensure the system meets the customer needs.

An example of an MBSE solutions used successfully in both industry and government-led projects is Cameo Systems Modeler from (Dassault Systems CATIA 2023a).

Need for Good Upfront Systems Engineering

Closely related to the requirements decomposition is the need to conduct an adequate upfront job of systems engineering. This is particularly true for systems that employ agile practices where, as with requirements decomposition, there is a tendency to get an initial architecture in place and then refine the design as the project progresses. The problem with taking this approach is that often, dependencies between various subsystems (both internal and external to the project) are often overlooked. This can result in schedule delays if certain dependencies are not available when the project needs them.

Data Rights, Licensing, and Intellectual Property

It is recommended that during initial mission engineering and the establishment of the initial system model, every effort be taken to ensure that all external COTS and FOSS licenses are in place and that intellectual property (IP) issues with all third-party propriety systems have been addressed.

Often, establishing these rights can be lengthy and, if not planned early on, can delay system development. This is particularly a challenge for projects where the actual software code was developed by one contractor and the subsequent contractor must then access and modify the same code. Having a complete understanding upfront on who owns the software, what modifications that will be made and which entities can use the resulting product (e.g., free use, licensed use, restricted use, etc.) is critical to ensuring a successful project.

Reserve Margin in the Timebox

When planning work for an upcoming period of performance (e.g., sprint or increment), leaving about 20% of labor capacity unassigned allows for the unexpected introduction of new requirements or technology, or to account for situations where the development team is not able to complete the assigned tasks in the allotted time (more on this in the chapter covering scaled agile). This 20% unassigned capacity is not only needed to incorporate unexpected requirements or technology injection, but in many cases, members of the scrum team may be temporarily reassigned to address potential upcoming requirements insertion (e.g., team members may be pulled to write a proposal to a request for proposal (RFP) or request for change (RFC)). These unexpected events happen frequently in large enterprise systems.

Conclusion

There are many unique challenges in applying mission engineering and technology integration in developing systems that are part of a much larger enterprise. Pitfalls include not fully identifying and understanding the system requirements, not identifying internal and external dependencies, and not completing an adequate job upfront on the system architecture, and not building sufficient flexibility into the system to adapt to a changing environment.

In many cases, taking a combined Agile and DevSecOps approach enhances technology integration. However, without adequate upfront engineering, the project may suffer from a design that is inflexible to change and may require subsequent replanning. A key method for building flexibility into the system is the use of digital engineering tools such as MBSE. These MBSE tools rely on models (almost all software-based) of the system being developed and deployed and are ideal for tracking how anticipated changes to system requirements will impact the system, prior to any new code being written. The challenge with MBSE is that if the model(s) are not kept up to date, the model(s) will be of little value to systems engineers when introducing new requirements into the product baseline.

As for the future, there is active research into extending the MBSE concept to areas such as digital twins or digital shadows. These (mostly) software-based systems mirror an exact product deployed in the field. For example, there may be several different versions of a product in operation – each with its unique configuration and use. Having a digital twin of each version will be useful in tracking how the product is used and how changes to the baseline system may or may not benefit the end-user.

References

Aparicio, M. (2023). Why is MVP Development So Important. https://www.wearecapicua.com/blog/mvp-development (accessed 18 June 2023).

Dassault Systems CATIA (2023a). Cameo systems modeler. https://www.3ds.com/products-services/catia/products/no-magic/cameo-systems-modeler/ (accessed 17 June 2023).

Dassault Systems CATIA (2023b). MagicDraw. https://www.3ds.com/products-services/catia/products/no-magic/solutions/model-based-system-engineering/ (accessed 17 June 2023).

InfoWorld (2022). What is CI/CD? Continuous integration and continuous delivery explained. https://www.infoworld.com/article/3271126/what-is-cicd-continuous-integration-and-continuous-delivery-explained.html (accessed 25 May 2022).

MicroFocus (2023). Fortify. https://www.microfocus.com/en-us/cyberres/application-security (accessed 18 June 2023).

Office of the Under Secretary of Defense for Acquisition and Sustainment (OUSD(A&S)) (2020). DoD instruction 5000.02: operation of the adaptive acquisition framework. https://www.esd.whs.mil/Portals/54/Documents/DD/issuances/dodi/500002p.pdf (accessed 18 June 2023).

Office of the Under Secretary of Defense for Research and Engineering (OUSD(R&E)) (2022). Mission engineering guide. https://ac.cto.mil/wp-content/uploads/2020/12/MEG-v40_20201130_shm.pdf (accessed 18 June 2023).

Productfolio (2020). MVP vs MMP. https://productfolio.com/mvp-vs-mmp/ (accessed 18 June 2023).

Tanner, M. (2023). DevStar. https://software.af.mil/dsop/dsop-devstar/ (accessed 18 June 2023).

The Manifesto Authors (2001). Manifesto for Agile software development. https://agilemanifesto.org/ (accessed 18 June 2023).

Wiegers, K.E. and Beatty, J. (2013). *Software Requirements*. Microsoft Press. ISBN: 9780735679658.

Wikipedia (2023a). V-Model. https://en.wikipedia.org/wiki/V-Model_(software_development) (accessed 18 June 2023).

Wikipedia (2023b). Timeboxing. https://en.wikipedia.org/wiki/Timeboxing (accessed 18 June 2023).

Biographical Sketches

Michael Orosz directs the Decision Systems Group at the University of Southern California's InforWmation Sciences Institute (USC/ISI) and is a research associate professor in USC's Sonny Astani Department of Civil and Environmental Engineering. Dr. Orosz has over 30 years' experience in government and commercial software development, systems engineering and acquisition, applied research and development, and project management and has developed several successful products in both the government and commercial sectors. Dr. Orosz received his BS in engineering from the Colorado School of Mines, an MS in computer science from the University of Colorado, and a PhD in computer science from UCLA.

Brian Duffy is a senior systems engineer with the University of Southern California Information Sciences Institute (USC/ISI). He conducts research and analysis to determine system engineering methods and metrics necessary to transition Major Défense Acquisition Programs from a traditional waterfall development to Agile/DevSecOps processes. Prior to USC/ISI, Mr. Duffy retired from the United States Air Force with multiple assignments related to National Security Space acquisition programs and command and control systems. Mr. Duffy holds a masters of aeronautical science degree from Embry-Riddle Aeronautical University and a bachelors of aeronautical and astronautical engineering degree from the University of Washington.

Craig Charlton is a senior systems engineer at USC's Information Sciences Institute (USC-ISI) and has provided acquisition support at Space Systems Command (SSC) at the Los Angeles Air Force Base during the past 20 years on a number of leading-edge satellite systems. Prior to his position at SSC, Mr. Charlton acquired more than 25 years of experience as a software engineer and in managing software projects in the commercial world, primarily in the fields of engineering and of law enforcement. Mr. Charlton received a BA in mathematics from California State University, Long Beach.

Captain Hector Saunders has served in the Department of the Air Force for six years. He served four years in the U.S. Air Force as a Cyber Warfare Operations Officer, and two years in the U.S. Space Force as a Program Manager. He currently leads an Integrated Product Team at Los Angeles Air Force Base to develop and modernize the GPS ground control system's software factory, which supports Agile/DevSecOps software development. Prior to his current assignment, he led cyber warfare missions and trained operators to defend the Department of the Air Forces' wide area network systems. He holds a BS in physics with minors in applied mathematics and computer science from the CUNY's The City College of New York.

Ellins Thomas is a lead senior DevOps engineer with Booz Allen Hamilton in Los Angeles. Ellins has 19 years of experience in government software development, DevSecOps, systems engineering, and project management. He has worked with large and small companies that worked in domains within the Army (Geospatial) and Space Force (DevSecOps) prototyping, developing, integrating, analyzing, and testing several products for both organizations and recently provided acquisition support at Space Systems Command (SSC). Ellins Thomas received his BS in Computer Science from Virginia Polytechnic Institute and State University (Virginia Tech).

Chapter 32

Reference Architecture: An Integration and Interoperability-Driven Framework

Joel S. Patton[1] *and James D. Moreland*[2]

[1] *Virginia Tech Applied Research Corporation, Arlington, VA, USA*
[2] *Virginia Tech University, Grado Department of Industrial and Systems, Blacksburg, VA, USA*

Definition of a Reference Architecture

Why Is an Architecture Useful?

As the complexity of systems and systems-of-systems (SoS) has grown so have the challenges for organizations to create, share, and utilize them effectively. To better address these challenges, it is necessary to create and apply ideas, principles, procedures, and modeling tools to support decision-making in an architectural framework. The use of architecture-based processes is now a common practice in commercial, government, civil, and military domains.

The application of an "architecture" is now applied to systems and to other entities that are not traditionally considered to be systems, such as enterprises, services, data, business functions, mission areas, product lines, families of systems, software items, etc. The concept of an architecture used in this document refers to a "reference architecture," which extends beyond the traditional use where the architecture entity can be considered a system or SoS. When the word architecture is used without any reference to its particular role the word refers to a more general case where the architecture entails the fundamental concepts and properties of an architecture entity. The following represents a broader interpretation of architecture entities and processes: enterprise, organization, solution, system (including software systems), subsystem, business, data (as a data element or data structure), application, information technology (as a collection), mission, product, service, software item, hardware item, etc. The kind of entity can also be expanded to include a product line, family of systems, SoS, collection of systems, collection of applications, etc. A "Reference Architecture" for the purposes of this document possesses several common qualities that distinguish it from other architectural constructs. It can be considered a Reference Architecture if it possesses the following attributes:

- Provides common lexicon across various stakeholders.
- Provides a consistent methodology for the implementation of technology intended to solve specific challenges.
- Supports the validation of solutions against proven system model.
- Based upon adherence to a common set of standards, specifications, and patterns.

Systems Engineering for the Digital Age: Practitioner Perspectives, First Edition. Edited by Dinesh Verma.

Companion website: www.wiley.com/go/verma/systemsengineering

- Illustrates and improves understanding of the various Mission Engineering components, processes, and systems, in the context of a vendor- and technology-agnostic conceptual models;
- Provides a technical reference for US government departments, agencies, and other consumers to understand, discuss, categorize, and compare Mission Engineering solutions; and
- Facilitates analysis of candidate standards for interoperability, portability, reusability, and extendibility.

What Is a Reference Architecture?

While there are a number of definitions for "Reference Architecture," a commonly accepted and often used definition (Reference Architecture 2022) is that it provides a methodology and/or set of practices and templates that are based on the generalization of a set of successful solutions for a particular category of solutions. Reference architectures provide guidance to conduct repeatable and reproducible studies designed to solve a particular class of problems. In this way, it serves as a "reference" for the specific architectures that organizations can implement to solve their problems unique to their operating environment. It is important to note it is never intended that a reference architecture would be implemented as-is, but rather used either as a point of reference or as a starting point for an organization's architectural efforts. It is, therefore, critical that development of reference architecture be as accurate a representation of the process or system being studied as possible to ensue viable solutions are developed.

It is important to understand that others in various field will refine the definition of reference architecture to match their unique demands. An example of such a definition is a service-oriented architecture (SOA), which is a description of how to build a class of artifacts. These artifacts can be embodied in many forms including design patterns, methodologies, standards, metadata, and documents of all sorts. It is important when developing a specific architecture based on best practices, one should begin by defining the scope of the architecture that is to be built.

It is also worth briefly reviewing common attributes between the concepts of reference architectures and architecture frameworks. Architecture frameworks, such as the Zachman Framework (Zachman 1987), the Open Group Architecture Framework (TOGAF) (*The TOGAF Library* 2023), and Department of Defense Architecture Framework (DoDAF) ("DoD Reference Architecture Description" 2010), provide approaches to describe and identify necessary inputs to a particular architecture as well as means to describe that architecture. While these architecture frameworks differs from a reference architecture, they can be used to inform and serve as a starting point. The reference architecture however then takes the architecture framework to the next step by accelerating the process for a particular architecture type, helping to identify which architectural approaches will satisfy particular requirements, and determine what a minimally acceptable set of architectural artifacts are needed to meet the "best practices" requirements. In fact, multiple reference architectures for the same domain are allowable and quite useful. Reference architectures can be complementary providing guidance for a single architecture, such as SOA, from multiple viewpoints.

As stated previously an examination and analysis of numerous existing Reference Architecture definitions within DoD, other Federal Agencies, and Industry revealed common points among them. A common theme among the definitions is that the primary purpose of a Reference Architecture is to guide and constrain the solution architectures as depicted in Figure 32.1 ("DoD Reference Architecture Description" 2010).

Another example is a government reference architecture (GRA), which is a high-level conceptual model crafted to serve as a tool to facilitate open discussion of the requirements, design

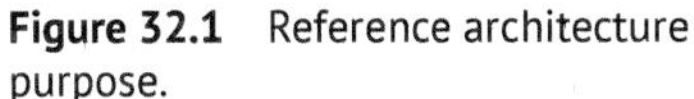

Figure 32.1 Reference architecture purpose.

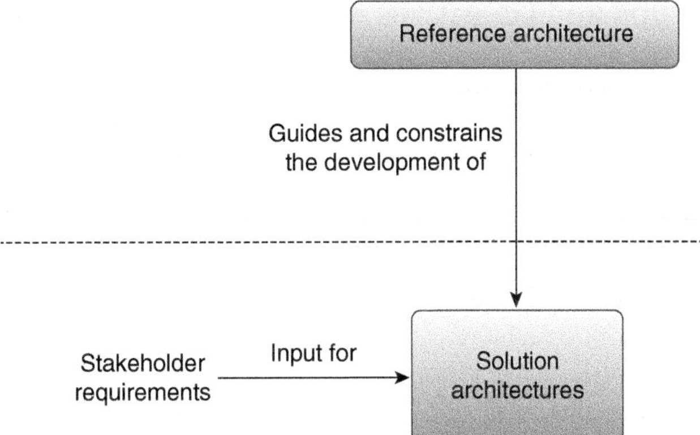

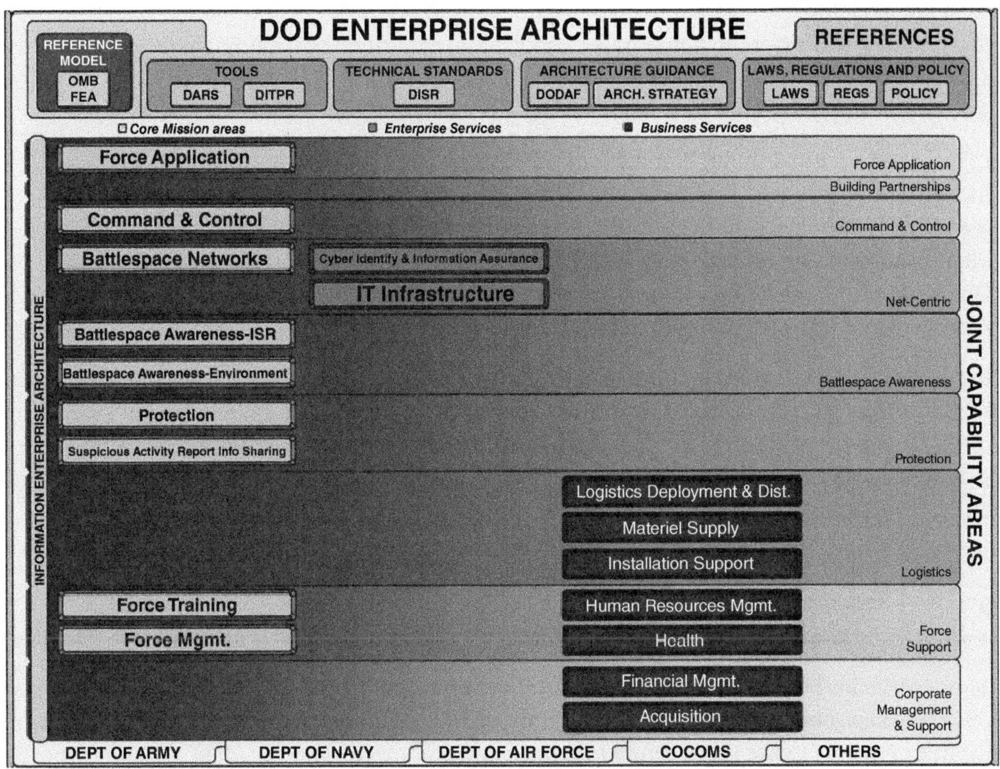

Figure 32.2 Example of DOD-wide government reference architecture.

structures, and operations inherent in Mission Engineering. The GRA is intended to facilitate the understanding of the operational intricacies in critical mission threads. It does not represent the system architecture of a specific Mission Engineering SoS, but rather is a tool for describing, discussing, and developing system-specific architectures using a common framework of reference. The model is not tied to any specific vendor products, services, or reference implementation, nor does it define prescriptive solutions that inhibit innovation. An example of a DoD-wide GRA is presented in Figure 32.2 ("DoD Reference Architecture Description" 2010).

Scope and Objectives of a Reference Architecture

Scope and Objective

Establishing a reference architecture is necessary to define minimum requirements on any framework, as one means of ensuring a consistent process in support of standards. A fundamental goal of an architecture framework is to codify a common set of practices within a community. This is typically done, especially in the DoD to promote interoperability and to enhance comprehensibility, commonality, and most importantly reducing unnecessary redundancies among individual architects. To achieve this goal, it is necessary to establish baseline requirements on architecture frameworks in terms of their content and presentation.

In the context of a DoD SoS approach, the development of a government reference architecture (GRA) provides a baseline for the future direction of Mission Engineering to address integration and interoperability by establishing a vendor-neutral, technology-, and infrastructure-agnostic conceptual model.

The requirements on architecture frameworks in the context of Mission Engineering are briefly outlined below.

An architecture framework shall include:

- Information identifying the architecture framework mapped back to an integrated capability framework.
- Identification of one or more gaps or deficiencies related to the systems performance.
- Identification of one or more stakeholders that have responsibility for that system(s).
- One or more architecture viewpoints that frame those gaps/deficiencies.
- Any corresponding operational or system limitations.

In constructing a reference architecture, an authoritative source of information about a specific subject area that guides and constrains the instantiations of multiple architectures and solutions is established. Reference architectures generally serve as a foundation for solution architectures and may also be used for comparison and alignment of instantiations of architectures and solutions.

High-Level Requirements

The development of a reference architecture requires a thorough understanding of current techniques, issues, and concerns in the context of the organization and environment. A reference architecture captures previous experience, for instance by adapting existing architectures. To be of value however, a reference architecture must be based on proven concepts. This validation of concepts in reference architectures is often derived from preceding architectures. This is especially important in situations where disruptive technologies or innovative applications are encountered. In these cases, reference implementations will need to accommodate prototyping and an incremental approach as part of the validation process. It is important to note that flaws in reference architectures will propagate to multiple architectures and actual systems, which will hinder the efficiencies you are trying to achieve as well as result in increased cost.

The value of a reference architecture is in its application to address the future which in terms depends on the vision and scope necessary in developing it. Therefore, the reference architecture is based on (future) customer behavior and business needs. These behaviors and needs are explored and analyzed and ultimately transformed into future requirements for the product portfolio.

In the case of government reference architectures, use cases are typically used to gain an understanding of current applications of integration and interoperability. This is accomplished by conducting a survey of reference architectures to understand commonalities. In the Government, this is performed within Mission Engineering Threads in use by developing a taxonomy to understand and organize the information collected and review existing technologies and trends relevant to mission tasks necessary to address new and emerging threats.

Developing a Reference Architecture

The purpose of a reference architecture is reflected in the set of requirements that the reference architecture must satisfy ("Reference Architectures; Why, What and How" 2007). The requirements must be structured into a set of goals, a set of critical success factors associated with these goals, and a set of requirements that are connected to the critical success factors that ensure their satisfaction. Creating your reference architecture is a challenging process as no predefined steps or processes exist. However, below are seven steps to help guide the implementation process.

Step 1. Identify the Purpose of Your Reference Architecture

Define the purpose of your reference architecture by asking the following questions:

- What information is important for the reference architecture?
- How much detail is needed to support analysis and decision making?
- Who will produce or use the reference architecture?
- How does the reference architecture drive Mission Integration Management?
- How is the reference architecture going to be used to evaluate capabilities to determine budget investments?

By knowing the purpose of the reference architecture, you can scope the necessary models and data that are needed to ensure people use your reference architecture for analysis and business decisions. A reference architecture should be strongly linked to company or organization's mission, vision, and strategy. The strategy determines what multi-dimensions have to be addressed, what the scope of the reference architecture is, what means, such as collaboration, are available to realize mission and vision. In fact, a reference architecture is an elaboration of mission, vision, and strategy as illustrated in Figure 32.3.

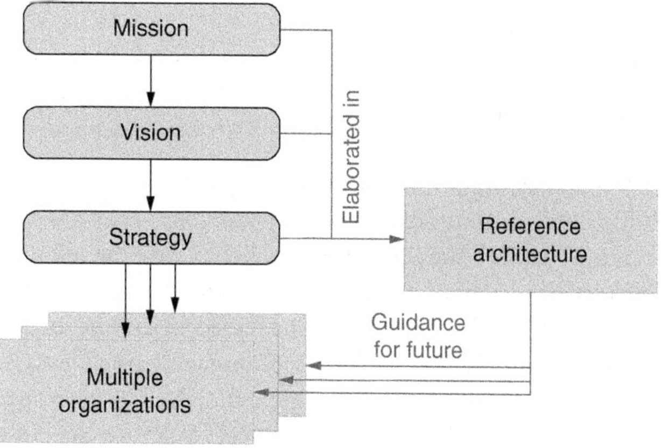

Figure 32.3 A reference architecture is an elaboration of mission, vision and strategy to provide guidance across multiple organizations.

Step 2. Identify Your Business Questions

Discuss the questions that are critical to the stakeholders' business; and then help them identify the questions that are hard for them to answer. The following questions are ones that many stakeholders need answered:

- What is the impact of retiring a system?
- What is the impact of changing key management elements?
- What applications are needed to support a business and requirements process?
- What is the impact of replacing systems?
- What processes need to be developed to support a new strategy?
- Where are the gaps or redundancies in our mission capability portfolio?

These questions should drive the content of your reference architecture. If most questions concern your mission capability portfolio, then focus on defining the mission area and the corresponding mission essential tasks. If you need to understand how your processes support a new strategy, then focus on the operational mission area. Then begin to expand the scope of your reference architecture with new mission questions to drive business decisions.

Step 3. Identify Assumptions and Business Rules

Now that you have identified the audience, purpose, and questions, you should identify the business rules that constrain or explain the area of interest to include all Corporate Planning Guidance. Every business has rules. For example, if you are capturing information about critical business processes, you must also capture any regulations or standards for the process.

An example of a regulation is the Health Insurance Portability and Accountability Act (HIPAA), which protects health insurance coverage for people who change jobs. A corporate regulation would then be created to show that the company is meeting the requirements of HIPAA.

You should capture assumptions about your reference architecture, such as "New application information will be uploaded on Friday" or "Every business unit is responsible for documenting business processes."

Step 4. Identify Your Framework

The following industry standard frameworks can be used as a starting point to create a reference architecture:

TOGAF: The Open Group Architecture Framework: Is the de facto industry standard framework, offering a methodological approach to Enterprise Architecture design, planning, implementation, and governance.

Zachman Framework: Is an IT-driven process that is based on a model-based approach that includes specifying deliverables, categorizes various aspects of enterprise system subsets into a matrix form and associates them with the decision choices of the business-I environment.

Using a standard framework gives your reference architecture a "skeleton" that you can then build out with your models. A framework also provides guidance on what information you need to capture based on the stakeholders who will use the reference architecture. It provides guidance on organizing information but does not suggest a specific implementation for mission specific architectures that will be built from this reference architecture.

As an example, the Zachman Framework provides a structured way for any organization to acquire the necessary knowledge about itself with respect to the Enterprise Architecture. The Zachman approach proposes a logical structure for classifying and organizing the descriptive

Interrogative perspectives →	What	How	Where	Who	When	Why	Target contributers
Scope							Strategists
Business							Executive leaders
System							Architects
Technology							Engineers
Component							Technicians
Operations		The enterprise					Workers
Audience perspectives	Inventory	Process	Network	Organization	Timing	Motivations	Targetdomains

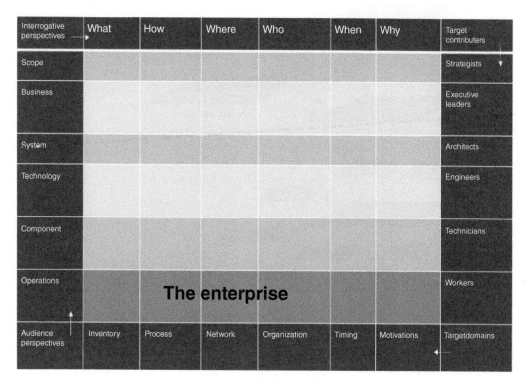

Figure 32.4 Zachman Framework applied to an enterprise.

representations of an enterprise, in different dimensions, and each dimension can be perceived in different perspectives as seen in Figure 32.4.

How you choose the method depends on the goal of your reference architecture, the experience of your team, and whether you want to follow a defined process like ToGAF, or just need help identifying which model to use for what purpose as in Zachman. You can also combine frameworks like the combination of ToGAF and Zachman, which are often used together. The framework in Figure 32.5 illustrates the linkages between approaches.

A framework provides guidance on what to model. Methodologies are then used to create models. A methodology is a rule set that explains how to model something. For example, the Business Process Modeling Notation (BPMN) methodology gives precise rules and symbols to model a business process.

The DoD Architecture Framework (DoDAF): Defines the solution architecture as a framework or structure that portrays the relationships among all the elements of something that answers a problem. It describes the fundamental organization of a system, embodied in its components, their relationships with each other and the environment, and the principles governing its design and evolution. Solution architecture instantiations are guided and constrained by all or part of a reference architecture where the generalized and logical abstract elements of the reference architecture are replaced by real world, physical elements according to the specified rules, principles, standards, and specifications. Reference architectures may also be complimentary in guiding architectures and solutions. Table 32.1 shows that reference architecture may guide and constrain various types of views and models depending on the purpose and scope. DoD-wide reference architecture, due to its broader purpose and scope, may guide and constrain Enterprise, Segment, Capability, and Solution Architectures.

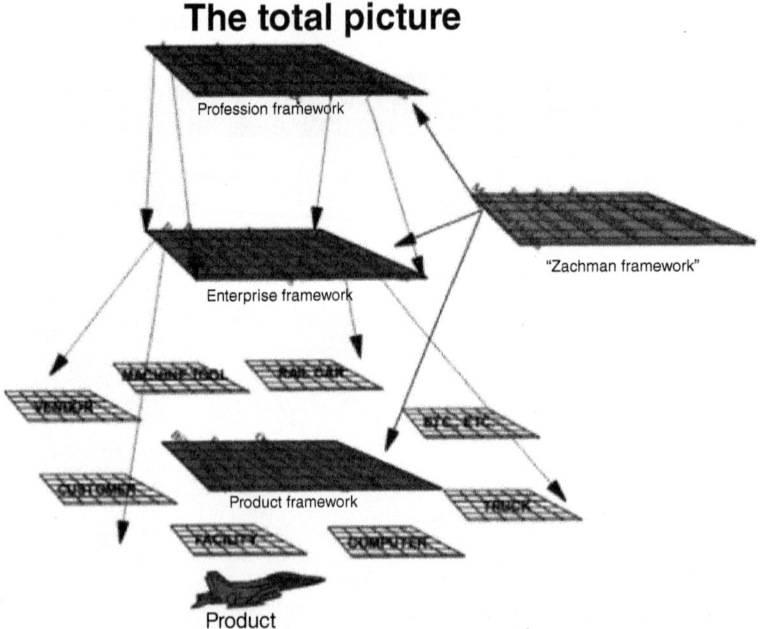

Figure 32.5 Linkages between TOGAF and Zachman Frameworks.

Table 32.1 DODAF models utilized in reference architectures.

Content	DoDAF 2.0 views/models	
Purpose: introduction, overview, context, scope, goals, purpose, why needed, and when and how used	• **AV-1 Overview and Summary Information** • **CV-1: Vision** – overall strategic concept and high-level scope • **OV-1 High Level Operational Concept Graphic** – executive operational summary level of what solution architectures are intended to do and how they are supposed to do it	
Principles: foundational organizational rules, culture, and values that drive technical positions and patterns	• **OV-6a Operational Rules Model** • **OV-6b Operational State Description** • **SvcV-10a Services Rules Model** • **SV-10a Systems Rules Model** • **OV-4 Organizational Relationships Chart** – architectural stakeholders	
Technical positions and policies	• **StdV-1 Standards Profile** – standards, specifications, guidance, policy applying to elements of the solution architectures	
Architectural patterns: generalized patterns of activities, service functionality and system functionality and their resources, providers and information/data resource flows Generalized scenario patterns of sequenced (sequential)	**Operational patterns** • **OV-2 (multiple) Operational Resource Flows** • **OV-5 {a, b} Activity diagrams** • **OV-6c Event-Trace Description** **Service patterns** • **SvcV-1 (multiple) Service Interfaces** • **SvcV-2 Service Resource Flows** • **SvcV-4 Service Functionality** • **SvcV-10b Service State Transitions** • **SvcV-10c Services Event-Trace**	**System patterns** • **SV-1 (multiple) System Interfaces** • **SV-2 System Resource Flows** • **SV-4 System Functionality** • **SV-10b System State Transitions** • **SV-10c Systems Event-Trace Description**

Step 5. Create a Meta-Model

A meta-model is an abstract view of your reference architecture. It shows the data you are trying to capture, and the relationships among the data. This is where you realize alignment, which is based on answers to your business questions. For example, if you need to know the application that supports a certain business process, there must be a relationship between those two things in your meta-model. Otherwise, there is no connection between the data, you cannot answer your business question, and the reference architecture is not functional.

Note that you do not want a direct relationship between everything in your meta-model, and you should only link things together that have logical relationships. For example, linking an organizational department to a technology does not make sense, but linking a technology to an application does. A good modeling tool such as Rational System Architect supports traversing the meta-model to create complex reports. So, in this meta-model example, you can report on the hardware that supports a business function even though there is not a direct relationship in the meta-model. In a meta-model you can potentially traverse from a business function to a business process owned by that function, to a location of the business process, to a supporting application the process needs, and finally to the technologies that support that application.

Your meta-model should include the following features:

- Relationships between the architecture elements. For example, a business process to an application.
- Definitions of the elements. For example, the meaning of the term "application" and what properties you will capture.
- Traceability to business questions. For example, if your question is "What applications support what business processes?" You know you need a business process and an application in your meta-model, with a direct or indirect relationship between them.

Step 6. Identify the Models Needed in the Reference Architecture

Now that you have identified your business questions, your framework, and the meta-model you need to answer your questions, you need to figure out what models to draw. Using a business process as an example, there are many industry standards that support modeling business processes, such as business capture processes and flow charts. Choose your modeling methodology based on the following criteria:

- **The audience for the information**: Managers understand simple diagrams like BPMN; software developers normally prefer UML sequence diagrams or use cases.
- **The elements of the meta-model**: If in your meta-model you need to understand data as it relates to business processes, consider using BPMN to model that. If instead you are just worried about the sequence of process steps, consider creating a flow chart.
- After knowing the audience and the content you want to model, you can then identify the diagrams you need to create. In the above example, since you needed information about business processes and system interfaces, you could select the following models:
  - Business Process Model and Notation (BPMN) – captures business processes.
  - System architecture (SA) – captures applications.
  - Model-based Systems Engineering (MBSE) – captures language, methodology, and tools.

It is important to remember that you cannot use a single diagram to model everything in your reference architecture. Further, separation of the architectural views, such as the application view from business view, is a best practice. If you try to model two views in the same diagram, it often creates confusion and does not capture information in a meaningful way.

Step 7. Integrate the Reference Architecture Domains

Link the data that you captured together based on the relationships you identified earlier. If you have existing reference architectures for programs or mission areas, and you want to create an overarching reference architecture, the easiest approach is to populate your reference architecture from the bottom up. Take existing reference architectures and pull common elements into a repository. Moving forward, try to standardize the models and terminology that is used across the organization, because everyone uses the same name for an organization, such as standardizing on "Finance" rather than having variations such as "Finance dept.," Accounting, or Accounting and Finance.

If this is your first reference architecture, use a common blueprint across your organization so that a reference architecture is created to reflect a series of integrated end-to-end tasks required to successfully achieve a desired outcome. This is referred to as a mission thread. Mission threads are designed to use the same framework, terminology, and models as the other mission-specific architectures that can be integrated. This creates an ability to link and report across the entire enterprise.

Application of Reference Architecture

The application of reference architectures provides guidance for the implementation of mandated data specifications and mission essential tasks required for the operation and execution of mission engineering threads in support of other enterprise, mission, organization, and community of interest specific reference or integrated solution architectures. It exists to pull together in a single document the logic flow of references and requirements to correctly and sufficiently implement the mandated specifications, tactics, techniques, and procedures to support operational mission execution to realize a fielded capability. Operational timelines require the reference architecture to monitor and dynamically refine data and mission essential tasks processing and access timelines in support of mission requirements. End-to-end operational timelines are decomposed into time budgets allocated to specific performance budgets for specific processes and activities.

As an example of application, the Department of the Navy (DON) produced a government reference architecture for the mission engineering data structure, see Figure 32.6.

This reference architecture provides assistance to DON program managers in the development of "solution architectures," as mandated by the Joint Capabilities and Integration Development System and Defense Acquisition System processes. This is done by providing program managers a "plug-and-play" integrated architecture reference to be used as a foundation for their program-specific solution architectures. It helps to minimize the need for solution architects to recreate portions of the enterprise architecture that are not specific to their individual program. This reference architecture also provides authoritative requirements and a common framework, which program managers must comply with to ensure their particular solution is aligned with achieving departmental goals and objectives.

The use of this reference architecture leads to more specific implementations as shown in Figure 32.7 with an Integrated Capability Framework (ICF) model used to develop specific products for the DON Mission Engineering effort.

The ICF model provides the data structure, taxonomy, and relationships needed to accurately and fully define a mission. It ensures commonality among products built and helps enable communication and collaboration by using a common taxonomy from authoritative sources. In addition, this ICF model captures required operational capabilities and utilizes the identified

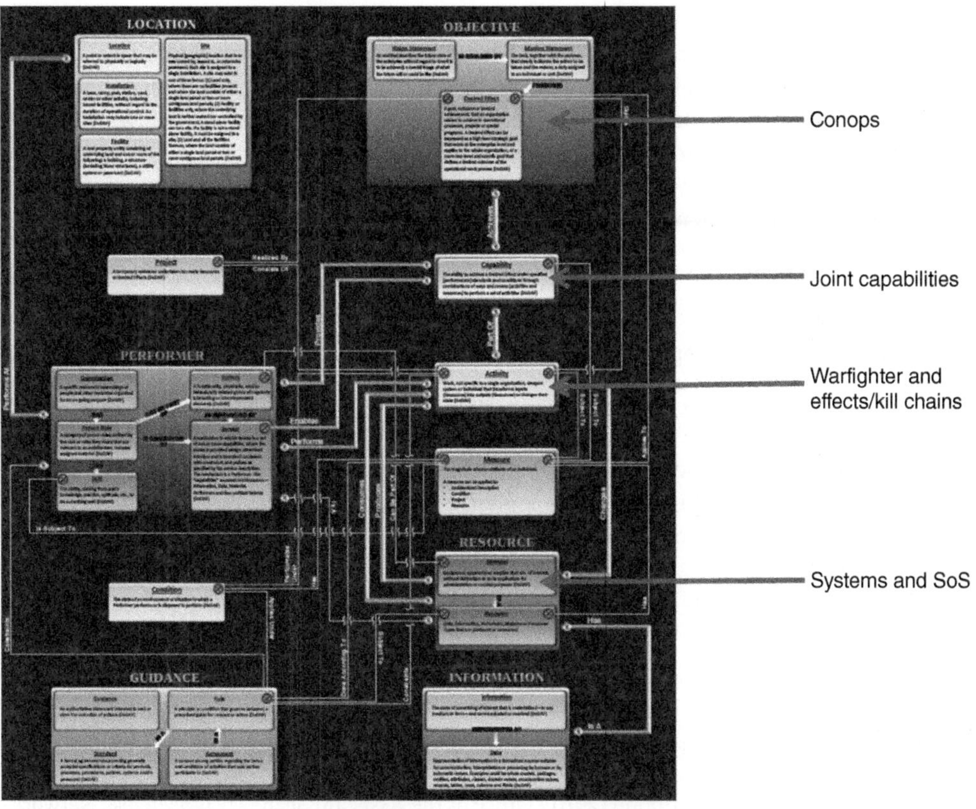

Figure 32.6 Reference architecture for DON Mission Engineering.

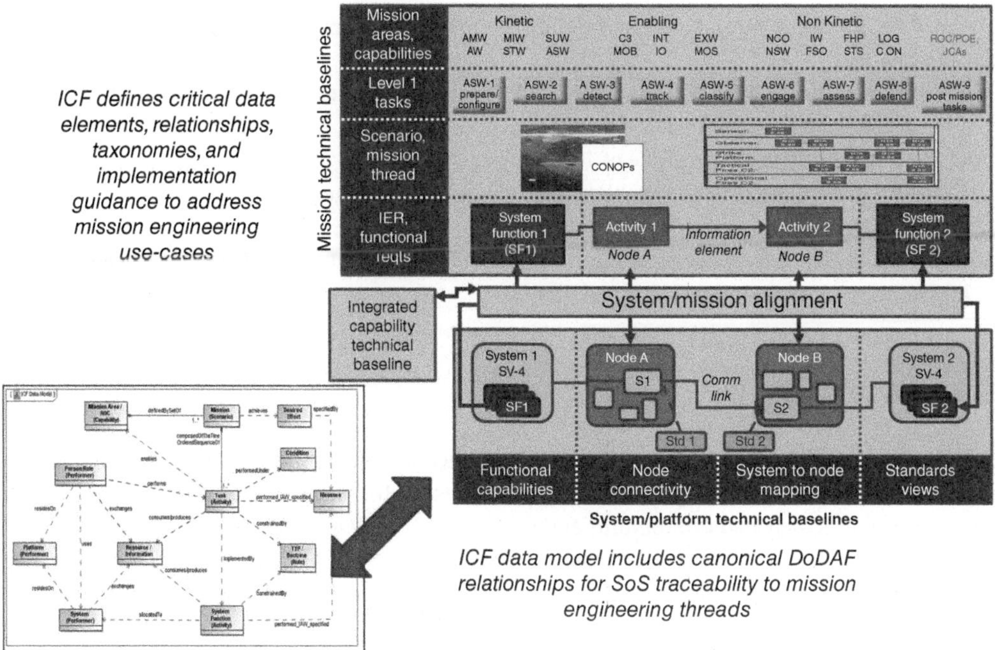

Figure 32.7 Integrated Capability Framework model for DON Mission Engineering.

relationships to identify system and platform functional, interface, and performance requirements. These relationships are identified in mission engineering threads by considering associated Rules of Engagement, Tactics, Techniques, Procedures (TTPs), as well as measures of performance, information exchange requirements, and required system functions. The output of System/Mission alignment is a linked set of products that provide the context needed for definition, analysis, and evaluation of mission capability, definition of mission need, and identification of potential integration and interoperability challenges.

Report Structure

The organization of the GRA report should roughly correspond to the following process:

- Section on high-level, system requirements in support of integration and interoperability relevant to the design of the GRA and discussion on the development of these requirements:
- Section presenting the generic, technology-independent GRA conceptual model.
- Section to discuss the main functional components of the GRA.
- Section to describe the system and life-cycle management considerations related to the GRA management fabric.
- Section to introduce security and privacy topics related to the security and privacy fabric of the GRA.

References

DoD Reference Architecture Description (2010). Office of the Assistant Secretary of Defense, Networks and Information Integration (OASD/NII), Reference Architecture Description, Prepared by the Office of the DoD CIO.

Reference Architecture (2022). https://en.wikipedia.org/wiki/Reference_architecture (accessed 31 May 2023).

Reference Architectures; Why, What and How (2007). White Paper Resulting from Architecture Forum Meeting March 12 & 13, 2007. Edited by: Dr. Gerrit Muller, Embedded Systems Institute and Mr. Eirik Hole, Stevens Institute of Technology.

The TOGAF Library (2023). https://publications.opengroup.org/togaf-library (accessed 31 May 2023).

Zachman, J.A. (1987). A framework for information systems architecture. *IBM Systems Journal* 26 (3): 276–292.

Biographical Sketches

Joel S. Patton Mr. Patton currently serves as a senior leader and subject matter expert in the field of Mission Engineering at Virginia Tech Applied Research Corporation. His prior experience includes serving as the Head of Programs at the Naval Surface Warfare Center Carderock Division. He has held similar position at the Pennsylvania State University Applied Research Lab, General Atomics' Electromagnetic Systems, and Siemens USA supporting the development and implementation of next-generation military systems.

Dr. James D. Moreland, Jr. Dr. Moreland currently serves as the Vice President, Mission Integration for Raytheon Missiles and Defense in the Air Power Mission Area. He entered the Senior Executive Service in September 2014 as the Director, Naval Warfare within the OUSD for Acquisition, Technology, and Logistics and served in this position until 2018. Dr. Moreland served as the Chief Engineer for the Naval Surface Warfare Center Dahlgren Division (NSWCDD) from 2010 to 2014. He is recognized as a world-renowned expert in Mission Engineering and served as the DoD lead and technical expert in this area. In addition, he served as a Senior Executive Science Advisor to the White House Office of Science & Technology Policy providing advice on all scientific, engineering, and technological aspects of national security.

Chapter 33

Mission Engineering Competency Framework

Gregg Vesonder and Nicole Hutchison

Stevens Institute of Technology, Hoboken, NJ, USA

Introduction

This chapter discusses mission engineering and what skills are needed to do it well. In order to understand the discipline of mission engineering, it is first necessary to define what a mission is.

DoD Joint Publication 3-0 (Joint Operations) defines a mission as the "task, together with the purpose, that clearly indicates the action to be taken and the reason thereby." (2018) In this context, mission definitions are the basis for evaluating solutions and trades – any proposal that does not advance the mission should not be considered. Broadening the context, NASA defines a mission as "a major activity required to accomplish [a NASA] goal or to effectively pursue a scientific, technological, or engineering opportunity directly related to [a NASA] goal. Mission needs are independent of any particular system or technological solution" (2017). In the colloquial sense a mission is a "specific task with which a person or a group is charged" and this can include a definite military, naval, or aerospace task (Miriam Webster Dictionary 2023). Mission engineering is the term commonly used in the US government. In industry, the term business analysis or engineering might be used. In the UK Ministry of Defense (MOD), the term "capability" is used. Regardless of the terminology, the concept is the same: missions are critical, major activities that an organization is trying to achieve. Mission engineering, then, is the discipline that defines and develops those activities and determines how potential solutions fit into the mission space.

There is no single definition of mission engineering. The US Department of Defense (DoD) defines it as "the deliberate planning, analyzing, organizing, and integrating of current and emerging operational and system capabilities to achieve desired warfighting mission effects" (Defense Acquisition Guide 2017). The office of the Deputy CTO for Mission Capabilities (DCTO(MC)) refines this definition, adding, "ME (mission engineering) is an analytical and data-driven approach to decompose and analyze the constituent parts of a mission in order to identify measurable tradeoffs and draw conclusions".[1] Mission engineering analyzes the mission goals and threads, analyzes the available as well as emerging operational and system capabilities, and designs a mission architecture to achieve the mission goal (Gold 2016). A more industrial definition of mission engineering is "the deliberate planning, analyzing, organizing, and integrating of current and emerging operational and system capabilities to achieve desired mission effects." For example, NASA defines mission engineering as an end-to-end, multi-mission development methodology

Systems Engineering for the Digital Age: Practitioner Perspectives, First Edition. Edited by Dinesh Verma.

Companion website: www.wiley.com/go/verma/systemsengineering

that seeks to integrate the development processes between the space, ground, science, and operations segments of a mission. It thereby promotes more mission-oriented system solutions, within and across missions (Ondrus and Fatig 1993). Other nations discuss it in terms of capabilities engineering (UK) or force design (Australia). In the United States DoD, mission engineering is a separate discipline, whereas in other agencies such as NASA and the Federal Aviation Administration (FAA), it is embedded within systems engineering.

Mission engineering applies the mission context to complicated and complex system-of-systems (SoS). Most current systems engineering practices do not fully address the unique characteristics of mission engineering, i.e., addressing the end-to-end mission as "the system" and extending further beyond data exchange between individual systems to enable cross-cutting functions, controls, and trades across systems.

Mission engineering differs from mission analysis in that the latter only addresses the examination of current operational and system capabilities, and not the design and engineering to assure the mission. Mission engineering within the US DoD applies an operational mission context to the complex SoS. The SoS approach has arisen in response to needs for capabilities requiring multiple linked systems that are greater than the sum of the capabilities of the constituent parts. Mission engineering is differentiated from traditional systems engineering because, from the mission engineering perspective, the individual systems that comprise the military capability are inherently flexible, functionally overlapping, multi-mission platforms supported by a complex backbone of information communication networks.

Mission Engineering Competency

Given the importance of mission engineering, the Office of the Deputy Assistant Secretary of Defense for Systems Engineering (ODASD(SE)) directed the Systems Engineering Research Center (SERC) to identify the critical skills required to successfully accomplish and shepherd mission engineering. The specific tasks were to:

- Identify competencies for mission engineering that are truly unique, showing where there is separation from the generally demanded acquisition competencies or systems engineering competencies.
- Identify critical overlaps between mission engineering and systems engineering competencies
- Identify aspects of mission engineering that are general enough to be considered critical by the broader acquisition workforce, yet specific enough to support building interdisciplinary mission engineering knowledge and abilities.
- Develop a mission engineering competency model that supports the US DoD engineering community but also provides input to each acquisition career field such as program management, and test and evaluation, unique to their responsibilities to support and manage mission engineering.

Competency, colloquially, is the ability to do something successfully or effectively, and this definition broadly applies when talking about workforce competencies. A competency model further lays out the knowledge (required facts and information), skills (technical proficiencies), abilities (capacity to apply knowledge and skills to complete a task), and behaviors (common responses to situations). Often the acronym "KSAB" is used to capture these aspects of a competency model that are critical to a discipline. In some organizations, "attitudes" is used instead of "abilities." The acronym KSA (knowledge, skills, and abilities) may be used instead of KSAB. In some organizations, KSABC is used, where the "C" stands for cognitions. Whatever terminology is used, the core purpose of competencies is to set expectations for what needs to be known; the skills and activities required to meet needs; and any expected responses to different types of situations.

Exploring Mission Engineering Competencies

The intent of the mission engineering competency work was to identify the skills necessary to perform critical mission engineering activities across complicated and complex systems and SoSs. These activities include but are not limited to mission analysis and synthesis; trade-off analyses; technology management; resource management; architecture development and modeling; mission modeling; addressing supporting capabilities such as communications and overarching mission functions; synchronization of testing; and individual system implementation. An effective competency model also reflects industry approaches and best practices.

The MECF was generated using the methodology illustrated in Figure 33.1. Researchers used a combination of literature review and in-depth interviews with practicing mission engineers and thought leaders. Interviewees included individuals across the US DoD (Army, Navy, Air Force, Marine Corps), other US government organizations (e.g., NASA, FAA), and industry. Overall, 32 in-depth (60+ minute) interviews were conducted.

There is a rich body of work in the open-source literature over the last several decades describing mission engineering applications, methods, and tooling. This literature includes articles in peer-reviewed journals and conferences as well as education courses and in-house publications and training. The mission engineering and SoS engineering literature include the following defense topical areas: anti-submarine warfare, electronic warfare, ballistic missile defense, theater missile defense, counter-air, brigade combat team air and missile defense, joint fires, strike, "always on" battlespace integration, and common operating picture. Non-defense topical areas include space exploration and the US national airspace system (NAS) (Deiotte and Garrett 2013; Garrett et al. 2011; Marvin and Garrett 2014; Marvin et al. 2014, 2016; Mindell 2015; Moreland 2009, 2014, 2015; Neaga et al. 2009; Parnell et al. 2016; Powers 2014; Rebovich 2014; Richards et al. 2009; Spaulding et al. 2011).

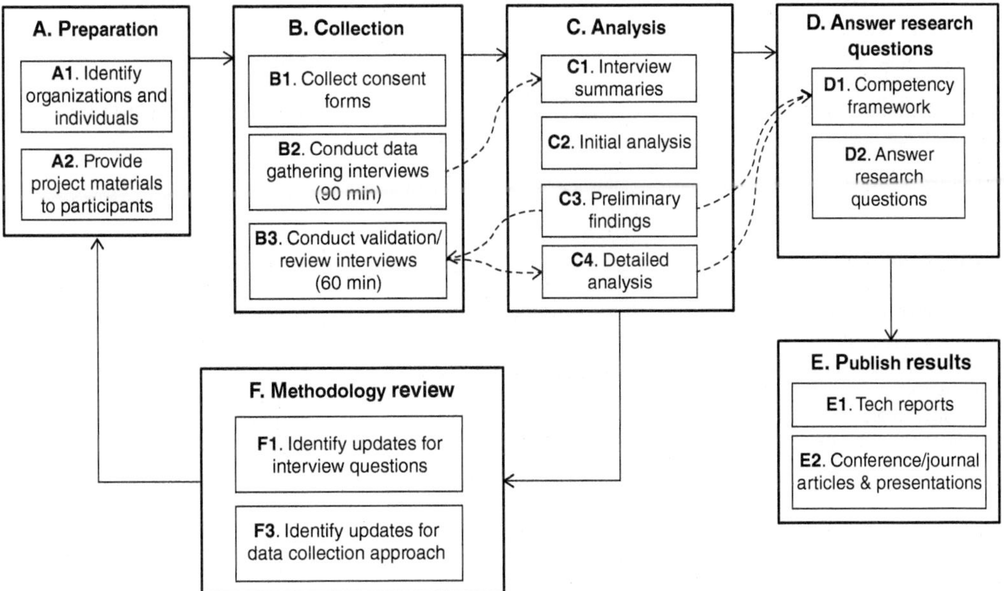

Figure 33.1 Mission engineering competency model development methodology. *Source:* Adapted from Vesonder et al. (2018).

Finally existing competency models were examined to identify any mission-related competencies that had already been identified (e.g., Hutchison et al. 2020; INCOSE 2018).

The Mission Engineering Competency Framework (MECF)

The analysis of existing documents and the in-depth interviews led to the development of the Mission Engineering Competency Framework (MECF), as illustrated in Figure 33.2.

There are 33 competencies included in the MECF. These are divided into six competencies areas (the colored labels in Figure 33.2):

- **Discipline and domain foundations:** This area focuses on the foundational understanding of the systems that will be required to support a given mission.
- **Mission concept:** This area focuses on an individual's ability to understand and work within the context of a given mission, including understanding the overall concept, scenarios, and relevant mission threads as well as understanding the factors that may influence the mission in addition to technology (doctrine, processes, training, etc.).
- **Systems engineering:** Mission engineering and systems engineering share critical overlaps (see Section 1). This area provides clarity on the specific systems engineering KSABCs that are most critical for mission engineering.

Discipline and domain foundations
- Principle and relevant disciplines
- Relevant domains
- System characteristics
- Relevant systems
- Relevant technologies
- Acquisition context

Technical leadership
- Guiding diverse stakeholders
- Team building
- Political savvy
- Decision making
- Workforce development

Mission concept
- Operational context
- Mission concept of operation
- Mission scenarios/threads
- DOTMLPF considerations

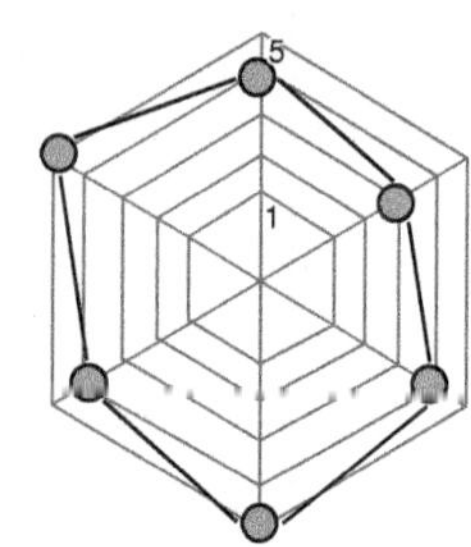

Interpersonal skills
- Communication
- Translation
- Enterprise context
- Building and utilizing a SME network
- Coordination
- Influence, persuasion, and negotiation

Systems engineering
- System of systems engineering
- Analysis
- Architecture
- Modeling and simulation
- Requirements
- Integration
- Gap analysis

—◯— An **example** of a mission engineer's profile

Systems mindset
- Big picture thinking
- Adaptability
- Paradoxical mindset
- Multi-scale abstraction
- Critical thinking

Figure 33.2 Initial mission engineering competency framework. *Source:* Adapted from Vesonder et al. (2018).

- **Systems mindset:** This area is analogous to the systems mindset in Helix (Hutchison et al. 2018) and includes the cognitive abilities around thinking holistically as well as being able to identify the right levels of detail and integrate these perspectives.
- **Interpersonal skills:** This area includes the skills and behaviors associated with the ability to work effectively in a multi-team environment and to coordinate across the mission scope.
- **Technical leadership:** Skills and behaviors associated with the ability to guide a diverse team of experts toward a specific technical goal.

In short, the MECF outlines technical competencies, operational understanding, and critical "nontechnical" skills. It is the combination of these different types of KSABs that will enable mission engineers to be successful. Appendix A contains a detailed description of all competencies contained within the MECF.

The radar chart in the center of Figure 33.2 provides an example of an individual mission engineer's competency profile. This is not intended to be a standard but an illustration – every mission engineer's profile will be a bit different. The main thing to keep in mind for building mission engineering teams is that the team overall needs to cover the competency space. Several interviewees stated that their best mission engineer was either a former operator (e.g., soldier, sailor, marine, or airman) who then got a degree in engineering or someone whose undergraduate degree was in an engineering or technical field who served in the military (gaining operational experience). Some teams use a combination of engineers and operators to cover the space, and several interviewees reported that this is highly successful as long as all on the team have sufficient nontechnical skills (the blue areas in Figure 33.2).

The study had several key findings:

- "Understanding the operational context" was considered *the* most important aspect of mission engineering among interviewees.
- Understanding current capabilities and their constraints is critical to enable mission engineers to identify gaps in developing or performing the overall mission.
- Applying the systems engineering technical process – such as architecture, modeling and simulation, interoperability, and feasibility analysis – with continuous feedback loops is a critical aspect of mission engineering.

Mission engineering is broader in scope than systems engineering, but cannot be successful without solid systems engineering practices.

The Path to Mission Engineering

As warfare, space travel, product development, and sustainment become more integrated and networked, mission engineering is becoming an increasingly vital skill. Technological artifacts are no longer islands but rather part of the networked, technological sprawl. To achieve success, engineers must not only understand the system but also the operational context in which it will operate. The system must be proven to effectively integrate with extant and future components in the field and be able to provide feedback on its effectiveness.

System engineering knowledge is necessary but not sufficient for mission engineering. Mission engineering requires the system engineer to be competent in digital engineering. As technologies in the field become increasingly networked, integrated, and dependent, it is necessary to develop and maintain an evolving architecture. This architecture must be supported by modeling and simulation and tested through applying case studies in simulations supported by a NearOps[2] environment not only in development and production but also in operation.

Appendix A: MECF Detailed Competency Descriptions by Area

This appendix provides a detailed discussion of each competency outlined in the MECF (Figure 33.2). These descriptions are copied from the original MECF document (Vesonder et al. 2018) with permission.

Discipline and Domain Foundations

Any given mission will cross multiple domains and require the support and integration of many engineering disciplines to be realized. Because a basic understanding of math, sciences, and fundamentals of engineering is assumed, the foundational building block for mission engineering, then, is an individual's capabilities in the most critical disciplines, domains, and technologies for a given mission as well as a grasp of complexity and the acquisition context in which they operate. This **Area** provides a grounding in the critical *systems* that will enable a mission.

Discipline and Domain Foundations

Competency (or Sub-competency)	Description
Principle and Relevant Disciplines	*Disciplines* are fundamental areas of education or expertise that are foundational to a system. For example, for a communications system, electrical engineering will be an important discipline to understand, while civil engineering will be less relevant. Specialties are disciplines that support mission engineering by applying cross-cutting knowledge. Specialties include Reliability, Availability, and Maintainability (RAM), Human Systems Integration, Safety Engineering, Affordability and other related topics. A mission engineer needs to understand which of these disciplines are most critical to support a given mission.
Relevant Domains	*Domain* refers to the overarching area of application for a given system; this includes things such as space, aerospace, marine, communication, finance, etc. Competency in relevant domains may enable an individual to be more effective.
System Characteristics	For mission systems, several specific characteristics were listed as critical and prevalent for mission systems. The most commonly-cited critical characteristics are provided below. However, when using this framework, the characteristics should be tailored to reflect the characteristics of the mission system.
Complexity	While systems will have a variety of characteristics that mission engineers must handle, for mission engineers, complexity is critical. In general, an individual will work on a spectrum of complexity, ranging from simple to complicated to complex to chaotic. (Adapted from the Cynefin framework (Snowden and Boone 2007)) Complexity is generally not measured by the number of parts of a system – which would be a measure of how complicated a system is – but of the interactions between system elements, disciplines, or technologies, and the properties that emerge out of these interactions that are not present in the individual elements. Mission engineering involves multiple systems and/or systems of systems, each with their own inherent complexity and, therefore, mission engineers consistently work in the "complex" space. One categorization of complexity includes structural complexity, dynamic complexity, and socio-political complexity (Collins et al. 2017); while another identifies two kinds of complexity: disorganized complexity and organized complexity (SEBoK authors, "Complexity", 2016). For mission engineering, this includes not the complexities of an individual system, but of the interactions between multiple systems or systems of systems that will enable a given mission. In many ways, complexity is increased by the following three characteristics:

Discipline and Domain Foundations	
Competency (or Sub-competency)	**Description**
Uncertainty	Uncertainty is the result of not having accurate or sufficient knowledge of a situation. (ISO/IEC 2009) Because mission engineers are trying to integrate across a variety of systems or SoS's, it is not possible for them to have every detail on each mission system. Because mission engineers also may be from outside the organizations or even services where a system is being developed, it may also be difficult to obtain required information. Being able to function, and make decisions under uncertainty is a critical skill for mission engineers.
Asynchrony	As with many SoS's, mission systems have to deal with issues of asynchrony: the quality or state of being out of concurrence in time. When individual systems are acquired, they each have their own lifecycle. When these systems then need to be combined into a larger SoS to support a mission, their acquisition lifecycles may not change. As one interviewee stated, "But the challenge is all the bits are delivered asynchronously. If [a contractor] delivers a new system, they select technology that will be mature at CDR and then deal with obsolescence once the system is in service. In that scenario, I wouldn't expect my subsystems to change that often. But in mission engineering, you have to expect that different bits will be delivered at different times. Some [are] nearly obsolete, some very new and untested." To be successful, mission engineers must be able to navigate asynchronous systems.
Legacy Systems	Though this is not unique to mission systems, especially in the defense context, it is critical to understand what many legacy systems may be expected to integrate into the broader mission. One interviewee stated, "[We] do leap in technology but [we are also] dealing with systems that entered the fleet in 1952 or 1953 - need to account for these." Another interviewee provided a specific example of this, "For example, aircrafts like the B-52 have been on the 'on and off' status for ongoing efforts. Due to funding cost or program office service, it shifted into monitoring and weren't active – more for knowledge management." Mission engineers have to be able to deal with these issues.
Relevant Systems in the Mission Space	The two categories above define the systems in the mission space and how they are expected to interact. This category, however, is focused on the mission engineer's understanding of these critical systems. This is not to say that mission engineers must be experts in every system, but they should understand the context of each system, including the mission and organizational context in which it is being developed. This familiarity will better enable the mission engineer to anticipate potential problems when integrating systems to perform a specific mission.
Relevant Technologies	Within the context of a mission, there are specific technologies that are relevant. For example, on a marine system, these may be technologies such as gas turbine, radar, and sonar systems; and each technology has its own terminology, challenges, etc. A mission engineer must be aware of the most critical technologies for the systems that are included within a given mission.
Acquisition Context	The ways that systems are developed and acquired provides critical context and boundaries for the ways in which missions can be addressed. Particularly, government acquisition systems have standard processes and rules that constrain the ways in which mission engineers can impact the systems of systems that must integrate to achieve a mission goal. Without understanding the constraints of the acquisition system, a mission engineer cannot effectively enable a given mission. This includes understanding the acquisition context of the programs that support the mission and where each of these programs fits within the acquisition process. This asynchronicity contributes to the complexity of mission engineering. It is important to note that this skillset around working in an acquisition environment must also be paired with understanding the mission environment (described in Section 3.1.2 below). One of the challenges in mission engineering is the dichotomy between the acquisition view of systems and the mission view of systems of systems (described in Section 2 above). It is mission engineers with both skillsets that were reported to be move effective.

Mission Concept

The first **Area**, above, provides the foundations for a mission engineer to be able to understand the systems which make up a mission. This **Area** defines the skills required for the mission engineer to understand the mission itself and how constituent systems are expected to support and enable the mission.

Mission Concept	
Competency (or Sub-competency)	**Description**
Operational Context	The operational context is the combination of the conditions, circumstances, and influences which will determine which systems will be used. In a Defense context, this includes the use of military forces. One of the most consistent themes heard throughout the interviews was the criticality of being able to understand all the systems that support a mission not just theoretically, but also in terms of how they function and the environment(s) in which they are expected to function. Several interviewees stated that they hired individuals with operational expertise such as Navy Seals or Army Special Forces. Their operational understanding is critical to successful mission engineering. These individuals often did not have a background in mission engineering, so their organizations trained them in mission engineering or paired them with experienced mission engineers.
Mission Concept of Operations	A system concept of operations (ConOps) is a lens through which to view a system, specifically of how key users will interact with the key systems within a mission. A mission ConOps is the view of the critical systems required to complete a mission which highlights how these systems will interact at a high-level to produce the desired mission effects.
Mission Scenarios/ Threads	Related to the Mission ConOps, mission threads define the end-to-end execution of a mission and enable individuals to understand how all the systems of systems work together. However, as opposed to the mission ConOps, which is intentionally at a high level, *mission threads include multiple levels of abstraction and are designed to enable each team working on a system or system of systems in the mission space to understand how to integrate with and support the overall mission*. This should help engineering teams understand the critical mission constraints of their systems and incorporate these constraints into their designs.
DOTMLPF Considerations	A critical aspect of mission engineering is understanding not only the systems required but also any areas where non-technical changes must be made to enable a mission. Many participants cited the DOTMLPF (Doctrine, Operations, Training, Materiel, Logistics, Personnel, Facility) considerations as a critical piece of the mission engineer's toolkit. Specifically, when working within the constraints of the acquisition system, where mission engineers may influence but not control systems, understanding when a mission need can be met with non-technical solutions or where an existing policy or practice must change in order to meet that need, is critical to successfully implementing mission approaches.

Systems Engineering

As described in Section 1, there are critical overlaps between mission engineering and systems engineering. To that end, participants described which systems engineering skills are particularly critical for mission engineers, which are reflected in this **Area**.

Systems Engineering	
Competency (or Sub-competency)	**Description**
System of Systems Engineering	Missions require a complex set of systems to work together to achieve a task that, likely, many of these systems were not designed to do. This is a classic system of systems problem. Maier (1998) defines a system of systems as an assemblage of components which individually may be regarded as systems, and which possess two additional properties: (a) operational independence of the components, so that if a system-of-systems is disassembled into its component systems, the component systems must be able to usefully operate independently, and (b) managerial independence of the components meaning the component systems not only can operate independently, they do operate independently. (p. 267-284) As stated throughout the research interviews, system of systems engineering is highly synonymous with mission engineering. Therefore, mastery of system of systems principles is critical for mission engineers to be effective.
Analysis	While synthesis – understanding how systems of systems can combine to deliver an overarching mission – is important and reflected above, another critical skill for mission engineers is that of analysis. Analysis is the use of data, simulations and theory to understand how something works, which may require breaking problems or systems down into smaller parts.
Architecture	Mission engineers consistently stated in their interviews that architecting was a critical skillset for mission engineers. Being able to develop architecture at the mission level is necessary to enable the diverse set of engineering teams to understand their role in the broader mission context. Likewise, an understanding of architecture is important; mission engineers must be able to understand the high-level architecture of the systems of systems with which they interact and how these architectures may or may not be compatible with the desired mission architecture.
Modeling and Simulation:	Today, most individual systems are sufficiently complicated and complex that no one person can understand the entire system holistically. Complexity increases exponentially as individual systems come together to complete a goal as a system of systems. For most missions, multiple systems of systems may be required to produce the desired mission effect. Clearly, no human being can fully understand and control this level of complexity. This is why modeling is a critical supporting discipline for mission engineering. Models allow appropriate simplifications to be made so that a human being can grasp the big picture of a mission. Rigorous models can also enable mission engineers to make trade-offs between the systems of multiple programs. Effective models can also be used as communication tools to enable a clear mission vision across the various systems and systems of systems involved.
Requirements	Finally, requirements engineering is a critical skill for mission engineers. One of the biggest challenges mission engineers cited in their interviews was the fact that current mission-level requirements are seldom generated and, when they are, they are often generated after many of the critical systems and systems of systems are already under development. The ability to clearly identify the most critical requirements for a mission and coordinate with the supporting systems and systems of systems to help them understand how to meet these mission level requirements is critical for mission engineering success.
Integration	IEEE 12207 defines integration as “a process that combines system elements to form complete or partial system configurations in order to create a product specified in the system requirements.” (ISO/IEEE 2008) In the context of mission engineering the concept expands to include combining not just system elements but entire systems or systems of systems with operational context and processes and procedures to help bring a mission together. Because missions tend to include many disparate systems or SoS’s, the ability to understand how the pieces can and must fit together to enable a mission is critical.

(*Continued*)

Systems Engineering	
Competency (or Sub-competency)	**Description**
Gap Analysis	Gap analysis is traditionally the comparison of actual performance with potential or desired performance. This definition applies in a mission context, but gap analysis for mission engineering can also include the comparison of planned performance for an individual system vs. the planned performance for that system in a mission context. A key example of this in a SoS is that of the Joint Tactical Radio System (JTRS) and the Army's Future Combat System (FCS). JTRS was intended to provide the critical communications infrastructure for FCS. There were many complications in the integration between JTRS and FCS. However, as noted in a CFS report, "The inability to meet . . . fundamental design and performance standards raised concerns that [JTRS] may not be able to accommodate," some of the critical uses for FCS. (CRS 2005) A comparison of what JTRS was required to provide – requirements that were set in the mid-1990s – and where its capabilities were actually evolving, was a critical activity for FCS engineers.

There are many other systems engineering KSABs that could support mission engineering. However, the critical skills that were consistently cited by interviewees are the ones highlighted here. The competency framework is expected to be tailored and additional systems engineering skills could certainly be added as part of this tailoring.

Systems Mindset

Systems Mindset is primarily focused on patterns of thinking, perceiving, and approaching a task that are particularly relevant to mission engineers, including holism and integration. The categories included in this area are:

Systems Mindset	
Competency (or Sub-competency)	**Description**
Big-Picture Thinking	Also referred to as 'systems thinking' and 'holistic thinking', this includes the ability to step back and take a broader view of the problem at hand; this is an important and essential characteristic of mission engineers. 'Big picture' could refer to a broader perspective along many different dimensions: the system as a whole including interfaces and integration, and not limited to any sub-system or component; the system while in operation, and its interactions with other systems and the operating environment; the entire lifecycle of the system, and not limited to the current stage of the system; the development program in the context of the organization and all its other development programs; the end goal or solution to the problem at hand; the perspectives of different stakeholders; and the technical as well as the human and business perspectives. A mission engineer is usually *the* person to bring this broader perspective, while classic engineers and subject matter experts often tend to be narrowly focused on their area of interest. Mission engineers are not only called to provide this big-picture perspective themselves, but also to enable others to see this bigger picture.

Systems Mindset	
Competency (or Sub-competency)	**Description**
Adaptability	The overall ability to deal with ambiguity and uncertainty, this involves the abilities to be open-minded, understand multiple disciplines, deal with challenges, and the ability to take rational risks. By definition, experts possess competency in a specific area, which is their 'comfort zone'; and they typically do not prefer going outside that circle or comfort zone. Such experts provide value to the organization by contributing their expertise in those focused areas. However, mission engineers tend to show an ability to broaden their comfort zones, and go beyond their current boundaries and they are also comfortable doing this.
Paradoxical Mindset	*The ability to hold and balance seemingly opposed views, and being able to move from one perspective to another appropriately.* Typically, an engineer may hold one view or the other, but rarely *both*. By having this paradoxical mindset, a mission engineer contributes value that is not usually expected from others. The opposing-concept pairs are:
Big-Picture Thinking and *Attention to Detail*	Big-picture thinking provides the broader higher-level perspective; at the same time, a mission engineer is also required to pay attention to the details of how things work and how they come together in a system.
Strategic and *Operational*	Mission engineers need to be strategic, focused on the end result of 'vision' for the mission, but also need to handle the tactical day-to-day activities and decisions required to reach that vision. They must also be able to appreciate "how what is done today is going to affect things downstream". A related concept pair is the ability to envision long-term issues but at the same time, have the drive for closure with the current situation in order to move on.
Analytic and *Synthetic*	A big-picture perspective may be associated with the ability to be synthetic, and to be able to bring together and integrate different pieces of a puzzle. However, a mission engineer also needs to be analytic and to be able to break down the big picture into smaller pieces on which others can focus and work. To do this effectively, a mission engineer needs to be able to operate at multiple levels (e.g., system, system-of-systems, and mission) and multiple dimensions (e.g., various technical disciplines and stakeholder perspectives).
Multi-Scale Abstraction	*The ability to filter out and understand the critical bits of information at the right level and to make relevant inferences.* Even with that filtered information, mission engineers using their mission engineering skills need to know when to use or not use pieces of information. Such abstraction also enables mission engineers to connect and extract meaning from different streams of information; for example, to tie together information that subject matter experts of two different disciplines are providing.
Critical Thinking	Critical thinking is the intellectually disciplined process of actively and skillfully conceptualizing, applying, analyzing, synthesizing, and/or evaluating information gathered from, or generated by, observation, experience, reflection, reasoning, or communication, as a guide to belief and action. (Paul and Elder 2008)

In addition to the broad category of "big picture" thinking, there are specific techniques and mindsets that are critical for thinking holistically about systems of systems. Keating and Gheorghe (2016) state that for systems of systems thinking, the focus is on system behaviors and specifically on the synergies created by interactions of specific systems, rather than from the specific systems themselves. Because missions require systems of systems working together, this focus on synergistic behavior is crucially important for mission engineers. This is not out of scope for general "systems thinking" but is important to highlight in a mission context.

Interpersonal Skills

The fifth competency area is *Interpersonal Skills*. Mission engineers can not work by themselves; they must interact with a variety of teams to affect change at the mission level. A mission engineer is expected to be proficient in a number of interpersonal skills. The specific categories contained within this competency area are listed below:

Interpersonal Skills	
Competency (or *Sub-competency*)	**Description**
Communication	Communication is critical for mission engineers since they interact with a variety of people, and this is a broad category covering a wide variety of related skills and abilities. Often they are an important link between individuals and groups, both internal and external to the organization – most importantly, the customers and end-users of the system being developed. Mission engineers need the ability to clearly express their thoughts and perspectives to establish a shared common understanding.
Audience	Mission engineers need to communicate with a variety of direct and indirect audiences: customers; subject matter experts; program managers; vice presidents; directors; specialty engineers; problem owners; technical teams; contractors; decision makers; system testers; and others working on or with the project.
Content	The variety of content that mission engineers need to communicate can be broadly divided into three types, based on the audience they are communicating with: 1) *Technical*: Communications with disciplinary and specialty engineers and subject matter experts involve high technical content. But communications of technical issues to managers, end-users, and others who may not be interested in or who may be confused by all the technical detail, involves adequate abstraction of the technical content. 2) *Managerial*: Mission engineers often provide project status to managers and supervisors and cost-schedule constraints and expectations to technical personnel. 3) *Social*: Mission engineers need to maintain an amicable environment within a team and to interact with others in a courteous manner. Such interactions involve communications that are neither technical nor managerial in nature.
Mode	Communicating the intended content to the target audience is done through a number of different modes: 1) *Oral*: This takes various forms, depending on the audience and context. It could be one-on-one, or as part of a team, in person, or remotely. 2) *Presentation*: A special form of communications is the ability to stand in front of an audience and to deliver a presentation using appropriate aids. Further, during presentations, mission engineers tend to represent others who may not be in the room: they present customer needs and requirements to others in the absence of customers, and they present design decisions and system related issues to customers in the absence of designers. 3) *Writing and Documentation*: Written communication skills are equally critical for mission engineers; the scale, audience, and objective of the written artifact also matter. It could range from a short email to communicate status, to a detailed test plan, to internal documentation supporting a project decision, to design documents being submitted for review.

Interpersonal Skills	
Competency (or *Sub-competency*)	**Description**
Translation	Building on the skills described under *Communication*, mission engineers serve a critical role as translators. They must help engineers understand the operational and mission context, operators understand engineering constraints, leadership at all levels understand the constraints of these environments, and help ensure that all stakeholders understand the impacts of the acquisition systems constraints on the art of what is possible for a given mission. Many mission engineers stated that this was one of the most critical benefits mission engineers provide.
Enterprise Context	The enterprise context is important in any system effort. As mission engineers try to influence multiple programs and projects to align with relevant missions, they must also have an understanding of the power structures and processes – 'how work gets done' – in each of the associated organizations. Without this skill set, individuals will struggle to bring about critical changes or garner support from decisions makers to enable mission development. These skills are critical to enabling mission engineers to influence stakeholders throughout the mission space.
Building & Utilizing a Subject Matter Expert (SME) Network	A mission engineer needs to be a 'people person', and build a social network of professional acquaintances. Such a network becomes a valuable resource for mission engineers to tap into, because they are not expected to know answers to all problems, but rather be able to find someone who has the expertise and ability to solve the problem.
Coordination	In this context, coordination is the organization of the different elements of a complex body or activity so as to enable them to work together effectively. A mission engineer must bring together and bring to agreement a broad set of individuals or groups who help to resolve mission related issues. This is an enabler to the *Guiding Diverse Stakeholders* competency in the next area. (Modified from the definition of "coordinator", Sheard 1996 and Hutchison et al. 2018.)
Influence, Persuasion, and Negotiation	It is critical for every mission engineer to have the skills needed to make a point and to successfully obtain buy-in. In many situations, mission engineers contribute a perspective that is different from that of others: a focus on the overall mission and directly on the strategic Defense needs. In such situations, it requires influence, persuasion, and negotiating skills for mission engineers to enable others to see the bigger picture on which they need to focus. As described in Section 1, mission engineers are often not empowered with any authority to enact the changes required to move a system already in development to be aligned with a mission need. They must therefore be persuasive and try to influence programs over which they have no direct control or authority.

Conflicts are bound to rise in a variety of scenarios – across teams, organizations, and services – between the technical side and business side of the organization; as well as outside of the organization. The mission engineer must resolve these conflicts while keeping the system goals in mind. In some cases, conflicts arise due to the existence of barriers, which may be related to the organizational culture, processes, team personalities, or other situations that could prevent an individual or team from getting their work done. The mission engineer needs the ability to break these barriers.

Technical Leadership

The sixth and final competency area is *Technical Leadership*. It is common and natural for mission engineers to play leadership roles at many levels within an organization. The specific categories contained within *Technical Leadership* are listed below.

Technical Leadership	
Competency (or Sub-competency)	**Description**
Guiding Diverse Stakeholders	This includes the ability to manage all the internal and external stakeholders, and to keep all teams engaged in the mission focused on the variety of stakeholder needs, especially those of the end user or customer. The mission engineer is uniquely positioned to interact with many stakeholders of the system – both external and internal to the organization. Being this 'touch point' person, the mission engineer needs to deal with multiple personalities, behaviors, organizations, and cultures. A key activity in this area is helping to manage expectations on the mission needs vs. the individual system needs.
Team Building	*The ability to identify, build, and effectively guide or coach a team comprising individuals with diverse expertise, perspectives, and personalities.* A mission engineer is charged with coordinating across programs and projects to deliver a mission engineering capability. The mission engineer needs to fully know each of the team members: their strengths, weaknesses, capacities, capabilities, limitations, personalities, expertise, and working styles. The mission engineer plays the roles of coach, guide, and teacher to develop the team's capabilities and to orchestrate it to perform the required tasks. Individual leadership styles could vary, but the overall objective of is to empower the team, to instill confidence, and to help them to deliver the solution and to be successful. Another key aspect of handling a team is the ability to delegate – the leader needs to build enough trust in the team to be able to delegate with confidence.
Political Savvy	Political savvy is the "ability to understand different people's agendas, and use this knowledge" to enable progress and influencing more effectively and with more sensitivity to different viewpoints. (Expert Program Management, 2017) Because mission engineers work with teams that span multiple organizations – across services, domains, and a combination of government and industry – understanding the political landscape is an important skill. This is a critical piece of context for the ability to *Guide Diverse Stakeholders* and provides critical understanding required for *Influence, Persuasion, and Negotiation*.
Decision Making	Specifically, the skillset that enables individuals to make decisions, especially with a group of people and when limited information is available. Though decision making requires interpersonal skills, in a mission engineering context, it is also critical for building consensus building between multipole stakeholding, requiring leadership and influence skills. For example, mission engineers may need to help systems engineers and program managers accept the trade-offs for their systems in favor of the mission; i.e. perhaps suboptimizing an individual system to enable the system to support the mission.
Workforce Development	Because mission engineering is an emerging discipline, one of the competencies that mission engineers are currently concerned with is how to build teams and develop individuals so that they can perform in mission engineering. As the discipline matures, individual mission engineers may be less concerned with this, but at the present time, this is an important aspect of how mission engineers function

Notes

1 CDTO(MC). 2023. "Mission Engineering." Office of the Deputy CTO for Mission Capabilities (DCTO(MC). https://ac.cto.mil/mission-engineering

2 NearOps describes a simulated operational environment that can be used for testing and evaluation and is as close to the operational environment as is reasonable.

References

Collins, B., Doskey, S. and Moreland, J. 2017. *Modeling the Convergence of Collaborative Systems of Systems: A Qualitative Case Study.* Systems Engineering, Wiley Periodicals, Hoboken, US-NJ.

Defense Acquisition University (2017). *Defense Acquisitio,n Guidebook*. Washington, DC: Pentagon.

Deiotte, R. and Garrett, R.K. Jr. (2013). *A Novel Approach to Mission-Level Engineering of Complex Systems of Systems: Addressing Integration and Interoperability Shortfalls by Interrogating the Interstitials*. Missile Defense Agency 13-MDA-7269, 29 April.

Expert Program Management (2017). Minute Tools Content Team, Political Awareness Skills, Minute Tools, Feb, 2017 https://expertprogrammanagement.com/2017/02/political-awareness-skills/.

Garrett, R.K. Jr., Anderson, S., Baron, N.T., and Moreland, J.D. Jr. (2011). Managing the interstitials, a system of systems framework suited for the ballistic missile defense system. *Systems Engineering* 14 (1): 87–109.

Glaser, B. and Strauss, A. (1967). *The Discovery of Grounded Theory: Strategies for Qualitative Research*. Chicago, IL: Aldine.

Gold, R. (2016). Mission engineering. *Proceedings of the 19th Annual National Defense Industrial Association (NDIA) Systems Engineering Conference*, Springfield, VA, USA (24–27 October 2016).

Hutchison, N., Henry, D., Pyster, A., and Pineda, R. (2014). Early findings from interviewing systems engineers who support the U.S. Department of Defense. In: *Proceedings of the International Council on Systems Engineering (INCOSE) 24th International Symposium*, Las Vegas, NV (30 June–3 July).

Hutchison, N., Verma, D., Burke, P. et al. (2017). *Atlas 1.1: The Theory of Effective Systems Engineers Revised*. Hoboken, NJ: Stevens Institute of Technology. Systems Engineering Research Center. SERC-2018-TR-101-A.

Hutchison, N., Verma, D., Burke, P. et al. (2018). *Atlas 1.1: An update to the Theory of Effective Systems Engineers*. Hoboken, NJ: Systems Engineering Research Center, Stevens Institute of Technology.

Hutchison, N., Verma, D., Burke, P. et al. (2020). *Atlas: Effective Systems Engineers and Systems Engineering*. SERC-2020-TR-007-A. Hoboken, NJ: Systems Engineering Research Center (SERC), Stevens Institute of Technology.

INCOSE (2018). *Systems Engineering Competency Framework*. INCOSE-TP-2018-002-01.0. San Diego, CA, USA: International Council on Systems Engineering (INCOSE).

Joint Chiefs of Staff (2018). *Joint Operations*, Joint Publication 3-0. Arlington, VA: U.S. Department of Defense. https://irp.fas.org/doddir/dod/jp3_0.pdf (accessed 22 October 2018).

Keating, C.B., and Gheorghe, A.V. (2016). *Systems thinking: Foundations for enhancing system of systems engineering*. IEEE International Conference on Systems of Systems Engineering. Norway, 1–6.

Maier, M. (1998). *Architecting principles for system-of-systems. Systems Engineering*, 1 (4): 267–284.

Marvin, J.W. and Garrett, R.K. Jr. (2014). Quantitative SoS architecture modeling. complex adaptive systems conference. *Procedia Computer Science* 36: 41–48.

Marvin, J., Whalen, T., Morantz, B. et al. (2014). Uncertainty quantification (UQ) in complex system of systems (SoS) modeling and simulation (M&S) environments. *Proceedings of the 24th International Council on Systems Engineering (INCOSE) International Symposium*, Las Vegas, NV, USA (30 June–3 July 2014).

Marvin, J.W., Schmitz, J.T., and Reed, R.A. (2016). Modeling-simulation-analysis-looping: 21st century game changer. In: *26th INCOSE International Symposium*. Edinburgh, Scotland, UK.

Mindell, D.A. (2015). *Our Robots, Ourselves: Robotics and the Myth of Autonomy*. New York, NY: Viking Chapter 4 War.

mission. (2023). Merriam-Webster.com. https://www.merriam-webster.com/dictionary/mission (accessed 31 January 2023).

Moreland, J. (2015). *Mission Engineering Integration and Interoperability (I&I)*. Dahlgren, VA (US): Leading Edge, Naval Surface Warfare Center, Dahlgren Division, January.

Moreland, J.D. Jr. (2009). Structuring a flexible, affordable naval force to meet strategic demand in the 21st century. *Naval Engineers Journal* (1): 35–51.

Moreland, J.D. Jr. (2014). Experimental research and future approach on evaluating service-oriented architecture (SOA) challenges in a hard real-time combat system environment. *Systems Engineering* 17 (1): 52–61.

NASA (2017). *Mission Engineering and Systems Analysis Division (MESA)*. Greenbelt, MD (US): https://aetd.gsfc.nasa.gov/590/index.php.

Neaga, E.I., Hensaw, M.J., and Yue, Y. (2009). The Influence of the Concept of Capability-Based Management on the Development of the Systems Engineering Discipline. In: *Proceedings of the 7th Annual Conference on Systems Engineering Research*. (April 20–23), Loughborough University, Loughborough, England, UK.

Office of the Deputy Under Secretary of Defense for Acquisition and Technology, Systems and Software Engineering (2008). *Systems Engineering Guide for Systems of Systems*, Version 1.0. Washington, DC (US): ODUSD(A&T)SSE.

Ondrus, P. and Fatig, M. (1993). *Mission Engineering*. Greenbelt, MD: NASA Goddard Space Flight Center https://ntrs.nasa.gov/search.jsp?R=19940019408.

Parnell, G., Miller, W.D., Michealson, K. et al. (2016). Industry support to mission analysis and mission engineering. *Prodceedings of the MORS Affordability Conference 2016*. https://content.ndia.org/-/media/sites/ndia/meetings-and-events/divisions/systems-engineering/past-events/division-meetings/2016-june/defense-industry-and-me-mission-analysis-v9-160518-annotated.ashx.

Paul, R., and Elder, L. (2008). *Critical thinking: The nuts and bolts of education*. Optometric Education 33: 88–91.

Powers, B. (2014). Normalizing digital close air support (DCAS): needs, challenges, and the role of unmanned systems. In: *United States Joint Forces Command, presented at the Western UAS Conference*, 11 September. San Diego, CA (US).

Rebovich, G. Jr. (2014). *Systems Engineering Guide*. Bedford, MA (US) and McLean, VA (US): MITRE.

Richards, M.G., Ross, A.M., Hastings, D.E., and Rhodes, D.H. (2009). Multi-attribute tradespace exploration for survivability. In: *7th Annual Conference on Systems Engineering Research*, Loughborough, England, UK (April 20–23). Loughborough University.

Snowden, D.J. and Bone, M.E. (2007). *A leader's framework for decision making*. Harvard Business Review 85 (11): 68–76.

Spaulding, C.R., Gibson, W.S., Schreurs, S.F. et al. (2011). Systems engineering for complex information systems in a federated, rapid development environment. *Johns Hopkins APL Technical Digest* 29 (4): 310–326.

Verma, D., Collopy, P., and Pallas, S. (2018). *Pathfinder Project Systems Engineering Research Needs and Workforce Development Assessment, Technical Report SERC-2018-TR-102*, 31 January. Hoboken, NJ (US): Stevens Institute of Technology, Systems Engineering Research Center.

Vesonder, G., Verma, D., Hutchison, N. et al. (2018). *Mission Engineering Competencies*. Hoboken, NJ: Stevens Institute of Technology, Systems Engineering Research Center.

Biographical Sketches

Dr. Gregg Vesonder is a Research Scientist at Stevens Institute of Technology and serves as a Senior Research Associate and Principal Investigator at SERC and AIRC. His projects include Agile Principles for Modern Acquisition and Competency Frameworks in AI, Machine Learning, Mission Engineering and Data Analytics. Gregg was previously Director of the Software Engineering Program at Stevens. Before Stevens Gregg was Executive Director at AT&T and Bell Labs Research and was awarded both a Bell Labs and AT&T Labs Fellow for his work in AI.

Dr. Nicole Hutchison is a research engineer with the SERC and serves as the Digital Engineering Competency Framework Principal Investigator. Other research projects include Helix, the Mission Engineering Competency Model, Digital Engineering Metrics. She is the Managing Editor of the Systems Engineering Body of Knowledge (SEBoK). Before coming to Stevens, Dr. Hutchison worked for Analytic Services Inc. primarily focused on public health, biodefense, and full-scale disaster exercises.

Part VII

Applying Systems Engineering to Enterprise Systems and Portfolio Management

Dan DeLaurenits

Chapter 34

Central Challenges in Modeling and Analyzing Enterprises as Systems

William B. Rouse

McCourt School of Public Policy, Georgetown University, Washington, DC, USA

Introduction

The technology and organizational trends outlined in this chapter pose substantial challenges for systems engineering and management. Many of these challenges involve information and knowledge management. Addressing these challenges will require many of the concepts, principles, methods, and tools discussed in other handbook chapters. It will also require new paradigms not discussed in these chapters.

These challenges were first outlined in the 1st edition of the ***Handbook of Systems Engineering and Management*** (Sage and Rouse, 1999) and then extended in the 2nd edition (Sage and Rouse, 2009). This chapter represents the elaboration of the 3rd generation of these challenges, with an additional 12th challenge. It is clear that progress has been made, but that elements of these challenges remain. Indeed, some of the challenges have worsened.

Systems Modeling

Our methods and tools for modeling, optimization, and control depend heavily on exploiting problem structure. Understanding the relationships and constraints underlying problem structure enables predicting system behaviors, as well as potentially controlling these behaviors. Decomposing problem structure, associating first principles with the elements resulting from this decomposition, and then recomposing these principles into an overall mathematical or computational model are typical steps of systems modeling.

For the loosely structured systems emerging due to information technology (IT), however, behavior does not emerge from fixed structures. Instead, structure emerges from collective behaviors. The distributed, collaborative, and virtual organizations that IT enables are such that the system elements are quite fluid. Distinctions between what is inside and outside the system depend on time-varying behaviors and consequences.

With such systems, modeling must be done on at least two levels. One level concerns the first principles for inclusion of elements in the system. The other level concerns the behaviors of elements if they are included in the system, which depend on what other elements are also included.

Systems Engineering for the Digital Age: Practitioner Perspectives, First Edition. Edited by Dinesh Verma.

Companion website: www.wiley.com/go/verma/systemsengineering

Thus, we need methods for modeling the inclinations of elements to become parts of systems, the behaviors that tend to result when included, and the ways in which these behaviors affect inclinations to be included. Satisfying this need significantly challenges typical modeling methods and tools.

Emergent and Complex Phenomena

Meeting the modeling challenge is complicated by the fact that not all critical phenomena can be fully understood, or even anticipated, based on analysis of the decomposed elements of the overall system. Complexity not only arises from there being many elements of the system but also from the possibility of collective behaviors that even the participants in the system could not have anticipated.

An excellent example is the process of technological innovation. Despite the focused intentions and immense efforts of the plethora of inventors and investors attracted by new technologies, the ultimate market success of these technologies almost always is other than what these people expect. It was not envisioned, for example, that the primary use of telephones or computer networks would be personal communication. Also, new technologies often cause great firms to fail.

Thus, many critical phenomena can only be studied once they emerge. In other words, the only way to identify such phenomena is to let them happen. The challenge is to create ways to recognize the emergence of unanticipated phenomena and be able to manage their consequences, especially in situations where likely consequences are highly undesirable.

Complexity theory is an emerging field of study that has evolved from five major knowledge areas: mathematics, physics, biology, organizational science, and computational intelligence and engineering. Fundamentally, a system is complex when we cannot understand it through simple cause-and-effect relationships or other standard methods of systems analysis. In a complex system, we cannot reduce the interplay of individual elements to the study of individual elements considered in isolation. Often, several different models of the complete system, each at a different level of abstraction, are needed.

There are several sciences of complexity, and they generally deal with approaches to understanding the dynamic behavior of units that range from individual organisms to the largest technical, economic, social, and political organizations. Often, such studies involve complex adaptive systems and hierarchical systems, are multidisciplinary in nature, and include or are at the limits of scientific knowledge.

Complexity studies attempt to pursue knowledge and discover features shared by systems described as complex. These include studies such as complex adaptive systems, complex systems theory, complexity theory, dynamic systems theory, complex nonlinear systems, and computational intelligence. Many scientific studies, prior to the development of simulation models and complexity theory, involved the use of linear models. When a study resulted in anomalous behavior, the failure was often incorrectly blamed on noise or experimental error. It is now recognized that such "errors" may reflect inherent inappropriateness of linear models—and linear thinking.

One measure of system complexity is the complexity of the simulation model necessary to effectively predict system behavior. The more the simulation model must embody the actual system to yield the same behavior, the more complex the system. In other words, outputs of complex systems cannot be predicted accurately based on models with typical types of simplifying assumptions. Consequently, creating models that will accurately predict the outcomes of complex systems is very difficult. We can, however, create a model that will accurately simulate the processes the system will use to create a given output.

This awareness has profound impacts for organizational efforts. For example, it raises concerns related to the real value of creating organizational mission statements and plans with expectations that these plans will be inexorably executed and missions thereby realized. It may be more valuable to create a model of an organization's planning processes themselves, subject this model to various input scenarios, and use the results to generate alternative output scenarios. The question then becomes one of how to manage an organization where this range of outputs is possible.

Interestingly, most studies of complex systems often run completely counter to the trend toward increasing fragmentation and specialization in most disciplines. The current trend in complexity studies is to reintegrate the fragmented interests of most disciplines into a common pathway. This transdisciplinarity provides the basis for creating a cohesive systems ecology to guide the use of IT for managing complex systems. Whether they be human-made systems, human systems, or organizational systems, the use of systems ecology could more quickly lead to organizing for complexity, and associated knowledge and enterprise integration.

An important aspect of complex systems is path dependence. The essence of this phenomenon begins with a supposedly minor advantage or inconsequential head start in the marketplace for some technology, product, or standard. This minor advantage can have important and irreversible influences on the ultimate market allocation of resources even if market participants make voluntary decisions and attempt to maximize their individual benefits. Such a result is not plausible with classic economic models that assume that the maximization of individual gain leads to market optimization unless the market is imperfect due to the existence of such effects as monopolies. Path dependence is a failure of traditional market mechanisms and suggests that users are "locked" into a suboptimal product, even though they are aware of the situation and may know that there is a superior alternative.

This type of path lock-in is generally attributed to two underlying drivers: (1) network effects, and (2) increasing returns of scale. Both of these drivers produce the same result, namely that the value of a product increases with the number of users. Network effects, or "network externalities," occur because the value of a product for an individual consumer may increase with increased adoption of that product by other consumers. This, in turn, raises the potential value for additional users. An example is the telephone, which is only useful if at least one other person has one as well, and becomes increasingly beneficial as the number of potential users of the telephone increases.

Increasing returns of scale imply that the average cost of a product decreases as higher volumes are manufactured. This effect is a feature of many knowledge-based products where high initial development costs dominate low marginal production and distribution costs. Thus, the average cost per unit decreases as the sales volume increases and the producing company is able to continuously reduce the price of the product. The increasing returns to scale, associated with high initial development costs and the low sales price create barriers against market entry by new potential competitors, even though they may have a superior product. If there is no competition, this phenomenon results in increasing profits due to the lack of incentives to decrease prices.

The controversy in the late 1990s over the integration of the Microsoft Internet Explorer with the Windows Operating System may be regarded as a potential example of path dependence, and appropriate models of this phenomenon can potentially be developed using complexity theory. These would allow exploration of whether network effects and increasing returns of scale can potentially reinforce the market dominance of an established but inferior product in the face of other superior products, or whether a given product is successful because its engineers have carefully and foresightedly integrated it with associated products such as to provide a seamless interface between several applications. To some extent, the more recent success of Google and social networking websites indicates how technology changes and market forces eventually emerge in such situations.

Uncertainties and Control

IT enables systems where the interactions of many loosely structured elements can produce unpredictable and uncertain responses that may be difficult to control. If critical phenomena are unpredictable—and elements can be inside or outside of the system—it can be very difficult to know what variables to control and where to exert control. Prediction and control can be akin to herding cats when you are not sure which cats are part of your herd.

To illustrate, consider a virtual enterprise configured to rapidly exploit an opportunity to create a new consumer product and quickly distribute it nationally. The speed needed to gain competitive advantage may be such that it is not clear who has joined the venture, who is still being wooed, and who has declined. Consequently, the availability of resources (i.e., capabilities for marketing, distribution, etc.) is uncertain, as is the sense of how all the pieces will play together, or potentially conflict. How can you predict and control the behaviors of this system?

The challenge is to understand such systems at a higher level. Control is likely to involve design and manipulation of incentives to participate and rewards for collaborative behaviors. It may be impossible, and probably undesirable, to control behaviors directly. The needed type of control is similar to policy formulation. Success depends on efficient experimentation much more than possibilities for mathematical optimization due to the inherent complexities that are involved. Again, insights from complexity theory may be brought to bear on these situations.

Access and Utilization of Information and Knowledge

Information access and utilization, as well as management of the knowledge resulting from this process, are complicated in a world with high levels of connectivity and a wealth of data, information, and knowledge. Ubiquitous networks and data warehouses or "data lakes" are touted as the means to taking advantage of this situation. However, providers of such "solutions" seldom address the basic issue of what information users really need, how this information should be processed and presented, and how it should be subsumed into knowledge that reflects context and experiential awareness of related issues.

The underlying problem is the usually tacit assumption that more information is inherently good to have. What users should do with this information and how value is provided by this usage are seldom clear. The result can be large investments in IT with negligible improvements of productivity. One of the major needs in this regard is for organizations to develop the capacity to become learning organizations and to support bilateral transformations between tacit and explicit knowledge.

Addressing these dilemmas should begin with the recognition that information is only a means to gaining knowledge and that information must be associated with a contingency-task structure to become knowledge. This knowledge is the source of desired advantages in the marketplace or vs. adversaries. Thus, understanding and supporting the transformations from information to knowledge to advantage are central challenges to enhancing information access and utilization in organizations.

For a period of time, organizations addressed the access and utilization challenge by investing in comprehensive and very expensive IT solutions. In recent years, however, investments in solutions have become highly driven by business issues. An organization will consider a particular business issue, such as demand forecasting, and determine the business value of resolving this issue. For example, how would revenues increase or costs decrease if forecasts were more accurate? They next look at the knowledge needed to provide this improvement, and then consider the information that must be accessed and utilized to generate this knowledge. Finally, they determine the life-cycle costs of this information. If these costs do not compare favorably with projected benefits, they do not make this investment. In this way, the challenge has become one of understanding information and knowledge needs in particular rather than in general.

Information and Knowledge Requirements

Beyond adopting a knowledge management perspective, one must deal with the tremendous challenge of specifying information requirements in a highly information-rich environment. Users can have access to virtually any information they want, regardless of whether they know what to do with it or how to utilize the information as knowledge. Users' natural tendencies to hedge and overspecify information needs present problems when these inflated requirements can easily be met.

The difficulty is users' limited abilities to consume information and transform it to the knowledge that will provide them desired advantages. While IT has evolved quite rapidly in recent decades, human information-processing abilities have not evolved significantly. These limited abilities become bottlenecks as users attempt to digest the wealth of information they have requested. The result is sluggish and hesitant decision-making, which in a sense is due to being overinformed.

The challenge is to develop methods for requirements analysis that are able to take into account the consequences of having met requirements. Trade-offs between meeting one requirement vs. another are often central aspects of requirements analysis. In an information-rich world, all requirements can be met and the key trade-offs are between meeting and perhaps exceeding these requirements and potentially overwhelming users. As an analogy, if energy were to become free, we then would be faced with trade-offs between energy usage and the negative consequences of usage, namely, pollution and global warming.

A key to dealing with these trade-offs is recognition that the traditional, crisp view of requirements is much too rigid. The old "procurement model" involved buying products and services from the lowest-price provider that meets requirements. When requirements can commonly be exceeded, the buyer has to decide how to attach value to these product or service attributes. "Nailing down" requirements is replaced by stakeholder-driven creation of innovative value propositions. This can involve many of the complications of multiattribute models. Nevertheless, this approach provides many more opportunities for gaining market advantage.

Information and Knowledge Support Systems

Information support systems, including decision support, can provide the means for helping users cope with information-rich environments. Such support systems are difficult to design and develop for loosely structured organizations due to inherently less well-defined tasks and decisions. It is also complicated by continually evolving information sources and organizational needs to respond to new opportunities and threats.

These difficulties can be overcome, at least in part, by adopting a human-centered approach to information and knowledge support system design. This approach begins with understanding the goals, needs, and preferences of system users and other stakeholders, for example, system maintainers. In particular, stakeholders' perceptions of the validity, acceptability, and viability of potential support functions drive trade-offs and decisions.

This approach focuses on stakeholders' abilities, limitations, and preferences, and attempts to synthesize solutions that enhance abilities, overcome limitations, and foster acceptance. From this perspective, IT and alternative sources of information and knowledge are enablers rather than ends in themselves. Consequently, design and development of support-system concepts can be premised on accessing and utilizing a wider range of resources. Nevertheless, design and development of such systems is challenging for loosely structured environments.

Inductive Reasoning

Prior to the development of simulation models and complexity theory, most studies involved use of linear models and assumed time-invariant processes (i.e., ergodicity). Most studies also assumed that humans use deductive reasoning and technoeconomic rationality to reach conclusions. But information imperfections, and limits on available time, often suggest that rationality must be bounded. Other forms of rationality and inductive reasoning are necessary.

There are a number of descriptive models of human problem-solving and decision-making. Generally, the appropriate model depends upon the contingency task structure, characteristics of the environment, and the experiential familiarity of humans with tasks and environment. Thus, the context surrounding information and the experiential familiarity of users of the information is most important. In fact, it is the use of information within the context of contingency task structures and the environment that results in the transformation from information to knowledge.

We interpret knowledge in terms of context and experience by sensing situations and recognizing patterns. We recognize features similar to previously recognized situations. We simplify the problem by using these to construct internal models, hypotheses, or schemata to use on a temporary basis. We attempt simplified deductions based on these hypotheses and act accordingly. Feedback of results from these interactions enables us to learn more about the environment and the nature of the task at hand. We revise our hypotheses, reinforcing appropriate ones and discarding poor ones. This use of simplified models is a central part of inductive behavior.

Models of inductive processes can be constructed in the following way. We first set up a collection of, generally heterogeneous, agents. We assume that they are able to form hypotheses based on mental models or subjective beliefs. Each agent is assumed to monitor performance relative to a personal set of belief models. These models are based on the results of actions, as well as prior beliefs and hypotheses. Through this iterative procedure, learning takes place as agents learn which hypotheses are most appropriate. Hypotheses, or models, are retained not because that are "correct" but because they have worked in the past. Agents differ in their approach to problems and the way in which they subjectively converge to a set of "useful" hypotheses. The question of where these mental models come from is interesting and a subject of much study in the learning literature.

This process may be modeled as a complex adaptive system. As noted, we cannot create models that will accurately predict the outcomes of many complex systems. We can, however, often create a model that will accurately simulate the processes the system uses to create outputs. The major constructs associated with such models are the interactions and feedback relations between the various agents whose choices depend upon the decisions of others, and linearity and return to scale considerations. There are many implications associated with these models. Among them are questions of steady state vs. continued evolutionary behavior, the nature and possibility of time-invariant processes (ergodicity), and questions of path dependence discussed earlier.

Machine learning models provide a form of inductive learning. Multi-layer statistical models are able to digest enormous data to create models that tend to be highly successful in pattern recognition, for example, of visual information and language. These models can be used to generate recommendations, for instance to radiologists. Unfortunately, the state of the art is such that these models can rarely explain their recommendations, which many humans find very limiting.

Misinformation and Disinformation

The challenges of misinformation and disinformation are increasingly pervasive due to numerous online information sources and social media platforms. Of course, people have long spread rumors and bought snake oil, health potions, and other shams, so the behavioral and social phenomena are not new. However, the connectivity and speed with which it now happens are unprecedented.

Social psychologists have observed that human beings evolved to gossip, preen, manipulate, and ostracize, and are now more connected to one another on platforms that have been designed to make outrage contagious. Social media does not make most people more aggressive but enables a minority to troll the majority. Social media deputizes everyone to administer justice with no due process.

A thorough review of relevant research suggests that reigning civil liberties in the United States will make it very difficult, perhaps even impossible, to the stem the flow of misinformation and disinformation on social media. However, we can mitigate the impacts of misinformation and disinformation on targeted audiences. Through education, policy, and technology, we can foster a societal immune system.

Misinformation and disinformation represent failures of societal information systems in the sense that information that can cause harm is available and widely spread to potential victims of this harm. There are four tasks associated with mitigating such failures:

- **Detection**: Determination that instances of social media postings represent misinformation and disinformation
- **Diagnosis**: Determination of the sources, targets, and acceptability of these instances
- **Compensation**: Controlling the system to achieve acceptable levels of misinformation and disinformation
- **Remediation**: Repairing or countering the cause(s) of the misinformation and disinformation

An integrated approach to successfully addressing these tasks is currently being pursued.

Learning Organizations

Realizing the full value of information and knowledge is strongly related to organizations' abilities to learn and become learning organizations. Learning involves the use of observations of the relationships between activities and outcomes, often obtained in an experiential manner, to improve behavior through the incorporation of appropriate changes in processes and products. Thus, learning represents acquired wisdom in the form of abilities for skill-based, rule-based, and knowledge-based - or formal – reasoning. It may involve know-how, in the form of skills or rules, or know-why, in terms of formal knowledge.

Learning generally involves several processes:

- Situation assessment
- Detection of a problem

- Synthesis of a potential solution
- Implementation of the solution
- Evaluation of the outcome
- Discovery of patterns among the preceding processes

This is a formal description of the learning process. It describes typical problem-solving processes and involves the basic steps of systems engineering and management in an inductive fashion.

While learning appears highly desirable, much of individual and organizational learning that occurs in practice is not necessarily beneficial or appropriate. For example, there is much literature that shows that individuals and organizations use improperly simplified and often distorted models of causal and diagnostic inferences. Similarly, they often employ improperly simplified and distorted models of the contingency task structure and the environment.

Organizational learning results when members of the organization react to changes in the internal or external environment of the organization by detection and correction of errors. A theory of reasoning, learning, and action for individual and organizational learning has been developed. In this model, learning is fundamentally associated with detection, diagnosis, and correction of errors.

The notion of error is singularly important in this theory. Errors are features of behavior that make actions ineffective. Detection and correction of errors produce learning. Individuals in an organization are agents of organizational action and organizational learning. Two information-related factors that inhibit organizational learning have been found: (1) information distortion such that its value in influencing quality decisions is lessened, and (2) lack of receptivity to corrective feedback.

Two types of organizational learning have been defined. Single-loop learning is learning that does not question the fundamental objectives or actions of an organization. This type of learning is essential to acting quickly, which is often needed. Such actions are usually based on rule-based or skill-based reasoning. When errors occur, members of the organization may discover sources of these errors and identify new strategic activities that may correct the errors. These activities may be identified either through use of a different rule or experientially based skill or through the application of formal reasoning to the situation at hand. The activities are then analyzed and evaluated, and one or more is selected for implementation. Single-loop learning enables the use of present policies to achieve present objectives. The organization may well improve, but this will be with respect to the current way of doing things. Organizational purpose, and perhaps even process, are seldom questioned.

In many cases, this approach is quite valid. Occasionally, however, this approach is not appropriate. For example, environmental control and self-protection through control over others, primarily by imposition of power, are typical managerial strategies. The consequences of this approach may include defensive group dynamics and low production of valid information. This lack of information does not result in disturbances to prevailing values. However, the resulting inefficiencies in decision-making encourage frustration and an increase in secrecy and loyalty demands from decision-makers. All of this is mutually self-reinforcing. It results in a stable autocratic organization and a self-fulfilling prophecy with respect to the need for organizational control. So, while there are many desirable features associated with single-loop learning, there are a number of potentially debilitating aspects as well. These are quite closely related to notions of organizational culture.

Double-loop learning involves identification of potential changes in organizational goals and approaches to inquiry that allow confrontation with and resolution of conflicts, rather than continued pursuit of incompatible objectives, which usually leads to increased conflict. Double-loop learning is the result of organizational inquiry that resolves incompatible organizational objectives through the setting of new priorities and objectives. New understanding is developed that results in updated cognitive maps and scripts of organizational behavior. Studies show that poorly performing organizations learn primarily on the basis of single-loop learning and rarely engage in double-loop learning in which the underlying organizational purposes and objectives are questioned.

Double-loop learning is particularly useful in the case when people's espoused theories of action – the "official" theories that people claim as a basis for action – conflict with their theories in use, which are the theories of action underlying actual behaviors. While people are often adept at identifying discrepancies between other people's espoused theories of action and theories in use, they are not equally capable of self-diagnosis. Further, the dictates of tactfulness normally prevent us from calling to the attention of others the observed inconsistencies between their espoused and actual theories of action. The result of this failure is inhibition of double-loop learning.

Two major inhibitions to learning, distancing and disconnectedness, denote the art of not accepting responsibility for either problems or solutions; disconnectedness occurs when individuals are not fully aware of the theories in use and the relationships between these theories and associated actions. Several other factors can interact to result in conflicting and intolerable pressures, including:

- Incongruity between espoused theory and theory in use that is recognized but not corrected.
- Inconsistency between theories in use of different members of the organization.
- Ineffectiveness, as objectives associated with theories in use become less and less achievable over time.
- Disutility, as theories in use become less valued over time.
- Unobservability, as theories in use result in suppression of information by others such that evaluation of effectiveness becomes impossible.

Detection and correction of conflicts between espoused theories of action and theories in use can lead to reductions of inhibitions of double-loop learning.

The result of double-loop learning is a new set of goals and operating policies that become part of the organization's knowledge base. It is when the environment and the contingency task structure changes that double-loop learning is called for. Learning organizations have abilities to accommodate double-loop learning and successfully integrate and utilize the appropriate blend of single- and double-loop learning.

Learning organizations are described as "organizations where people continually expand their capacity to create the results they truly desire, where new and expansive patterns of thinking are nurtured, where collective aspiration is set free, and where people are continually learning how to learn together." Five component technologies, or disciplines, enable this type of learning:

- Systems thinking.
- Personal mastery through proficiency and commitment to lifelong learning.
- Shared mental models of the organization markets, and competitors.
- Shared vision for the future of the organization.
- Team learning.

Systems thinking is denoted as the "fifth discipline." It is the catalyst and cornerstone of the learning organization that enables success through the other four disciplines.

Lack of organizational capacity in any of these disciplines is called a learning disability. One of the major disabilities is associated with implicit mental models that result in people having deeply rooted mental models without being aware of the cause-effect consequences that result from use of these models. Another is the tendency of people to envision themselves in terms of their position in an organization rather than in terms of their aptitudes and abilities. This often results in people becoming dislocated when organizational changes are necessary.

Each of the five learning disciplines can exist at three levels:

- **Principles**: The guiding ideas and insights that guide practices.
- **Practices**: The existing theories of action in practice.
- **Essences**: The wholistic and future-oriented understandings associated with each particular discipline.

These correspond very closely with the principles, practices, and perspectives that can be used to describe approaches to systems engineering and management. Based primarily on works in system dynamics, an approach to the study and modeling of systems of large scale and scope, and on efforts a variety of eminent thinkers, 11 laws of the fifth discipline can be stated:

- Contemporary and future problems often come about because of what were presumed to be past solutions.
- For every action, there is a reaction.
- Short-term improvements often lead to long-term difficulties.
- The easy solution may be no solution at all.
- The solution may be worse than the problem.
- Quick solutions, especially at the level of symptoms, often lead to more problems than existed initially and, hence, may be counterproductive.
- Cause and effect are not necessarily related closely, either in time or in space.
- The actions that will produce the most effective results are not necessarily obvious at first glance.
- Low cost and high effectiveness do not have to be subject to compensatory tradeoffs over all time.
- The entirety of an issue is often more than the simple aggregation of the components of the issue.
- The entire system, which comprises the organization and its environment, must be considered together.

Neglect of these laws can lead to any number of problems, most of which are relatively evident from their description. For example, failure to understand the last law leads to the fundamental attribution error in which we credit ourselves for successes and blame others for our failures.

On the basis of these laws, several leadership facets emerge. Leaders in learning organizations become designers, stewards, and teachers. Each of these leadership characteristics enables everyone in the organization to improve their understanding and use of the five important dimensions of organizational learning. This results in creative tension throughout the organization. This tension can be addressed in planning, which is one of the major activities of learning organizations, and it is through planning that much learning occurs. Organizational learning is one of the major contemporary thrusts in systems management today.

It is important to emphasize that the extended discussion of learning organizations in this section is central to understanding how to create organizations that can gain full benefits of information technology and knowledge management. This is crucial if we are to transform data to information to insights to programs of action.

Planning and Design

Dealing successfully with the challenges just discussed requires that approaches to planning and design be reconsidered. Traditionally, planning and design are activities that occur before solutions are placed into operation. In contrast, for loosely structured systems, planning and design must be transformed to something done in parallel with system operation rather than beforehand.

In the context of traditional systems engineering methods and tools, loosely structured systems pose needs for online identification, estimation, and control of time-varying, distributed knowledge-based systems. This is a tall order. Further, as noted earlier, control for such systems may mean manipulating the cost functional rather than directly affecting state variables. In other words, in the absence of being able to control behaviors, the key may be to control the incentives and rewards that determine behaviors.

From this perspective, planning and design become less a problem of specifying tasks, milestones, and schedules, and more an issue of influencing formulation and structuring of goals, priorities, and context. Defining and varying the agenda in light of current trends and events may be the essence of controlling loosely structured systems. This implies the need for assuring influence in the absence of possibilities for control.

An example related to the military scenario discussed earlier provides an excellent illustration of this concept. Traditionally, military operations have focused on threatening the use of, and perhaps employing, decisive and sometimes massive forces to deter, thwart, or defeat adversaries. This tactic is premised on having an adversary that is well defined, and, hence, can be located, identified, and engaged. Loosely structured and potentially virtual adversaries do not satisfy these criteria.

How can you combat virtual adversaries? One answer is to attack the decision to form covert coalitions. One can also attack the decisions of such coalitions to employ military forces. In other words, rather than trying to outwit an elusive adversary on the playing field, you create strong incentives for them to not play the game. Better yet, you can create compelling incentives to not even entertain entering the game. This requires that you understand their motivations and value systems much more than their abilities to execute military operations. Planning and design, in this case, must focus on abstractions somewhat removed from traditional command and control considerations and must give due consideration to the possibilities for cooperation and associated complexities.

Optimization vs. Agility

Over the past couple of decades, there have been huge investments in enterprise information systems and knowledge management. Enterprises have focused on becoming "lean," in part facilitated by back office efficiencies. Customer facing systems have been transformed to "self service" via web-based systems. The resulting savings have been substantial and added significantly to corporate profits.

However, sustaining such savings depends on continued generation of revenues which, in turn, depends on the continued relevance and competitiveness of the business models and processes upon which these enterprise information systems are premised. The optimal allocation of resources, for example, across supply chains, is only optimal if the business assumptions upon which the design is based remain valid.

However, the world has not remained static. Business models and processes have to change, and the associated information systems have to be sufficiently agile to support these changes. This has led to a surge of research into agile information systems and the inherent tradeoff between rigorous optimization for a given business model and agility to adapt to new business models.

The real issue, of course, is agile decision-making supported by agile information systems. Desires for decision-making agility is the driver for information system agility. This raises a variety of issues associated with the nature of decision-making and how decision-makers can be supported to be agile. Several other chapters in this handbook address these issues.

Measurement and Evaluation

Successfully addressing and resolving the many issues associated with the challenges described in this chapter requires that a variety of measurement challenges be understood and resolved. Approaches to modeling loosely structured systems, representing emergent and complex phenomena, and dealing with uncertainties and control all involve measurement.

Systems associated with access and utilization of information and knowledge management present particular measurement difficulties because the ways in which information and knowledge affect behaviors are often rather indirect. These inputs to humans often do not produce outputs in terms of observable behaviors. In fact, an important value of information and knowledge is their use in deciding not to act.

For this and a variety of related reasons, it can be quite difficult to evaluate the impact of information technology and knowledge management. Numerous studies have failed to identify measurable productivity improvements as the result of investments in these technologies. The difficulty is that the impact of information and knowledge is not usually directly related to numbers of products sold, manufactured, or shipped. Successful measurement requires understanding the often extended causal chain from information to actions and results.

Transformations from information to knowledge also present measurement problems. Information about the physical properties of a phenomenon are usually constant across applications. In contrast, knowledge about the context-specific implications of these properties depends on human intentions relative to these implications. Consequently, the ways in which information is best transformed to knowledge depends on the intentions of the humans involved. The overall measurement problem involves inferring, or otherwise determining, the intentions of users of information systems.

Both measuring the impact of information and inferring the intentions that underlie the transformations of information to knowledge present difficulties for evaluation, as well as creation of information and knowledge support systems. Many of the concepts discussed in this chapter are only implementable in the context of valid measurement systems. Otherwise, the promises of information technology and knowledge management mainly amount to hand waving.

Summary

In this chapter, we have posed 12 major systems engineering and management challenges. These challenges need to be addressed successfully if the promises of information technology and knowledge management are to be realized. These challenges concern our abilities to deal with:

- Systems modeling
- Emergent and complex phenomena
- Uncertainties and control
- Access and utilization of information and knowledge
- Information and knowledge requirements
- Information and knowledge support systems
- Inductive reasoning
  - Misinformation and disinformation
- Learning organizations

- Planning and design
  - Optimization vs. agility
- Measurement and evaluation

References

Sage, A.P. and Rouse, W.B. (1999). *Handbook of Systems Engineering & Management*, 1e (Chapter 30.4). New York: Wiley.

Sage, A.P. and Rouse, W.B. (2009). *Handbook of Systems Engineering & Management*, 2e (Chapter 34.4). New York: Wiley.

Biographical Sketches

William B. Rouse is research professor in the McCourt School of Public Policy at Georgetown University and professor emeritus and former chair of the School of Industrial and Systems Engineering at Georgia Institute of Technology. His research focuses on mathematical and computational modeling for policy design and analysis in complex public-private ecosystems, with particular emphasis on healthcare, education, energy, transportation, and national security. He is a member of the US National Academy of Engineering and Fellow of IEEE, INCOSE, INFORMS, and HFES. He is a member of the Research Council of the Systems Engineering Research Center and the Acquisition Innovation Research Council. His recent books include *Beyond Quick Fixes* (Oxford, 2014), *Bigger Pictures for Innovation* (Routledge, 2023), *Transforming Public-Private Ecosystems* (Oxford, 2022), *Failure Management* (Oxford, 2021), and *Computing Possible Futures* (Oxford, 2019). Rouse lives in Washington, DC.

Chapter 35

Methods for Integrating Dynamic Requirements and Emerging Technologies

William B. Rouse[1] *and Dinesh Verma*[2]

[1] *McCourt School of Public Policy, Georgetown University, Washington DC, USA*
[2] *Stevens Institute of Technology, School of Systems & Enterprises, Hoboken, NJ, USA*

Introduction

Much of engineering involves designing solutions to meet the needs of markets, or perhaps military missions or societal sector needs such as water, power, and transportation. These needs are often uncertain, especially if solutions are intended to operate far into the future.

There is also often uncertainty in how best to meet needs. New technologies may be needed and their likely performance and cost may be uncertain. Budgets may be insufficient to achieve what is needed. Competitors or adversaries may be creating competing solutions that are similar or superior.

Organizations would like to have the flexibility and agility to address both uncertain needs and uncertain technologies due to performance challenges, organizational experience, supply chains, etc. This is likely to require ways of thinking and allocating resources that are foreign to many organizations. This chapter outlines and illustrates these ways of thinking.

To illustrate how companies address uncertainties, consider two experiences at General Motors (GM). Both illustrations involved Ford surprising GM. The first led to a major failure and the second to a substantial success (Hanawalt and Rouse 2010).

In 1981, General Motors began planning for a complete refresh of its intermediate size vehicles: the front wheel drive A-Cars and the older rear wheel drive G-cars. The GM10 program would yield vehicles badged as Chevrolets, Pontiacs, Oldsmobiles, and Buicks. This program was to be the biggest R&D program in automotive history and with a US$5 billion dollar budget, the most ambitious new car program in GM's history.

The introduction of the Ford Taurus in 1985 was a huge market and business success, and a complete surprise to GM. It was one of the first projects in the United States to fully utilize the concept of cross-functional teams and concurrent engineering practices. The car and the process used to develop it were designed and engineered at the same time, ensuring higher quality and more efficient production. The revolutionary design of the Taurus coupled with its outstanding quality, created a new trend in the US automobile industry, and customers simply loved the car.

The Taurus forced GM to redesign the exterior sheet metal of the GM10 because senior executives thought the vehicles would look too similar. Many additional running changes were incorporated

Systems Engineering for the Digital Age: Practitioner Perspectives, First Edition. Edited by Dinesh Verma.

Companion website: www.wiley.com/go/verma/systemsengineering

into the design in an attempt to increase customer appeal. The first vehicles reached the market in 1988, ~$2 billion over budget and two years behind schedule.

All of the first GM10 entries were coupes, a GM tradition for the first year of any new platform. However, this market segment had moved overwhelmingly to a four-door sedan style. Two-door midsize family cars were useless to the largest group of customers in the segment – members of the Baby Boomer generation were now well into their child rearing years and needed four doors for their children. GM completely missed the target segment of the market. From 1985 to 1995, GM's share of new midsize cars tumbled from 51% to 36%.

The Lincoln Navigator is a full-size luxury SUV marketed and sold by the Lincoln brand of Ford Motor Company since the 1998 model year. Sold primarily in North America, the Navigator is the Lincoln counterpart of the Ford Expedition. While not the longest vehicle ever sold by the brand, it is the heaviest production Lincoln ever built. It is also the Lincoln with the greatest cargo capacity and the first non-limousine Lincoln to offer seating for more than six people.

GM was completely surprised by the Navigator. They had not imagined that customers would want luxurious large SUVs. GM responded with the Cadillac Escalade in 1999, intended to compete with the Navigator and other upscale SUVs. The Escalade went into production only 10 months after it was approved. The 1999 Escalade was nearly identical to the 1999 GMC Yukon Denali, except for the Cadillac badge and leather upholstery. It was redesigned for the 2002 model year to make its appearance and features fall more in line with Cadillac's image.

In 2019, 18,656 Navigators were sold, while 35,244 Escalades were sold. Escalade has outsold Navigator every year since 2002. GM had clearly adapted to the surprise of the Navigator. One can reasonably infer that the company learned from the GM10 debacle. Surprises happen. Be prepared.

We recently studied 12 cars withdrawn from the market in the 1930s, 1960s, and 2000s (Liu et al. 2015). We leveraged hundreds of historical accounts of these decisions, as well as production data for these cars and the market more broadly. We found that only one vehicle was withdrawn because of the nature of the car. People were unwilling to pay Packard prices for Studebaker quality, the two companies having merged in 1954.

The failure of the other 11 cars could be attributed to company decisions, market trends, and economic situations. For example, decisions by the Big Three companies to focus on cost reduction resulted in each manufacturer's car brands looking identical, effectively debadging them. Mercury, Oldsmobile, Plymouth, and Pontiac were the casualties. Honda and Toyota were the beneficiaries.

This chapter presents and illustrates a framework for addressing such scenarios. We first consider how uncertainties arise, contrasting the automotive and defense domains. We then propose an approach to managing uncertainties. This leads to consideration of how to represent alternative solutions and to estimate the value of these alternative solutions. We present a detailed case study of applying this framework and methodology.

Sources of Uncertainties

Table 35.1 portrays two domains where addressing uncertainties are often central and important aspects of decision-making. The primary domain emphasized in the case study in this article is automotive. However, we also want to emphasize the relevance of our line of thinking to the defense domain. The parallels are reasonably self-explanatory, but a few differences are worth elaborating.

Table 35.1 Multi-level comparison of automotive and defense domains.

Automotive domain	Defense domain
Economy	**Geopolitics**
• Geopolitics (e.g., regulations, tariffs, war)	• Military conflict (i.e., hot war)
• GDP and inflation (e.g., recession)	• Geopolitical tension (e.g., grey zone conflicts)
• Financial markets (e.g., interest rates)	• Civil wars (e.g., migration)
• Energy markets (e.g., fuel prices)	• Soft power (e.g., alliances)
Market	**Economics**
• Market growth/decline (e.g., consumers)	• GDP growth/decline
• Segment market saturation (e.g., sedans)	• Inflation/deflation
• External competitors (companies)	• Domestic and allies' defense budgets
• Internal competitors (brands)	• Congressional priorities (e.g., jobs)
Company priorities	**Defense priorities**
• Market strategy (e.g., positioning, pricing)	• Engagement strategies
• Product management (e.g., processes)	• Missions envisioned
• Dealer management (e.g., incentives)	• Adversary capabilities
• Financial management (e.g., investments)	• Capabilities required
• Brand management (e.g., rebadging)	• Emerging technologies
Vehicle	**Platform**
• Price	• Performance
• Design	• Schedule
• Quality	• Cost

In the auto domain, there are multiple providers of competing vehicles. In defense, there is typically one provider of each platform. Many customers make purchase decisions in the auto domain while, in defense, there is one (primary) customer making the purchase decision. The lack of competitive forces can lead to requirements being locked in prematurely.

In the auto domain, vehicles are used frequently. In defense, platforms are used when missions need them which, beyond training, may never occur. Competitors' relative market positions in the auto domain change with innovations, for example, in the powertrain. In defense, adversaries' positions change with strategic innovations, for instance, pursuits of asymmetric warfare. As former Defense Secretary James Mattis has said, "The enemy gets a vote on defense planning" (Mattis 2019).

Automobiles have model year changes, usually 3-year refreshes, and life spans of up to 10 years, typically 6–7. The B-52 bomber has been in use for almost 70 years and the F-15 fighter aircraft has been in use for almost 50 years. There are block upgrades of military aircraft every few years, typically for changes of avionics and weapon systems – rather than body style.

There are similarities that can be seen in Table 35.1. Uncertainties associated with market needs or mission requirements typically flow down in Table 35.1. Uncertainties associated with technology typically flow up, for example, when the engineering organization (at the company or vehicle level) is not sure of how to provide a function or whether performance or cost objectives can be met. New technologies enable new military capabilities. The most important weapons

Table 35.2 Comparison of automotive and defense domains.

Automotive domain uncertainties	Defense domain uncertainties
Customer future preferences	Mission plans will remain relevant
Customers future purchases will favor our offerings vs. competitors	Mission platforms will remain superior to adversaries' capabilities
Performance of our offerings after development	Performance of mission platforms after development
Affordability over the coming years	Affordability over the coming years
Budgets for our offerings across a range of future needs	Budgets for mission platforms across a range of future needs
Supply chains will be economical, efficient and secure	Supply chains will be economical, efficient and secure
Competitors' capabilities will not perceived to be superior	Adversaries' capabilities will be inferior, and certainly not superior
Our enterprise will continue to support our endeavors	Ensuring that sponsors, e.g., Congress, will continue to provide support

transforming warfare in the twentieth century, such as airplane, atomic weapons, the jet engine, electronic computers, did not emerge as a response to doctrinal requirement of the military (Chambers 1997).

Automobile companies are currently wrestling with pursuits of battery electric vehicles and the uncertain rate of market adoption (Liu et al. 2018). Just over the horizon is the opportunity to compete in the driverless car market (Liu et al. 2020), with significant uncertainties about the regulatory environment (Laris 2020). The case study later in this chapter addresses this opportunity.

There are also uncertainties associated with where to manufacture vehicles (Hanawalt and Rouse 2017). Labor costs used to dominate location decisions, but other economic, legal, and political factors are now being considered. Decisions to withdraw from manufacturing in Australia, Canada, and South Korea have resulted.

Product line or program managers in these two domains often have similar questions regarding common uncertainties. A comparison of these questions is shown in Table 35.2. It is often socially unacceptable to verbalize such questions. Unfortunately, uncertainties not verbalized are seldom well managed (Rouse 1998).

Managing Uncertainties

In both the automotive and defense domains, there are usually uncertainties about market or mission requirements as well as uncertainties about technologies and abilities needed to meet these requirements. This section outlines an approach to thinking about managing these uncertainties.

Consider a couple of extremes. You are absolutely sure a function will be required and you are absolutely sure of how to deliver it. In other words, you are not at all uncertain. You should invest to create a solution to meet this need; assuming that you are confident the necessary human and financial resources are available.

At the other extreme, you are absolutely sure a function will not be required. Regardless of your ability to deliver this function, you should not invest in creating this solution. Between these two extremes, there are several strategies a company might adopt. The choice depends on enterprises' abilities to predict their futures, as well as their anticipated abilities to respond to these futures. What strategies might enterprise decision-makers adopt to address alternative futures? As shown in Figure 35.1, we have found that there are four basic strategies that decision-makers can use: optimize, adapt, hedge, and accept.

If the phenomena of interest are highly predictable, then there is little chance that the enterprise will be pushed into unanticipated territory. Consequently, it is in the best interest of the enterprise to optimize its products and services to be as efficient as possible. In other words, if the unexpected cannot happen, then there is no reason to expend resources beyond process refinement and improvement.

If the phenomena of interest are not highly predictable, but products and services can be appropriately adapted when necessary, it may be in the best interest for the enterprise to plan to adapt. For example, agile capacities can be designed to enable their use in multiple ways to adapt to changing demands, e.g., the way Honda adjusted production capacity but other automakers could not in response to the Great Recession. Their planning was more efficient in the long run; even so, efficiency may have to be traded for the ability to adapt.

For this approach to work, the enterprise must be able to identify and respond to potential issues faster than the ecosystem changes. For example, consider unexpected increased customer demands that tax capacities beyond their designed limits. Design and building of new or expanded facilities can take considerable time. On the other hand, reconfiguration of agile capacities should be much faster, as Honda demonstrated.

The value of this approach is widely known in the military. As renown fighter pilot Robert Boyd, inventor of the OODA (observe–orient–decide–act) loop, noted that whoever can handle the quickest rate of change is the one who survives (Coram 2002). Similarly, Arie De Gues, head of Strategic Planning for Royal Dutch Shell, stated that the ability to learn faster than your competitors might be the only sustainable advantage (Senge 1990).

If the phenomena of interest are not very predictable and the enterprise has a limited ability to adapt and respond, it may be in the best interest of the enterprise to hedge its position. In this case, it can explore scenarios where the enterprise may not be able to handle sudden changes without prior investment. For example, an enterprise concerned about potential obsolescence of existing products and services may choose to invest in multiple, potential new offerings. Such investments might be pilot projects that enable the enterprise to learn how to deliver products and services differently or perhaps deliver different products and services.

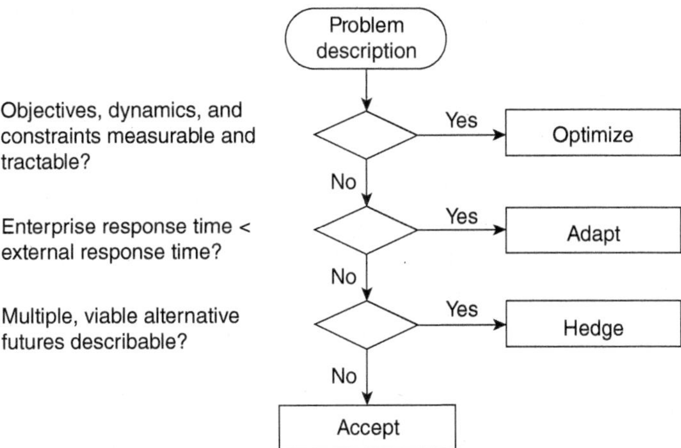

Figure 35.1 Strategy framework for enterprise decision-makers. *Source:* Pennock and Rouse 2016/Acquisition Research Program.

Over time, it will become clear which of these options makes most sense and the enterprise can exercise the best option by scaling up these offerings based on what they have learned during the pilot projects. In contrast, if the enterprise were to take a wait and see approach, it might not be able to respond quickly enough, and it might lose out to its competitors.

If the phenomena of interest are totally unpredictable and there is no viable way to respond, then the enterprise has no choice but to accept the risk. Accept is not so much a strategy as a default condition. If one is attempting to address a strategic challenge where there is little ability to optimize the efficacy of offerings, limited ability to adapt offerings, and no viable hedges against the uncertainties associated with these offerings, the enterprise must accept the conditions that emerge.

There is another version of acceptance that deserves mention – stay with the status quo. Yu et al. (2011) developed a computational theory of enterprise transformation, elaborating on a qualitative theory developed earlier (Rouse 2005). They employed this computational theory to assess when investing in change is attractive and unattractive. Investing in change is likely to be attractive when one is currently underperforming and the circumstances are such that investments will likely improve enterprise performance. In contrast, if one is already performing well, investments in change will be difficult to justify. Similarly, if performance cannot be predictably improved – due to noisy markets and/or highly discriminating customers – then investments may not be warranted despite current underperformance.

Lucero (2018) proposed that these four strategies would be differentially relevant for different areas of an uncertainty space with axes involving uncertainties around the requirements, and the ability to meet those requirements. We extended his thinking to formulate Figure 35.2, focusing on uncertainties in developing technologies. Figure 35.2 depicts the space as having nine discrete cells, which makes it easier to explain, but there are unlikely to be crisp borders between areas where the different strategies are applicable.

There are three types of hedges in Figure 35.2. The upper two cells of the middle column represent company or agency investments in creating technology options to meet possible requirements. The upper two cells of the left column represent licensing, joint development, or other arrangements to buy technology options from partners. The lower cell of the right column represents selling options to others so they can hedge their uncertainties.

The criteria on the left of Figure 35.1 constrain choices of strategies as well as positions in the uncertainty space. If, for example, the objectives, dynamics, and constraints are not measurable

Requirements uncertainty			
Definitely required	Hedge via partnership	Hedge via larger R&D investment	Optimize technology capability
Possibly required	Hedge via partnership	Hedge via smaller R&D investment	Adapt if requirement emerges
Not required	Accept current situation	Accept current situation	License patents to others
Technology uncertainty	Not feasible	Possibly feasible	Fully feasible

Figure 35.2 Strategies versus uncertainties.

and tractable, then optimization may lead to an inappropriate or at least fragile solution (Carlson and Doyle 2000).

At this point, we have strategies for addressing uncertainties. We now need to address the characteristics of the alternative solutions of interest, and then the projected expected utility of each alternative.

Representing Solutions

Whose preferences should guide decisions? While there may be one ultimate decision-maker, success usually depends on understanding all stakeholders. Human-centered design addresses the concerns, values, and perceptions of all stakeholders in designing, developing, manufacturing, buying, operating, and maintaining products and systems. The basic idea is to delight primary stakeholders and gain the support of the secondary stakeholders.

The human-centered design construct and an associated methodology has been elaborated in a book, ***Design for Success*** (Rouse 1991). Two other books soon followed (Rouse 1992, 1993). The human-centered design methodology has been applied many times and continually refined (Rouse 2007, 2015).

The premise of human-centered design is that the major stakeholders need to perceive products and services to be valid, acceptable, and viable. Valid products and services demonstrably help solve the problems for which they are intended. Acceptable products and services solve problems in ways that stakeholders prefer. Viable products and services provide benefits that are worth the costs of use. Costs here include the efforts needed to learn and use products and services, not just the purchase price.

Figure 35.3 embodies the principles of human-centered design, built around Set-Based Design (Sobek et al. 1999), Quality Function Deployment (Hauser and Clausing 1988), and Design

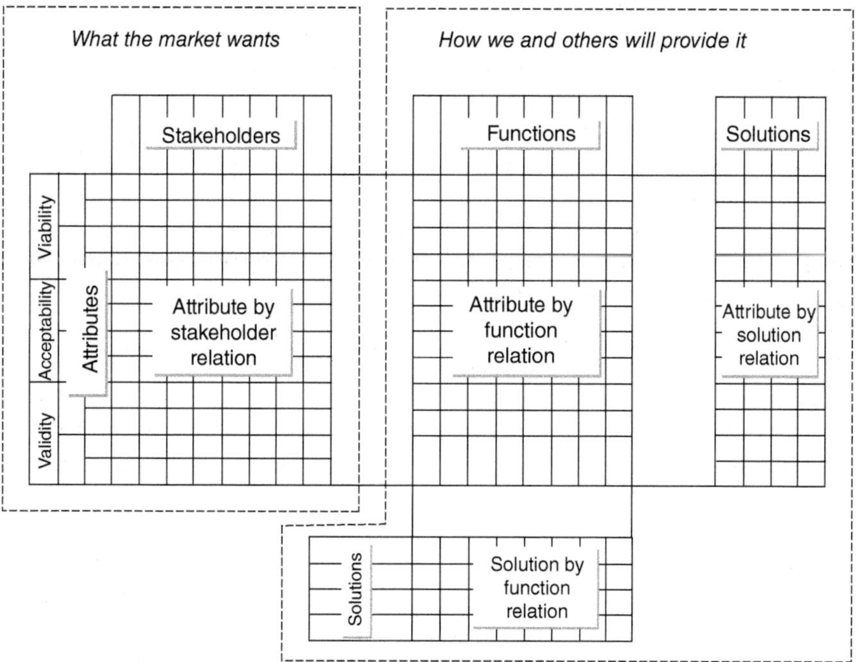

Figure 35.3 Model structure for technology platforms.

Structure Matrices (Eppinger and Browning 2012). As later discussed, multi-stakeholder, multi-attribute utility theory (Keeney and Raiffa 1993) is used to project the value of alternatives. Note that validity, acceptability, and viability in Figure 35.3 are defined in the above discussion of human-centered design.

Sobek et al. (1999) contrast set-based design (SBD) with point-based design. Developed by Toyota, SBD considers a broader range of possible designs and delays certain decisions longer. They argue that, "Taking time up front to explore and document feasible solutions from design and manufacturing perspectives leads to tremendous gains in efficiency and product integration later in the process and for subsequent development cycles." Al-Ashaab et al. (2013) and Singer et al. (2017) report on interesting applications of SBD to helicopter engines and surface combatant ships, respectively.

SBD is reflected in Figure 35.3 in terms of defining and elaborating multiple solutions, including those of competitors or adversaries. Quality Function Deployment (Hauser and Clausing 1988) translates the "voice of the customer" into engineering characteristics. For Figure 35.3, this translates into "voices of the stakeholders." Design Structure Matrices (Eppinger and Browning 2012) are used to model the structure of complex systems or processes. In Figure 35.3, multiple models are maintained to represent alternative offerings as well as current and anticipated competitors' offerings.

The "What the Market Wants" section of Figure 35.3 characterizes the stakeholders in the product or service and their utility functions associated with context-specific attributes clustered in terms of validity, acceptability, and viability. The section of Figure 35.3 labeled "How We and Others Will Provide It" specifies, on the right, the attribute values associated with each solution. The functions associated with each solution are defined on the left of this section. Functions are things like steering, accelerating, and braking, as well as functions that may not be available in all solutions, e.g., backup camera.

Attribute to function relationships in Figure 35.3 are expressed on a somewhat arbitrary scale from −3 to +3. Positive numbers indicate that improving a function increases the attribute. Negative numbers indicate that improving a function decreases an attribute. For example, a backup camera may increase the price of the vehicle but decrease insurance costs.

Solutions on the bottom of Figure 35.3 are composed of functions, which are related to attributes of interest to stakeholders. In keeping with the principles of Set-Based Design, multiple solutions are pursued in parallel, including potential offerings by competitors. While it is typical for one solution to eventually be selected for major investment, the representations of all solutions are retained, quite often being reused for subsequent opportunities.

There are additional considerations beyond SBD, QFD, and DSM. Uncertain or volatile requirements can be due to evolving performance targets, e.g., Ferreira et al. (2009), or surprises by competitors or adversaries, e.g., the Ford Taurus. Both causes tend to result in expensive rework. In the realm of defense, the end of the Cold War ended the need for a 70-ton self-propelled howitzer (Myers 2001). Advances in anti-ship cruise missiles and a challenging performance envelope doomed the Expeditionary Fighting Vehicle (Feickert 2009).

Decision-making may involve more than one epoch (Ross and Rhodes 2008) including both near-term and later decisions. For example, at GM, Epoch 1 involved creating an Escalade as a rebadged GMC in 1999. Epoch 2 involved offering an Escalade as a unique upscale SUV in 2002.

Another issue is the costs of switching from one solution to another (Silver and de Weck 2007). A surveillance and reconnaissance mission adopted an initial solution of a manned aircraft with an option to replace this solution with an unmanned air vehicle (UAV) several years later (Rouse 2010). A deterrent to switching was the very expensive manned aircraft, which would no longer be needed. This problem was resolved by negotiating, in advance, the sale of the aircraft to another agency, effectively taking it “off the books.” Thus, there can be significant value in flexibility. “A system is flexible to the extent that it can be cost-effectively modified to meet new needs or to capitalize on new opportunities” (Deshmukh et al. 2010).

Identifying options can be difficult (Mikaelian et al. 2012). What can you do when and what will it cost? Rouse et al. (2000) discuss case studies from the semiconductor industry. Rouse and Boff (2004) summarize 14 case studies from automotive, computing, defense, materials, and semiconductor industries.

Projecting Value

Using the framework provided by Figure 35.3, and principles from SBD, QFD, DSM, etc., one can create multi-attribute models of how alternatives address the concerns, values, and perceptions of all the stakeholders in designing, developing, manufacturing, buying, and using products and systems. The next issue of importance is the likely uncertainties associated with the attributes of the alternatives. These uncertainties involve what the market or mission needs – or will need – and how well solutions, in terms of functions and underlying technologies, will be able to meet these needs.

The expected value of an alternative can be defined as the value of the outcomes a solution provides times the probability that these outcomes will result. The probability may be discrete or it may be represented as a probability density function. For the former, the calculation involves multiplication and summation; for the latter, the calculation involves integration.

Following Keeney and Raiffa (1993), we will approach this problem using multi-stakeholder, multi-attribute utility theory. We can define the utility function of stakeholder i across the N attributes by

$$u_i = u(x_{1i}, x_{2i}, \ldots, x_{Ni}) = u(\mathbf{x}_i) \tag{35.1}$$

where the bold **x** denotes the vector of attributes. The utility of an alternative across all M stakeholders is given by

$$U = U\left[u(\mathbf{x}_1), u(\mathbf{x}_2), \ldots, u(\mathbf{x}_M)\right] \tag{35.2}$$

The appropriate forms of these functions vary by the assumptions one is willing to make. When there are many attributes, a weighted linear form is usually the most practical. The weights in Eq. (35.1) reflect how much a particular stakeholder cares about the attributed being weighted. It is quite common for most stakeholders to only care about a small subset of the overall set of attributes. Those for which they do not care receive weights of zero.

The weights in Eq. (35.2) reflect the extent to which the overall decision-maker or decision process cares about particular stakeholders. For example, is the customer the most important stakeholder or do corporate finances drive the decision? These weights are usually subject to considerable sensitivity analyses.

Who are typically the stakeholders? We have found that the concerns, values, and perceptions of the following entities are typically of interest:

- Market/Mission
- Customers/Users/Warfighters

- Operators/Maintainers
- Technologists/R&D
- Finance/Budgets
- Competitors – Current
- Competitors – Possible
- Investors
- Governments, e.g., Regulatory Authorities

For the case study presented in the section titled "Case Study: Driverless Cars for Disabled and Older Adults", we focus solely on the investor stakeholder. Investors in driverless cars are interested in three primary attributes:

- **Competitive Advantage (CA)**: To what extent will the investment of interest enable value-added pricing, reduce production costs, reduce operating costs, and leverage existing capacities?
- **Strategic Fit (SF)**: To what extent will the investment of interest leverage technology competencies, exploit current delivery architectures, complement existing value propositions, exploit current partnerships and infrastructure, and provide other opportunities for exploitation?
- **Return on Investment (ROI)**: What capital expenditures, technology acquisition costs, and labor expenses will be needed? What revenue and profits will likely result?

We will return to these attributes in the case study next.

Case Study: Driverless Cars for Disabled and Older Adults

Background

Assistive technologies (AT) hold enormous promise for the 100 million disabled and older adults in the United States (Rouse and McBride 2019). Driverless cars have the potential to greatly enhance the mobility of this population with attractive pricing. Note that the platforms of interest are autonomous vehicles, while the market or mission is to provide enhanced mobility to disabled and older adults.

The Auto Alliance hosted a series of three workshops on "AVs & Increased Accessibility" (Auto Alliance 2019). We focused on physical, sensory, and cognitive disabilities. Approximately 200 people participated in the three workshops from a wide range of advocacy groups, automobile manufacturers, and federal agencies. Workshop participants suggested a large number of needs, as well as approaches to meeting these needs. We clustered these needs into 20 categories. Eight categories covered 70% of the suggestions. Definitions of these categories are as follows:

- ***Displays and controls*** concern information that users can see, hear, touch, etc. and actions they can take.
- ***Locating and identifying vehicle*** concerns users knowing where their ride is waiting and recognizing the particular vehicle.
- ***Passenger profiles*** include secure access to information about passengers, in particular their specific needs.
- ***Emergencies*** concern events inside and outside the vehicle that may require off-normal operations and user support.
- ***Adaptation to passengers*** involves adjusting the human–machine interface to best support particular users with specific needs.

- ***Easy and safe entry and egress*** concerns getting into and out of the vehicle as well as safety relative to the vehicle's external environment
- ***Trip monitoring and progress*** relates to providing information as the trip proceeds, particularly with regard to route and schedule disruptions
- ***Onboard safety*** concerns what happens in the vehicle as the trip proceeds, assuring minimal passenger stress and injury avoidance

An example mapping from needs to technologies is shown in Table 35.3. Technologies required include hardware, software, sensing, networks, and especially enhanced human–machine interfaces. Human–machine interfaces need to enable requesting vehicle services, locating and accessing vehicles, monitoring trip progress, and egressing at destinations to desired locations. The content of Table 35.3 provides a starting point for filling in the framework in Figure 35.3.

Table 35.3 Market needs vs. enabling technologies.

	Technologies				
Needs	**Hardware**	**Software**	**Sensors**	**Networks**	**HMI**
Displays and controls	Hardware for displays and controls	Tutoring system for HMI use	Use and misuse of displays and controls	Access to device failure information	Auditory, braille, haptic, tactile, and visual displays
Locating and identifying vehicle	Vehicle-mounted sensors	Recognition software	Integration of sensed information	Sensors of external networks	Portrayal of vehicle and location
Passenger profiles, privacy	Phone or smart phones, tablets	App to securely provide profile information	Recognition of passenger	Access to baseline info. on disabilities	Portrayal to assure recognition
Emergencies	Controls to stop vehicle and move to safe space	Recognition and prediction of situation	Surrounding vehicles, people, and built environ	External services – police, fire, health	Portrayal of vehicle situation
Adaptation to passengers	Adjusting entry, egress, seating	Learning passenger preferences	Sensing reactions to adaptations	Access to baseline info. on adaptations	Portrayal to enable change confirmations
Easy and safe entry and egress	Sufficient space to maneuver	Capturing data on space conflicts	Surrounding vehicles, people and built environ	Networked access to, e.g., bldg. Directions	Portrayal of surrounding objects
Trip monitoring and progress	Speedometer, GPS, maps	Predictions of progress, points of interest	Surrounding vehicles, people and built environ	Access to traffic information, e.g., accidents	Portrayal of trip and progress
Onboard safety	Securement of wheelchairs and occupants	Capturing data on securement conflicts	Sensing and recording safety risks	Access to best practices on safety risks	Portrayal of securement status

Source: Auto Alliance (2019)/Acquisition Research Program.

The wealth of AT and supporting technologies in Table 35.3 suggest a substantial need for seamless technology integration to avoid overwhelming disabled and older adults, or indeed anybody (Duckmanton 2019, Mellody 2014, Moggridge 2007, Rouse 2014, Yu et al. 2016). We expect that AI-based cognitive assistants may be central to such integration. The question of who might provide which pieces of an overall integrated solution is addressed in this case study.

Investment Scenarios

The question of interest in this case study concerns how an automotive original equipment manufacturer (OEM) should position itself relative to this immense market opportunity (Rouse et al. 2021). We begin with SBD. The hypothetical OEM wants to consider five alternative solutions, indicated as scenarios in Table 35.4 because each includes a market strategy as well as a solution.

Predominant uncertainties include competitors' strategies, technologies (particularly software), abilities to execute, and time. The third scenario, ally with advocacy groups, merits elaboration. The key idea is an AARP branded vehicle, for example, similar to the Eddie Bauer branding of the Ford Explorer, with better paint job, leather seats, heated seats optional, and interior accents. This co-branding alliance with Ford lasted 20 years.

Table 35.4 Set of scenarios considered.

Scenario	Examples	Uncertainties	Confidence in requirements	Ability to respond
Provide total vehicle package	OEM itself or acquisition of autonomous vehicle player	Can OEM really compete against the tech companies?	Hardware is high; software has some unknowns	Strength in integration; easier when OEM controls
Provide vehicle platform to host intelligent software	Alliance with Amazon, Apple, Google, Microsoft or Uber	Why will intelligent platform players source OEM's vehicles?	Basic vehicle platform design is known, but can OEM do this at lowest cost?	Time to integrate software, which will evolve faster than hardware
Provide vehicle platform to host user-centered HMI	Alliance with advocacy groups for disabled and older adults	Why will user-centered HMI players source OEM's vehicles?	How will HMI requirements impact vehicle design?	Time to integrate software, which will evolve faster than hardware
Provide vehicle platform without alliance	OEM will manufacture desired platforms	Why will major players source OEM's vehicles?	Basic vehicle platform design is known; can OEM do this at the lowest cost?	Time to integrate software; design in modularity
Provide integrated mobility services	OEM will provide total mobility experiences	Can OEM competitively manage an end-to-end service?	Auto OEMs do not really understand business model, but does anyone?	Longer time to build out entire ecosystem

Multi-attribute Utility Model

The next step in applying the methodology outlined in this chapter is characterization of Competitive Advantage (CA), Strategic Fit (SF), and Return on Investment (ROI) for the set of five scenarios. We then want to consider uncertainties associated with each scenario, which for this case study will be characterized using discrete probabilities.

The expected utility of each scenario $E[U_S]$ can then be calculated using

$$E\left[U_S\right] = W_{CA} \times P_{CA} \times U_{CA} + W_{SF} \times P_{SF} \times U_{SF} + W_{ROI} \times P_{ROI} \times U_{ROI} \tag{35.3}$$

where $W_{CA} + W_{SF} + W_{ROI} = 1$ and P_{CA}, P_{SF}, and P_{ROI} are the probabilities of achieving U_{CA}, U_{SF}, and U_{ROI}, respectively. As noted earlier, in many situations, probability density functions are needed rather than discrete probabilities. The calculation then involves integration, rather than multiplication and summation.

Once we have the scenarios ranked by $E[U_S]$, we will return to consideration of the optimize, adapt, hedge, and accept strategies from Figure 35.1.

Table 35.5 summarizes assumed probabilities and utilities for the five scenarios. The risk associated with CA is primarily a requirements risk, i.e., the market risk of not having the right offering or best offering. The risk associated with SF is primarily a technology risk, i.e., the risk of not creating, or being able to create, a competitive technology platform. The risk associated with ROI includes both requirements and technology risks.

The reasoning underlying the assumptions in Table 35.5 is as follows:

- **Competitive Advantage**: U_{CA} is high if providing total solution, moderate if only providing vehicle; P_{CA} is low without strong partners, not just branding partners
- **Strategic Fit**: U_{SF} is high if only providing vehicle, moderate if also providing intelligent software; P_{SF} is high if only providing vehicle, moderate if integrating partners' intelligent software
- **Return on Investment**: U_{ROI} is high if providing total solution, moderate if partnering, low if only providing vehicle; P_{ROI} is low if providing total solution, moderate if partnering or only providing vehicle

Table 35.5 Assumed probabilities and utilities for the five scenarios.

	Competitive Advantage		Strategic Fit		Return on Investment	
Scenario	P_{CA}	U_{CA}	P_{SF}	U_{SF}	P_{ROI}	U_{ROI}
Provide total vehicle package	Low ($P = 0.1$)	High ($U = 0.9$)	Moderate ($P = 0.7$)	Moderate ($U = 0.5$)	Low ($P = 0.1$)	High ($U = 0.9$)
Provide vehicle platform as host	Moderate ($P = 0.3$)	High ($U = 0.9$)	Moderate ($P = 0.7$)	High ($U = 0.9$)	Moderate ($P = 0.3$)	Moderate ($U = 0.5$)
Provide vehicle platform to host HMI	Low ($P = 0.1$)	High ($U = 0.9$)	Moderate ($P = 0.7$)	High ($U = 0.9$)	Moderate ($P = 0.3$)	Moderate ($U = 0.5$)
Provide vehicle platform only	Low ($P = 0.1$)	Moderate ($U = 0.5$)	High ($P = 0.9$)	High ($U = 0.9$)	Moderate ($P = 0.3$)	Low ($U = 0.1$)
Provide integrated mobility services	Low ($P = 0.1$)	High ($U = 0.9$)	Moderate ($P = 0.7$)	Moderate ($U = 0.5$)	Low ($P = 0.1$)	High ($U = 0.9$)

The scenarios differ significantly in terms of probabilities of success and utilities if successful. The scenarios also differ significantly in terms of costs of success. Scenarios 1 and 5 represent total up-front commitments and the net present value (NPV) of financial projections would underlie ROI calculations. Scenarios 2 and 3 represent hedges against the risks of not being a player. For these scenarios, net option value (NOV) would be the metric in ROI calculations. Scenario 4 represents an accept strategy as it exploits existing capabilities and will require the least investment.

Boer (2008) suggests how to value a portfolio that includes some investments characterized by NPV and others by NOV. He argues for strategic value (SV), which is given by

$$SV = NPV + NOV \tag{35.4}$$

The NPV component represents the value associated with commitments already made, while the NOV component represents contingent opportunities for further investments should the options be "in the money" at a later time.

Expected Utilities vs. Weightings

Figure 35.4 provides results for $E[U_S]$ with varying assumptions regarding the relative importance (weighting) of CA, SF, and ROI. The overall results are as follows:

- Scenario 2 has the highest $E[U_S]$ unless SF dominates
- Scenarios 2 and 3 have the highest $E[U_S]$ if ROI and/or CA dominate
- Scenario 4, followed by 2 and 3, has the highest $E[U_S]$ if SF dominates
- Scenarios 1 and 5 have the lowest $E[U_S]$ across all weighting assumptions

Discussion

These results reflect, of course, the assumptions in Table 35.5. These assumptions could be varied to assess their impact, but given that $W \times P \times U$ occurs in all the underlying equations, the variations of W in Figure 35.4 reasonably reflects the range of possibilities.

Scenario 1 embodies a significant technology risk in a very competitive market, while Scenario 5 involves a significant requirements risk in attempting to provide services not typical for an OEM. Both of these risks could be hedged with acquisitions of a software company (Scenario 1) or

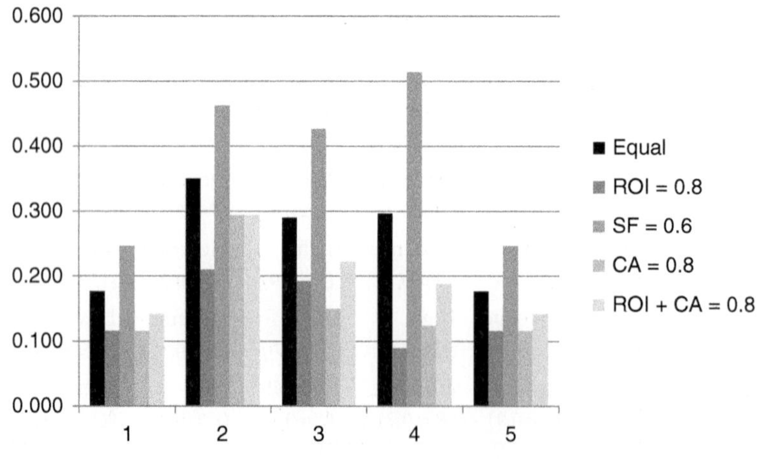

Figure 35.4 Expected utilities for the five scenarios with varying weights.

a service company (Scenario 5). This might be difficult as the market capitalizations of the auto OEMs are much lower than the capitalizations of likely and attractive acquisition targets.

Scenarios 2 and 3 represent hedges against these risks as well, but result in dividing the share of the vehicle that the OEM will provide and, hence, its revenues and profits. Nevertheless, they are attractive because they decrease the competition and provide key technologies. These scenarios also allow the freedom to pursue other strategies as uncertainties resolve themselves.

Scenario 4 focuses on leveraging Strategic Fit. It represents acceptance by the OEM of whatever leverage is provided by its core competencies. This also involves acceptance that they will have to compete with the other auto OEMs that want to provide the vehicle platform. They are quite familiar with this type of competition.

Overall Investment Strategy

The resulting overall strategy involves a portfolio of three investments:

- Substantial investment in Scenario 2 – a hedge against market and technology risks
- Moderate investment in Scenario 3 – a hedge against Scenario 2 not resulting in a partner
- Baseline investment in Scenario 4 – acceptance of a traditional role in the automotive marketplace

With the strategies decided, one is ready to apply the QFD and DSM aspects of Figure 35.3 to the functionality in Table 35.3. This requires that the set of stakeholders be expanded to include:

- OEM
- Partners
- Suppliers
- Car Service Providers
- Car Service Customers

It also requires characterizing competing offerings, whose likely functions, features, and pricing will have been sleuthed via business intelligence.

Discussion

This illustrates the multi-level nature of the methodology. The first question is which of the business scenarios make sense and, for those that make sense, determining the appropriate strategy for pursuing each scenario. The idea is to iteratively refine the chosen scenarios and strategies, which will influence the nature of investments, e.g., whether one makes a total commitment up front (NPV), hedges uncertainties with smaller investments (NOV), or simply accepts one's current position and waits to see how the market develops.

Conclusions

Engineering involves designing solutions to meet the needs of markets or missions. Organizations would like to have the flexibility and agility to address both uncertain needs and uncertain technologies for meeting these needs. This report has presented and illustrated a framework that provides this flexibility and agility. We considered how uncertainties arise, contrasting the automotive and defense domains. We proposed an approach to managing uncertainties. We considered how to represent alternative solutions and project the value of each alternative.

A detailed case study of autonomous vehicles to enhance the mobility of disabled and older adults was presented. We did not consider but need to acknowledge broader risks. It is quite imaginable that driverless car technologies, once deployed, will lead to inadvertent failures with substantial consequences (Danzig 2018). It is also possible that sweeping organizational and societal trends will substantially disrupt this seemingly immense market opportunity (Rouse 2019, 2021, 2022). The current pandemic is a case in point. The impacts of climate change are on the horizon.

Understanding and managing uncertainties need to be core competencies in companies, agencies, and institutions. As this report has argued, uncertainties need to be rigorously and systematically addressed. Managing for success must also include forecasting and managing potential failures (Rouse 2021).

References

Al-Ashaab, A., Golob, M., Attia, U.M. et al. (2013). The transformation of the product development process into lean environment using set-based concurrent engineering. *International Journal of Concurrent Engineering: Research and Applications* 18 (1): 41–53.

Auto Alliance (2019). *Assessing Transportation Needs of People with Disabilities and Older Adults: Report of Workshop 1*. Washington, DC: Auto Alliance.

Boer, F.P. (2008). *The Valuation of Technology: Business and Financial Issues in R&D*. New York: Wiley.

Carlson, J.M. and Doyle, J. (2000). Highly optimized tolerance: robustness and design in complex systems. *Physical Review Letters* 84 (11): 2529–2532.

Chambers, J. (ed.) (1997). *The Oxford Companion to American Military History*, 791. New York: Oxford University Press.

Coram, R.C. (2002). *Boyd: The Fighter Pilot Who Changed the Art of War*. Boston, MA: Little, Brown.

Danzig, R. (2018). *Technology Roulette: Managing Loss of Control as Many Militaries Pursue Technological Superiority*. Washington, DC: Center for a New American Security.

Deshmukh, A., Wortman, M., Boehm, B., et al. (2010). Valuing flexibility. *Proceedings of the 2nd Annual SERC Research Review Conference*, College Park, MD (9–10 November). Hoboken, NJ: Stevens Institute of Technology.

Duckmanton, D. (2019). Why UX and UI should remain separate. *UX Collective* (uxdesign.cc), 21 February. https://uxdesign.cc/why-ux-and-ui-should-remain-separate-7d6e3addb46f

Eppinger, S.D. and Browning, T.R. (2012). *Design Structure Matrix Methods and Applications*. Cambridge, MA: MIT Press.

Feickert, A. (2009). The Marines Expeditionary Fighting Vehicle (EFV): Background and Issues for Congress. *Congressional Research Service*, DTIC Accession Number: ADA498364 (3 August).

Ferreira, S., Collofello, J., Shunk, D., and Mackulak, G. (2009). Understanding the effects of requirements volatility in software engineering by using analytical modeling and software process simulation. *Journal of Systems and Software* 82 (10): 1568–1577.

Hanawalt, E.S. and Rouse, W.B. (2010). Car wars: factors underlying the success or failure of new car programs. *Journal of Systems Engineering* 13 (4): 389–404.

Hanawalt, E.S. and Rouse, W.B. (2017). Assessing location attractiveness for manufacturing automobiles. *Journal of Industrial Engineering and Management* 10 (3): 73–89.

Hauser, J.R. and Clausing, D. (1988). The house of quality. *Harvard Business Review* (May–June), pp. 63–73.

Keeney, R.L. and Raiffa, H. (1993). *Decisions with Multiple Objectives: Preference and Value Tradeoffs*. Cambridge, UK: Cambridge University Press.

Laris, M. (2020). Downsides of self-driving cars could swamp benefits if DC region fails to act, study says. *The Washington Post* (17 April).

Liu, C., Rouse, W.B., and Yu, X. (2015). When transformation fails: twelve case studies in the automobile industry. *Journal of Enterprise Transformation* 5 (2): 71–112.

Liu, C., Rouse, W.B., and Hanawalt, E. (2018). Adoption of powertrain technologies in automobiles: a system dynamics model of technology diffusion in the American market. *IEEE Transactions on Vehicular Technology* 67 (7): 5621–5634.

Liu, C., Rouse, W.B., and Belanger, D. (2020). Understanding risks and opportunities of autonomous vehicle technology adoption through systems dynamic scenario modeling – the American insurance industry. *IEEE Systems Journal* 14 (1): 1365–1374.

Lucero, D.S. (2018). The mash-up rubric: strategies for integrating emerging technologies to address dynamic requirements. *Proceedings of 28th Annual INCOSE International Symposium*, Washington, DC (11 July). San Diego, CA: International Council of Systems Engineers.

Mattis, J. (2019). Interview, *Meet the Press* (13 October).

Mellody, M. (2014). *Can Earth's and Society's Systems Meet the Needs of 10 Billion People? Summary of a Workshop*. Washington, DC: National Academies Press.

Mikaelian, T., Rhodes, D.H., Nightingale, D.J., and Hastings, D.E. (2012). A logical approach to real options identification with application to UAV systems. *IEEE Transactions on Systems, Man, and Cybernetics Part A: Systems and Humans* 42 (1): 32–47.

Moggridge, B. (2007). *Designing Interactions*. Cambridge, MA: MIT Press.

Myers, S.L. (2001). Pentagon panel urges scuttling howitzer system. *New York Times* (23 April). Section A, p. 1.

Pennock, M.J. and Rouse, W.B. (2016). The epistemology of enterprises. *Journal of Systems Engineering* 19 (1): 24–43.

Ross, A.M. and Rhodes, D.H. (2008). Using natural value-centric time scales for conceptualizing system timelines through Epoch-Era Analysis. *Proceedings of INCOSE International Symposium* 18 (1): 1186–1201.

Rouse, W.B. (1991). *Design for Success: A Human-Centered Approach to Designing Successful Products and Systems*. New York: Wiley.

Rouse, W.B. (1992). *Strategies for Innovation: Creating Successful Products, Systems, and Organizations*. New York: Wiley.

Rouse, W.B. (1993). *Catalysts for Change: Concepts and Principles for Enabling Innovation*. New York: Wiley.

Rouse, W.B. (1998). *Don't Jump to Solutions: Thirteen Delusions that Undermine Strategic Thinking*. San Francisco, CA: Jossey-Bass.

Rouse, W.B. (2005). A theory of enterprise transformation. *Journal of Systems Engineering* 8 (4): 279–295.

Rouse, W.B. (2007). *People and Organizations: Explorations of Human-Centered Design*. New York: Wiley.

Rouse, W.B. (2010). Options for surveillance and reconnaissance. In: *The Economics of Human Systems Integration* (Chapter 15) (ed. W.B. Rouse). New York: Wiley.

Rouse, W.B. (2014). Human interaction with policy flight simulators. *Journal of Applied Ergonomics* 45 (1): 72–77.

Rouse, W.B. (2015). *Modeling and Visualization of Complex Systems and Enterprises: Explorations of Physical, Human, Economic, and Social Phenomena*. New York: Wiley.

Rouse, W.B. (2019). *Computing Possible Futures: Model Based Explorations of "What if?"*. Oxford: Oxford University Press.

Rouse, W.B. (2021). *Failure Management: Malfunctions of Technologies, Organizations, and Society*. Oxford: Oxford University Press.

Rouse, W.B. (2022). Designing policy portfolios. In: *Transforming Public-Private Enterprises: Understanding and Enabling Innovation in Complex Systems*. Oxford: Oxford University Press.

Rouse, W.B. and Boff, K.R. (2004). Value-centered R&D organizations: ten principles for characterizing, assessing & managing value. *Journal of Systems Engineering* 7 (2): 167–185.

Rouse, W.B. and McBride, D.K. (2019). A systems approach to assistive technologies for disabled and older adults. *The Bridge* 49 (1): 32–38.

Rouse, W.B., Howard, C.W., Carns, W.E., and Prendergast, E.J. (2000). Technology investment advisor: an options-based approach to technology strategy. *Information-Knowledge-Systems Management* 2 (1): 63–81.

Rouse, W.B., Verma, D., Lucero, D.S., and Hanawalt, E.S. (2021). Strategies for addressing uncertain markets and uncertain technologies. *Proceedings of Annual Acquisition Research Symposium*, Naval Postgraduate School (11–13 May). Monterrey, CA: Naval Postgraduate School.

Senge, P. (1990). *The Fifth Discipline*. New York: Doubleday/Currency.

Silver, M.R. and De Weck, O.L. (2007). Time-expanded decision networks: a framework for designing evolvable complex systems. *Journal of Systems Engineering* 10 (2): 167–188.

Singer, D., Strickland, J., Doerry, N. et al. (2017). *Set-Based Design*, 7–12. Alexandria, VA: Society of Naval Architects and Marine Engineers, Technical and Research Bulletin.

Sobek, D.K., Ward, A.C., and Lifer, J.K. (1999). Toyota's principles of set-based concurrent engineering. *Sloan Management Review* 40 (2): 67–83.

Yu, X., Rouse, W.B., and Serban, N. (2011). A computational theory of enterprise transformation. *Journal of Systems Engineering* 14 (4): 441–454.

Yu, Z., Rouse, W.B., Serban, N., and Veral, E. (2016). A data-rich agent-based decision support model for hospital consolidation. *Journal of Enterprise Transformation* 6 (3/4): 136–161.

Biographical Sketches

William B. Rouse is research professor in the McCourt School of Public Policy at Georgetown University and professor emeritus and former chair of the School of Industrial and Systems Engineering at Georgia Institute of Technology. His research focuses on mathematical and computational modeling for policy design and analysis in complex public-private ecosystems, with particular emphasis on healthcare, education, energy, transportation, and national security. He is a member of the US National Academy of Engineering and Fellow of IEEE, INCOSE, INFORMS, and HFES. He is a member of the Research Council of the Systems Engineering Research Center and the Acquisition Innovation Research Council. His recent books include *Beyond Quick Fixes* (Oxford, 2014), *Bigger Pictures for Innovation* (Routledge, 2023), *Transforming Public-Private Ecosystems* (Oxford, 2022), *Failure Management* (Oxford, 2021), and *Computing Possible Futures* (Oxford, 2019). Rouse lives in Washington, DC.

Dinesh Verma served as the Founding Dean of the School of Systems and Enterprises at Stevens Institute of Technology from 2007 through 2016. He currently serves as the Founding Executive Director of the Systems Engineering Research Center (SERC), a US Department of Defense sponsored University Affiliated Research Center (UARC) focused on systems engineering research; along with the Acquisition Innovation Research Center (AIRC). Dr. Verma has authored over 150 technical papers, book reviews, technical monographs, and co-authored three textbooks: Maintainability: A Key to Effective Serviceability and Maintenance Management (Wiley, 1995), Economic Decision Analysis (Prentice Hall, 1998), Applied Space Systems Engineering (McGraw Hill, 2009). Verma has received three patents in the areas of life-cycle costing and fuzzy logic techniques for evaluating design concepts. He was recognized with an Honorary Doctorate Degree (Honoris Causa) in Technology and Design from Linnaeus University (Sweden) in January 2007; and with an Honorary Master of Engineering Degree (Honoris Causa) from Stevens Institute of Technology in September 2008.

Chapter 36

Portfolio Management and Optimization for System of Systems

Frank Patterson, David Fullmer, Daniel Browne, and Santiago Balestrini-Robinson

Georgia Tech Research Institute, Systems Engineering Research Division, Smyrna, GA, USA

Overview and Motivation

Management of a portfolio of capabilities, generally realized in a suite of physical systems, is a common problem faced by the agencies and programs within the United States Department of Defense (DoD). Standardized methods of tackling these problems have been developed over time, with a focus on sound systems engineering (SE) principles and tools. Although a standard process is commonly executed in solving these portfolio management problems, a standardized suite of tools is nonexistent to support that process. Most often, new tools are developed by each SE team, sometimes leveraging a past tool utilized by the team in an ad hoc manner. Similarly, each engineering or design challenge has its own level of specificity which oftentimes warrants, or even requires, a customized approach. Therefore, the balance between reusable capabilities and flexibility in integrating those capabilities is necessary to advance the state of practice and produce timely assessments of optimal portfolios.

This limited reuse of tools is a primary motivator for research in this area. Building this toolbox of portfolio analysis and optimization tools is something that organizations do over time, but are seldom able to further develop them unless they directly support a project or activity. It is imperative to have more reusable and easier to integrate tools to maximize the time spent by analysts doing value-added work over basic tool development and maintenance.

The chapter is organized as follows. First, an overview of the generic Systems Engineering process for portfolio analysis is provided, covering the core elements necessary to realize an analysis framework that can support optimization. Next, a review of formal optimization methods is provided to include a recommendation for the current state of practice for high-dimensional complex tradespaces. Finally, a review of example visualization techniques is provided that help illustrate the types of products that can be used to assess a set of portfolios.

Systems Engineering for the Digital Age: Practitioner Perspectives, First Edition. Edited by Dinesh Verma.

Companion website: www.wiley.com/go/verma/systemsengineering

Systems Engineering Process for Portfolio Management Problems

Portfolio management problems are well addressed when following a solid set of systems engineering principles and practices. As such, it is important that a systems engineer works closely with the customer as a facilitator. The overall process, depicted in Figure 36.1, is described in the following subsections to provide context for the discussion of the toolset that follows in the Combinatorial Design Space Analysis Tool section.

Most portfolio analyses problems share many of the characteristics of Wicked Problems (Rittel and Webber 1973), as oftentimes the stakeholders' understanding of the problem evolves as the analysts produce new solutions to elucidate tacit needs and goals. Furthermore, it is common for this iterative process to have no clear stopping rule, as new alternatives and concepts are identified and better characterization of the needs are evolved. This often leads to iterating until a good enough solution is identified, or the team runs out of time or money for their analysis. It is not uncommon to conclude a portfolio analysis exercise with a lingering feeling that more can be done or should be done. Ultimately, any exercise should help stakeholders make better-informed and more defensible decisions.

Define the Problem and List Alternatives

The process begins with a careful consideration of the problem. This entails both developing the problem statement (i.e., high-level goal(s)) and compiling a list of alternatives that serve as potential solutions. Generally, that list of alternatives is captured within a Market Research Database (MRDB) that captures both currently available off-the-shelf solutions as well as potential low TRL (Technology Readiness Level) future solutions. Note that these are created iteratively, not sequentially.

Analysts and stakeholders should develop a rhythm for iterating between problem definition and solution identification. Decomposition and revision of the problem statement may highlight capability gaps in the alternatives and necessitate further consideration. Alternatives may also become irrelevant as the problem statement is refined. Similarly, the inclusion (and exclusion) of alternatives may lead to reconsideration of the problem statement. The goal is to make sure that all

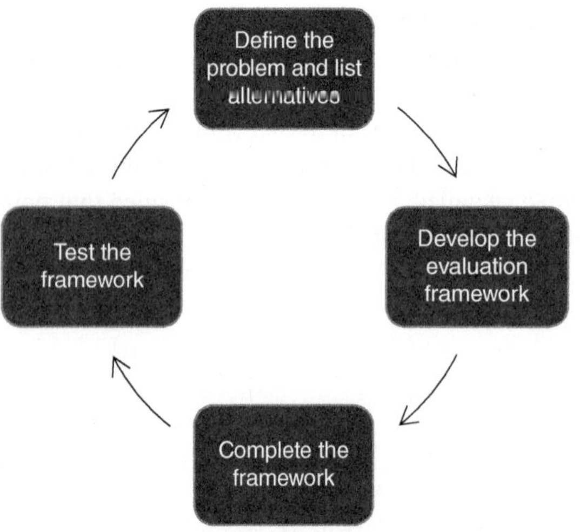

Figure 36.1 Overall process for portfolio management problems.

tacit needs are explicitly stated in the high-level goals and all relevant potential solutions are appropriately characterized and included in the candidate pool. Some of the questions that analysts and stakeholders should consider when doing this are included below.

This option seems relevant; why is it excluded?
How do its capabilities differ from the problem as stated?
Why are similar options included in the list of alternatives, i.e., what characteristics do they share that make them more suitable?
This alternative should not be preferred over this one, why is that? What is missing in the analysis, or what is being inappropriately assessed?

The provenance of both the problem statement and the list of alternatives must be traced to ensure that they are accurate, complete, and current. It is possible that the customer may already have an idea of the high-level goals of the project and potential alternatives. If not, the facilitator must be prepared to start from the ground up, potentially organizing a workshop or developing a market survey to elicit expertise from stakeholders and subject-matter experts (SMEs). When developing the list of alternatives, it is important to forecast future alternatives relative to the problem scope, accounting for options that may not be available now, but can be expected to be by a given implementation deadline. When including yet-to-be technology alternatives, consider that portfolio management exercises are often implemented in the not-so-near future, and that any analysis should account for the inherent risk of selecting unproven solutions.

Develop the Evaluation Framework

How are alternatives to be evaluated against the high-level goals of the project? The development of the evaluation framework requires extensive interaction with a small group of SMEs who possess a detailed understanding of the problem and the capabilities of the potential solutions under consideration. It is critical that these SMEs have sufficient availability to review the approach and the results produced by the framework, and help identify approaches to mitigate any shortfalls in the analysis. A successful portfolio analysis is generally predicated on the efficient incorporation of SME know-how.

Most portfolio analyses will have to consider a myriad of metrics from various stakeholders. These stakeholders have a mental model of the problem, what is required to solve it, and potentially how well different solutions will perform. In some cases, these models may be fairly flat and simple, e.g., they care about a few things that do not have to be broken down into more specific concerns. In most cases, there will be various concerns of interest, ranging from performance and effectiveness, to cost, risk, schedule, etc. If these concerns can be comparatively assessed with one another, and a model can be created for how the stakeholders perceive them, one can develop a model *a priori* of the analysis. If this cannot be done before solutions are generated, it will be necessary to assess the value of the solutions *a posteriori*. This will have implications for how the tradespace is explored (what methods are used to generate candidate solutions) and what affordances the interactive visualizations offer.

The high-level goals must be iteratively decomposed until they can be mapped to tangible alternatives. For example, a Logistics Information Technology (LogIT) scenario (e.g., high demand for bandwidth, low latency, common technology solutions across programs) might be decomposed into a series of activities (e.g., make supply request, update inventory). These activities might then be mapped directly to the capabilities of the alternatives. This is an iterative process that strives to reduce the number of levels required to complete the mapping, without a loss of traceability. Furthermore, the number of items in each level should increase monotonically. That is, a 10-item level should not map to a 5-item level and then up to a 15-item level. This is indicative of information loss and will tend to produce recommendations that do not agree with stakeholders' mental models.

The next step is to define the mapping between each level. The map is a general function, but in most cases, it takes the form of a two-dimensional matrix with its rows corresponding to the domain level and its columns corresponding to the range level. The matrix will be populated with values from a scale defined by the facilitator. The scale may be numeric (e.g., 0, 1, 3, 9) or semantic (e.g., None, Little, Some, Much), though a semantic scale generally has an underlying numeric equivalence. It is important to use clear and precise language when defining these levels. Avoid detailed scales with too much granularity; effort and time may be wasted in trying to differentiate between nearly equivalent values. Moreover, it may be necessary to hide the numeric meaning of the scale from the experts filling out the matrix. This information may affect their ratings either consciously or unconsciously. If there are few lower-level options (i.e., the domain level is limited), it may be possible to use more precise techniques to capture the relationship, e.g., pairwise comparison. Alternatively, a ranking method may be useful in some cases, whereby stakeholders do not have to provide a specific value, simply rank the items and from that ranking, a numeric value can be derived (e.g., using Rank Ordered Centroid). It is practically a necessity to have the framework support the ability to change these "mappings to numerical values" to conduct sensitivity analyses. Before any final recommendation is made, the analysts should ensure that if the preference of results changes as a function of the scaling/mappings used, stakeholders are made aware and a rationale for selecting a given approach is carefully crafted in coordination with stakeholders and SMEs.

Complete the Framework

The previous step developed the general framework for evaluating the alternatives against the high-level goals of the project. The next step is to have experts populate the details (e.g., enter data into the matrices that map between levels). This may be accomplished with a survey, a workshop, or ideally a combination of the two.

The usual biases that affect the quality of survey data apply here as well. It is important to make sure that responses are not influenced by superiors and that the loudest voice does not carry the day unquestioned. Conversely, non-productive discussions must be discouraged whenever possible, and participants reminded of the ultimate purpose of the overall activity. Challenges to the framework or the approach must be recognized, but if at all possible, not be allowed to disrupt the workshop.

When planning for this activity, the analysts must ensure that sufficient time is allotted for the data elicitation activity given the availability of the participants. It is not uncommon to hold a week-long workshop to elicit many 100s of inputs. Another important consideration when requesting large amounts of data are burnout and apathy. It is better to have a smaller pool of committed and knowledgeable stakeholders than a larger pool of uninterested or indifferent contributors. At the conclusion of the data elicitation process, the analysts must assess the accuracy of the data and explore it to identify any conspicuous patterns. There are no hard-set rules for doing this, but identifying clusters of stakeholders with similar views and comparing where they differ, or identifying which answers produced uniform or multimodal distributions can be critical to understanding how the stakeholders interpreted the questions asked.

Be efficient and effective when conducting a workshop. Distribute individual surveys at least two weeks before the workshop. Compile the data and identify disagreements for further discussion.

Use anonymous voting methods to minimize biases and intimidation tactics. Furthermore, work with a respected mediator capable of controlling the experts, keeping the discussion moving, and minimizing unproductive conflict. If possible, hold a mock workshop with one or two committed and knowledgeable participants to minimize the likelihood of finding issues with the framework during the larger workshop.

Pay attention to the proceedings. The validity of the framework may be assessed from the rate of progress. For example, slow progress may indicate that the map is too difficult to complete and further decomposition of the levels is necessary. Alternatively, rapid progress may indicate that the map is too obvious and there are too many levels. The framework may very well require further consideration.

It is not required that SMEs populate the entirety of the framework. In some cases, it may be suitable to forgo their qualitative assessments for more quantitative ones that leverage physics-based models or simulations. These hybrid approaches are not only feasible, but oftentimes desirable as the qualitative and quantitative data sources can be complementary of one another. With that said, it is seldom the case that a clear divide exists, and more often than not, the quantitative data available is limited by assumptions that may not be compatible with the problem at hand. Therefore, it may be necessary to provide the quantitative data to experts who can use it to inform their qualitative assessments. Determining the right mix between the two is as much a function of the problem of interest, as it is the time and resources available to conduct the analysis.

Test the Framework

Finally, test the framework with various inputs and examine the results. If the analysis team is surprised by the results, determine the reason. The framework (or its data) may require further iterations. If the framework does not behave as desired, it is important to fix the problem at its source and not just artificially manipulate the results. Once the analysis team is satisfied with the results, engage with the stakeholders to do additional assessments of the results produced by the framework. In some situations, it may be possible to formulate tests prior to the development of the framework by developing artificial alternatives with expected results (e.g., a highly capable but costly alternative vs. less costly and less capable ones). This can be a more thorough means to verify that the framework produces results that are congruent with the stakeholder mental models.

Expanding to Optimization

The process explained above should produce a framework for building and evaluating candidate portfolios. However, tradespaces with respect to portfolio analysis rapidly grow to intractable sizes, which leads to the need for using efficient algorithms to provide insight into these NP-Complete problem spaces. The "Portfolio Optimization and Analysis" section provides a review of formal optimization methods, as well as guidance to where each method is well suited with respect to problem construction.

Portfolio Optimization and Analysis

This section discusses logical and mathematical approaches to portfolio optimization in the context of mission engineering (ME) and systems-of-systems (SoS). There is a significant amount of variability in the potential problems captured here, but some common attributes are critical to help frame the discussion.

- Portfolio optimization refers to the process of selecting the best *portfolio* of options from a set of all portfolios possible.
- In the context of a mission or systems engineering problem, this often involves selecting from an available set of systems and subsystems to generate a holistic solution but does not necessarily exclude tuning system attributes.
- Portfolio optimization problems of this flavor extend beyond financial considerations (not to say those are not important) to include the capabilities of the possible portfolios.
- The nature of portfolio optimization problems in engineering are almost always multi-objective, with more than one competing criterion being considered.

In the case of multi-objective optimization, the concept of optimal is swapped with that of Pareto optimality. A Pareto optimal (non-dominated) solution is one that cannot be improved in one objective dimension without worsening in another. The collection of feasible Pareto points in a problem solution is often called a Pareto frontier.

Portfolio optimization is an important component of the DoD's shift from program-centric to portfolio-centric acquisition management (Section 809 Panel 2019) as well as the expanding field of digital mission engineering.

Portfolio Optimization Problems

Multi-objective optimization (MOO) is an expansive and growing mathematical field that has applications in many fields, and even many applications within systems engineering (Hwang and Masud 2012).

- Support decision-making in capturing and organizing stakeholder preferences.
- Support complex system or portfolio design decisions through multiple-criteria decision-making (MCDM)
- Enable multidisciplinary design optimization (MDO), sometimes known as multidisciplinary design analysis and optimization (MDAO).

The portfolio optimization problem discussed in this section is formulated as a multi-objective mixed integer nonlinear programming (MOMINLP). While variations of the problem might be formulated in different ways, in its most general sense, an ME or SoS optimization problem may have mixed design variables, is subject to multiple objectives for various stakeholders, and has multiple constraints. A MOMINLP problem has the following general format:

$$
\begin{aligned}
&\min \boldsymbol{F}(\boldsymbol{x}) - \left[f_1(\boldsymbol{x}), \ldots, f(\boldsymbol{x})_m \right]^T \\
&s.t.\ g_j(\boldsymbol{x}) \leq 0 \quad \forall j = 1, \ldots, n \\
&\quad h_k(\boldsymbol{x}) = 0 \quad \forall k = 1, \ldots, p \\
&\quad\quad x_i^L \leq x_i \leq x_i^U \quad x_i \varepsilon \mathbb{Z} \quad \forall i \varepsilon I
\end{aligned}
$$

In this specific portfolio case, decision variables in the set I are limited to integer values. These often denote subsystem portfolio quantities or presence of an item in the portfolio in the case of binary inclusion. It is up to the engineer to understand the problem at hand and formulate an optimization to guide design decisions as appropriate. Objectives in the case of a portfolio optimization problem are often the criteria by which stakeholders measure a solution, but can also include elements of risk or cost (to be minimized). Constraints for this optimization problem may include financial limits, system incompatibilities, or other limitations.

Similarities to "Portfolio Optimization"

It is worth noting that the portfolio optimization problems created by ME and SoS problems are tangential to the financial portfolio optimization (Best 2010), a process of selecting a "portfolio" of asset distributions to maximize a return. Areas of study like modern portfolio theory (mean-variance analysis) have inspired and informed approaches to discrete optimization that could be applied to problems being considered here (Bertsimas and Sim, 2004; Davendralingam and DeLaurentis 2013). However, these approaches often rely on assumptions (e.g., convex objectives, or linear constraints) that cannot always be met when addressing the types of problems considered in this chapter.

Potential Solutions Methods

It is important to understand that there exist a number of proposed methods for solving multi-objective optimization problems in general and MOMINLP specifically (Marler and Arora 2004). Given the nature or scale of a specific problem, a portfolio optimization-type problem can be solved through a number of methods. Often multiple approaches are hybridized together. In this section, we explore a number of potential solution methods, but then focus in on one of the most robust, evolutionary algorithms.

A Priori Approaches

One approach to optimizing a portfolio problem is to collapse any identified objective into a single dimension. These methods usually rely on *a priori* knowledge of stakeholder preferences, allowing expression of a single solution objective. This may be a less useful case in system of systems design problems due to a desire to explore a full solution tradespace, but can have applications in architecting to certain stakeholders.

There are a number of different *a priori* approaches to multi-objective optimization that can handle mixed decision variables and nonlinear cases. Some of the most commonly used ones are described below.

Decision-Making-Oriented Approaches A number of relatively simplistic approaches to *a priori* optimization have been developed that are particularly applicable to small decision spaces, where combinatorial alternatives are countable for individual analysis. These methods are sometimes referred to as multiple attribute decision making (MADM) (Zanakis et al. 1998). While often simple and not often applicable, these approaches often provide the benefit of being extremely fast and clearly traceable.

Weighted sum method: Also referred to as Simple Additive Weighting (SAW), assigns weights to each system attribute or criterion, weighting those attributes by stakeholder preference, and then summing up the weighted scores for each alternative to obtain a composite score.

AHP: The Analytical Hierarchy Process method involves constructing a hierarchy of attributes and criteria, and then using pairwise comparisons to determine the relative importance of each attribute. Alternatives can be ranked by their summed performance against each attribute.

TOPSIS: The Technique for Ordered Preference by Similarity to an Ideal Solution approach involves defining a composite ideal best and worst solutions, and then ranking preference weighted alternative scores based on their proximity to this ideal solution and distance from a worst solution. TOPSIS is often combined with AHP in order to weight criteria.

Goal Programming Goal programming involves defining a set of goals or targets for the objectives, and then formulating a mathematical model that minimizes the deviations between the actual and desired values. In this way, the optimal solutions obtained from the model inherently account for trade-offs between the objectives. There are a number of goal programming approaches that use various models for constructing goals and measuring how well they are satisfied as a whole (Azmi and Tamiz 2010). The majority of methods are relatively simple to perform and easy to setup, providing quick results for more basically constructed problems even at large scales of variables, objectives, or constraints. However, like all *a priori* methods, the results are based the constructed conditions and preferences between objectives, limiting the information that can be gleaned from results.

Utility Functions Utility function-based approaches utilize mappings that represent stakeholder satisfaction (or utility) based on the objective values. These utility functions often identify a critical value that an objective must meet, and the value at which improving an objective does not provide additional utility. In this way, the utility functions are a mathematical representation of stakeholder preferences (Rosenthal 1985) that normalizes all measured criteria between bounded values, e.g., 0% to 100%. The functions are often combined based on the relative importance of each objective or criteria, producing a single objective to drive optimization. This is often referred to as multi-attribute utility theory (MAUT) (von Winterfeldt and Fischer 1975).

Boolean Satisfiability Solvers If a heavily constrained portfolio optimization problem can be formulated entirely as a binary integer problem, Boolean Satisfiability Solvers (SAT Solvers) can be utilized to find results that meet all required constraints. SAT solvers excel at efficiently solving mathematical problems involving Boolean algebra. More explicitly, they are able to solve the Boolean Satisfiability Problem, determining if it is possible to find a combination of TRUE and FALSE values for the variables of a given Boolean formula such that the formula evaluates to TRUE. A portfolio problem can be formulated as a Boolean satisfiability problem by expressing the constraints of the portfolio as a logical expression. This expression can then be fed into a SAT Solver, which will attempt to find a set of values that satisfies the constraints. The SAT Solver can then be used to optimize the portfolio by finding the values that maximize the return while still satisfying the constraints.

SAT Solvers have the following benefits:

- They are able to provide a guaranteed optimal solution, as they are able to exhaustively search for the optimal portfolio.
- They can determine if no solution is feasible, i.e., the constraints cannot be simultaneously met.
- They are often faster than a global *a posteriori* search, as they are able to quickly evaluate a set of parameters and generate optimal portfolios.
- They also exhibit the following detriments:
- They can only optimize a single objective at a time and require explicit preferences upfront to balance multiple objectives.
- They can be computationally expensive, as they may require a significant amount of time to evaluate a large number of parameters and generate optimal portfolios.
- Not well suited to generate a diverse range of optimized portfolios, there is a high likelihood that the portfolios generated will be very similar with minor differences between them.
- Not able to provide a more detailed analysis of the trade-offs between multiple objectives, which can help stakeholders to make more informed portfolio decisions.
- More difficult to formulate a problem in the form required by most satisfiability solvers, and in some instances, it may not be possible to do so without a new conceptualization of the problem.

A Posteriori Methods

In contrast to *a priori* methods, *a posteriori* methods make no assumptions or have no access to stakeholder preference information. In many cases, stakeholders cannot explicitly express their preferences, or there is an inherent desire to see a range of solutions before making decisions. *A posteriori* methods are also useful when it is difficult or expensive to accurately model the objective functions or constraints, or when the problem is subject to significant uncertainty or variability. There are a number of *a posteriori* approaches for multi-objective optimization that can handle mixed decision variables, with many approaches borrowing from one another or even *a priori* or single objective approaches. The following subsections outline some common *a posteriori* methods used in solving portfolio or system-of-system optimization problems. Some of these general approaches can be combined or overlap with one another in specific applications.

Mathematical Programming Methods Mathematical programming approaches to multi-objective optimization are typically applied mathematical optimization techniques that are modified, often through repetition of the core algorithm, to solve multi-objective problems. An example might be the modification of the well-known linear programming (LP) method to multi-objective linear programming (MOLP). When problems are linear or convex, there are a number of vector optimization methods available, but nonlinear problems or those with unknown problem space shapes may require other approaches.

Some examples of mathematical programming approaches that may be potentially applicable depending on the nature of the Portfolio Optimization problem include:

- Successive Pareto Optimization (Mueller-Gritschneder et al. 2009)
- Normal Boundary Intersection (NBI) (Das and Dennis 1998)
- ε-Constraints Method (Mavrotas 2009)
- Multi-objective branch and bound (Przybylski and Gandibleux 2017)

Decomposition-Based Methods Decomposition is a general approach for solving a problem by breaking it up into smaller problems and solving those sequentially or in parallel. In multi-objective optimization decomposition methods, the problem is broken into a set of single objective subproblems, which are then solved and combined to form a final solution but can sometimes also involve decomposition of the decision variables (Santiago et al. 2014).

Machine Learning Methods Machine learning can be applied to multi-objective optimization in a number of ways. Traditionally, models can be trained with example data from the problem to approximate computationally (or otherwise) expensive objectives and constraints. The resulting models can then be operated on by any multi-objective method to predict a Pareto frontier of solutions.

More recently, however, deep learning techniques have been applied to directly learn where the Pareto frontier lies. An example is Pareto-Front Learning (PFL) (Navon et al. 2020), which uses hypernetworks to learn the entire Pareto Frontier and allows for run-time tuning of preferences.

Multi-objective Evolutionary Algorithms

The evolutionary algorithm (EA) approach discussed here is inherently an *a posteriori* method but is addressed separately here due to the ability of evolutionary algorithms to work as a global optimization technique across multiple objectives. The methods referred to previously as *a priori* and *a posteriori* are often traditionally expansions of single objective optimization approaches, but EAs can be employed to directly optimize multiple objectives at the same time. There are a growing variety of these methods, often referred to as multi-objective evolutionary algorithms (MOEAs) (Coello et al. 2007; Zhou et al. 2011).

Evolutionary algorithms also have a number of advantages that lend themselves widely to portfolio optimization. These advantages include but are not limited to:

- EAs support mixed integer problems through the flexible definition of decision variables as phenotypes and host of variable operations that have been developed to support various algorithms.
- EAs do not require gradient information or related data to proceed, and are generally acceptant of any time of problem space (discontinuous, non-convex, etc.).
- EAs can be hybridized with optimization methods for continuous and convex spaces, and have done so to great effect.
- A large variety of EAs have been proposed for various multi-objective problem types, with plentiful publications on various approaches.
- EAs are easily parallelizable and manageable where large computational resources are available.
- The computational cost of EAs (in particular genetic algorithms) grows linearly with problem size, as opposed to exponentially as many other methods tend to do.

Non-dominated Sorting Genetic Algorithms (NSGA) are evolutionary algorithms designed to solve multi-objective optimization problems (Deb et al. 2002). They work by sorting the population of potential solutions by their level of non-domination. Solutions with no better solutions in the same population are considered non-dominated. This method allows for an efficient sorting process with a minimal amount of computation, as the algorithm only needs to compare the solutions in the same population. The algorithm can then select the best solutions based on their non-domination level. NSGAs are useful for solving problems that require multiple objectives, as it can identify the best solutions based on their non-domination level.

Portfolio Optimization is often an exercise in helping stakeholders explore the effects of their preferences. NSGAs offer an opportunity to explore a frontier of stakeholder preferences, which can then be rapidly queried to gain insight into the underlying phenomena and common characteristic of what constitutes a "good" portfolio. Figure 36.2 illustrates this concept with a

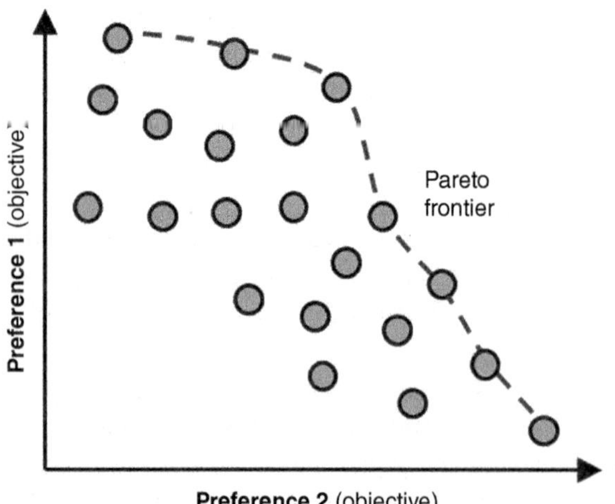

Figure 36.2 Notional tradespace with Pareto optimal solutions.

simple example, where each solution is depicted as a circle. The Pareto frontier is defined by the non-dominated solutions, i.e., those that to make any gains in one dimension (this example only includes two dimensions for illustration purposes) at least another dimension must see a reduction. In this case the ideal solution would be in the upper right quadrant, but due to physical, budgetary, or other considerations, the set of solutions is limited and a trade-off between the two objectives, Preference 1 and Preference 2, must be made. NSGAs are designed to push this frontier toward the ideal solution while maximizing its coverage, i.e., the spread between the points that defined the frontier.

The NSGA-II is a well-known and common approach to MOMINLP-type problems due to its flexibility, speed, and available implementations/documentation (Deb et al. 2002). Improvements to sorting speed and elitism (maintenance of the best solutions) from the original NSGA algorithm make this an excellent choice for solving many portfolio optimization problems. One weakness of the NSGA-II is it does not scale well to many-objective problems (traditionally performing best for 2–3 objectives). The NSGA-III (Deb and Jain 2013; Jain and Deb 2013) modifies aspects of the NSGA-II, primarily replacing the crowding distance operation with a reference point-based method to maintain solution diversity. The NSGA-III is shown to work well for problems beyond 10 objective dimensions.

Beyond the NSGA algorithms, several other MOEA algorithms have been developed with a widening availability of implementations. Some other notable examples include:

- Multi-objective evolutionary algorithm based on decomposition (MOEA/D) (Zhang and Li 2007)
- Adaptive evolutionary algorithm based on non-Euclidean geometry for many-objective optimization (AGE-MOEA) (Panichella 2019)
- S-Metric Selection Evolutionary Multi-objective Optimization Algorithm (SMS-EMOA) (Beume et al. 2007)
- Multi-objective hypervolume-based ant colony optimizer (MHACO)

It is up to the design engineer to carefully consider the problem they hope to solve and apply the best algorithm based on the problem, the available toolset, and any other constraints or limitations.

Application of an EA to a Mission Portfolio Optimization

There is more than one way to setup and solve a portfolio optimization problem for an ME or SoS problem. This section covers an example with some notes on how a different application might deviate from this example. The problem used is intended to be high-level, synthesizing many lower-level problems, to illustrate how it could be scaled and modified to fit a variety of situations.

In this general example, we consider a portfolio of options that consist of changes to the current force structure of a mission (Figure 36.3). These binary options may consist of investment or divestment in new or existing platforms, integration or removal of new or existing systems, and potentially programmatic changes that have nonmaterial effects on the force structure, its supporting elements, or their impacts. The inclusion or exclusion of each investment and divestment in a portfolio provides the binary decision variables ($x_i \epsilon [0, 1] \; \forall i = 1, \ldots, Q$ for Q investments + divestments).

In our example, every potential change to the force structure has some associated cost, some of which may recoup dollars. The first and most basic constraint in this problem is one where the

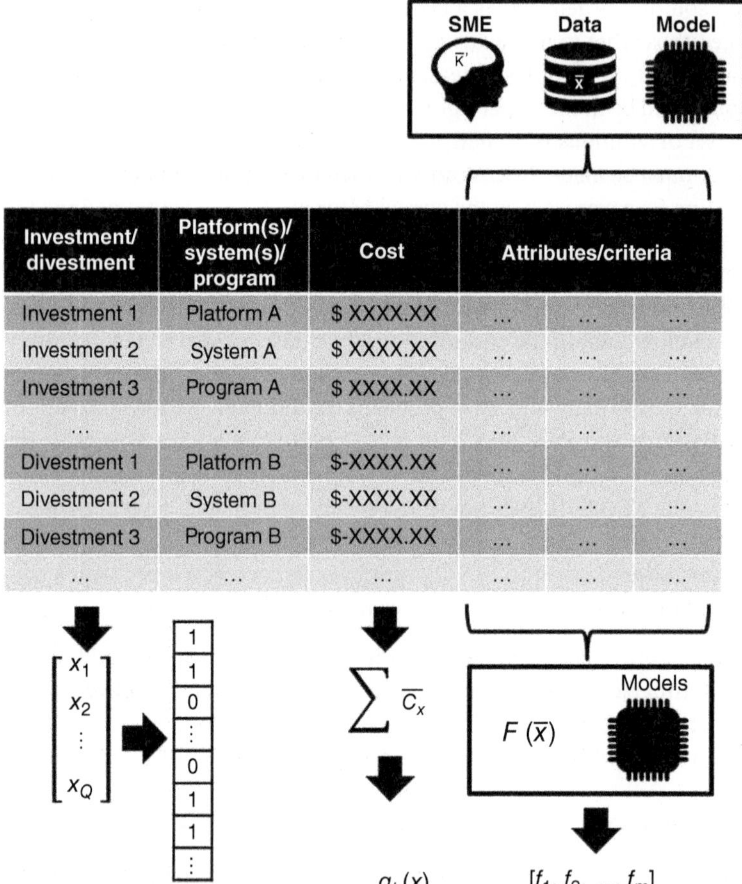

Figure 36.3 Example portfolio problem setup.

total cost of all selected changes must fall within some limits. Other constraints may be applied as specific to the problem at hand. Some example constraints may include:

- Constraints handling interoperability between platforms, systems, and programs.
- Constraints limiting selection of multiple investments that are scalar versions of one anther (e.g., invest in 5 new platforms of a given type or 10 new platforms, but not both).
- More detailed financial constraints concerning how investments are structured over time.
- Temporal constraints to reduce strain on simultaneous RDT&E and production of multiple investments.
- Constraints limiting the aggregate risk of combined investments/divestments.

Each change is associated with some attributes or criteria, which affect or measure that investment or divestment against stakeholder preferences. This may include individual mission capabilities, alignment to guidance, or any other metrics of interest. Given a subset of selected investments and divestments, each objective function may be calculated as a simple mathematical function of the known attributes, or they may require the use of a model (e.g., a campaign simulation). The exact setup would be particular to the specific problem being addressed as well as available tools, and the nature of the decision(s) being made.

A problem setup in this general manner, with available data, can then be solved utilizing an evolutionary algorithm. There are a number of different considerations that could be applied here based on the exact state of the problem, data, or how decision-makers want to utilize the results. Some examples of variations on the EA method are:

- Initial seeding of the algorithm could speed up results. These initial candidate portfolios could come from previous runs of the same (or a similar) problem. The candidates could also be specific portfolios that decision-makers want to explore.
- If the problem ends up heavily constrained, and solutions meeting all constraints are difficult to come by, a selection method that favors candidates that meet constraints could be utilized over a more exploratory random selection algorithm.
- If some objectives are exposed as being highly correlated, the number of objectives could be reduced to aid algorithm performance, reduce the population, and ease decision-making.

Other adjustments to the algorithm(s) utilized should be considered as results are analyzed. Careful consideration for "why" an algorithm produces the results it does is always a critical step in an optimization process. Providing detailed insight to stakeholders and decision-makers into the process and its results is one of the most important jobs of the analysts managing the portfolio analysis process. This leads to the importance of interactive visualizations to explore the results of these optimization algorithms, and give analysts and decision-makers various tools to understand why certain solutions are being presented by the algorithm(s) in addition to the proposed solutions themselves.

Combinatorial Design Space Analysis Tool

Once an evaluation framework is in place, and an optimization approach selected and implemented, it is possible to explore and analyze the combinatorial space in search for the right fit for the needs of the problem space. The "right" fit is determined based on a balance of capabilities, cost, and risk. Although the solution space is often unintuitive, proper visualization techniques can aid in understanding and facilitate analytical reasoning. The authors assert that no single toolbox or set of visualizations is universally applicable to portfolio optimization. However, there are common components and approaches that can be utilized in various integrated fashions to provide insight tailored to a given set of questions, stakeholders, and available data.

Visual Analytics

Visual analytics is the use of interactive visualizations to enable or advance the synthesis of information and derive insight from complex data (Thomas and Cook 2006). For example, in 1869 Charles Minard, a retired civil engineer, produced the following information graphic in order to tell the story of Napoleon's Russian Campaign of 1812 (see Figure 36.4).

Starting from the left in Figure 36.4 above the lighter-colored band depicts the size and location of Napoleon's army as Napoleon commenced his invasion of Russia. As the story unfolds Napoleon continues to rapidly push further into Russia in hopes of forcing the Russian army to battle. However, in each engagement the Russians retreat leaving the villages and crops burning, denying the French army the option of living off of the land. Without adequate food and supplies, the French army dwindles as they continue to march deeper into Russia. The dark band starting from the right and moving toward the left depicts the return trek of the French army and the continued suffering and losses.

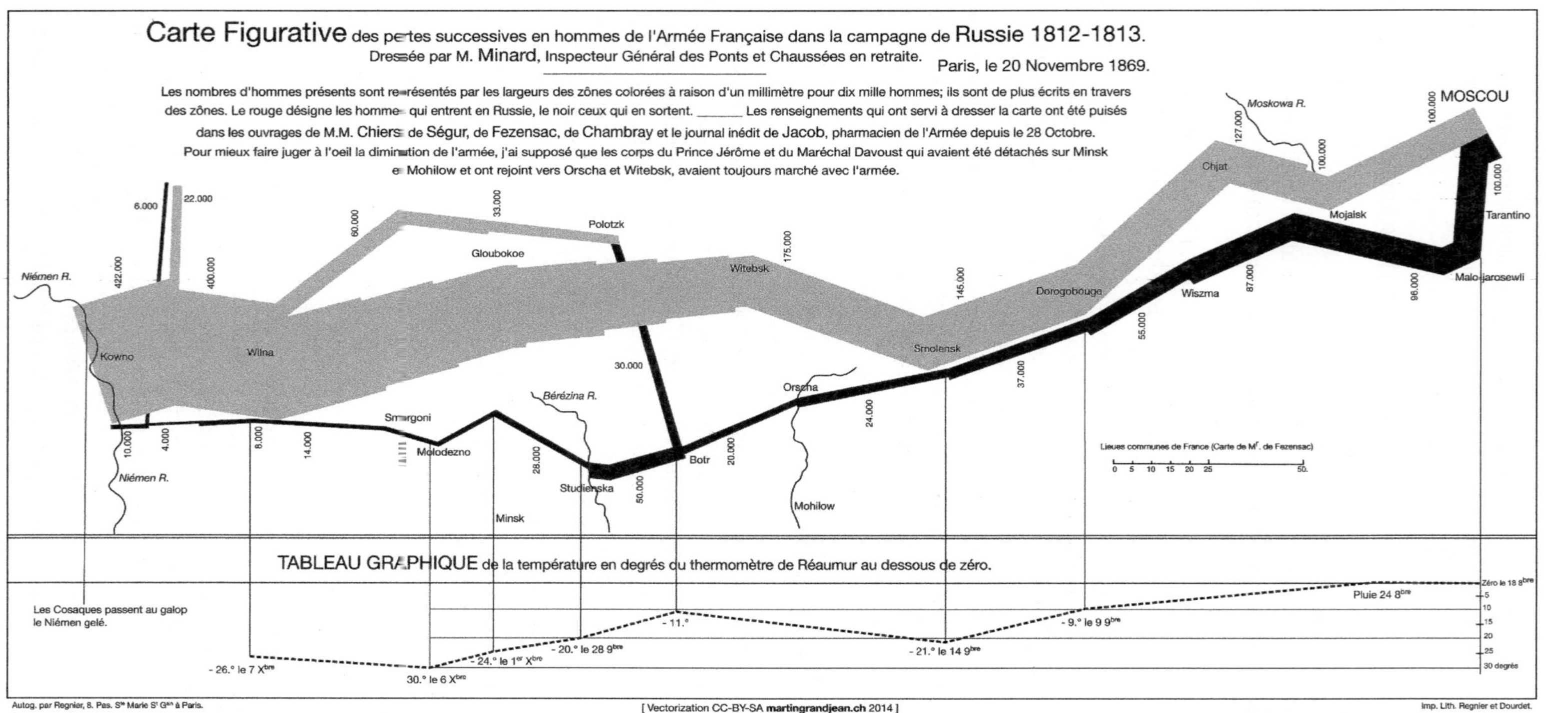

Figure 36.4 Modern redrawing of Napoleon's Russian Campaign of 1812 by Charles Minard. *Source:* Grandjean (2014).

The information graphic in this example is an effective storytelling visualization, and it is reasonable that a similar set of visualizations are able to tell the design and engineering story for the combinatorial design process. However, this narrative will require specific characteristics:

- It should provide insight into the interconnectivity (the relationships) of the components, functions, and capabilities;
- It should provide decision support;
- It should provide traceability through design decisions and decision impacts;
- It should provide a clear description of the final solution or set of candidate solutions.

Relationships

The first challenge in a narrative is to introduce the elements of the story in such a way that all of the relationships between the elements are understood. The fundamental elements in the combinatorial design process are components, functions, and capabilities as illustrated in Figure 36.5.

Components are the parts of a system that provide functionality. For example, a component could be a radio or an engine. Components also have attributes associated with them, such as mass or cost. Functionality is captured as a function, which is a specific process, action, or task that is performed. Capabilities represent a set of functions working together to meet needs. A system is an engineering solution to a set of needs. Systems are made up of interconnected components, which work together to provide capabilities. System attributes are derived from the aggregation of its component's attributes. A system architecture is a class of systems that utilize a specific set of components to achieve a set of capabilities. At times, systems must be combined together, into a System-of-Systems, in order to realize the end-goal capability set.

A problem formulation (e.g., LogIT) will have a specific set of needs, capabilities, functions, and components captured in machine- and human-friendly form (most often a spreadsheet). Checks are put in place to audit the data to help maintain integrity in the information being captured.

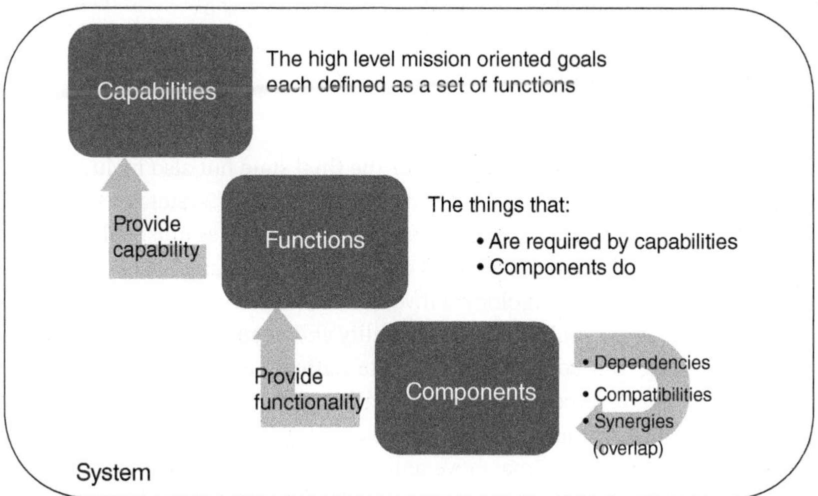

Figure 36.5 Elements of the SoS decomposition.

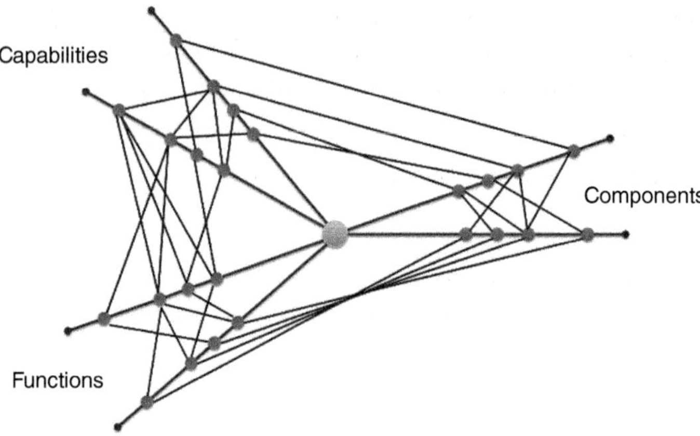

Figure 36.6 Example hive plot.

The resulting information will resemble a network graph with different types of nodes. While there are many layouts that could be used for drawing a graph it makes sense to utilize the position of the nodes with some intrinsic meaning in order to emphasize the relationships within the types of nodes and between the different types of nodes.

Figure 36.6 is an illustration of a hive plot, which groups the nodes by type and lays them out on a set of radial axes. The example illustrates how capabilities, functions, and components can be depicted. Each type of node (e.g., capabilities, components, and functions) has two axes assigned in order to allow the visualization of the relationships between the same types of node. Along each axis, a dot represents an item of this type; for this example, the location on the axis has no meaning. Lines between dots of the same element type could indicate a relationship such as one component requiring another or two components being incompatible (i.e., using a different color for the line). Lines between nodes of different types show relationships of one enabling the other. For example, links between components and functions identify the component(s), which satisfy that function, and similarly links between functions and capabilities identifies which functions are needed to realize a capability. The loop can be closed then, offering potential component solutions to meet a specific capability. As an interactive visualization, analysts can rapidly identify path dependencies between the desired capabilities and the specific components that can be integrated to realize that capability.

Notional Solution

The complete solution not only includes a representation of the final state but also includes each intermediate state in a clear plan of progression. Transformation of a complex system of systems is a gradual transition, and during each transition point, a core set of capabilities must still be available. As the duration of a project increases the likelihood of the solution changing also increases, particularly where relatively fast evolving technology is involved.

A true solution for such a project must embrace the dual reality and needs to map out a solution from beginning to end and enable the consideration of alternatives throughout its duration. Figure 36.7 is an illustration of how the process might be visualized.

The the darkest point on the right represents the currently desired system based on information collected from stakeholders. The adjacent points above and below represent uncertainty in that desired system. There is recognition that there is some uncertainty such as in the understanding of the requirements or that the need may change or evolve over time. The remaining light and dark

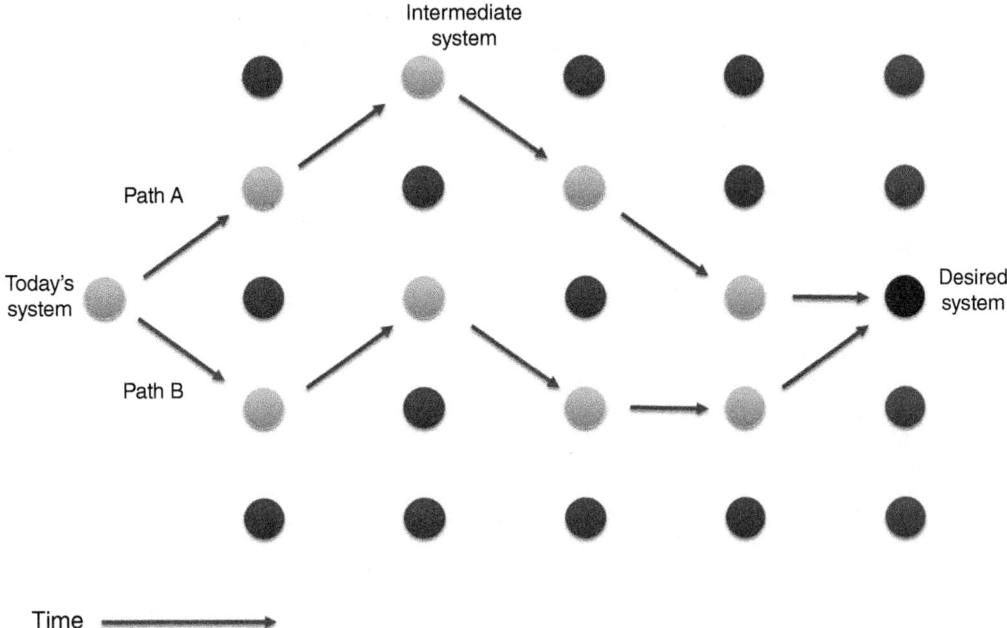

Figure 36.7 Portfolio design and change paths over time.

points, specifically those along Paths A and B, represent intermediate states of the portfolio. Because of the processes involved in DoD acquisition or limitations from funding cycles, it is often the case that a portfolio cannot be transformed in a single time step, but rather requires incremental modifications where each intermediate state is offering the base set of required capability while also moving toward the desired state. There are many intermediate states, which could be implemented; the light points represent those that fall along candidate paths of transition.

Visualizing Relationships

The implementation explored in this section is a framework designed for solving combinatorial design problems using a graph representation of the relationships between components, functions, and capabilities. The framework is developed in the Python programming language using a Jupyter Notebook. The notebook paradigm provides a convenient development and rapid prototyping environment with access to both the strong analytical capabilities of the programming language and a strong integration of a web-based visualization capability. The Notebook is an effective user interface allowing the developed software to be packaged as a Python module for easy portability and extensibility toward a broad range of problems. This provides a solution to the desired state of reusable components that can be rapidly integrated for a problem-specific portfolio management analysis capability.

Often at the system level, engineering problems can be represented as a graph of multiple types of nodes and edges. Where the nodes represent various parts (whether physical or abstract) of the system and the edges represent the relationships between the parts. While these graphs can be difficult to visualize and analyze for decision-making purposes, graphs are an effective representation for both computer and human consumption. The information the graphs are based on are the products of the research and experiences of SMEs.

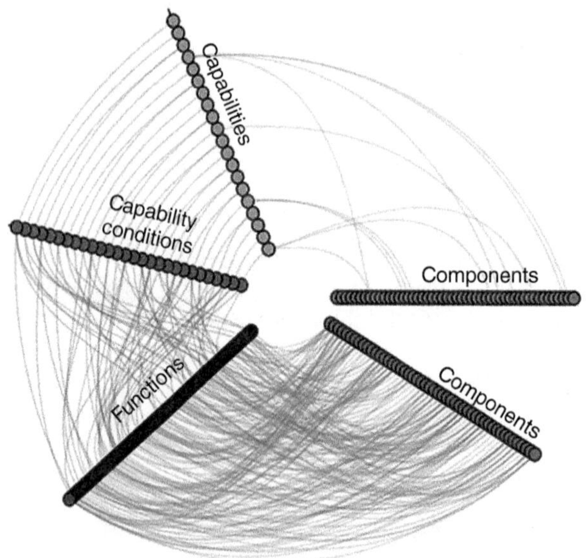

Figure 36.8 Example hive plot generated with D3 JavaScript library.

In the example graph in Figure 36.8, the different types of nodes (component, function, and capability nodes) are arranged in a hive chart as follows:

- The various nodes present components, functions, capabilies, and the various sets of conditions for which capabilites can be obtained.
- The edges between the component and function nodes represent the relationship between components and functions, meaning a component offers a function;
- The edges between the function and capability condition nodes represent the relationships between functions and the sets of conditions for obtaining capabilities;
- The edges between the capability condition and capability nodes represent the relationships between the sets of conditions and the actual capabilities;
- The edges between the capability and component nodes represent the cases where a capability can be directly obtained through the use of a single component;
- The unused space between the different axes of component nodes is a placeholder for capturing dependency and compatibility information between components.

The hive chart technique is an effective way of visually organizing the nodes in the graph in order to emphasize the types of relationships that exist between the different types of nodes. In addition, the hive technique coupled with interactive animations makes it a simple task to trace which combination of components produces which capabilities. However, the number of possible combinations is intractable for a typical number of components. Assuming that the system must have at least one component, then if there are 50 available components, there are $2^{50} - 1 = 1{,}125{,}899{,}906{,}842{,}623$ unique system combinations.

The data used in this example are notional, consisting of generic components, functions, and capabilities. The framework is written based on the assumption that all of the data are real and are accessible from an appropriate authoritative source (i.e., spreadsheets, data files, or a database). However, given the nature of research and the availability of data, it is advantageous to include support for generating generic placeholders. A few of the features of the generic data generator are as follows:

- Data are available to support the development of the framework;
- Data characteristics may be changed to design a more robust framework;
- Real and generic data may be mixed to support a partial MRDB during project execution.

Given the number of possible unique systems and the nonlinear nature of the domain space, a search algorithm (e.g., random combination generator, genetic algorithm, simulated annealing, etc.) cannot offer any guarantees as to the optimality of its results. However, rather than searching with the goal of finding the proverbial "needle in the haystack," this approach advocates searching for the purpose of learning and characterizing patterns with respect to design decisions associated with the components and component interactions.

Consider for a moment which sample points are the most valuable in terms of providing the information needed to answer the following questions:

- Which components complement each other?
- How close in terms of cost and lead-time is the next capability?
- Which components or combinations of components provide the most value?

A sample population of the over one quadrillion unique possible portfolios is necessary to gain insight useful for decision-making. Use of a value calculation to identify the relative "goodness" between two different portfolio solutions offers a means to understand a sampling of the vast tradespace. In this example, a modified genetic algorithm is utilized to search the options to produce a sample of 1000 good solutions. Since the goal is not to produce an optimal solution the ideal search results should show

- Good space exploration with visual variability
- Stable overall trends over multiple runs
- Quick run time

The "Visualizing Portfolio Solutions" section will provide examples of visualizations for comparing alternatives with the integration of a value calculation with a search or optimization algorithm.

Visualizing Portfolio Solutions

Distilling a vast tradespace into a digestible visualization or small set of visualizations is a key challenge in portfolio analysis. This section will present a few examples utilized in past efforts to provide ideas and insights into considerations for successful approaches.

In the first example provided in Figure 36.9, the general problem being explored was the development of a new system composed almost entirely of off-the-shelf components. The goal of the optimization was to understand what level of key requirements was achievable from a performance perspective across four key capabilities necessary for the system. In this visualization, the left side scatterplot provides thousands of individual alternative portfolio solutions, assessed against the estimated cost and an aggregated performance metric. The Pareto frontier is the upper-left portion, where solutions are maximizing performance while minimizing cost. The visualization on the right is a set of histograms that provide the quantity of use of specific component alternatives to meet a required function. For each function, there were anywhere from 2 to 20 potential component alternatives. These two visualizations were coordinated,

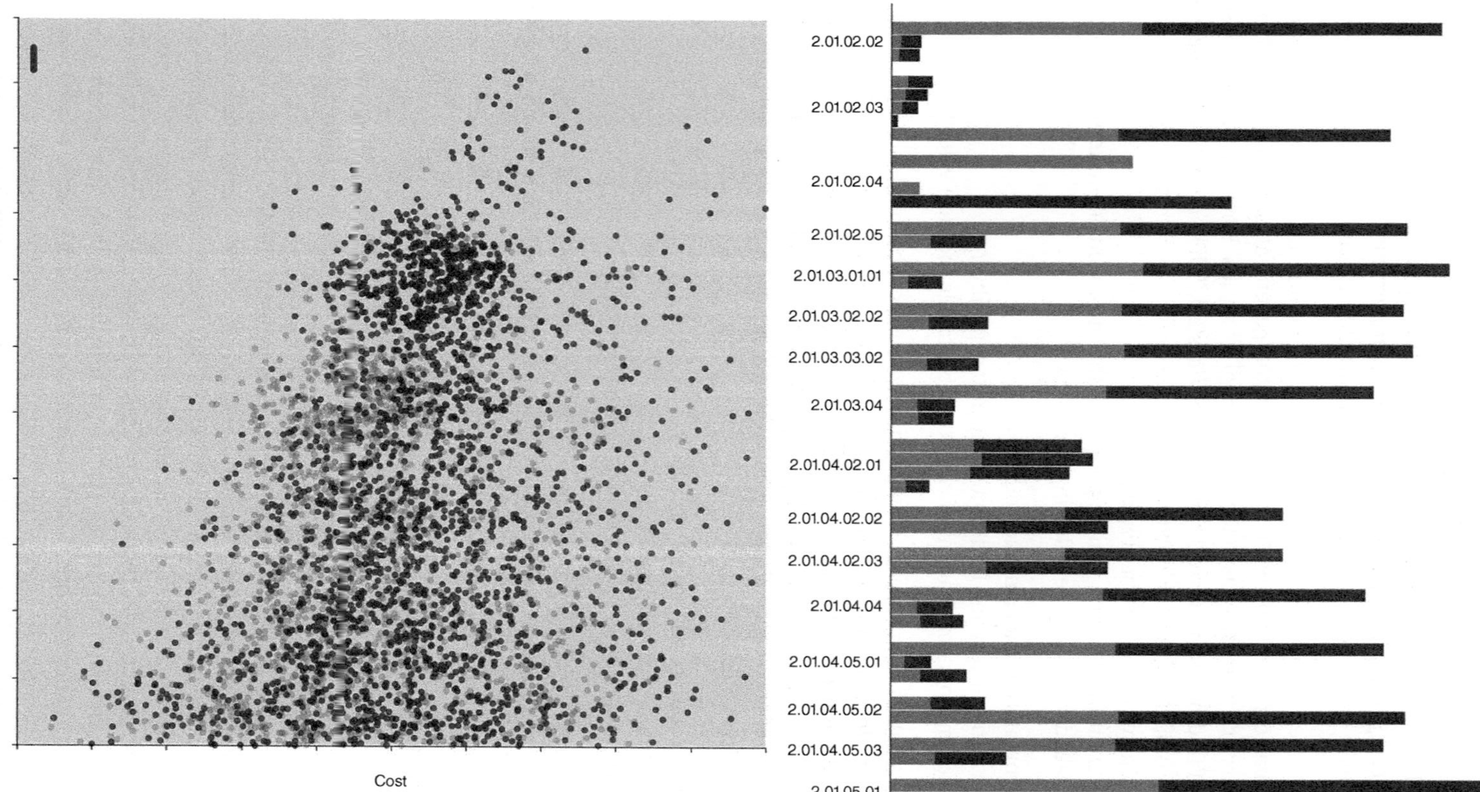

Figure 36.9 Coordinated scatterplot and histogram providing a combined perspective on cost, performance, and frequency of use of components.

meaning if a histogram bar were selected on the right, indicating the forced selection of a specific component, the scatterplot displayed solutions with that component as darker and other solutions without that component as lighter. Similarly, a lasso filter could be used on the left to highlight regions of interest (such as the Pareto frontier), which resulted in filtering the histograms into seperate regions indicating which components were utilized within the subset of selected alternatives.

With this visualization, decision-makers were able to assess the feasibility to achieve a desired performance for a given cost, while also understanding which individual components most frequently were utilized in the simulated solutions. Understanding the common use of components provided insight into the resilience of a design to survive a component change during integration. Meaning, if the specific component selected to meet a certain function could not be integrated, were other components available to replace it and what was the impact to cost and performance based on that change.

Another example of an integrated dashboard used to interactively explore sets of portfolios is presented in Figure 36.10. The dashboard presented is a sanitized version of real dashboard used to identify suitable portfolios of systems (generalized to Something 1 through 25) as they reduce the risk of failing to accomplish a task in four scenarios under two different conditions. At the center of the dashboard is the item selection controls. These allow users to select/deselect specific elements or items (i.e., Something 1 through 25) from the portfolio by clicking on the checkboxes. To the left of the item name is the quantity of items that the user expects to acquire by a given date. Although the details of how the quantity is used are not completely relevant to illustrate the concept, it is worthwhile to mention that they are used for two purposes: (1) to scale the impact of a selected item, and (2) to assess the cost of the portfolio. This was a means to include capacity concerns in the portfolio assessment. To the right of the checkboxes is the optimality indicator (i.e., the fraction on Pareto frontier) for each item. The details of what that is will be described below when the rest of the dashboard and its functionality is further explained.

To the left of the item selector is the risk matrix. The risk matrix contains eight risk items representing four scenarios (differentiated by color) and two conditions (differentiated by shape). If no item is selected, the risks default to their "baseline" value. When items are selected, capability gaps for a given scenario and condition change. This dashboard depicts this by reducing the impact of the consequence (shape moves down on the vertical axis). As even an unlimited budget cannot eliminate all risk, risk reduction is modeled as relative to initial risk position. As such, for a given improvement afforded by a portfolio of investments, change is more dramatic in scenarios where risk was initially higher. This reduction in risk can be quantitatively aggregated to a single number by assigning a total risk value that is correlated to a color depicted in the background of the risk cube (i.e., red, orange, yellow, green). For example, reducing the risk in Scenario 3 Condition 1 has a bigger impact than reducing the risk of Scenario 3 Condition 2 even though the distance is almost identical because Condition 2 is less likely to occur (i.e., is to the left of Condition 1). This aggregation into a single value as depicted in the "Total Risk Reduction" box on the lower right of the dashboard.

The final element of the dashboard is the portfolios scatterplot. The scatterplot depicts portfolio combinations created by the model. An ideal (theoretical) portfolio would reduce the risk by 100% and cost US\$0, and exist in the upper left corner of the scatterplot; conversely, the worst portfolio exists in the opposite corner. Between those corners exist more than 30 million portfolio options, with the optimal set being those that are not "dominated" by another portfolio. Dominated can be defined as not having another portfolio produce the same risk reduction for

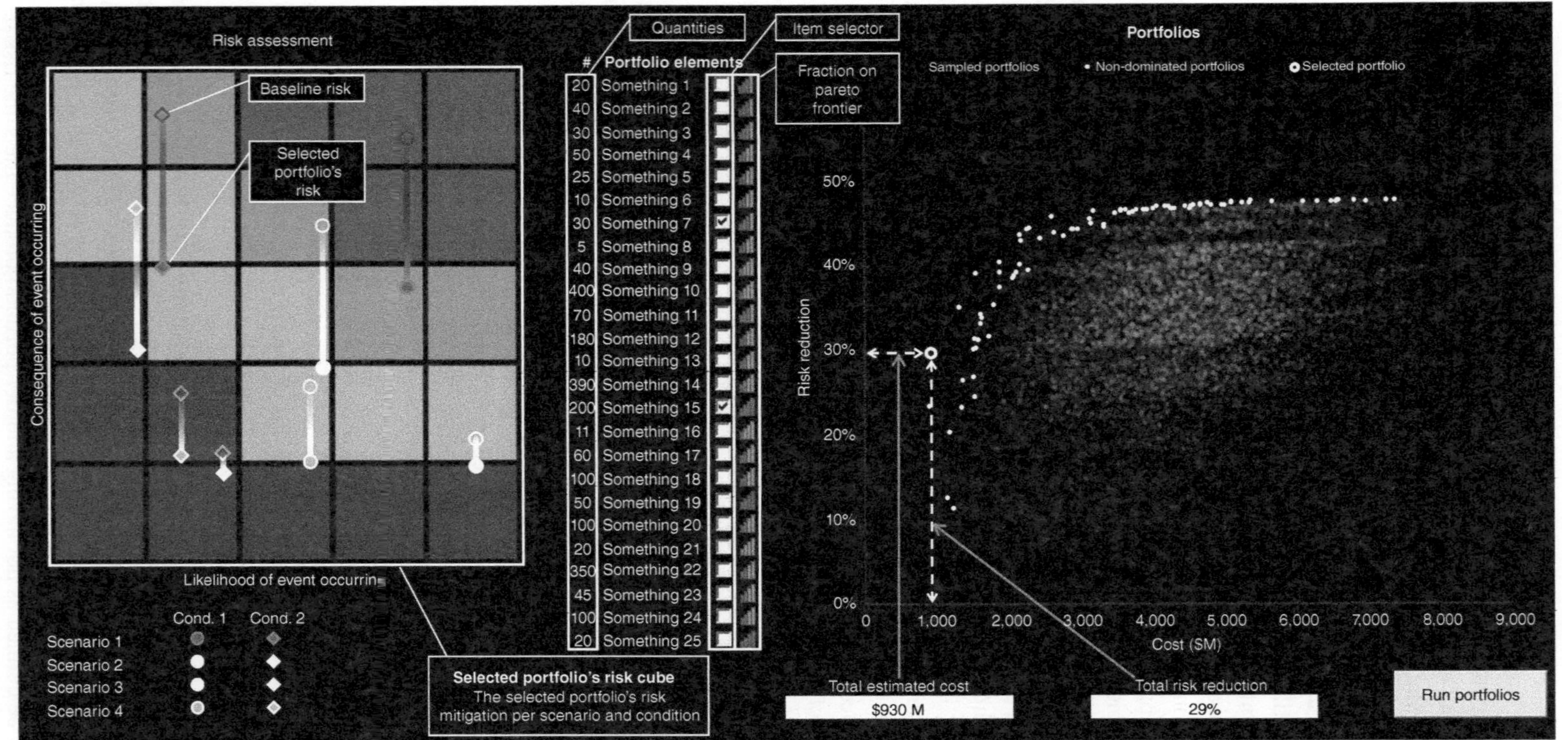

Figure 36.10 A sanitized risk-based assessment dashboard for portfolio analysis.

the same or lower cost. The brighter points in the scatterplot depict those non-dominated points, this set of non-dominated points is often referred to as the "Pareto frontier." The Pareto frontier allows a decision-maker to understand the trade-off between two or more requirements, in this case risk reduction and cost. It is important to highlight that the sampled portfolios represent a small set (~0.01%) of the possible portfolios that could be created. Nonetheless, this small sample can help identify items that are more prevalent in the set of non-dominated solutions. This is quantified by the bars (signal strength) next to the item selector check boxes (previously referred to as the "optimality indicator"). Full bars indicate that the majority of the Pareto optimal portfolios include that given item; conversely, empty bars indicate that the given item almost never exists in the set of Pareto optimal portfolios. In essence, this is a mechanism that supports portfolio generation using the principles of Set Based Design (SBD). By definition SBD does not mandate a specific solution, but provides information for a decision-maker to make an informed decision from a solution set of possibilities. Using this indicator, decision-makers can assemble their own portfolios and assess how the individual risks in a set of given scenarios and conditions are mitigated. For the purpose of comparison, the selected portfolio is depicted in the scatterplot as an empty orange circle. This portfolio may align with a sampled portfolio, or not, and it is entirely feasible that decision-makers can make portfolios that outperform the sampled Pareto frontier because the tool generates 4,096 (i.e., 2^{12}) random portfolios of a total of 33,554,432 (i.e., 2^{25}) possible portfolios, i.e., this represents a mere 0.012% of all possible portfolios. This dashboard is an example of how a stakeholder can use a portfolio modeling framework to explore portfolios, and an optimizer could have been integrated to generate solutions that pushed the Pareto frontier closer to the upper left corner.

A final example assessment dashboard is provided in Figure 36.11. This example is a sanitized version of a toolset, which supports the evaluation of changes to a five-year investment budget. The image on the left shows a parallel coordinates plot. Each of the vertical axes is a metric of interest to assess a portfolio against. For this example, the metrics are divided into three groups: Guidance (that is, how aligned is the portfolio with guidance from higher levels of leadership within an organization), Performance (how well does the portfolio evaluate against key capabilities it is intended to offer), and Cost (is the proposed portfolio increasing or decreasing the budget for each of the five years or in total across that period). Each of the lines is a complete portfolio: a combination of investments and divestments then evaluated against those dimensions of interest and plotted across the vertical axes. An ideal portfolio would be along the top of the Guidance and Performance regions, while at or below the $0 line in the cost region. Such a portfolio would indicate all possible improvement is achieved for no additional cost, or a cost savings.

The image on the right is a tornado chart. In essence, it is the same as a histogram; however, it provides the added benefit of quickly identifying a component as an investment (a bar to the right) or a divestment (a bar to the left). These two visualizations are coordinated, in that a user can filter the population of portfolios against any of the metrics displayed on the parallel axis chart, and the tornado chart automatically updates. Likewise, an element on the tornado chart can be "locked-in" or "locked-out" to reduce the space of solutions on the left. Like with the other examples, the tornado chart is inspired by the use of SBD, with the goal of understanding what components (here, investments, and divestments) are commonly found in "good" solutions. The last element of the visualization is the box-and-whisker plot above the tornado chart. This element shows the total variation in costs of the displayed portfolios. In this example, the portfolios range in cost from US$137M to US$216M, with respect to the total cost of the investments and divestments across the five-year period.

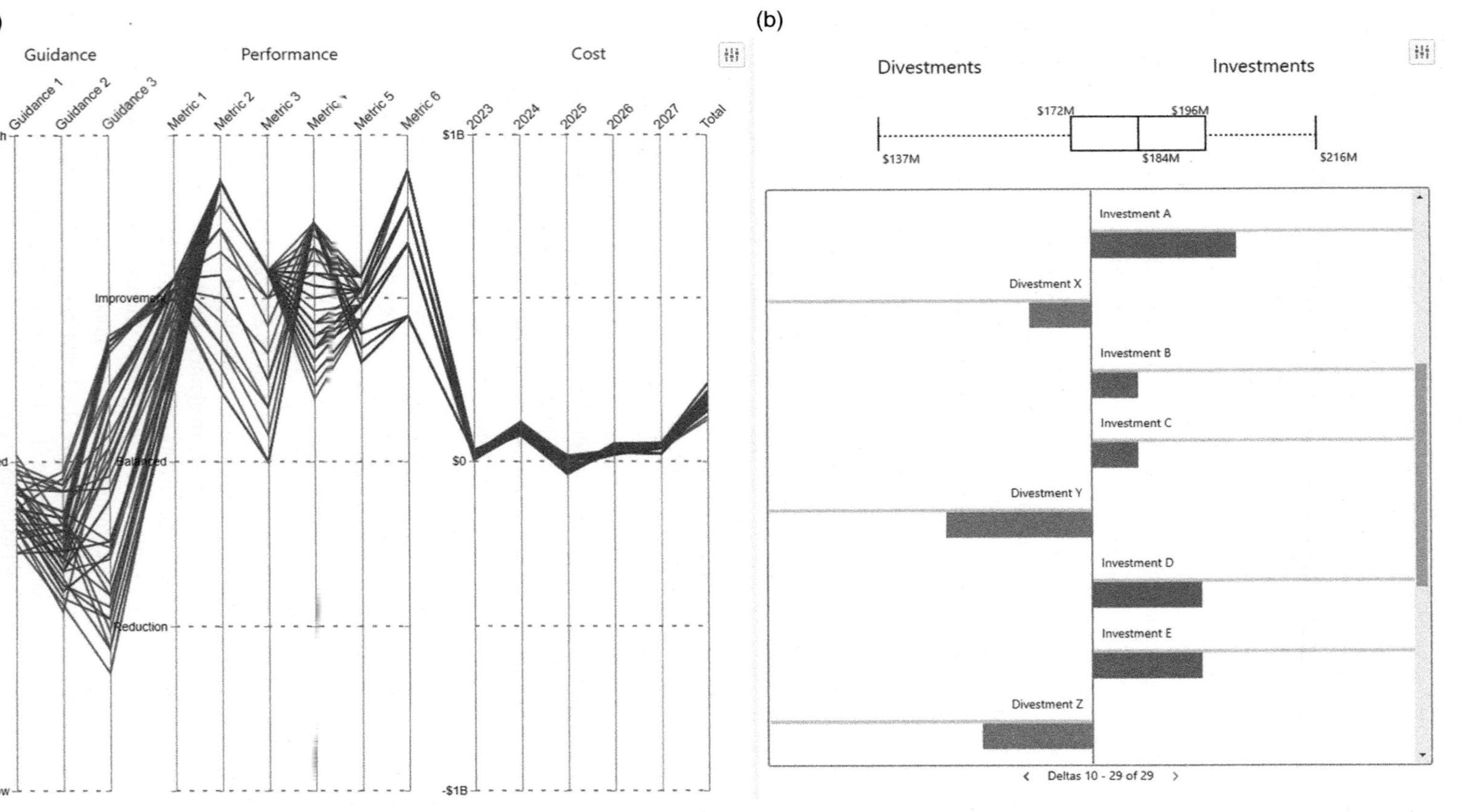

Figure 36.11 Coordinated parallel axis (a) and tornado chart (b) for assessing a portfolio.

Conclusions and Key Takeaways

This chapter provided an overview of the generic process for evaluating a portfolio or system of systems. Because portfolio analysis problems rapidly become intractable, this chapter provided a thorough review of search and optimization approaches and algorithms to explore these tradespaces in order to gain insight and help decision-makers better-informed and defensible decisions. There is no universal best approach to portfolio analysis, and depending on the specifics of the problem being addressed and the constraints imposed on the analysis team, different approaches will tend to provide a better portfolio analysis. Finally, this chapter highlighted the importance of visualization techniques to transform the vast amounts of data (that are often generated when assessing nontrivial portfolios) into actionable knowledge. Different examples of interactive visualization techniques were provided with descriptions of the problem sets they were used on to illustrate the art of the possible and inspire the practitioner to pursue their own hybridization of methods and techniques. The following subsections summarize the key takeaways from this chapter for the practicing systems engineer to consider.

Reusable Tools

Each problem set has its own unique set of requirements, therefore there is no one tool to be used across portfolio analysis problems. However, the goal is to ensure that analysts are afforded the opportunity to maximize value-added activities. Whenever possible, analysis should seek to have reusable components that can be integrated into different arrangements based on a problem's specific needs.

A critical set of tools to reuse are those used to generate feasible and desirable solutions. The authors have gravitated to using MOEAs, more explicitly NSGA-III, as they provide a high degree of flexibility and can be scaled to solve a wide variety of problems. That is not to say that other approaches may not be suitable to the reader's needs, and the authors encourage the reader to familiarize themselves with SAT solvers and MADM techniques if MOEAs are not proving to be fruitful for their needs.

Iterative Process

For most problems, a framework must be developed to (1) capture the stakeholders' needs, (2) characterize the potential solutions, and (3) integrate solution generation and exploration tools. The process must be iterative as portfolio analyses often share characteristics of wicked problems. Furthermore, stakeholders must be involved in this iterative process, so their mental models of the problem and potential solutions are improved as the analysis evolves. Conversely, the analysis should be directed by the stakeholders and SMEs' understanding; it is therefore critical that analysts, SMEs, and stakeholders work hand-in-hand through the process. Data elicitation must be done efficiently and effectively, it is possible to leverage SME inputs (qualitative data) as well as data from models and simulations (quantitative data) but the best way to integrate these is not only problem dependent, but heavily reliant on resources available (e.g., data, models, SMEs) and their applicability to the problem. Finally, the framework must be vetted, although formal VV&A processes are often not feasible within the scope of most problems, a vetting process should be established and documented.

Understand the Outputs

Ultimately, any portfolio analysis is intended to help decision-makers make better-informed decisions that they can defended. For this purpose, it is imperative that any portfolio analysis includes a significant level of effort to assess the solutions proposed, and that the patterns produced by the best performing solutions are understood. Sometimes the end result may be self-evident after the fact, other times the myriad of interactions may not be clearly traceable. At the very least, the portfolio analysis team should reserve sufficient time in the project's timeline perform a thorough final assessment of the best portfolio solutions identified.

References

Azmi, R. and Tamiz, M. (2010). A review of goal programming for portfolio selection. In: *New developments in Multiple Objective and Goal Programming* (ed. D. Jones, M. Tamiz, and J. Ries), 15–33. Berlin, Heidelberg: Springer.

Best, M.J. (2010). *Portfolio Optimization*. CRC Press.

Bertsimas, D. and Sim, M. (2004). The price of robustness. *Operations Research* 52 (1): 35–53.

Beume, N., Naujoks, B., and Emmerich, M. (2007). SMS-EMOA: multiobjective selection based on dominated hypervolume. *European Journal of Operational Research* 181 (3): 1653–1669.

Coello, C.A.C., Lamont, G.B., and Van Veldhuizen, D.A. (2007). *Evolutionary Algorithms for Solving Multi-objective Problems* (ed. D.E. Goldberg and J.R. Koza). New York: Springer.

Das, I. and Dennis, J.E. (1998). Normal-boundary intersection: a new method for generating the Pareto surface in nonlinear multicriteria optimization problems. *SIAM Journal on Optimization* 8 (3): 631–657.

Davendralingam, N. and DeLaurentis, D. (2013). A robust optimization framework to architecting system of systems. *Procedia Computer Science* 16: 255–264.

Deb, K., Pratap, A., Agarwal, S., and Meyarivan, T.A.M.T. (2002). A fast and elitist multiobjective genetic algorithm: NSGA-II. *IEEE Transactions on Evolutionary Computation* 6 (2): 182–197.

Deb, K. and Jain, H. (2013). An evolutionary many-objective optimization algorithm using reference-point-based nondominated sorting approach, part I: solving problems with box constraints. *IEEE Transactions on Evolutionary Computation* 18 (4): 577–601.

Grandjean, M. (2014). Historical Data Visualization: Minard's map vectorized and revisited. http://www.martingrandjean.ch/historical-data-visualization-minard-map (accessed 01 August 2023).

Jain, H. and Deb, K. (2013). An evolutionary many-objective optimization algorithm using reference-point based nondominated sorting approach, part II: handling constraints and extending to an adaptive approach. *IEEE Transactions on Evolutionary Computation* 18 (4): 602–622.

Hwang, C.L. and Masud, A.S.M. (2012). *Multiple Objective Decision Making—Methods and Applications: A State-of-the-Art Survey*, vol. 164. Springer Science & Business Media.

Navon, A., Shamsian, A., Chechik, G., and Fetaya, E. (2020). Learning the Pareto front with hypernetworks. *The Ninth International Conference on Learning Representations*. https://doi.org/10.48550/arXiv.2010.04104.

Marler, R.T. and Arora, J.S. (2004). Survey of multi-objective optimization methods for engineering. *Structural and Multidisciplinary Optimization* 26: 369–395.

Mavrotas, G. (2009). Effective implementation of the ε-constraint method in multi-objective mathematical programming problems. *Applied Mathematics and Computation* 213 (2): 455–465.

Mueller-Gritschneder, D., Graeb, H., and Schlichtmann, U. (2009). A successive approach to compute the bounded Pareto front of practical multiobjective optimization problems. *SIAM Journal on Optimization* 20 (2): 915–934.

Panichella, A. (2019). An adaptive evolutionary algorithm based on non-Euclidean geometry for many-objective optimization. In: *Proceedings of the Genetic and Evolutionary Computation Conference* (ed. M. López-Ibáñez), 595–603. New York: Association for Computing Machinery.

Przybylski, A. and Gandibleux, X. (2017). Multi-objective branch and bound. *European Journal of Operational Research* 260 (3): 856–872.

Rittel, H. and Webber, M. (1973). Dilemmas in a general theory of planning. In: *Policy Sciences*, vol. 4, 155–159. Amsterdam: Elsevier Scientific Publishing.

Rosenthal, R.E. (1985). Concepts, theory, and techniques principles of multiobjective optimization. *Decision Sciences* 16 (2): 133–152.

Santiago, A., Huacuja, H.J.F., Dorronsoro, B. et al. (2014). A survey of decomposition methods for multi-objective optimization. In: *Recent Advances on Hybrid Approaches for Designing Intelligent Systems* (ed. O. Castillo, P. Melin, W. Pedrycz, and J. Kacprzyk), 453–465. Cham: Springer. https://doi.org/10.1007/978-3-319-05170-3.

Section 809 Panel (2019). Report of the Advisory Panel on Streamlining and Codifying Acquisition Regulations. Streamlining and Codifying Acquisition, Volume 3 of 3.

Thomas, J.J. and Cook, K.A. (2006). A visual analytics agenda. *IEEE Computer Graphics and Applications* 26 (1): 10–13.

Von Winterfeldt, D. and Fischer, G.W. (1975). Multi-attribute utility theory: models and assessment procedures. In: *Utility, Probability, and Human Decision Making: Selected Proceedings of an Interdisciplinary Research Conference*, Rome (3–6 September 1973) (ed. D. Wendt and C. Vlek), 47–85. Dordrecht, Netherlands: Springer. https://doi.org/10.1007/978-94-010-1834-0.

Zanakis, S.H., Solomon, A., Wishart, N., and Dublish, S. (1998). Multi-attribute decision making: a simulation comparison of select methods. *European Journal of Operational Research* 107 (3): 507–529.

Zhang, Q. and Li, H. (2007). MOEA/D: a multiobjective evolutionary algorithm based on decomposition. *IEEE Transactions on Evolutionary Computation* 11 (6): 712–731.

Zhou, A., Qu, B.Y., Li, H. et al. (2011). Multiobjective evolutionary algorithms: a survey of the state of the art. *Swarm and Evolutionary Computation* 1 (1): 32–49.

Biographical Sketches

Dr. Frank Patterson is a senior research engineer at the Georgia Tech Research Institute (GTRI) in the Systems Engineering Research Division (SERD). His current research includes the application of state of the art of computational methods and tools to the design and analysis of complex systems. He is also experienced in the development, integration, and use of multi-disciplinary simulation, modeling, and analyses for the design of systems under uncertainty. He has over 15 years of experience supporting the DoD and warfighter as an engineer across various domains. Dr. Patterson earned his Bachelors, Masters, and PhD in Aerospace Engineering at Georgia Tech.

Dr. David Fullmer is a senior research engineer and head of the applied decision systems branch (ADSB) in the Systems Engineering Research Division at the Georgia Tech Research Institute. ADSB conducts DoD sponsored research in the design of decision frameworks, utilizing modern decision and analysis approaches including machine learning algorithm. He leads the Navy's Force Level Integration (FLINT) effort that provides a structured tradespace analysis approach for the Navy's integrated budgeting process, with custom developed software hosted on SIPRNet. His research focuses on enhancing the underlying models informing such tradespace analysis tools to provide better information to decision-makers.

Daniel C. Browne is a principal research engineer and chief of the Systems Engineering Research (SER) Division at the Georgia Tech Research Institute. SER Division conducts research in the design and application of best practices in model-based systems engineering (MBSE) and open architectures, transition of DoD acquisition to a digital engineering approach, decision methods development, and application of strategic decision systems. Danny's current research focuses on the application of MBSE and decision systems to support DoD acquisition and portfolio tradespace analysis.

Dr. Santiago Balestrini-Robinson is a senior research engineer in the Systems Engineering Research Division at the Georgia Tech Research Institute (GTRI). His primary area of research is the development of opinion-based and simulation-based decision support systems for large-scale system architectures. Dr. Balestrini-Robinson has led teams supporting multibillion-dollar military capital equipment acquisition programs. His interests in quantitative modeling and simulation span multiple domains, ranging from large-scale military campaign level analyses to the development of "peace-time" strategic-level cyber defense models, as well as humanitarian aid and disaster relief analyses. In conjunction with quantitative modeling and simulation-based analysis, he has developed novel opinion-based decision support tools. These tools enable decision-makers to collaboratively explore a decision space using an end-to-end, traceable, and defensible integrated and interactive visual decision support system to explore multi-level, multi-domain tradeoffs that leverage single or federated sources of truth.

Chapter 37

Assessing Benefits of Modularity in Missions and System of Systems

Navindran Davendralingam[1], Cesare Guariniello[2], and Lu Xiao[3]

[1] *Research Science, Amazon Inc., Seattle, WA, USA*
[2] *School of Aeronautics and Astronautics, Purdue University, West Lafayette, IN, USA*
[3] *School of Systems and Enterprise, Stevens Institute of Technology, Hoboken, NJ, USA*

Introduction

The US Department of Defense (DoD) has recognized the need to transition to a culture of performance and affordability. This includes prioritization of speed of delivery, continuous adaptation, and frequent upgrades. These objectives are supported by various DoD initiatives, including leveraging a modularization strategy via the modular open system approach (MOSA) – a defense acquisition initiative (and requirement in law for major programs) to encourage the adoption of modularization and of open architectures.

However, current MOSA guidelines provide limited insight into:

1) **The "What"**: The specific potential benefits of modularity and openness, the conflicting priorities of the diverse set of stakeholders involved, and the complex interdependencies that exist between the technical and business elements of modular system acquisitions;
2) **The "How"**: Which levers to play, and decision problems to solve, to realize the benefits of modularity and openness throughout the "MOSA ecosystem";
3) **The "Why"**: How can programs improve their evidence for specific MOSA implementation

While the above three points illustrate challenges relative to defense oriented MOSA implementation, the concerns are domain agnostic. Modularization is an intuitive approach for dividing a complex system into a manageable set of connectable components. However, as indicated above, relevant decision-makers require the means to rationalize modularization decisions across multiple verticals within a complex system (or system of systems). Prior research funded under auspices of the US Department of Defense Systems Engineering Research Center (US DoD SERC), were conducted to: (1) understand the benefits and drawbacks of modularity from prior literature and community driven knowledge, and (2) establish a decision support framework to enable objective assessment of modularity benefits using a multi-pronged approach that marries both quantitative and qualitative considerations of decision-making. Sections of this chapter are taken from a resultant series of publications from relevant SERC funded works. In the section entitled "Benefits of Modularity", we provide an overview on the motivating advantages of modularity, and potential drawbacks.

Systems Engineering for the Digital Age: Practitioner Perspectives, First Edition. Edited by Dinesh Verma.

Companion website: www.wiley.com/go/verma/systemsengineering

Benefits of Modularity

The following section on benefits of modularity is adapted from JDMS article.

Managing Complexity

Consider, for example, the development of vehicles by automobile manufacturers. The need to innovate and manufacture vehicles on short cycles has driven most large car manufacturers to a modularized strategy, where each module of a vehicle may be manufactured at a different geographic location. The modularization also involves compartmentalization of operations, logistics, and other such elements of the organization, much in accordance with Conway's Law (Conway 1968, 2019), which states that "any organization that designs a system will inevitably produce a design whose structure is a copy of the organization's communication structure." Such modularization makes the daunting task of designing, manufacturing, and delivering vehicles a manageable endeavor.

Parallelization of Development

The decomposition of a vehicle into subcomponents allows for more efficient innovations in the production phases as well, since teams can now concentrate on a more focused set of functions. In addition, the need for the mass production of vehicles has spurred new innovations in modular supply chain management strategies to keep up with the pace of both production and product innovations in the vehicle industry. This is true across many industries, especially in the case of computer software and hardware development. Modularization of computer components, even since the development of IBM's system/36024 has resulted in an exponential increase in computer technology innovations, due to the ability of both the consumer and producer to pick and choose different modules in both the use and development of parts of the computer system. The adaptability presented through modularization increased the likelihood of customer upgrades, which promoted further revenues and thus spurred further innovation in the computer industry. More recently, TARDEC has adopted an open architecture and modular approach toward the development of a Modular Active Protection Systems (MAPS), where the approach involves the use of a common controller in tandem with the best off-the-shelf active protection systems (APS) available. The common controller is developed by Lockheed Martin (Bethesda, MD) and utilizes a processor technology that is common across other Lockheed Martin products related to their airborne fixed-wing programs; the commonality of the product was described as a very pleasant surprise to the company (Judson, 2015), given the function of the processor as a new standalone product for TARDEC.

Hedge for Future Uncertainty

Much like the biology analogy provided earlier by Clune et al. (2013), the ability to adapt to uncertainty and change is made easier through modular design. For example, the Littoral Combat Ship (LCS), a US Navy program that produces ships with an interchangeable set of mission packages for operations in littoral water zones. The LCS modules consists primarily of Mine Counter Measures (MCM), Anti-Submarine Warfare (ASW), and Surface Warfare (SUW) packages; through an interchange of physical modules, with a common sea frame, the LCS can be refitted for various operational conditions, based on the modularity of both the physical and operational elements of the mission at hand. Modularity also accommodates uncertainty by enabling design resilient systems, which are robust to evolving environments, technological obsolescence, and uncertain financial conditions. In a modular architecture, updates may be necessary only in individual modules or clusters of modules, rather than involving the whole entity, with consequent economic and temporal advantages.

Promote Economic Benefits

The development of modular components, through competitive contracting across vendors, has brought about improved cost savings due to accelerated development and economies of scale. From the automotive industry (Takeishi and Takahiro 2001) to the computer industry (Baldwin and Clark 1997, 2000, 2006) and the military (O'Rourke 2017), the development of modular components, based on open standards, has prevented vendor lock-in and consequently dramatically reduced the costs of individual components. The US Army's VICTORY program (US Army 2015) for military ground vehicles was deemed to be successful due to the employment of a modular approach to how onboard data management systems are deployed within them. The interesting element is that this involved the participation of GE Intelligent Platforms (Charlottesville, VA), which supported the use of an open standards approach, seeing it as a key business opportunity since other prime vendors were focused on proprietary-based solutions. Practical benefits and drawbacks of modularity are also well documented in some studies that were done on the benefits vs. burdens of modularity (Dasch and Gorsich 2016). Three studies indicated net benefits, three found burdens of enforcing modularity, and one found that it could go either way. The complex relationship between variables within and across modules that affect modularity during various phases (e.g., design, engineering, production) give rise to the need for better means of identifying and objectively quantifying such interdependencies. As we examine the advantages that we have surveyed thus far, and contrast them with the perceived benefits of MOSA described in the section entitled "Benefits of modularity", it is evident there is a large degree of categorical overlap with the DoD's view of perceived benefits. Modularity/openness was adopted to fulfill a favorable localized objective, however, as opposed to an enforced "in hopes of" a perceived benefit.

Potential Drawbacks of Modularity

Organizational Structure Mismatch

For a commercial example, the Daimler–Chrysler merge (Ball and Miller 2000) was thought to be a good idea by both parties, due to the fulfillment of technical needs from both ends. Chrysler specialized in the modular designs of their vehicles, and so could rapidly innovate to meet customer trends each year. Daimler, on the other hand, specialized in highly integrated and care crafted vehicles that catered to the luxury vehicle segments of the market. The large differences in the supply-chain structures of each company (which mirrored their individual product designs), as well as development and production cycles, however, were significant enough to cause major problems in bringing products to market. Just as reflected in Conway's law (Conway 1968, 2019), the structure of an organization mirrors its product – in this case, the mirroring effect generated by organizations with different modes of operations (and supply chains) that did not create a single product line based on modularity.

Conway's law essentially suggests that there is a mirroring effect between the modular structure of a technical system and the respective organization structure. If the technical system is well

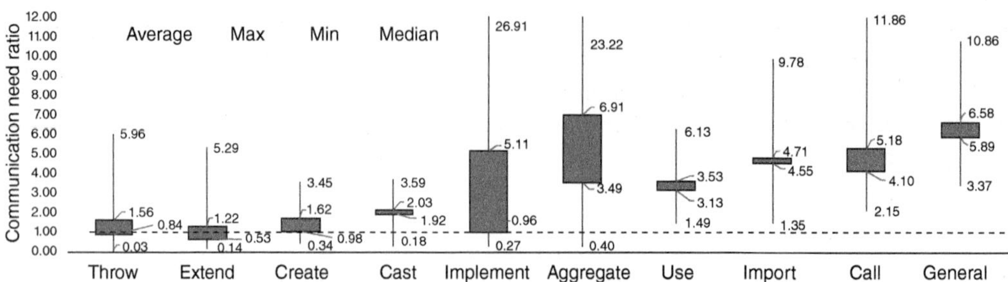

Figure 37.1 Communication ratio for different types of dependencies across software modules.

modularized, collaboration in the organization will productively mirror the modular structure. In contrast, if a system contains highly coupled modules, the collaboration among different parties requires more communication overhead and tends to be less efficient. Intensive research in the software engineering community has been conducted on related topics and led to a general consensus that highly congruent technical modular structure and organizational structure is beneficial for the quality of the software systems, as well as for the productivity of the team (Wang et al. 2018; Cataldo et al. 2008; Mauerer et al. 2021; Valetto et al. 2007).

In a case study of six open source software systems (Wang et al. 2018), the authors found that coupled software modules require 3–10 times more communication compared to software modules that are independent from each other. Different types of structural dependencies among modules require different levels of communication overhead as quantified in Figure 37.1. The x-axis shows the dependency type among software modules (i.e., source files); the y-axis measures the communication need ratio comparing the communication traffic for two independent modules. This is calculated from empirical data of these six projects. The authors also found that modules that are structurally independent from each other – but with excessive communication overhead among the developers who are in charge of these modules – are more problematic, e.g., more complicated and contain more bugs.

Limits Innovation

While modularity can be a means to achieve rapid innovations within the DoD acquisition strategy, there is a turning point toward such strategies. Work by Lee and Sorenson(2001), for example, has looked at the impact of modularization on innovation and found that most innovations occur with highly integrated product designs – as evidenced by an examination of patent data from the US Patent and Trademark Office. While it is true that modularity promotes rapid evolution of an architecture in a sequential, sub-component at-a-time manner, there is a loss of information on further potential innovation due to compartmentalization of the innovation to the domain space of the module itself.

Duplicated Efforts and Subsystems

Another pitfall of generating product variants through modularization is that it leads to duplication of support subsystems that are required to ensure interoperability; this duplication can increase overall size, weight, and power (SWaP) in complex systems. The development of modular architectures may also entail a lack of complete information about the impact that a developed module has on an overall system. This may cause problems due to the independent nature of the development of similar modules that only need to conform to a function, form, and interface standards, without explicit consideration of the impacts on the rest of the architecture. As in the case of integrated systems, however, this may be circumvented if appropriate testing mechanisms for operations and interfaces can be implemented.

Choice of Measure/Objectives

Numerous measures for determining modularity exist in the literature (Newcomb et al. 1998; Gershenson et al. 1999; Allen and Carlson-Skalak 1998; Mikkola 2000; Martin and Ishii 2002; Mattson and Magleby 2001; Sosa et al. 2000; Newman 2006). These metrics are primarily based on a variety of perspectives on interdependency analysis such as modular function deployment (MFD) (Erixon 1996) and Design Structure Matrix (DSM) (Eppinger and Browning 2012). The DSM describes the dependency between the various components of the system and can be analyzed to provide useful information related to information flow, coupling between design activities, and the choice or measure of modularity (including open standards) needed to reflect the desired outcomes. This can be a difficult task due to (1) the lack of *a priori* knowledge on the impact of selecting a measure of modularity; (2) the need for agreement to generate common open interface standards can be difficult due to vested interests across participants in the enterprise; and (3) the methods made available for assessing modularity, which are normally demonstrated for simplified systems and are challenged in more complex relational settings. In the case of openness, tools such as the Navy's Open Architecture Assessment Tool (US Navy 2009) provide guidance to program managers on assessing the level of a program's openness. The objective assessment of a program's openness and connection to modularity becomes more difficult, however, especially in view of many dependencies.

In addition, a phenomenon called modularity violations has been examined as a measurement of to what extent the modular structure of a system has been violated (Wong et al. 2011; Xiao et al. 2020). Modularity violations refer to the "latent" connections among structurally independent modules, reflected as the logical coupling among system modules with high maintenance costs. It is first investigated in pure software systems (Wong et al. 2011) and is later extended to examine systems of systems, such as cyber-physical systems (Xiao et al. 2020). Besides modularity violations, quantitative metrics, such as propagation cost (MacCormack et al. 2006) and decoupling level (Mo et al. 2016), are also used to measure the modular structure of complex systems. These two metrics measure to what extent the modules in a system are coupled with each other based on graph-theory.

Assessing Modularity Benefits in an SoS/Mission Context

In this section, we present a generalized decision support framework (DSF) that supports objective assessment of modularity, for both mission and system-of-systems contexts. The framework leverages both quantitative and qualitative perspectives and addresses key user (here, envisioned to be a program manager) needs in navigating modularity considerations. Prior to any consideration within the DSF, there naturally needs to be a problem definition phase that establishes the landscape for assessing modularity benefits. Furthermore, a user (Program/Project Manager, PM) would also bring to bear relevant knowledge and data artifacts that quantify and architecturally describe the landscape of where such decisions are to be developed and ultimately deployed. The following steps are prescribed in the analysis of assessing modularity.

Step 1: Problem Space Definition and Knowledge Gathering

The first step deals with determining the reason why modular benefits are being pursued in the first place – a step that requires prior insights and guidance. To this end, prior research under SERC has resulted in a PM Guidance Document (DeLaurentis 2017) that provides overarching guidance along four main categories:

1) What to measure and why?
2) Useful Strategies at Different Acquisition Lifecycle Phases
3) Emergent Phenomena
4) Technical Programmatic Pain Points

These topics in the PM guidance document can serve to help a PM determine:

1) Overarching Mission Objective(s) that should be pursued
2) Metrics, Key Performance Indices (KPIs), Risk Measures, "-ilities," and Criteria for Success
3) Organizational Disposition (i.e., what organizational construct operates at each phase of the system solution lifecycle)
4) Available resources for each phase of the system lifecycle, including acquisition of the final envisioned system solution.
5) Data and existing interfaces that exist currently

Step 2: Tradespace Analysis and Explorations via DSF

The DSF is a generalized framework that is two-pronged in nature. The first prong leverages a select set of tools to explore the quantitatively defined aspects of the modular Tradespace being evaluated. Typical analyses include seeking optimal solutions that balance cost (i.e., cost of acquisition and development), performance (end performance of the systems being developed), and risk (i.e., schedule development risk). The second prong relies on a series of cascading matrices, structured similarly to Quality Function Deployment (QFD) matrices, to help relate user inputs to relevant output, including available resources and organizational disposition. The DSF can now produce and evaluate architectures built upon different strategies, support trade-off between quantitative metrics of interest, including cost, performance, and risk, evaluated with SoS methods and tools, and indicate systems or architectures that can benefit from or require compliance with modularity need (i.e., in the case of defense, this would be MOSA principles). Figure 37.2 below illustrates the framework.

Step 3: Decision-Making for Modularity Benefits

Based on the knowledge gathered in the first phase of research, the user needs can be very diverse, and the projects that could benefit from the DSF belong to many different fields. In defense, the typical DoD PM user needs for MOSA have been assessed to be very diverse, but fall into some general categorizations as shown in Table 37.1 below.

The DSF structure permits the capture of elements caused by the possible complex interactions between the component systems that are often not readily captured in complex programs. The DSF also allows the user for comparison of systems architectures characterized by different strategies for modularity, including cases where it is imposed and cases where modular systems are not allowed. In the section entitled "Example Application: Amphibious Warfare Scenario", we provide an example scenario of application, and reference to a prototype DSF software platform for the results.

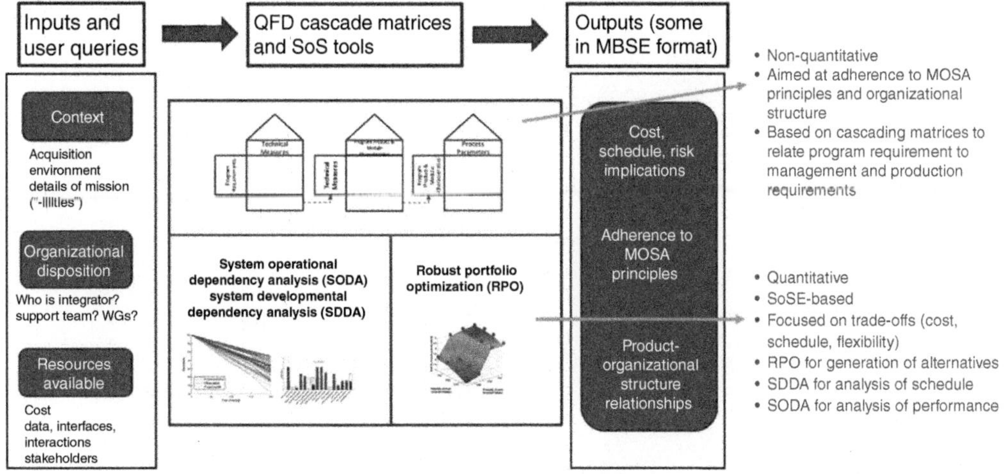

Figure 37.2 Generalized decision support framework (DSF) for MOSA.

Table 37.1 Mapping of categories of PM needs to outputs of the Decision Support Framework.

User need	DSF output
Rapid assessment of alternative systems architectures, modular, and not modular	Generation of Pareto-optimal architectures
Quantitative analysis of metrics of interest for comparison of architectures	Evaluation of cost, performance, and risk
Support to assess potential benefits of MOSA	Non-quantitative analysis of systems and sets of systems that can benefit from application of MOSA principles

Example Application: Amphibious Warfare Scenario

The synthetic problem is an Amphibious Warfare Scenario, which was chosen since it is a multi-domain problem involving air, ground, naval, and space systems. The systems in Amphibious Warfare interact to provide logistical support and system-level capabilities to achieve certain SoS-level capabilities. This high-level operational concept is illustrated in Figure 37.3, a Department of Defense Architecture Framework (DoDAF) OV-1 representation. A case study was developed for the synthetic problem that specifically defined systems from a World War II Amphibious Warfare Scenario. Use of World War II systems and context was chosen since many measures of system capabilities from this time period are public knowledge, which allowed the research team to create a case study with adequate detail.

The Mission System Library (MSL) was developed to be the key means to pass user inputs into the DSF. The MSL is created in an Excel workbook, where a series of eight sheets provide specific information on the problem:

1) **Main sheet**: System names, support capabilities (i.e., internal logistic requirement), system capabilities, and capability uncertainties
2) **SoS capabilities**: SoS capability names and sets of indices of the system capabilities that contribute to each SoS capability

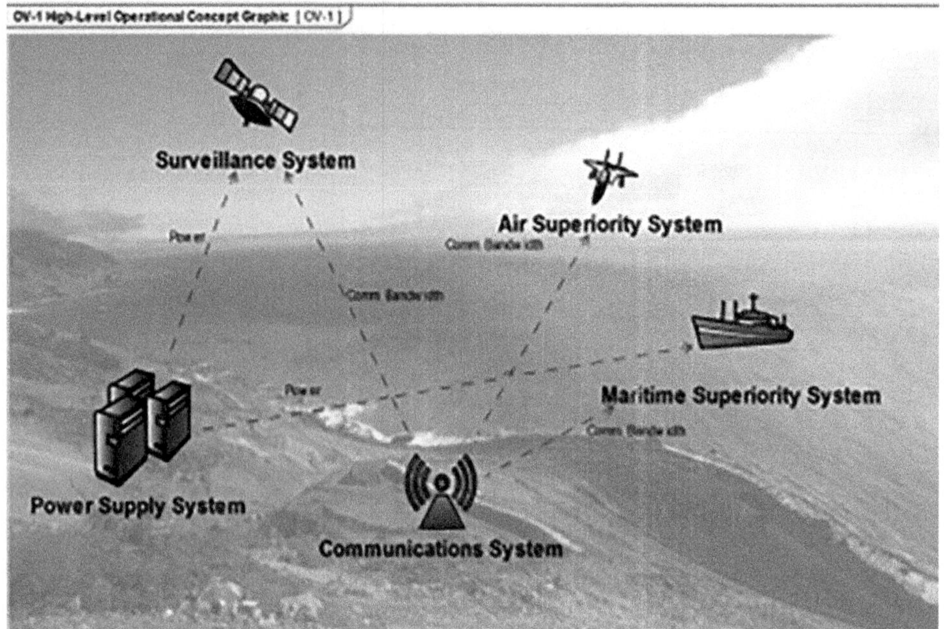

Figure 37.3 Concept of operations of a multi-domain amphibious warfare scenario.

3) **Compatibility constraints**: Matrices containing information on compatibility between systems, specification of maximum amount of specific systems allowed in a portfolio
4) **Must have systems**: The required systems in a portfolio
5) **Conditional must have systems**: The systems that are required to operate other systems
6) **Development covariance (MISDP)**: For alternative linear programming solvers (not active for this case study)
7) **Simulated capabilities (CVaR)**: For alternative linear programming solvers (not active for this case study)
8) **MOSA principles**: Empirical rules, based on case studies and PM Guidance, indicating systems and sets of systems that can benefit from application of specific MOSA principles

Each of these sheets can be read automatically by the DSF software to run the SoS analysis tools and create the outputs of the DSF. A DSF user is expected to create their own MSL for their specific problem. An example was created for the Amphibious Warfare Scenario (excerpts are shown in Figures 37.3 and 37.4) and is provided as a product of this research project along with the DSF software. Additionally, the User Guide to the Decision Support Framework developed for this project provides detailed guidance for future users on creating an MSL and each following step in the DSF.

In the Amphibious Warfare case study, systems were defined in each of the domains, including air, ground, naval, and space, as well as human systems (e.g., operators). On the Main Sheet of the MSL, systems are listed in each row. In total, 26 systems were defined, though only an excerpt is provided in Figures 37.4 and 37.5, and then each were evaluated for their support and system capabilities. Five support capabilities were defined for this case study: Transport Range (measured by range in miles), Transport Capacity (measured by capacity in pounds), Refuel (measured by fuel capacity in pounds), Communication Relay (measured using a constructed rating), and Operator

	A	B	C	D	E	F	G	H
1				Support Input Requirement				
2	No.	System Type	System Name	Transport Range	Transport Capacity	Refuel	Communication Relay	Operator
3	-	-	-	Range (mi)	Capacity (lb)	Fuel capacity (lb)	Rating (n.d.)	Number of Operators
4	1	Air Systems	P-51 Mustang	0	2000	2795	0	1
5	2		B-17 Flying Fortress	0	6000	18500	0	10
6	3		C-47	0	0	5369	0	4
7	4		B-52H Stratofortress	0	60000	321000	1	5
8	5		B-2 Spirit	0	40000	167000	1	2
9	6	Ground Systems	Infantry Platoon	10	1845	0	0	42
10	7		M114 155mm Howitzer	12480	12480	0	0	4
11	8		M-4 Sherman	150	1251	869	0	5
12	9		M8 Greyhound	175	274	353	0	4
13	10		Jeep Willis	0	0	95	0	1
14	11		"Deuce and a half" (supply truck)	0	0	378	0	1
15	12		Advanced Targeting Pod	0	0	0	0	0
16	13		TARDEC Chassis	0	0	378	0	1
17	14		TARDEC Anti Air Module	100	879	0	0	4
18	15		TARDEC Artillery Module	100	1750	0	0	4
19	16		TARDEC Personal Module	100	0	0	0	0
20	17		Bofors 40 mm gun (L60)	100	4800	0	0	4
21	18		Refuel Depot	0	0	0	0	0
22	19		Resupply Depot	0	0	0	0	0
23	20	Naval Systems	Allen M. Sumner Destroyer	0	0	0	0	336
24	21		Higgins Boat (LCVP)	0	0	0	0	3
25	22		Landing Ship, Tank (LST)	0	0	0	0	140
26	23		Battleship	0	0	0	0	2,220
27	24	Space Systems	Ultrahigh Frequency Follow-on (UFO) Communication Satellite	0	0	0	0	100
28	25		Wideband Global Satellite Communication Satellite (WGS)	0	0	0	0	100
	26	Human	General Personnel	0		0	0	0

1 Main Sheet | 2 SoS Capabilities | 3 Compatibility Constraints | 4 "Must Have" Systems | 5 Conditional Must Ha

Figure 37.4 List of available systems and support requirements.

(measured by number of operators). Each system might have one or more support input requirements, which must be fulfilled by a system that has a matching support output capability. Therefore, there are two sets of columns in the MSL for support capabilities: Support Input Requirement and Support Output Capability. It is worth mentioning that some systems might be only "support systems," if they only provide support output but do not provide system capabilities. Though the quantified SoS capabilities are evaluated using only the system capabilities, the Robust Portfolio Optimization tool can consider the support inputs and outputs by creating constraints that must be satisfied for any portfolio, making these interdependencies still critical to the architecture results.

The system capabilities are evaluated next. In the Amphibious Warfare case study, 23 system capabilities were defined, and some capabilities were assessed with multiple measures. For example, the Attack Air–Air capability is measured by Weapons Range, in miles, and Stopping Power, using a constructed rating. In this case study, there are nine variations of "Attack" system capabilities to represent the different combinations of each domain (air, sea, ground), and each uses the same measures. Additionally, there are nine "Defend" system capabilities, which use the measures Weapons Range, Stopping Power, and Robustness (a constructed rating). In several instances, constructed ratings were necessary where physical/natural measures were not possible or not consistent across types of technologies. For example, systems could receive a Stopping Power rating from zero to six, where zero indicated no stopping power and six indicated a high-explosive weight appropriate for large bombs/missiles. This rating allowed the research team to compare different types of technologies, where natural measures of stopping power – such as caliber (in inches) or

	A	B	C	AB	AC	AD	AE	AF	AG	AH	AI	AJ	AK	AL	AM	AN
1																
2	No.	System Type	System Name	SC15 = Defend Ground - Against Sea	SC16 = Defend Sea Against Air	SC17 = Defend Sea Against Ground	SC18 = Defend Sea Against Sea	SC19 = Mobility Air	SC20 = Mobility Ground	SC21 = Mobility Sea	SC22 = Surveillance	SC23 = Communication	Cost	Modular System?	Transport Range	Transpo Capaci
3	-	-	-	Weapons Range (mi), Stopping power (n.d.), Robustness (n.d.)	Weapons Range (mi), Stopping power (n.d.), Robustness (n.d.)	Weapons Range (mi), Stopping power (n.d.), Robustness (n.d.)	Weapons Range (mi), Stopping power (n.d.), Robustness (n.d.)	Combat Radius (mi), Operational Speed (mph)	Combat Radius (mi), Operational Speed (mph)	Combat Radius (nmi), Operational Speed (knots)	Detection rating (n.d.)	Communications Rating (n.d.)	($USD 2019)	Y/N	Uncertainty (+/- delta)	Uncertainty (+/- delta)
4	1	Air Systems	P-51 Mustang	0, 0, 0	0, 0, 0	0, 0, 0	0, 0, 0	1650, 360	0, 0	0, 0	2	1	$582,000.00	N	0	0
5	2		B-17 Flying Fortress	0, 0, 0	0, 0, 0	0, 0, 0	0, 0, 0	400, 150	0, 0	0, 0	1	1	$3,399,600.00	N	0	0
6	3		C-47	0, 0, 0	0, 0, 0	0, 0, 0	0, 0, 0	3800, 160	0, 0	0, 0	1	1	$2,173,800.00	N	0	200
7	4		B-52H Stratofortress	0, 0, 0	0, 0, 0	0, 0, 0	0, 0, 0	4400, 525	0, 0	0, 0	2	2	$78,900,000.00	N	0	0
8	5		B-2 Spirit	0, 0, 0	0, 0, 0	0, 0, 0	0, 0, 0	3450, 560	0, 0	0, 0	2	2	$3,423,000.00	N	0	0
9	6	Ground Systems	Infantry Platoon	1, 1, 1	0, 0, 0	0, 0, 0	0, 0, 0	0, 0	10, 3	0, 0	1	0	$8,876.91	N	0	0
10	7		M114 155mm Howitzer	9, 4, 1	0, 0, 0	0, 0, 0	0, 0, 0	0, 0	0, 0	0, 0	1	1	$182,581.95	N	0	0
11	8		M-4 Sherman	2, 3, 3	0, 0, 0	0, 0, 0	0, 0, 0	0, 0	150, 30	0, 0	1	1	$701,849.94	N	0	0
12	9		M8 Greyhound	1, 2, 2	0, 0, 0	0, 0, 0	0, 0, 0	0, 0	175, 55	0, 0	1	1	$359,147.56	N	0	0
13	10		Jeep Willis	0, 0, 0	0, 0, 0	0, 0, 0	0, 0, 0	0, 0	150, 65	0, 0	1	1	$1,575.21	N	0	70
14	11		"Deuce and a half" (supply truck)	0, 0, 0	0, 0, 0	0, 0, 0	0, 0, 0	0, 0	150, 45	0, 0	1	0	$19,879.13	N	0	600
15	12		Advanced Targeting Pod	0, 0, 0	0, 0, 0	0, 0, 0	0, 0, 0	0, 0	0, 0	0, 0	2	2	$1,000.00	Y	0	0
16	13		TARDEC Chassis	0, 0, 0	0, 0, 0	0, 0, 0	0, 0, 0	0, 0	100, 45	0, 0	0	0	$142,404.01	Y	0	200
17	14		TARDEC Anti Air Module	2, 2, 2	0, 0, 0	0, 0, 0	0, 0, 0	0, 0	0, 0	0, 0	1	1	$25,000.00	Y	0	0
18	15		TARDEC Artillery Module	5, 2, 3	0, 0, 0	0, 0, 0	0, 0, 0	0, 0	0, 0	0, 0	1	1	$25,000.00	Y	0	0
19	16		TARDEC Personal Module	0, 0, 0	0, 0, 0	0, 0, 0	0, 0, 0	0, 0	0, 0	0, 0	0	0	$20,000.00	Y	0	100
20	17		Bofors 40 mm gun (L60)	3, 2, 1	0, 0, 0	0, 0, 0	0, 0, 0	0, 0	0, 0	0, 0	0	0	$100,000.00	Y	0	0
21	18		Refuel Depot	0, 0, 0	0, 0, 0	0, 0, 0	0, 0, 0	0, 0	0, 0	0, 0	0	0	$0.00	N	0	0
22	19		Resupply Depot	0, 0, 0	0, 0, 0	0, 0, 0	0, 0, 0	0, 0	0, 0	0, 0	0	0	$0.00	N	0	100
23	20	Naval Systems	Allen M. Sumner Destroyer	0, 0, 0	4, 3, 3	4, 3, 3	4, 4, 4	0, 0	0, 0	3300, 20	2	1	$152,000,000.00	N	0	0
24	21		Higgins Boat (LCVP)	0, 0, 0	1, 1, 2	1, 1, 2	1, 1, 2	0, 0	0, 0	10, 9	0	1	$229,300.00	N	0	200
25	22		Landing Ship, Tank (LST)	0, 0, 0	4, 3, 3	4, 3, 3	4, 3, 3	0, 0	0, 0	10000, 12	2	1	$36,320,100.00	N	0	1000
26	23		Battleship	0, 0, 0	9, 3, 3	9, 3, 4	9, 4, 3	0, 0	0, 0	5900, 21	2	1	$100,238,000.00	N	0	0
27	24	Space Systems	Ultrahigh Frequency Follow-on (UFO) Communication Satellite	0, 0, 0	0, 0, 0	0, 0, 0	0, 0, 0	0, 0	0, 0	0, 0	2	2	$382,000,000.00	N	0	0
28	25		Wideband Global Satellite Communication Satellite (WGS)	0, 0, 0	0, 0, 0	0, 0, 0	0, 0, 0	0, 0	0, 0	0, 0	2	2	$300,000,000.00	N	0	0
	26	Human	General Personnel	0, 0, 0	0, 0, 0	0, 0, 0	0, 0, 0	0, 0	0, 0	0, 0	0	0	$170.00	N	0	0

1 Main Sheet | 2 SoS Capabilities | 3 Compatibility Constraints | 4 "Must Have" Systems | 5 Conditional Must Have S' ...

Figure 37.5 List of the same systems in Figure 37.3. Columns with "SC" in the header list the provided system capabilities; each system is also associated with cost and with an indication of modular vs. monolithic nature of the system; The last two columns of this excerpt are uncertainties on the capabilities of the systems.

explosive weight (in pounds) – changed depending on the type of technology – such as guns or bombs. The rating options for stopping power are listed in Table 37.2. Other constructed ratings were defined for Detection, Communication, and Communication Relay, and more information on all the constructed ratings are provided in the *User Guide to the Decision Support Framework*.

Each support and system capability may also be evaluated for uncertainty, where the possible change in capability range is defined in individual columns on the Main Sheet of the MSL. The uncertainty values together with a variable representing the user's risk aversion are considered by the Robust Portfolio Optimization tool, which is discussed in a later section of this report. The last columns of the Main Sheet are system cost (US$) and system modularity (yes/no).

Next, the SoS Capabilities sheet includes the names of all SoS Capabilities and the indices of the system capabilities that contribute to the SoS capability. The six SoS Capabilities for the case study are presented in Table 37.3. Each SoS capability is computed using a variation of the additive value model. Each of the subsequent sheets may optionally be filled out by the user to indicate desired strategies for the resulting portfolios (for example, whether certain systems are mandatory). For more information, please refer to the supplementary document, *User Guide to the Decision Support Framework*.

Table 37.2 Constructed ratings for stopping power measure of "Attack" and "Defend" capabilities.

Rating value	Natural measure: caliber or explosive weight range	Notes
0	0	No stopping power
1	[0.10, 0.45] in.	Dimension: caliber Appropriate for guns
2	[0.50, 1.00] in.	Dimension: caliber Appropriate for guns
3	>1.0 in. (but not explosive)	Cannons and other intermediate weapons
4	[500, 5,000] lb	Dimension: explosive weight Appropriate for bombs/missiles
5	[5,000, 10,000] lb	Dimension: explosive weight Appropriate for bombs/missiles
6	>10,000 lb	Dimension: explosive weight Appropriate for bombs/missiles

Table 37.3 SoS Capabilities for Amphibious Warfare case study, with system capability contributions.

No.	SoS-capability	System-capability indices						
1	Air superiority	1–6	19–27	46	47	52	53	
2	Naval superiority	13–18	37–45	50–53				
3	Tactical bombardment	3	4	46	47	52	53	
4	Land seizure	3	4	9	10	15	16	46–53
5	Land control	28–36	52	53				
6	Reconnaissance	46–53						

References

Allen, K.R. and Carlson-Skalak, S. (1998). Defining product architecture during conceptual design. In: *ASME Design Theory And Methodology '98*. DETC98/DTM-5650 (ed. K. Otto). New York: ASME.

Baldwin, C. and Clark, K.B. (1997). Managing in an age of modularity. *Harvard Business Review* 75 (5): 84–93.

Baldwin, C.Y. and Clark, K.B. (2000). *Design Rules: The Power of Modularity*, vol. 1. MIT Press.

Baldwin, C.Y. and Clark, K.B. (2006). Modularity in the design of complex engineering systems. In: *Complex Engineered Systems* (ed. D. Braha, A.A. Minai, and Y. Bar-Yam), 175–205. Berlin, Heidelberg: Springer.

Ball, J. and Miller, S. (2000). Daimler-Benz, Chrysler merger fails to live up to expectations. *The Wall Street Journal.*.

Cataldo, M., Herbsleb, J.D. & Carley, K.M. (2008). Socio-technical congruence: a framework for assessing the impact of technical and work dependencies on software development productivity. In Proceedings of the Second ACM-IEEE International Symposium on Empirical Software engineering and measurement, Kaiserslautern, Germany (pp. 2–11).

Clune, J., Mouret, J.B., and Lipson, H. (2013). The evolutionary origins of modularity. *Proceedings of the Royal Society B* 280: 1755.

Conway, M. (1968). How do committees invent? *Datamation* 14 (5): 28.

Conway, M. (2019). "Conway's Law". Mel Conway's Home Page. https://www.melconway.com/Home/Conways_Law.html (accessed 22 June 2023).

Dasch, J.M. and Gorsich, D.J. (2016). Modularity in military vehicles: benefits and burdens. *Defense Acquisition Research Journal* 23 (1): 2–27.

DeLaurentis, D. (2017). RT-163/185: Navigating in a MOSA Ecosystem: Guidance for Program Managers.

Eppinger, S. and Browning, T. (2012). *2012*. Design structure matrix methods and applications: MIT press.

Erixon, G. (1996). Modular function development (MFD), support for good product structure creation. In: *Proceedings of the 2nd WDK Workshop on Product Structuring*, vol. 3–4, 13–16. The Netherlands: University of Technology.

Gershenson, J.K., Prasad, G.J., and Allamneni, S. (1999). Modular product design: a lifecycle view. *Journal of Integrated Design & Process Science* 3 (4): 13–26.

Judson, J. (2015). Lockheed Martin developing open controller for Army's TARDEC. Orlando. http://www.defensenews.com/story/defense/land/vehicles/2015/12/08/lockheed-martin-developing-open-controller-for-armys-tardec/77004516/ (accessed 22 June 2017).

Lee, F. and Sorenson, O. (2001). The dangers of modularity. *Harvard Business Review* 79 (8): 20–21.

MacCormack, A., Rusnak, J., and Baldwin, C.Y. (2006). Exploring the structure of complex software designs: an empirical study of open source and proprietary code. *Management Science* 52 (7): 1015–1030.

Martin, M.V. and Ishii, K. (2002). Design for variety: developing standardized and modularized product platform architectures. *Research in Engineering Design* 13 (4): 213–235.

Mattson, C.A. and Magleby, S.P. (2001). The influence of product modularity during concept selection of consumer products. In: *ASME Design Engineering Technical Conference*, Pittsburgh, PA, 9–12. New York: ASME.

Mauerer, W., Joblin, M., Tamburri, D.A. et al. (2021). In search of socio-technical congruence: a large-scale longitudinal study. *IEEE Transactions on Software Engineering* 48 (8): 3159–3184.

Mikkola, J.H. (2000). Modularization assessment of product architecture. DRUID Winter Conference, Copenhagen Business School, Denmark.
Mo, R., Cai, Y., Kazman, R. et al. (2016). Decoupling level: a new metric for architectural maintenance complexity. In: *2016 IEEE/ACM 38th International Conference on Software Engineering (ICSE)*, 499–510. IEEE.
Newcomb, P.J., Bras, B., and Rosen, D.W. (1998). Implications of modularity on product design for the life cycle. *Journal of Mechanical Design* 120 (3): 483–490.
Newman, M.E. (2006). Modularity and community structure in networks. *Proceedings of the National Academy of Sciences of the United States of America* 103 (23): 8577–8582.
O'Rourke, R. (2017). Navy Littoral Combat Ship/Frigate (LCS/FF) Program: Background and Issues for Congress. *Congressional Research Service report RL33741*, Washington DC.
Sosa, M.E., Eppinger, S.D., and Rowles, C.M. (2000). Designing modular and integrative systems. In: *ASME Design Engineering Technical Conference Proceedings*, Baltimore, MD. New York: ASME.
Takeishi, A. and Takahiro, F. (2001). Modularisation in the auto industry: interlinked multiple hierarchies of product, production and supplier systems. *International Journal of Automotive Technology and Management* 1 (4): 379–396.
US Army (2015). Vehicular integration for C4ISR interoperability (VICTORY). https://www.army.mil/standto/archive/2015/07/06/#:~:text=What%20is%20it%3F,Vehicle%20and%20Ground%20Combat%20Systems (accessed 22 June 2023).
US Navy (August 2009). Open Architecture Assessment Tool 3.0.
Valetto, G., Helander, M., Ehrlich, K. et al. (2007). Using software repositories to investigate socio-technical congruence in development projects. In: *Fourth International Workshop on Mining Software Repositories (MSR'07: ICSE Workshops 2007)*, 25. IEEE.
Wang, X., Xiao, L., Yang, Y. et al. (2018). Identifying TraIn: a neglected form of socio-technical incongruence. In: *Proceedings of the 40th International Conference on Software Engineering: Companion Proceedings*, 358–359.
Wong, S., Cai, Y., Kim, M., and Dalton, M. (2011). Detecting software modularity violations. In: *Proceedings of the 33rd International Conference on Software Engineering*, 411–420.
Xiao, L., Pennock, M.J., Cardoso, J.L., and Wang, X. (2020). A case study on modularity violations in cyber-physical systems. *Systems Engineering* 23 (3): 338–349.

Biographical Sketches

Dr. Navindran Davendralingam is a senior research scientist at Amazon, and previously a research scientist at the Center for Integrated Systems in Aerospace (CISA) at Purdue University where he worked across a range of government and industry funded projects across areas of defense, agriculture, transportation, and aerospace technologies. Work in this book is a product of these prior works funded under the auspices of the Department of Defense (DoD) Systems Engineering Research Center (SERC).

Dr. Cesare Guariniello is a research scientist in the School of Aeronautics and Astronautics at Purdue University. He received his PhD in aeronautics and astronautics from Purdue University in 2016 and his master's degrees in astronautical engineering and computer and automation engineering from the University of Rome "La Sapienza." Dr. Guariniello works as part of the Center for Integrated Systems in Aerospace led by Dr. DeLaurentis and is currently engaged in projects funded by NASA, the DoD Systems Engineering Research Center (SERC), and the NSF. His main research interests include modeling and analysis of complex systems and SoS architectures – with particular focus on space mission architectures – aerospace technologies, and robotics. Dr. Guariniello is a senior member of IEEE and AIAA.

Dr. Lu Xiao is an assistant professor in the School of Systems and Enterprises at Stevens Institute of Technology. Her research interests lie in the broad area of software engineering, particularly in software architecture, software economics, cost estimation, and software ecosystems. She is an awardee of NSF CAREER project in 2021. She has published her work in different conferences and journals, including TSE, ICSE, FSE, and ICSA, etc. She completed her PhD in Computer Science at Drexel University in 2016, advised by Dr. Yuanfang Cai. She received the first-place prize at the ACM Student Research Competition in 2015. She earned her bachelor's degree in computer science from Beijing University of Posts and Telecommunications in 2009.

Part VIII

Systems Education and Competencies in the Age of Digital Engineering, Convergence, and AI

Nicole Hutchison

Chapter 38

Using the Systems Engineering Body of Knowledge (SEBoK)

Nicole Hutchison[1], Art Pyster[2], and Rob Cloutier[3]

[1] *Stevens Institute of Technology, Hoboken, NJ, USA*
[2] *George Mason University, Fairfax, VA, USA*
[3] *University of South Alabama, Mobile, AL, USA*

Development of the SEBoK

The Systems Engineering Body of Knowledge (SEBoK – pronounced "see-bach") is a global resource that is freely available. Beginning in 2009, the Systems Engineering Research Center (SERC) began pulling together this collaborative team. Initially led by Art Pyster (then Stevens Institute of Technology) and David Olwell (then Naval Postgraduate School), the initial author team consisted of over 70 subject-matter experts from around the world. The initial effort to create the SEBoK was a three-year endeavor funded by the Office of the Secretary of Defense (OSD) in the US Department of Defense (DoD). Though the funding was US DoD, the vision was that the SEBoK should be a resource that would be useful globally.

The SEBoK provides an overview of critical resources for systems engineering. It is not a compendium – something that has all of the resources fully incorporated – but provides references and links to critical information. It is maintained as a wiki (www.sebokwiki.org).

Starting in 2013, the SEBoK has been maintained by a Governing Board consisting of representatives from the International Council on Systems Engineering (INCOSE), IEEE (currently the IEEE Systems Council, previously the IEEE Computer Society), and the Stevens Institute of Technology, which runs the Systems Engineering Research Center (SERC).

What's Included in the SEBoK?

The SEBoK includes over 200 articles on topics that are relevant to systems engineers. These topics cover everything from the origins of SE to the latest emerging trends (see the next section for more information). Each article is a stand-alone wiki page about a given topic. The article will include a summary of the topic, relevant processes, examples, etc. Because the SEBoK is a wiki, each article

Systems Engineering for the Digital Age: Practitioner Perspectives, First Edition. Edited by Dinesh Verma.

Companion website: www.wiley.com/go/verma/systemsengineering

also contains links to critical related topics, references, and glossary terms. The structure is divided into three levels:

- **Articles** are the lowest level and are independent pages of the wiki that focus on a single topic;
- **Knowledge Areas** are collections of related topics which can be useful to helping an individual more thoroughly explore a space; and
- **Parts** are groupings of knowledge areas that reflect major themes of systems engineering knowledge.

In addition, each article includes a substantial reference section divided into three categories: works cited, primary references, and additional references. Works cited are just what they sound like – anything that is directly cited in the article is captured here (think standard good journal practice). The primary references, however, are a bit different. These are the handful of critical references that the authors and editors of the article have agreed are critical and foundational on the topic. Additional references provide either more depth or insights into a particular facet or approach for the topic. For example, an article on general systems architecture would include a resource on modular open systems architecture (MOSA) as an "additional" resource.

Think of it this way:

- **Articles** provide an introduction to the topic, helping an individual familiarize herself with the key tenants. Reading just the article would help you have basic awareness of the topic.
- **Primary References** provide a solid foundation of the topic and generally include the most-cited or most-critical references. Reading all the primary references would give you a solid foundational understanding of the topic. Each primary reference also has its own page showing you what articles (topics) cite it as primary, the full bibliographic information with a link to the source if possible, and an annotation from the SEBoK authors or editors explaining why that reference is considered important.
- **Additional References** provide deeper insight into a topic or a specialization in that topic. A more advanced practitioner would benefit from reviewing additional references to gain a more nuanced understanding of a topic.
- **Glossary** terms provide a basic definition. In some cases, more than one definition is provided if the definition is, for example, context-dependent or if the community has multiple accepted views on the term. References to the source(s) of the definition are provided and particularly when more than one definition is included, a discussion of the term's use in the community.

At the end of the day, the SEBoK covers the knowledge of SE but it is up to you as a systems engineer to apply that knowledge to advance as a practitioner.

How Is the SEBoK Organized?

The SEBoK consists primarily of articles. Articles are analogous to what you would find in Wikipedia – an overview of a given topic. Articles are grouped into knowledge areas and knowledge areas are conglomerated into Parts.

The SEBoK contains eight parts, each of which is intended to provide information and insight into a major theme of SE (Figure 38.1). Part 1 provides an introduction to the SEBoK itself, explaining the philosophy, layout, and concepts. It also sets the foundation for the SEBoK by explaining what systems engineering itself is. The SEBoK highlights multiple definitions of systems

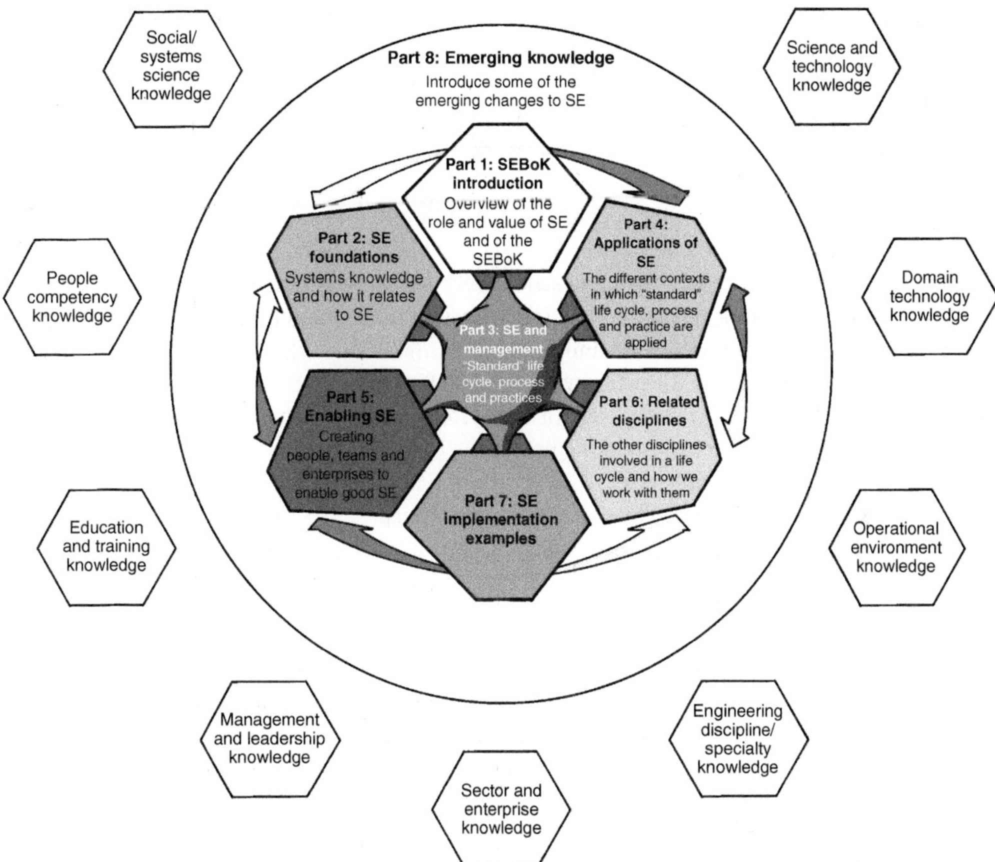

Figure 38.1 Overview of the SEBoK structure.

engineering but primarily focuses on the 2019 definition generated by the INCOSE Fellows: a transdisciplinary and integrative approach to enable the successful realization, use, and retirement of engineered systems, using systems principles and concepts, and scientific, technological, and management methods (INCOSE 2019). Part one provides insights into the history of SE and its origins, but also includes information that is very helpful to anyone trying to justify the use of SE, for example providing insight into the economic benefits of SE.

Part 2, "Systems Engineering Foundations," focuses on the systems science that underpins systems engineering. It provides overarching discussions of what systems are and what comprises them. This is where critical concepts such as systems thinking and models are introduced.

This section also highlights critical principles of systems engineering, including (Watson et al. 2023):

- SE in application is specific to stakeholder needs, solution space, resulting system solution(s), and context throughout the system life cycle.
- SE has a holistic system view that includes the system elements and the interactions among themselves, the enabling systems, and the system environment.

- SE influences and is influenced by internal and external resources, and political, economic, social, technological, environmental, and legal factors.
- Both policy and law must be properly understood to not overly constrain or under-constrain the system implementation.
- The real system is the perfect representation of the system. (models are only representations of real systems)
- A focus of SE is a progressively deeper understanding of the interactions, sensitivities, and behaviors of the system, stakeholder needs, and operational environment.
- Systems engineering addresses changing stakeholder needs over the system life cycle.
- SE addresses stakeholder needs, taking into consideration budget, schedule, and technical needs, along with other expectations and constraints.
- SE decisions are made under uncertainty accounting for risk.
- Decision quality depends on knowledge of the system, enabling system(s), and interoperating system(s) present in the decision-making process.
- SE spans the entire system life cycle.
- Complex systems are engineered by complex organizations.
- SE integrates engineering and scientific disciplines in an effective manner.
- SE is responsible for managing the discipline interactions within the organization.
- SE is based on a middle range set of theories.

Part 3 of the SEBoK, "Systems Engineering and Management," covers the topics that most people would expect to find in a resource on SE, covering different systems lifecycle models and all major activities across the systems lifecycle, both technical and non-technical. These articles provide exemplar processes and practices, which are tailorable for an engineering organization to satisfy strategic business goals and individual project objectives including:

- How engineering conducts system development
- The purpose of each engineering artifact generated
- How systems are integrated, and requirements verified
- How new product designs are transitioned to production operations
- How the resulting system is employed and sustained to satisfy customer needs (Carter and Singam 2022)

Unsurprisingly, Part 3 is the largest in the SEBoK, with over 50 individual articles (over totaling close to 500 pages in the PDF version). The materials in Part 3 are intended to be agnostic, meaning that they are intended to provide core principles, processes, and activities regardless of the type of system being engineered.

Part 4, "Systems Engineering Applications," builds on the information found in Part 3 and discusses how systems engineering approaches may differ depending on the type of system being developed. This is done in two ways: by type of system and by domain. For system types, Part 4 talks about products (which often closely align with "standard" SE practices), services, enterprise systems, and systems-of-systems (SoS).

For domains, the SEBoK currently has one example domain, that of healthcare. The articles in this area describe the terminology used in the healthcare sector and some of the ways SE is applied, such as in healthcare delivery and systems biology.

Part 5, "Enabling Systems Engineering," focuses on what is required to make SE happen and examines this through three lenses: individuals, teams, and organizations. For individuals, the SEBoK covers things like the most common roles for systems engineers, the different competency frameworks that are available and how they can be used, and special ethical considerations for systems engineers.

As you will hear if you ever attend an INCOSE event, SE is a "contact sport" – it cannot happen in isolation and, therefore, team considerations are also important. And as any practicing systems engineer knows, you will not be working on a team of systems engineers but a multidisciplinary team with individuals who have a variety of backgrounds. This part describes team vs. individual capability and discusses team dynamics and their importance: no matter how many skilled individual practitioners you have, you cannot be successful as a team if the individuals cannot form a cohesive whole. This part also includes special considerations for diversity, equity, and inclusion, building on the nontechnical skills of systems engineers to enable them to work effectively in this type of diverse team environment. Finally, the importance of technical leadership and the role that systems engineers often play as technical leaders is discussed.

Part 5 wraps with a discussion of organizational enablers to systems engineering, discussing strategy, required capabilities, and culture and how these can foster or inhibit successful SE.

Part 6, "Related Disciplines," describes the disciplines that systems engineers most commonly interact with and highlights critical relationships and interactions between these disciplines and SE. The main disciplines covered include: aerospace, electrical, mechanical, industrial, civil, software, environmental, and geospatial engineering, project management, economics, and enterprise IT.

Many articles are mature. Others are terse or even placeholders that will be fleshed out in a future release of the SEBoK. Where possible, Part 6 provides an overview of the discipline as well as a discussion of relationships with SE. For example, for project management and software, the articles discuss the discipline and provide an overview of the bodies of knowledge specifically associated with them (the PMBOK® and SWEBOK, respectively).

Part 6 closes with a series of articles on the relationship between systems engineering and a variety of quality attributes and the disciplines surrounding them. These are what are often called the "-ilities" or "nonfunctional" system attributes that, nevertheless, are crucial to successful system development and operation. The section includes an article on viewing these quality attributes from the lens of loss. The quality attributes included in the SEBoK are human systems integration, manufacturability and producibility, system affordability, system hardware assurance, system reliability, availability, and maintainability, system resilience, system resistance to electromagnetic interference, system safety, and system security.

Parts 2–6 represent the bulk of the information about systems engineering and its practice. Part 7, then, is where all of this is put into real-world context, presenting a variety of systems development examples (formerly called "case studies"). This includes a mix of both successful systems and systems failures and for each provides context of the system and links to the different knowledge elements of the SEBoK. The examples range across disciplines and countries. The case studies in the SEBoK (as of 2023) include:

- Defense System Examples
  - Submarine Warfare Federated Tactical Systems
  - Virginia Class Submarine
- Information System Examples
  - Complex Adaptive Taxi Service Scheduler
  - Successful Business Transformation within a Russian Information Technology Company
  - FBI Virtual Case File System

- Management System Examples
  - Project Management for a Complex Adaptive Operating System
- Medical System Examples
  - Next Generation Medical Infusion Pump
  - Medical Radiation
  - Design for Maintainability
- Space System Examples
  - Global Positioning System Case Study
  - Global Positioning System Case Study II
  - Russian Space Agency Project Management Systems
  - How Lack of Information Sharing Jeopardized the NASA/ESA Cassini/Huygens Mission to Saturn
  - Hubble Space Telescope
  - Applying a Model-Based Approach to Support Requirements Analysis on the Thirty-Meter Telescope
  - Miniature Seeker Technology Integration Spacecraft
  - Apollo 1 Disaster
- Transportation System Examples
  - Denver Airport Baggage Handling System
  - FAA Advanced Automation System (AAS)
  - Federal Aviation Administration Next Generation Air Transportation System
  - UK West Coast Route Modernisation Project
  - Standard Korean Light Transit System Vignette
- Utilities Examples
  - Northwest Hydro System
  - Singapore Water Management

Table 38.1 is from the SEBoK and illustrates the mapping between the examples and the different knowledge elements (though it should be noted that this provides the most common themes and is not exhaustive).

Each example also provides references to source materials so that you can learn about it in more depth.

Part 8 is the newest section of the SEBoK, added in 2020 to reflect emerging knowledge. This is where trends and new topics are identified, even if the literature related to systems engineering is not quite as mature as other areas of the SEBoK. As of 2023, the emerging topics include SE transformation, multiple articles on digital engineering and model-based systems engineering (MBSE), articles on the relationship between SE and artificial intelligence (AI), SoS and complexity, set-based design, and socio-technical systems. The intent with Part 8 is that as topics and articles mature, they will eventually be folded into the other parts of the SEBoK itself. For example, the Editorial Board, in addition to having articles in Part 8 around digital engineering, MBSE, and transformation also includes a discussion of these topics in relevant articles throughout Parts 3 and 4.

Part 8 also includes emerging doctoral research in SE, highlighting dissertations in SE as they are identified.

Table 38.1 Mapping of SEBoK examples and knowledge elements ("Topics").

SEBoK Topic	BT	ATC	NASA	HST	GPS	GPS II	Radiation	FBI VCF	MSTI	Infusion pump	DfM	CAS	PM	TS	SWFTS	NHS
Systems thinking					X	X		X	X	X	X	X	X	X		
Models and simulation		X	X			X							X	X	X	
Product systems engineering				X	X		X	X								X
Service systems engineering			X			X								X		X
Enterprise systems engineering	X	X	X			X							X	X		X
Systems of systems (SoS)	X	X				X								X		X
Life cycle models	X			X	X				X	X	X	X			X	
Business or mission analysis	X	X				X		X		X			X	X		
Stakeholder needs and requirements	X	X		X		X		X		X			X	X	X	
System requirements				X	X			X		X			X	X	X	
System architecture	X	X			X	X			X	X						
System analysis			X		X							X	X	X		
System implementation					X						X			X		
System integration	X		X	X	X	X			X		X		X	X		
System verification			X		X		X		X	X						
System validation	X				X	X	X		X	X						
System deployment					X	X					X					
Operation of the system					X	X				X	X			X		X
System maintenance																X
Logistics																
Planning	X	X		X				X		X			x			
Assessment and control							X	X	X				x	x		

(Continued)

Table 38.1 (Continued)

SEBoK Topic	BT	ATC	NASA	HST	GPS	GPS II	Radiation	FBI VCF	MSTI	Infusion pump	DfM	CAS	PM	TS	SWFTS	NHS
Risk management			X	X	X	X	X	X		X		X	x			
Measurement									X							
Decision management	X								X			X	x	X		
Configuration management					X	X			X		X	X				
Information management			X									X	x	X		
Quality management															X	
Enabling systems engineering			X		X	X			X	X	X	X	X		x	
Related disciplines						X	X	X		X		X				

BT - Business Transformation
ATC - FAA's NextGen Air Traffic Control
NASA - NASA's Mission to Saturn
HST - Hubble Space Telescope
GPS - Global Positioning System I
GPS II - Global Positioning System II
Radiation - Medical Radiation
FBI VCF - FBT Virtual Case File system
MSTI - Miniature Seeker Technology Integration
Infusion Pump - Next Generation Medical Infusion Pump
DfM - Design for Maintainability
CAS - Complex Adaptive Operating Systems
PM - Project Management
TS - Taxi Service
SWFTS - Submarine Warfare Federated Tactical Systems
Bag Handling - Denver Airport Baggage Handling System
VA Sub - Virginia Class Submarine
Route Mod - UK West Coast Route Modernization Project
Water Mgmt - Singapore Water Management
FAA AAS - FAA Advanced Automation System
Light Rail - Standard Korean Light Transit System
TMT - Thirty-Meter Telescope

How Can I Use the SEBoK?

The SEBoK is intended to be useful for anyone – from students to novice practitioners to experts. Below are a few examples of walking through the SEBoK with different personas. Find one like you to help you get started. But because it is delivered as a wiki, one of the best ways to get started is just to enter the SEBoK and navigate it yourself.

Jill – An Engineering Undergraduate

Jill is in her final year studying mechanical engineering at a public university. Her senior design project is a multi-disciplinary capstone, and she must work with students from the electrical and software engineering departments as well. At the first meeting of the capstone team, each student starts talking about their ideas to make their part of the project using the most advanced techniques and technologies available. They quickly start arguing about how to do this in the time and budget available. Her faculty advisor says that in order to be successful, they have to work together to create a useful product, which means that they can't each focus on their own parts in isolation. Her professor states that they all need to come to the next team meeting prepared to discuss their project as a system. Like all good undergraduate students, when she gets back to her dorm, the first thing she does is Google "system" and she quickly gets lost in the results (there are over 16 billion, after all).

For Jill, the SEBoK can provide a great way to get exposed to the idea of systems. In particular, Jill could benefit from reading the articles in the system fundamentals area, particularly "The Nature of Systems," which will introduce her to different views on and core definitions of a system. From there, the articles in the "Introduction to Systems Engineering" knowledge area would be useful. Finally, Jill should read the article "Systems Engineering and Mechanical Engineering," which relates to this new subject area what she has spent several years learning. In her reading, Jill might also identify a few references that she thinks will help her better understand this SE stuff and help per apply it.

This gives Jill a better understanding of what her professor meant – that all of the different elements of the project (or system) must work together in order to achieve their project goal (and their ultimate goal of graduating).

As the capstone progresses, Jill returns to the SEBoK to find additional resources as new issues come up.

Marcus – A New Systems Engineer

Marcus graduated with his master's in computer science three years ago and has since been working for a large company in their IT division. The team he is on has been operating and maintaining the existing customer support system, which has become increasingly cumbersome and costly. As part of this work, Marcus's manager noticed that he was always asking good questions, and several times Marcus's curiosity helped the team uncover problems that they had been unaware of. In addition, Marcus has exhibited "out of the box" thinking, which has helped create some novel solutions. Marcus's computer science and programming skills are competent. The company is about to kick off a project to build a whole new customer support system. Based on his experience with the current system and his inquisitiveness, Marcus's manager asked if he would consider taking a role as the systems software engineer, reporting directly to the chief systems engineer on the project. Marcus is excited to take on new responsibilities, but other than a few brief interactions with systems engineers, he's not sure what they do. He is set to start taking the company's introductory systems engineering course in a few weeks, but he wants to start understanding more about it now.

Marcus's manager recommends he take a look at the SEBoK in preparation for his upcoming course. Marcus, feeling a bit nervous, decides to search the SEBoK for "computer science" and finds the page, "*Exploring the Relationship between Systems Engineering and Software Engineering*." This is a page about a paper by the same title, which Marcus looks up to read later. In the SEBoK, he notices that this is used in something called, "Software Engineering in the Systems Engineering Life Cycle." Marcus reads this article and finds a number of related articles on the relationship between systems engineering and software engineering, which highlight the relationships between the two. Marcus is now feeling more confident that this "systems stuff" isn't completely different from what he has learned by applying his computer science degree on the job.

Marcus continues to explore the SEBoK, finding a collection of articles about systems thinking, which he finds useful and which leads him to articles about the systems engineering process. He learns about lifecycle approaches and finds an article on agile systems engineering. He is very familiar with agile approaches from his work over the last three years. While reading the SEBoK, Marcus looks at the references and identifies six articles and two books he wants to read to get more familiar with SE. He is feeling much more confident about his ability to tackle the upcoming course and his new role and bookmarks the SEBoK for future reference.

Yu Gin – An Experienced Systems Engineer

Yu Gin graduated 10 years ago with a degree in electrical engineering and was happy to join a large company in her area helping to design their medical devices. After a few years, Yu Gin was selected for an advancement program intended to help develop new systems talent in the company. As part of this, Yu Gin completed her graduate degree in systems engineering last year. She is currently working as a lead designer on one of the company's flagship products. She has taken all the courses her company offers around systems design and engineering and is looking for more resources to continue to develop her skills.

Yu Gin is aware of the SEBoK and has read a few articles but has not used it often. She decides to search for "knowledge skills abilities" (KSA) as her company uses this term around performance time. She finds articles on enabling individuals, ethical behavior, and roles and competencies. She reads the roles and competencies articles and learns about several different competency models. The primary references point her to these models, and she decides that the NASA SE competency model seems to align well with how her company views systems engineering. She decides to look through the competency model and see what competencies she has and which she might want to work on (see "Systems Engineering Workforce Competencies"). The competencies around the life cycle she is generally comfortable with, and she has had a few minor leadership roles, but some of the technical management competencies are less familiar. Yu Gin decides that learning more about technical management will improve her skillset and make her more valuable to the company.

She skims the articles on "SE Management" in the SEBoK, and the overview content is pretty familiar. Looking through the works cited and primary references, however, she finds a few really useful references. This leads to the definition of technical management in the SEBoK glossary, which points to a joint INCOSE/Project Management Institute (PMI) working group on the subject. Yu Gin reaches out to this group and identifies further resources for self-study.

After reviewing the resource, Yu Gin is now more familiar with the vocabulary of technical management and looks for related short courses to help her hone her skills. She, of course, is always looking for new ways to apply what she's learned on the job.

Sweta – A Chief Systems Engineer

Sweta is a chief systems engineer at a mid-size electronics company. She has led many programs to success, coordinating hundreds of engineers, project managers, and specialists. Sweta has recently been asked to step in and support the lead engineers in several smaller projects resolve some "knotty issues." She has noted that in almost every case, the issues had two root causes: lack of systems perspective and lack of coordination between teams of different disciplines. Sweta has seen several initiatives to move the workforce in one direction or another fail over her 18 years with the company, so is wary of top-down mandates saying that everyone needs to become a systems thinker or something similar.

Sweta is a long-time user of the SEBoK and remembers that there's a section somewhere about enabling systems engineers. She quickly finds the section and notices a knowledge area on "Enabling Businesses and Enterprises" and starts reading. The articles on determining what capabilities organizations need and developing those capabilities are particularly helpful and lead her to a Harvard Business Review article on change management that is really useful.

Armed with this information, Sweta gets a group of her peers together to discuss the challenges and some potential solutions.

Carlo – An Organizational Manager

The primary audience of the SEBoK is individuals working in the SE discipline. However, the articles are intended to be written in a way that will enable non-systems engineers to grasp the basics.

Carlo is a mid-level manager at a large firm that makes major weapons systems for the nation's defense department. The organization has recently launched an initiative that is intended to move the culture toward a more integrative and collaborative model. One aspect of this is a focus on systems and systems thinking. Leadership has emphasized how critical it is for all levels of management to really "get behind" this initiative for it to be successful. Carlo has never heard the term "systems thinking" before and the only connotation he has with "systems" are the IT systems that are always giving him such trouble.

Carlo Googles "systems thinking" and is immediately overwhelmed by the search results. He is inundated with definitions from companies trying to sell systems thinking services or training, universities advertising their programs, and blog posts, which could be by experts or by people who only think they know what systems thinking is. He stumbles upon something that links "systems thinking" to "systems engineering," which is a term he knows, though he has never been and has no desire to be an "engineer." He updates his search to include both "systems thinking" and "systems engineering" in Google Scholar to try to find more reputable sources. There are tens of thousands of results and, again, Carlo starts to become overwhelmed. He goes to a systems engineer who always says "hello" in the break room and asks him what to do. The systems engineer points him to the SEBoK.

In the SEBoK, Carlo quickly finds a knowledge area on systems thinking and several articles and papers that he believes he can trust to give authoritative information. After reading through the articles and references, Carlo now feels he has a basic understanding of what systems thinking is and he now understands why this is such a critical part of the company's strategy for change. Carlo now has the vocabulary and knowledge to support the initiative as a member of the management team.

Conclusions

The SEBoK is accessed, via the web, in excess of 80,000 times a month. Since version 1 was released in 2012, the SEBoK has been accessed more than six million times. When captured as a PDF, it contains over 1000 pages of content. This includes hundreds of critical references for the discipline organized by topic. It is a critical resource that has been compiled over more than a decade by dozens of subject matter experts and is carefully curated. It is a useful reference for all systems engineers, regardless of their level of expertise, and can be used as a resource to introduce nonsystems engineers to critical topics in the discipline.

References

Carter, J. and Singam, C. (2022). Systems engineering and management. In: *The Guide to the Systems Engineering Body of Knowledge (SEBoK)* (ed. R. Cloutier). Hoboken, NJ: The Trustees of the Stevens Institute of Technology. https://sebokwiki.org/wiki/Systems_Engineering_and_Management.

INCOSE (2019). *Fellows Briefing to the INCOSE Board of Directors*. San Diego, CA: International Council on Systems Engineering.

NASA (2009a). *NASA Competency Management Systems (CMS): Workforce Competency Dictionary, revision 7a*. Washington, DC, USA: U.S. National Aeronautics and Space Administration (NASA).

NASA (2009b). *Project Management and Systems Engineering Competency Model. Academy of Program/Project & Engineering Leadership (APPEL)*. Washington, DC, USA: US National Aeronautics and Space Administration (NASA). https://appel.nasa.gov/career-development/development-framework/ (accessed 3 June 2015).

SEBoK (eds.)(2022). *Guide to the Systems Engineering Body of Knowledge (SEBoK)*, v. 2.7. Robert Cloutier (Editor in Chief). Hoboken, NJ: Trustees of the Stevens Institute of Technology along with the International Council on Systems Engineering (INCOSE) and the IEEE Systems Council www.sebokwiki.org (accessed 31 January 2023).

Watson, M., Mesmer, B., Roedler, G. et al. (2023). Systems engineering principles. In: *The Guide to the Systems Engineering Body of Knowledge (SEBoK)* (ed. R. Cloutier). Hoboken, NJ: the Trustees of the Stevens Institute of Technology. https://sebokwiki.org/wiki/Systems_Engineering_Principles.

Biographical Sketches

Dr. Nicole Hutchison is a research engineer with the SERC and serves as the Digital Engineering Competency Framework principal investigator. Other research projects include Helix, the Mission Engineering Competency Model, Digital Engineering Metrics. She is the managing editor of the Systems Engineering Body of Knowledge (SEBoK). Before coming to Stevens, Dr. Hutchison worked for Analytic Services Inc. primarily focused on public health, biodefense, and full-scale disaster exercises.

Dr. Art Pyster has played leading technical, management, and executive roles in the telecommunications, aerospace, defense, air traffic control, education, and computing business sectors. Pyster's proudest professional accomplishments prior to coming to Mason are (1) standing up and operating the Systems Engineering Research Center, which is a University Affiliate Research Center and the largest and most impactful academic research program in systems engineering in the nation, (2) leading the development and operation of the Systems Engineering Body of Knowledge (SEBoK), whose articles are accessed 80,000 times monthly and is arguably the most widely read source on systems engineering in the world, (3) applying novel engineering processes at Digital Sound Corporation to develop extremely low-defect commercial telecommunications systems; and (4) as the Federal Aviation Administration's Deputy Chief Information Officer, leading its information security efforts, software research program, and IT policy development, and serving on the FAA's investment review board, which oversaw more than US $3 billion annually. At Mason, Pyster is a professor of systems engineering, but his primary role is associate dean for research in the College of Engineering and Computing (CEC). In that capacity, Pyster oversees the CEC research portfolio which has rapidly grown to an anticipated value of over $400M with annual expenditures approaching $70M. Pyster continues to oversee the SEBoK which is sponsored by the International Council on Systems Engineering (INCOSE), the IEEE Systems Council, and Stevens Institute of Technology. He is an INCOSE Fellow, Founders Award recipient, and Pioneer Award recipient. Pyster has authored more than 70 papers and two books.

Dr. Rob Cloutier is the program chair for Systems Engineering at the University of South Alabama. Dr. Cloutier's research interests include concept of operations, system architecting, model-based systems engineering, complex patterns for systems engineering, visualizing sociotechnical systems, and a systems engineering metaverse. He received his PhD in systems engineering in 2006 from the Stevens Institute of Technology in Hoboken, NJ. This thesis was entitled "Applicability of Patterns to Architecting Complex Systems." Dr. Cloutier has also served as the editor in chief of the Systems Engineering Body of Knowledge (SEBoK) since 2018.

Chapter 39

Understanding Critical Skills for Systems Engineers

Nicole Hutchison

Stevens Institute of Technology, Hoboken, NJ, USA

Understanding Critical Skills for Systems Engineers

Most individuals are not content to be mediocre at their jobs, and systems engineers are no exception. In over a decade of research on the skills and abilities that make systems engineers effective/successful/unique – choose your adjective – one of the common themes has been that systems engineers generally seeks to be better today than they were yesterday and still better tomorrow. This desire for continuous growth and improvement is a characteristic that drives systems engineers to be effective. We often say, "begin with the end in mind," so what does "better" look like? One of the most critical ways that systems engineers can improve their effectiveness is to understand the knowledge, skills, abilities, and behaviors (KSABs) that enable them to be *competent* in their jobs.

Competency is a loaded term. Often the term "competency" evokes immediate and strong reactions – we think of the competency models we know and what we love and hate about them. So let's unload the term. Competency is the ability of an individual to do something. Over this chapter, we will add layers and nuances and, of course, specifics to systems engineering to this definition. But we will come back to this simple definition.

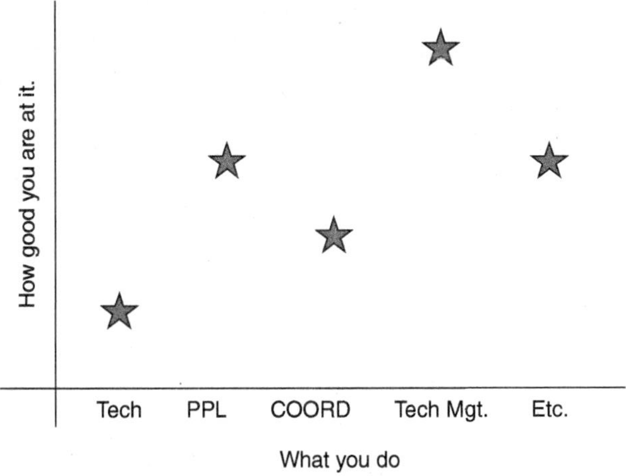

Systems Engineering for the Digital Age: Practitioner Perspectives, First Edition. Edited by Dinesh Verma.

Companion website: www.wiley.com/go/verma/systemsengineering

Individuals have a level of *proficiency* with respect to a competency. Again in simple terms, it's how well or to what degree an individual is able to do a given thing. There are different ways to assess proficiency and just like every competency model or framework has its chosen terms, so too does it have its own proficiency definitions. Some use a numeric scale, some verbal labels, but at the end of the day, the important thing to understand with proficiency is to use a consistent approach, with clearly defined terminology and repeatable steps for determining proficiency.

So let's recap: **competency** is the ability to *do* something and **proficiency** is *how well* you do that something.

Choose Your Competency

Anyone who grew up or had children in the 1980s or 1990s was likely to have at some point been exposed to "Choose Your Own Adventure" books. If you're unfamiliar with these, here's an overview: You read a story and at some point are faced with a choice. For example, perhaps you are alone at home and you hear a strange noise. Do you investigate the noise, or hide under the covers? You would be presented with each of these options and, based on your selection, you would turn to different pages to continue the story. There were many ways a story could go, but not an infinite number. Likewise, when you focus on your own competency development, you cannot make all the choices at once. You may want to improve your proficiency across all your skills, but as a human being with limited time and both a professional and personal life that requires you to distribute those resources, you will have to choose where to start your focus and adjust over time.

- If you already understand how to select what competencies you want to focus on, or better yet already have a personal development plan, go to Page 810. "Assess Your Competency".
- If you want some guidance on how to select where to focus, read on from here.

In order to understand where to focus your time and resources for growth, you must first assess your proficiency. There are many ways to do this. Some can be very rigorous – even too rigorous. If you have not had the mind-numbing experience of trying to assess yourself against 900+ KSAB statements as the newest form of annual evaluations, congratulations. For those of you who have, you will know that while these types of assessments are incredibly thorough, they take an inordinate amount of time, can be confusing, and are often demoralizing. Add in the layer of 360 evaluations, where you have to fill out these competency assessments not only for yourself but several colleagues, and mutiny is not far away.

Simpler competency assessments can be effective and less likely to result in masses fleeing your organization. These can be as simple as self-assessments based on competency and proficiency descriptions or can involve the creation of a more rigorous survey with questions designed to remove some of the biases implicit in self-assessment. Either way, without understanding the baseline of competency within yourself (or your organization), any efforts to improve on competency run the risk of not resulting in necessary changes.

Existing Competency Models and Frameworks

Many competency models currently exist to guide systems engineers. Some are focused on systems engineering directly and others are tangential, such as the mission engineering and Digital Engineering Competency Frameworks (MECF and DECF, respectively). A list of the most popular frameworks with brief descriptions is outlined below.[1]

- **INCOSE systems engineering competency framework:** The International Council on Systems Engineering (INCOSE) published its *Systems Engineering Competency Framework* in 2018. The framework was developed by INCOSE's competency working group over several years and includes a mix of technical and supporting "nontechnical" skills. Access the framework at: https://www.incose.org/docs/default-source/professional-development-portal/isecf.pdf?sfvrsn=dad06bc7_4.

- **Helix systems engineering proficiency model:** Starting in 2012, the Systems Engineering Research Center (SERC) was tasked by the Office of the Secretary of Defense (OSD) to provide insights on what makes systems engineers effective. Over the course of eight years, the research team conducted in-depth interviews with over 600 individuals across 32 organizations in 3 countries. About three-quarters of these individuals were practicing systems engineers, many of whom had been identified as key performers within their organizations, and the one-quarter individuals who work with systems engineers, including their leaders, managers, and colleagues on engineering teams. As part of this work, the team developed a model of the core proficiencies required for systems engineers to be effective. It includes technical, interpersonal, leadership, and mindset-related competencies. Access the framework at: https://sercproddata.s3.us-east-2.amazonaws.com/technical_reports/reports/1602166204-A013_SERC%20WRT%201004_Technical%20Report%20SERC-2020-TR-007.pdf.
- **NASA system engineering competency framework:** The US National Aeronautics and Space Administration's (NASA) SE competency framework was first published in 2007. It is focused primarily on the technical, though the current version does include considerations on professional and leadership development. Notably, it specifically calls out the "-ilities" of safety, security, and mission assurance. Access the Framework at: https://www.nasa.gov/pdf/303747main_Systems_Engineering_Competencies.pdf.
- **Mission engineering competency framework:** In 2017, the SERC published the *Mission Engineering Competency Framework*. It essentially defines mission engineering as systems engineering at the mission level – viewing missions as complex systems of systems. The framework includes technical foundations as well as competencies in cross-functional integration and organizational alacrity. Access the framework at: https://sercproddata.s3.us-east-2.amazonaws.com/technical_reports/reports/1602166204-A013_SERC%20WRT%201004_Technical%20Report%20SERC-2020-TR-007.pdf.
- **Digital engineering competency framework:** In 2021, the SERC published the *Digital Engineering Competency Framework*, which was focused on critical skills for enabling digital transformation. It includes foundational data skills such as governance and management; basics of software engineering; understanding of models and simulations; skills for developing, operating, and maintaining a digital engineering environment; and the systems engineering and analysis skills required in the digital space. Access the Framework at: https://sercproddata.s3.us-east-2.amazonaws.com/technical_reports/reports/1616668486.A013_SERC%20WRT%201006_Technical%20Report%20SERC-2021-TR-005_FINAL.pdf.

There is no "one framework to rule them all." The framework that works for you and your organization will depend on culture, values, and what you are already doing. We all have our favorites, but the important thing is that you chose one and try not to jump between multiple frameworks. By making a choice, you will have a consistent framework within and across your organization and you will have a consistent lens through which to view your own proficiencies.

T- or Pi-Shaped professionals

Systems engineers must be able to handle multiple systems and be transdisciplinary by nature. There are different perspectives on T-shaped skills (e.g., Waldawsky-Berger 2015; August et al. 2010; Michigan State University 2014), but all share two main characteristics: a deep dive of skills in one area (the vertical of the "T") and a lighter covering of skills across many different areas

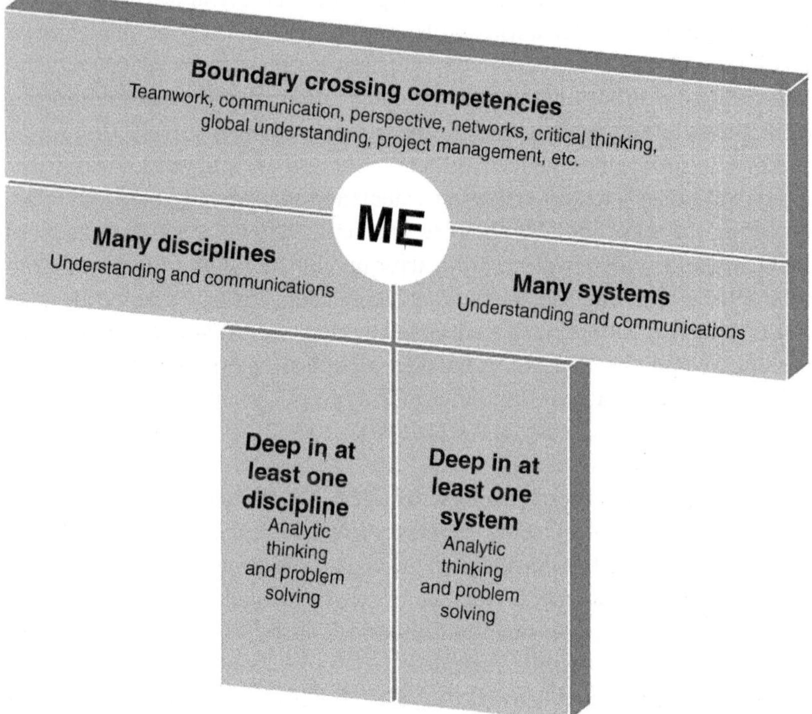

Figure 39.1 T-summit 2016 T-shaped skills model (Michigan State University 2014) fatter T's and Pi's over your career.

(the horizontal of the "T"). The details of the level of depth, the number of disciplines, and cross-cutting or multi-disciplinary aspects should be included, vary between models. One generic example from the T-Summit 2016 Conference (Figure 39.1, Michigan State University 2014) refers to disciplines and systems. This means that the approaches that prepare individuals to become systems engineers – the types of experiences, mentoring, and educational approaches – are also relevant to learning to be more "T-shaped."

A Pi-shaped skills model, similarly has two vertical areas of depth and can have more cross-cutting elements (resembling the Greek letter *p*, therefore, "pi"-shaped). We'll come back to the discussion of T- and Pi-shaped in a bit.

The Skills You Need

As outlined above, individuals and organizations will have to determine which skills are most critical for their work. While there are commonalities among all the frameworks highlighted above that apply to anyone with an interest in modern systems engineering, there are also competencies that are somewhat unique based on the context in which the framework was designed. Again, there is no one-size-fits-all competency model for all the different flavors of systems engineering being practiced around the world. However, any competency work should include:

- **Foundational skills:** There are basic foundations that are required for systems engineers. Generally, these reflect mathematics, an understanding of scientific methods with some grounding in natural science, an awareness of the fundamentals of engineering, and, more recently, basic computing foundations.

  - **Digital literacy:** The way engineering is practiced has become increasingly digital and this trend will continue to accelerate. Digital literacy here refers to an individual's ability to operate in a primarily digitally based environment as opposed to a document-centric one.
- **Domain and user context:** Systems are built for purpose with specific user needs and with an expected operating environment. Though systems principles apply regardless, it is critical for an individual to understand the terminology, considerations, and user expectations unique to their domain. Likewise, a given context will have its own constraints – safety regulations in the aviation and medical device industries, security requirements in IT systems, etc.
- **Foundational systems engineering:** These are the skills that most likely come to mind for anyone when they first think about systems engineering skills. They are the things you would think of when you discuss reflect on ISO 15288, the INCOSE *Systems Engineering Handbook* or the NASA *Systems Engineering Handbook*. These are the skills that support systems activities throughout the systems lifecycle – from initial problem statement to design and implementation, verification and validation, operation, and through to upgrades or disposal. These skills also include the orthogonal skillsets that align with project or engineering management – the critical activities that happen throughout the lifecycle like planning, configuration management, risk management, etc.
- **Modeling and simulation:** As systems engineering continues toward increasingly digital ways of working, modeling and simulation become critical foundations. Models are abstracted representations of the real world while simulations are models that mimic the operation of a system, allowing us to understand projected behaviors and test different scenarios. Model-based systems engineering (MBSE) has been a critical topic in the field for well over a decade and the current focus on digital transformation and digital engineering brings these skills even further to the forefront.
- **Data foundations:** The fundamental backbone of a digital environment, model, or simulation is data. Without authoritative data managed in a logical and understandable way, digitally based engineering will fail. It is imperative that systems engineers have at least a foundational understanding of what authoritative data is and how data should be managed.
- **Systems mindset:** Though many models mention systems thinking, systems mindset encompasses more than just this. Paradox mindset – the ability to hold two opposing but equally correct ideas in juxtaposition is a critical skillset of effective systems engineers. If you've heard jokes about how a systems engineer's answer to any given question is always "it depends," then you've heard some illustrations of this. Paradox mindset also enables systems engineers to focus on balance. An effective systems engineer will have a "30,000 foot" view of their system, but also understand enough of the details to facilitate balanced decision-making – and will have the judgement to know when to focus on one vs. the other. Finally, abstraction is a critical aspect of systems mindset. Abstraction here is the ability to understand specifics, but see the larger patterns that emerge from specific instances rather than focusing on the details. This pattern recognition is a critical way systems engineers provide value.
- **Interpersonal skills:** Systems engineering is by definition team-based activity with a clear social context. For systems engineers to be effective, they must have the ability to work within teams. Communication skills are a critical aspect of this – and these skills are diverse. Systems engineers must be able to share information efficiently in verbal or written formats and due to the transdisciplinary role they play, presentation skills are also critical. Another way systems engineers provide value is by applying their understanding of a team's skills and competencies to the problems at hand. One way this manifests is by being a resource who may not be able to solve all the problems but knows the right people to bring together to solve problems.
- **Technical leadership:** Systems engineers often have a leadership role on teams, whether formally or informally. In order to provide critical values such as leading teams toward a consistent system vision or making balanced decisions or taking rational risks at the systems level, systems

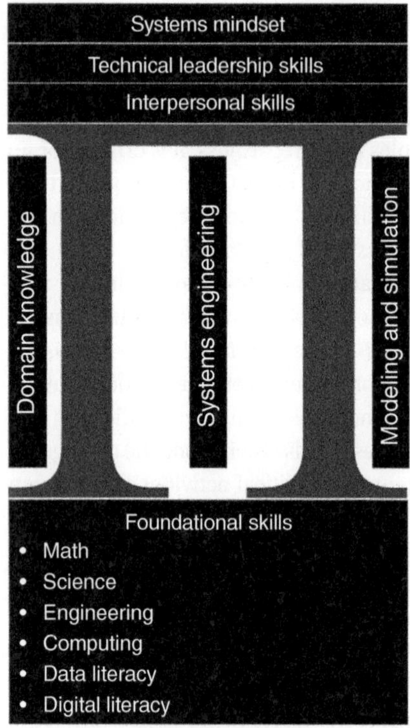

Figure 39.2 Systems skills in a Pi-shaped model.

engineers must have foundational leadership abilities that they can apply in the context of their technical tasks. A critical way systems engineers provide value in leadership roles is through conflict resolution and barrier-breaking – bringing their systems mindset skills forward to help a team come together to solve complex problems.

Let's come back to the context of T- or Pi-shaped skills. It should be easy to see that many aspects of systems engineering are cross-cutting. Technical leadership, interpersonal, and systems mindset skills all enable a systems engineer to span the boundaries of disciplines, domains, and even culture. And as stated previously, while systems engineers are in some ways jacks (or janes) of all trades, they also need to master a few areas to a sufficient level of depth to enable them to "speak engineer" and, bluntly, to have sufficient technical ability that engineering teams will trust their judgment. Conceptually, the various skills discussed above can be illustrated as shown in Figure 39.2.

You might be thinking to yourself, "there are more than two legs on that pi," and you would be right. And that is OK – the point of the Pi-shaped model is to think of skills in terms of whether they are largely cross-cussing or provide some depth of expertise. Obviously, no single systems engineer will be an expert at each one of these facets – nor should they be. And that is where the discussion of individuals and teams comes in (we'll get to this shortly).

Assess Your Competency

Now that you know the main skills required and have frameworks for reference, you have to decide which skills you want to focus on. There is no one person who could possibly master every single one of the competency areas described above, so choices must be made on where to focus. These decisions will depend largely on your strengths, your areas of interest, and the context in which you are working or hope to work. The other aspect to consider is the level of proficiency (depth) that you want to attain across these areas. For the sake of modeling the behavior we want to see, we will make a choice and use the Helix framework for our discussion of assessment.

The Helix framework has six competency areas:

- **STEM foundations:** foundational concepts from mathematics, physical sciences, and general engineering;
- **System's domain and operational context**: relevant domains, disciplines, and technologies for a given system and its operation;
- **Systems engineering discipline**: the foundation of systems science and systems engineering knowledge, which also includes technical skills such as modeling and simulation;
- **Systems mindset**: skills, behaviors, and cognition associated with being a systems engineer;

- **Interpersonal skills**: skills and behaviors associated with the ability to work effectively in a team environment and to coordinate across the problem domain and solution domain; and
- **Technical leadership**: skills and behaviors associated with the ability to guide a diverse team of experts toward a specific technical goal.

The proficiency levels used in Atlas can be found in Table 39.1.

With your framework and proficiency levels in hand, it is time to assess yourself. A few notes about this:

- **Be honest:** Self-assessments are notoriously biased. Most people will consistently rate their own skills higher than their managers or peers would rate them or than a more question-based rubric would. This does not mean self-assessment is useless – it just means that you have to be aware of this and be honest with yourself. The reason you are doing this is to improve your skills. This is impossible without clear self-reflection.
- **Consider experts:** One of the ways to keep yourself honest is to think of the best person you know in an area and think about how you compare to that person. This may be someone in your organization, a mentor, or just someone you admire. But if you think they are "expert" in an area and you know you are not at their level, then do not rate yourself as an "expert."

Table 39.1 Helix proficiency levels.

#	Level	Level description
1	***Fundamental awareness***	Individual has common knowledge or an understanding of basic techniques and concepts. Focus is on learning rather than doing
2	***Novice***	Individual has the level of experience gained in a classroom or as a trainee on-the-job. Individual can discuss terminology, concepts, principles, and issues related to this proficiency, and use the full range of reference and resource materials in this proficiency. Individual routinely needs help performing tasks that rely on this proficiency
3	***Intermediate***	Individual can successfully complete tasks relying on this proficiency. Help from an expert may be required from time to time, but the task is usually performed independently. The individual has applied this proficiency to situations occasionally while needing minimal guidance to perform it successfully. Individual understands and can discuss the application and implications of changes in tasks relying on the proficiency
4	***Advanced***	Individual can perform the actions associated with this proficiency without assistance. The individual has consistently provided practical and relevant ideas and perspectives on ways to improve the proficiency and its application and can coach others on this proficiency by translating complex nuances related to it into easy-to-understand terms. Individual participates in senior-level discussions regarding this proficiency and assists in the development of related reference and resource materials
5	***Expert***	Individual is known as an expert in this proficiency and provides guidance and troubleshooting and answers questions related to this proficiency and the roles where the proficiency is used. Focus is strategic. Individual has demonstrated consistent excellence in applying this proficiency across multiple projects and/or organizations. Individual can explain this proficiency to others in a commanding fashion, both inside and outside their organization

Source: Adapted from Pyster et al. (2018), used with permission.

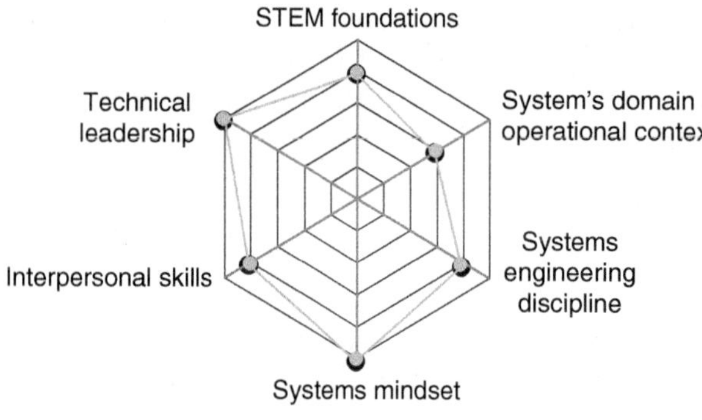

Figure 39.3 Helix proficiency model. *Source:* Hutchison et al. (2020, used with permission).

- **Consider details**: We can all have a quick reaction of our skills in an area. But looking at least one level down – in the Helix model the descriptions of the individual competencies. Often times, an area may include things you did not consider and it is only by considering the area holistically that you will get a realistic assessment.
- **Request review**: Our friends keep us honest. Ask someone you trust – peer, leader, mentor, whomever – and have them take a look at your self-assessment and give you some honest feedback. (*Hint*: Do not ask your subordinates. If you think self-assessment is biased, you should see the data on assessing leaders or managers when the feedback is not anonymous.)
- **Map It**: For many engineers, visualizations are more impactful than a lot of text. So map out your proficiency self-assessment. Figure 39.3 provides an example of this. Consider the middle of the radar chart to be "none" – i.e. "I can't even spell systems engineering" – and the outermost ring to be "expert." This quick visual and the insights you gained by working your way through the framework to assess yourself should give you some substantive insights.

Choose Your Path

You know where you are and have confirmed it with at least one trusted person. Now you need to choose where you're going. This is where having a visual map can be helpful because you cannot choose to improve in every area all at once. Let's assume that Figure 39.4 is your map. The solid line is where you are now, and you realize that you are doing well – functioning as "advanced" – in the systems domain and operational context area and interpersonal skills. So while you can still improve, these may not be the areas you want to focus on at the moment. You are an intermediate practitioner when it comes to your STEM foundations and application in the systems engineering domain. But you have rated yourself as a "novice" when it comes to technical leadership and systems mindset. You want to be great in everything – we all do – but realistically, you need to make some choices. So while you start with targets like the ones on the left, you (perhaps with the help of that trusted person) quickly decide that for the job you are currently doing, you really need to focus on improving your systems engineering mindset and systems engineering skills (the right). Why not focus on technical leadership? You are a junior person, so opportunities to hone leadership skills are hard to find currently, and in your organization honing your technical skills will make it more likely that you will have more opportunities to grow your leadership skills.

Now you know where you are and where you want to go. The decisions you need to make now focus on the how. How will you get there? What path(s) are readily available to you? The following chapter discusses career planning in more depth, but generally there are a few ways you can improve your skills:

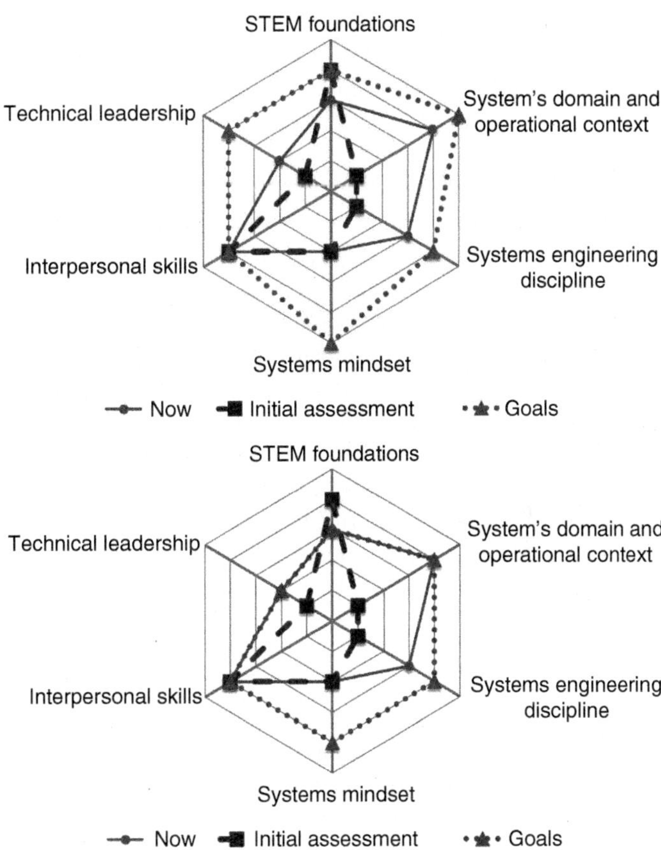

Figure 39.4 Example longitudinal competency maps.

Experiences

Experiences are considered the most critical factor contributing to the development of proficiencies and to the overall growth of systems engineers. Depending on the role you play in your organization, you may or may not have a lot of control over the experience opportunities you have. Sometimes the best way to gain experience is to volunteer for it. Taking on extra work is not always appealing, but if you are in an organization that does not allow you a lot of flexibility in choosing your assignments, it may be the easiest way to gain experience in needed areas. Sometimes that is not an option at work, and you will need to look for other options. Some people reported that they get the experiences they need through engaging with professional societies or groups like engineers and scientists acting locally, some volunteer with their children or niece's or nephew's schools to support competitions like Odyssey of the Mind or First Robotics, while others said that they simply became tinkerers, finding challenges they could to enhance their skills at home. There is no right way – which is why this is title "experiences" instead of "experience" – though there are some patterns.

The Helix project collected detailed career data from over 150 systems engineers, those who rose to the top – becoming chief systems engineers or chief engineers – tended to have a few things in common. The strongest commonality was variety – effective systems engineers tend to see variety in their roles, rather than staying in one area. Variety can mean different things: seeing different parts of the life cycle, trying a project in a new domain, or taking on a role at the systems level when you've previously worked on components. Doing the same thing repeatedly and expecting a different result is a definition of insanity – and it likely will not help you learn and grow the way expanding your boundaries will.

Mentoring/Coaching

We've talked already about getting feedback from others, and that is the essence of mentoring and coaching. **Mentoring** is a more general opportunity for a more experienced individual to provide guidance to a less-experienced individual. **Coaching** tends to be more focused on a specific problem – maybe you need to learn a new tool or you had been working on communications systems and now you need to focus more strongly on satellite systems – either way, a coach is someone who helps you through the specific on the job.

You can certainly ask someone to mentor you on your career, but when it comes to growing your proficiency, the main point here is to just engage someone outside yourself. Get feedback, get advice, seek a second opinion around what you are doing. There is no right way to mentor and, in fact, sometimes formal mentoring programs that are highly structured are uncomfortable for people. Find someone you trust to give you honest feedback and sit down with them in whatever way makes sense for you. Be like Nike: just do it.

Training and Education

In most organizations, if you tell someone you want to get better at something, their first instinct will be to tell you about the training available. New to systems engineering? We have a systems engineering 101 brown bag lunch once a month. Need to learn more about Systems Modeling Language (SysML) for your new project? We can send you to a short course on that. Training can be incredibly valuable, but can also be treated like a panacea. Be careful here: some things lend themselves well to training, others do not. In fact, in the Helix study, while training and education were found to be one of the drivers of systems engineers' growth, it was often cited as useful, but less helpful than experiences or mentoring. Training tends to be a shorter engagement around a particular topic or area. So if you are new to systems engineering, that SE101 brown bag may be a great first step, but training is only retained and valuable if it is applied. Pairing training with coaching, where someone helps you apply what you've learned on the job, is generally much more effective than just taking a course.

Education is generally more formal and broadly applicable, providing theoretical foundations, and falls the same way. It is valuable and has its place, but it may or may not be the right fit for what you are trying to learn.

Creating Your Path

As with most endeavors, it is useful to set SMART goals to your competency development approach:

- **Specific:** Clearly state your goals. Use action verbs and be specific.
- **Measurable:** As engineers, we sometimes can go a little overboard on this one. While we can really get into the details and create intricate metrics, when it comes to more qualitative goals, sometimes the best way to make sure it is "measurable" is to define it in a way that you can clearly say, "yes, I did that" or "no, I did not."

- **Achievable:** We are human. We balance career and home, sometimes school and other activities. While it is admirable to say we will go from a "novice" to an "expert" across the board in a year, it is not realistic. Don't set yourself up for failure by setting a goal that you cannot achieve.
- **Relevant:** This seems like common sense, but don't set a goal that won't actually help you reach your goal. If you are an advanced systems engineer, taking an "SE 101" course would not make a lot of sense; you could do it, but it is unlikely to help advance your practice. If you are brand new to systems engineering, jumping into a systems engineering PhD program probably is not the right step either.[2]
- **Time-bound:** Be specific about the time you will take to achieve this. Put a date on the calendar (with reminders leading up to it) and stick to it. Going back to the idea of "achievable," be realistic with your time frame. Most people can think 6 or 12 months ahead. People tend to struggle thinking a few years out. And a goal that will take 5 or more years to achieve? That might be your ultimate end goal, but you need to identify some waypoints on your journey.

In closing, there is no right or wrong way to work on your growth as a systems engineer. The only wrong thing is to not start. Pick any competency model that makes sense to you, spend some time reflecting honestly on your skills, and think about how you can do better. That's it. So get to work!

Notes

1 These are the frameworks of most relevance. Additional frameworks can be found in the references.

2 One caveat – sometimes an opportunity that seems tangentially but not directly aligned can be great. Almost every senior systems engineer interviewed in the Helix project said, "I came to systems engineering through a circuitous route." If a digital marketing short course might help you improve your presentation skills, especially when targeting specific stakeholders, go for it. Just make sure you can see the roadmap between the path and your goal.

References

Competency Models

DAU (2016a). *ENG: Engineering Career Field Competency Model*. Version 2.0. Fort Belvoir, VA: Defense Acquisition University, Department of Defense.

DAU (2016b). *PM: Project Management Career Field Competency Model*. Fort Belvoir, VA: Defense Acquisition University, Department of Defense.

Hutchison, N., Verma, D., Burke, P. et al. (2018). *Atlas 1.1: An update to the Theory of Effective Systems Engineers*. Hoboken, NJ: Systems Engineering Research Center, Stevens Institute of Technology.

Hutchison, N., Verma, D., Burke, P. et al. (2020). *Atlas: Effective Systems Engineers and Systems Engineering*. SERC-2020-TR-007-A. Hoboken, NJ: Systems Engineering Research Center (SERC), Stevens Institute of Technology.

IEEE (2014). *Software Engineering Competency Model (v. 1.0)*. IEEE Computer Society.

MITRE (2007). *Systems Engineering Competency Model*. McLean, VA: MITRE Corporation.

NASA (2009). *NASA's Systems Engineering Competencies*. Washington, DC: National Aeronautics and Space Administration http://www.nasa.gov/pdf/303747main_Systems_Engineering_Competencies.pdf.

NASA (2020). *aPPEL Knowledge Services: PM&SE Career Development Framework*. Washington, DC: National Aeronautics and Space Administration https://appel.nasa.gov/career-resources/development-framework/.

NAVAIR/NAVSEA Working Group (2020). Digital competency framework overview: NAVAIR/NAVSEA Working Group. Presentation.

Pyster, A.B., N.A.C. Hutchison, and Henry, D. (2018). *The Paradocixal Mindset of Systems Engineers: Uncommon Minds, Skills, and Careers*. Hoboken, NJ: John C. Wiley & Sons.

SFIA Foundation (2018). *Skills Framework for the Information Age 7: The Complete Reference. SFIA Foundation.*

Vesonder, G., Verma, D., Hutchison, N., et al. (2018). RT-171: mission engineering competencies technical report. *Systems Engineering Research Center (SERC), Stevens Institute of Technology, Hoboken, NJ. SERC-2018-TR-106.*

Whitcomb, C., Khan, R., and White, C. (2017). *The Systems Engineering Career Competency Model, Version 1.0*. Monterey, CA: Naval Postgraduate School.

T-Shaped Skills

August, P.V., Swift, J.M., Kellogg, D.Q. et al. (2010). The T assessment tool: a simple metric for assessing multidisciplinary graduate education. *Journal of Natural Resources & Life Sciences Education* 39: 15–21.

Michigan State University (2014). *"What is the 'T'?" Washington, DC, March 21–22, 2016, T-Summit 2016*. University of Michigan & IBM http://tsummit.org/t (accessed 3 June 2016).

Waldawsky-Berger, I. (2015). The rise of the T-shaped organization. *The Wall Street Journal*. https://www.wsj.com/articles/BL-CIOB-8754 (accessed 18 December 2015).

Additional Resources

Giffin, R., N. Hutchison, C. Lipizzi, et al. 2021. WRT-1018: DAU credential development. Hoboken, NJ: Stevens Institute of Technology, Systems Engineering Research Center (SERC). *SERC-2021-TR-008*.

Biographical Sketches

Dr. Nicole Hutchison is a research engineer with the SERC and serves as the Digital Engineering Competency Framework principal investigator. Other research projects include Helix, the Mission Engineering Competency Model, Digital Engineering Metrics. She is the managing editor of the Systems Engineering Body of Knowledge (SEBoK). Before coming to Stevens, Dr. Hutchison worked for Analytic Services Inc. primarily focused on public health, biodefense, and full-scale disaster exercises.

Chapter 40

Evolving University Programs on Systems Engineering

Paul T. Grogan

Stevens Institute of Technology, School of Systems and Enterprises, Hoboken, NJ, USA

Introduction

Systems engineering (SE) is an interdisciplinary field that manages technical processes within organizations undertaking complex engineering projects. The International Council on Systems Engineering (INCOSE) defines the practice of SE as "a transdisciplinary and integrative approach to enable the successful realization, use, and retirement of engineered systems, using systems principles and concepts, and scientific, technological, and management methods" (Sillitto et al. 2019). In addition to spanning multiple engineering disciplines, SE also incorporates concepts from adjacent fields like organizational science and management to provide a broad-based perspective on how to structure large-scale human activity to achieve technical objectives.

Historically, SE practice has relied upon significant practical experience accumulated over a career including disciplinary engineering roles undertaken during formative years. SE education programs thus seek to reduce the time required to effectively serve in a SE role by structuring and communicating key principles. University programs also build upon academic foundations that strive to understand and influence complex systems that exist at the boundaries of human perception. Most universities offering SE programs today target graduate education; however, broader calls for reform in engineering education (e.g., Dym 2004; Grasso and Burkins 2010; Crawley et al. 2014) show other opportunities for systems education including at the undergraduate level.

Background

University Programs

At the confluence of several domains and disciplines, SE educational programs take on many names and reside across various academic units. A review of more than 100 SE-affiliated programs at U.S. universities by the Systems Engineering Research Center, SERC and INCOSE (2021) shows educational programs reside in departments that cross various engineering disciplines including systems, industrial, manufacturing, mechanical, electrical, computer, and software and intersect

Systems Engineering for the Digital Age: Practitioner Perspectives, First Edition. Edited by Dinesh Verma.

Companion website: www.wiley.com/go/verma/systemsengineering

with additional topics like technology management, operations management, engineering management, engineering systems, and systems science.

While all SE programs are centered around common objectives of engineering complex systems, they tailor content to different audiences. For example, programs closely aligned with industrial engineering tend to emphasize manufacturing and operations research while those aligned with electrical and computer engineering tend to emphasize automatic control. Some programs specialize in individual application domains such as healthcare, aerospace, defense, and manufacturing through faculty expertise and selected industry and government partners.

Graduate Programs

Universities most frequently offer SE educational programs at the master's degree level, usually requiring a prior bachelor's degree in engineering, science, or mathematics. Some programs offer a master of science (MS), while others offer a master of engineering (MEng. or ME). While there are no formal criteria distinguishing between the two, MS degrees more often include a significant research component (e.g., a thesis), which can serve as an intermediate step to doctoral research. MEng. degrees are usually considered terminal degrees focused on professional preparation, consisting primarily or exclusively of coursework. Some universities offer MEng. degree programs in SE through a continuing education or professional education academic unit, rather than a traditional academic department, to specifically target and support students with significant working experience.

Master's degree programs vary by university for on-campus and online settings and full-time or part-time basis. Most programs consist of about 30 credit hours[1] of coursework and research, which can be completed in less than two years on a full-time basis. Completing a master's degree can take more than three years on a part-time basis. Some employers and other benefit programs provide tuition benefits to pursue a master's degree on a part-time basis.

Doctoral Programs

Advanced study of SE at the doctoral level is available at select institutions. A doctoral degree is required for some research-oriented career paths (e.g., university faculty, research scientist, research fellow, principal investigator, and some director-level positions) but is not generally required to practice SE. A doctor of philosophy (PhD) is the most common degree type; however, some programs alternatively provide a doctor of engineering (DEng.) which is more oriented toward industry professionals. The distinguishing feature of a doctoral program is an independent dissertation project through which the student, under guidance of an advisory committee of faculty and industry experts, demonstrates expertise in a topic by contributing new knowledge to a related research community. Publication of research results, e.g., in peer-reviewed conferences and academic journals, may be required for program completion.

Doctoral programs require a combination of coursework and research credits with coursework usually completed within the first two years. Most also require some amount of time spent in residence at the university or a research facility. Although labels and timing vary, doctoral program milestones include one or more oral or written examinations (qualifying, general, or comprehensive) on specific course material and general topics relevant to the area of study, a proposal defense (preliminary examination of the proposed research project), and dissertation defense (final examination) of the completed research project. The doctoral completion timeline is more uncertain than other degree programs because it depends on creative and novel contributions to the field. Full-time students usually plan to complete a doctoral program within four years; however, it is not uncommon to require five or more years.

Undergraduate Programs

Offering SE-focused education in undergraduate programs remains somewhat uncommon because students at this level often lack the practical experience to put more abstract SE learning objectives into context. Furthermore, the reduction of disciplinary engineering coursework is perceived to limit a student's ability to transition to disciplinary entry-level roles.

Given the challenges above, SE programs that offer a bachelor of science (BS) or bachelor of engineering (BEng. or BE) degree typically align it closely with a related disciplinary program or topic like industrial engineering, engineering management, or control systems engineering. Some engineering programs complement disciplinary degree programs with SE-focused introductory courses, capstone projects, minors, certificates, or concentrations to encourage pursuit of SE careers (e.g., see Muller and Bonnema 2013; Meyer et al. 2016; Muci-Kuchler et al. 2020).

University Program Structure

SE programs vary greatly from university to university, building on local strengths and expertise. Despite differences, recent efforts have been able to describe and categorize a generalizable SE program structure common to many programs.

The Graduate Reference Curriculum for Systems Engineering (GRCSE) provides recommendations for how to structure professional master's degree SE programs (Pyster et al. 2015). Objectives describe the purpose of the program with respect to what graduates can accomplish three-to-five years after program completion. Outcomes describe measurable student abilities at the time of graduation. Finally, a curriculum architecture discusses six knowledge components that comprise a curriculum: preparatory (knowledge required at the start of the program), foundation (common SE knowledge required of all graduates), concentration (specialized SE knowledge required of a selected concentration area), domain-specific (contextual knowledge about the intended SE application domain), program-specific (contextual knowledge from the program's institutional perspective), and capstone (knowledge developed through a final capstone experience).

GRCSE also provides a core body of knowledge (CorBoK) that outlines the specific SE knowledge students are expected to learn (Pyster et al. 2015). Topics consist of five parts: Foundations of SE, SE and Management, Applications of SE, Enabling SE, and Related Disciplines. Foundations of SE includes topics like systems fundamentals, systems science, systems thinking, modeling, and systems approaches in engineering. SE and Management covers lifecycle processes and models including concept definition, system definition, system realization, system development and use along with SE management, product and service life management, and SE standards. Applications of SE covers product SE, service SE, enterprise systems, and systems of systems. Enabling SE covers businesses and enterprises, teams, and individuals. Related Disciplines touches on software engineering, project management, industrial engineering, procurement and acquisition, and specialty engineering (e.g., safety, human factors, resilience).

The CorBoK topics closely align with results from El Khoury (2020), which identify six categories of SE education needs based on analysis of papers, handbooks, definitions, interviews, and job postings. Categories include requirements identification, generation, propagation, and analysis; system architecture and modeling; solution selection and implementation; system performance parameters analysis; verification, validation, and testing activities; and project management and planning, technical writing, and teamwork.

Professional Societies

While this chapter focuses on university educational programs, the applied nature of SE benefits from a surrounding ecosystem of organizations. Several professional societies serve the SE community by providing knowledge repositories and educational initiatives, as well as communities of practice for educators and practitioners alike.

Three societies are particularly focused on SE topics. INCOSE draws heavily on the community of SE practitioners to provide conferences, workshops, and cross-cutting products for working professionals and students alike. The Institute of Industrial and Systems Engineers (IISE) also serves student and professional communities, having formally added "Systems" to its name in 2016 to reflect the broader perspective expected of industrial engineers. Additionally, the Council of Engineering Systems Universities (CESUN) consists of member universities that work together to mature educational and research programs on interdisciplinary topics including SE and others like technology and policy, engineering management, innovation, entrepreneurship, operations research, manufacturing, product development, and industrial engineering.

Other professional societies also serve the SE community through focused divisions, technical committees, and program tracks at conferences. The American Society for Engineering Education (ASEE) maintains an SE division focused on educational efforts. The Institute of Electrical and Electronics Engineers (IEEE) maintains a Systems Council, which integrates activities with systems engineering specialty areas. The American Society of Mechanical Engineers (ASME) and American Institute of Aeronautics and Astronautics (AIAA) societies do not have dedicated technical committees or divisions focused on SE; however, special tracks on domain-specific SE topics are typically available at conferences and workshops.

Analysis of System Engineering Curricula

This section uses natural language processing (NLP) to analyze the contents of university program SE curricula through textual analysis of course titles and descriptions on publicly available course catalogs. Results show clusters of topics prevalent across SE programs that comprise the core of academic programs.

Data Source

The text corpora consist of titles and descriptions for 1561 graduate-level university courses. Data was collected from publicly available course catalogs for 47 university programs in the United States having an SE-affiliated master's degree program (listed in Table 40.1). These were identified from the Worldwide Directory of Systems and Industrial Engineering Programs (SERC and INCOSE 2021). Selected courses were either required for SE degree program completion or electives within the corresponding academic unit (department or school). All capstone, thesis, research, work experience, special topics, or similarly vaguely worded courses are censored from analysis.

Methods

The data processing pipeline constructs course associations though semantic similarity using open-source NLP methods in nltk and sklearn libraries available in the Python language.

Documents concatenate course titles and descriptions. A term frequency-inverse document frequency (TF-IDF) vectorizer from the nltk Python library transforms documents into vectors. To

Table 40.1 Selected university programs.

1) Air Force Institute of Technology	17) Johns Hopkins University	33) University of Arizona
2) Arizona State University	18) Kennesaw State University	34) University of Arkansas at Little Rock
3) Auburn University	19) Loyola Marymount University	35) University of Illinois at Urbana-Champaign
4) Binghamton University	20) Missouri University of Science and Technology	36) University of Maryland
5) Bradley University	21) Montana State University	37) University of Maryland, Baltimore County
6) California State Polytechnic University, Pomona	22) Oakland University	38) University of Michigan
7) California State University, Dominguez Hills	23) Ohio University	39) University of Minnesota Twin Cities
8) Colorado State University	24) Ohio State University	40) University of Pennsylvania
9) Cornell University	25) Old Dominion University	41) University of Rhode Island
10) Drexel University	26) Pennsylvania State University	42) University of South Alabama
11) Embry Riddle Aeronautics University	27) Portland State University	43) University of Texas at Arlington
12) Florida Institute of Technology	28) Rensselaer Polytechnic Institute	44) University of Texas at El Paso
13) George Mason University	29) Southern Methodist University	45) University of Virginia
14) George Washington University	30) Stevens Institute of Technology	46) Virginia Tech
15) Georgia Institute of Technology	31) University of Texas at Dallas	47) Worchester Polytechnic University
16) Iowa State University	32) University of Alabama – Huntsville	

avoid common terms, the maximum document frequency restricts vector components to words in less than 30% of documents (e.g., it would not be informative to know that the word "engineering" occurs in most of the course descriptions). To avoid infrequent terms, the minimum document frequency restricts vector components to words in at least 15 documents. Vector components consider n-grams of length 1-2 (1- or 2-word phrases) and strip out Unicode accents. The tokenizer uses the WordNet lemmatizer from the nltk Python library configured with English language and punctuation stop words to avoid duplicating vector components for each verb conjugations and closely related term. The resulting vector has 850 components (i.e., courses are described in an 850-dimension quantitative space).

Next, a k-means clustering algorithm from the sklearn.cluster Python library clusters the resulting vectorized documents into clusters. After trying between 5 and 14 clusters, $k = 12$ clusters were selected to maximize the silhouette score.

To visualize the vectorized documents, the multidimensional scaling (MDS) function from the sklearn.manifolds library projects the 850-dimensional data to two dimensions based on cosine similarity. In other words, two courses are close to each other if they share similar words within the 850 components selected by the vectorizer. The background color of the two-dimensional plane is computed based on concentration of similarly classified courses.

Results

Figure 40.1 shows the NLP analysis results as a two-dimensional projection of vectorized course titles and descriptions. Each dot shows the position of one of the 1561 courses and its color shows the best-fitting classification category. Nearby courses share common terms. The background is

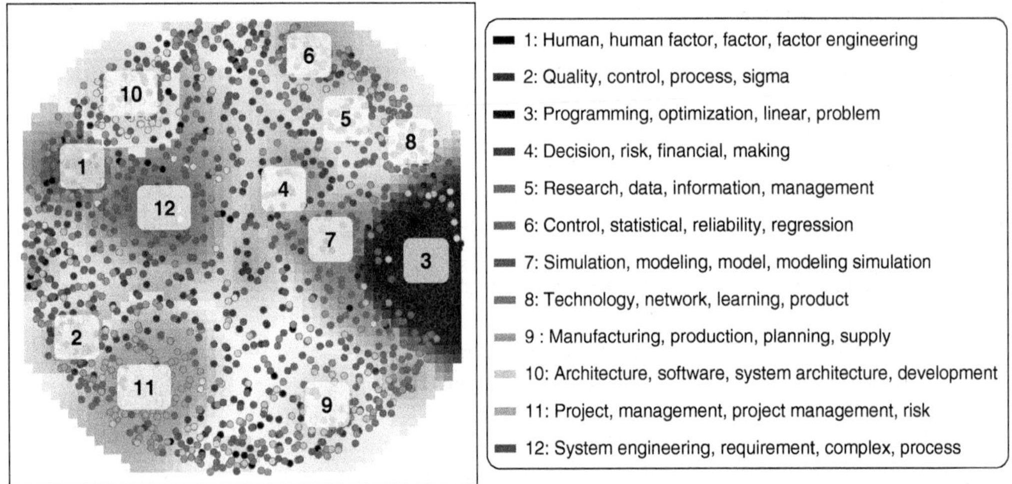

Figure 40.1 NLP analysis of titles and descriptions of 1561 graduate SE courses from 47 U.S. universities shows 12 clusters of closely related terms.

shaded based on the local prevalence of courses with the same classification. Numeric labels highlight the perceived centroid of each cluster.

The 12 classification groups listed in the figure legend list the top four *n*-grams (terms) representative of the classification. Some categories are very coherent (e.g., cluster 3, dark blue, captures core topics in operations research), while others (e.g., cluster 5, medium blue) are more diffuse due to smaller volumes of courses offered or cross-cutting topics.

Discussion

Results highlight several topic clusters within SE programs. While not every topic is represented in each university program and some programs focus only on a narrow set of topics based on local expertise and specialization, the results provide an overview of SE program breadth. Reviewing the overall landscape shows some interrelationships between topics.

The right side of the figure shows courses that traditionally are more quantitative in nature including optimization (cluster 3, dark blue), modeling and simulation (cluster 7, dark green), and learning technologies (cluster 8, medium green) including neural networks. These courses have a strong mathematical foundation and are emphasized from industrial engineering and data science perspectives.

The left side of the figure is dominated by topics aligned with broader perspectives on SE, such as requirements engineering and complex systems (cluster 12, red), system architecture (cluster 10, yellow), project management (cluster 11, orange), quality control (cluster 2, purple), and human factors (cluster 1, black). Many of these courses emphasize the role of humans, either as engineers or users of a system under consideration.

Some domain-specific applications are also visible within the results. Manufacturing, also encompassing production and supply chain planning, resides in the lower portion of the figure (cluster 9, lime green). Additionally, the software domain appears in the upper portion of the figure (cluster 10, system architecture). Other domains do not face sufficient volume across all programs to appear as separate clusters or highly referenced terms.

It should be noted that the results discussed in this section are based upon publicly available course descriptions that are somewhat terse in nature and may not be frequently updated due to required academic review and approval processes. This issue is mitigated somewhat by the large data set considered which comprises more than 1500 courses across 47 programs. Additional review of programmatic information, for example, course objectives and outcomes or more detailed syllabi, may help provide greater depth of analysis.

Evolving University Programs

Due to an intrinsic lag in academic program development, current SE curricular content cannot be expected to support all future SE needs. To help imagine the future, the *Systems Engineering Vision 2035* (INCOSE 2022) highlights themes that are expected to change the SE landscape over the next ten years. Broadly, the future of SE focuses on the design and management of increasingly complex cyber-physical systems (CPS) that blur the boundary between hardware and software. The following subsections look specifically at four features: digital and model-based practice, artificial intelligence and machine learning, data science, and human-systems integration.

Digital and Model-Based Practice

Building on the digital transformation that is already underway, SE will continue to move from document- to model-based activities, enabling more of the SE workflow to be completed in virtual environments with automation support. Digital models broadly facilitate the coordination and exchange of technical information across teams while enabling integrated testing prior to physical fabrication. Advanced visualizations help SE practitioners view large amounts of data and understand interdependencies between system components or across lifecycle phases. Computational models search a large design space for solutions, anticipate emergent system behavior, identify potential problems early on, and continuously monitor behavior throughout lifecycle processes from conception to retirement.

New educational opportunities significantly expand upon existing topics like systems modeling and simulation to consider a broader and more pervasive role for interdisciplinary modeling and computational simulation with particular attention to data standards and interoperability. Building on foundations in computer science and systems science, university programs can provide expertise in systems modeling activities using human- and machine-readable executable languages capable of generating simulated data to support SE lifecycle activities.

Digital and model-based practice provides another benefit to university programs: virtual SE environments can help align classroom and practitioner experiences. Simulated SE projects using computational models rather than physical components can help students experience an entire system lifecycle within the bounds of a one-semester class.

Artificial Intelligence and Machine Learning

Infusion of learning technologies including artificial intelligence (AI) and machine learning (ML) into engineered systems presents a fundamental challenge to SE because their behaviors cannot be fully tested in a closed setting. Rather, the field requires new technical management processes collectively described as SE4AI (McDermott et al. 2020) that can deal with the integration of adaptive or learning components within a larger system. Future SE practitioners must be aware of the technical foundations of AI/ML systems, as well as opportunities and challenges, and develop appropriate controls to specify and verify system requirements. Leveraging foundational coursework in computer science, university programs can provide opportunities to address AI and associated challenges in SE practice.

Application of AI/ML methods and tools within SE, described as AI4SE (McDermott et al. 2020), also presents a tantalizing opportunity to rapidly access and leverage knowledge across all stages of SE practice ranging from concept design to verification and validation. For example, coursework in generative design or NLP provides foundational methods and tools that could improve the efficiency and effectiveness of SE disciplinary work.

Data Science

Broader topics of data science provide new methods and tools to understand complex system behavior by creating, storing, accessing, and interpreting data. Future SE practitioners will curate large data sets and build innovative tools to leverage data across all lifecycle phases. Literacy in software development, data storage, cloud computing, and associated analytical and inferential platforms for computing at scale present significant opportunities for future SE.

University programs can provide advanced coursework for SE practitioners specializing in data science by drawing on adjacent topics like computer science, software engineering, management information systems, and business intelligence and analytics.

Human-Systems Integration

Finally, SE application domains increasingly consider the interface between engineered systems and humans as users, operators, partners, or citizens. Advances in human-system integration – at multiple scales ranging from individuals to society – will help understand the relationship between humans and intelligent, autonomous systems, enabling new levels of complexity.

Understanding and, to some extent, influencing human decision-making is a key component to broadening the scope of engineered systems to address global grand challenges such as climate change, resource management, and healthcare. Designing incentives or mechanisms into enterprise systems can encourage socially beneficent behaviors and position SE as a holistic field that reaches beyond the boundaries of traditional engineering.

To address major societal challenges at the interface with human systems, university programs must leverage other bodies of knowledge in economics, psychology, and organization management. Programs must also broaden interest beyond traditional SE domains like aerospace, defense, and manufacturing to attract and retain students who are eager to make an impact on the world.

Conclusion

In response to industry needs for systems engineers to coordinate and lead technical activities during all lifecycle phases of engineered systems, university programs provide SE education to prepare students for their future careers. While most programs focus on graduate education, ongoing program development considers opportunities to expose undergraduate students to SE concepts and nurture fundamental and applied research in doctoral programs.

NLP analysis of more than 1500 courses in current SE program curricula reveal a coherent structure of about a dozen topics. They range from mathematical solution frameworks like optimization and linear programming to process-oriented approaches to managing technical activity such as requirements engineering and systems architecture. Topics needed for future SE careers emphasizing model-based activities, applications of AI/ML technology, data science, and human-systems integration already exist within SE programs but are not as widespread as traditional topics like requirements engineering, project management, and system architecture.

Evolving university SE programs will require coordination with other academic disciplines including computational fields like computer science, software engineering, statistics, and data analytics, but also infusion from adjacent knowledge domains like organizational science, economics, and psychology. The future of SE mirrors the grand challenges facing society and, based on the solid foundation constructed based on complex engineered systems, the field is well-positioned to play a trans-disciplinary and integrative role over the coming decades.

Note

1 A credit hour is roughly equivalent to one hour of instruction per week over a 15-week semester.

References

Crawley, E.F., Malmqvist, J., Östlund, S. et al. (2014). *Rethinking Engineering Education: The CDIO Approach*. Springer.

Dym, C.L. (2004). Design, systems, and engineering education. *International Journal of Engineering Education* 20 (3): 305–312.

Grasso, D. and Burkins, M.B. (2010). *Holistic Engineering Education: Beyond Technology*. Springer.

INCOSE. Systems Engineering Vision 2035: engineering solutions for a better world. (2022). https://www.incose.org/about-systems-engineering/se-vision-2035 (accessed 19 September 2022).

Khoury, E.T. (2020). Are we teaching systems engineering students what they need to know? Master's thesis. School of Aeronautics and Astronautics, Purdue University.

McDermott, T., DeLaurentis, D., Beling, P. et al. (2020). AI4SE and SE4AI: a research roadmap. *Insight* 23 (1): 8–14.

Meyer, J., Nel, H., and van Rensburg, N.J. (2016). "Systems engineering education in an accredited undergraduate engineering program," *Proceedings of the ASME 2016 International Mechanical Engineering Conference and Exposition* (11–17 November 2016), Available online: 8 February 2017. Phoenix, AZ: ASME. https://doi.org/10.1115/IMECE2016-68038.

Muci-Kuchler, K.H., Birrenkott, C.M., Bedillion, M.D., et al. (2020). "Incorporating systems thinking and systems engineering concepts in a freshman-level mechanical engineering course," *2020 ASEE Virtual Annual Conference,* Virtual, Online (22–26 June 2020). ASEE. https://doi.org/10.18260/1-2--34813.

Muller, G. and Bonnema, G.M. (2013). Teaching systems engineering to undergraduates; experiences and considerations. *INCOSE International Symposium*, Philadelphia, PA (24–27 June 2013). Vol. 23(1), pp. 98–111. Wiley/INCOSE. https://doi.org/10.1002/j.2334-5837.2013.tb03006.x

Pyster, A., Olwell, D.H., Ferris, T.L.J. et al. (ed.) (2015). *Graduate Reference Curriculum for Systems Engineering (GRCSE™) V1.1*. Hoboken, NJ, USA: Trustees of the Stevens Institute of Technology. www.bkcase.org/grcse/.

SERC and INCOSE (2021). Worldwide directory of systems engineering and industrial engineering programs. https://wwdsie.com (accessed 19 September 2022).

Sillitto, H., Martin, J., McKinney, D., et al. (2019). Systems Engineering and Systems Definitions, version 1.0. INCOSE-TP-|22. INCOSE. https://www.incose.org/docs/default-source/default-document-library/final_-se-definition.pdf

Biographical Sketches

Paul Grogan is an associate professor with the School of Systems and Enterprises at Stevens Institute of Technology in Hoboken, New Jersey. He leads the Collective Design Lab, which develops and studies the use of information-based methods and tools for engineering design in systems with distributed architectures such as aerospace, defense, and critical infrastructure. He holds a PhD in engineering systems and an SM degree in aeronautics and astronautics from the Massachusetts Institute of Technology and a BS degree in engineering mechanics from the University of Wisconsin.

Chapter 41

Evolving University Programs for the Other 95% of Engineers: A Capstone Marketplace

William Shepherd

Stevens Institute of Technology, Hoboken, NJ, USA

Introduction to the Capstone Marketplace

In 2011, US Department of Defense's (DoD's) System Engineering Research Center (SERC) started an initiative to give undergraduate students opportunities to exercise basic system engineering (SE) principles in project-based coursework. DoD's then Undersecretary for Research and Acquisition[1] had expressed concerns that the rising complexity of defense systems would require increasing numbers of engineers and program managers trained in SE to create more individuals who could make effective trades between cost, schedule, risk, and technical performance in the next generation of advanced defense developments. DoD leaders felt that for SE to be effective, key aspects of it had to be introduced earlier in engineering education programs.

SERC was established to promote better tools and methods for SE and to promote human capital development in the SE domain. SERC proposed an effort to introduce undergraduate students to SE using a web-based "Capstone Marketplace" to enhance their senior engineering design courses. This Marketplace would identify valuable research topics from government and military organizations, market these needs and problems to universities, and attract academics and student teams to work on solutions. The Office of the Undersecretary of Defense for Research and Engineering (OUSD(R&E)) gave SERC a task order that created the Capstone Marketplace. The Marketplace soon attracted the participation of operational units in the military, who saw the forum as a way for enabling rapid technology insertions and filling capability shortfalls. This involvement of military "customers" provided university students with tangible project topics and connected them with the end users who would benefit from their work.

The Capstone Marketplace has helped to change the structure of typical university design projects. Traditionally, young undergraduate engineers followed a curriculum within a specialized discipline, with limited exposure to other types of engineering. SE education still is a subject mostly left to the graduate school environment (see "Evolving University Programs for on Systems Engineering"). Many universities now emphasize "project-based learning" and senior design capstone projects can be found in many different curriculums outside of engineering. The Capstone Marketplace is an authentic entrepreneurial environment – with real clients, real needs, contracts, money, schedules, and deliverables – and continues to be a unique way to introduce system thinking to undergraduates, enrich their education as engineers, and bring valuable research to government and military customers.

Systems Engineering for the Digital Age: Practitioner Perspectives, First Edition. Edited by Dinesh Verma.

Companion website: www.wiley.com/go/verma/systemsengineering

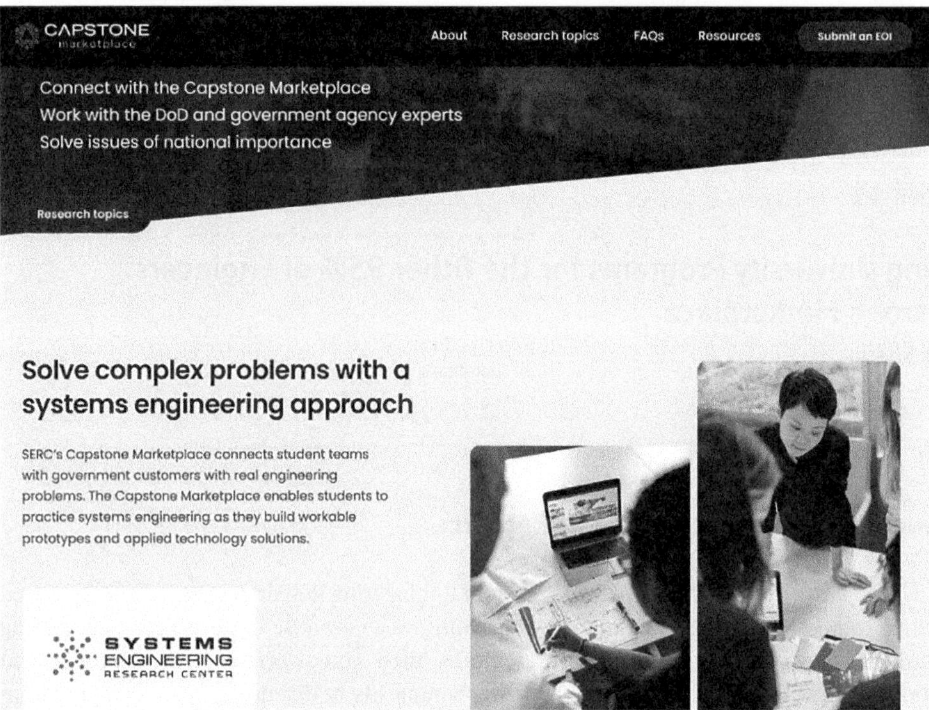

Capstone Marketplace. *Source:* capstonemarketplace.org/http://www.capstonemarketplace.org/last accessed under March 23, 2023.

Current Activity

The Capstone Marketplace has two overall objectives. First, it provides value to military and government customers, expanding available avenues of research and development for innovation and technology insertions. Second, it connects university students with real-world problems and common business practices and gives them exposure to the basics of SE in small development projects. The Marketplace is a portal that the military can use to advance unrealized ideas and to look at ways to meet their command's lower-priority needs. Students connect to the operators/end users, allowing the marketplace to strip away layers of R&D and acquisition hierarchy. This creates a very flat organizational structure linking those who have the "needs" and the technology developers who will work to meet them.

The Marketplace stimulates innovation both in students and within customer organizations. It helps to capture latent "human intellectual capital" – individuals who want to pursue an idea or create a path to their own solutions. In the Marketplace, military operators have can devise their own solutions to problems. This has long been a hallmark of the Special Operations culture, and units in the Special Operations Forces (SOF) have been the Marketplace's primary customers. The Marketplace teams present low-risk ways for units to evaluate new concepts that do not interfere with traditional R&D approaches; cost very little; and can provide quality analysis, prototypes, and technical data that would be considerably more expensive if produced by defense industry companies.

Capstone Marketplace teams completed 23 projects in the 2021–2022 academic year, and managed 17 teams in academic year 2022–2023. Figure 41.1 provides an overview of the activity for the Marketplace in the 2021–2022 academic year.

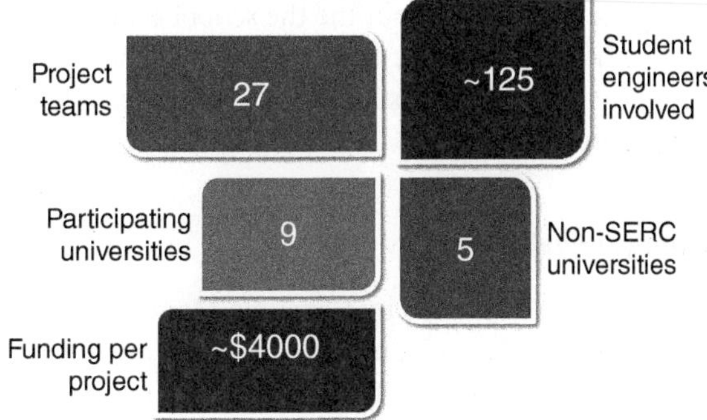

Figure 41.1 Overview of the 2020–2021 Capstone Marketplace activity.

How Do Capstone Projects Work?

Starting A Capstone

Capstone Marketplace managers meet with military units and staff to collect ideas and technical initiatives that are outside of their formal R&D processes. The Marketplace is able to harvest many good ideas that simply do not fit into the "bandwidth" of a military organization's current resources and processes. The benefits are many. The reception of "good ideas" and follow-through research encourages communication between an organization's innovators and its leadership so that the ideas are not rejected out of hand. The operator is the closest to the problem; he or she is the "end user," but in traditional R&D processes, layers of staff and managers can add bias to their ideas and potential solutions. The Marketplace flattens much of this interaction. The cost of a student project is zero to the operational commands (and they do not have research money anyway), as each student team receives a small amount of funding through the Capstone Marketplace itself.

Capstone Marketplace managers select project topics based on several considerations. The importance of the need is primary, followed by the complexity of the problem. Most real-world problems require multidisciplinary technical solutions, and the Marketplace looks for these. However, managers and their customers must consider that student teams have limited resources (time, money, skill, facilities, etc.) and expectations for the project outcomes must be reasonable. Complex problems are sometimes broken down into simpler tasks, each of which can be pursued as a separate Capstone development project. In all cases, the project tasks must provide academic value for students as well as potential benefits for the end user.

Capstone Marketplace research topics are posted on the website at the start of each US Fall semester (generally August or September). Topics are listed with short descriptions of the problems and become high-level "Statements of Work" in the subsequent university contracts. Universities may also submit their own ideas for research to the Marketplace. These are shared with government organizations that then choose to sponsor a Capstone project. Templates and additional guidance for these unsolicited research submissions are posted on the Marketplace website.

Universities submit Expressions of Interest (EOI) to the Marketplace through the website. After the Capstone Marketplace receives EOIs, Requests for Proposals (RFPs) are sent to each university. The Marketplace then coordinates direct contact between faculty advisors and the government customers and their subject-matter experts (SMEs). These SMEs provide the necessary details to assist faculty, administrators, and students in the preparation of their proposals. Proposals are purposely succinct and outline the school's interest, plans for organizing student teams, technical approaches for research, and an outline of the class schedule and the team's projected expenses. Capstone teams may also include graduate students who are receiving course credit for their participation.

Proposal Evaluations

Proposals are evaluated by Capstone Marketplace managers, SMEs, and other government representatives. Criteria for awards are based on faculty support of Capstone teams, the availability of labs, workshops, and test facilities, previous research, past performance, and funding requested. University collaboration is important; each university is expected to provide contributions to the resources that student teams need. SERC's Capstones are offered as an enrichment of the student design team experience, giving students real problems in a system engineering context.

Capstone awards are announced in the Fall; some awards are made after the Fall to start teams' Spring semester research.

Costs and Funding

The Marketplace provides resources for student design course activities. Formal agreements between the Marketplace and each university include incremental milestones and payments for each student team project. This is done in place of typical research grants so that each student team's work is tied to a series of specific deliverables. These contracts are collaborative agreements between SERC and each university. Schools are expected to provide academic instruction, classrooms, laboratories, shops, test facilities, and other resources on a collaborative basis.

The Capstone Marketplace does not pay for the labor of professors, graduate students, teaching assistants, or technicians. Funds are provided to universities for project material, services, and other student team support for needs that are typically outside normal class resources. University project proposals may include student team travel necessary to coordinate project activities, but this must be done within government regulations. Budget justifications are included in each proposal. The Capstone Marketplace is not a "work for hire" arrangement; it is a collaboration between SERC and each university, and it is expected that administrative fees for Capstone projects will be minimized. University overheads and fee structures are considerations in SERC Capstone awards.

Assignments of all research topics are normally completed by the end of September each year. SERC's awards for Capstone projects are limited to a maximum of US$5000 for two semesters of student design activity. Four design reviews are milestones in a typical Capstone project, two per semester. These reviews are conducted with government customers and Capstone Marketplace staff. At the end of the academic year, final design reviews include presentations of the prototype hardware and software, technical data, and any final reports. Demonstrations of the prototype with customers are usually included. At the completion of a student design project, each university's residual Capstone funds are retained for future student design projects.

Funding for SERC's Capstone Marketplace task has been shared by a number of government offices, including the Office of the Undersecretary of Defense for Research and Engineering (OUSD(R&E)), Office of the Undersecretary of Defense for Acquisition and Sustainment (OUSD(A&S)), US Special Operations Command, and the US Air Force Research Laboratory (AFRL).

Executing a Capstone Project

SERC provides resources to a university through Firm Fixed Price contract agreements. A Statement of Work (SOW) lists what is expected from the academic faculty and the student team they are supervising. The contract deliverables are typically split into four increments throughout the two-semester school year.

Getting Started

The first deliverables in early Fall are presented at a design review and the element for this review is the **Project Plan**. Each student team is required to write a detailed plan for their work in a format that mirrors a short industry research proposal. The Capstone Marketplace provides students guidance in developing their plans. Coordination between students and their military customers and SMEs is key to creating a plan that captures each party's expectations. The Capstone Marketplace Project Plan includes:

a) Statement of Work (generally more details than are in the university contract)
b) Project Organization (students and "customers")
c) Draft of a "Quad Chart"
d) Draft "Concept of Operations"
e) Schedules and Reporting Plan (including milestones and development path)
f) Facilities and Equipment
g) Communications Plan (including disclosures of information)
h) Budgets and Spend Plan
i) Intellectual Property
j) Closeout Plan for Project, including disposition of residual funds, materials, government property, etc.

The Project Plan becomes a living document central to subsequent design reviews. Students update this plan throughout the year-long project, providing a consistent way for reviewers to quickly assess team progress and identify new information or changes. The Kickoff review also has other important deliverables, especially a DOD-style "quad chart". This one-page document is used primarily for external communications from the team to others outside of the project. Quad charts are widely used in government projects and can have several different formats. The Marketplace chart has four panels with text and graphics (example shown in Figure 41.2). Panels show the customer's problem, what a proposed solution will look like, and the key technologies that will be pursued in a solution.

2020 SOF 19 traumatic brain injury (TBI) self-assessment

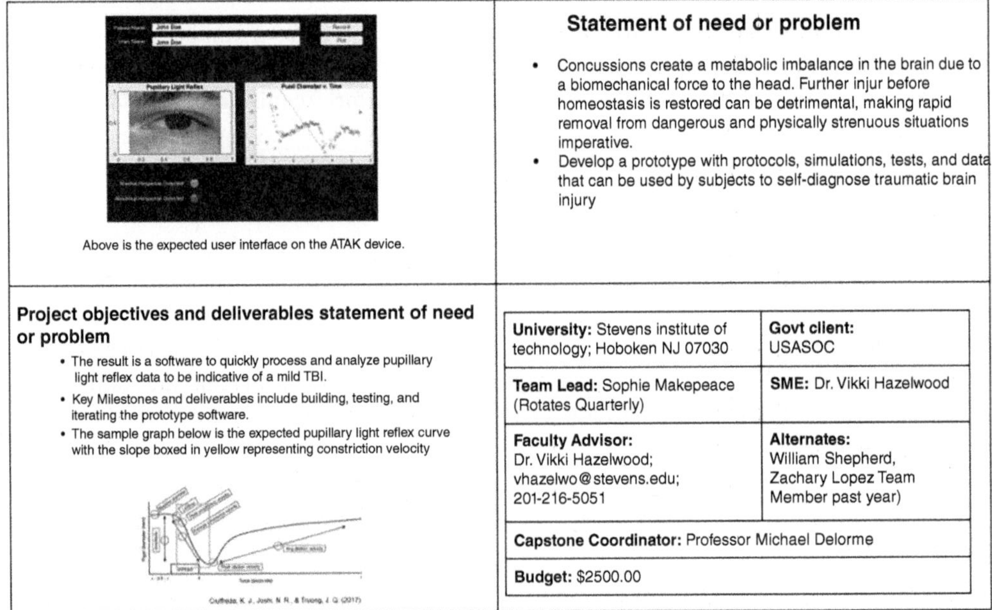

Above is the expected user interface on the ATAK device.
Statement of need or problem
• Concussions create a metabolic imbalance in the brain due to a biomechanical force to the head. Further injur before homeostasis is restored can be detrimental, making rapid removal from dangerous and physically strenuous situations imperative.
• Develop a prototype with protocols, simulations, tests, and data that can be used by subjects to self-diagnose traumatic brain injury
Project objectives and deliverables statement of need or problem
• The result is a software to quickly process and analyze pupillary light reflex data to be indicative of a mild TBI.
• Key Milestones and deliverables include building, testing, and iterating the prototype software.
• The sample graph below is the expected pupillary light reflex curve with the slope boxed in yellow representing constriction velocity
University: Stevens institute of technology; Hoboken NJ 07030
Govt client: USASOC
Team Lead: Sophie Makepeace (Rotates Quarterly)
SME: Dr. Vikki Hazelwood
Faculty Advisor: Dr. Vikki Hazelwood; vhazelwo@stevens.edu; 201-216-5051
Alternates: William Shepherd, Zachary Lopez Team Member past year)
Capstone Coordinator: Professor Michael Delorme
Budget: $2500.00

Figure 41.2 Example quad chart from the project "TBI Self Assessment." *Source:* Stevens Institute of Technology 2020.

System Engineering and Student Design

During the project, Capstone managers mentor student teams to discuss the basics of a system engineering flow. The Marketplace uses a lean process for its system engineering, steps that are less detailed than industry system engineering but tailored to general engineering students. The Marketplace's flow for Capstone work is:

- Concept of Operations
- Customer requirements
- Description of the complete system
- Research
- System Architectures
- Concepts
- Decomposition into subsystems and components, block diagrams
- Analysis, modeling, and simulations of concepts
- Concept down-selects
- Allocation of functions and requirements to various subsystems
- Selection of components, relevant technologies, arrangements
- Interface requirements for subsystems
- Test planning for subsystems and components
- Integration plans
- Preliminary design review(s) and baseline configuration

- Detailed subsystem design
- Acquisition of materials and components
- Fabrication of subsystems
- Subsystem testing
- Integration
- Integrated testing
- Full performance tests
- Demonstrations
- Reporting and Closeouts

(Some steps in the flow are recursive)

Concept of Operations (CONOPS)

Students are tasked with creating written and graphic descriptions of how end users will employ their prototypes and how prototype capabilities will solve customers' needs. The Concept of Operations (CONOPS) may address different user situations, phases of an operation, or other modes in which the solution will be needed. Students interview customer representatives and conduct exchanges with each project's Subject Matter Experts (SMEs). The CONOPS is an important document that describes the intended performance of the students' prototypes. The document can be modified during the prototype development, and prototype testing may reveal capabilities and ways of employment not initially envisioned when the CONOPS is first drafted. Figure 41.3 provides an example of a briefing chart reflecting the CONOPS for a student project from 2020.

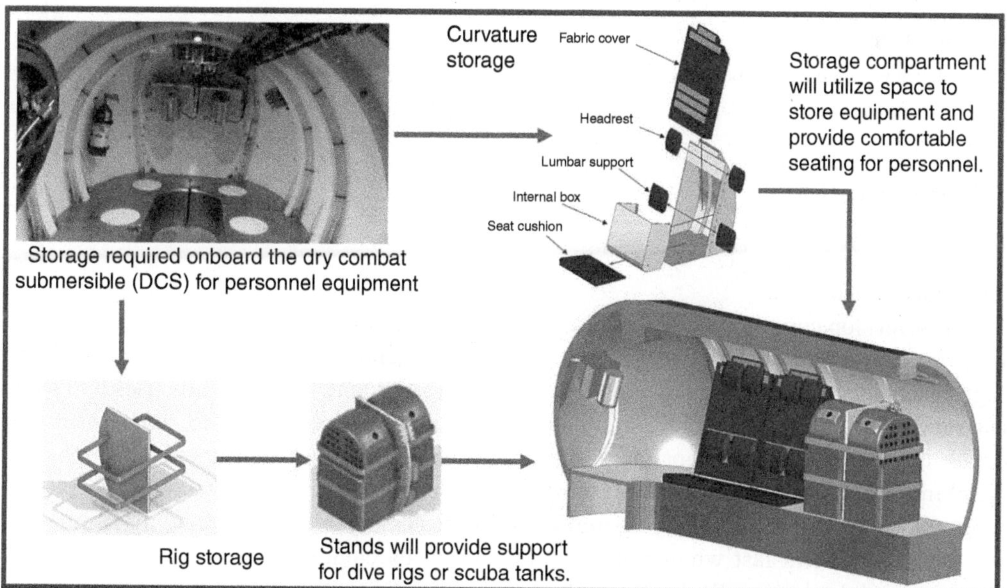

Figure 41.3 Concept of Operations for FWD Swimmer Compartment [4.1], part of the "DCS Internal Storage" Project. *Source:* N. Shelton/The University of Alabama in Huntsville.

Requirements, Functions, and the Voice of the Customer

During the initial stages of the project, students interview customers and SMEs to determine features that are necessary for their prototypes. Student teams are encouraged to use tools such as "N2" diagrams and computer applications that track the connectivity between intended functions, requirements, constraints, and possible conflicts.

Objective-Function-Constraint-Means (source Table 3.1) from the "DCS A DCS Atmospheric Monitoring" Project (University of South Alabama 2021).

Characteristics	O	C	F	M
Detect/measure carbon monoxide	X			
Display data			X	
Withstand pressure (1–4 atm)		X		
Industrial magnets (circular)				X
LCD screen				X
No ventilation		X		
Internally powered (battery)		X		
Carbon filter				X
Must operate for 72 h			X	
Remove humidity	X			
LED light				X
Detect/measure humidity	X			
Withstand temperature range (35—100 °F)		X		
Desiccants				X
Cannot exceed 100 cubic inches		X		
Alert passengers	X			
Mobile			X	
Remove carbon monoxide	X			

Conceptual Design

Students develop concept ideas by exploring the key technologies, form and general appearance, functions and performance, and top-level arrangements of their intended products. High-level block diagrams are included in concept designs, and are especially useful to show separations of subsystems and components. Students are encouraged to brainstorm and sketch to capture their concept ideas. Current engineering education does not emphasize sketching or hand drawing. Computer tools are quite evolved, and many designers bypass this hand work as (sketching) time-consuming and unnecessary. However, drawing simple figures on paper is still fast when compared to "design by computer." Students who develop sketching and drawing skills during such conceptualization work are likely to become more

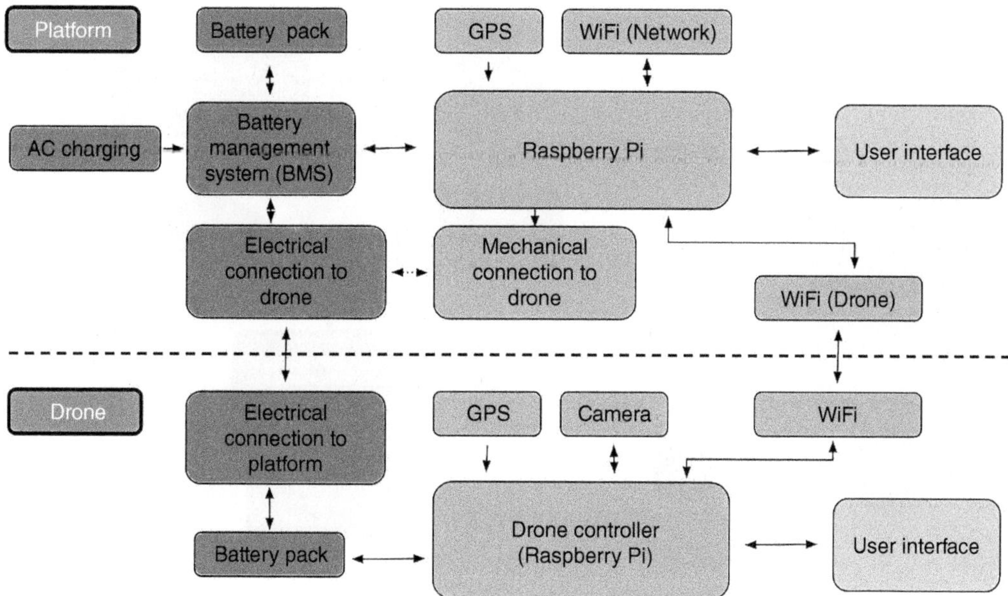

Figure 41.4 Block Diagram from the "Remote Drone Recharging" Project (University of Massachusetts Boston 2021).

perceptive and flexible designers. Figure 41.4 provides an example of a conceptual block diagram from a Capstone project.

Design Reviews and the Preliminary Design Review (PDR)

Project reviews are done throughout the academic year. In each review, the Capstone team presents updates to the Project Plan that include cost, schedule, and technical details. The student teams are also asked to assess various areas of project risk. A midstream milestone is the Preliminary Design Review (PDR). Students' PDR presentations show more mature designs, with key technology choices, subsystem and component divisions, arrangements, test planning, and system integration plans. The PDR is a major step in reducing project development risk. The teams show choices they have made – and that they are sufficient to provide the desired performance in the prototype end items. A successful PDR becomes a product "baseline" of the team design.

Most Capstone projects progress through two design reviews after PDR. At the end of each project, prototypes are intended to reach Technology Readiness Levels (TRLs) of 5 or 6 – items that can show intended utility in a laboratory or simulated operational environment. Final project reviews are conducted at the end of the academic year and may include product demonstrations to the customer. Critical Design Review (CDR) criteria that are typical for industry (e.g., 90% of all drawings complete) are not normally reached when Capstone teams are finished. Figure 41.5 shows a design artifact (software logic flow diagram) from one of the capstone projects.

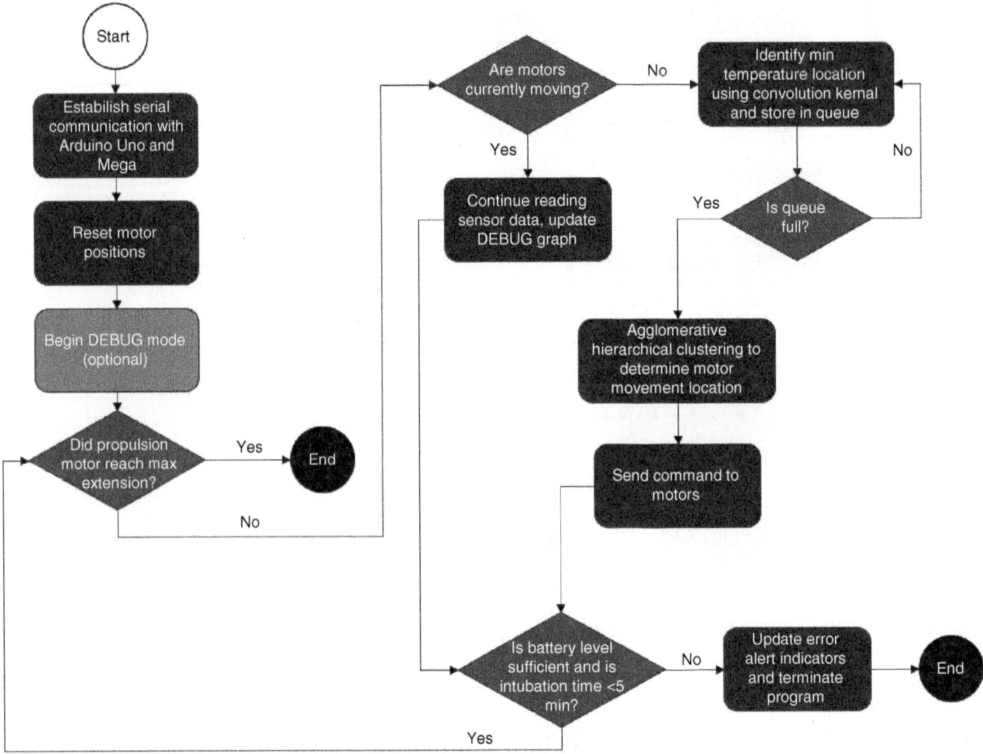

Figure 41.5 Software Logic Diagram from the "Self Intubating Airway" Project (University of Texas Austin 2021 – reformatted for inclusion).

Communications

Each student team is responsible for writing a communications plan for internal and external information exchange with team members, faculty advisors, customers, and Marketplace managers. Details of a communications plan include:

- File structure and share folders for documenting project details
- Networking applications used for both internal and external communications
- Calendars for meetings and videoconferencing with SMEs and Capstone managers
- Schedules and processes for design reviews
- Formats for reports and data
- Production of final video project summary

An example of the schedule and process materials included in the communications plan for a Capstone project is shown in Figure 41.6.

All student teams communicate with videoconferencing and file exchanges. However, in-person meetings are sometimes required, and student travel is allowed under the provisions of Capstone projects. Almost all SERC Capstones are done as unclassified projects; SERC does have the capability to do classified research at Service Academies and selected universities, but these instances are rare. Military and government customers may set restrictions on public disclosures of some information on their Capstone projects. Military customers generally require reviews of project

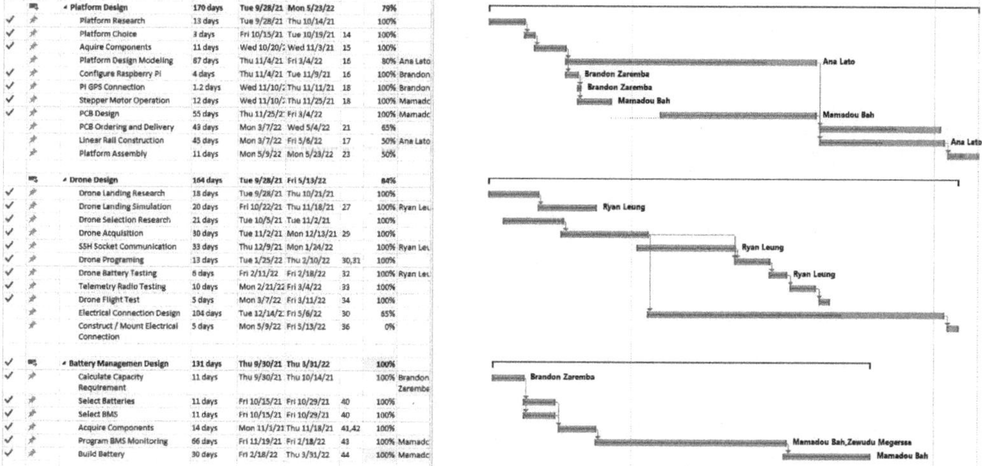

	Platform Design	170 days	Tue 9/28/21	Mon 5/23/22		79%	
✓	Platform Research	13 days	Tue 9/28/21	Thu 10/14/21		100%	
✓	Platform Choice	3 days	Fri 10/15/21	Tue 10/19/21	14	100%	
✓	Aquire Components	11 days	Wed 10/20/2	Wed 11/3/21	15	100%	
	Platform Design Modeling	87 days	Thu 11/4/21	Fri 3/4/22	16	80%	Ana Lato
✓	Configure Raspberry Pi	4 days	Thu 11/4/21	Tue 11/9/21	16	100%	Brandon
✓	PI GPS Connection	1.2 days	Wed 11/10/2	Thu 11/11/21	18	100%	Brandon
✓	Stepper Motor Operation	12 days	Wed 11/10/2	Thu 11/25/21	18	100%	Mamadc
✓	PCB Design	55 days	Thu 11/25/2	Fri 3/4/22		100%	Mamadc
	PCB Ordering and Delivery	43 days	Mon 3/7/22	Wed 5/4/22	21	65%	
	Linear Rail Construction	45 days	Mon 3/7/22	Fri 5/6/22	17	50%	Ana Lato
	Platform Assembly	11 days	Mon 5/9/22	Mon 5/23/22	23	50%	
	Drone Design	164 days	Tue 9/28/21	Fri 5/13/22		84%	
✓	Drone Landing Research	18 days	Tue 9/28/21	Thu 10/21/21		100%	
✓	Drone Landing Simulation	20 days	Fri 10/22/21	Thu 11/18/21	27	100%	Ryan Leu
✓	Drone Selection Research	21 days	Tue 10/5/21	Tue 11/2/21		100%	
✓	Drone Acquisition	30 days	Tue 11/2/21	Mon 12/13/21	29	100%	
✓	SSH Socket Communication	33 days	Thu 12/9/21	Mon 1/24/22		100%	Ryan Leu
✓	Drone Programing	13 days	Tue 1/25/22	Thu 2/10/22	30,31	100%	
✓	Drone Battery Testing	6 days	Fri 2/11/22	Fri 2/18/22	32	100%	Ryan Leu
✓	Telemetry Radio Testing	10 days	Mon 2/21/22	Fri 3/4/22	33	100%	
✓	Drone Flight Test	5 days	Mon 3/7/22	Fri 3/11/22	34	100%	
	Electrical Connection Design	104 days	Tue 12/14/2	Fri 5/6/22	30	65%	
	Construct / Mount Electrical Connection	5 days	Mon 5/9/22	Fri 5/13/22	36	0%	
✓	Battery Managemen Design	131 days	Thu 9/30/21	Thu 3/31/22		100%	
✓	Calculate Capacity Requirement	11 days	Thu 9/30/21	Thu 10/14/21		100%	Brandon Zarembe
✓	Select Batteries	11 days	Fri 10/15/21	Fri 10/29/21	40	100%	
✓	Select BMS	11 days	Fri 10/15/21	Fri 10/29/21	40	100%	
✓	Acquire Components	14 days	Mon 11/1/21	Thu 11/18/21	41,42	100%	
✓	Program BMS Monitoring	66 days	Fri 11/19/21	Fri 2/18/22	43	100%	Mamadc
✓	Build Battery	30 days	Fri 2/18/22	Thu 3/31/22	44	100%	Mamadc

Figure 41.6 MS Project Gantt Chart for the "Remote Drone Recharging" Project (University of Massachusetts Boston 2021).

material before any public dissemination is made. In most cases, customer reviews are mandatory before any briefings can be made to other school departments, publication of papers, media interviews, outside presentations, podcasts, etc.

Pedagogy of SERC's Capstones

The Capstone Marketplace introduces undergraduate engineers to real problems that belong to real people, bypassing many of the artificialities of design topics generated "in-house" without specific end-user needs. Capstone topics are selected that are suited for multidisciplinary research. Web-based "marketing" helps universities "flip the classroom" and gives students more choices for what they want to work on. The Marketplace also accepts students' inputs for project topics and assists them in forming research teams.

Marketplace managers work to balance the scope of project topics with the resources and abilities of the undergraduate students. The Marketplace creates an entrepreneurial environment and makes student teams responsible and accountable for total team performance. The style of team design reviews matches project reviews done in business. Team reviews are sequential, and presentations are done in standard formats. During the reviews, students are expected to quickly address schedules, costs, and risks, as well as their technical progress. Students explain rationales for technical decisions, changes in cost and schedule details, and present closeouts of action items from previous meetings. Customers and managers query students to probe the teams' knowledge and their confidence in the work done. Students are encouraged to provide direct and succinct answers and not to "wing it" if they are unsure.

In some cases, universities add graduate students to Capstone teams. They bring advanced skills to the teams not typically taught in undergraduate courses. Some of these graduate students have been from business schools, and they significantly enhance the teams' capabilities for business operations.

In contrast to typical government research and development efforts, occasional "failures" are accepted in Capstone Marketplace projects. Some projects just turn out to be dead ends. Research can prove intractable and beyond teams' capabilities. Even if students do not achieve desired project end states, the teams' results can have value to customers; knowledge of what not to do may be as important as finding the "right" answer.

Highlights of Capstone Marketplace's pedagogical approach are:

- Project Based Learning
- "Noble" work on real problems and needs
- Importance of the "voice of the customer"
- Brainstorming and free play for concepts. "Out of the Box" approaches welcome
- Direct and sustained communications with end users
- Total performance responsibility and accountability for teams – cost, schedule, technical performance, risk
- Decompositions of complex projects into subsets that can be more readily attacked
- Importance of subsystem and component interfaces, interface requirements, definitions, and testing
- Sequential developments to burn down project risks
- Importance of integration and test planning and execution
- Documentation of team technical data and design rationale
- Development of students' skills for presentations and customer interactions
- Tolerating failures

Capstone Highlights

Capstone Marketplace is a resource that allows organizations to research problems that are unaddressed in formal R&D programs. Some SERC Capstone student projects have proven very successful, providing DoD customers with new technical solutions of significant value, at very low cost. Capstone student teams have occasionally outperformed traditional industry and government organizations, providing their customers with fresh insights, novel approaches, and innovative solutions, which otherwise would have been overlooked. U.S. Special Operations Command has seen positive results from Capstone efforts. One Capstone project is illustrated in Figure 41.7: an

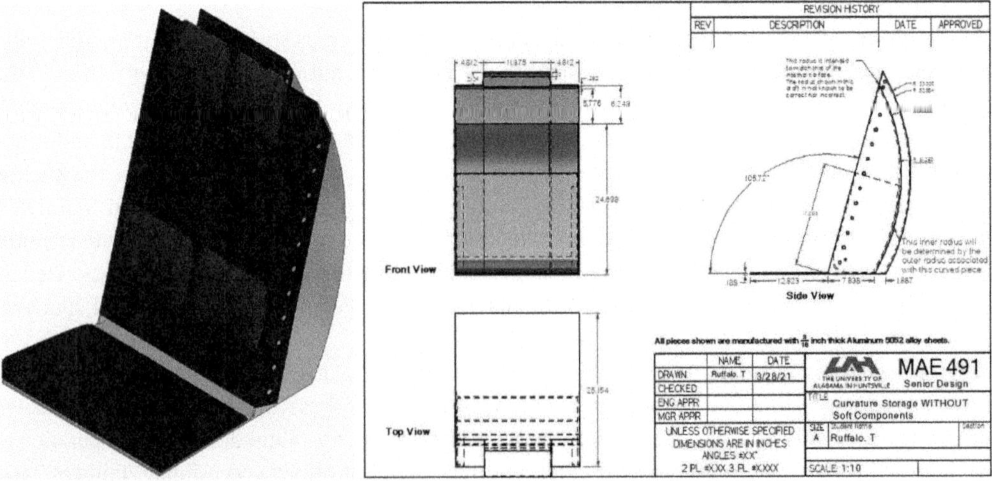

Figure 41.7 Curvature Storage System Draft and CAD (Credit: T. Ruffalo) from the "DCS Internal Storage" Project (University of Alabama Huntsville 2020).

internal stowage system design for a maritime platform that became a significant cost avoidance to a Navy customer. The student design is now a *fielded operational capability* that would have been unaffordable if procured through normal acquisition channels.

Additional successful Capstone projects have transitioned into operational systems for:

- Ship stopping

Example from a Stevens Institute of Technology Capstone. *Source:* Shane T. McCoy / Wikimedia Commons / Public Domain.

- Transparent Armor door mechanisms

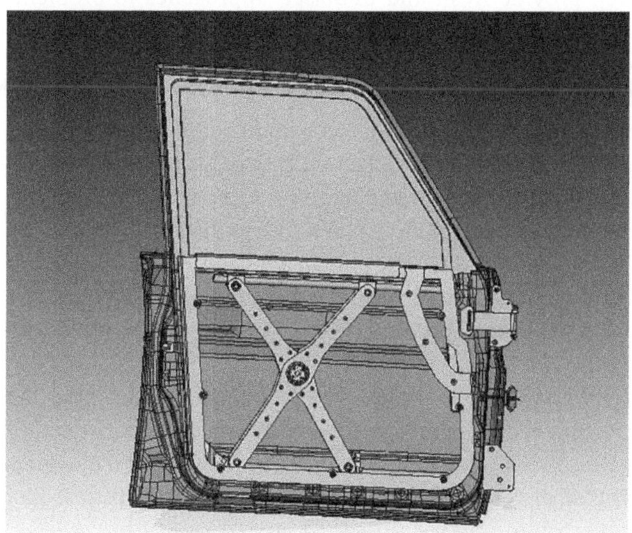

Example from a University of Alabama Huntsville Capstone.

- Advanced life jacket inflation system

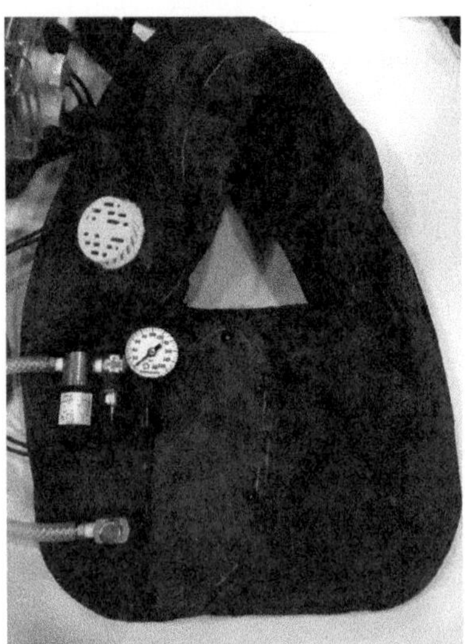

Example from a Stevens Institute of Technology Capstone.

Benefits and Challenges

Capstone Marketplace participation has several benefits. For government and military customers, the Marketplace does not require any organizational funds and provides increased bandwidth for research and development. Capstones provide multiple paths for problem-solving and help build tech databases that can be important for an organization's future programs and funding. Student teams bring knowledge of the latest tools and technologies and enthusiasm to their work; teams often provide fresh and unbiased approaches to problem-solving. The Marketplace is also a way to enhance an organization's culture. SMEs who work with Capstone teams get unique experiences as mentors, exposure to project management, and may become "MacGyvers" in their commands.

For universities and students, the Capstone Marketplace is an enhancement for students in their design classes. The Marketplace gives the student teams contact with real customers who have technology and capability needs, and who will use solutions that prove successful. This arrangement is often quite attractive to engineering students and displaces research topics that can be artificial. The Marketplace also provides students and faculty guidance, mentorship, and models for system engineering, project management, and common business operations. The Capstone Marketplace has also been successful in helping students transition to work as industry and government employees in the defense sector.

As with all worthwhile endeavors, there are challenges for the Marketplace. Matching the Marketplace's objectives, university academic requirements and constraints, and the performance desired by government customers can be difficult. Capstone managers work to adjust Statements of Work, descriptions of deliverables, and other details so that all parties involved in Capstones can reach agreement.

For the government and military customers, SMEs working with Capstone student teams have many other demands on their time. Their interactions with students teams are generally collateral duties. SMEs occasionally rotate, and new SMEs must be brought up to speed. Maintaining consistency in government requirements and expectations for Capstones projects is a challenge when key individuals change over the course of the academic year.

Military clients often have very limited access to open web services. This makes coordination of Capstone activities with SERC and other outside universities cumbersome. Government and military IT networks are increasing the allowable web browsers and collaboration tools that are most useful.

Occasionally, Capstone Marketplace processes and templates do not match a university's class syllabus. There can be situations where student teams are required to provide redundant reports and reviews. The Marketplace does its best to eliminate this duplication by consolidating schedules, events, and deliverables with students' academic requirements.

Future Directions: Get Involved

The Capstone Marketplace has several future objectives Achieving better transitions for Capstone projects into "Programs of Record" developments tops the list. The Marketplace has managed prototype transitions in a few instances, and the number of these transitions are increasing. Transitions are the most significant measure of Capstone effectiveness for our government and military customers. The Capstone Marketplace is expanding its participation among other government organizations outside of the DoD, such as to the National Aeronautics and Space Administration (NASA), the Department of Energy (DOE), and the Department of Homeland Security (DHS). To date, service components and national laboratories have had only minimum involvement with the Capstone Marketplace. A stronger transition record for student research and product development may increase these future Capstone collaborations.

There is great potential for industry to join the Capstone Marketplace and to assist student teams in prototype developments. Industry can contribute knowledge, expertise, hardware and software components, equipment, data, and other support to the students. In the future, industry partners can independently collaborate with universities and student teams. This makes them visible to potential government customers and gives companies contact with students who may become future employees. Under present guidelines, Capstone Marketplace is unable to solicit from or provide funds to industry for their participation in these projects. Despite these limitations, robust industry participation in the Capstone Marketplace remains a critical objective for future success.

There is no single reference in the Capstone Marketplace that can provide a "bare bones" guide for the system engineering processes that is suitable for an undergraduate Capstone project. There is a definite need in academia and the system engineering enterprise for such a text. The Capstone Marketplace uses the NASA *Systems Engineering Handbook* (2017) as a primary reference for descriptions of the system engineering process, definitions, and other explanations. The NASA *Handbook* is geared toward the management of very large engineering programs; however, it has been found to be a most useful source of information on the "how" and "why" of the system engineering art.

Note

1 Now Undersecretary of Defense for Research and Engineering (USD(R&E)).

References

INCOSE (2015). *System Engineering Handbook*, 4e. San Diego, CA: International Council on Systems Engineering (INCOSE). Published by Wiley. https://www.wiley.com/en-us/INCOSE+Systems+Engineering+Handbook%3A+A+Guide+for+System+Life+Cycle+Processes+and+Activities%2C+4th+Edition-p-9781118999400.

NASA (2017). System engineering handbook. Washington, DC: NASA/SP-2016-6105 Rev2, October 2017. https://www.nasa.gov/connect/ebooks/nasa-systems-engineering-handbook

Norman, D. (1988). *The Design of Everyday Things*. New York: Basic Books.

US DAU (2001). *Systems Engineering Fundamentals*. Fort Belvoir, VA: US Army Defense Acquisition University (DAU). https://acqnotes.com/wp-content/uploads/2017/07/DAU-Systems-Engineering-Fundamentals.pdf.

Wasson, C.S. (2016). *System Engineering Analysis, Design, and Development: Concepts, Principles, and Practices*, Wiley Series in Systems Engineering and Management, 2e. Hoboken, NJ: Wiley.

Biographical Sketches

Captain Shepherd is a retired Navy SEAL and NASA astronaut. He served in SEAL Teams on both coasts and was selected as one of 17 new NASA astronauts in 1984. He completed three space flights as a mission specialist flying on shuttles Atlantis, Discovery, and Columbia. In 1993, Captain Shepherd was the Program Manager for the International Space Station and in 2000, served as the commander of Expedition- 1 onboard the new station. He retired from active duty and was USSOCOM's first Science Advisor from 2008 to 2011, and managed the Special Operations Forces' science and technology portfolio. Capt. Shepherd is presently a Senior Researcher with Stevens Institute in Hoboken N.J.

Index

Systems Engineering for the Digital Age: Practitioner Perspectives, First Edition. Edited by Dinesh Verma.

Companion website: www.wiley.com/go/verma/systemsengineering

n

o

p

q

r

s